Table E (continued)

Cumulative Standard Normal Distribution

z	.00	.01	.02	.03	.04	.05	.06	.07	.08	.09
0.0	.5000	.5040	.5080	.5120	.5160	.5199	.5239	.5279	.5319	.5359
0.1	.5398	.5438	.5478	.5517	.5557	.5596	.5636	.5675	.5714	.5753
0.2	.5793	.5832	.5871	.5910	.5948	.5987	.6026	.6064	.6103	.6141
0.3	.6179	.6217	.6255	.6293	.6331	.6368	.6406	.6443	.6480	.6517
0.4	.6554	.6591	.6628	.6664	.6700	.6736	.6772	.6808	.6844	.6879
0.5	.6915	.6950	.6985	.7019	.7054	.7088	.7123	.7157	.7190	.7224
0.6	.7257	.7291	.7324	.7357	.7389	.7422	.7454	.7486	.7517	.7549
0.7	.7580	.7611	.7642	.7673	.7704	.7734	.7764	.7794	.7823	.7852
0.8	.7881	.7910	.7939	.7967	.7995	.8023	.8051	.8078	.8106	.8133
0.9	.8159	.8186	.8212	.8238	.8264	.8289	.8315	.8340	.8365	.8389
1.0	.8413	.8438	.8461	.8485	.8508	.8531	.8554	.8577	.8599	.8621
1.1	.8643	.8665	.8686	.8708	.8729	.8749	.8770	.8790	.8810	.8830
1.2	.8849	.8869	.8888	.8907	.8925	.8944	.8962	.8980	.8997	.9015
1.3	.9032	.9049	.9066	.9082	.9099	.9115	.9131	.9147	.9162	.9177
1.4	.9192	.9207	.9222	.9236	.9251	.9265	.9279	.9292	.9306	.9319
1.5	.9332	.9345	.9357	.9370	.9382	.9394	.9406	.9418	.9429	.9441
1.6	.9452	.9463	.9474	.9484	.9495	.9505	.9515	.9525	.9535	.9545
1.7	.9554	.9564	.9573	.9582	.9591	.9599	.9608	.9616	.9625	.9633
1.8	.9641	.9649	.9656	.9664	.9671	.9678	.9686	.9693	.9699	.9706
1.9	.9713	.9719	.9726	.9732	.9738	.9744	.9750	.9756	.9761	.9767
2.0	.9772	.9778	.9783	.9788	.9793	.9798	.9803	.9808	.9812	.9817
2.1	.9821	.9826	.9830	.9834	.9838	.9842	.9846	.9850	.9854	.9857
2.2	.9861	.9864	.9868	.9871	.9875	.9878	.9881	.9884	.9887	.9890
2.3	.9893	.9896	.9898	.9901	.9904	.9906	.9909	.9911	.9913	.9916
2.4	.9918	.9920	.9922	.9925	.9927	.9929	.9931	.9932	.9934	.9936
2.5	.9938	.9940	.9941	.9943	.9945	.9946	.9948	.9949	.9951	.9952
2.6	.9953	.9955	.9956	.9957	.9959	.9960	.9961	.9962	.9963	.9964
2.7	.9965	.9966	.9967	.9968	.9969	.9970	.9971	.9972	.9973	.9974
2.8	.9974	.9975	.9976	.9977	.9977	.9978	.9979	.9979	.9980	.9981
2.9	.9981	.9982	.9982	.9983	.9984	.9984	.9985	.9985	.9986	.9986
3.0	.9987	.9987	.9987	.9988	.9988	.9989	.9989	.9989	.9990	.9990
3.1	.9990	.9991	.9991	.9991	.9992	.9992	.9992	.9992	.9993	.9993
3.2	.9993	.9993	.9994	.9994	.9994	.9994	.9994	.9995	.9995	.9995
3.3	.9995	.9995	.9995	.9996	.9996	.9996	.9996	.9996	.9996	.9997
3.4	.9997	.9997	.9997	.9997	.9997	.9997	.9997	.9997	.9997	.9998

For z values greater than 3.49, use 0.9999.

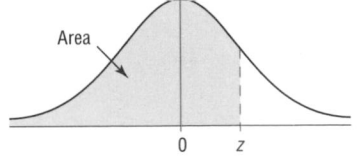

EIGHTH EDITION

Elementary Statistics

A Step by Step Approach

Allan G. Bluman

Professor Emeritus
Community College of Allegheny County

McGraw Hill

Connect
Learn
Succeed™

The McGraw-Hill Companies

Connect
Learn
Succeed™

ELEMENTARY STATISTICS: A STEP BY STEP APPROACH, EIGHTH EDITION

1 2 3 4 5 6 7 8 9 0 QDB/QDB 1 0 9 8 7 6 5 4 3 2 1

ISBN 978–0–07–338610–2
MHID 0–07–338610–3

ISBN 978–0–07–743858–6 (Annotated Instructor's Edition)
MHID 0–07–743858–2

Vice President, Editor-in-Chief: *Marty Lange*
Vice President, EDP: *Kimberly Meriwether David*
Senior Director of Development: *Kristine Tibbetts*
Editorial Director: *Stewart K. Mattson*
Sponsoring Editor: *John R. Osgood*
Developmental Editor: *Adam Fischer*
Marketing Manager: *Kevin M. Ernzen*
Senior Project Manager: *Vicki Krug*
Senior Buyer: *Sandy Ludovissy*
Designer: *Tara McDermott*
Cover Designer: *Ellen Pettengell*
Cover Image: *© Ric Ergenbright/CORBIS*
Senior Photo Research Coordinator: *Lori Hancock*
Compositor: *MPS Limited, a Macmillan Company*
Typeface: *10.5/12 Times Roman*
Printer: Quad/Graphics

Library of Congress Cataloging-in-Publication Data

Bluman, Allan G.
  Elementary statistics : a step by step approach / Allan Bluman. — 8th ed.
    p. cm.
  Includes bibliographical references and index.
  ISBN 978–0–07–338610–2 — ISBN 0–07–338610–3 (hard copy : alk. paper) 1. Statistics—Textbooks.
I. Title.
  QA276.12.B59 2012
  519.5—dc22

                                                                2010031466

www.mhhe.com

Allan G. Bluman

Allan G. Bluman is a professor emeritus at the Community College of Allegheny County, South Campus, near Pittsburgh, Pennsylvania. He has taught mathematics and statistics for over 35 years. He received an Apple for the Teacher award in recognition of his bringing excellence to the learning environment at South Campus. He has also taught statistics for Penn State University at the Greater Allegheny (McKeesport) Campus and at the Monroeville Center. He received his master's and doctor's degrees from the University of Pittsburgh.

He is also author of *Elementary Statistics: A Brief Version* and co-author of *Math in Our World*. In addition, he is the author of four mathematics books in the McGraw-Hill DeMystified Series. They are *Pre-Algebra, Math Word Problems, Business Math,* and *Probability.*

He is married and has two sons and a granddaughter.

Dedication: *To Betty Bluman, Earl McPeek, and Dr. G. Bradley Seager, Jr.*

STATISTICS

Hosted by **ALEKS Corp.**

Connect Statistics Hosted by ALEKS Corporation is an exciting, new assignment and assessment platform combining the strengths of McGraw-Hill Higher Education and ALEKS Corporation. Connect Statistics Hosted by ALEKS is the first platform on the market to combine an artificially-intelligent, diagnostic assessment with an intuitive ehomework platform designed to meet your needs.

Connect Statistics Hosted by ALEKS Corporation is the culmination of a one-of-a-kind market development process involving math full-time and adjunct statistics faculty at every step of the process. This process enables us to provide you with a solution that best meets your needs.

Connect Statistics Hosted by ALEKS Corporation is built by statistics educators for statistics educators!

1 *Your students want a well-organized homepage where key information is easily viewable.*

Modern Student Homepage

▶ This homepage provides a dashboard for students to immediately view their assignments, grades, and announcements for their course. (Assignments include HW, quizzes, and tests.)

▶ Students can access their assignments through the course Calendar to stay up-to-date and organized for their class.

Modern, intuitive, and simple interface.

2 *You want a way to identify the strengths and weaknesses of your class at the beginning of the term rather than after the first exam.*

Integrated ALEKS® Assessment

▶ This artificially-intelligent (AI), diagnostic assessment identifies precisely what a student knows and is ready to learn next.

▶ Detailed assessment reports provide instructors with specific information about where students are struggling most.

▶ This AI-driven assessment is the only one of its kind in an online homework platform.

Recommended to be used as the first assignment in any course.

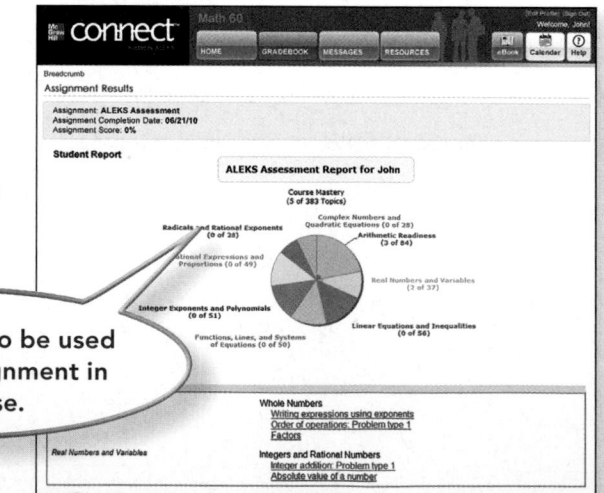

ALEKS is a registered trademark of ALEKS Corporation.

Built by Statistics Educators for Statistics Educators

3 **Your students want an assignment page that is easy to use and includes lots of extra help resources.**

Efficient Assignment Navigation

▶ Students have access to immediate feedback and help while working through assignments.

▶ Students have direct access to a media-rich eBook for easy referencing.

▶ Students can view detailed, step-by-step solutions written by instructors who teach the course, providing a unique solution to each and every exercise.

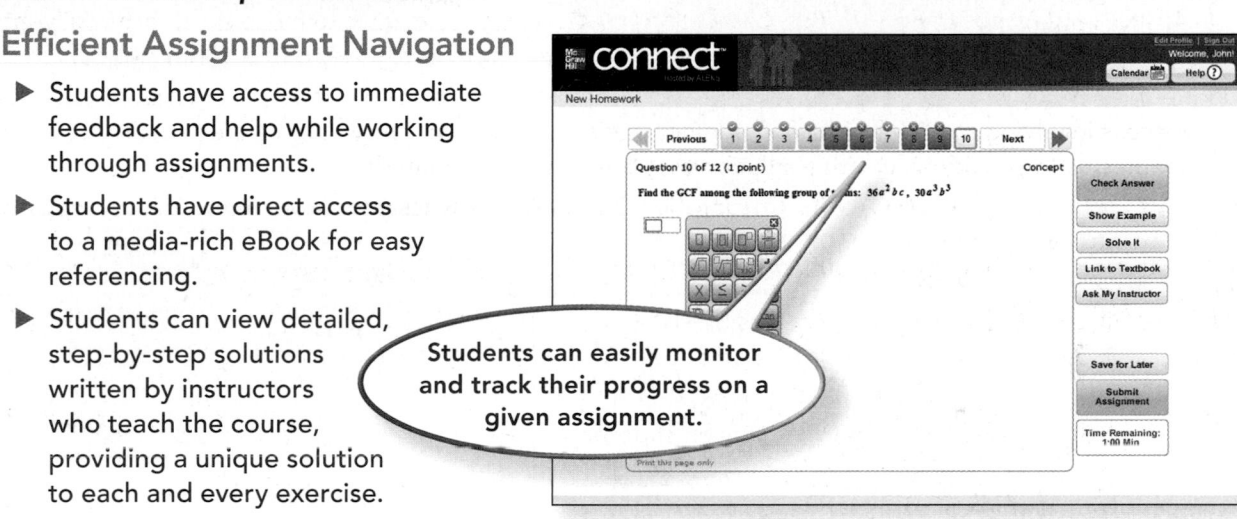

Students can easily monitor and track their progress on a given assignment.

4 **You want a more intuitive and efficient assignment creation process because of your busy schedule.**

Assignment Creation Process

▶ Instructors can select textbook-specific questions organized by chapter, section, and objective.

▶ Drag-and-drop functionality makes creating an assignment quick and easy.

▶ Instructors can preview their assignments for efficient editing.

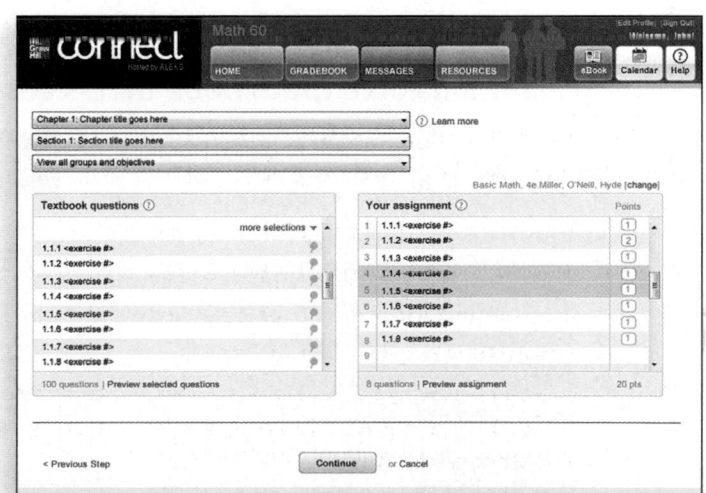

Connect Learn Succeed™

5 *Your students want an interactive eBook with rich functionality integrated into the product.*

Integrated Media-Rich eBook

▶ A Web-optimized eBook is seamlessly integrated within ConnectPlus Statistics Hosted by ALEKS Corp for ease of use.

▶ Students can access videos, images, and other media in context within each chapter or subject area to enhance their learning experience.

▶ Students can highlight, take notes, or even access shared instructor highlights/notes to learn the course material.

▶ The integrated eBook provides students with a cost-saving alternative to traditional textbooks.

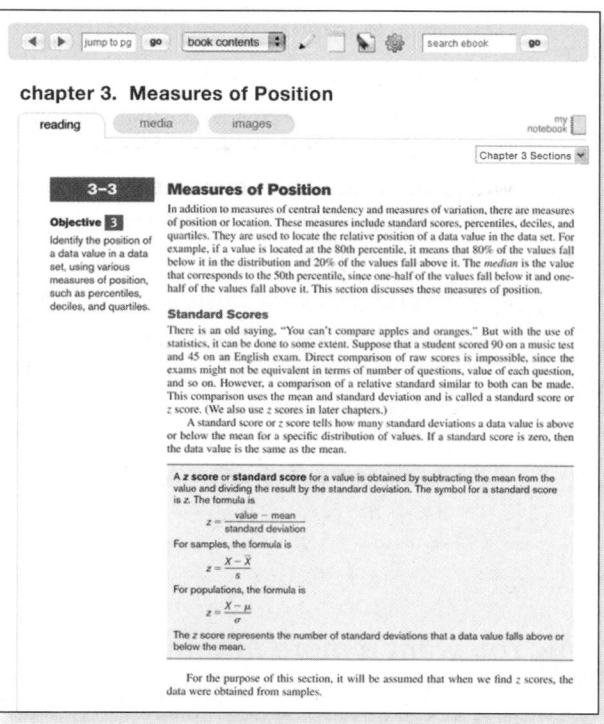

6 *You want a flexible gradebook that is easy to use.*

Flexible Instructor Gradebook

▶ Based on instructor feedback, Connect Statistics Hosted by ALEKS Corp's straightforward design creates an intuitive, visually pleasing grade management environment.

▶ Assignment types are color-coded for easy viewing.

▶ The gradebook allows instructors the flexibility to import and export additional grades.

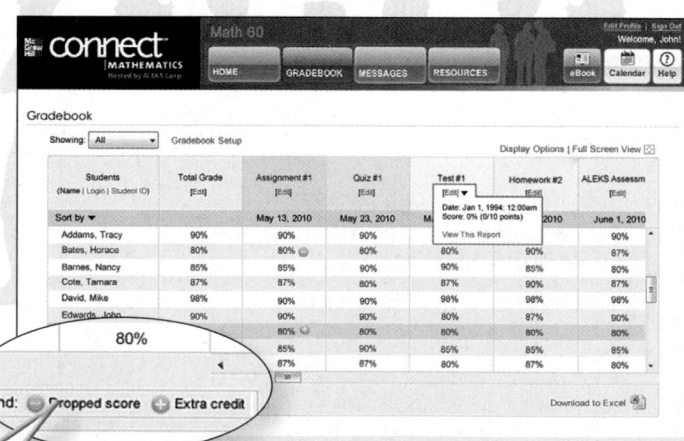

Instructors have the ability to drop grades as well as assign extra credit.

Built by Statistics Educators for Statistics Educators

 7 *You want algorithmic content that was developed by math faculty to ensure the content is pedagogically sound and accurate.*

Digital Content Development Story

The development of McGraw-Hill's Connect Statistics Hosted by ALEKS Corp. content involved collaboration between McGraw-Hill, experienced instructors, and ALEKS, a company known for its high-quality digital content. The result of this process, outlined below, is accurate content created with your students in mind. It is available in a simple-to-use interface with all the functionality tools needed to manage your course.

1. McGraw-Hill selected experienced instructors to work as Digital Contributors.
2. The Digital Contributors selected the textbook exercises to be included in the algorithmic content to ensure appropriate coverage of the textbook content.
3. The Digital Contributors created detailed, stepped-out solutions for use in the Guided Solution and Show Me features.
4. The Digital Contributors provided detailed instructions for authoring the algorithm specific to each exercise to maintain the original intent and integrity of each unique exercise.
5. Each algorithm was reviewed by the Contributor, went through a detailed quality control process by ALEKS Corporation, and was copyedited prior to being posted live.

Connect Statistics Hosted by ALEKS Corp.
Built by Statistics Educators for Statistics Educators

Lead Digital Contributors

Tim Chappell
Metropolitan Community College, Penn Valley

Jeremy Coffelt
Blinn College

Nancy Ikeda
Fullerton College

Amy Naughten

Digital Contributors

Al Bluman, *Community College of Allegheny County*

John Coburn, *St. Louis Community College, Florissant Valley*

Vanessa Coffelt, *Blinn College*

Donna Gerken, *Miami-Dade College*

Kimberly Graham

J.D. Herdlick, *St. Louis Community College, Meramec*

Vickie Flanders, *Baton Rouge Community College*

Nic LaHue, *Metropolitan Community College, Penn Valley*

Nicole Lloyd, *Lansing Community College*

Jackie Miller, *The Ohio State University*

Anne Marie Mosher, *St. Louis Community College, Florissant Valley*

Reva Narasimhan, *Kean University*

David Ray, *University of Tennessee, Martin*

Kristin Stoley, *Blinn College*

Stephen Toner, *Victor Valley College*

Paul Vroman, *St. Louis Community College, Florissant Valley*

Michelle Whitmer, *Lansing Community College*

www.connectmath.com

Contents

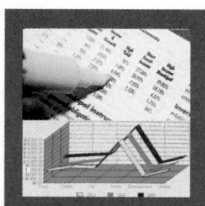

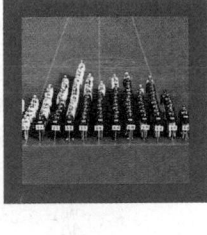

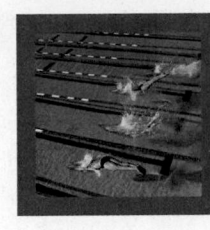

Preface

Approach

Elementary Statistics: A Step by Step Approach was written as an aid in the beginning statistics course to students whose mathematical background is limited to basic algebra. The book follows a nontheoretical approach without formal proofs, explaining concepts intuitively and supporting them with abundant examples. The applications span a broad range of topics certain to appeal to the interests of students of diverse backgrounds and include problems in business, sports, health, architecture, education, entertainment, political science, psychology, history, criminal justice, the environment, transportation, physical sciences, demographics, eating habits, and travel and leisure.

About This Book

While a number of important changes have been made in the eighth edition, the learning system remains untouched and provides students with a useful framework in which to learn and apply concepts. Some of the retained features include the following:

- **Over 1800** exercises are located at the end of major sections within each chapter.
- **Hypothesis-Testing Summaries** are found at the end of Chapter 9 (z, t, χ^2, and F tests for testing means, proportions, and variances), Chapter 12 (correlation, chi-square, and ANOVA), and Chapter 13 (nonparametric tests) to show students the different types of hypotheses and the types of tests to use.
- A **Data Bank** listing various attributes (educational level, cholesterol level, gender, etc.) for 100 people and several additional data sets using real data are included and referenced in various exercises and projects throughout the book.
- An updated **reference card** containing the formulas and the z, t, χ^2, and PPMC tables is included with this textbook.
- End-of-chapter **Summaries, Important Terms,** and **Important Formulas** give students a concise summary of the chapter topics and provide a good source for quiz or test preparation.
- **Review Exercises** are found at the end of each chapter.
- Special sections called **Data Analysis** require students to work with a data set to perform various statistical tests or procedures and then summarize the results. The data are included in the Data Bank in Appendix D and can be downloaded from the book's website at www.mhhe.com/bluman.
- **Chapter Quizzes,** found at the end of each chapter, include multiple-choice, true/false, and completion questions along with exercises to test students' knowledge and comprehension of chapter content.
- The **Appendixes** provide students with an essential algebra review, an outline for report writing, Bayes' theorem, extensive reference tables, a glossary, and answers to all quiz questions, all odd-numbered exercises, selected even-numbered exercises, and an alternate method for using the standard normal distribution.
- The **Applying the Concepts** feature is included in all sections and gives students an opportunity to think about the new concepts and apply them to hypothetical examples and scenarios similar to those found in newspapers, magazines, and radio and television news programs.

Changes in the Eighth Edition

Overall

- Added over 30 new Examples and 250 new Exercises throughout the book.
- Chapter summaries were revised into bulleted paragraphs representing each section from the chapter.
- New Historical Notes and Interesting facts have been added throughout the book.

Chapter 1

Updated and added new Speaking of Statistics. Revised the definition of nominal level of measurement.

Chapter 6

Revised presentation for finding areas under the standard normal distribution curve. New figures created to clarify explanations for steps in the Central Limit Theorem.

Chapter 7

Changed section 7.1 to Confidence Intervals for the Mean When σ is Known. Maximum error of the estimate has been updated to the margin of error. Updated the Formula for the Confidence Interval of the Mean for a Specific α to include when σ is Known. Added assumptions for Finding a Confidence Interval for a Mean When σ is Known. Revised the explanation for rounding up when determining sample size. Added assumptions for Finding a Confidence Interval for a Mean when σ is Unknown. Added assumptions for Finding a Confidence Interval for a Population Proportion. Added assumptions for Finding a Confidence Interval for a Variance or Standard Deviation.

Chapter 8

Added assumptions for the z Test for a Mean When σ Is Known. Added assumptions for the t Test for a Mean When σ Is Unknown. Added assumptions for Testing a Proportion.

Chapter 9

Revised the assumptions for the z Test to Determine the Difference Between Two Means. Added that it will be assumed that variances are not equal when using a t test to test the difference between means when the two samples are independent and when the samples are taken from two normally or approximately normally distributed populations. Added assumptions for the t Test for Two Independent Means When σ_1 and σ_2 Are Unknown. Added assumptions for the t Test for Two Means When the Samples Are Dependent. Added assumptions for the z Test for Two Proportions. Revised the assumptions for Testing the Difference Between Two Variables.

Chapter 10

Added assumptions for the Correlation Coefficient. Residuals, are now covered in full with figures illustrating different examples of Residual Plots.

Acknowledgments

It is important to acknowledge the many people whose contributions have gone into the Eighth Edition of *Elementary Statistics*. Very special thanks are due to Jackie Miller of The Ohio State University for her provision of the Index of Applications, her exhaustive accuracy check of the page proofs, and her general availability and advice concerning all matters statistical. The Technology Step by Step sections were provided by Gerry Moultine of Northwood University (MINITAB), John Thomas of College of Lake County (Excel), and Michael Keller of St. Johns River Community College (TI-83 Plus and TI-84 Plus).

I would also like to thank Diane P. Cope for providing the new exercises; Kelly Jackson for writing the new Data Projects; and Sally Robinson for error checking, adding technology-accurate answers to the answer appendix, and writing the Solutions Manuals.

Finally, at McGraw-Hill Higher Education, thanks to John Osgood, Sponsoring Editor; Adam Fischer, Developmental Editor; Kevin Ernzen, Marketing Manager; Vicki Krug, Project Manager; and Sandra Schnee, Senior Media Project Manager.

Allan G. Bluman

Special thanks for their advice and recommendations for revisions found in the Eighth Edition go to

Rosalie Abraham, *Florida State College, South Campus*

James Ball, *Indiana State University*

Luis Beltran, *Miami Dade College*

Abraham Biggs, *Broward College*

Melissa Bingham, *University of Wisconsin–Lacrosse*

Don Brown, *Macon State College*

Richard Carney, *Camden County College*

Joe Castillo, *Broward College*

James Cook, *Belmont University*

Rosemary Danaher, *Sacred Heart University*

Gregory Davis, *University of Wisconsin–Green Bay*

Hemangini Deshmukh, *Mercy Hurst College*

Abdulaziz Elfessi, *University of Wisconsin–Lacrosse*

Nancy Eschen, *Florida State College, South Campus*

Elaine Fitt, *Bucks County Community College*

David Gurney, *Southeastern Louisiana University*

John Todd Hammond, *Truman State University*

Willard Hannon, *Las Positas College*

James Helmreich, *Marist College*

Dr. James Hodge, *Mountain State University*

Kelly Jackson, *Camden County College*

Rose Jenkins, *Midlands Technical College*

June Jones, *Macon State College*

Grazyna Kamburowska, *State University College–Oneonta*

Jong Sung Kim, *Portland State University*

Janna Liberant, *Rockland Community College*

Scott McClintock, *West Chester University of Pennsylvania*

James Meyer, *University of Wisconsin–Green Bay*

David Milazzo, *Niagara County Community College–Sanborn*

Tommy Minton, *Seminole Community College*

Jason Molitierno, *Sacred Heart University*

Barry Monk, *Macon State College*

Carla Monticelli, *Camden County College*

Lyn Noble, *Florida State College, South Campus*

Jeanne Osborne, *Middlesex County College*

Ronald Persky, *Christopher Newport University*

Blanche Presley, *Macon State College*

William Radulovich, *Florida State College, South Campus*

Azar Raiszadeh, *Chattanooga State College*

Kandethody Ramachandran, *Hillsborough Community College–Brandon*

Dave Reineke, *University of Wisconsin–Lacrosse*

Vicki Schell, *Pensacola Junior College*

James Seibert, *Regis University*

Lee Seltzer, *Florida State College, South Campus*

Christine Tirella, *Niagara County Community College–Sanborn*

Christina Vertullo, *Marist College*

Jen-Ting Wang, *State University College–Oneonta*

Xubo Wang, *Macon State College*

Yajni Warnapala, *Roger Williams University*

Robert White, *Allan Hancock College*

Bridget Young, *Suffolk County Community College*

Bashar Zogheib, *Nova Southeastern College*

Guided Tour: Features and Supplements

Each chapter begins with an **outline** and a list of **learning objectives.** The objectives are repeated at the beginning of each section to help students focus on the concepts presented within that section.

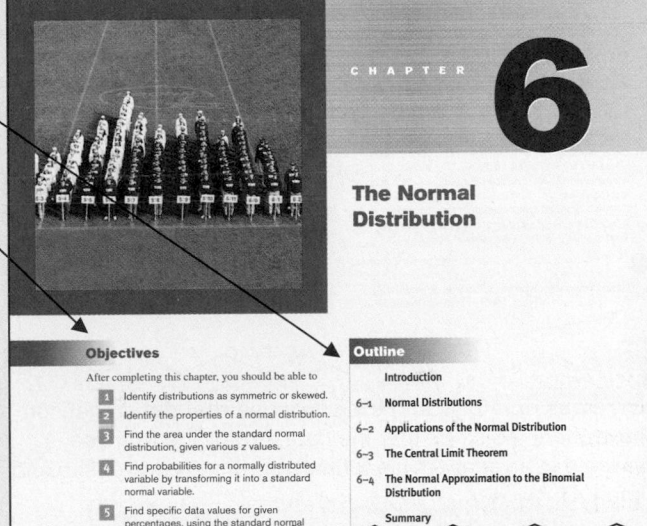

CHAPTER 6

The Normal Distribution

Objectives

After completing this chapter, you should be able to

1. Identify distributions as symmetric or skewed.
2. Identify the properties of a normal distribution.
3. Find the area under the standard normal distribution, given various z values.
4. Find probabilities for a normally distributed variable by transforming it into a standard normal variable.
5. Find specific data values for given percentages, using the standard normal ...

Outline

Introduction

6–1 Normal Distributions

6–2 Applications of the Normal Distribution

6–3 The Central Limit Theorem

6–4 The Normal Approximation to the Binomial Distribution

Summary

The outline and learning objectives are followed by a feature titled **Statistics Today,** in which a **real-life problem** shows students the relevance of the material in the chapter. This problem is subsequently solved near the end of the chapter by using the statistical techniques presented in the chapter.

592   Chapter 11 Other Chi-Square Tests

Statistics Today

Statistics and Heredity

An Austrian monk, Gregor Mendel (1822–1884), studied genetics, and its principles are the foundation for modern genetics. Mendel used his spare time to grow a variety of peas at the monastery. One of his many experiments involved crossbreeding peas that had smooth yellow seeds with peas that had wrinkled green seeds. He noticed that the results occurred with regularity. That is, some of the offspring had smooth yellow seeds, some had smooth green seeds, some had wrinkled yellow seeds, and some had wrinkled green seeds. Furthermore, after several experiments, the percentages of each type seemed to remain approximately the same. Mendel formulated his theory based on the assumption of dominant and recessive traits and tried to predict the results. He then crossbred his peas and examined 556 seeds over the next generation.

Finally, he compared the actual results with the theoretical results to see if his theory was correct. To do this, he used a "simple" chi-square test, which is explained in this chapter. See Statistics Today—Revisited at the end of this chapter.

Source: J. Hodges, Jr., D. Krech, and R. Crutchfield, *Stat Lab, An Empirical Introduction to Statistics* (New York: McGraw-Hill), pp. 228–229. Used with permission.

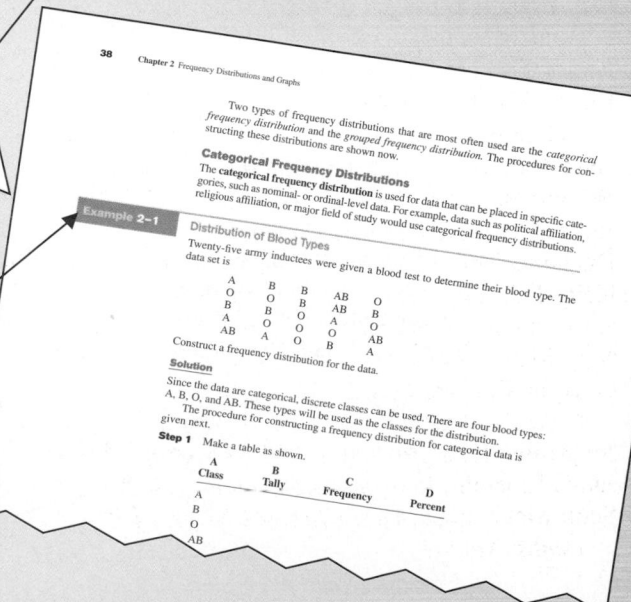

38   Chapter 2 Frequency Distributions and Graphs

Two types of frequency distributions that are most often used are the *categorical frequency distribution* and the *grouped frequency distribution.* The procedures for constructing these distributions are shown now.

Categorical Frequency Distributions

The **categorical frequency distribution** is used for data that can be placed in specific categories, such as nominal- or ordinal-level data. For example, data such as political affiliation, religious affiliation, or major field of study would use categorical frequency distributions.

Example 2–1

Distribution of Blood Types

Twenty-five army inductees were given a blood test to determine their blood type. The data set is

A	B	B	AB	O
O	O	B	AB	O
B	B	B	B	O
A	O	A	O	B
AB	A	O	A	AB
		O	B	A

Construct a frequency distribution for the data.

Solution

Since the data are categorical, discrete classes can be used. There are four blood types: A, B, O, and AB. These types will be used as the classes for the distribution.

The procedure for constructing a frequency distribution for categorical data is given next.

Step 1 Make a table as shown.

A Class	B Tally	C Frequency	D Percent
A			
B			
O			
AB			

Over 300 **examples** with detailed solutions serve as models to help students solve problems on their own. Examples are solved by using a step by step explanation, and illustrations provide a clear display of results for students.

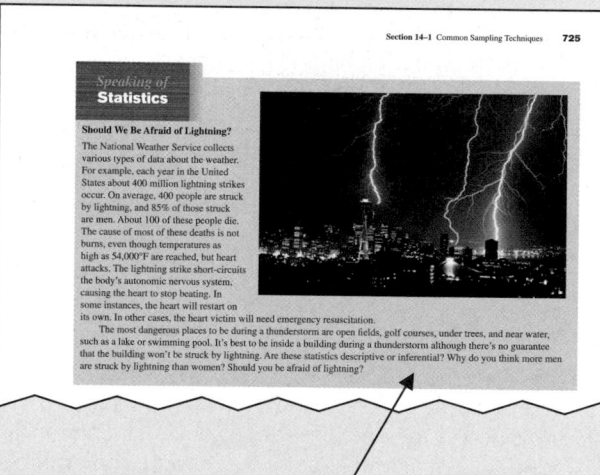

Exercises 8–2

For Exercises 1 through 13, perform each of the following steps.

a. State the hypotheses and identify the claim.
b. Find the critical value(s).
c. Compute the test value.
d. Make the decision.
e. Summarize the results.

Use diagrams to show the critical region (or regions), and use the traditional method of hypothesis testing unless otherwise specified.

1. **Warming and Ice Melt** The average depth of the Hudson Bay is 305 feet. Climatologists were interested in seeing if the effects of warming and ice melt were affecting the water level. Fifty-five measurements over a period of weeks yielded a sample mean of 306.2 feet. The population variance is known to be 3.57. Can it be concluded at the 0.05 level of significance that the average depth has increased? Is there evidence of what caused this to happen?

Source: *World Almanac and Book of Facts 2010.*

2. **Credit Card Debt** It has been reported that the average credit card debt for college seniors at the college book store for a specific college is $3262. The student senate at a large university feels that their seniors have a debt much less than this, so it conducts a study of 50 randomly selected seniors and finds that the average debt is $2995, and the population standard deviation is $1100. With $\alpha = 0.05$, is the student senate correct?

3. **Revenue of Large Businesses** A researcher estimates that the average revenue of the largest businesses in the United States is greater than $24 billion. A sample of 50 companies is selected, and the revenues (in billions of

8–24

dollars) are shown. At $\alpha = 0.05$, is there enough evidence to support the researcher's claim? Assume $\sigma = 28.7$.

178	122	91	44	35
61	56	46	20	32
30	28	28	20	27
29	16	16	19	15
41	38	36	15	25
31	30	19	19	19
24	16	15	15	19
25	25	18	14	15
24	23	17	17	22
22	21	20	17	20

Source: *New York Times Almanac.*

4. **Moviegoers** The average "moviegoer" sees 8.5 movies a year. A *moviegoer* is defined as a person who sees at least one movie in a theater in a 12-month period. A random sample of 40 moviegoers from a large university revealed that the average number of movies seen per person was 9.6. The population standard deviation is 3.2 movies. At the 0.05 level of significance, can it be concluded that this represents a difference from the national average?

Source: *MPAA Study.*

5. **Nonparental Care** According to the *Digest of Educational Statistics*, a certain group of preschool children under the age of one year each spends an average of 30.9 hours per week in nonparental care. A study of state university center-based programs indicated that a random sample of 32 infants spent an average of 32.1 hours per week in their care. The standard deviation of the population is 3.6 hours. At $\alpha = 0.01$ is there sufficient evidence to conclude that the sample mean differs from the national mean?

Source: www.nces.ed.gov

Numerous examples and exercises use **real data.** The icon shown here indicates that the data set for the exercise is available in a variety of file formats on the text's website and Data CD.

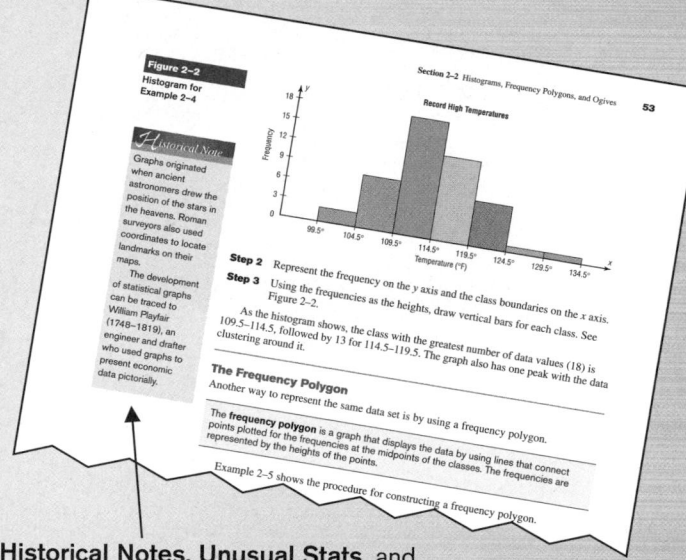

Procedure Table

Testing the Difference Between Means for Dependent Samples

Step 1 State the hypotheses and identify the claim.
Step 2 Find the critical value(s).
Step 3 Compute the test value.

a. Make a table, as shown.

X_1	X_2	A $D = X_1 - X_2$	B $D^2 = (X_1 - X_2)^2$
⋮	⋮	$\Sigma D =$ _____	$\Sigma D^2 =$ _____

b. Find the differences and place the results in column A.
$$D = X_1 - X_2$$

c. Find the mean of the differences.
$$\bar{D} = \frac{\Sigma D}{n}$$

d. Square the differences and place the results in column B. Complete the table.
$$D^2 = (X_1 - X_2)^2$$

e. Find the standard deviation of the differences.
$$s_D = \sqrt{\frac{n \Sigma D^2 - (\Sigma D)^2}{n(n - 1)}}$$

f. Find the test value.

Numerous **Procedure Tables** summarize processes for students' quick reference. All use the step by step method.

Section 14–1 Common Sampling Techniques **725**

Speaking of Statistics

Should We Be Afraid of Lightning?

The National Weather Service collects various types of data about the weather. For example, each year in the United States about 400 million lightning strikes occur. On average, 400 people are struck by lightning, and 85% of those struck are men. About 100 of these people die. The cause of most of these deaths is not burns, even though temperatures as high as 54,000°F are reached, but heart attacks. The lightning strike short-circuits the body's autonomic nervous system, causing the heart to stop beating. In some instances, the heart will restart on its own. In other cases, the heart victim will need emergency resuscitation.

The most dangerous places to be during a thunderstorm are open fields, golf courses, under trees, and near water, such as a lake or swimming pool. It's best to be inside a building during a thunderstorm although there's no guarantee that the building won't be struck by lightning. Are these statistics descriptive or inferential? Why do you think more men are struck by lightning than women? Should you be afraid of lightning?

The **Speaking of Statistics** sections invite students to think about poll results and other statistics-related news stories in another connection between statistics and the real world.

Rules and definitions are set off for easy referencing by the student.

Figure 2–2
Histogram for Example 2–4

Section 2–2 Histograms, Frequency Polygons, and Ogives **53**

Record High Temperatures

Step 2 Represent the frequency on the y axis and the class boundaries on the x axis.
Step 3 Using the frequencies as the heights, draw vertical bars for each class. See Figure 2–2.

As the histogram shows, the class with the greatest number of data values (18) is 109.5–114.5, followed by 13 for 114.5–119.5. The graph also has one peak with the data clustering around it.

The Frequency Polygon

Another way to represent the same data set is by using a frequency polygon.

The **frequency polygon** is a graph that displays the data by using lines that connect points plotted for the frequencies at the midpoints of the classes. The frequencies are represented by the heights of the points.

Example 2–5 shows the procedure for constructing a frequency polygon.

Historical Notes, Unusual Stats, and **Interesting Facts,** located in the margins, make statistics come alive for the reader.

---

418 Chapter 8 Hypothesis Testing

Again, remember that nothing is being proved true or false. The statistician is only stating that there is or is not enough evidence to say that a claim is *probably* true or false. As noted previously, the only way to prove something would be to use the entire population under study, and usually this cannot be done, especially when the population is large.

P-Value Method for Hypothesis Testing

Statisticians usually test hypotheses at the common α levels of 0.05 or 0.01 and sometimes at 0.10. Recall that the choice of the level depends on the seriousness of the type I error. Besides listing an α value, many computer statistical packages give a P-value for hypothesis tests.

The **P-value** (or probability value) is the probability of getting a sample statistic (such as the mean) or a more extreme sample statistic in the direction of the alternative hypothesis when the null hypothesis is true.

Critical Thinking sections at the end of each chapter challenge students to apply what they have learned to new situations. The problems presented are designed to deepen conceptual understanding and/or to extend topical coverage.

At the end of appropriate sections, **Technology Step by Step** boxes show students how to use MINITAB, the TI-83 Plus and TI-84 Plus graphing calculators, and Excel to solve the types of problems covered in the section. Instructions are presented in numbered steps, usually in the context of examples–including examples from the main part of the section. Numerous computer or calculator screens are displayed, showing intermediate steps as well as the final answer.

---

248 Chapter 4 Probability and Counting Rules

Critical Thinking Challenges

1. **Con Man Game** Consider this problem: A con man has 3 coins. One coin has been specially made and has a head on each side. A second coin has been specially made, and on each side it has a tail. Finally, a third coin has a head and a tail on it. All coins are of the same denomination. The con man places the 3 coins in his pocket, selects one, and shows you one side. It is heads. He is willing to bet you even money that it is the two-headed coin. His reasoning is that it can't be the two-tailed coin since a head is showing; therefore, there is a 50-50 chance of it being the two-headed coin. Would you take the bet? (*Hint:* See Exercise 1 in Data Projects.)

2. **de Méré Dice Game** Chevalier de Méré won money when he bet unsuspecting patrons that in 4 rolls of 1 die, he could get at least one 6; but he lost money when he bet that in 24 rolls of 2 dice, he could get at least a double 6. Using the probability rules, find the probability of each event and explain why he won the majority of the time on the first game but lost the majority of the time when playing the second game. (*Hint:* Find the probabilities of losing each game and subtract from 1.)

For example, suppose there were 3 people in the room. The probability that each had a different birthday would be

$$\frac{365}{365} \cdot \frac{364}{365} \cdot \frac{363}{365} = \frac{{}_{365}P_3}{365^3} = 0.992$$

Hence, the probability that at least 2 of the 3 people will have the same birthday will be

$$1 - 0.992 = 0.008$$

Hence, for k people, the formula is

$P(\text{at least 2 people have the same birthday})$
$= 1 - \frac{{}_{365}P_k}{365^k}$

Using your calculator, complete the table and verify that for at least a 50% chance of 2 people having the same birthday, 23 or more people will be needed.

Number of	Probability that at least 2 have the ...

Technology *Step by Step*

MINITAB
Step by Step

Data

5	29	34	44	45
63	68	74	74	81
88	91	97	98	113
110	151	158		

Determining Normality

There are several ways in which statisticians test a data set for normality. Four are shown here.

Construct a Histogram

Inspect the histogram for shape.

1. Enter the data in the first column of a new worksheet. Name the column Inventory.

2. Use **Stat>Basic Statistics>Graphical Summary** presented in Section 3–3 to create the histogram. Is it symmetric? Is there a single peak?

Check for Outliers

Inspect the boxplot for outliers. There are no outliers in this graph. Furthermore, the box is in the middle of the range, and the median is in the middle of the box. Most likely this is not a skewed distribution either.

Calculate The Pearson Coefficient of Skewness

The measure of skewness in the graphical summary is not the same as the Pearson coefficient. Use the calculator and the formula.

$$PC = \frac{3(\bar{X} - median)}{s}$$

3. Select **Calc>Calculator**, then type **PC** in the text box for Store result in:.
4. Enter the expression: **3*(MEAN(C1)−MEDI(C1))/(STDEV(C1))**. Make sure you get all the parentheses in the right place!
5. Click [OK]. The result, 0.148318, will be stored in the first row of C2 named PC. Since it is smaller than +1, the distribution is not skewed.

Construct a Normal Probability Plot

6. Select **Graph>Probability Plot**, then Single and click [OK].
7. D... click C1 ... to select ... to be gr...

Applying the Concepts **10–4**

More Math Means More Money

In a study to determine a person's yearly income 10 years after high school, it was found that the two biggest predictors are number of math courses taken and number of hours worked per week during a person's senior year of high school. The multiple regression equation generated from a sample of 20 individuals is

$$y' = 6000 + 4540x_1 + 1290x_2$$

Let x_1 represent the number of mathematics courses taken and x_2 represent hours worked. The correlation between income and mathematics courses is 0.63. The correlation between income and hours worked is 0.84, and the correlation between mathematics courses and hours worked is 0.31. Use this information to answer the following questions.

1. What is the dependent variable?
2. What are the independent variables?
3. What are the multiple regression assumptions?
4. Explain what 4540 and 1290 in the equation tell us.
5. What is the predicted income if a person took 8 math classes and worked 20 hours per week during her or his senior year in high school?
6. What does a multiple correlation coefficient of 0.77 mean?
7. Compute R^2.
8. Compute the adjusted R^2.
9. Would the equation be considered a good predictor of income?
10. What are your conclusions about the relationship among courses taken, hours worked, and yearly income?

See page 590 for the answers.

Applying the Concepts are exercises found at the end of each section to reinforce the concepts explained in the section. They give the student an opportunity to think about the concepts and apply them to hypothetical examples similar to real-life ones found in newspapers, magazines, and professional journals. Most contain open-ended questions–questions that require interpretation and may have more than one correct answer. These exercises can also be used as classroom discussion topics for instructors who like to use this type of teaching technique.

Data Projects

1. **Business and Finance** Use 30 stocks classified as the Dow Jones industrials as the sample. Note the amount each stock has gained or lost in the last quarter. Compute the mean and standard deviation for the data set. Compute the 95% confidence interval for the mean and the 95% confidence interval for the standard deviation. Compute the percentage of stocks that had a gain in the last quarter. Find a 95% confidence interval for the percentage of stocks with a gain.

2. **Sports and Leisure** Use the top home run hitter from each major league baseball team as the data set. Find the mean and the standard deviation for the number of home runs hit by the top hitter on each team. Find a 95% confidence interval for the mean number of home runs hit.

3. **Technology** Use the data collected in data project 3 of Chapter 2 regarding song lengths. Select a specific genre, and compute the percentage of songs in the sample that are of that genre. Create a 95% confidence interval for the true percentage. Use the entire music library, and find the population percentage of the library with that genre. Does the population percentage fall within the confidence interval?

4. **Health and Wellness** Use your class as the sample. Have each student take her or his temperature on a healthy day. Compute the mean and standard deviation for the sample. Create a 95% confidence interval for the mean temperature. Does the confidence interval obtained support the long-held belief that the average body temperature is 98.6°F?

5. **Politics and Economics** Select five political polls and note the margin of error, sample size, and percent favoring the candidate for each. For each poll, determine the level of confidence that must have been used to obtain the margin of error given, knowing the percent favoring the candidate and number of participants. Is there a pattern that emerges?

6. **Your Class** Have each student compute his or her body mass index (BMI) (703 times weight in pounds, divided by the quantity height in inches squared). Find the mean and standard deviation for the data set. Compute a 95% confidence interval for the mean BMI of a student. A BMI score over 30 is considered obese. Does the confidence interval indicate that the mean for BMI could be in the obese range?

Data Projects, which appear at the end of each chapter, further challenge students' understanding and application of the material presented in the chapter. Many of these require the student to gather, analyze, and report on real data.

Multimedia Supplements

Connect—www.connectstatistics.com

McGraw-Hill's Connect is a complete online homework system for mathematics and statistics. Instructors can assign textbook-specific content from over 40 McGraw-Hill titles as well as customize the level of feedback students receive, including the ability to have students show their work for any given exercise. Assignable content includes an array of videos and other multimedia tools along with algorithmic exercises, providing study tools for students with many different learning styles.

Within Connect, a diagnostic assessment tool powered by ALEKS™ is available to measure student preparedness and provide detailed reporting and personalized remediation. Connect also helps ensure consistent assignment delivery across several sections through a course administration function and makes sharing courses with other instructors easy.

For more information, visit the book's website (www.mhhe.com/bluman) or contact your local McGraw-Hill sales representative (www.mhhe.com/rep).

ALEKS®

ALEKS—www.aleks.com

ALEKS (**A**ssessment and **LE**arning in **K**nowledge **S**paces) is a dynamic online learning system for mathematics education, available over the Web 24/7. ALEKS assesses students, accurately determines their knowledge, and then guides them to the material that they are most ready to learn. With a variety of reports, Textbook Integration Plus, quizzes, and homework assignment capabilities, ALEKS offers flexibility and ease of use for instructors.

- ALEKS uses artificial intelligence to determine exactly what each student knows and is ready to learn. ALEKS remediates student gaps and provides highly efficient learning and improved learning outcomes.

- ALEKS is a comprehensive curriculum that aligns with syllabi or specified textbooks. When it is used in conjunction with McGraw-Hill texts, students also receive links to text-specific videos, multimedia tutorials, and textbook pages.

- Textbook Integration Plus allows ALEKS to be automatically aligned with syllabi or specified McGraw-Hill textbooks with instructor-chosen dates, chapter goals, homework, and quizzes.

- ALEKS with AI-2 gives instructors increased control over the scope and sequence of student learning. Students using ALEKS demonstrate a steadily increasing mastery of the content of the course.

- ALEKS offers a dynamic classroom management system that enables instructors to monitor and direct student progress toward mastery of course objectives.

ALEKS Prep for Statistics

ALEKS Prep for Statistics can be used during the beginning of the course to prepare students for future success and to increase retention and pass rates. Backed by two decades of National Science Foundation–funded research, ALEKS interacts with students much as a human tutor, with the ability to precisely assess a student's preparedness and provide instruction on the topics the student is ready to learn.

ALEKS Prep for Statistics

- Assists students in mastering core concepts that should have been learned prior to entering the present course.

- Frees up lecture time for instructors, allowing more time to focus on current course material and not review material.

- Provides up to six weeks of remediation and intelligent tutorial help to fill in students' individual knowledge gaps.

TEGRITY–http://tegritycampus.mhhe.com

Tegrity Campus is a service that makes class time available all the time by automatically capturing every lecture in a searchable format for students to review when they study and complete assignments. With a simple one-click start and stop process, you capture all computer screens and corresponding audio. Students replay any part of any class with easy-to-use browser-based viewing on a PC or Mac.

Educators know that the more students can see, hear, and experience class resources, the better they learn. With Tegrity Campus, students quickly recall key moments by using Tegrity Campus's unique search feature. This search helps students efficiently find what they need, when they need it across an entire semester of class recordings. Help turn all your students' study time into learning moments immediately supported by your lecture.

To learn more about Tegrity watch a 2 minute Flash demo at http://tegritycampus.mhhe.com

Electronic Textbook

CourseSmart is a new way for faculty to find and review eTextbooks. It's also a great option for students who are interested in accessing their course materials digitally and saving money. CourseSmart offers thousands of the most commonly adopted textbooks across hundreds of courses from a wide variety of higher education publishers. It is the only place for faculty to review and compare the full text of a textbook online, providing immediate access without the environmental impact of requesting a print exam copy. At CourseSmart, students can save up to 50% off the cost of a print book, reduce the impact on the environment, and gain access to powerful Web tools for learning including full text search, notes and highlighting, and e-mail tools for sharing notes between classmates. www.CourseSmart.com

MegaStat®

MegaStat® is a statistical add-in for Microsoft Excel, handcrafted by J. B. Orris of Butler University. When MegaStat is installed it appears as a menu item on the Excel menu bar and allows you to perform statistical analysis on data in an Excel workbook. ELEMENTARY STATISTICS: A BRIEF VERSION requires the use of this MegaStat add-in for Excel only for those Excel Technology Step by Step operations in the text that Excel would otherwise not have been able to perform. The MegaStat plug-in can be found at www.mhhe.com/bluman.

Computerized Test Bank (CTB) Online (instructors only)

The computerized test bank contains a variety of questions, including true/false, multiple-choice, short-answer, and short problems requiring analysis and written answers. The testing material is coded by type of question and level of difficulty. The Brownstone Diploma® system enables you to efficiently select, add, and organize questions, such as by type of question or by level of difficulty. It also allows for printing tests along with answer keys as well as editing the original questions, and it is available for Windows and Macintosh systems. Printable tests and a print version of the test bank can also be found on the website.

Lecture Videos

Lecture videos introduce concepts, definitions, theorems, formulas, and problem-solving procedures to help students better comprehend the topic at hand. These videos are closed-captioned for the hearing-impaired, are subtitled in Spanish, and meet the Americans with Disabilities Act Standards for Accessible Design. They can be found online at www.mhhe.com/bluman.

Exercise Videos

In these videos the instructor works through selected exercises, following the solution methodology employed in the text. Also included are tutorials for using the TI-83 Plus and TI-84 Plus calculators, Excel, and MINITAB, presented in an engaging format for students. These videos are closed-captioned for the hearing-impaired, are subtitled in Spanish, and meet the Americans with Disabilities Act Standards for Accessible Design. They can be found online at www.mhhe.com/bluman.

MINITAB Student Release 14

The student version of MINITAB statistical software is available with copies of the text. Ask your McGraw-Hill representative for details.

SPSS Student Version for Windows

A student version of SPSS statistical software is available with copies of this text. Consult your McGraw-Hill representative for details.

Print Supplements

Annotated Instructor's Edition (instructors only)

The Annotated Instructor's Edition contains answers to all exercises and tests. The answers to most questions are printed in red next to each problem. Answers not appearing on the page can be found in the Answer Appendix at the end of the book.

Instructor's Solutions Manual (instructors only)

By Sally Robinson of South Plains College, this manual includes worked-out solutions to all the exercises in the text and answers to all quiz questions. This manual can be found online at www.mhhe.com/bluman.

Student's Solutions Manual

By Sally Robinson of South Plains College, this manual contains detailed solutions to all odd-numbered text problems and answers to all quiz questions.

MINITAB 14 Manual

This manual provides the student with how-to information on data and file management, conducting various statistical analyses, and creating presentation-style graphics while following each text chapter.

TI-83 Plus and TI-84 Plus Graphing Calculator Manual

This friendly, practical manual teaches students to learn about statistics and solve problems by using these calculators while following each text chapter.

Excel Manual

This workbook, specially designed to accompany the text, provides additional practice in applying the chapter concepts while using Excel.

Index of Applications

CHAPTER **7**

Confidence Intervals and Sample Size

CHAPTER **14**

Sampling and Simulation

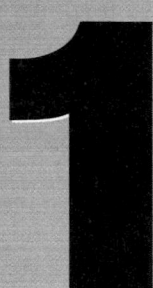

CHAPTER

The Nature of Probability and Statistics

Objectives

After completing this chapter, you should be able to

1 Demonstrate knowledge of statistical terms.

2 Differentiate between the two branches of statistics.

3 Identify types of data.

4 Identify the measurement level for each variable.

5 Identify the four basic sampling techniques.

6 Explain the difference between an observational and an experimental study.

7 Explain how statistics can be used and misused.

8 Explain the importance of computers and calculators in statistics.

Outline

Are We Improving Our Diet?

It has been determined that diets rich in fruits and vegetables are associated with a lower risk of chronic diseases such as cancer. Nutritionists recommend that Americans consume five or more servings of fruits and vegetables each day. Several researchers from the Division of Nutrition, the National Center for Chronic Disease Control and Prevention, the National Cancer Institute, and the National Institutes of Health decided to use statistical procedures to see how much progress is being made toward this goal.

The procedures they used and the results of the study will be explained in this chapter. See Statistics Today—Revisited at the end of this chapter.

Introduction

You may be familiar with probability and statistics through radio, television, newspapers, and magazines. For example, you may have read statements like the following found in newspapers.

- Nearly one in seven U.S. families are struggling with bills from medical expenses even though they have health insurance. (Source: *Psychology Today.*)
- Eating 10 grams of fiber a day reduces the risk of heart attack by 14%. (Source: *Archives of Internal Medicine, Reader's Digest.*)
- Thirty minutes of exercise two or three times each week can raise HDLs by 10% to 15%. (Source: *Prevention.*)
- In 2008, the average credit card debt for college students was $3173. (Source: Newser.com.)
- About 15% of men in the United States are left-handed and 9% of women are left-handed. (Source: Scripps Survey Research Center.)
- The median age of people who watch the *Tonight Show* with Jay Leno is 48.1. (Source: Nielsen Media Research.)

Statistics is used in almost all fields of human endeavor. In sports, for example, a statistician may keep records of the number of yards a running back gains during a football game, or the number of hits a baseball player gets in a season. In other areas, such as public health, an administrator might be concerned with the number of residents who contract a new strain of flu virus during a certain year. In education, a researcher might want to know if new methods of teaching are better than old ones. These are only a few examples of how statistics can be used in various occupations.

Furthermore, statistics is used to analyze the results of surveys and as a tool in scientific research to make decisions based on controlled experiments. Other uses of statistics include operations research, quality control, estimation, and prediction.

Statistics is the science of conducting studies to collect, organize, summarize, analyze, and draw conclusions from data.

Students study statistics for several reasons:

1. Like professional people, you must be able to read and understand the various statistical studies performed in your fields. To have this understanding, you must be knowledgeable about the vocabulary, symbols, concepts, and statistical procedures used in these studies.

2. You may be called on to conduct research in your field, since statistical procedures are basic to research. To accomplish this, you must be able to design experiments; collect, organize, analyze, and summarize data; and possibly make reliable predictions or forecasts for future use. You must also be able to communicate the results of the study in your own words.

3. You can also use the knowledge gained from studying statistics to become better consumers and citizens. For example, you can make intelligent decisions about what products to purchase based on consumer studies, about government spending based on utilization studies, and so on.

These reasons can be considered some of the goals for studying statistics.

It is the purpose of this chapter to introduce the goals for studying statistics by answering questions such as the following:

What are the branches of statistics?

What are data?

How are samples selected?

1–1   Descriptive and Inferential Statistics

Objective 1

Demonstrate knowledge of statistical terms.

To gain knowledge about seemingly haphazard situations, statisticians collect information for *variables,* which describe the situation.

A **variable** is a characteristic or attribute that can assume different values.

Data are the values (measurements or observations) that the variables can assume. Variables whose values are determined by chance are called **random variables.**

Suppose that an insurance company studies its records over the past several years and determines that, on average, 3 out of every 100 automobiles the company insured were involved in accidents during a 1-year period. Although there is no way to predict the specific automobiles that will be involved in an accident (random occurrence), the company can adjust its rates accordingly, since the company knows the general pattern over the long run. (That is, on average, 3% of the insured automobiles will be involved in an accident each year.)

A collection of data values forms a **data set.** Each value in the data set is called a **data value** or a **datum.**

Objective 2

Differentiate between the two branches of statistics.

Data can be used in different ways. The body of knowledge called statistics is sometimes divided into two main areas, depending on how data are used. The two areas are

1. Descriptive statistics

2. Inferential statistics

Descriptive statistics consists of the collection, organization, summarization, and presentation of data.

In *descriptive statistics* the statistician tries to describe a situation. Consider the national census conducted by the U.S. government every 10 years. Results of this census give you the average age, income, and other characteristics of the U.S. population. To obtain this information, the Census Bureau must have some means to collect relevant data. Once data are collected, the bureau must organize and summarize them. Finally, the bureau needs a means of presenting the data in some meaningful form, such as charts, graphs, or tables.

The second area of statistics is called *inferential statistics*.

Inferential statistics consists of generalizing from samples to populations, performing estimations and hypothesis tests, determining relationships among variables, and making predictions.

Here, the statistician tries to make inferences from *samples* to *populations*. Inferential statistics uses **probability,** i.e., the chance of an event occurring. You may be familiar with the concepts of probability through various forms of gambling. If you play cards, dice, bingo, or lotteries, you win or lose according to the laws of probability. Probability theory is also used in the insurance industry and other areas.

It is important to distinguish between a sample and a population.

A **population** consists of all subjects (human or otherwise) that are being studied.

Most of the time, due to the expense, time, size of population, medical concerns, etc., it is not possible to use the entire population for a statistical study; therefore, researchers use samples.

A **sample** is a group of subjects selected from a population.

If the subjects of a sample are properly selected, most of the time they should possess the same or similar characteristics as the subjects in the population. The techniques used to properly select a sample will be explained in Section 1–3.

An area of inferential statistics called **hypothesis testing** is a decision-making process for evaluating claims about a population, based on information obtained from samples. For example, a researcher may wish to know if a new drug will reduce the number of heart attacks in men over 70 years of age. For this study, two groups of men over 70 would be selected. One group would be given the drug, and the other would be given a placebo (a substance with no medical benefits or harm). Later, the number of heart attacks occurring in each group of men would be counted, a statistical test would be run, and a decision would be made about the effectiveness of the drug.

Statisticians also use statistics to determine *relationships* among variables. For example, relationships were the focus of the most noted study in the 20th century, "Smoking and Health," published by the Surgeon General of the United States in 1964. He stated that after reviewing and evaluating the data, his group found a definite relationship between smoking and lung cancer. He did not say that cigarette smoking actually causes lung cancer, but that there is a relationship between smoking and lung cancer. This conclusion was based on a study done in 1958 by Hammond and Horn. In this study, 187,783 men were observed over a period of 45 months. The death rate from

Speaking of
Statistics

Statistics and the New Planet

In the summer of 2005, astronomers announced the discovery of a new planet in our solar system. Astronomers have dubbed it Xena. They also discovered that it has a moon that is larger than Pluto.[1] Xena is about 9 billion miles from the Sun. (Some sources say 10 billion.) Its diameter is about 4200 miles. Its surface temperature has been estimated at −400°F, and it takes 560 years to circle the Sun.

How does Xena compare to the other planets? Let's look at the statistics.

Planet	Diameter (miles)	Distance from the Sun (millions of miles)	Orbital period (days)	Mean temperature (°F)	Number of moons
Mercury	3,032	36	88	333	0
Venus	7,521	67.2	224.7	867	0
Earth	7,926	93	365.2	59	1
Mars	4,222	141.6	687	−85	2
Jupiter	88,846	483.8	4,331	−166	63
Saturn	74,897	890.8	10,747	−220	47
Uranus	31,763	1,784.8	30,589	−320	27
Neptune	30,775	2,793.1	59,800	−330	13
Pluto[1]	1,485	3,647.2	90,588	−375	1

Source: NASA.
[1]Some astronomers no longer consider Pluto a planet.

With these statistics, we can make some comparisons. For example, Xena is about the size of the planet Mars, but it is over 21 times the size of Pluto. (Compare the volumes.) It takes about twice as long to circle the Sun as Pluto. What other comparisons can you make?

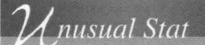
Unusual Stat

Twenty-nine percent of Americans want their boss's job.

lung cancer in this group of volunteers was 10 times as great for smokers as for nonsmokers.

Finally, by studying past and present data and conditions, statisticians try to make predictions based on this information. For example, a car dealer may look at past sales records for a specific month to decide what types of automobiles and how many of each type to order for that month next year.

Applying the Concepts **1–1**

Attendance and Grades

Read the following on attendance and grades, and answer the questions.

A study conducted at Manatee Community College revealed that students who attended class 95 to 100% of the time usually received an A in the class. Students who attended class

80 to 90% of the time usually received a B or C in the class. Students who attended class less than 80% of the time usually received a D or an F or eventually withdrew from the class.

Based on this information, attendance and grades are related. The more you attend class, the more likely it is you will receive a higher grade. If you improve your attendance, your grades will probably improve. Many factors affect your grade in a course. One factor that you have considerable control over is attendance. You can increase your opportunities for learning by attending class more often.

1. What are the variables under study?
2. What are the data in the study?
3. Are descriptive, inferential, or both types of statistics used?
4. What is the population under study?
5. Was a sample collected? If so, from where?
6. From the information given, comment on the relationship between the variables.

See page 33 for the answers.

1–2

Objective **3**

Identify types of data.

Variables and Types of Data

As stated in Section 1–1, statisticians gain information about a particular situation by collecting data for random variables. This section will explore in greater detail the nature of variables and types of data.

Variables can be classified as qualitative or quantitative. **Qualitative variables** are variables that can be placed into distinct categories, according to some characteristic or attribute. For example, if subjects are classified according to gender (male or female), then the variable *gender* is qualitative. Other examples of qualitative variables are religious preference and geographic locations.

Quantitative variables are numerical and can be ordered or ranked. For example, the variable *age* is numerical, and people can be ranked in order according to the value of their ages. Other examples of quantitative variables are heights, weights, and body temperatures.

Quantitative variables can be further classified into two groups: discrete and continuous. *Discrete variables* can be assigned values such as 0, 1, 2, 3 and are said to be countable. Examples of discrete variables are the number of children in a family, the number of students in a classroom, and the number of calls received by a switchboard operator each day for a month.

Discrete variables assume values that can be counted.

Continuous variables, by comparison, can assume an infinite number of values in an interval between any two specific values. Temperature, for example, is a continuous variable, since the variable can assume an infinite number of values between any two given temperatures.

Continuous variables can assume an infinite number of values between any two specific values. They are obtained by measuring. They often include fractions and decimals.

The classification of variables can be summarized as follows:

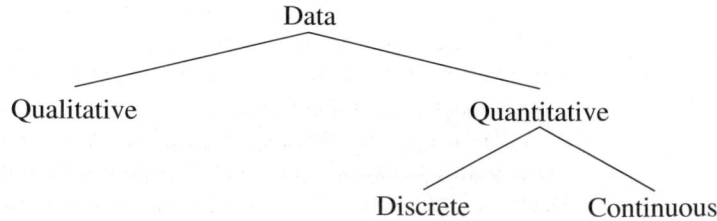

Since continuous data must be measured, answers must be rounded because of the limits of the measuring device. Usually, answers are rounded to the nearest given unit. For example, heights might be rounded to the nearest inch, weights to the nearest ounce, etc. Hence, a recorded height of 73 inches could mean any measure from 72.5 inches up to but not including 73.5 inches. Thus, the boundary of this measure is given as 72.5–73.5 inches. *Boundaries are written for convenience as 72.5–73.5 but are understood to mean all values up to but not including 73.5.* Actual data values of 73.5 would be rounded to 74 and would be included in a class with boundaries of 73.5 up to but not including 74.5, written as 73.5–74.5. As another example, if a recorded weight is 86 pounds, the exact boundaries are 85.5 up to but not including 86.5, written as 85.5–86.5 pounds. Table 1–1 helps to clarify this concept. The boundaries of a continuous variable are given in one additional decimal place and always end with the digit 5.

Table 1–1	Recorded Values and Boundaries	
Variable	**Recorded value**	**Boundaries**
Length	15 centimeters (cm)	14.5–15.5 cm
Temperature	86 degrees Fahrenheit (°F)	85.5–86.5°F
Time	0.43 second (sec)	0.425–0.435 sec
Mass	1.6 grams (g)	1.55–1.65 g

In addition to being classified as qualitative or quantitative, variables can be classified by how they are categorized, counted, or measured. For example, can the data be organized into specific categories, such as area of residence (rural, suburban, or urban)? Can the data values be ranked, such as first place, second place, etc.? Or are the values obtained from measurement, such as heights, IQs, or temperature? This type of classification—i.e., how variables are categorized, counted, or measured—uses **measurement scales,** and four common types of scales are used: nominal, ordinal, interval, and ratio.

Objective 4

Identify the measurement level for each variable.

The first level of measurement is called the *nominal level* of measurement. A sample of college instructors classified according to subject taught (e.g., English, history, psychology, or mathematics) is an example of nominal-level measurement. Classifying survey subjects as male or female is another example of nominal-level measurement. No ranking or order can be placed on the data. Classifying residents according to zip codes is also an example of the nominal level of measurement. Even though numbers are assigned as zip codes, there is no meaningful order or ranking. Other examples of nominal-level data are political party (Democratic, Republican, Independent, etc.), religion (Christianity, Judaism, Islam, etc.), and marital status (single, married, divorced, widowed, separated).

The **nominal level of measurement** classifies data into mutually exclusive (nonoverlapping) categories in which no order or ranking can be imposed on the data.

The next level of measurement is called the *ordinal level.* Data measured at this level can be placed into categories, and these categories can be ordered, or ranked. For example, from student evaluations, guest speakers might be ranked as superior, average, or poor. Floats in a homecoming parade might be ranked as first place, second place, etc. *Note that precise measurement of differences in the ordinal level of measurement* does not *exist.* For instance, when people are classified according to their build (small, medium, or large), a large variation exists among the individuals in each class.

Other examples of ordinal data are letter grades (A, B, C, D, F).

> The **ordinal level of measurement** classifies data into categories that can be ranked; however, precise differences between the ranks do not exist.

The third level of measurement is called the *interval level*. This level differs from the ordinal level in that precise differences do exist between units. For example, many standardized psychological tests yield values measured on an interval scale. IQ is an example of such a variable. There is a meaningful difference of 1 point between an IQ of 109 and an IQ of 110. Temperature is another example of interval measurement, since there is a meaningful difference of 1°F between each unit, such as 72 and 73°F. *One property is lacking in the interval scale: There is no true zero.* For example, IQ tests do not measure people who have no intelligence. For temperature, 0°F does not mean no heat at all.

> The **interval level of measurement** ranks data, and precise differences between units of measure do exist; however, there is no meaningful zero.

The final level of measurement is called the *ratio level*. Examples of ratio scales are those used to measure height, weight, area, and number of phone calls received. Ratio scales have differences between units (1 inch, 1 pound, etc.) and a true zero. In addition, the ratio scale contains a true ratio between values. For example, if one person can lift 200 pounds and another can lift 100 pounds, then the ratio between them is 2 to 1. Put another way, the first person can lift twice as much as the second person.

> The **ratio level of measurement** possesses all the characteristics of interval measurement, and there exists a true zero. In addition, true ratios exist when the same variable is measured on two different members of the population.

There is not complete agreement among statisticians about the classification of data into one of the four categories. For example, some researchers classify IQ data as ratio data rather than interval. Also, data can be altered so that they fit into a different category. For instance, if the incomes of all professors of a college are classified into the three categories of low, average, and high, then a ratio variable becomes an ordinal variable. Table 1–2 gives some examples of each type of data.

Table 1–2   Examples of Measurement Scales

Nominal-level data	Ordinal-level data	Interval-level data	Ratio-level data
Zip code	Grade (A, B, C, D, F)	SAT score	Height
Gender (male, female)	Judging (first place, second place, etc.)	IQ	Weight
Eye color (blue, brown, green, hazel)	Rating scale (poor, good, excellent)	Temperature	Time
Political affiliation	Ranking of tennis players		Salary
Religious affiliation			Age
Major field (mathematics, computers, etc.)			
Nationality			

Applying the Concepts **1–2**

Safe Travel

Read the following information about the transportation industry and answer the questions.

Transportation Safety

The chart shows the number of job-related injuries for each of the transportation industries for 1998.

Industry	Number of injuries
Railroad	4520
Intercity bus	5100
Subway	6850
Trucking	7144
Airline	9950

1. What are the variables under study?

2. Categorize each variable as quantitative or qualitative.

3. Categorize each quantitative variable as discrete or continuous.

4. Identify the level of measurement for each variable.

5. The railroad is shown as the safest transportation industry. Does that mean railroads have fewer accidents than the other industries? Explain.

6. What factors other than safety influence a person's choice of transportation?

7. From the information given, comment on the relationship between the variables.

See page 33 for the answers.

1–3

Objective 5

Identify the four basic sampling techniques.

Data Collection and Sampling Techniques

In research, statisticians use data in many different ways. As stated previously, data can be used to describe situations or events. For example, a manufacturer might want to know something about the consumers who will be purchasing his product so he can plan an effective marketing strategy. In another situation, the management of a company might survey its employees to assess their needs in order to negotiate a new contract with the employees' union. Data can be used to determine whether the educational goals of a school district are being met. Finally, trends in various areas, such as the stock market, can be analyzed, enabling prospective buyers to make more intelligent decisions concerning what stocks to purchase. These examples illustrate a few situations where collecting data will help people make better decisions on courses of action.

Data can be collected in a variety of ways. One of the most common methods is through the use of surveys. Surveys can be done by using a variety of methods. Three of the most common methods are the telephone survey, the mailed questionnaire, and the personal interview.

Telephone surveys have an advantage over personal interview surveys in that they are less costly. Also, people may be more candid in their opinions since there is no face-to-face contact. A major drawback to the telephone survey is that some people in the population will not have phones or will not answer when the calls are made; hence, not all people have a chance of being surveyed. Also, many people now have unlisted numbers and cell phones, so they cannot be surveyed. Finally, even the tone of the voice of the interviewer might influence the response of the person who is being interviewed.

Mailed questionnaire surveys can be used to cover a wider geographic area than telephone surveys or personal interviews since mailed questionnaire surveys are less expensive to conduct. Also, respondents can remain anonymous if they desire. Disadvantages

of mailed questionnaire surveys include a low number of responses and inappropriate answers to questions. Another drawback is that some people may have difficulty reading or understanding the questions.

Personal interview surveys have the advantage of obtaining in-depth responses to questions from the person being interviewed. One disadvantage is that interviewers must be trained in asking questions and recording responses, which makes the personal interview survey more costly than the other two survey methods. Another disadvantage is that the interviewer may be biased in his or her selection of respondents.

Data can also be collected in other ways, such as *surveying records* or *direct observation* of situations.

As stated in Section 1–1, researchers use samples to collect data and information about a particular variable from a large population. Using samples saves time and money and in some cases enables the researcher to get more detailed information about a particular subject. Samples cannot be selected in haphazard ways because the information obtained might be biased. For example, interviewing people on a street corner during the day would not include responses from people working in offices at that time or from people attending school; hence, not all subjects in a particular population would have a chance of being selected.

To obtain samples that are unbiased—i.e., that give each subject in the population an equally likely chance of being selected—statisticians use four basic methods of sampling: random, systematic, stratified, and cluster sampling.

Random Sampling

Random samples are selected by using chance methods or random numbers. One such method is to number each subject in the population. Then place numbered cards in a bowl, mix them thoroughly, and select as many cards as needed. The subjects whose numbers are selected constitute the sample. Since it is difficult to mix the cards

The Worst Day for Weight Loss

Many overweight people have difficulty losing weight. *Prevention* magazine reported that researchers from Washington University of Medicine studied the diets of 48 adult weight loss participants. They used food diaries, exercise monitors, and weigh-ins. They found that the participants ate an average of 236 more calories on Saturdays than they did on the other weekdays. This would amount to a weight gain of 9 pounds per year. So if you are watching your diet, be careful on Saturdays.

Are the statistics reported in this study descriptive or inferential in nature? What type of variables are used here?

thoroughly, there is a chance of obtaining a biased sample. For this reason, statisticians use another method of obtaining numbers. They generate random numbers with a computer or calculator. Before the invention of computers, random numbers were obtained from tables.

Some two-digit random numbers are shown in Table 1–3. To select a random sample of, say, 15 subjects out of 85 subjects, it is necessary to number each subject from 01 to 85. Then select a starting number by closing your eyes and placing your finger on a number in the table. (Although this may sound somewhat unusual, it enables us to find a starting number at random.) In this case suppose your finger landed on the number 12 in the second column. (It is the sixth number down from the top.) Then proceed downward until you have selected 15 different numbers between 01 and 85. When you reach the bottom of the column, go to the top of the next column. If you select a number greater than 85 or the number 00 or a duplicate number, just omit it. In our example, we will use the subjects numbered 12, 27, 75, 62, 57, 13, 31, 06, 16, 49, 46, 71, 53, 41, and 02. A more detailed procedure for selecting a random sample using a table of random numbers is given in Chapter 14, using Table D in Appendix C.

Systematic Sampling   *volunteer bias*

Researchers obtain **systematic samples** by numbering each subject of the population and then selecting every kth subject. For example, suppose there were 2000 subjects in the population and a sample of 50 subjects were needed. Since $2000 \div 50 = 40$, then $k = 40$, and every 40th subject would be selected; however, the first subject (numbered between 1 and 40) would be selected at random. Suppose subject 12 were the first subject selected; then the sample would consist of the subjects whose numbers were 12, 52, 92, etc., until 50 subjects were obtained. When using systematic sampling, you must be careful about how the subjects in the population are numbered. If subjects were arranged in a manner

Table 1–3	Random Numbers											
79	41	71	93	60	35	04	67	96	04	79	10	86
26	52	53	13	43	50	92	09	87	21	83	75	17
18	13	41	30	56	20	37	74	49	56	45	46	83
19	82	02	69	34	27	77	34	24	93	16	77	00
14	57	44	30	93	76	32	13	55	29	49	30	77
29	12	18	50	06	33	15	79	50	28	50	45	45
01	27	92	67	93	31	97	55	29	21	64	27	29
55	75	65	68	65	73	07	95	66	43	43	92	16
84	95	95	96	62	30	91	64	74	83	47	89	71
62	62	21	37	82	62	19	44	08	64	34	50	11
66	57	28	69	13	99	74	31	58	19	47	66	89
48	13	69	97	29	01	75	58	05	40	40	18	29
94	31	73	19	75	76	33	18	05	53	04	51	41
00	06	53	98	01	55	08	38	49	42	10	44	38
46	16	44	27	80	15	28	01	64	27	89	03	27
77	49	85	95	62	93	25	39	63	74	54	82	85
81	96	43	27	39	53	85	61	12	90	67	96	02
40	46	15	73	23	75	96	68	13	99	49	64	11

such as wife, husband, wife, husband, and every 40th subject were selected, the sample would consist of all husbands. Numbering is not always necessary. For example, a researcher may select every tenth item from an assembly line to test for defects.

Stratified Sampling

Researchers obtain **stratified samples** by dividing the population into groups (called strata) according to some characteristic that is important to the study, then sampling from each group. Samples within the strata should be randomly selected. For example, suppose the president of a two-year college wants to learn how students feel about a certain issue. Furthermore, the president wishes to see if the opinions of the first-year students differ from those of the second-year students. The president will randomly select students from each group to use in the sample.

Cluster Sampling

Researchers also use **cluster samples.** Here the population is divided into groups called clusters by some means such as geographic area or schools in a large school district, etc. Then the researcher randomly selects some of these clusters and uses all members of the selected clusters as the subjects of the samples. Suppose a researcher wishes to survey apartment dwellers in a large city. If there are 10 apartment buildings in the city, the researcher can select at random 2 buildings from the 10 and interview all the residents of these buildings. Cluster sampling is used when the population is large or when it involves subjects residing in a large geographic area. For example, if one wanted to do a study involving the patients in the hospitals in New York City, it would be very costly and time-consuming to try to obtain a random sample of patients since they would be spread over a large area. Instead, a few hospitals could be selected at random, and the patients in these hospitals would be interviewed in a cluster.

The four basic sampling methods are summarized in Table 1–4.

Other Sampling Methods

In addition to the four basic sampling methods, researchers use other methods to obtain samples. One such method is called a **convenience sample.** Here a researcher uses

Historical Note

In 1936, the *Literary Digest,* on the basis of a biased sample of its subscribers, predicted that Alf Landon would defeat Franklin D. Roosevelt in the upcoming presidential election. Roosevelt won by a landslide. The magazine ceased publication the following year.

Table 1–4	Summary of Sampling Methods
Random	Subjects are selected by random numbers.
Systematic	Subjects are selected by using every kth number after the first subject is randomly selected from 1 through k.
Stratified	Subjects are selected by dividing up the population into groups (strata), and subjects are randomly selected within groups.
Cluster	Subjects are selected by using an intact group that is representative of the population.

Interesting Facts

Older Americans are less likely to sacrifice happiness for a higher-paying job. According to one survey, 38% of those aged 18–29 said they would choose more money over happiness, while only 3% of those over 65 would.

subjects that are convenient. For example, the researcher may interview subjects entering a local mall to determine the nature of their visit or perhaps what stores they will be patronizing. This sample is probably not representative of the general customers for several reasons. For one thing, it was probably taken at a specific time of day, so not all customers entering the mall have an equal chance of being selected since they were not there when the survey was being conducted. But convenience samples can be representative of the population. If the researcher investigates the characteristics of the population and determines that the sample is representative, then it can be used.

Other sampling techniques, such as *sequential sampling, double sampling,* and *multistage sampling,* are explained in Chapter 14, along with a more detailed explanation of the four basic sampling techniques.

Applying the Concepts 1–3

American Culture and Drug Abuse

Assume you are a member of the Family Research Council and have become increasingly concerned about the drug use by professional sports players. You set up a plan and conduct a survey on how people believe the American culture (television, movies, magazines, and popular music) influences illegal drug use. Your survey consists of 2250 adults and adolescents from around the country. A consumer group petitions you for more information about your survey. Answer the following questions about your survey.

1. What type of survey did you use (phone, mail, or interview)?
2. What are the advantages and disadvantages of the surveying methods you did not use?
3. What type of scores did you use? Why?
4. Did you use a random method for deciding who would be in your sample?
5. Which of the methods (stratified, systematic, cluster, or convenience) did you use?
6. Why was that method more appropriate for this type of data collection?
7. If a convenience sample were obtained consisting of only adolescents, how would the results of the study be affected?

See page 33 for the answers.

1–4 Observational and Experimental Studies

Objective 6

Explain the difference between an observational and an experimental study.

There are several different ways to classify statistical studies. This section explains two types of studies: *observational studies* and *experimental studies.*

In an **observational study,** the researcher merely observes what is happening or what has happened in the past and tries to draw conclusions based on these observations.

For example, data from the Motorcycle Industry Council (*USA TODAY*) stated that "Motorcycle owners are getting older and richer." Data were collected on the ages and incomes of motorcycle owners for the years 1980 and 1998 and then compared. The findings showed considerable differences in the ages and incomes of motorcycle owners for the two years.

In this study, the researcher merely observed what had happened to the motorcycle owners over a period of time. There was no type of research intervention.

In an **experimental study,** the researcher manipulates one of the variables and tries to determine how the manipulation influences other variables.

For example, a study conducted at Virginia Polytechnic Institute and presented in *Psychology Today* divided female undergraduate students into two groups and had the students perform as many sit-ups as possible in 90 sec. The first group was told only to "Do your best," while the second group was told to try to increase the actual number of sit-ups done each day by 10%. After four days, the subjects in the group who were given the vague instructions to "Do your best" averaged 43 sit-ups, while the group that was given the more specific instructions to increase the number of sit-ups by 10% averaged 56 sit-ups by the last day's session. The conclusion then was that athletes who were given specific goals performed better than those who were not given specific goals.

This study is an example of a statistical experiment since the researchers intervened in the study by manipulating one of the variables, namely, the type of instructions given to each group.

In a true experimental study, the subjects should be assigned to groups randomly. Also, the treatments should be assigned to the groups at random. In the sit-up study, the article did not mention whether the subjects were randomly assigned to the groups.

Sometimes when random assignment is not possible, researchers use intact groups. These types of studies are done quite often in education where already intact groups are available in the form of existing classrooms. When these groups are used, the study is said to be a **quasi-experimental study.** The treatments, though, should be assigned at random. Most articles do not state whether random assignment of subjects was used.

Statistical studies usually include one or more *independent variables* and one *dependent variable.*

The **independent variable** in an experimental study is the one that is being manipulated by the researcher. The independent variable is also called the **explanatory variable.** The resultant variable is called the **dependent variable** or the **outcome variable.**

The outcome variable is the variable that is studied to see if it has changed significantly due to the manipulation of the independent variable. For example, in the sit-up study, the researchers gave the groups two different types of instructions, general and specific. Hence, the independent variable is the type of instruction. The dependent variable, then, is the resultant variable, that is, the number of sit-ups each group was able to perform after four days of exercise. If the differences in the dependent or outcome variable are large and other factors are equal, these differences can be attributed to the manipulation of the independent variable. In this case, specific instructions were shown to increase athletic performance.

In the sit-up study, there were two groups. The group that received the special instruction is called the **treatment group** while the other is called the **control group.** The treatment group receives a specific treatment (in this case, instructions for improvement) while the control group does not.

Both types of statistical studies have advantages and disadvantages. Experimental studies have the advantage that the researcher can decide how to select subjects and how to assign them to specific groups. The researcher can also control or manipulate the independent variable. For example, in studies that require the subjects to consume a certain amount of medicine each day, the researcher can determine the precise dosages and, if necessary, vary the dosage for the groups.

There are several disadvantages to experimental studies. First, they may occur in unnatural settings, such as laboratories and special classrooms. This can lead to several problems. One such problem is that the results might not apply to the natural setting. The age-old question then is, "This mouthwash may kill 10,000 germs in a test tube, but how many germs will it kill in my mouth?"

Another disadvantage with an experimental study is the **Hawthorne effect.** This effect was discovered in 1924 in a study of workers at the Hawthorne plant of the Western Electric Company. In this study, researchers found that the subjects who knew they were participating in an experiment actually changed their behavior in ways that affected the results of the study.

Another problem is called *confounding of variables.*

A **confounding variable** is one that influences the dependent or outcome variable but was not separated from the independent variable.

Researchers try to control most variables in a study, but this is not possible in some studies. For example, subjects who are put on an exercise program might also improve their diet unbeknownst to the researcher and perhaps improve their health in other ways not due to exercise alone. Then diet becomes a confounding variable.

Observational studies also have advantages and disadvantages. One advantage of an observational study is that it usually occurs in a natural setting. For example, researchers can observe people's driving patterns on streets and highways in large cities. Another advantage of an observational study is that it can be done in situations where it would be unethical or downright dangerous to conduct an experiment. Using observational studies, researchers can study suicides, rapes, murders, etc. In addition, observational studies can be done using variables that cannot be manipulated by the researcher, such as drug users versus nondrug users and right-handedness versus left-handedness.

Observational studies have disadvantages, too. As mentioned previously, since the variables are not controlled by the researcher, a definite cause-and-effect situation cannot be shown since other factors may have had an effect on the results. Observational studies can be expensive and time-consuming. For example, if one wanted to study the habitat of lions in Africa, one would need a lot of time and money, and there would be a certain amount of danger involved. Finally, since the researcher may not be using his or her own measurements, the results could be subject to the inaccuracies of those who collected the data. For example, if the researchers were doing a study of events that occurred in the 1800s, they would have to rely on information and records obtained by others from a previous era. There is no way to ensure the accuracy of these records.

When you read the results of statistical studies, decide if the study was observational or experimental. Then see if the conclusion follows logically, based on the nature of these studies.

No matter what type of study is conducted, two studies on the same subject sometimes have conflicting conclusions. Why might this occur? An article entitled "Bottom Line: Is It Good for You?" (*USA TODAY Weekend*) states that in the 1960s studies suggested that margarine was better for the heart than butter since margarine contains less saturated fat and users had lower cholesterol levels. In a 1980 study, researchers found that butter was better than margarine since margarine contained trans-fatty acids, which are worse for the heart than butter's saturated fat. Then in a 1998 study, researchers found that margarine was better for a person's health. Now, what is to be believed? Should one use butter or margarine?

The answer here is that you must take a closer look at these studies. Actually, it is not a choice between butter or margarine that counts, but the type of margarine used. In the 1980s, studies showed that solid margarine contains trans-fatty acids, and scientists believe that they are worse for the heart than butter's saturated fat. In the 1998 study, liquid margarine was used. It is very low in trans-fatty acids, and hence it is more healthful than butter because trans-fatty acids have been shown to raise cholesterol. Hence, the conclusion is to use liquid margarine instead of solid margarine or butter.

Before decisions based on research studies are made, it is important to get all the facts and examine them in light of the particular situation.

Applying the Concepts 1-4

Just a Pinch Between Your Cheek and Gum

As the evidence on the adverse effects of cigarette smoke grew, people tried many different ways to quit smoking. Some people tried chewing tobacco or, as it was called, smokeless tobacco. A small amount of tobacco was placed between the cheek and gum. Certain chemicals from the tobacco were absorbed into the bloodstream and gave the sensation of smoking cigarettes. This prompted studies on the adverse effects of smokeless tobacco. One study in particular used 40 university students as subjects. Twenty were given smokeless tobacco to chew, and twenty given a substance that looked and tasted like smokeless tobacco, but did not contain any of the harmful substances. The students were randomly assigned to one of the groups. The students' blood pressure and heart rate were measured before they started chewing and 20 minutes after they had been chewing. A significant increase in heart rate occurred in the group that chewed the smokeless tobacco. Answer the following questions.

1. What type of study was this (observational, quasi-experimental, or experimental)?
2. What are the independent and dependent variables?
3. Which was the treatment group?
4. Could the students' blood pressures be affected by knowing that they are part of a study?
5. List some possible confounding variables.
6. Do you think this is a good way to study the effect of smokeless tobacco?

See page 33 for the answers.

1-5

Objective 7

Explain how statistics can be used and misused.

Uses and Misuses of Statistics

As explained previously, statistical techniques can be used to describe data, compare two or more data sets, determine if a relationship exists between variables, test hypotheses, and make estimates about population characteristics. However, there is another aspect of statistics, and that is the misuse of statistical techniques to sell products that don't work properly, to attempt to prove something true that is really not true, or to get our attention by using statistics to evoke fear, shock, and outrage.

There are two sayings that have been around for a long time that illustrate this point:

"There are three types of lies—lies, damn lies, and statistics."

"Figures don't lie, but liars figure."

Just because we read or hear the results of a research study or an opinion poll in the media, this does not mean that these results are reliable or that they can be applied to any and all situations. For example, reporters sometimes leave out critical details such as the size of the sample used or how the research subjects were selected. Without this information, you cannot properly evaluate the research and properly interpret the conclusions of the study or survey.

It is the purpose of this section to show some ways that statistics can be misused. You should not infer that all research studies and surveys are suspect, but that there are many factors to consider when making decisions based on the results of research studies and surveys. Here are some ways that statistics can be misrepresented.

Suspect Samples

The first thing to consider is the sample that was used in the research study. Sometimes researchers use very small samples to obtain information. Several years ago, advertisements contained such statements as "Three out of four doctors surveyed recommend brand such and such." If only 4 doctors were surveyed, the results could have been obtained by chance alone; however, if 100 doctors were surveyed, the results might be quite different.

Not only is it important to have a sample size that is large enough, but also it is necessary to see how the subjects in the sample were selected. Studies using volunteers sometimes have a built-in bias. Volunteers generally do not represent the population at large. Sometimes they are recruited from a particular socioeconomic background, and sometimes unemployed people volunteer for research studies to get a stipend. Studies that require the subjects to spend several days or weeks in an environment other than their home or workplace automatically exclude people who are employed and cannot take time away from work. Sometimes only college students or retirees are used in studies. In the past, many studies have used only men, but have attempted to generalize the results to both men and women. Opinion polls that require a person to phone or mail in a response most often are not representative of the population in general, since only those with strong feelings for or against the issue usually call or respond by mail.

Another type of sample that may not be representative is the convenience sample. Educational studies sometimes use students in intact classrooms since it is convenient. Quite often, the students in these classrooms do not represent the student population of the entire school district.

When results are interpreted from studies using small samples, convenience samples, or volunteer samples, care should be used in generalizing the results to the entire population.

Ambiguous Averages

In Chapter 3, you will learn that there are four commonly used measures that are loosely called *averages.* They are the *mean, median, mode,* and *midrange.* For the same data set, these averages can differ markedly. People who know this can, without lying, select the one measure of average that lends the most evidence to support their position.

Changing the Subject

Another type of statistical distortion can occur when different values are used to represent the same data. For example, one political candidate who is running for reelection might say, "During my administration, expenditures increased a mere 3%." His opponent, who is trying to unseat him, might say, "During my opponent's administration, expenditures have increased a whopping $6,000,000." Here both figures are correct; however, expressing a 3% increase as $6,000,000 makes it sound like a very large increase. Here again, ask yourself, Which measure better represents the data?

Detached Statistics

Safe?

A claim that uses a detached statistic is one in which no comparison is made. For example, you may hear a claim such as "Our brand of crackers has one-third fewer calories." Here, no comparison is made. One-third fewer calories than what? Another example is a claim that uses a detached statistic such as "Brand A aspirin works four times faster." Four times faster than what? When you see statements such as this, always ask yourself, Compared to what?

Implied Connections

Many claims attempt to imply connections between variables that may not actually exist. For example, consider the following statement: "Eating fish may help to reduce your cholesterol." Notice the words *may help.* There is no guarantee that eating fish will definitely help you reduce your cholesterol.

"Studies suggest that using our exercise machine will reduce your weight." Here the word *suggest* is used; and again, there is no guarantee that you will lose weight by using the exercise machine advertised.

Another claim might say, "Taking calcium will lower blood pressure in some people." Note the word *some* is used. You may not be included in the group of "some" people. Be careful when you draw conclusions from claims that use words such as *may, in some people,* and *might help.*

Misleading Graphs

Statistical graphs give a visual representation of data that enables viewers to analyze and interpret data more easily than by simply looking at numbers. In Chapter 2, you will see how some graphs are used to represent data. However, if graphs are drawn inappropriately, they can misrepresent the data and lead the reader to draw false conclusions. The misuse of graphs is also explained in Chapter 2.

Faulty Survey Questions

When analyzing the results of a survey using questionnaires, you should be sure that the questions are properly written since the way questions are phrased can often influence the way people answer them. For example, the responses to a question such as "Do you feel that the North Huntingdon School District should build a new football stadium?" might be answered differently than a question such as "Do you favor increasing school taxes so that the North Huntingdon School District can build a new football stadium?" Each question asks something a little different, and the responses could be radically different. When you read and interpret the results obtained from questionnaire surveys, watch out for some of these common mistakes made in the writing of the survey questions.

In Chapter 14, you will find some common ways that survey questions could be misinterpreted by those responding and could therefore result in incorrect conclusions.

To restate the premise of this section, statistics, when used properly, can be beneficial in obtaining much information, but when used improperly, can lead to much misinformation. It is like your automobile. If you use your automobile to get to school or work or to go on a vacation, that's good. But if you use it to run over your neighbor's dog because it barks all night long and tears up your flower garden, that's not so good!

<table>
<tr><td>

1–6

Objective 8

Explain the importance of computers and calculators in statistics.

</td><td>

Computers and Calculators

In the past, statistical calculations were done with pencil and paper. However, with the advent of calculators, numerical computations became much easier. Computers do all the numerical calculation. All one does is to enter the data into the computer and use the appropriate command; the computer will print the answer or display it on the screen. Now the TI-83 Plus or TI-84 Plus graphing calculator accomplishes the same thing.

There are many statistical packages available; this book uses MINITAB and Microsoft Excel. Instructions for using MINITAB, the TI-83 Plus or TI-84 Plus graphing calculator, and Excel have been placed at the end of each relevant section, in subsections entitled Technology Step by Step.

You should realize that the computer and calculator merely give numerical answers and save the time and effort of doing calculations by hand. You are still responsible for understanding and interpreting each statistical concept. In addition, you should realize that the results come from the data and do not appear magically on the computer. Doing calculations using the procedure tables will help you reinforce this idea.

The author has left it up to instructors to choose how much technology they will incorporate into the course.

</td></tr>
</table>

Technology *Step by Step*

MINITAB
Step by Step

General Information

MINITAB statistical software provides a wide range of statistical analysis and graphing capabilities.

Take Note

In this text you will see captured screen images from computers running MINITAB Release 14. If you are using an earlier or later release of MINITAB, the screens you see on your computer may bear slight visual differences from the screens pictured in this text. **But don't be alarmed!** All the Step by Step operations described in this text, including the commands, the menu options, and the functionality, will work just fine on your computer.

Start the Program

1. Click the Windows Start Menu, then All Programs.
2. Click the MINITAB folder and then click ▄▄ MINITAB, the program icon. The program screen will look similar to the one shown here. You will see the Session Window, the Worksheet Window, and perhaps the Project Manager Window.
3. Click the Project Manager icon on the toolbar to bring the project manager to the front.

For Vista, click the Start button, then "All Programs." Next click "MINITAB Solutions" and then "MINITAB Statistical Software English."

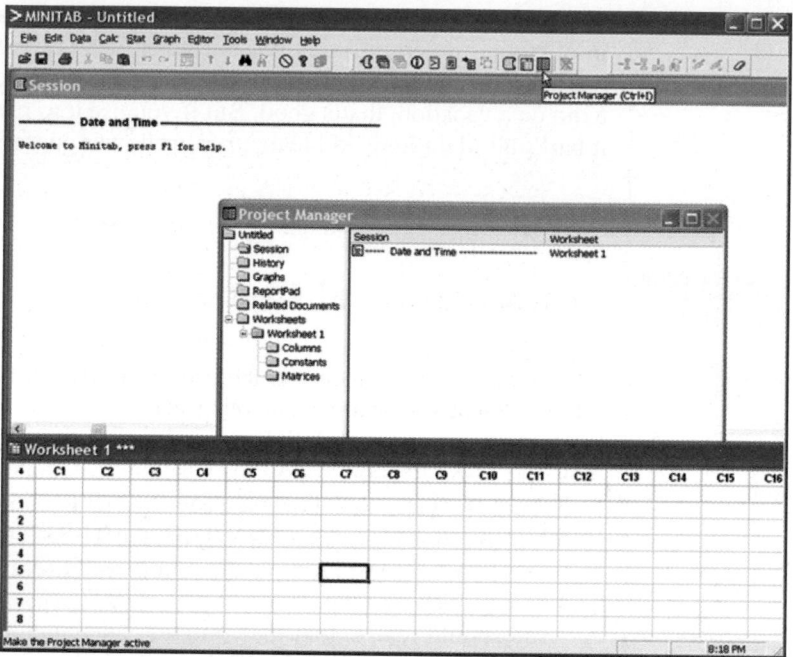

To use the program, data must be entered from the keyboard or from a file.

Entering Data in MINITAB

In MINITAB, all the data for one variable are stored in a column. Step by step instructions for entering these data follow.

 Data
213   208   203   215   222

1. Click in row 1 of Worksheet 1***. This makes the worksheet the active window and puts the cursor in the first cell. The small data entry arrow in the upper left-hand corner of the worksheet should be pointing down. If it is not, click it to change the direction in which the cursor will move when you press the [Enter] key.

2. Type in each number, pressing [Enter] after each entry, including the last number typed.

3. *Optional:* Click in the space above row 1 to type in **Weight,** the column label.

Save a Worksheet File

4. Click on the **File Menu.** *Note:* This is *not* the same as clicking the disk icon 💾 .

5. Click **Save Current Worksheet As . . .**

6. In the dialog box you will need to verify three items:

   a) Save in: Click on or type in the disk drive and directory where you will store your data. For a CD this might be **A:.**

   b) File Name: Type in the name of the file, such as **MyData.**

   c) Save as Type: The default here is MINITAB. An extension of mtw is added to the name.

      Click [Save]. The name of the worksheet will change from Worksheet 1*** to MyData.MTW.

Open the Databank File

The raw data are shown in Appendix D. There is a row for each person's data and a column for each variable. MINITAB data files comprised of data sets used in this book, including the

Databank, are available on the accompanying CD-ROM or at the Online Learning Center (www.mhhe.com/bluman). Here is how to get the data from a file into a worksheet.

1. Click **File>Open Worksheet.** A sequence of menu instructions will be shown this way.

   Note: This is *not* the same as clicking the file icon ☞ . If the dialog box says Open Project instead of Open Worksheet, click [Cancel] and use the correct menu item. The Open Worksheet dialog box will be displayed.

2. You must check three items in this dialog box.

   a) The Look In: dialog box should show the directory where the file is located.

   b) Make sure the Files of Type: shows the correct type, MINITAB [*.mtw].

   c) Double-click the file name in the list box Databank.mtw. A dialog box may inform you that a copy of this file is about to be added to the project. Click on the checkbox if you do not want to see this warning again.

3. Click the [OK] button. The data will be copied into a second worksheet. Part of the worksheet is shown here.

	C1	C2	C3	C4	C5	C6	C7	C8	C9	C10	C11-T	C12-T
	ID-NUMBER	AGE	ED-LEVEL	SMOKING	EXERCISE	WEIGHT	SERUM-	SYSTOLIC	IQ	SODIUM	GENDER	MARITAL-ST
1	1	27	2	1	1	120	193	126	118	136	F	M
2	2	18	1	0	1	145	210	120	105	137	M	S
3	3	32	2	0	0	118	196	128	115	135	F	M

   a) You may maximize the window and scroll if desired.

   b) C12-T Marital Status has a T appended to the label to indicate alphanumeric data. MyData.MTW is not erased or overwritten. Multiple worksheets can be available; however, only the active worksheet is available for analysis.

4. To switch between the worksheets, select **Window>MyData.MTW.**

5. Select **File>Exit** to quit. To save the project, click [Yes].

6. Type in the name of the file, **Chapter01.** The Data Window, the Session Window, and settings are all in one file called a project. Projects have an extension of mpj instead of mtw.

   Clicking the disk icon 💾 on the menu bar is the same as selecting **File>Save Project.**

   Clicking the file icon ☞ is the same as selecting **File>Open Project.**

7. Click [Save]. The mpj extension will be added to the name. The computer will return to the Windows desktop. The two worksheets, the Session Window results, and settings are saved in this project file. When a project file is opened, the program will start up right where you left off.

TI-83 Plus or TI-84 Plus
Step by Step

The TI-83 Plus or TI-84 Plus graphing calculator can be used for a variety of statistical graphs and tests.

General Information

To turn calculator on:
Press **ON** key.
To turn calculator off:
Press **2nd [OFF].**

To reset defaults only:

1. Press **2nd,** then [MEM].

2. Select **7,** then **2,** then **2.**

Optional. To reset settings on calculator and clear memory: (*Note:* This will clear all settings and programs in the calculator's memory.)
Press **2nd,** then [MEM]. Then press **7,** then **1,** then **2.**
(Also, the contrast may need to be adjusted after this.)

To adjust contrast (if necessary):

Press **2nd.** Then press and hold ▲ to darken or ▼ to lighten contrast.

To clear screen:

Press **CLEAR.**

(*Note:* This will return you to the screen you were using.)

To display a menu:

Press appropriate menu key. Example: **STAT.**

To return to home screen:

Press **2nd,** then **[QUIT].**

To move around on the screens:

Use the arrow keys.

To select items on the menu:

Press the corresponding number or move the cursor to the item, using the arrow keys. Then press **ENTER.**

(*Note:* In some cases, you do not have to press **ENTER,** and in other cases you may need to press **ENTER** twice.)

Entering Data

To enter single-variable data (if necessary, clear the old list):

1. Press **STAT** to display the Edit menu.

2. Press **ENTER** to select 1:Edit.

3. Enter the data in L_1 and press **ENTER** after each value.

4. After all data values are entered, press **STAT** to get back to the Edit menu or **2nd [QUIT]** to end.

Example TI1–1

Enter the following data values in L_1: **213, 208, 203, 215, 222.**

To enter multiple-variable data:

The TI-83 Plus or TI-84 Plus will take up to six lists designated L_1, L_2, L_3, L_4, L_5, and L_6.

1. To enter more than one set of data values, complete the preceding steps. Then move the cursor to L_2 by pressing the ▶ key.

2. Repeat the steps in the preceding part.

Output

L1	L2	L3	1
213	------	------	
208			
203			
215			
222			

L1(6)=

Editing Data

To correct a data value before pressing **ENTER,** use ◀ and retype the value and press **ENTER.**

To correct a data value in a list after pressing **ENTER,** move cursor to incorrect value in list and type in the correct value. Then press **ENTER.**

To delete a data value in a list:

Move cursor to value and press **DEL.**

To insert a data value in a list:

1. Move cursor to position where data value is to be inserted, then press **2nd [INS].**

2. Type data value; then press **ENTER.**

To clear a list:

1. Press **STAT,** then **4.**

2. Enter list to be cleared. Example: To clear L_1, press **2nd [L₁].** Then press **ENTER.**

   (*Note:* To clear several lists, follow STEP 1, but enter each list to be cleared, separating them with commas. To clear all lists at once, follow STEP 1; then press **ENTER.**)

Sorting Data

To sort the data in a list:

1. Enter the data in L_1.
2. Press **STAT 2** to get SortA to sort the list in ascending order.
3. Then press **2nd [L₁] ENTER.**

The calculator will display Done.

4. Press **STAT ENTER** to display sorted list.

(*Note:* The SortD or **3** sorts the list in descending order.)

Example TI1–2

Sort in ascending order the data values entered in Example TI1–1.

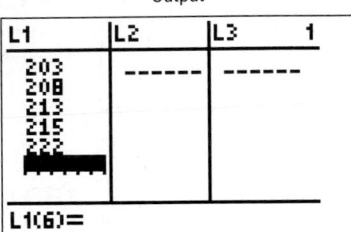

Output

Excel
Step by Step

Excel's Analysis ToolPak Add-In

General Information

Microsoft Excel 2007 has two different ways to solve statistical problems. First, there are built-in functions, such as STDEV and CHITEST, available from the standard toolbar by clicking Formulas, then selecting the Insert Function icon [fx Insert Function]. Another feature of Excel that is useful for calculating multiple statistical measures and performing statistical tests for a set of data is the Data Analysis command found in the Analysis Tool-Pak Add-in.

To load the Analysis Tool-Pak:

Click the Microsoft Office button [icon], then select Excel Options.

1. Click **Add-Ins,** and select **Add-ins** from the list of options on the left side of the options box.
2. Select the Analysis Tool-Pak, then click the **Go** button at the bottom of the options box.

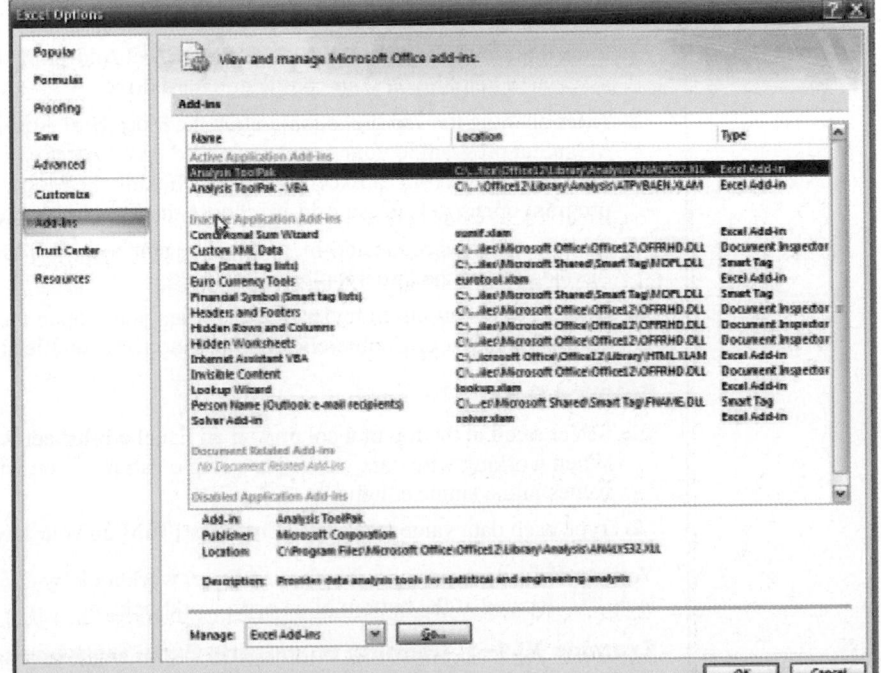

3. After loading the Analysis Tool-Pak, the Data Analysis command is available in the Analysis group on the Data tab.

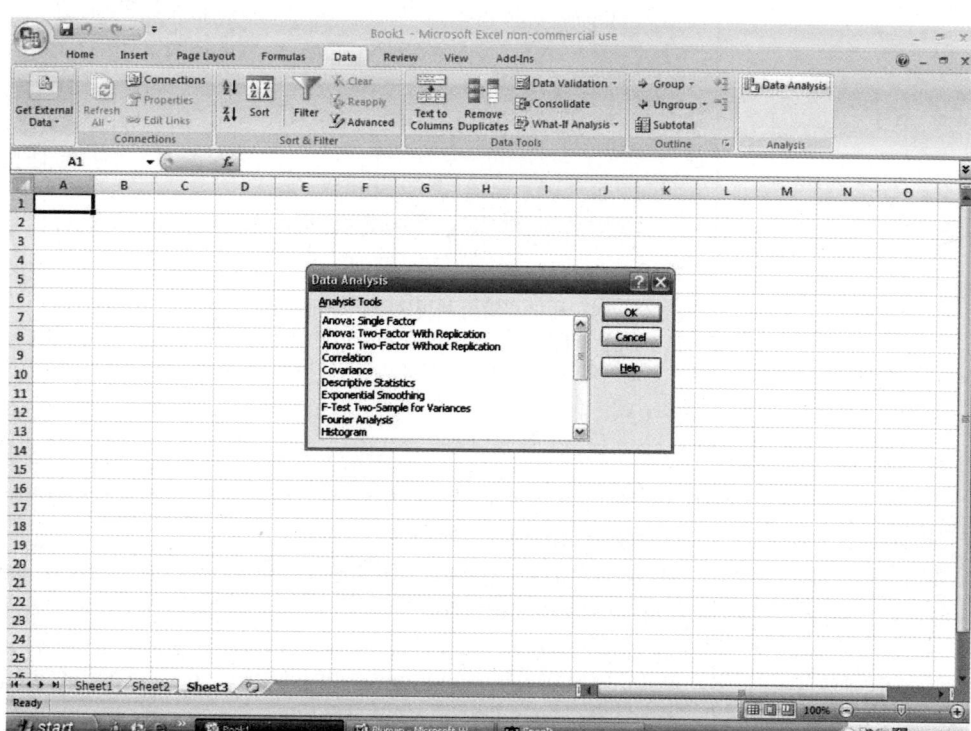

MegaStat

Later in this text you will encounter a few Excel Technology Step by Step operations that will require the use of the MegaStat Add-in for Excel. MegaStat can be downloaded from the CD that came with your textbook as well as from the text's Online Learning Center at www.mhhe.com/bluman.

1. Save the Zip file containing the MegaStat Excel Add-in file (MegaStat.xls) and the associated help file on your computer's hard drive.

2. After opening the Zip file, double-click the MegaStat Add-in file, then Extract the MegaStat program to your computer's hard drive. After extracting the file, you can load the MegaStat Add-in to Excel by double-clicking the MegaStat.xls file. When the Excel program opens to load the Add-in, choose the Enable Macros option.

3. After installation of the add-in, you will be able to access MegaStat by selecting the Add-ins tab on the Excel toolbar.

4. If MegaStat is not listed under Add-ins when you reopen the Excel program, then you can access MegaStat by double-clicking the MegaStat.xls file at any time.

Entering Data

1. Select a cell at the top of a column on an Excel worksheet where you want to enter data. When working with data values for a single variable, you will usually want to enter the values into a single column.

2. Type each data value and press **[Enter]** or **[Tab]** on your keyboard.

You can also add more worksheets to an Excel workbook by clicking the Insert Worksheet icon located at the bottom of an open workbook.

Example XL1–1: Opening an existing Excel workbook/worksheet

1. Open the Microsoft Office Excel 2007 program.

2. Click the Microsoft Click Office button , then click the Open file function. The Open dialog box will be displayed.
3. In the Look in box, click the folder where the Excel workbook file is located.
4. Double-click the file name in the list box. The selected workbook file will be opened in Excel for editing.

Summary*

- The two major areas of statistics are descriptive and inferential. Descriptive statistics includes the collection, organization, summarization, and presentation of data. Inferential statistics includes making inferences from samples to populations, estimations and hypothesis testing, determining relationships, and making predictions. Inferential statistics is based on *probability theory*. (1–1)

- Data can be classified as qualitative or quantitative. Quantitative data can be either discrete or continuous, depending on the values they can assume. Data can also be measured by various scales. The four basic levels of measurement are nominal, ordinal, interval, and ratio. (1–2)

- Since in most cases the populations under study are large, statisticians use subgroups called samples to get the necessary data for their studies. There are four basic methods used to obtain samples: random, systematic, stratified, and cluster. (1–3)

- There are two basic types of statistical studies: observational studies and experimental studies. When conducting observational studies, researchers observe what is happening or what has happened and then draw conclusions based on these observations. They do not attempt to manipulate the variables in any way. (1–4)

- When conducting an experimental study, researchers manipulate one or more of the independent or explanatory variables and see how this manipulation influences the dependent or outcome variable. (1–4)

- Finally, the applications of statistics are many and varied. People encounter them in everyday life, such as in reading newspapers or magazines, listening to the radio, or watching television. Since statistics is used in almost every field of endeavor, the educated individual should be knowledgeable about the vocabulary, concepts, and procedures of statistics. Also, everyone should be aware that statistics can be misused. (1–5)

- Today, computers and calculators are used extensively in statistics to facilitate the computations. (1–6)

Unusual Stat

The chance that someone will attempt to burglarize your home in any given year is 1 in 20.

LAFF - A - DAY

Bill Whitehead 2·11

"We've polled the entire populace, Your Majesty, and we've come up with exactly the results you ordered!"

© Dave Whitehead. King Features Syndicate.

*The numbers in parentheses indicate the chapter section where the material is explained.

Important Terms

cluster sample 12	experimental study 14	observational study 13	random variable 3
confounding variable 15	explanatory variable 14	ordinal level of measurement 8	ratio level of measurement 8
continuous variables 6	Hawthorne effect 15	outcome variable 14	sample 4
control group 14	hypothesis testing 4	population 4	statistics 3
convenience sample 12	independent variable 14	probability 4	stratified sample 12
data 3	inferential statistics 4	qualitative variables 6	systematic sample 11
data set 3	interval level of measurement 8	quantitative variables 6	treatment group 14
data value or datum 3	measurement scales 7	quasi-experimental study 14	variable 3
dependent variable 14	nominal level of measurement 7	random sample 10	
descriptive statistics 4			
discrete variables 6			

Answers not appearing on the page can be found in the answers appendix.

Review Exercises

Note: **All odd-numbered problems and even-numbered problems marked with "ans" are included in the answer section at the end of this book. The numbers in parentheses indicate the chapter section where the process to arrive at a solution is explained.**

1. Name and define the two areas of statistics. (1–1)

2. What is probability? Name two areas where probability is used. (1–1) Probability deals with events that occur by chance. It is used in gambling and insurance.

3. Suggest some ways statistics can be used in everyday life. (1–1) Answers will vary.

4. Explain the differences between a sample and a population. (1–1) A population is the totality of all subjects possessing certain common characteristics that are being studied.

5. Why are samples used in statistics? (1–1)

6. (ans) In each of these statements, tell whether descriptive or inferential statistics have been used.

    a. By 2040 at least 3.5 billion people will run short of water (World Future Society). Inferential
    b. Nine out of ten on-the-job fatalities are men (Source: *USA TODAY Weekend*). Descriptive
    c. Expenditures for the cable industry were $5.66 billion in 1996 (Source: *USA TODAY*). Descriptive
    d. The median household income for people aged 25–34 is $35,888 (Source: *USA TODAY*). Descriptive
    e. Allergy therapy makes bees go away (Source: *Prevention*). Inferential

    f. Drinking decaffeinated coffee can raise cholesterol levels by 7% (Source: American Heart Association).
    g. The national average annual medicine expenditure per person is $1052 (Source: *The Greensburg Tribune Review*). Descriptive
    h. Experts say that mortgage rates may soon hit bottom (Source: *USA TODAY*). (1–1) Inferential

7. Classify each as nominal-level, ordinal-level, interval-level, or ratio-level measurement.

    a. Pages in the 25 best-selling mystery novels. Ratio
    b. Rankings of golfers in a tournament. Ordinal
    c. Temperatures inside 10 pizza ovens. Interval
    d. Weights of selected cell phones. Ratio
    e. Salaries of the coaches in the NFL. Ratio
    f. Times required to complete a chess game. Ratio
    g. Ratings of textbooks (poor, fair, good, excellent). Ordinal
    h. Number of amps delivered by battery chargers. Ratio
    i. Ages of children in a day care center. Ratio
    j. Categories of magazines in a physician's office (sports, women's, health, men's, news). (1–2) Normal

8. Classify each variable as qualitative or quantitative.

    a. Marital status of nurses in a hospital. Qualitative
    b. Time it takes to run a marathon. Quantitative
    c. Weights of lobsters in a tank in a restaurant. Quantitative
    d. Colors of automobiles in a shopping center parking lot. Qualitative
    e. Ounces of ice cream in a large milkshake. Quantitative
    f. Capacity of the NFL football stadiums. Quantitative
    g. Ages of people living in a personal care home. (1–2) Quantitative

9. Classify each variable as discrete or continuous.

   a. Number of pizzas sold by Pizza Express each day. Discrete
   b. Relative humidity levels in operating rooms at local hospitals. Continuous
   c. Number of bananas in a bunch at several local supermarkets. Discrete
   d. Lifetimes (in hours) of 15 iPod batteries. Continuous
   e. Weights of the backpacks of first graders on a school bus. Continuous
   f. Number of students each day who make appointments with a math tutor at a local college. Discrete
   g. Blood pressures of runners in a marathon. (1–2) Continuous

10. Give the boundaries of each value.

   a. 36 inches. 35.5–36.5
   b. 105.4 miles. 105.35–105.45
   c. 72.6 tons. 72.55–72.65
   d. 5.27 centimeters. 5.265–5.275
   e. 5 ounces. (1–2) 4.5–5.5

11. Name and define the four basic sampling methods. (1–3) Random, systematic, stratified, cluster

12. (ans) Classify each sample as random, systematic, stratified, or cluster.

   a. In a large school district, all teachers from two buildings are interviewed to determine whether they believe the students have less homework to do now than in previous years. Cluster
   b. Every seventh customer entering a shopping mall is asked to select her or his favorite store. Systematic
   c. Nursing supervisors are selected using random numbers to determine annual salaries. Random
   d. Every 100th hamburger manufactured is checked to determine its fat content. Systematic
   e. Mail carriers of a large city are divided into four groups according to gender (male or female) and according to whether they walk or ride on their routes. Then 10 are selected from each group and interviewed to determine whether they have been bitten by a dog in the last year. (1–3) Stratified

13. Give three examples each of nominal, ordinal, interval, and ratio data. (1–2) Answers will vary.

14. For each of these statements, define a population and state how a sample might be obtained. Answers will vary.

   a. The average cost of an airline meal is $4.55 (Source: *Everything Has Its Price,* Richard E. Donley, Simon and Schuster).
   b. More than 1 in 4 United States children have cholesterol levels of 180 milligrams or higher (Source: The American Health Foundation).
   c. Every 10 minutes, 2 people die in car crashes and 170 are injured (Source: National Safety Council estimates).

   d. When older people with mild to moderate hypertension were given mineral salt for 6 months, the average blood pressure reading dropped by 8 points systolic and 3 points diastolic (Source: *Prevention*).
   e. The average amount spent per gift for Mom on Mother's Day is $25.95 (Source: The Gallup Organization). (1–3)

15. Select a newspaper or magazine article that involves a statistical study, and write a paper answering these questions. Answers will vary.

   a. Is this study descriptive or inferential? Explain your answer.
   b. What are the variables used in the study? In your opinion, what level of measurement was used to obtain the data from the variables?
   c. Does the article define the population? If so, how is it defined? If not, how could it be defined?
   d. Does the article state the sample size and how the sample was obtained? If so, determine the size of the sample and explain how it was selected. If not, suggest a way it could have been obtained.
   e. Explain *in your own words* what procedure (survey, comparison of groups, etc.) might have been used to determine the study's conclusions.
   f. Do you agree or disagree with the conclusions? State your reasons.

16. Information from research studies is sometimes taken out of context. Explain why the claims of these studies might be suspect. Answers will vary.

   a. Based on a recent telephone survey, 72% of those contacted shop online.
   b. In Greenville County there are 8324 deer.
   c. Nursing school graduates from Fairview University earn on average $33,456.
   d. Only 5% of the men surveyed wash the dishes after dinner.
   e. A recent study shows that high school dropouts spend less time on the Internet than those who graduated; therefore, the Internet raises your IQ.
   f. Most shark attacks occur in ocean water that is 3 feet deep; therefore, it is safer to swim in deep water. (1–5)

17. Identify each study as being either observational or experimental.

   a. Subjects were randomly assigned to two groups, and one group was given an herb and the other group a placebo. After 6 months, the numbers of respiratory tract infections each group had were compared. Experimental
   b. A researcher stood at a busy intersection to see if the color of the automobile that a person drives is related to running red lights. Observational

c. A researcher finds that people who are more hostile have higher total cholesterol levels than those who are less hostile. Observational

d. Subjects are randomly assigned to four groups. Each group is placed on one of four special diets—a low-fat diet, a high-fish diet, a combination of low-fat diet and high-fish diet, and a regular diet. After 6 months, the blood pressures of the groups are compared to see if diet has any effect on blood pressure. (1–4) Experimental

18. Identify the independent variable(s) and the dependent variable for each of the studies in Exercise 17. (1–4)

19. For each of the studies in Exercise 17, suggest possible confounding variables. (1–4)

20. Beneficial Bacteria According to a pilot study of 20 people conducted at the University of Minnesota, daily doses of a compound called arabinogalactan over a period of 6 months resulted in a significant increase in the beneficial lactobacillus species of bacteria. Why can't it be concluded that the compound is beneficial for the majority of people? (1–5) Only 20 people were used in the study.

21. Comment on the following statement, taken from a magazine advertisement: "In a recent clinical study, Brand ABC (actual brand will not be named) was proved to be 1950% better than creatine!" (1–5) The only time claims can be proved is when the entire population is used.

22. In an ad for women, the following statement was made: "For every 100 women, 91 have taken the road less traveled." Comment on this statement. (1–5)

23. In many ads for weight loss products, under the product claims and in small print, the following statement is made: "These results are not typical." What does this say about the product being advertised? (1–5)

24. In an ad for moisturizing lotion, the following claim is made: ". . . it's the number 1 dermatologist-recommended brand." What is misleading about this claim? (1–5) There is no mention of how this conclusion was obtained.

25. An ad for an exercise product stated: "Using this product will burn 74% more calories." What is misleading about this statement? (1–5) "74% more calories" than what? No comparison group is stated.

26. "Vitamin E is a proven antioxidant and may help in fighting cancer and heart disease." Is there anything ambiguous about this claim? Explain. (1–5) Since the word *may* is used, there is no guarantee that the product will help fight cancer.

27. "Just 1 capsule of Brand X can provide 24 hours of acid control." (Actual brand will not be named.) What needs to be more clearly defined in this statement? (1–5) What is meant by "24 hours of acid control"?

28. ". . . Male children born to women who smoke during pregnancy run a risk of violent and criminal behavior that lasts well into adulthood." Can we infer that smoking during pregnancy is responsible for criminal behavior in people? (1–5) No. There are many other factors that contribute to criminal behavior.

29. Caffeine and Health In the 1980s, a study linked coffee to a higher risk of heart disease and pancreatic cancer. In the early 1990s, studies showed that drinking coffee posed minimal health threats. However, in 1994, a study showed that pregnant women who drank 3 or more cups of tea daily may be at risk for spontaneous abortion. In 1998, a study claimed that women who drank more than a half-cup of caffeinated tea every day may actually increase their fertility. In 1998, a study showed that over a lifetime, a few extra cups of coffee a day can raise blood pressure, heart rate, and stress (Source: "Bottom Line: Is It Good for You? Or Bad?" by Monika Guttman, *USA TODAY Weekend*). Suggest some reasons why these studies appear to be conflicting. (1–5) Possible answer: It could be the amount of caffeine in the coffee or tea. It could have been the brewing method.

Extending the Concepts

30. Find an article that describes a statistical study, and identify the study as observational or experimental. Answers will vary.

31. For the article that you used in Exercise 30, identify the independent variable(s) and dependent variable for the study. Answers will vary.

32. For the article that you selected in Exercise 30, suggest some confounding variables that may have an effect on the results of the study. Answers will vary.

Are We Improving Our Diet?—Revisited

Researchers selected a *sample* of 23,699 adults in the United States, using phone numbers selected at *random*, and conducted a *telephone survey*. All respondents were asked six questions:

1. How often do you drink juices such as orange, grapefruit, or tomato?

2. Not counting juice, how often do you eat fruit?

3. How often do you eat green salad?

4. How often do you eat potatoes (not including french fries, fried potatoes, or potato chips)?

5. How often do you eat carrots?

6. Not counting carrots, potatoes, or salad, how many servings of vegetables do you usually eat?

Researchers found that men consumed fewer servings of fruits and vegetables per day (3.3) than women (3.7). Only 20% of the population consumed the recommended 5 or more daily servings. In addition, they found that youths and less-educated people consumed an even lower amount than the average.

Based on this study, they recommend that greater educational efforts be undertaken to improve fruit and vegetable consumption by Americans and to provide environmental and institutional support to encourage increased consumption.

Source: Mary K. Serdula, M.D., et al., "Fruit and Vegetable Intake Among Adults in 16 States: Results of a Brief Telephone Survey," *American Journal of Public Health* 85, no. 2. Copyright by the American Public Health Association.

Chapter Quiz

Determine whether each statement is true or false. If the statement is false, explain why.

1. Probability is used as a basis for inferential statistics. True

2. The heights of the mountains in the state of Alaska are an example of a variable. True

3. The lowest level of measurement is the nominal level. True

4. When the population of college professors is divided into groups according to their rank (instructor, assistant professor, etc.) and then several are selected from each group to make up a sample, the sample is called a cluster sample. False

5. The variable temperature is an example of a quantitative variable. True

6. The height of basketball players is considered a continuous variable. True

7. The boundary of a value such as 6 inches would be 5.9–6.1 inches. False

Select the best answer.

8. The number of ads on a one-hour television show is what type of data?
   a. Nominal
   b. Qualitative
   c. Discrete
   d. Continuous

9. What are the boundaries of 25.6 ounces?
   a. 25–26 ounces
   b. 25.55–25.65 ounces
   c. 25.5–25.7 ounces
   d. 20–39 ounces

10. A researcher divided subjects into two groups according to gender and then selected members from each group for her sample. What sampling method was the researcher using?
   a. Cluster
   b. Random
   c. Systematic
   d. Stratified

11. Data that can be classified according to color are measured on what scale?

   (a.) Nominal
   b. Ratio
   c. Ordinal
   d. Interval

12. A study that involves no researcher intervention is called

   a. An experimental study.
   b. A noninvolvement study.
   (c.) An observational study.
   d. A quasi-experimental study.

13. A variable that interferes with other variables in the study is called

   (a.) A confounding variable.
   b. An explanatory variable.
   c. An outcome variable.
   d. An interfering variable.

Use the best answer to complete these statements.

14. Two major branches of statistics are _____ and _____.
   Descriptive, inferential

15. Two uses of probability are _____ and _____.
   Gambling, insurance

16. The group of all subjects under study is called a(n) _____. Population

17. A group of subjects selected from the group of all subjects under study is called a(n) _____. Sample

18. Three reasons why samples are used in statistics are
   a. _____ b. _____ c. _____.
   a. Saves time   b. Saves money   c. Use when population is infinite

19. The four basic sampling methods are
   a. _____ b. _____ c. _____ d. _____.
   a. Random   b. Systematic   c. Cluster   d. Stratified

20. A study that uses intact groups when it is not possible to randomly assign participants to the groups is called a(n) _____ study. Quasi-experimental

21. In a research study, participants should be assigned to groups using _____ methods, if possible. Random

22. For each statement, decide whether descriptive or inferential statistics is used.

   a. The average life expectancy in New Zealand is 78.49 years (Source: *World Factbook*). Descriptive
   b. A diet high in fruits and vegetables will lower blood pressure (Source: Institute of Medicine). Inferential
   c. The total amount of estimated losses for Hurricane Katrina was $125 billion (Source: *The World Almanac and Book of Facts*). Descriptive
   d. Researchers stated that the shape of a person's ears is relative to the person's aggression (Source: *American Journal of Human Biology*). Inferential
   e. In 2013, the number of high school graduates will be 3.2 million students (Source: National Center for Education). Inferential

23. Classify each as nominal level, ordinal level, interval level, or ratio level of measurement.

   a. Rating of movies as G, PG, and R   Nominal
   b. Number of candy bars sold on a fund drive   Ratio
   c. Classification of automobiles as subcompact, compact, standard, and luxury   Ordinal
   d. Temperatures of hair dryers   Interval
   e. Weights of suitcases on a commercial airliner   Ratio

24. Classify each variable as discrete or continuous.

   a. Ages of people working in a large factory   Continuous
   b. Number of cups of coffee served at a restaurant   Discrete
   c. The amount of drug injections into a guinea pig   Continuous
   d. The time it takes a student to drive to school   Continuous
   e. The number of gallons of milk sold each day at a grocery store   Discrete

25. Give the boundaries of each.

   a. 32 minutes   31.5–32.5 minutes
   b. 0.48 millimeter   0.475–0.485 millimeter
   c. 6.2 inches   6.15–6.25 inches
   d. 19 pounds   18.5–19.5 pounds
   e. 12.1 quarts   12.05–12.15 quarts

Critical Thinking Challenges

1. **World's Busiest Airports** A study of the world's busiest airports was conducted by *Airports Council International.* Describe three variables that one could use to determine which airports are the busiest. What *units* would one use to measure these variables? Are these variables categorical, discrete, or continuous?

2. **Smoking and Criminal Behavior** The results of a study published in *Archives of General Psychiatry* stated that male children born to women who smoke during pregnancy run a risk of violent and criminal behavior that lasts into adulthood. The results of this study were challenged by some people in the media. Give several reasons why the results of this study would be challenged.

3. **Piano Lessons Improve Math Ability** The results of a study published in *Neurological Research* stated that second-graders who took piano lessons and played a computer math game more readily grasped math problems in fractions and proportions than a similar group who took an English class and played the same math game. What type of inferential study was this? Give several reasons why the piano lessons could improve a student's math ability.

4. **ACL Tears in Collegiate Soccer Players** A study of 2958 collegiate soccer players showed that in 46 anterior cruciate ligament (ACL) tears, 36 were in women. Calculate the percentages of tears for each gender.

   a. Can it be concluded that female athletes tear their knees more often than male athletes?
   b. Comment on how this study's conclusion might have been reached.

5. **Anger and Snap Judgments** Read the article entitled "Anger Can Cause Snap Judgments" and answer the following questions.

   a. Is the study experimental or observational?
   b. What is the independent variable?
   c. What is the dependent variable?
   d. Do you think the sample sizes are large enough to merit the conclusion?
   e. Based on the results of the study, what changes would you recommend to persons to help them reduce their anger?

6. **Hostile Children Fight Unemployment** Read the article entitled "Hostile Children Fight Unemployment" and answer the following questions.

   a. Is the study experimental or observational?
   b. What is the independent variable?
   c. What is the dependent variable?
   d. Suggest some confounding variables that may have influenced the results of the study.
   e. Identify the three groups of subjects used in the study.

ANGER CAN CAUSE SNAP JUDGMENTS

Anger can make a normally unbiased person act with prejudice, according to a forthcoming study in the journal *Psychological Science.*

Assistant psychology professors David DeSteno at Northeastern University in Boston and Nilanjana Dasgupta at the University of Massachusetts, Amherst, randomly divided 81 study participants into two groups and assigned them a writing task designed to induce angry, sad or neutral feelings. In a subsequent test to uncover nonconscious associations, angry subjects were quicker to connect negatively charged words—like war, death and vomit—with members of the opposite group—even though the groupings were completely arbitrary.

"These automatic responses guide our behavior when we're not paying attention," says DeSteno, and they can lead to discriminatory acts when there is pressure to make a quick decision. "If you're aware that your emotions might be coloring these gut reactions," he says, "you should take time to consider that possibility and adjust your actions accordingly."

—*Eric Strand*

Source: Reprinted with permission from *Psychology Today,* Copyright © (2004) Sussex Publishers, Inc.

UNEMPLOYMENT

Hostile Children Fight Unemployment

Aggressive children may be destined for later long-term unemployment. In a study that began in 1968, researchers at the University of Jyvaskyla in Finland examined about 300 participants at ages 8, 14, 27, and 36. They looked for aggressive behaviors like hurting other children, kicking objects when angry, or attacking others without reason.

Their results, published recently in the *International Journal of Behavioral Development*, suggest that children with low self-control of emotion —especially aggression—were significantly more prone to long-term unemployment. Children with behavioral inhibitions—such as passive and anxious behaviors—were also indirectly linked to unemployment

as they lacked the preliminary initiative needed for school success. And while unemployment rates were high in Finland during the last data collection, jobless participants who were aggressive as children were less likely to have a job two years later than their nonaggressive counterparts.

Ongoing unemployment can have serious psychological consequences, including depression, anxiety and stress. But lead researcher Lea Pulkkinen, Ph.D., a Jyvaskyla psychology professor, does have encouraging news for parents: Aggressive children with good social skills and child-centered parents were significantly less likely to be unemployed for more than two years as adults.

—*Tanya Zimbardo*

Source: Reprinted with permission from *Psychology Today,* Copyright © (2001) Sussex Publishers, Inc.

Data Projects

1. **Business and Finance** Investigate the types of data that are collected regarding stock and bonds, for example, price, earnings ratios, and bond ratings. Find as many types of data as possible. For each, identify the level of measure as nominal, ordinal, interval, or ratio. For any quantitative data, also note if they are discrete or continuous.

2. **Sports and Leisure** Select a professional sport. Investigate the types of data that are collected about that sport, for example, in baseball, the level of play (A, AA, AAA, Major League), batting average, and home-run hits. For each, identify the level of measure as nominal, ordinal, interval, or ratio. For any quantitative data, also note if they are discrete or continuous.

3. **Technology** Music organization programs on computers and music players maintain information about a song, such as the writer, song length, genre, and your personal rating. Investigate the types of data collected about a song. For each, identify the level of measure as nominal, ordinal, interval, or ratio. For any quantitative data, also note if they are discrete or continuous.

4. **Health and Wellness** Think about the types of data that can be collected about your health and wellness, things such as blood type, cholesterol level, smoking status, and BMI. Find as many data items as you can. For each, identify the level of measure as nominal, ordinal, interval, or ratio. For any quantitative data, also note if they are discrete or continuous.

5. **Politics and Economics** Every 10 years since 1790, the federal government has conducted a census of U.S. residents. Investigate the types of data that were collected in the 2010 census. For each, identify the level of measure as nominal, ordinal, interval, or ratio. For any quantitative data, also note if they are discrete or continuous. Use the library or a genealogy website to find a census form from 1860. What types of data were collected? How do the types of data differ?

6. **Your Class** Your school probably has a database that contains information about each student, such as age, county of residence, credits earned, and ethnicity. Investigate the types of student data that your college collects and reports. For each, identify the level of measure as nominal, ordinal, interval, or ratio. For any quantitative data, also note if they are discrete or continuous.

Answers to Applying the Concepts

Section 1–1 Attendance and Grades

1. The variables are grades and attendance.

2. The data consist of specific grades and attendance numbers.

3. These are descriptive statistics; however, if an inference were made to all students, then that would be inferential statistics.

4. The population under study is students at Manatee Community College (MCC).

5. While not specified, we probably have data from a sample of MCC students.

6. Based on the data, it appears that, in general, the better your attendance, the higher your grade.

Section 1–2 Safe Travel

1. The variables are industry and number of job-related injuries.

2. The type of industry is a qualitative variable, while the number of job-related injuries is quantitative.

3. The number of job-related injuries is discrete.

4. Type of industry is nominal, and the number of job-related injuries is ratio.

5. The railroads do show fewer job-related injuries; however, there may be other things to consider. For example, railroads employ fewer people than the other transportation industries in the study.

6. A person's choice of transportation might also be affected by convenience issues, cost, service, etc.

7. Answers will vary. One possible answer is that the railroads have the fewest job-related injuries, while the airline industry has the most job-related injuries (more than twice those of the railroad industry). The numbers of job-related injuries in the subway and trucking industries are fairly comparable.

Section 1–3 American Culture and Drug Abuse

Answers will vary, so this is one possible answer.

1. I used a telephone survey. The advantage to my survey method is that this was a relatively inexpensive survey method (although more expensive than using the mail) that could get a fairly sizable response. The disadvantage to my survey method is that I have not included anyone without a telephone. (*Note:* My survey used a random dialing method to include unlisted numbers and cell phone exchanges.)

2. A mail survey also would have been fairly inexpensive, but my response rate may have been much lower than what I got with my telephone survey. Interviewing would have allowed me to use follow-up questions and to clarify any questions of the respondents at the time of the interview. However, interviewing is very labor- and cost-intensive.

3. I used ordinal data on a scale of 1 to 5. The scores were 1 = strongly disagree, 2 = disagree, 3 = neutral, 4 = agree, 5 = strongly agree.

4. The random method that I used was a random dialing method.

5. To include people from each state, I used a stratified random sample, collecting data randomly from each of the area codes and telephone exchanges available.

6. This method allowed me to make sure that I had representation from each area of the United States.

7. Convenience samples may not be representative of the population, and a convenience sample of adolescents would probably differ greatly from the general population with regard to the influence of American culture on illegal drug use.

Section 1–4 Just a Pinch Between Your Cheek and Gum

1. This was an experiment, since the researchers imposed a treatment on each of the two groups involved in the study.

2. The independent variable is whether the participant chewed tobacco or not. The dependent variables are the students' blood pressures and heart rates.

3. The treatment group is the tobacco group—the other group was used as a control.

4. A student's blood pressure might not be affected by knowing that he or she was part of a study. However, if the student's blood pressure were affected by this knowledge, all the students (in both groups) would be affected similarly. This might be an example of the placebo effect.

5. Answers will vary. One possible answer is that confounding variables might include the way that the students chewed the tobacco, whether or not the students smoked (although this would hopefully have been evened out with the randomization), and that all the participants were university students.

6. Answers will vary. One possible answer is that the study design was fine, but that it cannot be generalized beyond the population of university students (or people around that age).

Frequency Distributions and Graphs

Objectives

After completing this chapter, you should be able to

1. Organize data using a frequency distribution.

2. Represent data in frequency distributions graphically using histograms, frequency polygons, and ogives.

3. Represent data using bar graphs, Pareto charts, time series graphs, and pie graphs.

4. Draw and interpret a stem and leaf plot.

Outline

Statistics Today

How Your Identity Can Be Stolen

Identity fraud is a big business today. The total amount of the fraud in 2006 was $56.6 billion. The average amount of the fraud for a victim is $6383, and the average time to correct the problem is 40 hours. The ways in which a person's identity can be stolen are presented in the following table:

Lost or stolen wallet, checkbook, or credit card	38%
Friends, acquaintances	15
Corrupt business employees	15
Computer viruses and hackers	9
Stolen mail or fraudulent change of address	8
Online purchases or transactions	4
Other methods	11

Source: Javelin Strategy & Research; Council of Better Business Bureau, Inc.

Looking at the numbers presented in a table does not have the same impact as presenting numbers in a well-drawn chart or graph. The article did not include any graphs. This chapter will show you how to construct appropriate graphs to represent data and help you to get your point across to your audience.

See Statistics Today—Revisited at the end of the chapter for some suggestions on how to represent the data graphically.

Introduction

When conducting a statistical study, the researcher must gather data for the particular variable under study. For example, if a researcher wishes to study the number of people who were bitten by poisonous snakes in a specific geographic area over the past several years, he or she has to gather the data from various doctors, hospitals, or health departments.

To describe situations, draw conclusions, or make inferences about events, the researcher must organize the data in some meaningful way. The most convenient method of organizing data is to construct a *frequency distribution.*

After organizing the data, the researcher must present them so they can be understood by those who will benefit from reading the study. The most useful method of presenting the data is by constructing *statistical charts* and *graphs.* There are many different types of charts and graphs, and each one has a specific purpose.

This chapter explains how to organize data by constructing frequency distributions and how to present the data by constructing charts and graphs. The charts and graphs illustrated here are histograms, frequency polygons, ogives, pie graphs, Pareto charts, and time series graphs. A graph that combines the characteristics of a frequency distribution and a histogram, called a stem and leaf plot, is also explained.

2–1

Objective 1

Organize data using a frequency distribution.

Organizing Data
Wealthy People

Suppose a researcher wished to do a study on the ages of the top 50 wealthiest people in the world. The researcher first would have to get the data on the ages of the people. In this case, these ages are listed in *Forbes Magazine*. When the data are in original form, they are called **raw data** and are listed next.

49	57	38	73	81
74	59	76	65	69
54	56	69	68	78
65	85	49	69	61
48	81	68	37	43
78	82	43	64	67
52	56	81	77	79
85	40	85	59	80
60	71	57	61	69
61	83	90	87	74

Since little information can be obtained from looking at raw data, the researcher organizes the data into what is called a *frequency distribution*. A frequency distribution consists of *classes* and their corresponding *frequencies*. Each raw data value is placed into a quantitative or qualitative category called a **class.** The **frequency** of a class then is the number of data values contained in a specific class. A frequency distribution is shown for the preceding data set.

Class limits	Tally	Frequency
35–41	///	3
42–48	///	3
49–55	////	4
56–62	ЖḤ ЖḤ	10
63–69	ЖḤ ЖḤ	10
70–76	ЖḤ	5
77–83	ЖḤ ЖḤ	10
84–90	ЖḤ	5
		Total 50

Now some general observations can be made from looking at the frequency distribution. For example, it can be stated that the majority of the wealthy people in the study are over 55 years old.

A **frequency distribution** is the organization of raw data in table form, using classes and frequencies.

The classes in this distribution are 35–41, 42–48, etc. These values are called *class limits.* The data values 35, 36, 37, 38, 39, 40, 41 can be tallied in the first class; 42, 43, 44, 45, 46, 47, 48 in the second class; and so on.

Two types of frequency distributions that are most often used are the *categorical frequency distribution* and the *grouped frequency distribution.* The procedures for constructing these distributions are shown now.

Categorical Frequency Distributions

The **categorical frequency distribution** is used for data that can be placed in specific categories, such as nominal- or ordinal-level data. For example, data such as political affiliation, religious affiliation, or major field of study would use categorical frequency distributions.

| Example 2–1 | Distribution of Blood Types |

Twenty-five army inductees were given a blood test to determine their blood type. The data set is

A	B	B	AB	O
O	O	B	AB	B
B	B	O	A	O
A	O	O	O	AB
AB	A	O	B	A

Construct a frequency distribution for the data.

Solution

Since the data are categorical, discrete classes can be used. There are four blood types: A, B, O, and AB. These types will be used as the classes for the distribution.

The procedure for constructing a frequency distribution for categorical data is given next.

Step 1 Make a table as shown.

A Class	B Tally	C Frequency	D Percent
A			
B			
O			
AB			

Step 2 Tally the data and place the results in column B.

Step 3 Count the tallies and place the results in column C.

Step 4 Find the percentage of values in each class by using the formula

$$\% = \frac{f}{n} \cdot 100\%$$

where f = frequency of the class and n = total number of values. For example, in the class of type A blood, the percentage is

$$\% = \frac{5}{25} \cdot 100\% = 20\%$$

Percentages are not normally part of a frequency distribution, but they can be added since they are used in certain types of graphs such as pie graphs. Also, the decimal equivalent of a percent is called a *relative frequency.*

Step 5 Find the totals for columns C (frequency) and D (percent). The completed table is shown.

A Class	B Tally	C Frequency	D Percent
A	ⅣⅣ	5	20
B	ⅣⅣ //	7	28
O	ⅣⅣ ////	9	36
AB	////	4	16
		Total 25	100

For the sample, more people have type O blood than any other type.

Grouped Frequency Distributions

When the range of the data is large, the data must be grouped into classes that are more than one unit in width, in what is called a **grouped frequency distribution.** For example, a distribution of the number of hours that boat batteries lasted is the following.

Class limits	Class boundaries	Tally	Frequency
24–30	23.5–30.5	///	3
31–37	30.5–37.5	/	1
38–44	37.5–44.5	ⅣⅣ	5
45–51	44.5–51.5	ⅣⅣ ////	9
52–58	51.5–58.5	ⅣⅣ /	6
59–65	58.5–65.5	/	1
			25

The procedure for constructing the preceding frequency distribution is given in Example 2–2; however, several things should be noted. In this distribution, the values 24 and 30 of the first class are called *class limits.* The **lower class limit** is 24; it represents the smallest data value that can be included in the class. The **upper class limit** is 30; it represents the largest data value that can be included in the class. The numbers in the second column are called **class boundaries.** These numbers are used to separate the classes so that there are no gaps in the frequency distribution. The gaps are due to the limits; for example, there is a gap between 30 and 31.

Students sometimes have difficulty finding class boundaries when given the class limits. The basic rule of thumb is that *the class limits should have the same decimal place value as the data, but the class boundaries should have one additional place value and end in a 5.* For example, if the values in the data set are whole numbers, such as 24, 32, and 18, the limits for a class might be 31–37, and the boundaries are 30.5–37.5. Find the boundaries by subtracting 0.5 from 31 (the lower class limit) and adding 0.5 to 37 (the upper class limit).

Lower limit − 0.5 = 31 − 0.5 = 30.5 = lower boundary
Upper limit + 0.5 = 37 + 0.5 = 37.5 = upper boundary

If the data are in tenths, such as 6.2, 7.8, and 12.6, the limits for a class hypothetically might be 7.8–8.8, and the boundaries for that class would be 7.75–8.85. Find these values by subtracting 0.05 from 7.8 and adding 0.05 to 8.8.

Finally, the **class width** for a class in a frequency distribution is found by subtracting the lower (or upper) class limit of one class from the lower (or upper) class limit of the next class. For example, the class width in the preceding distribution on the duration of boat batteries is 7, found from 31 − 24 = 7.

The class width can also be found by subtracting the lower boundary from the upper boundary for any given class. In this case, $30.5 - 23.5 = 7$.

Note: Do not subtract the limits of a single class. It will result in an incorrect answer.

The researcher must decide how many classes to use and the width of each class. To construct a frequency distribution, follow these rules:

1. *There should be between 5 and 20 classes.* Although there is no hard-and-fast rule for the number of classes contained in a frequency distribution, it is of the utmost importance to have enough classes to present a clear description of the collected data.

2. *It is preferable but not absolutely necessary that the class width be an odd number.* This ensures that the midpoint of each class has the same place value as the data. The **class midpoint** X_m is obtained by adding the lower and upper boundaries and dividing by 2, or adding the lower and upper limits and dividing by 2:

$$X_m = \frac{\text{lower boundary} + \text{upper boundary}}{2}$$

or

$$X_m = \frac{\text{lower limit} + \text{upper limit}}{2}$$

For example, the midpoint of the first class in the example with boat batteries is

$$\frac{24 + 30}{2} = 27 \qquad \text{or} \qquad \frac{23.5 + 30.5}{2} = 27$$

The midpoint is the numeric location of the center of the class. Midpoints are necessary for graphing (see Section 2–2). If the class width is an even number, the midpoint is in tenths. For example, if the class width is 6 and the boundaries are 5.5 and 11.5, the midpoint is

$$\frac{5.5 + 11.5}{2} = \frac{17}{2} = 8.5$$

Rule 2 is only a suggestion, and it is not rigorously followed, especially when a computer is used to group data.

3. *The classes must be mutually exclusive.* Mutually exclusive classes have nonoverlapping class limits so that data cannot be placed into two classes. Many times, frequency distributions such as

Age
10–20
20–30
30–40
40–50

are found in the literature or in surveys. If a person is 40 years old, into which class should she or he be placed? A better way to construct a frequency distribution is to use classes such as

Age
10–20
21–31
32–42
43–53

4. *The classes must be continuous.* Even if there are no values in a class, the class must be included in the frequency distribution. There should be no gaps in a

frequency distribution. The only exception occurs when the class with a zero frequency is the first or last class. A class with a zero frequency at either end can be omitted without affecting the distribution.

5. *The classes must be exhaustive.* There should be enough classes to accommodate all the data.

6. *The classes must be equal in width.* This avoids a distorted view of the data.

One exception occurs when a distribution has a class that is open-ended. That is, the class has no specific beginning value or no specific ending value. A frequency distribution with an open-ended class is called an **open-ended distribution.** Here are two examples of distributions with open-ended classes.

Age	Frequency	Minutes	Frequency
10–20	3	Below 110	16
21–31	6	110–114	24
32–42	4	115–119	38
43–53	10	120–124	14
54 and above	8	125–129	5

The frequency distribution for age is open-ended for the last class, which means that anybody who is 54 years or older will be tallied in the last class. The distribution for minutes is open-ended for the first class, meaning that any minute values below 110 will be tallied in that class.

Example 2–2 shows the procedure for constructing a grouped frequency distribution, i.e., when the classes contain more than one data value.

Example 2–2

Record High Temperatures

These data represent the record high temperatures in degrees Fahrenheit (°F) for each of the 50 states. Construct a grouped frequency distribution for the data using 7 classes.

112	100	127	120	134	118	105	110	109	112
110	118	117	116	118	122	114	114	105	109
107	112	114	115	118	117	118	122	106	110
116	108	110	121	113	120	119	111	104	111
120	113	120	117	105	110	118	112	114	114

Source: *The World Almanac and Book of Facts.*

Solution

The procedure for constructing a grouped frequency distribution for numerical data follows.

Step 1 Determine the classes.

Find the highest value and lowest value: $H = 134$ and $L = 100$.

Find the range: R = highest value − lowest value = $H − L$, so

$R = 134 − 100 = 34$

Select the number of classes desired (usually between 5 and 20). In this case, 7 is arbitrarily chosen.

Find the class width by dividing the range by the number of classes.

$$\text{Width} = \frac{R}{\text{number of classes}} = \frac{34}{7} = 4.9$$

Round the answer up to the nearest whole number if there is a remainder: $4.9 \approx 5$. (Rounding *up* is different from rounding *off*. A number is rounded up if there is any decimal remainder when dividing. For example, $85 \div 6 = 14.167$ and is rounded up to 15. Also, $53 \div 4 = 13.25$ and is rounded up to 14. Also, after dividing, if there is no remainder, you will need to add an extra class to accommodate all the data.)

Select a starting point for the lowest class limit. This can be the smallest data value or any convenient number less than the smallest data value. In this case, 100 is used. Add the width to the lowest score taken as the starting point to get the lower limit of the next class. Keep adding until there are 7 classes, as shown, 100, 105, 110, etc.

Subtract one unit from the lower limit of the second class to get the upper limit of the first class. Then add the width to each upper limit to get all the upper limits.

$$105 - 1 = 104$$

The first class is 100–104, the second class is 105–109, etc.

Find the class boundaries by subtracting 0.5 from each lower class limit and adding 0.5 to each upper class limit:

99.5–104.5, 104.5–109.5, etc.

Step 2   Tally the data.

Step 3   Find the numerical frequencies from the tallies.

The completed frequency distribution is

Class limits	Class boundaries	Tally	Frequency
100–104	99.5–104.5	//	2
105–109	104.5–109.5	ﾐ ///	8
110–114	109.5–114.5	ﾐ ﾐ ﾐ ///	18
115–119	114.5–119.5	ﾐ ﾐ ///	13
120–124	119.5–124.5	ﾐ //	7
125–129	124.5–129.5	/	1
130–134	129.5–134.5	/	1
			$n = \Sigma f = 50$

The frequency distribution shows that the class 109.5–114.5 contains the largest number of temperatures (18) followed by the class 114.5–119.5 with 13 temperatures. Hence, most of the temperatures (31) fall between 109.5 and 119.5°F.

Sometimes it is necessary to use a *cumulative frequency distribution*. A **cumulative frequency distribution** is a distribution that shows the number of data values less than or equal to a specific value (usually an upper boundary). The values are found by adding the frequencies of the classes less than or equal to the upper class boundary of a specific class. This gives an ascending cumulative frequency. In this example, the cumulative frequency for the first class is $0 + 2 = 2$; for the second class it is $0 + 2 + 8 = 10$; for the third class it is $0 + 2 + 8 + 18 = 28$. Naturally, a shorter way to do this would be to just add the cumulative frequency of the class below to the frequency of the given class. For

Historical Note

Florence Nightingale, a nurse in the Crimean War in 1854, used statistics to persuade government officials to improve hospital care of soldiers in order to reduce the death rate from unsanitary conditions in the military hospitals that cared for the wounded soldiers.

example, the cumulative frequency for the number of data values less than 114.5 can be found by adding $10 + 18 = 28$. The cumulative frequency distribution for the data in this example is as follows:

	Cumulative frequency
Less than 99.5	0
Less than 104.5	2
Less than 109.5	10
Less than 114.5	28
Less than 119.5	41
Less than 124.5	48
Less than 129.5	49
Less than 134.5	50

Cumulative frequencies are used to show how many data values are accumulated up to and including a specific class. In Example 2–2, 28 of the total record high temperatures are less than or equal to 114°F. Forty-eight of the total record high temperatures are less than or equal to 124°F.

After the raw data have been organized into a frequency distribution, it will be analyzed by looking for peaks and extreme values. The peaks show which class or classes have the most data values compared to the other classes. Extreme values, called outliers, show large or small data values that are relative to other data values.

When the range of the data values is relatively small, a frequency distribution can be constructed using single data values for each class. This type of distribution is called an **ungrouped frequency distribution** and is shown next.

Example 2–3   ### MPGs for SUVs

The data shown here represent the number of miles per gallon (mpg) that 30 selected four-wheel-drive sports utility vehicles obtained in city driving. Construct a frequency distribution, and analyze the distribution.

12	17	12	14	16	18
16	18	12	16	17	15
15	16	12	15	16	16
12	14	15	12	15	15
19	13	16	18	16	14

Source: *Model Year Fuel Economy Guide.* United States Environmental Protection Agency.

Solution

Step 1   Determine the classes. Since the range of the data set is small ($19 - 12 = 7$), classes consisting of a single data value can be used. They are 12, 13, 14, 15, 16, 17, 18, 19.

Note: If the data are continuous, class boundaries can be used. Subtract 0.5 from each class value to get the lower class boundary, and add 0.5 to each class value to get the upper class boundary.

Step 2   Tally the data.

Step 3   Find the numerical frequencies from the tallies, and find the cumulative frequencies.

The completed ungrouped frequency distribution is

Class limits	Class boundaries	Tally	Frequency
12	11.5–12.5	〃〃 /	6
13	12.5–13.5	/	1
14	13.5–14.5	///	3
15	14.5–15.5	〃〃 /	6
16	15.5–16.5	〃〃 ///	8
17	16.5–17.5	//	2
18	17.5–18.5	///	3
19	18.5–19.5	/	1

In this case, almost one-half (14) of the vehicles get 15 or 16 miles per gallon. The cumulative frequencies are

	Cumulative frequency
Less than 11.5	0
Less than 12.5	6
Less than 13.5	7
Less than 14.5	10
Less than 15.5	16
Less than 16.5	24
Less than 17.5	26
Less than 18.5	29
Less than 19.5	30

The steps for constructing a grouped frequency distribution are summarized in the following Procedure Table.

Procedure Table

Constructing a Grouped Frequency Distribution

Step 1   Determine the classes.

Find the highest and lowest values.

Find the range.

Select the number of classes desired.

Find the width by dividing the range by the number of classes and rounding up.

Select a starting point (usually the lowest value or any convenient number less than the lowest value); add the width to get the lower limits.

Find the upper class limits.

Find the boundaries.

Step 2   Tally the data.

Step 3   Find the numerical frequencies from the tallies, and find the cumulative frequencies.

When you are constructing a frequency distribution, the guidelines presented in this section should be followed. However, you can construct several different but correct frequency distributions for the same data by using a different class width, a different number of classes, or a different starting point.

Furthermore, the method shown here for constructing a frequency distribution is not unique, and there are other ways of constructing one. Slight variations exist, especially in computer packages. But regardless of what methods are used, classes should be mutually exclusive, continuous, exhaustive, and of equal width.

In summary, the different types of frequency distributions were shown in this section. The first type, shown in Example 2–1, is used when the data are categorical (nominal), such as blood type or political affiliation. This type is called a categorical frequency distribution. The second type of distribution is used when the range is large and classes several units in width are needed. This type is called a grouped frequency distribution and is shown in Example 2–2. Another type of distribution is used for numerical data and when the range of data is small, as shown in Example 2–3. Since each class is only one unit, this distribution is called an ungrouped frequency distribution.

All the different types of distributions are used in statistics and are helpful when one is organizing and presenting data.

The reasons for constructing a frequency distribution are as follows:

1. To organize the data in a meaningful, intelligible way.
2. To enable the reader to determine the nature or shape of the distribution.
3. To facilitate computational procedures for measures of average and spread (shown in Sections 3–1 and 3–2).
4. To enable the researcher to draw charts and graphs for the presentation of data (shown in Section 2–2).
5. To enable the reader to make comparisons among different data sets.

The factors used to analyze a frequency distribution are essentially the same as those used to analyze histograms and frequency polygons, which are shown in Section 2–2.

Applying the Concepts **2–1**

Ages of Presidents at Inauguration

The data represent the ages of our Presidents at the time they were first inaugurated.

57	61	57	57	58	57	61	54	68
51	49	64	50	48	65	52	56	46
54	49	51	47	55	55	54	42	51
56	55	51	54	51	60	62	43	55
56	61	52	69	64	46	54	47	

1. Were the data obtained from a population or a sample? Explain your answer.
2. What was the age of the oldest President?
3. What was the age of the youngest President?
4. Construct a frequency distribution for the data. (Use your own judgment as to the number of classes and class size.)
5. Are there any peaks in the distribution?

6. Identify any possible outliers.

7. Write a brief summary of the nature of the data as shown in the frequency distribution.

See page 101 for the answers.

Answers not appearing on the page can be found in the answers appendix.

Exercises 2–1

1. List five reasons for organizing data into a frequency distribution.

2. Name the three types of frequency distributions, and explain when each should be used. Categorical, ungrouped, grouped

3. Find the class boundaries, midpoints, and widths for each class.

   a. 32–38 31.5–38.5, 35, 7

   b. 86–104 85.5–104.5, 95, 19

   c. 895–905 894.5–905.5, 900, 11

   d. 12.3–13.5 12.25–13.55, 12.9, 1.3

   e. 3.18–4.96 3.175–4.965, 4.07, 1.79

4. How many classes should frequency distributions have? Why should the class width be an odd number?

5. Shown here are four frequency distributions. Each is incorrectly constructed. State the reason why.

   a. | Class | Frequency |
   |---|---|
   | 27–32 | 1 |
   | 33–38 | 0 |
   | 39–44 | 6 |
   | 45–49 | 4 |
   | 50–55 | 2   Class width is not uniform. |

   b. | Class | Frequency |
   |---|---|
   | 5–9 | 1 |
   | 9–13 | 2 |
   | 13–17 | 5 |
   | 17–20 | 6   Class limits overlap, and class |
   | 20–24 | 3   width is not uniform. |

   c. | Class | Frequency |
   |---|---|
   | 123–127 | 3 |
   | 128–132 | 7 |
   | 138–142 | 2 |
   | 143–147 | 19   A class has been omitted. |

   d. | Class | Frequency |
   |---|---|
   | 9–13 | 1 |
   | 14–19 | 6 |
   | 20–25 | 2 |
   | 26–28 | 5 |
   | 29–32 | 9   Class width is not uniform. |

6. What are open-ended frequency distributions? Why are they necessary?

7. **Trust in Internet Information** A survey was taken on how much trust people place in the information they read on the Internet. Construct a categorical frequency distribution for the data. A = trust in everything they read, M = trust in most of what they read, H = trust in about one-half of what they read, S = trust in a small portion of what they read. (Based on information from the *UCLA Internet Report.*)

M	M	M	A	H	M	S	M	H	M
S	M	M	M	M	A	M	M	A	M
M	M	H	M	M	M	H	M	H	M
A	M	M	M	H	M	M	M	M	M

8. **Grams per Food Serving** The data shown are the number of grams per serving of 30 selected brands of cakes. Construct a frequency distribution using 5 classes.

32	47	51	41	46	30
46	38	34	34	52	48
48	38	43	41	21	24
25	29	33	45	51	32
32	27	23	23	34	35

Source: *The Complete Food Counts.*

9. **Weights of the NBA's Top 50 Players** Listed are the weights of the NBA's top 50 players. Construct a grouped frequency distribution and a cumulative frequency distribution with 8 classes. Analyze the results in terms of peaks, extreme values, etc.

240	210	220	260	250	195	230	270	325	225
165	295	205	230	250	210	220	210	230	202
250	265	230	210	240	245	225	180	175	215
215	235	245	250	215	210	195	240	240	225
260	210	190	260	230	190	210	230	185	260

Source: www.msn.foxsports.com

10. **Stories in the World's Tallest Buildings** The number of stories in each of the world's 30 tallest buildings follows. Construct a grouped frequency distribution and a cumulative frequency distribution with 7 classes.

88	88	110	88	80	69	102	78	70	55
79	85	80	100	60	90	77	55	75	55
54	60	75	64	105	56	71	70	65	72

Source: *New York Times Almanac.*

11. GRE Scores at Top-Ranked Engineering Schools The average quantitative GRE scores for the top 30 graduate schools of engineering are listed. Construct a grouped frequency distribution and a cumulative frequency distribution with 5 classes.

767	770	761	760	771	768	776	771	756	770
763	760	747	766	754	771	771	778	766	762
780	750	746	764	769	759	757	753	758	746

Source: *U.S. News & World Report, Best Graduate Schools.*

12. Airline Passengers The number of passengers (in thousands) for the leading U.S. passenger airlines in 2004 is indicated below. Use the data to construct a grouped frequency distribution and a cumulative frequency distribution with a reasonable number of classes, and comment on the shape of the distribution.

91,570	86,755	81,066	70,786	55,373	42,400
40,551	21,119	16,280	14,869	13,659	13,417
13,170	12,632	11,731	10,420	10,024	9,122
7,041	6,954	6,406	6,362	5,930	5,585
5,427					

Source: *The World Almanac and Book of Facts.*

13. Ages of Declaration of Independence Signers The ages of the signers of the Declaration of Independence are shown. (Age is approximate since only the birth year appeared in the source, and one has been omitted since his birth year is unknown.) Construct a grouped frequency distribution and a cumulative frequency distribution for the data using 7 classes. (The data in this exercise will be used in Exercise 23 in Section 3–1.)

41	54	47	40	39	35	50	37	49	42	70	32
44	52	39	50	40	30	34	69	39	45	33	42
44	63	60	27	42	34	50	42	52	38	36	45
35	43	48	46	31	27	55	63	46	33	60	62
35	46	45	34	53	50	50					

Source: *The Universal Almanac.*

14. Unclaimed Expired Prizes The number of unclaimed expired prizes (in millions of dollars) for lottery tickets bought in a sample of states as shown. Construct a frequency distribution for the data using 5 classes. (The data in this exercise will be used for Exercise 22 in Section 3–1.)

28.5	51.7	19	5
2	1.2	14	14.6
0.8	11.6	3.5	30.1
1.7	1.3	13	14

15. Presidential Vetoes The number of total vetoes exercised by the past 20 Presidents is listed below. Use the data to construct a grouped frequency distribution and a cumulative frequency distribution with 5 classes. What is challenging about this set of data?

| 44 | 39 | 37 | 21 | 31 | 170 | 44 | 635 | 30 | 78 |
| 42 | 6 | 250 | 43 | 10 | 82 | 50 | 181 | 66 | 37 |

16. Salaries of College Coaches The data are the salaries (in hundred thousands of dollars) of a sample of 30 colleges and university coaches in the United States. Construct a frequency distribution for the data using 8 classes. (The data in this exercise will be used for Exercise 11 in Section 2–2.)

164	225	225	140	188
210	238	146	201	544
550	188	415	261	164
478	684	330	307	435
857	183	381	275	578
450	385	297	390	515

17. NFL Payrolls The data show the NFL team payrolls (in millions of dollars) for a specific year. Construct a frequency distribution for the payroll using 7 classes. (The data in this exercise will be used in Exercise 17 in Section 3–2.)

99	105	106	102
102	93	109	106
77	91	103	118
97	100	107	103
94	109	100	98
84	92	98	110
94	104	98	123
102	99	100	107

Source: *NFL.*

18. State Gasoline Tax The state gas tax in cents per gallon for 25 states is given below. Construct a grouped frequency distribution and a cumulative frequency distribution with 5 classes.

7.5	16	23.5	17	22
21.5	19	20	27.1	20
22	20.7	17	28	20
23	18.5	25.3	24	31
14.5	25.9	18	30	31.5

Source: *The World Almanac and Book of Facts.*

Extending the Concepts

19. JFK Assassination A researcher conducted a survey asking people if they believed more than one person was involved in the assassination of John F. Kennedy. The results were as follows: 73% said yes, 19% said no, and 9% had no opinion. Is there anything suspicious about the results?

Technology *Step by Step*

MINITAB
Step by Step

Make a Categorical Frequency Table
(Qualitative or Discrete Data)

1. Type in all the blood types from Example 2–1 down C1 of the worksheet.

    A B B AB O O O B AB B B B O A O A O O O AB AB A O B A

2. Click above row 1 and name the column **BloodType.**

3. Select **Stat>Tables>Tally Individual Values.**

    The cursor should be blinking in the Variables dialog box. If not, click inside the dialog box.

4. Double-click C1 in the Variables list.

5. Check the boxes for the statistics: Counts, Percents, and Cumulative percents.

6. Click [OK]. The results will be displayed in the Session Window as shown.

Tally for Discrete Variables: BloodType

BloodType	Count	Percent	CumPct
A	5	20.00	20.00
AB	4	16.00	36.00
B	7	28.00	64.00
O	9	36.00	100.00
N=	25		

Make a Grouped Frequency Distribution
(Quantitative Variable)

1. Select **File>New>New Worksheet.** A new worksheet will be added to the project.

2. Type the data used in Example 2–2 into C1. Name the column **TEMPERATURES.**

3. Use the instructions in the textbook to determine the class limits.

    In the next step you will create a new column of data, converting the numeric variable to text categories that can be tallied.

4. Select **Data>Code>Numeric to Text.**

    a) The cursor should be blinking in Code data from columns. If not, click inside the box, then double-click C1 Temperatures in the list. Only quantitative variables will be shown in this list.

    b) Click in the Into columns: then type the name of the new column, **TempCodes.**

    c) Press [Tab] to move to the next dialog box.

    d) Type in the first interval **100:104.**

    Use a colon to indicate the interval from 100 to 104 with no spaces before or after the colon.

    e) Press [Tab] to move to the New: column, and type the text category **100–104.**

    f) Continue to tab to each dialog box, typing the interval and then the category until the last category has been entered.

The dialog box should look like the one shown.

Code - Numeric to Text

Code data from columns:

TEMPERATURES

Into columns:

TempCodes

Original values (eg. 1:4 12):   **New:**

100:104	100 – 104
105:109	105 – 109
110:114	110 – 114
115:119	115 – 119
120:124	120 – 124
125:129	125 – 129
130:134	130 – 134

Select

Help   OK   Cancel

5. Click [OK]. In the worksheet, a new column of data will be created in the first empty column, C2. This new variable will contain the category for each value in C1. The column C2-T contains alphanumeric data.

6. Click **Stat>Tables>Tally Individual Values,** then double-click TempCodes in the Variables list.

  a) Check the boxes for the desired statistics, such as Counts, Percents, and Cumulative percents.

  b) Click [OK].

The table will be displayed in the Session Window. Eighteen states have high temperatures between 110 and 114°F. Eighty-two percent of the states have record high temperatures less than or equal to 119°F.

Tally for Discrete Variables: TempCodes

TempCodes	Count	Percent	CumPct
100–104	2	4.00	4.00
105–109	8	16.00	20.00
110–114	18	36.00	56.00
115–119	13	26.00	82.00
120–124	7	14.00	96.00
125–129	1	2.00	98.00
130–134	1	2.00	100.00
N=	50		

7. Click **File>Save Project As** . . . , and type the name of the project file, **Ch2-2.** This will save the two worksheets and the Session Window.

Excel
Step by Step

Categorical Frequency Table (Qualitative or Discrete Data)

1. In an open workbook select cell A1 and type in all the blood types from Example 2–1 down column A.

2. Type in the variable name **Blood Type** in cell B1.

3. Select cell B2 and type in the four different blood types down the column.

4. Type in the name **Count** in cell C1.

5. Select cell C2. From the toolbar, select the Formulas tab on the toolbar.

6. Select the Insert Function icon ⨍, then select the Statistical category in the Insert Function dialog box.

7. Select the Countif function from the function name list.

8. In the dialog box, type **A1:A25** in the **Range** box. Type in the blood type "A" in quotes in the **Criteria** box. The count or frequency of the number of data corresponding to the blood type should appear below the input. Repeat for the remaining blood types.

9. After all the data have been counted, select cell C6 in the worksheet.

10. From the toolbar select Formulas, then AutoSum and type in C2:C5 to insert the total frequency into cell C6.

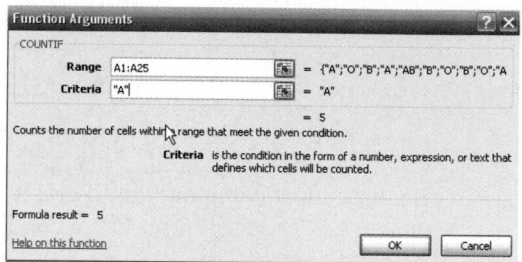

After entering data or a heading into a worksheet, you can change the width of a column to fit the input. To automatically change the width of a column to fit the data:

1. Select the column or columns that you want to change.

2. On the Home tab, in the Cells group, select Format.

3. Under Cell Size, click Autofit Column Width.

Making a Grouped Frequency Distribution (Quantitative Data)

1. Press **[Ctrl]-N** for a new workbook.

2. Enter the raw data from Example 2–2 in column A, one number per cell.

3. Enter the upper class boundaries in column B.

4. From the toolbar select the Data tab, then click Data Analysis.

5. In the Analysis Tools, select Histogram and click [OK].

6. In the Histogram dialog box, type **A1:A50** in the Input Range box and type **B1:B7** in the Bin Range box.

7. Select New Worksheet Ply, and check the Cumulative Percentage option. Click [OK].

8. You can change the label for the column containing the upper class boundaries and expand the width of the columns automatically after relabeling:

Select the Home tab from the toolbar.

Highlight the columns that you want to change.

Select Format, then AutoFit Column Width.

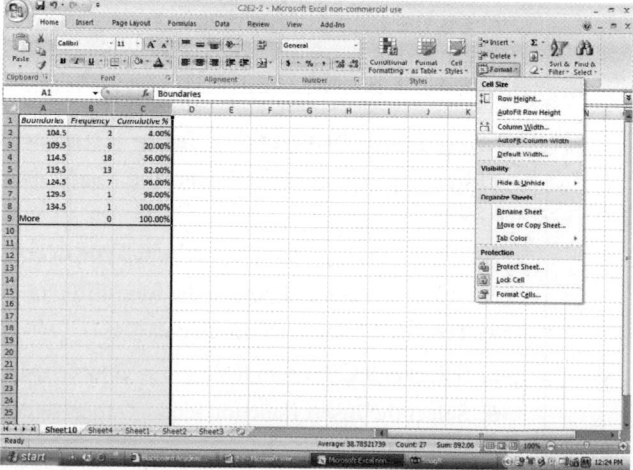

Note: By leaving the Chart Output unchecked, a new worksheet will display the table only.

2–2

Objective 2

Represent data in frequency distributions graphically using histograms, frequency polygons, and ogives.

Histograms, Frequency Polygons, and Ogives

After you have organized the data into a frequency distribution, you can present them in graphical form. The purpose of graphs in statistics is to convey the data to the viewers in pictorial form. It is easier for most people to comprehend the meaning of data presented graphically than data presented numerically in tables or frequency distributions. This is especially true if the users have little or no statistical knowledge.

Statistical graphs can be used to describe the data set or to analyze it. Graphs are also useful in getting the audience's attention in a publication or a speaking presentation. They can be used to discuss an issue, reinforce a critical point, or summarize a data set. They can also be used to discover a trend or pattern in a situation over a period of time.

The three most commonly used graphs in research are

1. The histogram.
2. The frequency polygon.
3. The cumulative frequency graph, or ogive (pronounced o-jive).

An example of each type of graph is shown in Figure 2–1. The data for each graph are the distribution of the miles that 20 randomly selected runners ran during a given week.

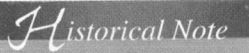

Historical Note

Karl Pearson introduced the histogram in 1891. He used it to show time concepts of various reigns of Prime Ministers.

The Histogram

The **histogram** is a graph that displays the data by using contiguous vertical bars (unless the frequency of a class is 0) of various heights to represent the frequencies of the classes.

Example 2–4

Record High Temperatures

Construct a histogram to represent the data shown for the record high temperatures for each of the 50 states (see Example 2–2).

Class boundaries	Frequency
99.5–104.5	2
104.5–109.5	8
109.5–114.5	18
114.5–119.5	13
119.5–124.5	7
124.5–129.5	1
129.5–134.5	1

Solution

Step 1 Draw and label the x and y axes. The x axis is always the horizontal axis, and the y axis is always the vertical axis.

Figure 2–1

Examples of Commonly Used Graphs

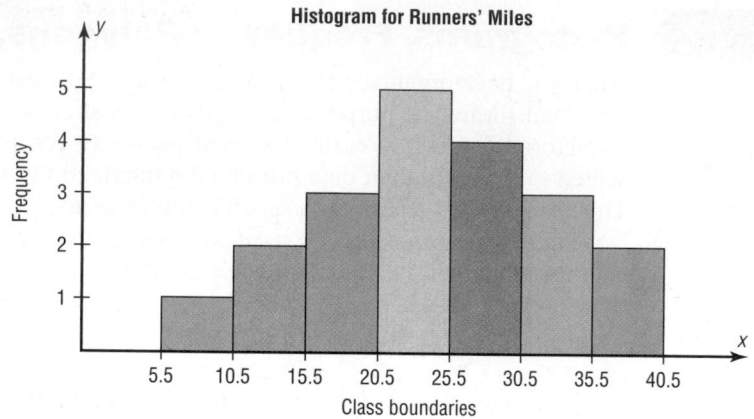

(a) Histogram

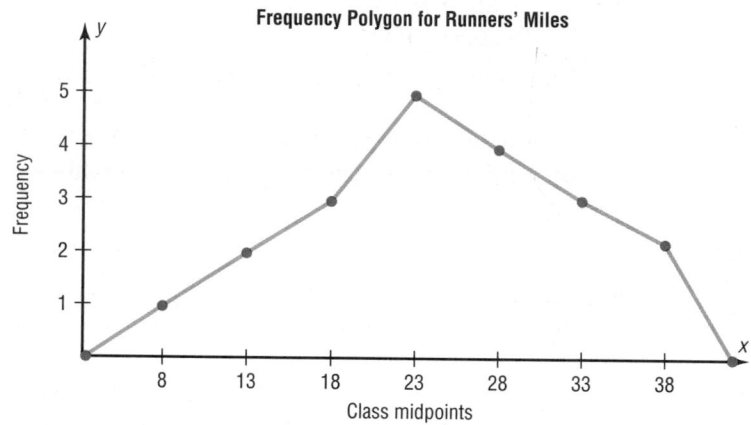

(b) Frequency polygon

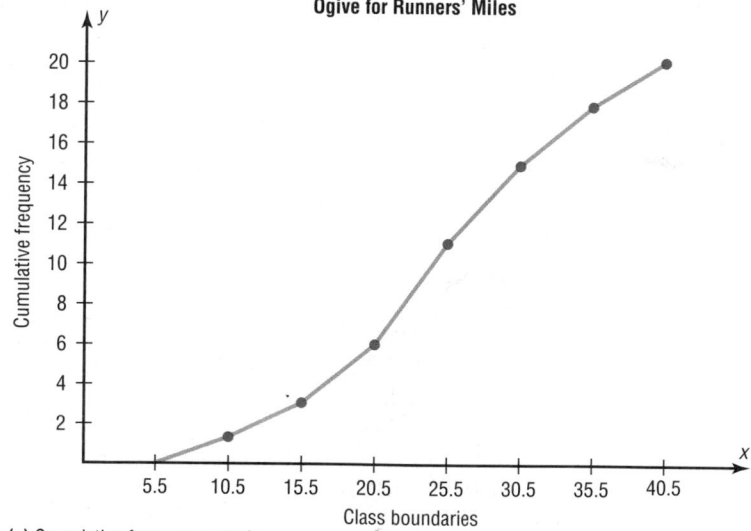

(c) Cumulative frequency graph

Figure 2–2

Histogram for
Example 2–4

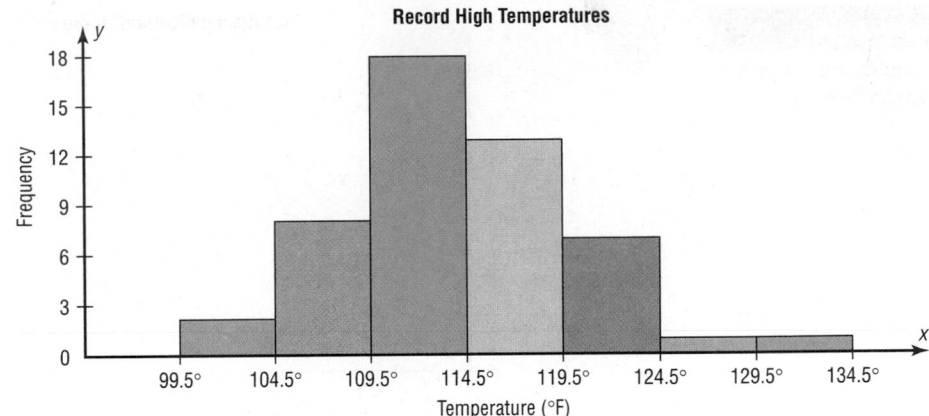

Record High Temperatures

Step 2   Represent the frequency on the y axis and the class boundaries on the x axis.

Step 3   Using the frequencies as the heights, draw vertical bars for each class. See
Figure 2–2.

As the histogram shows, the class with the greatest number of data values (18) is
109.5–114.5, followed by 13 for 114.5–119.5. The graph also has one peak with the data
clustering around it.

The Frequency Polygon

Another way to represent the same data set is by using a frequency polygon.

The **frequency polygon** is a graph that displays the data by using lines that connect
points plotted for the frequencies at the midpoints of the classes. The frequencies are
represented by the heights of the points.

Example 2–5 shows the procedure for constructing a frequency polygon.

Example 2–5

Record High Temperatures

Using the frequency distribution given in Example 2–4, construct a frequency polygon.

Solution

Step 1   Find the midpoints of each class. Recall that midpoints are found by adding
the upper and lower boundaries and dividing by 2:

$$\frac{99.5 + 104.5}{2} = 102 \qquad \frac{104.5 + 109.5}{2} = 107$$

and so on. The midpoints are

Class boundaries	Midpoints	Frequency
99.5–104.5	102	2
104.5–109.5	107	8
109.5–114.5	112	18
114.5–119.5	117	13
119.5–124.5	122	7
124.5–129.5	127	1
129.5–134.5	132	1

Figure 2–3

Frequency Polygon for
Example 2–5

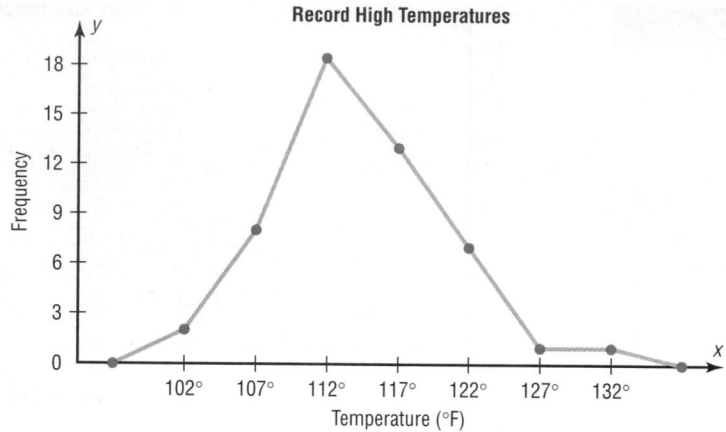

Record High Temperatures

Step 2 Draw the *x* and *y* axes. Label the *x* axis with the midpoint of each class, and then use a suitable scale on the *y* axis for the frequencies.

Step 3 Using the midpoints for the *x* values and the frequencies as the *y* values, plot the points.

Step 4 Connect adjacent points with line segments. Draw a line back to the *x* axis at the beginning and end of the graph, at the same distance that the previous and next midpoints would be located, as shown in Figure 2–3.

The frequency polygon and the histogram are two different ways to represent the same data set. The choice of which one to use is left to the discretion of the researcher.

The Ogive

The third type of graph that can be used represents the cumulative frequencies for the classes. This type of graph is called the *cumulative frequency graph,* or *ogive.* The **cumulative frequency** is the sum of the frequencies accumulated up to the upper boundary of a class in the distribution.

> The **ogive** is a graph that represents the cumulative frequencies for the classes in a frequency distribution.

Example 2–6 shows the procedure for constructing an ogive.

Example 2–6

Record High Temperatures

Construct an ogive for the frequency distribution described in Example 2–4.

Solution

Step 1 Find the cumulative frequency for each class.

	Cumulative frequency
Less than 99.5	0
Less than 104.5	2
Less than 109.5	10
Less than 114.5	28
Less than 119.5	41
Less than 124.5	48
Less than 129.5	49
Less than 134.5	50

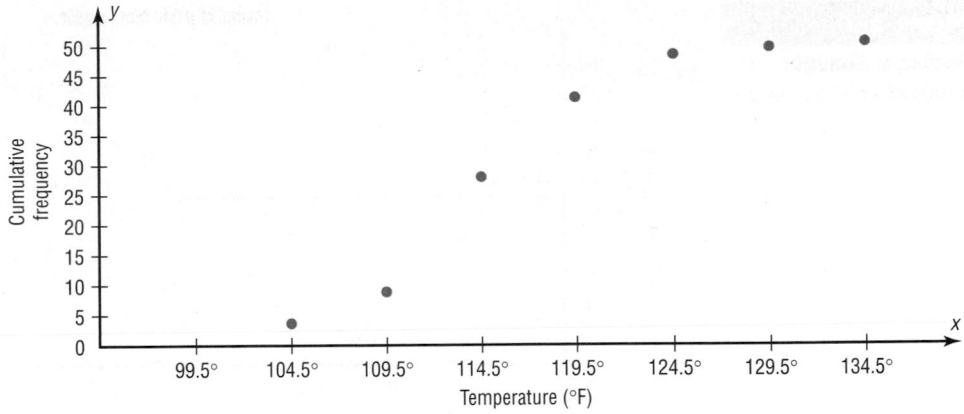

Figure 2–4

Plotting the Cumulative
Frequency for
Example 2–6

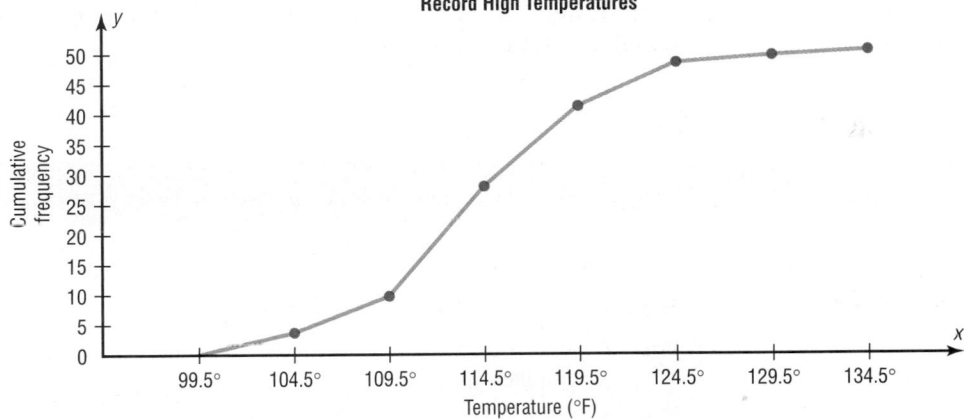

Figure 2–5

Ogive for Example 2–6

Step 2 Draw the x and y axes. Label the x axis with the class boundaries. Use an appropriate scale for the y axis to represent the cumulative frequencies. (Depending on the numbers in the cumulative frequency columns, scales such as 0, 1, 2, 3, . . . , or 5, 10, 15, 20, . . . , or 1000, 2000, 3000, . . . can be used. Do *not* label the y axis with the numbers in the cumulative frequency column.) In this example, a scale of 0, 5, 10, 15, . . . will be used.

Step 3 Plot the cumulative frequency at each upper class boundary, as shown in Figure 2–4. Upper boundaries are used since the cumulative frequencies represent the number of data values accumulated up to the upper boundary of each class.

Step 4 Starting with the first upper class boundary, 104.5, connect adjacent points with line segments, as shown in Figure 2–5. Then extend the graph to the first lower class boundary, 99.5, on the x axis.

Cumulative frequency graphs are used to visually represent how many values are below a certain upper class boundary. For example, to find out how many record high temperatures are less than 114.5°F, locate 114.5°F on the x axis, draw a vertical line up until it intersects the graph, and then draw a horizontal line at that point to the y axis. The y axis value is 28, as shown in Figure 2–6.

Figure 2–6

Finding a Specific
Cumulative Frequency

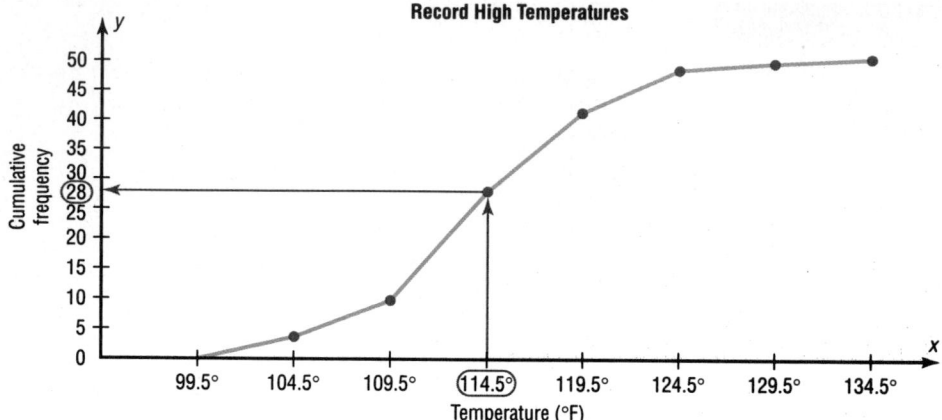

The steps for drawing these three types of graphs are shown in the following Procedure Table.

Procedure Table

Constructing Statistical Graphs

Step 1   Draw and label the x and y axes.

Step 2   Choose a suitable scale for the frequencies or cumulative frequencies, and label it on the y axis.

Step 3   Represent the class boundaries for the histogram or ogive, or the midpoint for the frequency polygon, on the x axis.

Step 4   Plot the points and then draw the bars or lines.

Relative Frequency Graphs

The histogram, the frequency polygon, and the ogive shown previously were constructed by using frequencies in terms of the raw data. These distributions can be converted to distributions using *proportions* instead of raw data as frequencies. These types of graphs are called **relative frequency graphs.**

Graphs of relative frequencies instead of frequencies are used when the proportion of data values that fall into a given class is more important than the actual number of data values that fall into that class. For example, if you wanted to compare the age distribution of adults in Philadelphia, Pennsylvania, with the age distribution of adults of Erie, Pennsylvania, you would use relative frequency distributions. The reason is that since the population of Philadelphia is 1,478,002 and the population of Erie is 105,270, the bars using the actual data values for Philadelphia would be much taller than those for the same classes for Erie.

To convert a frequency into a proportion or relative frequency, divide the frequency for each class by the total of the frequencies. The sum of the relative frequencies will always be 1. These graphs are similar to the ones that use raw data as frequencies, but the values on the y axis are in terms of proportions. Example 2–7 shows the three types of relative frequency graphs.

Example 2–7	**Miles Run per Week**

Construct a histogram, frequency polygon, and ogive using relative frequencies for the distribution (shown here) of the miles that 20 randomly selected runners ran during a given week.

Class boundaries	Frequency
5.5–10.5	1
10.5–15.5	2
15.5–20.5	3
20.5–25.5	5
25.5–30.5	4
30.5–35.5	3
35.5–40.5	2
	20

Solution

Step 1   Convert each frequency to a proportion or relative frequency by dividing the frequency for each class by the total number of observations.

For class 5.5–10.5, the relative frequency is $\frac{1}{20} = 0.05$; for class 10.5–15.5, the relative frequency is $\frac{2}{20} = 0.10$; for class 15.5–20.5, the relative frequency is $\frac{3}{20} = 0.15$; and so on.

Place these values in the column labeled Relative frequency.

Class boundaries	Midpoints	Relative frequency
5.5–10.5	8	0.05
10.5–15.5	13	0.10
15.5–20.5	18	0.15
20.5–25.5	23	0.25
25.5–30.5	28	0.20
30.5–35.5	33	0.15
35.5–40.5	38	0.10
		1.00

Step 2   Find the cumulative relative frequencies. To do this, add the frequency in each class to the total frequency of the preceding class. In this case, $0 + 0.05 = 0.05, 0.05 + 0.10 = 0.15, 0.15 + 0.15 = 0.30, 0.30 + 0.25 = 0.55$, etc. Place these values in the column labeled Cumulative relative frequency.

An alternative method would be to find the cumulative frequencies and then convert each one to a relative frequency.

	Cumulative frequency	Cumulative relative frequency
Less than 5.5	0	0.00
Less than 10.5	1	0.05
Less than 15.5	3	0.15
Less than 20.5	6	0.30
Less than 25.5	11	0.55
Less than 30.5	15	0.75
Less than 35.5	18	0.90
Less than 40.5	20	1.00

Step 3 Draw each graph as shown in Figure 2–7. For the histogram and ogive, use the class boundaries along the *x* axis. For the frequency polygon, use the midpoints on the *x* axis. The scale on the *y* axis uses proportions.

Figure 2–7

Graphs for
Example 2–7

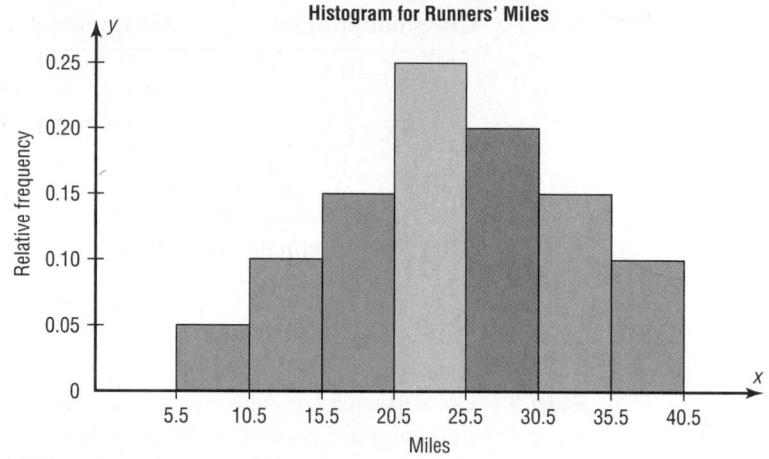

(a) Histogram

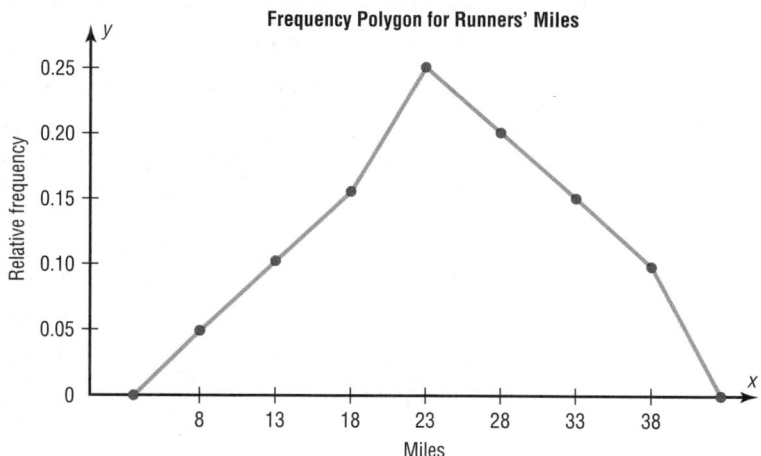

(b) Frequency polygon

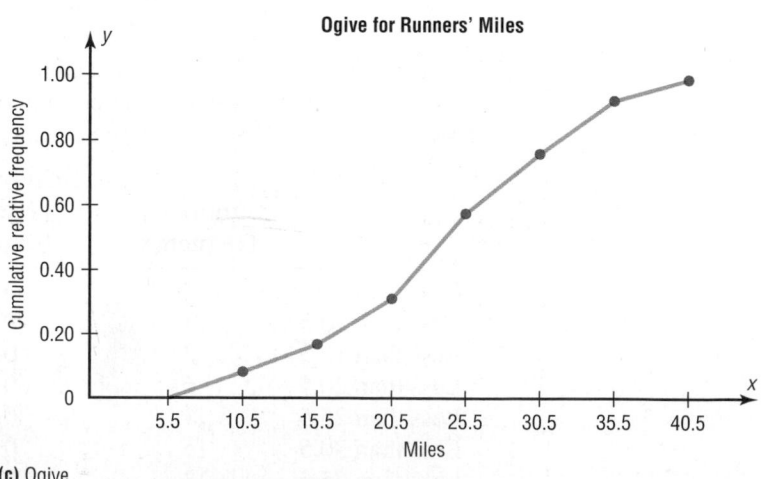

(c) Ogive

Distribution Shapes

When one is describing data, it is important to be able to recognize the shapes of the distribution values. In later chapters you will see that the shape of a distribution also determines the appropriate statistical methods used to analyze the data.

A distribution can have many shapes, and one method of analyzing a distribution is to draw a histogram or frequency polygon for the distribution. Several of the most common shapes are shown in Figure 2–8: *the bell-shaped or mound-shaped, the uniform-shaped, the J-shaped, the reverse J-shaped, the positively or right-skewed shape, the negatively or left-skewed shape, the bimodal-shaped, and the U-shaped.*

Distributions are most often not perfectly shaped, so it is not necessary to have an exact shape but rather to identify an overall pattern.

A *bell-shaped distribution* shown in Figure 2–8(a) has a single peak and tapers off at either end. It is approximately symmetric; i.e., it is roughly the same on both sides of a line running through the center.

Figure 2–8

Distribution Shapes

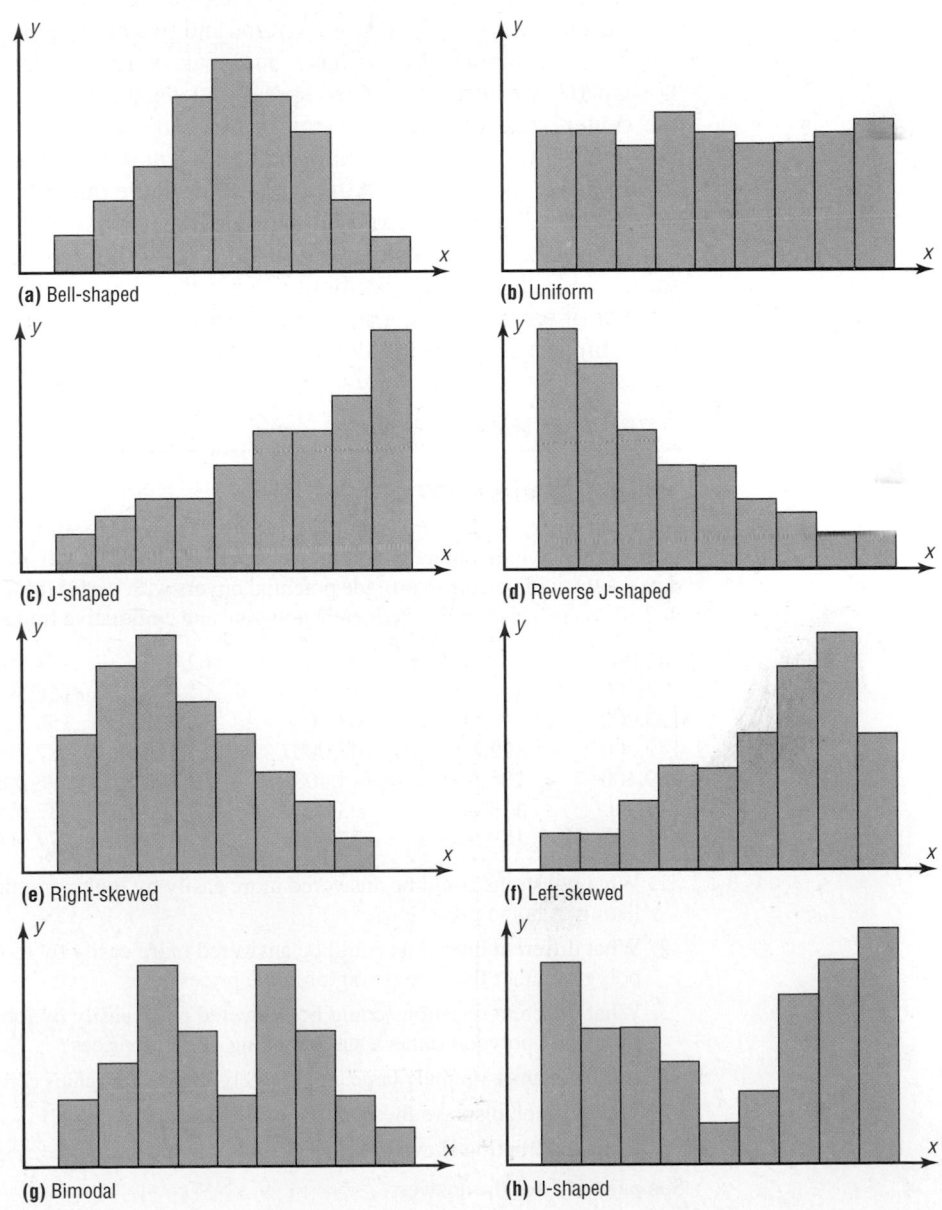

(a) Bell-shaped

(b) Uniform

(c) J-shaped

(d) Reverse J-shaped

(e) Right-skewed

(f) Left-skewed

(g) Bimodal

(h) U-shaped

A *uniform distribution* is basically flat or rectangular. See Figure 2–8(b).

A *J-shaped distribution* is shown in Figure 2–8(c), and it has a few data values on the left side and increases as one moves to the right. A *reverse J-shaped distribution* is the opposite of the J-shaped distribution. See Figure 2–8(d).

When the peak of a distribution is to the left and the data values taper off to the right, a distribution is said to be *positively or right-skewed*. See Figure 2–8(e). When the data values are clustered to the right and taper off to the left, a distribution is said to be *negatively or left-skewed*. See Figure 2–8(f). Skewness will be explained in detail in Chapter 3. Distributions with one peak, such as those shown in Figure 2–8(a), (e), and (f), are said to be *unimodal*. (The highest peak of a distribution indicates where the mode of the data values is. The mode is the data value that occurs more often than any other data value. Modes are explained in Chapter 3.) When a distribution has two peaks of the same height, it is said to be *bimodal*. See Figure 2–8(g). Finally, the graph shown in Figure 2–8(h) is a *U-shaped* distribution.

Distributions can have other shapes in addition to the ones shown here; however, these are some of the more common ones that you will encounter in analyzing data.

When you are analyzing histograms and frequency polygons, look at the shape of the curve. For example, does it have one peak or two peaks? Is it relatively flat, or is it U-shaped? Are the data values spread out on the graph, or are they clustered around the center? Are there data values in the extreme ends? These may be *outliers*. (See Section 3–3 for an explanation of outliers.) Are there any gaps in the histogram, or does the frequency polygon touch the x axis somewhere other than at the ends? Finally, are the data clustered at one end or the other, indicating a *skewed distribution*?

For example, the histogram for the record high temperatures shown in Figure 2–2 shows a single peaked distribution, with the class 109.5–114.5 containing the largest number of temperatures. The distribution has no gaps, and there are fewer temperatures in the highest class than in the lowest class.

Applying the Concepts **2–2**

Selling Real Estate

Assume you are a realtor in Bradenton, Florida. You have recently obtained a listing of the selling prices of the homes that have sold in that area in the last 6 months. You wish to organize those data so you will be able to provide potential buyers with useful information. Use the following data to create a histogram, frequency polygon, and cumulative frequency polygon.

142,000	127,000	99,600	162,000	89,000	93,000	99,500
73,800	135,000	119,500	67,900	156,300	104,500	108,650
123,000	91,000	205,000	110,000	156,300	104,000	133,900
179,000	112,000	147,000	321,550	87,900	88,400	180,000
159,400	205,300	144,400	163,000	96,000	81,000	131,000
114,000	119,600	93,000	123,000	187,000	96,000	80,000
231,000	189,500	177,600	83,400	77,000	132,300	166,000

1. What questions could be answered more easily by looking at the histogram rather than the listing of home prices?

2. What different questions could be answered more easily by looking at the frequency polygon rather than the listing of home prices?

3. What different questions could be answered more easily by looking at the cumulative frequency polygon rather than the listing of home prices?

4. Are there any extremely large or extremely small data values compared to the other data values?

5. Which graph displays these extremes the best?

6. Is the distribution skewed?

See page 101 for the answers.

1. Do Students Need Summer Development? For 108 randomly selected college applicants, the following frequency distribution for entrance exam scores was obtained. Construct a histogram, frequency polygon, and ogive for the data. (The data for this exercise will be used for Exercise 13 in this section.)

Class limits	Frequency
90–98	6
99–107	22
108–116	43
117–125	28
126–134	9

Applicants who score above 107 need not enroll in a summer developmental program. In this group, how many students do not have to enroll in the developmental program?

2. Number of College Faculty The number of faculty listed for a variety of private colleges that offer only bachelor's degrees is listed below. Use these data to construct a frequency distribution with 7 classes, a histogram, a frequency polygon, and an ogive. Discuss the shape of this distribution. What proportion of schools have 180 or more faculty?

165	221	218	206	138	135	224	204
70	210	207	154	155	82	120	116
176	162	225	214	93	389	77	135
221	161	128	310				

Source: *World Almanac and Book of Facts.*

3. Counties, Divisions, or Parishes for 50 States The number of counties, divisions, or parishes for each of the 50 states is given below. Use the data to construct a grouped frequency distribution with 6 classes, a histogram, a frequency polygon, and an ogive. Analyze the distribution. (The data in this exercise will be used for Exercise 24 in Section 2-2.)

67	27	15	75	58	64	8	67	159	5
102	44	92	99	105	120	64	16	23	14
83	87	82	114	56	93	16	10	21	33
62	100	53	88	77	36	67	5	46	66
95	254	29	14	95	39	55	72	23	3

Source: *World Almanac and Book of Facts.*

4. NFL Salaries The salaries (in millions of dollars) for 31 NFL teams for a specific season are given in this frequency distribution.

Class limits	Frequency
39.9–42.8	2
42.9–45.8	2
45.9–48.8	5
48.9–51.8	5
51.9–54.8	12
54.9–57.8	5

Source: NFL.com

Construct a histogram, a frequency polygon, and an ogive for the data; and comment on the shape of the distribution.

5. Railroad Crossing Accidents The data show the number of railroad crossing accidents for the 50 states of the United States for a specific year. Construct a histogram, frequency polygon, and ogive for the data. Comment on the skewness of the distribution. (The data in this exercise will be used for Exercise 14 in this section.)

Class limits	Frequency
1–43	24
44–86	17
87–129	3
130–172	4
173–215	1
216–258	0
259–301	0
302–344	1

Source: Federal Railroad Administration.

6. Costs of Utilities The frequency distribution represents the cost (in cents) for the utilities of states that supply much of their own power. Construct a histogram, frequency polygon, and ogive for the data. Is the distribution skewed?

Class limits	Frequency
6–8	12
9–11	16
12–14	3
15–17	1
18–20	0
21–23	0
24–26	1

7. Air Quality Standards The number of days that selected U.S. metropolitan areas failed to meet acceptable air quality standards is shown below for 1998 and 2003. Construct a grouped frequency distribution with 7 classes and a histogram for each set of data, and compare your results.

1998						2003					
43	76	51	14	0	10	10	11	14	20	15	6
20	0	5	17	67	25	17	0	5	19	127	4
38	0	56	8	0	9	31	5	88	1	1	16
14	5	37	14	95	20	14	19	20	9	138	22
23	12	33	0	3	45	13	10	20	20	20	12

Source: *World Almanac.*

8. How Quick Are Dogs? In a study of reaction times of dogs to a specific stimulus, an animal trainer obtained the following data, given in seconds. Construct a histogram, a frequency polygon, and an ogive for the data; analyze the results. (The histogram in this exercise will be used for Exercise 18 in this section, Exercise 16 in Section 3–1, and Exercise 26 in Section 3–2.)

Class limits	Frequency
2.3–2.9	10
3.0–3.6	12
3.7–4.3	6
4.4–5.0	8
5.1–5.7	4
5.8–6.4	2

9. Quality of Health Care The scores of health care quality as calculated by a professional risk management company are listed for selected states. Use the data to construct a frequency distribution with 6 classes, a histogram, a frequency polygon, and an ogive.

118.2 114.6 113.1 111.9 110.0 108.8 108.3 107.7 107.0 106.7
105.3 103.7 103.2 102.8 101.6 99.8 98.1 96.6 95.7 93.6
92.5 91.0 90.0 87.1 83.1

Source: *New York Times Almanac.*

10. Making the Grade The frequency distributions shown indicate the percentages of public school students in fourth-grade reading and mathematics who performed at or above the required proficiency levels for the 50 states in the United States. Draw histograms for each, and decide if there is any difference in the performance of the students in the subjects.

Class	Reading frequency	Math frequency
17.5–22.5	7	5
22.5–27.5	6	9
27.5–32.5	14	11
32.5–37.5	19	16
37.5–42.5	3	8
42.5–47.5	1	1

Source: *National Center for Educational Statistics.*

11. Construct a histogram, frequency polygon, and ogive for the data in Exercise 16 in Section 2–1 and analyze the results.

12. For the data in Exercise 18 in Section 2–1, construct a histogram for the state gasoline taxes.

13. For the data in Exercise 1 in this section, construct a histogram, a frequency polygon, and an ogive, using relative frequencies. What proportion of the applicants needs to enroll in the summer development program?

14. For the data in Exercise 5 in this section, construct a histogram, frequency polygon, and ogive using relative frequencies. What proportion of the railroad crossing accidents are less than 87?

15. Cereal Calories The number of calories per serving for selected ready-to-eat cereals is listed here. Construct a frequency distribution using 7 classes. Draw a histogram, a frequency polygon, and an ogive for the data, using relative frequencies. Describe the shape of the histogram.

130 190 140 80 100 120 220 220 110 100
210 130 100 90 210 120 200 120 180 120
190 210 120 200 130 180 260 270 100 160
190 240 80 120 90 190 200 210 190 180
115 210 110 225 190 130

Source: *The Doctor's Pocket Calorie, Fat, and Carbohydrate Counter.*

16. Protein Grams in Fast Food The amount of protein (in grams) for a variety of fast-food sandwiches is reported here. Construct a frequency distribution using 6 classes. Draw a histogram, a frequency polygon, and an ogive for the data, using relative frequencies. Describe the shape of the histogram.

23 30 20 27 44 26 35 20 29 29
25 15 18 27 19 22 12 26 34 15
27 35 26 43 35 14 24 12 23 31
40 35 38 57 22 42 24 21 27 33

Source: *The Doctor's Pocket Calorie, Fat, and Carbohydrate Counter.*

17. For the data for year 2003 in Exercise 7 in this section, construct a histogram, a frequency polygon, and an ogive, using relative frequencies.

18. How Quick Are Older Dogs? The animal trainer in Exercise 8 in this section selected another group of dogs who were much older than the first group and measured their reaction times to the same stimulus. Construct a histogram, a frequency polygon, and an ogive for the data.

Class limits	Frequency
2.3–2.9	1
3.0–3.6	3
3.7–4.3	4
4.4–5.0	16
5.1–5.7	14
5.8–6.4	4

Analyze the results and compare the histogram for this group with the one obtained in Exercise 8 in this section. Are there any differences in the histograms? (The data in this exercise will be used for Exercise 16 in Section 3–1 and Exercise 26 in Section 3–2.)

Extending the Concepts

19. Using the histogram shown here, do the following.

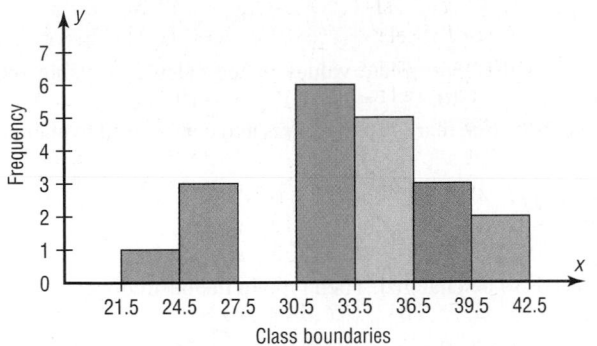

20. Using the results from Exercise 19, answer these questions.

a. Construct a frequency distribution; include class limits, class frequencies, midpoints, and cumulative frequencies.

b. Construct a frequency polygon.

c. Construct an ogive.

a. How many values are in the class 27.5–30.5? 0
b. How many values fall between 24.5 and 36.5? 14
c. How many values are below 33.5? 10
d. How many values are above 30.5? 16

Technology *Step by Step*

MINITAB
Step by Step

Construct a Histogram

1. Enter the data from Example 2–2, the high temperatures for the 50 states.
2. Select **Graph>Histogram.**
3. Select [Simple], then click [OK].
4. Click C1 TEMPERATURES in the Graph variables dialog box.
5. Click [Labels]. There are two tabs, Title/Footnote and Data Labels.

   a) Click in the box for Title, and type in Your Name and Course Section.

   b) Click [OK]. The Histogram dialog box is still open.

6. Click [OK]. A new graph window containing the histogram will open.

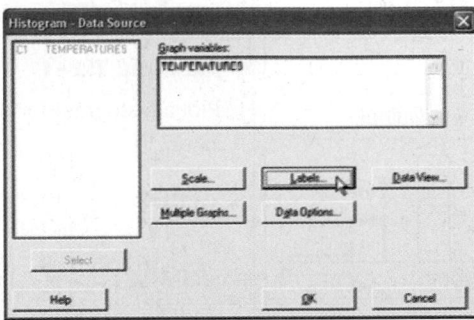

7. Click the **File** menu to print or save the graph.

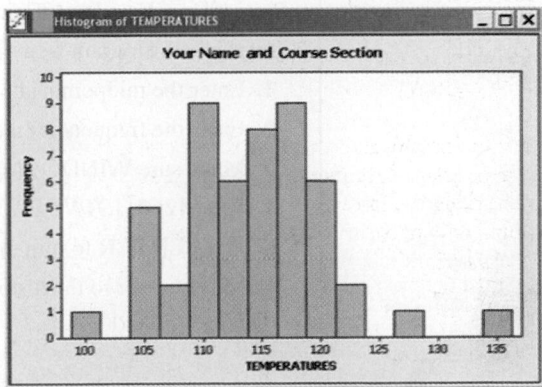

8. Click **File>Exit.**

9. Save the project as **Ch2-3.mpj.**

TI-83 Plus or TI-84 Plus
Step by Step

Constructing a Histogram

To display the graphs on the screen, enter the appropriate values in the calculator, using the WINDOW menu. The default values are $X_{min} = -10$, $X_{max} = +10$, $Y_{min} = -10$, and $Y_{max} = +10$. The X_{scl} changes the distance between the tick marks on the x axis and can be used to change the class width for the histogram.

To change the values in the WINDOW:

1. Press **WINDOW.**

2. Move the cursor to the value that needs to be changed. Then type in the desired value and press **ENTER.**

3. Continue until all values are appropriate.

4. Press **[2nd] [QUIT]** to leave the WINDOW menu.

To plot the histogram from raw data:

1. Enter the data in L_1.

Input

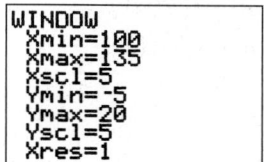

2. Make sure WINDOW values are appropriate for the histogram.

3. Press **[2nd] [STAT PLOT] ENTER.**

4. Press **ENTER** to turn the plot on, if necessary.

Input

5. Move cursor to the Histogram symbol and press **ENTER,** if necessary.

6. Make sure Xlist is L_1.

7. Make sure Freq is 1.

8. Press **GRAPH** to display the histogram.

9. To obtain the number of data values in each class, press the **TRACE** key, followed by ◄ or ► keys.

Output

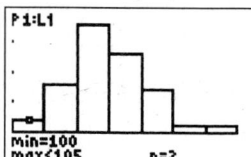

Example TI2–1

Plot a histogram for the following data from Examples 2–2 and 2–4.

112	100	127	120	134	118	105	110	109	112
110	118	117	116	118	122	114	114	105	109
107	112	114	115	118	117	118	122	106	110
116	108	110	121	113	120	119	111	104	111
120	113	120	117	105	110	118	112	114	114

Press **TRACE** and use the arrow keys to determine the number of values in each group.

To graph a histogram from grouped data:

1. Enter the midpoints into L_1.

2. Enter the frequencies into L_2.

3. Make sure WINDOW values are appropriate for the histogram.

4. Press **[2nd] [STAT PLOT] ENTER.**

5. Press **ENTER** to turn the plot on, if necessary.

6. Move cursor to the histogram symbol, and press **ENTER,** if necessary.

7. Make sure Xlist is L_1.

8. Make sure Freq is L_2.

9. Press **GRAPH** to display the histogram.

Example TI2–2

Plot a histogram for the data from Examples 2–4 and 2–5.

Class boundaries	Midpoints	Frequency
99.5–104.5	102	2
104.5–109.5	107	8
109.5–114.5	112	18
114.5–119.5	117	13
119.5–124.5	122	7
124.5–129.5	127	1
129.5–134.5	132	1

Output

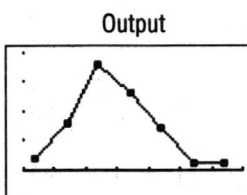

Output

Input

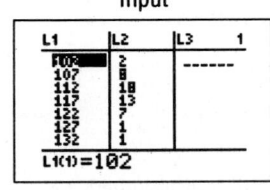

Input

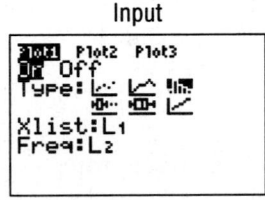

Output

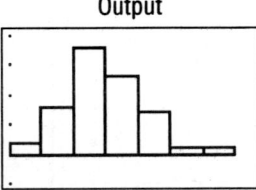

To graph a frequency polygon from grouped data, follow the same steps as for the histogram except change the graph type from histogram (third graph) to a line graph (second graph).

To graph an ogive from grouped data, modify the procedure for the histogram as follows:

1. Enter the upper class boundaries into L_1.
2. Enter the cumulative frequencies into L_2.
3. Change the graph type from histogram (third graph) to line (second graph).
4. Change the Y_{max} from the WINDOW menu to the sample size.

Excel
Step by Step

Constructing a Histogram

1. Press **[Ctrl]-N** for a new workbook.
2. Enter the data from Example 2–2 in column A, one number per cell.
3. Enter the upper boundaries into column B.
4. From the toolbar, select the Data tab, then select Data Analysis.
5. In Data Analysis, select Histogram and click [OK].
6. In the Histogram dialog box, type **A1:A50** in the Input Range box and type **B1:B7** in the Bin Range box.

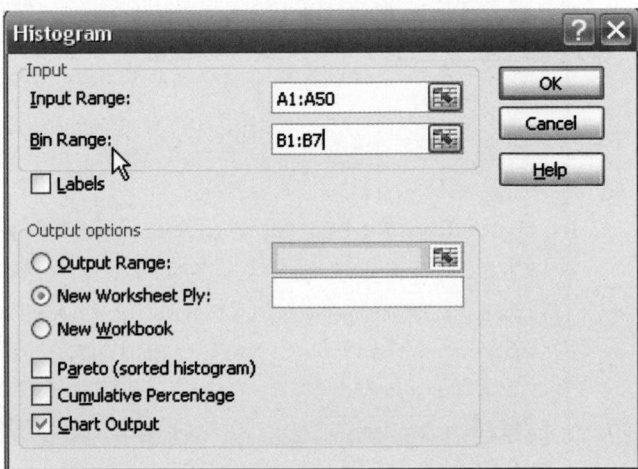

7. Select New Worksheet Ply and Chart Output. Click [OK].

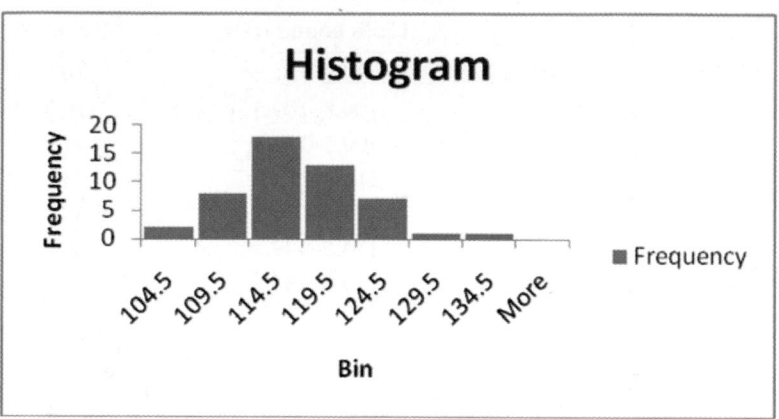

Editing the Histogram

To move the vertical bars of the histogram closer together:

1. Right-click one of the bars of the histogram, and select Format Data Series.

2. Move the Gap Width bar to the left to narrow the distance between bars.

To change the label for the horizontal axis:

1. Left-click the mouse over any region of the histogram.

2. Select the Chart Tools tab from the toolbar.

3. Select the Layout tab, Axis Titles and Primary Horizontal Axis Title.

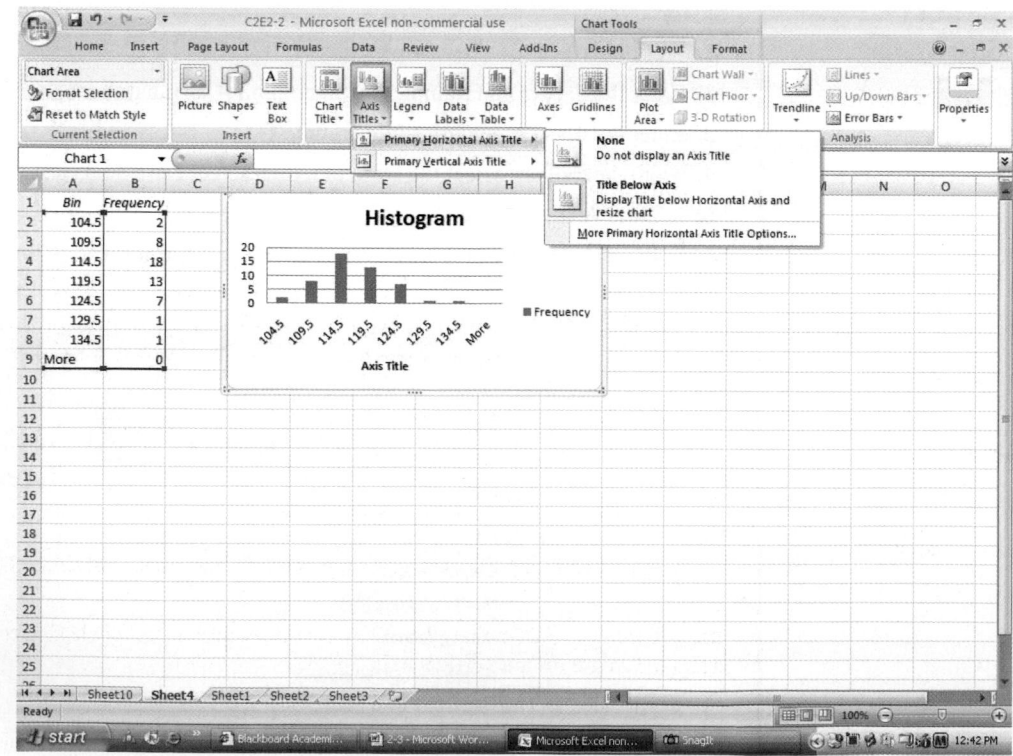

Once the Axis Title text box is selected, you can type in the name of the variable represented on the horizontal axis.

Constructing a Frequency Polygon

1. Press **[Ctrl]-N** for a new workbook.

2. Enter the midpoints of the data from Example 2–2 into column A. Enter the frequencies into column B.

	A	B
1	Midpoints	Frequencies
2	102	2
3	107	8
4	112	18
5	117	13
6	122	7
7	127	1
8	132	1

3. Highlight the Frequencies (including the label) from column B.

4. Select the Insert tab from the toolbar and the Line Chart option.

5. Select the 2-D line chart type.

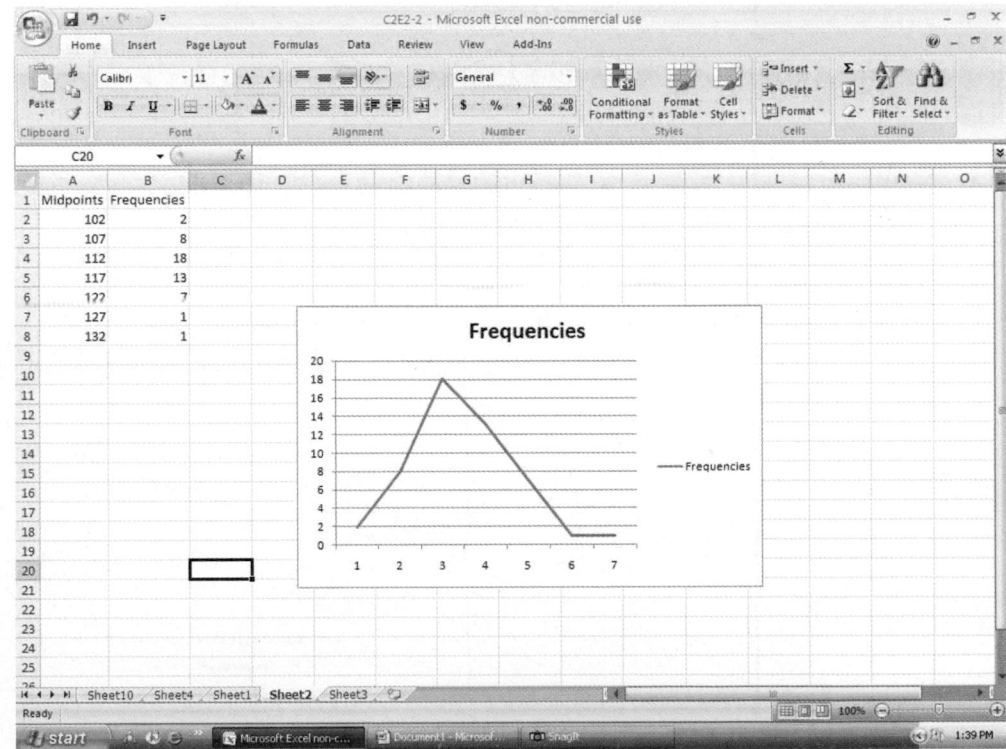

We will need to edit the graph so that the midpoints are on the horizontal axis and the frequencies are on the vertical axis.

1. Right-click the mouse on any region of the graph.

2. Select the Select Data option.

   3. Select Edit from the Horizontal Axis Labels and highlight the midpoints from column A, then click [OK].

   4. Click [OK] on the Select Data Source box.

Inserting Labels on the Axes

   1. Click the mouse on any region of the graph.

   2. Select Chart Tools and then Layout on the toolbar.

   3. Select Axis Titles to open the horizontal and vertical axis text boxes. Then manually type in labels for the axes.

Changing the Title

   1. Select Chart Tools, Layout from the toolbar.

   2. Select Chart Title.

   3. Choose one of the options from the Chart Title menu and edit.

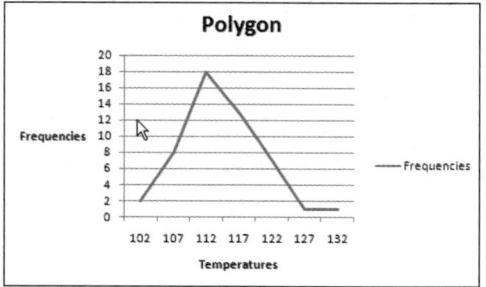

Constructing an Ogive

To create an ogive, you can use the upper class boundaries (horizontal axis) and cumulative frequencies (vertical axis) from the frequency distribution.

   1. Type the upper class boundaries and cumulative frequencies into adjacent columns of an Excel worksheet.

   2. Highlight the cumulative frequencies (including the label) and select the Insert tab from the toolbar.

   3. Select Line Chart, then the 2-D Line option.

As with the frequency polygon, you can insert labels on the axes and a chart title for the ogive.

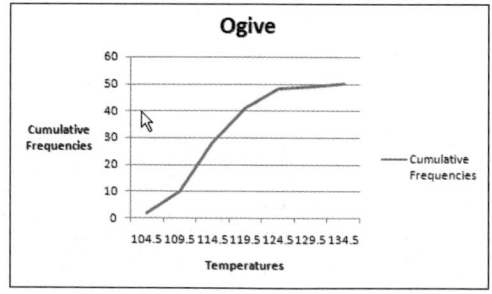

2–3   Other Types of Graphs

In addition to the histogram, the frequency polygon, and the ogive, several other types of graphs are often used in statistics. They are the bar graph, Pareto chart, time series graph, and pie graph. Figure 2–9 shows an example of each type of graph.

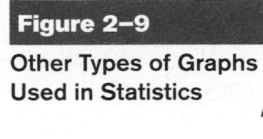

Figure 2–9

Other Types of Graphs
Used in Statistics

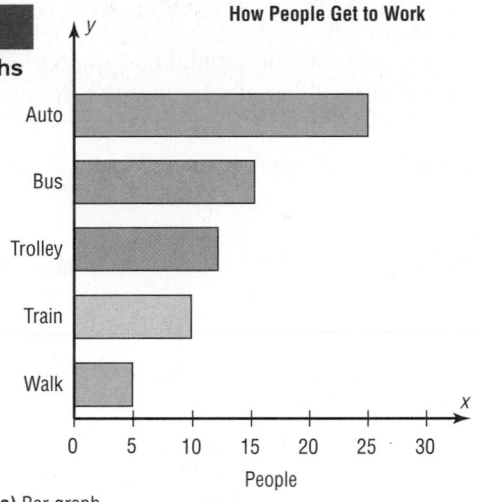

(a) Bar graph

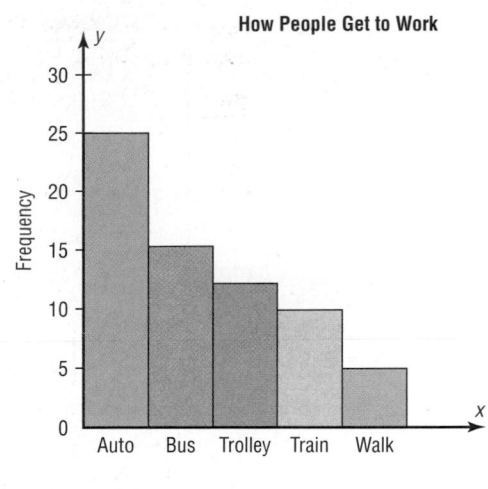

(b) Pareto chart

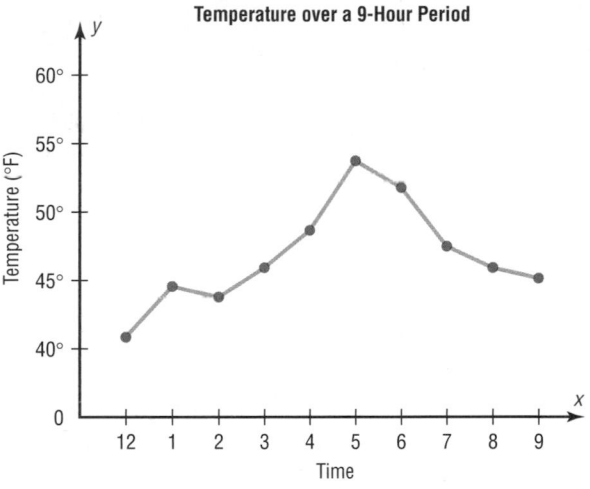

(c) Time series graph

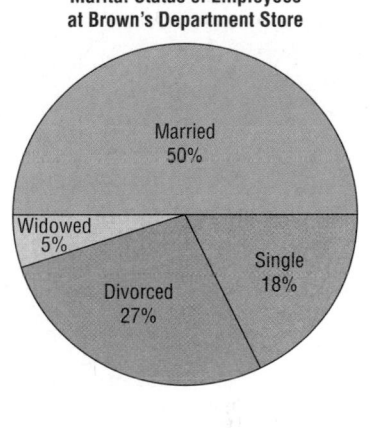

(d) Pie graph

Objective 3

Represent data using
bar graphs, Pareto
charts, time series
graphs, and
pie graphs.

Bar Graphs

When the data are qualitative or categorical, bar graphs can be used to represent the data.
A bar graph can be drawn using either horizontal or vertical bars.

> A **bar graph** represents the data by using vertical or horizontal bars whose heights or
> lengths represent the frequencies of the data.

Example 2–8

College Spending for First-Year Students

The table shows the average money spent by first-year college students. Draw a
horizontal and vertical bar graph for the data.

Electronics	$728
Dorm decor	344
Clothing	141
Shoes	72

Source: The National Retail Federation.

Solution

1. Draw and label the *x* and *y* axes. For the horizontal bar graph place the frequency scale on the *x* axis, and for the vertical bar graph place the frequency scale on the *y* axis.

2. Draw the bars corresponding to the frequencies. See Figure 2–10.

Figure 2–10

Bar Graphs for
Example 2–8

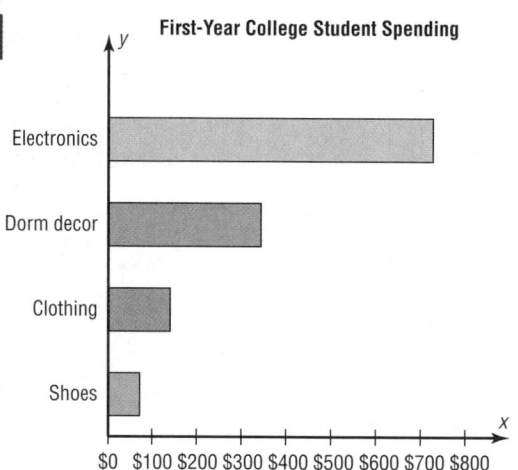

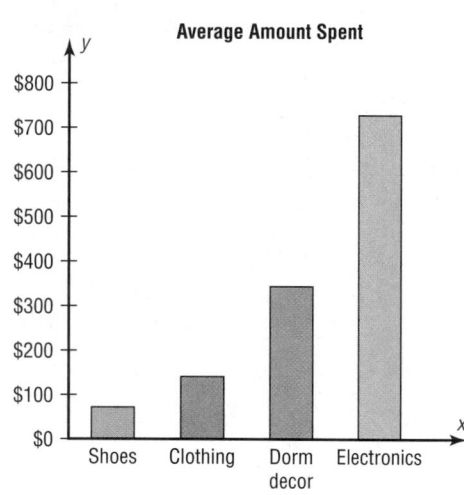

The graphs show that first-year college students spend the most on electronic equipment including computers.

Pareto Charts

When the variable displayed on the horizontal axis is qualitative or categorical, a *Pareto chart* can also be used to represent the data.

A **Pareto chart** is used to represent a frequency distribution for a categorical variable, and the frequencies are displayed by the heights of vertical bars, which are arranged in order from highest to lowest.

Example 2–9

Homeless People

The data shown here consist of the number of homeless people for a sample of selected cities. Construct and analyze a Pareto chart for the data.

City	Number
Atlanta	6832
Baltimore	2904
Chicago	6680
St. Louis	1485
Washington	5518

Source: U.S. Department of Housing and Urban Development.

Solution

Step 1 Arrange the data from the largest to smallest according to frequency.

City	Number
Atlanta	6832
Chicago	6680
Washington	5518
Baltimore	2904
St. Louis	1485

Step 2 Draw and label the x and y axes.

Step 3 Draw the bars corresponding to the frequencies. See Figure 2–11.

The graph shows that the number of homeless people is about the same for Atlanta and Chicago and a lot less for Baltimore and St. Louis.

Suggestions for Drawing Pareto Charts

1. Make the bars the same width.
2. Arrange the data from largest to smallest according to frequency.
3. Make the units that are used for the frequency equal in size.

When you analyze a Pareto chart, make comparisons by looking at the heights of the bars.

The Time Series Graph

When data are collected over a period of time, they can be represented by a time series graph.

Figure 2–11

Pareto Chart for Example 2–9

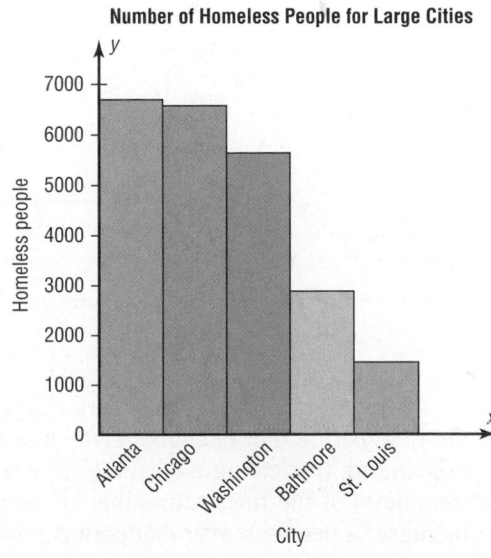

Number of Homeless People for Large Cities

A **time series graph** represents data that occur over a specific period of time.

Example 2–10 shows the procedure for constructing a time series graph.

Example 2–10

Workplace Homicides

The number of homicides that occurred in the workplace for the years 2003 to 2008 is shown. Draw and analyze a time series graph for the data.

Year	'03	'04	'05	'06	'07	'08
Number	632	559	567	540	628	517

Source: Bureau of Labor Statistics.

Solution

Step 1 Draw and label the x and y axes.

Step 2 Label the x axis for years and the y axis for the number.

Step 3 Plot each point according to the table.

Step 4 Draw line segments connecting adjacent points. Do not try to fit a smooth curve through the data points. See Figure 2–12.

There was a slight decrease in the years '04, '05, and '06, compared to '03, and again an increase in '07. The largest decrease occurred in '08.

Historical Note

Time series graphs are over 1000 years old. The first ones were used to chart the movements of the planets and the sun.

Figure 2–12

Time Series Graph for Example 2–10

There was a slight decrease in the years '04, '05, and '06, compared to '03, and again an increase in '07. The largest decrease occurred in '08.

When you analyze a time series graph, look for a trend or pattern that occurs over the time period. For example, is the line ascending (indicating an increase over time) or descending (indicating a decrease over time)? Another thing to look for is the slope, or steepness, of the line. A line that is steep over a specific time period indicates a rapid increase or decrease over that period.

Figure 2–13

Two Time Series
Graphs for Comparison

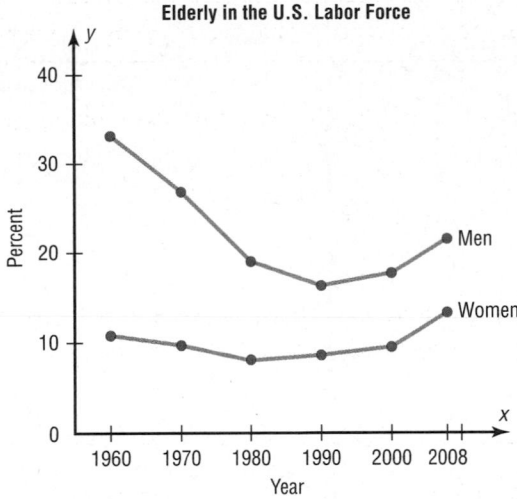

Elderly in the U.S. Labor Force

Source: Bureau of Census, U.S. Department of Commerce.

Two or more data sets can be compared on the same graph called a *compound time series graph* if two or more lines are used, as shown in Figure 2–13. This graph shows the percentage of elderly males and females in the United States labor force from 1960 to 2008. It shows that the percent of elderly men decreased significantly from 1960 to 1990 and then increased slightly after that. For the elderly females, the percent decreased slightly from 1960 to 1980 and then increased from 1980 to 2008.

The Pie Graph

Pie graphs are used extensively in statistics. The purpose of the pie graph is to show the relationship of the parts to the whole by visually comparing the sizes of the sections. Percentages or proportions can be used. The variable is nominal or categorical.

> A **pie graph** is a circle that is divided into sections or wedges according to the percentage of frequencies in each category of the distribution.

Example 2–11 shows the procedure for constructing a pie graph.

Example 2–11

Super Bowl Snack Foods

This frequency distribution shows the number of pounds of each snack food eaten during the Super Bowl. Construct a pie graph for the data.

Snack	Pounds (frequency)
Potato chips	11.2 million
Tortilla chips	8.2 million
Pretzels	4.3 million
Popcorn	3.8 million
Snack nuts	2.5 million
Total $n =$	30.0 million

Source: *USA TODAY Weekend.*

Cell Phone Usage

The graph shows the estimated number (in millions) of cell phone subscribers since 2000. How do you think the growth will affect our way of living? For example, emergencies can be handled faster since people are using their cell phones to call 911.

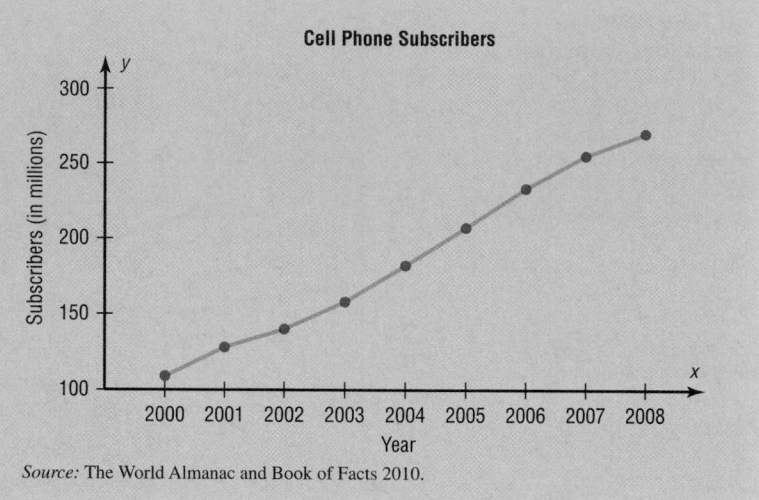

Cell Phone Subscribers

Source: The World Almanac and Book of Facts 2010.

Solution

Step 1 Since there are 360° in a circle, the frequency for each class must be converted into a proportional part of the circle. This conversion is done by using the formula

$$\text{Degrees} = \frac{f}{n} \cdot 360°$$

where f = frequency for each class and n = sum of the frequencies. Hence, the following conversions are obtained. The degrees should sum to 360°.*

Potato chips	$\frac{11.2}{30} \cdot 360°$ =	134°
Tortilla chips	$\frac{8.2}{30} \cdot 360°$ =	98°
Pretzels	$\frac{4.3}{30} \cdot 360°$ =	52°
Popcorn	$\frac{3.8}{30} \cdot 360°$ =	46°
Snack nuts	$\frac{2.5}{30} \cdot 360°$ =	30°
Total		360°

Step 2 Each frequency must also be converted to a percentage. Recall from Example 2–1 that this conversion is done by using the formula

$$\% = \frac{f}{n} \cdot 100$$

Hence, the following percentages are obtained. The percentages should sum to 100%.†

Potato chips	$\frac{11.2}{30} \cdot 100$	= 37.3%
Tortilla chips	$\frac{8.2}{30} \cdot 100$	= 27.3%

*Note: The degrees column does not always sum to 360° due to rounding.
†Note: The percent column does not always sum to 100% due to rounding.

Pretzels $\dfrac{4.3}{30} \cdot 100 = 14.3\%$

Popcorn $\dfrac{3.8}{30} \cdot 100 = 12.7\%$

Snack nuts $\dfrac{2.5}{30} \cdot 100 = 8.3\%$

Total   99.9%

Step 3   Next, using a protractor and a compass, draw the graph using the appropriate degree measures found in step 1, and label each section with the name and percentages, as shown in Figure 2–14.

Figure 2–14

Pie Graph for
Example 2–11

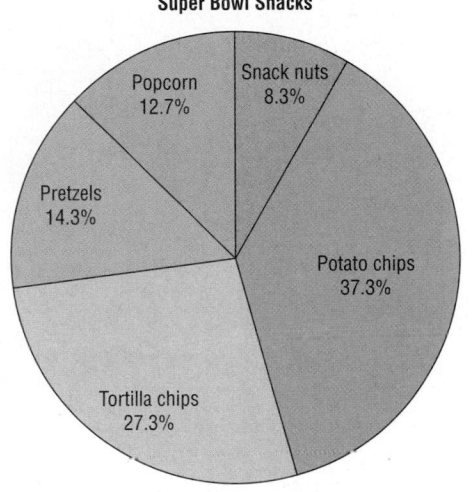

Super Bowl Snacks

Example 2–12   Construct a pie graph showing the blood types of the army inductees described in Example 2–1. The frequency distribution is repeated here.

Class	Frequency	Percent
A	5	20
B	7	28
O	9	36
AB	4	16
	25	100

Solution

Step 1   Find the number of degrees for each class, using the formula

$$\text{Degrees} = \frac{f}{n} \cdot 360°$$

For each class, then, the following results are obtained.

A $\dfrac{5}{25} \cdot 360° = 72°$

B $\dfrac{7}{25} \cdot 360° = 100.8°$

$$O \qquad \frac{9}{25} \cdot 360° = 129.6°$$

$$AB \qquad \frac{4}{25} \cdot 360° = 57.6°$$

Step 2   Find the percentages. (This was already done in Example 2–1.)

Step 3   Using a protractor, graph each section and write its name and corresponding percentage, as shown in Figure 2–15.

Figure 2–15

Pie Graph for
Example 2–12

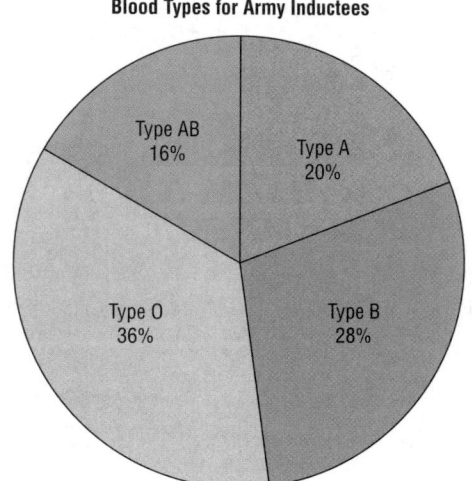

Blood Types for Army Inductees

The graph in Figure 2–15 shows that in this case the most common blood type is type O.

To analyze the nature of the data shown in the pie graph, look at the size of the sections in the pie graph. For example, are any sections relatively large compared to the rest?

Figure 2–15 shows that among the inductees, type O blood is more prevalent than any other type. People who have type AB blood are in the minority. More than twice as many people have type O blood as type AB.

Misleading Graphs

Graphs give a visual representation that enables readers to analyze and interpret data more easily than they could simply by looking at numbers. However, inappropriately drawn graphs can misrepresent the data and lead the reader to false conclusions. For example, a car manufacturer's ad stated that 98% of the vehicles it had sold in the past 10 years were still on the road. The ad then showed a graph similar to the one in Figure 2–16. The graph shows the percentage of the manufacturer's automobiles still on the road and the percentage of its competitors' automobiles still on the road. Is there a large difference? Not necessarily.

Notice the scale on the vertical axis in Figure 2–16. It has been cut off (or truncated) and starts at 95%. When the graph is redrawn using a scale that goes from 0 to 100%, as in Figure 2–17, there is hardly a noticeable difference in the percentages. Thus, changing the units at the starting point on the *y* axis can convey a very different visual representation of the data.

Figure 2–16

Graph of Automaker's
Claim Using a Scale
from 95 to 100%

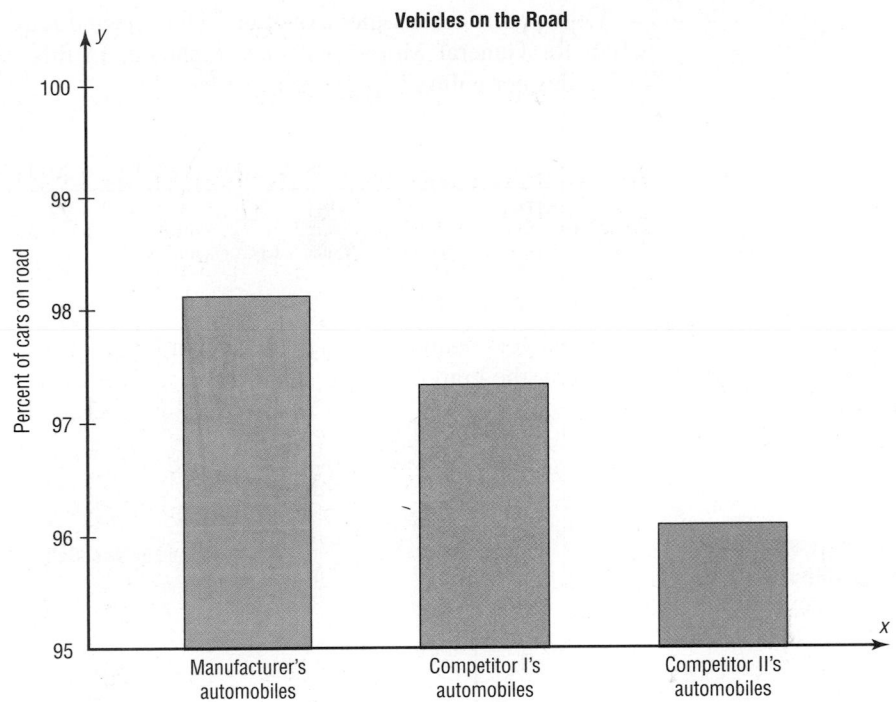

Vehicles on the Road

Figure 2–17

Graph in Figure 2–16
Redrawn Using a Scale
from 0 to 100%

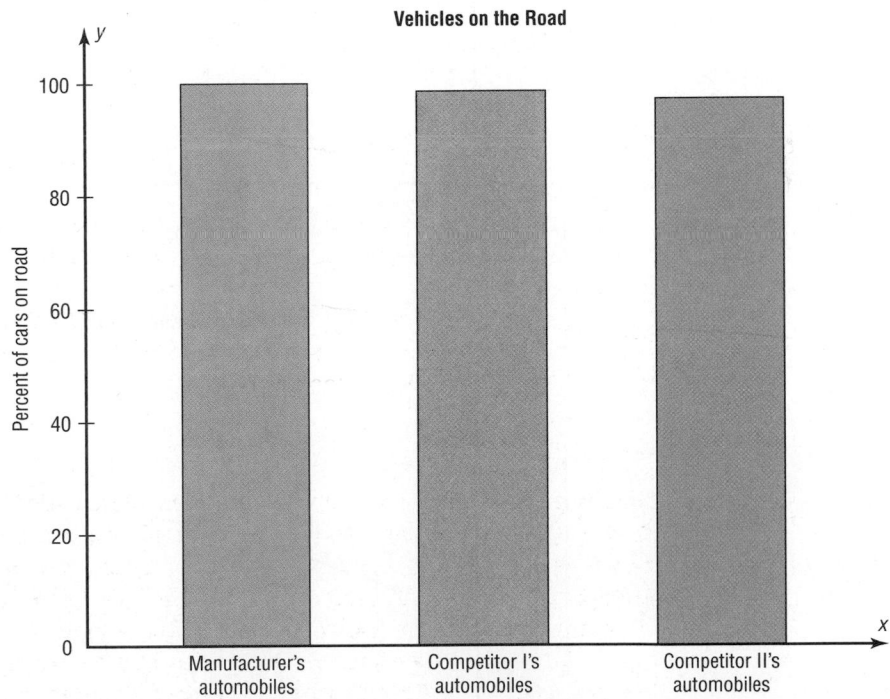

Vehicles on the Road

It is not wrong to truncate an axis of the graph; many times it is necessary to do so. However, the reader should be aware of this fact and interpret the graph accordingly. Do not be misled if an inappropriate impression is given.

Let us consider another example. The projected required fuel economy in miles per gallon for General Motors vehicles is shown. In this case, an increase from 21.9 to 23.2 miles per gallon is projected.

Year	2008	2009	2010	2011
MPG	21.9	22.6	22.9	23.2

Source: National Highway Traffic Safety Administration.

When you examine the graph shown in Figure 2–18(a) using a scale of 0 to 25 miles per gallon, the graph shows a slight increase. However, when the scale is changed to 21

Figure 2–18

Projected Miles per Gallon

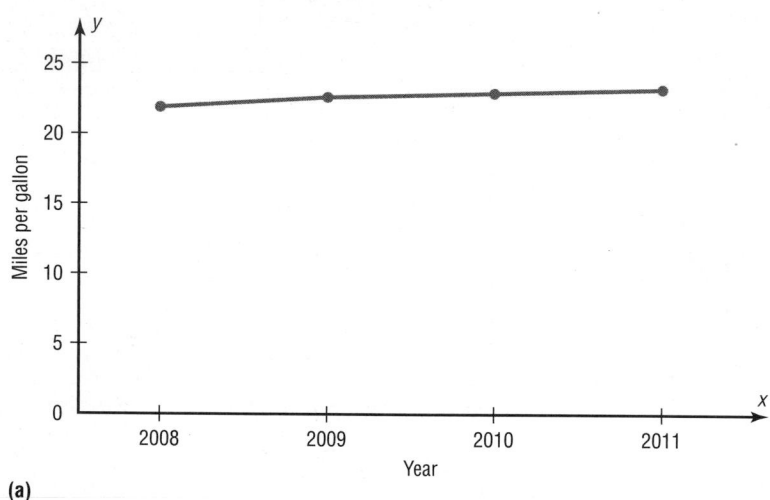

(a)

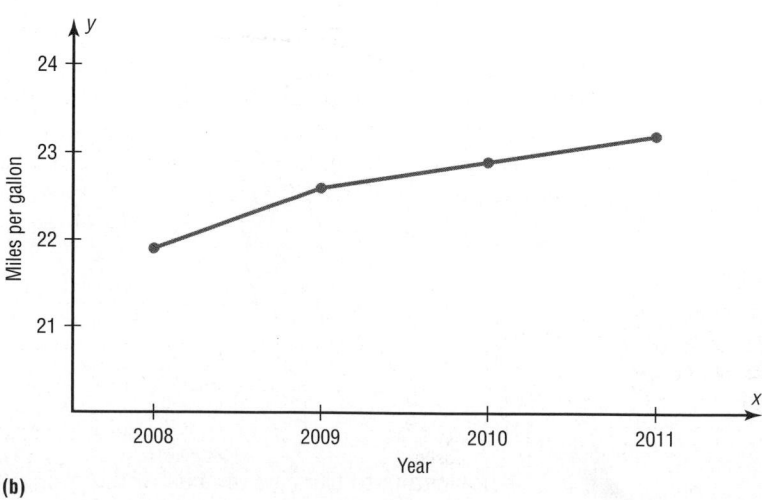

(b)

to 24 miles per gallon, the graph shows a much larger increase even though the data remain the same. See Figure 2–18(b). Again, by changing the units or starting point on the y axis, one can change the visual representation.

Another misleading graphing technique sometimes used involves exaggerating a one-dimensional increase by showing it in two dimensions. For example, the average cost of a 30-second Super Bowl commercial has increased from $42,000 in 1967 to $3 million in 2010 (Source: *USA TODAY*).

The increase shown by the graph in Figure 2–19(a) represents the change by a comparison of the heights of the two bars in one dimension. The same data are shown two-dimensionally with circles in Figure 2–19(b). Notice that the difference seems much larger because the eye is comparing the areas of the circles rather than the lengths of the diameters.

Note that it is not wrong to use the graphing techniques of truncating the scales or representing data by two-dimensional pictures. But when these techniques are used, the reader should be cautious of the conclusion drawn on the basis of the graphs.

Figure 2–19

Comparison of Costs for a 30-Second Super Bowl Commercial

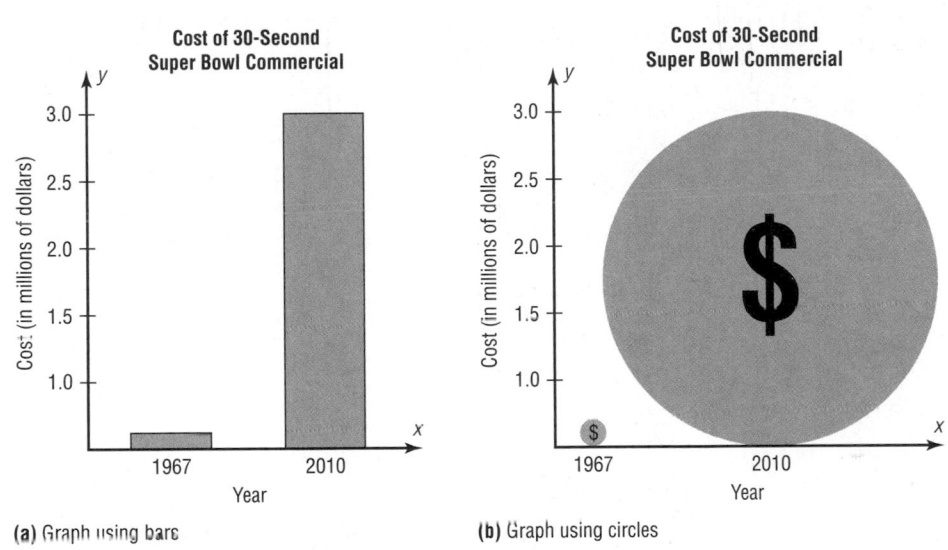

(a) Graph using bars

(b) Graph using circles

Another way to misrepresent data on a graph is by omitting labels or units on the axes of the graph. The graph shown in Figure 2–20 compares the cost of living, economic growth, population growth, etc., of four main geographic areas in the United States. However, since there are no numbers on the y axis, very little information can be gained from this graph, except a crude ranking of each factor. There is no way to decide the actual magnitude of the differences.

Figure 2–20

A Graph with No Units on the y Axis

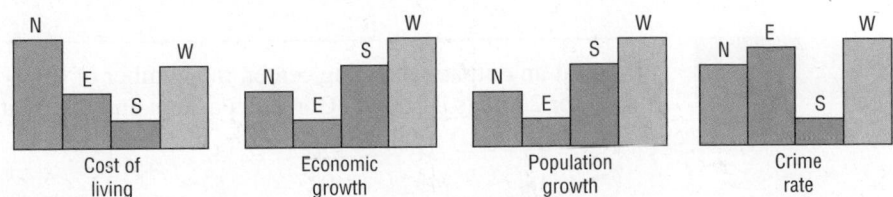

Finally, all graphs should contain a source for the information presented. The inclusion of a source for the data will enable you to check the reliability of the organization presenting the data. A summary of the types of graphs and their uses is shown in Figure 2–21.

Figure 2–21

Summary of Graphs and Uses of Each

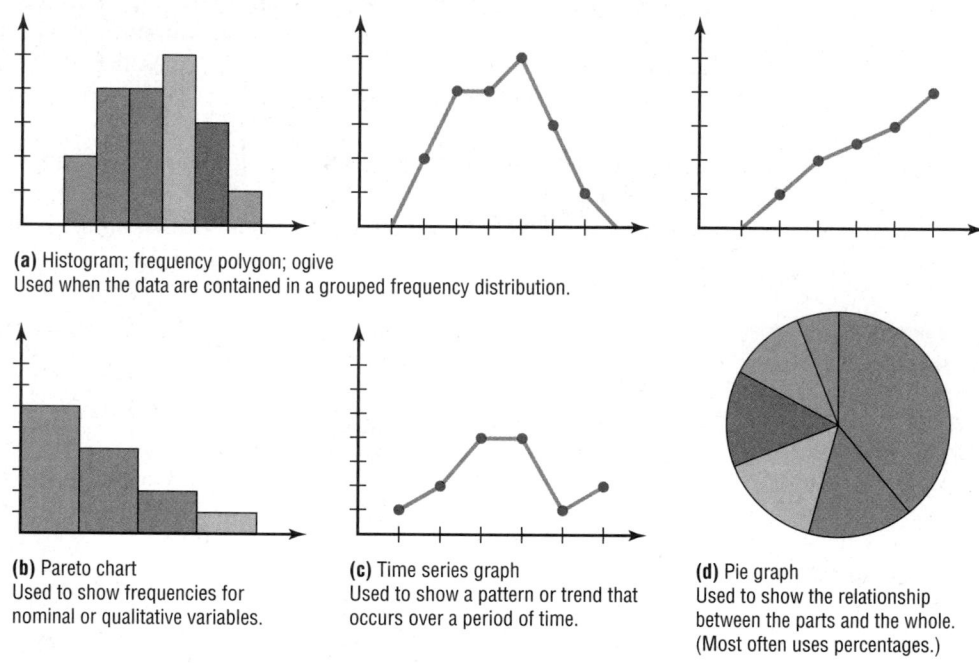

(a) Histogram; frequency polygon; ogive
Used when the data are contained in a grouped frequency distribution.

(b) Pareto chart
Used to show frequencies for nominal or qualitative variables.

(c) Time series graph
Used to show a pattern or trend that occurs over a period of time.

(d) Pie graph
Used to show the relationship between the parts and the whole. (Most often uses percentages.)

Stem and Leaf Plots

The stem and leaf plot is a method of organizing data and is a combination of sorting and graphing. It has the advantage over a grouped frequency distribution of retaining the actual data while showing them in graphical form.

Objective **4**

Draw and interpret a stem and leaf plot.

A **stem and leaf plot** is a data plot that uses part of the data value as the stem and part of the data value as the leaf to form groups or classes.

Example 2–13 shows the procedure for constructing a stem and leaf plot.

Example 2–13

At an outpatient testing center, the number of cardiograms performed each day for 20 days is shown. Construct a stem and leaf plot for the data.

25	31	20	32	13
14	43	02	57	23
36	32	33	32	44
32	52	44	51	45

How Much Paper Money Is in Circulation Today?

The Federal Reserve estimated that during a recent year, there were 22 billion bills in circulation. About 35% of them were $1 bills, 3% were $2 bills, 8% were $5 bills, 7% were $10 bills, 23% were $20 bills, 5% were $50 bills, and 19% were $100 bills. It costs about 3¢ to print each $1 bill.

  The average life of a $1 bill is 22 months, a $10 bill 3 years, a $20 bill 4 years, a $50 bill 9 years, and a $100 bill 9 years. What type of graph would you use to represent the average lifetimes of the bills?

Solution

Step 1   Arrange the data in order:

02, 13, 14, 20, 23, 25, 31, 32, 32, 32,
32, 33, 36, 43, 44, 44, 45, 51, 52, 57

Note: Arranging the data in order is not essential and can be cumbersome when the data set is large; however, it is helpful in constructing a stem and leaf plot. The leaves in the final stem and leaf plot should be arranged in order.

Step 2   Separate the data according to the first digit, as shown.

02   13, 14   20, 23, 25   31, 32, 32, 32, 32, 33, 36
43, 44, 44, 45   51, 52, 57

Step 3   A display can be made by using the leading digit as the *stem* and the trailing digit as the *leaf.* For example, for the value 32, the leading digit, 3, is the stem and the trailing digit, 2, is the leaf. For the value 14, the 1 is the stem and the 4 is the leaf. Now a plot can be constructed as shown in Figure 2–22.

Figure 2–22

Stem and Leaf Plot for Example 2–13

0	2
1	3 4
2	0 3 5
3	1 2 2 2 2 3 6
4	3 4 4 5
5	1 2 7

Leading digit (stem)	Trailing digit (leaf)
0	2
1	3 4
2	0 3 5
3	1 2 2 2 2 3 6
4	3 4 4 5
5	1 2 7

Figure 2–22 shows that the distribution peaks in the center and that there are no gaps in the data. For 7 of the 20 days, the number of patients receiving cardiograms was between 31 and 36. The plot also shows that the testing center treated from a minimum of 2 patients to a maximum of 57 patients in any one day.

If there are no data values in a class, you should write the stem number and leave the leaf row blank. Do not put a zero in the leaf row.

Example 2–14

An insurance company researcher conducted a survey on the number of car thefts in a large city for a period of 30 days last summer. The raw data are shown. Construct a stem and leaf plot by using classes 50–54, 55–59, 60–64, 65–69, 70–74, and 75–79.

52	62	51	50	69
58	77	66	53	57
75	56	55	67	73
79	59	68	65	72
57	51	63	69	75
65	53	78	66	55

Solution

Step 1   Arrange the data in order.

50, 51, 51, 52, 53, 53, 55, 55, 56, 57, 57, 58, 59, 62, 63,
65, 65, 66, 66, 67, 68, 69, 69, 72, 73, 75, 75, 77, 78, 79

Step 2   Separate the data according to the classes.

50, 51, 51, 52, 53, 53   55, 55, 56, 57, 57, 58, 59
62, 63   65, 65, 66, 66, 67, 68, 69, 69   72, 73
75, 75, 77, 78, 79

Figure 2–23

Stem and Leaf Plot for Example 2–14

5	0 1 1 2 3 3
5	5 5 6 7 7 8 9
6	2 3
6	5 5 6 6 7 8 9 9
7	2 3
7	5 5 7 8 9

Step 3   Plot the data as shown here.

Leading digit (stem)	Trailing digit (leaf)
5	0 1 1 2 3 3
5	5 5 6 7 7 8 9
6	2 3
6	5 5 6 6 7 8 9 9
7	2 3
7	5 5 7 8 9

The graph for this plot is shown in Figure 2–23.

When the data values are in the hundreds, such as 325, the stem is 32 and the leaf is 5. For example, the stem and leaf plot for the data values 325, 327, 330, 332, 335, 341, 345, and 347 looks like this.

32	5 7
33	0 2 5
34	1 5 7

When you analyze a stem and leaf plot, look for peaks and gaps in the distribution. See if the distribution is symmetric or skewed. Check the variability of the data by looking at the spread.

Related distributions can be compared by using a back-to-back stem and leaf plot. The back-to-back stem and leaf plot uses the same digits for the stems of both distributions, but the digits that are used for the leaves are arranged in order out from the stems on both sides. Example 2–15 shows a back-to-back stem and leaf plot.

Example 2–15

The number of stories in two selected samples of tall buildings in Atlanta and Philadelphia is shown. Construct a back-to-back stem and leaf plot, and compare the distributions.

Atlanta					Philadelphia				
55	70	44	36	40	61	40	38	32	30
63	40	44	34	38	58	40	40	25	30
60	47	52	32	32	54	40	36	30	30
50	53	32	28	31	53	39	36	34	33
52	32	34	32	50	50	38	36	39	32
26	29								

Source: *The World Almanac and Book of Facts.*

Solution

Step 1   Arrange the data for both data sets in order.

Step 2   Construct a stem and leaf plot using the same digits as stems. Place the digits for the leaves for Atlanta on the left side of the stem and the digits for the leaves for Philadelphia on the right side, as shown. See Figure 2–24.

Figure 2–24

Back-to-Back Stem and Leaf Plot for Example 2–15

Atlanta		Philadelphia
9 8 6	2	5
8 6 4 4 2 2 2 2 2 1	3	0 0 0 0 2 2 3 4 6 6 6 8 8 9 9
7 4 4 0 0	4	0 0 0 0
5 3 2 2 0 0	5	0 3 4 8
3 0	6	1
0	7	

Step 3   Compare the distributions. The buildings in Atlanta have a large variation in the number of stories per building. Although both distributions are peaked in the 30- to 39-story class, Philadelphia has more buildings in this class. Atlanta has more buildings that have 40 or more stories than Philadelphia does.

Stem and leaf plots are part of the techniques called *exploratory data analysis.* More information on this topic is presented in Chapter 3.

Applying the Concepts **2–3**

Leading Cause of Death

The following shows approximations of the leading causes of death among men ages 25–44 years. The rates are per 100,000 men. Answer the following questions about the graph.

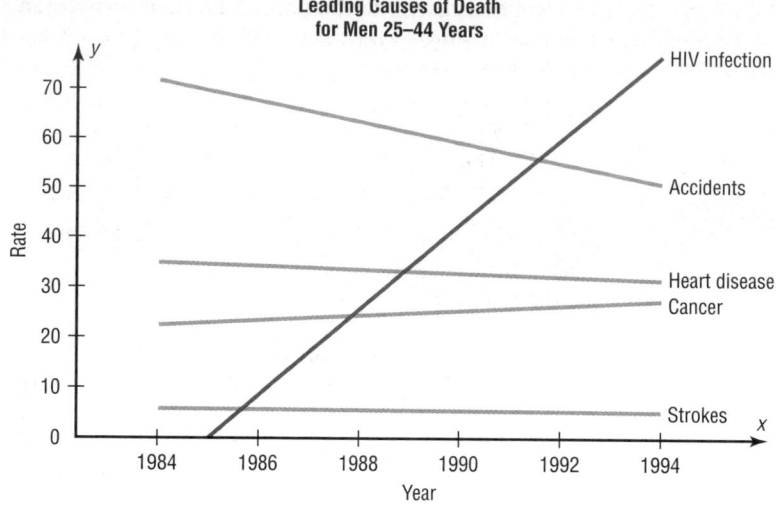

**Leading Causes of Death
for Men 25–44 Years**

1. What are the variables in the graph?
2. Are the variables qualitative or quantitative?
3. Are the variables discrete or continuous?
4. What type of graph was used to display the data?
5. Could a Pareto chart be used to display the data?
6. Could a pie chart be used to display the data?
7. List some typical uses for the Pareto chart.
8. List some typical uses for the time series chart.

See page 101 for the answers.

Exercises 2–3

1. **Number of Hurricanes** Construct a vertical bar chart for the total number of hurricanes by month from 1851 to 2008.

May	18
June	79
July	101
August	344
September	459
October	280
November	61

Source: National Hurricane Center.

2. **Worldwide Sales of Fast Foods** The worldwide sales (in billions of dollars) for several fast-food franchises for a specific year are shown. Construct a horizontal bar graph and a Pareto chart for the data.

Wendy's	$ 8.7
KFC	14.2
Pizza Hut	9.3
Burger King	12.7
Subway	10.0

Source: *Franchise Times*.

3. **Calories Burned While Exercising** Construct a Pareto chart for the following data on exercise.

	Calories burned per minute
Walking, 2 mph	2.8
Bicycling, 5.5 mph	3.2
Golfing	5.0
Tennis playing	7.1
Skiing, 3 mph	9.0
Running, 7 mph	14.5

Source: Physiology of Exercise.

4. **Roller Coaster Mania** The World Roller Coaster Census Report lists the following number of roller coasters on each continent. Represent the data graphically, using a Pareto chart and a horizontal bar graph.

Africa	17
Asia	315
Australia	22
Europe	413
North America	643
South America	45

Source: www.rcdb.com

5. Instruction Time The average weekly instruction time in schools for 5 selected countries is shown. Construct a vertical bar graph and a Pareto chart for the data.

Thailand	30.5 hours
China	26.9 hours
France	24.8 hours
United States	22.2 hours
Brazil	19 hours

Source: Organization for Economic Cooperation and Development.

6. Sales of Coffee The data show the total retail sales (in billions of dollars) of coffee for 6 years. Over the years, are the sales increasing or decreasing?

Year	2001	2002	2003	2004	2005	2006
Sales	$8.3	$8.4	$9.0	$9.6	$11.1	$12.3

Source: Specialty Coffee Association of America.

7. Safety Record of U.S. Airlines The safety record of U.S. airlines for 10 years is shown. Construct a time series graph for the data.

Year	Major Accidents
1997	2
1998	0
1999	2
2000	3
2001	1
2002	1
2003	2
2004	4
2005	2
2006	2
2007	0

Source: National Transportation Safety Board.

8. Average Global Temperatures The average global temperatures for the following years are shown. Draw a time series graph and comment on the trend.

Year	2004	2005	2006	2007	2008
Temperature	57.98	58.11	57.99	58.01	57.88

Source: National Oceanic and Atmospheric Administration.

9. Carbon Dioxide Concentrations The following data for the atmospheric concentration of carbon dioxide (in ppm^2) are shown. Draw a time series graph and comment on the trend.

Year	2004	2005	2006	2007	2008
Concentration	375	377	379	381	383

Source: U.S. Department of Energy.

10. Reasons We Travel The following data are based on a survey from American Travel Survey on why people travel. Construct a pie graph for the data and analyze the results.

Purpose	Number
Personal business	146
Visit friends or relatives	330
Work-related	225
Leisure	299

Source: *USA TODAY.*

11. Characteristics of the Population 65 and Over Two characteristics of the population aged 65 and over are shown below for 2004. Illustrate each characteristic with a pie graph.

Marital status		Educational attainment	
Never married	3.9%	Less than ninth grade	13.9%
Married	57.2	Completed grades 9–12	
Widowed	30.8	but no diploma	13.0
Divorced	8.1	H.S. graduate	36.0
		Some college/	
		associates degree	18.4
		Bachelor's/advanced	
		degree	18.7

Source: *New York Times Almanac.*

12. Colors of Automobiles The popular vehicle car colors are shown. Construct a pie graph for the data.

White	19%
Silver	18
Black	16
Red	13
Blue	12
Gray	12
Other	10

Source: Dupont Automotive Color Popularity Report.

13. Workers Switch Jobs In a recent survey, 3 in 10 people indicated that they are likely to leave their jobs when the economy improves. Of those surveyed, 34% indicated that they would make a career change, 29% want a new job in the same industry, 21% are going to start a business, and 16% are going to retire. Make a pie chart and a Pareto chart for the data. Which chart do you think better represents the data?

Source: National Survey Institute.

14. State which graph (Pareto chart, time series graph, or pie graph) would most appropriately represent the given situation.

a. The number of students enrolled at a local college for each year during the last 5 years.

b. The budget for the student activities department at a certain college for a specific year.

c. The means of transportation the students use to get to school.

d. The percentage of votes each of the four candidates received in the last election.

e. The record temperatures of a city for the last 30 years.

f. The frequency of each type of crime committed in a city during the year.

15. Presidents' Ages at Inauguration The age at inauguration for each U.S. President is shown. Construct a stem and leaf plot and analyze the data.

57	54	52	55	51	56	47
61	68	56	55	54	61	51
57	51	46	54	51	52	
57	49	54	42	60	69	
58	64	49	51	62	64	
57	48	51	56	43	46	
61	65	47	55	55	54	

Source: *New York Times Almanac.*

16. Calories in Salad Dressings A listing of calories per one ounce of selected salad dressings (not fat-free) is given below. Construct a stem and leaf plot for the data.

100	130	130	130	110	110	120	130	140	100
140	170	160	130	160	120	150	100	145	145
145	115	120	100	120	160	140	120	180	100
160	120	140	150	190	150	180	160		

17. Twenty Days of Plant Growth The growth (in centimeters) of two varieties of plant after 20 days is shown in this table. Construct a back-to-back stem and leaf plot for the data, and compare the distributions.

Variety 1				Variety 2			
20	12	39	38	18	45	62	59
41	43	51	52	53	25	13	57
59	55	53	59	42	55	56	38
50	58	35	38	41	36	50	62
23	32	43	53	45	55		

18. Math and Reading Achievement Scores The math and reading achievement scores from the National Assessment of Educational Progress for selected states are listed below. Construct a back-to-back stem and leaf plot with the data and compare the distributions.

Math					Reading				
52	66	69	62	61	65	76	76	66	67
63	57	59	59	55	71	70	70	66	61
55	59	74	72	73	61	69	78	76	77
68	76	73			77	77	80		

Source: *World Almanac.*

19. The sales of recorded music in 2004 by genre are listed below. Represent the data with an appropriate graph. Answers will vary.

Rock	23.9	Jazz	2.7
Country	13.0	Classical	2.0
Rap/hip-hop	12.1	Oldies	1.4
R&B/urban	11.3	Soundtracks	1.1
Pop	10.0	New age	1.0
Religious	6.0	Other	8.9
Children's	2.8		

Source: *World Almanac.*

Extending the Concepts

20. Successful Space Launches The number of successful space launches by the United States and Japan for the years 1993–1997 is shown here. Construct a compound time series graph for the data. What comparison can be made regarding the launches?

Year	1993	1994	1995	1996	1997
United States	29	27	24	32	37
Japan	1	4	2	1	2

Source: *The World Almanac and Book of Facts.*

21. Meat Production Meat production for veal and lamb for the years 1960–2000 is shown here. (Data are in millions of pounds.) Construct a compound time series graph for the data. What comparison can be made regarding meat production?

Year	1960	1970	1980	1990	2000
Veal	1109	588	400	327	225
Lamb	769	551	318	358	234

Source: *The World Almanac and Book of Facts.*

22. Top 10 Airlines During a recent year the top 10 airlines with the most aircraft are listed. Represent these data with an appropriate graph.

American	714	Continental	364
United	603	Southwest	327
Delta	600	British Airways	268
Northwest	424	American Eagle	245
U.S. Airways	384	Lufthansa (Ger.)	233

Source: *Top 10 of Everything.*

23. Nobel Prizes in Physiology or Medicine The top prize-winning countries for Nobel Prizes in Physiology or Medicine are listed here. Represent the data with an appropriate graph.

United States	80	Denmark	5
United Kingdom	24	Austria	4
Germany	16	Belgium	4
Sweden	8	Italy	3
France	7	Australia	3
Switzerland	6		

Source: *Top 10 of Everything.*

Source: Cartoon by Bradford Veley, Marquette, Michigan. Used with permission.

24. Cost of Milk The graph shows the increase in the price of a quart of milk. Why might the increase appear to be larger than it really is?

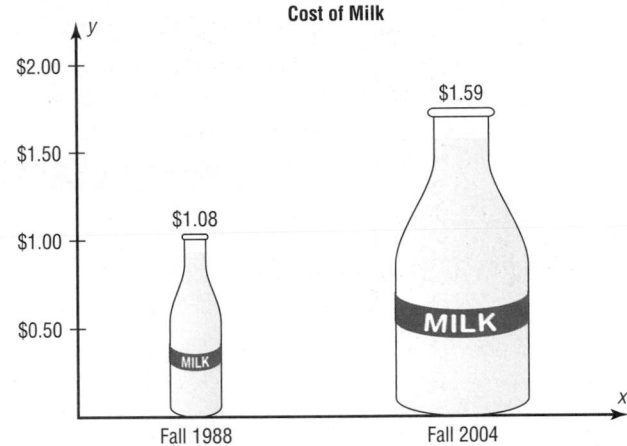

25. Boom in Number of Births The graph shows the projected boom (in millions) in the number of births. Cite several reasons why the graph might be misleading.

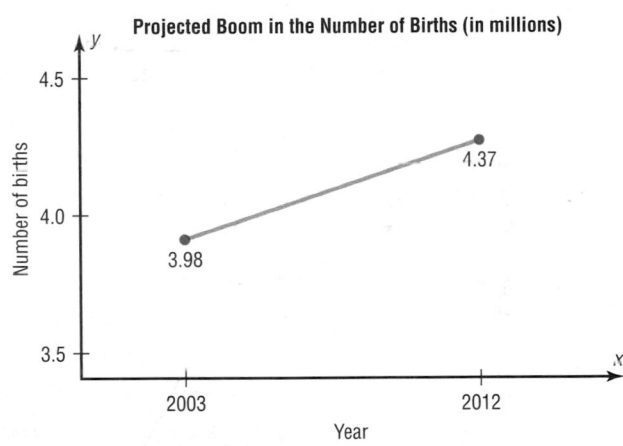

Technology *Step by Step*

MINITAB
Step by Step

Construct a Pie Chart

1. Enter the summary data for snack foods and frequencies from Example 2–11 into C1 and C2.

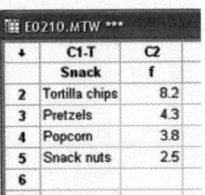

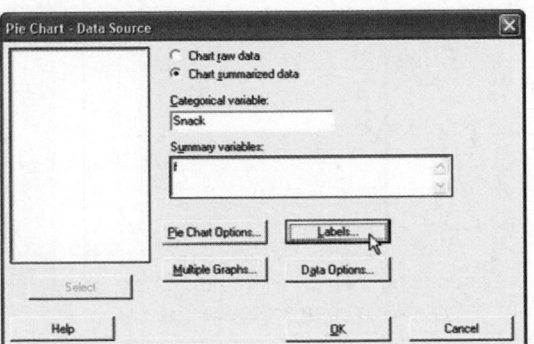

2. Name them **Snack** and **f.**
3. Select **Graph>Pie Chart.**
   a) Click the option for Chart summarized data.
   b) Press [Tab] to move to Categorical variable, then double-click C1 to select it.
   c) Press [Tab] to move to Summary variables, and select the column with the frequencies f.

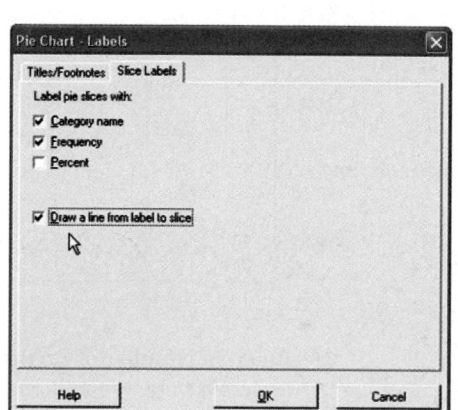
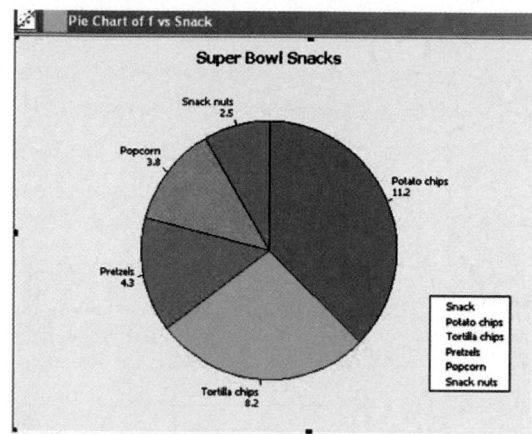

4. Click the [Labels] tab, then Titles/Footnotes.
   a) Type in the title: **Super Bowl Snacks.**
   b) Click the Slice Labels tab, then the options for Category name and Frequency.
   c) Click the option to Draw a line from label to slice.
   d) Click [OK] twice to create the chart.

Construct a Bar Chart

The procedure for constructing a bar chart is similar to that for the pie chart.

1. Select **Graph>Bar Chart.**
   a) Click on the drop-down list in Bars Represent: then select values from a table.
   b) Click on the Simple chart, then click [OK]. The dialog box will be similar to the Pie Chart Dialog Box.
2. Select the frequency column C2 f for Graph variables: and Snack for the Categorical variable.

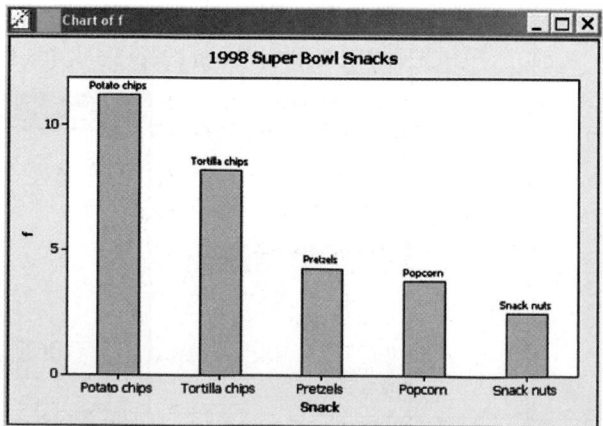

3. Click on [Labels], then type the title in the Titles/Footnote tab: **1998 Super Bowl Snacks.**

4. Click the tab for Data Labels, then click the option to Use labels from column: and select C1 Snacks.

5. Click [OK] twice.

Construct a Pareto Chart

Pareto charts are a quality control tool. They are similar to a bar chart with no gaps between the bars, and the bars are arranged by frequency.

1. Select **Stat>Quality Tools>Pareto.**

2. Click the option to Chart defects table.

3. Click in the box for the Labels in: and select Snack.

4. Click on the frequencies column C2 f.

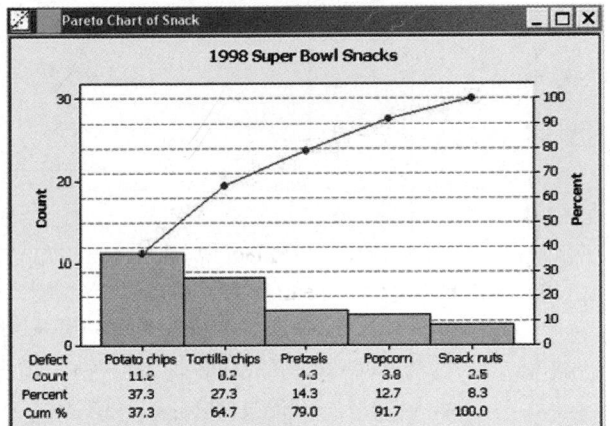

5. Click on [Options].

   a) Check the box for Cumulative percents.

   b) Type in the title, **1998 Super Bowl Snacks.**

6. Click [OK] twice. The chart is completed.

Construct a Time Series Plot

The data used are for the number of vehicles that used the Pennsylvania Turnpike.

Year	1999	2000	2001	2002	2003
Number	156.2	160.1	162.3	172.8	179.4

1. Add a blank worksheet to the project by selecting **File>New>New Worksheet.**

2. To enter the dates from 1999 to 2003 in C1, select **Calc>Make Patterned Data>Simple Set of Numbers.**

   a) Type **Year** in the text box for Store patterned data in.

   b) From first value: should be **1999.**

   c) To Last value: should be **2003.**

   d) In steps of should be **1** (for every other year). The last two boxes should be 1, the default value.

   e) Click [OK]. The sequence from 1999 to 2003 will be entered in C1 whose label will be Year.

3. Type **Vehicles (in millions)** for the label row above row 1 in C2.

4. Type **156.2** for the first number, then press [Enter]. Never enter the commas for large numbers!

5. Continue entering the value in each row of C2.

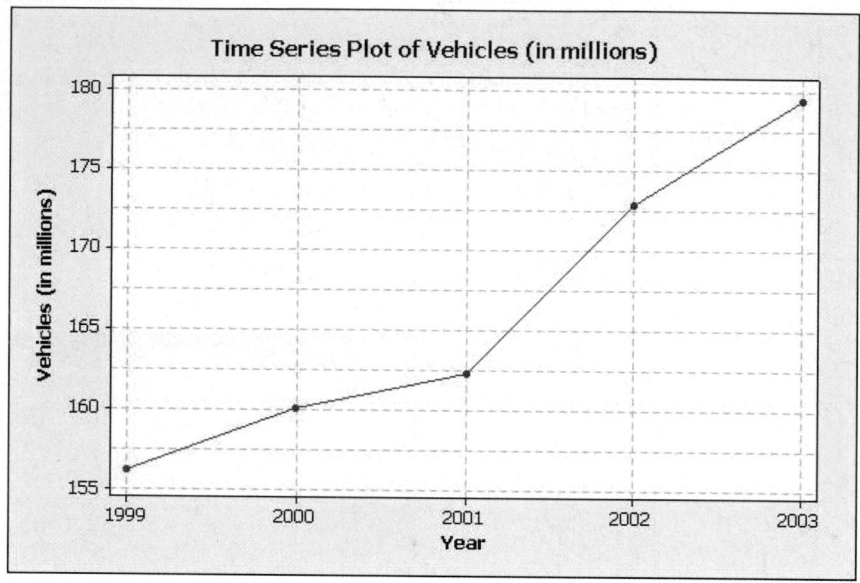

6. To make the graph, select **Graph>Time series plot,** then Simple, and press [OK].

   a) For Series select Vehicles (in millions), then click [Time/scale].

   b) Click the Stamp option and select Year for the Stamp column.

   c) Click the Gridlines tab and select all three boxes, Y major, Y minor, and X major.

   d) Click [OK] twice. A new window will open that contains the graph.

   e) To change the title, double-click the title in the graph window. A dialog box will open, allowing you to edit the text.

Construct a Stem and Leaf Plot

1. Type in the data for Example 2–14. Label the column **CarThefts.**

2. Select **STAT>EDA>Stem-and-Leaf.** This is the same as **Graph>Stem-and-Leaf.**

3. Double-click on C1 CarThefts in the column list.

4. Click in the Increment text box, and enter the class width of **5.**

5. Click [OK]. This character graph will be displayed in the session window.

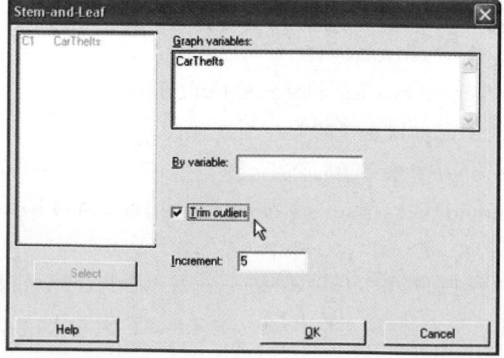

Stem-and-Leaf Display: CarThefts
Stem-and-leaf of CarThefts   N = 30
Leaf Unit = 1.0

6	5	011233
13	5	5567789
15	6	23
15	6	55667899
7	7	23
5	7	55789

TI-83 Plus or TI-84 Plus
Step by Step

To graph a time series, follow the procedure for a frequency polygon from Section 2–2, using the following data for the number of outdoor drive-in theaters

Year	1988	1990	1992	1994	1996	1998	2000
Number	1497	910	870	859	826	750	637

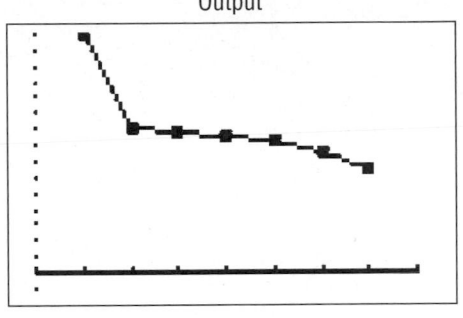

Output

Excel
Step by Step

Constructing a Pie Chart

To make a pie chart:

1. Enter the blood types from Example 2–12 into column A of a new worksheet.

2. Enter the frequencies corresponding to each blood type in column B.

3. Highlight the data in columns A and B and select Insert from the toolbar, then select the Pie chart type.

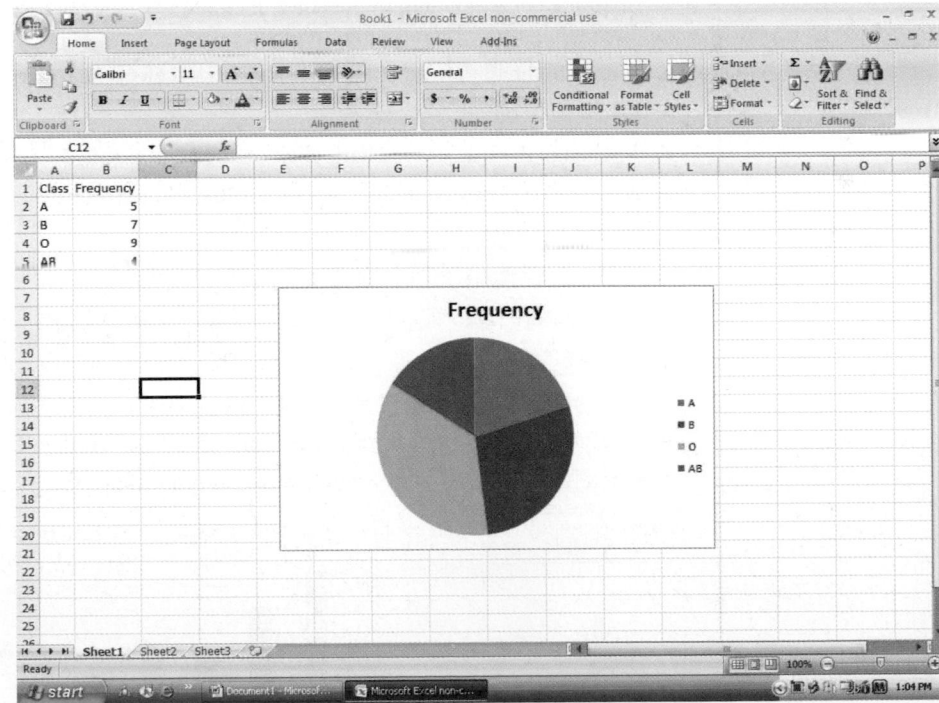

4. Click on any region of the chart. Then select Design from the Chart Tools tab on the toolbar.

5. Select Formulas from the chart Layouts tab on the toolbar.

6. To change the title of the chart, click on the current title of the chart.

7. When the text box containing the title is highlighted, click the mouse in the text box and change the title.

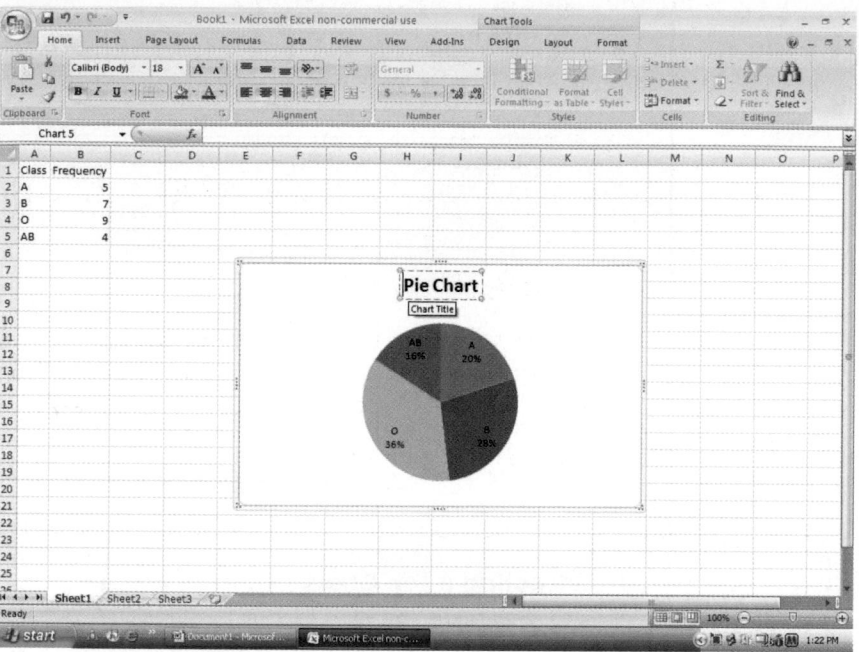

Constructing a Pareto Chart

To make a Pareto chart:

1. Enter the snack food categories from Example 2–11 into column A of a new worksheet.

2. Enter the corresponding frequencies in column B. The data should be entered in descending order according to frequency.

3. Highlight the data from columns A and B and select the Insert tab from the toolbar.

4. Select the Column Chart type.

5. To change the title of the chart, click on the current title of the chart.

6. When the text box containing the title is highlighted, click the mouse in the text box and change the title.

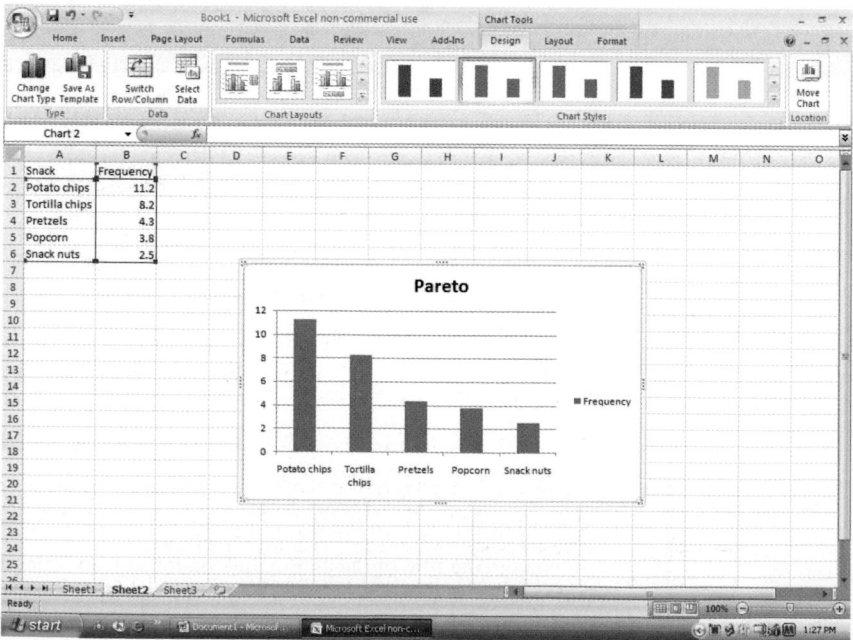

Constructing a Time Series Chart

Example

Year	1999	2000	2001	2002	2003
Vehicles*	156.2	160.1	162.3	172.8	179.4

*Vehicles (in millions) that used the Pennsylvania Turnpike.
Source: *Tribune Review*.

To make a time series chart:

1. Enter the years 1999 through 2003 from the example in column A of a new worksheet.
2. Enter the corresponding frequencies in column B.
3. Highlight the data from column B and select the Insert tab from the toolbar.
4. Select the Line chart type.

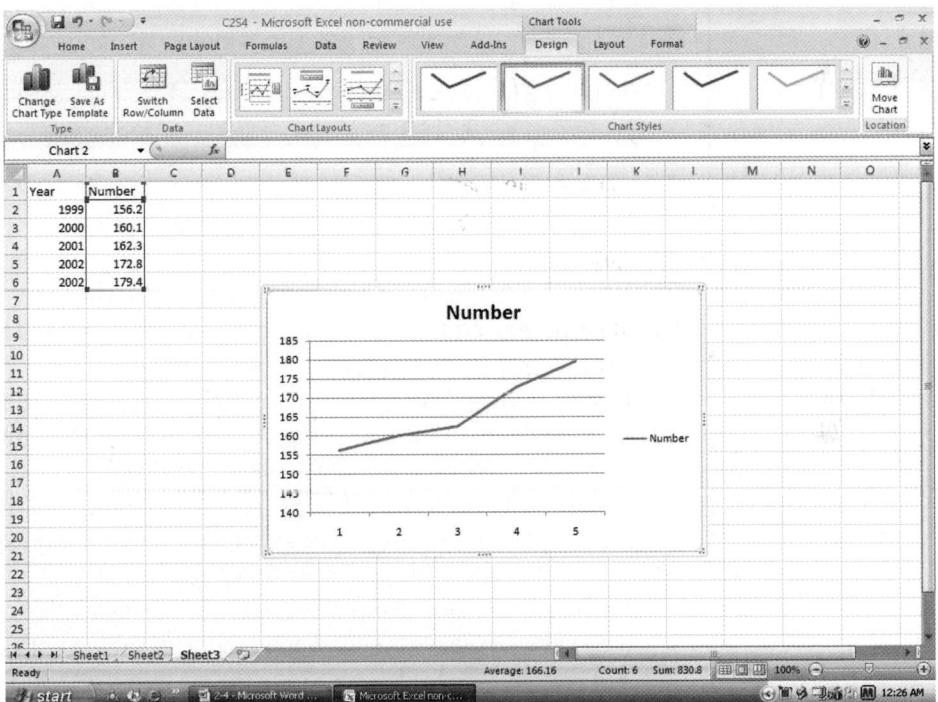

5. Right-click the mouse on any region of the graph.
6. Select the Select Data option.
7. Select Edit from the Horizontal Axis Labels and highlight the years from column A, then click [OK].
8. Click [OK] on the Select Data Source box.
9. Create a title for your chart, such as Number of Vehicles Using the Pennsylvania Turnpike Between 1999 and 2003. Right-click the mouse on any region of the chart. Select the Chart Tools tab from the toolbar, then Layout.
10. Select Chart Title and highlight the current title to change the title.
11. Select Axis Titles to change the horizontal and vertical axis labels.

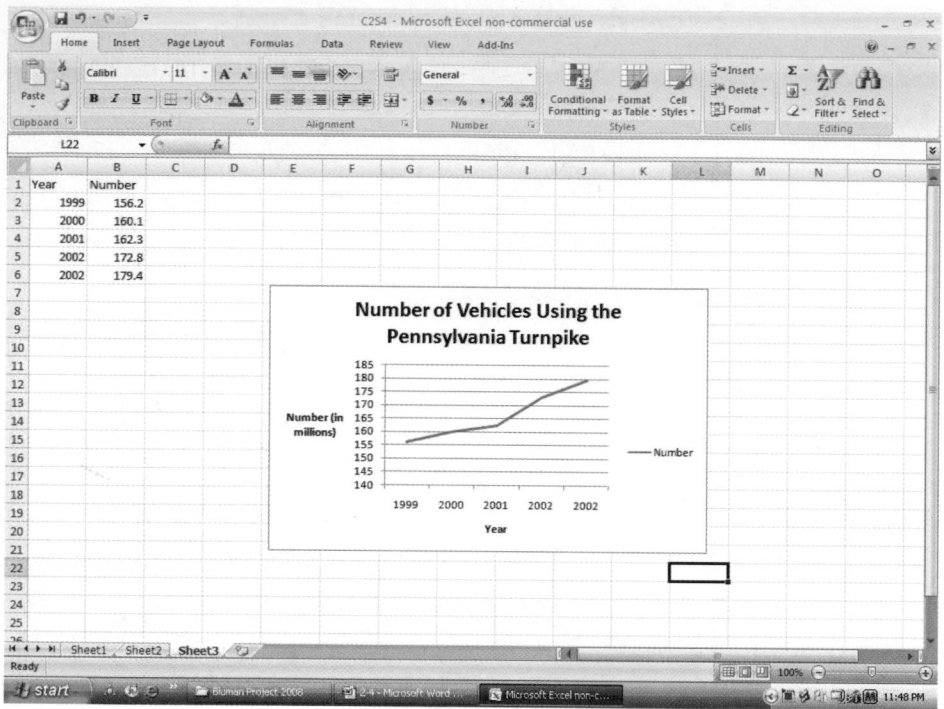

Summary

- When data are collected, the values are called raw data. Since very little knowledge can be obtained from raw data, they must be organized in some meaningful way. A frequency distribution using classes is the common method that is used. (2–1)

- Once a frequency distribution is constructed, graphs can be drawn to give a visual representation of the data. The most commonly used graphs in statistics are the histogram, frequency polygon, and ogive. (2–2)

- Other graphs such as the bar graph, Pareto chart, time series graph, and pie graph can also be used. Some of these graphs are frequently seen in newspapers, magazines, and various statistical reports. (2–3)

- Finally, a stem and leaf plot uses part of the data values as stems and part of the data values as leaves. This graph has the advantage of a frequency distribution and a histogram. (2–3)

Important Terms

bar graph 69

categorical frequency
distribution 38

class 37

class boundaries 39

class midpoint 40

class width 39

cumulative frequency 54

cumulative frequency
distribution 42

frequency 37

frequency distribution 37

frequency polygon 53

grouped frequency
distribution 39

histogram 51

lower class limit 39

ogive 54

open-ended distribution 41

Pareto chart 70

pie graph 73

raw data 37

relative frequency graph 56

stem and leaf plot 80

time series graph 72

ungrouped frequency
distribution 43

upper class limit 39

Important Formulas

Formula for the percentage of values in each class:

$$\% = \frac{f}{n} \cdot 100$$

where

  f = frequency of class
  n = total number of values

Formula for the range:

  R = **highest value − lowest value**

Formula for the class width:

  Class width = upper boundary − lower boundary

Formula for the class midpoint:

$$X_m = \frac{\text{lower boundary} + \text{upper boundary}}{2}$$

or

$$X_m = \frac{\text{lower limit} + \text{upper limit}}{2}$$

Formula for the degrees for each section of a pie graph:

$$\text{Degrees} = \frac{f}{n} \cdot 360°$$

Review Exercises

1. **How People Get Their News** The Brunswick Research Organization surveyed 50 randomly selected individuals and asked them the primary way they received the daily news. Their choices were via newspaper (N), television (T), radio (R), or Internet (I). Construct a categorical frequency distribution for the data and interpret the results. The data in this exercise will be used for Exercise 2 in this section. (2–1)

N	N	T	T	T	I	R	R	I	T
I	N	R	R	I	N	N	I	T	N
I	R	T	T	T	T	N	R	R	I
R	R	I	N	T	R	T	I	I	T
T	I	N	T	T	I	R	N	R	T

2. Construct a pie graph for the data in Exercise 1, and analyze the results. (2–3)

3. **Ball Sales** A sporting goods store kept a record of sales of five items for one randomly selected hour during a recent sale. Construct a frequency distribution for the data (B = baseballs, G = golf balls, T = tennis balls, S = soccer balls, F = footballs). (The data for this exercise will be used for Exercise 4 in this section.) (2–1)

F	B	B	B	G	T	F
G	G	F	S	G	T	
F	T	T	T	S	T	
F	S	S	G	S	B	

4. Draw a pie graph for the data in Exercise 3 showing the sales of each item, and analyze the results. (2–3)

5. **BUN Count** The blood urea nitrogen (BUN) count of 20 randomly selected patients is given here in

milligrams per deciliter (mg/dl). Construct an ungrouped frequency distribution for the data. (The data for this exercise will be used for Exercise 6.) (2–1)

17	18	13	14
12	17	11	20
13	18	19	17
14	16	17	12
16	15	19	22

6. Construct a histogram, a frequency polygon, and an ogive for the data in Exercise 5 in this section, and analyze the results. (2–2)

7. The percentage (rounded to the nearest whole percent) of persons from each state completing 4 years or more of college is listed below. Organize the data into a grouped frequency distribution with 5 classes. (2–1)

Percentage of persons completing 4 years of college

23	25	24	34	22	24	27	37	33	24
26	23	38	24	24	17	28	23	30	25
30	22	33	24	28	36	24	19	25	31
34	31	27	24	29	28	21	25	26	15
26	22	27	21	25	28	24	21	25	26

Source: *New York Times Almanac.*

8. Using the data in Exercise 7, construct a histogram, a frequency polygon, and an ogive. (2–2)

9. **NFL Franchise Values** The data shown (in millions of dollars) are the values of the 30 National Football League franchises. Construct a frequency distribution for the data using 8 classes. (The data for

this exercise will be used for Exercises 10 and 12 in this section.) (2–1)

170	191	171	235	173	187	181	191
200	218	243	200	182	320	184	239
186	199	186	210	209	240	204	193
211	186	197	204	188	242		

Source: *Pittsburgh Post-Gazette.*

10. Construct a histogram, a frequency polygon, and an ogive for the data in Exercise 9 in this section, and analyze the results. (2–2)

11. Ages of the Vice Presidents at the Time of Their Death The ages of the Vice Presidents of the United States at the time of their death are listed below. Use the data to construct a frequency distribution, histogram, frequency polygon, and ogive, using relative frequencies. Use 6 classes. (2–1, 2–2)

90	83	80	73	70	51	68	79	70	71
72	74	67	54	81	66	62	63	68	57
66	96	78	55	60	66	57	71	60	85
76	98	77	88	78	81	64	66	77	70

Source: *New York Times Almanac.*

12. Construct a histogram, frequency polygon, and ogive by using relative frequencies for the data in Exercise 9 in this section. (2–2)

13. Activities While Driving A survey of 1200 drivers showed the percentage of respondents who did the following while driving. Construct a horizontal bar graph and a Pareto chart for the data. (2–3)

Drink beverage	80%
Talk on cell phone	73
Eat a meal	41
Experience road rage	23
Smoke	21

Source: Nationwide Mutual Insurance Company.

14. Air Quality The following data show the number of days the air quality for Atlanta, Georgia, was below the accepted standards. Draw a time series graph for the data. (2–3)

Year	2005	2006	2007	2008
Days	5	14	15	4

Source: U.S. Environmental Protection Agency.

15. Bank Failures The following data show the number of bank failures for recent years. Draw a time series graph and comment on the trend. (2–3)

Year	'01	'02	'03	'04	'05	'06	'07	'08	'09
Number	4	11	3	4	0	0	3	26	98

Source: Federal Deposit Insurance Corporation.

16. Public Debt The following data show the public debt in billions of dollars for recent years. Draw a time series graph for the data. (2–3)

Year	'03	'04	'05	'06	'07	'08	'09
Debt	6783.2	7379.1	7932.7	8507.0	9007.7	10,025.0	11,956.6

Source: U.S. Department of the Treasury.

17. Gold Production in Colombia The following data show the amount of gold production in thousands of troy ounces for Colombia for recent years. Draw a time series graph and comment on the trend. (2–3)

Year	'03	'04	'05	'06	'07	'08
Amount	656	701	976	1250	1270	1620

Source: U.S. Department of the Interior.

18. Spending of College Freshmen The average amounts spent by college freshmen for school items are shown. Construct a pie graph for the data. (2–3)

Electronics/computers	$728
Dorm items	344
Clothing	141
Shoes	72

Source: National Retail Federation.

19. Career Changes A survey asked if people would like to spend the rest of their careers with their present employers. The results are shown. Construct a pie graph for the data and analyze the results. (2–3)

Answer	Number of people
Yes	660
No	260
Undecided	80

20. Museum Visitors The number of visitors to the Railroad Museum during 24 randomly selected hours is shown here. Construct a stem and leaf plot for the data. (2–3)

67	62	38	73	34	43	72	35
53	55	58	63	47	42	51	62
32	29	47	62	29	38	36	41

21. Public Libraries The numbers of public libraries in operation for selected states are listed below. Organize the data with a stem and leaf plot. (2–3)

102	176	210	142	189	176	108	113	205
209	184	144	108	192	176			

Source: *World Almanac.*

22. Job Aptitude Test A special aptitude test is given to job applicants. The data shown here represent the scores of 30 applicants. Construct a stem and leaf plot for the data and summarize the results. (2–3)

204	210	227	218	254
256	238	242	253	227
251	243	233	251	241
237	247	211	222	231
218	212	217	227	209
260	230	228	242	200

<table>
<tr><td>**Statistics Today**</td><td>

How Your Identity Can Be Stolen—Revisited

Data presented in numerical form do not convey an easy-to-interpret conclusion; however, when data are presented in graphical form, readers can see the visual impact of the numbers. In the case of identity fraud, the reader can see that most of the identity frauds are due to lost or stolen wallets, checkbooks, or credit cards, and very few identity frauds are caused by online purchases or transactions.

</td></tr>
</table>

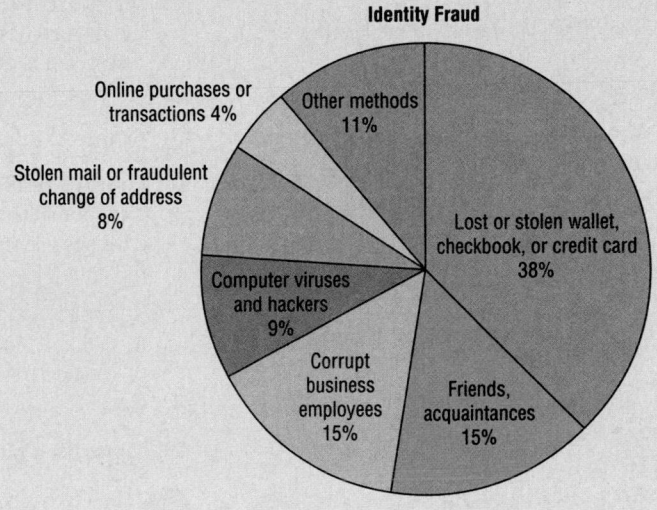

Data Analysis

A Data Bank is found in Appendix D, or on the World Wide Web by following links from

www.mhhe.com/math/stat/bluman

1. From the Data Bank located in Appendix D, choose one of the following variables: age, weight, cholesterol level, systolic pressure, IQ, or sodium level. Select at least 30 values. For these values, construct a grouped frequency distribution. Draw a histogram, frequency polygon, and ogive for the distribution. Describe briefly the shape of the distribution.

2. From the Data Bank, choose one of the following variables: educational level, smoking status, or exercise. Select at least 20 values. Construct an ungrouped frequency distribution for the data. For the distribution, draw a Pareto chart and describe briefly the nature of the chart.

3. From the Data Bank, select at least 30 subjects and construct a categorical distribution for their marital status. Draw a pie graph and describe briefly the findings.

4. Using the data from Data Set IV in Appendix D, construct a frequency distribution and draw a histogram. Describe briefly the shape of the distribution of the tallest buildings in New York City.

5. Using the data from Data Set XI in Appendix D, construct a frequency distribution and draw a frequency polygon. Describe briefly the shape of the distribution for the number of pages in statistics books.

6. Using the data from Data Set IX in Appendix D, divide the United States into four regions, as follows:

Northeast	CT ME MA NH NJ NY PA RI VT
Midwest	IL IN IA KS MI MN MS NE ND OH SD WI
South	AL AR DE DC FL GA KY LA MD NC OK SC TN TX VA WV
West	AK AZ CA CO HI ID MT NV NM OR UT WA WY

Find the total population for each region, and draw a Pareto chart and a pie graph for the data. Analyze the results. Explain which chart might be a better representation for the data.

7. Using the data from Data Set I in Appendix D, make a stem and leaf plot for the record low temperatures in the United States. Describe the nature of the plot.

Chapter Quiz

Determine whether each statement is true or false. If the statement is false, explain why.

1. In the construction of a frequency distribution, it is a good idea to have overlapping class limits, such as 10–20, 20–30, 30–40. False

2. Histograms can be drawn by using vertical or horizontal bars. False

3. It is not important to keep the width of each class the same in a frequency distribution. False

4. Frequency distributions can aid the researcher in drawing charts and graphs. True

5. The type of graph used to represent data is determined by the type of data collected and by the researcher's purpose. True

6. In construction of a frequency polygon, the class limits are used for the x axis. False

7. Data collected over a period of time can be graphed by using a pie graph. False

Select the best answer.

8. What is another name for the ogive?
   a. Histogram
   b. Frequency polygon
   c. Cumulative frequency graph
   d. Pareto chart

9. What are the boundaries for 8.6–8.8?
   a. 8–9
   b. 8.5–8.9
   c. 8.55–8.85
   d. 8.65–8.75

10. What graph should be used to show the relationship between the parts and the whole?
    a. Histogram
    b. Pie graph
    c. Pareto chart
    d. Ogive

11. Except for rounding errors, relative frequencies should add up to what sum?
    a. 0
    b. 1
    c. 50
    d. 100

Complete these statements with the best answers.

12. The three types of frequency distributions are _____, _____, and _____. Categorical, ungrouped, grouped

13. In a frequency distribution, the number of classes should be between _____ and _____. 5, 20

14. Data such as blood types (A, B, AB, O) can be organized into a(n) _____ frequency distribution. Categorical

15. Data collected over a period of time can be graphed using a(n) _____ graph. Time series

16. A statistical device used in exploratory data analysis that is a combination of a frequency distribution and a histogram is called a(n) _____. Stem and leaf plot

17. On a Pareto chart, the frequencies should be represented on the _____ axis. Vertical or y

18. **Housing Arrangements** A questionnaire on housing arrangements showed this information obtained from 25 respondents. Construct a frequency distribution for the data (H = house, A = apartment, M = mobile home, C = condominium).

H	C	H	M	H	A	C	A	M
C	M	C	A	M	A	C	C	M
C	C	H	A	H	H	M		

19. Construct a pie graph for the data in Exercise 18.

20. **Items Purchased at a Convenience Store** When 30 randomly selected customers left a convenience store, each was asked the number of items he or she purchased. Construct an ungrouped frequency distribution for the data.

2	9	4	3	6
6	2	8	6	5
7	5	3	8	6
6	2	3	2	4
6	9	9	8	9
4	2	1	7	4

21. Construct a histogram, a frequency polygon, and an ogive for the data in Exercise 20.

22. **Murders in Selected Cities** For a recent year, the number of murders in 25 selected cities is shown. Construct a frequency distribution using 9 classes, and analyze the nature of the data in terms of shape, extreme values, etc. (The information in this exercise will be used for Exercise 23 in this section.)

248	348	74	514	597
270	71	226	41	39
366	73	241	46	34
149	68	73	63	65
109	598	278	69	27

    Source: *Pittsburgh Tribune Review.*

23. Construct a histogram, frequency polygon, and ogive for the data in Exercise 22. Analyze the histogram.

24. **Recycled Trash** Construct a Pareto chart and a horizontal bar graph for the number of tons (in millions)

of trash recycled per year by Americans based on an Environmental Protection Agency study.

Type	Amount
Paper	320.0
Iron/steel	292.0
Aluminum	276.0
Yard waste	242.4
Glass	196.0
Plastics	41.6

Source: *USA TODAY.*

25. Identity Thefts The results of a survey of 84 people whose identities were stolen using various methods are shown. Draw a pie chart for the information.

Lost or stolen wallet, checkbook, or credit card	38
Retail purchases or telephone transactions	15
Stolen mail	9
Computer viruses or hackers	8
Phishing	4
Other	10
	84

Source: Javelin Strategy and Research.

26. Needless Deaths of Children *The New England Journal of Medicine* predicted the number of needless deaths due to childhood obesity. Draw a time series graph for the data.

Year	2020	2025	2030	2035
Deaths	130	550	1500	3700

 27. Museum Visitors The number of visitors to the Historic Museum for 25 randomly selected hours is shown. Construct a stem and leaf plot for the data.

15	53	48	19	38
86	63	98	79	38
62	89	67	39	26
28	35	54	88	76
31	47	53	41	68

Critical Thinking Challenges

1. Water Usage The graph shows the average number of gallons of water a person uses for various activities.

Can you see anything misleading about the way the graph is drawn?

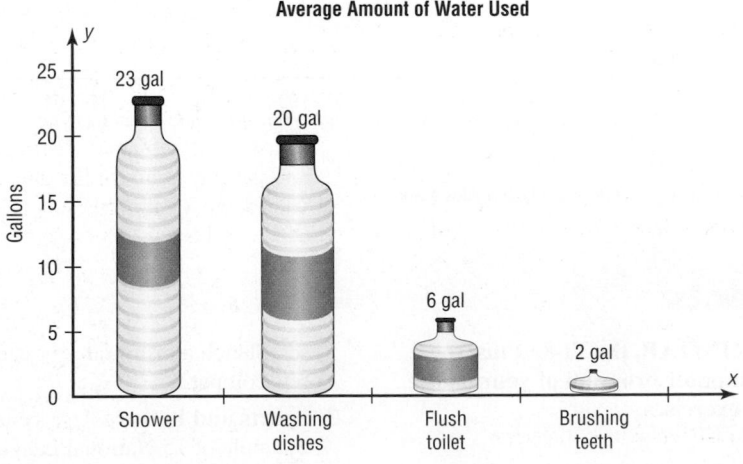

Average Amount of Water Used

2. **The Great Lakes** Shown are various statistics about the Great Lakes. Using appropriate graphs (your choice) and summary statements, write a report analyzing the data.

	Superior	Michigan	Huron	Erie	Ontario
Length (miles)	350	307	206	241	193
Breadth (miles)	160	118	183	57	53
Depth (feet)	1,330	923	750	210	802
Volume (cubic miles)	2,900	1,180	850	116	393
Area (square miles)	31,700	22,300	23,000	9,910	7,550
Shoreline (U.S., miles)	863	1,400	580	431	300

Source: *The World Almanac and Book of Facts.*

3. **Teacher Strikes** In Pennsylvania there were more teacher strikes in 2004 than there were in all other states combined. Because of the disruptions, state legislators want to pass a bill outlawing teacher strikes and submitting contract disputes to binding arbitration. The graph shows the number of teacher strikes in Pennsylvania for the school years 1992 to 2004. Use the graph to answer these questions.

a. In what year did the largest number of strikes occur? How many were there?

b. In what year did the smallest number of teacher strikes occur? How many were there?

c. In what year was the average duration of the strikes the longest? What was it?

d. In what year was the average duration of the strikes the shortest? What was it?

e. In what year was the number of teacher strikes the same as the average duration of the strikes?

f. Find the difference in the number of strikes for the school years 1992–1993 and 2004–2005.

g. Do you think teacher strikes should be outlawed? Justify your conclusions.

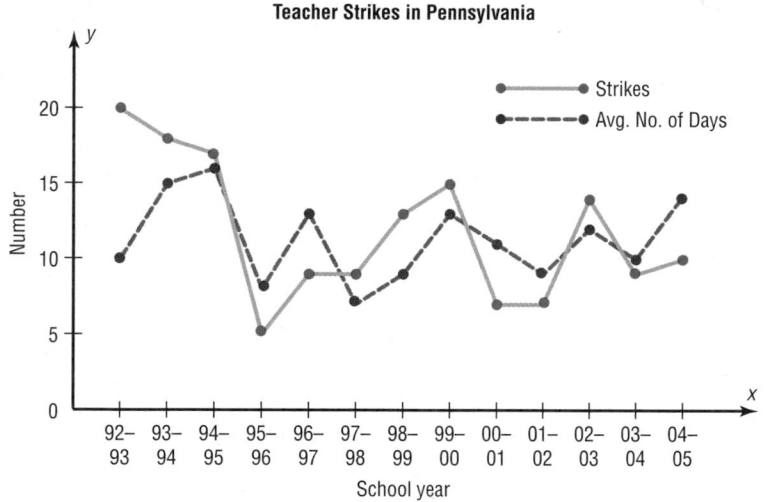

Source: Pennsylvania School Boards Associations.

Data Projects

Where appropriate, use MINITAB, the TI-83 Plus, the TI-84 Plus, Excel, or a computer program of your choice to complete the following exercises.

1. **Business and Finance** Consider the 30 stocks listed as the Dow Jones Industrials. For each, find their earnings per share. Randomly select 30 stocks traded on the NASDAQ. For each, find their earnings per share. Create a frequency table with 5 categories for each data set. Sketch a histogram for each. How do the two data sets compare?

2. **Sports and Leisure** Use systematic sampling to create a sample of 25 National League and 25 American League baseball players from the most recently completed season. Find the number of home runs for each player. Create a frequency table with 5 categories for each data set. Sketch a histogram for each. How do the two leagues compare?

3. **Technology** Randomly select 50 songs from your music player or music organization program. Find the length (in seconds) for each song. Use these data to create a frequency table with 6 categories. Sketch a frequency polygon for the frequency table. Is the shape of the distribution of times uniform, skewed, or bell-shaped? Also note the genre of each song. Create a Pareto chart showing the frequencies of the various categories. Finally, note the year each song was released. Create a pie chart organized by decade to show the percentage of songs from various time periods.

4. **Health and Wellness** Use information from the Red Cross to create a pie chart depicting the percentages of Americans with various blood types. Also find information about blood donations and the percentage

of each type donated. How do the charts compare? Why is the collection of type O blood so important?

5. **Politics and Economics** Consider the U.S. Electoral College System. For each of the 50 states, determine the number of delegates received. Create a frequency table with 8 classes. Is this distribution uniform, skewed, or bell-shaped?

6. **Your Class** Have each person in class take his or her pulse and determine the heart rate (beats in one minute). Use the data to create a frequency table with 6 classes. Then have everyone in the class do 25 jumping jacks and immediately take the pulse again after the activity. Create a frequency table for those data as well. Compare the two results. Are they similarly distributed? How does the range of scores compare?

Answers to Applying the Concepts

Section 2–1 Ages of Presidents at Inauguration

1. The data were obtained from the population of all Presidents at the time this text was written.

2. The oldest inauguration age was 69 years old.

3. The youngest inauguration age was 42 years old.

4. Answers will vary. One possible answer is

Age at inauguration	Frequency
42–45	2
46–49	7
50–53	8
54–57	16
58–61	5
62–65	4
66–69	2

5. Answers will vary. For the frequency distribution given in Question 4, there is a peak for the 54–57 bin.

6. Answers will vary. This frequency distribution shows no outliers. However, if we had split our frequency into 14 bins instead of 7, then the ages 42, 43, 68, and 69 might appear as outliers.

7. Answers will vary. The data appear to be unimodal and fairly symmetric, centering on 55 years of age.

Section 2–2 Selling Real Estate

1. A histogram of the data gives price ranges and the counts of homes in each price range. We can also talk about how the data are distributed by looking at a histogram.

2. A frequency polygon shows increases or decreases in the number of home prices around values.

3. A cumulative frequency polygon shows the number of homes sold at or below a given price.

4. The house that sold for $321,550 is an extreme value in this data set.

5. Answers will vary. One possible answer is that the histogram displays the outlier well since there is a gap in the prices of the homes sold.

6. The distribution of the data is skewed to the right.

Section 2–3 Leading Cause of Death

1. The variables in the graph are the year, cause of death, and rate of death per 100,000 men.

2. The cause of death is qualitative, while the year and death rates are quantitative.

3. Year is a discrete variable, and death rate is continuous. Since cause of death is qualitative, it is neither discrete nor continuous.

4. A line graph was used to display the data.

5. No, a Pareto chart could not be used to display the data, since we can only have one quantitative variable and one categorical variable in a Pareto chart.

6. We cannot use a pie chart for the same reasons as given for the Pareto chart.

7. A Pareto chart is typically used to show a categorical variable listed from the highest-frequency category to the category with the lowest frequency.

8. A time series chart is used to see trends in the data. It can also be used for forecasting and predicting.

Data Description

Objectives

After completing this chapter, you should be able to

1 Summarize data, using measures of central tendency, such as the mean, median, mode, and midrange.

2 Describe data, using measures of variation, such as the range, variance, and standard deviation.

3 Identify the position of a data value in a data set, using various measures of position, such as percentiles, deciles, and quartiles.

4 Use the techniques of exploratory data analysis, including boxplots and five-number summaries, to discover various aspects of data.

Outline

Introduction

3–1 **Measures of Central Tendency**

3–2 **Measures of Variation**

3–3 **Measures of Position**

3–4 **Exploratory Data Analysis**

Summary

Statistics Today

How Long Are You Delayed by Road Congestion?

No matter where you live, at one time or another, you have been stuck in traffic. To see whether there are more traffic delays in some cities than in others, statisticians make comparisons using descriptive statistics. A statistical study by the Texas Transportation Institute found that a driver is delayed by road congestion an average of 36 hours per year. To see how selected cities compare to this average, see Statistics Today—Revisited at the end of the chapter.

This chapter will show you how to obtain and interpret descriptive statistics such as measures of average, measures of variation, and measures of position.

Introduction

Chapter 2 showed how you can gain useful information from raw data by organizing them into a frequency distribution and then presenting the data by using various graphs. This chapter shows the statistical methods that can be used to summarize data. The most familiar of these methods is the finding of averages.

For example, you may read that the average speed of a car crossing midtown Manhattan during the day is 5.3 miles per hour or that the average number of minutes an American father of a 4-year-old spends alone with his child each day is 42.[1]

In the book *American Averages* by Mike Feinsilber and William B. Meed, the authors state:

"Average" when you stop to think of it is a funny concept. Although it describes all of us it describes none of us. . . . While none of us wants to be the average American, we all want to know about him or her.

The authors go on to give examples of averages:

The average American man is five feet, nine inches tall; the average woman is five feet, 3.6 inches.
The average American is sick in bed seven days a year missing five days of work.
On the average day, 24 million people receive animal bites.
By his or her 70th birthday, the average American will have eaten 14 steers, 1050 chickens, 3.5 lambs, and 25.2 hogs.[2]

[1]"Harper's Index," *Harper's* magazine.

[2]Mike Feinsilber and William B. Meed, *American Averages* (New York: Bantam Doubleday Dell).

In these examples, the word *average* is ambiguous, since several different methods can be used to obtain an average. Loosely stated, the average means the center of the distribution or the most typical case. Measures of average are also called *measures of central tendency* and include the *mean, median, mode,* and *midrange.*

Knowing the average of a data set is not enough to describe the data set entirely. Even though a shoe store owner knows that the average size of a man's shoe is size 10, she would not be in business very long if she ordered only size 10 shoes.

As this example shows, in addition to knowing the average, you must know how the data values are dispersed. That is, do the data values cluster around the mean, or are they spread more evenly throughout the distribution? The measures that determine the spread of the data values are called *measures of variation,* or *measures of dispersion.* These measures include the *range, variance,* and *standard deviation.*

Finally, another set of measures is necessary to describe data. These measures are called *measures of position.* They tell where a specific data value falls within the data set or its relative position in comparison with other data values. The most common position measures are *percentiles, deciles,* and *quartiles.* These measures are used extensively in psychology and education. Sometimes they are referred to as *norms.*

The measures of central tendency, variation, and position explained in this chapter are part of what is called *traditional statistics.*

Section 3–4 shows the techniques of what is called *exploratory data analysis.* These techniques include the *boxplot* and the *five-number summary.* They can be used to explore data to see what they show (as opposed to the traditional techniques, which are used to confirm conjectures about the data).

3-1   Measures of Central Tendency

Chapter 1 stated that statisticians use samples taken from populations; however, when populations are small, it is not necessary to use samples since the entire population can be used to gain information. For example, suppose an insurance manager wanted to know the average weekly sales of all the company's representatives. If the company employed a large number of salespeople, say, nationwide, he would have to use a sample and make

Objective 1

Summarize data, using measures of central tendency, such as the mean, median, mode, and midrange.

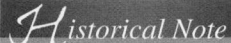
an inference to the entire sales force. But if the company had only a few salespeople, say, only 87 agents, he would be able to use all representatives' sales for a randomly chosen week and thus use the entire population.

Measures found by using all the data values in the population are called *parameters.* Measures obtained by using the data values from samples are called *statistics;* hence, the average of the sales from a sample of representatives is a *statistic,* and the average of sales obtained from the entire population is a *parameter.*

A **statistic** is a characteristic or measure obtained by using the data values from a sample.

A **parameter** is a characteristic or measure obtained by using all the data values from a specific population.

These concepts as well as the symbols used to represent them will be explained in detail in this chapter.

General Rounding Rule In statistics the basic rounding rule is that when computations are done in the calculation, rounding should not be done until the final answer is calculated. When rounding is done in the intermediate steps, it tends to increase the difference between that answer and the exact one. But in the textbook and solutions manual, it is not practical to show long decimals in the intermediate calculations; hence, the values in the examples are carried out to enough places (usually three or four) to obtain the same answer that a calculator would give after rounding on the last step.

The Mean

The *mean,* also known as the *arithmetic average,* is found by adding the values of the data and dividing by the total number of values. For example, the mean of 3, 2, 6, 5, and 4 is found by adding $3 + 2 + 6 + 5 + 4 = 20$ and dividing by 5; hence, the mean of the data is $20 \div 5 = 4$. The values of the data are represented by X's. In this data set, $X_1 = 3$, $X_2 = 2$, $X_3 = 6$, $X_4 = 5$, and $X_5 = 4$. To show a sum of the total X values, the symbol Σ (the capital Greek letter sigma) is used, and ΣX means to find the sum of the X values in the data set. The summation notation is explained in Appendix A.

The **mean** is the sum of the values, divided by the total number of values. The symbol $\overline{X}$ represents the sample mean.

$$\overline{X} = \frac{X_1 + X_2 + X_3 + \cdots + X_n}{n} = \frac{\Sigma X}{n}$$

where n represents the total number of values in the sample.

For a population, the Greek letter μ (mu) is used for the mean.

$$\mu = \frac{X_1 + X_2 + X_3 + \cdots + X_N}{N} = \frac{\Sigma X}{N}$$

where N represents the total number of values in the population.

In statistics, Greek letters are used to denote parameters, and Roman letters are used to denote statistics. Assume that the data are obtained from samples unless otherwise specified.

Example 3–1

Days Off per Year

The data represent the number of days off per year for a sample of individuals selected from nine different countries. Find the mean.

20, 26, 40, 36, 23, 42, 35, 24, 30

Source: World Tourism Organization.

Solution

$$\bar{X} = \frac{\Sigma X}{n} = \frac{20 + 26 + 40 + 36 + 23 + 42 + 35 + 24 + 30}{9} = \frac{276}{9} = 30.7 \text{ days}$$

Hence, the mean of the number of days off is 30.7 days.

Example 3–2

Hospital Infections

The data show the number of patients in a sample of six hospitals who acquired an infection while hospitalized. Find the mean.

| 110 | 76 | 29 | 38 | 105 | 31 |

Source: Pennsylvania Health Care Cost Containment Council.

Solution

$$\bar{X} = \frac{\Sigma X}{n} = \frac{110 + 76 + 29 + 38 + 105 + 31}{6} = \frac{389}{6} = 64.8$$

The mean of the number of hospital infections for the six hospitals is 64.8.

The mean, in most cases, is not an actual data value.

Rounding Rule for the Mean The mean should be rounded to one more decimal place than occurs in the raw data. For example, if the raw data are given in whole numbers, the mean should be rounded to the nearest tenth. If the data are given in tenths, the mean should be rounded to the nearest hundredth, and so on.

The procedure for finding the mean for grouped data uses the midpoints of the classes. This procedure is shown next.

Example 3–3

Miles Run per Week

Using the frequency distribution for Example 2–7, find the mean. The data represent the number of miles run during one week for a sample of 20 runners.

Solution

The procedure for finding the mean for grouped data is given here.

Step 1 Make a table as shown.

A Class	B Frequency f	C Midpoint X_m	D $f \cdot X_m$
5.5–10.5	1		
10.5–15.5	2		
15.5–20.5	3		
20.5–25.5	5		
25.5–30.5	4		
30.5–35.5	3		
35.5–40.5	2		
	$n = 20$		

Interesting Fact

The average time it takes a person to find a new job is 5.9 months.

Step 2 Find the midpoints of each class and enter them in column C.

$$X_m = \frac{5.5 + 10.5}{2} = 8 \qquad \frac{10.5 + 15.5}{2} = 13 \qquad \text{etc.}$$

Step 3 For each class, multiply the frequency by the midpoint, as shown, and place the product in column D.

$1 \cdot 8 = 8 \qquad 2 \cdot 13 = 26 \qquad$ etc.

The completed table is shown here.

A Class	B Frequency f	C Midpoint X_m	D $f \cdot X_m$
5.5–10.5	1	8	8
10.5–15.5	2	13	26
15.5–20.5	3	18	54
20.5–25.5	5	23	115
25.5–30.5	4	28	112
30.5–35.5	3	33	99
35.5–40.5	2	38	76
	$n = 20$		$\Sigma f \cdot X_m = 490$

Step 4 Find the sum of column D.

Step 5 Divide the sum by n to get the mean.

$$\overline{X} = \frac{\Sigma f \cdot X_m}{n} = \frac{490}{20} = 24.5 \text{ miles}$$

The procedure for finding the mean for grouped data assumes that the mean of all the raw data values in each class is equal to the midpoint of the class. In reality, this is not true, since the average of the raw data values in each class usually will not be exactly equal to the midpoint. However, using this procedure will give an acceptable approximation of the mean, since some values fall above the midpoint and other values fall below the midpoint for each class, and the midpoint represents an estimate of all values in the class.

The steps for finding the mean for grouped data are summarized in the next Procedure Table.

Procedure Table

Finding the Mean for Grouped Data

Step 1 Make a table as shown.

A Class	B Frequency f	C Midpoint X_m	D $f \cdot X_m$

Step 2 Find the midpoints of each class and place them in column C.

Step 3 Multiply the frequency by the midpoint for each class, and place the product in column D.

Step 4 Find the sum of column D.

Step 5 Divide the sum obtained in column D by the sum of the frequencies obtained in column B.

The formula for the mean is

$$\overline{X} = \frac{\Sigma f \cdot X_m}{n}$$

[*Note:* The symbols $\Sigma f \cdot X_m$ mean to find the sum of the product of the frequency (f) and the midpoint (X_m) for each class.]

Ages of the Top 50 Wealthiest People

The histogram shows the ages of the top 50 wealthiest individuals according to *Forbes Magazine* for a recent year. The mean age is 66.04 years. The median age is 68 years. Explain why these two statistics are not enough to adequately describe the data.

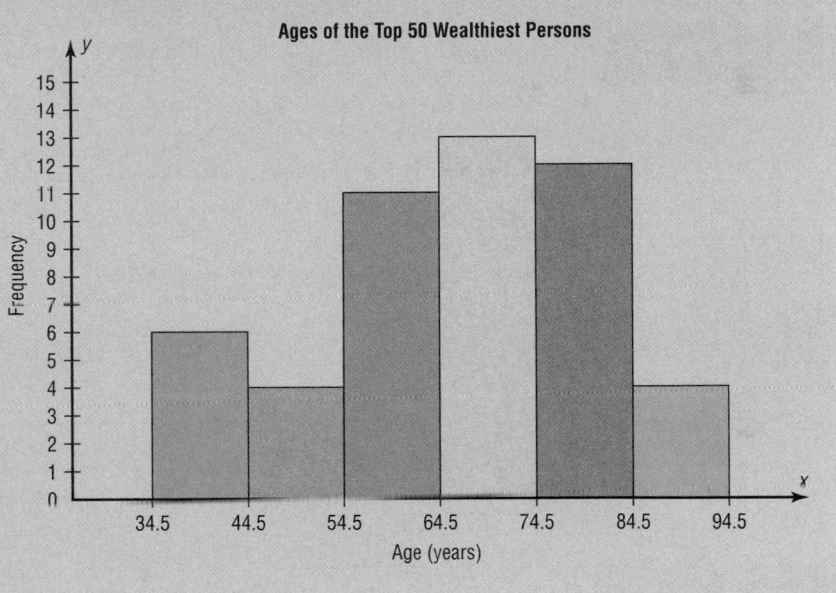

Ages of the Top 50 Wealthiest Persons

The Median

An article recently reported that the median income for college professors was $43,250. This measure of central tendency means that one-half of all the professors surveyed earned more than $43,250, and one-half earned less than $43,250.

The *median* is the halfway point in a data set. Before you can find this point, the data must be arranged in order. When the data set is ordered, it is called a **data array.** The median either will be a specific value in the data set or will fall between two values, as shown in Examples 3–4 through 3–8.

The **median** is the midpoint of the data array. The symbol for the median is MD.

Steps in computing the median of a data array

Step 1   Arrange the data in order.

Step 2   Select the middle point.

Example 3–4	Hotel Rooms

 The number of rooms in the seven hotels in downtown Pittsburgh is 713, 300, 618, 595, 311, 401, and 292. Find the median.

Source: Interstate Hotels Corporation.

Solution

Step 1   Arrange the data in order.

292, 300, 311, 401, 595, 618, 713

Step 2   Select the middle value.

292, 300, 311, 401, 595, 618, 713
$\uparrow$
Median

Hence, the median is 401 rooms.

Example 3–5	National Park Vehicle Pass Costs

Find the median for the daily vehicle pass charge for five U.S. National Parks. The costs are $25, $15, $15, $20, and $15.

Source: National Park Service.

Solution

$15 $15 $15 $20 $25
$\uparrow$
Median

The median cost is $15.

Examples 3–4 and 3–5 each had an odd number of values in the data set; hence, the median was an actual data value. When there are an even number of values in the data set, the median will fall between two given values, as illustrated in Examples 3–6, 3–7, and 3–8.

Example 3–6	Tornadoes in the United States

 The number of tornadoes that have occurred in the United States over an 8-year period follows. Find the median.

684, 764, 656, 702, 856, 1133, 1132, 1303

Source: *The Universal Almanac.*

Solution

656, 684, 702, 764, 856, 1132, 1133, 1303
$\uparrow$
Median

Since the middle point falls halfway between 764 and 856, find the median MD by adding the two values and dividing by 2.

$$MD = \frac{764 + 856}{2} = \frac{1620}{2} = 810$$

The median number of tornadoes is 810.

Example 3–7

Asthma Cases

The number of children with asthma during a specific year in seven local districts is shown. Find the median.

253, 125, 328, 417, 201, 70, 90

Source: Pennsylvania Department of Health.

Solution

70, 90, 125, 201, 253, 328, 417
           ↑
        Median

Since the number 201 is at the center of the distribution, the median is 201.

Example 3–8

Magazines Purchased

Six customers purchased these numbers of magazines: 1, 7, 3, 2, 3, 4. Find the median.

Solution

$$1, 2, 3, 3, 4, 7 \qquad MD = \frac{3+3}{2} = 3$$
    ↑
  Mcdian

Hence, the median number of magazines purchased is 3.

The Mode

The third measure of average is called the *mode*. The mode is the value that occurs most often in the data set. It is sometimes said to be the most typical case.

> The value that occurs most often in a data set is called the **mode.**

A data set that has only one value that occurs with the greatest frequency is said to be **unimodal.**

If a data set has two values that occur with the same greatest frequency, both values are considered to be the mode and the data set is said to be **bimodal.** If a data set has more than two values that occur with the same greatest frequency, each value is used as the mode, and the data set is said to be **multimodal.** When no data value occurs more than once, the data set is said to have *no mode*. A data set can have more than one mode or no mode at all. These situations will be shown in some of the examples that follow.

Example 3–9

NFL Signing Bonuses

Find the mode of the signing bonuses of eight NFL players for a specific year. The bonuses in millions of dollars are

18.0, 14.0, 34.5, 10, 11.3, 10, 12.4, 10

Source: *USA TODAY.*

Solution

It is helpful to arrange the data in order although it is not necessary.

10, 10, 10, 11.3, 12.4, 14.0, 18.0, 34.5

Since $10 million occurred 3 times—a frequency larger than any other number—the mode is $10 million.

Example 3–10

Branches of Large Banks

 Find the mode for the number of branches that six banks have.

401, 344, 209, 201, 227, 353

Source: SNL Financial.

Solution

Since each value occurs only once, there is no mode.
Note: Do not say that the mode is zero. That would be incorrect, because in some data, such as temperature, zero can be an actual value.

Example 3–11

Licensed Nuclear Reactors

 The data show the number of licensed nuclear reactors in the United States for a recent 15-year period. Find the mode.

Source: *The World Almanac and Book of Facts.*

104	104	104	104	104
107	109	109	109	110
109	111	112	111	109

Solution

Since the values 104 and 109 both occur 5 times, the modes are 104 and 109. The data set is said to be bimodal.

The mode for grouped data is the modal class. The **modal class** is the class with the largest frequency.

Example 3–12

Miles Run per Week

Find the modal class for the frequency distribution of miles that 20 runners ran in one week, used in Example 2–7.

Class	Frequency	
5.5–10.5	1	
10.5–15.5	2	
15.5–20.5	3	
20.5–25.5	5	← Modal class
25.5–30.5	4	
30.5–35.5	3	
35.5–40.5	2	

Solution

The modal class is 20.5–25.5, since it has the largest frequency. Sometimes the midpoint of the class is used rather than the boundaries; hence, the mode could also be given as 23 miles per week.

The mode is the only measure of central tendency that can be used in finding the most typical case when the data are nominal or categorical.

Example 3–13	Area Boat Registrations

The data show the number of boats registered for six counties in southwestern Pennsylvania. Find the mode.

Westmoreland	11,008
Butler	9,002
Washington	6,843
Beaver	6,367
Fayette	4,208
Armstrong	3,782

Source: Pennsylvania Fish and Boat Commission.

Solution

Since the category with the highest frequency is Westmoreland, the most typical case is Westmoreland. Hence the mode is 11,008.

An extremely high or extremely low data value in a data set can have a striking effect on the mean of the data set. These extreme values are called *outliers*. This is one reason why when analyzing a frequency distribution, you should be aware of any of these values. For the data set shown in Example 3–14, the mean, median, and mode can be quite different because of extreme values. A method for identifying outliers is given in Section 3–3.

Example 3–14	Salaries of Personnel

A small company consists of the owner, the manager, the salesperson, and two technicians, all of whose annual salaries are listed here. (Assume that this is the entire population.)

Staff	Salary
Owner	$50,000
Manager	20,000
Salesperson	12,000
Technician	9,000
Technician	9,000

Find the mean, median, and mode.

Solution

$$\mu = \frac{\Sigma X}{N} = \frac{50,000 + 20,000 + 12,000 + 9000 + 9000}{5} = \$20,000$$

Hence, the mean is $20,000, the median is $12,000, and the mode is $9,000.

In Example 3–14, the mean is much higher than the median or the mode. This is so because the extremely high salary of the owner tends to raise the value of the mean. In this and similar situations, the median should be used as the measure of central tendency.

The Midrange

The *midrange* is a rough estimate of the middle. It is found by adding the lowest and highest values in the data set and dividing by 2. It is a very rough estimate of the average and can be affected by one extremely high or low value.

The **midrange** is defined as the sum of the lowest and highest values in the data set, divided by 2. The symbol MR is used for the midrange.

$$MR = \frac{\text{lowest value} + \text{highest value}}{2}$$

Example 3–15

Water-Line Breaks

 In the last two winter seasons, the city of Brownsville, Minnesota, reported these numbers of water-line breaks per month. Find the midrange.

2, 3, 6, 8, 4, 1

Solution

$$MR = \frac{1 + 8}{2} = \frac{9}{2} = 4.5$$

Hence, the midrange is 4.5.

If the data set contains one extremely large value or one extremely small value, a higher or lower midrange value will result and may not be a typical description of the middle.

Example 3–16

NFL Signing Bonuses

Find the midrange of data for the NFL signing bonuses in Example 3–9. The bonuses in millions of dollars are

18.0, 14.0, 34.5, 10, 11.3, 10, 12.4, 10

Solution

The smallest bonus is $10 million and the largest bonus is $34.5 million.

$$MR = \frac{10 + 34.5}{2} = \frac{44.5}{2} = \$22.25 \text{ million}$$

Notice that this amount is larger than seven of the eight amounts and is not typical of the average of the bonuses. The reason is that there is one very high bonus, namely, $34.5 million.

In statistics, several measures can be used for an average. The most common measures are the mean, median, mode, and midrange. Each has its own specific purpose and use. Exercises 39 through 41 show examples of other averages, such as the harmonic mean, the geometric mean, and the quadratic mean. Their applications are limited to specific areas, as shown in the exercises.

The Weighted Mean

Sometimes, you must find the mean of a data set in which not all values are equally represented. Consider the case of finding the average cost of a gallon of gasoline for three taxis. Suppose the drivers buy gasoline at three different service stations at a cost of $3.22, $3.53, and $3.63 per gallon. You might try to find the average by using the formula

$$\bar{X} = \frac{\Sigma X}{n}$$

$$= \frac{3.22 + 3.53 + 3.63}{3} = \frac{10.38}{3} = \$3.46$$

But not all drivers purchased the same number of gallons. Hence, to find the true average cost per gallon, you must take into consideration the number of gallons each driver purchased.

The type of mean that considers an additional factor is called the *weighted mean,* and it is used when the values are not all equally represented.

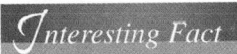

Interesting Fact

The average American drives about 10,000 miles a year.

Find the **weighted mean** of a variable X by multiplying each value by its corresponding weight and dividing the sum of the products by the sum of the weights.

$$\bar{X} = \frac{w_1 X_1 + w_2 X_2 + \cdots + w_n X_n}{w_1 + w_2 + \cdots + w_n} = \frac{\Sigma w X}{\Sigma w}$$

where $w_1, w_2, \ldots, w_n$ are the weights and $X_1, X_2, \ldots, X_n$ are the values.

Example 3–17 shows how the weighted mean is used to compute a grade point average. Since courses vary in their credit value, the number of credits must be used as weights.

Example 3–17   Grade Point Average

A student received an A in English Composition I (3 credits), a C in Introduction to Psychology (3 credits), a B in Biology I (4 credits), and a D in Physical Education (2 credits). Assuming A = 4 grade points, B = 3 grade points, C = 2 grade points, D = 1 grade point, and F = 0 grade points, find the student's grade point average.

Solution

Course	Credits (w)	Grade (X)
English Composition I	3	A (4 points)
Introduction to Psychology	3	C (2 points)
Biology I	4	B (3 points)
Physical Education	2	D (1 point)

$$\bar{X} = \frac{\Sigma w X}{\Sigma w} = \frac{3 \cdot 4 + 3 \cdot 2 + 4 \cdot 3 + 2 \cdot 1}{3 + 3 + 4 + 2} = \frac{32}{12} = 2.7$$

The grade point average is 2.7.

Table 3–1 summarizes the measures of central tendency.

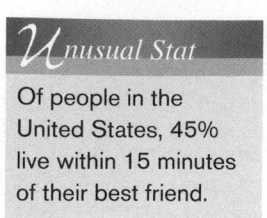

Table 3–1	Summary of Measures of Central Tendency	
Measure	**Definition**	**Symbol(s)**
Mean	Sum of values, divided by total number of values	$\mu, \overline{X}$
Median	Middle point in data set that has been ordered	MD
Mode	Most frequent data value	None
Midrange	Lowest value plus highest value, divided by 2	MR

Researchers and statisticians must know which measure of central tendency is being used and when to use each measure of central tendency. The properties and uses of the four measures of central tendency are summarized next.

Properties and Uses of Central Tendency

The Mean
1. The mean is found by using all the values of the data.
2. The mean varies less than the median or mode when samples are taken from the same population and all three measures are computed for these samples.
3. The mean is used in computing other statistics, such as the variance.
4. The mean for the data set is unique and not necessarily one of the data values.
5. The mean cannot be computed for the data in a frequency distribution that has an open-ended class.
6. The mean is affected by extremely high or low values, called outliers, and may not be the appropriate average to use in these situations.

The Median
1. The median is used to find the center or middle value of a data set.
2. The median is used when it is necessary to find out whether the data values fall into the upper half or lower half of the distribution.
3. The median is used for an open-ended distribution.
4. The median is affected less than the mean by extremely high or extremely low values.

The Mode
1. The mode is used when the most typical case is desired.
2. The mode is the easiest average to compute.
3. The mode can be used when the data are nominal or categorical, such as religious preference, gender, or political affiliation.
4. The mode is not always unique. A data set can have more than one mode, or the mode may not exist for a data set.

The Midrange
1. The midrange is easy to compute.
2. The midrange gives the midpoint.
3. The midrange is affected by extremely high or low values in a data set.

Figure 3–1

Types of Distributions

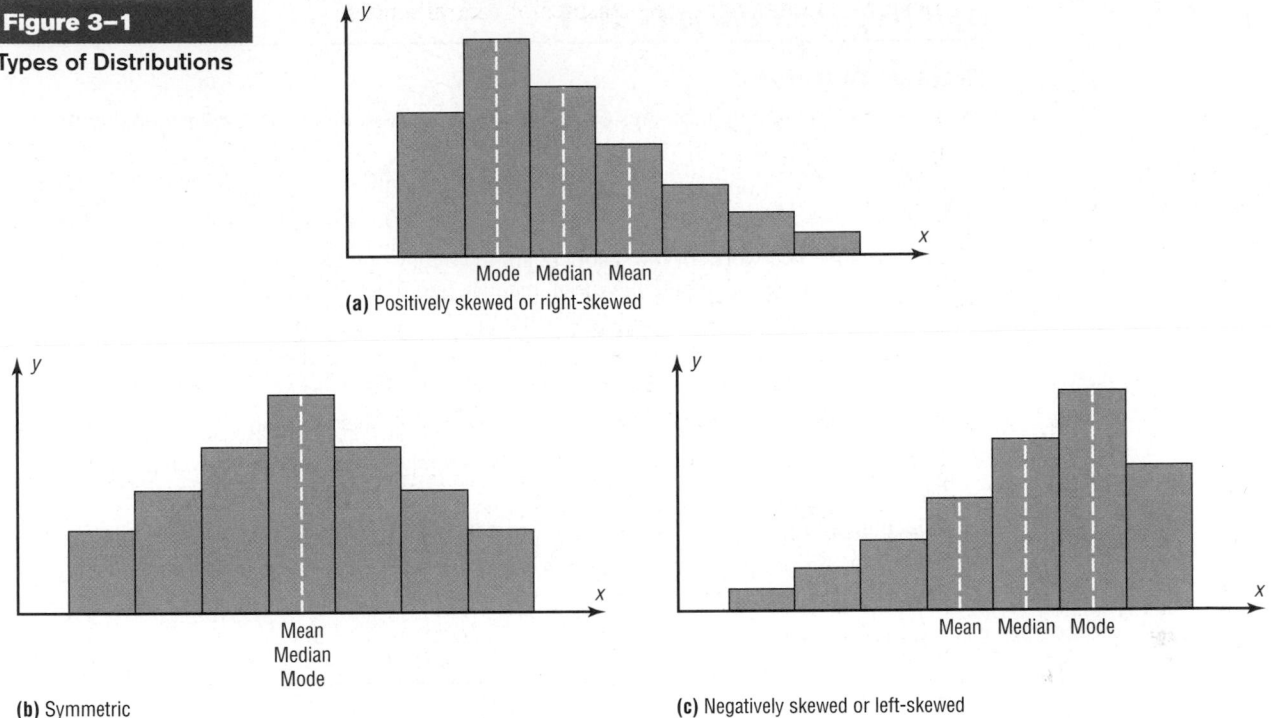

(a) Positively skewed or right-skewed

(b) Symmetric

(c) Negatively skewed or left-skewed

Distribution Shapes

Frequency distributions can assume many shapes. The three most important shapes are positively skewed, symmetric, and negatively skewed. Figure 3–1 shows histograms of each.

In a **positively skewed** or **right-skewed distribution,** the majority of the data values fall to the left of the mean and cluster at the lower end of the distribution; the "tail" is to the right. Also, the mean is to the right of the median, and the mode is to the left of the median.

For example, if an instructor gave an examination and most of the students did poorly, their scores would tend to cluster on the left side of the distribution. A few high scores would constitute the tail of the distribution, which would be on the right side. Another example of a positively skewed distribution is the incomes of the population of the United States. Most of the incomes cluster about the low end of the distribution; those with high incomes are in the minority and are in the tail at the right of the distribution.

In a **symmetric distribution,** the data values are evenly distributed on both sides of the mean. In addition, when the distribution is unimodal, the mean, median, and mode are the same and are at the center of the distribution. Examples of symmetric distributions are IQ scores and heights of adult males.

When the majority of the data values fall to the right of the mean and cluster at the upper end of the distribution, with the tail to the left, the distribution is said to be **negatively skewed** or **left-skewed.** Also, the mean is to the left of the median, and the mode is to the right of the median. As an example, a negatively skewed distribution results if the majority of students score very high on an instructor's examination. These scores will tend to cluster to the right of the distribution.

When a distribution is extremely skewed, the value of the mean will be pulled toward the tail, but the majority of the data values will be greater than the mean or less than the mean (depending on which way the data are skewed); hence, the median rather than the mean is a more appropriate measure of central tendency. An extremely skewed distribution can also affect other statistics.

A measure of skewness for a distribution is discussed in Exercise 48 in Section 3–2.

Applying the Concepts **3–1**

Teacher Salaries

The following data represent salaries (in dollars) from a school district in Greenwood, South Carolina.

10,000	11,000	11,000	12,500	14,300	17,500
18,000	16,600	19,200	21,560	16,400	107,000

1. First, assume you work for the school board in Greenwood and do not wish to raise taxes to increase salaries. Compute the mean, median, and mode, and decide which one would best support your position to not raise salaries.

2. Second, assume you work for the teachers' union and want a raise for the teachers. Use the best measure of central tendency to support your position.

3. Explain how outliers can be used to support one or the other position.

4. If the salaries represented every teacher in the school district, would the averages be parameters or statistics?

5. Which measure of central tendency can be misleading when a data set contains outliers?

6. When you are comparing the measures of central tendency, does the distribution display any skewness? Explain.

See page 180 for the answers.

Exercises 3–1

For Exercises 1 through 9, find (*a*) the mean, (*b*) the median, (*c*) the mode, and (*d*) the midrange.

1. Grade Point Averages The average undergraduate grade point average (GPA) for the 25 top-ranked medical schools is listed below. *a.* 3.724   *b.* 3.73
c. 3.74 and 3.70   *d.* 3.715

3.80	3.77	3.70	3.74	3.70
3.86	3.76	3.68	3.67	3.57
3.83	3.70	3.80	3.74	3.67
3.78	3.74	3.73	3.65	3.66
3.75	3.64	3.78	3.73	3.64

Source: *U.S. News & World Report Best Graduate Schools.*

2. Airport Parking The number of short-term parking spaces at 15 airports is shown.   *a.* 3174.6
b. 1479   *c.* No mode   *d.* 5012.5

750	3400	1962	700	203
900	8662	260	1479	5905
9239	690	9822	1131	2516

Source: *USA Today.*

3. High Temperatures The reported high temperatures (in degrees Fahrenheit) for selected world cities on an October day are shown below. Which measure of central tendency do you think best describes these data?

62 72 66 79 83 61 62 85 72 64 74 71
42 38 91 66 77 90 74 63 64 68 42

Source: www.accuweather.com   *a.* 68.1   *b.* 68
   c. 42, 62, 64, 66, 72, 74   *d.* 64.5

4. Observers in the Frogwatch Program The number of observers in the Frogwatch USA program (a wildlife conservation program dedicated to helping conserve frogs and toads) for the top 10 states with the most observers is 484, 483, 422, 396, 378, 352, 338, 331, 318, and 302. The top 10 states with the most active watchers list these numbers of visits: 634, 464, 406, 267, 219, 194, 191, 150, 130, and 114. Compare the measures of central tendency for these two groups of data.

Source: www.nwf.org/frogwatch

5. Expenditures per Pupil for Selected States The expenditures per pupil for selected states are listed below. Based on these data, what do you think of the claim that the average expenditure per pupil in the United States exceeds $10,000?   *a.* 9422.2   *b.* 8988
c. 7552, 12,568, 8632   *d.* 9434. Claim seems a little high.

6,300	11,847	8,319	9,344	9,870
10,460	7,491	7,552	12,568	8,632
7,552	12,568	8,632	11,057	10,454
8,109				

Source: *New York Times Almanac.*

6. Earnings of Nonliving Celebrities *Forbes* magazine prints an annual Top-Earning Nonliving Celebrities list (based on royalties and estate earnings). Find the measures of central tendency for these data and comment on the skewness. Figures represent millions of dollars.   *a.* 19   *b.* 10   *c.* 7   *d.* 28.5 (Isn't it cool that Albert Einstein is on this list?)

Kurt Cobain	50	Ray Charles	10
Elvis Presley	42	Marilyn Monroe	8
Charles M. Schulz	35	Johnny Cash	8
John Lennon	24	J.R.R. Tolkien	7
Albert Einstein	20	George Harrison	7
Andy Warhol	19	Bob Marley	7
Theodore Geisel (Dr. Seuss)	10		

Source: articles.moneycentral.msn.com

7. Earthquake Strengths Twelve major earthquakes had Richter magnitudes shown here.

7.0, 6.2, 7.7, 8.0, 6.4, 6.2,
7.2, 5.4, 6.4, 6.5, 7.2, 5.4

Which would you consider the best measure of average?

Source: *The Universal Almanac.*
a. 6.63   *b.* 6.45   *c.* 5.4, 6.2, 6.4, 7.2   *d.* 6.7; answers will vary

8. Top-Paid CEOs The data shown are the total compensation (in millions of dollars) for the 50 top-paid CEOs for a recent year. Compare the averages, and state which one you think is the best measure.

17.5	18.0	36.8	31.7	31.7
17.3	24.3	47.7	38.5	17.0
23.7	16.5	25.1	17.4	18.0
37.6	19.7	21.4	28.6	21.6
19.3	20.0	16.9	25.2	19.8
25.0	17.2	20.4	20.1	29.1
19.1	25.2	23.2	25.9	24.0
41.7	24.0	16.8	26.8	31.4
16.9	17.2	24.1	35.2	19.1
22.9	18.2	25.4	35.4	25.5

Source: *USA TODAY.* 24.42; 23.45; 16.9, 17.2, 18, 19.1, 24, 25.2, 31.7; 32.1. It appears that the mean and median are good measures of the average.

9. Garbage Collection The amount of garbage in millions of tons collected over a 16-year period is shown. *a.* 46.78   *b.* 47.65   *c.* None   *d.* 44.05

29.7	47.3	32.9	36
48	57.2	53.7	52.8
58.4	55.8	46.1	46.4
37.9	43.5	50.1	52.7

Source: *Environmental Protection Agency.*

10. Foreign Workers The number of foreign workers' certificates for the New England states and the northwestern states is shown. Find the mean, median, and mode for both areas and compare the results.

New England States	Northwest States
6768	1870
3196	622
1112	620
819	23
1019	172
1795	112

Source: Department of Labor.

11. Populations of Selected Cities Populations for towns and cities of 5000 or more (based on the 2004 figures) in the 15XXX zip code area are listed here for two different years. Find the mean, median, mode, and midrange for each set of data. What do your findings suggest?

2004			1990		
11,270	8,825	7,439	13,374	9,200	8,133
8,220	5,132	8,395	9,278	4,768	9,135
5,463	8,174	5,044	6,113	9,656	5,784
8,739	5,282	7,869	9,229	21,923	8,286
6,199	5,307	10,493	10,687	5,319	9,126
10,309	14,925	8,397	11,221	15,174	9,901
9,964	14,849	5,094	10,823	15,864	5,445
14,340	5,707	6,672	14,292	5,748	6,961

Source: *World Almanac.*

For Exercises 12 through 21, find the (*a*) mean and (*b*) modal class.

12. Executive Bonuses A random sample of bonuses (in millions of dollars) paid by large companies to their executives is shown. These data will be used for Exercise 18 in Section 3–2. *a.* 5   *b.* 3.5–6.5

Class boundaries	Frequency
0.5–3.5	11
3.5–6.5	12
6.5–9.5	4
9.5–12.5	2
12.5–15.5	1

13. Hourly Compensation for Production Workers The hourly compensation costs (in U.S. dollars) for production workers in selected countries are represented below.

Class	Frequency	
2.48–7.48	7	
7.49–12.49	3	
12.50–17.50	1	
17.51–22.51	7	
22.52–27.52	5	*a.* 17.68 *b.* 2.48–7.48 and
27.53–32.53	5	17.51–22.51. Group mean is less.

Compare the mean of these grouped data to the U.S. mean of $21.97.

Source: *New York Times Almanac.*

14. Automobile Fuel Efficiency Thirty automobiles were tested for fuel efficiency (in miles per gallon). This frequency distribution was obtained. (The data in this exercise will be used in Exercise 20 in Section 3–2.)
a. 19.7 *b.* 17.5–22.5

Class boundaries	Frequency
7.5–12.5	3
12.5–17.5	5
17.5–22.5	15
22.5–27.5	5
27.5–32.5	2

15. Percentage of Foreign-Born People The percentage of foreign-born population for each of the 50 states is represented below. Do you think the mean is the best average for this set of data? Explain. *a.* 6.5 *b.* 0.8–4.4. Probably not—data are "top heavy."

Percentage	Frequency
0.8–4.4	26
4.5–8.1	11
8.2–11.8	4
11.9–15.5	5
15.6–19.2	2
19.3–22.9	1
23.0–26.6	1

Source: *World Almanac.*

16. Find the mean and modal class for each set of data in Exercises 8 and 18 in Section 2–2. Is the average about the same for both sets of data?

17. Percentage of College-Educated Population over 25 Below are the percentages of the population over 25 years of age who have completed 4 years of college or more for the 50 states and the District of Columbia. Find the mean and modal class. *a.* 26.7 *b.* 24.2–28.6

Percentage	Frequency
15.2–19.6	3
19.7–24.1	15
24.2–28.6	19
28.7–33.1	6
33.2–37.6	7
37.7–42.1	0
42.2–46.6	1

Source: *New York Times Almanac.*

18. Net Worth of Corporations These data represent the net worth (in millions of dollars) of 45 national corporations. *a.* 42.9 *b.* 32–42

Class limits	Frequency
10–20	2
21–31	8
32–42	15
43–53	7
54–64	10
65–75	3

19. Specialty Coffee Shops A random sample of 30 states shows the number of specialty coffee shops for a specific company. *a.* 34.1 *b.* 0.5–19.5

Class boundaries	Frequency
0.5–19.5	12
19.5–38.5	7
38.5–57.5	5
57.5–76.5	3
76.5–95.5	3

20. Commissions Earned This frequency distribution represents the commission earned (in dollars) by 100 salespeople employed at several branches of a large chain store. *a.* 180.3 *b.* 177–185

Class limits	Frequency
150–158	5
159–167	16
168–176	20
177–185	21
186–194	20
195–203	15
204–212	3

21. Copier Service Calls This frequency distribution represents the data obtained from a sample of 75 copying machine service technicians. The values represent the days between service calls for various copying machines. *a.* 23.7 *b.* 21.5–24.5

Class boundaries	Frequency
15.5–18.5	14
18.5–21.5	12
21.5–24.5	18
24.5–27.5	10
27.5–30.5	15
30.5–33.5	6

22. Use the data from Exercise 14 in Section 2–1 and find the mean and modal class. *a.* 14.6 *b.* 0–10

23. Find the mean and modal class for the data in Exercise 13 in Section 2–1. 44.8; 40.5–47.5

24. Use the data from Exercise 3 in Section 2–2 and find the mean and modal class. *a.* 64.4 *b.* 3–45 and 46–88

25. Enrollments for Selected Independent Religiously Controlled 4-Year Colleges Listed below are the enrollments for selected independent religiously controlled 4-year colleges that offer bachelor's degrees only. Construct a grouped frequency distribution with six classes and find the mean and modal class. *a.* 1804.6 *b.* 1013–1345

1013 1867 1268 1666 2309 1231 3005 2895 2166 1136
1532 1461 1750 1069 1723 1827 1155 1714 2391 2155
1412 1688 2471 1759 3008 2511 2577 1082 1067 1062
1319 1037 2400

Source: *World Almanac.*

26. Find the weighted mean price of three models of automobiles sold. The number and price of each model sold are shown in this list. $9866.67

Model	Number	Price
A	8	$10,000
B	10	12,000
C	12	8,000

27. **Fat Grams** Using the weighted mean, find the average number of grams of fat per ounce of meat or fish that a person would consume over a 5-day period if he ate these:

Meat or fish	Fat (g/oz)
3 oz fried shrimp	3.33
3 oz veal cutlet (broiled)	3.00
2 oz roast beef (lean)	2.50
2.5 oz fried chicken drumstick	4.40
4 oz tuna (canned in oil)	1.75
Source: *The World Almanac and Book of Facts.*	2.896

28. **Diet Cola Preference** A recent survey of a new diet cola reported the following percentages of people who liked the taste. Find the weighted mean of the percentages. 35.4%

Area	% Favored	Number surveyed
1	40	1000
2	30	3000
3	50	800

29. **Costs of Helicopters** The costs of three models of helicopters are shown here. Find the weighted mean of the costs of the models. $545,666.67

Model	Number sold	Cost
Sunscraper	9	$427,000
Skycoaster	6	365,000
High-flyer	12	725,000

30. **Final Grade** An instructor grades exams, 20%; term paper, 30%; final exam, 50%. A student had grades of 83, 72, and 90, respectively, for exams, term paper, and final exam. Find the student's final average. Use the weighted mean. 83.2

31. **Final Grade** Another instructor gives four 1-hour exams and one final exam, which counts as two 1-hour exams. Find a student's grade if she received 62, 83, 97, and 90 on the 1-hour exams and 82 on the final exam. 82.7

32. For these situations, state which measure of central tendency—mean, median, or mode—should be used.

    a. The most typical case is desired. Mode
    b. The distribution is open-ended. Median
    c. There is an extreme value in the data set. Median
    d. The data are categorical. Mode
    e. Further statistical computations will be needed. Mean
    f. The values are to be divided into two approximately equal groups, one group containing the larger values and one containing the smaller values. Median

33. Describe which measure of central tendency—mean, median, or mode—was probably used in each situation.

    a. One-half of the factory workers make more than $5.37 per hour, and one-half make less than $5.37 per hour. Median
    b. The average number of children per family in the Plaza Heights Complex is 1.8. Mean
    c. Most people prefer red convertibles over any other color. Mode
    d. The average person cuts the lawn once a week. Mode
    e. The most common fear today is fear of speaking in public. Mode
    f. The average age of college professors is 42.3 years. Mean

34. What types of symbols are used to represent sample statistics? Give an example. What types of symbols are used to represent population parameters? Give an example. Roman letters, $\bar{X}$; Greek letters, μ

35. A local fast-food company claims that the average salary of its employees is $13.23 per hour. An employee states that most employees make minimum wage. If both are being truthful, how could both be correct? Both could be true since one may be using the mean for the average salary and the other may be using the mode for the average.

Extending the Concepts

36. If the mean of five values is 64, find the sum of the values. 320

37. If the mean of five values is 8.2 and four of the values are 6, 10, 7, and 12, find the fifth value. 6

38. Find the mean of 10, 20, 30, 40, and 50.

    a. Add 10 to each value and find the mean. 40
    b. Subtract 10 from each value and find the mean. 20
    c. Multiply each value by 10 and find the mean. 300
    d. Divide each value by 10 and find the mean. 3
    e. Make a general statement about each situation. The results will be the same as if you add, subtract, multiply, and divide the mean by 10.

39. The *harmonic mean* (HM) is defined as the number of values divided by the sum of the reciprocals of each value. The formula is

$$HM = \frac{n}{\Sigma(1/X)}$$

For example, the harmonic mean of 1, 4, 5, and 2 is

$$\text{HM} = \frac{4}{1/1 + 1/4 + 1/5 + 1/2} = 2.05$$

This mean is useful for finding the average speed. Suppose a person drove 100 miles at 40 miles per hour and returned driving 50 miles per hour. The average miles per hour is *not* 45 miles per hour, which is found by adding 40 and 50 and dividing by 2. The average is found as shown.

Since

$$\text{Time} = \text{distance} \div \text{rate}$$

then

$$\text{Time 1} = \frac{100}{40} = 2.5 \text{ hours to make the trip}$$

$$\text{Time 2} = \frac{100}{50} = 2 \text{ hours to return}$$

Hence, the total time is 4.5 hours, and the total miles driven are 200. Now, the average speed is

$$\text{Rate} = \frac{\text{distance}}{\text{time}} = \frac{200}{4.5} = 44.44 \text{ miles per hour}$$

This value can also be found by using the harmonic mean formula

$$\text{HM} = \frac{2}{1/40 + 1/50} = 44.44$$

Using the harmonic mean, find each of these.

a. A salesperson drives 300 miles round trip at 30 miles per hour going to Chicago and 45 miles per hour returning home. Find the average miles per hour. 36 mph

b. A bus driver drives the 50 miles to West Chester at 40 miles per hour and returns driving 25 miles per hour. Find the average miles per hour. 30.77 mph

c. A carpenter buys $500 worth of nails at $50 per pound and $500 worth of nails at $10 per pound. Find the average cost of 1 pound of nails. $16.67

40. The *geometric mean* (GM) is defined as the *n*th root of the product of *n* values. The formula is

$$\text{GM} = \sqrt[n]{(X_1)(X_2)(X_3)\cdots(X_n)}$$

The geometric mean of 4 and 16 is

$$\text{GM} = \sqrt{(4)(16)} = \sqrt{64} = 8$$

The geometric mean of 1, 3, and 9 is

$$\text{GM} = \sqrt[3]{(1)(3)(9)} = \sqrt[3]{27} = 3$$

The geometric mean is useful in finding the average of percentages, ratios, indexes, or growth rates. For example, if a person receives a 20% raise after 1 year of service and a 10% raise after the second year of service, the average percentage raise per year is not 15 but 14.89%, as shown.

$$\text{GM} = \sqrt{(1.2)(1.1)} = 1.1489$$

or

$$\text{GM} = \sqrt{(120)(110)} = 114.89\%$$

His salary is 120% at the end of the first year and 110% at the end of the second year. This is equivalent to an average of 14.89%, since 114.89% − 100% = 14.89%.

This answer can also be shown by assuming that the person makes $10,000 to start and receives two raises of 20 and 10%.

$$\text{Raise 1} = 10{,}000 \cdot 20\% = \$2000$$
$$\text{Raise 2} = 12{,}000 \cdot 10\% = \$1200$$

His total salary raise is $3200. This total is equivalent to

$$\$10{,}000 \cdot 14.89\% = \$1489.00$$
$$\$11{,}489 \cdot 14.89\% = \underline{\;\;1710.71}$$
$$\$3199.71 \approx \$3200$$

Find the geometric mean of each of these.

a. The growth rates of the Living Life Insurance Corporation for the past 3 years were 35, 24, and 18%. 25.5%

b. A person received these percentage raises in salary over a 4-year period: 8, 6, 4, and 5%. 5.7%

c. A stock increased each year for 5 years at these percentages: 10, 8, 12, 9, and 3%. 8.4%

d. The price increases, in percentages, for the cost of food in a specific geographic region for the past 3 years were 1, 3, and 5.5%. 3.2%

41. A useful mean in the physical sciences (such as voltage) is the *quadratic mean* (QM), which is found by taking the square root of the average of the squares of each value. The formula is

$$\text{QM} = \sqrt{\frac{\Sigma X^2}{n}}$$

The quadratic mean of 3, 5, 6, and 10 is

$$\text{QM} = \sqrt{\frac{3^2 + 5^2 + 6^2 + 10^2}{4}}$$
$$= \sqrt{42.5} = 6.52$$

Find the quadratic mean of 8, 6, 3, 5, and 4. 5.48

42. An approximate median can be found for data that have been grouped into a frequency distribution. First it is necessary to find the median class. This is the class that contains the median value. That is the *n*/2 data value. Then it is assumed that the data values are evenly distributed throughout the median class. The formula is

$$\text{MD} = \frac{n/2 - \text{cf}}{f}(w) + L_m$$

where
n = sum of frequencies
cf = cumulative frequency of class immediately preceding the median class
w = width of median class
f = frequency of median class
L_m = lower boundary of median class

Using this formula, find the median for data in the frequency distribution of Exercise 15. 4.31

Excel
Step by Step

Finding Measures of Central Tendency

Example XL3–1

Find the mean, mode, and median of the data from Example 3–11. The data represent the population of licensed nuclear reactors in the United States for a recent 15-year period.

104	104	104	104	104
107	109	109	109	110
109	111	112	111	109

1. On an Excel worksheet enter the numbers in cells A2–A16. Enter a label for the variable in cell A1.

On the same worksheet as the data:

2. Compute the mean of the data: key in **=AVERAGE(A2:A16)** in a blank cell.

3. Compute the mode of the data: key in **=MODE(A2:A16)** in a blank cell.

4. Compute the median of the data: key in **=MEDIAN(A2:A16)** in a blank cell.

These and other statistical functions can also be accessed without typing them into the worksheet directly.

1. Select the Formulas tab from the toolbar and select the Insert Function Icon 📊.

2. Select the Statistical category for statistical functions.

3. Scroll to find the appropriate function and click [OK].

	A	B	C
1	Number of Reactors		
2	104	107.7333	mean
3	104	104	mode
4	104	109	median
5	104		
6	104		
7	107		
8	109		
9	109		
10	109		
11	110		
12	109		
13	111		
14	112		
15	111		
16	109		

3–2   Measures of Variation

In statistics, to describe the data set accurately, statisticians must know more than the measures of central tendency. Consider Example 3–18.

Example 3–18

Objective 2

Describe data, using measures of variation, such as the range, variance, and standard deviation.

Comparison of Outdoor Paint

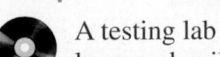
A testing lab wishes to test two experimental brands of outdoor paint to see how long each will last before fading. The testing lab makes 6 gallons of each paint to test. Since different chemical agents are added to each group and only six cans are involved, these two groups constitute two small populations. The results (in months) are shown. Find the mean of each group.

Brand A	Brand B
10	35
60	45
50	30
30	35
40	40
20	25

Solution

The mean for brand A is

$$\mu = \frac{\Sigma X}{N} = \frac{210}{6} = 35 \text{ months}$$

The mean for brand B is

$$\mu = \frac{\Sigma X}{N} = \frac{210}{6} = 35 \text{ months}$$

Since the means are equal in Example 3–18, you might conclude that both brands of paint last equally well. However, when the data sets are examined graphically, a somewhat different conclusion might be drawn. See Figure 3–2.

As Figure 3–2 shows, even though the means are the same for both brands, the spread, or variation, is quite different. Figure 3–2 shows that brand B performs more consistently; it is less variable. For the spread or variability of a data set, three measures are commonly used: *range, variance,* and *standard deviation.* Each measure will be discussed in this section.

Range

The range is the simplest of the three measures and is defined now.

The **range** is the highest value minus the lowest value. The symbol R is used for the range.

R = highest value − lowest value

Figure 3–2

Examining Data Sets Graphically

Variation of paint (in months)

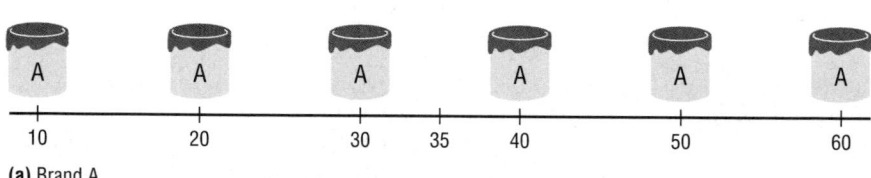

(a) Brand A

Variation of paint (in months)

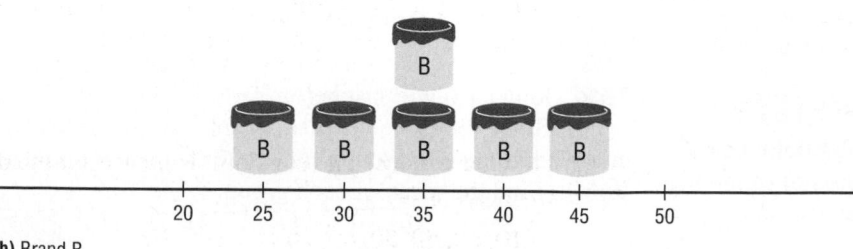

(b) Brand B

Example 3–19	**Comparison of Outdoor Paint**

Find the ranges for the paints in Example 3–18.

Solution

For brand A, the range is

$$R = 60 - 10 = 50 \text{ months}$$

For brand B, the range is

$$R = 45 - 25 = 20 \text{ months}$$

Make sure the range is given as a single number.

    The range for brand A shows that 50 months separate the largest data value from the smallest data value. For brand B, 20 months separate the largest data value from the smallest data value, which is less than one-half of brand A's range.

    One extremely high or one extremely low data value can affect the range markedly, as shown in Example 3–20.

Example 3–20	**Employee Salaries**

 The salaries for the staff of the XYZ Manufacturing Co. are shown here. Find the range.

Staff	Salary
Owner	$100,000
Manager	40,000
Sales representative	30,000
Workers	25,000
	15,000
	18,000

Solution

The range is $R = \$100,000 - \$15,000 = \$85,000$.

    Since the owner's salary is included in the data for Example 3–20, the range is a large number. To have a more meaningful statistic to measure the variability, statisticians use measures called the *variance* and *standard deviation*.

Population Variance and Standard Deviation

Before the variance and standard deviation are defined formally, the computational procedure will be shown, since the definition is derived from the procedure.

Rounding Rule for the Standard Deviation The rounding rule for the standard deviation is the same as that for the mean. The final answer should be rounded to one more decimal place than that of the original data.

Example 3–21	**Comparison of Outdoor Paint**

 Find the variance and standard deviation for the data set for brand A paint in Example 3–18.

    10, 60, 50, 30, 40, 20

Solution

Step 1 Find the mean for the data.

$$\mu = \frac{\Sigma X}{N} = \frac{10 + 60 + 50 + 30 + 40 + 20}{6} = \frac{210}{6} = 35$$

Step 2 Subtract the mean from each data value.

$$10 - 35 = -25 \qquad 50 - 35 = +15 \qquad 40 - 35 = +5$$
$$60 - 35 = +25 \qquad 30 - 35 = -5 \qquad 20 - 35 = -15$$

Step 3 Square each result.

$$(-25)^2 = 625 \qquad (+15)^2 = 225 \qquad (+5)^2 = 25$$
$$(+25)^2 = 625 \qquad (-5)^2 = 25 \qquad (-15)^2 = 225$$

Step 4 Find the sum of the squares.

$$625 + 625 + 225 + 25 + 25 + 225 = 1750$$

Step 5 Divide the sum by N to get the variance.

$$\text{Variance} = 1750 \div 6 = 291.7$$

Step 6 Take the square root of the variance to get the standard deviation. Hence, the standard deviation equals $\sqrt{291.7}$, or 17.1. It is helpful to make a table.

A Values X	B $X - \mu$	C $(X - \mu)^2$
10	-25	625
60	$+25$	625
50	$+15$	225
30	-5	25
40	$+5$	25
20	-15	225
		1750

Column A contains the raw data X. Column B contains the differences $X - \mu$ obtained in step 2. Column C contains the squares of the differences obtained in step 3.

Karl Pearson in 1892 and 1893 introduced the statistical concepts of the range and standard deviation.

The preceding computational procedure reveals several things. First, the square root of the variance gives the standard deviation; and vice versa, squaring the standard deviation gives the variance. Second, the variance is actually the average of the square of the distance that each value is from the mean. Therefore, if the values are near the mean, the variance will be small. In contrast, if the values are far from the mean, the variance will be large.

You might wonder why the squared distances are used instead of the actual distances. One reason is that the sum of the distances will always be zero. To verify this result for a specific case, add the values in column B of the table in Example 3–21. When each value is squared, the negative signs are eliminated.

Finally, why is it necessary to take the square root? The reason is that since the distances were squared, the units of the resultant numbers are the squares of the units of the original raw data. Finding the square root of the variance puts the standard deviation in the same units as the raw data.

When you are finding the square root, always use its positive value, since the variance and standard deviation of a data set can never be negative.

The **variance** is the average of the squares of the distance each value is from the mean. The symbol for the population variance is σ^2 (σ is the Greek lowercase letter sigma). The formula for the population variance is

$$\sigma^2 = \frac{\Sigma(X - \mu)^2}{N}$$

where

X = individual value
μ = population mean
N = population size

The **standard deviation** is the square root of the variance. The symbol for the population standard deviation is σ.
The corresponding formula for the population standard deviation is

$$\sigma = \sqrt{\sigma^2} = \sqrt{\frac{\Sigma(X - \mu)^2}{N}}$$

| **Example 3–22** | **Comparison of Outdoor Paint** |

Find the variance and standard deviation for brand B paint data in Example 3–18. The months were

35, 45, 30, 35, 40, 25

Solution

Step 1 Find the mean.

$$\mu = \frac{\Sigma X}{N} = \frac{35 + 45 + 30 + 35 + 40 + 25}{6} = \frac{210}{6} = 35$$

Step 2 Subtract the mean from each value, and place the result in column B of the table.

Step 3 Square each result and place the squares in column C of the table.

A	B	C
X	$X - \mu$	$(X - \mu)^2$
35	0	0
45	10	100
30	−5	25
35	0	0
40	5	25
25	−10	100

Step 4 Find the sum of the squares in column C.

$$\Sigma(X - \mu)^2 = 0 + 100 + 25 + 0 + 25 + 100 = 250$$

Step 5 Divide the sum by N to get the variance.

$$\sigma^2 = \frac{\Sigma(X - \mu)^2}{N} = \frac{250}{6} = 41.7$$

Step 6 Take the square root to get the standard deviation.

$$\sigma = \sqrt{\frac{\Sigma(X - \mu)^2}{N}} = \sqrt{41.7} = 6.5$$

Hence, the standard deviation is 6.5.

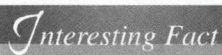

Interesting Fact

Each person receives on average 598 pieces of mail per year.

Since the standard deviation of brand A is 17.1 (see Example 3–21) and the standard deviation of brand B is 6.5, the data are more variable for brand A. *In summary, when the means are equal, the larger the variance or standard deviation is, the more variable the data are.*

Sample Variance and Standard Deviation

When computing the variance for a sample, one might expect the following expression to be used:

$$\frac{\Sigma(X - \overline{X})^2}{n}$$

where $\overline{X}$ is the sample mean and n is the sample size. *This formula is not usually used, however, since in most cases the purpose of calculating the statistic is to estimate the corresponding parameter.* For example, the sample mean $\overline{X}$ is used to estimate the population mean μ. The expression

$$\frac{\Sigma(X - \overline{X})^2}{n}$$

does not give the best estimate of the population variance because when the population is large and the sample is small (usually less than 30), the variance computed by this formula usually underestimates the population variance. Therefore, instead of dividing by n, find the variance of the sample by dividing by $n - 1$, giving a slightly larger value and an *unbiased* estimate of the population variance.

The formula for the sample variance, denoted by s^2, is

$$s^2 = \frac{\Sigma(X - \overline{X})^2}{n - 1}$$

where
$\overline{X}$ = sample mean
n = sample size

To find the standard deviation of a sample, you must take the square root of the sample variance, which was found by using the preceding formula.

Formula for the Sample Standard Deviation

The standard deviation of a sample (denoted by s) is

$$s = \sqrt{s^2} = \sqrt{\frac{\Sigma(X - \overline{X})^2}{n - 1}}$$

where
X = individual value
$\overline{X}$ = sample mean
n = sample size

Shortcut formulas for computing the variance and standard deviation are presented next and will be used in the remainder of the chapter and in the exercises. These formulas are mathematically equivalent to the preceding formulas and do not involve using the mean. They save time when repeated subtracting and squaring occur in the original formulas. They are also more accurate when the mean has been rounded.

Shortcut or Computational Formulas for s^2 and s

The shortcut formulas for computing the variance and standard deviation for data obtained from samples are as follows.

Variance	**Standard deviation**

$$s^2 = \frac{n(\Sigma X^2) - (\Sigma X)^2}{n(n-1)} \qquad s = \sqrt{\frac{n(\Sigma X^2) - (\Sigma X)^2}{n(n-1)}}$$

Examples 3–23 and 3–24 explain how to use the shortcut formulas.

Example 3–23

European Auto Sales

Find the sample variance and standard deviation for the amount of European auto sales for a sample of 6 years shown. The data are in millions of dollars.

11.2, 11.9, 12.0, 12.8, 13.4, 14.3

Source: *USA TODAY.*

Solution

Step 1 Find the sum of the values.

$\Sigma X = 11.2 + 11.9 + 12.0 + 12.8 + 13.4 + 14.3 = 75.6$

Step 2 Square each value and find the sum.

$\Sigma X^2 = 11.2^2 + 11.9^2 + 12.0^2 + 12.8^2 + 13.4^2 + 14.3^2 = 958.94$

Step 3 Substitute in the formulas and solve.

$$s^2 = \frac{n(\Sigma X^2) - (\Sigma X)^2}{n(n-1)}$$
$$= \frac{6(958.94) - 75.6^2}{6(6-1)}$$
$$= \frac{5753.64 - 5715.36}{6(5)}$$
$$= \frac{38.28}{30}$$
$$= 1.276$$

The variance is 1.28 rounded.

$s = \sqrt{1.28} = 1.13$

Hence, the sample standard deviation is 1.13.

Note that ΣX^2 is not the same as $(\Sigma X)^2$. The notation ΣX^2 means to square the values first, then sum; $(\Sigma X)^2$ means to sum the values first, then square the sum.

Variance and Standard Deviation for Grouped Data

The procedure for finding the variance and standard deviation for grouped data is similar to that for finding the mean for grouped data, and it uses the midpoints of each class.

Example 3–24	Miles Run per Week

Find the variance and the standard deviation for the frequency distribution of the data in Example 2–7. The data represent the number of miles that 20 runners ran during one week.

Class	Frequency	Midpoint
5.5–10.5	1	8
10.5–15.5	2	13
15.5–20.5	3	18
20.5–25.5	5	23
25.5–30.5	4	28
30.5–35.5	3	33
35.5–40.5	2	38

Solution

Step 1   Make a table as shown, and find the midpoint of each class.

A	B	C	D	E
	Frequency	Midpoint		
Class	f	X_m	$f \cdot X_m$	$f \cdot X_m^2$
5.5–10.5	1	8		
10.5–15.5	2	13		
15.5–20.5	3	18		
20.5–25.5	5	23		
25.5–30.5	4	28		
30.5–35.5	3	33		
35.5–40.5	2	38		

Step 2   Multiply the frequency by the midpoint for each class, and place the products in column D.

$1 \cdot 8 = 8 \qquad 2 \cdot 13 = 26 \qquad \ldots \qquad 2 \cdot 38 = 76$

Step 3   Multiply the frequency by the square of the midpoint, and place the products in column E.

$1 \cdot 8^2 = 64 \qquad 2 \cdot 13^2 = 338 \qquad \ldots \qquad 2 \cdot 38^2 = 2888$

Step 4   Find the sums of columns B, D, and E. The sum of column B is n, the sum of column D is $\Sigma f \cdot X_m$, and the sum of column E is $\Sigma f \cdot X_m^2$. The completed table is shown.

A Class	B Frequency	C Midpoint	D $f \cdot X_m$	E $f \cdot X_m^2$
5.5–10.5	1	8	8	64
10.5–15.5	2	13	26	338
15.5–20.5	3	18	54	972
20.5–25.5	5	23	115	2,645
25.5–30.5	4	28	112	3,136
30.5–35.5	3	33	99	3,267
35.5–40.5	2	38	76	2,888
	$n = 20$		$\Sigma f \cdot X_m = 490$	$\Sigma f \cdot X_m^2 = 13,310$

Step 5   Substitute in the formula and solve for s^2 to get the variance.

$$s^2 = \frac{n(\Sigma f \cdot X_m^2) - (\Sigma f \cdot X_m)^2}{n(n-1)}$$

$$= \frac{20(13,310) - 490^2}{20(20-1)}$$

$$= \frac{266,200 - 240,100}{20(19)}$$

$$= \frac{26,100}{380}$$

$$= 68.7$$

Step 6   Take the square root to get the standard deviation.

$$s = \sqrt{68.7} = 8.3$$

Be sure to use the number found in the sum of column B (i.e., the sum of the frequencies) for n. Do not use the number of classes.

The steps for finding the variance and standard deviation for grouped data are summarized in this Procedure Table.

Procedure Table

Finding the Sample Variance and Standard Deviation for Grouped Data

Step 1   Make a table as shown, and find the midpoint of each class.

A	B	C	D	E
Class	Frequency	Midpoint	$f \cdot X_m$	$f \cdot X_m^2$

Step 2   Multiply the frequency by the midpoint for each class, and place the products in column D.

Step 3   Multiply the frequency by the square of the midpoint, and place the products in column E.

Step 4   Find the sums of columns B, D, and E. (The sum of column B is n. The sum of column D is $\Sigma f \cdot X_m$. The sum of column E is $\Sigma f \cdot X_m^2$.)

Step 5   Substitute in the formula and solve to get the variance.

$$s^2 = \frac{n(\Sigma f \cdot X_m^2) - (\Sigma f \cdot X_m)^2}{n(n-1)}$$

Step 6   Take the square root to get the standard deviation.

The three measures of variation are summarized in Table 3–2.

Table 3–2	Summary of Measures of Variation	
Measure	**Definition**	**Symbol(s)**
Range	Distance between highest value and lowest value	R
Variance	Average of the squares of the distance that each value is from the mean	σ^2, s^2
Standard deviation	Square root of the variance	σ, s

Uses of the Variance and Standard Deviation

1. As previously stated, variances and standard deviations can be used to determine the spread of the data. If the variance or standard deviation is large, the data are more dispersed. This information is useful in comparing two (or more) data sets to determine which is more (most) variable.

2. The measures of variance and standard deviation are used to determine the consistency of a variable. For example, in the manufacture of fittings, such as nuts and bolts, the variation in the diameters must be small, or the parts will not fit together.

3. The variance and standard deviation are used to determine the number of data values that fall within a specified interval in a distribution. For example, Chebyshev's theorem (explained later) shows that, for any distribution, at least 75% of the data values will fall within 2 standard deviations of the mean.

4. Finally, the variance and standard deviation are used quite often in inferential statistics. These uses will be shown in later chapters of this textbook.

Coefficient of Variation

Historical Note

Karl Pearson devised the coefficient of variation to compare the deviations of two different groups such as the heights of men and women.

Whenever two samples have the same units of measure, the variance and standard deviation for each can be compared directly. For example, suppose an automobile dealer wanted to compare the standard deviation of miles driven for the cars she received as trade-ins on new cars. She found that for a specific year, the standard deviation for Buicks was 422 miles and the standard deviation for Cadillacs was 350 miles. She could say that the variation in mileage was greater in the Buicks. But what if a manager wanted to compare the standard deviations of two different variables, such as the number of sales per salesperson over a 3-month period and the commissions made by these salespeople?

A statistic that allows you to compare standard deviations when the units are different, as in this example, is called the *coefficient of variation*.

The **coefficient of variation,** denoted by CVar, is the standard deviation divided by the mean. The result is expressed as a percentage.

For samples,

$$CVar = \frac{s}{\overline{X}} \cdot 100$$

For populations,

$$CVar = \frac{\sigma}{\mu} \cdot 100$$

Example 3–25

Sales of Automobiles

The mean of the number of sales of cars over a 3-month period is 87, and the standard deviation is 5. The mean of the commissions is $5225, and the standard deviation is $773. Compare the variations of the two.

Solution

The coefficients of variation are

$$CVar = \frac{s}{\overline{X}} = \frac{5}{87} \cdot 100 = 5.7\% \qquad \text{sales}$$

$$CVar = \frac{773}{5225} \cdot 100 = 14.8\% \qquad \text{commissions}$$

Since the coefficient of variation is larger for commissions, the commissions are more variable than the sales.

Example 3–26	Pages in Women's Fitness Magazines

The mean for the number of pages of a sample of women's fitness magazines is 132, with a variance of 23; the mean for the number of advertisements of a sample of women's fitness magazines is 182, with a variance of 62. Compare the variations.

Solution

The coefficients of variation are

$$\text{CVar} = \frac{\sqrt{23}}{132} \cdot 100 = 3.6\% \qquad \text{pages}$$

$$\text{CVar} = \frac{\sqrt{62}}{182} \cdot 100 = 4.3\% \qquad \text{advertisements}$$

The number of advertisements is more variable than the number of pages since the coefficient of variation is larger for advertisements.

Range Rule of Thumb

The range can be used to approximate the standard deviation. The approximation is called the **range rule of thumb.**

The Range Rule of Thumb

A rough estimate of the standard deviation is

$$s \approx \frac{\text{range}}{4}$$

In other words, if the range is divided by 4, an approximate value for the standard deviation is obtained. For example, the standard deviation for the data set 5, 8, 8, 9, 10, 12, and 13 is 2.7, and the range is $13 - 5 = 8$. The range rule of thumb is $s \approx 2$. The range rule of thumb in this case underestimates the standard deviation somewhat; however, it is in the ballpark.

A note of caution should be mentioned here. The range rule of thumb is only an *approximation* and should be used when the distribution of data values is unimodal and roughly symmetric.

The range rule of thumb can be used to estimate the largest and smallest data values of a data set. The smallest data value will be approximately 2 standard deviations below the mean, and the largest data value will be approximately 2 standard deviations above the mean of the data set. The mean for the previous data set is 9.3; hence,

$$\text{Smallest data value} = \overline{X} - 2s = 9.3 - 2(2.8) = 3.7$$

$$\text{Largest data value} = \overline{X} + 2s = 9.3 + 2(2.8) = 14.9$$

Notice that the smallest data value was 5, and the largest data value was 13. Again, these are rough approximations. For many data sets, almost all data values will fall within 2 standard deviations of the mean. Better approximations can be obtained by using Chebyshev's theorem and the empirical rule. These are explained next.

Chebyshev's Theorem

As stated previously, the variance and standard deviation of a variable can be used to determine the spread, or dispersion, of a variable. That is, the larger the variance or standard deviation, the more the data values are dispersed. For example, if two variables measured in the same units have the same mean, say, 70, and the first variable has a standard deviation of 1.5 while the second variable has a standard deviation of 10, then the data for the second variable will be more spread out than the data for the first variable. *Chebyshev's theorem,* developed by the Russian mathematician Chebyshev (1821–1894), specifies the proportions of the spread in terms of the standard deviation.

Chebyshev's theorem   The proportion of values from a data set that will fall within k standard deviations of the mean will be at least $1 - 1/k^2$, where k is a number greater than 1 (k is not necessarily an integer).

This theorem states that at least three-fourths, or 75%, of the data values will fall within 2 standard deviations of the mean of the data set. This result is found by substituting $k = 2$ in the expression.

$$1 - \frac{1}{k^2} \qquad \text{or} \qquad 1 - \frac{1}{2^2} = 1 - \frac{1}{4} = \frac{3}{4} = 75\%$$

For the example in which variable 1 has a mean of 70 and a standard deviation of 1.5, at least three-fourths, or 75%, of the data values fall between 67 and 73. These values are found by adding 2 standard deviations to the mean and subtracting 2 standard deviations from the mean, as shown:

$$70 + 2(1.5) = 70 + 3 = 73$$

and

$$70 - 2(1.5) = 70 - 3 = 67$$

For variable 2, at least three-fourths, or 75%, of the data values fall between 50 and 90. Again, these values are found by adding and subtracting, respectively, 2 standard deviations to and from the mean.

$$70 + 2(10) = 70 + 20 = 90$$

and

$$70 - 2(10) = 70 - 20 = 50$$

Furthermore, the theorem states that at least eight-ninths, or 88.89%, of the data values will fall within 3 standard deviations of the mean. This result is found by letting $k = 3$ and substituting in the expression.

$$1 - \frac{1}{k^2} \qquad \text{or} \qquad 1 - \frac{1}{3^2} = 1 - \frac{1}{9} = \frac{8}{9} = 88.89\%$$

For variable 1, at least eight-ninths, or 88.89%, of the data values fall between 65.5 and 74.5, since

$$70 + 3(1.5) = 70 + 4.5 = 74.5$$

and

$$70 - 3(1.5) = 70 - 4.5 = 65.5$$

For variable 2, at least eight-ninths, or 88.89%, of the data values fall between 40 and 100.

Figure 3–3

Chebyshev's Theorem

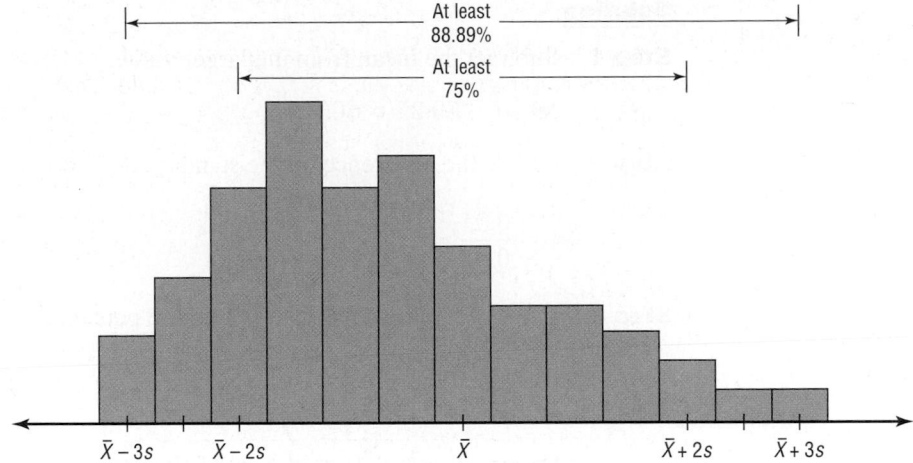

This theorem can be applied to any distribution regardless of its shape (see Figure 3–3).

Examples 3–27 and 3–28 illustrate the application of Chebyshev's theorem.

Example 3–27

Prices of Homes

The mean price of houses in a certain neighborhood is $50,000, and the standard deviation is $10,000. Find the price range for which at least 75% of the houses will sell.

Solution

Chebyshev's theorem states that three-fourths, or 75%, of the data values will fall within 2 standard deviations of the mean. Thus,

$$\$50,000 + 2(\$10,000) = \$50,000 + \$20,000 = \$70,000$$

and

$$\$50,000 - 2(\$10,000) = \$50,000 - \$20,000 = \$30,000$$

Hence, at least 75% of all homes sold in the area will have a price range from $30,000 to $70,000.

Chebyshev's theorem can be used to find the minimum percentage of data values that will fall between any two given values. The procedure is shown in Example 3–28.

Example 3–28

Travel Allowances

A survey of local companies found that the mean amount of travel allowance for executives was $0.25 per mile. The standard deviation was $0.02. Using Chebyshev's theorem, find the minimum percentage of the data values that will fall between $0.20 and $0.30.

Solution

Step 1 Subtract the mean from the larger value.

$$\$0.30 - \$0.25 = \$0.05$$

Step 2 Divide the difference by the standard deviation to get k.

$$k = \frac{0.05}{0.02} = 2.5$$

Step 3 Use Chebyshev's theorem to find the percentage.

$$1 - \frac{1}{k^2} = 1 - \frac{1}{2.5^2} = 1 - \frac{1}{6.25} = 1 - 0.16 = 0.84 \qquad \text{or} \qquad 84\%$$

Hence, at least 84% of the data values will fall between $0.20 and $0.30.

The Empirical (Normal) Rule

Chebyshev's theorem applies to any distribution regardless of its shape. However, when a distribution is *bell-shaped* (or what is called *normal*), the following statements, which make up the **empirical rule,** are true.

> Approximately 68% of the data values will fall within 1 standard deviation of the mean.
>
> Approximately 95% of the data values will fall within 2 standard deviations of the mean.
>
> Approximately 99.7% of the data values will fall within 3 standard deviations of the mean.

For example, suppose that the scores on a national achievement exam have a mean of 480 and a standard deviation of 90. If these scores are normally distributed, then approximately 68% will fall between 390 and 570 (480 + 90 = 570 and 480 − 90 = 390). Approximately 95% of the scores will fall between 300 and 660 (480 + 2 · 90 = 660 and 480 − 2 · 90 = 300). Approximately 99.7% will fall between 210 and 750 (480 + 3 · 90 = 750 and 480 − 3 · 90 = 210). See Figure 3–4. (The empirical rule is explained in greater detail in Chapter 6.)

Figure 3–4

The Empirical Rule

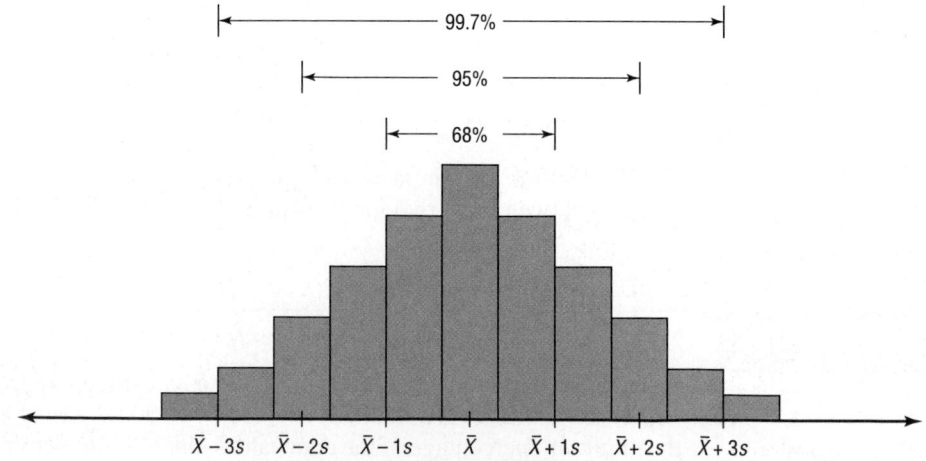

Applying the Concepts 3–2

Blood Pressure

The table lists means and standard deviations. The mean is the number before the plus/minus, and the standard deviation is the number after the plus/minus. The results are from a study attempting to find the average blood pressure of older adults. Use the results to answer the questions.

	Normotensive		Hypertensive	
	Men ($n = 1200$)	Women ($n = 1400$)	Men ($n = 1100$)	Women ($n = 1300$)
Age	55 ± 10	55 ± 10	60 ± 10	64 ± 10
Blood pressure (mm Hg)				
Systolic	123 ± 9	121 ± 11	153 ± 17	156 ± 20
Diastolic	78 ± 7	76 ± 7	91 ± 10	88 ± 10

1. Apply Chebyshev's theorem to the systolic blood pressure of normotensive men. At least how many of the men in the study fall within 1 standard deviation of the mean?

2. At least how many of those men in the study fall within 2 standard deviations of the mean?

Assume that blood pressure is normally distributed among older adults. Answer the following questions, using the empirical rule instead of Chebyshev's theorem.

3. Give ranges for the diastolic blood pressure (normotensive and hypertensive) of older women.

4. Do the normotensive, male, systolic blood pressure ranges overlap with the hypertensive, male, systolic blood pressure ranges?

See page 180 for the answers.

Exercises 3–2

1. What is the relationship between the variance and the standard deviation? The square root of the variance is the standard deviation.

2. Why might the range *not* be the best estimate of variability? One extremely high or one extremely low data value will influence the range.

3. What are the symbols used to represent the population variance and standard deviation? σ^2; σ

4. What are the symbols used to represent the sample variance and standard deviation? s^2; s

5. Why is the unbiased estimator of variance used?

6. The three data sets have the same mean and range, but is the variation the same? Prove your answer by computing the standard deviation. Assume the data were obtained from samples.

a. 5, 7, 9, 11, 13, 15, 17
b. 5, 6, 7, 11, 15, 16, 17
c. 5, 5, 5, 11, 17, 17, 17   No, *a* has the smallest variation; *c* has the biggest variation.

For Exercises 7–17, find the range, variance, and standard deviation unless the question asks for something different. Assume the data represent samples, and use the shortcut formula for the unbiased estimator to compute the variance and standard deviation.

7. **Police Calls in Schools** The number of incidents in which police were needed for a sample of 10 schools in Allegheny County is 7, 37, 3, 8, 48, 11, 6, 0, 10, 3. Are the data consistent or do they vary? Explain your answer. 48; 254.7; 15.9 (rounded to 16)   The data vary widely.
Source: U.S. Department of Education.

8. **Cigarette Taxes** The increases (in cents) in cigarette taxes for 17 states in a 6-month period are

60, 20, 40, 40, 45, 12, 34, 51, 30, 70, 42, 31, 69, 32, 8, 18, 50

Use the range rule of thumb to estimate the standard deviation. Compare the estimate to the actual standard deviation. 62; 332.4; 18.2; using the range rule of thumb, $s \approx 15.5$. This is close to the actual standard deviation of 18.2.
Source: Federation of Tax Administrators.

9. Precipitation and High Temperatures The normal daily high temperatures (in degrees Fahrenheit) in January for 10 selected cities are as follows.

50, 37, 29, 54, 30, 61, 47, 38, 34, 61

The normal monthly precipitation (in inches) for these same 10 cities is listed here.

4.8, 2.6, 1.5, 1.8, 1.8, 3.3, 5.1, 1.1, 1.8, 2.5

Which set is more variable?

Source: *New York Times Almanac.*

10. Size of U.S. States The total surface area (in square miles) for each of six selected Eastern states is listed here.

28,995	PA	37,534	FL
31,361	NY	27,087	VA
20,966	ME	37,741	GA

The total surface area for each of six selected Western states is listed (in square miles).

72,964	AZ	70,763	NV
101,510	CA	62,161	OR
66,625	CO	54,339	UT

Which set is more variable?

Source: *New York Times Almanac.*

11. Stories in the Tallest Buildings The number of stories in the 13 tallest buildings for two different cities is listed below. Which set of data is more variable?

Houston: 75, 71, 64, 56, 53, 55, 47, 55, 52, 50, 50, 50, 47

Pittsburgh: 64, 54, 40, 32, 46, 44, 42, 41, 40, 40, 34, 32, 30

Source: *World Almanac.*

12. Starting Teachers' Salaries Starting teachers' salaries (in equivalent U.S. dollars) for upper secondary education in selected countries are listed below. Which set of data is more variable? (The U.S. average starting salary at this time was $29,641.)

Europe		Asia	
Sweden	$48,704	Korea	$26,852
Germany	41,441	Japan	23,493
Spain	32,679	India	18,247
Finland	32,136	Malaysia	13,647
Denmark	30,384	Philippines	9,857
Netherlands	29,326	Thailand	5,862
Scotland	27,789		

Source: *World Almanac.*

13. The average age of U.S. astronaut candidates in the past has been 34, but candidates have ranged in age from 26 to 46. Use the range rule of thumb to estimate the standard deviation of the applicants' ages.

Source: www.nasa.gov $s \approx R/4$ so $s \approx 5$ years.

14. Times Spent in Rush-Hour Traffic A sample of 12 drivers shows the time that they spent (in minutes) stopped in rush-hour traffic on a specific snowy day last winter. *a.* 22 *b.* 35.5 *c.* 5.96

52	56	53
61	49	51
53	58	53
60	71	58

15. Football Playoff Statistics The number of yards gained in NFL playoff games by rookie quarterbacks is shown. *a.* 160 *b.* 1984.5 *c.* 44.5

193	66	136	140
157	163	181	226
135	199		

16. Passenger Vehicle Deaths The number of people killed in each state from passenger vehicle crashes for a specific year is shown. *a.* 2721 *b.* 355,427.6 *c.* 596.2

778	309	1110	324	705
1067	826	76	205	152
218	492	65	186	712
193	262	452	875	82
730	1185	2707	1279	390
305	123	948	343	602
69	451	951	104	985
155	450	2080	565	875
414	981	2786	82	793
214	130	396	620	797

Source: National Highway Traffic Safety Administration.

17. Find the range, variance, and standard deviation for the data in Exercise 17 of Section 2–1. *a.* 46 *b.* 77.48 *c.* 8.8

For Exercises 18 through 27, find the variance and standard deviation.

18. Baseball Team Batting Averages Team batting averages for major league baseball in 2005 are represented below. Find the variance and standard deviation for each league. Compare the results.

NL		AL	
0.252–0.256	4	0.256–0.261	2
0.257–0.261	6	0.262–0.267	5
0.262–0.266	1	0.268–0.273	4
0.267–0.271	4	0.274–0.279	2
0.272–0.276	1	0.280–0.285	1

Source: *World Almanac.* NL: $s^2 = 0.00004$, $s = 0.0066$
AL: $s^2 = 0.0000476$, $s = 0.0069$

19. Cost per Load of Laundry Detergents The costs per load (in cents) of 35 laundry detergents tested by a consumer organization are shown here. 133.6; 11.6

Class limits	Frequency
13–19	2
20–26	7
27–33	12
34–40	5
41–47	6
48–54	1
55–61	0
62–68	2

20. **Automotive Fuel Efficiency** Thirty automobiles were tested for fuel efficiency (in miles per gallon). This frequency distribution was obtained. 25.7; 5.1

Class boundaries	Frequency
7.5–12.5	3
12.5–17.5	5
17.5–22.5	15
22.5–27.5	5
27.5–32.5	2

21. **Murders in Cities** The data show the number of murders in 25 selected cities. 27,941.46; 167.2

Class limits	Frequency
34–96	13
97–159	2
160–222	0
223–285	5
286–348	1
349–411	1
412–474	0
475–537	1
538–600	2

22. **Reaction Times** In a study of reaction times to a specific stimulus, a psychologist recorded these data (in seconds).

Class limits	Frequency
2.1–2.7	12
2.8–3.4	13
3.5–4.1	7
4.2–4.8	5
4.9–5.5	2
5.6–6.2	1

0.847; 0.920

23. **FM Radio Stations** A random sample of 30 states shows the number of low-power FM radio stations for each state.

Class limits	Frequency
1–9	5
10–18	7
19–27	10
28–36	3
37–45	3
46–54	2

Source: Federal Communications Commission. 167.2; 12.93

24. **Murder Rates** The data represent the murder rate per 100,000 individuals in a sample of selected cities in the United States. 134.3; 11.6

Class	Frequency
5–11	8
12–18	5
19–25	7
26–32	1
33–39	1
40–46	3

Source: FBI and U.S. Census Bureau.

25. **Battery Lives** Eighty randomly selected batteries were tested to determine their lifetimes (in hours). The following frequency distribution was obtained.

Class boundaries	Frequency
62.5–73.5	5
73.5–84.5	14
84.5–95.5	18
95.5–106.5	25
106.5–117.5	12
117.5–128.5	6

Can it be concluded that the lifetimes of these brands of batteries are consistent? 211.2; 14.5; no, the variability of the lifetimes of the batteries is quite large.

26. Find the variance and standard deviation for the two distributions in Exercises 8 and 18 in Section 2–2. Compare the variation of the data sets. Decide if one data set is more variable than the other.

27. **Word Processor Repairs** This frequency distribution represents the data obtained from a sample of word processor repairers. The values are the days between service calls on 80 machines. 11.7; 3.4

Class boundaries	Frequency
25.5–28.5	5
28.5–31.5	9
31.5–34.5	32
34.5–37.5	20
37.5–40.5	12
40.5–43.5	2

28. **Missing Work** The average number of days construction workers miss per year is 11. The standard deviation is 2.3. The average number of days factory workers miss per year is 8 with a standard deviation of 1.8. Which class is more variable in terms of days missed?

29. **Suspension Bridges** The lengths (in feet) of the main span of the longest suspension bridges in the United States and the rest of the world are shown below. Which set of data is more variable?

United States: 4205, 4200, 3800, 3500, 3478, 2800, 2800, 2310
World: 6570, 5538, 5328, 4888, 4626, 4544, 4518, 3970

Source: *World Almanac.*

30. **Hospital Emergency Waiting Times** The mean of the waiting times in an emergency room is 80.2 minutes with a standard deviation of 10.5 minutes for people who are admitted for additional treatment. The mean waiting time for patients who are discharged after receiving treatment is 120.6 minutes with a standard deviation of 18.3 minutes. Which times are more variable?

31. **Ages of Accountants** The average age of the accountants at Three Rivers Corp. is 26 years, with a standard deviation of 6 years; the average salary of the accountants is $31,000, with a standard deviation of $4000. Compare the variations of age and income. 23.1%; 12.9%; age is more variable.

32. Using Chebyshev's theorem, solve these problems for a distribution with a mean of 80 and a standard deviation of 10.

    a. At least what percentage of values will fall between 60 and 100? 75%

    b. At least what percentage of values will fall between 65 and 95? 56%

33. The mean of a distribution is 20 and the standard deviation is 2. Use Chebyshev's theorem.

    a. At least what percentage of the values will fall between 10 and 30? 96%

    b. At least what percentage of the values will fall between 12 and 28? 93.75%

34. In a distribution of 160 values with a mean of 72, at least 120 fall within the interval 67–77. Approximately what percentage of values should fall in the interval 62–82? Use Chebyshev's theorem. At least 93.75%

35. **Calories** The average number of calories in a regular-size bagel is 240. If the standard deviation is 38 calories, find the range in which at least 75% of the data will lie. Use Chebyshev's theorem. Between 164 and 316 calories

36. **Time Spent Online** Americans spend an average of 3 hours per day online. If the standard deviation is 32 minutes, find the range in which at least 88.89% of the data will lie. Use Chebyshev's theorem.

    Source: www.cs.cmu.edu Between 84 and 276 minutes

37. **Solid Waste Production** The average college student produces 640 pounds of solid waste each year. If the standard deviation is approximately 85 pounds, within what weight limits will at least 88.89% of all students' garbage lie? Between 385 and 895 pounds

    Source: Environmental Sustainability Committee, www.esc.mtu.edu

38. **Sale Price of Homes** The average sale price of new one-family houses in the United States for 2003 was $246,300. Find the range of values in which at least 75% of the sale prices will lie if the standard deviation is $48,500. Between $149,300 and $343,300

    Source: *New York Times Almanac.*

39. **Trials to Learn a Maze** The average of the number of trials it took a sample of mice to learn to traverse a maze was 12. The standard deviation was 3. Using Chebyshev's theorem, find the minimum percentage of data values that will fall in the range of 4–20 trials. 86%

40. **Farm Sizes** The average farm in the United States in 2004 contained 443 acres. The standard deviation is 42 acres. Use Chebyshev's theorem to find the minimum percentage of data values that will fall in the range of 338–548 acres. At least 84%

    Source: *World Almanac.*

41. **Citrus Fruit Consumption** The average U.S. yearly per capita consumption of citrus fruit is 26.8 pounds. Suppose that the distribution of fruit amounts consumed is bell-shaped with a standard deviation equal to 4.2 pounds. What percentage of Americans would you expect to consume more than 31 pounds of citrus fruit per year? 16%

    Source: USDA/Economic Research Service.

42. **Work Hours for College Faculty** The average full-time faculty member in a post-secondary degree-granting institution works an average of 53 hours per week.

    a. If we assume the standard deviation is 2.8 hours, what percentage of faculty members work more than 58.6 hours a week? No more than 12.5%

    b. If we assume a bell-shaped distribution, what percentage of faculty members work more than 58.6 hours a week? 2.5%

    Source: National Center for Education Statistics.

Extending the Concepts

43. **Serum Cholesterol Levels** For this data set, find the mean and standard deviation of the variable. The data represent the serum cholesterol levels of 30 individuals. Count the number of data values that fall within 2 standard deviations of the mean. Compare this with the number obtained from Chebyshev's theorem. Comment on the answer.

211	240	255	219	204
200	212	193	187	205
256	203	210	221	249
231	212	236	204	187
201	247	206	187	200
237	227	221	192	196

All the data values fall within 2 standard deviations of the mean.

44. **Ages of Consumers** For this data set, find the mean and standard deviation of the variable. The data represent the ages of 30 customers who ordered a product advertised on television. Count the number of data values that fall within 2 standard deviations of the mean. Compare this with the number obtained from Chebyshev's theorem. Comment on the answer. 93.3%; All but two data values fall within 2 standard deviations of the mean.

42	44	62	35	20
30	56	20	23	41
55	22	31	27	66
21	18	24	42	25
32	50	31	26	36
39	40	18	36	22

45. Using Chebyshev's theorem, complete the table to find the minimum percentage of data values that fall within k standard deviations of the mean.

k	1.5	2	2.5	3	3.5
Percent	56	75	84	88.89	92

46. Use this data set: 10, 20, 30, 40, 50
 a. Find the standard deviation. 15.81
 b. Add 5 to each value, and then find the standard deviation. 15.81
 c. Subtract 5 from each value and find the standard deviation. 15.81
 d. Multiply each value by 5 and find the standard deviation. 79.06
 e. Divide each value by 5 and find the standard deviation. 3.16
 f. Generalize the results of parts *b* through *e*.
 g. Compare these results with those in Exercise 38 of Exercises 3–1.

 47. The mean deviation is found by using this formula:

$$\text{Mean deviation} = \frac{\Sigma |X - \overline{X}|}{n}$$

where
 X = value
 $\overline{X}$ = mean
 n = number of values
 $|\ |$ = absolute value

Find the mean deviation for these data.

5, 9, 10, 11, 11, 12, 15, 18, 20, 22   4.36

48. A measure to determine the skewness of a distribution is called the *Pearson coefficient of skewness* (*PC*). The formula is

$$PC = \frac{3(\overline{X} - MD)}{s}$$

The values of the coefficient usually range from -3 to $+3$. When the distribution is symmetric, the coefficient is zero; when the distribution is positively skewed, it is positive; and when the distribution is negatively skewed, it is negative.

Using the formula, find the coefficient of skewness for each distribution, and describe the shape of the distribution.

 a. Mean = 10, median = 8, standard deviation = 3.
 b. Mean = 42, median = 45, standard deviation = 4.
 c. Mean = 18.6, median = 18.6, standard deviation = 1.5.
 d. Mean = 98, median = 97.6, standard deviation = 4.

49. All values of a data set must be within $s\sqrt{n-1}$ of the mean. If a person collected 25 data values that had a mean of 50 and a standard deviation of 3 and you saw that one data value was 67, what would you conclude?

Technology *Step by Step*

Excel
Step by Step

Finding Measures of Variation

Example XL3–2

Find the variance, standard deviation, and range of the data from Example 3–23. The data represent the amount (in millions of dollars) of European auto sales for a sample of 6 years.

  11.2 11.9 12.0 12.8 13.4 14.3

1. On an Excel worksheet enter the data in cells A2–A7. Enter a label for the variable in cell A1.

2. For the sample variance, enter **=VAR(A2:A7)**.

3. For the sample standard deviation, enter **=STDEV(A2:A7)**.

4. For the range, compute the difference between the maximum and the minimum values by entering **=MAX(A2:A7) − MIN(A2:A7)**.

These and other statistical functions can also be accessed without typing them into the worksheet directly.

1. Select the Formulas tab from the toolbar and select the Insert Function Icon .

2. Select the Statistical category for statistical functions.

3. Scroll to find the appropriate function and click [OK].

	A	B	C	D
1	Sales (in millions $)	1.276	variance	
2	11.2	1.129602	standard deviation	
3	11.9	3.1	range	
4	12			
5	12.8			
6	13.4			
7	14.3			
8				
9				

3-3

Measures of Position

Objective 3

Identify the position of a data value in a data set, using various measures of position, such as percentiles, deciles, and quartiles.

In addition to measures of central tendency and measures of variation, there are measures of position or location. These measures include standard scores, percentiles, deciles, and quartiles. They are used to locate the relative position of a data value in the data set. For example, if a value is located at the 80th percentile, it means that 80% of the values fall below it in the distribution and 20% of the values fall above it. The *median* is the value that corresponds to the 50th percentile, since one-half of the values fall below it and one-half of the values fall above it. This section discusses these measures of position.

Standard Scores

There is an old saying, "You can't compare apples and oranges." But with the use of statistics, it can be done to some extent. Suppose that a student scored 90 on a music test and 45 on an English exam. Direct comparison of raw scores is impossible, since the exams might not be equivalent in terms of number of questions, value of each question, and so on. However, a comparison of a relative standard similar to both can be made. This comparison uses the mean and standard deviation and is called a standard score or *z* score. (We also use *z* scores in later chapters.)

A standard score or *z* score tells how many standard deviations a data value is above or below the mean for a specific distribution of values. If a standard score is zero, then the data value is the same as the mean.

A **z score** or **standard score** for a value is obtained by subtracting the mean from the value and dividing the result by the standard deviation. The symbol for a standard score is *z*. The formula is

$$z = \frac{\text{value} - \text{mean}}{\text{standard deviation}}$$

For samples, the formula is

$$z = \frac{X - \bar{X}}{s}$$

For populations, the formula is

$$z = \frac{X - \mu}{\sigma}$$

The *z* score represents the number of standard deviations that a data value falls above or below the mean.

For the purpose of this section, it will be assumed that when we find *z* scores, the data were obtained from samples.

Example 3–29

Test Scores

Interesting Fact

The average number of faces that a person learns to recognize and remember during his or her lifetime is 10,000.

A student scored 65 on a calculus test that had a mean of 50 and a standard deviation of 10; she scored 30 on a history test with a mean of 25 and a standard deviation of 5. Compare her relative positions on the two tests.

Solution

First, find the *z* scores. For calculus the *z* score is

$$z = \frac{X - \bar{X}}{s} = \frac{65 - 50}{10} = 1.5$$

For history the z score is

$$z = \frac{30 - 25}{5} = 1.0$$

Since the z score for calculus is larger, her relative position in the calculus class is higher than her relative position in the history class.

Note that if the z score is positive, the score is above the mean. If the z score is 0, the score is the same as the mean. And if the z score is negative, the score is below the mean.

Example 3-30

Test Scores

Find the z score for each test, and state which is higher.

Test A	$X = 38$	$\overline{X} = 40$	$s = 5$
Test B	$X = 94$	$\overline{X} = 100$	$s = 10$

Solution

For test A,

$$z = \frac{X - \overline{X}}{s} = \frac{38 - 40}{5} = -0.4$$

For test B,

$$z = \frac{94 - 100}{10} = -0.6$$

The score for test A is relatively higher than the score for test B.

When all data for a variable are transformed into z scores, the resulting distribution will have a mean of 0 and a standard deviation of 1. A z score, then, is actually the number of standard deviations each value is from the mean for a specific distribution. In Example 3–29, the calculus score of 65 was actually 1.5 standard deviations above the mean of 50. This will be explained in greater detail in Chapter 6.

Percentiles

Percentiles are position measures used in educational and health-related fields to indicate the position of an individual in a group.

Percentiles divide the data set into 100 equal groups.

In many situations, the graphs and tables showing the percentiles for various measures such as test scores, heights, or weights have already been completed. Table 3–3 shows the percentile ranks for scaled scores on the Test of English as a Foreign Language. If a student had a scaled score of 58 for section 1 (listening and comprehension), that student would have a percentile rank of 81. Hence, that student did better than 81% of the students who took section 1 of the exam.

Table 3–3 **Percentile Ranks and Scaled Scores on the Test of English as a Foreign Language***

Scaled score	Section 1: Listening comprehension	Section 2: Structure and written expression	Section 3: Vocabulary and reading comprehension	Total scaled score	Percentile rank
68	99	98			
66	98	96	98	660	99
64	96	94	96	640	97
62	92	90	93	620	94
60	87	84	88	600	89
→58	81	76	81	580	82
56	73	68	72	560	73
54	64	58	61	540	62
52	54	48	50	520	50
50	42	38	40	500	39
48	32	29	30	480	29
46	22	21	23	460	20
44	14	15	16	440	13
42	9	10	11	420	9
40	5	7	8	400	5
38	3	4	5	380	3
36	2	3	3	360	1
34	1	2	2	340	1
32		1	1	320	
30		1	1	300	
Mean	51.5	52.2	51.4	517	Mean
S.D.	7.1	7.9	7.5	68	S.D.

*Based on the total group of 1,178,193 examinees tested from July 1989 through June 1991.

Source: Reprinted by permission of Educational Testing Service, the copyright owner. However, the test question and any other testing information are provided in their entirety by McGraw-Hill Companies, Inc. No endorsement of this publication by Educational Testing Service should be inferred.

Figure 3–5 shows percentiles in graphical form of weights of girls from ages 2 to 18. To find the percentile rank of an 11-year-old who weighs 82 pounds, start at the 82-pound weight on the left axis and move horizontally to the right. Find 11 on the horizontal axis and move up vertically. The two lines meet at the 50th percentile curved line; hence, an 11-year-old girl who weighs 82 pounds is in the 50th percentile for her age group. If the lines do not meet exactly on one of the curved percentile lines, then the percentile rank must be approximated.

Percentiles are also used to compare an individual's test score with the national norm. For example, tests such as the National Educational Development Test (NEDT) are taken by students in ninth or tenth grade. A student's scores are compared with those of other students locally and nationally by using percentile ranks. A similar test for elementary school students is called the California Achievement Test.

Percentiles are not the same as percentages. That is, if a student gets 72 correct answers out of a possible 100, she obtains a percentage score of 72. There is no indication of her position with respect to the rest of the class. She could have scored the highest, the lowest, or somewhere in between. On the other hand, if a raw score of 72 corresponds to the 64th percentile, then she did better than 64% of the students in her class.

Figure 3–5

Weights of Girls by Age and Percentile Rankings

Source: Distributed by Mead Johnson Nutritional Division. Reprinted with permission.

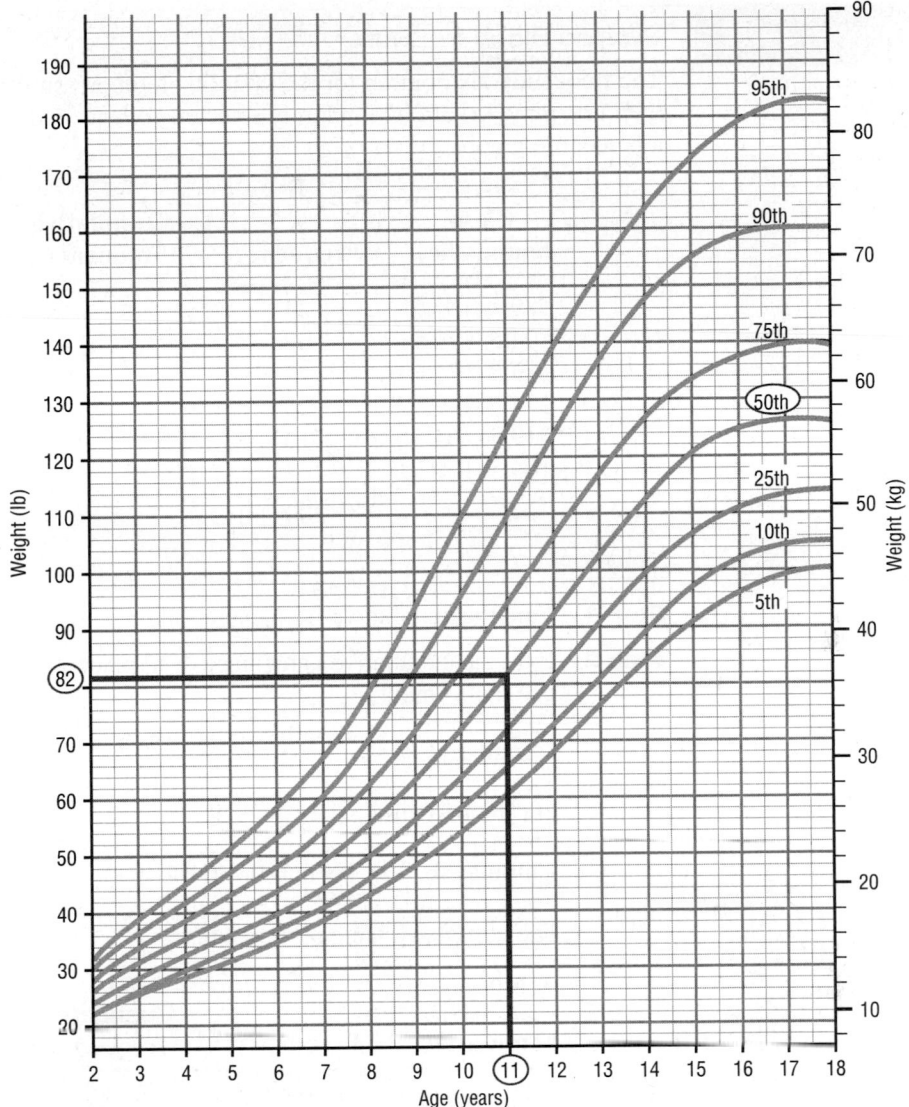

Percentiles are symbolized by

$$P_1, P_2, P_3, \ldots, P_{99}$$

and divide the distribution into 100 groups.

Percentile graphs can be constructed as shown in Example 3–31. Percentile graphs use the same values as the cumulative relative frequency graphs described in Section 2–2, except that the proportions have been converted to percents.

Example 3–31   Systolic Blood Pressure

The frequency distribution for the systolic blood pressure readings (in millimeters of mercury, mm Hg) of 200 randomly selected college students is shown here. Construct a percentile graph.

A Class boundaries	B Frequency	C Cumulative frequency	D Cumulative percent
89.5–104.5	24		
104.5–119.5	62		
119.5–134.5	72		
134.5–149.5	26		
149.5–164.5	12		
164.5–179.5	4		
	200		

Solution

Step 1   Find the cumulative frequencies and place them in column C.

Step 2   Find the cumulative percentages and place them in column D. To do this step, use the formula

$$\text{Cumulative \%} = \frac{\text{cumulative frequency}}{n} \cdot 100$$

For the first class,

$$\text{Cumulative \%} = \frac{24}{200} \cdot 100 = 12\%$$

The completed table is shown here.

A Class boundaries	B Frequency	C Cumulative frequency	D Cumulative percent
89.5–104.5	24	24	12
104.5–119.5	62	86	43
119.5–134.5	72	158	79
134.5–149.5	26	184	92
149.5–164.5	12	196	98
164.5–179.5	4	200	100
	200		

Step 3   Graph the data, using class boundaries for the x axis and the percentages for the y axis, as shown in Figure 3–6.

Once a percentile graph has been constructed, one can find the approximate corresponding percentile ranks for given blood pressure values and find approximate blood pressure values for given percentile ranks.

For example, to find the percentile rank of a blood pressure reading of 130, find 130 on the x axis of Figure 3–6, and draw a vertical line to the graph. Then move horizontally to the value on the y axis. Note that a blood pressure of 130 corresponds to approximately the 70th percentile.

If the value that corresponds to the 40th percentile is desired, start on the y axis at 40 and draw a horizontal line to the graph. Then draw a vertical line to the x axis and read

Figure 3–6

Percentile Graph for
Example 3–31

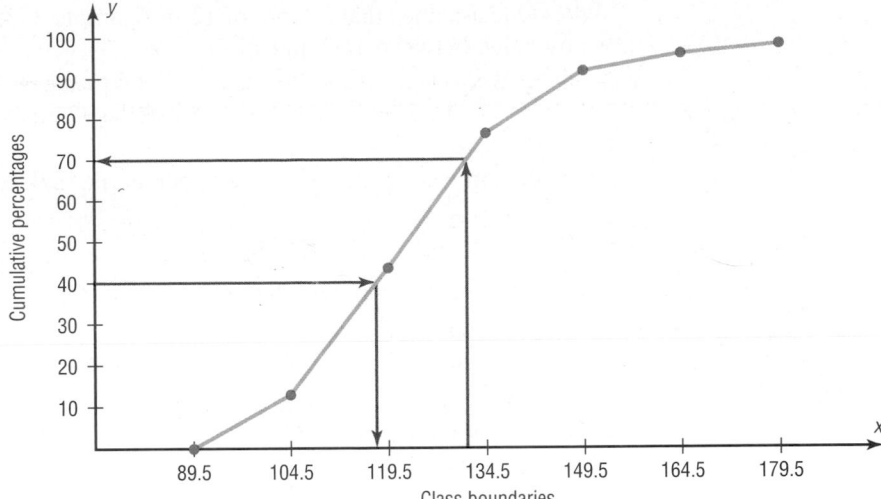

the value. In Figure 3–6, the 40th percentile corresponds to a value of approximately 118. Thus, if a person has a blood pressure of 118, he or she is at the 40th percentile.

Finding values and the corresponding percentile ranks by using a graph yields only approximate answers. Several mathematical methods exist for computing percentiles for data. These methods can be used to find the approximate percentile rank of a data value or to find a data value corresponding to a given percentile. When the data set is large (100 or more), these methods yield better results. Examples 3–32 through 3–35 show these methods.

Percentile Formula

The percentile corresponding to a given value X is computed by using the following formula:

$$\text{Percentile} = \frac{(\text{number of values below } X) + 0.5}{\text{total number of values}} \cdot 100$$

Example 3–32

Test Scores

 A teacher gives a 20-point test to 10 students. The scores are shown here. Find the percentile rank of a score of 12.

18, 15, 12, 6, 8, 2, 3, 5, 20, 10

Solution

Arrange the data in order from lowest to highest.

2, 3, 5, 6, 8, 10, 12, 15, 18, 20

Then substitute into the formula.

$$\text{Percentile} = \frac{(\text{number of values below } X) + 0.5}{\text{total number of values}} \cdot 100$$

Since there are six values below a score of 12, the solution is

$$\text{Percentile} = \frac{6 + 0.5}{10} \cdot 100 = 65\text{th percentile}$$

Thus, a student whose score was 12 did better than 65% of the class.

Note: One assumes that a score of 12 in Example 3–32, for instance, means theoretically any value between 11.5 and 12.5.

Example 3–33

Test Scores

Using the data in Example 3–32, find the percentile rank for a score of 6.

Solution

There are three values below 6. Thus

$$\text{Percentile} = \frac{3 + 0.5}{10} \cdot 100 = 35\text{th percentile}$$

A student who scored 6 did better than 35% of the class.

Examples 3–34 and 3–35 show a procedure for finding a value corresponding to a given percentile.

Example 3–34

Test Scores

Using the scores in Example 3–32, find the value corresponding to the 25th percentile.

Solution

Step 1   Arrange the data in order from lowest to highest.

2, 3, 5, 6, 8, 10, 12, 15, 18, 20

Step 2   Compute

$$c = \frac{n \cdot p}{100}$$

where
   n = total number of values
   p = percentile

Thus,

$$c = \frac{10 \cdot 25}{100} = 2.5$$

Step 3   If c is not a whole number, round it up to the next whole number; in this case, $c = 3$. (If c is a whole number, see Example 3–35.) Start at the lowest value and count over to the third value, which is 5. Hence, the value 5 corresponds to the 25th percentile.

Example 3–35

Using the data set in Example 3–32, find the value that corresponds to the 60th percentile.

Solution

Step 1   Arrange the data in order from smallest to largest.

2, 3, 5, 6, 8, 10, 12, 15, 18, 20

Step 2   Substitute in the formula.

$$c = \frac{n \cdot p}{100} = \frac{10 \cdot 60}{100} = 6$$

Step 3   If c is a whole number, use the value halfway between the c and $c + 1$ values when counting up from the lowest value—in this case, the 6th and 7th values.

2, 3, 5, 6, 8, 10, 12, 15, 18, 20
　　　　　　　↗　↖
　　　　6th value　　7th value

The value halfway between 10 and 12 is 11. Find it by adding the two values and dividing by 2.

$$\frac{10 + 12}{2} = 11$$

Hence, 11 corresponds to the 60th percentile. Anyone scoring 11 would have done better than 60% of the class.

The steps for finding a value corresponding to a given percentile are summarized in this Procedure Table.

Procedure Table

Finding a Data Value Corresponding to a Given Percentile

Step 1   Arrange the data in order from lowest to highest.

Step 2   Substitute into the formula

$$c = \frac{n \cdot p}{100}$$

where
　　　n = total number of values
　　　p = percentile

Step 3A   If c is not a whole number, round up to the next whole number. Starting at the lowest value, count over to the number that corresponds to the rounded-up value.

Step 3B   If c is a whole number, use the value halfway between the cth and $(c + 1)$st values when counting up from the lowest value.

Quartiles and Deciles

Quartiles divide the distribution into four groups, separated by Q_1, Q_2, Q_3.

Note that Q_1 is the same as the 25th percentile; Q_2 is the same as the 50th percentile, or the median; Q_3 corresponds to the 75th percentile, as shown:

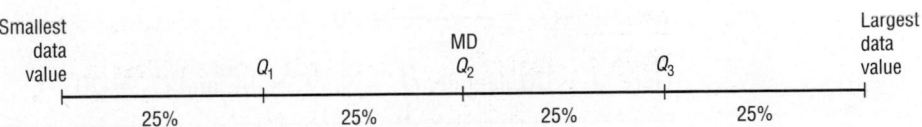

Quartiles can be computed by using the formula given for computing percentiles on page 147. For Q_1 use $p = 25$. For Q_2 use $p = 50$. For Q_3 use $p = 75$. However, an easier method for finding quartiles is found in this Procedure Table.

Procedure Table

Finding Data Values Corresponding to Q_1, Q_2, and Q_3

Step 1   Arrange the data in order from lowest to highest.

Step 2   Find the median of the data values. This is the value for Q_2.

Step 3   Find the median of the data values that fall below Q_2. This is the value for Q_1.

Step 4   Find the median of the data values that fall above Q_2. This is the value for Q_3.

Example 3–36 shows how to find the values of Q_1, Q_2, and Q_3.

Example 3–36   Find Q_1, Q_2, and Q_3 for the data set 15, 13, 6, 5, 12, 50, 22, 18.

Solution

Step 1   Arrange the data in order.

5, 6, 12, 13, 15, 18, 22, 50

Step 2   Find the median (Q_2).

5, 6, 12, 13, 15, 18, 22, 50
                ↑
               MD

$$MD = \frac{13 + 15}{2} = 14$$

Step 3   Find the median of the data values less than 14.

5, 6, 12, 13
       ↑
      Q_1

$$Q_1 = \frac{6 + 12}{2} = 9$$

So Q_1 is 9.

Step 4   Find the median of the data values greater than 14.

15, 18, 22, 50
        ↑
       Q_3

$$Q_3 = \frac{18 + 22}{2} = 20$$

Here Q_3 is 20. Hence, $Q_1 = 9$, $Q_2 = 14$, and $Q_3 = 20$.

In addition to dividing the data set into four groups, quartiles can be used as a rough measurement of variability. The **interquartile range (IQR)** is defined as the difference between Q_1 and Q_3 and is the range of the middle 50% of the data.

The interquartile range is used to identify outliers, and it is also used as a measure of variability in exploratory data analysis, as shown in Section 3–4.

Deciles divide the distribution into 10 groups, as shown. They are denoted by D_1, D_2, etc.

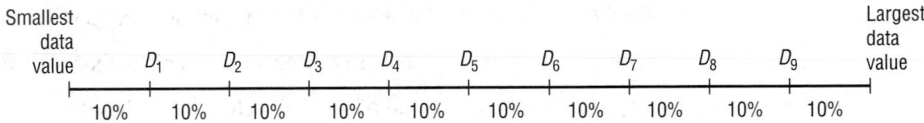

Note that D_1 corresponds to P_{10}; D_2 corresponds to P_{20}; etc. Deciles can be found by using the formulas given for percentiles. Taken altogether then, these are the relationships among percentiles, deciles, and quartiles.

Deciles are denoted by $D_1, D_2, D_3, \ldots, D_9$, and they correspond to $P_{10}, P_{20}, P_{30}, \ldots, P_{90}$.

Quartiles are denoted by Q_1, Q_2, Q_3 and they correspond to P_{25}, P_{50}, P_{75}.

The median is the same as P_{50} or Q_2 or D_5.

The position measures are summarized in Table 3–4.

Table 3–4	Summary of Position Measures	
Measure	**Definition**	**Symbol(s)**
Standard score or z score	Number of standard deviations that a data value is above or below the mean	z
Percentile	Position in hundredths that a data value holds in the distribution	P_n
Decile	Position in tenths that a data value holds in the distribution	D_n
Quartile	Position in fourths that a data value holds in the distribution	Q_n

Outliers

A data set should be checked for extremely high or extremely low values. These values are called *outliers*.

An **outlier** is an extremely high or an extremely low data value when compared with the rest of the data values.

An outlier can strongly affect the mean and standard deviation of a variable. For example, suppose a researcher mistakenly recorded an extremely high data value. This value would then make the mean and standard deviation of the variable much larger than they really were. Outliers can have an effect on other statistics as well.

There are several ways to check a data set for outliers. One method is shown in this Procedure Table.

Procedure Table

Procedure for Identifying Outliers

Step 1 Arrange the data in order and find Q_1 and Q_3.

Step 2 Find the interquartile range: IQR = $Q_3 - Q_1$.

Step 3 Multiply the IQR by 1.5.

Step 4 Subtract the value obtained in step 3 from Q_1 and add the value to Q_3.

Step 5 Check the data set for any data value that is smaller than $Q_1 - 1.5(\text{IQR})$ or larger than $Q_3 + 1.5(\text{IQR})$.

This procedure is shown in Example 3–37.

Example 3–37

 Check the following data set for outliers.

5, 6, 12, 13, 15, 18, 22, 50

Solution

The data value 50 is extremely suspect. These are the steps in checking for an outlier.

Step 1 Find Q_1 and Q_3. This was done in Example 3–36; Q_1 is 9 and Q_3 is 20.

Step 2 Find the interquartile range (IQR), which is $Q_3 - Q_1$.

$$\text{IQR} = Q_3 - Q_1 = 20 - 9 = 11$$

Step 3 Multiply this value by 1.5.

$$1.5(11) = 16.5$$

Step 4 Subtract the value obtained in step 3 from Q_1, and add the value obtained in step 3 to Q_3.

$$9 - 16.5 = -7.5 \qquad \text{and} \qquad 20 + 16.5 = 36.5$$

Step 5 Check the data set for any data values that fall outside the interval from −7.5 to 36.5. The value 50 is outside this interval; hence, it can be considered an outlier.

There are several reasons why outliers may occur. First, the data value may have resulted from a measurement or observational error. Perhaps the researcher measured the variable incorrectly. Second, the data value may have resulted from a recording error. That is, it may have been written or typed incorrectly. Third, the data value may have been obtained from a subject that is not in the defined population. For example, suppose test scores were obtained from a seventh-grade class, but a student in that class was actually in the sixth grade and had special permission to attend the class. This student might have scored extremely low on that particular exam on that day. Fourth, the data value might be a legitimate value that occurred by chance (although the probability is extremely small).

There are no hard-and-fast rules on what to do with outliers, nor is there complete agreement among statisticians on ways to identify them. Obviously, if they occurred as a result of an error, an attempt should be made to correct the error or else the data value should be omitted entirely. When they occur naturally by chance, the statistician must make a decision about whether to include them in the data set.

When a distribution is normal or bell-shaped, data values that are beyond 3 standard deviations of the mean can be considered suspected outliers.

Applying the Concepts 3–3

Determining Dosages

In an attempt to determine necessary dosages of a new drug (HDL) used to control sepsis, assume you administer varying amounts of HDL to 40 mice. You create four groups and label them *low dosage, moderate dosage, large dosage,* and *very large dosage.* The dosages also vary within each group. After the mice are injected with the HDL and the sepsis bacteria, the time until the onset of sepsis is recorded. Your job as a statistician is to effectively communicate the results of the study.

1. Which measures of position could be used to help describe the data results?
2. If 40% of the mice in the top quartile survived after the injection, how many mice would that be?
3. What information can be given from using percentiles?
4. What information can be given from using quartiles?
5. What information can be given from using standard scores?

See page 180 for the answers.

Exercises 3–3

1. What is a z score? A *z* score tells how many standard deviations the data value is above or below the mean.

2. Define *percentile rank.* A percentile rank indicates the percentage of data values that fall below the specific rank.

3. What is the difference between a percentage and a percentile? A percentile is a relative measurement of position; a percentage is an absolute measure of the part to the total.

4. Define *quartile.* A quartile is a relative measure of position obtained by dividing the data set into quarters.

5. What is the relationship between quartiles and percentiles? $Q_1 = P_{25}; Q_2 = P_{50}; Q_3 = P_{75}$

6. What is a decile? A decile is a relative measure of position obtained by dividing the data set into tenths.

7. How are deciles related to percentiles? $D_1 = P_{10}; D_2 = P_{20}; D_3 = P_{30};$ etc.

8. To which percentile, quartile, and decile does the median correspond? $P_{50}; Q_2; D_5$

9. Vacation Days If the average number of vacation days for a selection of various countries has a mean of 29.4 days and a standard deviation of 8.6, find the z scores for the average number of vacation days in each of these countries.

Canada	26 days	−0.40
Italy	42 days	1.47
United States	13 days	−1.91

Source: www.infoplease.com

10. Age of Senators The average age of Senators in the 108th Congress was 59.5 years. If the standard deviation was 11.5 years, find the z scores corresponding to the oldest and youngest Senators: Robert C. Byrd (D, WV), 86, and John Sununu (R, NH), 40. Byrd: $z = 2.30$ Sununu: $z = -1.70$

Source: *CRS Report for Congress.*

11. Driver's License Exam Scores The average score on a state CDL license exam is 76 with a standard deviation of 5. Find the corresponding z score for each raw score.

a. 79 0.6
b. 70 −1.2
c. 88 2.4
d. 65 −2.2
e. 77 0.2

12. Teacher's Salary The average teacher's salary in a particular state is $54,166. If the standard deviation is

$10,200, find the salaries corresponding to the following z scores.

a. 2 $74,566   d. 2.5 $79,666
b. −1 $43,966   e. −1.6 $37,846
c. 0 $54,166

13. Which has a better relative position: a score of 75 on a statistics test with a mean of 60 and a standard deviation of 10 or a score of 36 on an accounting test with a mean of 30 and a variance of 16? Neither; z = 1.5 for each

14. **College and University Debt** A student graduated from a 4-year college with an outstanding loan of $9650 where the average debt is $8455 with a standard deviation of $1865. Another student graduated from a university with an outstanding loan of $12,360 where the average of the outstanding loans was $10,326 with a standard deviation of $2143. Which student had a higher debt in relationship to his or her peers? 0.64; 0.95. The student from the university has a higher relative debt.

15. Which score indicates the highest relative position?

a. A score of 3.2 on a test with $\overline{X} = 4.6$ and $s = 1.5$ −0.93
b. A score of 630 on a test with $\overline{X} = 800$ and $s = 200$ −0.85
c. A score of 43 on a test with $\overline{X} = 50$ and $s = 5$ −1.4; score in part b is highest

16. **College Room and Board Costs** Room and board costs for selected schools are summarized in this distribution. Find the approximate cost of room and board corresponding to each of the following percentiles.

Costs (in dollars)	Frequency
3000.5–4000.5	5
4000.5–5000.5	6
5000.5–6000.5	18
6000.5–7000.5	24
7000.5–8000.5	19
8000.5–9000.5	8
9000.5–10,000.5	5

a. 30th percentile $5806
b. 50th percentile $6563
c. 75th percentile $7566
d. 90th percentile $8563

Source: *World Almanac.*

17. Using the data in Exercise 16, find the approximate percentile rank of each of the following costs.

a. 5500 24th
b. 7200 67th
c. 6500 48th
d. 8300 88th

18. **Achievement Test Scores (ans)** The data shown represent the scores on a national achievement test for a group of 10th-grade students. Find the approximate

percentile ranks of these scores by constructing a percentile graph.

a. 220 6   d. 280 76
b. 245 24   e. 300 94
c. 276 68

Score	Frequency
196.5–217.5	5
217.5–238.5	17
238.5–259.5	22
259.5–280.5	48
280.5–301.5	22
301.5–322.5	6

19. For the data in Exercise 18, find the approximate scores that correspond to these percentiles.

a. 15th 234   d. 65th 274
b. 29th 251   e. 80th 284
c. 43rd 263

20. **Airplane Speeds (ans)** The airborne speeds in miles per hour of 21 planes are shown. Find the approximate values that correspond to the given percentiles by constructing a percentile graph.

Class	Frequency
366–386	4
387–407	2
408–428	3
429–449	2
450–470	1
471–491	2
492–512	3
513–533	4
	21

Source: *The World Almanac and Book of Facts.*

a. 9th 375   d. 60th 477
b. 20th 389   e. 75th 504
c. 45th 433

21. Using the data in Exercise 20, find the approximate percentile ranks of the following miles per hour (mph).

a. 380 mph 13th   d. 505 mph 76th
b. 425 mph 40th   e. 525 mph 92nd
c. 455 mph 54th

22. **Average Weekly Earnings** The average weekly earnings in dollars for various industries are listed below. Find the percentile rank of each value.

804 736 659 489 777 623 597 524 228
94th; 72nd; 61st; 17th; 83rd; 50th; 39th; 28th; 6th

Source: *New York Times Almanac.*

23. For the data from Exercise 22, what value corresponds to the 40th percentile? 597

24. Test Scores Find the percentile rank for each test score in the data set. 7th; 21st; 36th; 50th; 64th; 79th; 93rd

12, 28, 35, 42, 47, 49, 50

25. In Exercise 24, what value corresponds to the 60th percentile? 47

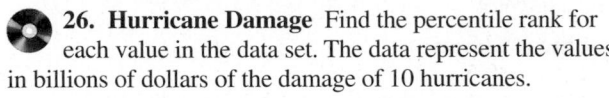

26. Hurricane Damage Find the percentile rank for each value in the data set. The data represent the values in billions of dollars of the damage of 10 hurricanes. 5th; 15th; 25th; 35th; 45th; 55th; 65th; 75th; 85th; 95th

1.1, 1.7, 1.9, 2.1, 2.2, 2.5, 3.3, 6.2, 6.8, 20.3

Source: Insurance Services Office.

27. What value in Exercise 26 corresponds to the 40th percentile? 2.1

28. Test Scores Find the percentile rank for each test score in the data set. 8th; 25th; 42nd; 58th; 75th; 92nd

5, 12, 15, 16, 20, 21

29. What test score in Exercise 28 corresponds to the 33rd percentile? 12

30. Using the procedure shown in Example 3–37, check each data set for outliers.

a. 16, 18, 22, 19, 3, 21, 17, 20 3
b. 24, 32, 54, 31, 16, 18, 19, 14, 17, 20 54
c. 321, 343, 350, 327, 200 None
d. 88, 72, 97, 84, 86, 85, 100 None
e. 145, 119, 122, 118, 125, 116 145
f. 14, 16, 27, 18, 13, 19, 36, 15, 20 None

31. Another measure of average is called the *midquartile;* it is the numerical value halfway between Q_1 and Q_3, and the formula is

$$\text{Midquartile} = \frac{Q_1 + Q_3}{2}$$

Using this formula and other formulas, find Q_1, Q_2, Q_3, the midquartile, and the interquartile range for each data set.

a. 5, 12, 16, 25, 32, 38 12; 20.5; 32; 22; 20
b. 53, 62, 78, 94, 96, 99, 103 62; 94; 99; 80.5; 37

MINITAB
Step by Step

Calculate Descriptive Statistics from Data

Example MT3–1

1. Enter the data from Example 3–23 into C1 of MINITAB. Name the column **AutoSales.**
2. Select **Stat>Basic Statistics>Display Descriptive Statistics.**
3. The cursor will be blinking in the Variables text box. Double-click C1 AutoSales.
4. Click [Statistics] to view the statistics that can be calculated with this command.
   a) Check the boxes for Mean, Standard deviation, Variance, Coefficient of variation, Median, Minimum, Maximum, and N nonmissing.

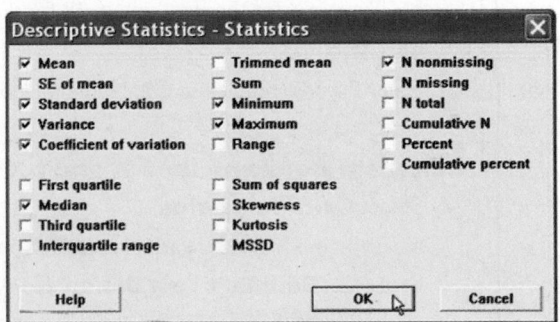

   b) Remove the checks from other options.

   5. Click [OK] twice. The results will be displayed in the session window as shown.

Descriptive Statistics: AutoSales

Variable	N	Mean	Median	StDev	Variance	CoefVar	Minimum	Maximum
AutoSales	6	12.6	12.4	1.12960	1.276	8.96509	11.2	14.3

   Session window results are in text format. A high-resolution graphical window displays the descriptive statistics, a histogram, and a boxplot.

   6. Select **Stat>Basic Statistics>Graphical Summary.**

   7. Double-click C1 AutoSales.

   8. Click [OK].

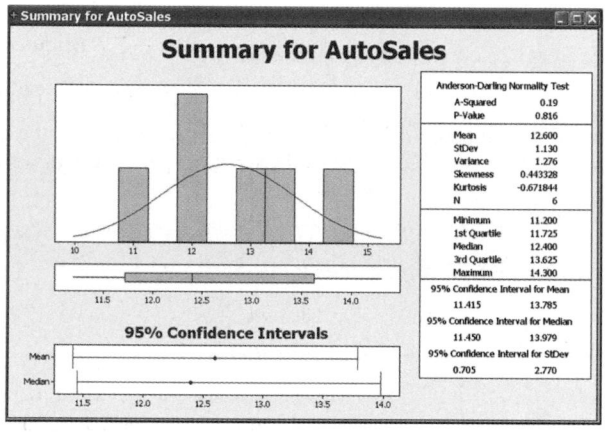

The graphical summary will be displayed in a separate window as shown.

Calculate Descriptive Statistics from a Frequency Distribution

Multiple menu selections must be used to calculate the statistics from a table. We will use data given in Example 3–24.

Enter Midpoints and Frequencies

   1. Select **File>New>New Worksheet** to open an empty worksheet.

   2. To enter the midpoints into C1, select **Calc>Make Patterned Data>Simple Set of Numbers.**

      a) Type **X** to name the column.

      b) Type in **8** for the First value, **38** for the Last value, and **5** for Steps.

      c) Click [OK].

   3. Enter the frequencies in C2. Name the column **f.**

Calculate Columns for f·X and f·X²

   4. Select **Calc>Calculator.**

      a) Type in **fX** for the variable and **f*X** in the Expression dialog box. Click [OK].

      b) Select **Edit>Edit Last Dialog** and type in **fX2** for the variable and **f*X**2** for the expression.

      c) Click [OK]. There are now four columns in the worksheet.

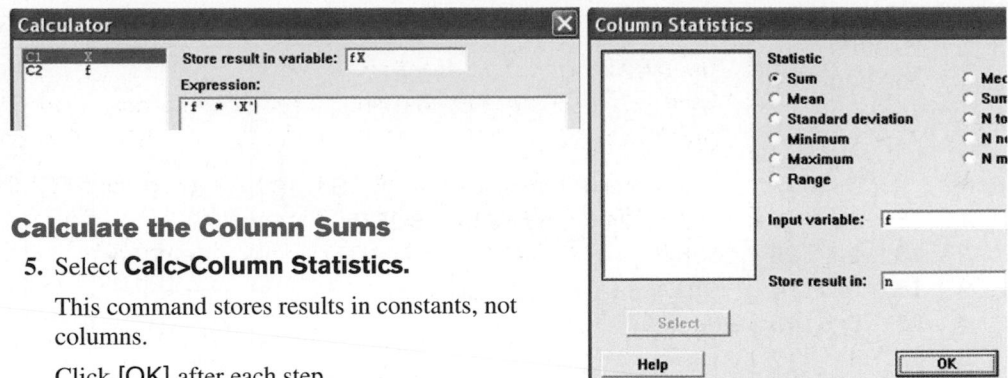

Calculate the Column Sums

5. Select **Calc>Column Statistics.**

   This command stores results in constants, not columns.

   Click [OK] after each step.

   a) Click the option for Sum; then select C2 f for the Input column, and type **n** for Store result in.

   b) Select **Edit>Edit Last Dialog;** then select C3 fX for the column and type **sumX** for storage.

   c) Edit the last dialog box again. This time select C4 fX2 for the column, then type **sumX2** for storage.

To verify the results, navigate to the Project Manager window, then the constants folder of the worksheet. The sums are 20, 490, and 13,310.

Calculate the Mean, Variance, and Standard Deviation

6. Select **Calc>Calculator.**

   a) Type **Mean** for the variable, then click in the box for the Expression and type **sumX/n.** Click [OK]. If you double-click the constants instead of typing them, single quotes will surround the names. The quotes are not required unless the column name has spaces.

   b) Click the **EditLast Dialog** icon and type **Variance** for the variable.

   c) In the expression box type in

   (sumX2-sumX**2/n)/(n-1)

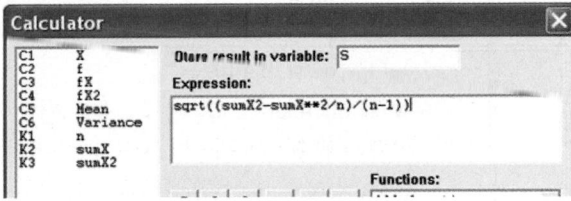

   d) Edit the last dialog box and type **S** for the variable. In the expression box, drag the mouse over the previous expression to highlight it.

   e) Click the button in the keypad for parentheses. Type **SQRT** at the beginning of the line, upper- or lowercase will work. The expression should be SQRT((sumX2-sumX**2/n)/(n-1)).

   f) Click [OK].

Display Results

   g) Select **Data>Display Data,** then highlight all columns and constants in the list.

   h) Click [Select] then [OK].

The session window will display all our work! Create the histogram with instructions from Chapter 2.

Data Display

n   20.0000

sumX   490.000

sumX2 13310.0

Row	X	f	fX	fX2	Mean	Variance	S
1	8	1	8	64	24.5	68.6842	8.28759
2	13	2	26	338			
3	18	3	54	972			
4	23	5	115	2645			
5	28	4	112	3136			
6	33	3	99	3267			
7	38	2	76	2888			

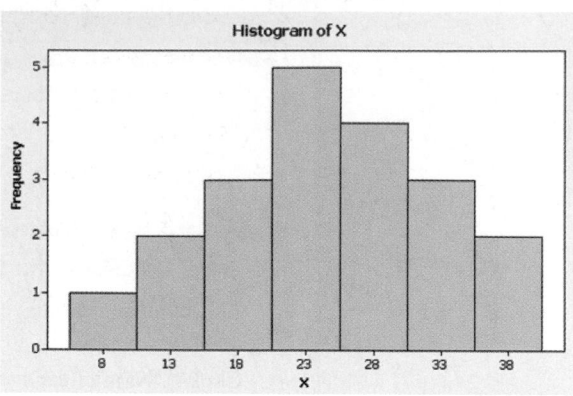

TI-83 Plus or TI-84 Plus
Step by Step

Calculating Descriptive Statistics

To calculate various descriptive statistics:

1. Enter data into L_1.
2. Press **STAT** to get the menu.
3. Press ▶ to move cursor to CALC; then press **1** for 1-Var Stats.
4. Press **2nd [L_1]**, then **ENTER**.

The calculator will display

$\bar{x}$ sample mean

Σx sum of the data values

Σx^2 sum of the squares of the data values

S_x sample standard deviation

σ_x population standard deviation

n number of data values

minX   smallest data value

Q_1 lower quartile

Med   median

Q_3 upper quartile

maxX   largest data value

Example TI3–1

Find the various descriptive statistics for the auto sales data from Example 3–23:

     11.2, 11.9, 12.0, 12.8, 13.4, 14.3

Output

```
1-Var Stats
 x̄=12.6
 Σx=75.6
 Σx²=958.94
 Sx=1.1296017
 σx=1.031180553
↓n=6
```

Output

```
1-Var Stats
↑n=6
 minX=11.2
 Q₁=11.9
 Med=12.4
 Q₃=13.4
 maxX=14.3
```

Following the steps just shown, we obtain these results, as shown on the screen:

The mean is 12.6.

The sum of x is 75.6.

The sum of x^2 is 958.94.

The sample standard deviation S_x is 1.1296017.

The population standard deviation σ_x is 1.031180553.

The sample size n is 6.

The smallest data value is 11.2.

Q_1 is 11.9.

The median is 12.4.

Q_3 is 13.4.

The largest data value is 14.3.

To calculate the mean and standard deviation from grouped data:

1. Enter the midpoints into L_1.
2. Enter the frequencies into L_2.
3. Press **STAT** to get the menu.
4. Use the arrow keys to move the cursor to CALC; then press **1** for 1-Var Stats.
5. Press **2nd [L1], 2nd [L2]**, then **ENTER.**

Example TI3–2

Calculate the mean and standard deviation for the data given in Examples 3–3 and 3–24.

Class	Frequency	Midpoint
5.5–10.5	1	8
10.5–15.5	2	13
15.5–20.5	3	18
20.5–25.5	5	23
25.5–30.5	4	28
30.5–35.5	3	33
35.5–40.5	2	38

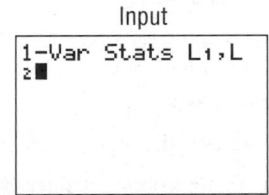

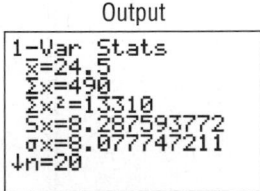

The sample mean is 24.5, and the sample standard deviation is 8.287593772.

To graph a percentile graph, follow the procedure for an ogive but use the cumulative percent in L_2, 100 for Y_{max}, and the data from Example 3–31.

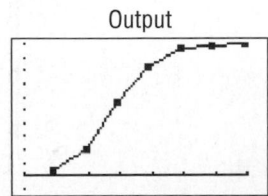

Excel

Step by Step

Measures of Position

Example XL3–3

Find the *z* scores for each value of the data from Example 3–23. The data represent the amount (in millions of dollars) of European auto sales for a sample of 6 years.

$$11.2 \quad 11.9 \quad 12.0 \quad 12.8 \quad 13.4 \quad 14.3$$

1. On an Excel worksheet enter the data in cells A2–A7. Enter a label for the variable in cell A1.
2. Label cell B1 as *z* score.
3. Select cell B2.
4. Select the Formulas tab from the toolbar and Insert Function ![fx Insert Function].
5. Select the Statistical category for statistical functions and scroll in the function list to STANDARDIZE and click [OK].

In the STANDARDIZE dialog box:

6. Type A2 for the X value.
7. Type average(A2:A7) for the Mean.
8. Type stdev(A2:A7) for the Standard_dev. Then click [OK].
9. Repeat the procedure above for each data value in column A.

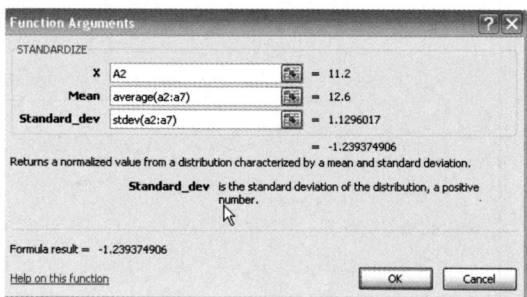

Example XL3–4

Find the percentile rank for each value of the data from Example 3–23. The data represent the amount (in millions of dollars) of European auto sales for a sample of 6 years.

$$11.2 \quad 11.9 \quad 12.0 \quad 12.8 \quad 13.4 \quad 14.3$$

1. On an Excel worksheet enter the data in cells A2–A7. Enter a label for the variable in cell A1.
2. Label cell B1 as *z* score.
3. Select cell B2.
4. Select the Formulas tab from the toolbar and Insert Function ![fx Insert Function].
5. Select the Statistical category for statistical functions and scroll in the function list to PERCENTRANK and click [OK].

In the PERCENTRANK dialog box:

6. Type A2:A7 for the Array.

7. Type A2 for the X value, then click [OK].

8. Repeat the procedure above for each data value in column A.

The PERCENTRANK function returns the percentile rank as a decimal. To convert this to a percentage, multiply the function output by 100. Make sure to select a new column before multiplying the percentile rank by 100.

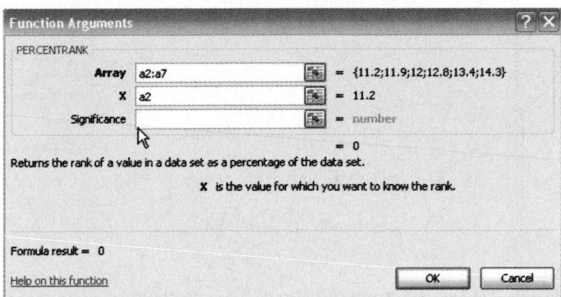

Descriptive Statistics in Excel

Example XL3–5

 Excel Analysis Tool-Pak Add-in Data Analysis includes an item called Descriptive Statistics that reports many useful measures for a set of data.

1. Enter the data set shown in cells A1 to A9 of a new worksheet.

   12   17   15   16   16   14   18   13   10

See the Excel Step by Step in Chapter 1 for the instructions on loading the Analysis Tool-Pak Add-in.

2. Select the Data tab on the toolbar and select Data Analysis.

3. In the Analysis Tools dialog box, scroll to Descriptive Statistics, then click [OK].

4. Type A1:A9 in the Input Range box and check the Grouped by Columns option.

5. Select the Output Range option and type in cell C1.

6. Check the Summary statistics option and click [OK].

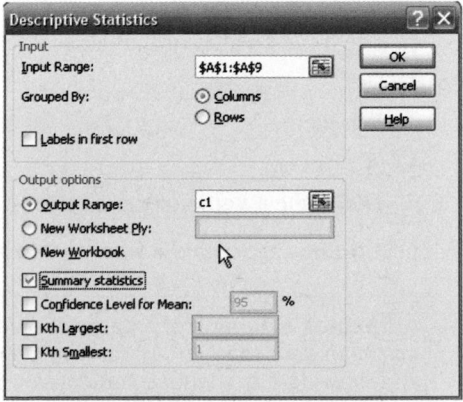

Below is the summary output for this data set.

Column1	
Mean	14.55555556
Standard Error	0.85165054
Median	15
Mode	16
Standard Deviation	2.554951619
Sample Variance	6.527777778
Kurtosis	-0.3943866
Skewness	-0.51631073
Range	8
Minimum	10
Maximum	18
Sum	131
Count	9

3–4

Objective 4

Use the techniques of exploratory data analysis, including boxplots and five-number summaries, to discover various aspects of data.

Exploratory Data Analysis

In traditional statistics, data are organized by using a frequency distribution. From this distribution various graphs such as the histogram, frequency polygon, and ogive can be constructed to determine the shape or nature of the distribution. In addition, various statistics such as the mean and standard deviation can be computed to summarize the data.

The purpose of traditional analysis is to confirm various conjectures about the nature of the data. For example, from a carefully designed study, a researcher might want to know if the proportion of Americans who are exercising today has increased from 10 years ago. This study would contain various assumptions about the population, various definitions such as of exercise, and so on.

In **exploratory data analysis (EDA),** data can be organized using a *stem and leaf plot.* (See Chapter 2.) The measure of central tendency used in EDA is the *median.* The measure of variation used in EDA is the *interquartile range $Q_3 - Q_1$.* In EDA the data are represented graphically using a *boxplot* (sometimes called a box-and-whisker plot). The purpose of exploratory data analysis is to examine data to find out what information can be discovered about the data such as the center and the spread. Exploratory data analysis was developed by John Tukey and presented in his book *Exploratory Data Analysis* (Addison-Wesley, 1977).

The Five-Number Summary and Boxplots

A **boxplot** can be used to graphically represent the data set. These plots involve five specific values:

1. The lowest value of the data set (i.e., minimum)
2. Q_1
3. The median
4. Q_3
5. The highest value of the data set (i.e., maximum)

These values are called a **five-number summary** of the data set.

A **boxplot** is a graph of a data set obtained by drawing a horizontal line from the minimum data value to Q_1, drawing a horizontal line from Q_3 to the maximum data value, and drawing a box whose vertical sides pass through Q_1 and Q_3 with a vertical line inside the box passing through the median or Q_2.

Procedure for constructing a boxplot

1. Find the five-number summary for the data values, that is, the maximum and minimum data values, Q_1 and Q_3, and the median.
2. Draw a horizontal axis with a scale such that it includes the maximum and minimum data values.
3. Draw a box whose vertical sides go through Q_1 and Q_3, and draw a vertical line though the median.
4. Draw a line from the minimum data value to the left side of the box and a line from the maximum data value to the right side of the box.

Example 3–38   ## Number of Meteorites Found

The number of meteorites found in 10 states of the United States is 89, 47, 164, 296, 30, 215, 138, 78, 48, 39. Construct a boxplot for the data.

Source: Natural History Museum.

Solution

Step 1   Arrange the data in order:

30, 39, 47, 48, 78, 89, 138, 164, 215, 296

Step 2   Find the median.

30, 39, 47, 48, 78, 89, 138, 164, 215, 296
↑
Median

$$\text{Median} = \frac{78 + 89}{2} = 83.5$$

Step 3   Find Q_1.

30, 39, 47, 48, 78
↑
Q_1

Step 4   Find Q_3.

89, 138, 164, 215, 296
↑
Q_3

Step 5   Draw a scale for the data on the x axis.

Step 6   Locate the lowest value, Q_1, median, Q_3, and the highest value on the scale.

Step 7   Draw a box around Q_1 and Q_3, draw a vertical line through the median, and connect the upper value and the lower value to the box. See Figure 3–7.

Figure 3–7

Boxplot for
Example 3–38

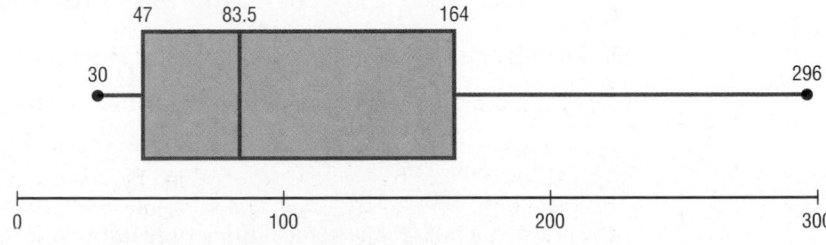

The distribution is somewhat positively skewed.

Information Obtained from a Boxplot

1. *a.* If the median is near the center of the box, the distribution is approximately symmetric.
    b. If the median falls to the left of the center of the box, the distribution is positively skewed.
    c. If the median falls to the right of the center, the distribution is negatively skewed.

2. *a.* If the lines are about the same length, the distribution is approximately symmetric.
    b. If the right line is larger than the left line, the distribution is positively skewed.
    c. If the left line is larger than the right line, the distribution is negatively skewed.

The boxplot in Figure 3–7 indicates that the distribution is slightly positively skewed.

If the boxplots for two or more data sets are graphed on the same axis, the distributions can be compared. To compare the averages, use the location of the medians. To compare the variability, use the interquartile range, i.e., the length of the boxes. Example 3–39 shows this procedure.

Example 3–39

Sodium Content of Cheese

A dietitian is interested in comparing the sodium content of real cheese with the sodium content of a cheese substitute. The data for two random samples are shown. Compare the distributions, using boxplots.

Real cheese				**Cheese substitute**			
310	420	45	40	270	180	250	290
220	240	180	90	130	260	340	310

Source: *The Complete Book of Food Counts.*

Solution

Step 1 Find Q_1, MD, and Q_3 for the real cheese data.

$$40 \quad 45 \quad 90 \quad 180 \quad 220 \quad 240 \quad 310 \quad 420$$

$$\qquad\quad \uparrow \qquad\qquad \uparrow \qquad\qquad \uparrow$$

$$\qquad\quad Q_1 \qquad\qquad MD \qquad\qquad Q_3$$

$$Q_1 = \frac{45 + 90}{2} = 67.5 \qquad MD = \frac{180 + 220}{2} = 200$$

$$Q_3 = \frac{240 + 310}{2} = 275$$

Step 2 Find Q_1, MD, and Q_3 for the cheese substitute data.

$$130 \quad 180 \quad 250 \quad 260 \quad 270 \quad 290 \quad 310 \quad 340$$

$$\qquad\quad \uparrow \qquad\qquad \uparrow \qquad\qquad \uparrow$$

$$\qquad\quad Q_1 \qquad\qquad MD \qquad\qquad Q_3$$

$$Q_1 = \frac{180 + 250}{2} = 215 \qquad MD = \frac{260 + 270}{2} = 265$$

$$Q_3 = \frac{290 + 310}{2} = 300$$

Step 3 Draw the boxplots for each distribution on the same graph. See Figure 3–8.

Step 4 Compare the plots. It is quite apparent that the distribution for the cheese substitute data has a higher median than the median for the distribution for the real cheese data. The variation or spread for the distribution of the real cheese data is larger than the variation for the distribution of the cheese substitute data.

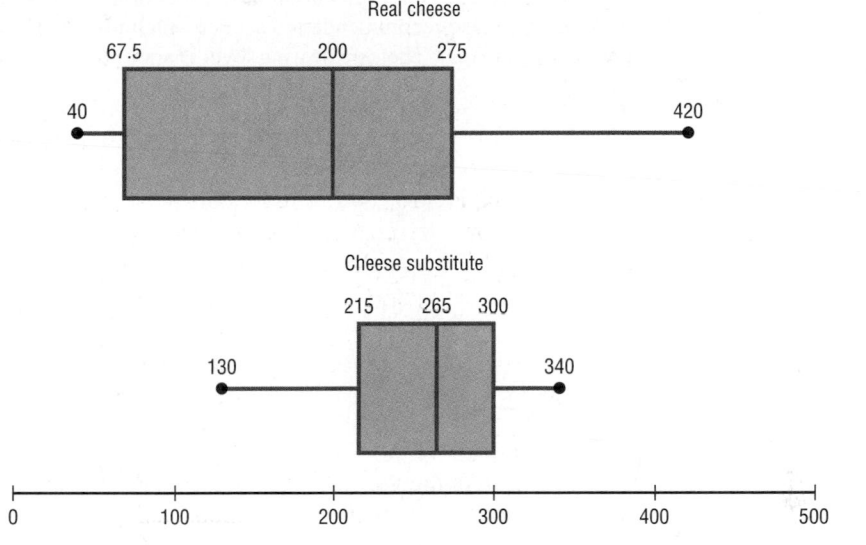

Figure 3–8

Boxplots for
Example 3–39

A *modified boxplot* can be drawn and used to check for outliers. See Exercise 18 in Extending the Concepts in this section.

In exploratory data analysis, *hinges* are used instead of quartiles to construct boxplots. When the data set consists of an even number of values, hinges are the same as quartiles. Hinges for a data set with an odd number of values differ somewhat from quartiles. However, since most calculators and computer programs use quartiles, they will be used in this textbook.

Another important point to remember is that the summary statistics (median and interquartile range) used in exploratory data analysis are said to be *resistant statistics*. A **resistant statistic** is relatively less affected by outliers than a *nonresistant statistic*. The mean and standard deviation are nonresistant statistics. Sometimes when a distribution is skewed or contains outliers, the median and interquartile range may more accurately summarize the data than the mean and standard deviation, since the mean and standard deviation are more affected in this case.

Table 3–5 shows the correspondence between the traditional and the exploratory data analysis approach.

Table 3–5	Traditional versus EDA Techniques	
Traditional		**Exploratory data analysis**
Frequency distribution		Stem and leaf plot
Histogram		Boxplot
Mean		Median
Standard deviation		Interquartile range

Applying the Concepts **3–4**

The Noisy Workplace

Assume you work for OSHA (Occupational Safety and Health Administration) and have complaints about noise levels from some of the workers at a state power plant. You charge the power plant with taking decibel readings at six different areas of the plant at different times of the day and week. The results of the data collection are listed. Use boxplots to initially explore the data and make recommendations about which plant areas workers must be provided with protective ear wear. The safe hearing level is approximately 120 decibels.

Area 1	Area 2	Area 3	Area 4	Area 5	Area 6
30	64	100	25	59	67
12	99	59	15	63	80
35	87	78	30	81	99
65	59	97	20	110	49
24	23	84	61	65	67
59	16	64	56	112	56
68	94	53	34	132	80
57	78	59	22	145	125
100	57	89	24	163	100
61	32	88	21	120	93
32	52	94	32	84	56
45	78	66	52	99	45
92	59	57	14	105	80
56	55	62	10	68	34
44	55	64	33	75	21

See page 180 for the answers.

See page 180 for the answers.

Exercises 3–4

For Exercises 1–6, identify the five-number summary and find the interquartile range.

 1. 8, 12, 32, 6, 27, 19, 54   6, 8, 19, 32, 54; 24

 2. 19, 16, 48, 22, 7   7, 11.5, 19, 35, 48; 23.5

 3. 362, 589, 437, 316, 192, 188
188, 192, 339, 437, 589; 245

 4. 147, 243, 156, 632, 543, 303
147, 156, 273, 543, 632; 387

 5. 14.6, 19.8, 16.3, 15.5, 18.2
14.6, 15.05, 16.3, 19, 19.8; 3.95

 6. 9.7, 4.6, 2.2, 3.7, 6.2, 9.4, 3.8   2.2, 3.7, 4.6, 9.4, 9.7; 5.7

For Exercises 7–10, use each boxplot to identify the maximum value, minimum value, median, first quartile, third quartile, and interquartile range.

7.   11, 3, 8, 5, 9, 4

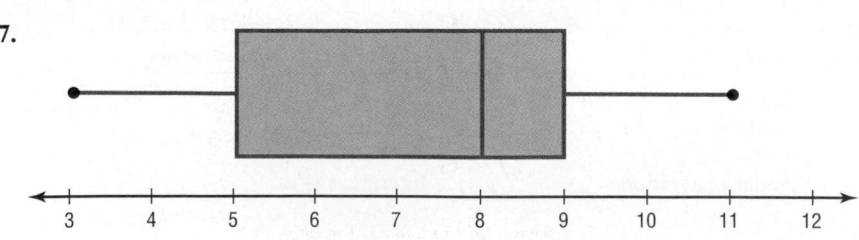

8.

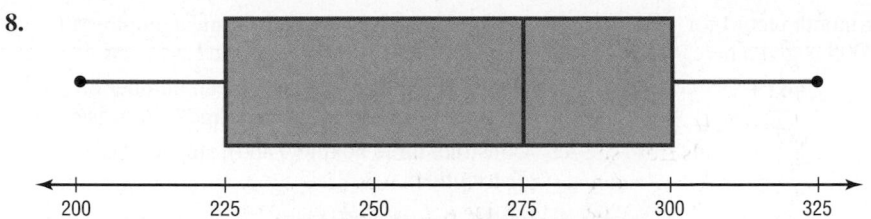

325, 200, 275, 225, 300, 75

9.

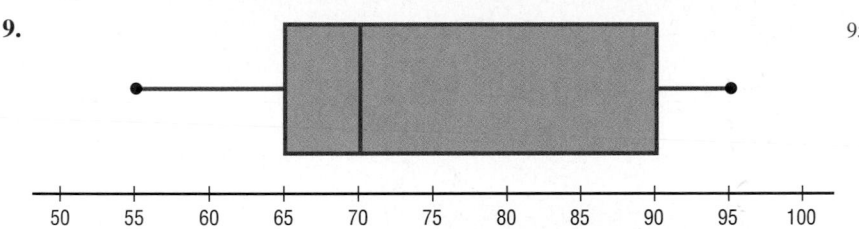

95, 55, 70, 65, 90, 25

10.

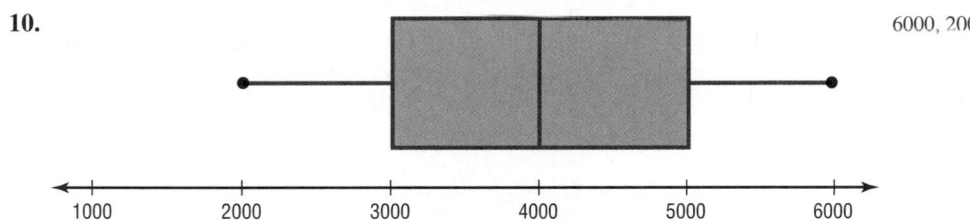

6000, 2000, 4000, 3000, 5000; 2000

11. Earned Run Average—Number of Games Pitched Construct a boxplot for the following data and comment on the shape of the distribution representing the number of games pitched by major league baseball's earned run average (ERA) leaders for the past few years.

30 34 29 30 34 29 31 33 34 27
30 27 34 32

Source: *World Almanac.*

12. Innings Pitched Construct a boxplot for the following data which represent the number of innings pitched by the ERA leaders for the past few years. Comment on the shape of the distribution.

192 228 186 199 238 217 213 234 264 187
214 115 238 246

Source: *World Almanac.*

13. Teacher Strikes The number of teacher strikes over a 13-year period in Pennsylvania is shown. Construct a boxplot for the data. Is the distribution symmetric?

20	18	7	13
7	14	5	9
9	9	10	17
15			

Source: Pennsylvania School Boards Association.

14. Visitors Who Travel to Foreign Countries Construct a boxplot for the number (in millions) of visitors who traveled to a foreign country each year for a random selection of years. Comment on the skewness of the distribution.

4.3	0.5	0.6	0.8	0.5
0.4	3.8	1.3	0.4	0.3

15. Tornadoes in 2005 Construct a boxplot and comment on its skewness for the number of tornadoes recorded each month in 2005.

33 10 62 132 123 316 138 123 133 18
150 26

Source: Storm Prediction Center.

16. Size of Dams These data represent the volumes in cubic yards of the largest dams in the United States and in South America. Construct a boxplot of the data for each region and compare the distributions.

United States	South America
125,628	311,539
92,000	274,026
78,008	105,944
77,700	102,014
66,500	56,242
62,850	46,563
52,435	
50,000	

Source: *New York Times Almanac.*

17. Number of Tornadoes A four-month record for the number of tornadoes in 2003–2005 is given here.

	2005	2004	2003
April	132	125	157
May	123	509	543
June	316	268	292
July	138	124	167

a. Which month had the highest mean number of tornadoes for this 3-year period? May: 391.7

b. Which year has the highest mean number of tornadoes for this 4-month period? 2003: 289.8

c. Construct three boxplots and compare the distributions.

Source: NWS, Storm Prediction Center.

Extending the Concepts

 18. Unhealthful Smog Days A *modified boxplot* can be drawn by placing a box around Q_1 and Q_3 and then extending the whiskers to the largest and/or smallest values within 1.5 times the interquartile range (that is, $Q_3 - Q_1$). *Mild outliers* are values between 1.5(IQR) and 3(IQR). *Extreme outliers* are data values beyond 3(IQR).

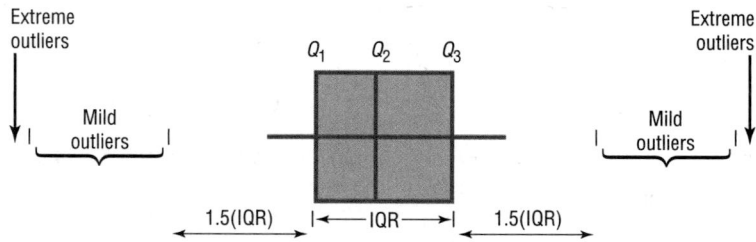

For the data shown here, draw a modified boxplot and identify any mild or extreme outliers. The data represent the number of unhealthful smog days for a specific year for the highest 10 locations.

| 97 | 39 | 43 | 66 | 91 |
| 43 | 54 | 42 | 53 | 39 |

Source: U.S. Public Interest Research Group and Clean Air Network.

Technology *Step by Step*

MINITAB
Step by Step

Construct a Boxplot

1. Type in the data 33, 38, 43, 30, 29, 40, 51, 27, 42, 23, 31. Label the column **Clients.**

2. Select **Stat>EDA>Boxplot.**

3. Double-click Clients to select it for the Y variable.

4. Click on [Labels].

   a) In the Title 1: of the Title/Footnotes folder, type **Number of Clients.**

   b) Press the [Tab] key and type **Your Name** in the text box for Subtitle 1:.

5. Click [OK] twice. The graph will be displayed in a graph window.

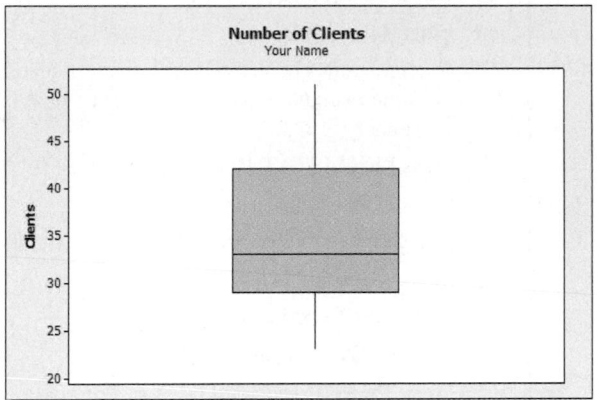

Example MT3–2

The number of car thefts in a large city over a 30-day period is shown.

52	62	51	50	69
58	77	66	53	57
75	56	65	67	73
79	59	68	65	72
57	51	63	69	75
65	53	78	66	55

1. Enter the data for this example. Label the column **CARS-THEFT.**

2. Select **Stat>EDA>Boxplot.**

3. Double-click CARS-THEFT to select it for the Y variable.

4. Click on the drop-down arrow for Annotation.

5. Click on Title, then enter an appropriate title such as **Car Thefts for Large City, U.S.A.**

6. Click [OK] twice.

A high-resolution graph will be displayed in a graph window.

Boxplot Dialog Box and Boxplot

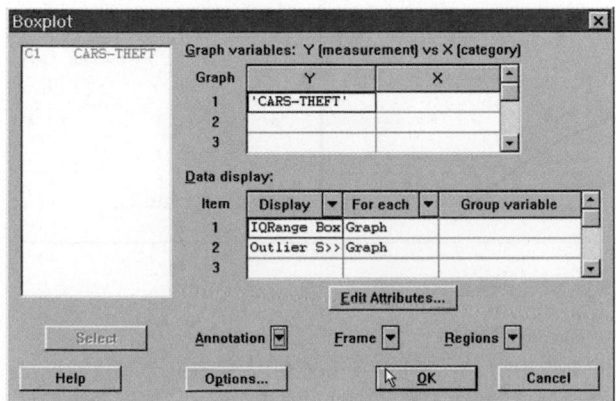

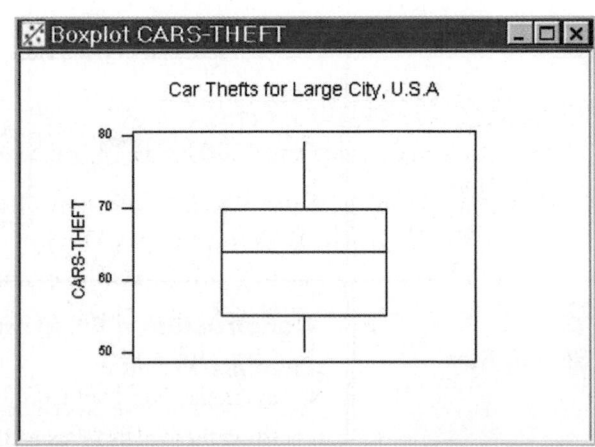

TI-83 Plus or TI-84 Plus
Step by Step

Constructing a Boxplot

To draw a boxplot:

1. Enter data into L_1.
2. Change values in WINDOW menu, if necessary. (*Note:* Make X_{min} somewhat smaller than the smallest data value and X_{max} somewhat larger than the largest data value.) Change Y_{min} to 0 and Y_{max} to 1.
3. Press **[2nd] [STAT PLOT]**, then **1** for Plot 1.
4. Press **ENTER** to turn Plot 1 on.
5. Move cursor to Boxplot symbol (fifth graph) on the Type: line, then press **ENTER.**
6. Make sure Xlist is L_1.
7. Make sure Freq is 1.
8. Press **GRAPH** to display the boxplot.
9. Press **TRACE** followed by ◄ or ► to obtain the values from the five-number summary on the boxplot.

To display two boxplots on the same display, follow the above steps and use the 2: Plot 2 and L_2 symbols.

Example TI3–3

Construct a boxplot for the data values:

$$33, 38, 43, 30, 29, 40, 51, 27, 42, 23, 31$$

Input	Input
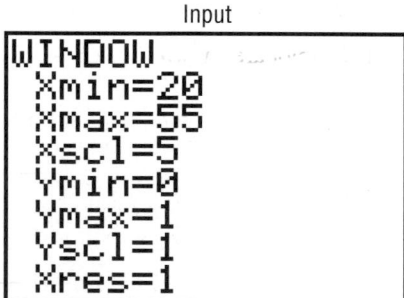	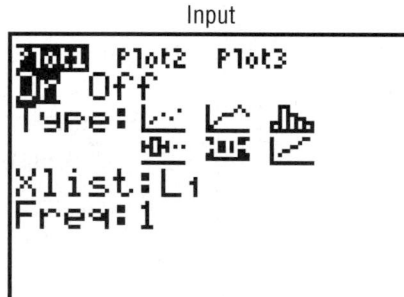

Using the **TRACE** key along with the ◄ and ► keys, we obtain the five-number summary. The minimum value is 23; Q_1 is 29; the median is 33; Q_3 is 42; the maximum value is 51.

Output

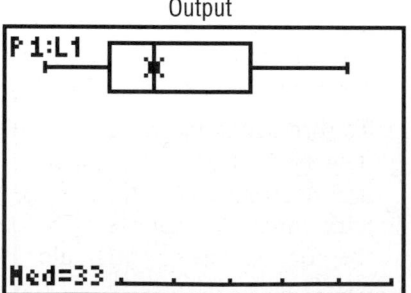

Excel
Step by Step

Constructing a Stem and Leaf Plot and a Boxplot

Example XL3–6

Excel does not have procedures to produce stem and leaf plots or boxplots. However, you may construct these plots by using the MegaStat Add-in available on your CD or from the Online

Learning Center. If you have not installed this add-in, refer to the instructions in the Excel Step by Step section of Chapter 1.

To obtain a boxplot and stem and leaf plot:

1. Enter the data values 33, 38, 43, 30, 29, 40, 51, 27, 42, 23, 31 into column A of a new Excel worksheet.

2. Select the Add-Ins tab, then MegaStat from the toolbar.

3. Select Descriptive Statistics from the MegaStat menu.

4. Enter the cell range A1:A11 in the Input range.

5. Check both Boxplot and Stem and Leaf Plot. *Note:* You may leave the other output options unchecked for this example. Click [OK].

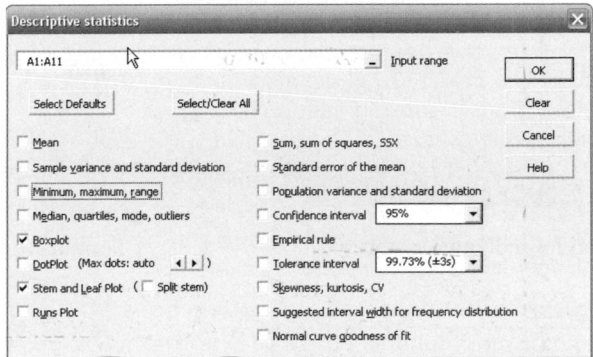

The stem and leaf plot and the boxplot are shown below.

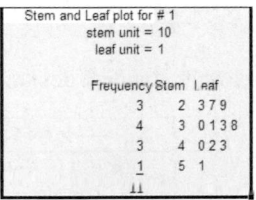

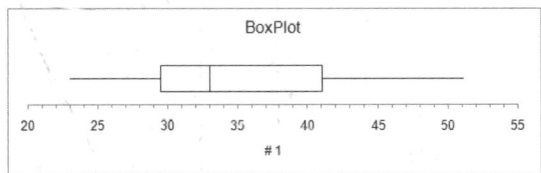

Summary

- This chapter explains the basic ways to summarize data. These include measures of central tendency. They are the mean, median, mode, and midrange. The weighted mean can also be used. (3–1)

- To summarize the variation of data, statisticians use measures of variation or dispersion. The three most common measures of variation are the range, variance, and standard deviation. The coefficient of variation can be used to compare the variation of two data sets. The data values are distributed according to Chebyshev's theorem on the empirical rule. (3–2)

- There are several measures of the position of data values in the set. There are standard scores, percentiles, quartiles, and deciles. Sometimes a data set contains an extremely high or extremely low data value, called an outlier. (3–3)

- Other methods can be used to describe a data set. These methods are the five-number summary and boxplots. These methods are called exploratory data analysis. (3–4)

The techniques explained in Chapter 2 and this chapter are the basic techniques used in descriptive statistics.

Important Terms

bimodal 111

boxplot 162

Chebyshev's theorem 134

coefficient of variation 132

data array 109

decile 151

empirical rule 136

exploratory data
analysis (EDA) 162

five-number summary 162

interquartile range (IQR) 151

mean 106

median 109

midrange 114

modal class 112

mode 111

multimodal 111

negatively skewed or left-
skewed distribution 117

outlier 151

parameter 106

percentile 143

positively skewed or right-
skewed distribution 117

quartile 149

range 124

range rule of thumb 133

resistant statistic 165

standard deviation 127

statistic 106

symmetric
distribution 117

unimodal 111

variance 127

weighted mean 115

z score or standard
score 142

Important Formulas

Formula for the mean for individual data:

$$\bar{X} = \frac{\Sigma X}{n} \qquad \mu = \frac{\Sigma X}{N}$$

Formula for the mean for grouped data:

$$\bar{X} = \frac{\Sigma f \cdot X_m}{n}$$

Formula for the weighted mean:

$$\bar{X} = \frac{\Sigma wX}{\Sigma w}$$

Formula for the midrange:

$$\text{MR} = \frac{\text{lowest value} + \text{highest value}}{2}$$

Formula for the range:

$$R = \text{highest value} - \text{lowest value}$$

Formula for the variance for population data:

$$\sigma^2 = \frac{\Sigma(X - \mu)^2}{N}$$

Formula for the variance for sample data (shortcut formula
for the unbiased estimator):

$$s^2 = \frac{n(\Sigma X^2) - (\Sigma X)^2}{n(n - 1)}$$

Formula for the variance for grouped data:

$$s^2 = \frac{n(\Sigma f \cdot X_m^2) - (\Sigma f \cdot X_m)^2}{n(n - 1)}$$

Formula for the standard deviation for population data:

$$\sigma = \sqrt{\frac{\Sigma(X - \mu)^2}{N}}$$

Formula for the standard deviation for sample data
(shortcut formula):

$$s = \sqrt{\frac{n(\Sigma X^2) - (\Sigma X)^2}{n(n - 1)}}$$

Formula for the standard deviation for grouped data:

$$s = \sqrt{\frac{n(\Sigma f \cdot X_m^2) - (\Sigma f \cdot X_m)^2}{n(n - 1)}}$$

Formula for the coefficient of variation:

$$\text{CVar} = \frac{s}{\bar{X}} \cdot 100 \qquad \text{or} \qquad \text{CVar} = \frac{\sigma}{\mu} \cdot 100$$

Range rule of thumb:

$$s \approx \frac{\text{range}}{4}$$

Expression for Chebyshev's theorem: The proportion of
values from a data set that will fall within k standard
deviations of the mean will be at least

$$1 - \frac{1}{k^2}$$

where k is a number greater than 1.

Formula for the z score (standard score):

$$z = \frac{X - \mu}{\sigma} \qquad \text{or} \qquad z = \frac{X - \bar{X}}{s}$$

Formula for the cumulative percentage:

$$\text{Cumulative \%} = \frac{\text{cumulative}\atop\text{frequency}}{n} \cdot 100$$

Formula for the percentile rank of a value X:

$$\text{Percentile} = \frac{\text{number of values}\atop\text{below } X + 0.5}{\text{total number}\atop\text{of values}} \cdot 100$$

Formula for finding a value corresponding to a given percentile:

$$c = \frac{n \cdot p}{100}$$

Formula for interquartile range:

$$\text{IQR} = Q_3 - Q_1$$

Review Exercises

1. Net Worth of Wealthy People The net worth (in billions of dollars) of a sample of the richest people in the United States is shown. Find the mean, median, mode, midrange, variance, and standard deviation for the data. (3–1) (3–2)

59	52	28	26	19
19	18	17	17	17

Source: *Forbes Magazine.*

2. Shark Attacks The number of shark attacks and deaths over a recent 5-year period is shown. Find the mean, median, mode, midrange, variance, and standard deviation for the data. Which data set is more variable? (3–1) (3–2)

Attacks	71	64	61	65	57
Deaths	1	4	4	7	4

3. Battery Lives Twelve batteries were tested to see how many hours they would last. The frequency distribution is shown here.

Hours	Frequency
1–3	1
4–6	4
7–9	5
10–12	1
13–15	1

Find each of these. (3–1) (3–2)

a. Mean 7.3   *c.* Variance 10.0
b. Modal class 7–9   *d.* Standard deviation 3.2

4. SAT Scores The mean SAT math scores for selected states are represented below. Find the mean class, modal class, variance, and standard deviation, and comment on the shape of the data. (3–1) (3–2)

Score	Frequency
478–504	4
505–531	6
532–558	2
559–585	2
586–612	2

Source: *World Almanac.*

5. Rise in Tides Shown here is a frequency distribution for the rise in tides at 30 selected locations in the United States.

Rise in tides (inches)	Frequency
12.5–27.5	6
27.5–42.5	3
42.5–57.5	5
57.5–72.5	8
72.5–87.5	6
87.5–102.5	2

Find each of these. (3–1) (3–2)

a. Mean 55.5   *c.* Variance 566.1
b. Modal class 57.5–72.5   *d.* Standard deviation 23.8

6. Fuel Capacity The fuel capacity in gallons of 50 randomly selected cars is shown here.

Class	Frequency
10–12	6
13–15	4
16–18	14
19–21	15
22–24	8
25–27	2
28–30	1
	50

Find each of these. (3–1) (3–2)

a. Mean 18.5   *c.* Variance 17.7
b. Modal class 19–21   *d.* Standard deviation 4.2

7. Households with Four Television Networks A survey showed the number of viewers and number of households of four television networks. Find the average number of viewers, using the weighted mean. (3–1) 1.43 viewers

Households	1.4	0.8	0.3	1.6
Viewers (in millions)	1.6	0.8	0.4	1.8

Source: Nielsen Media Research.

8. Investment Earnings An investor calculated these percentages of each of three stock investments with payoffs as shown. Find the average payoff. Use the weighted mean. (3–1) $4700.00

Stock	Percent	Payoff
A	30	$10,000
B	50	3,000
C	20	1,000

9. Years of Service of Employees In an advertisement, a transmission service center stated that the average years of service of its employees were 13. The distribution is shown here. Using the weighted mean, calculate the correct average. (3–1) 6

Number of employees	Years of service
8	3
1	6
1	30

10. Textbooks in Professors' Offices If the average number of textbooks in professors' offices is 16, the standard deviation is 5, and the average age of the professors is 43, with a standard deviation of 8, which data set is more variable? (3–2) 31.25%; 18.6%; the number of books is more variable

11. Magazines in Bookstores A survey of bookstores showed that the average number of magazines carried is 56, with a standard deviation of 12. The same survey showed that the average length of time each store had been in business was 6 years, with a standard deviation of 2.5 years. Which is more variable, the number of magazines or the number of years? (3–2) Magazine variance: 0.214; year variance: 0.417; years are more variable

12. Years of Service of Supreme Court Members The number of years served by selected past members of the U.S. Supreme Court is listed below. Find the percentile rank for each value. Which value corresponds to the 40th percentile? Construct a boxplot for the data and comment on their shape. (3–3) (3–4)

19, 15, 16, 24, 17, 4, 3, 31, 23, 5, 33

Source: World Almanac.

13. NFL Salaries The salaries (in millions of dollars) for 29 NFL teams for the 1999–2000 season are given in this frequency distribution. (3–3)

Class limits	Frequency
39.9–42.8	2
42.9–45.8	2
45.9–48.8	5
48.9–51.8	5
51.9–54.8	12
54.9–57.8	3

Source: www.NFL.com

a. Construct a percentile graph.
b. Find the values that correspond to the 35th, 65th, and 85th percentiles. 50, 53, 55
c. Find the percentile of values 44, 48, and 54. 10th; 26th; 78th

14. Check each data set for outliers. (3–3)

a. 506, 511, 517, 514, 400, 521 400
b. 3, 7, 9, 6, 8, 10, 14, 16, 20, 12 None
c. 14, 18, 27, 26, 19, 13, 5, 25 None
d. 112, 157, 192, 116, 153, 129, 131 None

15. Cost of Car Rentals A survey of car rental agencies shows that the average cost of a car rental is $0.32 per mile. The standard deviation is $0.03. Using Chebyshev's theorem, find the range in which at least 75% of the data values will fall. (3–2) $0.26–$0.38

16. Average Earnings of Workers The average earnings of year-round full-time workers 25–34 years old with a bachelor's degree or higher were $58,500 in 2003. If the standard deviation is $11,200, what can you say about the percentage of these workers who earn (3–2)

a. Between $47,300 and $69,700? Nothing because $k = 1$
b. More than $80,900? At most ¼ or 25%
c. How likely is it that someone earns more than $100,000? At most 7.3%

Source: New York Times Almanac.

17. Labor Charges The average labor charge for automobile mechanics is $54 per hour. The standard deviation is $4. Find the minimum percentage of data values that will fall within the range of $48 to $60. Use Chebyshev's theorem. (3–2) 56%

18. Costs to Train Employees For a certain type of job, it costs a company an average of $231 to train an employee to perform the task. The standard deviation is $5. Find the minimum percentage of data values that will fall in the range of $219 to $243. Use Chebyshev's theorem. (3–2) 83%

19. Delivery Charges The average delivery charge for a refrigerator is $32. The standard deviation is $4. Find the minimum percentage of data values that will fall in the range of $20 to $44. Use Chebyshev's theorem. (3–2) 88.89%

20. Exam Grades Which of these exam grades has a better relative position? (3–3)

  a. A grade of 82 on a test with $\overline{X} = 85$ and $s = 6$ –0.5
  b. A grade of 56 on a test with $\overline{X} = 60$ and $s = 5$ –0.8

  The test in part *a* is better.

21. Top Movie Sites The number of sites at which the top nine movies (based on the daily gross earnings) opened in a particular week is indicated below.

3017	3687	2525
2516	2820	2579
3211	3044	2330

Construct a boxplot for the data.

The 10th movie on the list opened at only 909 theaters. Add this number to the above set of data and comment on the changes that occur. (3–4)

Source: www.showbizdata.com The range is much larger.

22. Hours Worked The data shown here represent the number of hours that 12 part-time employees at a toy store worked during the weeks before and after Christmas. Construct two boxplots and compare the distributions. (3–4)

Before	38	16	18	24	12	30	35	32	31	30	24	35
After	26	15	12	18	24	32	14	18	16	18	22	12

23. Commuter Times The mean of the times it takes a commuter to get to work in Baltimore is 29.7 minutes. If the standard deviation is 6 minutes, within what limits would you expect approximately 68% of the times to fall? Assume the distribution is approximately bell-shaped. (3–3) 23.7–35.7

Statistics Today

How Long Are You Delayed by Road Congestion?—Revisited

The average number of hours per year that a driver is delayed by road congestion is listed here.

Los Angeles	56
Atlanta	53
Seattle	53
Houston	50
Dallas	46
Washington	46
Austin	45
Denver	45
St. Louis	44
Orlando	42
U.S. average	36

Source: Texas Transportation Institute.

By making comparisons using averages, you can see that drivers in these 10 cities are delayed by road congestion more than the national average.

Data Analysis

A Data Bank is found in Appendix D, or on the World Wide Web by following links from www.mhhe.com/math/stat/bluman/

1. From the Data Bank, choose one of the following variables: age, weight, cholesterol level, systolic pressure, IQ, or sodium level. Select at least 30 values, and find the mean, median, mode, and midrange. State which measurement of central tendency best describes the average and why.

2. Find the range, variance, and standard deviation for the data selected in Exercise 1.

3. From the Data Bank, choose 10 values from any variable, construct a boxplot, and interpret the results.

4. Randomly select 10 values from the number of suspensions in the local school districts in southwestern Pennsylvania in Data Set V in Appendix D. Find the mean, median, mode, range, variance, and standard deviation of the number of suspensions by using the Pearson coefficient of skewness.

5. Using the data from Data Set VII in Appendix D, find the mean, median, mode, range, variance, and standard deviation of the acreage owned by the municipalities. Comment on the skewness of the data, using the Pearson coefficient of skewness.

Chapter Quiz

Determine whether each statement is true or false. If the statement is false, explain why.

1. When the mean is computed for individual data, all values in the data set are used. True

2. The mean cannot be found for grouped data when there is an open class. True

3. A single, extremely large value can affect the median more than the mean. False

4. One-half of all the data values will fall above the mode, and one-half will fall below the mode. False

5. In a data set, the mode will always be unique. False

6. The range and midrange are both measures of variation. False

7. One disadvantage of the median is that it is not unique. False

8. The mode and midrange are both measures of variation. False

9. If a person's score on an exam corresponds to the 75th percentile, then that person obtained 75 correct answers out of 100 questions. False

Select the best answer.

10. What is the value of the mode when all values in the data set are different?

    a. 0
    b. 1
    c. There is no mode.
    d. It cannot be determined unless the data values are given.

11. When data are categorized as, for example, places of residence (rural, suburban, urban), the most appropriate measure of central tendency is the

    a. Mean   c. Mode
    b. Median   d. Midrange

12. P_{50} corresponds to   *a and b*

    a. Q_2
    b. D_5
    c. IQR
    d. Midrange

13. Which is not part of the five-number summary?

    a. Q_1 and Q_3
    b. The mean
    c. The median
    d. The smallest and the largest data values

14. A statistic that tells the number of standard deviations a data value is above or below the mean is called

    a. A quartile
    b. A percentile

    c. A coefficient of variation
    d. A z score

15. When a distribution is bell-shaped, approximately what percentage of data values will fall within 1 standard deviation of the mean?

    a. 50%
    b. 68%
    c. 95%
    d. 99.7%

Complete these statements with the best answer.

16. A measure obtained from sample data is called a(n) _____. Statistic

17. Generally, Greek letters are used to represent _____, and Roman letters are used to represent _____. Parameters, statistics

18. The positive square root of the variance is called the _____. Standard deviation

19. The symbol for the population standard deviation is _____. σ

20. When the sum of the lowest data value and the highest data value is divided by 2, the measure is called _____. Midrange

21. If the mode is to the left of the median and the mean is to the right of the median, then the distribution is _____ skewed. Positively

22. An extremely high or extremely low data value is called a(n) _____. Outlier

 23. Miles per Gallon The number of highway miles per gallon of the 10 worst vehicles is shown.

12 15 13 14 15 16 17 16 17 18

Source: *Pittsburgh Post Gazette.*

Find each of these.

    a. Mean 15.3
    b. Median 15.5
    c. Mode 15, 16, and 17
    d. Midrange 15
    e. Range 6
    f. Variance 3.57
    g. Standard deviation 1.9

24. **Errors on a Typing Test** The distribution of the number of errors that 10 students made on a typing test is shown.

Errors	Frequency
0–2	1
3–5	3
6–8	4
9–11	1
12–14	1

Find each of these.

a. Mean 6.4
b. Modal class 6–8
c. Variance 11.6
d. Standard deviation 3.4

25. **Inches of Rain** Shown here is a frequency distribution for the number of inches of rain received in 1 year in 25 selected cities in the United States.

Number of inches	Frequency
5.5–20.5	2
20.5–35.5	3
35.5–50.5	8
50.5–65.5	6
65.5–80.5	3
80.5–95.5	3

Find each of these.

a. Mean 51.4
b. Modal class 35.5–50.5
c. Variance 451.5
d. Standard deviation 21.2

26. **Shipment Times** A survey of 36 selected recording companies showed these numbers of days that it took to receive a shipment from the day it was ordered.

Days	Frequency
1–3	6
4–6	8
7–9	10
10–12	7
13–15	0
16–18	5

Find each of these.

a. Mean 8.2
b. Modal class 7–9
c. Variance 21.6
d. Standard deviation 4.6

27. **Best Friends of Students** In a survey of third-grade students, this distribution was obtained for the number of "best friends" each had. 1.6

Number of students	Number of best friends
8	1
6	2
5	3
3	0

Find the average number of best friends for the class. Use the weighted mean.

28. **Employee Years of Service** In an advertisement, a retail store stated that its employees averaged 9 years of service. The distribution is shown here. 4.5

Number of employees	Years of service
8	2
2	6
3	10

Using the weighted mean, calculate the correct average.

29. **Newspapers for Sale** The average number of newspapers for sale in an airport newsstand is 12, and the standard deviation is 4. The average age of the pilots is 37 years, with a standard deviation of 6 years. Which data set is more variable? 0.33; 0.162; newspapers

30. **Brands of Toothpaste Carried** A survey of grocery stores showed that the average number of brands of toothpaste carried was 16, with a standard deviation of 5. The same survey showed the average length of time each store was in business was 7 years, with a standard deviation of 1.6 years. Which is more variable, the number of brands or the number of years? 0.3125; 0.229; brands

31. **Test Scores** A student scored 76 on a general science test where the class mean and standard deviation were 82 and 8, respectively; he also scored 53 on a psychology test where the class mean and standard deviation were 58 and 3, respectively. In which class was his relative position higher? -0.75; -1.67; science

32. Which score has the highest relative position?

a. $X = 12$ $\overline{X} = 10$ $s = 4$ 0.5
b. $X = 170$ $\overline{X} = 120$ $s = 32$ 1.6
c. $X = 180$ $\overline{X} = 60$ $s = 8$ 15, c is highest

33. **Sizes of Malls** The number of square feet (in millions) of eight of the largest malls in southwestern Pennsylvania is shown.

| 1 | 0.9 | 1.3 | 0.8 |
| 1.4 | 0.77 | 0.7 | 1.2 |

Source: International Council of Shopping Centers.

a. Find the percentile for each value.
b. What value corresponds to the 40th percentile?
c. Construct a boxplot and comment on the nature of the distribution.

34. **Exam Scores** On a philosophy comprehensive exam, this distribution was obtained from 25 students.

Score	Frequency
40.5–45.5	3
45.5–50.5	8
50.5–55.5	10
55.5–60.5	3
60.5–65.5	1

a. Construct a percentile graph.
b. Find the values that correspond to the 22nd, 78th, and 99th percentiles. 47; 55; 64
c. Find the percentiles of the values 52, 43, and 64. 56th, 6th, 99th percentiles

35. **Gas Prices for Rental Cars** The first column of these data represents the prebuy gas price of a rental car, and the second column represents the price charged if the car is returned without refilling the gas tank for a selected car rental company. Draw two boxplots for the data and compare the distributions. (*Note:* The data were collected several years ago.)

Prebuy cost	No prebuy cost
$1.55	$3.80
1.54	3.99
1.62	3.99
1.65	3.85
1.72	3.99
1.63	3.95
1.65	3.94
1.72	4.19
1.45	3.84
1.52	3.94

Source: *USA TODAY.*

36. SAT Scores The average national SAT score is 1019. If we assume a bell-shaped distribution and a standard deviation equal to 110, what percentage of scores will you expect to fall above 1129? Above 799? 16%, 97.5%

Source: *New York Times Almanac,* 2002.

Critical Thinking Challenges

1. Average Cost of Weddings Averages give us information to help us to see where we stand and enable us to make comparisons. Here is a study on the average cost of a wedding. What type of average—mean, median, mode, or midrange—might have been used for each category?

OTHER PEOPLE'S MONEY

Question: What is the hottest wedding month? **Answer:** It's a tie. September now ranks as high as June in U.S. nuptials. The average attendence is 186 guests. And what kind of tabs are people running up for these affairs? Well, the next time a bride is throwing a bouquet, single women might want to . . . duck!

Reception	**$7246**
Rings	**4042**
Photos/videography.	**1263**
Bridal gown	**790**
Flowers	**775**
Music	**745**
Invitations	**374**
Mother of the bride's dress	**198**
Other (veil, limo, fees, etc.)	**3441**
Average cost of a wedding	**$18,874**

Stats: Bride's 2000 State of the Union Report

Source: Reprinted with permission from the September 2001 Reader's Digest. Copyright © 2001 by The Reader's Digest Assn., Inc.

2. Average Cost of Smoking This article states that the average yearly cost of smoking a pack of cigarettes a day is $1190. Find the average cost of a pack of cigarettes in your area, and compute the cost per day for 1 year. Compare your answer with the one in the article.

Burning Through the Cash

Everyone knows the health-related reasons to quit smoking, so heres an economic ar gument: A pack a day adds up to $1190 a year on average; it's more in states that have higher taxes on tobacco. To calculate what you or a loved one spends, visit ashline.org/ASH/quit/contemplation/index.html and try out their smoker's calculator. You'll be stunned.

Source: Reprinted with permission from the April 2002 Reader's Digest. Copyright © 2002 by The Reader's Digest Assn., Inc.

3. Ages of U.S. Residents The table shows the median ages of residents for the 10 oldest states and the 10 youngest states of the United States including Washington, D.C. Explain why the median is used instead of the mean.

	10 Oldest			10 Youngest	
Rank	**State**	**Median age**	**Rank**	**State**	**Median age**
1	West Virginia	38.9	51	Utah	27.1
2	Florida	38.7	50	Texas	32.3
3	Maine	38.6	49	Alaska	32.4
4	Pennsylvania	38.0	48	Idaho	33.2
5	Vermont	37.7	47	California	33.3
6	Montana	37.5	46	Georgia	33.4
7	Connecticut	37.4	45	Mississippi	33.8
8	New Hampshire	37.1	44	Louisiana	34.0
9	New Jersey	36.7	43	Arizona	34.2
10	Rhode Island	36.7	42	Colorado	34.3

Source: U.S. Census Bureau.

 **Data Projects**

Where appropriate, use MINITAB, the TI-83 Plus, the TI-84 Plus, or a computer program of your choice to complete the following exercises.

1. **Business and Finance** Use the data collected in data project 1 of Chapter 2 regarding earnings per share. Determine the mean, mode, median, and midrange for the two data sets. Is one measure of center more appropriate than the other for these data? Do the measures of center appear similar? What does this say about the symmetry of the distribution?

2. **Sports and Leisure** Use the data collected in data project 2 of Chapter 2 regarding home runs. Determine the mean, mode, median, and midrange for the two data sets. Is one measure of center more appropriate than the

other for these data? Do the measures of center appear similar? What does this say about the symmetry of the distribution?

3. **Technology** Use the data collected in data project 3 of Chapter 2. Determine the mean for the frequency table created in that project. Find the actual mean length of all 50 songs. How does the grouped mean compare to the actual mean?

4. **Health and Wellness** Use the data collected in data project 6 of Chapter 2 regarding heart rates. Determine the mean and standard deviation for each set of data. Do the means seem very different from one another? Do the standard deviations appear very different from one another?

5. Politics and Economics Use the data collected in data project 5 of Chapter 2 regarding delegates. Use the formulas for population mean and standard deviation to compute the parameters for all 50 states. What is the z score associated with California? Delaware? Ohio? Which states are more than 2 standard deviations from the mean?

6. Your Class Use your class as a sample. Determine the mean, median, and standard deviation for the age of students in your class. What z score would a 40-year-old have? Would it be unusual to have an age of 40? Determine the skew of the data, using the Pearson coefficient of skewness. (See Exercise 48, page 141.)

Answers to Applying the Concepts

Section 3–1 Teacher Salaries

1. The sample mean is $22,921.67, the sample median is $16,500, and the sample mode is $11,000. If you work for the school board and do not want to raise salaries, you could say that the average teacher salary is $22,921.67.

2. If you work for the teachers' union and want a raise for the teachers, either the sample median of $16,500 or the sample mode of $11,000 would be a good measure of center to report.

3. The outlier is $107,000. With the outlier removed, the sample mean is $15,278.18, the sample median is $16,400, and the sample mode is still $11,000. The mean is greatly affected by the outlier and allows the school board to report an average teacher salary that is not representative of a "typical" teacher salary.

4. If the salaries represented every teacher in the school district, the averages would be parameters, since we have data from the entire population.

5. The mean can be misleading in the presence of outliers, since it is greatly affected by these extreme values.

6. Since the mean is greater than both the median and the mode, the distribution is skewed to the right (positively skewed).

Section 3–2 Blood Pressure

1. Chebyshev's theorem does not work for scores within 1 standard deviation of the mean.

2. At least 75% (900) of the normotensive men will fall in the interval 105–141 mm Hg.

3. About 95% (1330) of the normotensive women have diastolic blood pressures between 62 and 90 mm Hg. About 95% (1235) of the hypertensive women have diastolic blood pressures between 68 and 108 mm Hg.

4. About 95% (1140) of the normotensive men have systolic blood pressures between 105 and 141 mm Hg. About 95% (1045) of the hypertensive men have systolic blood pressures between 119 and 187 mm Hg. These two ranges do overlap.

Section 3–3 Determining Dosages

1. The quartiles could be used to describe the data results.

2. Since there are 10 mice in the upper quartile, this would mean that 4 of them survived.

3. The percentiles would give us the position of a single mouse with respect to all other mice.

4. The quartiles divide the data into four groups of equal size.

5. Standard scores would give us the position of a single mouse with respect to the mean time until the onset of sepsis.

Section 3–4 The Noisy Workplace

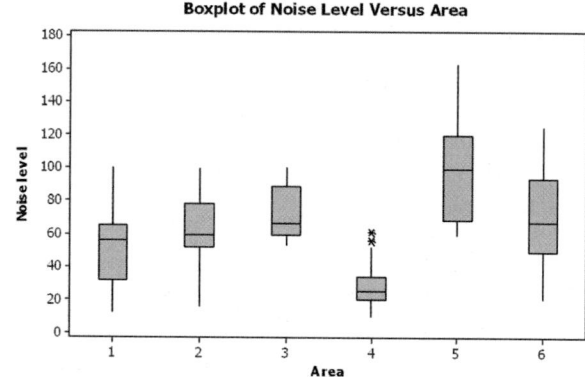

Boxplot of Noise Level Versus Area

From this boxplot, we see that about 25% of the readings in area 5 are above the safe hearing level of 120 decibels. Those workers in area 5 should definitely have protective ear wear. One of the readings in area 6 is above the safe hearing level. It might be a good idea to provide protective ear wear to those workers in area 6 as well. Areas 1–4 appear to be "safe" with respect to hearing level, with area 4 being the safest.

Probability and Counting Rules

Objectives

After completing this chapter, you should be able to

1 Determine sample spaces and find the probability of an event, using classical probability or empirical probability.

2 Find the probability of compound events, using the addition rules.

3 Find the probability of compound events, using the multiplication rules.

4 Find the conditional probability of an event.

5 Find the total number of outcomes in a sequence of events, using the fundamental counting rule.

6 Find the number of ways that *r* objects can be selected from *n* objects, using the permutation rule.

7 Find the number of ways that *r* objects can be selected from *n* objects without regard to order, using the combination rule.

8 Find the probability of an event, using the counting rules.

Outline

Would You Bet Your Life?

Humans not only bet money when they gamble, but also bet their lives by engaging in unhealthy activities such as smoking, drinking, using drugs, and exceeding the speed limit when driving. Many people don't care about the risks involved in these activities since they do not understand the concepts of probability. On the other hand, people may fear activities that involve little risk to health or life because these activities have been sensationalized by the press and media.

In his book *Probabilities in Everyday Life* (Ivy Books, p. 191), John D. McGervey states

When people have been asked to estimate the frequency of death from various causes, the most overestimated categories are those involving pregnancy, tornadoes, floods, fire, and homicide. The most underestimated categories include deaths from diseases such as diabetes, strokes, tuberculosis, asthma, and stomach cancer (although cancer in general is overestimated).

The question then is, Would you feel safer if you flew across the United States on a commercial airline or if you drove? How much greater is the risk of one way to travel over the other? See Statistics Today—Revisited at the end of the chapter for the answer.

In this chapter, you will learn about probability—its meaning, how it is computed, and how to evaluate it in terms of the likelihood of an event actually happening.

Introduction

A cynical person once said, "The only two sure things are death and taxes." This philosophy no doubt arose because so much in people's lives is affected by chance. From the time you awake until you go to bed, you make decisions regarding the possible events that are governed at least in part by chance. For example, should you carry an umbrella to work today? Will your car battery last until spring? Should you accept that new job?

Probability as a general concept can be defined as the chance of an event occurring. Many people are familiar with probability from observing or playing games of chance, such as card games, slot machines, or lotteries. In addition to being used in games of chance, probability theory is used in the fields of insurance, investments, and weather forecasting and in various other areas. Finally, as stated in Chapter 1, probability is the basis of inferential statistics. For example, predictions are based on probability, and hypotheses are tested by using probability.

The basic concepts of probability are explained in this chapter. These concepts include *probability experiments, sample spaces,* the *addition* and *multiplication rules,* and the *probabilities of complementary events.* Also in this chapter, you will learn the rule for counting, the differences between permutations and combinations, and how to figure out how many different combinations for specific situations exist. Finally, Section 4–5 explains how the counting rules and the probability rules can be used together to solve a wide variety of problems.

4–1

Sample Spaces and Probability

The theory of probability grew out of the study of various games of chance using coins, dice, and cards. Since these devices lend themselves well to the application of concepts of probability, they will be used in this chapter as examples. This section begins by explaining some basic concepts of probability. Then the types of probability and probability rules are discussed.

Basic Concepts

Processes such as flipping a coin, rolling a die, or drawing a card from a deck are called *probability experiments.*

Objective 1

Determine sample spaces and find the probability of an event, using classical probability or empirical probability.

> A **probability experiment** is a chance process that leads to well-defined results called outcomes.
>
> An **outcome** is the result of a single trial of a probability experiment.

A trial means flipping a coin once, rolling one die once, or the like. When a coin is tossed, there are two possible outcomes: head or tail. (*Note:* We exclude the possibility of a coin landing on its edge.) In the roll of a single die, there are six possible outcomes: 1, 2, 3, 4, 5, or 6. In any experiment, the set of all possible outcomes is called the *sample space.*

> A **sample space** is the set of all possible outcomes of a probability experiment.

Some sample spaces for various probability experiments are shown here.

Experiment	Sample space
Toss one coin	Head, tail
Roll a die	1, 2, 3, 4, 5, 6
Answer a true/false question	True, false
Toss two coins	Head-head, tail-tail, head-tail, tail-head

It is important to realize that when two coins are tossed, there are *four* possible outcomes, as shown in the fourth experiment above. Both coins could fall heads up. Both coins could fall tails up. Coin 1 could fall heads up and coin 2 tails up. Or coin 1 could fall tails up and coin 2 heads up. Heads and tails will be abbreviated as H and T throughout this chapter.

Example 4–1

Rolling Dice

Find the sample space for rolling two dice.

Solution

Since each die can land in six different ways, and two dice are rolled, the sample space can be presented by a rectangular array, as shown in Figure 4–1. The sample space is the list of pairs of numbers in the chart.

Figure 4–1

Sample Space for
Rolling Two Dice
(Example 4–1)

	Die 2					
Die 1	1	2	3	4	5	6
1	(1, 1)	(1, 2)	(1, 3)	(1, 4)	(1, 5)	(1, 6)
2	(2, 1)	(2, 2)	(2, 3)	(2, 4)	(2, 5)	(2, 6)
3	(3, 1)	(3, 2)	(3, 3)	(3, 4)	(3, 5)	(3, 6)
4	(4, 1)	(4, 2)	(4, 3)	(4, 4)	(4, 5)	(4, 6)
5	(5, 1)	(5, 2)	(5, 3)	(5, 4)	(5, 5)	(5, 6)
6	(6, 1)	(6, 2)	(6, 3)	(6, 4)	(6, 5)	(6, 6)

Example 4–2

Drawing Cards

Find the sample space for drawing one card from an ordinary deck of cards.

Solution

Since there are 4 suits (hearts, clubs, diamonds, and spades) and 13 cards for each suit (ace through king), there are 52 outcomes in the sample space. See Figure 4–2.

Figure 4–2

Sample Space for
Drawing a Card
(Example 4–2)

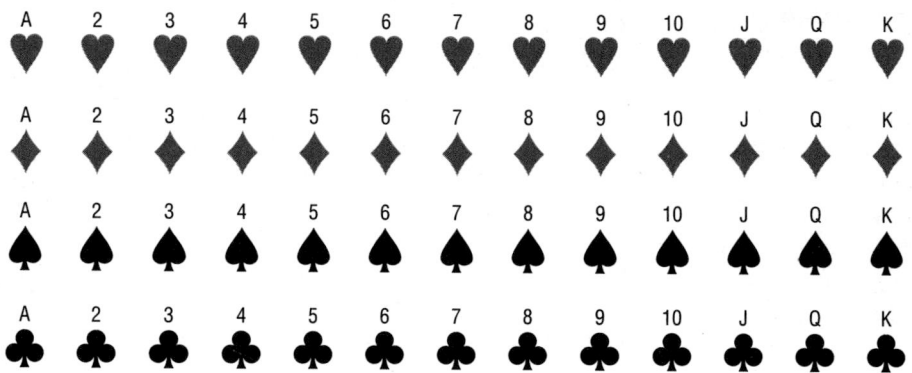

Example 4–3

Gender of Children

Find the sample space for the gender of the children if a family has three children. Use B for boy and G for girl.

Solution

There are two genders, male and female, and each child could be either gender. Hence, there are eight possibilities, as shown here.

    BBB   BBG   BGB   GBB   GGG   GGB   GBG   BGG

In Examples 4–1 through 4–3, the sample spaces were found by observation and reasoning; however, another way to find all possible outcomes of a probability experiment is to use a *tree diagram.*

A **tree diagram** is a device consisting of line segments emanating from a starting point and also from the outcome point. It is used to determine all possible outcomes of a probability experiment.

Example 4–4

Gender of Children

Use a tree diagram to find the sample space for the gender of three children in a family, as in Example 4–3.

Solution

Since there are two possibilities (boy or girl) for the first child, draw two branches from a starting point and label one B and the other G. Then if the first child is a boy, there are two possibilities for the second child (boy or girl), so draw two branches from B and label one B and the other G. Do the same if the first child is a girl. Follow the same procedure for the third child. The completed tree diagram is shown in Figure 4–3. To find the outcomes for the sample space, trace through all the possible branches, beginning at the starting point for each one.

Figure 4–3

Tree Diagram for Example 4–4

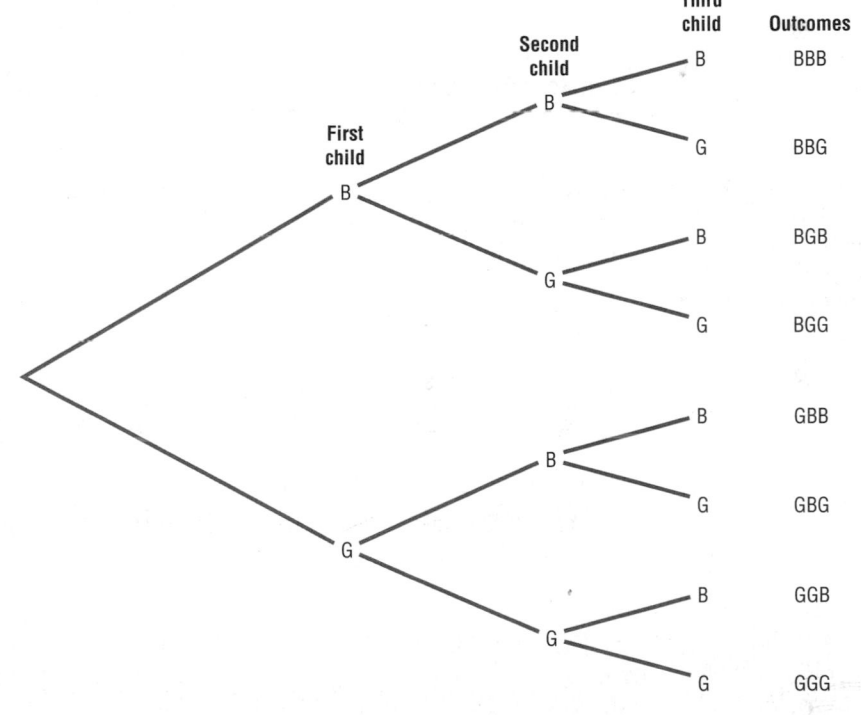

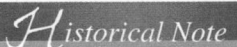
An outcome was defined previously as the result of a single trial of a probability experiment. In many problems, one must find the probability of two or more outcomes. For this reason, it is necessary to distinguish between an outcome and an event.

An **event** consists of a set of outcomes of a probability experiment.

An event can be one outcome or more than one outcome. For example, if a die is rolled and a 6 shows, this result is called an *outcome,* since it is a result of a single trial. An event with one outcome is called a **simple event.** The event of getting an odd number when a die is rolled is called a **compound event,** since it consists of three outcomes or three simple events. In general, a compound event consists of two or more outcomes or simple events.

There are three basic interpretations of probability:

1. Classical probability
2. Empirical or relative frequency probability
3. Subjective probability

Classical Probability

Classical probability uses sample spaces to determine the numerical probability that an event will happen. You do not actually have to perform the experiment to determine that probability. Classical probability is so named because it was the first type of probability studied formally by mathematicians in the 17th and 18th centuries.

Classical probability assumes that all outcomes in the sample space are equally likely to occur. For example, when a single die is rolled, each outcome has the same probability of occurring. Since there are six outcomes, each outcome has a probability of $\frac{1}{6}$. When a card is selected from an ordinary deck of 52 cards, you assume that the deck has been shuffled, and each card has the same probability of being selected. In this case, it is $\frac{1}{52}$.

Equally likely events are events that have the same probability of occurring.

Formula for Classical Probability

The probability of any event E is

$$\frac{\text{Number of outcomes in } E}{\text{Total number of outcomes in the sample space}}$$

This probability is denoted by

$$P(E) = \frac{n(E)}{n(S)}$$

This probability is called *classical probability,* and it uses the sample space S.

Probabilities can be expressed as fractions, decimals, or—where appropriate—percentages. If you ask, "What is the probability of getting a head when a coin is tossed?" typical responses can be any of the following three.

"One-half."

"Point five."

"Fifty percent."[1]

These answers are all equivalent. In most cases, the answers to examples and exercises given in this chapter are expressed as fractions or decimals, but percentages are used where appropriate.

[1]Strictly speaking, a percent is not a probability. However, in everyday language, probabilities are often expressed as percents (i.e., there is a 60% chance of rain tomorrow). For this reason, some probabilities will be expressed as percents throughout this book.

Rounding Rule for Probabilities Probabilities should be expressed as reduced fractions or rounded to two or three decimal places. When the probability of an event is an extremely small decimal, it is permissible to round the decimal to the first nonzero digit after the point. For example, 0.0000587 would be 0.00006. When obtaining probabilities from one of the tables in Appendix C, use the number of decimal places given in the table. If decimals are converted to percentages to express probabilities, move the decimal point two places to the right and add a percent sign.

Example 4–5	**Drawing Cards**

Find the probability of getting a black 10 when drawing a card from a deck.

Solution

There are 52 cards in a deck, and there are two black 10s—the 10 of spades and the 10 of clubs. Hence the probability of getting a black 10 is $P(\text{black } 10) = \frac{2}{52} = \frac{1}{26}$.

Example 4–6	**Gender of Children**

If a family has three children, find the probability that two of the three children are girls.

Solution

The sample space for the gender of the children for a family that has three children has eight outcomes, that is, BBB, BBG, BGB, GBB, GGG, GGB, GBG, and BGG. (See Examples 4–3 and 4–4.) Since there are three ways to have two girls, namely, GGB, GBG, and BGG, $P(\text{two girls}) = \frac{3}{8}$.

Historical Note

Ancient Greeks and Romans made crude dice from animal bones, various stones, minerals, and ivory. When the dice were tested mathematically, some were found to be quite accurate.

In probability theory, it is important to understand the meaning of the words *and* and *or*. For example, if you were asked to find the probability of getting a queen *and* a heart when you were drawing a single card from a deck, you would be looking for the queen of hearts. Here the word *and* means "at the same time." The word *or* has two meanings. For example, if you were asked to find the probability of selecting a queen *or* a heart when one card is selected from a deck, you would be looking for one of the 4 queens or one of the 13 hearts. In this case, the queen of hearts would be included in both cases and counted twice. So there would be $4 + 13 - 1 = 16$ possibilities.

On the other hand, if you were asked to find the probability of getting a queen *or* a king, you would be looking for one of the 4 queens or one of the 4 kings. In this case, there would be $4 + 4 = 8$ possibilities. In the first case, both events can occur at the same time; we say that this is an example of the *inclusive or*. In the second case, both events cannot occur at the same time, and we say that this is an example of the *exclusive or*.

Example 4–7	**Drawing Cards**

A card is drawn from an ordinary deck. Find these probabilities.

  a. Of getting a jack
  b. Of getting the 6 of clubs (i.e., a 6 and a club)
  c. Of getting a 3 or a diamond
  d. Of getting a 3 or a 6

Solution

a. Refer to the sample space in Figure 4–2. There are 4 jacks so there are 4 outcomes in event E and 52 possible outcomes in the sample space. Hence,

$$P(\text{jack}) = \tfrac{4}{52} = \tfrac{1}{13}$$

b. Since there is only one 6 of clubs in event E, the probability of getting a 6 of clubs is

$$P(\text{6 of clubs}) = \tfrac{1}{52}$$

c. There are four 3s and 13 diamonds, but the 3 of diamonds is counted twice in this listing. Hence, there are 16 possibilities of drawing a 3 or a diamond, so

$$P(\text{3 or diamond}) = \tfrac{16}{52} = \tfrac{4}{13}$$

This is an example of the inclusive or.

d. Since there are four 3s and four 6s,

$$P(\text{3 or 6}) = \tfrac{8}{52} = \tfrac{2}{13}$$

This is an example of the exclusive or.

There are four basic probability rules. These rules are helpful in solving probability problems, in understanding the nature of probability, and in deciding if your answers to the problems are correct.

Probability Rule 1

The probability of any event E is a number (either a fraction or decimal) between and including 0 and 1. This is denoted by $0 \le P(E) \le 1$.

Rule 1 states that probabilities cannot be negative or greater than 1.

Probability Rule 2

If an event E cannot occur (i.e., the event contains no members in the sample space), its probability is 0.

Example 4–8

Rolling a Die

When a single die is rolled, find the probability of getting a 9.

Solution

Since the sample space is 1, 2, 3, 4, 5, and 6, it is impossible to get a 9. Hence, the probability is $P(9) = \tfrac{0}{6} = 0$.

Probability Rule 3

If an event E is certain, then the probability of E is 1.

In other words, if $P(E) = 1$, then the event E is certain to occur. This rule is illustrated in Example 4–9.

Example 4–9

Rolling a Die

When a single die is rolled, what is the probability of getting a number less than 7?

Solution

Since all outcomes—1, 2, 3, 4, 5, and 6—are less than 7, the probability is

$$P(\text{number less than 7}) = \tfrac{6}{6} = 1$$

The event of getting a number less than 7 is certain.

In other words, probability values range from 0 to 1. When the probability of an event is close to 0, its occurrence is highly unlikely. When the probability of an event is near 0.5, there is about a 50-50 chance that the event will occur; and when the probability of an event is close to 1, the event is highly likely to occur.

Probability Rule 4

The sum of the probabilities of all the outcomes in the sample space is 1.

For example, in the roll of a fair die, each outcome in the sample space has a probability of $\tfrac{1}{6}$. Hence, the sum of the probabilities of the outcomes is as shown.

Outcome	1	2	3	4	5	6
Probability	$\tfrac{1}{6}$	$\tfrac{1}{6}$	$\tfrac{1}{6}$	$\tfrac{1}{6}$	$\tfrac{1}{6}$	$\tfrac{1}{6}$
Sum	$\tfrac{1}{6}$ +	$\tfrac{1}{6}$ +	$\tfrac{1}{6}$ +	$\tfrac{1}{6}$ +	$\tfrac{1}{6}$ +	$\tfrac{1}{6}$ = $\tfrac{6}{6}$ = 1

Complementary Events

Another important concept in probability theory is that of *complementary events*. When a die is rolled, for instance, the sample space consists of the outcomes 1, 2, 3, 4, 5, and 6. The event E of getting odd numbers consists of the outcomes 1, 3, and 5. The event of not getting an odd number is called the *complement* of event E, and it consists of the outcomes 2, 4, and 6.

The **complement of an event** E is the set of outcomes in the sample space that are not included in the outcomes of event E. The complement of E is denoted by $\overline{E}$ (read "E bar").

Example 4–10 further illustrates the concept of complementary events.

Example 4–10

Finding Complements

Find the complement of each event.

a. Rolling a die and getting a 4

b. Selecting a letter of the alphabet and getting a vowel

    c. Selecting a month and getting a month that begins with a J

    d. Selecting a day of the week and getting a weekday

Solution

    a. Getting a 1, 2, 3, 5, or 6

    b. Getting a consonant (assume y is a consonant)

    c. Getting February, March, April, May, August, September, October, November, or December

    d. Getting Saturday or Sunday

The outcomes of an event and the outcomes of the complement make up the entire sample space. For example, if two coins are tossed, the sample space is HH, HT, TH, and TT. The complement of "getting all heads" is not "getting all tails," since the event "all heads" is HH, and the complement of HH is HT, TH, and TT. Hence, the complement of the event "all heads" is the event "getting at least one tail."

Since the event and its complement make up the entire sample space, it follows that the sum of the probability of the event and the probability of its complement will equal 1. That is, $P(E) + P(\overline{E}) = 1$. For example, let $E =$ all heads, or HH, and let $\overline{E} =$ at least one tail, or HT, TH, TT. Then $P(E) = \frac{1}{4}$ and $P(\overline{E}) = \frac{3}{4}$; hence, $P(E) + P(\overline{E}) = \frac{1}{4} + \frac{3}{4} = 1$.

The rule for complementary events can be stated algebraically in three ways.

Rule for Complementary Events

$$P(\overline{E}) = 1 - P(E) \qquad \text{or} \qquad P(E) = 1 - P(\overline{E}) \qquad \text{or} \qquad P(E) + P(\overline{E}) = 1$$

Stated in words, the rule is: *If the probability of an event or the probability of its complement is known, then the other can be found by subtracting the probability from 1.* This rule is important in probability theory because at times the best solution to a problem is to find the probability of the complement of an event and then subtract from 1 to get the probability of the event itself.

Example 4–11

Residence of People

If the probability that a person lives in an industrialized country of the world is $\frac{1}{5}$, find the probability that a person does not live in an industrialized country.

Source: *Harper's Index.*

Solution

$P(\text{not living in an industrialized country}) = 1 - P(\text{living in an industrialized country})$
$$= 1 - \tfrac{1}{5} = \tfrac{4}{5}$$

Probabilities can be represented pictorially by **Venn diagrams.** Figure 4–4(a) shows the probability of a simple event *E.* The area inside the circle represents the probability of event *E,* that is, $P(E)$. The area inside the rectangle represents the probability of all the events in the sample space $P(S)$.

Figure 4–4

Venn Diagram for the Probability and Complement

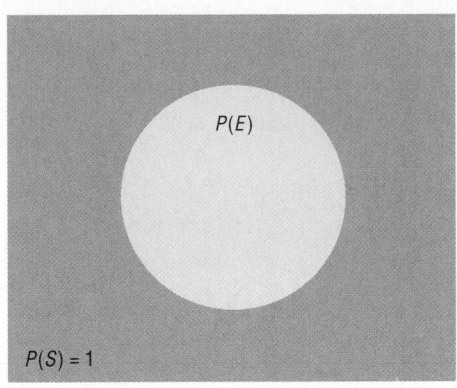

$P(E)$

$P(S) = 1$

(a) Simple probability

$P(E)$

$P(\bar{E})$

(b) $P(\bar{E}) = 1 - P(E)$

The Venn diagram that represents the probability of the complement of an event $P(\bar{E})$ is shown in Figure 4–4(b). In this case, $P(\bar{E}) = 1 - P(E)$, which is the area inside the rectangle but outside the circle representing $P(E)$. Recall that $P(S) = 1$ and $P(E) = 1 - P(\bar{E})$. The reasoning is that $P(E)$ is represented by the area of the circle and $P(\bar{E})$ is the probability of the events that are outside the circle.

Empirical Probability

The difference between classical and **empirical probability** is that classical probability assumes that certain outcomes are equally likely (such as the outcomes when a die is rolled), while empirical probability relies on actual experience to determine the likelihood of outcomes. In empirical probability, one might actually roll a given die 6000 times, observe the various frequencies, and use these frequencies to determine the probability of an outcome.

Suppose, for example, that a researcher for the American Automobile Association (AAA) asked 50 people who plan to travel over the Thanksgiving holiday how they will get to their destination. The results can be categorized in a frequency distribution as shown.

Method	Frequency
Drive	41
Fly	6
Train or bus	3
	50

Now probabilities can be computed for various categories. For example, the probability of selecting a person who is driving is $\frac{41}{50}$, since 41 out of the 50 people said that they were driving.

Formula for Empirical Probability

Given a frequency distribution, the probability of an event being in a given class is

$$P(E) = \frac{\text{frequency for the class}}{\text{total frequencies in the distribution}} = \frac{f}{n}$$

This probability is called *empirical probability* and is based on observation.

Example 4–12	**Travel Survey**

In the travel survey just described, find the probability that a person will travel by airplane over the Thanksgiving holiday.

Solution

$$P(E) = \frac{f}{n} = \frac{6}{50} = \frac{3}{25}$$

Note: These figures are based on an AAA survey.

Example 4–13	**Distribution of Blood Types**

In a sample of 50 people, 21 had type O blood, 22 had type A blood, 5 had type B blood, and 2 had type AB blood. Set up a frequency distribution and find the following probabilities.

   a. A person has type O blood.

   b. A person has type A or type B blood.

   c. A person has neither type A nor type O blood.

   d. A person does not have type AB blood.

Source: The American Red Cross.

Solution

Type	Frequency
A	22
B	5
AB	2
O	21
Total	50

a. $P(\text{O}) = \dfrac{f}{n} = \dfrac{21}{50}$

b. $P(\text{A or B}) = \dfrac{22}{50} + \dfrac{5}{50} = \dfrac{27}{50}$
(Add the frequencies of the two classes.)

c. $P(\text{neither A nor O}) = \dfrac{5}{50} + \dfrac{2}{50} = \dfrac{7}{50}$
(Neither A nor O means that a person has either type B or type AB blood.)

d. $P(\text{not AB}) = 1 - P(\text{AB}) = 1 - \dfrac{2}{50} = \dfrac{48}{50} = \dfrac{24}{25}$
(Find the probability of not AB by subtracting the probability of type AB from 1.)

Example 4–14   Hospital Stays for Maternity Patients

Hospital records indicated that knee replacement patients stayed in the hospital for the number of days shown in the distribution.

Number of days stayed	Frequency
3	15
4	32
5	56
6	19
7	5
	127

Find these probabilities.

    a. A patient stayed exactly 5 days.   *c.* A patient stayed at most 4 days.

    b. A patient stayed less than 6 days.   *d.* A patient stayed at least 5 days.

Solution

    a. $P(5) = \dfrac{56}{127}$

    b. $P(\text{fewer than 6 days}) = \dfrac{15}{127} + \dfrac{32}{127} + \dfrac{56}{127} = \dfrac{103}{127}$

        (Less than 6 days means 3, 4, or 5 days.)

    c. $P(\text{at most 4 days}) = \dfrac{15}{127} + \dfrac{32}{127} = \dfrac{47}{127}$

        (At most 4 days means 3 or 4 days.)

    d. $P(\text{at least 5 days}) = \dfrac{56}{127} + \dfrac{19}{127} + \dfrac{5}{127} = \dfrac{80}{127}$

        (At least 5 days means 5, 6, or 7 days.)

Empirical probabilities can also be found by using a relative frequency distribution, as shown in Section 2–2. For example, the relative frequency distribution of the travel survey shown previously is

Method	Frequency	Relative frequency
Drive	41	0.82
Fly	6	0.12
Train or bus	3	0.06
	50	1.00

These frequencies are the same as the relative frequencies explained in Chapter 2.

Law of Large Numbers

When a coin is tossed one time, it is common knowledge that the probability of getting a head is $\frac{1}{2}$. But what happens when the coin is tossed 50 times? Will it come up heads

25 times? Not all the time. You should expect about 25 heads if the coin is fair. But due to chance variation, 25 heads will not occur most of the time.

If the empirical probability of getting a head is computed by using a small number of trials, it is usually not exactly $\frac{1}{2}$. However, as the number of trials increases, the empirical probability of getting a head will approach the theoretical probability of $\frac{1}{2}$, if in fact the coin is fair (i.e., balanced). This phenomenon is an example of the **law of large numbers.**

You should be careful to not think that the number of heads and number of tails tend to "even out." As the number of trials increases, the proportion of heads to the total number of trials will approach $\frac{1}{2}$. This law holds for any type of gambling game—tossing dice, playing roulette, and so on.

It should be pointed out that the probabilities that the proportions steadily approach may or may not agree with those theorized in the classical model. If not, it can have important implications, such as "the die is not fair." Pit bosses in Las Vegas watch for empirical trends that do not agree with classical theories, and they will sometimes take a set of dice out of play if observed frequencies are too far out of line with classical expected frequencies.

Subjective Probability

The third type of probability is called *subjective probability.* **Subjective probability** uses a probability value based on an educated guess or estimate, employing opinions and inexact information.

In subjective probability, a person or group makes an educated guess at the chance that an event will occur. This guess is based on the person's experience and evaluation of a solution. For example, a sportswriter may say that there is a 70% probability that the Pirates will win the pennant next year. A physician might say that, on the basis of her diagnosis, there is a 30% chance the patient will need an operation. A seismologist might say there is an 80% probability that an earthquake will occur in a certain area. These are only a few examples of how subjective probability is used in everyday life.

All three types of probability (classical, empirical, and subjective) are used to solve a variety of problems in business, engineering, and other fields.

Probability and Risk Taking

An area in which people fail to understand probability is risk taking. Actually, people fear situations or events that have a relatively small probability of happening rather than those events that have a greater likelihood of occurring. For example, many people think that the crime rate is increasing every year. However, in his book entitled *How Risk Affects Your Everyday Life,* author James Walsh states: "Despite widespread concern about the number of crimes committed in the United States, FBI and Justice Department statistics show that the national crime rate has remained fairly level for 20 years. It even dropped slightly in the early 1990s."

He further states, "Today most media coverage of risk to health and well-being focuses on shock and outrage." Shock and outrage make good stories and can scare us about the wrong dangers. For example, the author states that if a person is 20% overweight, the loss of life expectancy is 900 days (about 3 years), but loss of life expectancy from exposure to radiation emitted by nuclear power plants is 0.02 day. As you can see, being overweight is much more of a threat than being exposed to radioactive emission.

Many people gamble daily with their lives, for example, by using tobacco, drinking and driving, and riding motorcycles. When people are asked to estimate the probabilities or frequencies of death from various causes, they tend to overestimate causes such as accidents, fires, and floods and to underestimate the probabilities of death from diseases (other than cancer), strokes, etc. For example, most people think that their chances of dying of a heart attack are 1 in 20, when in fact they are almost 1 in 3; the chances of dying by pesticide poisoning are 1 in 200,000 (*True Odds* by James Walsh). The reason people think this way is that the news media sensationalize deaths resulting from catastrophic events and rarely mention deaths from disease.

When you are dealing with life-threatening catastrophes such as hurricanes, floods, automobile accidents, or texting while driving, it is important to get the facts. That is, get the actual numbers from accredited statistical agencies or reliable statistical studies, and then compute the probabilities and make decisions based on your knowledge of probability and statistics.

In summary, then, when you make a decision or plan a course of action based on probability, make sure that you understand the true probability of the event occurring. Also, find out how the information was obtained (i.e., from a reliable source). Weigh the cost of the action and decide if it is worth it. Finally, look for other alternatives or courses of action with less risk involved.

Applying the Concepts **4–1**

Tossing a Coin

Assume you are at a carnival and decide to play one of the games. You spot a table where a person is flipping a coin, and since you have an understanding of basic probability, you believe that the odds of winning are in your favor. When you get to the table, you find out that all you have to do is to guess which side of the coin will be facing up after it is tossed. You are assured that the coin is fair, meaning that each of the two sides has an equally likely chance of occurring. You think back about what you learned in your statistics class about probability before you decide what to bet on. Answer the following questions about the coin-tossing game.

1. What is the sample space?
2. What are the possible outcomes?
3. What does the classical approach to probability say about computing probabilities for this type of problem?

You decide to bet on heads, believing that it has a 50% chance of coming up. A friend of yours, who had been playing the game for awhile before you got there, tells you that heads has come up the last 9 times in a row. You remember the law of large numbers.

4. What is the law of large numbers, and does it change your thoughts about what will occur on the next toss?
5. What does the empirical approach to probability say about this problem, and could you use it to solve this problem?
6. Can subjective probabilities be used to help solve this problem? Explain.
7. Assume you could win $1 million if you could guess what the results of the next toss will be. What would you bet on? Why?

See page 249 for the answers.

Exercises 4–1

1. What is a probability experiment? A probability experiment is a chance process that leads to well-defined outcomes.

2. Define *sample space*. The set of all possible outcomes of a probability experiment is called a sample space.

3. What is the difference between an outcome and an event? An outcome is the result of a single trial of a probability experiment, but an event can consist of more than one outcome.

4. What are equally likely events? Equally likely events have the same probability of occurring.

5. What is the range of the values of the probability of an event? The range of values is 0 to 1 inclusive.

6. When an event is certain to occur, what is its probability? 1

7. If an event cannot happen, what value is assigned to its probability? 0

8. What is the sum of the probabilities of all the outcomes in a sample space? 1

9. If the probability that it will rain tomorrow is 0.20, what is the probability that it won't rain tomorrow? Would you recommend taking an umbrella? 0.80 Since the probability that it won't rain is 80%, you could leave your umbrella at home and be fairly safe.

10. A probability experiment is conducted. Which of these cannot be considered a probability outcome?

   a. $\frac{2}{3}$ *d.* 1.65   *g.* 1

   b. 0.63   *e.* −0.44   *h.* 125%

   c. −$\frac{3}{5}$ *f.* 0   *i.* 24%

11. Classify each statement as an example of classical probability, empirical probability, or subjective probability.

   a. The probability that a person will watch the 6 o'clock evening news is 0.15. Empirical

   b. The probability of winning at a Chuck-a-Luck game is $\frac{5}{36}$. Classical

   c. The probability that a bus will be in an accident on a specific run is about 6%. Empirical

   d. The probability of getting a royal flush when five cards are selected at random is $\frac{1}{649,740}$. Classical

   e. The probability that a student will get a C or better in a statistics course is about 70%. Empirical

   f. The probability that a new fast-food restaurant will be a success in Chicago is 35%. Empirical

   g. The probability that interest rates will rise in the next 6 months is 0.50. Subjective

12. (ans) **Rolling a Die** If a die is rolled one time, find these probabilities.

   a. Getting a 2 $\frac{1}{6}$

   b. Getting a number greater than 6 0

   c. Getting an odd number $\frac{1}{2}$

   d. Getting a 4 or an odd number $\frac{2}{3}$

   e. Getting a number less than 7 1

   f. Getting a number greater than or equal to 3 $\frac{2}{3}$

   g. Getting a number greater than 2 and an even number $\frac{1}{3}$

13. **Rolling Two Dice** If two dice are rolled one time, find the probability of getting these results.

   a. A sum of 9 $\frac{1}{9}$

   b. A sum of 7 or 11 $\frac{2}{9}$

   c. Doubles $\frac{1}{6}$

   d. A sum less than 9 $\frac{13}{18}$

   e. A sum greater than or equal to 10 $\frac{1}{6}$

14. (ans) **Drawing a Card** If one card is drawn from a deck, find the probability of getting these results.

   a. A queen $\frac{1}{13}$

   b. A club $\frac{1}{4}$

   c. A queen of clubs $\frac{1}{52}$

   d. A 3 or an 8 $\frac{2}{13}$

   e. A 6 or a spade $\frac{4}{13}$

   f. A 6 and a spade $\frac{1}{52}$

   g. A black king $\frac{1}{26}$

   h. A red card and a 7 $\frac{1}{26}$

   i. A diamond or a heart $\frac{1}{2}$

   j. A black card $\frac{1}{2}$

15. **Shopping Mall Promotion** A shopping mall has set up a promotion as follows. With any mall purchase of $50 or more, the customer gets to spin the wheel shown here. If a number 1 comes up, the customer wins $10. If the number 2 comes up, the customer wins $5; and if the number 3 or 4 comes up, the customer wins a discount coupon. Find the following probabilities.

   a. The customer wins $10. 0.1

   b. The customer wins money. 0.2

   c. The customer wins a coupon. 0.8

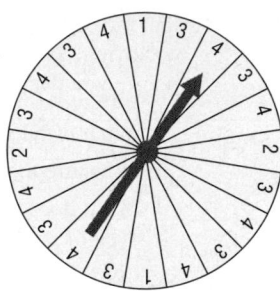

16. **Selecting a State** Choose one of the 50 states at random.

   a. What is the probability that it begins with M? $\frac{4}{25}$

   b. What is the probability that it doesn't begin with a vowel? $\frac{19}{25}$

17. **Human Blood Types** Human blood is grouped into four types. The percentages of Americans with each type are listed below.

   O 43%   A 40%   B 12%   AB 5%

   Choose one American at random. Find the probability that this person

   a. Has type O blood 0.43

   b. Has type A or B 0.52

   c. Does not have type O or A 0.17

   Source: www.infoplease.com

18. **Gender of College Students** In 2004, 57.2% of all enrolled college students were female. Choose one enrolled student at random. What is the probability that the student was male? 0.428

   Source: www.nces.ed.gov

19. Prime Numbers A prime number is a number that is evenly divisible only by 1 and itself. The prime numbers less than 100 are listed below.

2   3   5   7   11   13   17   19   23   29   31
37   41   43   47   53   59   61   67   71   73   79
83   89   97

Choose one of these numbers at random. Find the probability that

a.   The number is even 0.04

b.   The sum of the number's digits is even 0.52

c.   The number is greater than 50 0.4

20. Rural Speed Limits Rural speed limits for all 50 states are indicated below.

60 mph	65 mph	70 mph	75 mph
1 (HI)	18	18	13

Choose one state at random. Find the probability that its speed limit is

a.   60 or 70 miles per hour 0.38

b.   Greater than 65 miles per hour 0.62

c.   70 miles per hour or less 0.74

Source: *World Almanac.*

21. Gender of Children A couple has three children. Find each probability.

a.   All boys $\frac{1}{8}$

b.   All girls or all boys $\frac{1}{4}$

c.   Exactly two boys or two girls $\frac{3}{4}$

d.   At least one child of each gender $\frac{3}{4}$

22. Craps Game In the game of craps using two dice, a person wins on the first roll if a 7 or an 11 is rolled. Find the probability of winning on the first roll. $\frac{2}{9}$

23. Craps Game In a game of craps, a player loses on the roll if a 2, 3, or 12 is tossed on the first roll. Find the probability of losing on the first roll. $\frac{1}{9}$

24. Computers in Elementary Schools Elementary and secondary schools were classified by the number of computers they had. Choose one of these schools at random.

Computers	1–10	11–20	21–50	51–100	100+
Schools	3170	4590	16,741	23,753	34,803

Choose one school at random. Find the probability that it has

a.   50 or fewer computers 0.295

b.   More than 100 computers 0.419

c.   No more than 20 computers 0.093

Source: *World Almanac.*

25. College Debt The following information shows the amount of debt students who graduated from college incur.

$1 to $5000	$5001 to $20,000	$20,001 to $50,000	$50,000+
27%	40%	19%	14%

If a person who graduates has some debt, find the probability that

a.   It is less than $5001 27%

b.   It is more than $20,000 33%

c.   It is between $1 and $20,000 67%

d.   It is more than $50,000 14%

Source: *USA Today.*

26. Gasoline Mileage for Autos and Trucks Of the top 10 cars and trucks based on gas mileage, 4 are Hondas, 3 are Toyotas, and 3 are Volkswagens. Choose one at random. Find the probability that it is

a.   Japanese 0.7

b.   Japanese or German 1

c.   Not foreign 0

Source: www.autobytel.com

27. Large Monetary Bills in Circulation There are 1,765,000 five thousand dollar bills in circulation and 3,460,000 ten thousand dollar bills in circulation. Choose one bill at random (wouldn't that be nice!). What is the probability that it is a ten thousand dollar bill? 0.662

Source: *World Almanac.*

28. Sources of Energy Uses in the United States A breakdown of the sources of energy used in the United States is shown below. Choose one energy source at random. Find the probability that it is

a.   Not oil 0.61

b.   Natural gas or oil 0.63

c.   Nuclear 0.08

Oil 39%	Natural gas 24%	Coal 23%
Nuclear 8%	Hydropower 3%	Other 3%

Source: www.infoplease.com

29. Rolling Dice Roll two dice and multiply the numbers.

a.   Write out the sample space.

b.   What is the probability that the product is a multiple of 6? $\frac{5}{12}$

c.   What is the probability that the product is less than 10? $\frac{17}{36}$

30. Federal Government Revenue The source of federal government revenue for a specific year is

50% from individual income taxes

32% from social insurance payroll taxes

10% from corporate income taxes

3% from excise taxes

5% other

If a revenue source is selected at random, what is the probability that it comes from individual or corporate income taxes? 0.6

Source: *New York Times Almanac.*

31. Selecting a Bill A box contains a $1 bill, a $5 bill, a $10 bill, and a $20 bill. A bill is selected at random, and it is not replaced; then a second bill is selected at random. Draw a tree diagram and determine the sample space.

32. Tossing Coins Draw a tree diagram and determine the sample space for tossing four coins.

33. Selecting Numbered Balls Four balls numbered 1 through 4 are placed in a box. A ball is selected at random, and its number is noted; then it is replaced. A second ball is selected at random, and its number

is noted. Draw a tree diagram and determine the sample space.

34. Family Dinner Combinations A family special at a neighborhood restaurant offers dinner for four for $39.99. There are 3 appetizers available, 4 entrees, and 3 desserts from which to choose. The special includes one of each. Represent the possible dinner combinations with a tree diagram.

35. Required First-Year College Courses First-year students at a particular college must take one English class, one class in mathematics, a first-year seminar, and an elective. There are 2 English classes to choose from, 3 mathematics classes, 5 electives, and everyone takes the same first-year seminar. Represent the possible schedules, using a tree diagram.

36. Tossing a Coin and Rolling a Die A coin is tossed; if it falls heads up, it is tossed again. If it falls tails up, a die is rolled. Draw a tree diagram and determine the outcomes.

Extending the Concepts

37. Distribution of CEO Ages The distribution of ages of CEOs is as follows:

Age	Frequency
21–30	1
31–40	8
41–50	27
51–60	29
61–70	24
71–up	11

Source: Information based on *USA TODAY* Snapshot.

If a CEO is selected at random, find the probability that his or her age is

a. Between 31 and 40 0.08

b. Under 31 0.01

c. Over 30 and under 51 0.35

d. Under 31 or over 60 0.36

38. Tossing a Coin A person flipped a coin 100 times and obtained 73 heads. Can the person conclude that the coin was unbalanced? Probably

39. Medical Treatment A medical doctor stated that with a certain treatment, a patient has a 50% chance of recovering without surgery. That is, "Either he will get well or he won't get well." Comment on this statement. The statement is probably not based on empirical probability, and is probably not true.

40. Wheel Spinner The wheel spinner shown here is spun twice. Find the sample space, and then determine the probability of the following events.

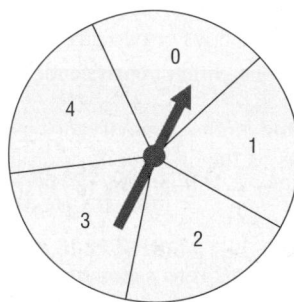

a. An odd number on the first spin and an even number on the second spin (*Note:* 0 is considered even.) $\frac{6}{25}$

b. A sum greater than 4 $\frac{2}{5}$

c. Even numbers on both spins $\frac{9}{25}$

d. A sum that is odd $\frac{12}{25}$

e. The same number on both spins $\frac{1}{5}$

41. Tossing Coins Toss three coins 128 times and record the number of heads (0, 1, 2, or 3); then record your results with the theoretical probabilities. Compute the empirical probabilities of each. Answers will vary.

42. Tossing Coins Toss two coins 100 times and record the number of heads (0, 1, 2). Compute the probabilities of each outcome, and compare these probabilities with the theoretical results. Approximately $\frac{1}{4}, \frac{1}{2}$, and $\frac{1}{4}$, respectively

43. Odds Odds are used in gambling games to make them fair. For example, if you rolled a die and won every time you rolled a 6, then you would win on average once every 6 times. So that the game is fair, the odds of 5 to 1 are given. This means that if you bet $1 and won, you could win $5. On average, you would win $5 once in 6 rolls and lose $1 on the other 5 rolls—hence the term *fair game.*

In most gambling games, the odds given are not fair. For example, if the odds of winning are really 20 to 1, the house might offer 15 to 1 in order to make a profit.

Odds can be expressed as a fraction or as a ratio, such as $\frac{5}{1}$, 5:1, or 5 to 1. Odds are computed in favor of the event or against the event. The formulas for odds are

$$\text{Odds in favor} = \frac{P(E)}{1 - P(E)}$$

$$\text{Odds against} = \frac{P(\overline{E})}{1 - P(\overline{E})}$$

In the die example,

$$\text{Odds in favor of a 6} = \frac{\frac{1}{6}}{\frac{5}{6}} = \frac{1}{5} \text{ or } 1:5$$

$$\text{Odds against a 6} = \frac{\frac{5}{6}}{\frac{1}{6}} = \frac{5}{1} \text{ or } 5:1$$

Find the odds in favor of and against each event.

a. Rolling a die and getting a 2 1:5, 5:1
b. Rolling a die and getting an even number 1:1, 1:1
c. Drawing a card from a deck and getting a spade 1:3, 3:1
d. Drawing a card and getting a red card 1:1, 1:1
e. Drawing a card and getting a queen 1:12, 12:1
f. Tossing two coins and getting two tails 1:3, 3:1
g. Tossing two coins and getting one tail 1:1, 1:1

The Addition Rules for Probability

Objective 2

Find the probability of compound events, using the addition rules.

Many problems involve finding the probability of two or more events. For example, at a large political gathering, you might wish to know, for a person selected at random, the probability that the person is a female or is a Republican. In this case, there are three possibilities to consider:

1. The person is a female.
2. The person is a Republican.
3. The person is both a female and a Republican.

Consider another example. At the same gathering there are Republicans, Democrats, and Independents. If a person is selected at random, what is the probability that the person is a Democrat or an Independent? In this case, there are only two possibilities:

1. The person is a Democrat.
2. The person is an Independent.

The difference between the two examples is that in the first case, the person selected can be a female and a Republican at the same time. In the second case, the person selected cannot be both a Democrat and an Independent at the same time. In the second case, the two events are said to be *mutually exclusive;* in the first case, they are not mutually exclusive.

Two events are **mutually exclusive events** if they cannot occur at the same time (i.e., they have no outcomes in common).

In another situation, the events of getting a 4 and getting a 6 when a single card is drawn from a deck are mutually exclusive events, since a single card cannot be both a 4 and a 6. On the other hand, the events of getting a 4 and getting a heart on a single draw are not mutually exclusive, since you can select the 4 of hearts when drawing a single card from an ordinary deck.

| Example 4–15 | **Rolling a Die** |

Determine which events are mutually exclusive and which are not, when a single die is rolled.

    a. Getting an odd number and getting an even number

    b. Getting a 3 and getting an odd number

    c. Getting an odd number and getting a number less than 4

    d. Getting a number greater than 4 and getting a number less than 4

Solution

    a. The events are mutually exclusive, since the first event can be 1, 3, or 5 and the second event can be 2, 4, or 6.

    b. The events are not mutually exclusive, since the first event is a 3 and the second can be 1, 3, or 5. Hence, 3 is contained in both events.

    c. The events are not mutually exclusive, since the first event can be 1, 3, or 5 and the second can be 1, 2, or 3. Hence, 1 and 3 are contained in both events.

    d. The events are mutually exclusive, since the first event can be 5 or 6 and the second event can be 1, 2, or 3.

| Example 4–16 | **Drawing a Card** |

Determine which events are mutually exclusive and which are not when a single card is drawn from a deck.

    a. Getting a 7 and getting a jack

    b. Getting a club and getting a king

    c. Getting a face card and getting an ace

    d. Getting a face card and getting a spade

Solution

Only the events in parts *a* and *c* are mutually exclusive.

     The probability of two or more events can be determined by the *addition rules.* The first addition rule is used when the events are mutually exclusive.

Addition Rule 1

When two events A and B are mutually exclusive, the probability that A or B will occur is

$$P(A \text{ or } B) = P(A) + P(B)$$

| Example 4–17 | **Coffee Shop Selection** |

A city has 9 coffee shops: 3 Starbuck's, 2 Caribou Coffees, and 4 Crazy Mocho Coffees. If a person selects one shop at random to buy a cup of coffee, find the probability that it is either a Starbuck's or Crazy Mocho Coffees.

Solution

Since there are 3 Starbuck's and 4 Crazy Mochos, and a total of 9 coffee shops,
$P(\text{Starbuck's or Crazy Mocho}) = P(\text{Starbuck's}) + P(\text{Crazy Mocho}) = \frac{3}{9} + \frac{4}{9} = \frac{7}{9}$.
The events are mutually exclusive.

Example 4–18

Research and Development Employees

The corporate research and development centers for three local companies have the following number of employees:

U.S. Steel	110
Alcoa	750
Bayer Material Science	250

If a research employee is selected at random, find the probability that the employee is employed by U.S. Steel or Alcoa.

Source: *Pittsburgh Tribune Review.*

Solution

$$P(\text{U.S. Steel or Alcoa}) = P(\text{U.S. Steel}) + P(\text{Alcoa})$$

$$= \frac{110}{1110} + \frac{750}{1110} = \frac{860}{1110} = \frac{86}{111}$$

Example 4–19

Selecting a Day of the Week

A day of the week is selected at random. Find the probability that it is a weekend day.

Solution

$$P(\text{Saturday or Sunday}) = P(\text{Saturday}) + P(\text{Sunday}) = \frac{1}{7} + \frac{1}{7} = \frac{2}{7}$$

When two events are not mutually exclusive, we must subtract one of the two probabilities of the outcomes that are common to both events, since they have been counted twice. This technique is illustrated in Example 4–20.

Example 4–20

Drawing a Card

A single card is drawn at random from an ordinary deck of cards. Find the probability that it is either an ace or a black card.

Solution

Since there are 4 aces and 26 black cards (13 spades and 13 clubs), 2 of the aces are black cards, namely, the ace of spades and the ace of clubs. Hence the probabilities of the two outcomes must be subtracted since they have been counted twice.

$$P(\text{ace or black card}) = P(\text{ace}) + P(\text{black card}) - P(\text{black aces}) = \frac{4}{52} + \frac{26}{52} - \frac{2}{52} = \frac{28}{52} = \frac{7}{13}$$

When events are not mutually exclusive, addition rule 2 can be used to find the probability of the events.

Addition Rule 2

If A and B are *not* mutually exclusive, then

$$P(A \text{ or } B) = P(A) + P(B) - P(A \text{ and } B)$$

Note: This rule can also be used when the events are mutually exclusive, since $P(A$ and $B)$ will always equal 0. However, it is important to make a distinction between the two situations.

Example 4–21   Selecting a Medical Staff Person

In a hospital unit there are 8 nurses and 5 physicians; 7 nurses and 3 physicians are females. If a staff person is selected, find the probability that the subject is a nurse or a male.

Solution

The sample space is shown here.

Staff	Females	Males	Total
Nurses	7	1	8
Physicians	3	2	5
Total	10	3	13

The probability is

$$P(\text{nurse or male}) = P(\text{nurse}) + P(\text{male}) - P(\text{male nurse})$$
$$= \tfrac{8}{13} + \tfrac{3}{13} - \tfrac{1}{13} = \tfrac{10}{13}$$

Example 4–22   Driving While Intoxicated

On New Year's Eve, the probability of a person driving while intoxicated is 0.32, the probability of a person having a driving accident is 0.09, and the probability of a person having a driving accident while intoxicated is 0.06. What is the probability of a person driving while intoxicated or having a driving accident?

Solution

$$P(\text{intoxicated or accident}) = P(\text{intoxicated}) + P(\text{accident})$$
$$- P(\text{intoxicated and accident})$$
$$= 0.32 + 0.09 - 0.06 = 0.35$$

In summary, then, when the two events are mutually exclusive, use addition rule 1. When the events are not mutually exclusive, use addition rule 2.

The probability rules can be extended to three or more events. For three mutually exclusive events A, B, and C,

$$P(A \text{ or } B \text{ or } C) = P(A) + P(B) + P(C)$$

Figure 4–5

Venn Diagrams for the Addition Rules

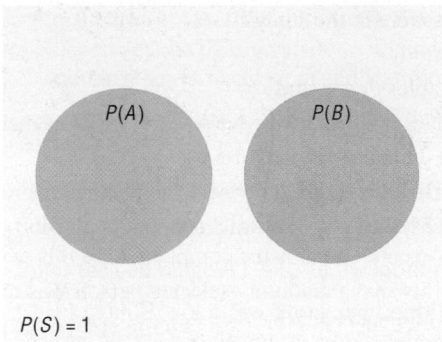

$P(S) = 1$

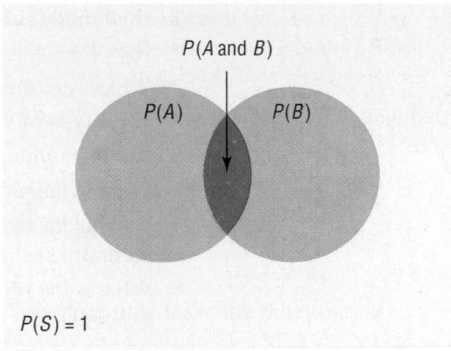

$P(A \text{ and } B)$

$P(S) = 1$

(a) Mutually exclusive events
$P(A \text{ or } B) = P(A) + P(B)$

(b) Nonmutually exclusive events
$P(A \text{ or } B) = P(A) + P(B) - P(A \text{ and } B)$

For three events that are *not* mutually exclusive,

$$P(A \text{ or } B \text{ or } C) = P(A) + P(B) + P(C) - P(A \text{ and } B) - P(A \text{ and } C) - P(B \text{ and } C) + P(A \text{ and } B \text{ and } C)$$

See Exercises 23 and 24 in this section.

Figure 4–5(a) shows a Venn diagram that represents two mutually exclusive events A and B. In this case, $P(A \text{ or } B) = P(A) + P(B)$, since these events are mutually exclusive and do not overlap. In other words, the probability of occurrence of event A or event B is the sum of the areas of the two circles.

Figure 4–5(b) represents the probability of two events that are *not* mutually exclusive. In this case, $P(A \text{ or } B) = P(A) + P(B) - P(A \text{ and } B)$. The area in the intersection or overlapping part of both circles corresponds to $P(A \text{ and } B)$; and when the area of circle A is added to the area of circle B, the overlapping part is counted twice. It must therefore be subtracted once to get the correct area or probability.

Note: Venn diagrams were developed by mathematician John Venn (1834–1923) and are used in set theory and symbolic logic. They have been adapted to probability theory also. In set theory, the symbol $\cup$ represents the *union* of two sets, and $A \cup B$ corresponds to A or B. The symbol $\cap$ represents the *intersection* of two sets, and $A \cap B$ corresponds to A and B. Venn diagrams show only a general picture of the probability rules and do not portray all situations, such as $P(A) = 0$, accurately.

Applying the Concepts **4–2**

Which Pain Reliever Is Best?

Assume that following an injury you received from playing your favorite sport, you obtain and read information on new pain medications. In that information you read of a study that was conducted to test the side effects of two new pain medications. Use the following table to answer the questions and decide which, if any, of the two new pain medications you will use.

Side effect	Number of side effects in 12-week clinical trial		
	Placebo $n = 192$	Drug A $n = 186$	Drug B $n = 188$
Upper respiratory congestion	10	32	19
Sinus headache	11	25	32
Stomach ache	2	46	12
Neurological headache	34	55	72
Cough	22	18	31
Lower respiratory congestion	2	5	1

1. How many subjects were in the study?
2. How long was the study?
3. What were the variables under study?
4. What type of variables are they, and what level of measurement are they on?
5. Are the numbers in the table exact figures?
6. What is the probability that a randomly selected person was receiving a placebo?
7. What is the probability that a person was receiving a placebo or drug A? Are these mutually exclusive events? What is the complement to this event?
8. What is the probability that a randomly selected person was receiving a placebo or experienced a neurological headache?
9. What is the probability that a randomly selected person was not receiving a placebo or experienced a sinus headache?

See page 249 for the answers.

Exercises 4–2

1. **Define mutually exclusive events, and give an example of two events that are mutually exclusive and two events that are not mutually exclusive.** Two events are mutually exclusive if they cannot occur at the same time (i.e., they have no outcomes in common). Examples will vary.

2. **Determine whether these events are mutually exclusive.**

   a. Roll a die: Get an even number, and get a number less than 3. No

   b. Roll a die: Get a prime number (2, 3, 5), and get an odd number. No

   c. Roll a die: Get a number greater than 3, and get a number less than 3. Yes

   d. Select a student in your class: The student has blond hair, and the student has blue eyes. No

   e. Select a student in your college: The student is a sophomore, and the student is a business major. No

   f. Select any course: It is a calculus course, and it is an English course. Yes

   g. Select a registered voter: The voter is a Republican, and the voter is a Democrat. Yes

3. **College Degrees Awarded** The table below represents the college degrees awarded in a recent academic year by gender.

	Bachelor's	Master's	Doctorate
Men	573,079	211,381	24,341
Women	775,424	301,264	21,683

   Choose a degree at random. Find the probability that it is

   a. A bachelor's degree 0.707

   b. A doctorate or a degree awarded to a woman 0.589

   c. A doctorate awarded to a woman 0.011

   d. Not a master's degree 0.731

   Source: www.nces.ed.gov

4. **Selecting a Fish** In a fish tank, there are 24 goldfish, 2 angel fish, and 5 guppies. If a fish is selected at random, find the probability that it is a goldfish or an angel fish. $\frac{26}{31}$

5. **Selecting an Instructor** At a convention there are 7 mathematics instructors, 5 computer science instructors, 3 statistics instructors, and 4 science instructors. If an instructor is selected, find the probability of getting a science instructor or a math instructor. $\frac{11}{19}$

6. **Selecting a Movie** A media rental store rented the following number of movie titles in each of these categories: 170 horror, 230 drama, 120 mystery, 310 romance, and 150 comedy. If a person selects a movie to rent, find the probability that it is a romance or a comedy. Is this event likely or unlikely to occur? Explain your answer. $\frac{23}{49}$; the probability of the event is slightly less than 0.5, which makes it about equally likely to occur or not to occur.

7. **Hospital Staff** On a hospital staff, there are 4 dermatologists, 7 surgeons, 5 general practitioners, 3 psychiatrists, and 3 orthopedic specialists. If a doctor is selected at random, find the probability that the doctor is

   a. A psychiatrist, surgeon, or dermatologist $\frac{7}{11}$

   b. A general practitioner or surgeon $\frac{6}{11}$

   c. An orthopedic specialist, a surgeon, or a dermatologist $\frac{7}{11}$

   d. A surgeon or dermatologist $\frac{1}{2}$

8. **Tourist Destinations** The probability that a given tourist goes to the amusement park is 0.47, and the probability that she goes to the water park is 0.58. If the probability that she goes to either the water park or the amusement park is 0.95, what is the probability that she visits both of the parks on vacation? 0.10 or 10%

9. **Sports Participation** At a particular school with 200 male students, 58 play football, 40 play basketball, and 8 play both. What is the probability that a randomly selected male student plays neither sport? 0.55

10. **Selecting a Card** A single card is drawn from a deck. Find the probability of selecting the following.

    a. A 4 or a diamond $\frac{4}{13}$

    b. A club or a diamond $\frac{1}{2}$

    c. A jack or a black card $\frac{7}{13}$

11. **Selecting a Student** In a statistics class there are 18 juniors and 10 seniors; 6 of the seniors are females, and 12 of the juniors are males. If a student is selected at random, find the probability of selecting the following.

    a. A junior or a female $\frac{6}{7}$

    b. A senior or a female $\frac{4}{7}$

    c. A junior or a senior 1

12. **Selecting a Book** At a used-book sale, 100 books are adult books and 160 are children's books. Of the adult books, 70 are nonfiction while 60 of the children's books are nonfiction. If a book is selected at random, find the probability that it is

    a. Fiction 0.5

    b. Not a children's nonfiction book 0.7692

    c. An adult book or a children's nonfiction book 0.6154

13. **Young Adult Residences** According to the Bureau of the Census, the following statistics describe the number (in thousands) of young adults living at home or in a dormitory in the year 2004.

	Ages 18–24	Ages 25–34
Male	7922	2534
Female	5779	995

Source: *World Almanac.*

Choose one student at random. Find the probability that the student is

    a. A female student aged 25–34 0.058

    b. Male or aged 18–24 0.942

    c. Under 25 years of age and not male 0.335

14. **Endangered Species** The chart below shows the numbers of endangered and threatened species both here in the United States and abroad.

	Endangered		Threatened	
	United States	Foreign	United States	Foreign
Mammals	68	251	10	20
Birds	77	175	13	6
Reptiles	14	64	22	16
Amphibians	11	8	10	1

Source: www.infoplease.com

Choose one species at random. Find the probability that it is

    a. Threatened and in the United States 0.072

    b. An endangered foreign bird 0.229

    c. A mammal or a threatened foreign species 0.4856

15. **Multiple Births** The number of multiple births in the United States for a recent year indicated that there were 128,665 sets of twins, 7110 sets of triplets, 468 sets of quadruplets, and 85 sets of quintuplets. Choose one set of siblings at random. Find the probability that it

    a. Represented more than two babies 0.056

    b. Represented quads or quints 0.004

    c. Now choose one baby from these multiple births. What is the probability that the baby was a triplet? 0.076

16. **Licensed Drivers in the United States** In a recent year there were the following numbers (in thousands) of licensed drivers in the United States.

	Male	Female
Age 19 and under	4746	4517
Age 20	1625	1553
Age 21	1679	1627

Source: *World Almanac.*

Choose one driver at random. Find the probability that the driver is

    a. Male and 19 or under 0.301

    b. Age 20 or female 0.592

    c. At least 20 years old 0.412

17. **Student Survey** In a recent survey, the following data were obtained in response to the question, "If the number of summer classes were increased, would you be more likely to enroll in one or more of them?"

Class	Yes	No	No opinion
Freshmen	15	8	5
Sophomores	24	4	2

If a student is selected at random, find the probability that the student

    a. Has no opinion $\frac{7}{58}$

    b. Is a freshman or is against the issue $\frac{16}{29}$

    c. Is a sophomore and favors the issue $\frac{12}{29}$

18. **Mail Delivery** A local postal carrier distributes first-class letters, advertisements, and magazines. For a certain day, she distributed the following numbers of each type of item.

Delivered to	First-class letters	Ads	Magazines
Home	325	406	203
Business	732	1021	97

If an item of mail is selected at random, find these probabilities.

a. The item went to a home. $\frac{467}{1392}$

b. The item was an ad, or it went to a business. $\frac{47}{58}$

c. The item was a first-class letter, or it went to a home. $\frac{833}{1392}$

19. **Medical Tests on Emergency Patients** The frequency distribution shown here illustrates the number of medical tests conducted on 30 randomly selected emergency patients.

Number of tests performed	Number of patients
0	12
1	8
2	2
3	3
4 or more	5

If a patient is selected at random, find these probabilities.

a. The patient has had exactly 2 tests done. $\frac{1}{15}$

b. The patient has had at least 2 tests done. $\frac{1}{3}$

c. The patient has had at most 3 tests done. $\frac{5}{6}$

d. The patient has had 3 or fewer tests done. $\frac{5}{6}$

e. The patient has had 1 or 2 tests done. $\frac{1}{3}$

20. A social organization of 32 members sold college sweatshirts as a fundraiser. The results of their sale are shown below.

No. of sweatshirts	No. of students
0	2
1–5	13
6–10	8
11–15	4
16–20	4
20+	1

Choose one student at random. Find the probability that the student sold

a. More than 10 sweatshirts 0.2813

b. At least one sweatshirt 0.9375

c. 1–5 or more than 15 sweatshirts 0.5625

21. **Door-to-Door Sales** A sales representative who visits customers at home finds she sells 0, 1, 2, 3, or 4 items according to the following frequency distribution.

Items sold	Frequency
0	8
1	10
2	3
3	2
4	1

Find the probability that she sells the following.

a. Exactly 1 item $\frac{5}{12}$

b. More than 2 items $\frac{1}{8}$

c. At least 1 item $\frac{2}{3}$

d. At most 3 items $\frac{23}{24}$

22. **Medical Patients** A recent study of 300 patients found that of 100 alcoholic patients, 87 had elevated cholesterol levels, and of 200 nonalcoholic patients, 43 had elevated cholesterol levels. If a patient is selected at random, find the probability that the patient is the following.

a. An alcoholic with elevated cholesterol level $\frac{29}{100}$

b. A nonalcoholic $\frac{2}{3}$

c. A nonalcoholic with nonelevated cholesterol level $\frac{157}{300}$

23. **Selecting a Card** If one card is drawn from an ordinary deck of cards, find the probability of getting the following.

a. A king or a queen or a jack $\frac{3}{13}$

b. A club or a heart or a spade $\frac{3}{4}$

c. A king or a queen or a diamond $\frac{19}{52}$

d. An ace or a diamond or a heart $\frac{7}{13}$

e. A 9 or a 10 or a spade or a club $\frac{15}{26}$

24. **Rolling Die** Two dice are rolled. Find the probability of getting

a. A sum of 8, 9, or 10 $\frac{1}{3}$

b. Doubles or a sum of 7 $\frac{1}{3}$

c. A sum greater than 9 or less than 4 $\frac{1}{4}$

d. Based on the answers to a, b, and c, which is least likely to occur? Choice c is least likely to occur.

25. **Corn Products** U.S. growers harvested 11 billion bushels of corn in 2005. About 1.9 billion bushels were exported, and 1.6 billion bushels were used for ethanol. Choose one bushel of corn at random. What is the probability that it was used either for export or for ethanol? 0.318

Source: www.census.gov

26. **Rolling Dice** Three dice are rolled. Find the probability of getting

a. Triples $\frac{1}{36}$ b. A sum of 5 $\frac{1}{36}$

Extending the Concepts

27. Purchasing a Pizza The probability that a customer selects a pizza with mushrooms or pepperoni is 0.55, and the probability that the customer selects only mushrooms is 0.32. If the probability that he or she selects only pepperoni is 0.17, find the probability of the customer selecting both items. 0.06

28. Building a New Home In building new homes, a contractor finds that the probability of a home buyer selecting a two-car garage is 0.70 and of selecting a one-car garage is 0.20. Find the probability that the buyer will select no garage. The builder does not build houses with three-car or more garages. 0.10

29. In Exercise 28, find the probability that the buyer will not want a two-car garage. 0.30

30. Suppose that $P(A) = 0.42$, $P(B) = 0.38$, and $P(A \cup B) = 0.70$. Are A and B mutually exclusive? Explain. No. $P(A \cap B) \neq 0$

"I know you haven't had an accident in thirteen years. We're raising your rates because you're about due one."

© Bob Schochet. King Features Syndicate.

Technology *Step by Step*

MINITAB
Step by Step

Calculate Relative Frequency Probabilities

The random variable X represents the number of days patients stayed in the hospital from Example 4–14.

1. In C1 of a worksheet, type in the values of X. Name the column **X.**
2. In C2 enter the frequencies. Name the column **f.**
3. To calculate the relative frequencies and store them in a new column named Px:
   a) Select **Calc>Calculator.**
   b) Type **Px** in the box for Store result in variable:.
   c) Click in the Expression box, then double-click C2 f.
   d) Type or click the division operator.
   e) Scroll down the function list to Sum, then click [Select].
   f) Double-click C2 f to select it.
   g) Click [OK].

The dialog box and completed worksheet are shown.

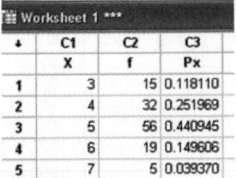

	C1	C2	C3
	X	f	Px
1	3	15	0.118110
2	4	32	0.251969
3	5	56	0.440945
4	6	19	0.149606
5	7	5	0.039370

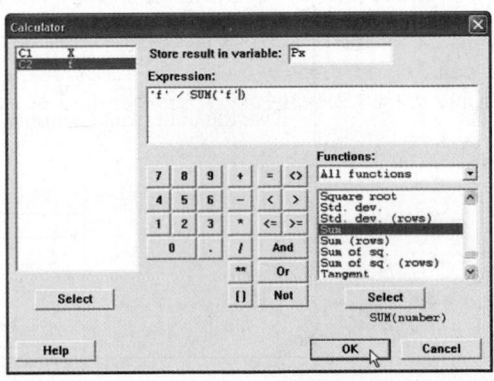

If the original data, rather than the table, are in a worksheet, use **Stat>Tables>Tally** to make the tables with percents (Section 2–1).

MINITAB can also make a two-way classification table.

Construct a Contingency Table

1. Select **File>Open Worksheet** to open the Databank.mtw file.

2. Select **Stat>Tables>Crosstabulation ...**

   a) Double-click C4 SMOKING STATUS to select it For rows:.

   b) Select C11 GENDER for the For Columns: Field.

   c) Click on option for Counts and then [OK].

The session window and completed dialog box are shown.

Tabulated statistics: SMOKING STATUS, GENDER

Rows: SMOKING STATUS   Columns: GENDER

	F	M	All
0	25	22	47
1	18	19	37
2	7	9	16
All	50	50	100

Cell Contents: Count

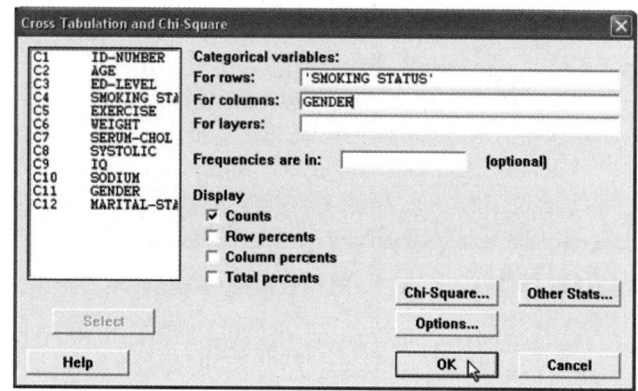

In this sample of 100 there are 25 females who do not smoke compared to 22 men. Sixteen individuals smoke 1 pack or more per day.

TI-83 Plus or TI-84 Plus
Step by Step

To construct a relative frequency table:

1. Enter the data values in L_1 and the frequencies in L_2.

2. Move the cursor to the top of the L_3 column so that L_3 is highlighted.

3. Type L_2 divided by the sample size, then press **ENTER.**

Use the data from Example 4–14.

L1	L2	L3	3
3	15	------	
4	32		
5	56		
6	19		
7	5		
------	------		

L3 =L2/127▪

L1	L2	L3	3
3	15	.11811	
4	32	.25197	
5	56	.44094	
6	19	.14961	
7	5	.03937	
------	------	------	

L3(1)=.1181102362...

Excel
Step by Step

Constructing a Relative Frequency Distribution

Use the data from Example 4–14.

1. In a new worksheet, type the label **DAYS** in cell A1. Beginning in cell A2, type in the data for the variable representing the number of days maternity patients stayed in the hospital.

2. In cell B1, type the label for the frequency, **COUNT**. Beginning in cell B2, type in the frequencies.

3. In cell B7, compute the total frequency by selecting the sum icon Σ from the toolbar and press **Enter.**

4. In cell C1, type a label for the relative frequencies, **Rf**. In cell C2, type **=(B2)/(B7)** and **Enter.** In cell C2, type **=(B3)/(B7)** and **Enter.** Repeat this for each of the remaining frequencies.

5. To find the total relative frequency, select the sum icon Σ from the toolbar and **Enter.** This sum should be 1.

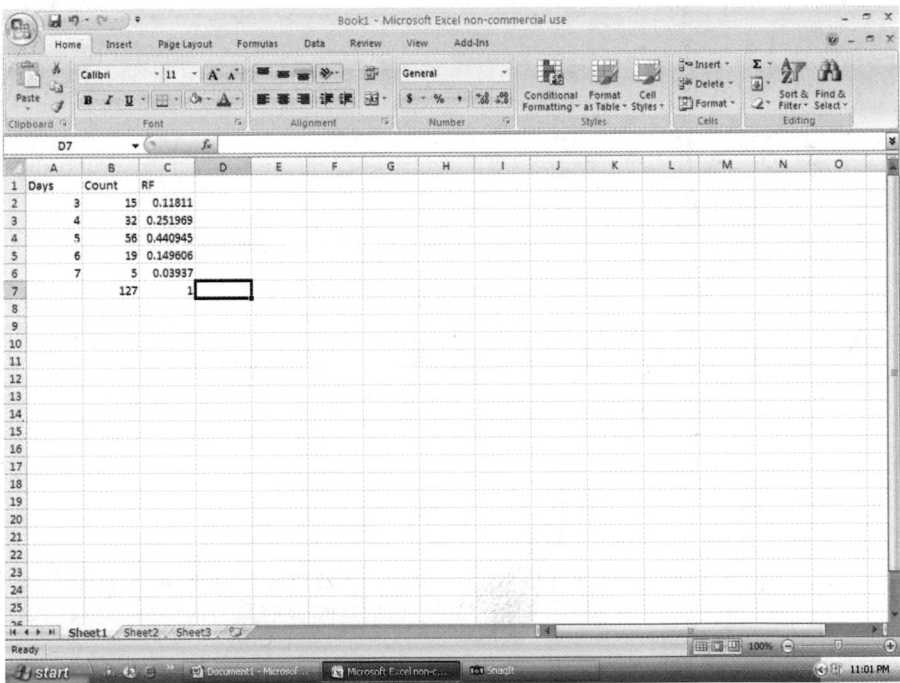

Constructing a Contingency Table
Example XL4–1

For this example, you will need to have the MegaStat Add-In installed on Excel (refer to Chapter 1, Excel Step by Step instructions for instructions on installing MegaStat).

1. Open the Databank.xls file from the CD-ROM that came with your text. To do this:

   Double-click My Computer on the Desktop.

   Double-click the Bluman CD-ROM icon in the CD drive holding the disk.

   Double-click the datasets folder. Then double-click the all_data-sets folder.

   Double-click the bluman_es_data-sets_excel-windows folder. In this folder double-click the Databank.xls file. The Excel program will open automatically once you open this file.

2. Highlight the column labeled SMOKING STATUS to copy these data onto a new Excel worksheet.

3. Click the Microsoft Office Button , select New Blank Workbook, then Create.

4. With cell A1 selected, click the Paste icon on the toolbar to paste the data into the new workbook.

5. Return to the Databank.xls file. Highlight the column labeled Gender. Copy and paste these data into column B of the worksheet containing the SMOKING STATUS data.

6. Type in the categories for SMOKING STATUS, **0, 1,** and **2** into cells C2–C4. In cell D2, type M for male and in cell D3, type F for female.

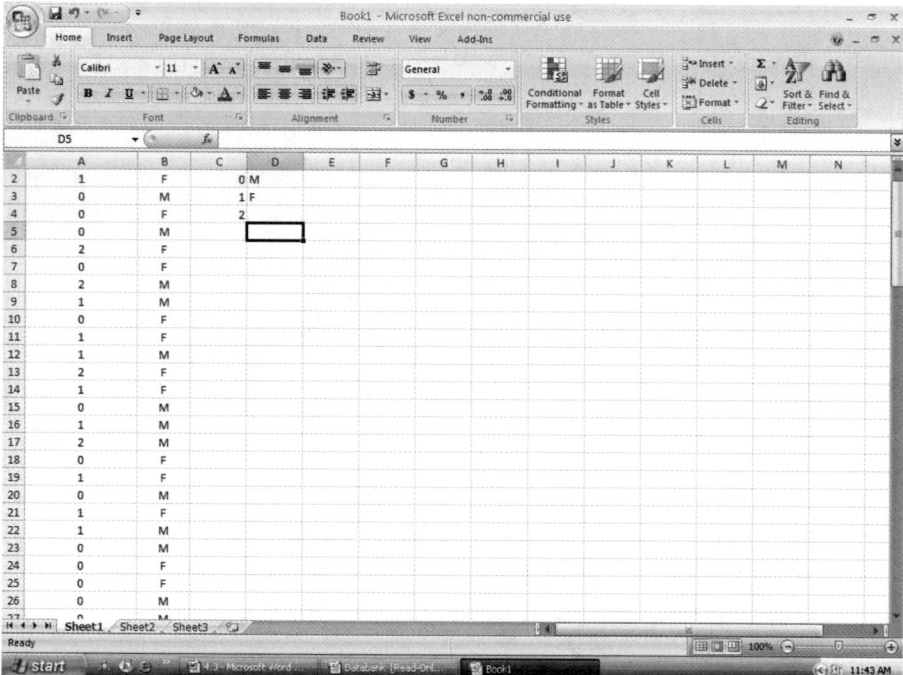

7. On the toolbar, select Add-Ins. Then select MegaStat. *Note:* You may need to open MegaStat from the file MegaStat.xls saved on your computer's hard drive.

8. Select **Chi-Square/Crosstab>Crosstabulation.**

9. In the Row variable Data range box, type A1:A101. In the Row variable Specification range box, type C2:C4. In the Column variable Data range box, type B1:B101. In the Column variable Specification range box, type D2:D3. Remove any checks from the Output Options. Then click [OK].

Crosstabulation

		GENDER		
		M	F	Total
SMOKING STATUS	0	21	25	46
	1	19	18	37
	2	9	8	17
	Total	49	51	100

4–3	# The Multiplication Rules and Conditional Probability

Section 4–2 showed that the addition rules are used to compute probabilities for mutually exclusive and non-mutually exclusive events. This section introduces the multiplication rules.

Objective 3

Find the probability of compound events, using the multiplication rules.

The Multiplication Rules

The *multiplication rules* can be used to find the probability of two or more events that occur in sequence. For example, if you toss a coin and then roll a die, you can find the probability of getting a head on the coin *and* a 4 on the die. These two events are said to be *independent* since the outcome of the first event (tossing a coin) does not affect the probability outcome of the second event (rolling a die).

Two events A and B are **independent events** if the fact that A occurs does not affect the probability of B occurring.

Here are other examples of independent events:

Rolling a die and getting a 6, and then rolling a second die and getting a 3.

Drawing a card from a deck and getting a queen, replacing it, and drawing a second card and getting a queen.

To find the probability of two independent events that occur in sequence, you must find the probability of each event occurring separately and then multiply the answers. For example, if a coin is tossed twice, the probability of getting two heads is $\frac{1}{2} \cdot \frac{1}{2} = \frac{1}{4}$. This result can be verified by looking at the sample space HH, HT, TH, TT. Then $P(\text{HH}) = \frac{1}{4}$.

Multiplication Rule 1

When two events are independent, the probability of both occurring is

$$P(A \text{ and } B) = P(A) \cdot P(B)$$

Example 4–23	Tossing a Coin

A coin is flipped and a die is rolled. Find the probability of getting a head on the coin and a 4 on the die.

Solution

$$P(\text{head and 4}) = P(\text{head}) \cdot P(4) = \frac{1}{2} \cdot \frac{1}{6} = \frac{1}{12}$$

Note that the sample space for the coin is H, T; and for the die it is 1, 2, 3, 4, 5, 6.

The problem in Example 4–23 can also be solved by using the sample space

H1   H2   H3   H4   H5   H6   T1   T2   T3   T4   T5   T6

The solution is $\frac{1}{12}$, since there is only one way to get the head-4 outcome.

Example 4–24

Drawing a Card

A card is drawn from a deck and replaced; then a second card is drawn. Find the probability of getting a queen and then an ace.

Solution

The probability of getting a queen is $\frac{4}{52}$, and since the card is replaced, the probability of getting an ace is $\frac{4}{52}$. Hence, the probability of getting a queen and an ace is

$$P(\text{queen and ace}) = P(\text{queen}) \cdot P(\text{ace}) = \frac{4}{52} \cdot \frac{4}{52} = \frac{16}{2704} = \frac{1}{169}$$

Example 4–25

Selecting a Colored Ball

An urn contains 3 red balls, 2 blue balls, and 5 white balls. A ball is selected and its color noted. Then it is replaced. A second ball is selected and its color noted. Find the probability of each of these.

  a. Selecting 2 blue balls
  b. Selecting 1 blue ball and then 1 white ball
  c. Selecting 1 red ball and then 1 blue ball

Solution

  a. $P(\text{blue and blue}) = P(\text{blue}) \cdot P(\text{blue}) = \frac{2}{10} \cdot \frac{2}{10} = \frac{4}{100} = \frac{1}{25}$

  b. $P(\text{blue and white}) = P(\text{blue}) \cdot P(\text{white}) = \frac{2}{10} \cdot \frac{5}{10} = \frac{10}{100} = \frac{1}{10}$

  c. $P(\text{red and blue}) = P(\text{red}) \cdot P(\text{blue}) = \frac{3}{10} \cdot \frac{2}{10} = \frac{6}{100} = \frac{3}{50}$

Multiplication rule 1 can be extended to three or more independent events by using the formula

$$P(A \text{ and } B \text{ and } C \text{ and } \ldots \text{ and } K) = P(A) \cdot P(B) \cdot P(C) \cdots P(K)$$

When a small sample is selected from a large population and the subjects are not replaced, the probability of the event occurring changes so slightly that for the most part, it is considered to remain the same. Examples 4–26 and 4–27 illustrate this concept.

Example 4–26

Survey on Stress

A Harris poll found that 46% of Americans say they suffer great stress at least once a week. If three people are selected at random, find the probability that all three will say that they suffer great stress at least once a week.

Source: *100% American.*

Solution

Let S denote stress. Then

$$P(S \text{ and } S \text{ and } S) = P(S) \cdot P(S) \cdot P(S)$$
$$= (0.46)(0.46)(0.46) \approx 0.097$$

| Example 4–27 | Male Color Blindness |

Approximately 9% of men have a type of color blindness that prevents them from distinguishing between red and green. If 3 men are selected at random, find the probability that all of them will have this type of red-green color blindness.

Source: *USA TODAY*.

Solution

Let C denote red-green color blindness. Then

$$P(C \text{ and } C \text{ and } C) = P(C) \cdot P(C) \cdot P(C)$$
$$= (0.09)(0.09)(0.09)$$
$$= 0.000729$$

Hence, the rounded probability is 0.0007.

In Examples 4–23 through 4–27, the events were independent of one another, since the occurrence of the first event in no way affected the outcome of the second event. On the other hand, when the occurrence of the first event changes the probability of the occurrence of the second event, the two events are said to be *dependent*. For example, suppose a card is drawn from a deck and *not* replaced, and then a second card is drawn. What is the probability of selecting an ace on the first card and a king on the second card?

Before an answer to the question can be given, you must realize that the events are dependent. The probability of selecting an ace on the first draw is $\frac{4}{52}$. If that card is *not* replaced, the probability of selecting a king on the second card is $\frac{4}{51}$, since there are 4 kings and 51 cards remaining. The outcome of the first draw has affected the outcome of the second draw.

Dependent events are formally defined now.

When the outcome or occurrence of the first event affects the outcome or occurrence of the second event in such a way that the probability is changed, the events are said to be **dependent events**.

Here are some examples of dependent events:

Drawing a card from a deck, not replacing it, and then drawing a second card.

Selecting a ball from an urn, not replacing it, and then selecting a second ball.

Being a lifeguard and getting a suntan.

Having high grades and getting a scholarship.

Parking in a no-parking zone and getting a parking ticket.

To find probabilities when events are dependent, use the multiplication rule with a modification in notation. For the problem just discussed, the probability of getting an ace on the first draw is $\frac{4}{52}$, and the probability of getting a king on the second draw is $\frac{4}{51}$. By the multiplication rule, the probability of both events occurring is

$$\frac{4}{52} \cdot \frac{4}{51} = \frac{16}{2652} = \frac{4}{663}$$

The event of getting a king on the second draw *given* that an ace was drawn the first time is called a *conditional probability*.

The **conditional probability** of an event B in relationship to an event A is the probability that event B occurs after event A has already occurred. The notation for conditional probability is $P(B|A)$. This notation does not mean that B is divided by A; rather, it means the probability that event B occurs given that event A has already occurred. In the card example, $P(B|A)$ is the probability that the second card is a king given that the first card is an ace, and it is equal to $\frac{4}{51}$ since the first card was *not* replaced.

Multiplication Rule 2

When two events are dependent, the probability of both occurring is

$$P(A \text{ and } B) = P(A) \cdot P(B|A)$$

Example 4–28   ## University Crime

At a university in western Pennsylvania, there were 5 burglaries reported in 2003, 16 in 2004, and 32 in 2005. If a researcher wishes to select at random two burglaries to further investigate, find the probability that both will have occurred in 2004.

Source: IUP Police Department.

Solution

In this case, the events are dependent since the researcher wishes to investigate two distinct cases. Hence the first case is selected and not replaced.

$$P(C_1 \text{ and } C_2) = P(C_1) \cdot P(C_2|C_1) = \frac{16}{53} \cdot \frac{15}{52} = \frac{60}{689}$$

Example 4–29   ## Homeowner's and Automobile Insurance

World Wide Insurance Company found that 53% of the residents of a city had homeowner's insurance (H) with the company. Of these clients, 27% also had automobile insurance (A) with the company. If a resident is selected at random, find the probability that the resident has both homeowner's and automobile insurance with World Wide Insurance Company.

Solution

$$P(H \text{ and } A) = P(H) \cdot P(A|H) = (0.53)(0.27) = 0.1431$$

This multiplication rule can be extended to three or more events, as shown in Example 4–30.

Example 4–30   ## Drawing Cards

Three cards are drawn from an ordinary deck and not replaced. Find the probability of these events.

   a. Getting 3 jacks
   b. Getting an ace, a king, and a queen in order
   c. Getting a club, a spade, and a heart in order
   d. Getting 3 clubs

Solution

$$a.\ P(3 \text{ jacks}) = \frac{4}{52} \cdot \frac{3}{51} \cdot \frac{2}{50} = \frac{24}{132{,}600} = \frac{1}{5525}$$

$b.$ $P(\text{ace and king and queen}) = \dfrac{4}{52} \cdot \dfrac{4}{51} \cdot \dfrac{4}{50} = \dfrac{64}{132,600} = \dfrac{8}{16,575}$

$c.$ $P(\text{club and spade and heart}) = \dfrac{13}{52} \cdot \dfrac{13}{51} \cdot \dfrac{13}{50} = \dfrac{2197}{132,600} = \dfrac{169}{10,200}$

$d.$ $P(\text{3 clubs}) = \dfrac{13}{52} \cdot \dfrac{12}{51} \cdot \dfrac{11}{50} = \dfrac{1716}{132,600} = \dfrac{11}{850}$

Tree diagrams can be used as an aid to finding the solution to probability problems when the events are sequential. Example 4–31 illustrates the use of tree diagrams.

Example 4–31

Selecting Colored Balls

Box 1 contains 2 red balls and 1 blue ball. Box 2 contains 3 blue balls and 1 red ball. A coin is tossed. If it falls heads up, box 1 is selected and a ball is drawn. If it falls tails up, box 2 is selected and a ball is drawn. Find the probability of selecting a red ball.

Solution

The first two branches designate the selection of either box 1 or box 2. Then from box 1, either a red ball or a blue ball can be selected. Likewise, a red ball or blue ball can be selected from box 2. Hence a tree diagram of the example is shown in Figure 4–6.

Next determine the probabilities for each branch. Since a coin is being tossed for the box selection, each branch has a probability of $\frac{1}{2}$, that is, heads for box 1 or tails for box 2. The probabilities for the second branches are found by using the basic probability rule. For example, if box 1 is selected and there are 2 red balls and 1 blue ball, the probability of selecting a red ball is $\frac{2}{3}$ and the probability of selecting a blue ball is $\frac{1}{3}$. If box 2 is selected and it contains 3 blue balls and 1 red ball, then the probability of selecting a red ball is $\frac{1}{4}$ and the probability of selecting a blue ball is $\frac{3}{4}$.

Next multiply the probability for each outcome, using the rule $P(A \text{ and } B) = P(A) \cdot P(B|A)$. For example, the probability of selecting box 1 and selecting a red ball is $\frac{1}{2} \cdot \frac{2}{3} = \frac{2}{6}$. The probability of selecting box 1 and a blue ball is $\frac{1}{2} \cdot \frac{1}{3} = \frac{1}{6}$. The probability of selecting box 2 and selecting a red ball is $\frac{1}{2} \cdot \frac{1}{4} = \frac{1}{8}$. The probability of selecting box 2 and a blue ball is $\frac{1}{2} \cdot \frac{3}{4} = \frac{3}{8}$. (Note that the sum of these probabilities is 1.)

Finally a red ball can be selected from either box 1 or box 2 so $P(\text{red}) = \frac{2}{6} + \frac{1}{8} = \frac{8}{24} + \frac{3}{24} = \frac{11}{24}$.

Figure 4–6

Tree Diagram for
Example 4–31

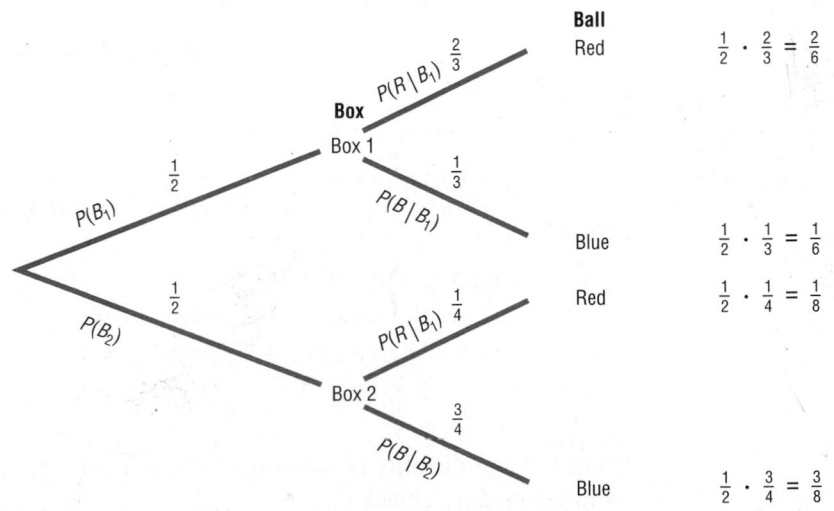

Tree diagrams can be used when the events are independent or dependent, and they can also be used for sequences of three or more events.

Conditional Probability

The conditional probability of an event B in relationship to an event A was defined as the probability that event B occurs after event A has already occurred.

The conditional probability of an event can be found by dividing both sides of the equation for multiplication rule 2 by $P(A)$, as shown:

$$P(A \text{ and } B) = P(A) \cdot P(B|A)$$

$$\frac{P(A \text{ and } B)}{P(A)} = \frac{\cancel{P(A)} \cdot P(B|A)}{\cancel{P(A)}}$$

$$\frac{P(A \text{ and } B)}{P(A)} = P(B|A)$$

Formula for Conditional Probability

The probability that the second event B occurs given that the first event A has occurred can be found by dividing the probability that both events occurred by the probability that the first event has occurred. The formula is

$$P(B|A) = \frac{P(A \text{ and } B)}{P(A)}$$

Examples 4–32, 4–33, and 4–34 illustrate the use of this rule.

Example 4–32

Selecting Colored Chips

A box contains black chips and white chips. A person selects two chips without replacement. If the probability of selecting a black chip *and* a white chip is $\frac{15}{56}$, and the probability of selecting a black chip on the first draw is $\frac{3}{8}$, find the probability of selecting the white chip on the second draw, *given* that the first chip selected was a black chip.

Solution

Let

$$B = \text{selecting a black chip} \qquad W = \text{selecting a white chip}$$

Then

$$P(W|B) = \frac{P(B \text{ and } W)}{P(B)} = \frac{15/56}{3/8}$$

$$= \frac{15}{56} \div \frac{3}{8} = \frac{15}{56} \cdot \frac{8}{3} = \frac{\overset{5}{\cancel{15}}}{\underset{7}{\cancel{56}}} \cdot \frac{\overset{1}{\cancel{8}}}{\underset{1}{\cancel{3}}} = \frac{5}{7}$$

Hence, the probability of selecting a white chip on the second draw given that the first chip selected was black is $\frac{5}{7}$.

Example 4–33

Parking Tickets

The probability that Sam parks in a no-parking zone *and* gets a parking ticket is 0.06, and the probability that Sam cannot find a legal parking space and has to park in the no-parking zone is 0.20. On Tuesday, Sam arrives at school and has to park in a no-parking zone. Find the probability that he will get a parking ticket.

Solution

Let

$$N = \text{parking in a no-parking zone} \qquad T = \text{getting a ticket}$$

Then

$$P(T|N) = \frac{P(N \text{ and } T)}{P(N)} = \frac{0.06}{0.20} = 0.30$$

Hence, Sam has a 0.30 probability of getting a parking ticket, given that he parked in a no-parking zone.

The conditional probability of events occurring can also be computed when the data are given in table form, as shown in Example 4–34.

Example 4–34

Survey on Women in the Military

A recent survey asked 100 people if they thought women in the armed forces should be permitted to participate in combat. The results of the survey are shown.

Gender	Yes	No	Total
Male	32	18	50
Female	8	42	50
Total	40	60	100

Find these probabilities.

  a. The respondent answered yes, given that the respondent was a female.

  b. The respondent was a male, given that the respondent answered no.

Solution

Let

$$M = \text{respondent was a male} \qquad Y = \text{respondent answered yes}$$
$$F = \text{respondent was a female} \qquad N = \text{respondent answered no}$$

a. The problem is to find $P(Y|F)$. The rule states

$$P(Y|F) = \frac{P(F \text{ and } Y)}{P(F)}$$

The probability $P(F \text{ and } Y)$ is the number of females who responded yes, divided by the total number of respondents:

$$P(F \text{ and } Y) = \frac{8}{100}$$

The probability $P(F)$ is the probability of selecting a female:

$$P(F) = \frac{50}{100}$$

Then

$$P(Y|F) = \frac{P(F \text{ and } Y)}{P(F)} = \frac{8/100}{50/100}$$

$$= \frac{8}{100} \div \frac{50}{100} = \frac{\overset{4}{\cancel{8}}}{\cancel{100}} \cdot \frac{\overset{1}{\cancel{100}}}{\underset{25}{\cancel{50}}} = \frac{4}{25}$$

b. The problem is to find $P(M|N)$.

$$P(M|N) = \frac{P(N \text{ and } M)}{P(N)} = \frac{18/100}{60/100}$$

$$= \frac{18}{100} \div \frac{60}{100} = \frac{\overset{3}{\cancel{18}}}{\underset{1}{\cancel{100}}} \cdot \frac{\overset{1}{\cancel{100}}}{\underset{10}{\cancel{60}}} = \frac{3}{10}$$

The Venn diagram for conditional probability is shown in Figure 4–7. In this case,

$$P(B|A) = \frac{P(A \text{ and } B)}{P(A)}$$

which is represented by the area in the intersection or overlapping part of the circles A and B, divided by the area of circle A. The reasoning here is that if you assume A has occurred, then A becomes the sample space for the next calculation and is the denominator of the probability fraction $\dfrac{P(A \text{ and } B)}{P(A)}$. The numerator $P(A \text{ and } B)$ represents the probability of the part of B that is contained in A. Hence, $P(A \text{ and } B)$ becomes the numerator of the probability fraction $\dfrac{P(A \text{ and } B)}{P(A)}$. Imposing a condition reduces the sample space.

Probabilities for "At Least"

The multiplication rules can be used with the complementary event rule (Section 4–1) to simplify solving probability problems involving "at least." Examples 4–35, 4–36, and 4–37 illustrate how this is done.

Figure 4–7

Venn Diagram for Conditional Probability

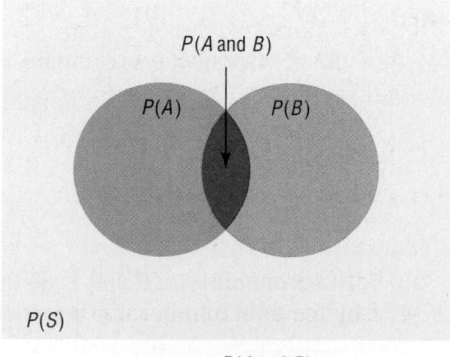

$$P(B|A) = \frac{P(A \text{ and } B)}{P(A)}$$

| Example 4–35 | Drawing Cards |

A game is played by drawing 4 cards from an ordinary deck and replacing each card after it is drawn. Find the probability that at least 1 ace is drawn.

Solution

It is much easier to find the probability that no aces are drawn (i.e., losing) and then subtract that value from 1 than to find the solution directly, because that would involve finding the probability of getting 1 ace, 2 aces, 3 aces, and 4 aces and then adding the results.

Let E = at least 1 ace is drawn and $\overline{E}$ = no aces drawn. Then

$$P(\overline{E}) = \frac{48}{52} \cdot \frac{48}{52} \cdot \frac{48}{52} \cdot \frac{48}{52}$$

$$= \frac{12}{13} \cdot \frac{12}{13} \cdot \frac{12}{13} \cdot \frac{12}{13} = \frac{20{,}736}{28{,}561}$$

Hence,

$$P(E) = 1 - P(\overline{E})$$

$$P(\text{winning}) = 1 - P(\text{losing}) = 1 - \frac{20{,}736}{28{,}561} = \frac{7825}{28{,}561} \approx 0.27$$

or a hand with at least 1 ace will occur about 27% of the time.

| Example 4–36 | Tossing Coins |

A coin is tossed 5 times. Find the probability of getting at least 1 tail.

Solution

It is easier to find the probability of the complement of the event, which is "all heads," and then subtract the probability from 1 to get the probability of at least 1 tail.

$$P(E) = 1 - P(\overline{E})$$

$$P(\text{at least 1 tail}) = 1 - P(\text{all heads})$$

$$P(\text{all heads}) = \left(\frac{1}{2}\right)^5 = \frac{1}{32}$$

Hence,

$$P(\text{at least 1 tail}) = 1 - \frac{1}{32} = \frac{31}{32}$$

| Example 4–37 | |

The Neckware Association of America reported that 3% of ties sold in the United States are bow ties. If 4 customers who purchased a tie are randomly selected, find the probability that at least 1 purchased a bow tie.

Solution

Let E = at least 1 bow tie is purchased and $\overline{E}$ = no bow ties are purchased. Then

$$P(E) = 0.03 \qquad \text{and} \qquad P(\overline{E}) = 1 - 0.03 = 0.97$$

$P(\text{no bow ties are purchased}) = (0.97)(0.97)(0.97)(0.97) \approx 0.885$; hence,
$P(\text{at least one bow tie is purchased}) = 1 - 0.885 = 0.115$.

Similar methods can be used for problems involving "at most."

Applying the Concepts **4–3**

Guilty or Innocent?

In July 1964, an elderly woman was mugged in Costa Mesa, California. In the vicinity of the crime a tall, bearded man sat waiting in a yellow car. Shortly after the crime was committed, a young, tall woman, wearing her blond hair in a ponytail, was seen running from the scene of the crime and getting into the car, which sped off. The police broadcast a description of the suspected muggers. Soon afterward, a couple fitting the description was arrested and convicted of the crime. Although the evidence in the case was largely circumstantial, the two people arrested were nonetheless convicted of the crime. The prosecutor based his entire case on basic probability theory, showing the unlikeness of another couple being in that area while having all the same characteristics that the elderly woman described. The following probabilities were used.

Characteristic	Assumed probability
Drives yellow car	1 out of 12
Man over 6 feet tall	1 out of 10
Man wearing tennis shoes	1 out of 4
Man with beard	1 out of 11
Woman with blond hair	1 out of 3
Woman with hair in a ponytail	1 out of 13
Woman over 6 feet tall	1 out of 100

1. Compute the probability of another couple being in that area with the same characteristics.

2. Would you use the addition or multiplication rule? Why?

3. Are the characteristics independent or dependent?

4. How are the computations affected by the assumption of independence or dependence?

5. Should any court case be based solely on probabilities?

6. Would you convict the couple who was arrested even if there were no eyewitnesses?

7. Comment on why in today's justice system no person can be convicted solely on the results of probabilities.

8. In actuality, aren't most court cases based on uncalculated probabilities?

See page 249 for the answers.

Exercises 4–3

1. State which events are independent and which are dependent.

   a. Tossing a coin and drawing a card from a deck Independent

   b. Drawing a ball from an urn, not replacing it, and then drawing a second ball Dependent

   c. Getting a raise in salary and purchasing a new car Dependent

   d. Driving on ice and having an accident Dependent

   e. Having a large shoe size and having a high IQ Independent

   f. A father being left-handed and a daughter being left-handed Dependent

   g. Smoking excessively and having lung cancer Dependent

   h. Eating an excessive amount of ice cream and smoking an excessive amount of cigarettes Independent

2. **Exercise** If 37% of high school students said that they exercise regularly, find the probability that 5 randomly selected high school students will say that they exercise regularly. Would you consider this event likely or unlikely to occur? Explain your answer. 0.007; the event is very unlikely to occur since its probability is very small.

3. **Video and Computer Games** Sixty-nine percent of U.S. heads of households play video or computer games. Choose 4 heads of households at random. Find the probability that

   a. None play video or computer games 0.009

   b. All four do 0.227

   Source: www.theesa.com

4. Seat Belt Use The Gallup Poll reported that 52% of Americans used a seat belt the last time they got into a car. If 4 people are selected at random, find the probability that they all used a seat belt the last time they got into a car. 7.3%

Source: *100% American.*

5. Automobile Sales An automobile salesperson finds the probability of making a sale is 0.21. If she talks to 4 customers, find the probability that she will make 4 sales. Is the event likely or unlikely to occur? Explain your answer. 0.00194 The event is highly unlikely since the probability is small.

6. Prison Populations If 25% of U.S. federal prison inmates are not U.S. citizens, find the probability that 2 randomly selected federal prison inmates will not be U.S. citizens. 6.3%

Source: *Harper's Index.*

7. MLS Players Of the 216 players on major league soccer rosters, 80.1% are U.S. citizens. If 3 players are selected at random for an exhibition, what is the probability that all are U.S. citizens? 0.5139

Source: *USA Today.*

8. Working Women and Computer Use It is reported that 72% of working women use computers at work. Choose 5 working women at random. Find

   a. The probability that at least 1 doesn't use a computer at work 0.807

   b. The probability that all 5 use a computer in their jobs 0.194

Source: www.infoplease.com

9. Text Messages via Cell Phones Thirty-five percent of people who own cell phones use their phones to send and receive text messages. Choose 4 cell phone owners at random. What is the probability that none use their phones for texting? 0.179

10. Cards If 2 cards are selected from a standard deck of 52 cards without replacement, find these probabilities.

   a. Both are spades. $\frac{1}{17}$

   b. Both are the same suit. $\frac{4}{17}$

   c. Both are kings. $\frac{1}{221}$

11. Cable Television In 2006, 86% of U.S. households had cable TV. Choose 3 households at random. Find the probability that

   a. None of the 3 households had cable TV 0.003

   b. All 3 households had cable TV 0.636

   c. At least 1 of the 3 households had cable TV 0.997

Source: www.infoplease.com

12. Flashlight Batteries A flashlight has 6 batteries, 2 of which are defective. If 2 are selected at random without replacement, find the probability that both are defective. $\frac{1}{15}$

13. Drawing a Card Four cards are drawn from a deck *without* replacement. Find these probabilities.

   a. All are kings. $\frac{1}{270,725}$

   b. All are diamonds. $\frac{11}{4165}$

   c. All are red cards. $\frac{46}{833}$

14. Scientific Study In a scientific study there are 8 guinea pigs, 5 of which are pregnant. If 3 are selected at random without replacement, find the probability that all are pregnant. $\frac{5}{28}$

15. In Exercise 14, find the probability that none are pregnant. $\frac{1}{56}$

16. Winning a Door Prize At a gathering consisting of 10 men and 20 women, two door prizes are awarded. Find the probability that both prizes are won by men. The winning ticket is not replaced. Would you consider this event likely or unlikely to occur? $\frac{3}{29}$ unlikely

17. In Exercise 16, find the probability that both prizes are won by women. Which event (Exercise 16 or 17) is most likely to occur? $\frac{38}{87}$ Number 20 is more likely to occur.

18. Sales A manufacturer makes two models of an item: model I, which accounts for 80% of unit sales, and model II, which accounts for 20% of unit sales. Because of defects, the manufacturer has to replace (or exchange) 10% of its model I and 18% of its model II. If a model is selected at random, find the probability that it will be defective. 0.116

19. Student Financial Aid In a recent year 8,073,000 male students and 10,980,000 female students were enrolled as undergraduates. Receiving aid were 60.6% of the male students and 65.2% of the female students. Of those receiving aid, 44.8% of the males got federal aid and 50.4% of the females got federal aid. Choose 1 student at random. (*Hint:* Make a tree diagram.) Find the probability that the student is

   a. A male student without aid 0.167

   b. A male student, given that the student has aid 0.406

   c. A female student or a student who receives federal aid 0.691

Source: www.nces.gov

20. Selecting Colored Balls Urn 1 contains 5 red balls and 3 black balls. Urn 2 contains 3 red balls and 1 black ball. Urn 3 contains 4 red balls and 2 black balls. If an urn is selected at random and a ball is drawn, find the probability it will be red. $\frac{49}{72}$

21. Automobile Insurance An insurance company classifies drivers as low-risk, medium-risk, and high-risk. Of those insured, 60% are low-risk, 30% are medium-risk, and 10% are high-risk. After a study, the company finds that during a 1-year period, 1% of the low-risk drivers had an accident, 5% of the medium-risk drivers had an accident, and 9% of the high-risk drivers had an accident. If a driver is selected at random, find the probability that the driver will have had an accident during the year. 0.03

22. **Defective Items** A production process produces an item. On average, 15% of all items produced are defective. Each item is inspected before being shipped, and the inspector misclassifies an item 10% of the time. What proportion of the items will be "classified as good"? What is the probability that an item is defective given that it was classified as good? 0.78 0.0192

23. **Prison Populations** For a recent year, 0.99 of the incarcerated population is adults and 0.07 of these are female. If an incarcerated person is selected at random, find the probability that the person is a female given that the person is an adult. 0.071

    Source: Bureau of Justice.

24. **Rolling Dice** Roll two standard dice and add the numbers. What is the probability of getting a number larger than 9 for the first time on the third roll? 0.1157

25. **Model Railroad Circuit** A circuit to run a model railroad has 8 switches. Two are defective. If you select 2 switches at random and test them, find the probability that the second one is defective, given that the first one is defective. $\frac{1}{7}$

26. **Country Club Activities** At the Avonlea Country Club, 73% of the members play bridge and swim, and 82% play bridge. If a member is selected at random, find the probability that the member swims, given that the member plays bridge. 89%

27. **College Courses** At a large university, the probability that a student takes calculus and is on the dean's list is 0.042. The probability that a student is on the dean's list is 0.21. Find the probability that the student is taking calculus, given that he or she is on the dean's list. 0.2

28. **Country Club Members** At the Coulterville Country Club, 72% of the members play golf and are college graduates, and 80% of the members play golf. If a member is selected at random, find the probability that the member is a college graduate given that the member plays golf. 0.9

29. **Pizza and Salads** In a pizza restaurant, 95% of the customers order pizza. If 65% of the customers order pizza and a salad, find the probability that a customer who orders pizza will also order a salad. 68.4%

30. **Gift Baskets** The Gift Basket Store had the following premade gift baskets containing the following combinations in stock.

	Cookies	Mugs	Candy
Coffee	20	13	10
Tea	12	10	12

Choose 1 basket at random. Find the probability that it contains

a. Coffee or candy 0.7143

b. Tea given that it contains mugs 0.4348

c. Tea and cookies 0.1558

Source: www.infoplease.com

31. **Blood Types and Rh Factors** In addition to being grouped into four types, human blood is grouped by its Rhesus (Rh) factor. Consider the figures below which show the distributions of these groups for Americans.

	O	A	B	AB
Rh+	37%	34%	10%	4%
Rh−	6%	6%	2%	1%

Choose 1 American at random. Find the probability that the person

a. Is a universal donor, i.e., has O negative blood 0.06

b. Has type O blood given that the person is Rh+ 0.4353

c. Has A+ or AB− blood 0.35

d. Has Rh− given that the person has type B 0.1667

Source: www.infoplease.com

32. **Doctor Specialties** Below are listed the numbers of doctors in various specialties by gender.

	Pathology	Pediatrics	Psychiatry
Male	12,575	33,020	27,803
Female	5,604	33,351	12,292

Choose 1 doctor at random.

a. Find $P(\text{male}|\text{pediatrician})$. 0.498

b. Find $P(\text{pathologist}|\text{female})$. 0.109

c. Are the characteristics "female" and "pathologist" independent? Explain. No. $P(\text{path}|\text{female}) \neq P(\text{path})$

Source: *World Almanac.*

33. **Olympic Medals** The medal distribution from the 2008 Summer Olympic Games for the top 23 countries is shown below.

	Gold	Silver	Bronze
United States	36	38	36
Russia	23	21	28
China	51	21	28
Great Britain	19	13	15
Others	173	209	246

Choose 1 medal winner at random.

a. Find the probability that the winner won the gold medal, given that the winner was from the United States. 0.327

b. Find the probability that the winner was from the United States, given that she or he won a gold medal. 0.119

c. Are the events "medal winner is from United States" and "gold medal won" independent? Explain. No. $P(G|U.S.) \neq P(G)$

34. Computer Ownership At a local university 54.3% of incoming first-year students have computers. If 3 students are selected at random, find the following probabilities.

a. None have computers. 0.0954

b. At least one has a computer. 0.9046

c. All have computers. 0.1601

35. Leisure Time Exercise Only 27% of U.S. adults get enough leisure time exercise to achieve cardiovascular fitness. Choose 3 adults at random. Find the probability that

a. All 3 get enough daily exercise 0.0197

b. At least 1 of the 3 gets enough exercise 0.611

Source: www.infoplease.com

36. Customer Purchases In a department store there are 120 customers, 90 of whom will buy at least 1 item. If 5 customers are selected at random, one by one, find the probability that all will buy at least 1 item. 0.231

37. Marital Status of Women According to the *Statistical Abstract of the United States,* 70.3% of females ages 20 to 24 have never been married. Choose 5 young women in this age category at random. Find the probability that

a. None has ever been married 0.1717

b. At least 1 has been married 0.8283

Source: *New York Times Almanac.*

38. Fatal Accidents The American Automobile Association (AAA) reports that of the fatal car and truck accidents, 54% are caused by car driver error. If 3 accidents are chosen at random, find the probability that

a. All are caused by car driver error 0.157

b. None is caused by car driver error 0.097

c. At least 1 is caused by car driver error 0.903

Source: AAA quoted on CNN.

39. On-Time Airplane Arrivals The greater Cincinnati airport led major U.S. airports in on-time arrivals in the last quarter of 2005 with an 84.3% on-time rate. Choose 5 arrivals at random and find the probability that at least 1 was not on time. 0.574

Source: www.census.gov

40. Online Electronic Games Fifty-six percent of electronic gamers play games online, and sixty-four percent of those gamers are female. What is the probability that a randomly selected gamer plays games online and is male? 0.202

Source: www.tech.msn.com

41. Reading to Children Fifty-eight percent of American children (ages 3 to 5) are read to every day by someone at home. Suppose 5 children are randomly selected. What is the probability that at least 1 is read to every day by someone at home? 0.9869

Source: Federal Interagency Forum on Child and Family Statistics.

42. Doctoral Assistantships Of Ph.D. students, 60% have paid assistantships. If 3 students are selected at random, find the probabilities

a. All have assistantships 0.216

b. None has an assistantship 0.064

c. At least 1 has an assistantship 0.936

Source: U.S. Department of Education, *Chronicle of Higher Education.*

43. Selecting Cards If 4 cards are drawn from a deck of 52 and not replaced, find the probability of getting at least 1 club. $\frac{14,498}{20,825}$

44. Full-Time College Enrollment The majority (69%) of undergraduate students were enrolled in a 4-year college in a recent year. Eighty-one percent of those enrolled attended full-time. Choose 1 enrolled undergraduate student at random. What is the probability that she or he is a part-time student at a 4-year college? 0.131

Source: www.census.gov

45. Family and Children's Computer Games It was reported that 19.8% of computer games sold in 2005 were classified as "family and children's." Choose 5 purchased computer games at random. Find the probability that

a. None of the 5 was family and children's 0.332

b. At least 1 of the 5 was family and children's 0.668

Source: www.theesa.com

46. Medication Effectiveness A medication is 75% effective against a bacterial infection. Find the probability that if 12 people take the medication, at least 1 person's infection will not improve. 96.8%

47. Tossing a Coin A coin is tossed 5 times; find the probability of getting at least 1 tail. Would you consider this event likely to happen? Explain your answer. $\frac{31}{32}$

48. Selecting a Letter of the Alphabet If 3 letters of the alphabet are selected at random, find the probability of getting at least 1 letter x. Letters can be used more than once. Would you consider this event likely to happen? Explain your answer. 0.111; the event is very unlikely to occur since the probability is only about 11%.

49. Rolling a Die A die is rolled 6 times. Find the probability of getting at least one 4. Would you consider this event likely or unlikely? Explain your answer. 0.665 It will happen almost 67% of the time. It's somewhat likely.

50. **High School Grades of First-Year College Students**
Forty-seven percent of first-year college students enrolled in 2005 had an average grade of A in high school compared to 20% of first-year college students in 1970. Choose 6 first-year college students at random enrolled in 2005. Find the probability that

 a. All had an A average in high school 0.011
 b. None had an A average in high school 0.022
 c. At least 1 had an A average in high school 0.978

Source: www.census.gov

51. **Rolling a Die** If a die is rolled 3 times, find the probability of getting at least 1 even number. $\frac{7}{8}$

52. **Selecting a Flower** In a large vase, there are 8 roses, 5 daisies, 12 lilies, and 9 orchids. If 4 flowers are selected at random, find the probability that at least 1 of the flowers is a rose. Would you consider this event likely to occur? Explain your answer. 0.678; yes the event is a little more likely to occur than not since the probability is about 68%.

Extending the Concepts

53. Let A and B be two mutually exclusive events. Are A and B independent events? Explain your answer. No, since $P(A \cap B) = 0$ and does not equal $P(A) \cdot P(B)$.

54. **Types of Vehicles** The Bargain Auto Mall has the following cars in stock.

	SUV	Compact	Mid-sized
Foreign	20	50	20
Domestic	65	100	45

Are the events "compact" and "domestic" independent? Explain. No, since $P(C|D) \neq P(C)$.

55. **College Enrollment** An admissions director knows that the probability a student will enroll after a campus visit is 0.55, or $P(E) = 0.55$. While students are on campus visits, interviews with professors are arranged.

The admissions director computes these conditional probabilities for students enrolling after visiting three professors, DW, LP, and MH.

$$P(E|DW) = 0.95 \qquad P(E|LP) = 0.55 \qquad P(E|MH) = 0.15$$

Is there something wrong with the numbers? Explain.

56. **Commercials** Event A is the event that a person remembers a certain product commercial. Event B is the event that a person buys the product. If $P(B) = 0.35$, comment on each of these conditional probabilities if you were vice president for sales.

 a. $P(B|A) = 0.20$
 b. $P(B|A) = 0.35$
 c. $P(B|A) = 0.55$

4–4   Counting Rules

Many times a person must know the number of all possible outcomes for a sequence of events. To determine this number, three rules can be used: the *fundamental counting rule,* the *permutation rule,* and the *combination rule.* These rules are explained here, and they will be used in Section 4–5 to find probabilities of events.

The first rule is called the **fundamental counting rule.**

The Fundamental Counting Rule

Objective 5

Find the total number of outcomes in a sequence of events, using the fundamental counting rule.

Fundamental Counting Rule

In a sequence of n events in which the first one has k_1 possibilities and the second event has k_2 and the third has k_3, and so forth, the total number of possibilities of the sequence will be

$$k_1 \cdot k_2 \cdot k_3 \cdots k_n$$

Note: In this case *and* means to multiply.

Examples 4–38 through 4–41 illustrate the fundamental counting rule.

Example 4–38

Tossing a Coin and Rolling a Die

A coin is tossed and a die is rolled. Find the number of outcomes for the sequence of events.

Figure 4–8

Complete Tree Diagram for Example 4–38

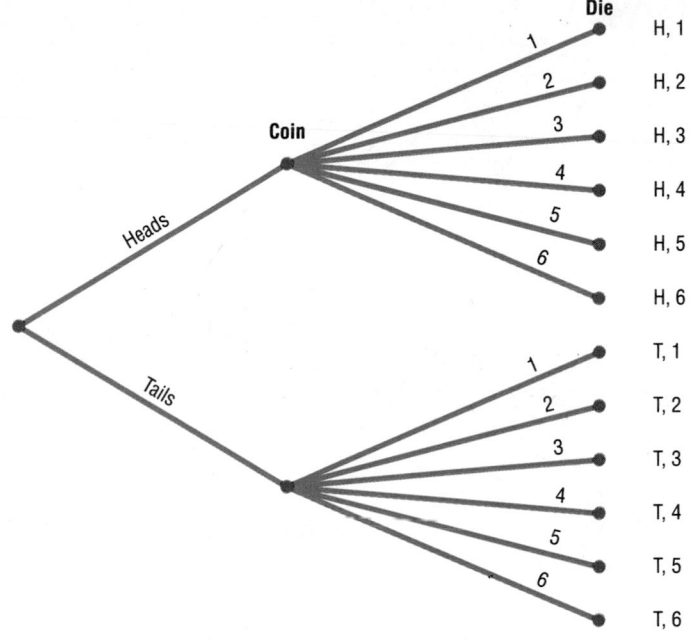

Solution

Since the coin can land either heads up or tails up and since the die can land with any one of six numbers showing face up, there are $2 \cdot 6 = 12$ possibilities. A tree diagram can also be drawn for the sequence of events. See Figure 4–8.

Interesting Fact

Possible games of chess: 25×10^{115}.

Example 4–39

Types of Paint

A paint manufacturer wishes to manufacture several different paints. The categories include

Color	Red, blue, white, black, green, brown, yellow
Type	Latex, oil
Texture	Flat, semigloss, high gloss
Use	Outdoor, indoor

How many different kinds of paint can be made if you can select one color, one type, one texture, and one use?

Solution

You can choose one color and one type and one texture and one use. Since there are 7 color choices, 2 type choices, 3 texture choices, and 2 use choices, the total number of possible different paints is

Color	Type	Texture	Use	
7	· 2	· 3	· 2	= 84

Example 4–40

Distribution of Blood Types

There are four blood types, A, B, AB, and O. Blood can also be Rh+ and Rh−. Finally, a blood donor can be classified as either male or female. How many different ways can a donor have his or her blood labeled?

Figure 4–9

Complete Tree
Diagram for
Example 4–40

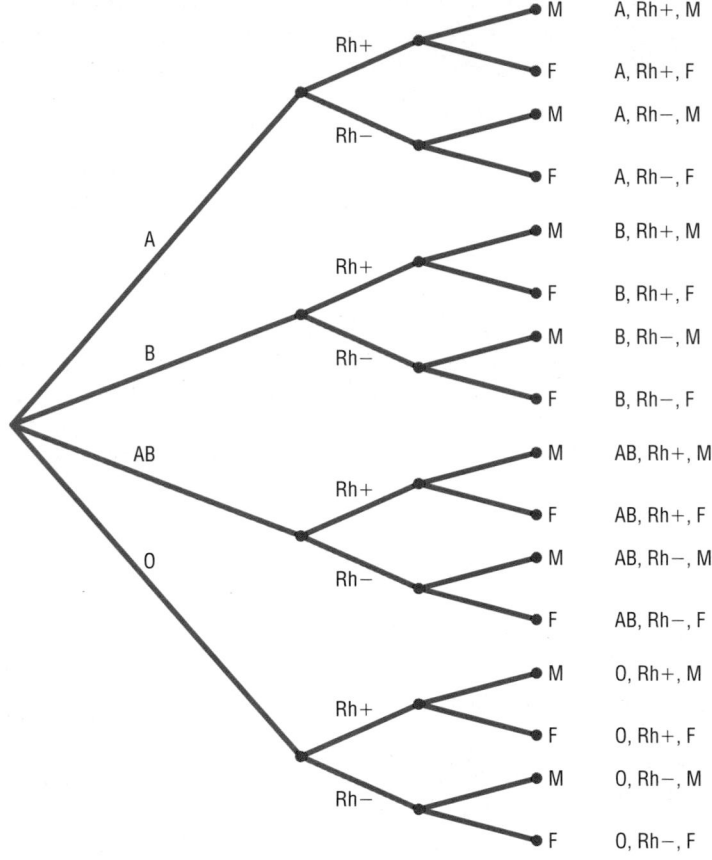

Solution

Since there are 4 possibilities for blood type, 2 possibilities for Rh factor, and 2 possibilities for the gender of the donor, there are 4 · 2 · 2, or 16, different classification categories, as shown.

Blood type		Rh		Gender	
4	·	2	·	2	= 16

A tree diagram for the events is shown in Figure 4–9.

   When determining the number of different possibilities of a sequence of events, you must know whether repetitions are permissible.

Example 4–41

Identification Cards

The manager of a department store chain wishes to make four-digit identification cards for her employees. How many different cards can be made if she uses the digits 1, 2, 3, 4, 5, and 6 and repetitions are permitted?

Solution

Since there are 4 spaces to fill on each card and there are 6 choices for each space, the total number of cards that can be made is $6 \cdot 6 \cdot 6 \cdot 6 = 1296$.

Now, what if repetitions are not permitted? For Example 4–41, the first digit can be chosen in 6 ways. But the second digit can be chosen in only 5 ways, since there are only five digits left, etc. Thus, the solution is

$$6 \cdot 5 \cdot 4 \cdot 3 = 360$$

The same situation occurs when one is drawing balls from an urn or cards from a deck. If the ball or card is replaced before the next one is selected, then repetitions are permitted, since the same one can be selected again. But if the selected ball or card is not replaced, then repetitions are not permitted, since the same ball or card cannot be selected the second time.

These examples illustrate the fundamental counting rule. In summary: *If repetitions are permitted, then the numbers stay the same going from left to right. If repetitions are not permitted, then the numbers decrease by 1 for each place left to right.*

Two other rules that can be used to determine the total number of possibilities of a sequence of events are the permutation rule and the combination rule.

Factorial Notation

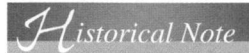

These rules use *factorial notation*. The factorial notation uses the exclamation point.

$$5! = 5 \cdot 4 \cdot 3 \cdot 2 \cdot 1$$
$$9! = 9 \cdot 8 \cdot 7 \cdot 6 \cdot 5 \cdot 4 \cdot 3 \cdot 2 \cdot 1$$

To use the formulas in the permutation and combination rules, a special definition of $0!$ is needed. $0! = 1$.

Factorial Formulas

For any counting n

$$n! = n(n - 1)(n - 2) \cdots 1$$
$$0! = 1$$

Permutations

A **permutation** is an arrangement of n objects in a specific order.

Examples 4–42 and 4–43 illustrate permutations.

Example 4–42   Business Location

Suppose a business owner has a choice of 5 locations in which to establish her business. She decides to rank each location according to certain criteria, such as price of the store and parking facilities. How many different ways can she rank the 5 locations?

Solution

There are

$$5! = 5 \cdot 4 \cdot 3 \cdot 2 \cdot 1 = 120$$

different possible rankings. The reason is that she has 5 choices for the first location, 4 choices for the second location, 3 choices for the third location, etc.

In Example 4–42 all objects were used up. But what happens when not all objects are used up? The answer to this question is given in Example 4–43.

Example 4–43

Business Location

Suppose the business owner in Example 4–42 wishes to rank only the top 3 of the 5 locations. How many different ways can she rank them?

Solution

Using the fundamental counting rule, she can select any one of the 5 for first choice, then any one of the remaining 4 locations for her second choice, and finally, any one of the remaining locations for her third choice, as shown.

First choice		**Second choice**		**Third choice**	
5	$\cdot$	4	$\cdot$	3	$= 60$

The solutions in Examples 4–42 and 4–43 are permutations.

Objective 6

Find the number of ways that r objects can be selected from n objects, using the permutation rule.

Permutation Rule

The arrangement of n objects in a specific order using r objects at a time is called a *permutation of n objects taking r objects at a time*. It is written as $_nP_r$, and the formula is

$$_nP_r = \frac{n!}{(n-r)!}$$

The notation $_nP_r$ is used for permutations.

$$_6P_4 \text{ means } \frac{6!}{(6-4)!} \quad \text{or} \quad \frac{6!}{2!} = \frac{6 \cdot 5 \cdot 4 \cdot 3 \cdot \cancel{2} \cdot \cancel{1}}{\cancel{2} \cdot \cancel{1}} = 360$$

Although Examples 4–42 and 4–43 were solved by the multiplication rule, they can now be solved by the permutation rule.

In Example 4–42, 5 locations were taken and then arranged in order; hence,

$$_5P_5 = \frac{5!}{(5-5)!} = \frac{5!}{0!} = \frac{5 \cdot 4 \cdot 3 \cdot 2 \cdot 1}{1} = 120$$

(Recall that $0! = 1$.)

In Example 4–43, 3 locations were selected from 5 locations, so $n = 5$ and $r = 3$; hence

$$_5P_3 = \frac{5!}{(5 - 3)!} = \frac{5!}{2!} = \frac{5 \cdot 4 \cdot 3 \cdot \cancel{2} \cdot \cancel{1}}{\cancel{2} \cdot \cancel{1}} = 60$$

Examples 4–44 and 4–45 illustrate the permutation rule.

Example 4–44

Television Ads

The advertising director for a television show has 7 ads to use on the program. If she selects 1 of them for the opening of the show, 1 for the middle of the show, and 1 for the ending of the show, how many possible ways can this be accomplished?

Solution

Since order is important, the solution is

$$_7P_3 = \frac{7!}{(7 - 3)!} = \frac{7!}{4!} = 210$$

Hence, there would be 210 ways to show 3 ads.

Example 4–45

School Musical Plays

A school musical director can select 2 musical plays to present next year. One will be presented in the fall, and one will be presented in the spring. If she has 9 to pick from, how many different possibilities are there?

Solution

Order is important since one play can be presented in the fall and the other play in the spring.

$$_9P_2 = \frac{9!}{(9 - 2)!} = \frac{9!}{7!} = \frac{9 \cdot 8 \cdot 7!}{7!} = 72$$

There are 72 different possibilities.

Combinations

Suppose a dress designer wishes to select two colors of material to design a new dress, and she has on hand four colors. How many different possibilities can there be in this situation?

Objective 7

Find the number of ways that *r* objects can be selected from *n* objects without regard to order, using the combination rule.

This type of problem differs from previous ones in that the order of selection is not important. That is, if the designer selects yellow and red, this selection is the same as the selection red and yellow. This type of selection is called a *combination*. The difference between a permutation and a combination is that in a combination, the order or arrangement of the objects is not important; by contrast, order *is* important in a permutation. Example 4–46 illustrates this difference.

A selection of distinct objects without regard to order is called a **combination**.

| Example 4–46 | **Letters** |

Given the letters A, B, C, and D, list the permutations and combinations for selecting two letters.

Solution

The permutations are

AB	BA	CA	DA
AC	BC	CB	DB
AD	BD	CD	DC

In permutations, AB is different from BA. But in combinations, AB is the same as BA since the order of the objects does not matter in combinations. Therefore, if duplicates are removed from a list of permutations, what is left is a list of combinations, as shown.

AB	B̶A̶	C̶A̶	D̶A̶
AC	BC	C̶B̶	D̶B̶
AD	BD	CD	D̶C̶

Hence the combinations of A, B, C, and D are AB, AC, AD, BC, BD, and CD. (Alternatively, BA could be listed and AB crossed out, etc.) The combinations have been listed alphabetically for convenience, but this is not a requirement.

Combinations are used when the order or arrangement is not important, as in the selecting process. Suppose a committee of 5 students is to be selected from 25 students. The 5 selected students represent a combination, since it does not matter who is selected first, second, etc.

Combination Rule

The number of combinations of r objects selected from n objects is denoted by ${}_nC_r$ and is given by the formula

$$_nC_r = \frac{n!}{(n-r)!\,r!}$$

Example 4–47	**Combinations**

How many combinations of 4 objects are there, taken 2 at a time?

Solution

Since this is a combination problem, the answer is

$$_4C_2 = \frac{4!}{(4-2)!2!} = \frac{4!}{2!2!} = \frac{\overset{2}{\cancel{4}} \cdot 3 \cdot \cancel{2!}}{\cancel{2} \cdot 1 \cdot \cancel{2!}} = 6$$

This is the same result shown in Example 4–46.

Notice that the expression for $_nC_r$ is

$$\frac{n!}{(n-r)!r!}$$

which is the formula for permutations with $r!$ in the denominator. In other words,

$$_nC_r = \frac{_nP_r}{r!}$$

This $r!$ divides out the duplicates from the number of permutations, as shown in Example 4–46. For each two letters, there are two permutations but only one combination. Hence, dividing the number of permutations by $r!$ eliminates the duplicates. This result can be verified for other values of n and r. *Note:* $_nC_n = 1$.

Example 4–48	**Book Reviews**

A newspaper editor has received 8 books to review. He decides that he can use 3 reviews in his newspaper. How many different ways can these 3 reviews be selected?

Solution

$$_8C_3 = \frac{8!}{(8-3)!3!} = \frac{8!}{5!3!} = \frac{8 \cdot 7 \cdot 6}{3 \cdot 2 \cdot 1} = 56$$

There are 56 possibilities.

Example 4–49	**Committee Selection**

In a club there are 7 women and 5 men. A committee of 3 women and 2 men is to be chosen. How many different possibilities are there?

Solution

Here, you must select 3 women from 7 women, which can be done in $_7C_3$, or 35, ways. Next, 2 men must be selected from 5 men, which can be done in $_5C_2$, or 10, ways. Finally, by the fundamental counting rule, the total number of different ways is $35 \cdot 10 = 350$, since you are choosing both men and women. Using the formula gives

$$_7C_3 \cdot {_5C_2} = \frac{7!}{(7-3)!3!} \cdot \frac{5!}{(5-2)!2!} = 350$$

Table 4–1 summarizes the counting rules.

Table 4–1	Summary of Counting Rules	
Rule	**Definition**	**Formula**
Fundamental counting rule	The number of ways a sequence of n events can occur if the first event can occur in k_1 ways, the second event can occur in k_2 ways, etc.	$k_1 \cdot k_2 \cdot k_3 \cdots k_n$
Permutation rule	The number of permutations of n objects taking r objects at a time (order is important)	$_nP_r = \dfrac{n!}{(n-r)!}$
Combination rule	The number of combinations of r objects taken from n objects (order is not important)	$_nC_r = \dfrac{n!}{(n-r)!r!}$

Applying the Concepts **4–4**

Garage Door Openers

Garage door openers originally had a series of four on/off switches so that homeowners could personalize the frequencies that opened their garage doors. If all garage door openers were set at the same frequency, anyone with a garage door opener could open anyone else's garage door.

1. Use a tree diagram to show how many different positions 4 consecutive on/off switches could be in.

After garage door openers became more popular, another set of 4 on/off switches was added to the systems.

2. Find a pattern of how many different positions are possible with the addition of each on/off switch.
3. How many different positions are possible with 8 consecutive on/off switches?
4. Is it reasonable to assume, if you owned a garage door opener with 8 switches, that someone could use his or her garage door opener to open your garage door by trying all the different possible positions?

In 1989 it was reported that the ignition keys for 1988 Dodge Caravans were made from a single blank that had five cuts on it. Each cut was made at one out of five possible levels. In 1988, assume there were 420,000 Dodge Caravans sold in the United States.

5. How many different possible keys can be made from the same key blank?
6. How many different 1988 Dodge Caravans could any one key start?

Look at the ignition key for your car and count the number of cuts on it. Assume that the cuts are made at one of any of five possible levels. Most car companies use one key blank for all their makes and models of cars.

7. Conjecture how many cars your car company sold over recent years, and then figure out how many other cars your car key could start. What would you do to decrease the odds of someone being able to open another vehicle with his or her key?

See page 250 for the answers.

Exercises 4–4

1. **Zip Codes** How many 5-digit zip codes are possible if digits can be repeated? If there cannot be repetitions? 100,000; 30,240

2. **Batting Order** How many ways can a baseball manager arrange a batting order of 9 players? 362,880

3. **Video Games** How many different ways can 6 different video game cartridges be arranged on a shelf? 720

4. **Visiting Nurses** How many different ways can a visiting nurse visit 9 patients if she wants to visit them all in one day? 362,880

5. **Laundry Soap Display** A store manager wishes to display 7 different kinds of laundry soap in a row. How many different ways can this be done? 5040 ways

6. **Show Programs** Three bands and two comics are performing for a student talent show. How many different programs (in terms of order) can be arranged? How many if the comics must perform between bands? 120; 12

7. **Campus Tours** Student volunteers take visitors on a tour of 10 campus buildings. How many different tours are possible? (Assume order is important.) 3,628,000

8. **Radio Station Call Letters** The call letters of a radio station must have 4 letters. The first letter must be a K or a W. How many different station call letters can be made if repetitions are not allowed? If repetitions are allowed? 27,600; 35,152

9. **Identification Tags** How many different 3-digit identification tags can be made if the digits can be used more than once? If the first digit must be a 5 and repetitions are not permitted? 1000; 72

10. **Secret Code Word** How many 4-letter code words can be made using the letters in the word *pencil* if repetitions are permitted? If repetitions are not permitted? 1296; 360

11. **Selection of Officers** Six students are running for the positions of president and vice-president, and five students are running for secretary and treasurer. If the two highest vote getters in each of the two contests are elected, how many winning combinations can there be? 600

12. **Automobile Trips** There are 2 major roads from city X to city Y and 4 major roads from city Y to city Z. How many different trips can be made from city X to city Z passing through city Y? 8

13. Evaluate each of these.

    a. $8!$ 40,320   e. $_7P_5$ 2520   i. $_5P_5$ 120
    b. $10!$ 3,628,800   f. $_{12}P_4$ 11,880 j. $_6P_2$ 30
    c. $0!$ 1   g. $_5P_3$ 60
    d. $1!$ 1   h. $_6P_0$ 1

14. **County Assessments** The County Assessment Bureau decides to reassess homes in 8 different areas. How many different ways can this be accomplished? 40,320

15. **Sports Car Stripes** How many different 4-color code stripes can be made on a sports car if each code consists of the colors green, red, blue, and white? All colors are used only once. 24

16. **Manufacturing Tests** An inspector must select 3 tests to perform in a certain order on a manufactured part. He has a choice of 7 tests. How many ways can he perform 3 different tests? 210

17. **Threatened Species of Reptiles** There are 22 threatened species of reptiles in the United States. In how many ways can you choose 4 to write about? (Order is not important.) 7315

    Source: www.infoplease.com

18. **Inspecting Restaurants** How many different ways can a city health department inspector visit 5 restaurants in a city with 10 restaurants? 30,240

19. How many different 4-letter permutations can be formed from the letters in the word *decagon*? 840

20. **Cell Phone Models** A particular cell phone company offers 4 models of phones, each in 6 different colors and each available with any one of 5 calling plans. How many combinations are possible? 120

21. **ID Cards** How many different ID cards can be made if there are 6 digits on a card and no digit can be used more than once? 151,200

22. **Free-Sample Requests** An online coupon service has 13 offers for free samples. How may different requests are possible if a customer must request exactly 3 free samples? How many are possible if the customer may request up to 3 free samples? 286; 378 (count 0)

23. **Ticket Selection** How many different ways can 4 tickets be selected from 50 tickets if each ticket wins a different prize? 5,527,200

24. **Movie Selections** The Foreign Language Club is showing a four-movie marathon of subtitled movies. How many ways can they choose 4 from the 11 available? 330

25. **Task Assignments** How many ways can an adviser choose 4 students from a class of 12 if they are all assigned the same task? How many ways can the students be chosen if they are each given a different task? 495; 11,880

26. **Agency Cases** An investigative agency has 7 cases and 5 agents. How many different ways can the cases be assigned if only 1 case is assigned to each agent? 2520

27. **(ans)** Evaluate each expression.

    a. $_5C_2$ 10   d. $_6C_2$ 15   g. $_3C_3$ 1   j. $_4C_3$ 4

    b. $_8C_3$ 56   e. $_6C_4$ 15   h. $_9C_7$ 36

    c. $_7C_4$ 35   f. $_3C_0$ 1   i. $_{12}C_2$ 66

28. **Selecting Cards** How many ways can 3 cards be selected from a standard deck of 52 cards, disregarding the order of selection? 22,100

29. **Selecting Coins** How many ways can a person select 3 coins from a box consisting of a penny, a nickel, a dime, a quarter, a half-dollar, and a one-dollar coin? 120

30. **Selecting Players** How many ways can 4 baseball players and 3 basketball players be selected from 12 baseball players and 9 basketball players? 41,580

31. **Selecting a Committee** How many ways can a committee of 4 people be selected from a group of 10 people? 210

32. **Selecting Christmas Presents** If a person can select 3 presents from 10 presents under a Christmas tree, how many different combinations are there? 120

33. **Questions for a Test** How many different tests can be made from a test bank of 20 questions if the test consists of 5 questions? 15,504

34. **Promotional Program** The general manager of a fast-food restaurant chain must select 6 restaurants from 11 for a promotional program. How many different possible ways can this selection be done? 462

35. **Music Program Selections** A jazz band has prepared 18 selections for a concert tour. At each stop they will perform 10. How many different programs are possible? How many programs are possible if they always begin with the same song and end with the same song? 43,758; 12,870

36. **Freight Train Cars** In a train yard there are 4 tank cars, 12 boxcars, and 7 flatcars. How many ways can a train be made up consisting of 2 tank cars, 5 boxcars, and 3 flatcars? (In this case, order is not important.) 166,320

37. **Selecting a Committee** There are 7 women and 5 men in a department. How many ways can a committee of 4 people be selected? How many ways can this committee be selected if there must be 2 men and 2 women on the committee? How many ways can this committee be selected if there must be at least 2 women on the committee? 495; 210; 420

38. **Selecting Cereal Boxes** Wake Up cereal comes in 2 types, crispy and crunchy. If a researcher has 10 boxes of each, how many ways can she select 3 boxes of each for a quality control test? 14,400

39. **Hawaiian Words** The Hawaiian alphabet consists of 7 consonants and 5 vowels. How many three-letter "words" are possible if there are never two consonants together and if a word must always end in a vowel? 475

40. **Selecting a Jury** How many ways can a jury of 6 women and 6 men be selected from 10 women and 12 men? 194,040

41. **Selecting a Golf Foursome** How many ways can a foursome of 2 men and 2 women be selected from 10 men and 12 women in a golf club? 2970

42. **Investigative Team** The state narcotics bureau must form a 5-member investigative team. If it has 25 agents from which to choose, how many different possible teams can be formed? 53,130

43. **Dominoes** A domino is a flat rectangular block the face of which is divided into two square parts, each part showing from zero to six pips (or dots). Playing a game consists of playing dominoes with a matching number of pips. Explain why there are 28 dominoes in a complete set. $_7C_2$ is 21 combinations + 7 double tiles = 28

44. **Charity Event Participants** There are 16 seniors and 15 juniors in a particular social organization. In how many ways can 4 seniors and 2 juniors be chosen to participate in a charity event? 191,100

45. **Selecting Commercials** How many ways can a person select 7 television commercials from 11 television commercials? 330

46. **DVD Selection** How many ways can a person select 8 DVDs from a display of 13 DVDs? 1287

47. **Candy Bar Selection** How many ways can a person select 6 candy bars from a list of 10 and 6 salty snacks from a list of 12 to put in a vending machine? 194,040

48. **Selecting a Location** An advertising manager decides to have an ad campaign in which 8 special calculators will be hidden at various locations in a shopping mall. If he has 17 locations from which to pick, how many different possible combinations can he choose? 24,310

Permutations and Combinations

49. **Selecting Posters** A buyer decides to stock 8 different posters. How many ways can she select these 8 if there are 20 from which to choose? 125,970

50. **Test Marketing Products** Anderson Research Company decides to test-market a product in 6 areas. How many different ways can 3 areas be selected in a certain order for the first test? 120

51. **Selecting Rats** How many different ways can a researcher select 5 rats from 20 rats and assign each to a different test? 1,860,480

52. **Selecting Musicals** How many different ways can a theatrical group select 2 musicals and 3 dramas from 11 musicals and 8 dramas to be presented during the year? 3080

53. **Textbook Selection** How many different ways can an instructor select 2 textbooks from a possible 17? 136

54. **DVD Selection** How many ways can a person select 8 DVDs from 10 DVDs? 45

55. **Public Service Announcements** How many different ways can 5 public service announcements be run during 1 hour? 120

56. **Signal Flags** How many different signals can be made by using at least 3 different flags if there are 5 different flags from which to select? 300

57. **Dinner Selections** How many ways can a dinner patron select 3 appetizers and 2 vegetables if there are 6 appetizers and 5 vegetables on the menu? 200

58. **Air Pollution** The Environmental Protection Agency must investigate 9 mills for complaints of air pollution. How many different ways can a representative select 5 of these to investigate this week? 126

59. **Selecting Officers** In a board of directors composed of 8 people, how many ways can one chief executive officer, one director, and one treasurer be selected? 336

Extending the Concepts

60. **Selecting Coins** How many different ways can you select one or more coins if you have 2 nickels, 1 dime, and 1 half-dollar? 15

61. **People Seated in a Circle** In how many ways can 3 people be seated in a circle? 4? n? (*Hint:* Think of them standing in a line before they sit down and/or draw diagrams.) 2; 6; $(n-1)!$

62. **Seating in a Movie Theater** How many different ways can 5 people—A, B, C, D, and E—sit in a row at a movie theater if (*a*) A and B must sit together; (*b*) C must sit to the right of, but not necessarily next to, B; (*c*) D and E will not sit next to each other? *a.* 48 *b.* 60 *c.* 72

63. **Poker Hands** Using combinations, calculate the number of each poker hand in a deck of cards. (A poker hand consists of 5 cards dealt in any order.)
  a. Royal flush 4
  b. Straight flush 36
  c. Four of a kind 624
  d. Full house 3744

Technology *Step by Step*

TI-83 Plus or TI-84 Plus
Step by Step

Factorials, Permutations, and Combinations

Factorials $n!$
1. Type the value of n.
2. Press **MATH** and move the cursor to PRB, then press **4** for !.
3. Press **ENTER**.

Permutations $_nP_r$
1. Type the value of n.
2. Press **MATH** and move the cursor to PRB, then press **2** for $_nP_r$.
3. Type the value of r.
4. Press **ENTER**.

Combinations $_nC_r$
1. Type the value of n.
2. Press **MATH** and move the cursor to PRB, then press **3** for $_nC_r$.
3. Type the value of r.
4. Press **ENTER**.

```
5!
           120
8 nPr 3
           336
12 nCr 5
           792
```

Calculate $5!$, $_8P_3$, and $_{12}C_5$ (Examples 4–42, 4–44, and 4–48 from the text).

Excel
Step by Step

Permutations, Combinations, and Factorials

To find a value of a permutation, for example, $_5P_3$:

1. In an open cell in an Excel worksheet, select the Formulas tab on the toolbar. Then click the Insert function icon .

2. Select the Statistical function category, then the PERMUT function, and click [OK].

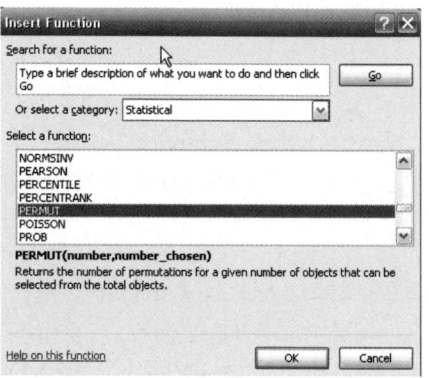

3. Type **5** in the Number box.

4. Type **3** in the Number_chosen box and click [OK].

The selected cell will display the answer: 60.

    To find a value of a combination, for example, $_5C_3$:

1. In an open cell, select the Formulas tab on the toolbar. Click the Insert function icon.

2. Select the All function category, then the COMBIN function, and click [OK].

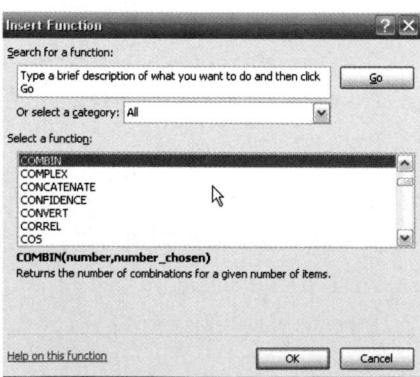

3. Type **5** in the Number box.

4. Type **3** in the Number_chosen box and click [OK].

The selected cell will display the answer: 10.

    To find a factorial of a number, for example, 7!:

1. In an open cell, select the Formulas tab on the toolbar. Click the Insert function icon.

2. Select the Math & Trig function category, then the FACT function, and click [OK].

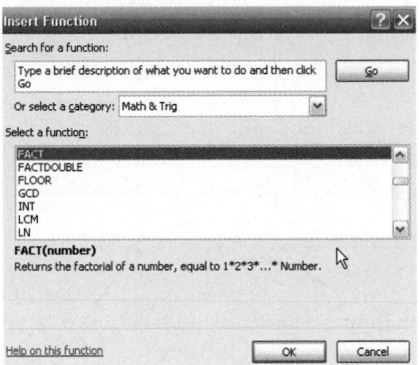

3. Type **7** in the Number box and click [OK].

The selected cell will display the answer: 5040.

4–5 Probability and Counting Rules

Objective **8**

Find the probability of an event, using the counting rules.

The counting rules can be combined with the probability rules in this chapter to solve many types of probability problems. By using the fundamental counting rule, the permutation rules, and the combination rule, you can compute the probability of outcomes of many experiments, such as getting a full house when 5 cards are dealt or selecting a committee of 3 women and 2 men from a club consisting of 10 women and 10 men.

Example 4–50 **Four Aces**

Find the probability of getting 4 aces when 5 cards are drawn from an ordinary deck of cards.

Solution

There are $_{52}C_5$ ways to draw 5 cards from a deck. There is only 1 way to get 4 aces (that is, $_4C_4$), but there are 48 possibilities to get the fifth card. Therefore, there are 48 ways to get 4 aces and 1 other card. Hence,

$$P(4 \text{ aces}) = \frac{_4C_4 \cdot 48}{_{52}C_5} = \frac{1 \cdot 48}{2{,}598{,}960} = \frac{48}{2{,}598{,}960} = \frac{1}{54{,}145}$$

Example 4–51   Defective Transistors

A box contains 24 transistors, 4 of which are defective. If 4 are sold at random, find the following probabilities.

    a. Exactly 2 are defective.   *c.* All are defective.
    b. None is defective.   *d.* At least 1 is defective.

Solution

There are $_{24}C_4$ ways to sell 4 transistors, so the denominator in each case will be 10,626.

    a. Two defective transistors can be selected as $_4C_2$ and two nondefective ones as $_{20}C_2$. Hence,

$$P(\text{exactly 2 defectives}) = \frac{_4C_2 \cdot _{20}C_2}{_{24}C_4} = \frac{1140}{10{,}626} = \frac{190}{1771}$$

    b. The number of ways to choose no defectives is $_{20}C_4$. Hence,

$$P(\text{no defectives}) = \frac{_{20}C_4}{_{24}C_4} = \frac{4845}{10{,}626} = \frac{1615}{3542}$$

    c. The number of ways to choose 4 defectives from 4 is $_4C_4$, or 1. Hence,

$$P(\text{all defective}) = \frac{1}{_{24}C_4} = \frac{1}{10{,}626}$$

    d. To find the probability of at least 1 defective transistor, find the probability that there are no defective transistors, and then subtract that probability from 1.

$$P(\text{at least 1 defective}) = 1 - P(\text{no defectives})$$

$$= 1 - \frac{_{20}C_4}{_{24}C_4} = 1 - \frac{1615}{3542} = \frac{1927}{3542}$$

Example 4–52   Magazines

A store has 6 *TV Graphic* magazines and 8 *Newstime* magazines on the counter. If two customers purchased a magazine, find the probability that one of each magazine was purchased.

Solution

$$P(1 \text{ } TV \text{ } Graphic \text{ and } 1 \text{ } Newstime) = \frac{_6C_1 \cdot _8C_1}{_{14}C_2} = \frac{6 \cdot 8}{91} = \frac{48}{91}$$

Example 4–53	Combination Lock

A combination lock consists of the 26 letters of the alphabet. If a 3-letter combination is needed, find the probability that the combination will consist of the letters ABC in that order. The same letter can be used more than once. (*Note:* A combination lock is really a permutation lock.)

Solution

Since repetitions are permitted, there are $26 \cdot 26 \cdot 26 = 17{,}576$ different possible combinations. And since there is only one ABC combination, the probability is $P(\text{ABC}) = 1/26^3 = 1/17{,}576$.

Example 4–54	Tennis Tournament

There are 8 married couples in a tennis club. If 1 man and 1 woman are selected at random to plan the summer tournament, find the probability that they are married to each other.

Solution

Since there are 8 ways to select the man and 8 ways to select the woman, there are $8 \cdot 8$, or 64, ways to select 1 man and 1 woman. Since there are 8 married couples, the solution is $\frac{8}{64} = \frac{1}{8}$.

As indicated at the beginning of this section, the counting rules and the probability rules can be used to solve a large variety of probability problems found in business, gambling, economics, biology, and other fields.

Applying the Concepts **4–5**

Counting Rules and Probability

One of the biggest problems for students when doing probability problems is to decide which formula or formulas to use. Another problem is to decide whether two events are independent or dependent. Use the following problem to help develop a better understanding of these concepts.

Assume you are given a 5-question multiple-choice quiz. Each question has 5 possible answers: A, B, C, D, and E.

1. How many events are there?
2. Are the events independent or dependent?
3. If you guess at each question, what is the probability that you get all of them correct?
4. What is the probability that a person would guess answer A for each question?

Assume that you are given a test in which you are to match the correct answers in the right column with the questions in the left column. You can use each answer only once.

5. How many events are there?
6. Are the events independent or dependent?
7. What is the probability of getting them all correct if you are guessing?
8. What is the difference between the two problems?

See page 250 for the answers.

Speaking of
Statistics

The Mathematics of Gambling

Gambling is big business. There are state lotteries, casinos, sports betting, and church bingos. It seems that today everybody is either watching or playing Texas Hold 'Em Poker.

Using permutations, combinations, and the probability rules, mathematicians can find the probabilities of various gambling games. Here are the probabilities of the various 5-card poker hands.

Hand	Number of ways	Probability
Straight flush	40	0.000015
Four of a kind	624	0.000240
Full house	3,744	0.001441
Flush	5,108	0.001965
Straight	10,200	0.003925
Three of a kind	54,912	0.021129
Two pairs	123,552	0.047539
One pair	1,098,240	0.422569
Less than one pair	1,302,540	0.501177
Total	2,598,960	1.000000

The chance of winning at gambling games can be compared by using what is called the house advantage, house edge, or house percentage. For example, the house advantage for roulette is about 5.26%, which means in the long run, the house wins 5.26 cents on every $1 bet; or you will lose, on average, 5.26 cents on every $1 you bet. The lower the house advantage, the more favorable the game is to you.

For the game of craps, the house advantage is anywhere between 1.4 and 15%, depending on what you bet on. For the game called keno, the house advantage is 29.5%. The house of advantage for Chuck-a-Luck is 7.87%, and for baccarat, it is either 1.36 or 1.17% depending on your bet.

Slot machines have a house advantage anywhere from about 4 to 10% depending on the geographic location, such as Atlantic City, Las Vegas, and Mississippi, and the amount put in the machine, such as 5¢, 25¢, and $1.

Actually, gamblers found winning strategies for the game blackjack or 21 such as card counting. However, the casinos retaliated by using multiple decks and by banning card counters.

Exercises 4–5

1. **Selecting Cards** Find the probability of getting 2 face cards (king, queen, or jack) when 2 cards are drawn from a deck without replacement. $\frac{11}{221}$

2. **Selecting a Committee** A parent-teacher committee consisting of 4 people is to be formed from 20 parents and 5 teachers. Find the probability that the committee will consist of these people. (Assume that the selection will be random.)

    a. All teachers $\frac{1}{2530}$

    b. 2 teachers and 2 parents $\frac{38}{253}$

    c. All parents $\frac{969}{2530}$

    d. 1 teacher and 3 parents $\frac{114}{253}$

3. **Management Seminar** In a company there are 7 executives: 4 women and 3 men. Three are selected to attend a management seminar. Find these probabilities.

    a. All 3 selected will be women. $\frac{4}{35}$

    b. All 3 selected will be men. $\frac{1}{35}$

    c. 2 men and 1 woman will be selected. $\frac{12}{35}$

    d. 1 man and 2 women will be selected. $\frac{18}{35}$

4. Senate Partisanship The composition of the Senate of the 111th Congress is

41 Republicans   2 Independent   57 Democrats

A new committee is being formed to study ways to benefit the arts in education. If 3 Senators are selected at random to head the committee, what is the probability that they will all be Republicans? What is the probability that they will all be Democrats? What is the probability that there will be 1 from each party, including the Independent? 0.0659; 0.1810; 0.0289

Source: *New York Times Almanac.*

5. Congressional Committee Memberships The composition of the 108th Congress was 51 Republicans, 48 Democrats, and 1 Independent. A committee on aid to higher education is to be formed with 3 Senators to be chosen at random to head the committee. Find the probability that the group of 3 consists of

a. All Republicans 0.129

b. All Democrats 0.107

c. 1 Democrat, 1 Republican, and 1 Independent 0.0908

6. Defective Resistors A package contains 12 resistors, 3 of which are defective. If 4 are selected, find the probability of getting

a. 0 defective resistors $\frac{14}{55}$

b. 1 defective resistor $\frac{28}{55}$

c. 3 defective resistors $\frac{1}{55}$

7. Winning Tickets If 50 tickets are sold and 2 prizes are to be awarded, find the probability that one person will win 2 prizes if that person buys 2 tickets. $\frac{1}{1225}$

8. Getting a Full House Find the probability of getting a full house (3 cards of one denomination and 2 of another) when 5 cards are dealt from an ordinary deck. $\frac{18}{12,495} = \frac{6}{4165}$

9. Flight School Graduation At a recent graduation at a naval flight school, 18 Marines, 10 members of the Navy, and 3 members of the Coast Guard got their wings. Choose 3 pilots at random to feature on a training brochure. Find the probability that there will be

a. 1 of each 0.120

b. 0 members of the Navy 0.296

c. 3 Marines 0.182

10. Selecting Cards The red face cards and the black cards numbered 2–9 are put into a bag. Four cards are drawn at random without replacement. Find the following probabilities:

a. All 4 cards are red. 0.002

b. 2 cards are red and 2 cards are black. 0.246

c. At least 1 of the cards is red. 0.751

d. All 4 cards are black. 0.249

11. Socks in a Drawer A drawer contains 11 identical red socks and 8 identical black socks. Suppose that you choose 2 socks at random in the dark.

a. What is the probability that you get a pair of red socks? 0.3216

b. What is the probability that you get a pair of black socks? 0.1637

c. What is the probability that you get 2 unmatched socks? 0.5146

d. Where did the other red sock go? It probably got lost in the wash!

12. Selecting Books Find the probability of selecting 3 science books and 4 math books from 8 science books and 9 math books. The books are selected at random. $\frac{882}{2431}$

13. Rolling Three Dice When 3 dice are rolled, find the probability of getting a sum of 7. $\frac{5}{72}$

14. Football Team Selection A football team consists of 20 each freshmen and sophomores, 15 juniors, and 10 seniors. Four players are selected at random to serve as captains. Find the probability that

a. All 4 are seniors 0.0003

b. There is 1 each: freshman, sophomore, junior, and senior 0.089

c. There are 2 sophomores and 2 freshmen 0.053

d. At least 1 of the students is a senior 0.496

15. Arrangement of Washers Find the probability that if 5 different-sized washers are arranged in a row, they will be arranged in order of size. $\frac{1}{60}$

16. Using the information in Exercise 63 in Section 4–4, find the probability of each poker hand.

a. Royal flush $\frac{4}{2,598,960}$

b. Straight flush $\frac{36}{2,598,960}$

c. Four of a kind $\frac{624}{2,598,960}$

17. Plant Selection All holly plants are dioecious—a male plant must be planted within 30 to 40 feet of the female plants in order to yield berries. A home improvement store has 12 unmarked holly plants for sale, 8 of which are female. If a homeowner buys 3 plants at random, what is the probability that berries will be produced? 0.727

Summary

In this chapter, the basic concepts of probability are explained.

- There are three basic types of probability. They are classical probability, empirical probability, and subjective probability. Classical probability uses samples spaces. Empirical probability uses frequency distributions, and subjective probability uses an educated guess to determine the probability of an event. The probability of any event is a number from 0 to 1. If an event cannot occur, the probability is 0. If an event is certain, the probability is 1. The sum of the probability of all the events in the sample space is 1. To find the probability of the complement of an event, subtract the probability of the event from 1. (4–1)

- Two events are mutually exclusive if they cannot occur at the same time; otherwise, the events are not mutually exclusive. To find the probability of two mutually exclusive events occurring, add the probability of each event. To find the probability of two events when they are not mutually exclusive, add the possibilities of the individual events and then subtract the probability that both events occur at the same time. These types of probability problems can be solved by using the addition rules. (4–2)

- Two events are independent if the occurrence of the first event does not change the probability of the second event occurring. Otherwise, the events are dependent. To find the probability of two independent events occurring, multiply the probabilities of each event. To find the probability that two dependent events occur, multiply the probability that the first event occurs by the probability that the second event occurs given that the first event has already occurred. The complement of an event is found by selecting the outcomes in the sample space that are not involved in the outcomes of the event. These types of problems can be solved by using the multiplication rules and the complementary event rules. (4–3)

- Finally, when a large number of events can occur, the fundamental counting rule, the permutation rule, and the combination rule can be used to determine the number of ways that these events can occur. (4–4)

- The counting rules and the probability rules can be used to solve more-complex probability problems. (4–5)

Important Terms

classical probability 186

combination 229

complement of an event 189

compound event 186

conditional probability 213

dependent events 213

empirical probability 191

equally likely events 186

event 185

fundamental counting rule 224

independent events 211

law of large numbers 194

mutually exclusive events 199

outcome 183

permutation 227

probability 182

probability experiment 183

sample space 183

simple event 185

subjective probability 194

tree diagram 185

Venn diagrams 190

Important Formulas

Formula for classical probability:

$$P(E) = \frac{\text{number of outcomes in } E}{\text{total number of outcomes in sample space}} = \frac{n(E)}{n(S)}$$

Formula for empirical probability:

$$P(E) = \frac{\text{frequency for class}}{\text{total frequencies in distribution}} = \frac{f}{n}$$

Addition rule 1, for two mutually exclusive events:

$$P(A \text{ or } B) = P(A) + P(B)$$

Addition rule 2, for events that are not mutually exclusive:

$$P(A \text{ or } B) = P(A) + P(B) - P(A \text{ and } B)$$

Multiplication rule 1, for independent events:

$$P(A \text{ and } B) = P(A) \cdot P(B)$$

Multiplication rule 2, for dependent events:

$$P(A \text{ and } B) = P(A) \cdot P(B|A)$$

Formula for conditional probability:

$$P(B|A) = \frac{P(A \text{ and } B)}{P(A)}$$

Formula for complementary events:

$$P(\overline{E}) = 1 - P(E) \quad \text{or} \quad P(E) = 1 - P(\overline{E})$$
$$\text{or} \quad P(E) + P(\overline{E}) = 1$$

Fundamental counting rule: In a sequence of n events in which the first one has k_1 possibilities, the second event has k_2 possibilities, the third has k_3 possibilities, etc., the total number of possibilities of the sequence will be

$$k_1 \cdot k_2 \cdot k_3 \cdots k_n$$

Permutation rule: The number of permutations of n objects taking r objects at a time when order is important is

$$_nP_r = \frac{n!}{(n-r)!}$$

Combination rule: The number of combinations of r objects selected from n objects when order is not important is

$$_nC_r = \frac{n!}{(n-r)!r!}$$

Review Exercises

1. When a standard die is rolled, find the probability of getting
   a. A 5   0.167
   b. A number larger than 2   0.667
   c. An odd number (4–1)   0.5

2. **Selecting a Card** When a card is selected from a deck, find the probability of getting
   a. A club $\frac{1}{4}$
   b. A face card or a heart $\frac{11}{26}$
   c. A 6 and a spade $\frac{1}{52}$
   d. A king $\frac{1}{13}$
   e. A red card (4–1) $\frac{1}{2}$

3. **Software Selection** The top-10 selling computer software titles last year consisted of 3 for doing taxes, 5 antivirus or security programs, and 2 "other." Choose one title at random.
   a. What is the probability that it is not used for doing taxes?   0.7
   b. What is the probability that it is used for taxes or is one of the "other" programs? (4–1)   0.5

   Source: www.infoplease.com

4. A six-sided die is printed with the numbers 1, 2, 3, 5, 8, and 13. Roll the die once—what is the probability of getting an even number? Roll the die twice and add the numbers. What is the probability of getting an odd sum on the dice? (4–1)   0.333; 0.444

5. **Breakfast Drink** In a recent survey, 18 people preferred milk, 29 people preferred coffee, and 13 people preferred juice as their primary drink for breakfast. If a person is selected at random, find the probability that the person preferred juice as her or his primary drink. (4–1) $\frac{13}{60}$

6. **Purchasing Sweaters** During a sale at a men's store, 16 white sweaters, 3 red sweaters, 9 blue sweaters, and 7 yellow sweaters were purchased. If a customer is selected at random, find the probability that he bought.
   a. A blue sweater $\frac{9}{35}$
   b. A yellow or a white sweater $\frac{23}{35}$
   c. A red, a blue, or a yellow sweater $\frac{19}{35}$
   d. A sweater that was not white (4–2) $\frac{19}{35}$

7. **Budget Rental Cars** Cheap Rentals has nothing but budget cars for rental. The probability that a car has air conditioning is 0.5, and the probability that a car has a CD player is 0.37. The probability that a car has both air conditioning and a CD player is 0.06. What is the probability that a randomly selected car has neither air conditioning nor a CD player? (4–2)   0.19

8. **Rolling Two Dice** When two dice are rolled, find the probability of getting
   a. A sum of 5 or 6 $\frac{1}{4}$
   b. A sum greater than 9 $\frac{1}{6}$
   c. A sum less than 4 or greater than 9 $\frac{1}{4}$
   d. A sum that is divisible by 4 $\frac{1}{4}$
   e. A sum of 14   0
   f. A sum less than 13 (4–1)   1

9. **Car and Boat Ownership** The probability that a person owns a car is 0.80, that a person owns a boat is 0.30, and that a person owns both a car and a boat is 0.12. Find the probability that a person owns either a boat or a car. (4–2)   0.98

10. **Car Purchases** There is a 0.39 probability that John will purchase a new car, a 0.73 probability that Mary will purchase a new car, and a 0.36 probability that both will purchase a new car. Find the probability that neither will purchase a new car. (4–2)   0.24

11. **Online Course Selection** Roughly 1 in 6 students enrolled in higher education took at least one online course last fall. Choose 5 enrolled students at random. Find the probability that
    a. All 5 took online courses   0.0001
    b. None of the 5 took a course online   0.402
    c. At least 1 took an online course (4–2)   0.598

    Source: www.encarta.msn.com

12. **Borrowing Books** Of Americans using library services, 67% borrow books. If 5 patrons are chosen at random, what is the probability that all borrowed books? That none borrowed books? (4–3)

    Source: American Library Association.   0.1350; 0.0039

13. **Drawing Cards** Three cards are drawn from an ordinary deck *without* replacement. Find the probability of getting

    a. All black cards $\frac{2}{17}$

    b. All spades $\frac{11}{850}$

    c. All queens (4–3) $\frac{1}{5525}$

14. **Coin Toss and Card Drawn** A coin is tossed and a card is drawn from a deck. Find the probability of getting

    a. A head and a 6 $\frac{1}{26}$

    b. A tail and a red card $\frac{1}{4}$

    c. A head and a club (4–3) $\frac{1}{8}$

15. **Movie Releases** The top five countries for movie releases so far this year are the United States with 471 releases, United Kingdom with 386, Japan with 79, Germany with 316, and France with 132. Choose 1 new release at random. Find the probability that it is

    a. European 0.603

    b. From the United States 0.340

    c. German or French 0.324

    d. German given that it is European (4–2) 0.379

    Source: www.showbizdata.com

16. **Factory Output** A manufacturing company has three factories: X, Y, and Z. The daily output of each is shown here.

Product	Factory X	Factory Y	Factory Z
TVs	18	32	15
Stereos	6	20	13

    If one item is selected at random, find these probabilities.

    a. It was manufactured at factory X or is a stereo. $\frac{57}{104}$

    b. It was manufactured at factory Y or factory Z. $\frac{10}{13}$

    c. It is a TV or was manufactured at factory Z. (4–3) $\frac{3}{4}$

17. **Effectiveness of Vaccine** A vaccine has a 90% probability of being effective in preventing a certain disease. The probability of getting the disease if a person is not vaccinated is 50%. In a certain geographic region, 25% of the people get vaccinated. If a person is selected at random, find the probability that he or she will contract the disease. (4–3) 0.4

18. **Television Models** A manufacturer makes three models of a television set, models A, B, and C. A store sells 40% of model A sets, 40% of model B sets, and 20% of model C sets. Of model A sets, 3% have stereo sound; of model B sets, 7% have stereo sound; and of model C sets, 9% have stereo sound. If a set is sold at random, find the probability that it has stereo sound. (4–3) 5.8%

19. **Car Purchase** The probability that Sue will live on campus and buy a new car is 0.37. If the probability

that she will live on campus is 0.73, find the probability that she will buy a new car, given that she lives on campus. (4–3) 0.51

20. **Applying Shipping Labels** Four unmarked packages have lost their shipping labels, and you must reapply them. What is the probability that you apply the labels and get all 4 of them correct? Exactly 3 correct? Exactly 2? At least 1 correct? (4–3) 0.0417; impossible; 0.25; 0.625

21. **Health Club Membership** Of the members of the Blue River Health Club, 43% have a lifetime membership and exercise regularly (three or more times a week). If 75% of the club members exercise regularly, find the probability that a randomly selected member is a life member, given that he or she exercises regularly. (4–3) 57.3%

22. **Bad Weather** The probability that it snows and the bus arrives late is 0.023. José hears the weather forecast, and there is a 40% chance of snow tomorrow. Find the probability that the bus will be late, given that it snows. (4–3) 0.058

23. **Education Level and Smoking** At a large factory, the employees were surveyed and classified according to their level of education and whether they smoked. The data are shown in the table.

	Educational level		
Smoking habit	Not high school graduate	High school graduate	College graduate
Smoke	6	14	19
Do not smoke	18	7	25

    If an employee is selected at random, find these probabilities.

    a. The employee smokes, given that he or she graduated from college. $\frac{19}{44}$

    b. Given that the employee did not graduate from high school, he or she is a smoker. (4–3) $\frac{1}{4}$

24. **War Veterans** Approximately 11% of the civilian population are veterans. Choose 5 civilians at random. What is the probability that none are veterans? What is the probability that at least 1 is a veteran? (4–3) 0.558; 0.442

    Source: www.factfinder.census.gov

25. **DVD Players** Eighty-one percent of U.S. households have DVD players. Choose 6 households at random. What is the probability that at least 1 does not have a DVD player? (4–3) 0.718

    Source: www.infoplease.com

26. **Chronic Sinusitis** The U.S. Department of Health and Human Services reports that 15% of Americans have chronic sinusitis. If 5 people are selected at random, find the probability that at least 1 has chronic sinusitis. (4–3) 55.6%

    Source: *100% American.*

27. **Automobile License Plate** An automobile license plate consists of 3 letters followed by 4 digits. How many different plates can be made if repetitions are allowed? If repetitions are not allowed? If repetitions are allowed in the letters but not in the digits? (4–4) 175,760,000; 78,624,000; 88,583,040

28. **Types of Copy Paper** White copy paper is offered in 5 different strengths and 11 different degrees of brightness, recycled or not, and acid-free or not. How many different types of paper are available for order? (4–4) 220

29. **Baseball Players** How many ways can 3 outfielders and 4 infielders be chosen from 5 outfielders and 7 infielders? (4–4) 350

30. **Computer Operators** How many different ways can 8 computer operators be seated in a row? (4–4) 40,320

31. **Student Representatives** How many ways can a student select 2 electives from a possible choice of 10 electives? (4–4) 45

32. **Committee Representation** There are 6 Republican, 5 Democrat, and 4 Independent candidates. How many different ways can a committee of 3 Republicans, 2 Democrats, and 1 Independent be selected? (4–4) 800

33. **Song Selections** A promotional MP3 player is available with the capacity to store 100 songs which can be reordered at the push of a button. How many different arrangements of these songs are possible? (*Note:* Factorials get very big, very fast! How large a factorial will your calculator calculate?) (4–4) 100! (Answers may vary regarding calculator.)

34. **Employee Health Care Plans** A new employee has a choice of 5 health care plans, 3 retirement plans, and 2 different expense accounts. If a person selects 1 of each option, how many different options does he or she have? (4–4) 30

35. **Course Enrollment** There are 12 students who wish to enroll in a particular course. There are only 4 seats left in the classroom. How many different ways can 4 students be selected to attend the class? (4–4) 495

36. **Candy Selection** A candy store allows customers to select 3 different candies to be packaged and mailed. If there are 13 varieties available, how many possible selections can be made? (4–4) 286

37. **Book Selection** If a student can select 5 novels from a reading list of 20 for a course in literature, how many different possible ways can this selection be done? (4–4) 15,504

38. **Course Selection** If a student can select one of 3 language courses, one of 5 mathematics courses, and one of 4 history courses, how many different schedules can be made? (4–4) 60

39. **License Plates** License plates are to be issued with 3 letters followed by 4 single digits. How many such license plates are possible? If the plates are issued at random, what is the probability that the license plate says USA followed by a number that is divisible by 5? (4–5) 175,760,000; 0.0000114

40. **Leisure Activities** A newspaper advertises 5 different movies, 3 plays, and 2 baseball games for the weekend. If a couple selects 3 activities, find the probability that they attend 2 plays and 1 movie. (4–5) $\frac{1}{8}$

41. **Territorial Selection** Several territories and colonies today are still under the jurisdiction of another country. France holds the most with 16 territories, the United Kingdom has 15, the United States has 14, and several other countries have territories as well. Choose 3 territories at random from those held by France, the United Kingdom, and the United States. What is the probability that all 3 belong to the same country? (4–5)

Source: www.infoplease.com 0.097

42. **Yahtzee** Yahtzee is a game played with 5 dice. Players attempt to score points by rolling various combinations. When all 5 dice show the same number, it is called a *Yahtzee* and scores 50 points for the first one and 100 points for each subsequent Yahtzee in the same game. What is the probability that a person throws a Yahtzee on the very first roll? What is the probability that a person throws two Yahtzees on two successive turns? (4–5) 0.000772; 0.0000006

43. **Personnel Classification** For a survey, a subject can be classified as follows:
Gender: male or female
Marital status: single, married, widowed, divorced
Occupation: administration, faculty, staff
Draw a tree diagram for the different ways a person can be classified. (4–4)

Statistics Today

Would You Bet Your Life?–Revisited

In his book *Probabilities in Everyday Life*, John D. McGervey states that the chance of being killed on any given commercial airline flight is almost 1 in 1 million and that the chance of being killed during a transcontinental auto trip is about 1 in 8000. The corresponding probabilities are 1/1,000,000 = 0.000001 as compared to 1/8000 = 0.000125. Since the second number is 125 times greater than the first number, you have a much higher risk driving than flying across the United States.

Chapter Quiz

Determine whether each statement is true or false. If the statement is false, explain why.

1. Subjective probability has little use in the real world. False

2. Classical probability uses a frequency distribution to compute probabilities. False

3. In classical probability, all outcomes in the sample space are equally likely. True

4. When two events are not mutually exclusive, $P(A \text{ or } B) = P(A) + P(B)$. False

5. If two events are dependent, they must have the same probability of occurring. False

6. An event and its complement can occur at the same time. False

7. The arrangement ABC is the same as BAC for combinations. True

8. When objects are arranged in a specific order, the arrangement is called a combination. False

Select the best answer.

9. The probability that an event happens is 0.42. What is the probability that the event won't happen?
   - a. −0.42
   - b. 0.58
   - c. 0
   - d. 1

10. When a meteorologist says that there is a 30% chance of showers, what type of probability is the person using?
   - a. Classical
   - b. Empirical
   - c. Relative
   - d. Subjective

11. The sample space for tossing 3 coins consists of how many outcomes?
   - a. 2
   - b. 4
   - c. 6
   - d. 8

12. The complement of guessing 5 correct answers on a 5-question true/false exam is
   - a. Guessing 5 incorrect answers
   - b. Guessing at least 1 incorrect answer
   - c. Guessing at least 1 correct answer
   - d. Guessing no incorrect answers

13. When two dice are rolled, the sample space consists of how many events?
   - a. 6
   - b. 12
   - c. 36
   - d. 54

14. What is $_nP_0$?
   - a. 0
   - b. 1
   - c. n
   - d. It cannot be determined.

15. What is the number of permutations of 6 different objects taken all together?
   - a. 0
   - b. 1
   - c. 36
   - d. 720

16. What is 0!?
   - a. 0
   - b. 1
   - c. Undefined
   - d. 10

17. What is $_nC_n$?
   - a. 0
   - b. 1
   - c. n
   - d. It cannot be determined.

Complete the following statements with the best answer.

18. The set of all possible outcomes of a probability experiment is called the _____. Sample space

19. The probability of an event can be any number between and including _____ and _____. 0, 1

20. If an event cannot occur, its probability is _____. 0

21. The sum of the probabilities of the events in the sample space is _____. 1

22. When two events cannot occur at the same time, they are said to be _____. Mutually exclusive

23. When a card is drawn, find the probability of getting
   - a. A jack $\frac{1}{13}$
   - b. A 4 $\frac{1}{13}$
   - c. A card less than 6 (an ace is considered above 6) $\frac{4}{13}$

24. **Selecting a Card** When a card is drawn from a deck, find the probability of getting
   - a. A diamond $\frac{1}{4}$
   - b. A 5 or a heart $\frac{4}{13}$
   - c. A 5 and a heart $\frac{1}{52}$
   - d. A king $\frac{1}{13}$
   - e. A red card $\frac{1}{2}$

25. **Selecting a Sweater** At a men's clothing store, 12 men purchased blue golf sweaters, 8 purchased green sweaters, 4 purchased gray sweaters, and 7 bought black sweaters. If a customer is selected at random, find the probability that he purchased
   - a. A blue sweater $\frac{12}{31}$
   - b. A green or gray sweater $\frac{12}{31}$
   - c. A green or black or blue sweater $\frac{27}{31}$
   - d. A sweater that was not black $\frac{24}{31}$

26. **Rolling Dice** When 2 dice are rolled, find the probability of getting
   - a. A sum of 6 or 7 $\frac{11}{36}$
   - b. A sum greater than 8 $\frac{5}{18}$
   - c. A sum less than 3 or greater than 8 $\frac{11}{36}$
   - d. A sum that is divisible by 3 $\frac{1}{3}$
   - e. A sum of 16 0
   - f. A sum less than 11 $\frac{11}{12}$

27. Appliance Ownership The probability that a person owns a microwave oven is 0.75, that a person owns a compact disk player is 0.25, and that a person owns both a microwave and a CD player is 0.16. Find the probability that a person owns either a microwave or a CD player, but not both. 0.68

28. Starting Salaries Of the physics graduates of a university, 30% received a starting salary of $30,000 or more. If 5 of the graduates are selected at random, find the probability that all had a starting salary of $30,000 or more. 0.002

29. Selecting Cards Five cards are drawn from an ordinary deck *without* replacement. Find the probability of getting

   a. All red cards $\frac{253}{9996}$
   b. All diamonds $\frac{33}{66,640}$
   c. All aces 0

30. Scholarships The probability that Samantha will be accepted by the college of her choice and obtain a scholarship is 0.35. If the probability that she is accepted by the college is 0.65, find the probability that she will obtain a scholarship given that she is accepted by the college. 0.54

31. New Car Warranty The probability that a customer will buy a car and an extended warranty is 0.16. If the probability that a customer will purchase a car is 0.30, find the probability that the customer will also purchase the extended warranty. 0.53

32. Bowling and Club Membership Of the members of the Spring Lake Bowling Lanes, 57% have a lifetime membership and bowl regularly (three or more times a week). If 70% of the club members bowl regularly, find the probability that a randomly selected member is a lifetime member, given that he or she bowls regularly. 0.81

33. Work and Weather The probability that Mike has to work overtime and it rains is 0.028. Mike hears the weather forecast, and there is a 50% chance of rain. Find the probability that he will have to work overtime, given that it rains. 0.056

34. Education of Factory Employees At a large factory, the employees were surveyed and classified according to their level of education and whether they attend a sports event at least once a month. The data are shown in the table.

	Educational level		
Sports event	**High school graduate**	**Two-year college degree**	**Four-year college degree**
Attend	16	20	24
Do not attend	12	19	25

If an employee is selected at random, find the probability that

   a. The employee attends sports events regularly, given that he or she graduated from college (2- or 4-year degree) $\frac{1}{2}$
   b. Given that the employee is a high school graduate, he or she does not attend sports events regularly $\frac{3}{7}$

35. Heart Attacks In a certain high-risk group, the chances of a person having suffered a heart attack are 55%. If 6 people are chosen, find the probability that at least 1 will have had a heart attack. 0.99

36. Rolling a Die A single die is rolled 4 times. Find the probability of getting at least one 5. 0.518

37. Eye Color If 85% of all people have brown eyes and 6 people are selected at random, find the probability that at least 1 of them has brown eyes. 0.9999886

38. Singer Selection How many ways can 5 sopranos and 4 altos be selected from 7 sopranos and 9 altos? 2646

39. Speaker Selection How many different ways can 8 speakers be seated on a stage? 40,320

40. Stocking Machines A soda machine servicer must restock and collect money from 15 machines, each one at a different location. How many ways can she select 4 machines to service in 1 day? 1365

41. ID Cards One company's ID cards consist of 5 letters followed by 2 digits. How many cards can be made if repetitions are allowed? If repetitions are not allowed? 1,188,137,600; 710,424,000

42. How many different arrangements of the letters in the word *number* can be made? 720

43. Physics Test A physics test consists of 25 true/false questions. How many different possible answer keys can be made? 33,554,432

44. Cellular Telephones How many different ways can 3 cellular telephones be selected from 8 cellular phones? 56

45. Fruit Selection On a lunch counter, there are 3 oranges, 5 apples, and 2 bananas. If 3 pieces of fruit are selected, find the probability that 1 orange, 1 apple, and 1 banana are selected. $\frac{1}{4}$

46. Cruise Ship Activities A cruise director schedules 4 different movies, 2 bridge games, and 3 tennis games for a two-day period. If a couple selects 3 activities, find the probability that they attend 2 movies and 1 tennis game. $\frac{3}{14}$

47. Committee Selection At a sorority meeting, there are 6 seniors, 4 juniors, and 2 sophomores. If a committee of 3 is to be formed, find the probability that 1 of each will be selected. $\frac{12}{55}$

48. Banquet Meal Choices For a banquet, a committee can select beef, pork, chicken, or veal; baked potatoes or mashed potatoes; and peas or green beans for a vegetable. Draw a tree diagram for all possible choices of a meat, a potato, and a vegetable.

Critical Thinking Challenges

1. **Con Man Game** Consider this problem: A con man has 3 coins. One coin has been specially made and has a head on each side. A second coin has been specially made, and on each side it has a tail. Finally, a third coin has a head and a tail on it. All coins are of the same denomination. The con man places the 3 coins in his pocket, selects one, and shows you one side. It is heads. He is willing to bet you even money that it is the two-headed coin. His reasoning is that it can't be the two-tailed coin since a head is showing; therefore, there is a 50-50 chance of it being the two-headed coin. Would you take the bet? (*Hint:* See Exercise 1 in Data Projects.)

2. **de Méré Dice Game** Chevalier de Méré won money when he bet unsuspecting patrons that in 4 rolls of 1 die, he could get at least one 6; but he lost money when he bet that in 24 rolls of 2 dice, he could get at least a double 6. Using the probability rules, find the probability of each event and explain why he won the majority of the time on the first game but lost the majority of the time when playing the second game. (*Hint:* Find the probabilities of losing each game and subtract from 1.)

3. **Classical Birthday Problem** How many people do you think need to be in a room so that 2 people will have the same birthday (month and day)? You might think it is 366. This would, of course, guarantee it (excluding leap year), but how many people would need to be in a room so that there would be a 90% probability that 2 people would be born on the same day? What about a 50% probability?

   Actually, the number is much smaller than you may think. For example, if you have 50 people in a room, the probability that 2 people will have the same birthday is 97%. If you have 23 people in a room, there is a 50% probability that 2 people were born on the same day!

   The problem can be solved by using the probability rules. It must be assumed that all birthdays are equally likely, but this assumption will have little effect on the answers. The way to find the answer is by using the complementary event rule as $P(2$ people having the same birthday$) = 1 - P($all have different birthdays$)$.

For example, suppose there were 3 people in the room. The probability that each had a different birthday would be

$$\frac{365}{365} \cdot \frac{364}{365} \cdot \frac{363}{365} = \frac{_{365}P_3}{365^3} = 0.992$$

Hence, the probability that at least 2 of the 3 people will have the same birthday will be

$$1 - 0.992 = 0.008$$

Hence, for k people, the formula is

$P($at least 2 people have the same birthday$)$

$$= 1 - \frac{_{365}P_k}{365^k}$$

Using your calculator, complete the table and verify that for at least a 50% chance of 2 people having the same birthday, 23 or more people will be needed.

Number of people	Probability that at least 2 have the same birthday
1	0.000
2	0.003
5	0.027
10	
15	
20	
21	
22	
23	

4. We know that if the probability of an event happening is 100%, then the event is a certainty. Can it be concluded that if there is a 50% chance of contracting a communicable disease through contact with an infected person, there would be a 100% chance of contracting the disease if 2 contacts were made with the infected person? Explain your answer.

 Data Projects

1. **Business and Finance** Select a pizza restaurant and a sandwich shop. For the pizza restaurant look at the menu to determine how many sizes, crust types, and toppings are available. How many different pizza types are possible? For the sandwich shop determine how many breads, meats, veggies, cheeses, sauces, and condiments are available. How many different sandwich choices are possible?

2. **Sports and Leisure** When poker games are shown on television, there are often percentages displayed that show how likely it is that a certain hand will win. Investigate how these percentages are determined. Show an example with two competing hands in a Texas Hold 'Em game. Include the percentages that each hand will win after the deal, the flop, the turn, and the river.

3. **Technology** A music player or music organization program can keep track of how many different artists are in a library. First note how many different artists are in your music library. Then find the probability that if 25 songs are selected at random, none will have the same artist.

4. **Health and Wellness** Assume that the gender distribution of babies is such that one-half the time females are born and one-half the time males are born. In a family of 3 children, what is the probability that all are girls? In a family of 4? Is it unusual that in a family with 4 children all would be girls? In a family of 5?

5. **Politics and Economics** Consider the U.S. Senate. Find out about the composition of any three of the Senate's standing committees. How many different committees of Senators are possible, knowing the party composition of the Senate and the number of committee members from each party for each committee?

6. **Your Class** Research the famous Monty Hall probability problem. Conduct a simulation of the Monty Hall problem online using a simulation program or in class using live "contestants." After 50 simulations compare your results to those stated in the research you did. Did your simulation support the conclusions?

Answers to Applying the Concepts

Section 4–1   Tossing a Coin

1. The sample space is the listing of all possible outcomes of the coin toss.

2. The possible outcomes are heads or tails.

3. Classical probability says that a fair coin has a 50-50 chance of coming up heads or tails.

4. The law of large numbers says that as you increase the number of trials, the overall results will approach the theoretical probability. However, since the coin has no "memory," it still has a 50-50 chance of coming up heads or tails on the next toss. Knowing what has already happened should not change your opinion on what will happen on the next toss.

5. The empirical approach to probability is based on running an experiment and looking at the results. You cannot do that at this time.

6. Subjective probabilities could be used if you believe the coin is biased.

7. Answers will vary; however, they should address that a fair coin has a 50-50 chance of coming up heads or tails on the next flip.

Section 4–2   Which Pain Reliever Is Best?

1. There were $192 + 186 + 188 = 566$ subjects in the study.

2. The study lasted for 12 weeks.

3. The variables are the type of pain reliever and the side effects.

4. Both variables are qualitative and nominal.

5. The numbers in the table are exact figures.

6. The probability that a randomly selected person was receiving a placebo is $192/566 = 0.3392$ (about 34%).

7. The probability that a randomly selected person was receiving a placebo or drug A is $(192 + 186)/566 = 378/566 = 0.6678$ (about 67%). These are mutually

exclusive events. The complement is that a randomly selected person was receiving drug B.

8. The probability that a randomly selected person was receiving a placebo or experienced a neurological headache is $(192 + 55 + 72)/566 = 319/566 = 0.5636$ (about 56%).

9. The probability that a randomly selected person was not receiving a placebo or experienced a sinus headache is $(186 + 188)/566 + 11/566 = 385/566 = 0.6802$ (about 68%).

Section 4–3   Guilty or Innocent?

1. The probability of another couple with the same characteristics being in that area is $\frac{1}{12} \cdot \frac{1}{10} \cdot \frac{1}{4} \cdot \frac{1}{11} \cdot \frac{1}{3} \cdot \frac{1}{13} \cdot \frac{1}{100} = \frac{1}{20,592,000}$, assuming the characteristics are independent of one another.

2. You would use the multiplication rule, since you are looking for the probability of multiple events happening together.

3. We do not know if the characteristics are dependent or independent, but we assumed independence for the calculation in question 1.

4. The probabilities would change if there were dependence among two or more events.

5. Answers will vary. One possible answer is that probabilities can be used to explain how unlikely it is to have a set of events occur at the same time (in this case, how unlikely it is to have another couple with the same characteristics in that area).

6. Answers will vary. One possible answer is that if the only eyewitness was the woman who was mugged and the probabilities are accurate, it seems very unlikely that a couple matching these characteristics would be in that area at that time. This might cause you to convict the couple.

7. Answers will vary. One possible answer is that our probabilities are theoretical and serve a purpose when appropriate, but that court cases are based on much more than impersonal chance.

8. Answers will vary. One possible answer is that juries decide whether to convict a defendant if they find evidence "beyond a reasonable doubt" that the person is guilty. In probability terms, this means that if the defendant was actually innocent, then the chance of seeing the events that occurred is so unlikely as to have occurred by chance. Therefore, the jury concludes that the defendant is guilty.

Section 4–4   Garage Door Openers

1. Four on/off switches lead to 16 different settings.

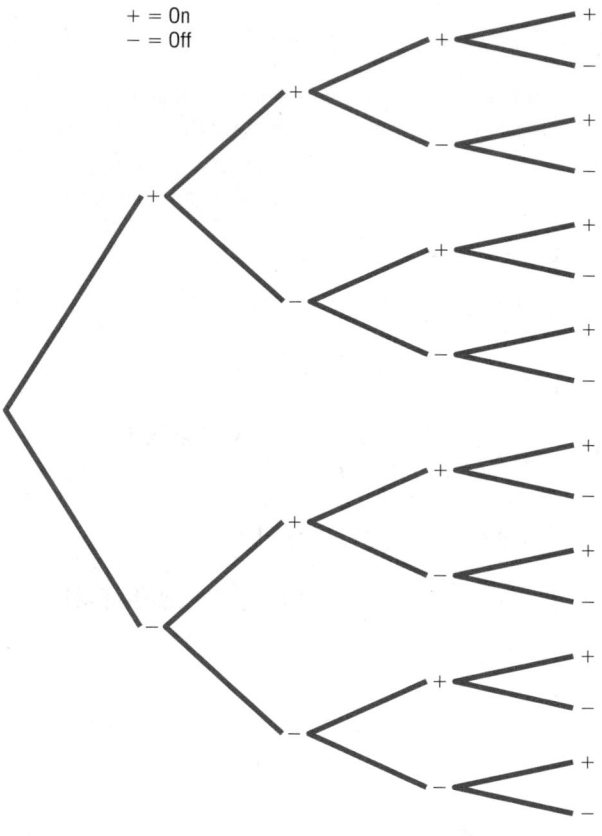

+ = On
− = Off

2. With 5 on/off switches, there are $2^5 = 32$ different settings. With 6 on/off switches, there are $2^6 = 64$ different settings. In general, if there are k on/off switches, there are 2^k different settings.

3. With 8 consecutive on/off switches, there are $2^8 = 256$ different settings.

4. It is less likely for someone to be able to open your garage door if you have 8 on/off settings (probability about 0.4%) than if you have 4 on/off switches (probability about 6.0%). Having 8 on/off switches in the opener seems pretty safe.

5. Each key blank could be made into $5^5 = 3125$ possible keys.

6. If there were 420,000 Dodge Caravans sold in the United States, then any one key could start about $420,000/3125 = 134.4$, or about 134, different Caravans.

7. Answers will vary.

Section 4–5   Counting Rules and Probability

1. There are five different events: each multiple-choice question is an event.

2. These events are independent.

3. If you guess on 1 question, the probability of getting it correct is 0.20. Thus, if you guess on all 5 questions, the probability of getting all of them correct is $(0.20)^5 = 0.00032$.

4. The probability that a person would guess answer A for a question is 0.20, so the probability that a person would guess answer A for each question is $(0.20)^5 = 0.00032$.

5. There are five different events: each matching question is an event.

6. These are dependent events.

7. The probability of getting them all correct if you are guessing is $\frac{1}{5} \cdot \frac{1}{4} \cdot \frac{1}{3} \cdot \frac{1}{2} \cdot \frac{1}{1} = \frac{1}{120} = 0.0083$.

8. The difference between the two problems is that we are sampling without replacement in the second problem, so the denominator changes in the event probabilities.

CHAPTER

5

Discrete Probability Distributions

Objectives

After completing this chapter, you should be able to

1 Construct a probability distribution for a random variable.

2 Find the mean, variance, standard deviation, and expected value for a discrete random variable.

3 Find the exact probability for X successes in n trials of a binomial experiment.

4 Find the mean, variance, and standard deviation for the variable of a binomial distribution.

5 Find probabilities for outcomes of variables, using the Poisson, hypergeometric, and multinomial distributions.

Outline

Is Pooling Worthwhile?

Blood samples are used to screen people for certain diseases. When the disease is rare, health care workers sometimes combine or pool the blood samples of a group of individuals into one batch and then test it. If the test result of the batch is negative, no further testing is needed since none of the individuals in the group has the disease. However, if the test result of the batch is positive, each individual in the group must be tested.

Consider this hypothetical example: Suppose the probability of a person having the disease is 0.05, and a pooled sample of 15 individuals is tested. What is the probability that no further testing will be needed for the individuals in the sample? The answer to this question can be found by using what is called the *binomial distribution*. See Statistics Today—Revisited at the end of the chapter.

This chapter explains probability distributions in general and a specific, often used distribution called the binomial distribution. The Poisson, hypergeometric, and multinomial distributions are also explained.

Introduction

Many decisions in business, insurance, and other real-life situations are made by assigning probabilities to all possible outcomes pertaining to the situation and then evaluating the results. For example, a saleswoman can compute the probability that she will make 0, 1, 2, or 3 or more sales in a single day. An insurance company might be able to assign probabilities to the number of vehicles a family owns. A self-employed speaker might be able to compute the probabilities for giving 0, 1, 2, 3, or 4 or more speeches each week. Once these probabilities are assigned, statistics such as the mean, variance, and standard deviation can be computed for these events. With these statistics, various decisions can be made. The saleswoman will be able to compute the average number of sales she makes per week, and if she is working on commission, she will be able to approximate her weekly income over a period of time, say, monthly. The public speaker will be able to plan ahead and approximate his average income and expenses. The insurance company can use its information to design special computer forms and programs to accommodate its customers' future needs.

This chapter explains the concepts and applications of what is called a *probability distribution*. In addition, special probability distributions, such as the *binomial, multinomial, Poisson,* and *hypergeometric* distributions, are explained.

5–1 Probability Distributions

Objective 1

Construct a probability distribution for a random variable.

Before probability distribution is defined formally, the definition of a variable is reviewed. In Chapter 1, a *variable* was defined as a characteristic or attribute that can assume different values. Various letters of the alphabet, such as X, Y, or Z, are used to represent variables. Since the variables in this chapter are associated with probability, they are called *random variables.*

For example, if a die is rolled, a letter such as X can be used to represent the outcomes. Then the value that X can assume is 1, 2, 3, 4, 5, or 6, corresponding to the outcomes of rolling a single die. If two coins are tossed, a letter, say Y, can be used to represent the number of heads, in this case 0, 1, or 2. As another example, if the temperature at 8:00 A.M. is 43° and at noon it is 53°, then the values T that the temperature assumes are said to be random, since they are due to various atmospheric conditions at the time the temperature was taken.

A **random variable** is a variable whose values are determined by chance.

Also recall from Chapter 1 that you can classify variables as discrete or continuous by observing the values the variable can assume. If a variable can assume only a specific number of values, such as the outcomes for the roll of a die or the outcomes for the toss of a coin, then the variable is called a *discrete variable.*

Discrete variables have a finite number of possible values or an infinite number of values that can be counted. The word *counted* means that they can be enumerated using the numbers 1, 2, 3, etc. For example, the number of joggers in Riverview Park each day and the number of phone calls received after a TV commercial airs are examples of discrete variables, since they can be counted.

Variables that can assume all values in the interval between any two given values are called *continuous variables.* For example, if the temperature goes from 62 to 78° in a 24-hour period, it has passed through every possible number from 62 to 78. *Continuous random variables are obtained from data that can be measured rather than counted.* Continuous random variables can assume an infinite number of values and can be decimal and fractional values. On a continuous scale, a person's weight might be exactly 183.426 pounds if a scale could measure weight to the thousandths place; however, on a digital scale that measures only to tenths of pounds, the weight would be 183.4 pounds. Examples of continuous variables are heights, weights, temperatures, and time. In this chapter only discrete random variables are used; Chapter 6 explains continuous random variables.

The procedure shown here for constructing a probability distribution for a discrete random variable uses the probability experiment of tossing three coins. Recall that when three coins are tossed, the sample space is represented as TTT, TTH, THT, HTT, HHT, HTH, THH, HHH; and if X is the random variable for the number of heads, then X assumes the value 0, 1, 2, or 3.

Probabilities for the values of X can be determined as follows:

No heads	One head			Two heads			Three heads
TTT	TTH	THT	HTT	HHT	HTH	THH	HHH
$\frac{1}{8}$	$\frac{1}{8}$	$\frac{1}{8}$	$\frac{1}{8}$	$\frac{1}{8}$	$\frac{1}{8}$	$\frac{1}{8}$	$\frac{1}{8}$
$\frac{1}{8}$	$\frac{3}{8}$			$\frac{3}{8}$			$\frac{1}{8}$

Hence, the probability of getting no heads is $\frac{1}{8}$, one head is $\frac{3}{8}$, two heads is $\frac{3}{8}$, and three heads is $\frac{1}{8}$. From these values, a probability distribution can be constructed by listing the outcomes and assigning the probability of each outcome, as shown here.

Number of heads X	0	1	2	3
Probability $P(X)$	$\frac{1}{8}$	$\frac{3}{8}$	$\frac{3}{8}$	$\frac{1}{8}$

> A **discrete probability distribution** consists of the values a random variable can assume and the corresponding probabilities of the values. The probabilities are determined theoretically or by observation.

Discrete probability distributions can be shown by using a graph or a table. Probability distributions can also be represented by a formula. See Exercises 31–36 at the end of this section for examples.

Example 5–1

Rolling a Die

Construct a probability distribution for rolling a single die.

Solution

Since the sample space is 1, 2, 3, 4, 5, 6 and each outcome has a probability of $\frac{1}{6}$, the distribution is as shown.

Outcome X	1	2	3	4	5	6
Probability $P(X)$	$\frac{1}{6}$	$\frac{1}{6}$	$\frac{1}{6}$	$\frac{1}{6}$	$\frac{1}{6}$	$\frac{1}{6}$

Probability distributions can be shown graphically by representing the values of X on the x axis and the probabilities $P(X)$ on the y axis.

Example 5–2

Tossing Coins

Represent graphically the probability distribution for the sample space for tossing three coins.

Number of heads X	0	1	2	3
Probability $P(X)$	$\frac{1}{8}$	$\frac{3}{8}$	$\frac{3}{8}$	$\frac{1}{8}$

Solution

The values that X assumes are located on the x axis, and the values for $P(X)$ are located on the y axis. The graph is shown in Figure 5–1.

Note that for visual appearances, it is not necessary to start with 0 at the origin.

Examples 5–1 and 5–2 are illustrations of *theoretical* probability distributions. You did not need to actually perform the experiments to compute the probabilities. In contrast, to construct actual probability distributions, you must observe the variable over a period of time. They are empirical, as shown in Example 5–3.

Figure 5–1

Probability Distribution for Example 5–2

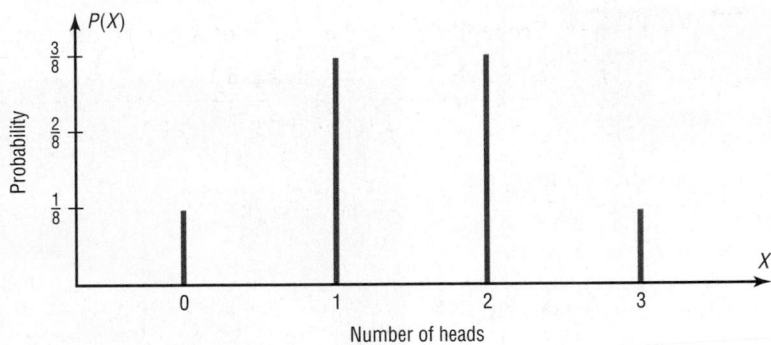

Example 5–3

Baseball World Series

The baseball World Series is played by the winner of the National League and the American League. The first team to win four games wins the World Series. In other words, the series will consist of four to seven games, depending on the individual victories. The data shown consist of 40 World Series events. The number of games played in each series is represented by the variable X. Find the probability $P(X)$ for each X, construct a probability distribution, and draw a graph for the data.

X	Number of games played
4	8
5	7
6	9
7	16
	40

Solution

The probability $P(X)$ can be computed for each X by dividing the number of games X by the total.

For 4 games, $\frac{8}{40} = 0.200$ For 6 games, $\frac{9}{40} = 0.225$

For 5 games, $\frac{7}{40} = 0.175$ For 7 games, $\frac{16}{40} = 0.400$

The probability distribution is

Number of games X	4	5	6	7
Probability $P(X)$	0.200	0.175	0.225	0.400

The graph is shown in Figure 5–2.

Figure 5–2

Probability Distribution for Example 5–3

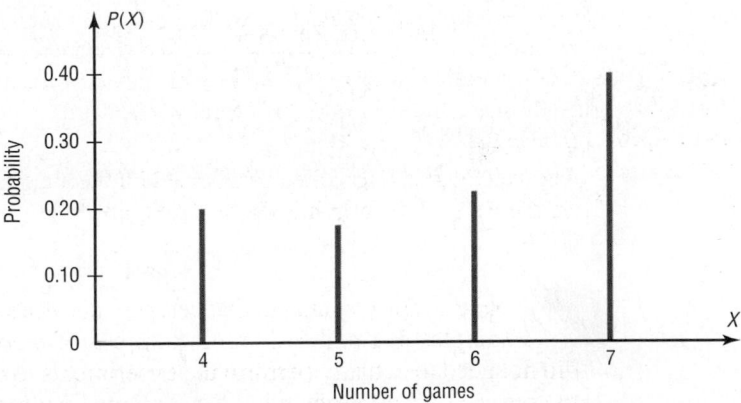

Coins, Births, and Other Random (?) Events

Examples of random events such as tossing coins are used in almost all books on probability. But is flipping a coin really a random event?

Tossing coins dates back to ancient Roman times when the coins usually consisted of the Emperor's head on one side (i.e., heads) and another icon such as a ship on the other side (i.e., ships). Tossing coins was used in both fortune telling and ancient Roman games.

A Chinese form of divination called the *I-Ching* (pronounced E-Ching) is

thought to be at least 4000 years old. It consists of 64 hexagrams made up of six horizontal lines. Each line is either broken or unbroken, representing the yin and the yang. These 64 hexagrams are supposed to represent all possible situations in life. To consult the I-Ching, a question is asked and then three coins are tossed six times. The way the coins fall, either heads up or heads down, determines whether the line is broken (yin) or unbroken (yang). Once the hexagon is determined, its meaning is consulted and interpreted to get the answer to the question. (*Note:* Another method used to determine the hexagon employs yarrow sticks.)

In the 16th century, a mathematician named Abraham DeMoivre used the outcomes of tossing coins to study what later became known as the normal distribution; however, his work at that time was not widely known.

Mathematicians usually consider the outcomes of a coin toss a random event. That is, each probability of getting a head is $\frac{1}{2}$, and the probability of getting a tail is $\frac{1}{2}$. Also, it is not possible to predict with 100% certainty which outcome will occur. But new studies question this theory. During World War II a South African mathematician named John Kerrich tossed a coin 10,000 times while he was interned in a German prison camp. Unfortunately, the results of his experiment were never recorded, so we don't know the number of heads that occurred.

Several studies have shown that when a coin-tossing device is used, the probability that a coin will land on the same side on which it is placed on the coin-tossing device is about 51%. It would take about 10,000 tosses to become aware of this bias. Furthermore, researchers showed that when a coin is spun on its edge, the coin falls tails up about 80% of the time since there is more metal on the heads side of a coin. This makes the coin slightly heavier on the heads side than on the tails side.

Another assumption commonly made in probability theory is that the number of male births is equal to the number of female births and that the probability of a boy being born is $\frac{1}{2}$ and the probability of a girl being born is $\frac{1}{2}$. We know this is not exactly true.

In the later 1700s, a French mathematician named Pierre Simon Laplace attempted to prove that more males than females are born. He used records from 1745 to 1770 in Paris and showed that the percentage of females born was about 49%. Although these percentages vary somewhat from location to location, further surveys show they are generally true worldwide. Even though there are discrepancies, we generally consider the outcomes to be 50-50 since these discrepancies are relatively small.

Based on this article, would you consider the coin toss at the beginning of a football game fair?

Two Requirements for a Probability Distribution

1. The sum of the probabilities of all the events in the sample space must equal 1; that is, $\Sigma P(X) = 1$.
2. The probability of each event in the sample space must be between or equal to 0 and 1. That is, $0 \leq P(X) \leq 1$.

   The first requirement states that the sum of the probabilities of all the events must be equal to 1. This sum cannot be less than 1 or greater than 1 since the sample space includes *all* possible outcomes of the probability experiment. The second requirement states that the probability of any individual event must be a value from 0 to 1. The reason (as stated in Chapter 4) is that the range of the probability of any individual value can be 0, 1, or any value between 0 and 1. A probability cannot be a negative number or greater than 1.

Example 5–4

Probability Distributions

Determine whether each distribution is a probability distribution.

a.
X	4	6	8	10
P(X)	−0.6	0.2	0.7	1.5

c.
X	8	9	12
P(X)	$\frac{2}{3}$	$\frac{1}{6}$	$\frac{1}{6}$

b.
X	1	2	3	4
P(X)	$\frac{1}{4}$	$\frac{1}{4}$	$\frac{1}{4}$	$\frac{1}{4}$

d.
X	1	3	5	7	9
P(X)	0.3	0.1	0.2	0.4	−0.7

Solution

a. No. It is not a probability distribution since $P(X)$ cannot be negative or greater than 1.

b. Yes. It is a probability distribution.

c. Yes. It is a probability distribution.

d. No, since $P(X) \neq -0.7$.

   Many variables in business, education, engineering, and other areas can be analyzed by using probability distributions. Section 5–2 shows methods for finding the mean and standard deviation for a probability distribution.

Applying the Concepts **5–1**

Dropping College Courses

Use the following table to answer the questions.

Reason for Dropping a College Course	Frequency	Percentage
Too difficult	45	
Illness	40	
Change in work schedule	20	
Change of major	14	
Family-related problems	9	
Money	7	
Miscellaneous	6	
No meaningful reason	3	

1. What is the variable under study? Is it a random variable?

2. How many people were in the study?

3. Complete the table.

4. From the information given, what is the probability that a student will drop a class because of illness? Money? Change of major?

5. Would you consider the information in the table to be a probability distribution?

6. Are the categories mutually exclusive?

7. Are the categories independent?

8. Are the categories exhaustive?

9. Are the two requirements for a discrete probability distribution met?

See page 297 for the answers.

Exercises 5–1

1. Define and give three examples of a random variable. A random variable is a variable whose values are determined by chance. Examples will vary.

2. Explain the difference between a discrete and a continuous random variable.

3. Give three examples of a discrete random variable.

4. Give three examples of a continuous random variable.

5. What is a probability distribution? Give an example.

For Exercises 6 through 11, determine whether the distribution represents a probability distribution. If it does not, state why.

6.
X	3	7	9	12	14
P(X)	$\frac{4}{13}$	$\frac{1}{13}$	$\frac{3}{13}$	$\frac{1}{13}$	$\frac{2}{13}$

7.
X	3	6	8	12
P(X)	0.3	0.5	0.7	−0.8

8.
X	5	7	9
P(X)	0.6	0.8	−0.4

No. Probabilities cannot be negative.

9.
X	1	2	3	4	5
P(X)	$\frac{3}{10}$	$\frac{1}{10}$	$\frac{1}{10}$	$\frac{2}{10}$	$\frac{3}{10}$

Yes

10.
X	20	30	40	50
P(X)	0.05	0.35	0.4	0.2

Yes

11.
X	7	14	21
P(X)	0.3	0.1	1.7

No. A probability cannot be greater than 1.

For Exercises 12 through 18, state whether the variable is discrete or continuous.

12. The speed of a jet airplane   Continuous

13. The number of cheeseburgers a fast-food restaurant serves each day   Discrete

14. The number of people who play the state lottery each day   Discrete

15. The weight of an automobile.   Continuous

16. The time it takes to have a medical physical exam.   Continuous

17. The number of mathematics majors in your school   Discrete

18. The blood pressures of all patients admitted to a hospital on a specific day   Continuous

For Exercises 19 through 28, construct a probability distribution for the data and draw a graph for the distribution.

19. **Medical Tests** The probabilities that a patient will have 0, 1, 2, or 3 medical tests performed on entering a hospital are $\frac{6}{15}$, $\frac{5}{15}$, $\frac{3}{15}$, and $\frac{1}{15}$, respectively.

20. **Investment Return** The probabilities of a return on an investment of $5,000, $7,000, and $9,000 are $\frac{1}{2}$, $\frac{3}{8}$, and $\frac{1}{8}$.

21. **Birthday Cake Sales** The probabilities that a bakery has a demand for 2, 3, 5, or 7 birthday cakes on any given day are 0.35, 0.41, 0.15, and 0.09, respectively.

22. **DVD Rentals** The probabilities that a customer will rent 0, 1, 2, 3, or 4 DVDs on a single visit to the rental store are 0.15, 0.25, 0.3, 0.25, and 0.05, respectively.

23. **Loaded Die** A die is loaded in such a way that the probabilities of getting 1, 2, 3, 4, 5, and 6 are $\frac{1}{2}$, $\frac{1}{6}$, $\frac{1}{12}$, $\frac{1}{12}$, $\frac{1}{12}$, and $\frac{1}{12}$, respectively.

24. **Item Selection** The probabilities that a customer selects 1, 2, 3, 4, and 5 items at a convenience store are 0.32, 0.12, 0.23, 0.18, and 0.15, respectively.

25. **Student Classes** The probabilities that a student is registered for 2, 3, 4, or 5 classes are 0.01, 0.34, 0.62, and 0.03, respectively.

26. **Garage Space** The probabilities that a randomly selected home has garage space for 0, 1, 2, or 3 cars are 0.22, 0.33, 0.37, and 0.08, respectively.

27. **Selecting a Monetary Bill** A box contains three $1 bills, two $5 bills, five $10 bills, and one $20 bill. Construct a probability distribution for the data if x represents the value of a single bill drawn at random and then replaced.

28. **Family with Children** Construct a probability distribution for a family with 4 children. Let X be the number of girls.

29. **Drawing a Card** Construct a probability distribution for drawing a card from a deck of 40 cards consisting 10 cards numbered 1, 10 cards numbered 2, 15 cards numbered 3, and 5 cards numbered 4.

30. **Rolling Two Dice** Using the sample space for tossing two dice, construct a probability distribution for the sums 2 through 12.

Extending the Concepts

A probability distribution can be written in formula notation such as $P(X) = 1/X$, where $X = 2, 3, 6$. The distribution is shown as follows:

X	2	3	6
$P(X)$	$\frac{1}{2}$	$\frac{1}{3}$	$\frac{1}{6}$

For Exercises 31 through 36, write the distribution for the formula and determine whether it is a probability distribution.

31. $P(X) = X/6$ for $X = 1, 2, 3$

32. $P(X) = X$ for $X = 0.2, 0.3, 0.5$

33. $P(X) = X/6$ for $X = 3, 4, 7$

34. $P(X) = X + 0.1$ for $X = 0.1, 0.02, 0.04$

35. $P(X) = X/7$ for $X = 1, 2, 4$

36. $P(X) = X/(X + 2)$ for $X = 0, 1, 2$

5–2   Mean, Variance, Standard Deviation, and Expectation

Objective 2

Find the mean, variance, standard deviation, and expected value for a discrete random variable.

The mean, variance, and standard deviation for a probability distribution are computed differently from the mean, variance, and standard deviation for samples. This section explains how these measures—as well as a new measure called the *expectation*—are calculated for probability distributions.

Mean

In Chapter 3, the mean for a sample or population was computed by adding the values and dividing by the total number of values, as shown in these formulas:

$$\overline{X} = \frac{\Sigma X}{n} \qquad \mu = \frac{\Sigma X}{N}$$

But how would you compute the mean of the number of spots that show on top when a die is rolled? You could try rolling the die, say, 10 times, recording the number of spots, and finding the mean; however, this answer would only approximate the true mean. What about 50 rolls or 100 rolls? Actually, the more times the die is rolled, the better the approximation. You might ask, then, How many times must the die be rolled to get the exact answer? *It must be rolled an infinite number of times.* Since this task is impossible, the previous formulas cannot be used because the denominators would be infinity. Hence, a new method of computing the mean is necessary. This method gives the exact theoretical value of the mean as if it were possible to roll the die an infinite number of times.

Before the formula is stated, an example will be used to explain the concept. Suppose two coins are tossed repeatedly, and the number of heads that occurred is recorded. What will be the mean of the number of heads? The sample space is

HH, HT, TH, TT

Historical Note

A professor, Augustin Louis Cauchy (1789–1857), wrote a book on probability. While he was teaching at the Military School of Paris, one of his students was Napoleon Bonaparte.

and each outcome has a probability of $\frac{1}{4}$. Now, in the long run, you would *expect* two heads (HH) to occur approximately $\frac{1}{4}$ of the time, one head to occur approximately $\frac{1}{2}$ of the time (HT or TH), and no heads (TT) to occur approximately $\frac{1}{4}$ of the time. Hence, on average, you would expect the number of heads to be

$$\frac{1}{4} \cdot 2 + \frac{1}{2} \cdot 1 + \frac{1}{4} \cdot 0 = 1$$

That is, if it were possible to toss the coins many times or an infinite number of times, the *average* of the number of heads would be 1.

Hence, to find the mean for a probability distribution, you must multiply each possible outcome by its corresponding probability and find the sum of the products.

Formula for the Mean of a Probability Distribution

The mean of a random variable with a discrete probability distribution is

$$\mu = X_1 \cdot P(X_1) + X_2 \cdot P(X_2) + X_3 \cdot P(X_3) + \cdots + X_n \cdot P(X_n)$$
$$= \Sigma X \cdot P(X)$$

where $X_1, X_2, X_3, \ldots, X_n$ are the outcomes and $P(X_1), P(X_2), P(X_3), \ldots, P(X_n)$ are the corresponding probabilities.

Note: $\Sigma X \cdot P(X)$ means to sum the products.

Rounding Rule for the Mean, Variance, and Standard Deviation for a Probability Distribution The rounding rule for the mean, variance, and standard deviation for variables of a probability distribution is this: The mean, variance, and standard deviation should be rounded to one more decimal place than the outcome X. When fractions are used, they should be reduced to lowest terms.

Examples 5–5 through 5–8 illustrate the use of the formula.

Example 5–5

Rolling a Die

Find the mean of the number of spots that appear when a die is tossed.

Solution

In the toss of a die, the mean can be computed thus.

Outcome X	1	2	3	4	5	6
Probability $P(X)$	$\frac{1}{6}$	$\frac{1}{6}$	$\frac{1}{6}$	$\frac{1}{6}$	$\frac{1}{6}$	$\frac{1}{6}$

$$\mu = \Sigma X \cdot P(X) = 1 \cdot \frac{1}{6} + 2 \cdot \frac{1}{6} + 3 \cdot \frac{1}{6} + 4 \cdot \frac{1}{6} + 5 \cdot \frac{1}{6} + 6 \cdot \frac{1}{6}$$
$$= \frac{21}{6} = 3\frac{1}{2} \text{ or } 3.5$$

That is, when a die is tossed many times, the theoretical mean will be 3.5. Note that even though the die cannot show a 3.5, the theoretical average is 3.5.

The reason why this formula gives the theoretical mean is that in the long run, each outcome would occur approximately $\frac{1}{6}$ of the time. Hence, multiplying the outcome by its corresponding probability and finding the sum would yield the theoretical mean. In other words, outcome 1 would occur approximately $\frac{1}{6}$ of the time, outcome 2 would occur approximately $\frac{1}{6}$ of the time, etc.

Example 5–6	**Children in a Family**

In a family with two children, find the mean of the number of children who will be girls.

Solution

The probability distribution is as follows:

Number of girls X	0	1	2
Probability $P(X)$	$\frac{1}{4}$	$\frac{1}{2}$	$\frac{1}{4}$

Hence, the mean is

$$\mu = \Sigma X \cdot P(X) = 0 \cdot \tfrac{1}{4} + 1 \cdot \tfrac{1}{2} + 2 \cdot \tfrac{1}{4} = 1$$

Example 5–7	**Tossing Coins**

If three coins are tossed, find the mean of the number of heads that occur. (See the table preceding Example 5–1.)

Solution

The probability distribution is

Number of heads X	0	1	2	3
Probability $P(X)$	$\frac{1}{8}$	$\frac{3}{8}$	$\frac{3}{8}$	$\frac{1}{8}$

The mean is

$$\mu = \Sigma X \cdot P(X) = 0 \cdot \tfrac{1}{8} + 1 \cdot \tfrac{3}{8} + 2 \cdot \tfrac{3}{8} + 3 \cdot \tfrac{1}{8} = \tfrac{12}{8} = 1\tfrac{1}{2} \text{ or } 1.5$$

The value 1.5 cannot occur as an outcome. Nevertheless, it is the long-run or theoretical average.

Example 5–8	**Number of Trips of Five Nights or More**

The probability distribution shown represents the number of trips of five nights or more that American adults take per year. (That is, 6% do not take any trips lasting five nights or more, 70% take one trip lasting five nights or more per year, etc.) Find the mean.

Number of trips X	0	1	2	3	4
Probability $P(X)$	0.06	0.70	0.20	0.03	0.01

Solution

$$\mu = \Sigma X \cdot P(X)$$
$$= (0)(0.06) + (1)(0.70) + (2)(0.20) + (3)(0.03) + (4)(0.01)$$
$$= 0 + 0.70 + 0.40 + 0.09 + 0.04$$
$$= 1.23 \approx 1.2$$

Hence, the mean of the number of trips lasting five nights or more per year taken by American adults is 1.2.

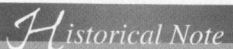

Fey Manufacturing
Co., located in San
Francisco, invented
the first three-reel,
automatic payout slot
machine in 1895.

Variance and Standard Deviation

For a probability distribution, the mean of the random variable describes the measure of the so-called long-run or theoretical average, but it does not tell anything about the spread of the distribution. Recall from Chapter 3 that to measure this spread or variability, statisticians use the variance and standard deviation. These formulas were used:

$$\sigma^2 = \frac{\Sigma(X - \mu)^2}{N} \qquad \text{or} \qquad \sigma = \sqrt{\frac{\Sigma(X - \mu)^2}{N}}$$

These formulas cannot be used for a random variable of a probability distribution since N is infinite, so the variance and standard deviation must be computed differently.

To find the variance for the random variable of a probability distribution, subtract the theoretical mean of the random variable from each outcome and square the difference. Then multiply each difference by its corresponding probability and add the products. The formula is

$$\sigma^2 = \Sigma[(X - \mu)^2 \cdot P(X)]$$

Finding the variance by using this formula is somewhat tedious. So for simplified computations, a shortcut formula can be used. This formula is algebraically equivalent to the longer one and is used in the examples that follow.

Formula for the Variance of a Probability Distribution

Find the variance of a probability distribution by multiplying the square of each outcome by its corresponding probability, summing those products, and subtracting the square of the mean. The formula for the variance of a probability distribution is

$$\sigma^2 = \Sigma[X^2 \cdot P(X)] - \mu^2$$

The standard deviation of a probability distribution is

$$\sigma = \sqrt{\sigma^2} \qquad \text{or} \qquad \sqrt{\Sigma[X^2 \cdot P(X)] - \mu^2}$$

Remember that the variance and standard deviation cannot be negative.

Example 5–9

Rolling a Die

Compute the variance and standard deviation for the probability distribution in Example 5–5.

Solution

Recall that the mean is $\mu = 3.5$, as computed in Example 5–5. Square each outcome and multiply by the corresponding probability, sum those products, and then subtract the square of the mean.

$$\sigma^2 = (1^2 \cdot \tfrac{1}{6} + 2^2 \cdot \tfrac{1}{6} + 3^2 \cdot \tfrac{1}{6} + 4^2 \cdot \tfrac{1}{6} + 5^2 \cdot \tfrac{1}{6} + 6^2 \cdot \tfrac{1}{6}) - (3.5)^2 = 2.9$$

To get the standard deviation, find the square root of the variance.

$$\sigma = \sqrt{2.9} = 1.7$$

| Example 5–10 | Selecting Numbered Balls |

A box contains 5 balls. Two are numbered 3, one is numbered 4, and two are numbered 5. The balls are mixed and one is selected at random. After a ball is selected, its number is recorded. Then it is replaced. If the experiment is repeated many times, find the variance and standard deviation of the numbers on the balls.

Solution

Let X be the number on each ball. The probability distribution is

Number on ball X	3	4	5
Probability $P(X)$	$\frac{2}{5}$	$\frac{1}{5}$	$\frac{2}{5}$

The mean is

$$\mu = \Sigma X \cdot P(X) = 3 \cdot \tfrac{2}{5} + 4 \cdot \tfrac{1}{5} + 5 \cdot \tfrac{2}{5} = 4$$

The variance is

$$\sigma = \Sigma[X^2 \cdot P(X)] - \mu^2$$
$$= 3^2 \cdot \tfrac{2}{5} + 4^2 \cdot \tfrac{1}{5} + 5^2 \cdot \tfrac{2}{5} - 4^2$$
$$= 16\tfrac{4}{5} - 16$$
$$= \tfrac{4}{5}$$

The standard deviation is

$$\sigma = \sqrt{\frac{4}{5}} = \sqrt{0.8} = 0.894$$

The mean, variance, and standard deviation can also be found by using vertical columns, as shown.

X	$P(X)$	$X \cdot P(X)$	$X^2 \cdot P(X)$
3	0.4	1.2	3.6
4	0.2	0.8	3.2
5	0.4	2.0	10
		$\Sigma X \cdot P(X) = 4.0$	16.8

Find the mean by summing the $\Sigma X \cdot P(X)$ column, and find the variance by summing the $X^2 \cdot P(X)$ column and subtracting the square of the mean.

$$\sigma^2 = 16.8 - 4^2 = 16.8 - 16 = 0.8$$

and

$$\sigma = \sqrt{0.8} = 0.894$$

| Example 5–11 | On Hold for Talk Radio |

A talk radio station has four telephone lines. If the host is unable to talk (i.e., during a commercial) or is talking to a person, the other callers are placed on hold. When all lines are in use, others who are trying to call in get a busy signal. The probability that 0, 1, 2, 3, or 4 people will get through is shown in the distribution. Find the variance and standard deviation for the distribution.

X	0	1	2	3	4
$P(X)$	0.18	0.34	0.23	0.21	0.04

Should the station have considered getting more phone lines installed?

Solution

The mean is

$$\mu = \Sigma X \cdot P(X)$$
$$= 0 \cdot (0.18) + 1 \cdot (0.34) + 2 \cdot (0.23) + 3 \cdot (0.21) + 4 \cdot (0.04)$$
$$= 1.6$$

The variance is

$$\sigma^2 = \Sigma[X^2 \cdot P(X)] - \mu^2$$
$$= [0^2 \cdot (0.18) + 1^2 \cdot (0.34) + 2^2 \cdot (0.23) + 3^2 \cdot (0.21) + 4^2 \cdot (0.04)] - 1.6^2$$
$$= [0 + 0.34 + 0.92 + 1.89 + 0.64] - 2.56$$
$$= 3.79 - 2.56 = 1.23$$
$$= 1.2 \text{ (rounded)}$$

The standard deviation is $\sigma = \sqrt{\sigma^2}$, or $\sigma = \sqrt{1.2} = 1.1$.

No. The mean number of people calling at any one time is 1.6. Since the standard deviation is 1.1, most callers would be accommodated by having four phone lines because $\mu + 2\sigma$ would be $1.6 + 2(1.1) = 1.6 + 2.2 = 3.8$. Very few callers would get a busy signal since at least 75% of the callers would either get through or be put on hold. (See Chebyshev's theorem in Section 3–2.)

Expectation

Another concept related to the mean for a probability distribution is that of expected value or expectation. Expected value is used in various types of games of chance, in insurance, and in other areas, such as decision theory.

The **expected value** of a discrete random variable of a probability distribution is the theoretical average of the variable. The formula is

$$\mu = E(X) = \Sigma X \cdot P(X)$$

The symbol $E(X)$ is used for the expected value.

The formula for the expected value is the same as the formula for the theoretical mean. The expected value, then, is the theoretical mean of the probability distribution. That is, $E(X) = \mu$.

When expected value problems involve money, it is customary to round the answer to the nearest cent.

Example 5–12

Winning Tickets

One thousand tickets are sold at $1 each for a color television valued at $350. What is the expected value of the gain if you purchase one ticket?

Solution

The problem can be set up as follows:

	Win	Lose
Gain X	$349	−$1
Probability $P(X)$	$\frac{1}{1000}$	$\frac{999}{1000}$

Two things should be noted. First, for a win, the net gain is $349, since you do not get the cost of the ticket ($1) back. Second, for a loss, the gain is represented by a negative number, in this case −$1. The solution, then, is

$$E(X) = \$349 \cdot \frac{1}{1000} + (-\$1) \cdot \frac{999}{1000} = -\$0.65$$

Expected value problems of this type can also be solved by finding the overall gain (i.e., the value of the prize won or the amount of money won, not considering the cost of the ticket for the prize or the cost to play the game) and subtracting the cost of the tickets or the cost to play the game, as shown:

$$E(X) = \$350 \cdot \frac{1}{1000} - \$1 = -\$0.65$$

Here, the overall gain ($350) must be used.

Note that the expectation is −$0.65. This does not mean that you lose $0.65, since you can only win a television set valued at $350 or lose $1 on the ticket. What this expectation means is that the average of the losses is $0.65 for each of the 1000 ticket holders. Here is another way of looking at this situation: If you purchased one ticket each week over a long time, the average loss would be $0.65 per ticket, since theoretically, on average, you would win the set once for each 1000 tickets purchased.

Example 5–13

Special Die

A special six-sided die is made in which 3 sides have 6 spots, 2 sides have 4 spots, and 1 side has 1 spot. If the die is rolled, find the expected value of the number of spots that will occur.

Solution

Since there are 3 sides with 6 spots, the probability of getting a 6 is $\frac{3}{6} = \frac{1}{2}$. Since there are 2 sides with 4 spots, the probability of getting 4 spots is $\frac{2}{6} = \frac{1}{3}$. The probability of getting 1 spot is $\frac{1}{6}$ since 1 side has 1 spot.

Gain X	1	4	6
Probability $P(X)$	$\frac{1}{6}$	$\frac{1}{3}$	$\frac{1}{2}$

$$E(X) = 1 \cdot \tfrac{1}{6} + 4 \cdot \tfrac{1}{3} + 6 \cdot \tfrac{1}{2} = 4\tfrac{1}{2}$$

Notice you can only get 1, 4, or 6 spots; but if you rolled the die a large number of times and found the average, it would be about $4\frac{1}{2}$.

Example 5–14

Bond Investment

A financial adviser suggests that his client select one of two types of bonds in which to invest $5000. Bond X pays a return of 4% and has a default rate of 2%. Bond Y has a $2\frac{1}{2}$% return and a default rate of 1%. Find the expected rate of return and decide which bond would be a better investment. When the bond defaults, the investor loses all the investment.

Solution

The return on bond X is $\$5000 \cdot 4\% = \$200$. The expected return then is

$$E(X) = \$200(0.98) - \$5000(0.02) = \$96$$

The return on bond Y is $\$5000 \cdot 2\frac{1}{2}\% = \$125$. The expected return then is

$$E(X) = \$125(0.99) - \$5000(0.01) = \$73.75$$

Hence, bond X would be a better investment since the expected return is higher.

In gambling games, if the expected value of the game is zero, the game is said to be fair. If the expected value of a game is positive, then the game is in favor of the player. That is, the player has a better than even chance of winning. If the expected value of the game is negative, then the game is said to be in favor of the house. That is, in the long run, the players will lose money.

In his book *Probabilities in Everyday Life* (Ivy Books, 1986), author John D. McGervy gives the expectations for various casino games. For keno, the house wins $\$0.27$ on every $\$1.00$ bet. For Chuck-a-Luck, the house wins about $\$0.52$ on every $\$1.00$ bet. For roulette, the house wins about $\$0.90$ on every $\$1.00$ bet. For craps, the house wins about $\$0.88$ on every $\$1.00$ bet. The bottom line here is that if you gamble long enough, sooner or later you will end up losing money.

Applying the Concepts **5–2**

Expected Value

On March 28, 1979, the nuclear generating facility at Three Mile Island, Pennsylvania, began discharging radiation into the atmosphere. People exposed to even low levels of radiation can experience health problems ranging from very mild to severe, even causing death. A local newspaper reported that 11 babies were born with kidney problems in the three-county area surrounding the Three Mile Island nuclear power plant. The expected value for that problem in infants in that area was 3. Answer the following questions.

1. What does *expected value* mean?
2. Would you expect the exact value of 3 all the time?
3. If a news reporter stated that the number of cases of kidney problems in newborns was nearly four times as much as was usually expected, do you think pregnant mothers living in that area would be overly concerned?
4. Is it unlikely that 11 occurred by chance?
5. Are there any other statistics that could better inform the public?
6. Assume that 3 out of 2500 babies were born with kidney problems in that three-county area the year before the accident. Also assume that 11 out of 2500 babies were born with kidney problems in that three-county area the year after the accident. What is the real percent of increase in that abnormality?
7. Do you think that pregnant mothers living in that area should be overly concerned after looking at the results in terms of rates?

See page 298 for the answers.

Exercises 5–2

1. **Defective DVDs** From past experience, a company found that in cartons of DVDs, 90% contain no defective DVDs, 5% contain one defective DVD, 3% contain two defective DVDs, and 2% contain three defective DVDs. Find the mean, variance, and standard deviation for the number of defective DVDs. 0.17; 0.321; 0.567

2. **Suit Sales** The number of suits sold per day at a retail store is shown in the table, with the corresponding probabilities. Find the mean, variance, and standard deviation of the distribution. 20.8; 1.6; 1.3

Number of suits sold X	19	20	21	22	23
Probability P(X)	0.2	0.2	0.3	0.2	0.1

If the manager of the retail store wants to be sure that he has enough suits for the next 5 days, how many should the manager purchase? 104 suits

3. **Number of Credit Cards** A bank vice president feels that each savings account customer has, on average, three credit cards. The following distribution represents the number of credit cards people own. Find the mean, variance, and standard deviation. Is the vice president correct? 1.3, 0.9, 1. No, on average, each person has about 1 credit card.

Number of cards X	0	1	2	3	4
Probability P(X)	0.18	0.44	0.27	0.08	0.03

4. **Trivia Quiz** The probabilities that a player will get 5 to 10 questions right on a trivia quiz are shown below. Find the mean, variance, and standard deviation for the distribution. 7.4; 1.84; 1.356

X	5	6	7	8	9	10
P(X)	0.05	0.2	0.4	0.1	0.15	0.1

5. **Cellular Phone Sales** The probability that a cellular phone company kiosk sells X number of new phone contracts per day is shown below. Find the mean, variance, and standard deviation for this probability distribution. 5.4; 2.94; 1.71

X	4	5	6	8	10
P(X)	0.4	0.3	0.1	0.15	0.05

What is the probability that they will sell 6 or more contracts three days in a row? 0.027

6. **Traffic Accidents** The county highway department recorded the following probabilities for the number of accidents per day on a certain freeway for one month. The number of accidents per day and their corresponding probabilities are shown. Find the mean, variance, and standard deviation. 1.3; 1.81; 1.35

Number of accidents X	0	1	2	3	4
Probability P(X)	0.4	0.2	0.2	0.1	0.1

7. **Commercials During Children's TV Programs** A concerned parents group determined the number of commercials shown in each of five children's programs over a period of time. Find the mean, variance, and standard deviation for the distribution shown. 6.6; 1.3; 1.1

Number of commercials X	5	6	7	8	9
Probability P(X)	0.2	0.25	0.38	0.10	0.07

8. **Number of Televisions per Household** A study conducted by a TV station showed the number of televisions per household and the corresponding probabilities for each. Find the mean, variance, and standard deviation. 1.9; 0.6; 0.8

Number of televisions X	1	2	3	4
Probability P(X)	0.32	0.51	0.12	0.05

If you were taking a survey on the programs that were watched on television, how many program diaries would you send to each household in the survey? 2 diaries

9. **Students Using the Math Lab** The number of students using the Math Lab per day is found in the distribution below. Find the mean, variance, and standard deviation for this probability distribution. 9.4; 5.24; 2.289

X	6	8	10	12	14
P(X)	0.15	0.3	0.35	0.1	0.1

What is the probability that fewer than 8 or more than 12 use the lab in a given day? 0.25

10. **Pizza Deliveries** A pizza shop owner determines the number of pizzas that are delivered each day. Find the mean, variance, and standard deviation for the distribution shown. If the manager stated that 45 pizzas were delivered on one day, do you think that this is a believable claim? 37.1; 1.3; 1.1; it could happen (perhaps on a Super Bowl Sunday), but it is highly unlikely.

Number of deliveries X	35	36	37	38	39
Probability P(X)	0.1	0.2	0.3	0.3	0.1

11. **Insurance** An insurance company insures a person's antique coin collection worth $20,000 for an annual premium of $300. If the company figures that the probability of the collection being stolen is 0.002, what will be the company's expected profit? $260

12. **Job Bids** A landscape contractor bids on jobs where he can make $3000 profit. The probabilities of getting 1, 2, 3, or 4 jobs per month are shown.

Number of jobs	1	2	3	4
Probability	0.2	0.3	0.4	0.1

Find the contractor's expected profit per month. $7200

13. **Rolling Dice** If a person rolls doubles when she tosses two dice, she wins $5. For the game to be fair, how much should she pay to play the game? $0.83

14. **Dice Game** A person pays $2 to play a certain game by rolling a single die once. If a 1 or a 2 comes up, the person wins nothing. If, however, the player rolls a 3, 4, 5, or 6, he or she wins the difference between the number rolled and $2. Find the expectation for this game. Is the game fair? −33.3 cents; no

15. **Lottery Prizes** A lottery offers one $1000 prize, one $500 prize, and five $100 prizes. One thousand tickets are sold at $3 each. Find the expectation if a person buys one ticket. −$1.00

16. In Exercise 15, find the expectation if a person buys two tickets. Assume that the player's ticket is replaced after each draw and that the same ticket can win more than one prize. −$2.00

17. **Winning the Lottery** For a daily lottery, a person selects a three-digit number. If the person plays for $1, she can win $500. Find the expectation. In the same daily lottery, if a person boxes a number, she will win $80. Find the expectation if the number 123 is played for $1 and boxed. (When a number is "boxed," it can win when the digits occur in any order.) −$0.50, −$0.52

18. **Life Insurance** A 35-year-old woman purchases a $100,000 term life insurance policy for an annual payment of $360. Based on a period life table for the U.S. government, the probability that she will survive the year is 0.999057. Find the expected value of the policy for the insurance company. $265.70

19. **Roulette** A roulette wheel has 38 numbers, 1 through 36, 0, and 00. One-half of the numbers from 1 through 36 are red, and the other half are black; 0 and 00 are green. A ball is rolled, and it falls into one of the 38 slots, giving a number and a color. The payoffs (winnings) for a $1 bet are as follows:?

Red or black	$1	0	$35
Odd or even	$1	00	$35
1–18	$1	Any single number	$35
9–36	$1	0 or 00	$17

If a person bets $1, find the expected value for each.

a. Red −5.26 cents   d. Any single number −5.26 cents
b. Even −5.26 cents   e. 0 or 00 −5.26 cents
c. 00 −5.26 cents

Extending the Concepts

20. **Rolling Dice** Construct a probability distribution for the sum shown on the faces when two dice are rolled. Find the mean, variance, and standard deviation of the distribution. 7; 5.8; 2.4

21. **Rolling a Die** When one die is rolled, the expected value of the number of spots is 3.5. In Exercise 20, the mean number of spots was found for rolling two dice. What is the mean number of spots if three dice are rolled? 10.5

22. The formula for finding the variance for a probability distribution is

$$\sigma^2 = \Sigma[(X - \mu)^2 \cdot P(X)]$$

Verify algebraically that this formula gives the same result as the shortcut formula shown in this section.

23. **Rolling a Die** Roll a die 100 times. Compute the mean and standard deviation. How does the result compare with the theoretical results of Example 5–5? Answers will vary.

24. **Rolling Two Dice** Roll two dice 100 times and find the mean, variance, and standard deviation of the sum of the spots. Compare the result with the theoretical results obtained in Exercise 20. Answers will vary.

25. **Extracurricular Activities** Conduct a survey of the number of extracurricular activities your classmates are enrolled in. Construct a probability distribution and find the mean, variance, and standard deviation. Answers will vary.

26. **Promotional Campaign** In a recent promotional campaign, a company offered these prizes and the corresponding probabilities. Find the expected value of winning. The tickets are free.

Number of prizes	Amount	Probability
1	$100,000	$\frac{1}{1,000,000}$
2	10,000	$\frac{1}{50,000}$
5	1,000	$\frac{1}{10,000}$
10	100	$\frac{1}{1000}$

If the winner has to mail in the winning ticket to claim the prize, what will be the expectation if the cost of the stamp is considered? Use the current cost of a stamp for a first-class letter. $1.56 with the cost of a stamp = $0.44

Speaking of
Statistics

This study shows that a part of the brain reacts to the impact of losing, and it might explain why people tend to increase their bets after losing when gambling. Explain how this type of split decision making may influence fighter pilots, firefighters, or police officers, as the article states.

THE GAMBLER'S FALLACY

WHY WE EXPECT TO STRIKE IT RICH AFTER A LOSING STREAK

A GAMBLER USUALLY WAGERS more after taking a loss, in the misguided belief that a run of bad luck increases the probability of a win. We tend to cling to the misconception that past events can skew future odds. "On some level, you're thinking, 'If I just lost, it's going to even out.' The extent to which you're disturbed by a loss seems to go along with risky behavior," says University of Michigan psychologist William Gehring, Ph.D., co-author of a new study linking dicey decision-making to neurological activity originating in the medial frontal cortex, long thought to be an area of the brain used in error detection.

Because people are so driven to up the ante after a loss, Gehring believes that the medial frontal cortex unconsciously influences future decisions based on the impact of the loss, in addition to registering the loss itself.

Gehring drew this conclusion by asking 12 subjects fitted with electrode caps to choose either the number 5 or 25, with the larger number representing the riskier bet.

On any given round, both numbers could amount to a loss, both could amount to a gain or the results could split, one number signifying a loss, the other a gain.

The medial frontal cortex responded to the outcome of a gamble within a quarter of a second, registering sharp electrical impulses only after a loss. Gehring points out that if the medial frontal cortex simply detected errors it would have reacted after participants chose the lesser of two possible gains. In other words, choosing "5" during a round in which both numbers paid off and betting on "25" would have yielded a larger profit.

After the study appeared in *Science*, Gehring received several e-mails from stock traders likening the "gambler's fallacy" to impulsive trading decisions made directly after off-loading a losing security. Researchers speculate that such risky, split-second decision-making could extend to fighter pilots, firemen and policemen—professions in which rapid-fire decisions are crucial and frequent.

—*Dan Schulman*

Reprinted with permission from *Psychology Today* magazine (copyright © 2002 Sussex Publishers, LLC).

Technology *Step by Step*

TI-83 Plus or TI-84 Plus

Step by Step

To calculate the mean and variance for a discrete random variable by using the formulas:

1. Enter the x values into L_1 and the probabilities into L_2.
2. Move the cursor to the top of the L_3 column so that L_3 is highlighted.
3. Type L_1 multiplied by L_2, then press **ENTER**.
4. Move the cursor to the top of the L_4 column so that L_4 is highlighted.
5. Type L_1 followed by the x^2 key multiplied by L_2, then press **ENTER**.
6. Type **2nd QUIT** to return to the home screen.
7. Type **2nd LIST,** move the cursor to MATH, type **5** for sum, then type L_3, then press **ENTER**.
8. Type **2nd ENTER,** move the cursor to L_3, type L_4, then press **ENTER**.

Example TI5–1

Number on ball X	0	2	4	6	8
Probability $P(X)$	$\frac{1}{5}$	$\frac{1}{5}$	$\frac{1}{5}$	$\frac{1}{5}$	$\frac{1}{5}$

Using the data from Example TI5–1 gives the following:

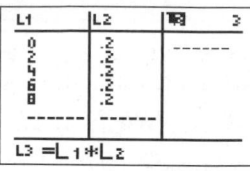 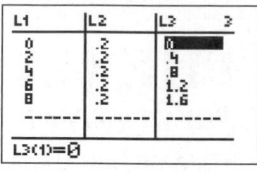

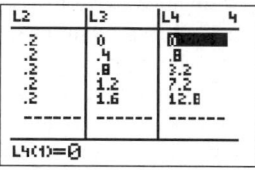 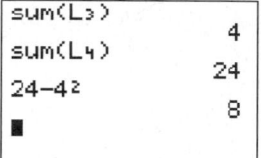

To calculate the mean and standard deviation for a discrete random variable without using the formulas, modify the procedure to calculate the mean and standard deviation from grouped data (Chapter 3) by entering the x values into L_1 and the probabilities into L_2.

```
1-Var Stats L₁,L
2█
```

```
1-Var Stats
 x̄=4
 Σx=4
 Σx²=24
 Sx=
 σx=2.828427125
↓n=1
```

5–3   The Binomial Distribution

Many types of probability problems have only two outcomes or can be reduced to two outcomes. For example, when a coin is tossed, it can land heads or tails. When a baby is born, it will be either male or female. In a basketball game, a team either wins or loses. A true/false item can be answered in only two ways, true or false. Other situations can be

Objective 3

Find the exact probability for *X* successes in *n* trials of a binomial experiment.

reduced to two outcomes. For example, a medical treatment can be classified as effective or ineffective, depending on the results. A person can be classified as having normal or abnormal blood pressure, depending on the measure of the blood pressure gauge. A multiple-choice question, even though there are four or five answer choices, can be classified as correct or incorrect. Situations like these are called *binomial experiments.*

A **binomial experiment** is a probability experiment that satisfies the following four requirements:

1. There must be a fixed number of trials.
2. Each trial can have only two outcomes or outcomes that can be reduced to two outcomes. These outcomes can be considered as either success or failure.
3. The outcomes of each trial must be independent of one another.
4. The probability of a success must remain the same for each trial.

A binomial experiment and its results give rise to a special probability distribution called the *binomial distribution.*

The outcomes of a binomial experiment and the corresponding probabilities of these outcomes are called a **binomial distribution.**

In binomial experiments, the outcomes are usually classified as successes or failures. For example, the correct answer to a multiple-choice item can be classified as a success, but any of the other choices would be incorrect and hence classified as a failure. The notation that is commonly used for binomial experiments and the binomial distribution is defined now.

Notation for the Binomial Distribution

$P(S)$	The symbol for the probability of success
$P(F)$	The symbol for the probability of failure
p	The numerical probability of a success
q	The numerical probability of a failure

$$P(S) = p \quad \text{and} \quad P(F) = 1 - p = q$$

n	The number of trials
X	The number of successes in n trials

Note that $0 \leq X \leq n$ and $X = 0, 1, 2, 3, \ldots, n$.

The probability of a success in a binomial experiment can be computed with this formula.

Binomial Probability Formula

In a binomial experiment, the probability of exactly X successes in n trials is

$$P(X) = \frac{n!}{(n - X)!X!} \cdot p^X \cdot q^{n-X}$$

An explanation of why the formula works is given following Example 5–15.

| Example 5–15 | Tossing Coins |

A coin is tossed 3 times. Find the probability of getting exactly two heads.

Solution

This problem can be solved by looking at the sample space. There are three ways to get two heads.

HHH, <u>HHT, HTH, THH,</u> TTH, THT, HTT, TTT

The answer is $\frac{3}{8}$, or 0.375.

Looking at the problem in Example 5–15 from the standpoint of a binomial experiment, one can show that it meets the four requirements.

1. There are a fixed number of trials (three).
2. There are only two outcomes for each trial, heads or tails.
3. The outcomes are independent of one another (the outcome of one toss in no way affects the outcome of another toss).
4. The probability of a success (heads) is $\frac{1}{2}$ in each case.

In this case, $n = 3$, $X = 2$, $p = \frac{1}{2}$, and $q = \frac{1}{2}$. Hence, substituting in the formula gives

$$P(2 \text{ heads}) = \frac{3!}{(3-2)!2!} \cdot \left(\frac{1}{2}\right)^2 \left(\frac{1}{2}\right)^1 = \frac{3}{8} = 0.375$$

which is the same answer obtained by using the sample space.

The same example can be used to explain the formula. First, note that there are three ways to get exactly two heads and one tail from a possible eight ways. They are HHT, HTH, and THH. In this case, then, the number of ways of obtaining two heads from three coin tosses is $_3C_2$, or 3, as shown in Chapter 4. In general, the number of ways to get X successes from n trials without regard to order is

$$_nC_X = \frac{n!}{(n-X)!X!}$$

This is the first part of the binomial formula. (Some calculators can be used for this.)

Next, each success has a probability of $\frac{1}{2}$ and can occur twice. Likewise, each failure has a probability of $\frac{1}{2}$ and can occur once, giving the $(\frac{1}{2})^2(\frac{1}{2})^1$ part of the formula. To generalize, then, each success has a probability of p and can occur X times, and each failure has a probability of q and can occur $n - X$ times. Putting it all together yields the binomial probability formula.

| Example 5–16 | Survey on Doctor Visits |

A survey found that one out of five Americans say he or she has visited a doctor in any given month. If 10 people are selected at random, find the probability that exactly 3 will have visited a doctor last month.

Source: *Reader's Digest.*

Solution

In this case, $n = 10$, $X = 3$, $p = \frac{1}{5}$, and $q = \frac{4}{5}$. Hence,

$$P(3) = \frac{10!}{(10-3)!3!}\left(\frac{1}{5}\right)^3\left(\frac{4}{5}\right)^7 = 0.201$$

| Example 5–17 | Survey on Employment |

A survey from Teenage Research Unlimited (Northbrook, Illinois) found that 30% of teenage consumers receive their spending money from part-time jobs. If 5 teenagers are selected at random, find the probability that at least 3 of them will have part-time jobs.

Solution

To find the probability that at least 3 have part-time jobs, it is necessary to find the individual probabilities for 3, or 4, or 5 and then add them to get the total probability.

$$P(3) = \frac{5!}{(5-3)!3!}(0.3)^3(0.7)^2 = 0.132$$

$$P(4) = \frac{5!}{(5-4)!4!}(0.3)^4(0.7)^1 = 0.028$$

$$P(5) = \frac{5!}{(5-5)!5!}(0.3)^5(0.7)^0 = 0.002$$

Hence,

P(at least three teenagers have part-time jobs)
$= 0.132 + 0.028 + 0.002 = 0.162$

Computing probabilities by using the binomial probability formula can be quite tedious at times, so tables have been developed for selected values of n and p. Table B in Appendix C gives the probabilities for individual events. Example 5–18 shows how to use Table B to compute probabilities for binomial experiments.

| Example 5–18 | Tossing Coins |

Solve the problem in Example 5–15 by using Table B.

Solution

Since $n = 3$, $X = 2$, and $p = 0.5$, the value 0.375 is found as shown in Figure 5–3.

| Figure 5–3 |

Using Table B for Example 5–18

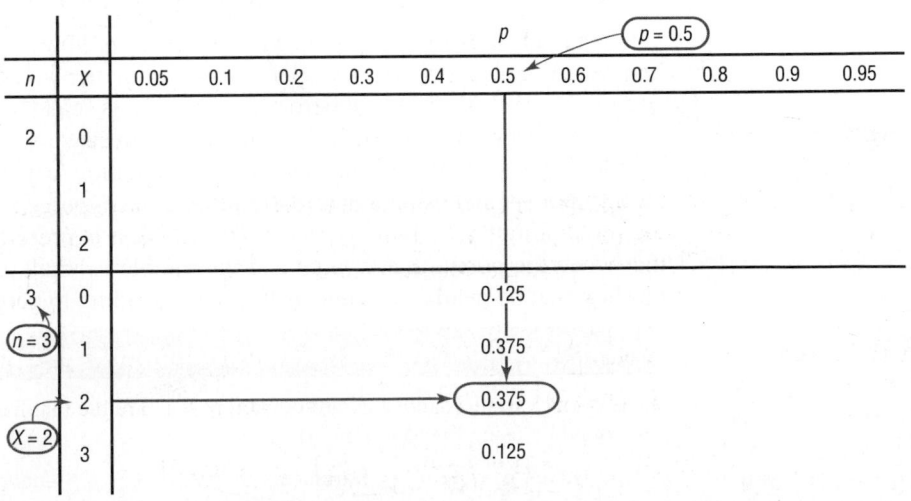

| Example 5–19 | Survey on Fear of Being Home Alone at Night |

Public Opinion reported that 5% of Americans are afraid of being alone in a house at night. If a random sample of 20 Americans is selected, find these probabilities by using the binomial table.

   a. There are exactly 5 people in the sample who are afraid of being alone at night.
   b. There are at most 3 people in the sample who are afraid of being alone at night.
   c. There are at least 3 people in the sample who are afraid of being alone at night.

Source: *100% American* by Daniel Evan Weiss.

Solution

   a. $n = 20$, $p = 0.05$, and $X = 5$. From the table, we get 0.002.
   b. $n = 20$ and $p = 0.05$. "At most 3 people" means 0, or 1, or 2, or 3. Hence, the solution is

$$P(0) + P(1) + P(2) + P(3) = 0.358 + 0.377 + 0.189 + 0.060$$
$$= 0.984$$

   c. $n = 20$ and $p = 0.05$. "At least 3 people" means 3, 4, 5, . . . , 20. This problem can best be solved by finding $P(0) + P(1) + P(2)$ and subtracting from 1.

$$P(0) + P(1) + P(2) = 0.358 + 0.377 + 0.189 = 0.924$$
$$1 - 0.924 = 0.076$$

| Example 5–20 | Driving While Intoxicated |

A report from the Secretary of Health and Human Services stated that 70% of single-vehicle traffic fatalities that occur at night on weekends involve an intoxicated driver. If a sample of 15 single-vehicle traffic fatalities that occur at night on a weekend is selected, find the probability that exactly 12 involve a driver who is intoxicated.

Source: *100% American* by Daniel Evan Weiss.

Solution

Now, $n = 15$, $p = 0.70$, and $X = 12$. From Table B, $P(12) = 0.170$. Hence, the probability is 0.17.

Remember that in the use of the binomial distribution, the outcomes must be independent. For example, in the selection of components from a batch to be tested, each component must be replaced before the next one is selected. Otherwise, the outcomes are not independent. However, a dilemma arises because there is a chance that the same component could be selected again. This situation can be avoided by not replacing the component and using a distribution called the hypergeometric distribution to calculate the probabilities. The hypergeometric distribution is presented later in this chapter. Note that when the population is large and the sample is small, the binomial probabilities can be shown to be nearly the same as the corresponding hypergeometric probabilities.

Objective 4

Find the mean, variance, and standard deviation for the variable of a binomial distribution.

Mean, Variance, and Standard Deviation for the Binomial Distribution

The mean, variance, and standard deviation of a variable that has the *binomial distribution* can be found by using the following formulas.

Mean: $\mu = n \cdot p$ Variance: $\sigma^2 = n \cdot p \cdot q$ Standard deviation: $\sigma = \sqrt{n \cdot p \cdot q}$

These formulas are algebraically equivalent to the formulas for the mean, variance, and standard deviation of the variables for probability distributions, but because they are for variables of the binomial distribution, they have been simplified by using algebra. The algebraic derivation is omitted here, but their equivalence is shown in Example 5–21.

Example 5–21

Tossing a Coin

A coin is tossed 4 times. Find the mean, variance, and standard deviation of the number of heads that will be obtained.

Solution

With the formulas for the binomial distribution and $n = 4$, $p = \frac{1}{2}$, and $q = \frac{1}{2}$, the results are

$$\mu = n \cdot p = 4 \cdot \tfrac{1}{2} = 2$$
$$\sigma^2 = n \cdot p \cdot q = 4 \cdot \tfrac{1}{2} \cdot \tfrac{1}{2} = 1$$
$$\sigma = \sqrt{1} = 1$$

From Example 5–21, when four coins are tossed many, many times, the average of the number of heads that appear is 2, and the standard deviation of the number of heads is 1. Note that these are theoretical values.

As stated previously, this problem can be solved by using the formulas for expected value. The distribution is shown.

No. of heads X	0	1	2	3	4
Probability $P(X)$	$\frac{1}{16}$	$\frac{4}{16}$	$\frac{6}{16}$	$\frac{4}{16}$	$\frac{1}{16}$

$$\mu = E(X) = \Sigma X \cdot P(X) = 0 \cdot \tfrac{1}{16} + 1 \cdot \tfrac{4}{16} + 2 \cdot \tfrac{6}{16} + 3 \cdot \tfrac{4}{16} + 4 \cdot \tfrac{1}{16} = \tfrac{32}{16} = 2$$
$$\sigma^2 = \Sigma X^2 \cdot P(X) - \mu^2$$
$$= 0^2 \cdot \tfrac{1}{16} + 1^2 \cdot \tfrac{4}{16} + 2^2 \cdot \tfrac{6}{16} + 3^2 \cdot \tfrac{4}{16} + 4^2 \cdot \tfrac{1}{16} - 2^2 = \tfrac{80}{16} - 4 = 1$$
$$\sigma = \sqrt{1} = 1$$

Hence, the simplified binomial formulas give the same results.

Example 5–22

Rolling a Die

A die is rolled 480 times. Find the mean, variance, and standard deviation of the number of 3s that will be rolled.

Solution

This is a binomial experiment since getting a 3 is a success and not getting a 3 is considered a failure.

Hence $n = 480$, $p = \frac{1}{6}$, and $q = \frac{5}{6}$.

$$\mu = n \cdot p = 480 \cdot \tfrac{1}{6} = 80$$
$$\sigma^2 = n \cdot p \cdot q = 480 \cdot \tfrac{1}{6} \cdot \tfrac{5}{6} = 66.67$$
$$\sigma = \sqrt{66.67} = 8.16$$

| Example 5–23 | **Likelihood of Twins** |

The *Statistical Bulletin* published by Metropolitan Life Insurance Co. reported that 2% of all American births result in twins. If a random sample of 8000 births is taken, find the mean, variance, and standard deviation of the number of births that would result in twins.

Source: *100% American* by Daniel Evan Weiss.

Solution

This is a binomial situation, since a birth can result in either twins or not twins (i.e., two outcomes).

$$\mu = n \cdot p = (8000)(0.02) = 160$$
$$\sigma^2 = n \cdot p \cdot q = (8000)(0.02)(0.98) = 156.8$$
$$\sigma = \sqrt{n \cdot p \cdot q} = \sqrt{156.8} = 12.5$$

For the sample, the average number of births that would result in twins is 160, the variance is 156.8, or 157, and the standard deviation is 12.5, or 13 if rounded.

Applying the Concepts **5–3**

Unsanitary Restaurants

Health officials routinely check sanitary conditions of restaurants. Assume you visit a popular tourist spot and read in the newspaper that in 3 out of every 7 restaurants checked, there were unsatisfactory health conditions found. Assuming you are planning to eat out 10 times while you are there on vacation, answer the following questions.

1. How likely is it that you will eat at three restaurants with unsanitary conditions?
2. How likely is it that you will eat at four or five restaurants with unsanitary conditions?
3. Explain how you would compute the probability of eating in at least one restaurant with unsanitary conditions. Could you use the complement to solve this problem?
4. What is the most likely number to occur in this experiment?
5. How variable will the data be around the most likely number?
6. How do you know that this is a binomial distribution?
7. If it is a binomial distribution, does that mean that the likelihood of a success is always 50% since there are only two possible outcomes?

Check your answers by using the following computer-generated table.

Mean = 4.29   **Std. dev. = 1.56492**

X	P(X)	Cum. Prob.
0	0.00371	0.00371
1	0.02784	0.03155
2	0.09396	0.12552
3	0.18793	0.31344
4	0.24665	0.56009
5	0.22199	0.78208
6	0.13874	0.92082
7	0.05946	0.98028
8	0.01672	0.99700
9	0.00279	0.99979
10	0.00021	1.00000

See page 298 for the answers.

Exercises 5–3

1. Which of the following are binomial experiments or can be reduced to binomial experiments?
   a. Surveying 100 people to determine if they like Sudsy Soap Yes
   b. Tossing a coin 100 times to see how many heads occur Yes
   c. Drawing a card with replacement from a deck and getting a heart Yes
   d. Asking 1000 people which brand of cigarettes they smoke No
   e. Testing four different brands of aspirin to see which brands are effective No
   f. Testing one brand of aspirin by using 10 people to determine whether it is effective Yes
   g. Asking 100 people if they smoke Yes
   h. Checking 1000 applicants to see whether they were admitted to White Oak College Yes
   i. Surveying 300 prisoners to see how many different crimes they were convicted of No
   j. Surveying 300 prisoners to see whether this is their first offense Yes

2. (ans) Compute the probability of X successes, using Table B in Appendix C.
   a. $n = 2, p = 0.30, X = 1$ 0.420
   b. $n = 4, p = 0.60, X = 3$ 0.346
   c. $n = 5, p = 0.10, X = 0$ 0.590
   d. $n = 10, p = 0.40, X = 4$ 0.251
   e. $n = 12, p = 0.90, X = 2$ 0.000
   f. $n = 15, p = 0.80, X = 12$ 0.250
   g. $n = 17, p = 0.05, X = 0$ 0.418
   h. $n = 20, p = 0.50, X = 10$ 0.176
   i. $n = 16, p = 0.20, X = 3$ 0.246

3. Compute the probability of X successes, using the binomial formula.
   a. $n = 6, X = 3, p = 0.03$ 0.0005
   b. $n = 4, X = 2, p = 0.18$ 0.131
   c. $n = 5, X = 3, p = 0.63$ 0.342
   d. $n = 9, X = 0, p = 0.42$ 0.007
   e. $n = 10, X = 5, p = 0.37$ 0.173

For Exercises 4 through 13, assume all variables are binomial. (*Note:* If values are not found in Table B of Appendix C, use the binomial formula.)

4. **Guidance Missile System** A missile guidance system has five fail-safe components. The probability of each failing is 0.05. Find these probabilities.
   a. Exactly 2 will fail. 0.021 (TI 0.0214)
   b. More than 2 will fail. 0.001 (TI 0.001158)
   c. All will fail. 0 (TI 0.0000003)
   d. Compare the answers for parts a, b, and c, and explain why these results are reasonable. Since the probability of each event becomes less likely, the probabilities become smaller.

5. **True/False Exam** A student takes a 20-question, true/false exam and guesses on each question. Find the probability of passing if the lowest passing grade is 15 correct out of 20. Would you consider this event likely to occur? Explain your answer. 0.021; no, it's only about a 2% chance.

6. **Multiple-Choice Exam** A student takes a 20-question, multiple-choice exam with five choices for each question and guesses on each question. Find the probability of guessing at least 15 out of 20 correctly. Would you consider this event likely or unlikely to occur? Explain your answer. 0.000; the probability is extremely small.

7. **Driving to Work Alone** It is reported that 77% of workers aged 16 and over drive to work alone. Choose 8 workers at random. Find the probability that
   a. All drive to work alone 0.124
   b. More than one-half drive to work alone 0.912
   c. Exactly 3 drive to work alone 0.017
   Source: www.factfinder.census.gov

8. **High School Dropouts** Approximately 10.3% of American high school students drop out of school before graduation. Choose 10 students entering high school at random. Find the probability that
   a. No more than two drop out 0.925
   b. At least 6 graduate 0.998
   c. All 10 stay in school and graduate 0.337
   Source: www.infoplease.com

9. **Survey on Concern for Criminals** In a survey, 3 of 4 students said the courts show "too much concern" for criminals. Find the probability that at most 3 out of 7 randomly selected students will agree with this statement. Source: *Harper's Index.* 0.071

10. **Labor Force Couples** The percentage of couples where both parties are in the labor force is 52.1. Choose 5 couples at random. Find the probability that
    a. None of the couples have both persons working 0.025
    b. More than 3 of the couples have both persons in the labor force 0.215
    c. Fewer than 2 of the couples have both parties working 0.162
    Source: www.bls.gov

11. **College Education and Business World Success** R. H. Bruskin Associates Market Research found that 40% of Americans do not think that having a college education is important to succeed in the business world. If a random sample of five Americans is selected, find these probabilities.
    a. Exactly 2 people will agree with that statement. 0.346
    b. At most 3 people will agree with that statement. 0.913
    c. At least 2 people will agree with that statement. 0.663
    d. Fewer than 3 people will agree with that statement. Source: *100% American* by Daniel Evans Weiss. 0.683

12. **Destination Weddings** Twenty-six percent of couples who plan to marry this year are planning destination weddings. In a random sample of 12 couples who plan to marry, find the probability that

 a. Exactly 6 couples will have a destination wedding
 b. At least 6 couples will have a destination wedding
 c. Fewer than 5 couples will have a destination wedding   *a.* 0.047 *b.* 0.065 *c.* 0.821

 Source: *Time* magazine.

13. **People Who Have Some College Education** Fifty-three percent of all persons in the U.S. population have at least some college education. Choose 10 persons at random. Find the probability that

 a. Exactly one-half have some college education 0.242
 b. At least 5 do not have any college education 0.547
 c. Fewer than 5 have some college education 0.306

 Source: *New York Times Almanac.*

14. **(ans)** Find the mean, variance, and standard deviation for each of the values of n and p when the conditions for the binomial distribution are met.

 a. $n = 100, p = 0.75$ 75; 18.8; 4.3
 b. $n = 300, p = 0.3$ 90; 63; 7.9
 c. $n = 20, p = 0.5$ 10; 5; 2.2
 d. $n = 10, p = 0.8$ 8; 1.6; 1.3
 e. $n = 1000, p = 0.1$ 100; 90; 9.5
 f. $n = 500, p = 0.25$ 125; 93.8; 9.7
 g. $n = 50, p = \frac{2}{5}$ 20; 12; 3.5
 h. $n = 36, p = \frac{1}{6}$ 6; 5; 2.2

15. **Social Security Recipients** A study found that 1% of Social Security recipients are too young to vote. If 800 Social Security recipients are randomly selected, find the mean, variance, and standard deviation of the number of recipients who are too young to vote. 8; 7.9; 2.8

 Source: *Harper's Index.*

16. **Tossing Coins** Find the mean, variance, and standard deviation for the number of heads when ten coins are tossed. 5; 2.5; 1.58

17. **Defective Calculators** If 3% of calculators are defective, find the mean, variance, and standard deviation of a lot of 300 calculators. 9; 8.73; 2.95

18. **Federal Government Employee E-mail Use** It has been reported that 83% of federal government employees use e-mail. If a sample of 200 federal government employees is selected, find the mean, variance, and standard deviation of the number who use e-mail.

 Source: *USA TODAY.* 166; 28.2; 5.3

19. **Watching Fireworks** A survey found that 21% of Americans watch fireworks on television on July 4. Find the mean, variance, and standard deviation of the number of individuals who watch fireworks on television on July 4 if a random sample of 1000 Americans is selected.

 Source: USA Snapshot, *USA TODAY.* 210; 165.9; 12.9

20. **Alternate Sources of Fuel** Eighty-five percent of Americans favor spending government money to develop alternative sources of fuel for automobiles. For a random sample of 120 Americans, find the mean, variance, and standard deviation for the number who favor government spending for alternative fuels.

 Source: www.pollingreport.com 102; 15.3; 3.912

21. **Survey on Bathing Pets** A survey found that 25% of pet owners had their pets bathed professionally rather than do it themselves. If 18 pet owners are randomly selected, find the probability that exactly 5 people have their pets bathed professionally. 0.199

 Source: USA Snapshot, *USA TODAY.*

22. **Survey on Answering Machine Ownership** In a survey, 63% of Americans said they own an answering machine. If 14 Americans are selected at random, find the probability that exactly 9 own an answering machine. 0.217

 Source: USA Snapshot, *USA TODAY.*

23. **Poverty and the Federal Government** One out of every three Americans believes that the U.S. government should take "primary responsibility" for eliminating poverty in the United States. If 10 Americans are selected, find the probability that at most 3 will believe that the U.S. government should take primary responsibility for eliminating poverty. 0.559

 Source: *Harper's Index.*

24. **Internet Purchases** Thirty-two percent of adult Internet users have purchased products or services online. For a random sample of 200 adult Internet users, find the mean, variance, and standard deviation for the number who have purchased goods or services online. 64; 43.52; 6.597

 Source: www.infoplease.com

25. **Survey on Internet Awareness** In a survey, 58% of American adults said they had never heard of the Internet. If 20 American adults are selected at random, find the probability that exactly 12 will say they have never heard of the Internet. 0.177

 Source: *Harper's Index.*

26. **Job Elimination** In the past year, 13% of businesses have eliminated jobs. If 5 businesses are selected at random, find the probability that at least 3 have eliminated jobs during the last year. 0.018

 Source: *USA TODAY.*

27. **Survey of High School Seniors** Of graduating high school seniors, 14% said that their generation will be remembered for their social concerns. If 7 graduating seniors are selected at random, find the probability that either 2 or 3 will agree with that statement. 0.246

 Source: *USA TODAY.*

28. Is this a binomial distribution? Explain.

X	0	1	2	3
$P(X)$	0.064	0.288	0.432	0.216

Extending the Concepts

29. Children in a Family The graph shown here represents the probability distribution for the number of girls in a family of three children. From this graph, construct a probability distribution.

30. Construct a binomial distribution graph for the number of defective computer chips in a lot of 4 if $p = 0.3$.

$P(X)$

0.375	
0.250	
0.125	

Probability

0   1   2   3 X

Number of girls

Technology *Step by Step*

MINITAB
Step by Step

The Binomial Distribution

Calculate a Binomial Probability

From Example 5–19, it is known that 5% of the population is afraid of being alone at night. If a random sample of 20 Americans is selected, what is the probability that exactly 5 of them are afraid?

$$n = 20 \qquad p = 0.05 \ (5\%) \qquad \text{and} \qquad X = 5 \ (5 \text{ out of } 20)$$

No data need to be entered in the worksheet.

1. Select **Calc>Probability Distributions>Binomial.**

2. Click the option for Probability.

3. Click in the text box for Number of trials:.

4. Type in **20,** then Tab to Probability of success, then type **.05.**

5. Click the option for Input constant, then type in **5.** Leave the text box for Optional storage empty. If the name of a constant such as K1 is entered here, the results are stored but not displayed in the session window.

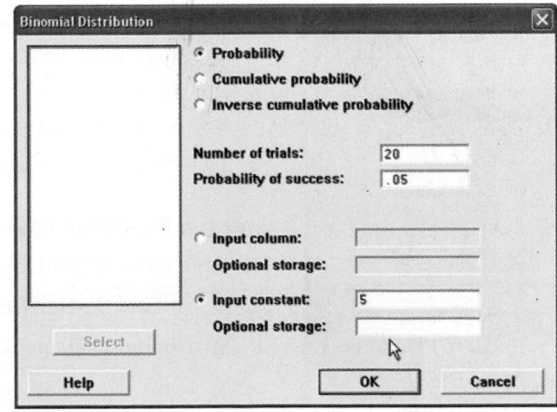

6. Click [OK]. The results are visible in the session window.

Probability Density Function
Binomial with n = 20 and p = 0.05
x   f(x)
5 0.0022446

Construct a Binomial Distribution

These instructions will use $n = 20$ and $p = 0.05$.

1. Select **Calc>Make Patterned Data>Simple Set of Numbers.**
2. You must enter three items:
   a) Enter **X** in the box for Store patterned data in:. MINITAB will use the first empty column of the active worksheet and name it X.
   b) Press Tab. Enter the value of **0** for the first value. Press Tab.
   c) Enter **20** for the last value. This value should be n. In steps of:, the value should be 1.
3. Click [OK].
4. Select **Calc>Probability Distributions>Binomial.**
5. In the dialog box you must enter five items.
   a) Click the button for Probability.
   b) In the box for Number of trials enter **20**.
   c) Enter **.05** in the Probability of success.

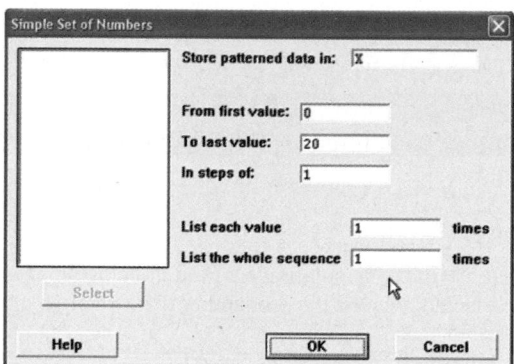

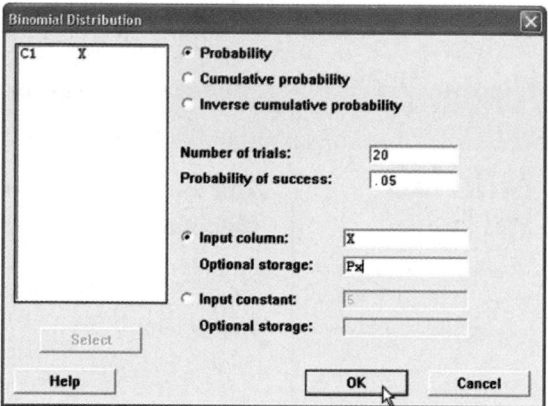

   d) Check the button for Input columns, then type the column name, **X,** in the text box.
   e) Click in the box for Optional storage, then type **Px.**
6. Click [OK]. The first available column will be named Px, and the calculated probabilities will be stored in it.
7. To view the completed table, click the worksheet icon on the toolbar.

Graph a Binomial Distribution

The table must be available in the worksheet.

1. Select **Graph>Scatterplot,** then Simple.
   a) Double-click on C2 Px for the Y variable and C1 X for the X variable.
   b) Click [Data view], then Project lines, then [OK]. Deselect any other type of display that may be selected in this list.
   c) Click on [Labels], then Title/Footnotes.
   d) Type an appropriate title, such as **Binomial Distribution n = 20, p = .05.**
   e) Press Tab to the Subtitle 1, then type in Your Name.
   f) Optional: Click [Scales] then [Gridlines] then check the box for Y major ticks.
   g) Click [OK] twice.

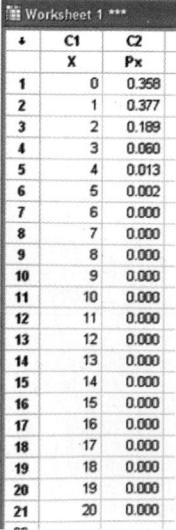

	C1	C2
	X	Px
1	0	0.358
2	1	0.377
3	2	0.189
4	3	0.060
5	4	0.013
6	5	0.002
7	6	0.000
8	7	0.000
9	8	0.000
10	9	0.000
11	10	0.000
12	11	0.000
13	12	0.000
14	13	0.000
15	14	0.000
16	15	0.000
17	16	0.000
18	17	0.000
19	18	0.000
20	19	0.000
21	20	0.000

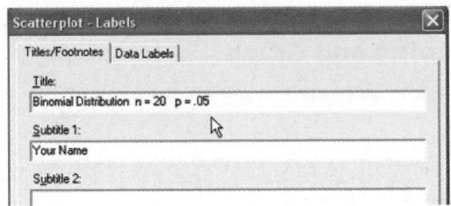

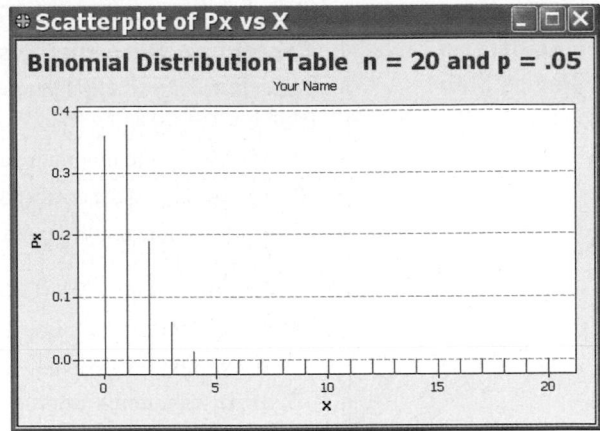

The graph will be displayed in a window. Right-click the control box to save, print, or close the graph.

TI-83 Plus or TI-84 Plus
Step by Step

Binomial Random Variables

To find the probability for a binomial variable:
Press **2nd [DISTR]** then **0** for binomial pdf((Note: On the TI-84 Plus Use A)
The form is binompdf(n,p,X).

Example: $n = 20$, $X = 5$, $p = .05$. (Example 5–19a from the text) binompdf(20,.05,5)

Example: $n = 20$, $X = 0, 1, 2, 3$, $p = .05$. (Example 5–19b from the text)
binompdf(20,.05,{0,1,2,3})
The calculator will display the probabilities in a list. Use the arrow keys to view entire display.

To find the cumulative probability for a binomial random variable:
Press **2nd [DISTR]** then **A (ALPHA MATH)** for binomcdf((Note: On the TI-84 Plus Use B)
The form is binomcdf(n,p,X). This will calculate the cumulative probability for values from 0 to X.

Example: $n = 20$, $X = 0, 1, 2, 3$, $p = .05$ (Example 5–19b from the text)
binomcdf(20,.05,3)

```
binompdf(20,.05,
5)
           .002244646
binompdf(20,.05,
{0,1,2,3})
{.3584859224 .3…
```

```
binomcdf(20,.05,
3)
           .984098474
```

To construct a binomial probability table:

1. Enter the X values 0 through n into L₁.

2. Move the cursor to the top of the L₂ column so that L₂ is highlighted.

3. Type the command binompdf($n,p,$L₁), then press **ENTER.**

Example: $n = 20$, $p = .05$ (Example 5–19 from the text)

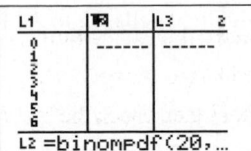

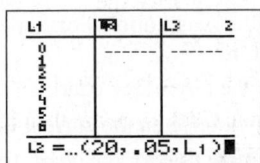

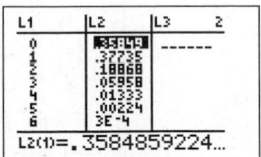

Excel
Step by Step

**Random Number
Generation Dialog Box**

Creating a Binomial Distribution and Graph

These instructions will demonstrate how Excel can be used to construct a binomial distribution table for $n = 20$ and $p = 0.35$.

1. Type **X** for the binomial variable label in cell A1 of an Excel worksheet.
2. Type **P(X)** for the corresponding probabilities in cell B1.
3. Enter the integers from 0 to 20 in column A starting at cell A2. Select the Data tab from the toolbar. Then select Data Analysis. Under Analysis Tools, select Random Number Generation and click [OK].
4. In the Random Number Generation dialog box, enter the following:
   a) Number of Variables: **1**
   b) Distribution: Patterned
   c) Parameters: From **0** to **20** in steps of **1,** repeating each number: **1** times and repeating each sequence **1** times
   d) Output range: **A2:A21**
5. Then click [OK].

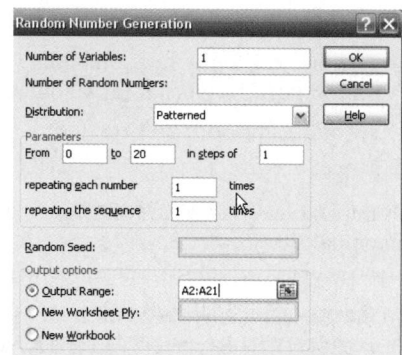

6. To determine the probability corresponding to the first value of the binomial random variable, select cell B2 and type: **=BINOMDIST(0,20,.35,FALSE).** This will give the probability of obtaining 0 successes in 20 trials of a binomial experiment for which the probability of success is 0.35.
7. Repeat step **6,** changing the first parameter, for each of the values of the random variable from column A.

Note: If you wish to obtain the cumulative probabilities for each of the values in column A, you can type: **=BINOMDIST(0,20,.35,TRUE)** and repeat for each of the values in column A.

To create the graph:

1. Select the Insert tab from the toolbar and the Column Chart.
2. Select the Clustered Column (the first column chart under the 2-D Column selections).
3. You will need to edit the data for the chart.
   a) Right-click the mouse on any location of the chart. Click the Select Data option. The Select Data Source dialog box will appear.
   b) Click **X** in the Legend Entries box and click Remove.
   c) Click the Edit button under Horizontal Axis Labels to insert a range for the variable X.
   d) When the Axis Labels box appears, highlight cells A2 to A21 on the worksheet, then click [OK].
4. To change the title of the chart:
   a) Left-click once on the current title.
   b) Type a new title for the chart, for example, Binomial Distribution (20, .35, .65).

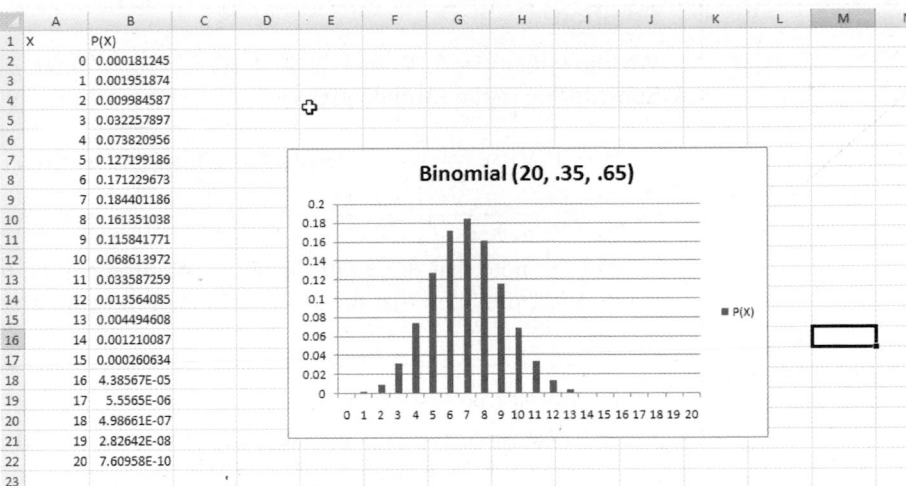

	A	B
1	X	P(X)
2	0	0.000181245
3	1	0.001951874
4	2	0.009984587
5	3	0.032257897
6	4	0.073820956
7	5	0.127199186
8	6	0.171229673
9	7	0.184401186
10	8	0.161351038
11	9	0.115841771
12	10	0.068613972
13	11	0.033587259
14	12	0.013564085
15	13	0.004494608
16	14	0.001210087
17	15	0.000260634
18	16	4.38567E-05
19	17	5.5565E-06
20	18	4.98661E-07
21	19	2.82642E-08
22	20	7.60958E-10
23		

5–4 Other Types of Distributions (Optional)

In addition to the binomial distribution, other types of distributions are used in statistics. Three of the most commonly used distributions are the multinomial distribution, the Poisson distribution, and the hypergeometric distribution. They are described next.

The Multinomial Distribution

Objective 5

Find probabilities for outcomes of variables, using the Poisson, hypergeometric, and multinomial distributions.

Recall that in order for an experiment to be binomial, two outcomes are required for each trial. But if each trial in an experiment has more than two outcomes, a distribution called the **multinomial distribution** must be used. For example, a survey might require the responses of "approve," "disapprove," or "no opinion." In another situation, a person may have a choice of one of five activities for Friday night, such as a movie, dinner, baseball game, play, or party. Since these situations have more than two possible outcomes for each trial, the binomial distribution cannot be used to compute probabilities.

The multinomial distribution can be used for such situations if the probabilities for each trial remain constant and the outcomes are independent for a fixed number of trials. The events must also be mutually exclusive.

Formula for the Multinomial Distribution

If X consists of events $E_1, E_2, E_3, \ldots, E_k$, which have corresponding probabilities $p_1, p_2, p_3, \ldots, p_k$ of occurring, and X_1 is the number of times E_1 will occur, X_2 is the number of times E_2 will occur, X_3 is the number of times E_3 will occur, etc., then the probability that X will occur is

$$P(X) = \frac{n!}{X_1! \cdot X_2! \cdot X_3! \cdots X_k!} \cdot p_1^{X_1} \cdot p_2^{X_2} \cdots p_k^{X_k}$$

where $X_1 + X_2 + X_3 + \cdots + X_k = n$ and $p_1 + p_2 + p_3 + \cdots + p_k = 1$.

Example 5–24   Leisure Activities

In a large city, 50% of the people choose a movie, 30% choose dinner and a play, and 20% choose shopping as a leisure activity. If a sample of 5 people is randomly selected, find the probability that 3 are planning to go to a movie, 1 to a play, and 1 to a shopping mall.

Solution

We know that $n = 5$, $X_1 = 3$, $X_2 = 1$, $X_3 = 1$, $p_1 = 0.50$, $p_2 = 0.30$, and $p_3 = 0.20$. Substituting in the formula gives

$$P(X) = \frac{5!}{3! \cdot 1! \cdot 1!} \cdot (0.50)^3(0.30)^1(0.20)^1 = 0.15$$

Again, note that the multinomial distribution can be used even though replacement is not done, provided that the sample is small in comparison with the population.

Example 5–25

Coffee Shop Customers

A small airport coffee shop manager found that the probabilities a customer buys 0, 1, 2, or 3 cups of coffee are 0.3, 0.5, 0.15, and 0.05, respectively. If 8 customers enter the shop, find the probability that 2 will purchase something other than coffee, 4 will purchase 1 cup of coffee, 1 will purchase 2 cups, and 1 will purchase 3 cups.

Solution

Let $n = 8$, $X_1 = 2$, $X_2 = 4$, $X_3 = 1$, and $X_4 = 1$.

$$p_1 = 0.3 \qquad p_2 = 0.5 \qquad p_3 = 0.15 \qquad \text{and} \qquad p_4 = 0.05$$

Then

$$P(X) = \frac{8!}{2!4!1!1!} \cdot (0.3)^2(0.5)^4(0.15)^1(0.05)^1 = 0.0354$$

Example 5–26

Selecting Colored Balls

A box contains 4 white balls, 3 red balls, and 3 blue balls. A ball is selected at random, and its color is written down. It is replaced each time. Find the probability that if 5 balls are selected, 2 are white, 2 are red, and 1 is blue.

Solution

We know that $n = 5$, $X_1 = 2$, $X_2 = 2$, $X_3 = 1$; $p_1 = \frac{4}{10}$, $p_2 = \frac{3}{10}$, and $p_3 = \frac{3}{10}$; hence,

$$P(X) = \frac{5!}{2!2!1!} \cdot \left(\frac{4}{10}\right)^2\left(\frac{3}{10}\right)^2\left(\frac{3}{10}\right)^1 = \frac{81}{625}$$

Thus, the multinomial distribution is similar to the binomial distribution but has the advantage of allowing you to compute probabilities when there are more than two outcomes for each trial in the experiment. That is, the multinomial distribution is a general distribution, and the binomial distribution is a special case of the multinomial distribution.

The Poisson Distribution

A discrete probability distribution that is useful when n is large and p is small and when the independent variables occur over a period of time is called the **Poisson distribution.** In addition to being used for the stated conditions (i.e., n is large, p is small, and the variables occur over a period of time), the Poisson distribution can be used when a density of items is distributed over a given area or volume, such as the number of plants growing per acre or the number of defects in a given length of videotape.

Historical Notes

Simeon D. Poisson (1781–1840) formulated the distribution that bears his name. It appears only once in his writings and is only one page long. Mathematicians paid little attention to it until 1907, when a statistician named W. S. Gosset found real applications for it.

Formula for the Poisson Distribution

The probability of X occurrences in an interval of time, volume, area, etc., for a variable where λ (Greek letter lambda) is the mean number of occurrences per unit (time, volume, area, etc.) is

$$P(X; \lambda) = \frac{e^{-\lambda}\lambda^X}{X!} \qquad \text{where } X = 0, 1, 2, \ldots$$

The letter e is a constant approximately equal to 2.7183.

Round the answers to four decimal places.

Example 5–27

Typographical Errors

If there are 200 typographical errors randomly distributed in a 500-page manuscript, find the probability that a given page contains exactly 3 errors.

Solution

First, find the mean number λ of errors. Since there are 200 errors distributed over 500 pages, each page has an average of

$$\lambda = \frac{200}{500} = \frac{2}{5} = 0.4$$

or 0.4 error per page. Since $X = 3$, substituting into the formula yields

$$P(X; \lambda) = \frac{e^{-\lambda}\lambda^X}{X!} = \frac{(2.7183)^{-0.4}(0.4)^3}{3!} = 0.0072$$

Thus, there is less than a 1% chance that any given page will contain exactly 3 errors.

Since the mathematics involved in computing Poisson probabilities is somewhat complicated, tables have been compiled for these probabilities. Table C in Appendix C gives P for various values for λ and X.

In Example 5–27, where X is 3 and λ is 0.4, the table gives the value 0.0072 for the probability. See Figure 5–4.

Figure 5–4

Using Table C

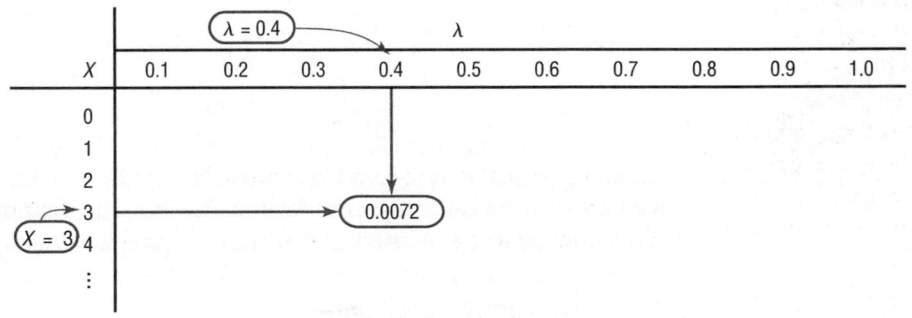

Example 5–28

Toll-Free Telephone Calls

A sales firm receives, on average, 3 calls per hour on its toll-free number. For any given hour, find the probability that it will receive the following.

a. At most 3 calls　　　*b.* At least 3 calls　　　*c.* 5 or more calls

Solution

a. "At most 3 calls" means 0, 1, 2, or 3 calls. Hence,

$$P(0; 3) + P(1; 3) + P(2; 3) + P(3; 3)$$
$$= 0.0498 + 0.1494 + 0.2240 + 0.2240$$
$$= 0.6472$$

b. "At least 3 calls" means 3 or more calls. It is easier to find the probability of 0, 1, and 2 calls and then subtract this answer from 1 to get the probability of at least 3 calls.

$$P(0; 3) + P(1; 3) + P(2; 3) = 0.0498 + 0.1494 + 0.2240 = 0.4232$$

and

$$1 - 0.4232 = 0.5768$$

c. For the probability of 5 or more calls, it is easier to find the probability of getting 0, 1, 2, 3, or 4 calls and subtract this answer from 1. Hence,

$$P(0; 3) + P(1; 3) + P(2; 3) + P(3; 3) + P(4; 3)$$
$$= 0.0498 + 0.1494 + 0.2240 + 0.2240 + 0.1680$$
$$= 0.8152$$

and

$$1 - 0.8152 = 0.1848$$

Thus, for the events described, the part *a* event is most likely to occur, and the part *c* event is least likely to occur.

The Poisson distribution can also be used to approximate the binomial distribution when the expected value $\lambda = n \cdot p$ is less than 5, as shown in Example 5–29. (The same is true when $n \cdot q < 5$.)

Example 5–29   Left-Handed People

If approximately 2% of the people in a room of 200 people are left-handed, find the probability that exactly 5 people there are left-handed.

Solution

Since $\lambda = n \cdot p$, then $\lambda = (200)(0.02) = 4$. Hence,

$$P(X; \lambda) = \frac{(2.7183)^{-4}(4)^5}{5!} = 0.1563$$

which is verified by the formula $_{200}C_5(0.02)^5(0.98)^{195} \approx 0.1579$. The difference between the two answers is based on the fact that the Poisson distribution is an approximation and rounding has been used.

The Hypergeometric Distribution

When sampling is done *without* replacement, the binomial distribution does not give exact probabilities, since the trials are not independent. The smaller the size of the population, the less accurate the binomial probabilities will be.

For example, suppose a committee of 4 people is to be selected from 7 women and 5 men. What is the probability that the committee will consist of 3 women and 1 man?

To solve this problem, you must find the number of ways a committee of 3 women and 1 man can be selected from 7 women and 5 men. This answer can be found by using combinations; it is

$$_7C_3 \cdot {}_5C_1 = 35 \cdot 5 = 175$$

Next, find the total number of ways a committee of 4 people can be selected from 12 people. Again, by the use of combinations, the answer is

$$_{12}C_4 = 495$$

Finally, the probability of getting a committee of 3 women and 1 man from 7 women and 5 men is

$$P(X) = \frac{175}{495} = \frac{35}{99}$$

The results of the problem can be generalized by using a special probability distribution called the hypergeometric distribution. The **hypergeometric distribution** is a distribution of a variable that has two outcomes when sampling is done without replacement.

The probabilities for the hypergeometric distribution can be calculated by using the formula given next.

Formula for the Hypergeometric Distribution

Given a population with only two types of objects (females and males, defective and nondefective, successes and failures, etc.), such that there are a items of one kind and b items of another kind and $a + b$ equals the total population, the probability $P(X)$ of selecting without replacement a sample of size n with X items of type a and $n - X$ items of type b is

$$P(X) = \frac{{}_aC_X \cdot {}_bC_{n-X}}{{}_{a+b}C_n}$$

The basis of the formula is that there are $_aC_X$ ways of selecting the first type of items, $_bC_{n-X}$ ways of selecting the second type of items, and $_{a+b}C_n$ ways of selecting n items from the entire population.

Example 5–30

Assistant Manager Applicants

Ten people apply for a job as assistant manager of a restaurant. Five have completed college and five have not. If the manager selects 3 applicants at random, find the probability that all 3 are college graduates.

Solution

Assigning the values to the variables gives

$$a = 5 \text{ college graduates} \qquad n = 3$$
$$b = 5 \text{ nongraduates} \qquad X = 3$$

and $n - X = 0$. Substituting in the formula gives

$$P(X) = \frac{{}_5C_3 \cdot {}_5C_0}{{}_{10}C_3} = \frac{10}{120} = \frac{1}{12}$$

Example 5–31

House Insurance

A recent study found that 2 out of every 10 houses in a neighborhood have no insurance. If 5 houses are selected from 10 houses, find the probability that exactly 1 will be uninsured.

Solution

In this example, $a = 2$, $b = 8$, $n = 5$, $X = 1$, and $n - X = 4$.

$$P(X) = \frac{{}_2C_1 \cdot {}_8C_4}{{}_{10}C_5} = \frac{2 \cdot 70}{252} = \frac{140}{252} = \frac{5}{9}$$

In many situations where objects are manufactured and shipped to a company, the company selects a few items and tests them to see whether they are satisfactory or defective. If a certain percentage is defective, the company then can refuse the whole shipment. This procedure saves the time and cost of testing every single item. To make the judgment about whether to accept or reject the whole shipment based on a small sample of tests, the company must know the probability of getting a specific number of defective items. To calculate the probability, the company uses the hypergeometric distribution.

Example 5–32

Defective Compressor Tanks

A lot of 12 compressor tanks is checked to see whether there are any defective tanks. Three tanks are checked for leaks. If 1 or more of the 3 is defective, the lot is rejected. Find the probability that the lot will be rejected if there are actually 3 defective tanks in the lot.

Solution

Since the lot is rejected if at least 1 tank is found to be defective, it is necessary to find the probability that none are defective and subtract this probability from 1.
    Here, $a = 3$, $b = 9$, $n = 3$, and $X = 0$; so

$$P(X) = \frac{{}_3C_0 \cdot {}_9C_3}{{}_{12}C_3} = \frac{1 \cdot 84}{220} = 0.38$$

Hence,

$$P(\text{at least 1 defective}) = 1 - P(\text{no defectives}) = 1 - 0.38 = 0.62$$

There is a 0.62, or 62%, probability that the lot will be rejected when 3 of the 12 tanks are defective.

A summary of the discrete distributions used in this chapter is shown in Table 5–1.

Interesting Fact

An IBM supercomputer set a world record in 2008 by performing 1.026 quadrillion calculations in 1 second.

Table 5–1	Summary of Discrete Distributions

1. Binomial distribution

$$P(X) = \frac{n!}{(n - X)!X!} \cdot p^X \cdot q^{n-X}$$

$$\mu = n \cdot p \qquad \sigma = \sqrt{n \cdot p \cdot q}$$

Used when there are only two outcomes for a fixed number of independent trials and the probability for each success remains the same for each trial.

2. Multinomial distribution

$$P(X) = \frac{n!}{X_1! \cdot X_2! \cdot X_3! \cdots X_k!} \cdot p_1^{X_1} \cdot p_2^{X_2} \cdots p_k^{X_k}$$

where

$$X_1 + X_2 + X_3 + \cdots + X_k = n \qquad \text{and} \qquad p_1 + p_2 + p_3 + \cdots + p_k = 1$$

Used when the distribution has more than two outcomes, the probabilities for each trial remain constant, outcomes are independent, and there are a fixed number of trials.

3. Poisson distribution

$$P(X; \lambda) = \frac{e^{-\lambda}\lambda^X}{X!} \qquad \text{where } X = 0, 1, 2, \ldots$$

Used when n is large and p is small, the independent variable occurs over a period of time, or a density of items is distributed over a given area or volume.

4. Hypergeometric distribution

$$P(X) = \frac{{}_aC_X \cdot {}_bC_{n-X}}{{}_{a+b}C_n}$$

Used when there are two outcomes and sampling is done without replacement.

Applying the Concepts 5–4

Rockets and Targets

During the latter days of World War II, the Germans developed flying rocket bombs. These bombs were used to attack London. Allied military intelligence didn't know whether these bombs were fired at random or had a sophisticated aiming device. To determine the answer, they used the Poisson distribution.

To assess the accuracy of these bombs, London was divided into 576 square regions. Each region was $\frac{1}{4}$ square kilometer in area. They then compared the number of actual hits with the theoretical number of hits by using the Poisson distribution. If the values in both distributions were close, then they would conclude that the rockets were fired at random. The actual distribution is as follows:

Hits	0	1	2	3	4	5
Regions	229	211	93	35	7	1

1. Using the Poisson distribution, find the theoretical values for each number of hits. In this case, the number of bombs was 535, and the number of regions was 576. So

$$\mu = \frac{535}{576} = 0.929$$

For 3 hits,

$$P(X) = \frac{\mu^X \cdot e^{-\mu}}{X!}$$

$$= \frac{(0.929)^3(2.7183)^{-0.929}}{3!} = 0.0528$$

Hence the number of hits is $(0.0528)(576) = 30.4128$.
Complete the table for the other number of hits.

Hits	0	1	2	3	4	5
Regions				30.4		

2. Write a brief statement comparing the two distributions.

3. Based on your answer to question 2, can you conclude that the rockets were fired at random?

See page 298 for the answer.

Exercises 5–4

1. Use the multinomial formula and find the probabilities for each.

   a. $n = 6, X_1 = 3, X_2 = 2, X_3 = 1, p_1 = 0.5, p_2 = 0.3, p_3 = 0.2$ 0.135

   b. $n = 5, X_1 = 1, X_2 = 2, X_3 = 2, p_1 = 0.3, p_2 = 0.6, p_3 = 0.1$ 0.0324

   c. $n = 4, X_1 = 1, X_2 = 1, X_3 = 2, p_1 = 0.8, p_2 = 0.1, p_3 = 0.1$ 0.0096

   d. $n = 3, X_1 = 1, X_2 = 1, X_3 = 1, p_1 = 0.5, p_2 = 0.3, p_3 = 0.2$ 0.18

   e. $n = 5, X_1 = 1, X_2 = 3, X_3 = 1, p_1 = 0.7, p_2 = 0.2, p_3 = 0.1$ 0.0112

2. **Firearm Sales** When people were asked if they felt that the laws covering the sale of firearms should be more strict, less strict, or kept as they are now, 54% responded more strict, 11% responded less, 34% said keep them as they are now, and 1% had no opinion. If 10 randomly selected people are asked the same question, what is the probability that 4 will respond more strict, 3 less, 2 keep them the same, and 1 have no opinion? 0.0016

   Source: www.pollingreport.com

3. **M&M Color Distribution** According to the manufacturer, M&M's are produced and distributed in the following proportions: 13% brown, 13% red, 14% yellow, 16% green, 20% orange, and 24% blue. In a random sample of 12 M&M's, what is the probability of having 2 of each color? 0.0025

4. **Truck Inspection Violations** The probabilities are 0.50, 0.40, and 0.10 that a trailer truck will have no violations, 1 violation, or 2 or more violations when it is given a safety inspection by state police. If 5 trailer trucks are inspected, find the probability that 3 will have no violations, 1 will have 1 violation, and 1 will have 2 or more violations. 0.1

5. **Rolling a Die** A die is rolled 4 times. Find the probability of two 1s, one 2, and one 3. $\frac{1}{108}$

6. **Mendel's Theory** According to Mendel's theory, if tall and colorful plants are crossed with short and colorless plants, the corresponding probabilities are $\frac{9}{16}, \frac{3}{16}, \frac{3}{16}$, and $\frac{1}{16}$ for tall and colorful, tall and colorless, short and colorful, and short and colorless, respectively. If 8 plants are selected, find the probability that 1 will be tall and colorful, 3 will be tall and colorless, 3 will be short and colorful, and 1 will be short and colorless. 0.002

7. Find each probability $P(X; \lambda)$, using Table C in Appendix C.

   a. $P(5; 4)$ 0.1563

   b. $P(2; 4)$ 0.1465

   c. $P(6; 3)$ 0.0504

   d. $P(10; 7)$ 0.071

   e. $P(9; 8)$ 0.1241

8. **Copy Machine Output** A copy machine randomly puts out 10 blank sheets per 500 copies processed. Find the probability that in a run of 300 copies, 5 sheets of paper will be blank. 0.1606

9. **Study of Robberies** A recent study of robberies for a certain geographic region showed an average of 1 robbery per 20,000 people. In a city of 80,000 people, find the probability of the following.

   a. 0 robberies 0.0183

   b. 1 robbery 0.0733

   c. 2 robberies 0.1465

   d. 3 or more robberies 0.7619

10. **Misprints on Manuscript Pages** In a 400-page manuscript, there are 200 randomly distributed misprints. If a page is selected, find the probability that it has 1 misprint. 0.3033

11. Telephone Soliciting A telephone soliciting company obtains an average of 5 orders per 1000 solicitations. If the company reaches 250 potential customers, find the probability of obtaining at least 2 orders. 0.3554

12. Mail Ordering A mail-order company receives an average of 5 orders per 500 solicitations. If it sends out 100 advertisements, find the probability of receiving at least 2 orders. 0.2642

13. Company Mailing Of a company's mailings 1.5% are returned because of incorrect or incomplete addresses. In a mailing of 200 pieces, find the probability that none will be returned. 0.0498

14. Emission Inspection Failures If 3% of all cars fail the emissions inspection, find the probability that in a sample of 90 cars, 3 will fail. Use the Poisson approximation. 0.2205

15. Phone Inquiries The average number of phone inquiries per day at the poison control center is 4. Find the probability it will receive 5 calls on a given day. Use the Poisson approximation. 0.1563

16. Defective Calculators In a batch of 2000 calculators, there are, on average, 8 defective ones. If a random sample of 150 is selected, find the probability of 5 defective ones. 0.0004

17. School Newspaper Staff A school newspaper staff is comprised of 5 seniors, 4 juniors, 5 sophomores, and 7 freshmen. If 4 staff members are chosen at random for a publicity photo, what is the probability that there will be 1 student from each class? 0.117

18. Missing Pages from Books A bookstore owner examines 5 books from each lot of 25 to check for missing pages. If he finds at least 2 books with missing pages, the entire lot is returned. If, indeed, there are 5 books with missing pages, find the probability that the lot will be returned. 0.252

19. Types of CDs A CD case contains 10 jazz albums, 4 classical albums, and 2 soundtracks. Choose 3 at random to put in a CD changer. What is the probability of selecting 2 jazz albums and 1 classical album? 0.321

20. Defective Computer Keyboards A shipment of 24 computer keyboards is rejected if 4 are checked for defects and at least 1 is found to be defective. Find the probability that the shipment will be returned if there are actually 6 defective keyboards. 0.712

21. Defective Electronics A shipment of 24 electric typewriters is rejected if 3 are checked for defects and at least 1 is found to be defective. Find the probability that the shipment will be returned if there are actually 6 typewriters that are defective. 0.597

Technology *Step by Step*

TI-83 Plus or TI-84 Plus
Step by Step

Poisson Random Variables

To find the probability for a Poisson random variable:
Press **2nd [DISTR]** then **B (ALPHA APPS)** for poissonpdf((Note: On the TI-84 Plus Use C)
The form is poissonpdf(λ,X).

Example: $\lambda = 0.4$, $X = 3$ (Example 5–27 from the text)
poissonpdf(.4,3)

Example: $\lambda = 3$, $X = 0, 1, 2, 3$ (Example 5–28a from the text)
poissonpdf(3,{0,1,2,3})
The calculator will display the probabilities in a list. Use the arrow keys to view the entire display.

To find the cumulative probability for a Poisson random variable:
Press **2nd [DISTR]** then **C (ALPHA PRGM)** for poissoncdf((Note: On the TI-84 Plus Use D)
The form is poissoncdf(λ,X). This will calculate the cumulative probability for values from 0 to X.

Example: $\lambda = 3$, $X = 0, 1, 2, 3$ (Example 5–28a from the text)
poissoncdf(3,3)

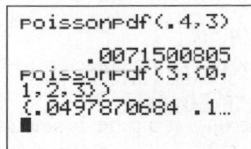

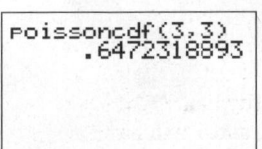

To construct a Poisson probability table:

1. Enter the X values 0 through a large possible value of X into L_1.

2. Move the cursor to the top of the L_2 column so that L_2 is highlighted.

3. Enter the command poissonpdf(λ,L_1) then press **ENTER**.

Example: $\lambda = 3$, $X = 0, 1, 2, 3, \ldots, 10$ (Example 5–28 from the text)

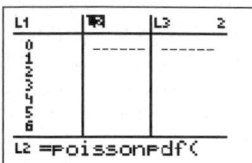 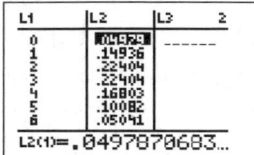

Summary

- A discrete probability distribution consists of the values a random variable can assume and the corresponding probabilities of these values. There are two requirements of a probability distribution: the sum of the probabilities of the events must equal 1, and the probability of any single event must be a number from 0 to 1. Probability distributions can be graphed. (5–1)

- The mean, variance, and standard deviation of a probability distribution can be found. The expected value of a discrete random variable of a probability distribution can also be found. This is basically a measure of the average. (5–2)

- A binomial experiment has four requirements. There must be a fixed number of trials. Each trial can have only two outcomes. The outcomes are independent of each other, and the probability of a success must remain the same for each trial. The probabilities of the outcomes can be found by using the binomial formula or the binomial table. (5–3)

- In addition to the binomial distribution, there are some other commonly used probability distributions. They are the multinomial distribution, the Poisson distribution, and the hypergeometric distribution. (5–4)

Important Terms

binomial
distribution 271

binomial
experiment 271

discrete probability
distribution 254

expected value 264

hypergeometric
distribution 287

multinomial
distribution 283

Poisson distribution 284

random variable 253

Important Formulas

Formula for the mean of a probability distribution:

$$\mu = \Sigma X \cdot P(X)$$

Formulas for the variance and standard deviation of a probability distribution:

$$\sigma^2 = \Sigma[X^2 \cdot P(X)] - \mu^2$$
$$\sigma = \sqrt{\Sigma[X^2 \cdot P(X)] - \mu^2}$$

Formula for expected value:

$$E(X) = \Sigma X \cdot P(X)$$

Binomial probability formula:

$$P(X) = \frac{n!}{(n-X)!X!} \cdot p^X \cdot q^{n-X} \quad \text{where } X = 0, 1, 2, 3, \ldots n$$

Formula for the mean of the binomial distribution:

$$\mu = n \cdot p$$

Formulas for the variance and standard deviation of the binomial distribution:

$$\sigma^2 = n \cdot p \cdot q \qquad \sigma = \sqrt{n \cdot p \cdot q}$$

Formula for the multinomial distribution:

$$P(X) = \frac{n!}{X_1! \cdot X_2! \cdot X_3! \cdots X_k!} \cdot p_1^{X_1} \cdot p_2^{X_2} \cdots p_k^{X_k}$$

(The Xs sum to n and the ps sum to one)

Formula for the Poisson distribution:

$$P(X; \lambda) = \frac{e^{-\lambda}\lambda^X}{X!} \qquad \text{where } X = 0, 1, 2, \ldots$$

Formula for the hypergeometric distribution:

$$P(X) = \frac{{}_aC_X \cdot {}_bC_{n-X}}{{}_{a+b}C_n}$$

Review Exercises

For Exercises 1 through 3, determine whether the distribution represents a probability distribution. If it does not, state why.

1. (5–1)

X	1	2	3	4	5
$P(X)$	$\frac{1}{10}$	$\frac{3}{10}$	$\frac{1}{10}$	$\frac{2}{10}$	$\frac{3}{10}$

Yes

2. (5–1) No. The sum of the probabilities does not equal 1.

X	5	10	15
$P(X)$	0.3	0.4	0.1

3. (5–1) No; the sum of the probabilities is greater than 1.

X	8	12	16	20
$P(X)$	$\frac{2}{6}$	$\frac{1}{12}$	$\frac{1}{12}$	$\frac{1}{12}$

4. Emergency Calls The number of emergency calls a local police department receives per 24-hour period is distributed as shown here. Construct a graph for the data. (5–1)

Number of calls X	10	11	12	13	14
Probability $P(X)$	0.02	0.12	0.40	0.31	0.15

5. Credit Cards A large retail company encourages its employees to get customers to apply for the store credit card. Below is the distribution for the number of credit card applications received per employee for an 8-hour shift.

X	0	1	2	3	4	5
$P(X)$	0.27	0.28	0.20	0.15	0.08	0.02

a. What is the probability that an employee will get 2 or 3 applications during any given shift? (5–1) 0.35
b. Find the mean, variance, and standard deviation for this probability distribution. (5–2) 1.55; 1.8075; 1.3444

6. Coins in a Box A box contains 5 pennies, 3 dimes, 1 quarter, and 1 half-dollar. Construct a probability distribution and draw a graph for the data. (5–1)

7. Tie Purchases At Tyler's Tie Shop, Tyler found the probabilities that a customer will buy 0, 1, 2, 3, or 4 ties, as shown. Construct a graph for the distribution. (5–1)

Number of ties X	0	1	2	3	4
Probability $P(X)$	0.30	0.50	0.10	0.08	0.02

8. Customers in a Bank A bank has a drive-through service. The number of customers arriving during a 15-minute period is distributed as shown. Find the mean, variance, and standard deviation for the distribution. (5–2) 2.1; 1.4; 1.2

Number of customers X	0	1	2	3	4
Probability $P(X)$	0.12	0.20	0.31	0.25	0.12

9. Arrivals at an Airport At a small rural airport, the number of arrivals per hour during the day has the distribution shown. Find the mean, variance, and standard deviation for the data. (5–2) 7.22; 2.1716; 1.47

Number X	5	6	7	8	9	10
Probability $P(X)$	0.14	0.21	0.24	0.18	0.16	0.07

10. Cans of Paint Purchased During a recent paint sale at Corner Hardware, the number of cans of paint purchased was distributed as shown. Find the mean, variance, and standard deviation of the distribution. (5–2) 2.1; 1.5; 1.2

Number of cans X	1	2	3	4	5
Probability $P(X)$	0.42	0.27	0.15	0.10	0.06

11. Inquiries Received The number of inquiries received per day for a college catalog is distributed as shown. Find the mean, variance, and standard deviation for the data. (5–2) 24.2; 1.5; 1.2

Number of inquiries X	22	23	24	25	26	27
Probability $P(X)$	0.08	0.19	0.36	0.25	0.07	0.05

12. Outdoor Regatta A producer plans an outdoor regatta for May 3. The cost of the regatta is $8000. This includes advertising, security, printing tickets, entertainment, etc. The producer plans to make $15,000 profit if all goes well. However, if it rains, the regatta will have to be canceled. According to the weather report, the probability of rain is 0.3. Find the producer's expected profit. (5–2) $8100

13. Card Game A game is set up as follows: All the diamonds are removed from a deck of cards, and these 13 cards are placed in a bag. The cards are mixed up, and then one card is chosen at random (and then replaced). The player wins according to the following rules.

If the ace is drawn, the player loses $20.
If a face card is drawn, the player wins $10.
If any other card (2–10) is drawn, the player wins $2.

How much should be charged to play this game in order for it to be fair? (5–2) $2.15

14. Using Exercise 13, how much should be charged if instead of winning $2 for drawing a 2–10, the player wins the amount shown on the card in dollars? (5–2) $4.92

15. Let x be a binomial random variable with $n = 12$ and $p = 0.3$. Find the following:

 a. $P(X = 8)$ 0.008
 b. $P(X < 5)$ 0.724
 c. $P(X \geq 10)$ 0.0002
 d. $P(4 < X \leq 9)$ (5–3) 0.276

16. **Internet Access via Cell Phone** Fourteen percent of cell phone users use their cell phones to access the Internet. In a random sample of 10 cell phone users, what is the probability that exactly 2 have used their phones to access the Internet? More than 2? (5–3) 0.2639; 0.155

Source: www.infoplease.com

17. **Computer Literacy Test** If 80% of job applicants are able to pass a computer literacy test, find the mean, variance, and standard deviation of the number of people who pass the examination in a sample of 150 applicants. (5–3) 120; 24; 4.9

18. **Flu Shots** It has been reported that 63% of adults aged 65 and over got their flu shots last year. In a random sample of 300 adults aged 65 and over, find the mean, variance, and standard deviation for the number who got their flu shots. (5–3) 189; 69.93; 8.3624

Source: U.S. Center for Disease Control and Prevention.

19. **U.S. Police Chiefs and the Death Penalty** The chance that a U.S. police chief believes the death penalty "significantly reduces the number of homicides" is 1 in 4. If a random sample of 8 police chiefs is selected, find the probability that at most 3 believe that the death penalty significantly reduces the number of homicides. (5–3) 0.886

Source: *Harper's Index.*

20. **Household Wood Burning** *American Energy Review* reported that 27% of American households burn wood. If a random sample of 500 American households is selected, find the mean, variance, and standard deviation of the number of households that burn wood. (5–3) 135; 98.6; 9.9

Source: *100% American* by Daniel Evan Weiss.

21. **Pizza for Breakfast** Three out of four American adults under age 35 have eaten pizza for breakfast. If a random sample of 20 adults under age 35 is selected, find the probability that exactly 16 have eaten pizza for breakfast. (5–3)

Source: *Harper's Index.* 0.190

22. **Unmarried Women** According to survey records, 75.4% of women aged 20–24 have never been married. In a random sample of 250 young women aged 20–24,

find the mean, variance, and standard deviation for the number who are or who have been married. (5–3)

Source: www.infoplease.com 61.5; 46.371; 6.8096

23. **(Opt.) Accuracy Count of Votes** After a recent national election, voters were asked how confident they were that votes in their state would be counted accurately. The results are shown below. 0.0193

46% Very confident   41% Somewhat confident
9% Not very confident   3% Not at all confident

If 10 voters are selected at random, find the probability that 5 would be very confident, 3 somewhat confident, 1 not very confident, and 1 not at all confident. (5–4)

Source: *New York Times.*

24. **(Opt.)** Before a DVD leaves the factory, it is given a quality control check. The probabilities that a DVD contains 0, 1, or 2 defects are 0.90, 0.06, and 0.04, respectively. In a sample of 12 recorders, find the probability that 8 have 0 defects, 3 have 1 defect, and 1 has 2 defects. (5–4) 0.007

25. **(Opt.)** In a Christmas display, the probability that all lights are the same color is 0.50; that 2 colors are used is 0.40; and that 3 or more colors are used is 0.10. If a sample of 10 displays is selected, find the probability that 5 have only 1 color of light, 3 have 2 colors, and 2 have 3 or more colors. (5–4) 0.050

26. **(Opt.) Lost Luggage in Airlines** Transportation officials reported that 8.25 out of every 1000 airline passengers lost luggage during their travels last year. If we randomly select 400 airline passengers, what is the probability that 5 lost some luggage? (5–4) 0.1203

Source: U.S. Department of Transportation.

27. **(Opt.)** Computer Help Hot Line receives, on average, 6 calls per hour asking for assistance. The distribution is Poisson. For any randomly selected hour, find the probability that the company will receive

 a. At least 6 calls 0.5543
 b. 4 or more calls 0.8488
 c. At most 5 calls (5–4) 0.4457

28. **(Opt.)** The number of boating accidents on Lake Emilie follows a Poisson distribution. The probability of an accident is 0.003. If there are 1000 boats on the lake during a summer month, find the probability that there will be 6 accidents. (5–4) 0.0504

29. **(Opt.)** If 5 cards are drawn from a deck, find the probability that 2 will be hearts. (5–4) 0.27

30. **(Opt.)** Of the 50 automobiles in a used-car lot, 10 are white. If 5 automobiles are selected to be sold at an auction, find the probability that exactly 2 will be white. (5–4) 0.21

31. **(Opt.) Items Donated to a Food Bank** At a food bank a case of donated items contains 10 cans of soup, 8 cans of vegetables, and 8 cans of fruit. If 3 cans are selected at random to distribute, find the probability of getting 1 vegetable and 2 cans of fruit. (5–4) 0.0862

Statistics Today	Is Pooling Worthwhile?–Revisited

In the case of the pooled sample, the probability that only one test will be needed can be determined by using the binomial distribution. The question being asked is, In a sample of 15 individuals, what is the probability that no individual will have the disease? Hence, $n = 15$, $p = 0.05$, and $X = 0$. From Table B in Appendix C, the probability is 0.463, or 46% of the time, only one test will be needed. For screening purposes, then, pooling samples in this case would save considerable time, money, and effort as opposed to testing every individual in the population.

Chapter Quiz

Determine whether each statement is true or false. If the statement is false, explain why.

1. The expected value of a random variable can be thought of as a long-run average. True

2. The number of courses a student is taking this semester is an example of a continuous random variable. False

3. When the binomial distribution is used, the outcomes must be dependent. False

4. A binomial experiment has a fixed number of trials. True

Complete these statements with the best answer.

5. Random variable values are determined by ___chance___.

6. The mean for a binomial variable can be found by using the formula ___$n \cdot p$___.

7. One requirement for a probability distribution is that the sum of all the events in the sample space must equal ___1___.

Select the best answer.

8. What is the sum of the probabilities of all outcomes in a probability distribution?
   a. 0
   b. $\frac{1}{2}$
   c. 1
   d. It cannot be determined.

9. How many outcomes are there in a binomial experiment?
   a. 0
   b. 1
   c. 2
   d. It varies.

10. The number of trials for a binomial experiment
   a. Can be infinite
   b. Is unchanged
   c. Is unlimited
   d. Must be fixed

For questions 11 through 14, determine if the distribution represents a probability distribution. If not, state why.

11.

X	1	2	3	4	5
P(X)	$\frac{1}{7}$	$\frac{2}{7}$	$\frac{2}{7}$	$\frac{3}{7}$	$\frac{2}{7}$

No, since $\Sigma P(X) > 1$

12.

X	3	6	9	12	15
P(X)	0.3	0.5	0.1	0.08	0.02

Yes

13.

X	50	75	100
P(X)	0.5	0.2	0.3

Yes

14.

X	4	8	12	16
P(X)	$\frac{1}{6}$	$\frac{3}{12}$	$\frac{1}{2}$	$\frac{1}{12}$

Yes

15. **Calls for a Fire Company** The number of fire calls the Conestoga Valley Fire Company receives per day is distributed as follows:

Number X	5	6	7	8	9
Probability P(X)	0.28	0.32	0.09	0.21	0.10

Construct a graph for the data.

16. **Telephones per Household** A study was conducted to determine the number of telephones each household has. The data are shown here.

Number of telephones	0	1	2	3	4
Frequency	2	30	48	13	7

Construct a probability distribution and draw a graph for the data.

17. **CD Purchases** During a recent CD sale at Matt's Music Store, the number of CDs customers purchased was distributed as follows:

Number X	0	1	2	3	4
Probability P(X)	0.10	0.23	0.31	0.27	0.09

Find the mean, variance, and standard deviation of the distribution. 2.0; 1.3; 1.1

18. **Calls for a Crisis Hot Line** The number of calls received per day at a crisis hot line is distributed as follows:

Number X	30	31	32	33	34
Probability P(X)	0.05	0.21	0.38	0.25	0.11

Find the mean, variance, and standard deviation of the distribution. 32.2; 1.1; 1.0

296   Chapter 5 Discrete Probability Distributions

19. Selecting a Card There are 6 playing cards placed face down in a box. They are the 4 of diamonds, the 5 of hearts, the 2 of clubs, the 10 of spades, the 3 of diamonds, and the 7 of hearts. A person selects a card. Find the expected value of the draw. 5.2

20. Selecting a Card A person selects a card from an ordinary deck of cards. If it is a black card, she wins $2. If it is a red card between or including 3 and 7, she wins $10. If it is a red face card, she wins $25; and if it is a black jack, she wins an extra $100. Find the expectation of the game. $9.65

21. Carpooling If 40% of all commuters ride to work in carpools, find the probability that if 8 workers are selected, 5 will ride in carpools. 0.124

22. Employed Women If 60% of all women are employed outside the home, find the probability that in a sample of 20 women,

  a. Exactly 15 are employed 0.075
  b. At least 10 are employed 0.872
  c. At most 5 are not employed outside the home 0.125

23. Driver's Exam If 80% of the applicants are able to pass a driver's proficiency road test, find the mean, variance, and standard deviation of the number of people who pass the test in a sample of 300 applicants. 240; 48; 6.9

24. Meeting Attendance A history class has 75 members. If there is a 12% absentee rate per class meeting, find the mean, variance, and standard deviation of the number of students who will be absent from each class. 9; 7.9; 2.8

25. Income Tax Errors (Optional) The probability that a person will make 0, 1, 2, or 3 errors on his or her income tax return is 0.50, 0.30, 0.15, and 0.05, respectively. If 30 claims are selected, find the probability that 15 will contain 0 errors, 8 will contain 1 error, 5 will contain 2 errors, and 2 will contain 3 errors. 0.008

26. Quality Control Check (Optional) Before a television set leaves the factory, it is given a quality control check. The probability that a television contains 0, 1, or 2 defects is 0.88, 0.08, and 0.04, respectively. In a sample of 16 televisions, find the probability that 9 will have 0 defects, 4 will have 1 defect, and 3 will have 2 defects. 0.0003

27. Bowling Team Uniforms (Optional) Among the teams in a bowling league, the probability that the uniforms are all 1 color is 0.45, that 2 colors are used is 0.35, and that 3 or more colors are used is 0.20. If a sample of 12 uniforms is selected, find the probability that 5 contain only 1 color, 4 contain 2 colors, and 3 contain 3 or more colors. 0.061

28. Elm Trees (Optional) If 8% of the population of trees are elm trees, find the probability that in a sample of 100 trees, there are exactly 6 elm trees. Assume the distribution is approximately Poisson. 0.122

29. Sports Score Hot Line Calls (Optional) Sports Scores Hot Line receives, on the average, 8 calls per hour requesting the latest sports scores. The distribution is Poisson in nature. For any randomly selected hour, find the probability that the company will receive

  a. At least 8 calls 0.5470
  b. 3 or more calls 0.9863
  c. At most 7 calls 0.4529

30. Color of Raincoats (Optional) There are 48 raincoats for sale at a local men's clothing store. Twelve are black. If 6 raincoats are selected to be marked down, find the probability that exactly 3 will be black. 0.128

31. Youth Group Officers (Optional) A youth group has 8 boys and 6 girls. If a slate of 4 officers is selected, find the probability that exactly

  a. 3 are girls 0.160
  b. 2 are girls 0.42
  c. 4 are boys 0.07

Critical Thinking Challenges

1. Lottery Numbers Pennsylvania has a lottery entitled "Big 4." To win, a player must correctly match four digits from a daily lottery in which four digits are selected. Find the probability of winning.

2. Lottery Numbers In the Big 4 lottery, for a bet of $100, the payoff is $5000. What is the expected value of winning? Is it worth it?

3. Lottery Numbers If you played the same four-digit number every day (or any four-digit number for that matter) in the Big 4, how often (in years) would you win, assuming you have average luck?

4. Chuck-a-Luck In the game Chuck-a-Luck, three dice are rolled. A player bets a certain amount (say $1.00) on a number from 1 to 6. If the number appears on 1 die, the person wins $1.00. If it appears on 2 dice, the person wins $2.00, and if it appears on all 3 dice, the person wins $3.00. What are the chances of winning $1.00? $2.00? $3.00?

5. Chuck-a-Luck What is the expected value of the game of Chuck-a-Luck if a player bets $1.00 on one number?

 **Data Projects**

1. **Business and Finance** Assume that a life insurance company would like to make a profit of $250 on a $100,000 policy sold to a person whose probability of surviving the year is 0.9985. What premium should the company charge the customer? If the company would like to make a $250 profit on a $100,000 policy at a premium of $500, what is the lowest life expectancy it should accept for a customer?

2. **Sports and Leisure** Baseball, hockey, and basketball all use a seven-game series to determine their championship. Find the probability that with two evenly matched teams a champion will be found in 4 games. Repeat for 5, 6, and 7 games. Look at the historical results for the three sports. How do the actual results compare to the theoretical?

3. **Technology** Use your most recent itemized phone bill for the data in this problem. Assume that incoming and outgoing calls are equal in the population (why is this a reasonable assumption?). This means assume $p = 0.5$. For the number of calls you made last month, what would be the mean number of outgoing calls in a random selection of calls? Also, compute the standard deviation. Was the number of outgoing calls you made an unusual amount given the above? In a selection of 12 calls, what is the probability that less than 3 were outgoing?

4. **Health and Wellness** Use Red Cross data to determine the percentage of the population with an Rh factor that is positive (A+, B+, AB+, or O+ blood types). Use that value for p. How many students in your class have a positive Rh factor? Is this an unusual amount?

5. **Politics and Economics** Find out what percentage of citizens in your state is registered to vote. Assuming that this is a binomial variable, what would be the mean number of registered voters in a random group of citizens with a sample size equal to the number of students in your class? Also determine the standard deviation. How many students in your class are registered to vote? Is this an unusual number, given the above?

6. **Your Class** Have each student in class toss 4 coins on her or his desk, and note how many heads are showing. Create a frequency table displaying the results. Compare the frequency table to the theoretical probability distribution for the outcome when 4 coins are tossed. Find the mean for the frequency table. How does it compare with the mean for the probability distribution?

Answers to Applying the Concepts

Section 5–1   Dropping College Courses

1. The random variable under study is the reason for dropping a college course.

2. There were a total of 144 people in the study.

3. The complete table is as follows:

Reason for Dropping a College Course	Frequency	Percentage
Too difficult	45	31.25
Illness	40	27.78
Change in work schedule	20	13.89
Change of major	14	9.72
Family-related problems	9	6.25
Money	7	4.86
Miscellaneous	6	4.17
No meaningful reason	3	2.08

4. The probability that a student will drop a class because of illness is about 28%. The probability that a student will drop a class because of money is about 5%. The probability that a student will drop a class because of a change of major is about 10%.

5. The information is not itself a probability distribution, but it can be used as one.

6. The categories are not necessarily mutually exclusive, but we treated them as such in computing the probabilities.

7. The categories are not independent.

8. The categories are exhaustive.

9. Since all the probabilities are between 0 and 1, inclusive, and the probabilities sum to 1, the requirements for a discrete probability distribution are met.

Section 5–2 Expected Value

1. The expected value is the mean in a discrete probability distribution.

2. We would expect variation from the expected value of 3.

3. Answers will vary. One possible answer is that pregnant mothers in that area might be overly concerned upon hearing that the number of cases of kidney problems in newborns was nearly 4 times what was usually expected. Other mothers (particularly those who had taken a statistics course!) might ask for more information about the claim.

4. Answers will vary. One possible answer is that it does seem unlikely to have 11 newborns with kidney problems when we expect only 3 newborns to have kidney problems.

5. The public might better be informed by percentages or rates (e.g., rate per 1000 newborns).

6. The increase of 8 babies born with kidney problems represents a 0.32% increase (less than $\frac{1}{2}$%).

7. Answers will vary. One possible answer is that the percentage increase does not seem to be something to be overly concerned about.

Section 5–3 Unsanitary Restaurants

1. The probability of eating at 3 restaurants with unsanitary conditions out of the 10 restaurants is 0.18793.

2. The probability of eating at 4 or 5 restaurants with unsanitary conditions out of the 10 restaurants is $(0.24665) + (0.22199) = 0.46864$.

3. To find this probability, you could add the probabilities for eating at 1, 2, . . . , 10 unsanitary restaurants. An easier way to compute the probability is to subtract the probability of eating at no unsanitary restaurants from 1 (using the complement rule).

4. The highest probability for this distribution is 4, but the expected number of unsanitary restaurants that you would eat at is $10 \cdot \frac{3}{7} = 4.29$.

5. The standard deviation for this distribution is $\sqrt{(10)(\frac{3}{7})(\frac{4}{7})} = 1.56$.

6. We have two possible outcomes: "success" is eating in an unsanitary restaurant; "failure" is eating in a sanitary restaurant. The probability that one restaurant is unsanitary is independent of the probability that any other restaurant is unsanitary. The probability that a restaurant is unsanitary remains constant at $\frac{3}{7}$. And we are looking at the number of unsanitary restaurants that we eat at out of 10 "trials."

7. The likelihood of success will vary from situation to situation. Just because we have two possible outcomes, this does not mean that each outcome occurs with probability 0.50.

Section 5–4 Rockets and Targets

1. The theoretical values for the number of hits are as follows:

Hits	0	1	2	3	4	5
Regions	227.5	211.3	98.2	30.4	7.1	1.3

2. The actual values are very close to the theoretical values.

3. Since the actual values are close to the theoretical values, it does appear that the rockets were fired at random.

CHAPTER 6

The Normal Distribution

Objectives

After completing this chapter, you should be able to

1. Identify distributions as symmetric or skewed.

2. Identify the properties of a normal distribution.

3. Find the area under the standard normal distribution, given various z values.

4. Find probabilities for a normally distributed variable by transforming it into a standard normal variable.

5. Find specific data values for given percentages, using the standard normal distribution.

6. Use the central limit theorem to solve problems involving sample means for large samples.

7. Use the normal approximation to compute probabilities for a binomial variable.

Outline

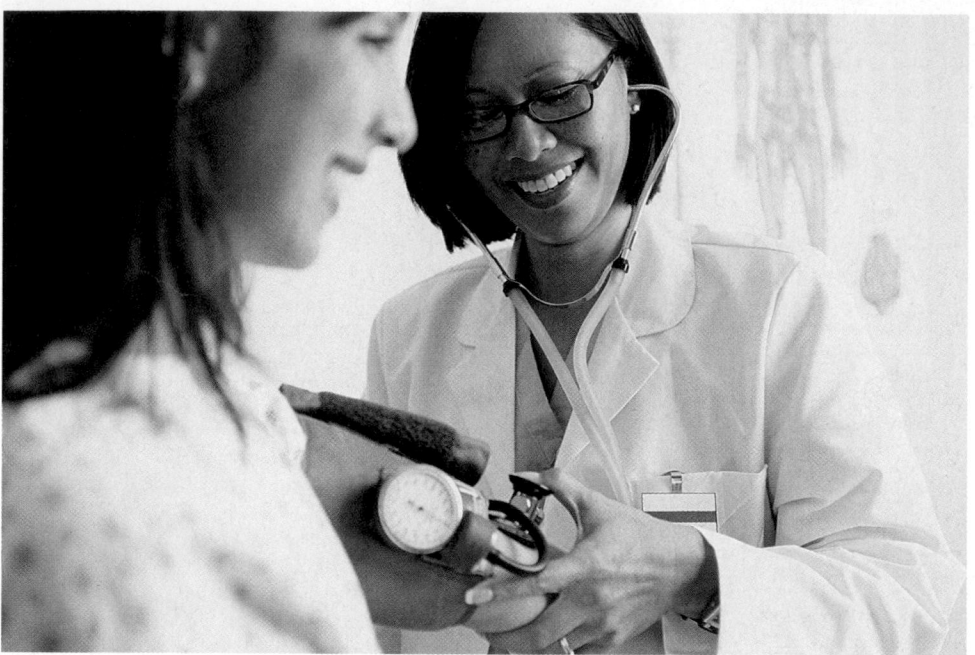

What Is Normal?

Medical researchers have determined so-called normal intervals for a person's blood pressure, cholesterol, triglycerides, and the like. For example, the normal range of systolic blood pressure is 110 to 140. The normal interval for a person's triglycerides is from 30 to 200 milligrams per deciliter (mg/dl). By measuring these variables, a physician can determine if a patient's vital statistics are within the normal interval or if some type of treatment is needed to correct a condition and avoid future illnesses. The question then is, How does one determine the so-called normal intervals? See Statistics Today—Revisited at the end of the chapter.

In this chapter, you will learn how researchers determine normal intervals for specific medical tests by using a normal distribution. You will see how the same methods are used to determine the lifetimes of batteries, the strength of ropes, and many other traits.

Introduction

Random variables can be either discrete or continuous. Discrete variables and their distributions were explained in Chapter 5. Recall that a discrete variable cannot assume all values between any two given values of the variables. On the other hand, a continuous variable can assume all values between any two given values of the variables. Examples of continuous variables are the heights of adult men, body temperatures of rats, and cholesterol levels of adults. Many continuous variables, such as the examples just mentioned, have distributions that are bell-shaped, and these are called *approximately normally distributed variables*. For example, if a researcher selects a random sample of 100 adult women, measures their heights, and constructs a histogram, the researcher gets a graph similar to the one shown in Figure 6–1(a). Now, if the researcher increases the sample size and decreases the width of the classes, the histograms will look like the ones shown in Figure 6–1(b) and (c). Finally, if it were possible to measure exactly the heights of all adult females in the United States and plot them, the histogram would approach what is called a *normal distribution,* shown in Figure 6–1(d). This distribution is also known as

Figure 6–1

Histograms for the Distribution of Heights of Adult Women

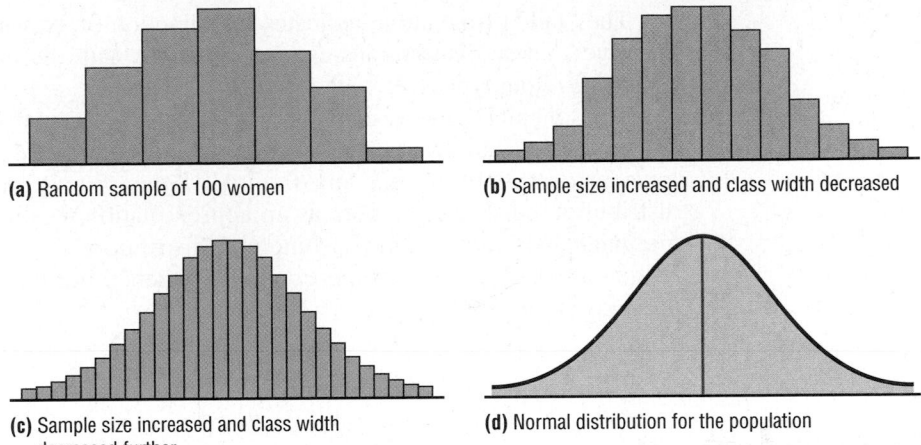

(a) Random sample of 100 women

(b) Sample size increased and class width decreased

(c) Sample size increased and class width decreased further

(d) Normal distribution for the population

Figure 6–2

Normal and Skewed Distributions

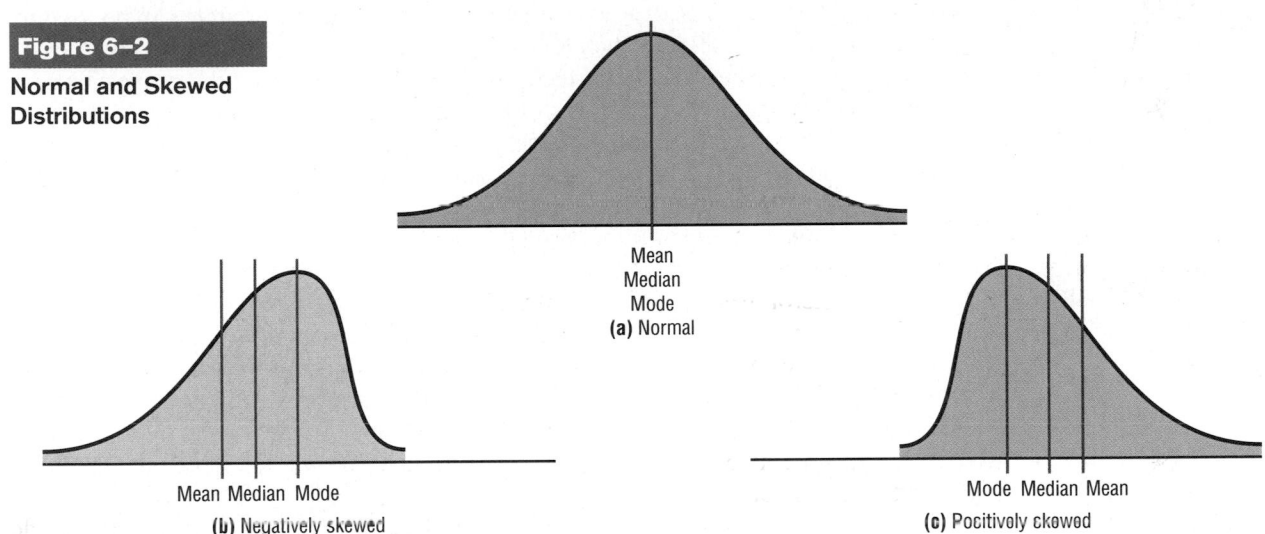

Mean
Median
Mode
(a) Normal

Mean Median Mode
(b) Negatively skewed

Mode Median Mean
(c) Positively skewed

a *bell curve* or a *Gaussian distribution,* named for the German mathematician Carl Friedrich Gauss (1777–1855), who derived its equation.

No variable fits a normal distribution perfectly, since a normal distribution is a theoretical distribution. However, a normal distribution can be used to describe many variables, because the deviations from a normal distribution are very small. This concept will be explained further in Section 6–1.

Objective 1

Identify distributions as symmetric or skewed.

When the data values are evenly distributed about the mean, a distribution is said to be a **symmetric distribution.** (A normal distribution is symmetric.) Figure 6–2(a) shows a symmetric distribution. When the majority of the data values fall to the left or right of the mean, the distribution is said to be *skewed.* When the majority of the data values fall to the right of the mean, the distribution is said to be a **negatively or left-skewed distribution.** The mean is to the left of the median, and the mean and the median are to the left of the mode. See Figure 6–2(b). When the majority of the data values fall to the left of the mean, a distribution is said to be a **positively or right-skewed distribution.** The mean falls to the right of the median, and both the mean and the median fall to the right of the mode. See Figure 6–2(c).

The "tail" of the curve indicates the direction of skewness (right is positive, left is negative). These distributions can be compared with the ones shown in Figure 3–1 in Chapter 3. Both types follow the same principles.

This chapter will present the properties of a normal distribution and discuss its applications. Then a very important fact about a normal distribution called the *central limit theorem* will be explained. Finally, the chapter will explain how a normal distribution curve can be used as an approximation to other distributions, such as the binomial distribution. Since a binomial distribution is a discrete distribution, a correction for continuity may be employed when a normal distribution is used for its approximation.

6–1

Normal Distributions

Objective 2

Identify the properties of a normal distribution.

In mathematics, curves can be represented by equations. For example, the equation of the circle shown in Figure 6–3 is $x^2 + y^2 = r^2$, where r is the radius. A circle can be used to represent many physical objects, such as a wheel or a gear. Even though it is not possible to manufacture a wheel that is perfectly round, the equation and the properties of a circle can be used to study many aspects of the wheel, such as area, velocity, and acceleration. In a similar manner, the theoretical curve, called a *normal distribution curve,* can be used to study many variables that are not perfectly normally distributed but are nevertheless approximately normal.

The mathematical equation for a normal distribution is

$$y = \frac{e^{-(X-\mu)^2/(2\sigma^2)}}{\sigma\sqrt{2\pi}}$$

where

$e \approx 2.718$ ($\approx$ means "is approximately equal to")

$\pi \approx 3.14$

μ = population mean

σ = population standard deviation

This equation may look formidable, but in applied statistics, tables or technology is used for specific problems instead of the equation.

Another important consideration in applied statistics is that the area under a normal distribution curve is used more often than the values on the y axis. Therefore, when a normal distribution is pictured, the y axis is sometimes omitted.

Circles can be different sizes, depending on their diameters (or radii), and can be used to represent wheels of different sizes. Likewise, normal curves have different shapes and can be used to represent different variables.

The shape and position of a normal distribution curve depend on two parameters, the *mean* and the *standard deviation.* Each normally distributed variable has its own normal distribution curve, which depends on the values of the variable's mean and standard deviation. Figure 6–4(a) shows two normal distributions with the same mean values but different standard deviations. The larger the standard deviation, the more dispersed, or spread out, the distribution is. Figure 6–4(b) shows two normal distributions with the same standard deviation but with different means. These curves have the same shapes but are located at different positions on the x axis. Figure 6–4(c) shows two normal distributions with different means and different standard deviations.

Figure 6–3

Graph of a Circle and an Application

Circle

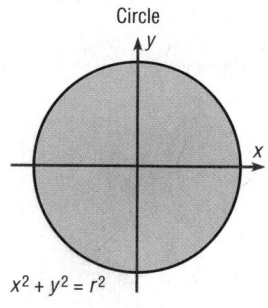

$x^2 + y^2 = r^2$

Wheel

Figure 6–4

Shapes of Normal Distributions

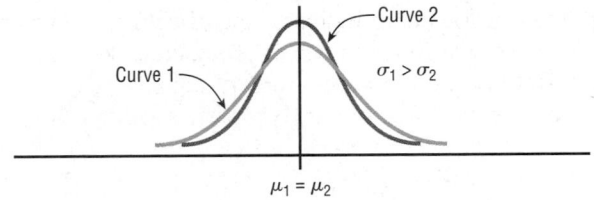

(a) Same means but different standard deviations

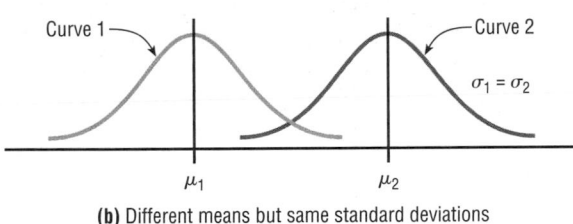

(b) Different means but same standard deviations

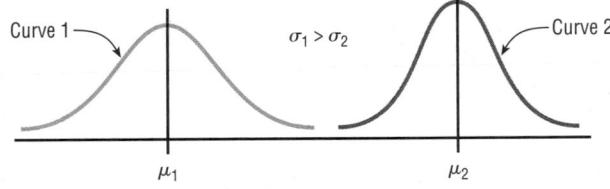

(c) Different means and different standard deviations

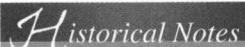

The discovery of the equation for a normal distribution can be traced to three mathematicians. In 1733, the French mathematician Abraham DeMoivre derived an equation for a normal distribution based on the random variation of the number of heads appearing when a large number of coins were tossed. Not realizing any connection with the naturally occurring variables, he showed this formula to only a few friends. About 100 years later, two mathematicians, Pierre Laplace in France and Carl Gauss in Germany, derived the equation of the normal curve independently and without any knowledge of DeMoivre's work. In 1924, Karl Pearson found that DeMoivre had discovered the formula before Laplace or Gauss.

A **normal distribution** is a continuous, symmetric, bell-shaped distribution of a variable.

The properties of a normal distribution, including those mentioned in the definition, are explained next.

Summary of the Properties of the Theoretical Normal Distribution

1. A normal distribution curve is bell-shaped.
2. The mean, median, and mode are equal and are located at the center of the distribution.
3. A normal distribution curve is unimodal (i.e., it has only one mode).
4. The curve is symmetric about the mean, which is equivalent to saying that its shape is the same on both sides of a vertical line passing through the center.
5. The curve is continuous; that is, there are no gaps or holes. For each value of X, there is a corresponding value of Y.
6. The curve never touches the x axis. Theoretically, no matter how far in either direction the curve extends, it never meets the x axis—but it gets increasingly closer.
7. The total area under a normal distribution curve is equal to 1.00, or 100%. This fact may seem unusual, since the curve never touches the x axis, but one can prove it mathematically by using calculus. (The proof is beyond the scope of this textbook.)
8. The area under the part of a normal curve that lies within 1 standard deviation of the mean is approximately 0.68, or 68%; within 2 standard deviations, about 0.95, or 95%; and within 3 standard deviations, about 0.997, or 99.7%. See Figure 6–5, which also shows the area in each region.

The values given in item 8 of the summary follow the *empirical rule* for data given in Section 3–2.

You must know these properties in order to solve problems involving distributions that are approximately normal.

Figure 6–5

Areas Under a Normal
Distribution Curve

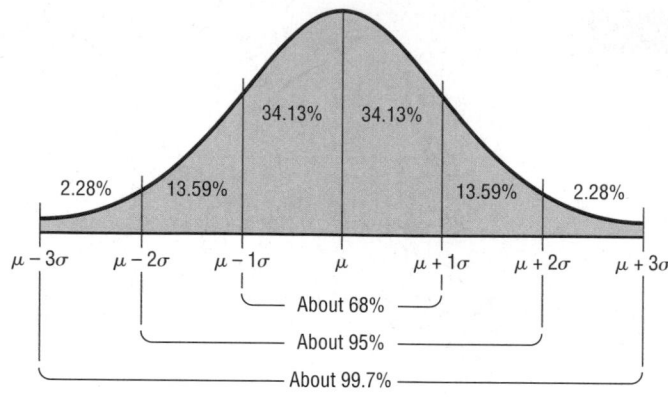

The Standard Normal Distribution

Since each normally distributed variable has its own mean and standard deviation, as
stated earlier, the shape and location of these curves will vary. In practical applications,
then, you would have to have a table of areas under the curve for each variable. To simplify this situation, statisticians use what is called the *standard normal distribution*.

Objective 3

Find the area under
the standard normal
distribution, given
various *z* values.

> The **standard normal distribution** is a normal distribution with a mean of 0 and a
> standard deviation of 1.

The standard normal distribution is shown in Figure 6–6.

The values under the curve indicate the proportion of area in each section. For example, the area between the mean and 1 standard deviation above or below the mean is
about 0.3413, or 34.13%.

The formula for the standard normal distribution is

$$y = \frac{e^{-z^2/2}}{\sqrt{2\pi}}$$

All normally distributed variables can be transformed into the standard normally distributed variable by using the formula for the standard score:

$$z = \frac{\text{value} - \text{mean}}{\text{standard deviation}} \qquad \text{or} \qquad z = \frac{X - \mu}{\sigma}$$

This is the same formula used in Section 3–3. The use of this formula will be explained
in Section 6–3.

As stated earlier, the area under a normal distribution curve is used to solve practical application problems, such as finding the percentage of adult women whose height is
between 5 feet 4 inches and 5 feet 7 inches, or finding the probability that a new battery
will last longer than 4 years. Hence, the major emphasis of this section will be to show
the procedure for finding the area under the standard normal distribution curve for any
z value. The applications will be shown in Section 6–2. Once the *X* values are transformed by using the preceding formula, they are called *z* values. The ***z*** **value** or ***z*** **score**
is actually the number of standard deviations that a particular *X* value is away from the
mean. Table E in Appendix C gives the area (to four decimal places) under the standard
normal curve for any *z* value from -3.49 to 3.49.

Figure 6–6

Standard Normal
Distribution

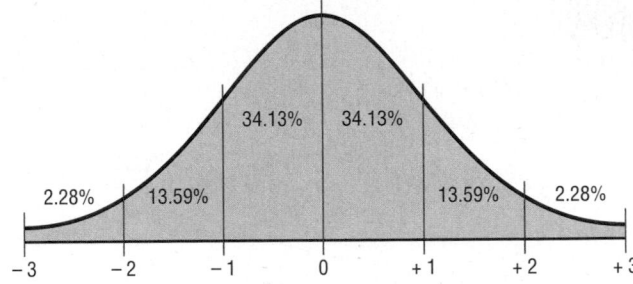

34.13%   34.13%

2.28%   13.59%   13.59%   2.28%

−3   −2   −1   0   +1   +2   +3

Finding Areas Under the Standard Normal Distribution Curve

For the solution of problems using the standard normal distribution, a two-step process is recommended with the use of the Procedure Table shown.

The two steps are

Step 1   Draw the normal distribution curve and shade the area.

Step 2   Find the appropriate figure in the Procedure Table and follow the directions given.

There are three basic types of problems, and all three are summarized in the Procedure Table. Note that this table is presented as an aid in understanding how to use the standard normal distribution table and in visualizing the problems. After learning the procedures, you should not find it necessary to refer to the Procedure Table for every problem.

Procedure Table

Finding the Area Under the Standard Normal Distribution Curve

1. To the left of any z value:
   Look up the z value in the table and use the area given.

2. To the right of any z value:
   Look up the z value and subtract the area from 1.

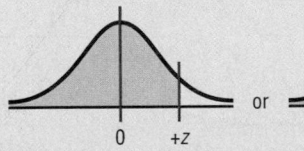 or 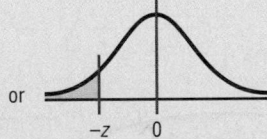   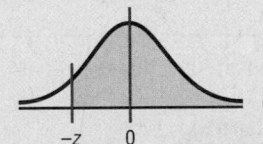 or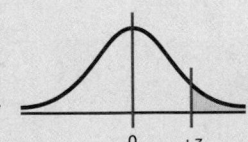

3. Between any two z values:
   Look up both z values and subtract the
   corresponding areas.

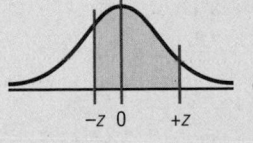 or 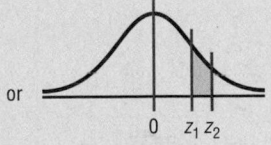 or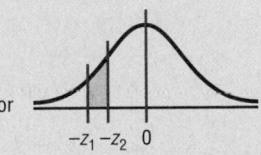

Figure 6–7

Table E Area Value for
z = 1.39

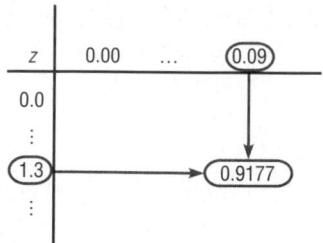

Table E in Appendix C gives the area under the normal distribution curve to the left of any z value given in two decimal places. For example, the area to the left of a z value of 1.39 is found by looking up 1.3 in the left column and 0.09 in the top row. Where the two lines meet gives an area of 0.9177. See Figure 6–7.

Example 6–1

Find the area to the left of $z = 2.06$.

Solution

Step 1 Draw the figure. The desired area is shown in Figure 6–8.

Figure 6–8

Area Under the
Standard Normal
Distribution Curve for
Example 6–1

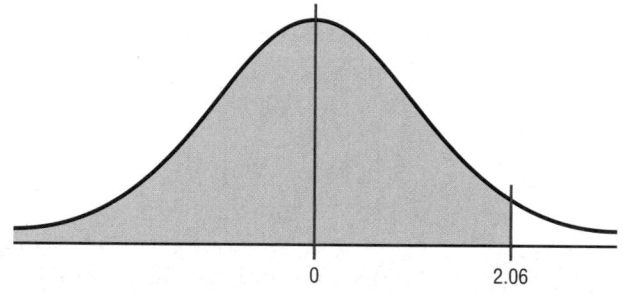

Step 2 We are looking for the area under the standard normal distribution to the left of $z = 2.06$. Since this is an example of the first case, look up the area in the table. It is 0.9803. Hence, 98.03% of the area is less than $z = 2.06$.

Example 6–2

Find the area to the right of $z = -1.19$.

Solution

Step 1 Draw the figure. The desired area is shown in Figure 6–9.

Figure 6–9

Area Under the
Standard Normal
Distribution Curve for
Example 6–2

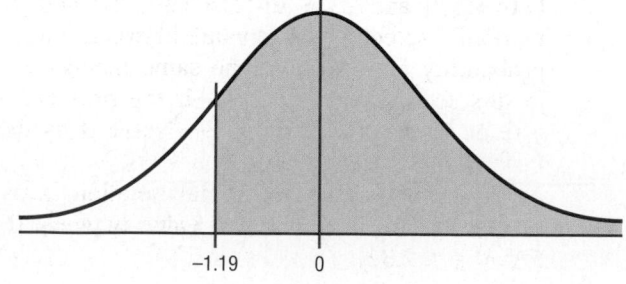

Step 2 We are looking for the area to the right of $z = -1.19$. This is an example of the second case. Look up the area for $z = -1.19$. It is 0.1170. Subtract it from 1.0000. $1.0000 - 0.1170 = 0.8830$. Hence, 88.30% of the area under the standard normal distribution curve is to the left of $z = -1.19$.

Example 6–3

Find the area between $z = +1.68$ and $z = -1.37$.

Solution

Step 1 Draw the figure as shown. The desired area is shown in Figure 6–10.

Figure 6–10

Area Under the
Standard Normal
Distribution Curve
for Example 6–3

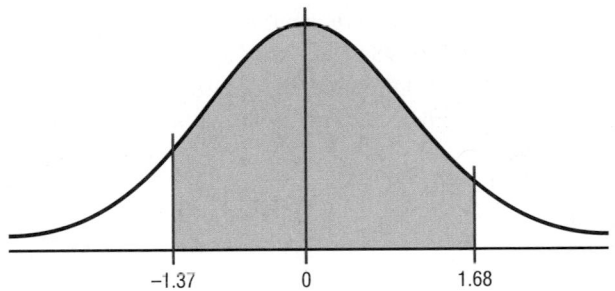

Step 2 Since the area desired is between two given z values, look up the areas corresponding to the two z values and subtract the smaller area from the larger area. (Do not subtract the z values.) The area for $z = +1.68$ is 0.9535, and the area for $z = -1.37$ is 0.0853. The area between the two z values is $0.9535 - 0.0853 = 0.8682$ or 86.82%.

A Normal Distribution Curve as a Probability Distribution Curve

A normal distribution curve can be used as a probability distribution curve for normally distributed variables. Recall that a normal distribution is a *continuous distribution,* as opposed to a discrete probability distribution, as explained in Chapter 5. The fact that it is continuous means that there are no gaps in the curve. In other words, for every z value on the x axis, there is a corresponding height, or frequency, value.

The area under the standard normal distribution curve can also be thought of as a probability. That is, if it were possible to select any z value at random, the probability of choosing one, say, between 0 and 2.00 would be the same as the area under the curve between 0 and 2.00. In this case, the area is 0.4772. Therefore, the probability of randomly selecting any z value between 0 and 2.00 is 0.4772. The problems involving probability are solved in the same manner as the previous examples involving areas in this section. For example, if the problem is to find the probability of selecting a z value between 2.25 and 2.94, solve it by using the method shown in case 3 of the Procedure Table.

For probabilities, a special notation is used. For example, if the problem is to find the probability of any z value between 0 and 2.32, this probability is written as $P(0 < z < 2.32)$.

Note: In a continuous distribution, the probability of any exact z value is 0 since the area would be represented by a vertical line above the value. But vertical lines in theory have no area. So $P(a \leq z \leq b) = P(a < z < b)$.

Example 6–4

Find the probability for each.

    a. $P(0 < z < 2.32)$
    b. $P(z < 1.65)$
    c. $P(z > 1.91)$

Solution

    a. $P(0 < z < 2.32)$ means to find the area under the standard normal distribution curve between 0 and 2.32. First look up the area corresponding to 2.32. It is 0.9898. Then look up the area corresponding to $z = 0$. It is 0.500. Subtract the two areas: $0.9898 - 0.5000 = 0.4898$. Hence the probability is 0.4898, or 48.98%. This is shown in Figure 6–11.

Figure 6–11

Area Under the Standard Normal Distribution Curve for Part *a* of Example 6–4

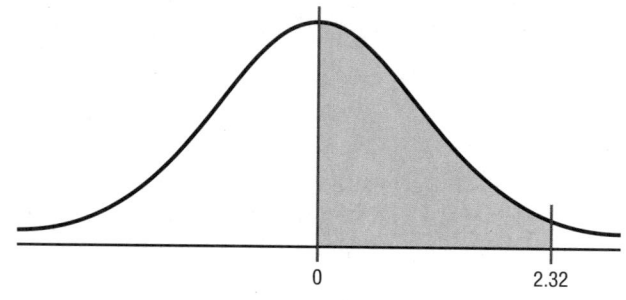

    b. $P(z < 1.65)$ is represented in Figure 6–12. Look up the area corresponding to $z = 1.65$ in Table E. It is 0.9505. Hence, $P(z < 1.65) = 0.9505$, or 95.05%.

Figure 6–12

Area Under the Standard Normal Distribution Curve for Part *b* of Example 6–4

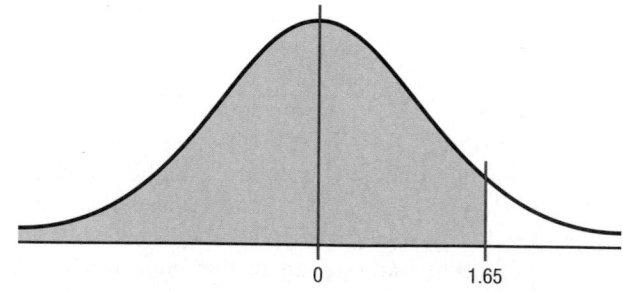

    c. $P(z > 1.91)$ is shown in Figure 6–13. Look up the area that corresponds to $z = 1.91$. It is 0.9719. Then subtract this area from 1.0000. $P(z > 1.91) = 1.0000 - 0.9719 = 0.0281$, or 2.81%.

Figure 6–13

Area Under the
Standard Normal
Distribution Curve
for Part *c* of
Example 6–4

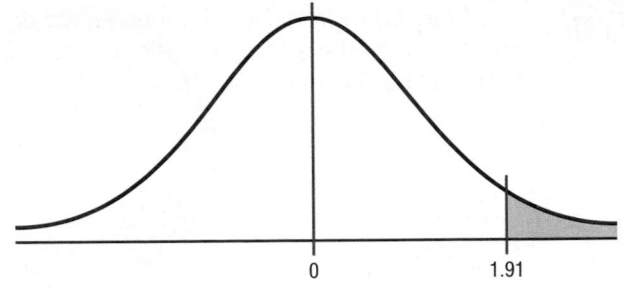

Sometimes, one must find a specific *z* value for a given area under the standard normal distribution curve. The procedure is to work backward, using Table E.

Since Table E is cumulative, it is necessary to locate the cumulative area up to a given *z* value. Example 6–5 shows this.

Example 6–5

Find the *z* value such that the area under the standard normal distribution curve between 0 and the *z* value is 0.2123.

Solution

Draw the figure. The area is shown in Figure 6–14.

Figure 6–14

Area Under the
Standard Normal
Distribution Curve for
Example 6–5

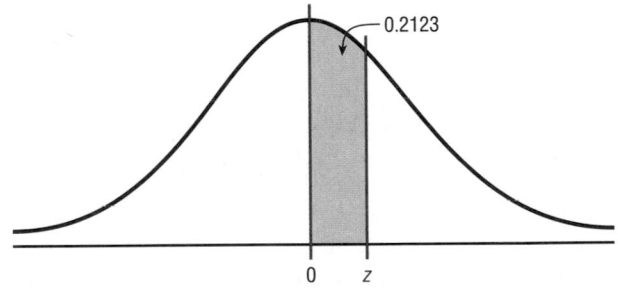

In this case it is necessary to add 0.5000 to the given area of 0.2123 to get the cumulative area of 0.7123. Look up the area in Table E. The value in the left column is 0.5, and the top value is 0.06. Add these two values to get $z = 0.56$. See Figure 6–15.

Figure 6–15

Finding the *z* Value
from Table E for
Example 6–5

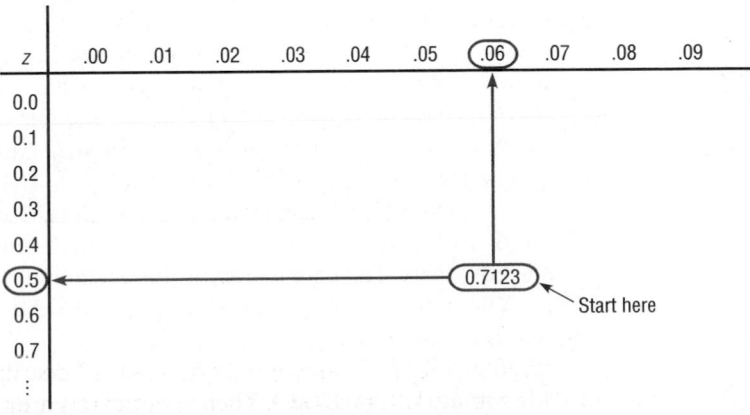

Figure 6–16

The Relationship
Between Area and
Probability

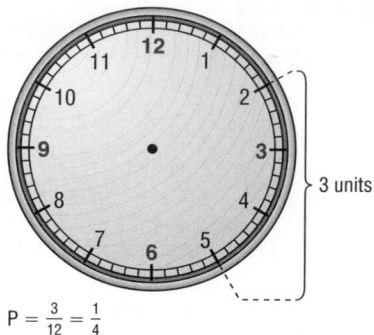

$$P = \frac{3}{12} = \frac{1}{4}$$

(a) Clock

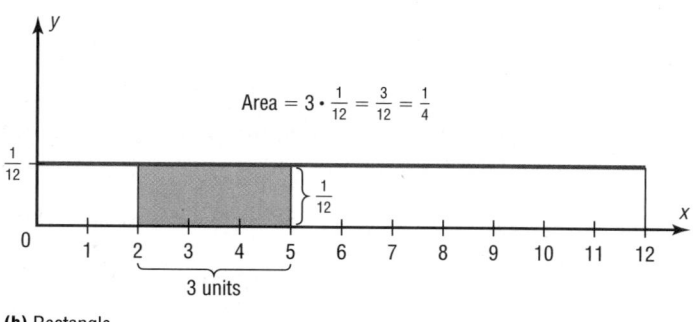

$$\text{Area} = 3 \cdot \frac{1}{12} = \frac{3}{12} = \frac{1}{4}$$

(b) Rectangle

If the exact area cannot be found, use the closest value. For example, if you wanted to find the z value for an area 0.9241, the closest area is 0.9236, which gives a z value of 1.43. See Table E in Appendix C.

The rationale for using an area under a continuous curve to determine a probability can be understood by considering the example of a watch that is powered by a battery. When the battery goes dead, what is the probability that the minute hand will stop somewhere between the numbers 2 and 5 on the face of the watch? In this case, the values of the variable constitute a continuous variable since the hour hand can stop anywhere on the dial's face between 0 and 12 (one revolution of the minute hand). Hence, the sample space can be considered to be 12 units long, and the distance between the numbers 2 and 5 is 5 − 2, or 3 units. Hence, the probability that the minute hand stops on a number between 2 and 5 is $\frac{3}{12} = \frac{1}{4}$. See Figure 6–16(a).

The problem could also be solved by using a graph of a continuous variable. Let us assume that since the watch can stop anytime at random, the values where the minute hand would land are spread evenly over the range of 0 through 12. The graph would then consist of a *continuous uniform distribution* with a range of 12 units. Now if we require the area under the curve to be 1 (like the area under the standard normal distribution), the height of the rectangle formed by the curve and the x axis would need to be $\frac{1}{12}$. The reason is that the area of a rectangle is equal to the base times the height. If the base is 12 units long, then the height has to be $\frac{1}{12}$ since $12 \cdot \frac{1}{12} = 1$.

The area of the rectangle with a base from 2 through 5 would be $3 \cdot \frac{1}{12}$, or $\frac{1}{4}$. See Figure 6–16(b). Notice that the area of the small rectangle is the same as the probability found previously. Hence the area of this rectangle corresponds to the probability of this event. The same reasoning can be applied to the standard normal distribution curve shown in Example 6–5.

Finding the area under the standard normal distribution curve is the first step in solving a wide variety of practical applications in which the variables are normally distributed. Some of these applications will be presented in Section 6–2.

Applying the Concepts **6–1**

Assessing Normality

Many times in statistics it is necessary to see if a set of data values is approximately normally distributed. There are special techniques that can be used. One technique is to draw a histogram for the data and see if it is approximately bell-shaped. (*Note:* It does not have to be exactly symmetric to be bell-shaped.)

The numbers of branches of the 50 top libraries are shown.

67	84	80	77	97	59	62	37	33	42
36	54	18	12	19	33	49	24	25	22
24	29	9	21	21	24	31	17	15	21
13	19	19	22	22	30	41	22	18	20
26	33	14	14	16	22	26	10	16	24

Source: *The World Almanac and Book of Facts.*

1. Construct a frequency distribution for the data.

2. Construct a histogram for the data.

3. Describe the shape of the histogram.

4. Based on your answer to question 3, do you feel that the distribution is approximately normal?

In addition to the histogram, distributions that are approximately normal have about 68% of the values fall within 1 standard deviation of the mean, about 95% of the data values fall within 2 standard deviations of the mean, and almost 100% of the data values fall within 3 standard deviations of the mean. (See Figure 6–5.)

5. Find the mean and standard deviation for the data.

6. What percent of the data values fall within 1 standard deviation of the mean?

7. What percent of the data values fall within 2 standard deviations of the mean?

8. What percent of the data values fall within 3 standard deviations of the mean?

9. How do your answers to questions 6, 7, and 8 compare to 68, 95, and 100%, respectively?

10. Does your answer help support the conclusion you reached in question 4? Explain.

(More techniques for assessing normality are explained in Section 6–2.)
See pages 353 and 354 for the answers.

Exercises 6–1

1. What are the characteristics of a normal distribution?

2. Why is the standard normal distribution important in statistical analysis? Many variables are normally distributed, and the distribution can be used to describe these variables.

3. What is the total area under the standard normal distribution curve? 1 or 100%

4. What percentage of the area falls below the mean? Above the mean? 50% of the area lies below the mean, and 50% of the area lies above the mean.

5. About what percentage of the area under the normal distribution curve falls within 1 standard deviation above and below the mean? 2 standard deviations? 3 standard deviations? 68%; 95%; 99.7%

For Exercises 6 through 25, find the area under the standard normal distribution curve.

6. Between $z = 0$ and $z = 1.77$ 0.4616

7. Between $z = 0$ and $z = 0.75$ 0.2734

8. Between $z = 0$ and $z = -0.32$ 0.1255

9. Between $z = 0$ and $z = -2.07$ 0.4808

10. To the right of $z = 2.01$ 0.0222

11. To the right of $z = 0.29$ 0.3859

12. To the left of $z = -0.75$ 0.2266

13. To the left of $z = -1.39$ 0.0823

14. Between $z = 1.23$ and $z = 1.90$ 0.0806

15. Between $z = 1.05$ and $z = 1.78$ 0.1094

16. Between $z = -0.96$ and $z = -0.36$ 0.1909

17. Between $z = -1.56$ and $z = -1.83$ 0.0258

18. Between $z = 0.24$ and $z = -1.12$ 0.4634

19. Between $z = -1.46$ and $z = -1.98$ 0.0482

20. To the left of $z = 1.31$ 0.9049

21. To the left of $z = 2.11$ 0.9826

22. To the right of $z = -1.92$ 0.9726

23. To the right of $z = -0.17$ 0.5675

24. To the left of $z = -2.15$ and to the right of $z = 1.62$
0.0684

25. To the right of $z = 1.92$ and to the left of $z = -0.44$
0.3574

In Exercises 26 through 39, find the probabilities for each, using the standard normal distribution.

26. $P(0 < z < 1.96)$ 0.4750

27. $P(0 < z < 0.67)$ 0.2486

28. $P(-1.23 < z < 0)$ 0.3907

29. $P(-1.43 < z < 0)$ 0.4236

30. $P(z > 0.82)$ 0.2061

31. $P(z > 2.83)$ 0.0023

32. $P(z < -1.77)$ 0.0384

33. $P(z < -1.32)$ 0.0934

34. $P(-0.20 < z < 1.56)$ 0.5199

35. $P(-2.46 < z < 1.74)$ 0.9522 (TI: 0.9521)

36. $P(1.12 < z < 1.43)$ 0.0550

37. $P(1.46 < z < 2.97)$ 0.0706 (TI: 0.0707)

38. $P(z > -1.43)$ 0.9236

39. $P(z < 1.42)$ 0.9222

For Exercises 40 through 45, find the z value that corresponds to the given area.

40.

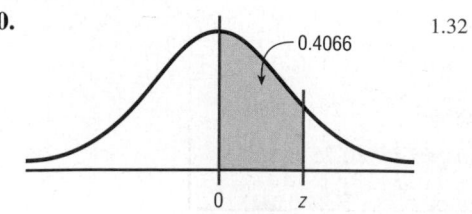

1.32

41.

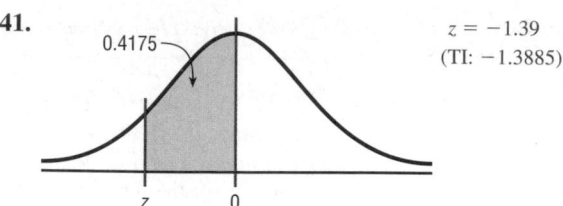

$z = -1.39$
(TI: -1.3885)

42.

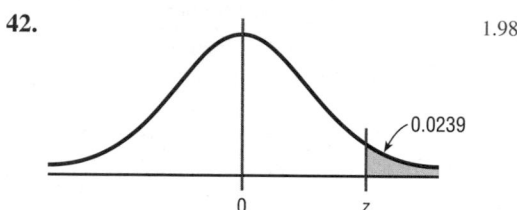

1.98

43.

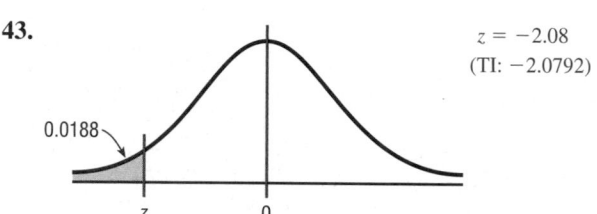

$z = -2.08$
(TI: -2.0792)

44.

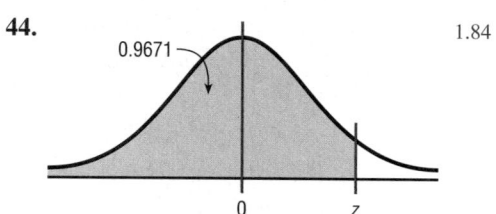

1.84

45.

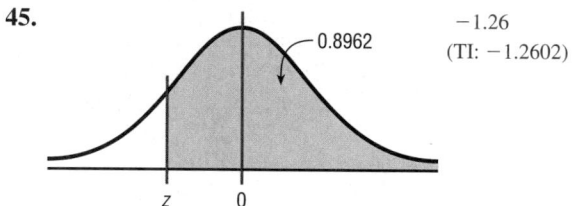

-1.26
(TI: -1.2602)

46. Find the z value to the right of the mean so that

  a. 54.78% of the area under the distribution curve lies to the left of it. 0.12

  b. 69.85% of the area under the distribution curve lies to the left of it. 0.52

  c. 88.10% of the area under the distribution curve lies to the left of it. 1.18

47. Find the z value to the left of the mean so that

  a. 98.87% of the area under the distribution curve lies to the right of it. -2.28 (TI: -2.2801)

  b. 82.12% of the area under the distribution curve lies to the right of it. -0.92 (TI: -0.91995)

  c. 60.64% of the area under the distribution curve lies to the right of it. -0.27 (TI: -0.26995)

48. Find two z values so that 48% of the middle area is bounded by them. $z = \pm 0.64$

49. Find two z values, one positive and one negative, that are equidistant from the mean so that the areas in the two tails total the following values.

a. 5% $z = +1.96$ and $z = -1.96$ (TI: ± 1.95996)

b. 10% $z = +1.65$ and $z = -1.65$, approximately (TI: ± 1.64485)

c. 1% $z = +2.58$ and $z = -2.58$, approximately (TI: ± 2.57583)

Extending the Concepts

50. In the standard normal distribution, find the values of z for the 75th, 80th, and 92nd percentiles. 0.6745; 0.8416; 1.41

51. Find $P(-1 < z < 1)$, $P(-2 < z < 2)$, and $P(-3 < z < 3)$. How do these values compare with the empirical rule? 0.6827; 0.9545; 0.9973; they are very close.

52. Find z_0 such that $P(z > z_0) = 0.1234$. 1.16

53. Find z_0 such that $P(-1.2 < z < z_0) = 0.8671$. 2.10

54. Find z_0 such that $P(z_0 < z < 2.5) = 0.7672$. -0.75

55. Find z_0 such that the area between z_0 and $z = -0.5$ is 0.2345 (two answers). -1.45 and 0.11

56. Find z_0 such that $P(-z_0 < z < z_0) = 0.76$. 1.175

57. Find the equation for the standard normal distribution by substituting 0 for μ and 1 for σ in the equation

$$y = \frac{e^{-(X-\mu)^2/(2\sigma^2)}}{\sigma\sqrt{2\pi}} \qquad y = \frac{e^{-X^2/2}}{\sqrt{2\pi}}$$

58. Graph by hand the standard normal distribution by using the formula derived in Exercise 57. Let $\pi \approx 3.14$ and $e \approx 2.718$. Use X values of $-2, -1.5, -1, -0.5, 0, 0.5, 1, 1.5,$ and 2. (Use a calculator to compute the y values.)

Technology *Step by Step*

MINITAB
Step by Step

The Standard Normal Distribution

It is possible to determine the height of the density curve given a value of z, the cumulative area given a value of z, or a z value given a cumulative area. Examples are from Table E in Appendix C.

Find the Area to the Left of $z = 1.39$

1. Select **Calc>Probability Distributions>Normal.** There are three options.
2. Click the button for Cumulative probability. In the center section, the mean and standard deviation for the standard normal distribution are the defaults. The mean should be **0,** and the standard deviation should be **1.**
3. Click the button for Input Constant, then click inside the text box and type in **1.39.** Leave the storage box empty.
4. Click [OK].

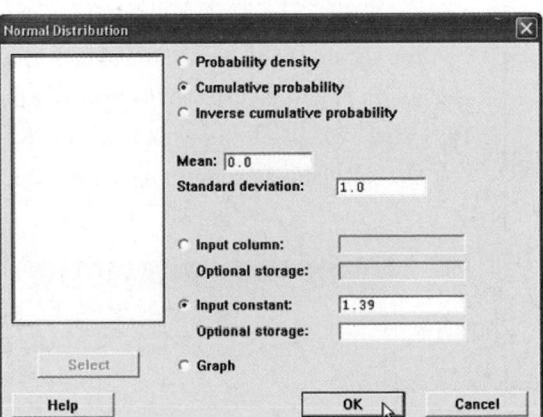

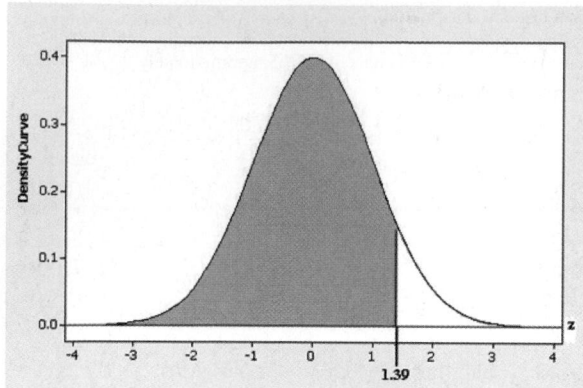

Cumulative Distribution Function

Normal with mean = 0 and standard deviation = 1

x P(X <= x)

1.39   0.917736

The graph is not shown in the output.

The session window displays the result, 0.917736. If you choose the optional storage, type in a variable name such as **K1.** The result will be stored in the constant and will not be in the session window.

Find the Area to the Right of −2.06

1. Select **Calc>Probability Distributions>Normal.**

2. Click the button for Cumulative probability.

3. Click the button for Input Constant, then enter **−2.06** in the text box. Do not forget the minus sign.

4. Click in the text box for Optional storage and type **K1.**

5. Click [OK]. The area to the left of −2.06 is stored in K1 but not displayed in the session window.

   To determine the area to the right of the z value, subtract this constant from 1, then display the result.

6. Select **Calc>Calculator.**

   a) Type **K2** in the text box for Store result in:.

   b) Type in the expression **1 − K1,** then click [OK].

7. Select **Data>Display Data.** Drag the mouse over K1 and K2, then click [Select] and [OK].

   The results will be in the session window and stored in the constants.

Data Display

K1   0.0196993

K2   0.980301

8. To see the constants and other information about the worksheet, click the Project Manager icon. In the left pane click on the green worksheet icon, and then click the constants folder. You should see all constants and their values in the right pane of the Project Manager.

9. For the third example calculate the two probabilities and store them in K1 and K2.

10. Use the calculator to subtract K1 from K2 and store in K3.

    The calculator and project manager windows are shown.

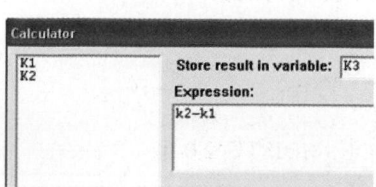

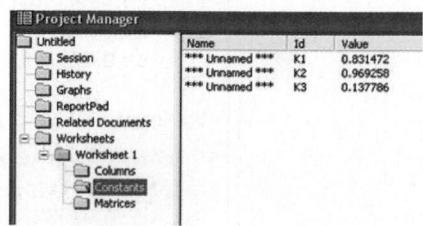

Calculate a z Value Given the Cumulative Probability

Find the z value for a cumulative probability of 0.025.

1. Select **Calc>Probability Distributions>Normal.**

2. Click the option for Inverse cumulative probability, then the option for Input constant.

3. In the text box type **.025,** the cumulative area, then click [OK].

4. In the dialog box, the z value will be returned, -1.960.

Inverse Cumulative Distribution Function
Normal with mean = 0 and standard deviation = 1
P (X <= x)   x
      0.025   −1.95996

In the session window z is -1.95996.

TI-83 Plus or TI-84 Plus
Step by Step

```
normalcdf(1.11,E
99)
      .1334995565
normalcdf( E99,-
1.93)
      .0268033499
■
```

```
normalcdf(2,2.47
)
      .0159944012
invNorm(.7123)
      .560116461
■
```

Standard Normal Random Variables

To find the probability for a standard normal random variable:
Press **2nd [DISTR],** then **2** for normalcdf(
The form is normalcdf(lower z score, upper z score).
Use E99 for ∞ (infinity) and −E99 for $-\infty$ (negative infinity). Press **2nd [EE]** to get E.

Example: Area to the right of $z = 1.11$
normalcdf(1.11,E99)

Example: Area to the left of $z = -1.93$
normalcdf(−E99,−1.93)

Example: Area between $z = 2.00$ and $z = 2.47$
normalcdf(2.00,2.47)

To find the percentile for a standard normal random variable:
Press **2nd [DISTR],** then **3** for the invNorm(
The form is invNorm(area to the left of z score)

Example: Find the z score such that the area under the standard normal curve to the left of it is 0.7123
invNorm(.7123)

Excel
Step by Step

The Standard Normal Distribution

Finding areas under the standard normal distribution curve

Example XL6-1

Find the area to the left of $z = 1.99$.
In a blank cell type: =NORMSDIST(1.99)
Answer: 0.976705

Example XL6-2

Find the area to the right of $z = -2.04$.
In a blank cell type: = 1-NORMSDIST(−2.04)
Answer: 0.979325

Example XL6–3

Find the area between $z = -2.04$ and $z = 1.99$.
In a blank cell type: =NORMSDIST(1.99) − NORMSDIST(−2.04)
Answer: 0.956029

Finding a *z* value given an area under the standard normal distribution curve

Example XL6–4

Find a z score given the cumulative area (area to the left of z) is 0.0250.
In a blank cell type: =NORMSINV(.025)
Answer: −1.95996

6–2 Applications of the Normal Distribution

Objective 4

Find probabilities for a normally distributed variable by transforming it into a standard normal variable.

The standard normal distribution curve can be used to solve a wide variety of practical problems. The only requirement is that the variable be normally or approximately normally distributed. There are several mathematical tests to determine whether a variable is normally distributed. See the Critical Thinking Challenges on page 352. For all the problems presented in this chapter, you can assume that the variable is normally or approximately normally distributed.

To solve problems by using the standard normal distribution, transform the original variable to a standard normal distribution variable by using the formula

$$z = \frac{\text{value} - \text{mean}}{\text{standard deviation}} \qquad \text{or} \qquad z = \frac{X - \mu}{\sigma}$$

This is the same formula presented in Section 3–3. This formula transforms the values of the variable into standard units or z values. Once the variable is transformed, then the Procedure Table and Table E in Appendix C can be used to solve problems.

For example, suppose that the scores for a standardized test are normally distributed, have a mean of 100, and have a standard deviation of 15. When the scores are transformed to z values, the two distributions coincide, as shown in Figure 6–17. (Recall that the z distribution has a mean of 0 and a standard deviation of 1.)

Figure 6–17

Test Scores and Their Corresponding *z* Values

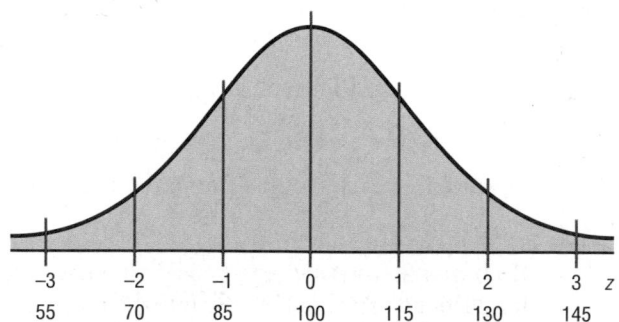

To solve the application problems in this section, transform the values of the variable to z values and then find the areas under the standard normal distribution, as shown in Section 6–1.

Example 6–6

Summer Spending

A survey found that women spend on average $146.21 on beauty products during the summer months. Assume the standard deviation is $29.44. Find the percentage of women who spend less than $160.00. Assume the variable is normally distributed.

Solution

Step 1   Draw the figure and represent the area as shown in Figure 6–18.

Figure 6–18

Area Under a
Normal Curve for
Example 6–6

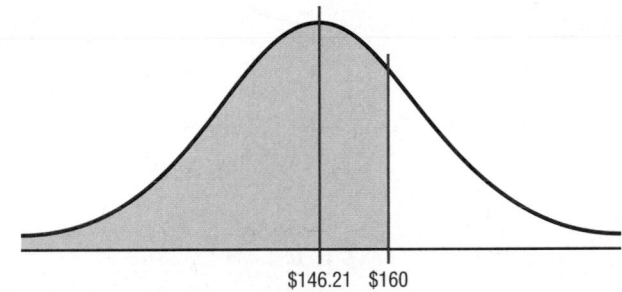

$146.21 $160

Step 2   Find the z value corresponding to $160.00.

$$z = \frac{X - \mu}{\sigma} = \frac{\$160.00 - \$146.21}{\$29.44} = 0.47$$

Hence $160.00 is 0.47 of a standard deviation above the mean of $146.21, as shown in the z distribution in Figure 6–19.

Figure 6–19

Area and z Values for
Example 6–6

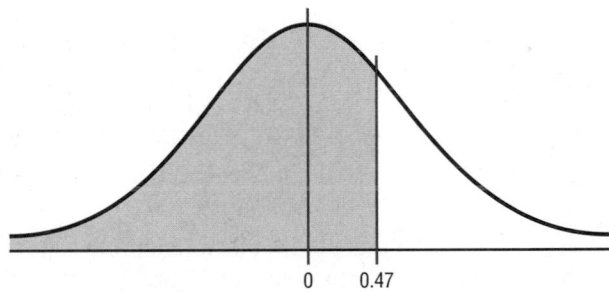

0   0.47

Step 3   Find the area, using Table E. The area under the curve to the left of $z = 0.47$ is 0.6808.

Therefore 0.6808, or 68.08%, of the women spend less than $160.00 on beauty products during the summer months.

Example 6–7

Monthly Newspaper Recycling

Each month, an American household generates an average of 28 pounds of newspaper for garbage or recycling. Assume the standard deviation is 2 pounds. If a household is selected at random, find the probability of its generating

   a. Between 27 and 31 pounds per month

   b. More than 30.2 pounds per month

Assume the variable is approximately normally distributed.

Source: Michael D. Shook and Robert L. Shook, *The Book of Odds.*

Solution *a*

Step 1   Draw the figure and represent the area. See Figure 6–20.

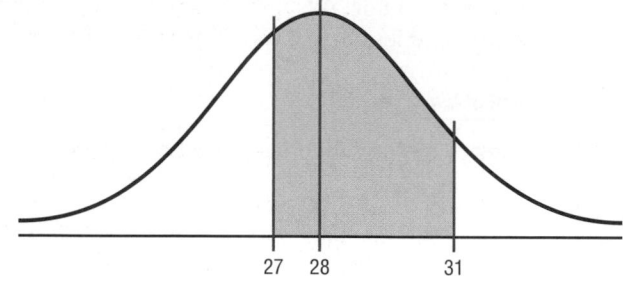

Step 2   Find the two *z* values.

$$z_1 = \frac{X - \mu}{\sigma} = \frac{27 - 28}{2} = -\frac{1}{2} = -0.5$$

$$z_2 = \frac{X - \mu}{\sigma} = \frac{31 - 28}{2} = \frac{3}{2} = 1.5$$

Step 3   Find the appropriate area, using Table E. The area to the left of z_2 is 0.9332, and the area to the left of z_1 is 0.3085. Hence the area between z_1 and z_2 is $0.9332 - 0.3085 = 0.6247$. See Figure 6–21.

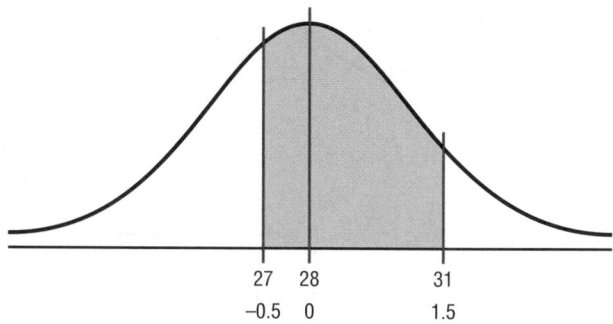

Hence, the probability that a randomly selected household generates between 27 and 31 pounds of newspapers per month is 62.47%.

Solution *b*

Step 1   Draw the figure and represent the area, as shown in Figure 6–22.

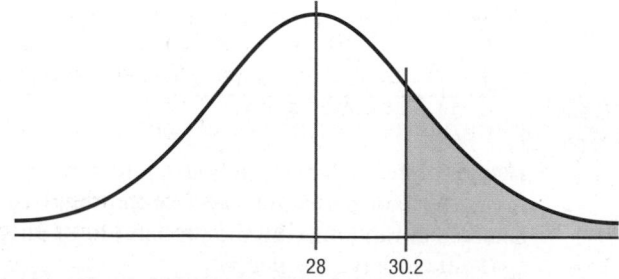

Step 2   Find the *z* value for 30.2.

$$z = \frac{X - \mu}{\sigma} = \frac{30.2 - 28}{2} = \frac{2.2}{2} = 1.1$$

Step 3 Find the appropriate area. The area to the left of $z = 1.1$ is 0.8643. Hence the area to the right of $z = 1.1$ is $1.0000 - 0.8643 = 0.1357$.

Hence, the probability that a randomly selected household will accumulate more than 30.2 pounds of newspapers is 0.1357, or 13.57%.

A normal distribution can also be used to answer questions of "How many?" This application is shown in Example 6–8.

Example 6–8	### Coffee Consumption

Americans consume an average of 1.64 cups of coffee per day. Assume the variable is approximately normally distributed with a standard deviation of 0.24 cup. If 500 individuals are selected, approximately how many will drink less than 1 cup of coffee per day?

Source: *Chicago Sun-Times.*

Solution

Step 1 Draw a figure and represent the area as shown in Figure 6–23.

Figure 6–23	
Area Under a Normal Curve for Example 6–8	

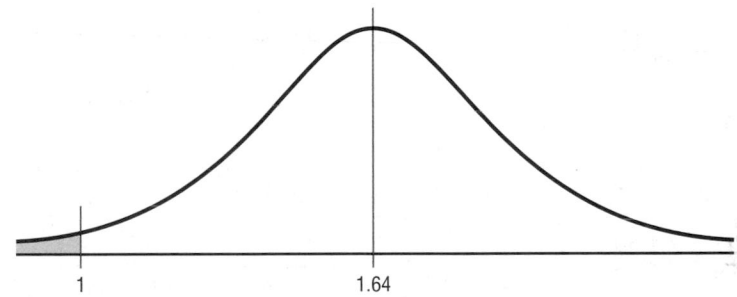

Step 2 Find the z value for 1.

$$z = \frac{X - \mu}{\sigma} = \frac{1 - 1.64}{0.24} = -2.67$$

Step 3 Find the area to the left of $z = -2.67$. It is 0.0038.

Step 4 To find how many people drank less than 1 cup of coffee, multiply the sample size 500 by 0.0038 to get 1.9. Since we are asking about people, round the answer to 2 people. Hence, approximately 2 people will drink less than 1 cup of coffee a day.

Note: For problems using percentages, be sure to change the percentage to a decimal before multiplying. Also, round the answer to the nearest whole number, since it is not possible to have 1.9 people.

Finding Data Values Given Specific Probabilities

A normal distribution can also be used to find specific data values for given percentages. This application is shown in Example 6–9.

Police Academy Qualifications

To qualify for a police academy, candidates must score in the top 10% on a general abilities test. The test has a mean of 200 and a standard deviation of 20. Find the lowest possible score to qualify. Assume the test scores are normally distributed.

Solution

Since the test scores are normally distributed, the test value X that cuts off the upper 10% of the area under a normal distribution curve is desired. This area is shown in Figure 6–24.

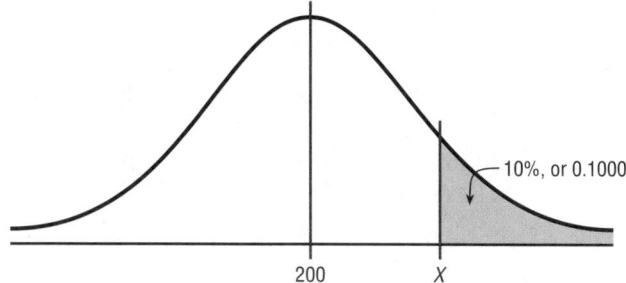

Work backward to solve this problem.

Step 1 Subtract 0.1000 from 1.000 to get the area under the normal distribution to the left of x: $1.0000 - 0.10000 = 0.9000$.

Step 2 Find the z value that corresponds to an area of 0.9000 by looking up 0.9000 in the area portion of Table E. If the specific value cannot be found, use the closest value—in this case 0.8997, as shown in Figure 6–25. The corresponding z value is 1.28. (If the area falls exactly halfway between two z values, use the larger of the two z values. For example, the area 0.9500 falls halfway between 0.9495 and 0.9505. In this case use 1.65 rather than 1.64 for the z value.)

z	.00	.01	.02	.03	.04	.05	.06	.07	.08	.09
0.0										
0.1										
0.2										
⋮										
1.1										0.9000
1.2									0.8997	0.9015
1.3										
1.4										
⋮										

Specific value → 0.9000

Closest value → 0.8997

Step 3 Substitute in the formula $z = (X - \mu)/\sigma$ and solve for X.

$$1.28 = \frac{X - 200}{20}$$

$$(1.28)(20) + 200 = X$$

$$25.60 + 200 = X$$

$$225.60 = X$$

$$226 = X$$

A score of 226 should be used as a cutoff. Anybody scoring 226 or higher qualifies.

Instead of using the formula shown in step 3, you can use the formula $X = z \cdot \sigma + \mu$. This is obtained by solving

$$z = \frac{X - \mu}{\sigma}$$

for X as shown.

$z \cdot \sigma = X - \mu$ Multiply both sides by σ.

$z \cdot \sigma + \mu = X$ Add μ to both sides.

$X = z \cdot \sigma + \mu$ Exchange both sides of the equation.

Formula for Finding X

When you must find the value of X, you can use the following formula:

$$X = z \cdot \sigma + \mu$$

Example 6–10

Systolic Blood Pressure

For a medical study, a researcher wishes to select people in the middle 60% of the population based on blood pressure. If the mean systolic blood pressure is 120 and the standard deviation is 8, find the upper and lower readings that would qualify people to participate in the study.

Solution

Assume that blood pressure readings are normally distributed; then cutoff points are as shown in Figure 6–26.

Figure 6–26

Area Under a
Normal Curve for
Example 6–10

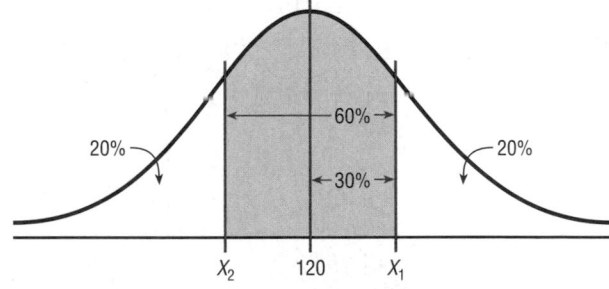

Figure 6–26 shows that two values are needed, one above the mean and one below the mean. To get the area to the left of the positive z value, add $0.5000 + 0.3000 = 0.8000$ ($30\% = 0.3000$). The z value with area to the left closest to 0.8000 is 0.84. Substituting in the formula $X = z\sigma + \mu$ gives

$$X_1 = z\sigma + \mu = (0.84)(8) + 120 = 126.72$$

The area to the left of the negative z value is 20%, or 0.2000. The area closest to 0.2000 is -0.84.

$$X_2 = (-0.84)(8) + 120 = 113.28$$

Therefore, the middle 60% will have blood pressure readings of $113.28 < X < 126.72$.

As shown in this section, a normal distribution is a useful tool in answering many questions about variables that are normally or approximately normally distributed.

Determining Normality

A normally shaped or bell-shaped distribution is only one of many shapes that a distribution can assume; however, it is very important since many statistical methods require that the distribution of values (shown in subsequent chapters) be normally or approximately normally shaped.

There are several ways statisticians check for normality. The easiest way is to draw a histogram for the data and check its shape. If the histogram is not approximately bell-shaped, then the data are not normally distributed.

Skewness can be checked by using the Pearson coefficient of skewness (PC) also called Pearson's index of skewness. The formula is

$$PC = \frac{3(\bar{X} - \text{median})}{s}$$

If the index is greater than or equal to $+1$ or less than or equal to -1, it can be concluded that the data are significantly skewed.

In addition, the data should be checked for outliers by using the method shown in Chapter 3. Even one or two outliers can have a big effect on normality.

Examples 6–11 and 6–12 show how to check for normality.

Example 6–11

Technology Inventories

A survey of 18 high-technology firms showed the number of days' inventory they had on hand. Determine if the data are approximately normally distributed.

| 5 | 29 | 34 | 44 | 45 | 63 | 68 | 74 | 74 |
| 81 | 88 | 91 | 97 | 98 | 113 | 118 | 151 | 158 |

Source: *USA TODAY.*

Solution

Step 1 Construct a frequency distribution and draw a histogram for the data, as shown in Figure 6–27.

Class	Frequency
5–29	2
30–54	3
55–79	4
80–104	5
105–129	2
130–154	1
155–179	1

Figure 6–27

Histogram for Example 6–11

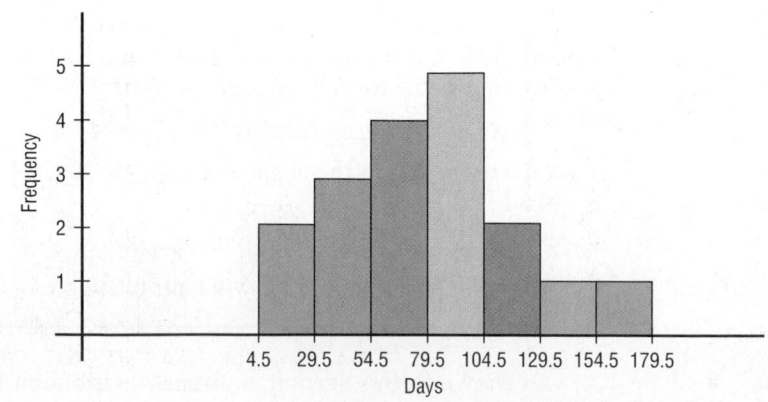

Since the histogram is approximately bell-shaped, we can say that the distribution is approximately normal.

Step 2 Check for skewness. For these data, $\bar{X} = 79.5$, median $= 77.5$, and $s = 40.5$. Using the Pearson coefficient of skewness gives

$$PC = \frac{3(79.5 - 77.5)}{40.5}$$
$$= 0.148$$

In this case, the PC is not greater than $+1$ or less than -1, so it can be concluded that the distribution is not significantly skewed.

Step 3 Check for outliers. Recall that an outlier is a data value that lies more than 1.5(IQR) units below Q_1 or 1.5(IQR) units above Q_3. In this case, $Q_1 = 45$ and $Q_3 = 98$; hence, IQR $= Q_3 - Q_1 = 98 - 45 = 53$. An outlier would be a data value less than $45 - 1.5(53) = -34.5$ or a data value larger than $98 + 1.5(53) = 177.5$. In this case, there are no outliers.

Since the histogram is approximately bell-shaped, the data are not significantly skewed, and there are no outliers, it can be concluded that the distribution is approximately normally distributed.

Example 6–12

Number of Baseball Games Played

The data shown consist of the number of games played each year in the career of Baseball Hall of Famer Bill Mazeroski. Determine if the data are approximately normally distributed.

81	148	152	135	151	152
159	142	34	162	130	162
163	143	67	112	70	

Source. *Greensburg Tribune Review.*

Solution

Step 1 Construct a frequency distribution and draw a histogram for the data. See Figure 6–28.

Figure 6–28

Histogram for Example 6–12

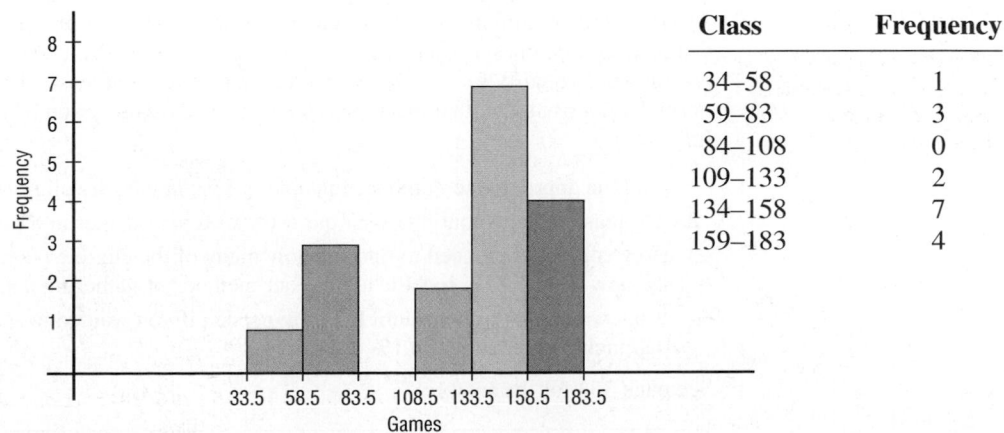

Class	Frequency
34–58	1
59–83	3
84–108	0
109–133	2
134–158	7
159–183	4

The histogram shows that the frequency distribution is somewhat negatively skewed.

Step 2 Check for skewness; $\overline{X} = 127.24$, median $= 143$, and $s = 39.87$.

$$PC = \frac{3(\overline{X} - \text{median})}{s}$$
$$= \frac{3(127.24 - 143)}{39.87}$$
$$= -1.19$$

Since the PC is less than -1, it can be concluded that the distribution is significantly skewed to the left.

Step 3 Check for outliers. In this case, $Q_1 = 96.5$ and $Q_3 = 155.5$. IQR $= Q_3 - Q_1 = 155.5 - 96.5 = 59$. Any value less than $96.5 - 1.5(59) = 8$ or above $155.5 + 1.5(59) = 244$ is considered an outlier. There are no outliers.

In summary, the distribution is somewhat negatively skewed.

Another method that is used to check normality is to draw a *normal quantile plot. Quantiles,* sometimes called *fractiles,* are values that separate the data set into approximately equal groups. Recall that quartiles separate the data set into four approximately equal groups, and deciles separate the data set into 10 approximately equal groups. A normal quantile plot consists of a graph of points using the data values for the x coordinates and the z values of the quantiles corresponding to the x values for the y coordinates. (*Note:* The calculations of the z values are somewhat complicated, and technology is usually used to draw the graph. The Technology Step by Step section shows how to draw a normal quantile plot.) If the points of the quantile plot do not lie in an approximately straight line, then normality can be rejected.

There are several other methods used to check for normality. A method using normal probability graph paper is shown in the Critical Thinking Challenge section at the end of this chapter, and the chi-square goodness-of-fit test is shown in Chapter 11. Two other tests sometimes used to check normality are the Kolmogorov-Smikirov test and the Lilliefors test. An explanation of these tests can be found in advanced textbooks.

Applying the Concepts 6–2

Smart People

Assume you are thinking about starting a Mensa chapter in your hometown of Visiala, California, which has a population of about 10,000 people. You need to know how many people would qualify for Mensa, which requires an IQ of at least 130. You realize that IQ is normally distributed with a mean of 100 and a standard deviation of 15. Complete the following.

1. Find the approximate number of people in Visiala who are eligible for Mensa.
2. Is it reasonable to continue your quest for a Mensa chapter in Visiala?
3. How could you proceed to find out how many of the eligible people would actually join the new chapter? Be specific about your methods of gathering data.
4. What would be the minimum IQ score needed if you wanted to start an Ultra-Mensa club that included only the top 1% of IQ scores?

See page 354 for the answers.

Exercises 6-2

1. **Admission Charge for Movies** The average early-bird special admission price for a movie is $5.81. If the distribution of movie admission charges is approximately normal with a standard deviation of $0.81, what is the probability that a randomly selected admission charge is less than $3.50? 0.0022

2. **Teachers' Salaries** The average annual salary for all U.S. teachers is $47,750. Assume that the distribution is normal and the standard deviation is $5680. Find the probability that a randomly selected teacher earns

   a. Between $35,000 and $45,000 a year 0.3031
   b. More than $40,000 a year 0.9131
   c. If you were applying for a teaching position and were offered $31,000 a year, how would you feel (based on this information)? Not too happy—it's really at the bottom of the heap! (prob. = 0.0016)

   Source: *New York Times Almanac.*

3. **Population in U.S. Jails** The average daily jail population in the United States is 706,242. If the distribution is normal and the standard deviation is 52,145, find the probability that on a randomly selected day, the jail population is

   a. Greater than 750,000 0.2005 (TI: 0.2007)
   b. Between 600,000 and 700,000 0.4315 (TI: 0.4316)

   Source: *New York Times Almanac.*

4. **SAT Scores** The national average SAT score (for Verbal and Math) is 1028. If we assume a normal distribution with $\sigma = 92$, what is the 90th percentile score? What is the probability that a randomly selected score exceeds 1200? 1146; 0.0307

   Source: *New York Times Almanac.*

5. **Chocolate Bar Calories** The average number of calories in a 1.5-ounce chocolate bar is 225. Suppose that the distribution of calories is approximately normal with $\sigma = 10$. Find the probability that a randomly selected chocolate bar will have

   a. Between 200 and 220 calories 0.3023
   b. Less than 200 calories 0.0062

   Source: *The Doctor's Pocket Calorie, Fat, and Carbohydrate Counter.*

6. **Monthly Mortgage Payments** The average monthly mortgage payment including principal and interest is $982 in the United States. If the standard deviation is approximately $180 and the mortgage payments are approximately normally distributed, find the probability that a randomly selected monthly payment is

   a. More than $1000 0.4602
   b. More than $1475 0.0031
   c. Between $800 and $1150 0.6676

   Source: *World Almanac.*

7. **Professors' Salaries** The average salary for a Queens College full professor is $85,900. If the average salaries are normally distributed with a standard deviation of $11,000, find these probabilities. *a.* 0.3557 (TI: 0.3547)

   a. The professor makes more than $90,000.
   b. The professor makes more than $75,000.

   Source: AAUP, *Chronicle of Higher Education. b.* 0.8389 (TI: 0.8391)

8. **Doctoral Student Salaries** Full-time Ph.D. students receive an average of $12,837 per year. If the average salaries are normally distributed with a standard deviation of $1500, find these probabilities.

   a. The student makes more than $15,000. 0.0749
   b. The student makes between $13,000 and $14,000. 0.2385

   Source: U.S. Education Dept., *Chronicle of Higher Education.*

9. **Miles Driven Annually** The mean number of miles driven per vehicle annually in the United States is 12,494 miles. Choose a randomly selected vehicle, and assume the annual mileage is normally distributed with a standard deviation of 1290 miles. What is the probability that the vehicle was driven more than 15,000 miles? Less than 8000 miles? Would you buy a vehicle if you had been told that it had been driven less than 6000 miles in the past year?

   Source: *World Almanac.*

10. **Commute Time to Work** The average commute to work (one way) is 25 minutes according to the 2005 American Community Survey. If we assume that commuting times are normally distributed and that the standard deviation is 6.1 minutes, what is the probability that a randomly selected commuter spends more than 30 minutes commuting one way? Less than 18 minutes? 0.2061; 0.1251

    Source: www.census.gov

11. **Credit Card Debt** The average credit card debt for college seniors is $3262. If the debt is normally distributed with a standard deviation of $1100, find these probabilities.

    a. That the senior owes at least $1000 0.9803 (TI: 0.9801)
    b. That the senior owes more than $4000
    c. That the senior owes between $3000 and $4000

    Source: *USA TODAY. b.* 0.2514 (TI: 0.2511) *c.* 0.3434 (TI: 0.3430)

12. **Price of Gasoline** The average retail price of gasoline (all types) for the first half of 2009 was 236.5 cents. What would the standard deviation have to be in order for a 15% probability that a gallon of gas costs less than $2.00?

    Source: *World Almanac.* 35.1 cents

13. **Waiting Time at a Bank Drive-in Window** The average waiting time at a drive-in window of a local bank is 10.3 minutes, with a standard deviation of 2.7 minutes. Assume the variable is normally distributed. If a customer arrives at the bank, find the probability that the customer will have to wait

a. Between 4 and 9 minutes 0.3057

b. Less than 5 minutes or more than 10 minutes 0.5688

c. How might a customer estimate his or her approximate waiting time?

14. Newborn Elephant Weights Newborn elephant calves usually weigh between 200 and 250 pounds—until October 2006, that is. An Asian elephant at the Houston (Texas) Zoo gave birth to a male calf weighing in at a whopping 384 pounds! Mack (like the truck) is believed to be the heaviest elephant calf ever born at a facility accredited by the Association of Zoos and Aquariums. If, indeed, the mean weight for newborn elephant calves is 225 pounds with a standard deviation of 45 pounds, what is the probability of a newborn weighing at least 384 pounds? Assume that the weights of newborn elephants are normally distributed. Less than 0.0001

Source: www.houstonzoo.org

15. Waiting to Be Seated The average waiting time to be seated for dinner at a popular restaurant is 23.5 minutes, with a standard deviation of 3.6 minutes. Assume the variable is normally distributed. When a patron arrives at the restaurant for dinner, find the probability that the patron will have to wait the following time.

a. Between 15 and 22 minutes 0.3281

b. Less than 18 minutes or more than 25 minutes 0.4002

c. Is it likely that a person will be seated in less than 15 minutes? Not usually

16. Salary of Full-Time Male Professors The average salary of a male full professor at a public four-year institution offering classes at the doctoral level is $99,685. For a female full professor at the same kind of institution, the salary is $90,330. If the standard deviation for the salaries of both genders is approximately $5200 and the salaries are normally distributed, find the 80th percentile salary for male professors and for female professors.

Source: *World Almanac.*   Men: $104,053   Women: $94,698

17. Lake Temperatures During September, the average temperature in Keystone Lake is 71.2°, and the standard deviation is 3.4°. For a randomly selected day, find the probability that the temperature of the lake is less than 63°. Based on your answer, would this be a likely or unlikely occurrence? Assume the variable is normally distributed. 0.0080 or 0.8%. A temperature of 63° is unlikely since the probability is about 0.8%.

18. Itemized Charitable Contributions The average charitable contribution itemized per income tax return in Pennsylvania is $792. Suppose that the distribution of contributions is normal with a standard deviation of $103. Find the limits for the middle 50% of contributions. $722.99 and $861.01

Source: IRS, *Statistics of Income Bulletin.*

19. New Home Sizes A contractor decided to build homes that will include the middle 80% of the market. If the average size of homes built is 1810 square feet,

find the maximum and minimum sizes of the homes the contractor should build. Assume that the standard deviation is 92 square feet and the variable is normally distributed.

Source: Michael D. Shook and Robert L. Shook, *The Book of Odds.*

20. New Home Prices If the average price of a new one-family home is $246,300 with a standard deviation of $15,000, find the minimum and maximum prices of the houses that a contractor will build to satisfy the middle 80% of the market. Assume that the variable is normally distributed. $227,100 to $265,500

Source: *New York Times Almanac.*

21. Cost of Personal Computers The average price of a personal computer (PC) is $949. If the computer prices are approximately normally distributed and $\sigma = \$100$, what is the probability that a randomly selected PC costs more than $1200? The least expensive 10% of personal computers cost less than what amount? 0.006; $821

Source: *New York Times Almanac.*

22. Reading Improvement Program To help students improve their reading, a school district decides to implement a reading program. It is to be administered to the bottom 5% of the students in the district, based on the scores on a reading achievement exam. If the average score for the students in the district is 122.6, find the cutoff score that will make a student eligible for the program. The standard deviation is 18. Assume the variable is normally distributed. 92.99 or 93

23. Used Car Prices An automobile dealer finds that the average price of a previously owned vehicle is $8256. He decides to sell cars that will appeal to the middle 60% of the market in terms of price. Find the maximum and minimum prices of the cars the dealer will sell. The standard deviation is $1150, and the variable is normally distributed. The maximum price is $9222, and the minimum price is $7290. (TI: $7288.14 minimum, $9223.86 maximum)

24. Ages of Amtrak Passenger Cars The average age of Amtrak passenger train cars is 19.4 years. If the distribution of ages is normal and 20% of the cars are older than 22.8 years, find the standard deviation. 4.05

Source: *New York Times Almanac.*

25. Lengths of Hospital Stays The average length of a hospital stay for all diagnoses is 4.8 days. If we assume that the lengths of hospital stays are normally distributed with a variance of 2.1, then 10% of hospital stays are longer than how many days? Thirty percent of stays are less than how many days?

Source: www.cdc.gov 6.7; 4.05 (TI: for 10%, 6.657; for 30%, 4.040)

26. High School Competency Test A mandatory competency test for high school sophomores has a normal distribution with a mean of 400 and a standard deviation of 100.

a. The top 3% of students receive $500. What is the minimum score you would need to receive this award? 588 *b.* The bottom 1.5% of students must go to summer school. What is the minimum score you would need to stay out of this group? 183

27. **Product Marketing** An advertising company plans to market a product to low-income families. A study states that for a particular area, the average income per family is $24,596 and the standard deviation is $6256. If the company plans to target the bottom 18% of the families based on income, find the cutoff income. Assume the variable is normally distributed. $18,840.48 (TI: $18,869.48)

28. **Bottled Drinking Water** Americans drank an average of 23.2 gallons of bottled water per capita in 2008. If the standard deviation is 2.7 gallons and the variable is normally distributed, find the probability that a randomly selected American drank more than 25 gallons of bottled water. What is the probability that the selected person drank between 22 and 30 gallons? 0.0968; 0.6641

Source: www.census.gov

29. **Wristwatch Lifetimes** The mean lifetime of a wristwatch is 25 months, with a standard deviation of 5 months. If the distribution is normal, for how many months should a guarantee be made if the manufacturer does not want to exchange more than 10% of the watches? Assume the variable is normally distributed. 18.6 months

30. **Police Academy Acceptance Exams** To qualify for a police academy, applicants are given a test of physical fitness. The scores are normally distributed with a mean of 64 and a standard deviation of 9. If only the top 20% of the applicants are selected, find the cutoff score. 71.6 or 72

31. In the distributions shown, state the mean and standard deviation for each. *Hint:* See Figures 6–5 and 6–6. Also the vertical lines are 1 standard deviation apart. *a.* $\mu = 120, \sigma = 20$; *b.* $\mu = 15, \sigma = 2.5$; *c.* $\mu = 30, \sigma = 5$

a.

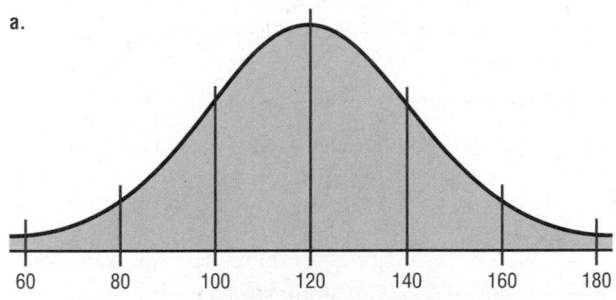

b.

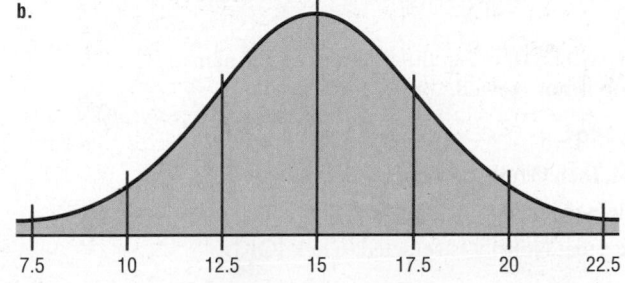

c.

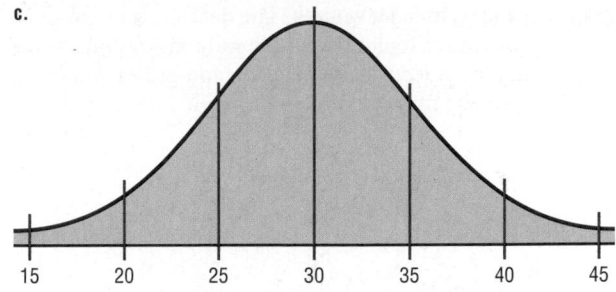

32. **SAT Scores** Suppose that the mathematics SAT scores for high school seniors for a specific year have a mean of 456 and a standard deviation of 100 and are approximately normally distributed. If a subgroup of these high school seniors, those who are in the National Honor Society, is selected, would you expect the distribution of scores to have the same mean and standard deviation? Explain your answer.

33. Given a data set, how could you decide if the distribution of the data was approximately normal?
There are several mathematics tests that can be used.

34. If a distribution of raw scores were plotted and then the scores were transformed to *z* scores, would the shape of the distribution change? Explain your answer.
No. The shape of the distribution would be the same.

35. In a normal distribution, find σ when $\mu = 105$ and 5.48% of the area lies to the right of 110. 3.125

36. In a normal distribution, find μ when σ is 6 and 3.75% of the area lies to the left of 85. 95.68

37. In a certain normal distribution, 1.25% of the area lies to the left of 42, and 1.25% of the area lies to the right of 48. Find μ and σ. $\mu = 45, \sigma = 1.34$

38. **Exam Scores** An instructor gives a 100-point examination in which the grades are normally distributed. The mean is 60 and the standard deviation is 10. If there are 5% A's and 5% F's, 15% B's and 15% D's, and 60% C's, find the scores that divide the distribution into those categories.

39. **Drive-in Movies** The data shown represent the number of outdoor drive-in movies in the United States for a 14-year period. Check for normality. Not normal

| 2084 | 1497 | 1014 | 910 | 899 | 870 | 837 | 859 |
| 848 | 826 | 815 | 750 | 637 | 737 | | |

Source: National Association of Theater Owners.

40. **Cigarette Taxes** The data shown represent the cigarette tax (in cents) for 50 selected states. Check for normality. Not normal

200	160	156	200	30	300	224	346	170	55
160	170	270	60	57	80	37	153	200	60
100	178	302	84	251	125	44	435	79	166
68	37	153	252	300	141	57	42	134	136
200	98	45	118	200	87	103	250	17	62

Source: http://www.tobaccofreekids.org

41. Box Office Revenues The data shown represent the box office total revenue (in millions of dollars) for a randomly selected sample of the top-grossing films in 2009. Check for normality. Not normal

37	32	155	277
146	80	66	113
71	29	166	36
28	72	32	32
30	32	52	84
37	402	42	109

Source: http://boxofficemojo.com

42. Number of Runs Made The data shown represent the number of runs made each year during Bill Mazeroski's career. Check for normality. Not normal

30	59	69	50	58	71	55	43	66	52	56	62
36	13	29	17	3							

Source: *Greensburg Tribune Review.*

Technology *Step by Step*

MINITAB
Step by Step

Data

5	29	34	44	45
63	68	74	74	81
88	91	97	98	113
118	151	158		

Determining Normality

There are several ways in which statisticians test a data set for normality. Four are shown here.

Construct a Histogram

Inspect the histogram for shape.

1. Enter the data in the first column of a new worksheet. Name the column Inventory.

2. Use **Stat>Basic Statistics>Graphical Summary** presented in Section 3–3 to create the histogram. Is it symmetric? Is there a single peak?

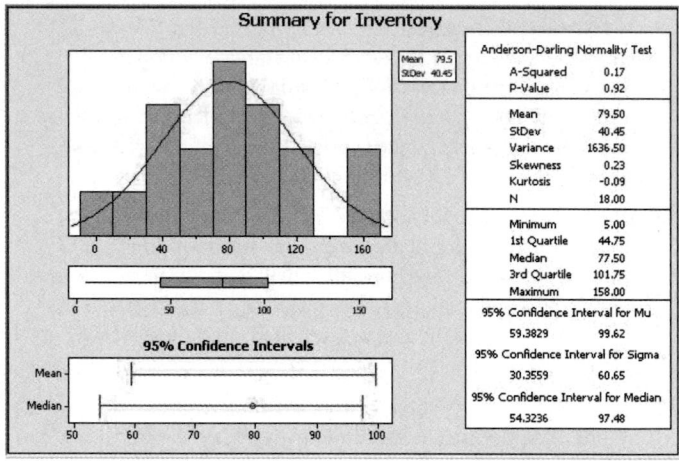

Check for Outliers

Inspect the boxplot for outliers. There are no outliers in this graph. Furthermore, the box is in the middle of the range, and the median is in the middle of the box. Most likely this is not a skewed distribution either.

Calculate The Pearson Coefficient of Skewness

The measure of skewness in the graphical summary is not the same as the Pearson coefficient. Use the calculator and the formula.

$$PC = \frac{3(\overline{X} - \text{median})}{s}$$

3. Select **Calc>Calculator,** then type **PC** in the text box for Store result in:.

4. Enter the expression: **3*(MEAN(C1)−MEDI(C1))/(STDEV(C1)).** Make sure you get all the parentheses in the right place!

5. Click [OK]. The result, 0.148318, will be stored in the first row of **C2** named **PC.** Since it is smaller than +1, the distribution is not skewed.

Construct a Normal Probability Plot

6. Select **Graph>Probability Plot,** then Single and click [OK].

7. Double-click C1 Inventory to select the data to be graphed.

8. Click [Distribution] and make sure that Normal is selected. Click [OK].

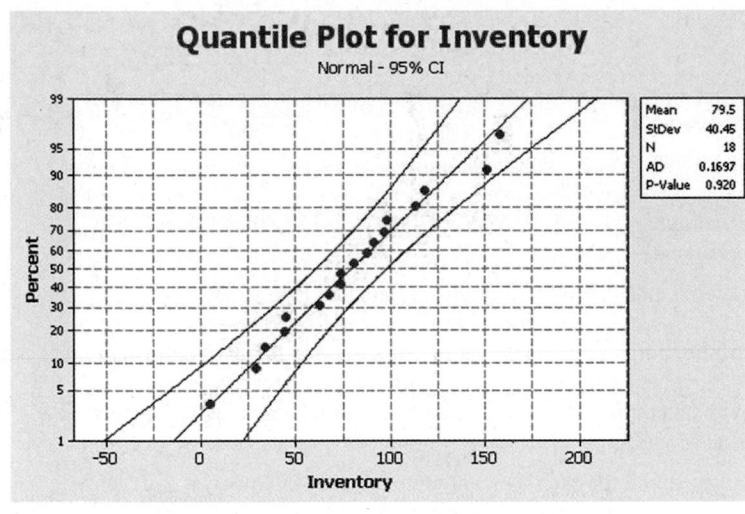

9. Click [Labels] and enter the title for the graph: **Quantile Plot for Inventory.** You may also put **Your Name** in the subtitle.

10. Click [OK] twice. Inspect the graph to see if the graph of the points is linear.

These data are nearly normal.

What do you look for in the plot?

a) An "S curve" indicates a distribution that is too thick in the tails, a uniform distribution, for example.

b) Concave plots indicate a skewed distribution.

c) If one end has a point that is extremely high or low, there may be outliers.

This data set appears to be nearly normal by every one of the four criteria!

TI-83 Plus or TI-84 Plus
Step by Step

Normal Random Variables

To find the probability for a normal random variable:
Press **2nd [DISTR]**, then **2** for normalcdf(
The form is normalcdf(lower x value, upper x value, μ, σ)
Use E99 for ∞ (infinity) and $-$E99 for $-\infty$ (negative infinity). Press **2nd [EE]** to get E.

Example: Find the probability that x is between 27 and 31 when $\mu = 28$ and $\sigma = 2$ (Example 6–7a from the text).
normalcdf(27,31,28,2)

To find the percentile for a normal random variable:
Press **2nd [DISTR]**, then **3** for invNorm(
The form is invNorm(area to the left of x value, μ, σ)

Example: Find the 90th percentile when $\mu = 200$ and $\sigma = 20$ (Example 6–9 from text).
invNorm(.9,200,20)

```
normalcdf(27,31,
20,2)
        .6246552391
invNorm(.9,200,2
0)
        225.6310313
```

To construct a normal quantile plot:

1. Enter the data values into L₁.

2. Press **2nd [STAT PLOT]** to get the STAT PLOT menu.

3. Press **1** for Plot 1.

4. Turn on the plot by pressing **ENTER** while the cursor is flashing over ON.

5. Move the cursor to the normal quantile plot (6th graph).

6. Make sure L₁ is entered for the Data List and X is highlighted for the Data Axis.

7. Press **WINDOW** for the Window menu. Adjust Xmin and Xmax according to the data values. Adjust Ymin and Ymax as well, Ymin = -3 and Ymax = 3 usually work fine.

8. Press **GRAPH.**

Using the data from the previous example gives

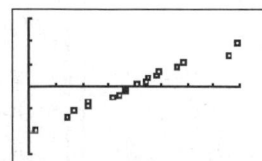

Since the points in the normal quantile plot lie close to a straight line, the distribution is approximately normal.

Excel
Step by Step

Normal Quantile Plot

Excel can be used to construct a normal quantile plot in order to examine if a set of data is approximately normally distributed.

1. Enter the data from the MINITAB example into column A of a new worksheet. The data should be sorted in ascending order. If the data are not already sorted in ascending order, highlight the data to be sorted and select the Sort & Filter icon from the toolbar. Then select Sort Smallest to Largest.

2. After all the data are entered and sorted in column A, select cell B1. Type: =NORMSINV(1/(2*18)). Since the sample size is 18, each score represents $\frac{1}{18}$, or approximately 5.6%, of the sample. Each data value is assumed to subdivide the data into equal intervals. Each data value corresponds to the midpoint of a particular subinterval. Thus, this procedure will standardize the data by assuming each data value represents the midpoint of a subinterval of width $\frac{1}{18}$.

3. Repeat the procedure from step 2 for each data value in column A. However, for each subsequent value in column A, enter the next odd multiple of $\frac{1}{36}$ in the argument for the NORMSINV function. For example, in cell B2, type: =NORMSINV(3/(2*18)). In cell B3, type: =NORMSINV(5/(2*18)), and so on until all the data values have corresponding z scores.

4. Highlight the data from columns A and B, and select Insert, then Scatter chart. Select the Scatter with only markers (the first Scatter chart).

5. To insert a title to the chart: Left-click on any region of the chart. Select Chart Tools and Layout from the toolbar. Then select Chart Title.

6. To insert a label for the variable on the horizontal axis: Left-click on any region of the chart. Select Chart Tools and Layout form the toolbar. Then select Axis Titles>Primary Horizontal Axis Title.

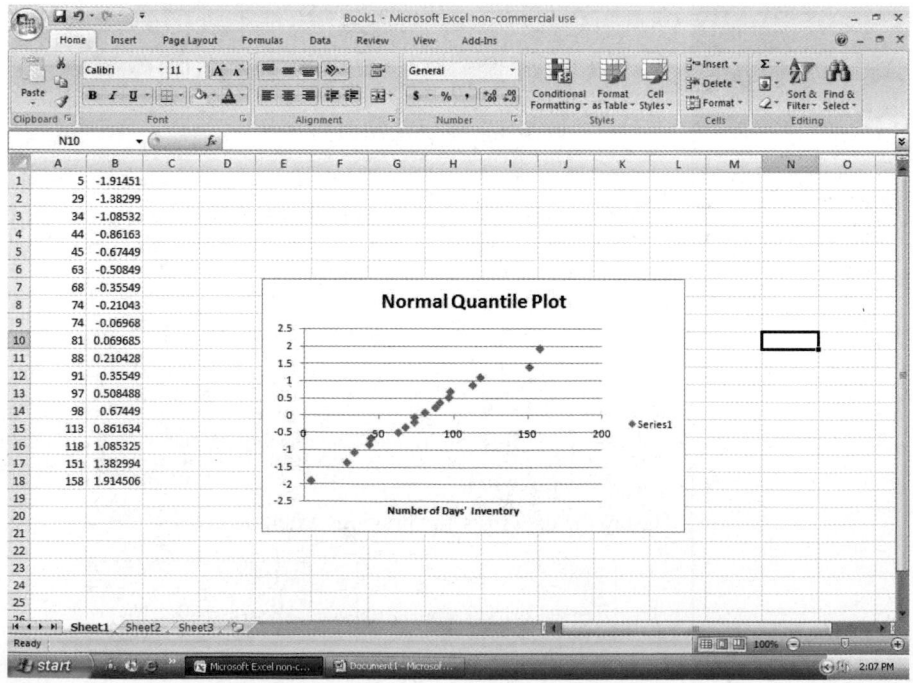

The points on the chart appear to lie close to a straight line. Thus, we deduce that the data are approximately normally distributed.

The Central Limit Theorem

Objective 6

Use the central limit theorem to solve problems involving sample means for large samples.

In addition to knowing how individual data values vary about the mean for a population, statisticians are interested in knowing how the means of samples of the same size taken from the same population vary about the population mean.

Distribution of Sample Means

Suppose a researcher selects a sample of 30 adult males and finds the mean of the measure of the triglyceride levels for the sample subjects to be 187 milligrams/deciliter. Then suppose a second sample is selected, and the mean of that sample is found to be 192 milligrams/deciliter. Continue the process for 100 samples. What happens then is that the mean becomes a random variable, and the sample means 187, 192, 184, . . . , 196 constitute a *sampling distribution of sample means*.

A **sampling distribution of sample means** is a distribution using the means computed from all possible random samples of a specific size taken from a population.

If the samples are randomly selected with replacement, the sample means, for the most part, will be somewhat different from the population mean μ. These differences are caused by sampling error.

Sampling error is the difference between the sample measure and the corresponding population measure due to the fact that the sample is not a perfect representation of the population.

When all possible samples of a specific size are selected with replacement from a population, the distribution of the sample means for a variable has two important properties, which are explained next.

Properties of the Distribution of Sample Means

1. The mean of the sample means will be the same as the population mean.
2. The standard deviation of the sample means will be smaller than the standard deviation of the population, and it will be equal to the population standard deviation divided by the square root of the sample size.

The following example illustrates these two properties. Suppose a professor gave an 8-point quiz to a small class of four students. The results of the quiz were 2, 6, 4, and 8. For the sake of discussion, assume that the four students constitute the population. The mean of the population is

$$\mu = \frac{2 + 6 + 4 + 8}{4} = 5$$

The standard deviation of the population is

$$\sigma = \sqrt{\frac{(2-5)^2 + (6-5)^2 + (4-5)^2 + (8-5)^2}{4}} = 2.236$$

The graph of the original distribution is shown in Figure 6–29. This is called a *uniform distribution*.

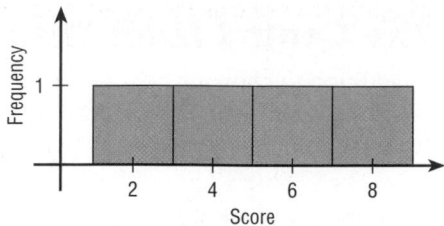

Now, if all samples of size 2 are taken with replacement and the mean of each sample is found, the distribution is as shown.

Sample	Mean	Sample	Mean
2, 2	2	6, 2	4
2, 4	3	6, 4	5
2, 6	4	6, 6	6
2, 8	5	6, 8	7
4, 2	3	8, 2	5
4, 4	4	8, 4	6
4, 6	5	8, 6	7
4, 8	6	8, 8	8

A frequency distribution of sample means is as follows.

$\overline{X}$	f
2	1
3	2
4	3
5	4
6	3
7	2
8	1

For the data from the example just discussed, Figure 6–30 shows the graph of the sample means. The histogram appears to be approximately normal.

The mean of the sample means, denoted by $\mu_{\overline{X}}$, is

$$\mu_{\overline{X}} = \frac{2 + 3 + \cdots + 8}{16} = \frac{80}{16} = 5$$

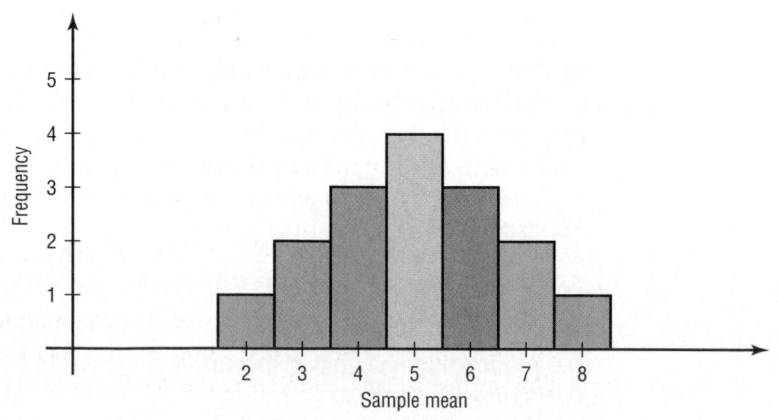

which is the same as the population mean. Hence,

$$\mu_{\bar{X}} = \mu$$

The standard deviation of sample means, denoted by $\sigma_{\bar{X}}$, is

$$\sigma_{\bar{X}} = \sqrt{\frac{(2-5)^2 + (3-5)^2 + \cdots + (8-5)^2}{16}} = 1.581$$

which is the same as the population standard deviation, divided by $\sqrt{2}$:

$$\sigma_{\bar{X}} = \frac{2.236}{\sqrt{2}} = 1.581$$

(*Note:* Rounding rules were not used here in order to show that the answers coincide.)

In summary, if all possible samples of size n are taken with replacement from the same population, the mean of the sample means, denoted by $\mu_{\bar{X}}$, equals the population mean μ; and the standard deviation of the sample means, denoted by $\sigma_{\bar{X}}$, equals $\sigma/\sqrt{n}$. The standard deviation of the sample means is called the **standard error of the mean.** Hence,

$$\sigma_{\bar{X}} = \frac{\sigma}{\sqrt{n}}$$

A third property of the sampling distribution of sample means pertains to the shape of the distribution and is explained by the **central limit theorem.**

The Central Limit Theorem

As the sample size n increases without limit, the shape of the distribution of the sample means taken with replacement from a population with mean μ and standard deviation σ will approach a normal distribution. As previously shown, this distribution will have a mean μ and a standard deviation $\sigma/\sqrt{n}$.

If the sample size is sufficiently large, the central limit theorem can be used to answer questions about sample means in the same manner that a normal distribution can be used to answer questions about individual values. The only difference is that a new formula must be used for the z values. It is

$$z = \frac{\bar{X} - \mu}{\sigma/\sqrt{n}}$$

Notice that $\bar{X}$ is the sample mean, and the denominator must be adjusted since means are being used instead of individual data values. The denominator is the standard deviation of the sample means.

If a large number of samples of a given size are selected from a normally distributed population, or if a large number of samples of a given size that is greater than or equal to 30 are selected from a population that is not normally distributed, and the sample means are computed, then the distribution of sample means will look like the one shown in Figure 6–31. Their percentages indicate the areas of the regions.

It's important to remember two things when you use the central limit theorem:

1. When the original variable is normally distributed, the distribution of the sample means will be normally distributed, for any sample size n.

2. When the distribution of the original variable might not be normal, a sample size of 30 or more is needed to use a normal distribution to approximate the distribution of the sample means. The larger the sample, the better the approximation will be.

Figure 6–31

Distribution of Sample
Means for a Large
Number of Samples

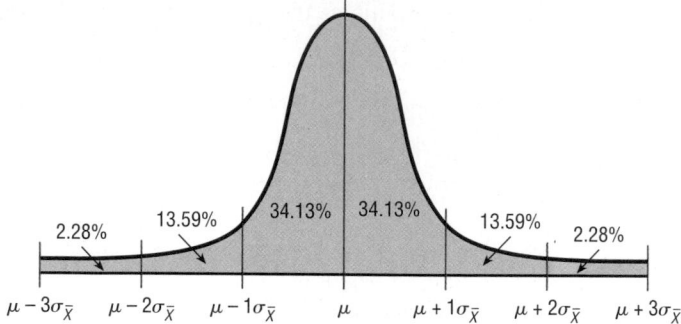

Examples 6–13 through 6–15 show how the standard normal distribution can be used
to answer questions about sample means.

Example 6–13   **Hours That Children Watch Television**

A. C. Neilsen reported that children between the ages of 2 and 5 watch an average of
25 hours of television per week. Assume the variable is normally distributed and the
standard deviation is 3 hours. If 20 children between the ages of 2 and 5 are randomly
selected, find the probability that the mean of the number of hours they watch television
will be greater than 26.3 hours.

Source: Michael D. Shook and Robert L. Shook, *The Book of Odds.*

Solution

Since the variable is approximately normally distributed, the distribution of sample
means will be approximately normal, with a mean of 25. The standard deviation of the
sample means is

$$\sigma_{\bar{X}} = \frac{\sigma}{\sqrt{n}} = \frac{3}{\sqrt{20}} = 0.671$$

The distribution of the means is shown in Figure 6–32, with the appropriate area
shaded.

Figure 6–32

Distribution of
the Means for
Example 6–13

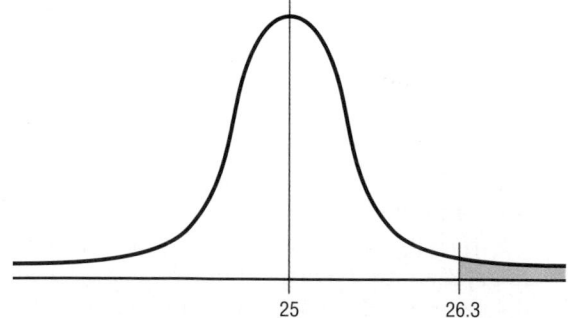

The z value is

$$z = \frac{\bar{X} - \mu}{\sigma/\sqrt{n}} = \frac{26.3 - 25}{3/\sqrt{20}} = \frac{1.3}{0.671} = 1.94$$

The area to the right of 1.94 is $1.000 - 0.9738 = 0.0262$, or 2.62%.
   One can conclude that the probability of obtaining a sample mean larger than
26.3 hours is 2.62% [i.e., $P(\bar{X} > 26.3) = 2.62\%$].

Example 6–14

The average age of a vehicle registered in the United States is 8 years, or 96 months. Assume the standard deviation is 16 months. If a random sample of 36 vehicles is selected, find the probability that the mean of their age is between 90 and 100 months.

Source: *Harper's Index.*

Solution

Since the sample is 30 or larger, the normality assumption is not necessary. The desired area is shown in Figure 6–33.

Figure 6–33

Area Under a
Normal Curve for
Example 6–14

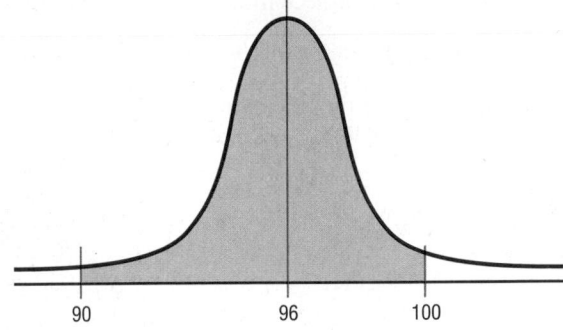

The two z values are

$$z_1 = \frac{90 - 96}{16/\sqrt{36}} = -2.25$$

$$z_2 = \frac{100 - 96}{16/\sqrt{36}} = 1.50$$

To find the area between the two z values of -2.25 and 1.50, look up the corresponding area in Table E and subtract one from the other. The area for $z = -2.25$ is 0.0122, and the area for $z = 1.50$ is 0.9332. Hence the area between the two values is $0.9332 - 0.0122 = 0.9210$, or 92.1%.

Hence, the probability of obtaining a sample mean between 90 and 100 months is 92.1%, that is, $P(90 < \overline{X} < 100) = 92.1\%$.

Students sometimes have difficulty deciding whether to use

$$z = \frac{\overline{X} - \mu}{\sigma/\sqrt{n}} \qquad \text{or} \qquad z = \frac{X - \mu}{\sigma}$$

The formula

$$z = \frac{\overline{X} - \mu}{\sigma/\sqrt{n}}$$

should be used to gain information about a sample mean, as shown in this section. The formula

$$z = \frac{X - \mu}{\sigma}$$

is used to gain information about an individual data value obtained from the population. Notice that the first formula contains $\overline{X}$, the symbol for the sample mean, while the second formula contains X, the symbol for an individual data value. Example 6–15 illustrates the uses of the two formulas.

| Example 6-15 | Meat Consumption |

The average number of pounds of meat that a person consumes per year is 218.4 pounds. Assume that the standard deviation is 25 pounds and the distribution is approximately normal.

Source: Michael D. Shook and Robert L. Shook, *The Book of Odds*.

    a. Find the probability that a person selected at random consumes less than 224 pounds per year.

    b. If a sample of 40 individuals is selected, find the probability that the mean of the sample will be less than 224 pounds per year.

Solution

    a. Since the question asks about an individual person, the formula $z = (X - \mu)/\sigma$ is used. The distribution is shown in Figure 6–34.

Figure 6-34

Area Under a Normal Curve for Part *a* of Example 6-15

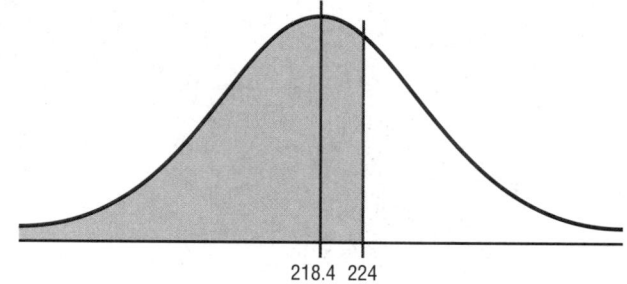

218.4 224
Distribution of individual data values for the population

The *z* value is

$$z = \frac{X - \mu}{\sigma} = \frac{224 - 218.4}{25} = 0.22$$

The area to the left of $z = 0.22$ is 0.5871. Hence, the probability of selecting an individual who consumes less than 224 pounds of meat per year is 0.5871, or 58.71% [i.e., $P(X < 224) = 0.5871$].

    b. Since the question concerns the mean of a sample with a size of 40, the formula $z = (\bar{X} - \mu)/(\sigma/\sqrt{n})$ is used. The area is shown in Figure 6–35.

Figure 6-35

Area Under a Normal Curve for Part *b* of Example 6-15

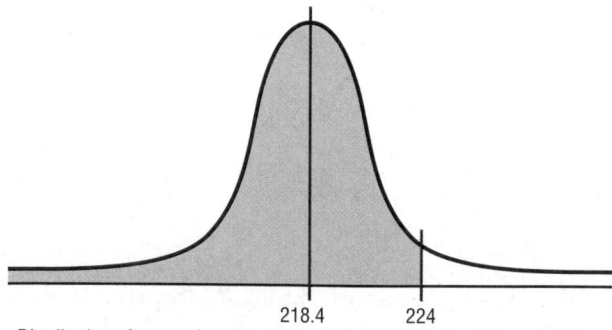

218.4   224
Distribution of means for all samples of size 40 taken from the population

The *z* value is

$$z = \frac{\bar{X} - \mu}{\sigma/\sqrt{n}} = \frac{224 - 218.4}{25/\sqrt{40}} = 1.42$$

The area to the left of $z = 1.42$ is 0.9222.

Hence, the probability that the mean of a sample of 40 individuals is less than 224 pounds per year is 0.9222, or 92.22%. That is, $P(\overline{X} < 224) = 0.9222$.

Comparing the two probabilities, you can see that the probability of selecting an individual who consumes less than 224 pounds of meat per year is 58.71%, but the probability of selecting a sample of 40 people with a mean consumption of meat that is less than 224 pounds per year is 92.22%. This rather large difference is due to the fact that the distribution of sample means is much less variable than the distribution of individual data values. (*Note:* An individual person is the equivalent of saying $n = 1$.)

Finite Population Correction Factor (Optional)

The formula for the standard error of the mean $\sigma/\sqrt{n}$ is accurate when the samples are drawn with replacement or are drawn without replacement from a very large or infinite population. Since sampling with replacement is for the most part unrealistic, a *correction factor* is necessary for computing the standard error of the mean for samples drawn without replacement from a finite population. Compute the correction factor by using the expression

$$\sqrt{\frac{N - n}{N - 1}}$$

where N is the population size and n is the sample size.

This correction factor is necessary if relatively large samples are taken from a small population, because the sample mean will then more accurately estimate the population mean and there will be less error in the estimation. Therefore, the standard error of the mean must be multiplied by the correction factor to adjust for large samples taken from a small population. That is,

$$\sigma_{\overline{X}} = \frac{\sigma}{\sqrt{n}} \cdot \sqrt{\frac{N - n}{N - 1}}$$

Finally, the formula for the z value becomes

$$z = \frac{\overline{X} - \mu}{\frac{\sigma}{\sqrt{n}} \cdot \sqrt{\frac{N - n}{N - 1}}}$$

When the population is large and the sample is small, the correction factor is generally not used, since it will be very close to 1.00.

The formulas and their uses are summarized in Table 6–1.

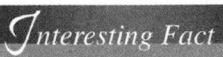

Interesting Fact

The bubonic plague killed more than 25 million people in Europe between 1347 and 1351.

Table 6–1	Summary of Formulas and Their Uses
Formula	**Use**
1. $z = \dfrac{X - \mu}{\sigma}$	Used to gain information about an individual data value when the variable is normally distributed.
2. $z = \dfrac{\overline{X} - \mu}{\sigma/\sqrt{n}}$	Used to gain information when applying the central limit theorem about a sample mean when the variable is normally distributed or when the sample size is 30 or more.

Applying the Concepts **6–3**

Central Limit Theorem

Twenty students from a statistics class each collected a random sample of times on how long it took students to get to class from their homes. All the sample sizes were 30. The resulting means are listed.

Student	Mean	Std. Dev.		Student	Mean	Std. Dev.
1	22	3.7		11	27	1.4
2	31	4.6		12	24	2.2
3	18	2.4		13	14	3.1
4	27	1.9		14	29	2.4
5	20	3.0		15	37	2.8
6	17	2.8		16	23	2.7
7	26	1.9		17	26	1.8
8	34	4.2		18	21	2.0
9	23	2.6		19	30	2.2
10	29	2.1		20	29	2.8

1. The students noticed that everyone had different answers. If you randomly sample over and over from any population, with the same sample size, will the results ever be the same?

2. The students wondered whose results were right. How can they find out what the population mean and standard deviation are?

3. Input the means into the computer and check to see if the distribution is normal.

4. Check the mean and standard deviation of the means. How do these values compare to the students' individual scores?

5. Is the distribution of the means a sampling distribution?

6. Check the sampling error for students 3, 7, and 14.

7. Compare the standard deviation of the sample of the 20 means. Is that equal to the standard deviation from student 3 divided by the square of the sample size? How about for student 7, or 14?

See page 354 for the answers.

Exercises 6–3

1. If samples of a specific size are selected from a population and the means are computed, what is this distribution of means called? The distribution is called the sampling distribution of sample means.

2. Why do most of the sample means differ somewhat from the population mean? What is this difference called? The sample is not a perfect representation of the population. The difference is due to what is called sampling error.

3. What is the mean of the sample means? The mean of the sample means is equal to the population mean.

4. What is the standard deviation of the sample means called? What is the formula for this standard deviation? The standard error of the mean: $\sigma_{\bar{x}} = \sigma/\sqrt{n}$.

5. What does the central limit theorem say about the shape of the distribution of sample means? The distribution will be approximately normal when the sample size is large.

6. What formula is used to gain information about an individual data value when the variable is normally distributed? $z = \dfrac{X - \mu}{\sigma}$

7. What formula is used to gain information about a sample mean when the variable is normally distributed or when the sample size is 30 or more? $z = \dfrac{\bar{X} - \mu}{\sigma/\sqrt{n}}$

For Exercises 8 through 25, assume that the sample is taken from a large population and the correction factor can be ignored.

8. **Glass Garbage Generation** A survey found that the American family generates an average of 17.2 pounds of glass garbage each year. Assume the standard deviation of the distribution is 2.5 pounds. Find the probability that the mean of a sample of 55 families will be between 17 and 18 pounds. 0.7135

Source: Michael D. Shook and Robert L. Shook, *The Book of Odds*.

9. **College Costs** The mean undergraduate cost for tuition, fees, room, and board for four-year institutions was $26,489 for a recent academic year. Suppose

that $\sigma = \$3204$ and that 36 four-year institutions are randomly selected. Find the probability that the sample mean cost for these 36 schools is

a. Less than $25,000 0.0026 (TI: 0.0026)

b. Greater than $26,000 0.8212 (TI: 0.8201)

c. Between $24,000 and $26,000 0.1787 (TI: 0.1799)

Source: www.nces.ed.gov

10. Teachers' Salaries in Connecticut The average teacher's salary in Connecticut (ranked first among states) is $57,337. Suppose that the distribution of salaries is normal with a standard deviation of $7500.

a. What is the probability that a randomly selected teacher makes less than $52,000 per year? 0.2389

b. If we sample 100 teachers' salaries, what is the probability that the sample mean is less than $56,000? 0.0375

Source: *New York Times Almanac.*

11. Serum Cholesterol Levels The mean serum cholesterol level of a large population of overweight children is 220 milligrams per deciliter (mg/dl), and the standard deviation is 16.3 mg/dl. If a random sample of 35 overweight children is selected, find the probability that the mean will be between 220 and 222 mg/dl. Assume the serum cholesterol level variable is normally distributed. 0.2673

12. Teachers' Salaries in North Dakota The average teacher's salary in North Dakota is $37,764. Assume a normal distribution with $\sigma = \$5100$.

a. What is the probability that a randomly selected teacher's salary is greater than $45,000? 0.0778

b. For a sample of 75 teachers, what is the probability that the sample mean is greater than $38,000? 0.3466

Source: *New York Times Almanac.*

13. Fuel Efficiency for U.S. Light Vehicles The average fuel efficiency of U.S. light vehicles (cars, SUVs, minivans, vans, and light trucks) for 2005 was 21 mpg. If the standard deviation of the population was 2.9 and the gas ratings were normally distributed, what is the probability that the mean mpg for a random sample of 25 light vehicles is under 20? Between 20 and 25?

Source: *World Almanac.* 0.0427; 0.9572 (TI: 0.0423; 0.9577)

14. SAT Scores The national average SAT score (for Verbal and Math) is 1028. Suppose that nothing is known about the shape of the distribution and that the standard deviation is 100. If a random sample of 200 scores were selected and the sample mean were calculated to be 1050, would you be surprised? Explain. Yes—the probability of such is less than 0.0001.

Source: *New York Times Almanac.*

15. Sodium in Frozen Food The average number of milligrams (mg) of sodium in a certain brand of low-salt microwave frozen dinners is 660 mg, and the standard deviation is 35 mg. Assume the variable is normally distributed. a. 0.3859 (TI: 0.3875)

a. If a single dinner is selected, find the probability that the sodium content will be more than 670 mg.

b. If a sample of 10 dinners is selected, find the probability that the mean of the sample will be larger than 670 mg. 0.1841 (TI: 0.1831)

c. Why is the probability for part *a* greater than that for part *b*? Individual values are more variable than means.

16. Cell Phone Lifetimes A recent study of the lifetimes of cell phones found the average is 24.3 months. The standard deviation is 2.6 months. If a company provides its 33 employees with a cell phone, find the probability that the mean lifetime of these phones will be less than 23.8 months. Assume cell phone life is a normally distributed variable. 0.1357

17. Water Use The *Old Farmer's Almanac* reports that the average person uses 123 gallons of water daily. If the standard deviation is 21 gallons, find the probability that the mean of a randomly selected sample of 15 people will be between 120 and 126 gallons. Assume the variable is normally distributed. 0.4176 (TI: 0.4199)

18. Medicare Hospital Insurance The average yearly Medicare Hospital Insurance benefit per person was $4064 in a recent year. If the benefits are normally distributed with a standard deviation of $460, find the probability that the mean benefit for a random sample of 20 patients is

a. Less than $3800 0.0051

b. More than $4100 0.3632

Source: *New York Times Almanac.*

19. Amount of Laundry Washed Each Year Procter & Gamble reported that an American family of four washes an average of 1 ton (2000 pounds) of clothes each year. If the standard deviation of the distribution is 187.5 pounds, find the probability that the mean of a randomly selected sample of 50 families of four will be between 1980 and 1990 pounds. 0.1254 (TI: 0.12769)

Source: *The Harper's Index Book.*

20. Per Capita Income of Delaware Residents In a recent year, Delaware had the highest per capita annual income with $51,803. If $\sigma = \$4850$, what is the probability that a random sample of 34 state residents had a mean income greater than $50,000? Less than $48,000?

Source: *New York Times Almanac.* 0.9850; Less than 0.0001

21. Annual Precipitation The average annual precipitation for a large Midwest city is 30.85 inches with a standard deviation of 3.6 inches. Assume the variable is normally distributed.

a. Find the probability that a randomly selected month will have less than 30 inches. 0.4052 or 40.52%

b. Find the probability that the mean of a random selection of 32 months will have a mean less than 30 inches. 0.0901 or 9.01%

c. Does it seem reasonable that one month could have a rainfall amount less than 30 inches? Yes, the probability is slightly more than 40%.

d. Does it seem reasonable that the mean of a sample of 32 months could be less than 30 inches? It's possible since the probability is about 9%.

22. **Systolic Blood Pressure** Assume that the mean systolic blood pressure of normal adults is 120 millimeters of mercury (mm Hg) and the standard deviation is 5.6. Assume the variable is normally distributed.

a. If an individual is selected, find the probability that the individual's pressure will be between 120 and 121.8 mm Hg. 0.1255

b. If a sample of 30 adults is randomly selected, find the probability that the sample mean will be between 120 and 121.8 mm Hg. 0.4608

c. Why is the answer to part *a* so much smaller than the answer to part *b*? Means are less variable than individual data.

23. **Cholesterol Content** The average cholesterol content of a certain brand of eggs is 215 milligrams, and the standard deviation is 15 milligrams. Assume the variable is normally distributed.

a. If a single egg is selected, find the probability that the cholesterol content will be greater than 220 milligrams. 0.3707 (TI: 0.3694)

b. If a sample of 25 eggs is selected, find the probability that the mean of the sample will be larger than 220 milligrams. 0.0475 (TI: 0.04779)

Source: *Living Fit.*

24. **Ages of Proofreaders** At a large publishing company, the mean age of proofreaders is 36.2 years, and the standard deviation is 3.7 years. Assume the variable is normally distributed.

a. If a proofreader from the company is randomly selected, find the probability that his or her age will be between 36 and 37.5 years. 0.1567

b. If a random sample of 15 proofreaders is selected, find the probability that the mean age of the proofreaders in the sample will be between 36 and 37.5 years. 0.4963

25. **Weekly Income of Private Industry Information Workers** The average weekly income of information workers in private industry is $777. If the standard deviation is $77, what is the probability that a random sample of 50 information workers will earn, on average, more than $800 per week? Do we need to assume a normal distribution? Explain.

Source: *World Almanac.* 0.0174 No—the central limit theorem applies.

Extending the Concepts

For Exercises 26 and 27, check to see whether the correction factor should be used. If so, be sure to include it in the calculations.

26. **Life Expectancies** In a study of the life expectancy of 500 people in a certain geographic region, the mean age at death was 72.0 years, and the standard deviation was 5.3 years. If a sample of 50 people from this region is selected, find the probability that the mean life expectancy will be less than 70 years. 0.0025

27. **Home Values** A study of 800 homeowners in a certain area showed that the average value of the homes was $82,000, and the standard deviation was $5000. If 50 homes are for sale, find the probability that the mean of the values of these homes is greater than $83,500. 0.0143

28. **Breaking Strength of Steel Cable** The average breaking strength of a certain brand of steel cable is 2000 pounds, with a standard deviation of 100 pounds. A sample of 20 cables is selected and tested. Find the sample mean that will cut off the upper 95% of all samples of size 20 taken from the population. Assume the variable is normally distributed. 1963.10 pounds

29. The standard deviation of a variable is 15. If a sample of 100 individuals is selected, compute the standard error of the mean. What size sample is necessary to double the standard error of the mean? $\sigma_{\bar{x}} = 1.5, n = 25$

30. In Exercise 29, what size sample is needed to cut the standard error of the mean in half? 400

6–4

The Normal Approximation to the Binomial Distribution

A normal distribution is often used to solve problems that involve the binomial distribution since when *n* is large (say, 100), the calculations are too difficult to do by hand using the binomial distribution. Recall from Chapter 5 that a binomial distribution has the following characteristics:

1. There must be a fixed number of trials.

2. The outcome of each trial must be independent.

3. Each experiment can have only two outcomes or outcomes that can be reduced to two outcomes.

4. The probability of a success must remain the same for each trial.

Also, recall that a binomial distribution is determined by n (the number of trials) and p (the probability of a success). When p is approximately 0.5, and as n increases, the shape of the binomial distribution becomes similar to that of a normal distribution. The larger n is and the closer p is to 0.5, the more similar the shape of the binomial distribution is to that of a normal distribution.

But when p is close to 0 or 1 and n is relatively small, a normal approximation is inaccurate. As a rule of thumb, statisticians generally agree that a normal approximation should be used only when $n \cdot p$ and $n \cdot q$ are both greater than or equal to 5. (*Note:* $q = 1 - p$.) For example, if p is 0.3 and n is 10, then $np = (10)(0.3) = 3$, and a normal distribution should not be used as an approximation. On the other hand, if $p = 0.5$ and $n = 10$, then $np = (10)(0.5) = 5$ and $nq = (10)(0.5) = 5$, and a normal distribution can be used as an approximation. See Figure 6–36.

Objective **7**

Use the normal approximation to compute probabilities for a binomial variable.

Figure 6–36

Comparison of the Binomial Distribution and a Normal Distribution

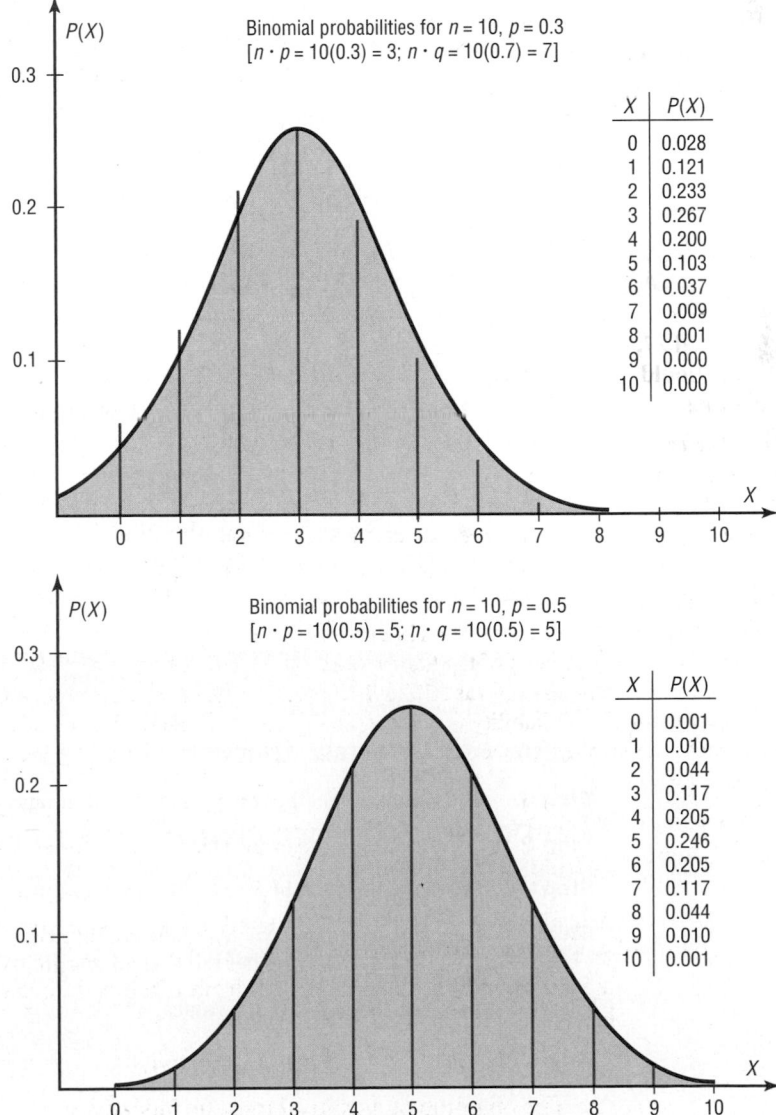

In addition to the previous condition of $np \geq 5$ and $nq \geq 5$, a correction for continuity may be used in the normal approximation.

A **correction for continuity** is a correction employed when a continuous distribution is used to approximate a discrete distribution.

The continuity correction means that for any specific value of X, say 8, the boundaries of X in the binomial distribution (in this case, 7.5 to 8.5) must be used. (See Section 1–2.) Hence, when you employ a normal distribution to approximate the binomial, you must use the boundaries of any specific value X as they are shown in the binomial distribution. For example, for $P(X = 8)$, the correction is $P(7.5 < X < 8.5)$. For $P(X \leq 7)$, the correction is $P(X < 7.5)$. For $P(X \geq 3)$, the correction is $P(X > 2.5)$.

Students sometimes have difficulty deciding whether to add 0.5 or subtract 0.5 from the data value for the correction factor. Table 6–2 summarizes the different situations.

Table 6–2	Summary of the Normal Approximation to the Binomial Distribution
Binomial	**Normal**
When finding:	Use:
1. $P(X = a)$	$P(a - 0.5 < X < a + 0.5)$
2. $P(X \geq a)$	$P(X > a - 0.5)$
3. $P(X > a)$	$P(X > a + 0.5)$
4. $P(X \leq a)$	$P(X < a + 0.5)$
5. $P(X < a)$	$P(X < a - 0.5)$
For all cases, $\mu = n \cdot p$, $\sigma = \sqrt{n \cdot p \cdot q}$, $n \cdot p \geq 5$, and $n \cdot q \geq 5$.	

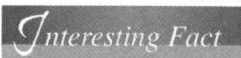

Interesting Fact

Of the 12 months, August ranks first in the number of births for Americans.

The formulas for the mean and standard deviation for the binomial distribution are necessary for calculations. They are

$$\mu = n \cdot p \qquad \text{and} \qquad \sigma = \sqrt{n \cdot p \cdot q}$$

The steps for using the normal distribution to approximate the binomial distribution are shown in this Procedure Table.

Procedure Table

Procedure for the Normal Approximation to the Binomial Distribution

Step 1   Check to see whether the normal approximation can be used.

Step 2   Find the mean μ and the standard deviation σ.

Step 3   Write the problem in probability notation, using X.

Step 4   Rewrite the problem by using the continuity correction factor, and show the corresponding area under the normal distribution.

Step 5   Find the corresponding z values.

Step 6   Find the solution.

Example 6–16	**Reading While Driving**

A magazine reported that 6% of American drivers read the newspaper while driving. If 300 drivers are selected at random, find the probability that exactly 25 say they read the newspaper while driving.

Source: USA Snapshot, *USA TODAY.*

Solution

Here, $p = 0.06$, $q = 0.94$, and $n = 300$.

Step 1   Check to see whether a normal approximation can be used.

$np = (300)(0.06) = 18 \qquad nq = (300)(0.94) = 282$

Since $np \geq 5$ and $nq \geq 5$, the normal distribution can be used.

Step 2   Find the mean and standard deviation.

$\mu = np = (300)(0.06) = 18$

$\sigma = \sqrt{npq} = \sqrt{(300)(0.06)(0.94)} = \sqrt{16.92} = 4.11$

Step 3   Write the problem in probability notation: $P(X = 25)$.

Step 4   Rewrite the problem by using the continuity correction factor. See approximation number 1 in Table 6–2: $P(25 - 0.5 < X < 25 + 0.5) = P(24.5 < X < 25.5)$. Show the corresponding area under the normal distribution curve. See Figure 6–37.

Figure 6–37

Area Under a Normal Curve and X Values for Example 6–16

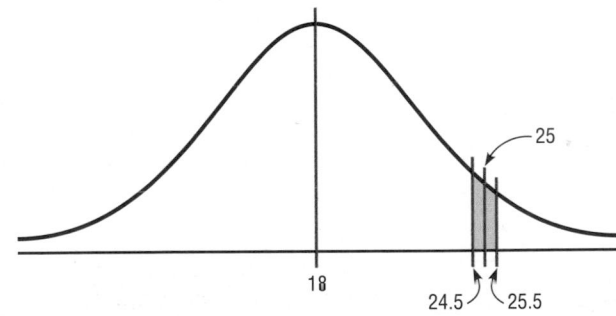

Step 5   Find the corresponding z values. Since 25 represents any value between 24.5 and 25.5, find both z values.

$$z_1 = \frac{25.5 - 18}{4.11} = 1.82 \qquad z_2 = \frac{24.5 - 18}{4.11} = 1.58$$

Step 6   The area to the left of $z = 1.82$ is 0.9656, and the area to the left of $z = 1.58$ is 0.9429. The area between the two z values is $0.9656 - 0.9429 = 0.0227$, or 2.27%. Hence, the probability that exactly 25 people read the newspaper while driving is 2.27%.

Example 6–17	**Widowed Bowlers**

Of the members of a bowling league, 10% are widowed. If 200 bowling league members are selected at random, find the probability that 10 or more will be widowed.

Solution

Here, $p = 0.10$, $q = 0.90$, and $n = 200$.

Step 1   Since $np = (200)(0.10) = 20$ and $nq = (200)(0.90) = 180$, the normal approximation can be used.

Step 2 $\mu = np = (200)(0.10) = 20$

$\sigma = \sqrt{npq} = \sqrt{(200)(0.10)(0.90)} = \sqrt{18} = 4.24$

Step 3 $P(X \geq 10)$

Step 4 See approximation number 2 in Table 6–2: $P(X > 10 - 0.5) = P(X > 9.5)$. The desired area is shown in Figure 6–38.

Figure 6–38

Area Under a Normal Curve and X Value for Example 6–17

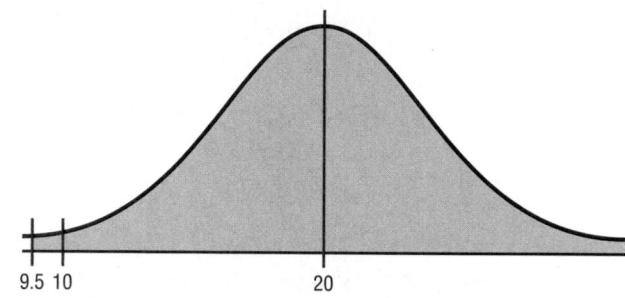

Step 5 Since the problem is to find the probability of 10 or more positive responses, a normal distribution graph is as shown in Figure 6–38.
The z value is

$$z = \frac{9.5 - 20}{4.24} = -2.48$$

Step 6 The area to the left of $z = -2.48$ is 0.0066. Hence the area to the right of $z = -2.48$ is $1.0000 - 0.0066 = 0.9934$, or 99.34%.

It can be concluded, then, that the probability of 10 or more widowed people in a random sample of 200 bowling league members is 99.34%.

Example 6–18

Batting Averages

If a baseball player's batting average is 0.320 (32%), find the probability that the player will get at most 26 hits in 100 times at bat.

Solution

Here, $p = 0.32$, $q = 0.68$, and $n = 100$.

Step 1 Since $np = (100)(0.320) = 32$ and $nq = (100)(0.680) = 68$, the normal distribution can be used to approximate the binomial distribution.

Step 2 $\mu = np = (100)(0.320) = 32$

$\sigma = \sqrt{npq} = \sqrt{(100)(0.32)(0.68)} = \sqrt{21.76} = 4.66$

Step 3 $P(X \leq 26)$

Step 4 See approximation number 4 in Table 6–2: $P(X < 26 + 0.5) = P(X < 26.5)$. The desired area is shown in Figure 6–39.

Step 5 The z value is

$$z = \frac{26.5 - 32}{4.66} = -1.18$$

Figure 6-39

Area Under a
Normal Curve for
Example 6-18

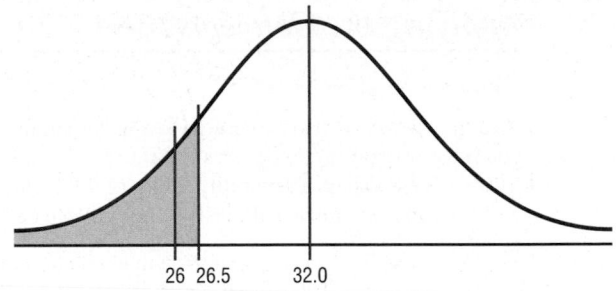

26 26.5   32.0

Step 6 The area to the left of $z = -1.18$ is 0.1190. Hence the probability is 0.1190, or 11.9%.

The closeness of the normal approximation is shown in Example 6-19.

Example 6-19

When $n = 10$ and $p = 0.5$, use the binomial distribution table (Table B in Appendix C) to find the probability that $X = 6$. Then use the normal approximation to find the probability that $X = 6$.

Solution

From Table B, for $n = 10$, $p = 0.5$, and $X = 6$, the probability is 0.205.
For a normal approximation,

$$\mu = np = (10)(0.5) = 5$$

$$\sigma = \sqrt{npq} = \sqrt{(10)(0.5)(0.5)} = 1.58$$

Now, $X = 6$ is represented by the boundaries 5.5 and 6.5. So the z values are

$$z_1 = \frac{6.5 - 5}{1.58} = 0.95 \qquad z_2 = \frac{5.5 - 5}{1.58} = 0.32$$

The corresponding area for 0.95 is 0.8289, and the corresponding area for 0.32 is 0.6255. The area between the two z values of 0.95 and 0.32 is $0.8289 - 0.6255 = 0.2034$, which is very close to the binomial table value of 0.205. See Figure 6-40.

Figure 6-40

Area Under a
Normal Curve for
Example 6-19

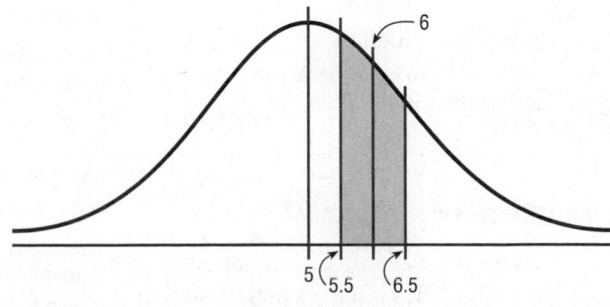

6

5   5.5   6.5

The normal approximation also can be used to approximate other distributions, such as the Poisson distribution (see Table C in Appendix C).

Applying the Concepts 6–4

How Safe Are You?

Assume one of your favorite activities is mountain climbing. When you go mountain climbing, you have several safety devices to keep you from falling. You notice that attached to one of your safety hooks is a reliability rating of 97%. You estimate that throughout the next year you will be using this device about 100 times. Answer the following questions.

1. Does a reliability rating of 97% mean that there is a 97% chance that the device will not fail any of the 100 times?
2. What is the probability of at least one failure?
3. What is the complement of this event?
4. Can this be considered a binomial experiment?
5. Can you use the binomial probability formula? Why or why not?
6. Find the probability of at least two failures.
7. Can you use a normal distribution to accurately approximate the binomial distribution? Explain why or why not.
8. Is correction for continuity needed?
9. How much safer would it be to use a second safety hook independently of the first?

See page 354 for the answers.

Exercises 6–4

1. Explain why a normal distribution can be used as an approximation to a binomial distribution. What conditions must be met to use the normal distribution to approximate the binomial distribution? Why is a correction for continuity necessary?

2. (ans) Use the normal approximation to the binomial to find the probabilities for the specific value(s) of X.

   a. $n = 30, p = 0.5, X = 18$ 0.0811
   b. $n = 50, p = 0.8, X = 44$ 0.0516
   c. $n = 100, p = 0.1, X = 12$ 0.1052
   d. $n = 10, p = 0.5, X \geq 7$ 0.1711
   e. $n = 20, p = 0.7, X \leq 12$ 0.2327
   f. $n = 50, p = 0.6, X \leq 40$ 0.9988

3. Check each binomial distribution to see whether it can be approximated by a normal distribution (i.e., are $np \geq 5$ and $nq \geq 5$?).

   a. $n = 20, p = 0.5$ Yes   d. $n = 50, p = 0.2$ Yes
   b. $n = 10, p = 0.6$ No   e. $n = 30, p = 0.8$ Yes
   c. $n = 40, p = 0.9$ No   f. $n = 20, p = 0.85$ No

4. **School Enrollment** Of all 3- to 5-year-old children, 56% are enrolled in school. If a sample of 500 such children is randomly selected, find the probability that at least 250 will be enrolled in school. 0.9970

   Source: *Statistical Abstract of the United States.*

5. **Youth Smoking** Two out of five adult smokers acquired the habit by age 14. If 400 smokers are randomly selected, find the probability that 170 or fewer acquired the habit by age 14. 0.8577

   Source: *Harper's Index.*

6. **Mail Order** A mail order company has an 8% success rate. If it mails advertisements to 600 people, find the probability of getting less than 40 sales. 0.1003

7. **Voter Preference** A political candidate estimates that 30% of the voters in her party favor her proposed tax reform bill. If there are 400 people at a rally, find the probability that at least 100 voters will favor her tax bill. Based on your answer, is it likely that 100 or more people will favor the bill? 0.9875

8. **Household Computers** According to recent surveys, 60% of households have personal computers. If a random sample of 180 households is selected, what is the probability that more than 60 but fewer than 100 have a personal computer?

   Source: *New York Times Almanac.* 0.0984

9. **Female Americans Who Have Completed 4 Years of College** The percentage of female Americans 25 years old and older who have completed 4 years of college or more is 26.1. In a random sample of 200 American women who are at least 25, what is the probability that at most 50 have completed 4 years of college or more?

   Source: *New York Times Almanac.* 0.3936

10. Population of College Cities College students often make up a substantial portion of the population of college cities and towns. State College, Pennsylvania, ranks first with 71.1% of its population made up of college students. What is the probability that in a random sample of 150 people from State College, more than 50 are not college students? 0.0985

Source: www.infoplease.com

11. Elementary School Teachers Women comprise 80.3% of all elementary school teachers. In a random sample of 300 elementary teachers, what is the probability that less than three-fourths are women? 0.0087

Source: *New York Times Almanac.*

12. Telephone Answering Devices Seventy-eight percent of U.S. homes have a telephone answering device. In a random sample of 250 homes, what is the probability

that fewer than 50 do not have a telephone answering device? 0.2005

Source: *New York Times Almanac.*

13. Parking Lot Construction The mayor of a small town estimates that 35% of the residents in the town favor the construction of a municipal parking lot. If there are 350 people at a town meeting, find the probability that at least 100 favor construction of the parking lot. Based on your answer, is it likely that 100 or more people would favor the parking lot? 0.9951; yes (TI: 0.9950)

14. Residences of U.S. Citizens According to the U.S. Census, 67.5% of the U.S. population were born in their state of residence. In a random sample of 200 Americans, what is the probability that fewer than 125 were born in their state of residence? 0.0559

Source: www.census.gov

Extending the Concepts

15. Recall that for use of a normal distribution as an approximation to the binomial distribution, the conditions $np \geq 5$ and $nq \geq 5$ must be met. For each given probability, compute the minimum sample size needed for use of the normal approximation.

a. $p = 0.1$ $n \geq 50$
b. $p = 0.3$ $n \geq 17$
c. $p = 0.5$ $n \geq 10$
d. $p = 0.8$ $n > 25$
e. $p = 0.9$ $n \geq 50$

Summary

- A normal distribution can be used to describe a variety of variables, such as heights, weights, and temperatures. A normal distribution is bell-shaped, unimodal, symmetric, and continuous; its mean, median, and mode are equal. Since each normally distributed variable has its own distribution with mean μ and standard deviation σ, mathematicians use the standard normal distribution, which has a mean of 0 and a standard deviation of 1. Other approximately normally distributed variables can be transformed to the standard normal distribution with the formula $z = (X - \mu)/\sigma$. (6–1)

- A normal distribution can be used to solve a variety of problems in which the variables are approximately normally distributed. (6–2)

- A sampling distribution of sample means is a distribution using the means computed from all possible random samples of a specific size taken from a population. The difference between a sample measure and the corresponding population measure is due to what is called *sampling error*. The mean of the sample means will be the same as the population mean. The standard deviation of the sample mean will be equal to the population standard deviation divided by the square root of the sample size. The central limit theorem states that as the sample size increases without limit, the shape of the distribution of the sample means taken with replacement from a population will approach a normal distribution. (6–3)

- A normal distribution can be used to approximate other distributions, such as a binomial distribution. For a normal distribution to be used as an approximation, the conditions $np \geq 5$ and $nq \geq 5$ must be met. Also, a correction for continuity may be used for more accurate results. (6–4)

Important Terms

central limit theorem 333

correction for
continuity 342

negatively or left-skewed
distribution 301

normal distribution 303

positively or right-skewed
distribution 301

sampling distribution of
sample means 331

sampling error 331

standard error of the
mean 333

standard normal
distribution 304

symmetric
distribution 301

z value (score) 304

Important Formulas

Formula for the z score (or standard score):

$$z = \frac{X - \mu}{\sigma}$$

Formula for finding a specific data value:

$$X = z \cdot \sigma + \mu$$

Formula for the mean of the sample means:

$$\mu_{\bar{X}} = \mu$$

Formula for the standard error of the mean:

$$\sigma_{\bar{x}} = \frac{\sigma}{\sqrt{n}}$$

Formula for the z value for the central limit theorem:

$$z = \frac{\bar{X} - \mu}{\sigma/\sqrt{n}}$$

Formulas for the mean and standard deviation for the binomial distribution:

$$\mu = n \cdot p \qquad \sigma = \sqrt{n \cdot p \cdot q}$$

Review Exercises

1. Find the area under the standard normal distribution curve for each. (6–1)

   a. Between $z = 0$ and $z = 1.95$ 0.4744
   b. Between $z = 0$ and $z = 0.37$ 0.1443
   c. Between $z = 1.32$ and $z = 1.82$ 0.0590
   d. Between $z = -1.05$ and $z = 2.05$ 0.8329 (TI: 0.8330)
   e. Between $z = -0.03$ and $z = 0.53$ 0.2139
   f. Between $z = +1.10$ and $z = -1.80$ 0.8284
   g. To the right of $z = 1.99$ 0.0233
   h. To the right of $z = -1.36$ 0.9131
   i. To the left of $z = -2.09$ 0.0183
   j. To the left of $z = 1.68$ 0.9535

2. Using the standard normal distribution, find each probability. (6–1)

   a. $P(0 < z < 2.07)$ 0.4808
   b. $P(-1.83 < z < 0)$ 0.0336
   c. $P(-1.59 < z < +2.01)$ 0.9219
   d. $P(1.33 < z < 1.88)$ 0.0617
   e. $P(-2.56 < z < 0.37)$ 0.6391
   f. $P(z > 1.66)$ 0.0485
   g. $P(z < -2.03)$ 0.0212
   h. $P(z > -1.19)$ 0.8830
   i. $P(z < 1.93)$ 0.9732
   j. $P(z > -1.77)$ 0.9616

3. **Per Capita Spending on Health Care** The average per capita spending on health care in the United States is $5274. If the standard deviation is $600 and the distribution of health care spending is approximately normal, what is the probability that a randomly selected person spends more than $6000? Find the limits of the middle 50% of individual health care expenditures. (6–2)

   Source: *World Almanac.* 0.1131; $4872 and $5676 (TI: $4869.31 minimum, $5678.69 maximum)

4. **Salaries for Actuaries** The average salary for graduates entering the actuarial field is $40,000. If the salaries are normally distributed with a standard deviation of $5000, find the probability that

   a. An individual graduate will have a salary over $45,000. 0.1587
   b. A group of nine graduates will have a group average over $45,000. (6–2) 0.0013

   Source: www.BeAnActuary.org

5. **Commuter Train Passengers** On a certain run of a commuter train, the average number of passengers is 476 and the standard deviation is 22. Assume the variable is normally distributed. If the train makes the run, find the probability that the number of passengers will be

a. Between 476 and 500 passengers 0.3621 or 36.21%
b. Less than 450 passengers 0.1190 or 11.9%
c. More than 510 passengers (6–2) 0.0606 or 6.06%

6. **Monthly Spending for Paging and Messaging Services** The average individual monthly spending in the United States for paging and messaging services is $10.15. If the standard deviation is $2.45 and the amounts are normally distributed, what is the probability that a randomly selected user of these services pays more than $15.00 per month? Between $12.00 and $14.00 per month? (6–2) 0.0239; 0.1654

Source: *New York Times Almanac.*

7. **Cost of iPod Repair** The average cost of repairing an iPod is $120 with a standard deviation of $10.50. The costs are normally distributed. If 15% of the costs are considered excessive, find the cost in dollars that would be considered excessive. (6–2) $130.92

8. **Heights of Active Volcanoes** The heights (in feet above sea level) of a random sample of the world's active volcanoes are shown here. Check for normality. (6–2) Not normal

13,435	5,135	11,339	12,224	7,470
9,482	12,381	7,674	5,223	5,631
3,566	7,113	5,850	5,679	15,584
5,587	8,077	9,550	8,064	2,686
5,250	6,351	4,594	2,621	9,348
6,013	2,398	5,658	2,145	3,038

Source: *New York Times Almanac.*

9. **Private Four-Year College Enrollment** A random sample of enrollments in Pennsylvania's private four-year colleges is listed here. Check for normality. (6–2) Not normal

1350	1886	1743	1290	1767
2067	1118	3980	1773	4605
1445	3883	1486	980	1217
3587				

Source: *New York Times Almanac.*

10. **Average Precipitation** For the first 7 months of the year, the average precipitation in Toledo, Ohio, is 19.32 inches. If the average precipitation is normally distributed with a standard deviation of 2.44 inches, find these probabilities.
a. A randomly selected year will have precipitation greater than 18 inches for the first 7 months.
b. Five randomly selected years will have an average precipitation greater than 18 inches for the first 7 months. (6–3)

Source: *Toledo Blade.* a. 0.7054 (TI: 0.7057) b. 0.8869 (TI: 0.8868)

11. **Confectionary Products** Americans ate an average of 25.7 pounds of confectionary products each last year and spent an average of $61.50 per person doing so. If the standard deviation for consumption is 3.75 pounds and the standard deviation for the amount spent is $5.89, find the following:
a. The probability that the sample mean confectionary consumption for a random sample of 40 American consumers was greater than 27 pounds
b. The probability that for a random sample of 50, the sample mean for confectionary spending exceeded $60.00 (6–3)

Source: www.census.gov a. 0.0143 (TI: 0.0142) b. 0.9641

12. **Portable CD Player Lifetimes** A recent study of the life span of portable compact disc players found the average to be 3.7 years with a standard deviation of 0.6 year. If a random sample of 32 people who own CD players is selected, find the probability that the mean lifetime of the sample will be less than 3.4 years. If the mean is less than 3.4 years, would you consider that 3.7 years might be incorrect? (6–3) 0.0023; yes, since the probability is less than 1%.

13. **Retirement Income** Of the total population of American households, including older Americans and perhaps some not so old, 17.3% receive retirement income. In a random sample of 120 households, what is the probability that more than 20 households but less than 35 households receive a retirement income? (6–4)

Source: www.bls.gov 0.5234

14. **Slot Machines** The probability of winning on a slot machine is 5%. If a person plays the machine 500 times, find the probability of winning 30 times. Use the normal approximation to the binomial distribution. (6–4) 0.0496

15. **Multiple-Job Holders** According to the government 5.3% of those employed are multiple-job holders. In a random sample of 150 people who are employed, what is the probability that fewer than 10 hold multiple jobs? What is the probability that more than 50 are not multiple-job holders? (6–4)

Source: www.bls.gov 0.7123; 0.9999 (TI: 0.7139; 0.9999)

16. **Enrollment in Personal Finance Course** In a large university, 30% of the incoming first-year students elect to enroll in a personal finance course offered by the university. Find the probability that of 800 randomly selected incoming first-year students, at least 260 have elected to enroll in the course. (6–4) 0.0668

17. **U.S. Population** Of the total population of the United States, 20% live in the northeast. If 200 residents of the United States are selected at random, find the probability that at least 50 live in the northeast. (6–4) 0.0465

Source: *Statistical Abstract of the United States.*

What Is Normal?–Revisited

Many of the variables measured in medical tests—blood pressure, triglyceride level, etc.—are approximately normally distributed for the majority of the population in the United States. Thus, researchers can find the mean and standard deviation of these variables. Then, using these two measures along with the z values, they can find normal intervals for healthy individuals. For example, 95% of the systolic blood pressures of healthy individuals fall within 2 standard deviations of the mean. If an individual's pressure is outside the determined normal range (either above or below), the physician will look for a possible cause and prescribe treatment if necessary.

Chapter Quiz

Determine whether each statement is true or false. If the statement is false, explain why.

1. The total area under a normal distribution is infinite. False

2. The standard normal distribution is a continuous distribution. True

3. All variables that are approximately normally distributed can be transformed to standard normal variables. True

4. The z value corresponding to a number below the mean is always negative. True

5. The area under the standard normal distribution to the left of $z = 0$ is negative. False

6. The central limit theorem applies to means of samples selected from different populations. False

Select the best answer.

7. The mean of the standard normal distribution is
   a. 0   c. 100
   b. 1   d. Variable

8. Approximately what percentage of normally distributed data values will fall within 1 standard deviation above or below the mean?
   a. 68%   c. 99.7%
   b. 95%   d. Variable

9. Which is not a property of the standard normal distribution?
   a. It's symmetric about the mean.
   b. It's uniform.
   c. It's bell-shaped.
   d. It's unimodal.

10. When a distribution is positively skewed, the relationship of the mean, median, and mode from left to right will be
   a. Mean, median, mode   c. Median, mode, mean
   b. Mode, median, mean   d. Mean, mode, median

11. The standard deviation of all possible sample means equals
   a. The population standard deviation
   b. The population standard deviation divided by the population mean

c. The population standard deviation divided by the square root of the sample size
   d. The square root of the population standard deviation

Complete the following statements with the best answer.

12. When one is using the standard normal distribution, $P(z < 0) =$ _____. 0.5

13. The difference between a sample mean and a population mean is due to _____. Sampling error

14. The mean of the sample means equals the population mean

15. The standard deviation of all possible sample means is called the _____. standard error of the mean

16. The normal distribution can be used to approximate the binomial distribution when $n \cdot p$ and $n \cdot q$ are both greater than or equal to _____. 5

17. The correction factor for the central limit theorem should be used when the sample size is greater than _____ of the size of the population. 5%

18. Find the area under the standard normal distribution for each.
   a. Between 0 and 1.50 0.4332
   b. Between 0 and −1.25 0.3944
   c. Between 1.56 and 1.96 0.0344
   d. Between −1.20 and −2.25 0.1029
   e. Between −0.06 and 0.73 0.2912
   f. Between 1.10 and −1.80 0.8284
   g. To the right of $z = 1.75$ 0.0401
   h. To the right of $z = -1.28$ 0.8997
   i. To the left of $z = -2.12$ 0.017
   j. To the left of $z = 1.36$ 0.9131

19. Using the standard normal distribution, find each probability.
   a. $P(0 < z < 2.16)$ 0.4846
   b. $P(-1.87 < z < 0)$ 0.4693
   c. $P(-1.63 < z < 2.17)$ 0.9334
   d. $P(1.72 < z < 1.98)$ 0.0188
   e. $P(-2.17 < z < 0.71)$ 0.7461
   f. $P(z > 1.77)$ 0.0384
   g. $P(z < -2.37)$ 0.0089
   h. $P(z > -1.73)$ 0.9582

i. $P(z < 2.03)$ 0.9788
j. $P(z > -1.02)$ 0.8461

20. Amount of Rain in a City The average amount of rain per year in Greenville is 49 inches. The standard deviation is 8 inches. Find the probability that next year Greenville will receive the following amount of rainfall. Assume the variable is normally distributed.

a. At most 55 inches of rain 0.7734
b. At least 62 inches of rain 0.0516
c. Between 46 and 54 inches of rain 0.3837
d. How many inches of rain would you consider to be an extremely wet year?

21. Heights of People The average height of a certain age group of people is 53 inches. The standard deviation is 4 inches. If the variable is normally distributed, find the probability that a selected individual's height will be

a. Greater than 59 inches 0.0668
b. Less than 45 inches 0.0228
c. Between 50 and 55 inches 0.4649
d. Between 58 and 62 inches 0.0934

22. Lemonade Consumption The average number of gallons of lemonade consumed by the football team during a game is 20, with a standard deviation of 3 gallons. Assume the variable is normally distributed. When a game is played, find the probability of using

a. Between 20 and 25 gallons 0.4525
b. Less than 19 gallons 0.3707
c. More than 21 gallons 0.3707
d. Between 26 and 28 gallons 0.019

23. Years to Complete a Graduate Program The average number of years a person takes to complete a graduate degree program is 3. The standard deviation is 4 months. Assume the variable is normally distributed. If an individual enrolls in the program, find the probability that it will take

a. More than 4 years to complete the program 0.0013
b. Less than 3 years to complete the program 0.5
c. Between 3.8 and 4.5 years to complete the program 0.0081
d. Between 2.5 and 3.1 years to complete the program 0.5511

24. Passengers on a Bus On the daily run of an express bus, the average number of passengers is 48. The standard deviation is 3. Assume the variable is normally distributed. Find the probability that the bus will have

a. Between 36 and 40 passengers 0.0037
b. Fewer than 42 passengers 0.0228
c. More than 48 passengers 0.5
d. Between 43 and 47 passengers 0.3232

25. Thickness of Library Books The average thickness of books on a library shelf is 8.3 centimeters. The standard deviation is 0.6 centimeter. If 20% of the books are oversized, find the minimum thickness of the oversized books on the library shelf. Assume the variable is normally distributed. 8.804 centimeters

26. Membership in an Organization Membership in an elite organization requires a test score in the upper 30% range. If $\mu = 115$ and $\sigma = 12$, find the lowest acceptable score that would enable a candidate to apply for membership. Assume the variable is normally distributed. 121.24 is the lowest acceptable score.

27. Repair Cost for Microwave Ovens The average repair cost of a microwave oven is $55, with a standard deviation of $8. The costs are normally distributed. If 12 ovens are repaired, find the probability that the mean of the repair bills will be greater than $60. 0.015

28. Electric Bills The average electric bill in a residential area is $72 for the month of April. The standard deviation is $6. If the amounts of the electric bills are normally distributed, find the probability that the mean of the bill for 15 residents will be less than $75. 0.9738

29. Sleep Survey According to a recent survey, 38% of Americans get 6 hours or less of sleep each night. If 25 people are selected, find the probability that 14 or more people will get 6 hours or less of sleep each night. Does this number seem likely? 0.0495; no

Source: *Amazing Almanac.*

30. Unemployment If 8% of all people in a certain geographic region are unemployed, find the probability that in a sample of 200 people, there are fewer than 10 people who are unemployed. 0.0455 or 4.55%

31. Household Online Connection The percentage of U.S. households that have online connections is 44.9%. In a random sample of 420 households, what is the probability that fewer than 200 have online connections? 0.8577

Source: *New York Times Almanac.*

32. Computer Ownership Fifty-three percent of U.S. households have a personal computer. In a random sample of 250 households, what is the probability that fewer than 120 have a PC? 0.0495

Source: *New York Times Almanac.*

33. Calories in Fast-Food Sandwiches The number of calories contained in a selection of fast-food sandwiches is shown here. Check for normality. Not normal

390	405	580	300	320
540	225	720	470	560
535	660	530	290	440
390	675	530	1010	450
320	460	290	340	610
430	530			

Source: *The Doctor's Pocket Calorie, Fat, and Carbohydrate Counter.*

34. GMAT Scores The average GMAT scores for the top-30 ranked graduate schools of business are listed here. Check for normality. Approximately normal

718	703	703	703	700	690	695	705	690	688
676	681	689	686	691	669	674	652	680	670
651	651	637	662	641	645	645	642	660	636

Source: *U.S. News & World Report Best Graduate Schools.*

Critical Thinking Challenges

Sometimes a researcher must decide whether a variable is normally distributed. There are several ways to do this. One simple but very subjective method uses special graph paper, which is called *normal probability paper*. For the distribution of systolic blood pressure readings given in Chapter 3 of the textbook, the following method can be used:

1. Make a table, as shown.

Boundaries	Frequency	Cumulative frequency	Cumulative percent frequency
89.5–104.5	24		
104.5–119.5	62		
119.5–134.5	72		
134.5–149.5	26		
149.5–164.5	12		
164.5–179.5	4		
	200		

2. Find the cumulative frequencies for each class, and place the results in the third column.

3. Find the cumulative percents for each class by dividing each cumulative frequency by 200 (the total frequencies) and multiplying by 100%. (For the first class, it would be $24/200 \times 100\% = 12\%$.) Place these values in the last column.

4. Using the normal probability paper shown in Table 6–3, label the x axis with the class boundaries as shown and plot the percents.

5. If the points fall approximately in a straight line, it can be concluded that the distribution is normal. Do you feel that this distribution is approximately normal? Explain your answer.

6. To find an approximation of the mean or median, draw a horizontal line from the 50% point on the y axis over to the curve and then a vertical line down to the x axis. Compare this approximation of the mean with the computed mean.

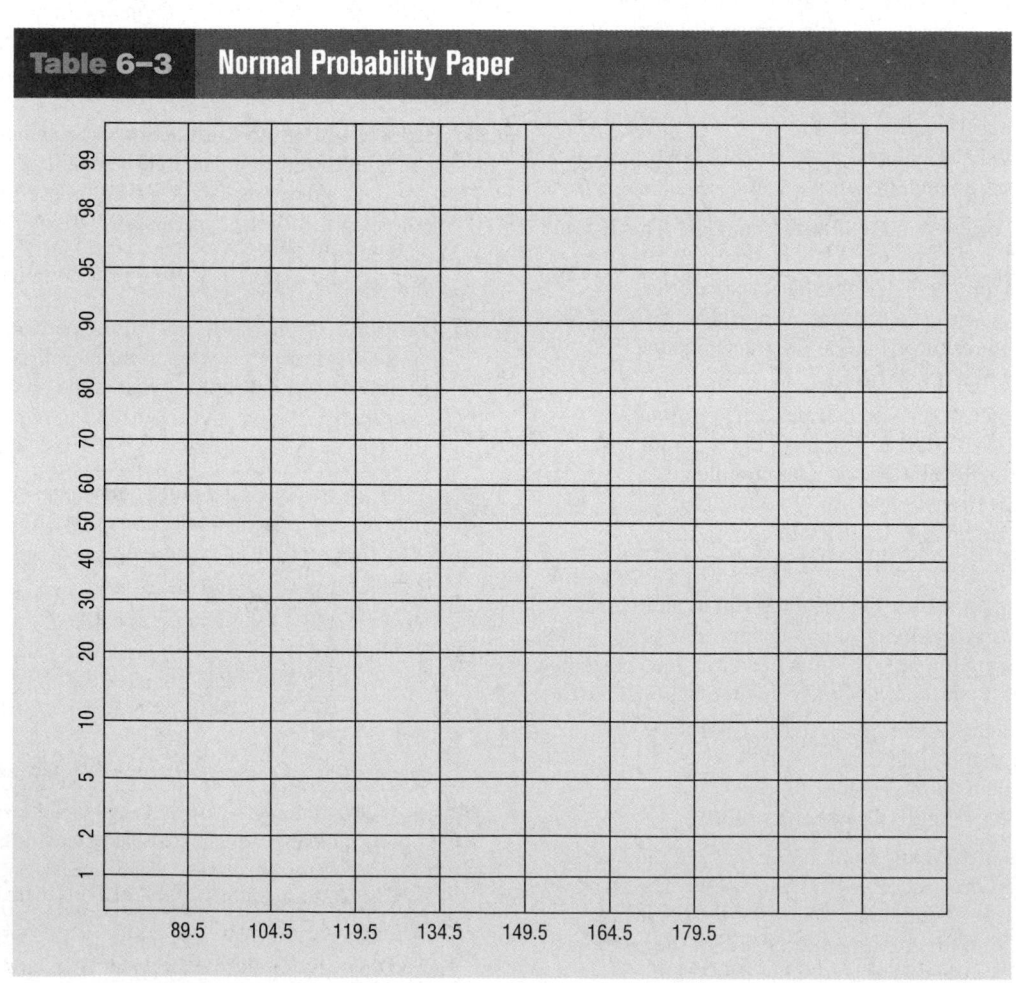

Table 6–3 **Normal Probability Paper**

7. To find an approximation of the standard deviation, locate the values on the *x* axis that correspond to the 16 and 84% values on the *y* axis. Subtract these two values and divide the result by 2. Compare this

approximate standard deviation to the computed standard deviation.

8. Explain why the method used in step 7 works.

Data Projects

1. Business and Finance Use the data collected in data project 1 of Chapter 2 regarding earnings per share to complete this problem. Use the mean and standard deviation computed in data project 1 of Chapter 3 as estimates for the population parameters. What value separates the top 5% of stocks from the others?

2. Sports and Leisure Find the mean and standard deviation for the batting average for a player in the most recently completed MBL season. What batting average would separate the top 5% of all hitters from the rest? What is the probability that a randomly selected player bats over 0.300? What is the probability that a team of 25 players has a mean that is above 0.275?

3. Technology Use the data collected in data project 3 of Chapter 2 regarding song lengths. If the sample estimates for mean and standard deviation are used as replacements for the population parameters for this data set, what song length separates the bottom 5% and top 5% from the other values?

4. Health and Wellness Use the data regarding heart rates collected in data project 4 of Chapter 2 for this problem. Use the sample mean and standard deviation as estimates of the population parameters. For the before-exercise data, what heart rate separates the top

10% from the other values? For the after-exercise data, what heart rate separates the bottom 10% from the other values? If a student was selected at random, what is the probability that her or his mean heart rate before exercise was less than 72? If 25 students were selected at random, what is the probability that their mean heart rate before exercise was less than 72?

5. Politics and Economics Use the data collected in data project 6 of Chapter 2 regarding Math SAT scores to complete this problem. What are the mean and standard deviation for statewide Math SAT scores? What SAT score separates the bottom 10% of states from the others? What is the probability that a randomly selected state has a statewide SAT score above 500?

6. Your Class Confirm the two formulas hold true for the central limit theorem for the population containing the elements {1, 5, 10}. First, compute the population mean and standard deviation for the data set. Next, create a list of all 9 of the possible two-element samples that can be created with replacement: {1, 1}, {1, 5}, etc. For each of the 9 compute the sample mean. Now find the mean of the sample means. Does it equal the population mean? Compute the standard deviation of the sample means. Does it equal the population standard deviation, divided by the square root of *n*?

Answers to Applying the Concepts

Section 6–1 Assessing Normality

1. Answers will vary. One possible frequency distribution is the following:

Branches	Frequency
0–9	1
10–19	14
20–29	17
30–39	7
40–49	3
50–59	2
60–69	2
70–79	1
80–89	2
90–99	1

2. Answers will vary according to the frequency distribution in question 1. This histogram matches the frequency distribution in question 1.

Histogram of Libraries

3. The histogram is unimodal and skewed to the right (positively skewed).

4. The distribution does not appear to be normal.

5. The mean number of branches is $\bar{x} = 31.4$, and the standard deviation is $s = 20.6$.

6. Of the data values, 80% fall within 1 standard deviation of the mean (between 10.8 and 52).

7. Of the data values, 92% fall within 2 standard deviations of the mean (between 0 and 72.6).

8. Of the data values, 98% fall within 3 standard deviations of the mean (between 0 and 93.2).

9. My values in questions 6–8 differ from the 68, 95, and 100% that we would see in a normal distribution.

10. These values support the conclusion that the distribution of the variable is not normal.

Section 6–2   Smart People

1. $z = \frac{130 - 100}{15} = 2$. The area to the right of 2 in the standard normal table is about 0.0228, so I would expect about $10,000(0.0228) = 228$ people in Visiala to qualify for Mensa.

2. It does seem reasonable to continue my quest to start a Mensa chapter in Visiala.

3. Answers will vary. One possible answer would be to randomly call telephone numbers (both home and cell phones) in Visiala, ask to speak to an adult, and ask whether the person would be interested in joining Mensa.

4. To have an Ultra-Mensa club, I would need to find the people in Visiala who have IQs that are at least 2.326 standard deviations above average. This means that I would need to recruit those with IQs that are at least 135:

$$2.326 = \frac{x - 100}{15} \Rightarrow x = 100 + 2.326(15) = 134.89$$

Section 6–3   Central Limit Theorem

1. It is very unlikely that we would ever get the same results for any of our random samples. While it is a remote possibility, it is highly unlikely.

2. A good estimate for the population mean would be to find the average of the students' sample means. Similarly, a good estimate for the population standard deviation would be to find the average of the students' sample standard deviations.

3. The distribution appears to be somewhat left-skewed (negatively skewed).

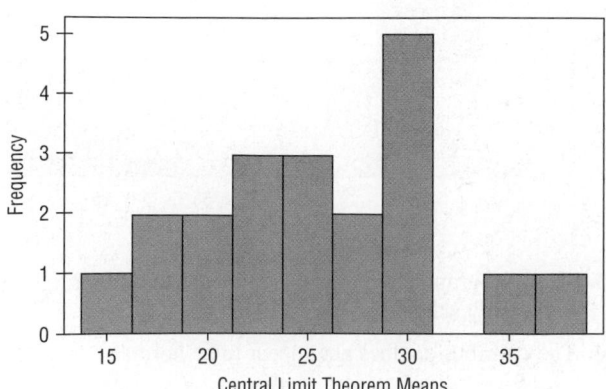

Histogram of Central Limit Theorem Means

4. The mean of the students' means is 25.4, and the standard deviation is 5.8.

5. The distribution of the means is not a sampling distribution, since it represents just 20 of all possible samples of size 30 from the population.

6. The sampling error for student 3 is $18 - 25.4 = -7.4$; the sampling error for student 7 is $26 - 25.4 = +0.6$; the sampling error for student 14 is $29 - 25.4 = +3.6$.

7. The standard deviation for the sample of the 20 means is greater than the standard deviations for each of the individual students. So it is not equal to the standard deviation divided by the square root of the sample size.

Section 6–4   How Safe Are You?

1. A reliability rating of 97% means that, on average, the device will not fail 97% of the time. We do not know how many times it will fail for any particular set of 100 climbs.

2. The probability of at least 1 failure in 100 climbs is $1 - (0.97)^{100} = 1 - 0.0476 = 0.9524$ (about 95%).

3. The complement of the event in question 2 is the event of "no failures in 100 climbs."

4. This can be considered a binomial experiment. We have two outcomes: success and failure. The probability of the equipment working (success) remains constant at 97%. We have 100 independent climbs. And we are counting the number of times the equipment works in these 100 climbs.

5. We could use the binomial probability formula, but it would be very messy computationally.

6. The probability of at least two failures *cannot* be estimated with the normal distribution (see below). So the probability is $1 - [(0.97)^{100} + 100(0.97)^{99}(0.03)] = 1 - 0.1946 = 0.8054$ (about 80.5%).

7. We *should not* use the normal approximation to the binomial since $nq < 10$.

8. If we had used the normal approximation, we would have needed a correction for continuity, since we would have been approximating a discrete distribution with a continuous distribution.

9. Since a second safety hook will be successful or fail independently of the first safety hook, the probability of failure drops from 3% to $(0.03)(0.03) = 0.0009$, or 0.09%.

CHAPTER 7

Confidence Intervals and Sample Size

Objectives

After completing this chapter, you should be able to

1 Find the confidence interval for the mean when σ is known.

2 Determine the minimum sample size for finding a confidence interval for the mean.

3 Find the confidence interval for the mean when σ is unknown.

4 Find the confidence interval for a proportion.

5 Determine the minimum sample size for finding a confidence interval for a proportion.

6 Find a confidence interval for a variance and a standard deviation.

Outline

Statistics Today

Would You Change the Channel?

A survey by the Roper Organization found that 45% of the people who were offended by a television program would change the channel, while 15% would turn off their television sets. The survey further stated that the margin of error is 3 percentage points, and 4000 adults were interviewed.

Several questions arise:

1. How do these estimates compare with the true population percentages?
2. What is meant by a margin of error of 3 percentage points?
3. Is the sample of 4000 large enough to represent the population of all adults who watch television in the United States?

See Statistics Today—Revisited at the end of the chapter for the answers.

After reading this chapter, you will be able to answer these questions, since this chapter explains how statisticians can use statistics to make estimates of parameters.

Source: The Associated Press.

Introduction

One aspect of inferential statistics is **estimation,** which is the process of estimating the value of a parameter from information obtained from a sample. For example, *The Book of Odds,* by Michael D. Shook and Robert L. Shook (New York: Penguin Putnam, Inc.), contains the following statements:

"One out of 4 Americans is currently dieting." (Calorie Control Council)
"Seventy-two percent of Americans have flown on commercial airlines." ("The Bristol Meyers Report: Medicine in the Next Century")
"The average kindergarten student has seen more than 5000 hours of television." (U.S. Department of Education)
"The average school nurse makes $32,786 a year." (National Association of School Nurses)
"The average amount of life insurance is $108,000 per household with life insurance." (American Council of Life Insurance)

Since the populations from which these values were obtained are large, these values are only *estimates* of the true parameters and are derived from data collected from samples.

The statistical procedures for estimating the population mean, proportion, variance, and standard deviation will be explained in this chapter.

An important question in estimation is that of sample size. How large should the sample be in order to make an accurate estimate? This question is not easy to answer since the size of the sample depends on several factors, such as the accuracy desired and the probability of making a correct estimate. The question of sample size will be explained in this chapter also.

Inferential statistical techniques have various **assumptions** that must be met before valid conclusions can be obtained. One common assumption is that the samples must be randomly selected. Chapter 1 explains how to obtain a random sample. The other common assumption is that either the sample size must be greater than or equal to 30 or the population must be normally or approximately normally distributed if the sample size is less than 30.

To check this assumption, you can use the methods explained in Chapter 6. Just for review, the methods are to check the histogram to see if it is approximately bell-shaped, check for outliers, and if possible, generate a normal quartile plot and see if the points fall close to a straight line. (*Note:* An area of statistics called nonparametric statistics does not require the variable to be normally distributed.)

Some statistical techniques are called **robust.** This means that the distribution of the variable can depart somewhat from normality, and valid conclusions can still be obtained.

7–1

Confidence Intervals for the Mean When σ Is Known

Objective 1

Find the confidence interval for the mean when σ is known.

Suppose a college president wishes to estimate the average age of students attending classes this semester. The president could select a random sample of 100 students and find the average age of these students, say, 22.3 years. From the sample mean, the president could infer that the average age of all the students is 22.3 years. This type of estimate is called a *point estimate.*

A **point estimate** is a specific numerical value estimate of a parameter. The best point estimate of the population mean μ is the sample mean $\overline{X}$.

You might ask why other measures of central tendency, such as the median and mode, are not used to estimate the population mean. The reason is that the means of samples vary less than other statistics (such as medians and modes) when many samples are selected from the same population. Therefore, the sample mean is the best estimate of the population mean.

Sample measures (i.e., statistics) are used to estimate population measures (i.e., parameters). These statistics are called **estimators.** As previously stated, the sample mean is a better estimator of the population mean than the sample median or sample mode.

A good estimator should satisfy the three properties described now.

Three Properties of a Good Estimator

1. The estimator should be an **unbiased estimator.** That is, the expected value or the mean of the estimates obtained from samples of a given size is equal to the parameter being estimated.
2. The estimator should be consistent. For a **consistent estimator,** as sample size increases, the value of the estimator approaches the value of the parameter estimated.
3. The estimator should be a **relatively efficient estimator.** That is, of all the statistics that can be used to estimate a parameter, the relatively efficient estimator has the smallest variance.

Confidence Intervals

As stated in Chapter 6, the sample mean will be, for the most part, somewhat different from the population mean due to sampling error. Therefore, you might ask a second question: How good is a point estimate? The answer is that there is no way of knowing how close a particular point estimate is to the population mean.

This answer places some doubt on the accuracy of point estimates. For this reason, statisticians prefer another type of estimate, called an *interval estimate*.

An **interval estimate** of a parameter is an interval or a range of values used to estimate the parameter. This estimate may or may not contain the value of the parameter being estimated.

In an interval estimate, the parameter is specified as being between two values. For example, an interval estimate for the average age of all students might be $21.9 < \mu < 22.7$, or 22.3 ± 0.4 years.

Either the interval contains the parameter or it does not. A degree of confidence (usually a percent) can be assigned before an interval estimate is made. For instance, you may wish to be 95% confident that the interval contains the true population mean. Another question then arises. Why 95%? Why not 99 or 99.5%?

If you desire to be more confident, such as 99 or 99.5% confident, then you must make the interval larger. For example, a 99% confidence interval for the mean age of college students might be $21.7 < \mu < 22.9$, or 22.3 ± 0.6. Hence, a tradeoff occurs. To be more confident that the interval contains the true population mean, you must make the interval wider.

The **confidence level** of an interval estimate of a parameter is the probability that the interval estimate will contain the parameter, assuming that a large number of samples are selected and that the estimation process on the same parameter is repeated.

A **confidence interval** is a specific interval estimate of a parameter determined by using data obtained from a sample and by using the specific confidence level of the estimate.

Intervals constructed in this way are called *confidence intervals*. Three common confidence intervals are used: the 90, the 95, and the 99% confidence intervals.

The algebraic derivation of the formula for determining a confidence interval for a mean will be shown later. A brief intuitive explanation will be given first.

The central limit theorem states that when the sample size is large, approximately 95% of the sample means taken from a population and same sample size will fall within ± 1.96 standard errors of the population mean, that is,

$$\mu \pm 1.96\left(\frac{\sigma}{\sqrt{n}}\right)$$

Now, if a specific sample mean is selected, say, $\overline{X}$, there is a 95% probability that the interval $\mu \pm 1.96(\sigma/\sqrt{n})$ contains $\overline{X}$. Likewise, there is a 95% probability that the interval specified by

$$\overline{X} \pm 1.96\left(\frac{\sigma}{\sqrt{n}}\right)$$

will contain μ, as will be shown later. Stated another way,

$$\overline{X} - 1.96\left(\frac{\sigma}{\sqrt{n}}\right) < \mu < \overline{X} + 1.96\left(\frac{\sigma}{\sqrt{n}}\right)$$

Historical Notes

Point and interval estimates were known as long ago as the late 1700s. However, it wasn't until 1937 that a mathematician, J. Neyman, formulated practical applications for them.

Hence, you can be 95% confident that the population mean is contained within that interval when the values of the variable are normally distributed in the population.

The value used for the 95% confidence interval, 1.96, is obtained from Table E in Appendix C. For a 99% confidence interval, the value 2.58 is used instead of 1.96 in the formula. This value is also obtained from Table E and is based on the standard normal distribution. Since other confidence intervals are used in statistics, the symbol $z_{\alpha/2}$ (read "zee sub alpha over two") is used in the general formula for confidence intervals. The Greek letter α (alpha) represents the total area in both tails of the standard normal distribution curve, and $\alpha/2$ represents the area in each one of the tails. More will be said after Examples 7–1 and 7–2 about finding other values for $z_{\alpha/2}$.

The relationship between α and the confidence level is that the stated confidence level is the percentage equivalent to the decimal value of $1 - \alpha$, and vice versa. When the 95% confidence interval is to be found, $\alpha = 0.05$, since $1 - 0.05 = 0.95$, or 95%. When $\alpha = 0.01$, then $1 - \alpha = 1 - 0.01 = 0.99$, and the 99% confidence interval is being calculated.

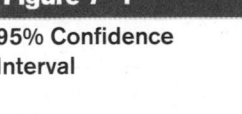

Interesting Fact

A postal worker who delivers mail walks on average 5.2 miles per day.

Formula for the Confidence Interval of the Mean for a Specific α When σ is Known

$$\bar{X} - z_{\alpha/2}\left(\frac{\sigma}{\sqrt{n}}\right) < \mu < \bar{X} + z_{\alpha/2}\left(\frac{\sigma}{\sqrt{n}}\right)$$

For a 90% confidence interval, $z_{\alpha/2} = 1.65$; for a 95% confidence interval, $z_{\alpha/2} = 1.96$; and for a 99% confidence interval, $z_{\alpha/2} = 2.58$.

The term $z_{\alpha/2}(\sigma/\sqrt{n})$ is called the *margin of error* (also called the *maximum error of the estimate*). For a specific value, say, $\alpha = 0.05$, 95% of the sample means will fall within this error value on either side of the population mean, as previously explained. See Figure 7–1.

Figure 7–1

95% Confidence Interval

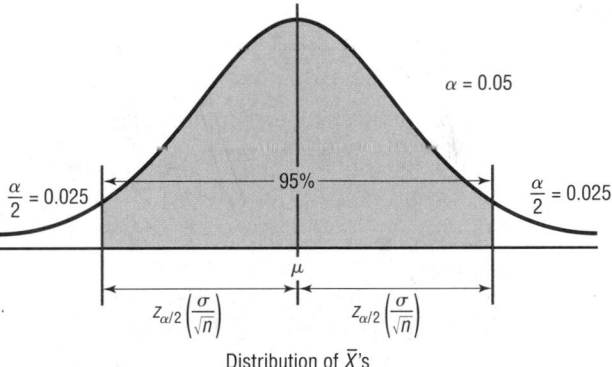

$\alpha = 0.05$

$\frac{\alpha}{2} = 0.025$ 95% $\frac{\alpha}{2} = 0.025$

μ

$z_{\alpha/2}\left(\frac{\sigma}{\sqrt{n}}\right)$ $z_{\alpha/2}\left(\frac{\sigma}{\sqrt{n}}\right)$

Distribution of $\bar{X}$'s

When $n \geq 30$, s can be substituted for σ, but a different distribution is used.

The **margin of error** also called the maximum error of the estimate is the maximum likely difference between the point estimate of a parameter and the actual value of the parameter.

A more detailed explanation of the margin of error follows Examples 7–1 and 7–2, which illustrate the computation of confidence intervals.

Assumptions for Finding a Confidence Interval for a Mean When σ Is Known

1. The sample is a random sample.
2. Either $n \geq 30$ or the population is normally distributed if $n < 30$.

Rounding Rule for a Confidence Interval for a Mean When you are computing a confidence interval for a population mean by using *raw data,* round off to one more decimal place than the number of decimal places in the original data. When you are computing a confidence interval for a population mean by using a sample mean and a standard deviation, round off to the same number of decimal places as given for the mean.

Example 7–1

Days It Takes to Sell an Aveo

A researcher wishes to estimate the number of days it takes an automobile dealer to sell a Chevrolet Aveo. A sample of 50 cars had a mean time on the dealer's lot of 54 days. Assume the population standard deviation to be 6.0 days. Find the best point estimate of the population mean and the 95% confidence interval of the population mean.

Source: Based on information obtained from Power Information Network.

Solution

The best point estimate of the mean is 54 days. For the 95% confidence interval use $z = 1.96$.

$$\overline{X} - z_{\alpha/2}\left(\frac{\sigma}{\sqrt{n}}\right) < \mu < \overline{X} + z_{\alpha/2}\left(\frac{\sigma}{\sqrt{n}}\right)$$

$$54 - 1.96\left(\frac{6.0}{\sqrt{50}}\right) < \mu < 54 + 1.96\left(\frac{6.0}{\sqrt{50}}\right)$$

$$54 - 1.7 < \mu < 54 + 1.7$$

$$52.3 < \mu < 55.7 \text{ or } 54 \pm 1.7$$

Hence one can say with 95% confidence that the interval between 52.3 and 55.7 days does contain the population mean, based on a sample of 50 automobiles.

Example 7–2

Waiting Times in Emergency Rooms

A survey of 30 emergency room patients found that the average waiting time for treatment was 174.3 minutes. Assuming that the population standard deviation is 46.5 minutes, find the best point estimate of the population mean and the 99% confidence of the population mean.

Source: Based on information from Press Ganey Associates Inc.

Solution

The best point estimate is 174.3 minutes. The 99% confidence is interval is

$$\overline{X} - z_{\alpha/2}\left(\frac{\sigma}{\sqrt{n}}\right) < \mu < \overline{X} + z_{\alpha/2}\left(\frac{\sigma}{\sqrt{n}}\right)$$

$$174.3 - 2.58\left(\frac{46.5}{\sqrt{30}}\right) < \mu < \overline{X} + 2.58\left(\frac{46.5}{\sqrt{30}}\right)$$

$$174.3 - 21.9 < \mu < 174.3 + 21.9$$

$$152.4 < \mu < 196.2$$

Hence, one can be 99% confident that the mean waiting time for emergency room treatment is between 152.4 and 196.2 minutes.

Another way of looking at a confidence interval is shown in Figure 7–2. According to the central limit theorem, approximately 95% of the sample means fall within 1.96 standard deviations of the population mean if the sample size is 30 or more, or if σ is known when n

Figure 7–2

95% Confidence
Interval for Sample
Means

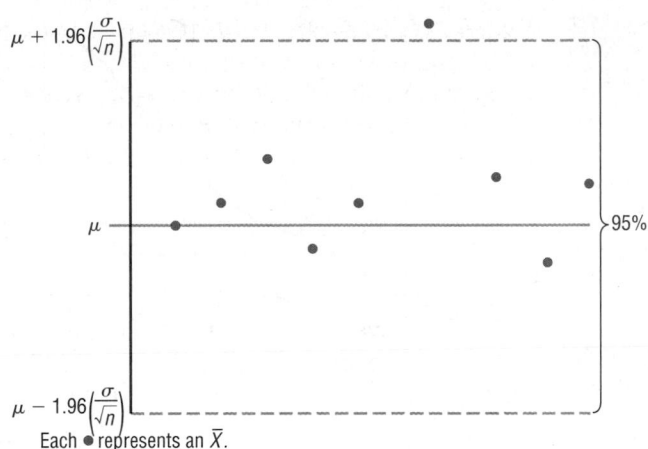

Each ● represents an $\overline{X}$.

Figure 7–3

95% Confidence
Intervals for Each
Sample Mean

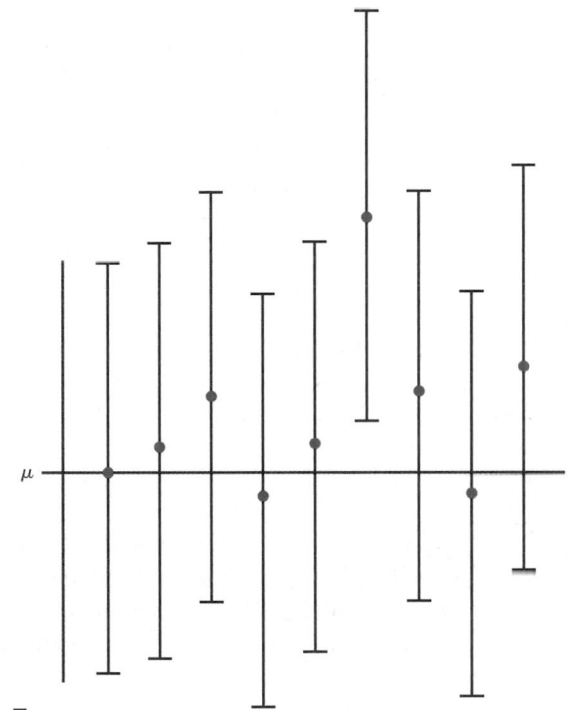

Each ⊥ represents an interval about a sample mean.

is less than 30 and the population is normally distributed. If it were possible to build a confidence interval about each sample mean, as was done in Examples 7–1 and 7–2 for μ, 95% of these intervals would contain the population mean, as shown in Figure 7–3. Hence, you can be 95% confident that an interval built around a specific sample mean would contain the population mean. If you desire to be 99% confident, you must enlarge the confidence intervals so that 99 out of every 100 intervals contain the population mean.

Since other confidence intervals (besides 90, 95, and 99%) are sometimes used in statistics, an explanation of how to find the values for $z_{\alpha/2}$ is necessary. As stated previously, the Greek letter α represents the total of the areas in both tails of the normal distribution. The value for α is found by subtracting the decimal equivalent for the desired confidence level from 1. For example, if you wanted to find the 98% confidence interval, you would change 98% to 0.98 and find $\alpha = 1 - 0.98$, or 0.02. Then $\alpha/2$ is obtained by dividing α by 2. So $\alpha/2$ is 0.02/2, or 0.01. Finally, $z_{0.01}$ is the z value that will give an area of 0.01 in the right tail of the standard normal distribution curve. See Figure 7–4.

Figure 7–4

Finding $\alpha/2$ for a 98% Confidence Interval

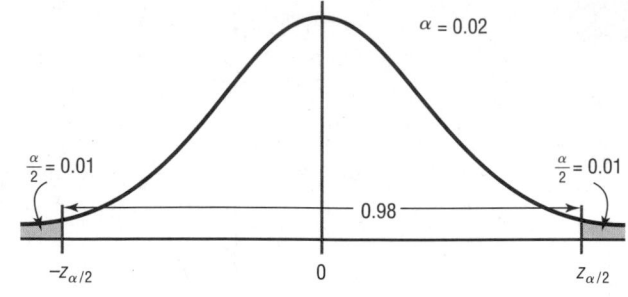

$\alpha = 0.02$

$\dfrac{\alpha}{2} = 0.01$

$\dfrac{\alpha}{2} = 0.01$

0.98

$-z_{\alpha/2}$ 0 $z_{\alpha/2}$

Figure 7–5

Finding $z_{\alpha/2}$ for a 98% Confidence Interval

Table E
The Standard Normal Distribution

z	.00	.01	.02	.03	...	.09
0.0						
0.1						
⋮						
2.3				0.9901		

Once $\alpha/2$ is determined, the corresponding $z_{\alpha/2}$ value can be found by using the procedure shown in Chapter 6, which is reviewed here. To get the $z_{\alpha/2}$ value for a 98% confidence interval, subtract 0.01 from 1.0000 to get 0.9900. Next, locate the area that is closest to 0.9900 (in this case, 0.9901) in Table E, and then find the corresponding z value. In this example, it is 2.33. See Figure 7–5.

For confidence intervals, only the positive z value is used in the formula.

When the original variable is normally distributed and σ is known, the standard normal distribution can be used to find confidence intervals regardless of the size of the sample. When $n \geq 30$, the distribution of means will be approximately normal even if the original distribution of the variable departs from normality.

When σ is unknown, s can be used as an estimate of σ, but a different distribution is used for the critical values. This method is explained in Section 7–2.

Example 7–3

Credit Union Assets

The following data represent a sample of the assets (in millions of dollars) of 30 credit unions in southwestern Pennsylvania. Find the 90% confidence interval of the mean.

12.23	16.56	4.39
2.89	1.24	2.17
13.19	9.16	1.42
73.25	1.91	14.64
11.59	6.69	1.06
8.74	3.17	18.13
7.92	4.78	16.85
40.22	2.42	21.58
5.01	1.47	12.24
2.27	12.77	2.76

Source: *Pittsburgh Post Gazette.*

Solution

Step 1 Find the mean and standard deviation for the data. Use the formulas shown in Chapter 3 or your calculator. The mean $\bar{X} = 11.091$. Assume the standard deviation of the population is 14.405.

Step 2 Find $\alpha/2$. Since the 90% confidence interval is to be used, $\alpha = 1 - 0.90 = 0.10$, and

$$\frac{\alpha}{2} = \frac{0.10}{2} = 0.05$$

Step 3 Find $z_{\alpha/2}$. Subtract 0.05 from 1.000 to get 0.9500. The corresponding z value obtained from Table E is 1.65. (*Note:* This value is found by using the z value for an area between 0.9495 and 0.9505. A more precise z value obtained mathematically is 1.645 and is sometimes used; however, 1.65 will be used in this textbook.)

Step 4 Substitute in the formula

$$\bar{X} - z_{\alpha/2}\left(\frac{\sigma}{\sqrt{n}}\right) < \mu < \bar{X} + z_{\alpha/2}\left(\frac{\sigma}{\sqrt{n}}\right)$$

$$11.091 - 1.65\left(\frac{14.405}{\sqrt{30}}\right) < \mu < 11.091 + 1.65\left(\frac{14.405}{\sqrt{30}}\right)$$

$$11.091 - 4.339 < \mu < 11.091 + 4.339$$

$$6.752 < \mu < 15.430$$

Hence, one can be 90% confident that the population mean of the assets of all credit unions is between $6.752 million and $15.430 million, based on a sample of 30 credit unions.

Comment to Computer and Statistical Calculator Users

This chapter and subsequent chapters include examples using raw data. If you are using computer or calculator programs to find the solutions, the answers you get may vary somewhat from the ones given in the textbook. This is so because computers and calculators do not round the answers in the intermediate steps and can use 12 or more decimal places for computation. Also, they use more-exact critical values than those given in the tables in the back of this book. These small discrepancies are part and parcel of statistics.

Sample Size

Objective 2

Determine the minimum sample size for finding a confidence interval for the mean.

Sample size determination is closely related to statistical estimation. Quite often you ask, How large a sample is necessary to make an accurate estimate? The answer is not simple, since it depends on three things: the margin of error, the population standard deviation, and the degree of confidence. For example, how close to the true mean do you want to be (2 units, 5 units, etc.), and how confident do you wish to be (90, 95, 99%, etc.)? For the purpose of this chapter, it will be assumed that the population standard deviation of the variable is known or has been estimated from a previous study.

The formula for sample size is derived from the margin of error formula

$$E = z_{\alpha/2}\left(\frac{\sigma}{\sqrt{n}}\right)$$

and this formula is solved for n as follows:

$$E\sqrt{n} = z_{\alpha/2}(\sigma)$$

$$\sqrt{n} = \frac{z_{\alpha/2} \cdot \sigma}{E}$$

Hence, $\qquad n = \left(\frac{z_{\alpha/2} \cdot \sigma}{E}\right)^2$

Formula for the Minimum Sample Size Needed for an Interval Estimate of the Population Mean

$$n = \left(\frac{z_{\alpha/2} \cdot \sigma}{E}\right)^2$$

where E is the margin of error. If necessary, round the answer up to obtain a whole number. That is, if there is any fraction or decimal portion in the answer, use the next whole number for sample size n.

Example 7–4

Depth of a River

A scientist wishes to estimate the average depth of a river. He wants to be 99% confident that the estimate is accurate within 2 feet. From a previous study, the standard deviation of the depths measured was 4.33 feet.

Solution

Since $\alpha = 0.01$ (or $1 - 0.99$), $z_{\alpha/2} = 2.58$ and $E = 2$. Substituting in the formula,

$$n = \left(\frac{z_{\alpha/2} \cdot \sigma}{E}\right)^2 = \left[\frac{(2.58)(4.33)}{2}\right]^2 = 31.2$$

Round the value 31.2 up to 32. Therefore, to be 99% confident that the estimate is within 2 feet of the true mean depth, the scientist needs at least a sample of 32 measurements.

In most cases in statistics, we round off. However, when determining sample size, we always round up to the next whole number.

Notice that when you are finding the sample size, the size of the population is irrelevant when the population is large or infinite or when sampling is done with replacement. In other cases, an adjustment is made in the formula for computing sample size. This adjustment is beyond the scope of this book.

The formula for determining sample size requires the use of the population standard deviation. What happens when σ is unknown? In this case, an attempt is made to estimate σ. One such way is to use the standard deviation s obtained from a sample taken previously as an estimate for σ. The standard deviation can also be estimated by dividing the range by 4.

Sometimes, interval estimates rather than point estimates are reported. For instance, you may read a statement: "On the basis of a sample of 200 families, the survey estimates that an American family of two spends an average of $84 per week for groceries. One

Interesting Fact

It has been estimated that the amount of pizza consumed every day in the United States would cover a farm consisting of 75 acres.

can be 95% confident that this estimate is accurate within \$3 of the true mean." This statement means that the 95% confidence interval of the true mean is

$$\$84 - \$3 < \mu < \$84 + \$3$$
$$\$81 < \mu < \$87$$

The algebraic derivation of the formula for a confidence interval is shown next. As explained in Chapter 6, the sampling distribution of the mean is approximately normal when large samples ($n \geq 30$) are taken from a population. Also,

$$z = \frac{\overline{X} - \mu}{\sigma/\sqrt{n}}$$

Furthermore, there is a probability of $1 - \alpha$ that a z will have a value between $-z_{\alpha/2}$ and $+z_{\alpha/2}$. Hence,

$$-z_{\alpha/2} < \frac{\overline{X} - \mu}{\sigma/\sqrt{n}} < z_{\alpha/2}$$

By using algebra, the formula can be rewritten as

$$-z_{\alpha/2} \cdot \frac{\sigma}{\sqrt{n}} < \overline{X} - \mu < z_{\alpha/2} \cdot \frac{\sigma}{\sqrt{n}}$$

Subtracting $\overline{X}$ from both sides and from the middle gives

$$-\overline{X} - z_{\alpha/2} \cdot \frac{\sigma}{\sqrt{n}} < -\mu < -\overline{X} + z_{\alpha/2} \cdot \frac{\sigma}{\sqrt{n}}$$

Multiplying by -1 gives

$$\overline{X} + z_{\alpha/2} \cdot \frac{\sigma}{\sqrt{n}} > \mu > \overline{X} - z_{\alpha/2} \cdot \frac{\sigma}{\sqrt{n}}$$

Reversing the inequality yields the formula for the confidence interval:

$$\overline{X} - z_{\alpha/2} \cdot \frac{\sigma}{\sqrt{n}} < \mu < \overline{X} + z_{\alpha/2} \cdot \frac{\sigma}{\sqrt{n}}$$

Applying the Concepts **7-1**

Making Decisions with Confidence Intervals

Assume you work for Kimberly Clark Corporation, the makers of Kleenex. The job you are presently working on requires you to decide how many Kleenexes are to be put in the new automobile glove compartment boxes. Complete the following.

1. How will you decide on a reasonable number of Kleenexes to put in the boxes?

2. When do people usually need Kleenexes?

3. What type of data collection technique would you use?

4. Assume you found out that from your sample of 85 people, on average about 57 Kleenexes are used throughout the duration of a cold, with a population standard deviation of 15. Use a confidence interval to help you decide how many Kleenexes will go in the boxes.

5. Explain how you decided how many Kleenexes will go in the boxes.

See page 398 for the answers.

Exercises 7–1

1. What is the difference between a point estimate and an interval estimate of a parameter? Which is better? Why?

2. What information is necessary to calculate a confidence interval?

3. What is the margin of error?

4. What is meant by the 95% confidence interval of the mean?

5. What are three properties of a good estimator? A good estimator should be unbiased, consistent, and relatively efficient.

6. What statistic best estimates μ? $\overline{X}$

7. What is necessary to determine the sample size?

8. In determining the sample size for a confidence interval, is the size of the population relevant? No, as long as it is much larger than the sample size needed.

9. Find each.

   a. $z_{\alpha/2}$ for the 99% confidence interval 2.58
   b. $z_{\alpha/2}$ for the 98% confidence interval 2.33
   c. $z_{\alpha/2}$ for the 95% confidence interval 1.96
   d. $z_{\alpha/2}$ for the 90% confidence interval 1.65
   e. $z_{\alpha/2}$ for the 94% confidence interval 1.88

10. **Number of Faculty** The numbers of faculty at 32 randomly selected state-controlled colleges and universities with enrollment under 12,000 students are shown below. Use these data to estimate the mean number of faculty at all state-controlled colleges and universities with enrollment under 12,000 with 92% confidence. Assume $\sigma = 165.1$.

211	384	396	211	224	337	395	121	356
621	367	408	515	280	289	180	431	176
318	836	203	374	224	121	412	134	539
471	638	425	159	324				

Source: *World Almanac.* $295.15 < \mu < 397.35$

11. **Playing Video Games** In a recent study of 35 ninth-grade students, the mean number of hours per week that they played video games was 16.6. The standard deviation of the population was 2.8.

   a. Find the best point estimate of the mean. 16.6 hours
   b. Find the 95% confidence interval of the mean of the time playing video games. $15.7 < \mu < 17.5$
   c. Find the 99% confidence interval of the mean time playing video games. $15.4 < \mu < 17.8$
   d. Which is larger? Explain why.

12. **Freshmen's GPA** First-semester GPAs for a random selection of freshmen at a large university are shown. Estimate the true mean GPA of the freshman class with 99% confidence. Assume $\sigma = 0.62$. $2.55 < \mu < 3.09$

1.9	3.2	2.0	2.9	2.7	3.3
2.8	3.0	3.8	2.7	2.0	1.9
2.5	2.7	2.8	3.2	3.0	3.8
3.1	2.7	3.5	3.8	3.9	2.7
2.0	2.8	1.9	4.0	2.2	2.8
2.1	2.4	3.0	3.4	2.9	2.1

13. **Workers' Distractions** A recent study showed that the modern working person experiences an average of 2.1 hours per day of distractions (phone calls, e-mails, impromptu visits, etc.). A random sample of 50 workers for a large corporation found that these workers were distracted an average of 1.8 hours per day and the population standard deviation was 20 minutes. Estimate the true mean population distraction time with 90% confidence, and compare your answer to the results of the study. $1.72 < \mu < 1.88$; lower

Source: *Time Almanac.*

14. **Number of Jobs** A sociologist found that in a sample of 50 retired men, the average number of jobs they had during their lifetimes was 7.2. The population standard deviation is 2.1.

   a. Find the best point estimate of the mean. 7.2 jobs
   b. Find the 95% confidence interval of the mean number of jobs. $6.6 < \mu < 7.8$
   c. Find the 99% confidence interval of the mean number of jobs. $6.4 < \mu < 8.0$
   d. Which is smaller? Explain why.

15. **Actuary Exams** A survey of 35 individuals who passed the seven exams and obtained the rank of Fellow in the actuarial field finds the average salary to be $150,000. If the standard deviation for the population is $15,000, construct a 95% confidence interval for all Fellows.

Source: www.BeAnActuary.org $145,030 < \mu < 154,970$

16. **Number of Farms** A random sample of the number of farms (in thousands) in various states follows. Estimate the mean number of farms per state with 90% confidence. Assume $\sigma = 31$.

47	95	54	33	64	4	8	57	9	80
8	90	3	49	4	44	79	80	48	16
68	7	15	21	52	6	78	109	40	50
29									

Source: *New York Times Almanac.* $34.3 < \mu < 52.7$

17. **Television Viewing** A study of 415 kindergarten students showed that they have seen on average 5000 hours of television. If the sample standard deviation of the population is 900, find the 95% confidence level of the mean for all students. If a parent claimed that his children watched 4000 hours, would the claim be believable?

Source: *U.S. Department of Education.*

18. **Day Care Tuition** A random sample of 50 four-year-olds attending day care centers provided a yearly tuition average of $3987 and the population standard deviation of $630. Find the 90% confidence interval of the true mean. If a day care center were starting up and wanted to keep tuition low, what would be a reasonable amount to charge? $3840 < μ < $4134; $3800

19. **Hospital Noise Levels** Noise levels at various area urban hospitals were measured in decibels. The mean of the noise levels in 84 corridors was 61.2 decibels, and the standard deviation of the population was 7.9. Find the 95% confidence interval of the true mean.

Source: M. Bayo, A. Garcia, and A. Garcia, "Noise Levels in an Urban Hospital and Workers' Subjective Responses," *Archives of Environmental Health* 50, no. 3, p. 249 (May–June 1995). Reprinted with permission of the Helen Dwight Reid Educational Foundation. Published by Heldref Publications, 1319 Eighteenth St. N.W., Washington, D.C. 20036-1802. Copyright © 1995. $59.5 < \mu < 62.9$

20. **Length of Growing Seasons** The growing seasons for a random sample of 35 U.S. cities were recorded, yielding a sample mean of 190.7 days and the population standard deviation of 54.2 days. Estimate the true mean population of the growing season with 95% confidence.

Source: *The Old Farmer's Almanac.* $172.74 < \mu < 208.66$

21. **Convenience Store Shoppers** A random sample of shoppers at a convenience store are selected to see how much they spent on that visit. The standard deviation of the population is $6.43. How large a sample must be selected if the researcher wants to be 99% confident of finding whether the true mean differs from the sample mean by $1.50? 123 subjects

22. In the hospital study cited in Exercise 19, the mean noise level in the 171 ward areas was 58.0 decibels, and

the population standard deviation is 4.8. Find the 90% confidence interval of the true mean. $57.4 < \mu < 58.6$

Source: M. Bayo, A. Garcia, and A. Garcia, "Noise Levels in an Urban Hospital and Workers' Subjective Responses," *Archives of Environmental Health* 50, no. 3, p. 249 (May–June 1995). Reprinted with permission of the Helen Dwight Reid Educational Foundation. Published by Heldref Publications, 1319 Eighteenth St. N.W., Washington, D.C. 20036-1802. Copyright © 1995.

23. **Birth Weights of Infants** A health care professional wishes to estimate the birth weights of infants. How large a sample must be obtained if she desires to be 90% confident that the true mean is within 2 ounces of the sample mean? Assume $\sigma = 8$ ounces. 44 subjects

24. **Cost of Pizzas** A pizza shop owner wishes to find the 95% confidence interval of the true mean cost of a large plain pizza. How large should the sample be if she wishes to be accurate to within $0.15? A previous study showed that the standard deviation of the price was $0.26. 12

25. **National Accounting Examination** If the variance of a national accounting examination is 900, how large a sample is needed to estimate the true mean score within 5 points with 99% confidence? 240 exams

26. **Commuting Times in New York** The 90% confidence interval for the mean one-way commuting time in New York City is $37.8 < \mu < 38.8$ minutes. Construct a 95% confidence interval based on the same data. Which interval provides more information?

Source: www.census.gov $37.71 < \mu < 38.89$; the 90% interval

---

MINITAB
Step by Step

Finding a z Confidence Interval for the Mean

For Example 7–3, find the 90% confidence interval estimate for the mean amount of assets for credit unions in southwestern Pennsylvania.

1. Maximize the worksheet, then enter the data into C1 of a MINITAB worksheet. If sigma is known, skip to step 3.

2. Calculate the standard deviation for the sample. It will be used as an estimate for sigma.

   a) Select **Calc>Column statistics.**

   b) Click the option for Standard deviation.

   c) Enter **C1 Assets** for the Input variable and **s** for Store in:.

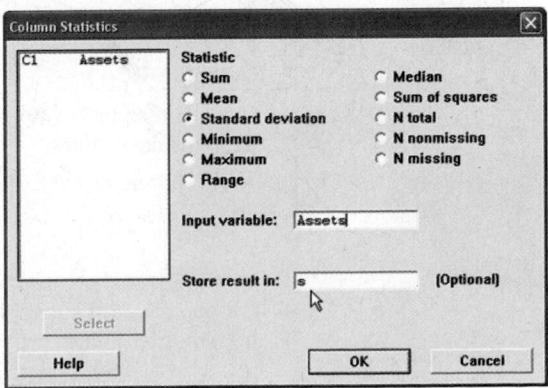

3. Select **Stat>Basic Statistics>1-Sample Z.**

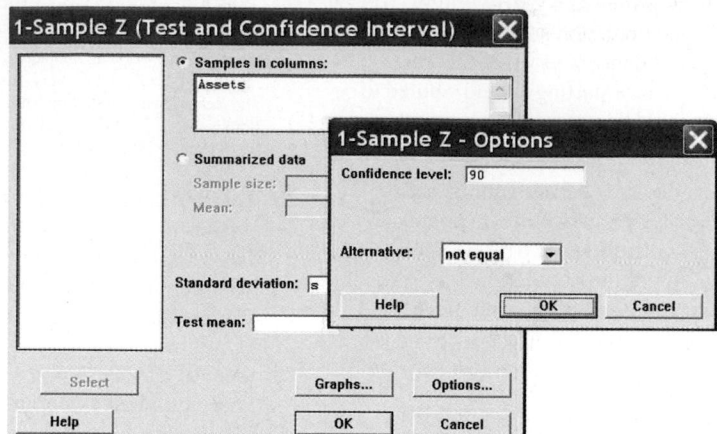

4. Select C1 Assets for the Samples in Columns.

5. Click in the box for Standard Deviation and enter **s.** Leave the box for Test mean empty.

6. Click the [Options] button. In the dialog box make sure the Confidence Level is 90 and the Alternative is not equal.

7. Optional: Click [Graphs], then select Boxplot of data. The boxplot of these data would clearly show the outliers!

8. Click [OK] twice. The results will be displayed in the session window.

One-Sample Z: Assets

```
The assumed sigma = 14.4054
Variable      N       Mean      StDev     SE Mean        90% CI
Assets       30     11.0907    14.4054    2.6301    (6.7646, 15.4167)
```

TI-83 Plus or TI-84 Plus
Step by Step

Finding a z Confidence Interval for the Mean (Data)

1. Enter the data into L_1.

2. Press **STAT** and move the cursor to TESTS.

3. Press **7** for ZInterval.

4. Move the cursor to Data and press **ENTER.**

5. Type in the appropriate values.

6. Move the cursor to Calculate and press **ENTER.**

Example TI7–1

This is Example 7–3 from the text. Find the 90% confidence interval for the population mean, given the data values.

12.23	2.89	13.19	73.25	11.59	8.74	7.92	40.22	5.01	2.27
16.56	1.24	9.16	1.91	6.69	3.17	4.78	2.42	1.47	12.77
4.39	2.17	1.42	14.64	1.06	18.13	16.85	21.58	12.24	2.76

The population standard deviation σ is unknown. Since the sample size is $n = 30$, you can use the sample standard deviation s as an approximation for σ. After the data values are entered in L_1 (step 1 above), press **STAT,** move the cursor to CALC, press **1** for 1-Var Stats, then press **ENTER.** The sample standard deviation of 14.40544747 will be one of the statistics listed. Then continue with step 2. At step 5 on the line for σ, press **VARS** for variables, press **5** for Statistics, press **3** for S_x.

The 90% confidence interval is $6.765 < \mu < 15.417$. The difference between these limits and the ones in Example 7–3 is due to rounding.

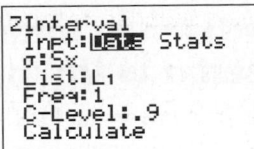
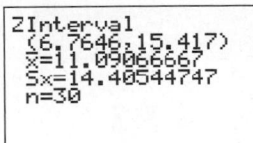

Finding a *z* Confidence Interval for the Mean (Statistics)

1. Press **STAT** and move the cursor to TESTS.
2. Press **7** for ZInterval.
3. Move the cursor to Stats and press **ENTER.**
4. Type in the appropriate values.
5. Move the cursor to Calculate and press **ENTER.**

Example TI7–2

Find the 95% confidence interval for the population mean, given $\sigma = 2$, $\overline{X} = 23.2$, and $n = 50$.

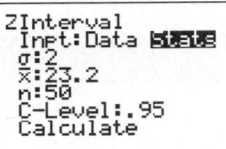
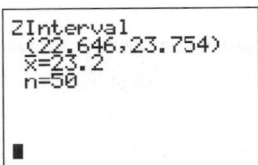

The 95% confidence interval is $22.6 < \mu < 23.8$.

Excel
Step by Step

Finding a *z* Confidence Interval for the Mean

Excel has a procedure to compute the margin of error. But it does not compute confidence intervals. However, you may determine confidence intervals for the mean by using the MegaStat Add-in available on your CD. If you have not installed this add-in, do so, following the instructions from the Chapter 1 Excel Step by Step.

Example XL7–1

 Find the 95% confidence interval for the mean if $\sigma = 11$, using this sample:

43	52	18	20	25	45	43	21	42	32	24	32	19	25	26
44	42	41	41	53	22	25	23	21	27	33	36	47	19	20

1. Enter the data into an Excel worksheet.
2. From the toolbar, select Add-Ins, **MegaStat>Confidence Intervals/Sample Size.**
   Note: You may need to open **MegaStat** from the **MegaStat.xls** file on your computer's hard drive.
3. Enter the mean of the data, **32.03.**
4. Select *z* for the standard normal distribution.
5. Enter **11** for the standard deviation and **30** for *n*, the sample size.
6. Either type in or scroll to 95% for the Confidence Level, then click [OK].

The result of the procedure is shown next.

Confidence interval—mean

```
   95% Confidence level
32.03 Mean
   11 Standard deviation
   30 n
1.960 z
3.936 Half-width
35.966 Upper confidence limit
28.094 Lower confidence limit
```

Confidence Intervals for the Mean When σ Is Unknown

Objective **3**

Find the confidence interval for the mean when σ is unknown.

When σ is known and the sample size is 30 or more, or the population is normally distributed if the sample size is less than 30, the confidence interval for the mean can be found by using the z distribution as shown in Section 7–1. However, most of the time, the value of σ is not known, so it must be estimated by using s, namely, the standard deviation of the sample. When s is used, especially when the sample size is small, critical values greater than the values for $z_{\alpha/2}$ are used in confidence intervals in order to keep the interval at a given level, such as the 95%. These values are taken from the *Student t distribution,* most often called the t **distribution.**

To use this method, the samples must be simple random samples, and the population from which the samples were taken must be normally or approximately normally distributed, or the sample size must be 30 or more.

Some important characteristics of the t distribution are described now.

Characteristics of the t Distribution

The t distribution shares some characteristics of the normal distribution and differs from it in others. The t distribution is similar to the standard normal distribution in these ways:

1. It is bell-shaped.
2. It is symmetric about the mean.
3. The mean, median, and mode are equal to 0 and are located at the center of the distribution.
4. The curve never touches the x axis.

The t distribution differs from the standard normal distribution in the following ways:

1. The variance is greater than 1.
2. The t distribution is actually a family of curves based on the concept of *degrees of freedom,* which is related to sample size.
3. As the sample size increases, the t distribution approaches the standard normal distribution. See Figure 7–6.

Many statistical distributions use the concept of degrees of freedom, and the formulas for finding the degrees of freedom vary for different statistical tests. The **degrees of freedom** are the number of values that are free to vary after a sample statistic has been computed, and they tell the researcher which specific curve to use when a distribution consists of a family of curves.

For example, if the mean of 5 values is 10, then 4 of the 5 values are free to vary. But once 4 values are selected, the fifth value must be a specific number to get a sum of 50, since $50 \div 5 = 10$. Hence, the degrees of freedom are $5 - 1 = 4$, and this value tells the researcher which t curve to use.

Figure 7–6

The t Family of Curves

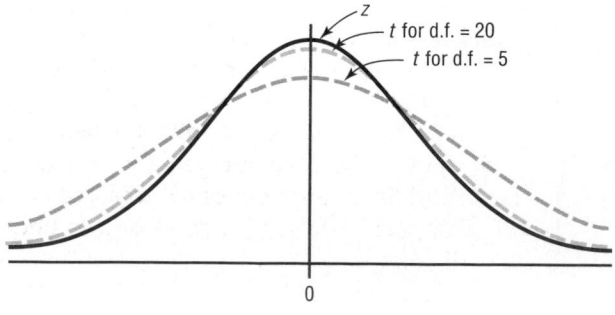

One can be 99% confident that the population mean number of home fires started by candles each year is between 4785.2 and 9297.6, based on a sample of home fires occurring over a period of 7 years.

Students sometimes have difficulty deciding whether to use $z_{\alpha/2}$ or $t_{\alpha/2}$ values when finding confidence intervals for the mean. As stated previously, when σ is known, $z_{\alpha/2}$ values can be used *no matter what the sample size is,* as long as the variable is normally distributed or $n \geq 30$. When σ is unknown and $n \geq 30$, then s can be used in the formula and $t_{\alpha/2}$ values can be used. Finally, when σ is unknown and $n < 30$, s is used in the formula and $t_{\alpha/2}$ values are used, as long as the variable is approximately normally distributed. These rules are summarized in Figure 7–8.

Figure 7–8

When to Use the z or t Distribution

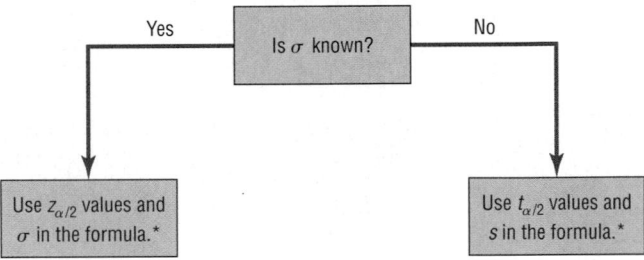

*If $n < 30$, the variable must be normally distributed.

Applying the Concepts **7–2**

Sport Drink Decision

Assume you get a new job as a coach for a sports team, and one of your first decisions is to choose the sports drink that the team will use during practices and games. You obtain a *Sports Report* magazine so you can use your statistical background to help you make the best decision. The following table lists the most popular sports drinks and some important information about each of them. Answer the following questions about the table.

Drink	Calories	Sodium	Potassium	Cost
Gatorade	60	110	25	$1.29
Powerade	68	77	32	1.19
All Sport	75	55	55	0.89
10-K	63	55	35	0.79
Exceed	69	50	44	1.59
1st Ade	58	58	25	1.09
Hydra Fuel	85	23	50	1.89

1. Would this be considered a small sample?

2. Compute the mean cost per container, and create a 90% confidence interval about that mean. Do all the costs per container fall inside the confidence interval? If not, which ones do not?

3. Are there any you would consider outliers?

4. How many degrees of freedom are there?

5. If cost is a major factor influencing your decision, would you consider cost per container or cost per serving?

6. List which drink you would recommend and why.

See page 398 for the answers.

Exercises 7–2

1. What are the properties of the *t* distribution?

2. What is meant by *degrees of freedom?*

3. When should the *t* distribution be used to find a confidence interval for the mean? The *t* distribution should be used when σ is unknown.

4. (**ans**) Find the values for each.

   a. $t_{\alpha/2}$ and $n = 18$ for the 99% confidence interval for the mean 2.898

   b. $t_{\alpha/2}$ and $n = 23$ for the 95% confidence interval for the mean 2.074

   c. $t_{\alpha/2}$ and $n = 15$ for the 98% confidence interval for the mean 2.624

   d. $t_{\alpha/2}$ and $n = 10$ for the 90% confidence interval for the mean 1.833

   e. $t_{\alpha/2}$ and $n = 20$ for the 95% confidence interval for the mean 2.093

For Exercises 5 through 20, assume that all variables are approximately normally distributed.

5. **Visits to Networking Sites** A sample of 10 networking sites for a specific month has a mean of 26.1 and a standard deviation of 4.2. Find the 99% confidence interval of the true mean. $21.8 < \mu < 30.4$

6. **Digital Camera Prices** The prices (in dollars) for a particular model of digital camera with 6.0 megapixels and an optical 3X zoom lens are shown below for 10 online retailers. Estimate the true mean price for this particular model with 95% confidence. $205.2 < \mu < 230.2$. Assume the variable is normally distributed.

   225 240 215 206 211 210 193 250 225 202

7. **Women Representatives in State Legislature** A state representative wishes to estimate the mean number of women representatives per state legislature. A random sample of 17 states is selected, and the number of women representatives is shown. Based on the sample, what is the point estimate of the mean? Find the 90% confidence interval of the mean population. (*Note:* The population mean is actually 31.72, or about 32.) Compare this value to the point estimate and the confidence interval. There is something unusual about the data. Describe it and state how it would affect the confidence interval.

5	33	35	37	24
31	16	45	19	13
18	29	15	39	18
58	132			

8. **State Gasoline Taxes** A random sample of state gasoline taxes (in cents) is shown here for 12 states. Use the data to estimate the true population mean gasoline tax with 90% confidence. Does your interval contain the national average of 44.7 cents? $38.70 < \mu < 48.28$. Assume normal distribution; yes.

38.4	40.9	67	32.5	51.5	43.4
38	43.4	50.7	35.4	39.3	41.4

   Source: http://www.api.org/statistics/fueltaxes/

9. **Workplace Homicides** A sample of six recent years had an average of 573.8 workplace homicides per year with a standard deviation of 46.8. Find the 99% confidence interval of the true mean of all workplace homicides per year. If in a certain year there were 625 homicides, would this be considered unusually high?

   Source: Based on statistics from the Bureau of Labor Statistics.

10. **Dance Company Students** The number of students who belong to the dance company at each of several randomly selected small universities is shown below. Estimate the true population mean size of a university dance company with 99% confidence. $25.8 < \mu < 33.9$. Assume normal distribution.

21	25	32	22	28	30	29	30
47	26	35	26	35	26	28	28
32	27	40					

11. **Distance Traveled to Work** A recent study of 28 employees of XYZ company showed that the mean of the distance they traveled to work was 14.3 miles. The standard deviation of the sample mean was 2 miles. Find the 95% confidence interval of the true mean. If a manager wanted to be sure that most of his employees would not be late, how much time would he suggest they allow for the commute if the average speed were 30 miles per hour? $13.5 < \mu < 15.1$; about 30 minutes.

12. **Thunderstorm Speeds** A meteorologist who sampled 13 thunderstorms found that the average speed at which they traveled across a certain state was 15 miles per hour. The standard deviation of the sample was 1.7 miles per hour. Find the 99% confidence interval of the mean. If a meteorologist wanted to use the highest speed to predict the times it would take storms to travel across the state in order to issue warnings, what figure would she likely use? $13.6 < \mu < 16.4$; 16.4 miles per hour

13. **Students per Teacher in U.S. Public Schools** The national average for the number of students per teacher for all U.S. public schools is 15.9. A random sample of 12 school districts from a moderately populated area showed that the mean number of students per teacher was 19.2 with a variance of 4.41. Estimate the true mean number of students per teacher with 95% confidence. How does your estimate compare with the national average?

    Source: *World Almanac.* $17.87 < \mu < 20.53$. Assume normal distribution; it's higher.

14. **Social Networking Sites** A recent survey of 8 social networking sites has a mean of 13.1 million visitors for a specific month. The standard deviation was 4.1 million. Find the 95% confidence interval of the true mean.

    Source: ComScore Media Matrix. $9.7 < \mu < 16.5$

15. **Chicago Commuters** A sample of 14 commuters in Chicago showed the average of the commuting times was 33.2 minutes. If the standard deviation was 8.3 minutes, find the 95% confidence interval of the true mean.

    Source: U.S. Census Bureau. $28.4 < \mu < 38.0$

16. **Hospital Noise Levels** For a sample of 24 operating rooms taken in the hospital study mentioned in Exercise 19 in Section 7–1, the mean noise level was 41.6 decibels, and the standard deviation was 7.5. Find the 95% confidence interval of the true mean of the noise levels in the operating rooms.

Source: M. Bayo, A. Garcia, and A. Garcia, "Noise Levels in an Urban Hospital and Workers' Subjective Responses," *Archives of Environmental Health* 50, no. 3, p. 249 (May–June 1995). Reprinted with permission of the Helen Dwight Reid Educational Foundation. Published by Heldref Publications, 1319 Eighteenth St. N.W., Washington, D.C. 20036-1802. Copyright © 1995. $38.4 < \mu < 44.8$

17. **Costs for a 30-Second Spot on Cable Television** The approximate costs for a 30-second spot for various cable networks in a random selection of cities are shown below. Estimate the true population mean cost for a 30-second advertisement on cable network with 90% confidence. $32.0 < \mu < 71$. Assume normal distribution.

| 14 | 55 | 165 | 9 | 15 | 66 | 23 | 30 | 150 |
| 22 | 12 | 13 | 54 | 73 | 55 | 41 | 78 | |

Source: www.spotrunner.com

18. **Football Player Heart Rates** For a group of 22 college football players, the mean heart rate after a morning workout session was 86 beats per minute, and the standard deviation was 5. Find the 90% confidence interval of the true mean for all college football players after a workout session. If a coach did not want to work his team beyond its capacity, what maximum value should he use for the mean number of heartbeats per minute? $84.2 < \mu < 87.8$. He probably used a maximum pulse rate of 88 on average.

19. **Grooming Times for Men and Women** It has been reported that 20- to 24-year-old men spend an average of 37 minutes per day grooming and 20- to 24-year-old women spend an average of 49 minutes per day grooming. Ask your classmates for their individual grooming time per day (unless you're in an 8:00 A.M. class), and use the data to estimate the true mean grooming time for your school with 95% confidence. Answers will vary.
Source: *Time* magazine, Oct. 2006.

20. **Unhealthy Days in Cities** The number of unhealthy days based on the AQI (Air Quality Index) for a random sample of metropolitan areas is shown. Construct a 98% confidence interval based on the data.

| 61 | 12 | 6 | 40 | 27 | 38 | 93 | 5 | 13 | 40 |

Source: *New York Times Almanac.* $8.8 < \mu < 58.2$

Extending the Concepts

21. A *one-sided confidence* interval can be found for a mean by using

$$\mu > \overline{X} - t_\alpha \frac{s}{\sqrt{n}} \quad \text{or} \quad \mu < \overline{X} + t_\alpha \frac{s}{\sqrt{n}}$$

where t_α is the value found under the row labeled One tail. Find two one-sided 95% confidence intervals of the population mean for the data shown, and interpret the answers. The data represent the daily revenues in dollars from 20 parking meters in a small municipality.

2.60	1.05	2.45	2.90
1.30	3.10	2.35	2.00
2.40	2.35	2.40	1.95
2.80	2.50	2.10	1.75
1.00	2.75	1.80	1.95

Technology *Step by Step*

MINITAB
Step by Step

Find a *t* Interval for the Mean

For Example 7–7, find the 99% confidence interval for the mean number of home fires started by candles each year.

1. Type the data into C1 of a MINITAB worksheet. Name the column **HomeFires**.

2. Select **Stat>Basic Statistics>1-Sample t.**

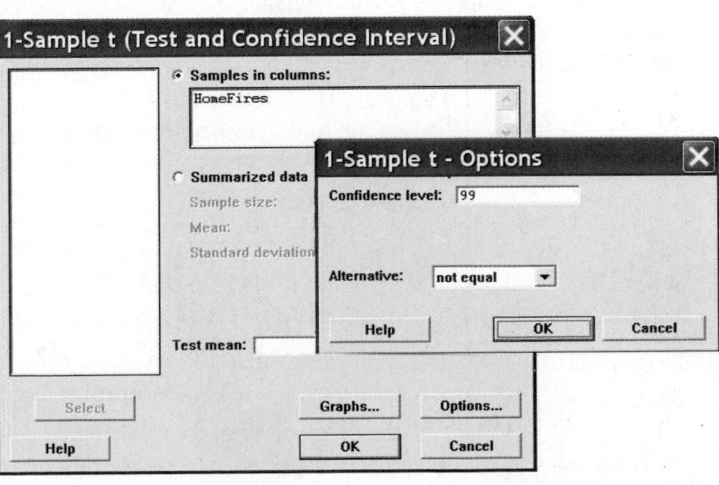

3. Double-click C1 HomeFires for the Samples in Columns.

4. Click on [Options] and be sure the Confidence Level is 99 and the Alternative is not equal.

5. Click [OK] twice.

6. Check for normality:

   a) Select **Graph>Probability Plot,** then Single.

   b) Select C1 HomeFires for the variable. The normal plot is concave, a skewed distribution.

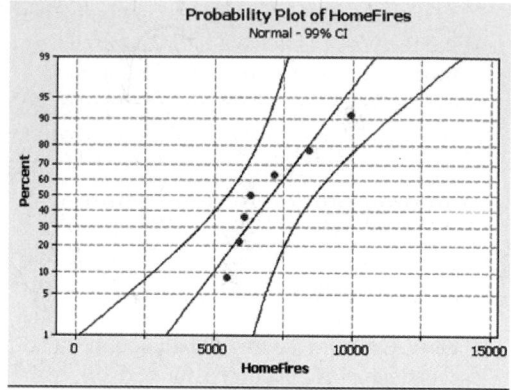

In the session window you will see the results. The 99% confidence interval estimate for μ is between 4784.99 and 9784.99. The sample size, mean, standard deviation, and standard error of the mean are also shown.

However, this small sample appears to have a nonnormal population. The interval is less likely to contain the true mean.

One-Sample T: HomeFires

```
Variable    N     Mean     StDev    SE Mean         99% CI
HomeFires   7   7041.43   1610.27    608.63    (4784.99, 9297.87)
```

TI-83 Plus or TI-84 Plus
Step by Step

Finding a *t* Confidence Interval for the Mean (Data)

1. Enter the data into L_1.
2. Press **STAT** and move the cursor to TESTS.
3. Press **8** for TInterval.
4. Move the cursor to Data and press **ENTER**.
5. Type in the appropriate values.
6. Move the cursor to Calculate and press **ENTER**.

Finding a *t* Confidence Interval for the Mean (Statistics)

1. Press **STAT** and move the cursor to TESTS.
2. Press **8** for TInterval.
3. Move the cursor to Stats and press **ENTER**.
4. Type in the appropriate values.
5. Move the cursor to Calculate and press **ENTER**.

Excel
Step by Step

Finding a *t* Confidence Interval for the Mean

Excel has a procedure to compute the margin of error. But it does not compute confidence intervals. However, you may determine confidence intervals for the mean by using the MegaStat Add-in available on your CD. If you have not installed this add-in, do so, following the instructions from the Chapter 1 Excel Step by Step.

Example XL7–2

Find the 95% confidence interval, using these sample data:

| 625 | 675 | 535 | 406 | 512 | 680 | 483 | 522 | 619 | 575 |

1. Enter the data into an Excel worksheet.
2. From the toolbar, select Add-Ins, **MegaStat>Confidence Intervals/Sample Size.** *Note:* You may need to open **MegaStat** from the **MegaStat.xls** file on your computer's hard drive.
3. Enter the mean of the data, **563.2.**

4. Select *t* for the *t* distribution.

5. Enter **87.9** for the standard deviation and **10** for *n*, the sample size.

6. Either type in or scroll to 95% for the Confidence Level, then click [OK].

The result of the procedure is shown next.

Confidence interval—mean

```
    95% Confidence level
 563.2 Mean
  87.9 Standard deviation
    10 n
 2.262 t (d.f. = 9)
62.880 Half-width
626.080 Upper confidence limit
500.320 Lower confidence limit
```

7–3

Confidence Intervals and Sample Size for Proportions

Objective **4**

Find the confidence interval for a proportion.

A *USA TODAY* Snapshots feature stated that 12% of the pleasure boats in the United States were named *Serenity*. The parameter 12% is called a **proportion.** It means that of all the pleasure boats in the United States, 12 out of every 100 are named *Serenity*. A proportion represents a part of a whole. It can be expressed as a fraction, decimal, or percentage. In this case, $12\% = 0.12 = \frac{12}{100}$ or $\frac{3}{25}$. Proportions can also represent probabilities. In this case, if a pleasure boat is selected at random, the probability that it is called *Serenity* is 0.12.

Proportions can be obtained from samples or populations. The following symbols will be used.

Symbols Used in Proportion Notation

p = population proportion

$\hat{p}$ (read "*p* hat") = sample proportion

For a sample proportion,

$$\hat{p} = \frac{X}{n} \quad \text{and} \quad \hat{q} = \frac{n - X}{n} \quad \text{or} \quad \hat{q} = 1 - \hat{p}$$

where X = number of sample units that possess the characteristics of interest and n = sample size.

For example, in a study, 200 people were asked if they were satisfied with their job or profession; 162 said that they were. In this case, $n = 200$, $X = 162$, and $\hat{p} = X/n = 162/200 = 0.81$. It can be said that for this sample, 0.81, or 81%, of those surveyed were satisfied with their job or profession. The sample proportion is $\hat{p} = 0.81$.

The proportion of people who did not respond favorably when asked if they were satisfied with their job or profession constituted $\hat{q}$, where $\hat{q} = (n - X)/n$. For this survey, $\hat{q} = (200 - 162)/200 = 38/200$, or 0.19, or 19%.

When $\hat{p}$ and $\hat{q}$ are given in decimals or fractions, $\hat{p} + \hat{q} = 1$. When $\hat{p}$ and $\hat{q}$ are given in percentages, $\hat{p} + \hat{q} = 100\%$. It follows, then, that $\hat{q} = 1 - \hat{p}$, or $\hat{p} = 1 - \hat{q}$, when $\hat{p}$ and $\hat{q}$ are in decimal or fraction form. For the sample survey on job satisfaction, $\hat{q}$ can also be found by using $\hat{q} = 1 - \hat{p}$, or $1 - 0.81 = 0.19$.

Similar reasoning applies to population proportions; that is, $p = 1 - q$, $q = 1 - p$, and $p + q = 1$, when p and q are expressed in decimal or fraction form. When p and q are expressed as percentages, $p + q = 100\%$, $p = 100\% - q$, and $q = 100\% - p$.

Example 7–8	Air Conditioned Households

In a recent survey of 150 households, 54 had central air conditioning. Find $\hat{p}$ and $\hat{q}$, where $\hat{p}$ is the proportion of households that have central air conditioning.

Solution

Since $X = 54$ and $n = 150$,

$$\hat{p} = \frac{X}{n} = \frac{54}{150} = 0.36 = 36\%$$

$$\hat{q} = \frac{n - X}{n} = \frac{150 - 54}{150} = \frac{96}{150} = 0.64 = 64\%$$

You can also find $\hat{q}$ by using the formula $\hat{q} = 1 - \hat{p}$. In this case, $\hat{q} = 1 - 0.36 = 0.64$.

As with means, the statistician, given the sample proportion, tries to estimate the population proportion. Point and interval estimates for a population proportion can be made by using the sample proportion. For a point estimate of p (the population proportion), $\hat{p}$ (the sample proportion) is used. On the basis of the three properties of a good estimator, $\hat{p}$ is unbiased, consistent, and relatively efficient. But as with means, one is not able to decide how good the point estimate of p is. Therefore, statisticians also use an interval estimate for a proportion, and they can assign a probability that the interval will contain the population proportion.

The confidence interval for a particular p is based on the sampling distribution of $\hat{p}$. When the sample size n is no more than 5% of the population size, the sampling distribution of $\hat{p}$ is approximately normal with a mean of p and a standard deviation of $\sqrt{pq/n}$, where $q = 1 - p$.

Confidence Intervals

To construct a confidence interval about a proportion, you must use the margin of error, which is

$$E = z_{\alpha/2}\sqrt{\frac{\hat{p}\hat{q}}{n}}$$

Confidence intervals about proportions must meet the criteria that $n\hat{p} \geq 5$ and $n\hat{q} \geq 5$.

Formula for a Specific Confidence Interval for a Proportion
$$\hat{p} - z_{\alpha/2}\sqrt{\frac{\hat{p}\hat{q}}{n}} < p < \hat{p} + z_{\alpha/2}\sqrt{\frac{\hat{p}\hat{q}}{n}}$$ when $n\hat{p}$ and $n\hat{q}$ are each greater than or equal to 5.

Assumptions for Finding a Confidence Interval for a Population Proportion
1. The sample is a random sample.   2. The conditions for a binomial experiment are satisfied (See Chapter 5).

Rounding Rule for a Confidence Interval for a Proportion Round off to three decimal places.

Example 7–9	Covering College Costs

A survey conducted by Sallie Mae and Gallup of 1404 respondents found that 323 students paid for their education by student loans. Find the 90% confidence of the true proportion of students who paid for their education by student loans.

Solution

Since $\alpha = 1 - 0.90 = 0.10$, $z_{\alpha/2} = 1.65$. Substitute in the formula

$$\hat{p} - z_{\alpha/2}\sqrt{\frac{\hat{p}\hat{q}}{n}} < p < \hat{p} + z_{\alpha/2}\sqrt{\frac{\hat{p}\hat{q}}{n}}$$

Find $\hat{p}$ and $\hat{q}$.

$$\hat{p} = \frac{323}{1404} = 0.23 \quad \text{and} \quad \hat{q} = 1 - \hat{p} = 1 - 0.23 = 0.77$$

$$0.23 - 1.65\sqrt{\frac{(0.23)(0.77)}{1404}} < p < 0.23 + 1.65\sqrt{\frac{(0.23)(0.77)}{1404}}$$

$$0.23 - 0.019 < p < 0.23 + 0.019$$

$$0.211 < p < 0.249$$

or

$$21.1\% < p < 24.9\%$$

Hence, you can be 90% confident that the percentage of students who pay for their college education by student loans is between 21.1 and 24.9%.

When a specific percentage is given, the percentage becomes $\hat{p}$ when it is changed to a decimal. For example, if the problem states that 12% of the applicants were men, then $\hat{p} = 0.12$.

Example 7–10	Religious Books

A survey of 1721 people found that 15.9% of individuals purchase religious books at a Christian bookstore. Find the 95% confidence interval of the true proportion of people who purchase their religious books at a Christian bookstore.

Source: Baylor University.

Solution

Here $\hat{p} = 0.159$ (i.e., 15.9%), and $\hat{q} = 1 - 0.159 = 0.841$. For the 95% confidence interval $z_{\alpha/2} = 1.96$.

$$\hat{p} - z_{\alpha/2}\sqrt{\frac{\hat{p}\hat{q}}{n}} < p < \hat{p} + z_{\alpha/2}\sqrt{\frac{\hat{p}\hat{q}}{n}}$$

$$0.159 - 1.96\sqrt{\frac{(0.159)(0.841)}{1721}} < p < 0.159 + 1.96\sqrt{\frac{(0.159)(0.841)}{1721}}$$

$$0.142 < p < 0.176$$

Hence, you can say with 95% confidence that the true percentage is between 14.2 and 17.6%.

Sample Size for Proportions

To find the sample size needed to determine a confidence interval about a proportion, use this formula:

Objective 5

Determine the minimum sample size for finding a confidence interval for a proportion.

> **Formula for Minimum Sample Size Needed for Interval Estimate of a Population Proportion**
>
> $$n = \hat{p}\hat{q}\left(\frac{z_{\alpha/2}}{E}\right)^2$$
>
> If necessary, round up to obtain a whole number.

This formula can be found by solving the margin of error value for n in the formula

$$E = z_{\alpha/2}\sqrt{\frac{\hat{p}\hat{q}}{n}}$$

There are two situations to consider. First, if some approximation of $\hat{p}$ is known (e.g., from a previous study), that value can be used in the formula.

Second, if no approximation of $\hat{p}$ is known, you should use $\hat{p} = 0.5$. This value will give a sample size sufficiently large to guarantee an accurate prediction, given the confidence interval and the error of estimate. The reason is that when $\hat{p}$ and $\hat{q}$ are each 0.5, the product $\hat{p}\hat{q}$ is at maximum, as shown here.

$\hat{p}$	$\hat{q}$	$\hat{p}\hat{q}$
0.1	0.9	0.09
0.2	0.8	0.16
0.3	0.7	0.21
0.4	0.6	0.24
0.5	**0.5**	**0.25**
0.6	0.4	0.24
0.7	0.3	0.21
0.8	0.2	0.16
0.9	0.1	0.09

Example 7–11

Home Computers

A researcher wishes to estimate, with 95% confidence, the proportion of people who own a home computer. A previous study shows that 40% of those interviewed had a computer at home. The researcher wishes to be accurate within 2% of the true proportion. Find the minimum sample size necessary.

Solution

Since $z_{\alpha/2} = 1.96$, $E = 0.02$, $\hat{p} = 0.40$, and $\hat{q} = 0.60$, then

$$n = \hat{p}\hat{q}\left(\frac{z_{\alpha/2}}{E}\right)^2 = (0.40)(0.60)\left(\frac{1.96}{0.02}\right)^2 = 2304.96$$

which, when rounded up, is 2305 people to interview.

Example 7–12

M&M Colors

A researcher wishes to estimate the percentage of M&M's that are brown. He wants to be 95% confident and be accurate within 3% of the true proportion. How large a sample size would be necessary?

Speaking of
Statistics

Does Success Bring Happiness?

W. C. Fields said, "Start every day off with a smile and get it over with."

Do you think people are happy because they are successful, or are they successful because they are just happy people? A recent survey conducted by *Money* magazine showed that 34% of the people surveyed said that they were happy because they were successful; however, 63% said that they were successful because they were happy individuals. The people surveyed had an average household income of $75,000 or more. The margin of error was ±2.5%. Based on the information in this article, what would be the confidence interval for each percent?

Solution

Since no prior knowledge of $\hat{p}$ is known, assign a value of 0.5 and then $\hat{q} = 1 - \hat{p} = 1 - 0.5 = 0.5$. Substitute in the formula, using $E = 0.03$.

$$n = \hat{p}\hat{q}\left(\frac{z_{\alpha/2}}{E}\right)^2 = (0.5)(0.5)\left(\frac{1.96}{0.03}\right)^2 = 1067.1$$

Hence, a sample size of 1068 would be needed.

In determining the sample size, the size of the population is irrelevant. Only the degree of confidence and the margin of error are necessary to make the determination.

Applying the Concepts **7–3**

Contracting Influenza

To answer the questions, use the following table describing the percentage of people who reported contracting influenza by gender and race/ethnicity.

Characteristic	Influenza Percent	(95% CI)
Gender		
Men	48.8	(47.1–50.5%)
Women	51.5	(50.2–52.8%)
Race/ethnicity		
Caucasian	52.2	(51.1–53.3%)
African American	33.1	(29.5–36.7%)
Hispanic	47.6	(40.9–54.3%)
Other	39.7	(30.8–48.5%)
Total	50.4	(49.3–51.5%)

Forty-nine states and the District of Columbia participated in the study. Weighted means were used. The sample size was 19,774. There were 12,774 women and 7000 men.

1. Explain what (95% CI) means.
2. How large is the error for men reporting influenza?
3. What is the sample size?
4. How does sample size affect the size of the confidence interval?
5. Would the confidence intervals be larger or smaller for a 90% CI, using the same data?
6. Where does the 51.5% under influenza for women fit into its associated 95% CI?

See page 398 for the answers.

Exercises 7–3

1. In each case, find $\hat{p}$ and $\hat{q}$.
   a. $n = 80$ and $X = 40$ 0.5, 0.5
   b. $n = 200$ and $X = 90$ 0.45, 0.55
   c. $n = 130$ and $X = 60$ 0.46, 0.54
   d. $n = 60$ and $X = 35$ 0.58, 0.42
   e. $n = 95$ and $X = 43$ 0.45, 0.55

2. (**ans**) Find $\hat{p}$ and $\hat{q}$ for each percentage. (Use each percentage for $\hat{p}$.)
   a. 25% $\hat{p} = 0.25$, $\hat{q} = 0.75$
   b. 42% $\hat{p} = 0.42$, $\hat{q} = 0.58$
   c. 68% $\hat{p} = 0.68$, $\hat{q} = 0.32$
   d. 55% $\hat{p} = 0.55$, $\hat{q} = 0.45$
   e. 12% $\hat{p} = 0.12$, $\hat{q} = 0.88$

3. **Vacations** A U.S. Travel Data Center survey conducted for *Better Homes and Gardens* of 1500 adults found that 39% said that they would take more vacations this year than last year. Find the 95% confidence interval for the true proportion of adults who said that they will travel more this year. $0.365 < p < 0.415$
   Source: *USA TODAY.*

4. **Regular Voters in America** Thirty-five percent of adult Americans are regular voters. A random sample of 250 adults in a medium-size college town were surveyed, and it was found that 110 were regular voters. Estimate the true proportion of regular voters with 90% confidence and comment on your results. $0.388 < p < 0.492$. It is probably higher because of increased awareness in a college town.
   Source: *Time* magazine, Oct. 2006.

5. **Private Schools** The proportion of students in private schools is around 11%. A random sample of 450 students from a wide geographic area indicated that 55 attended private schools. Estimate the true proportion of students attending private schools with 95% confidence. How does your estimate compare to 11%?
   $0.092 < p < 0.153$; 11% is contained in the confidence interval.
   Source: National Center for Education Statistics (www.nces.ed.gov).

6. **Belief in Haunted Places** A random sample of 205 college students were asked if they believed that places could be haunted, and 65 responded yes. Estimate the

true proportion of college students who believe in the possibility of haunted places with 99% confidence. According to *Time* magazine, 37% of Americans believe that places can be haunted.
   Source: *Time* magazine, Oct. 2006. $0.233 < p < 0.401$

7. **Work Interruptions** A survey found that out of 200 workers, 168 said they were interrupted three or more times an hour by phone messages, faxes, etc. Find the 90% confidence interval of the population proportion of workers who are interrupted three or more times an hour.
   Source: Based on information from *USA TODAY* Snapshot. $0.797 < p < 0.883$

8. **Travel to Outer Space** A CBS News/*New York Times* poll found that 329 out of 763 adults said they would travel to outer space in their lifetime, given the chance. Estimate the true proportion of adults who would like to travel to outer space with 92% confidence. $0.400 < p < 0.463$
   Source: www.pollingreport.com

9. **High School Graduates Who Take the SAT** The national average for the percentage of high school graduates taking the SAT is 49%, but the state averages vary from a low of 4% to a high of 92%. A random sample of 300 graduating high school seniors was polled across a particular tristate area, and it was found that 195 had taken the SAT. Estimate the true proportion of high school graduates in this region who take the SAT with 95% confidence. $0.596 < p < 0.704$
   Source: *World Almanac.*

10. **Educational Television** In a sample of 200 people, 154 said that they watched educational television. Find the 90% confidence interval of the true proportion of people who watched educational television. If the television company wanted to publicize the proportion of viewers, do you think it should use the 90% confidence interval?
   $0.721 < p < 0.819$

11. **Fruit Consumption** A nutritionist found that in a sample of 80 families, 25% indicated that they ate fruit at least 3 times a week. Find the 99% confidence interval of the true proportion of families who said that they ate fruit at least 3 times a week. Would a proportion of families equal to 28% be considered large? $0.125 < p < 0.375$. No, since 0.28 is contained in the interval.

12. Students Who Major in Business It has been reported that 20.4% of incoming freshmen indicate that they will major in business or a related field. A random sample of 400 incoming college freshmen was asked their preference, and 95 replied that they were considering business as a major. Estimate the true proportion of freshman business majors with 98% confidence. Does your interval contain 20.4? $0.188 < p < 0.288$; yes

Source: *New York Times Almanac.*

13. Financial Well-being In a Gallup Poll of 1005 individuals, 452 thought they were worse off financially than a year ago. Find the 95% confidence interval for the true proportion of individuals who feel they are worse off financially. $0.419 < p < 0.481$

Source: Gallup Poll.

14. Fighting U.S. Hunger In a poll of 1000 likely voters, 560 say that the United States spends too little on fighting hunger at home. Find a 95% confidence interval for the true proportion of voters who feel this way.

Source: Alliance to End Hunger. $0.529 < p < 0.591$

15. Overseas Travel A researcher wishes to be 95% confident that her estimate of the true proportion of individuals who travel overseas is within 4% of the true proportion. Find the sample necessary if in a prior study, a sample of 200 people showed that 40 traveled overseas last year. If no estimate of the sample proportion is available, how large should the sample be? 385; 601

16. Widows A recent study indicated that 29% of the 100 women over age 55 in the study were widows.

a. How large a sample must you take to be 90% confident that the estimate is within 0.05 of

the true proportion of women over age 55 who are widows? 225

b. If no estimate of the sample proportion is available, how large should the sample be? 273

17. Direct Satellite Television It is believed that 25% of U.S. homes have a direct satellite television receiver. How large a sample is necessary to estimate the true population of homes which do with 95% confidence and within 3 percentage points? How large a sample is necessary if nothing is known about the proportion?

Source: *New York Times Almanac.* 801 homes; 1068 homes

18. Obesity Obesity is defined as a *body mass index* (BMI) of 30 kg/m² or more. A 95% confidence interval for the percentage of U.S. adults aged 20 years and over who were obese was found to be 22.4 to 23.5%. What was the sample size? 318

Source: National Center for Health Statistics (www.cdc.gov/nchs).

19. Unmarried Americans Nearly one-half of Americans aged 25 to 29 are unmarried. How large a sample is necessary to estimate the true proportion of unmarried Americans in this age group within 2½ percentage points with 90% confidence? 1089

Source: *Time* magazine, Oct. 2006.

20. Diet Habits A federal report indicated that 27% of children ages 2 to 5 years had a good diet—an increase over previous years. How large a sample is needed to estimate the true proportion of children with good diets within 2% with 95% confidence? 1893

Source: Federal Interagency Forum on Child and Family Statistics, *Washington Observer-Reporter.*

Extending the Concepts

21. Gun Control If a sample of 600 people is selected and the researcher decides to have a margin of error of 4% on the specific proportion who favor gun control, find the degree of confidence. A recent study showed that 50% were in favor of some form of gun control. 95%

22. Survey on Politics In a study, 68% of 1015 adults said that they believe the Republicans favor the rich. If the margin of error was 3 percentage points, what was the confidence interval used for the proportion? 96%

Source: *USA TODAY.*

Technology *Step by Step*

MINITAB
Step by Step

Find a Confidence Interval for a Proportion

MINITAB will calculate a confidence interval, given the statistics from a sample *or* given the raw data. In a sample of 500 nursing applications 60 were from men. Find the 90% confidence interval estimate for the true proportion of male applicants.

1. Select **Stat>Basic Statistics>1 Proportion.**
2. Click on the button for Summarized data. No data will be entered in the worksheet.
3. Click in the box for Number of trials and enter **500**.
4. In the Number of events box, enter **60**.
5. Click on [Options].

6. Type **90** for the confidence level.

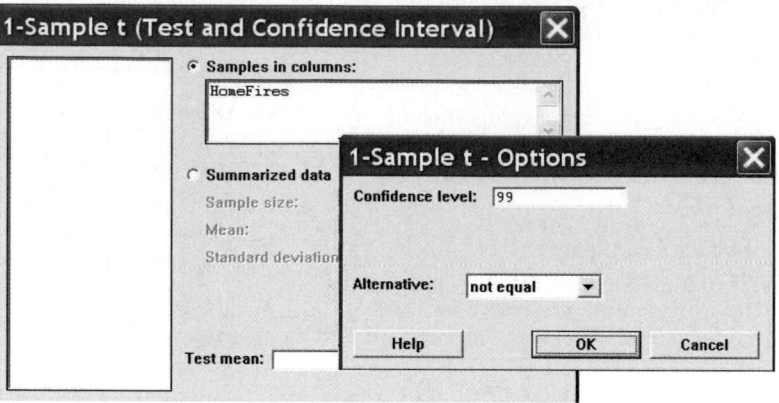

7. Check the box for Use test and interval based on normal distribution.
8. Click [OK] twice.

The results for the confidence interval will be displayed in the session window.

Test and CI for One Proportion

```
Test of p = 0.5 vs p not = 0.
Sample   X     N    Sample p            90% CI          Z-Value   P-Value
1       60   500   0.120000   (0.096096, 0.143904)     -16.99     0.000
```

TI-83 Plus or TI-84 Plus
Step by Step

Finding a Confidence Interval for a Proportion

1. Press **STAT** and move the cursor to TESTS.
2. Press **A (ALPHA, MATH)** for 1-PropZInt.
3. Type in the appropriate values.
4. Move the cursor to Calculate and press **ENTER.**

Example TI7–3

Find the 95% confidence interval of p when $X = 60$ and $n = 500$.
    The 95% confidence level for p is $0.09152 < p < 0.14848$.
Also $\hat{p}$ is given.

Input
```
1-PropZInt
 x:60
 n:500
 C-Level:.95
 Calculate
```

Output
```
1-PropZInt
 (.09152,.14848)
 p=.12
 n=500
```

Excel
Step by Step

Finding a Confidence Interval for a Proportion

Excel has a procedure to compute the margin of error. But it does not compute confidence intervals. However, you may determine confidence intervals for a proportion by using the MegaStat Add-in available on your CD. If you have not installed this add-in, do so, following the instructions from the Chapter 1 Excel Step by Step.

Example XL7–3

There were 500 nursing applications in a sample, including 60 from men. Find the 90% confidence interval for the true proportion of male applicants.

1. From the toolbar, select Add-Ins, **MegaStat>Confidence Intervals/Sample Size.**
   Note: You may need to open **MegaStat** from the **MegaStat.xls** file on your computer's hard drive.
2. In the dialog box, select Confidence interval–p.
3. Enter **60** in the box labeled p; p will automatically change to x.

Here is a survey about college students' credit card usage. Suggest several ways that the study could have been more meaningful if confidence intervals had been used.

OTHER PEOPLE'S MONEY

Undergrads love their plastic. That means—you guessed it—students are learning to become debtors. According to the Public Interest Research Groups, only half of all students pay off card balances in full each month, 36% sometimes do and 14% never do. Meanwhile, 48% have paid a late fee. Here's how undergrads stack up, according to Nellie Mae, a provider of college loans:

Undergrads with a credit card **78%**

Average number of cards owned . . **3**

Average student card debt **$1236**

Students with 4 or more cards **32%**

Balances of $3000 to $7000 . . . . . **13%**

Balances over $7000 . . . . **9%**

Reprinted with permission from the January 2002 Reader's Digest.
Copyright © 2002 by The Reader's Digest Assn. Inc.

4. Enter **500** in the box labeled n.

5. Either type in or scroll to 90% for the Confidence Level, then click [OK].

The result of the procedure is shown next.

Confidence interval—proportion

```
 90% Confidence level
0.12 Proportion
 500 n
1.645 z
0.024 Half-width
0.144 Upper confidence limit
0.096 Lower confidence limit
```

7–4

Confidence Intervals for Variances and Standard Deviations

Objective 6

Find a confidence interval for a variance and a standard deviation.

In Sections 7–1 through 7–3 confidence intervals were calculated for means and proportions. This section will explain how to find confidence intervals for variances and standard deviations. In statistics, the variance and standard deviation of a variable are as important as the mean. For example, when products that fit together (such as pipes) are manufactured, it is important to keep the variations of the diameters of the products as small as possible; otherwise, they will not fit together properly and will have to be scrapped. In the manufacture of medicines, the variance and standard deviation of the medication in the pills play an important role in making sure patients receive the proper dosage. For these reasons, confidence intervals for variances and standard deviations are necessary.

To calculate these confidence intervals, a new statistical distribution is needed. It is called the **chi-square distribution.**

The chi-square variable is similar to the t variable in that its distribution is a family of curves based on the number of degrees of freedom. The symbol for chi-square is χ^2 (Greek letter chi, pronounced "ki"). Several of the distributions are shown in Figure 7–9, along with the corresponding degrees of freedom. The chi-square distribution is obtained from the values of $(n - 1)s^2/\sigma^2$ when random samples are selected from a normally distributed population whose variance is σ^2.

A chi-square variable cannot be negative, and the distributions are skewed to the right. At about 100 degrees of freedom, the chi-square distribution becomes somewhat symmetric. The area under each chi-square distribution is equal to 1.00, or 100%.

Table G in Appendix C gives the values for the chi-square distribution. These values are used in the denominators of the formulas for confidence intervals. Two different values

Figure 7–9

**The Chi-Square Family
of Curves**

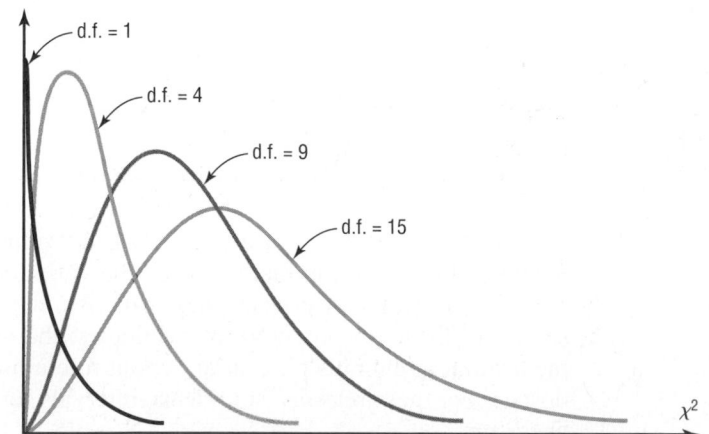

are used in the formula because the distribution is not symmetric. One value is found on the left side of the table, and the other is on the right. See Figure 7–10. For example, to find the table values corresponding to the 95% confidence interval, you must first change 95% to a decimal and subtract it from 1 ($1 - 0.95 = 0.05$). Then divide the answer by 2 ($\alpha/2 = 0.05/2 = 0.025$). This is the column on the right side of the table, used to get the values for χ^2_{right}. To get the value for χ^2_{left}, subtract the value of $\alpha/2$ from 1 ($1 - 0.05/2 = 0.975$). Finally, find the appropriate row corresponding to the degrees of freedom $n - 1$. A similar procedure is used to find the values for a 90 or 99% confidence interval.

Figure 7–10

Chi-Square Distribution for d.f. $= n - 1$

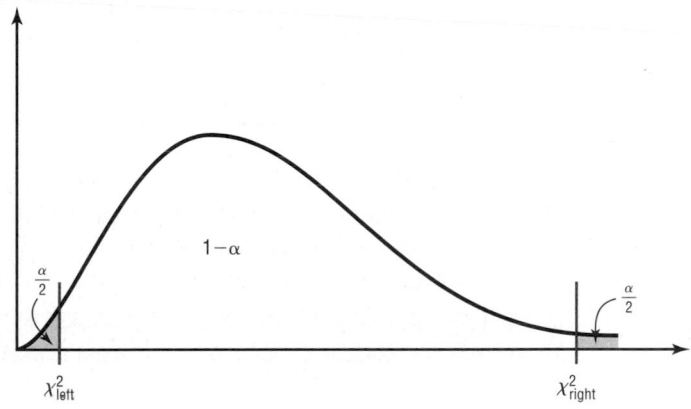

Example 7–13

Find the values for χ^2_{right} and χ^2_{left} for a 90% confidence interval when $n = 25$.

Solution

To find χ^2_{right}, subtract $1 - 0.90 = 0.10$ and divide by 2 to get 0.05.

To find χ^2_{left}, subtract $1 - 0.05$ to get 0.95. Hence, use the 0.95 and 0.05 columns and the row corresponding to 24 d.f. See Figure 7–11.

Figure 7–11

χ^2 **Table for Example 7–13**

					Table G					
				The Chi-square Distribution						
Degrees of					α					
freedom	0.995	0.99	0.975	0.95	0.90	0.10	0.05	0.025	0.01	0.005
1										
2										
⋮										
24				13.848			36.415			
				χ^2_{left}			χ^2_{right}			

The answers are

$$\chi^2_{\text{right}} = 36.415$$

$$\chi^2_{\text{left}} = 13.848$$

See Figure 7–12.

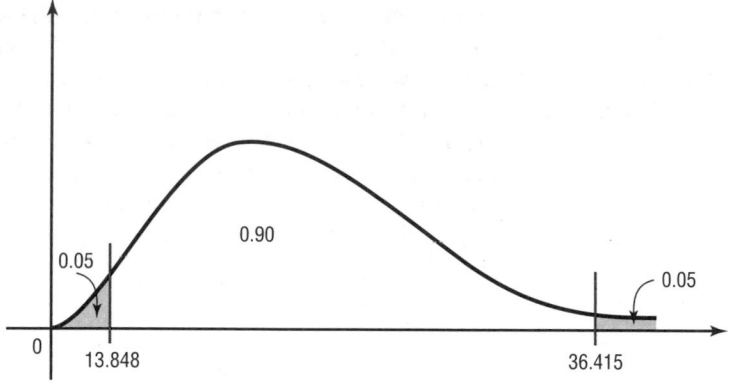

Useful estimates for σ^2 and σ are s^2 and s, respectively.

To find confidence intervals for variances and standard deviations, you must assume that the variable is normally distributed.

The formulas for the confidence intervals are shown here.

Formula for the Confidence Interval for a Variance

$$\frac{(n-1)s^2}{\chi^2_{\text{right}}} < \sigma^2 < \frac{(n-1)s^2}{\chi^2_{\text{left}}}$$

d.f. = $n - 1$

Formula for the Confidence Interval for a Standard Deviation

$$\sqrt{\frac{(n-1)s^2}{\chi^2_{\text{right}}}} < \sigma < \sqrt{\frac{(n-1)s^2}{\chi^2_{\text{left}}}}$$

d.f. = $n - 1$

Recall that s^2 is the symbol for the sample variance and s is the symbol for the sample standard deviation. If the problem gives the sample standard deviation s, be sure to *square* it when you are using the formula. But if the problem gives the sample variance s^2, *do not square it* when you are using the formula, since the variance is already in square units.

Assumptions for Finding a Confidence Interval for a Variance or Standard Deviation

1. The sample is a random sample.
2. The population must be normally distributed.

Rounding Rule for a Confidence Interval for a Variance or Standard Deviation When you are computing a confidence interval for a population variance or standard deviation by using raw data, round off to one more decimal place than the number of decimal places in the original data.

When you are computing a confidence interval for a population variance or standard deviation by using a sample variance or standard deviation, round off to the same number of decimal places as given for the sample variance or standard deviation.

Example 7–14 shows how to find a confidence interval for a variance and standard deviation.

Example 7–14 | **Nicotine Content**

Find the 95% confidence interval for the variance and standard deviation of the nicotine content of cigarettes manufactured if a sample of 20 cigarettes has a standard deviation of 1.6 milligrams.

Solution

Since $\alpha = 0.05$, the two critical values, respectively, for the 0.025 and 0.975 levels for 19 degrees of freedom are 32.852 and 8.907. The 95% confidence interval for the variance is found by substituting in the formula.

$$\frac{(n-1)s^2}{\chi^2_{right}} < \sigma^2 < \frac{(n-1)s^2}{\chi^2_{left}}$$

$$\frac{(20-1)(1.6)^2}{32.852} < \sigma^2 < \frac{(20-1)(1.6)^2}{8.907}$$

$$1.5 < \sigma^2 < 5.5$$

Hence, you can be 95% confident that the true variance for the nicotine content is between 1.5 and 5.5.

For the standard deviation, the confidence interval is

$$\sqrt{1.5} < \sigma < \sqrt{5.5}$$

$$1.2 < \sigma < 2.3$$

Hence, you can be 95% confident that the true standard deviation for the nicotine content of all cigarettes manufactured is between 1.2 and 2.3 milligrams based on a sample of 20 cigarettes.

Example 7–15 | **Cost of Ski Lift Tickets**

Find the 90% confidence interval for the variance and standard deviation for the price in dollars of an adult single-day ski lift ticket. The data represent a selected sample of nationwide ski resorts. Assume the variable is normally distributed.

59	54	53	52	51
39	49	46	49	48

Source: *USA TODAY.*

Solution

Step 1 Find the variance for the data. Use the formulas in Chapter 3 or your calculator. The variance $s^2 = 28.2$.

Step 2 Find χ^2_{right} and χ^2_{left} from Table G in Appendix C. Since $\alpha = 0.10$, the two critical values are 3.325 and 16.919, using d.f. = 9 and 0.95 and 0.05.

Step 3 Substitute in the formula and solve.

$$\frac{(n-1)s^2}{\chi^2_{right}} < \sigma^2 < \frac{(n-1)s^2}{\chi^2_{left}}$$

$$\frac{(10-1)(28.2)}{16.919} < \sigma^2 < \frac{(10-1)(28.2)}{3.325}$$

$$15.0 < \sigma^2 < 76.3$$

For the standard deviation

$$\sqrt{15} < \sigma < \sqrt{76.3}$$
$$3.87 < \sigma < 8.73$$

Hence you can be 90% confident that the standard deviation for the price of all single-day ski lift tickets of the population is between $3.87 and $8.73 based on a sample of 10 nationwide ski resorts. (Two decimal places are used since the data are in dollars and cents.)

Note: If you are using the standard deviation instead (as in Example 7–14) of the variance, be sure to square the standard deviation when substituting in the formula.

Applying the Concepts 7–4

Confidence Interval for Standard Deviation

Shown are the ages (in years) of the Presidents at the times of their deaths.

67	90	83	85	73	80	78	79
68	71	53	65	74	64	77	56
66	63	70	49	57	71	67	71
58	60	72	67	57	60	90	63
88	78	46	64	81	93	93	

1. Do the data represent a population or a sample?
2. Select a random sample of 12 ages and find the variance and standard deviation.
3. Find the 95% confidence interval of the standard deviation.
4. Find the standard deviation of all the data values.
5. Does the confidence interval calculated in question 3 contain the mean?
6. If it does not, give a reason why.
7. What assumption(s) must be considered for constructing the confidence interval in step 3?

See page 398 for the answers.

Exercises 7–4

1. What distribution must be used when computing confidence intervals for variances and standard deviations? Chi-square

2. What assumption must be made when computing confidence intervals for variances and standard deviations? The variable must be normally distributed.

3. Using Table G, find the values for χ^2_{left} and χ^2_{right}.

   a. $\alpha = 0.05, n = 12$ 3.816; 21.920
   b. $\alpha = 0.10, n = 20$ 10.117; 30.144
   c. $\alpha = 0.05, n = 27$ 13.844; 41.923
   d. $\alpha = 0.01, n = 6$ 0.412; 16.750
   e. $\alpha = 0.10, n = 41$ 26.509; 55.758

4. **Lifetimes of Wristwatches** Find the 90% confidence interval for the variance and standard deviation for the lifetimes of inexpensive wristwatches if a sample of 24 watches has a standard deviation of 4.8 months.

Assume the variable is normally distributed. Do you feel that the lifetimes are relatively consistent? $15.1 < \sigma^2 < 40.5$; $3.9 < \sigma < 6.4$

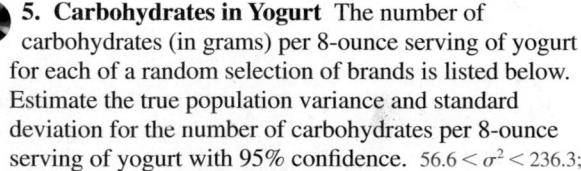 5. **Carbohydrates in Yogurt** The number of carbohydrates (in grams) per 8-ounce serving of yogurt for each of a random selection of brands is listed below. Estimate the true population variance and standard deviation for the number of carbohydrates per 8-ounce serving of yogurt with 95% confidence. $56.6 < \sigma^2 < 236.3$; $7.5 < \sigma < 15.4$

17	42	41	20	39	41	35	15	43
25	38	33	42	23	17	25	34	

6. **Carbon Monoxide Deaths** A study of generation-related carbon monoxide deaths showed that a sample of 6 recent years had a standard deviation of 4.1 deaths per year. Find the 99% confidence interval of the variance and standard distribution. Assume the variable is normally distributed. $5.0 < \sigma^2 < 204.0$; $2.2 < \sigma < 14.3$

Source: Based on information from Consumer Protection Safety Commission.

7. Cost of Knee Replacement Surgery U.S. insurers' costs for knee replacement surgery range from $17,627 to $25,462. Estimate the population variance (standard deviation) in cost with 98% confidence based on a random sample of 10 persons who have had this surgery. The retail costs (for uninsured persons) for the same procedure range from $40,640 to $58,702. Estimate the population variance and standard deviation in cost with 98% confidence based on a sample of 10 persons, and compare your two intervals.

Source: *Time Almanac.*

8. Age of College Students Find the 90% confidence interval for the variance and standard deviation of the ages of seniors at Oak Park College if a sample of 24 students has a standard deviation of 2.3 years. Assume the variable is normally distributed. $3.5 < \sigma^2 < 9.3; 1.9 < \sigma < 3.0$

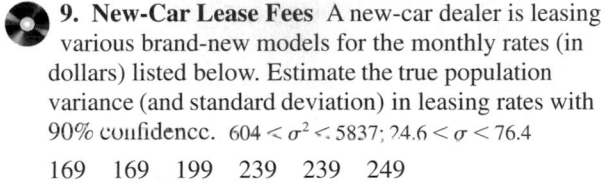 **9. New-Car Lease Fees** A new-car dealer is leasing various brand-new models for the monthly rates (in dollars) listed below. Estimate the true population variance (and standard deviation) in leasing rates with 90% confidence. $604 < \sigma^2 < 5837; 24.6 < \sigma < 76.4$

169   169   199   239   239   249

10. Stock Prices A random sample of stock prices per share (in dollars) is shown. Find the 90% confidence interval for the variance and standard deviation for the prices. Assume the variable is normally distributed.

26.69	13.88	28.37	12.00
75.37	7.50	47.50	43.00
3.81	53.81	13.62	45.12
6.94	28.25	28.00	60.50
40.25	10.87	46.12	14.75

Source: *Pittsburgh Tribune Review.*
$259.343 < \sigma^2 < 772.724; 16.104 < \sigma < 27.798$

11. Number of Homeless Individuals A researcher wishes to find the confidence interval of the population standard deviation for the number of homeless people in a large city. A sample of 25 months had a standard deviation of 462. Find the 95% confidence interval. $130,136 < \sigma^2 < 413,084; 361 < \sigma < 643$

12. Home Ownership Rates The percentage rates of home ownership for 8 randomly selected states are listed below. Estimate the population variance and standard deviation for the percentage rate of home ownership with 99% confidence.

66.0   75.8   70.9   73.9   63.4   68.5   73.3   65.9

Source: *World Almanac.* $6.8 < \sigma^2 < 140; 2.6 < \sigma < 11.8$

Extending the Concepts

13. Calculator Battery Lifetimes A confidence interval for a standard deviation for large samples taken from a normally distributed population can be approximated by

$$s - z_{\alpha/2}\frac{s}{\sqrt{2n}} < \sigma < s + z_{\alpha/2}\frac{s}{\sqrt{2n}}$$

Find the 95% confidence interval for the population standard deviation of calculator batteries. A sample of 200 calculator batteries has a standard deviation of 18 months. $16.2 < \sigma < 19.8$

Technology *Step by Step*

TI-83 Plus or TI-84 Plus
Step by Step

The TI-83 Plus and TI-84 Plus do not have a built-in confidence interval for the variance or standard deviation. However, the downloadable program named SDINT is available on your CD and Online Learning Center. Follow the instructions with your CD for downloading the program.

Finding a Confidence Interval for the Variance and Standard Deviation (Data)

1. Enter the data values into L_1.
2. Press **PRGM,** move the cursor to the program named SDINT, and press **ENTER** twice.
3. Press **1** for Data.
4. Type L_1 for the list and press **ENTER.**
5. Type the confidence level and press **ENTER.**
6. Press **ENTER** to clear the screen.

Example TI7–4

This refers to Example 7–15 in the text. Find the 90% confidence interval for the variance and standard deviation for the data:

59   54   53   52   51   39   49   46   49   48

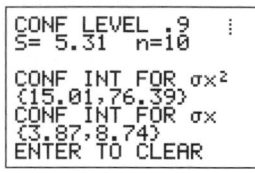

Finding a Confidence Interval for the Variance and Standard Deviation (Statistics)

1. Press **PRGM,** move the cursor to the program named SDINT, and press **ENTER** twice.

2. Press **2** for Stats.

3. Type the sample standard deviation and press **ENTER.**

4. Type the sample size and press **ENTER.**

5. Type the confidence level and press **ENTER.**

6. Press **ENTER** to clear the screen.

Example TI7–5

This refers to Example 7–14 in the text. Find the 95% confidence interval for the variance and standard deviation, given $n = 20$ and $s = 1.6$.

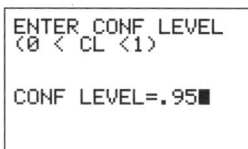

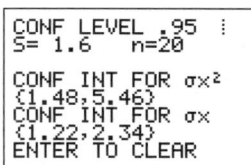

Summary

- An important aspect of inferential statistics is estimation. Estimations of parameters of populations are accomplished by selecting a random sample from that population and choosing and computing a statistic that is the best estimator of the parameter. A good estimator must be unbiased, consistent, and relatively efficient. The best estimate of μ is $\overline{X}$. (7–1)

- There are two types of estimates of a parameter: point estimates and interval estimates. A point estimate is a specific value. For example, if a researcher wishes to estimate the average length of a certain adult fish, a sample of the fish is selected and measured. The mean of this sample is computed, for example, 3.2 centimeters. From this sample mean, the researcher estimates the population mean to be 3.2 centimeters. The problem with point estimates is that the accuracy of the estimate cannot be determined. For this reason, statisticians prefer to use the interval estimate. By computing an interval about the sample value, statisticians can be 95 or 99% (or some other percentage) confident that their estimate contains the true parameter. The confidence level is determined by the researcher. The higher the confidence level, the wider the interval of the estimate must be. For example, a 95% confidence interval of the true mean length of a certain species of fish might be

$$3.17 < \mu < 3.23$$

whereas the 99% confidence interval might be

$$3.15 < \mu < 3.25 \quad (7\text{–}1)$$

- When the population standard deviation is known, the z value is used to compute the confidence interval. (7–1)
- Closely related to computing confidence intervals is the determination of the sample size to make an estimate of the mean. This information is needed to determine the minimum sample size necessary.

  1. The degree of confidence must be stated.
  2. The population standard deviation must be known or be able to be estimated.
  3. The margin of error must be stated. (7–1)

- If the population standard deviation is unknown, the t value is used. When the sample size is less than 30, the population must be normally distributed. (7–2)
- Confidence intervals and sample sizes can also be computed for proportions by using the normal distribution. (7–3)
- Finally, confidence intervals for variances and standard deviations can be computed by using the chi-square distribution. (7–4)

Important Terms

assumptions 357	degrees of freedom 370	margin of error 359	robust 357
chi-square distribution 386	estimation 356	point estimate 357	t distribution 370
confidence interval 358	estimator 357	proportion 377	unbiased estimator 357
confidence level 358	interval estimate 358	relatively efficient estimator 357	
consistent estimator 357			

Important Formulas

Formula for the confidence interval of the mean when σ is known (when $n \geq 30$, s can be used if σ is unknown):

$$\overline{X} - z_{\alpha/2}\left(\frac{\sigma}{\sqrt{n}}\right) < \mu < \overline{X} + z_{\alpha/2}\left(\frac{\sigma}{\sqrt{n}}\right)$$

Formula for the sample size for means:

$$n = \left(\frac{z_{\alpha/2} \cdot \sigma}{E}\right)^2$$

where E is the margin of error.

Formula for the confidence interval of the mean when σ is unknown:

$$\overline{X} - t_{\alpha/2}\left(\frac{s}{\sqrt{n}}\right) < \mu < \overline{X} + t_{\alpha/2}\left(\frac{s}{\sqrt{n}}\right)$$

Formula for the confidence interval for a proportion:

$$\hat{p} - z_{\alpha/2}\sqrt{\frac{\hat{p}\hat{q}}{n}} < p < \hat{p} + z_{\alpha/2}\sqrt{\frac{\hat{p}\hat{q}}{n}}$$

where $\hat{p} = X/n$ and $\hat{q} = 1 - \hat{p}$.

Formula for the sample size for proportions:

$$n = \hat{p}\hat{q}\left(\frac{z_{\alpha/2}}{E}\right)^2$$

Formula for the confidence interval for a variance:

$$\frac{(n-1)s^2}{\chi^2_{\text{right}}} < \sigma^2 < \frac{(n-1)s^2}{\chi^2_{\text{left}}}$$

Formula for confidence interval for a standard deviation:

$$\sqrt{\frac{(n-1)s^2}{\chi^2_{\text{right}}}} < \sigma < \sqrt{\frac{(n-1)s^2}{\chi^2_{\text{left}}}}$$

Review Exercises

1. Eight chemical elements do not have isotopes (different forms of the same element having the same atomic number but different atomic weights). A random sample of 30 of the elements that do have isotopes showed a mean number of 19.63 isotopes per element and the population a standard deviation of 18.73. Estimate the true mean number of isotopes for all elements with isotopes with 90% confidence. (7–1)

   Source: *Time Almanac.* $13.99 < \mu < 25.27$ (or $14 < \mu < 25$) (TI: $14.005 < \mu < 25.255$)

2. **Vacation Days** A U.S. Travel Data Center survey reported that Americans stayed an average of 7.5 nights when they went on vacation. The sample size was 1500. Find a point estimate of the population mean. Find the 95% confidence interval of the true mean. Assume the population standard deviation was 0.8. (7–1)

   Source: *USA TODAY.* 7.5; $7.46 < \mu < 7.54$

3. **Spending for Postage** A researcher wishes to estimate within $25 the average cost of postage a community college spends in one year. If she wishes to be 90% confident, how large of a sample would be necessary if the population standard deviation is $80? (7–1) 28

4. **Shopping Survey** A random sample of 49 shoppers showed that they spend an average of $23.45 per visit at the Saturday Mornings Bookstore. The standard deviation of the population is $2.80. Find a point estimate of the population mean. Find the 90% confidence interval of the true mean. (7–1) $23.45; $22.79 < \mu < 24.11

5. **Lengths of Children's Animated Films** The lengths (in minutes) of a random selection of popular children's animated films are listed below. Estimate the true mean length of all children's animated films with 95% confidence. (7–2) $76.9 < \mu < 88.3$. Assume normal distribution.

   | 93 | 83 | 76 | 92 | 77 | 81 | 78 | 100 | 78 | 76 | 75 |

6. **Dog Bites to Postal Workers** For a certain urban area, in a sample of 5 months, on average 28 mail carriers were bitten by dogs each month. The standard deviation of the sample was 3. Find the 90% confidence interval of the true mean number of mail carriers who are bitten by dogs each month. Assume the variable is normally distributed. (7–2) $25 < \mu < 31$

7. **Presidential Travel** In a survey of 1004 individuals, 442 felt that President George W. Bush spent too much time away from Washington. Find a 95% confidence interval for the true population proportion. (7–3)

   Source: *USA TODAY/CNN/Gallup Poll.* $0.409 < p < 0.471$

8. **Vacation Sites** A U.S. Travel Data Center's survey of 1500 adults found that 42% of respondents stated that they favor historical sites as vacations. Find the 95% confidence interval of the true proportion of

   all adults who favor visiting historical sites as vacations. (7–3)

   Source: *USA TODAY.* $0.395 < p < 0.445$

9. **Emergency Room Accidents** In a study of 200 accidents that required treatment in an emergency room, 80 occurred at work. Find the 90% confidence interval of the true proportion of accidents that occurred at work. (7–3) $0.343 < p < 0.457$

10. A local county has a very active adult education venue. A random sample of the population showed that 189 out of 400 persons 16 years old or older participated in some type of formal adult education activities, such as basic skills training, apprenticeships, personal interest courses, and part-time college or university degree programs. Estimate the true proportion of adults participating in some kind of formal education program with 98% confidence. (7–3) $0.414 < p < 0.531$

11. **Health Insurance Coverage for Children** A federal report stated that 88% of children under age 18 were covered by health insurance in 2000. How large a sample is needed to estimate the true proportion of covered children with 90% confidence with a confidence interval 0.05 wide? (7–3) 460

    Source: *Washington Observer-Reporter.*

12. **Child Care Programs** A study found that 73% of prekindergarten children ages 3 to 5 whose mothers had a bachelor's degree or higher were enrolled in center-based early childhood care and education programs. How large a sample is needed to estimate the true proportion within 3 percentage points with 95% confidence? How large a sample is needed if you had no prior knowledge of the proportion? (7–3) 842 children; 1068 children

13. **Baseball Diameters** The standard deviation of the diameter of 18 baseballs was 0.29 cm. Find the 95% confidence interval of the true standard deviation of the diameters of the baseballs. Do you think the manufacturing process should be checked for inconsistency? (7–4) $0.218 < \sigma < 0.435$. Yes. It seems that there is a large standard deviation.

14. **MPG for Lawn Mowers** A random sample of 22 lawn mowers was selected, and the motors were tested to see how many miles per gallon of gasoline each one obtained. The variance of the measurements was 2.6. Find the 95% confidence interval of the true variance. (7–4) $1.5 < \sigma^2 < 5.3$

15. **Lifetimes of Snowmobiles** A random sample of 15 snowmobiles was selected, and the lifetime (in months) of the batteries was measured. The variance of the sample was 8.6. Find the 90% confidence interval of the true variance. (7–4) $5.1 < \sigma^2 < 18.3$

16. **Length of Children's Animated Films** Use the data from Exercise 5 to estimate the population variance (standard deviation) in length of children's animated films with 99% confidence. (7–4) $28.6 < \sigma^2 < 334.2$; $5.3 < \sigma < 18.3$

Statistics Today

Would You Change the Channel?–Revisited

The estimates given in the survey are point estimates. However, since the margin of error is stated to be 3 percentage points, an interval estimate can easily be obtained. For example, if 45% of the people changed the channel, then the confidence interval of the true percentages of people who changed channels would be 42% $< p <$ 48%. The article fails to state whether a 90%, 95%, or some other percentage was used for the confidence interval.

Using the formula given in Section 7–3, a minimum sample size of 1068 would be needed to obtain a 95% confidence interval for p, as shown. Use $\hat{p}$ and $\hat{q}$ as 0.5, since no value is known for $\hat{p}$.

$$n = \hat{p}\hat{q}\left(\frac{z_{\alpha/2}}{E}\right)^2$$

$$= (0.5)(0.5)\left(\frac{1.96}{0.03}\right)^2 = 1067.1$$

$$= 1068$$

Data Analysis

The Data Bank is found in Appendix D, or on the World Wide Web by following links from www.mhhe.com/math/stat/bluman/.

1. From the Data Bank choose a variable, find the mean, and construct the 95 and 99% confidence intervals of the population mean. Use a sample of at least 30 subjects. Find the mean of the population, and determine whether it falls within the confidence interval.

2. Repeat Exercise 1, using a different variable and a sample of 15.

3. Repeat Exercise 1, using a proportion. For example, construct a confidence interval for the proportion of individuals who did not complete high school.

4. From Data Set III in Appendix D, select a sample of 30 values and construct the 95 and 99% confidence

intervals of the mean length in miles of major North American rivers. Find the mean of all the values, and determine if the confidence intervals contain the mean.

5. From Data Set VI in Appendix D, select a sample of 20 values and find the 90% confidence interval of the mean of the number of acres. Find the mean of all the values, and determine if the confidence interval contains the mean.

6. Select a random sample of 20 of the record high temperatures in the United States, found in Data Set I in Appendix D. Find the proportion of temperatures below 110°. Construct a 95% confidence interval for this proportion. Then find the true proportion of temperatures below 110°, using all the data. Is the true proportion contained in the confidence interval? Explain.

Chapter Quiz

Determine whether each statement is true or false. If the statement is false, explain why.

1. Interval estimates are preferred over point estimates since a confidence level can be specified. True

2. For a specific confidence interval, the larger the sample size, the smaller the margin of error will be. True

3. An estimator is consistent if as the sample size decreases, the value of the estimator approaches the value of the parameter estimated. False

4. To determine the sample size needed to estimate a parameter, you must know the margin of error. True

Select the best answer.

5. When a 99% confidence interval is calculated instead of a 95% confidence interval with n being the same, the margin of error will be
   a. Smaller
   b. Larger
   c. The same
   d. It cannot be determined.

6. The best point estimate of the population mean is
   a. The sample mean
   b. The sample median
   c. The sample mode
   d. The sample midrange

7. When the population standard deviation is unknown and the sample size is less than 30, what table value should be used in computing a confidence interval for a mean?
   a. z
   b. t
   c. Chi-square
   d. None of the above

Complete the following statements with the best answer.

8. A good estimator should be _____, _____, and _____. Unbiased, consistent, relatively efficient

9. The maximum difference between the point estimate of a parameter and the actual value of the parameter is called _____. Margin of error

10. The statement "The average height of an adult male is 5 feet 10 inches" is an example of a(n) _____ estimate. Point

11. The three confidence intervals used most often are the _____%, _____%, and _____%. 90; 95; 99

12. **Cost of Textbooks** An irate student complained that the cost of textbooks was too high. He randomly surveyed 36 other students and found that the mean amount of money spent for textbooks was $121.60. If the standard deviation of the population was $6.36, find the best point estimate and the 90% confidence interval of the true mean. $121.60; $119.85 < \mu < 123.35

13. **Doctor Visit Costs** An irate patient complained that the cost of a doctor's visit was too high. She randomly surveyed 20 other patients and found that the mean amount of money they spent on each doctor's visit was $44.80. The standard deviation of the sample was $3.53. Find a point estimate of the population mean. Find the 95% confidence interval of the population mean. Assume the variable is normally distributed. $44.80; $43.15 < \mu < 46.45

14. **Weights of Minivans** The average weight of 40 randomly selected minivans was 4150 pounds. The standard deviation was 480 pounds. Find a point estimate of the population mean. Find the 99% confidence interval of the true mean weight of the minivans. 4150; $3954 < \mu < 4346$

15. **Ages of Insurance Representatives** In a study of 10 insurance sales representatives from a certain large city, the average age of the group was 48.6 years and the standard deviation was 4.1 years. Assume the variable is normally distributed. Find the 95% confidence interval of the population mean age of all insurance sales representatives in that city. $45.7 < \mu < 51.5$

16. **Patients Treated in Hospital Emergency Rooms** In a hospital, a sample of 8 weeks was selected, and it was found that an average of 438 patients was treated in the emergency room each week. The standard deviation was 16. Find the 99% confidence interval of the true mean. Assume the variable is normally distributed. $418 < \mu < 458$

17. **Burglaries** For a certain urban area, it was found that in a sample of 4 months, an average of 31 burglaries occurred each month. The standard deviation was 4. Assume the variable is normally distributed. Find the 90% confidence interval of the true mean number of burglaries each month. $26 < \mu < 36$

18. **Hours Spent Studying** A university dean wishes to estimate the average number of hours that freshmen study each week. The standard deviation from a previous study is 2.6 hours. How large a sample must be selected if he wants to be 99% confident of finding whether the true mean differs from the sample mean by 0.5 hour? 180

19. **Money Spent on Road Repairs** A researcher wishes to estimate within $300 the true average amount of money a county spends on road repairs each year. If she wants to be 90% confident, how large a sample is necessary? The standard deviation is known to be $900. 25

20. **Political Survey** A political analyst found that 43% of 300 Republican voters feel that the federal government has too much power. Find the 95% confidence interval of the population proportion of Republican voters who feel this way. $0.374 < p < 0.486$

21. **Emergency Room Accidents** In a study of 150 accidents that required treatment in an emergency room, 36% involved children under 6 years of age. Find the 90% confidence interval of the true proportion of accidents that involve children under the age of 6. $0.295 < p < 0.425$

22. **Television Set Ownership** A survey of 90 families showed that 40 owned at least one television set. Find the 95% confidence interval of the true proportion of families who own at least one television set. $0.342 < p < 0.547$

23. **Skipping Lunch** A nutritionist wishes to determine, within 3%, the true proportion of adults who do not eat any lunch. If he wishes to be 95% confident that his estimate contains the population proportion, how large a sample will be necessary? A previous study found that 15% of the 125 people surveyed said they did not eat lunch. 545

24. **Novel Pages** A sample of 25 novels has a standard deviation of 9 pages. Find the 95% confidence interval of the population standard deviation. $7 < \sigma < 13$

25. **Truck Safety Check** Find the 90% confidence interval for the variance and standard deviation for the time it takes a state police inspector to check a truck for safety if a sample of 27 trucks has a standard deviation of 6.8 minutes. Assume the variable is normally distributed. $30.9 < \sigma^2 < 78.2; 5.6 < \sigma < 8.8$

26. **Automobile Pollution** A sample of 20 automobiles has a pollution by-product release standard deviation of 2.3 ounces when 1 gallon of gasoline is used. Find the 90% confidence interval of the population standard deviation. $1.8 < \sigma < 3.2$

Critical Thinking Challenges

A confidence interval for a median can be found by using these formulas

$$U = \frac{n+1}{2} + \frac{z_{\alpha/2}\sqrt{n}}{2}$$ (round up)

$$L = n - U + 1$$

to define positions in the set of ordered data values.

Suppose a data set has 30 values, and you want to find the 95% confidence interval for the median. Substituting in the formulas, you get

$$U = \frac{30+1}{2} + \frac{1.96\sqrt{30}}{2} = 21$$ (rounded up)

$$L = 30 - 21 + 1 = 10$$

when $n = 30$ and $z_{\alpha/2} = 1.96$.

Arrange the data in order from smallest to largest, and then select the 10th and 21st values of the data array; hence, $X_{10} <$ median $< X_{21}$.

Find the 90% confidence interval for the median for the given data.

84	49	3	133	85	4340	461	60	28	97
14	252	18	16	24	346	254	29	254	6
31	104	72	29	391	19	125	10	6	17
72	31	23	225	72	5	61	366	77	8
26	8	55	138	158	846	123	47	21	82

Data Projects

1. **Business and Finance** Use 30 stocks classified as the Dow Jones industrials as the sample. Note the amount each stock has gained or lost in the last quarter. Compute the mean and standard deviation for the data set. Compute the 95% confidence interval for the mean and the 95% confidence interval for the standard deviation. Compute the percentage of stocks that had a gain in the last quarter. Find a 95% confidence interval for the percentage of stocks with a gain.

2. **Sports and Leisure** Use the top home run hitter from each major league baseball team as the data set. Find the mean and the standard deviation for the number of home runs hit by the top hitter on each team. Find a 95% confidence interval for the mean number of home runs hit.

3. **Technology** Use the data collected in data project 3 of Chapter 2 regarding song lengths. Select a specific genre, and compute the percentage of songs in the sample that are of that genre. Create a 95% confidence interval for the true percentage. Use the entire music library, and find the population percentage of the library with that genre. Does the population percentage fall within the confidence interval?

4. **Health and Wellness** Use your class as the sample. Have each student take her or his temperature on a healthy day. Compute the mean and standard deviation for the sample. Create a 95% confidence interval for the mean temperature. Does the confidence interval obtained support the long-held belief that the average body temperature is 98.6°F?

5. **Politics and Economics** Select five political polls and note the margin of error, sample size, and percent favoring the candidate for each. For each poll, determine the level of confidence that must have been used to obtain the margin of error given, knowing the percent favoring the candidate and number of participants. Is there a pattern that emerges?

6. **Your Class** Have each student compute his or her body mass index (BMI) (703 times weight in pounds, divided by the quantity height in inches squared). Find the mean and standard deviation for the data set. Compute a 95% confidence interval for the mean BMI of a student. A BMI score over 30 is considered obese. Does the confidence interval indicate that the mean for BMI could be in the obese range?

Answers to Applying the Concepts

Section 7–1   Making Decisions with Confidence Intervals

1. Answers will vary. One possible answer is to find out the average number of Kleenexes that a group of randomly selected individuals use in a 2-week period.

2. People usually need Kleenexes when they have a cold or when their allergies are acting up.

3. If we want to concentrate on the number of Kleenexes used when people have colds, we select a random sample of people with colds and have them keep a record of how many Kleenexes they use during their colds.

4. Answers may vary. I will use a 95% confidence interval:

$$\bar{x} \pm 1.96\frac{\sigma}{\sqrt{n}} = 57 \pm 1.96\frac{15}{\sqrt{85}} = 57 \pm 3.2$$

I am 95% confident that the interval 53.8–60.2 contains the true mean number of Kleenexes used by people when they have colds. It seems reasonable to put 60 Kleenexes in the new automobile glove compartment boxes.

5. Answers will vary. Since I am 95% confident that the interval contains the true average, any number of Kleenexes between 54 and 60 would be reasonable. Sixty seemed to be the most reasonable answer, since it is close to 2 standard deviations above the sample mean.

Section 7–2   Sport Drink Decision

1. Answers will vary. One possible answer is that this is a small sample since we are only looking at seven popular sport drinks.

2. The mean cost per container is $1.25, with standard deviation of $0.39. The 90% confidence interval is

$$\overline{X} \pm t_{\alpha/2}\frac{s}{\sqrt{n}} = 1.25 \pm 1.943\frac{0.39}{\sqrt{7}} = 1.25 \pm 0.29$$

or $0.96 < \mu < 1.54$

The 10-K, All Sport, Exceed, and Hydra Fuel all fall outside of the confidence interval.

3. None of the values appear to be outliers.

4. There are $7 - 1 = 6$ degrees of freedom.

5. Cost per serving would impact my decision on purchasing a sport drink, since this would allow me to compare the costs on an equal scale.

6. Answers will vary.

Section 7–3   Contracting Influenza

1. (95% CI) means that these are the 95% confidence intervals constructed from the data.

2. The margin of error for men reporting influenza is $(50.5 - 47.1)/2 = 1.7\%$.

3. The total sample size was 19,774.

4. The larger the sample size, the smaller the margin of error (all other things being held constant).

5. A 90% confidence interval would be narrower (smaller) than a 95% confidence interval, since we need to include fewer values in the interval.

6. The 51.5% is the middle of the confidence interval, since it is the point estimate for the confidence interval.

Section 7–4   Confidence Interval for Standard Deviation

1. The data represent a population, since we have the age at death for all deceased Presidents (at the time of the writing of this book).

2. Answers will vary. One possible sample is 56, 67, 53, 46, 63, 77, 63, 57, 71, 57, 80, 65, which results in a standard deviation of 9.9 years and a variance of 98.0.

3. Answers will vary. The 95% confidence interval for the standard deviation is $\sqrt{\frac{(n-1)s^2}{\chi^2_{\text{right}}}}$ to $\sqrt{\frac{(n-1)s^2}{\chi^2_{\text{left}}}}$. In this case we have $\sqrt{\frac{(12-1)9.9^2}{21.920}} = \sqrt{49.1839} = 7.0$ to $\sqrt{\frac{(12-1)9.9^2}{3.8158}} = \sqrt{282.538} = 16.8$, or 7.0 to 16.8 years.

4. The standard deviation for all the data values is 12.0 years.

5. Answers will vary. Yes, the confidence interval does contain the population standard deviation.

6. Answers will vary.

7. We need to assume that the distribution of ages at death is normal.

CHAPTER

8

Hypothesis Testing

Objectives

After completing this chapter, you should be able to

1 Understand the definitions used in hypothesis testing.

2 State the null and alternative hypotheses.

3 Find critical values for the z test.

4 State the five steps used in hypothesis testing.

5 Test means when σ is known, using the z test.

6 Test means when σ is unknown, using the t test.

7 Test proportions, using the z test.

8 Test variances or standard deviations, using the chi-square test.

9 Test hypotheses, using confidence intervals.

10 Explain the relationship between type I and type II errors and the power of a test.

Outline

How Much Better Is Better?

Suppose a school superintendent reads an article which states that the overall mean score for the SAT is 910. Furthermore, suppose that, for a sample of students, the average of the SAT scores in the superintendent's school district is 960. Can the superintendent conclude that the students in his school district scored higher on average? At first glance, you might be inclined to say yes, since 960 is higher than 910. But recall that the means of samples vary about the population mean when samples are selected from a specific population. So the question arises, Is there a real difference in the means, or is the difference simply due to chance (i.e., sampling error)? In this chapter, you will learn how to answer that question by using statistics that explain hypothesis testing. See Statistics Today—Revisited for the answer. In this chapter, you will learn how to answer many questions of this type by using statistics that are explained in the theory of hypothesis testing.

Introduction

Researchers are interested in answering many types of questions. For example, a scientist might want to know whether the earth is warming up. A physician might want to know whether a new medication will lower a person's blood pressure. An educator might wish to see whether a new teaching technique is better than a traditional one. A retail merchant might want to know whether the public prefers a certain color in a new line of fashion. Automobile manufacturers are interested in determining whether seat belts will reduce the severity of injuries caused by accidents. These types of questions can be addressed through statistical **hypothesis testing,** which is a decision-making process for evaluating claims about a population. In hypothesis testing, the researcher must define the population under study, state the particular hypotheses that will be investigated, give the significance level, select a sample from the population, collect the data, perform the calculations required for the statistical test, and reach a conclusion.

Hypotheses concerning parameters such as means and proportions can be investigated. There are two specific statistical tests used for hypotheses concerning means: the *z test* and the *t test.* This chapter will explain in detail the hypothesis-testing procedure along with the *z* test and the *t* test. In addition, a hypothesis-testing procedure for testing a single variance or standard deviation using the chi-square distribution is explained in Section 8–5.

The three methods used to test hypotheses are

1. The traditional method
2. The *P*-value method
3. The confidence interval method

The *traditional method* will be explained first. It has been used since the hypothesis-testing method was formulated. A newer method, called the *P-value method,* has become popular with the advent of modern computers and high-powered statistical calculators. It will be explained at the end of Section 8–2. The third method, the *confidence interval method,* is explained in Section 8–6 and illustrates the relationship between hypothesis testing and confidence intervals.

8–1   Steps in Hypothesis Testing–Traditional Method

Every hypothesis-testing situation begins with the statement of a hypothesis.

A **statistical hypothesis** is a conjecture about a population parameter. This conjecture may or may not be true.

Objective 1

Understand the definitions used in hypothesis testing.

There are two types of statistical hypotheses for each situation: the null hypothesis and the alternative hypothesis.

The **null hypothesis,** symbolized by H_0, is a statistical hypothesis that states that there is no difference between a parameter and a specific value, or that there is no difference between two parameters.

The **alternative hypothesis,** symbolized by H_1, is a statistical hypothesis that states the existence of a difference between a parameter and a specific value, or states that there is a difference between two parameters.

(*Note:* Although the definitions of null and alternative hypotheses given here use the word *parameter,* these definitions can be extended to include other terms such as *distributions* and *randomness.* This is explained in later chapters.)

As an illustration of how hypotheses should be stated, three different statistical studies will be used as examples.

Objective 2

State the null and alternative hypotheses.

Situation A   A medical researcher is interested in finding out whether a new medication will have any undesirable side effects. The researcher is particularly concerned with the pulse rate of the patients who take the medication. Will the pulse rate increase, decrease, or remain unchanged after a patient takes the medication?

Since the researcher knows that the mean pulse rate for the population under study is 82 beats per minute, the hypotheses for this situation are

$$H_0: \mu = 82 \qquad \text{and} \qquad H_1: \mu \neq 82$$

The null hypothesis specifies that the mean will remain unchanged, and the alternative hypothesis states that it will be different. This test is called a *two-tailed test* (a term that will be formally defined later in this section), since the possible side effects of the medicine could be to raise or lower the pulse rate.

Situation B A chemist invents an additive to increase the life of an automobile battery. If the mean lifetime of the automobile battery without the additive is 36 months, then her hypotheses are

$$H_0: \mu = 36 \quad \text{and} \quad H_1: \mu > 36$$

In this situation, the chemist is interested only in increasing the lifetime of the batteries, so her alternative hypothesis is that the mean is greater than 36 months. The null hypothesis is that the mean is equal to 36 months. This test is called *right-tailed,* since the interest is in an increase only.

Situation C A contractor wishes to lower heating bills by using a special type of insulation in houses. If the average of the monthly heating bills is $78, her hypotheses about heating costs with the use of insulation are

$$H_0: \mu = \$78 \quad \text{and} \quad H_1: \mu < \$78$$

This test is a *left-tailed test,* since the contractor is interested only in lowering heating costs.

To state hypotheses correctly, researchers must translate the *conjecture* or *claim* from words into mathematical symbols. The basic symbols used are as follows:

Equal to	$=$	Greater than	$>$
Not equal to	$\neq$	Less than	$<$

The null and alternative hypotheses are stated together, and the null hypothesis contains the equals sign, as shown (where k represents a specified number).

Two-tailed test	Right-tailed test	Left-tailed test
$H_0: \mu = k$	$H_0: \mu = k$	$H_0: \mu = k$
$H_1: \mu \neq k$	$H_1: \mu > k$	$H_1: \mu < k$

The formal definitions of the different types of tests are given later in this section.

In this book, the null hypothesis is always stated using the equals sign. This is done because in most professional journals, and when we test the null hypothesis, the assumption is that the mean, proportion, or standard deviation is equal to a given specific value. Also, when a researcher conducts a study, he or she is generally looking for evidence to support a claim. Therefore, the claim should be stated as the alternative hypothesis, i.e., using $<$ or $>$ or $\neq$. Because of this, the alternative hypothesis is sometimes called the **research hypothesis.**

Table 8–1	Hypothesis-Testing Common Phrases	
$>$		**$<$**
Is greater than		Is less than
Is above		Is below
Is higher than		Is lower than
Is longer than		Is shorter than
Is bigger than		Is smaller than
Is increased		Is decreased or reduced from
$=$		**$\neq$**
Is equal to		Is not equal to
Is the same as		Is different from
Has not changed from		Has changed from
Is the same as		Is not the same as

A claim, though, can be stated as either the null hypothesis or the alternative hypothesis; however, the statistical evidence can only *support* the claim if it is the alternative hypothesis. Statistical evidence can be used to *reject* the claim if the claim is the null hypothesis. These facts are important when you are stating the conclusion of a statistical study.

Table 8–1 shows some common phrases that are used in hypotheses and conjectures, and the corresponding symbols. This table should be helpful in translating verbal conjectures into mathematical symbols.

Example 8–1

State the null and alternative hypotheses for each conjecture.

  a. A researcher thinks that if expectant mothers use vitamin pills, the birth weight of the babies will increase. The average birth weight of the population is 8.6 pounds.

  b. An engineer hypothesizes that the mean number of defects can be decreased in a manufacturing process of compact disks by using robots instead of humans for certain tasks. The mean number of defective disks per 1000 is 18.

  c. A psychologist feels that playing soft music during a test will change the results of the test. The psychologist is not sure whether the grades will be higher or lower. In the past, the mean of the scores was 73.

Solution

  a. H_0: $\mu = 8.6$ and H_1: $\mu > 8.6$

  b. H_0: $\mu = 18$ and H_1: $\mu < 18$

  c. H_0: $\mu = 73$ and H_1: $\mu \neq 73$

After stating the hypothesis, the researcher designs the study. The researcher selects the correct *statistical test,* chooses an appropriate *level of significance,* and formulates a plan for conducting the study. In situation A, for instance, the researcher will select a sample of patients who will be given the drug. After allowing a suitable time for the drug to be absorbed, the researcher will measure each person's pulse rate.

Recall that when samples of a specific size are selected from a population, the means of these samples will vary about the population mean, and the distribution of the sample means will be approximately normal when the sample size is 30 or more. (See Section 6–3.) So even if the null hypothesis is true, the mean of the pulse rates of the sample of patients will not, in most cases, be exactly equal to the population mean of 82 beats per minute. There are two possibilities. Either the null hypothesis is true, and the difference between the sample mean and the population mean is due to chance; *or* the null hypothesis is false, and the sample came from a population whose mean is not 82 beats per minute but is some other value that is not known. These situations are shown in Figure 8–1.

The farther away the sample mean is from the population mean, the more evidence there would be for rejecting the null hypothesis. The probability that the sample came from a population whose mean is 82 decreases as the distance or absolute value of the difference between the means increases.

If the mean pulse rate of the sample were, say, 83, the researcher would probably conclude that this difference was due to chance and would not reject the null hypothesis. But if the sample mean were, say, 90, then in all likelihood the researcher would conclude that the medication increased the pulse rate of the users and would reject the null hypothesis. The question is, Where does the researcher draw the line? This decision is not made on feelings or intuition; it is made statistically. That is, the difference must be significant and in all likelihood not due to chance. Here is where the concepts of statistical test and level of significance are used.

Figure 8–1

**Situations in
Hypothesis Testing**

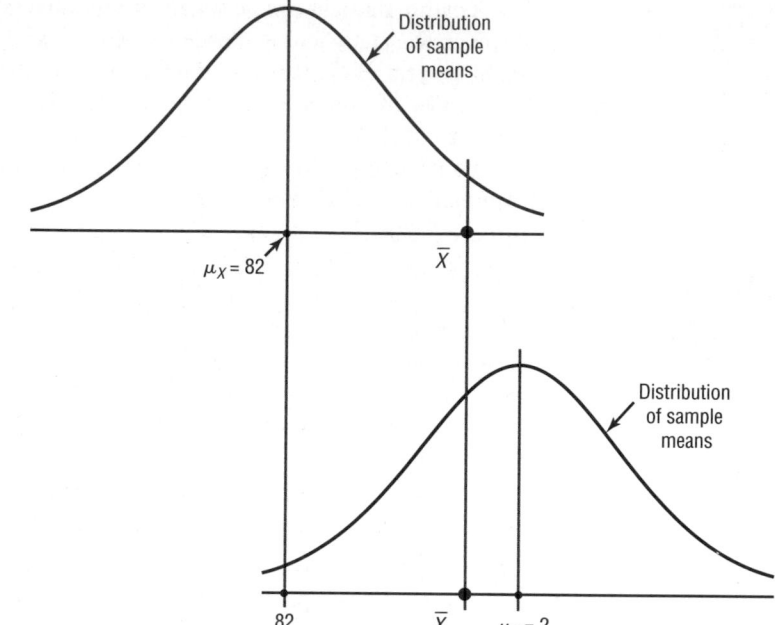

(a) H_0 is true

Distribution
of sample
means

$\mu_{\bar{X}} = 82$ $\bar{X}$

(b) H_0 is false

Distribution
of sample
means

82    $\bar{X}$ $\mu_{\bar{X}} = ?$

A **statistical test** uses the data obtained from a sample to make a decision about
whether the null hypothesis should be rejected.

The numerical value obtained from a statistical test is called the **test value.**

In this type of statistical test, the mean is computed for the data obtained from the
sample and is compared with the population mean. Then a decision is made to reject or
not reject the null hypothesis on the basis of the value obtained from the statistical test.
If the difference is significant, the null hypothesis is rejected. If it is not, then the null
hypothesis is not rejected.

In the hypothesis-testing situation, there are four possible outcomes. In reality, the
null hypothesis may or may not be true, and a decision is made to reject or not reject it
on the basis of the data obtained from a sample. The four possible outcomes are shown
in Figure 8–2. Notice that there are two possibilities for a correct decision and two possibilities for an incorrect decision.

Figure 8–2

**Possible Outcomes of
a Hypothesis Test**

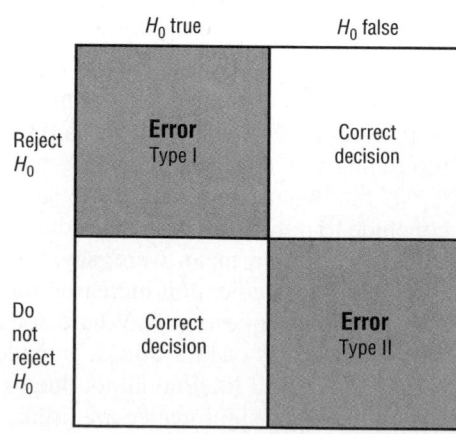

	H_0 true	H_0 false
Reject H_0	**Error** Type I	Correct decision
Do not reject H_0	Correct decision	**Error** Type II

If a null hypothesis is true and it is rejected, then a *type I error* is made. In situation A, for instance, the medication might not significantly change the pulse rate of all the users in the population; but it might change the rate, by chance, of the subjects in the sample. In this case, the researcher will reject the null hypothesis when it is really true, thus committing a type I error.

On the other hand, the medication might not change the pulse rate of the subjects in the sample, but when it is given to the general population, it might cause a significant increase or decrease in the pulse rate of users. The researcher, on the basis of the data obtained from the sample, will not reject the null hypothesis, thus committing a *type II error.*

In situation B, the additive might not significantly increase the lifetimes of automobile batteries in the population, but it might increase the lifetimes of the batteries in the sample. In this case, the null hypothesis would be rejected when it was really true. This would be a type I error. On the other hand, the additive might not work on the batteries selected for the sample, but if it were to be used in the general population of batteries, it might significantly increase their lifetimes. The researcher, on the basis of information obtained from the sample, would not reject the null hypothesis, thus committing a type II error.

A **type I error** occurs if you reject the null hypothesis when it is true.

A **type II error** occurs if you do not reject the null hypothesis when it is false.

The hypothesis-testing situation can be likened to a jury trial. In a jury trial, there are four possible outcomes. The defendant is either guilty or innocent, and he or she will be convicted or acquitted. See Figure 8–3.

Now the hypotheses are

H_0: The defendant is innocent

H_1: The defendant is not innocent (i.e., guilty)

Next, the evidence is presented in court by the prosecutor, and based on this evidence, the jury decides the verdict, innocent or guilty.

If the defendant is convicted but he or she did not commit the crime, then a type I error has been committed. See block 1 of Figure 8–3. On the other hand, if the defendant is convicted and he or she has committed the crime, then a correct decision has been made. See block 2.

If the defendant is acquitted and he or she did not commit the crime, a correct decision has been made by the jury. See block 3. However, if the defendant is acquitted and he or she did commit the crime, then a type II error has been made. See block 4.

Figure 8–3

Hypothesis Testing and a Jury Trial

H_0: The defendant is innocent.
H_1: The defendant is not innocent.

The results of a trial can be shown as follows:

	H_0 true (innocent)	H_0 false (not innocent)
Reject H_0 (convict)	Type I error 1.	Correct decision 2.
Do not reject H_0 (acquit)	Correct decision 3.	Type II error 4.

The decision of the jury does not prove that the defendant did or did not commit the crime. The decision is based on the evidence presented. If the evidence is strong enough, the defendant will be convicted in most cases. If the evidence is weak, the defendant will be acquitted in most cases. Nothing is proved absolutely. Likewise, the decision to reject or not reject the null hypothesis does not prove anything. *The only way to prove anything statistically is to use the entire population,* which, in most cases, is not possible. The decision, then, is made on the basis of probabilities. That is, when there is a large difference between the mean obtained from the sample and the hypothesized mean, the null hypothesis is probably not true. The question is, How large a difference is necessary to reject the null hypothesis? Here is where the level of significance is used.

> The **level of significance** is the maximum probability of committing a type I error. This probability is symbolized by α (Greek letter **alpha**). That is, $P(\text{type I error}) = \alpha$.

The probability of a type II error is symbolized by β, the Greek letter **beta.** That is, $P(\text{type II error}) = \beta$. In most hypothesis-testing situations, β cannot be easily computed; however, α and β are related in that decreasing one increases the other.

Statisticians generally agree on using three arbitrary significance levels: the 0.10, 0.05, and 0.01 levels. That is, if the null hypothesis is rejected, the probability of a type I error will be 10%, 5%, or 1%, depending on which level of significance is used. Here is another way of putting it: When $\alpha = 0.10$, there is a 10% chance of rejecting a true null hypothesis; when $\alpha = 0.05$, there is a 5% chance of rejecting a true null hypothesis; and when $\alpha = 0.01$, there is a 1% chance of rejecting a true null hypothesis.

In a hypothesis-testing situation, the researcher decides what level of significance to use. It does not have to be the 0.10, 0.05, or 0.01 level. It can be any level, depending on the seriousness of the type I error. After a significance level is chosen, a *critical value* is selected from a table for the appropriate test. If a z test is used, for example, the z table (Table E in Appendix C) is consulted to find the critical value. The critical value determines the critical and noncritical regions.

> The **critical value** separates the critical region from the noncritical region. The symbol for critical value is C.V.
>
> The **critical** or **rejection region** is the range of values of the test value that indicates that there is a significant difference and that the null hypothesis should be rejected.
>
> The **noncritical** or **nonrejection region** is the range of values of the test value that indicates that the difference was probably due to chance and that the null hypothesis should not be rejected.

The critical value can be on the right side of the mean or on the left side of the mean for a one-tailed test. Its location depends on the inequality sign of the alternative hypothesis. For example, in situation B, where the chemist is interested in increasing the average lifetime of automobile batteries, the alternative hypothesis is H_1: $\mu > 36$. Since the inequality sign is $>$, the null hypothesis will be rejected only when the sample mean is significantly greater than 36. Hence, the critical value must be on the right side of the mean. Therefore, this test is called a right-tailed test.

> A **one-tailed test** indicates that the null hypothesis should be rejected when the test value is in the critical region on one side of the mean. A one-tailed test is either a **right-tailed test** or **left-tailed test,** depending on the direction of the inequality of the alternative hypothesis.

Figure 8–4

Finding the Critical Value for $\alpha = 0.01$ (Right-Tailed Test)

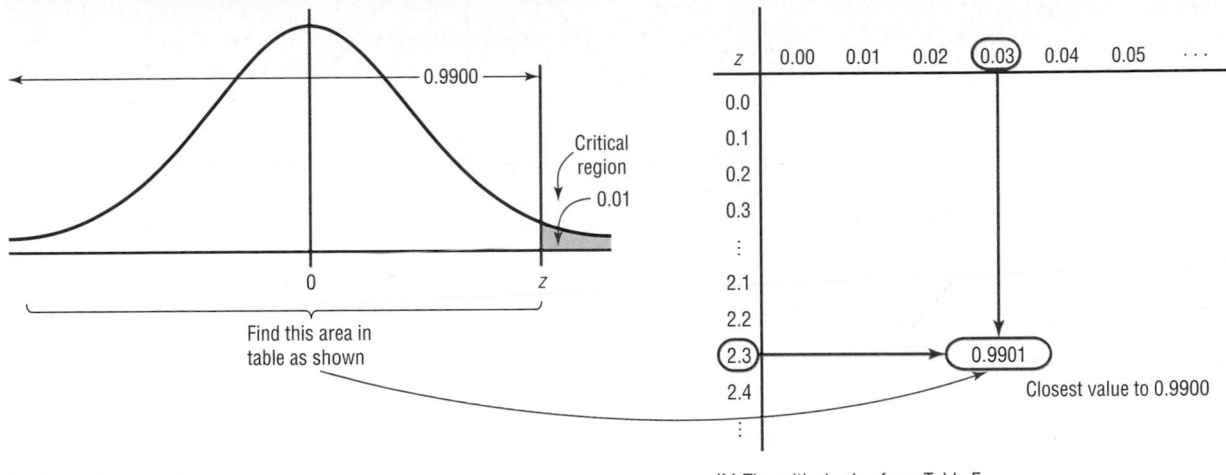

(a) The critical region

(b) The critical value from Table E

To obtain the critical value, the researcher must choose an alpha level. In situation B, suppose the researcher chose $\alpha = 0.01$. Then the researcher must find a z value such that 1% of the area falls to the right of the z value and 99% falls to the left of the z value, as shown in Figure 8–4(a).

Objective 3

Find critical values for the z test.

Next, the researcher must find the area value in Table E closest to 0.9900. The critical z value is 2.33, since that value gives the area closest to 0.9900 (that is, 0.9901), as shown in Figure 8–4(b).

The critical and noncritical regions and the critical value are shown in Figure 8–5.

Figure 8–5

Critical and Noncritical Regions for $\alpha = 0.01$ (Right-Tailed Test)

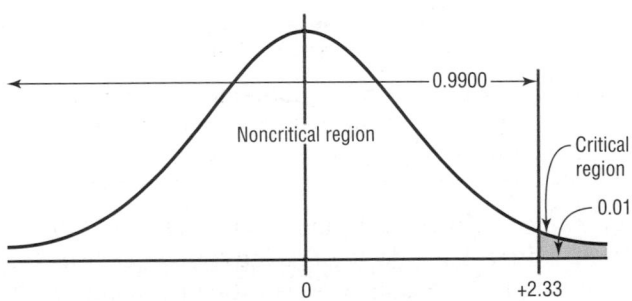

Now, move on to situation C, where the contractor is interested in lowering the heating bills. The alternative hypothesis is H_1: $\mu < \$78$. Hence, the critical value falls to the left of the mean. This test is thus a left-tailed test. At $\alpha = 0.01$, the critical value is -2.33, since 0.0099 is the closest value to 0.01. This is shown in Figure 8–6.

When a researcher conducts a two-tailed test, as in situation A, the null hypothesis can be rejected when there is a significant difference in either direction, above or below the mean.

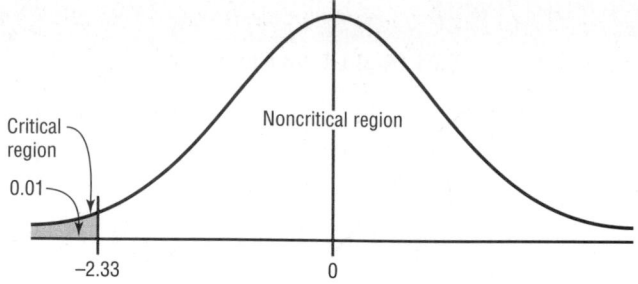

Figure 8–6

Critical and Noncritical
Regions for $\alpha = 0.01$
(Left-Tailed Test)

In a **two-tailed test,** the null hypothesis should be rejected when the test value is in either of the two critical regions.

For a two-tailed test, then, the critical region must be split into two equal parts. If $\alpha = 0.01$, then one-half of the area, or 0.005, must be to the right of the mean and one-half must be to the left of the mean, as shown in Figure 8–7.

In this case, the z value on the left side is found by looking up the z value corresponding to an area of 0.0050. The z value falls about halfway between -2.57 and -2.58 corresponding to the areas 0.0049 and 0.0051. The average of -2.57 and -2.58 is $[(-2.57) + (-2.58)] \div 2 = -2.575$ so if the z value is needed to three decimal places, -2.575 is used; however, if the z value is rounded to two decimal places, -2.58 is used.

On the right side, it is necessary to find the z value corresponding to $0.99 + 0.005$, or 0.9950. Again, the value falls between 0.9949 and 0.9951, so $+2.575$ or 2.58 can be used. See Figure 8–7.

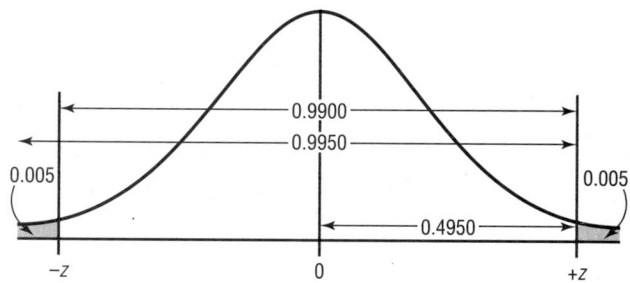

Figure 8–7

Finding the Critical
Values for $\alpha = 0.01$
(Two-Tailed Test)

The critical values are $+2.58$ and -2.58, as shown in Figure 8–8.

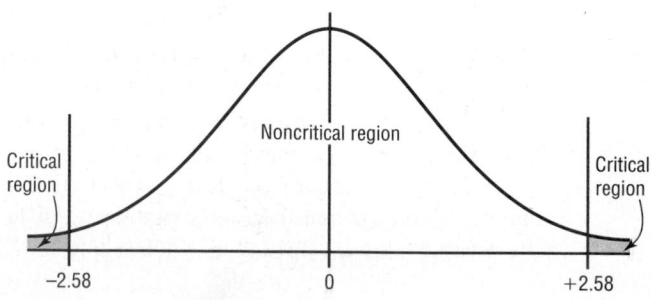

Figure 8–8

Critical and Noncritical
Regions for $\alpha = 0.01$
(Two-Tailed Test)

Similar procedures are used to find other values of α.

Figure 8–9 with rejection regions shaded shows the critical value (C.V.) for the three situations discussed in this section for values of $\alpha = 0.10$, $\alpha = 0.05$, and $\alpha = 0.01$. The procedure for finding critical values is outlined next (where k is a specified number).

Figure 8–9

Summary of Hypothesis Testing and Critical Values

$H_0: \mu = k$ $\begin{cases} \alpha = 0.10, \text{C.V.} = -1.28 \\ \alpha = 0.05, \text{C.V.} = -1.65 \\ \alpha = 0.01, \text{C.V.} = -2.33 \end{cases}$
$H_1: \mu < k$

(a) Left-tailed

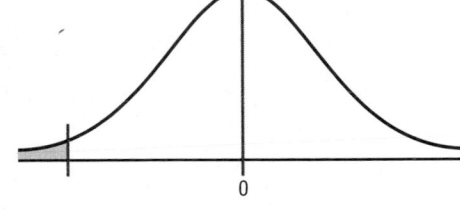

$H_0: \mu = k$ $\begin{cases} \alpha = 0.10, \text{C.V.} = +1.28 \\ \alpha = 0.05, \text{C.V.} = +1.65 \\ \alpha = 0.01, \text{C.V.} = +2.33 \end{cases}$
$H_1: \mu > k$

(b) Right-tailed

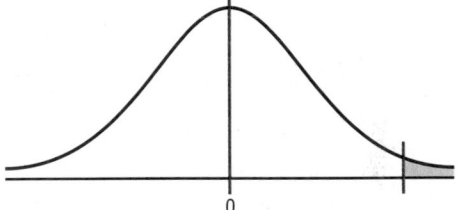

$H_0: \mu = k$ $\begin{cases} \alpha = 0.10, \text{C.V.} = \pm1.65 \\ \alpha = 0.05, \text{C.V.} = \pm1.96 \\ \alpha = 0.01, \text{C.V.} = \pm2.58 \end{cases}$
$H_1: \mu \neq k$

(c) Two-tailed

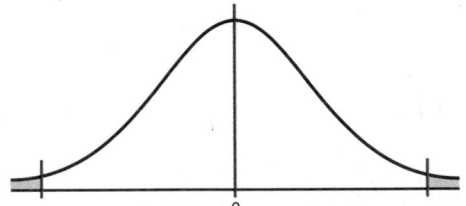

Procedure Table

Finding the Critical Values for Specific α Values, Using Table E

Step 1 Draw the figure and indicate the appropriate area.

   a. If the test is left-tailed, the critical region, with an area equal to α, will be on the left side of the mean.

   b. If the test is right-tailed, the critical region, with an area equal to α, will be on the right side of the mean.

   c. If the test is two-tailed, α must be divided by 2; one-half of the area will be to the right of the mean, and one-half will be to the left of the mean.

Step 2 *a.* For a left-tailed test, use the z value that corresponds to the area equivalent to α. in Table E.

   b. For a right-tailed test, use the z value that corresponds to the area equivalent to $1 - \alpha$.

   c. For a two-tailed test, use the z value that corresponds to $\alpha/2$ for the left value. It will be negative. For the right value, use the z value that corresponds to the area equivalent to $1 - \alpha/2$. It will be positive.

| Example 8–2 | Using Table E in Appendix C, find the critical value(s) for each situation and draw the appropriate figure, showing the critical region. |

    a. A left-tailed test with $\alpha = 0.10$.

    b. A two-tailed test with $\alpha = 0.02$.

    c. A right-tailed test with $\alpha = 0.005$.

Solution *a*

Step 1   Draw the figure and indicate the appropriate area. Since this is a left-tailed test, the area of 0.10 is located in the left tail, as shown in Figure 8–10.

Step 2   Find the area closest to 0.1000 in Table E. In this case, it is 0.1003. Find the z value that corresponds to the area 0.1003. It is -1.28. See Figure 8–10.

Figure 8–10	
Critical Value and Critical Region for part *a* of Example 8–2	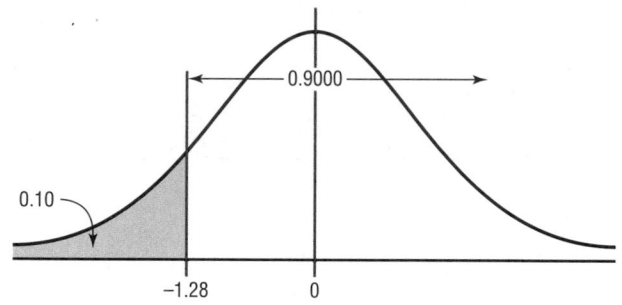

Solution *b*

Step 1   Draw the figure and indicate the appropriate area. In this case, there are two areas equivalent to $\alpha/2$, or $0.02/2 = 0.01$.

Step 2   For the left z critical value, find the area closest to $\alpha/2$, or $0.02/2 = 0.01$. In this case, it is 0.0099.

    For the right z critical value, find the area closest to $1 - \alpha/2$, or $1 - 0.02/2 = 0.9900$. In this case, it is 0.9901.

    Find the z values for each of the areas. For 0.0099, $z = -2.33$. For the area of 0.9901, $z = 0.9901$, $z = +2.33$. See Figure 8–11.

Figure 8–11	
Critical Values and Critical Regions for part *b* of Example 8–2	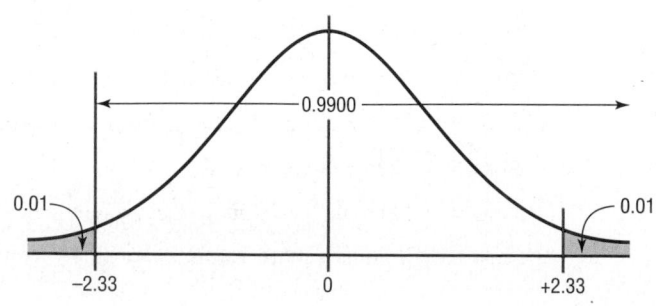

Solution *c*

Step 1   Draw the figure and indicate the appropriate area. Since this is a right-tailed test, the area 0.005 is located in the right tail, as shown in Figure 8–12.

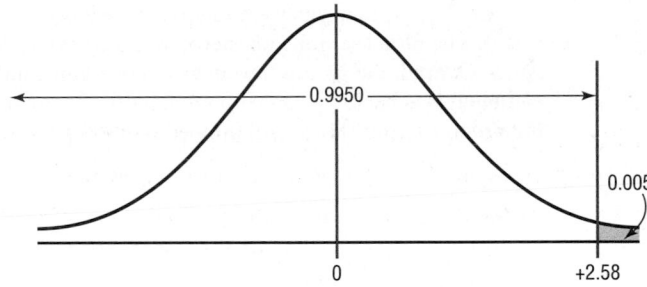

Step 2   Find the area closest to $1 - \alpha$, or $1 - 0.005 = 0.9950$. In this case, it is 0.9949 or 0.9951.

The two *z* values corresponding to 0.9949 and 0.9951 are $+2.57$ and $+2.58$. Since 0.9500 is halfway between these two values, find the average of the two values $(+2.57 + 2.58) \div 2 = +2.575$. However, 2.58 is most often used. See Figure 8–12.

Objective 4

State the five steps
used in hypothesis
testing.

In hypothesis testing, the following steps are recommended.

1. State the hypotheses. Be sure to state both the null and the alternative hypotheses.

2. Design the study. This step includes selecting the correct statistical test, choosing a level of significance, and formulating a plan to carry out the study. The plan should include information such as the definition of the population, the way the sample will be selected, and the methods that will be used to collect the data.

3. Conduct the study and collect the data.

4. Evaluate the data. The data should be tabulated in this step, and the statistical test should be conducted. Finally, decide whether to reject or not reject the null hypothesis.

5. Summarize the results.

For the purposes of this chapter, a simplified version of the hypothesis-testing procedure will be used, since designing the study and collecting the data will be omitted. The steps are summarized in the Procedure Table.

Procedure Table

Solving Hypothesis-Testing Problems (Traditional Method)

Step 1   State the hypotheses and identify the claim.

Step 2   Find the critical value(s) from the appropriate table in Appendix C.

Step 3   Compute the test value.

Step 4   Make the decision to reject or not reject the null hypothesis.

Step 5   Summarize the results.

Applying the Concepts **8–1**

Eggs and Your Health

The Incredible Edible Egg company recently found that eating eggs does not increase a person's blood serum cholesterol. Five hundred subjects participated in a study that lasted for 2 years. The participants were randomly assigned to either a no-egg group or a moderate-egg group. The blood serum cholesterol levels were checked at the beginning and at the end of the study. Overall, the groups' levels were not significantly different. The company reminds us that eating eggs is healthy if done in moderation. Many of the previous studies relating eggs and high blood serum cholesterol jumped to improper conclusions.

Using this information, answer these questions.

1. What prompted the study?
2. What is the population under study?
3. Was a sample collected?
4. What was the hypothesis?
5. Were data collected?
6. Were any statistical tests run?
7. What was the conclusion?

See page 469 for the answers.

Exercises 8–1

1. Define *null* and *alternative hypotheses,* and give an example of each.

2. What is meant by a type I error? A type II error? How are they related?

3. What is meant by a statistical test?

4. Explain the difference between a one-tailed and a two-tailed test.

5. What is meant by the critical region? The noncritical region?

6. What symbols are used to represent the null hypothesis and the alternative hypothesis? H_0 represents the null hypothesis; H_1 represents the alternative hypothesis.

7. What symbols are used to represent the probabilities of type I and type II errors? α, β

8. Explain what is meant by a significant difference.

9. When should a one-tailed test be used? A two-tailed test?

10. List the steps in hypothesis testing.

11. In hypothesis testing, why can't the hypothesis be proved true?

12. **(ans)** Using the z table (Table E), find the critical value (or values) for each.

    a. $\alpha = 0.05$, two-tailed test ± 1.96

    b. $\alpha = 0.01$, left-tailed test -2.33

    c. $\alpha = 0.005$, right-tailed test $+2.58$

    d. $\alpha = 0.01$, right-tailed test $+2.33$

    e. $\alpha = 0.05$, left-tailed test -1.65

    f. $\alpha = 0.02$, left-tailed test -2.05

    g. $\alpha = 0.05$, right-tailed test $+1.65$

    h. $\alpha = 0.01$, two-tailed test ± 2.58

    i. $\alpha = 0.04$, left-tailed test -1.75

    j. $\alpha = 0.02$, right-tailed test $+2.05$

13. For each conjecture, state the null and alternative hypotheses.

    a. The average age of community college students is 24.6 years. $H_0: \mu = 24.6$ and $H_1: \mu \neq 24.6$

    b. The average income of accountants is $51,497. $H_0: \mu = \$51,497$ and $H_1: \mu \neq \$51,497$

    c. The average age of attorneys is greater than 25.4 years. $H_0: \mu = 25.4$ and $H_1: \mu > 25.4$

    d. The average score of high school basketball games is less than 88. $H_0: \mu = 88$ and $H_1: \mu < 88$

    e. The average pulse rate of male marathon runners is less than 70 beats per minute. $H_0: \mu = 70$ and $H_1: \mu < 70$

    f. The average cost of a DVD player is $79.95. $H_0: \mu = \$79.95$ and $H_1: \mu \neq \$79.95$

    g. The average weight loss for a sample of people who exercise 30 minutes per day for 6 weeks is 8.2 pounds. $H_0: \mu = 8.2$ and $H_1: \mu \neq 8.2$

| 8–2 | **z Test for a Mean** |

Objective 5

Test means when σ is known, using the z test.

In this chapter, two statistical tests will be explained: the z test is used when σ is known, and the t test is used when σ is unknown. This section explains the z test, and Section 8–3 explains the t test.

Many hypotheses are tested using a statistical test based on the following general formula:

$$\text{Test value} = \frac{(\text{observed value}) - (\text{expected value})}{\text{standard error}}$$

The observed value is the statistic (such as the sample mean) that is computed from the sample data. The expected value is the parameter (such as the population mean) that you would expect to obtain if the null hypothesis were true—in other words, the hypothesized value. The denominator is the standard error of the statistic being tested (in this case, the standard error of the mean).

The z test is defined formally as follows.

> The **z test** is a statistical test for the mean of a population. It can be used when $n \geq 30$, or when the population is normally distributed and σ is known.
>   The formula for the z test is
>
> $$z = \frac{\bar{X} - \mu}{\sigma/\sqrt{n}}$$
>
> where
> $\bar{X}$ = sample mean
> μ = hypothesized population mean
> σ = population standard deviation
> n = sample size

For the z test, the observed value is the value of the sample mean. The expected value is the value of the population mean, assuming that the null hypothesis is true. The denominator $\sigma/\sqrt{n}$ is the standard error of the mean.

The formula for the z test is the same formula shown in Chapter 6 for the situation where you are using a distribution of sample means. Recall that the central limit theorem allows you to use the standard normal distribution to approximate the distribution of sample means when $n \geq 30$.

Note: Your first encounter with hypothesis testing can be somewhat challenging and confusing, since there are many new concepts being introduced at the same time. *To understand all the concepts, you must carefully follow each step in the examples and try each exercise that is assigned.* Only after careful study and patience will these concepts become clear.

> **Assumptions for the z Test for a Mean When σ Is Known**
>
> 1. The sample is a random sample.
> 2. Either $n \geq 30$ or the population is normally distributed if $n < 30$.

As stated in Section 8–1, there are five steps for solving *hypothesis-testing* problems:

Step 1   State the hypotheses and identify the claim.

Step 2   Find the critical value(s).

Speaking of
Statistics

This study found that people who used pedometers reported having increased energy, mood improvement, and weight loss. State possible null and alternative hypotheses for the study. What would be a likely population? What is the sample size? Comment on the sample size.

RD HEALTH

Step to It

IT FITS in your hand, costs less than $30, and will make you feel great. Give up? A pedometer. Brenda Rooney, an epidemiologist at Gundersen Lutheran Medical Center in LaCrosse, Wis., gave 500 people pedometers and asked them to take 10,000 steps—about five miles—a day. (Office workers typically average about 4000 steps a day.) By the end of eight weeks, 56 percent reported having more energy, 47 percent improved their mood and 50 percent lost weight. The subjects reported that seeing their total step-count motivated them to take more.

— JENNIFER BRAUNSCHWEIGER

Source: Reprinted with permission from the April 2002 Reader's Digest. Copyright © 2002 by The Reader's Digest Assn. Inc.

Step 3 Compute the test value.

Step 4 Make the decision to reject or not reject the null hypothesis.

Step 5 Summarize the results.

Example 8–3 illustrates these five steps.

Example 8–3

Days on Dealers' Lots

A researcher wishes to see if the mean number of days that a basic, low-price, small automobile sits on a dealer's lot is 29. A sample of 30 automobile dealers has a mean of 30.1 days for basic, low-price, small automobiles. At $\alpha = 0.05$, test the claim that the mean time is greater than 29 days. The standard deviation of the population is 3.8 days.

Source: Based on information from *Power Information Network*.

Solution

Step 1 State the hypotheses and identify the claim.
$H_0: \mu = 29$ and $H_1: \mu > 29$ (claim)

Step 2 Find the critical value. Since $\alpha = 0.05$ and the test is a right-tailed test, the critical value is $z = +1.65$.

Step 3 Compute the test value.

$$z = \frac{\overline{X} - \mu}{\sigma/\sqrt{n}} = \frac{30.1 - 29}{3.8/\sqrt{30}} = 1.59$$

Step 4 Make the decision. Since the test value, $+1.59$, is less than the critical value, $+1.65$, and is not in the critical region, the decision is to not reject the null hypothesis. This test is summarized in Figure 8–13.

Figure 8–13

Summary of the *z* Test of Example 8–3

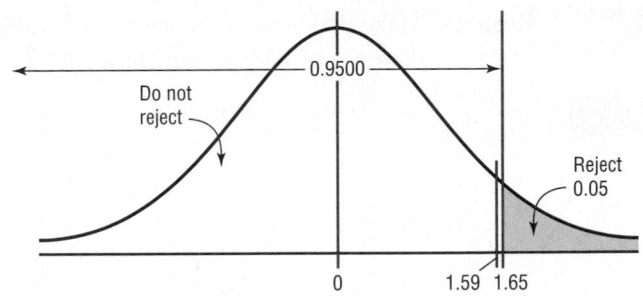

Step 5   Summarize the results. There is not enough evidence to support the claim that the mean time is greater than 29 days.

Comment: Even though in Example 8–3 the sample mean of 30.1 is higher than the hypothesized population mean of 29, it is not *significantly* higher. Hence, the difference may be due to chance. When the null hypothesis is not rejected, there is still a probability of a type II error, i.e., of not rejecting the null hypothesis when it is false.

The probability of a type II error is not easily ascertained. Further explanation about the type II error is given in Section 8–6. For now, it is only necessary to realize that the probability of type II error exists when the decision is not to reject the null hypothesis.

Also note that when the null hypothesis is not rejected, it cannot be accepted as true. There is merely not enough evidence to say that it is false. This guideline may sound a little confusing, but the situation is analogous to a jury trial. The verdict is either guilty or not guilty and is based on the evidence presented. If a person is judged not guilty, it does not mean that the person is proved innocent; it only means that there was not enough evidence to reach the guilty verdict.

Example 8–4

Costs of Men's Athletic Shoes

A researcher claims that the average cost of men's athletic shoes is less than $80. He selects a random sample of 36 pairs of shoes from a catalog and finds the following costs (in dollars). (The costs have been rounded to the nearest dollar.) Is there enough evidence to support the researcher's claim at $\alpha = 0.10$? Assume $\sigma = 19.2$.

60	70	75	55	80	55
50	40	80	70	50	95
120	90	75	85	80	60
110	65	80	85	85	45
75	60	90	90	60	95
110	85	45	90	70	70

Solution

Step 1   State the hypotheses and identify the claim

$$H_0: \mu = \$80 \quad \text{and} \quad H_1: \mu < \$80 \text{ (claim)}$$

Step 2   Find the critical value. Since $\alpha = 0.10$ and the test is a left-tailed test, the critical value is -1.28.

Step 3   Compute the test value. Since the exercise gives raw data, it is necessary to find the mean of the data. Using the formulas in Chapter 3 or your calculator gives $\overline{X} = 75.0$ and $\sigma = 19.2$. Substitute in the formula

$$z = \frac{\overline{X} - \mu}{\sigma/\sqrt{n}} = \frac{75 - 80}{19.2/\sqrt{36}} = -1.56$$

Step 4   Make the decision. Since the test value, -1.56, falls in the critical region, the decision is to reject the null hypothesis. See Figure 8–14.

Figure 8–14

Critical and Test Values for Example 8–4

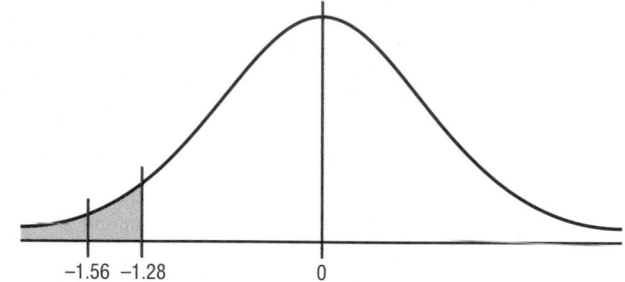

-1.56 -1.28 0

Step 5   Summarize the results. There is enough evidence to support the claim that the average cost of men's athletic shoes is less than $80.

Comment: In Example 8–4, the difference is said to be significant. However, when the null hypothesis is rejected, there is always a chance of a type I error. In this case, the probability of a type I error is at most 0.10, or 10%.

Example 8–5

Cost of Rehabilitation

The Medical Rehabilitation Education Foundation reports that the average cost of rehabilitation for stroke victims is $24,672. To see if the average cost of rehabilitation is different at a particular hospital, a researcher selects a random sample of 35 stroke victims at the hospital and finds that the average cost of their rehabilitation is $26,343. The standard deviation of the population is $3251. At $\alpha = 0.01$, can it be concluded that the average cost of stroke rehabilitation at a particular hospital is different from $24,672?

Source: Snapshot, *USA TODAY.*

Solution

Step 1   State the hypotheses and identify the claim.

$$H_0: \mu = \$24,672 \qquad \text{and} \qquad H_1: \mu \neq \$24,672 \text{ (claim)}$$

Step 2   Find the critical values. Since $\alpha = 0.01$ and the test is a two-tailed test, the critical values are $+2.58$ and -2.58.

Step 3   Compute the test value.

$$z = \frac{\bar{X} - \mu}{\sigma/\sqrt{n}} = \frac{26,343 - 24,672}{3251/\sqrt{35}} = 3.04$$

Step 4   Make the decision. Reject the null hypothesis, since the test value falls in the critical region, as shown in Figure 8–15.

Figure 8–15

Critical and Test Values for Example 8–5

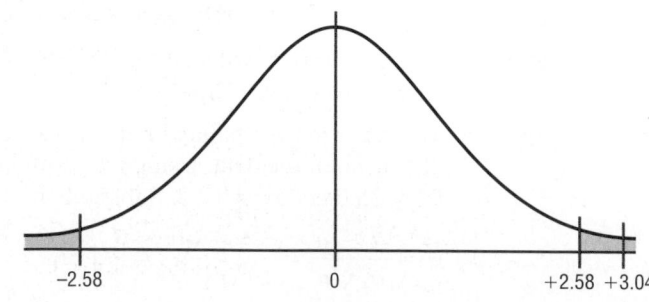

-2.58 0 $+2.58$ $+3.04$

Step 5 Summarize the results. There is enough evidence to support the claim that the average cost of rehabilitation at the particular hospital is different from $24,672.

Students sometimes have difficulty summarizing the results of a hypothesis test. Figure 8–16 shows the four possible outcomes and the summary statement for each situation.

Figure 8–16

Outcomes of a Hypothesis-Testing Situation

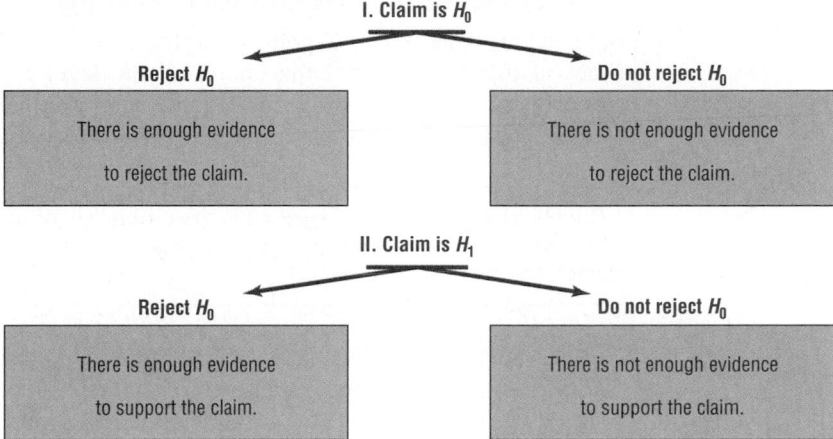

First, the claim can be either the null or alternative hypothesis, and one should identify which it is. Second, after the study is completed, the null hypothesis is either rejected or not rejected. From these two facts, the decision can be identified in the appropriate block of Figure 8–16.

For example, suppose a researcher claims that the mean weight of an adult animal of a particular species is 42 pounds. In this case, the claim would be the null hypothesis, H_0: $\mu = 42$, since the researcher is asserting that the parameter is a specific value. If the null hypothesis is rejected, the conclusion would be that there is enough evidence to reject the claim that the mean weight of the adult animal is 42 pounds. See Figure 8–17(a).

On the other hand, suppose the researcher claims that the mean weight of the adult animals is not 42 pounds. The claim would be the alternative hypothesis H_1: $\mu \neq 42$. Furthermore, suppose that the null hypothesis is not rejected. The conclusion, then, would be that there is not enough evidence to support the claim that the mean weight of the adult animals is not 42 pounds. See Figure 8–17(b).

Figure 8–17

Outcomes of a Hypothesis-Testing Situation for Two Specific Cases

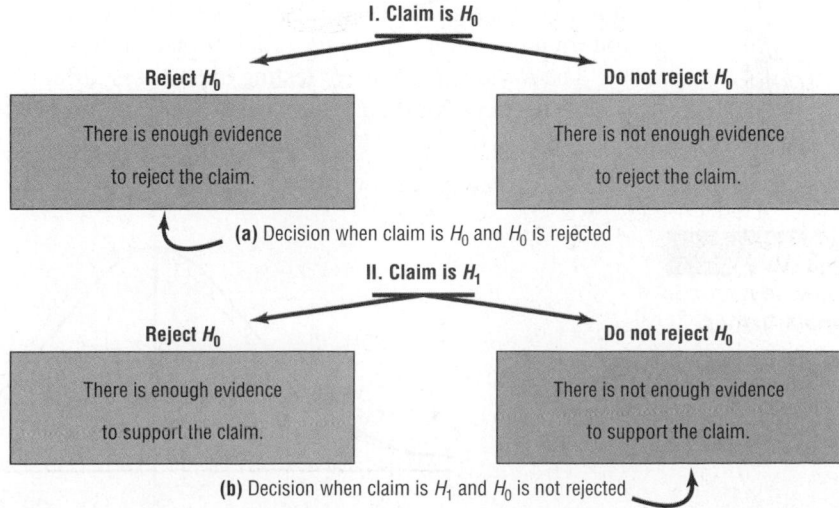

Again, remember that nothing is being proved true or false. The statistician is only stating that there is or is not enough evidence to say that a claim is *probably* true or false. As noted previously, the only way to prove something would be to use the entire population under study, and usually this cannot be done, especially when the population is large.

P-Value Method for Hypothesis Testing

Statisticians usually test hypotheses at the common α levels of 0.05 or 0.01 and sometimes at 0.10. Recall that the choice of the level depends on the seriousness of the type I error. Besides listing an α value, many computer statistical packages give a *P*-value for hypothesis tests.

> The ***P*-value** (or probability value) is the probability of getting a sample statistic (such as the mean) or a more extreme sample statistic in the direction of the alternative hypothesis when the null hypothesis is true.

In other words, the *P*-value is the actual area under the standard normal distribution curve (or other curve, depending on what statistical test is being used) representing the probability of a particular sample statistic or a more extreme sample statistic occurring if the null hypothesis is true.

For example, suppose that an alternative hypothesis is H_1: $\mu > 50$ and the mean of a sample is $\overline{X} = 52$. If the computer printed a *P*-value of 0.0356 for a statistical test, then the probability of getting a sample mean of 52 or greater is 0.0356 if the true population mean is 50 (for the given sample size and standard deviation). The relationship between the *P*-value and the α value can be explained in this manner. For $P = 0.0356$, the null hypothesis would be rejected at $\alpha = 0.05$ but not at $\alpha = 0.01$. See Figure 8–18.

When the hypothesis test is two-tailed, the area in one tail must be doubled. For a two-tailed test, if α is 0.05 and the area in one tail is 0.0356, the *P*-value will be $2(0.0356) = 0.0712$. That is, the null hypothesis should not be rejected at $\alpha = 0.05$, since 0.0712 is greater than 0.05. In summary, then, if the *P*-value is less than α, reject the null hypothesis. If the *P*-value is greater than α, do not reject the null hypothesis.

The *P*-values for the *z* test can be found by using Table E in Appendix C. First find the area under the standard normal distribution curve corresponding to the *z* test value. For a left-tailed test, use the area given in the table; for a right-tailed test, use 1.0000 minus the area given in the table. To get the *P*-value for a two-tailed test, double the area you found in the tail. This procedure is shown in step 3 of Examples 8–6 and 8–7.

The *P*-value method for testing hypotheses differs from the traditional method somewhat. The steps for the *P*-value method are summarized next.

Figure 8–18

Comparison of α
Values and *P*-Values

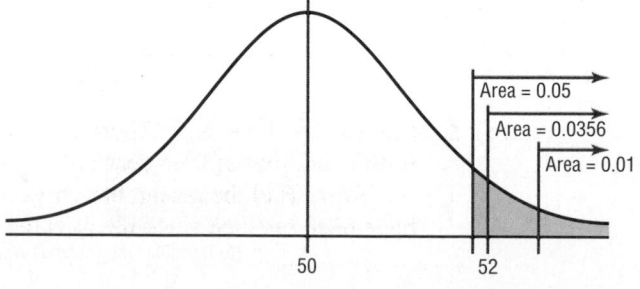

Procedure Table

Solving Hypothesis-Testing Problems (*P*-Value Method)

Step 1   State the hypotheses and identify the claim.

Step 2   Compute the test value.

Step 3   Find the *P*-value.

Step 4   Make the decision.

Step 5   Summarize the results.

Examples 8–6 and 8–7 show how to use the *P*-value method to test hypotheses.

Example 8–6

Cost of College Tuition

A researcher wishes to test the claim that the average cost of tuition and fees at a four-year public college is greater than $5700. She selects a random sample of 36 four-year public colleges and finds the mean to be $5950. The population standard deviation is $659. Is there evidence to support the claim at $\alpha = 0.05$? Use the *P*-value method.

Source: Based on information from the College Board.

Solution

Step 1   State the hypotheses and identify the claim. H_0: $\mu = \$5700$ and H_1: $\mu > \$5700$ (claim).

Step 2   Compute the test value.

$$z = \frac{\bar{X} - \mu}{\sigma/\sqrt{n}} = \frac{5950 - 5700}{659/\sqrt{36}} = 2.28$$

Step 3   Find the *P*-value. Using Table E in Appendix C, find the corresponding area under the normal distribution for $z = 2.28$. It is 0.9887. Subtract this value for the area from 1.0000 to find the area in the right tail.

$$1.0000 - 0.9887 = 0.0113$$

Hence the *P*-value is 0.0113.

Step 4   Make the decision. Since the *P*-value is less than 0.05, the decision is to reject the null hypothesis. See Figure 8–19.

Figure 8–19

P-Value and α Value for Example 8–6

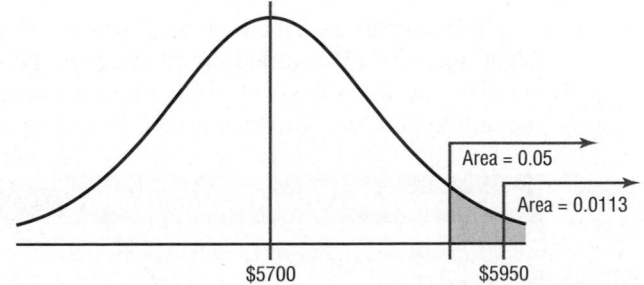

Step 5   Summarize the results. There is enough evidence to support the claim that the tuition and fees at four-year public colleges are greater than $5700.

    Note: Had the researcher chosen $\alpha = 0.01$, the null hypothesis would not have been rejected since the *P*-value (0.0113) is greater than 0.01.

| Example 8–7 | Wind Speed |

A researcher claims that the average wind speed in a certain city is 8 miles per hour. A sample of 32 days has an average wind speed of 8.2 miles per hour. The standard deviation of the population is 0.6 mile per hour. At $\alpha = 0.05$, is there enough evidence to reject the claim? Use the P-value method.

Solution

Step 1 State the hypotheses and identify the claim.

$$H_0: \mu = 8 \text{ (claim)} \qquad \text{and} \qquad H_1: \mu \neq 8$$

Step 2 Compute the test value.

$$z = \frac{8.2 - 8}{0.6/\sqrt{32}} = 1.89$$

Step 3 Find the P-value. Using Table E, find the corresponding area for $z = 1.89$. It is 0.9706. Subtract the value from 1.0000.

$$1.0000 - 0.9706 = 0.0294$$

Since this is a two-tailed test, the area of 0.0294 must be doubled to get the P-value.

$$2(0.0294) = 0.0588$$

Step 4 Make the decision. The decision is to not reject the null hypothesis, since the P-value is greater than 0.05. See Figure 8–20.

| Figure 8–20 |

P-Values and α Values for Example 8–7

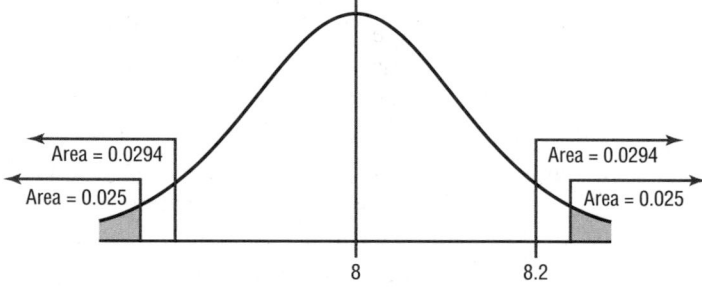

Step 5 Summarize the results. There is not enough evidence to reject the claim that the average wind speed is 8 miles per hour.

In Examples 8–6 and 8–7, the P-value and the α value were shown on a normal distribution curve to illustrate the relationship between the two values; however, it is not necessary to draw the normal distribution curve to make the decision whether to reject the null hypothesis. You can use the following rule:

| Decision Rule When Using a P-Value |

If P-value $\leq \alpha$, reject the null hypothesis.
If P-value $> \alpha$, do not reject the null hypothesis.

In Example 8–6, P-value $= 0.0113$ and $\alpha = 0.05$. Since P-value $\leq \alpha$, the null hypothesis was rejected. In Example 8–7, P-value $= 0.0588$ and $\alpha = 0.05$. Since P-value $> \alpha$, the null hypothesis was not rejected.

The *P*-values given on calculators and computers are slightly different from those found with Table E. This is so because *z* values and the values in Table E have been rounded. *Also, most calculators and computers give the exact P-value for two-tailed tests, so it should not be doubled (as it should when the area found in Table E is used).*

A clear distinction between the α value and the *P*-value should be made. The α value is chosen by the researcher *before* the statistical test is conducted. The *P*-value is computed after the sample mean has been found.

There are two schools of thought on *P*-values. Some researchers do not choose an α value but report the *P*-value and allow the reader to decide whether the null hypothesis should be rejected.

In this case, the following guidelines can be used, but be advised that these guidelines are not written in stone, and some statisticians may have other opinions.

Guidelines for *P*-Values

If *P*-value $\leq$ 0.01, reject the null hypothesis. The difference is highly significant.

If *P*-value $>$ 0.01 but *P*-value $\leq$ 0.05, reject the null hypothesis. The difference is significant.

If *P*-value $>$ 0.05 but *P*-value $\leq$ 0.10, consider the consequences of type I error before rejecting the null hypothesis.

If *P*-value $>$ 0.10, do not reject the null hypothesis. The difference is not significant.

Others decide on the α value in advance and use the *P*-value to make the decision, as shown in Examples 8–6 and 8–7. A note of caution is needed here: If a researcher selects $\alpha = 0.01$ and the *P*-value is 0.03, the researcher may decide to change the α value from 0.01 to 0.05 so that the null hypothesis will be rejected. This, of course, should not be done. If the α level is selected in advance, it should be used in making the decision.

One additional note on hypothesis testing is that the researcher should distinguish between *statistical significance* and *practical significance*. When the null hypothesis is rejected at a specific significance level, it can be concluded that the difference is probably not due to chance and thus is statistically significant. However, the results may not have any practical significance. For example, suppose that a new fuel additive increases the miles per gallon that a car can get by $\frac{1}{4}$ mile for a sample of 1000 automobiles. The results may be statistically significant at the 0.05 level, but it would hardly be worthwhile to market the product for such a small increase. Hence, there is no practical significance to the results. It is up to the researcher to use common sense when interpreting the results of a statistical test.

Applying the Concepts **8–2**

Car Thefts

You recently received a job with a company that manufactures an automobile antitheft device. To conduct an advertising campaign for the product, you need to make a claim about the number of automobile thefts per year. Since the population of various cities in the United States varies, you decide to use rates per 10,000 people. (The rates are based on the number of people living in the cities.) Your boss said that last year the theft rate per 10,000 people was 44 vehicles. You want to see if it has changed. The following are rates per 10,000 people for 36 randomly selected locations in the United States.

55	42	125	62	134	73
39	69	23	94	73	24
51	55	26	66	41	67
15	53	56	91	20	78
70	25	62	115	17	36
58	56	33	75	20	16

Source: Based on information from the National Insurance Crime Bureau.

Using this information, answer these questions.

1. What hypotheses would you use?
2. Is the sample considered small or large?
3. What assumption must be met before the hypothesis test can be conducted?
4. Which probability distribution would you use?
5. Would you select a one- or two-tailed test? Why?
6. What critical value(s) would you use?
7. Conduct a hypothesis test. Use $\sigma = 30.3$.
8. What is your decision?
9. What is your conclusion?
10. Write a brief statement summarizing your conclusion.
11. If you lived in a city whose population was about 50,000, how many automobile thefts per year would you expect to occur?

See page 469 for the answers.

Exercises 8–2

For Exercises 1 through 13, perform each of the following steps.

   a. State the hypotheses and identify the claim.
   b. Find the critical value(s).
   c. Compute the test value.
   d. Make the decision.
   e. Summarize the results.

Use diagrams to show the critical region (or regions), and use the traditional method of hypothesis testing unless otherwise specified.

1. **Warming and Ice Melt** The average depth of the Hudson Bay is 305 feet. Climatologists were interested in seeing if the effects of warming and ice melt were affecting the water level. Fifty-five measurements over a period of weeks yielded a sample mean of 306.2 feet. The population variance is known to be 3.57. Can it be concluded at the 0.05 level of significance that the average depth has increased? Is there evidence of what caused this to happen?

   Source: *World Almanac and Book of Facts 2010.*

2. **Credit Card Debt** It has been reported that the average credit card debt for college seniors at the college book store for a specific college is $3262. The student senate at a large university feels that their seniors have a debt much less than this, so it conducts a study of 50 randomly selected seniors and finds that the average debt is $2995, and the population standard deviation is $1100. With $\alpha = 0.05$, is the student senate correct?

3. **Revenue of Large Businesses** A researcher estimates that the average revenue of the largest businesses in the United States is greater than $24 billion. A sample of 50 companies is selected, and the revenues (in billions of dollars) are shown. At $\alpha = 0.05$, is there enough evidence to support the researcher's claim? Assume $\sigma = 28.7$.

178	122	91	44	35
61	56	46	20	32
30	28	28	20	27
29	16	16	19	15
41	38	36	15	25
31	30	19	19	19
24	16	15	15	19
25	25	18	14	15
24	23	17	17	22
22	21	20	17	20

Source: *New York Times Almanac.*

4. **Moviegoers** The average "moviegoer" sees 8.5 movies a year. A *moviegoer* is defined as a person who sees at least one movie in a theater in a 12-month period. A random sample of 40 moviegoers from a large university revealed that the average number of movies seen per person was 9.6. The population standard deviation is 3.2 movies. At the 0.05 level of significance, can it be concluded that this represents a difference from the national average?

   Source: *MPAA Study.*

5. **Nonparental Care** According to the *Digest of Educational Statistics*, a certain group of preschool children under the age of one year each spends an average of 30.9 hours per week in nonparental care. A study of state university center-based programs indicated that a random sample of 32 infants spent an average of 32.1 hours per week in their care. The standard deviation of the population is 3.6 hours. At $\alpha = 0.01$ is there sufficient evidence to conclude that the sample mean differs from the national mean?

   Source: *www.nces.ed.gov*

6. **Peanut Production in Virginia** The average production of peanuts in Virginia is 3000 pounds per acre. A new plant food has been developed and is tested on 60 individual plots of land. The mean yield with the new plant food is 3120 pounds of peanuts per acre, and the population standard deviation is 578 pounds. At $\alpha = 0.05$, can you conclude that the average production has increased?

Source: *The Old Farmer's Almanac.*

7. **Heights of 1-Year-Olds** The average 1-year-old (both genders) is 29 inches tall. A random sample of 30 1-year-olds in a large day care franchise resulted in the following heights. At $\alpha = 0.05$, can it be concluded that the average height differs from 29 inches? Assume $\sigma = 2.61$.

25	32	35	25	30	26.5	26	25.5	29.5	32
30	28.5	30	32	28	31.5	29	29.5	30	34
29	32	27	28	33	28	27	32	29	29.5

Source: www.healthepic.com

8. **Salaries of Government Employees** The mean salary of federal government employees on the General Schedule is $59,593. The average salary of 30 state employees who do similar work is $58,800 with $\sigma = \$1500$. At the 0.01 level of significance, can it be concluded that state employees earn on average less than federal employees?

Source: *New York Times Almanac.*

9. **Operating Costs of an Automobile** The average cost of owning and operating an automobile is $8121 per 15,000 miles including fixed and variable costs. A random survey of 40 automobile owners revealed an average cost of $8350 with a population standard deviation of $750. Is there sufficient evidence to conclude that the average is greater than $8121? Use $\alpha = 0.01$.

Source: *New York Times Almanac 2010.*

10. **Home Prices in Pennsylvania** A real estate agent claims that the average price of a home sold in Beaver County, Pennsylvania, is $60,000. A random sample of 36 homes sold in the county is selected, and the prices in dollars are shown. Is there enough evidence to reject the agent's claim at $\alpha = 0.05$? Assume $\sigma = \$76,025$.

9,500	54,000	99,000	94,000	80,000
29,000	121,500	184,750	15,000	164,450
6,000	13,000	188,400	121,000	308,000
42,000	7,500	32,900	126,900	25,225
95,000	92,000	38,000	60,000	211,000
15,000	28,000	53,500	27,000	21,000
76,000	85,000	25,225	40,000	97,000
284,000				

Source: *Pittsburgh Tribune-Review.*

11. **Use of Disposable Cups** The average college student goes through 500 disposable cups in a year. To raise environmental awareness, a student group at a large university volunteered to help count how many cups were used by students on their campus. A random sample of 50 students' results found that they used a mean of 476 cups with $\sigma = 42$ cups. At $\alpha = 0.01$, is there sufficient evidence to conclude that the mean differs from 500?

Source: www.esc.mtu.edu/SFES.php

12. **Student Expenditures** The average expenditure per student (based on average daily attendance) for a certain school year was $10,337 with a population standard deviation of $1560. A survey for the next school year of 150 randomly selected students resulted in a sample mean of $10,798. Do these results indicate that the average expenditure has changed? Choose your own level of significance.

Source: *World Almanac.*

13. **Ages of U.S. Senators** The mean age of Senators in the 109th Congress was 60.35 years. A random sample of 40 senators from various state senates had an average age of 55.4 years, and the population standard deviation is 6.5 years. At $\alpha = 0.05$, is there sufficient evidence that state senators are on average younger than the Senators in Washington?

Source: *CG Today.*

14. What is meant by a *P*-value? The *P*-value is the actual probability of getting the sample mean if the null hypothesis is true.

15. State whether the null hypothesis should be rejected on the basis of the given *P*-value.

 a. *P*-value = 0.258, $\alpha = 0.05$, one-tailed test   Do not reject.

 b. *P*-value = 0.0684, $\alpha = 0.10$, two-tailed test   Reject.

 c. *P*-value = 0.0153, $\alpha = 0.01$, one-tailed test   Do not reject.

 d. *P*-value = 0.0232, $\alpha = 0.05$, two-tailed test   Reject.

 e. *P*-value = 0.002, $\alpha = 0.01$, one-tailed test   Reject.

16. **Soft Drink Consumption** A researcher claims that the yearly consumption of soft drinks per person is 52 gallons. In a sample of 50 randomly selected people, the mean of the yearly consumption was 56.3 gallons. The standard deviation of the population is 3.5 gallons. Find the *P*-value for the test. On the basis of the *P*-value, is the researcher's claim valid?

Source: *U.S. Department of Agriculture.*

17. **Stopping Distances** A study found that the average stopping distance of a school bus traveling 50 miles per hour was 264 feet. A group of automotive engineers decided to conduct a study of its school buses and found that for 20 buses, the average stopping distance of buses traveling 50 miles per hour was 262.3 feet. The standard deviation of the population was 3 feet. Test the claim that the average stopping distance of the company's buses is actually less than 264 feet. Find the *P*-value. On the basis of the *P*-value, should the null hypothesis be rejected at $\alpha = 0.01$? Assume that the variable is normally distributed.

Source: *Snapshot, USA TODAY.*

18. **Copy Machine Use** A store manager hypothesizes that the average number of pages a person copies on the store's copy machine is less than 40. A sample of 50 customers' orders is selected. At $\alpha = 0.01$, is there enough evidence to support the claim? Use the *P*-value hypothesis-testing method. Assume $\sigma = 30.9$.

2	2	2	5	32
5	29	8	2	49
21	1	24	72	70
21	85	61	8	42
3	15	27	113	36
37	5	3	58	82
9	2	1	6	9
80	9	51	2	122
21	49	36	43	61
3	17	17	4	1

19. **Burning Calories by Playing Tennis** A health researcher read that a 200-pound male can burn an average of 546 calories per hour playing tennis. Thirty-six males were randomly selected and tested. The mean of the number of calories burned per hour was 544.8. Test the claim that the average number of calories burned is actually less than 546, and find the P-value. On the basis of the P-value, should the null hypothesis be rejected at $\alpha = 0.01$? The standard deviation of the population is 3. Can it be concluded that the average number of calories burned is less than originally thought?

20. **Breaking Strength of Cable** A special cable has a breaking strength of 800 pounds. The standard deviation of the population is 12 pounds. A researcher selects a sample of 20 cables and finds that the average breaking strength is 793 pounds. Can he reject the claim that the breaking strength is 800 pounds? Find the P-value. Should the null hypothesis be rejected at $\alpha = 0.01$? Assume that the variable is normally distributed.

21. **Farm Sizes** The average farm size in the United States is 444 acres. A random sample of 40 farms in Oregon indicated a mean size of 430 acres, and the population standard deviation is 52 acres. At $\alpha = 0.05$, can it be concluded that the average farm in Oregon differs from the national mean? Use the P-value method.

Source: *New York Times Almanac.*

22. **Farm Sizes** Ten years ago, the average acreage of farms in a certain geographic region was 65 acres. The standard deviation of the population was 7 acres. A recent study consisting of 22 farms showed that the average was 63.2 acres per farm. Test the claim, at $\alpha = 0.10$, that the average has not changed by finding the P-value for the test. Assume that σ has not changed and the variable is normally distributed.

23. **Transmission Service** A car dealer recommends that transmissions be serviced at 30,000 miles. To see whether her customers are adhering to this recommendation, the dealer selects a sample of 40 customers and finds that the average mileage of the automobiles serviced is 30,456. The standard deviation of the population is 1684 miles. By finding the P-value, determine whether the owners are having their transmissions serviced at 30,000 miles. Use $\alpha = 0.10$. Do you think the α value of 0.10 is an appropriate significance level?

24. **Speeding Tickets** A motorist claims that the South Boro Police issue an average of 60 speeding tickets per day. These data show the number of speeding tickets issued each day for a period of one month. Assume σ is 13.42. Is there enough evidence to reject the motorist's claim at $\alpha = 0.05$? Use the P-value method.

72	45	36	68	69	71	57	60
83	26	60	72	58	87	48	59
60	56	64	68	42	57	57	
58	63	49	73	75	42	63	

25. **Sick Days** A manager states that in his factory, the average number of days per year missed by the employees due to illness is less than the national average of 10. The following data show the number of days missed by 40 employees last year. Is there sufficient evidence to believe the manager's statement at $\alpha = 0.05$? $\sigma = 3.63$. Use the P-value method.

0	6	12	3	3	5	4	1
3	9	6	0	7	6	3	4
7	4	7	1	0	8	12	3
2	5	10	5	15	3	2	5
3	11	8	2	2	4	1	9

Extending the Concepts

26. Suppose a statistician chose to test a hypothesis at $\alpha = 0.01$. The critical value for a right-tailed test is $+2.33$. If the test value were 1.97, what would the decision be? What would happen if, after seeing the test value, she decided to choose $\alpha = 0.05$? What would the decision be? Explain the contradiction, if there is one.

27. **Hourly Wage** The president of a company states that the average hourly wage of her employees is $8.65. A sample of 50 employees has the distribution shown. At $\alpha = 0.05$, is the president's statement believable? Assume $\sigma = 0.105$.

Class	Frequency
8.35–8.43	2
8.44–8.52	6
8.53–8.61	12
8.62–8.70	18
8.71–8.79	10
8.80–8.88	2

Technology *Step by Step*

MINITAB
Step by Step

Hypothesis Test for the Mean and the *z* Distribution

MINITAB can be used to calculate the test statistic and its *P*-value. The *P*-value approach does not require a critical value from the table. If the *P*-value is smaller than α, the null hypothesis is rejected. For Example 8–4, test the claim that the mean shoe cost is less than $80.

1. Enter the data into a column of MINITAB. Do not try to type in the dollar signs! Name the column **ShoeCost.**
2. If sigma is known, skip to step 3; otherwise estimate sigma from the sample standard deviation *s*.

Calculate the Standard Deviation in the Sample

   a) Select **Calc>Column Statistics.**
   b) Check the button for Standard deviation.
   c) Select ShoeCost for the Input variable.
   d) Type **s** in the text box for Store the result in:.
   e) Click [OK].

Calculate the Test Statistic and *P*-Value

3. Select **Stat>Basic Statistics>1 Sample Z,** then select ShoeCost in the Variable text box.
4. Click in the text box and enter the value of sigma or type *s*, the sample standard deviation.
5. Click in the text box for Test mean, and enter the hypothesized value of **80.**
6. Click on [Options].
   a) Change the Confidence level to **90.**
   b) Change the Alternative to less than. This setting is crucial for calculating the *P*-value.
7. Click [OK] twice.

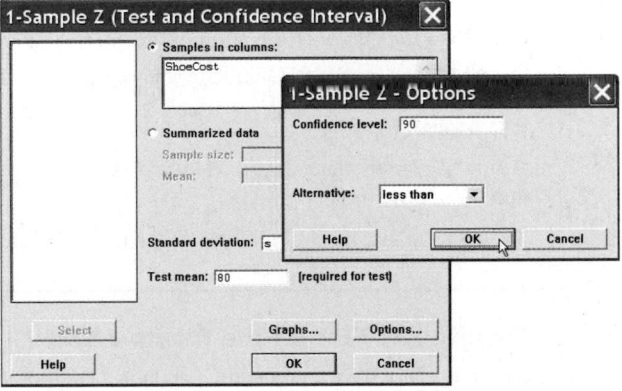

One-Sample Z: ShoeCost
```
Test of mu = 80 vs < 80
The assumed sigma 19.161
                                                    90%
                                                   Upper
Variable    N     Mean    StDev   SE Mean   Bound      Z      P
ShoeCost    36   75.0000  19.1610  3.1935   79.0926  -1.57  0.059
```

Since the *P*-value of 0.059 is less than α, reject the null hypothesis. There is enough evidence in the sample to conclude the mean cost is less than $80.

TI-83 Plus or TI-84 Plus
Step by Step

Hypothesis Test for the Mean and the *z* Distribution (Data)

1. Enter the data values into L₁.
2. Press **STAT** and move the cursor to TESTS.
3. Press 1 for ZTest.
4. Move the cursor to Data and press **ENTER.**
5. Type in the appropriate values.
6. Move the cursor to the appropriate alternative hypothesis and press **ENTER.**
7. Move the cursor to Calculate and press **ENTER.**

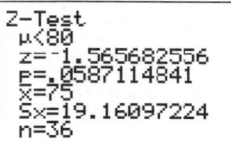

Example TI8–1

 This relates to Example 8–4 from the text. At the 10% significance level, test the claim that $\mu < 80$ given the data values.

60	70	75	55	80	55	50	40	80	70	50	95
120	90	75	85	80	60	110	65	80	85	85	45
75	60	90	90	60	95	110	85	45	90	70	70

The population standard deviation σ is unknown. Since the sample size $n = 36 \geq 30$, you can use the sample standard deviation s as an approximation for σ. After the data values are entered in L₁ (step 1), press **STAT,** move the cursor to CALC, press **1** for 1-Var Stats, then press **ENTER.** The sample standard deviation of 19.16097224 will be one of the statistics listed. Then continue with step 2. At step 5 on the line for σ press **VARS** for variables, press **5** for Statistics, press **3** for S_x.

The test statistic is $z = -1.565682556$, and the *P*-value is 0.0587114841.

Hypothesis Test for the Mean and the *z* Distribution (Statistics)

1. Press **STAT** and move the cursor to TESTS.
2. Press 1 for ZTest.
3. Move the cursor to Stats and press **ENTER.**
4. Type in the appropriate values.
5. Move the cursor to the appropriate alternative hypothesis and press **ENTER.**
6. Move the cursor to Calculate and press **ENTER.**

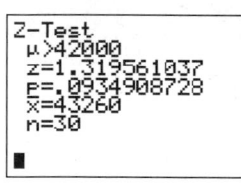

Example TI8–2

At the 5% significance level, test the claim that $\mu > 42,000$ given $\sigma = 5230$, $\overline{X} = 43,260$, and $n = 30$.

The test statistic is $z = 1.319561037$, and the *P*-value is 0.0934908728.

Excel
Step by Step

Hypothesis Test for the Mean: *z* Test

Excel does not have a procedure to conduct a hypothesis test for the mean. However, you may conduct the test of the mean by using the MegaStat Add-in available on your CD. If you have not installed this add-in, do so, following the instructions from the Chapter 1 Excel Step by Step.

Example XL8–1

This example relates to Example 8–4 from the text. At the 10% significance level, test the claim that $\mu < 80$. The MegaStat *z* test uses the *P*-value method. Therefore, it is not necessary to enter a significance level.

1. Enter the data into column A of a new worksheet.
2. From the toolbar, select Add-Ins, **MegaStat>Hypothesis Tests>Mean vs. Hypothesized Value.** *Note:* You may need to open MegaStat from the MegaStat.xls file on your computer's hard drive.

3. Select data input and type **A1:A36** as the Input Range.

4. Type **80** for the Hypothesized mean and select the "less than" Alternative.

5. Select *z* test and click [OK].

The result of the procedure is shown next.

Hypothesis Test: Mean vs. Hypothesized Value

```
 80.000  Hypothesized value
 75.000  Mean data
 19.161  Standard deviation
  3.193  Standard error
     36  n

 -1.57  z
 0.0587  P-value (one-tailed, lower)
```

8–3   *t* Test for a Mean

Objective 6

Test means when σ is unknown, using the *t* test.

When the population standard deviation is unknown, the *z* test is not normally used for testing hypotheses involving means. A different test, called the *t test,* is used. The distribution of the variable should be approximately normal.

As stated in Chapter 7, the *t* distribution is similar to the standard normal distribution in the following ways.

1. It is bell-shaped.

2. It is symmetric about the mean.

3. The mean, median, and mode are equal to 0 and are located at the center of the distribution.

4. The curve never touches the *x* axis.

The *t* distribution differs from the standard normal distribution in the following ways.

1. The variance is greater than 1.

2. The *t* distribution is a family of curves based on the *degrees of freedom,* which is a number related to sample size. (Recall that the symbol for degrees of freedom is d.f. See Section 7–2 for an explanation of degrees of freedom.)

3. As the sample size increases, the *t* distribution approaches the normal distribution.

The *t* test is defined next.

The **t test** is a statistical test for the mean of a population and is used when the population is normally or approximately normally distributed, and σ is unknown.

The formula for the *t* test is

$$t = \frac{\bar{X} - \mu}{s/\sqrt{n}}$$

The degrees of freedom are d.f. $= n - 1$.

The formula for the *t* test is similar to the formula for the *z* test. But since the population standard deviation σ is unknown, the sample standard deviation *s* is used instead.

The critical values for the *t* test are given in Table F in Appendix C. For a one-tailed test, find the α level by looking at the top row of the table and finding the appropriate column. Find the degrees of freedom by looking down the left-hand column.

Notice that the degrees of freedom are given for values from 1 through 30, then at intervals above 30. When the degrees of freedom are above 30, some textbooks will tell you to use the nearest table value; however, in this textbook, you should always round down to the nearest table value. For example, if d.f. = 59, use d.f. = 55 to find the critical value or values. This is a conservative approach.

As the degrees of freedom get larger, the critical values approach the *z* values. Hence the bottom values (large sample size) are the same as the *z* values that were used in the last section.

Example 8–8

Find the critical *t* value for $\alpha = 0.05$ with d.f. = 16 for a right-tailed *t* test.

Solution

Find the 0.05 column in the top row and 16 in the left-hand column. Where the row and column meet, the appropriate critical value is found; it is +1.746. See Figure 8–21.

Figure 8–21

Finding the Critical Value for the *t* Test in Table F (Example 8–8)

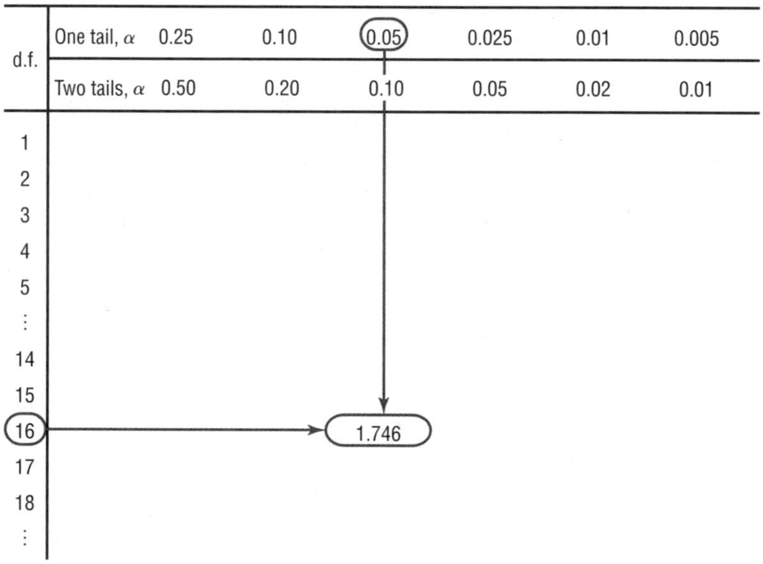

Example 8–9

Find the critical *t* value for $\alpha = 0.01$ with d.f. = 22 for a left-tailed test.

Solution

Find the 0.01 column in the row labeled One tail, and find 22 in the left column. The critical value is −2.508 since the test is a one-tailed left test.

Example 8–10

Find the critical values for $\alpha = 0.10$ with d.f. = 18 for a two-tailed *t* test.

Solution

Find the 0.10 column in the row labeled Two tails, and find 18 in the column labeled d.f. The critical values are +1.734 and −1.734.

Example 8–11

Find the critical value for $\alpha = 0.05$ with d.f. = 28 for a right-tailed *t* test.

Solution

Find the 0.05 column in the One-tail row and 28 in the left column. The critical value is +1.701.

Assumptions for the *t* Test for a Mean When σ Is Unknown

1. The sample is a random sample.
2. Either $n \geq 30$ or the population is normally distributed if $n < 30$.

When you test hypotheses by using the *t* test (traditional method), follow the same procedure as for the *z* test, except use Table F.

Step 1 State the hypotheses and identify the claim.

Step 2 Find the critical value(s) from Table F.

Step 3 Compute the test value.

Step 4 Make the decision to reject or not reject the null hypothesis.

Step 5 Summarize the results.

Remember that the t test should be used when the population is approximately normally distributed and the population standard deviation is unknown.
    Examples 8–12 through 8–14 illustrate the application of the *t* test.

Example 8–12

Hospital Infections

A medical investigation claims that the average number of infections per week at a hospital in southwestern Pennsylvania is 16.3. A random sample of 10 weeks had a mean number of 17.7 infections. The sample standard deviation is 1.8. Is there enough evidence to reject the investigator's claim at $\alpha = 0.05$?

Source: Based on information obtained from Pennsylvania Health Care Cost Containment Council.

Solution

Step 1 $H_0: \mu = 16.3$ (claim) and $H_1: \mu \neq 16.3$.

Step 2 The critical values are $+2.262$ and -2.262 for $\alpha = 0.05$ and d.f. $= 9$.

Step 3 The test value is

$$t = \frac{\bar{X} - \mu}{s/\sqrt{n}} = \frac{17.7 - 16.3}{1.8/\sqrt{10}} = 2.46$$

Step 4 Reject the null hypothesis since $2.46 > 2.262$. See Figure 8–22.

Figure 8–22

Summary of the *t* Test of Example 8–12

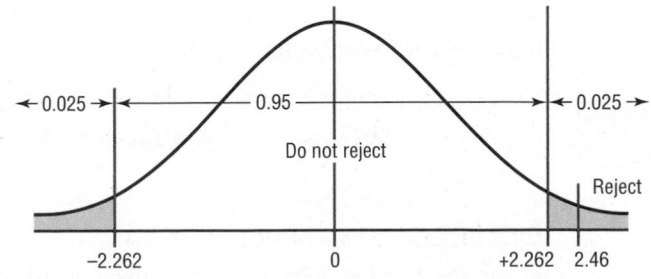

Step 5 There is enough evidence to reject the claim that the average number of infections is 16.3.

Example 8–13

Substitute Teachers' Salaries

An educator claims that the average salary of substitute teachers in school districts in Allegheny County, Pennsylvania, is less than $60 per day. A random sample of eight school districts is selected, and the daily salaries (in dollars) are shown. Is there enough evidence to support the educator's claim at $\alpha = 0.10$?

60 56 60 55 70 55 60 55

Source: *Pittsburgh Tribune-Review.*

Solution

Step 1 H_0: $\mu = \$60$ and H_1: $\mu < \$60$ (claim).

Step 2 At $\alpha = 0.10$ and d.f. = 7, the critical value is -1.415.

Step 3 To compute the test value, the mean and standard deviation must be found. Using either the formulas in Chapter 3 or your calculator, $\overline{X} = \$58.88$, and $s = 5.08$, you find

$$t = \frac{\overline{X} - \mu}{s/\sqrt{n}} = \frac{58.88 - 60}{5.08/\sqrt{8}} = -0.624$$

Step 4 Do not reject the null hypothesis since -0.624 falls in the noncritical region. See Figure 8–23.

Figure 8–23

Critical Value and Test Value for Example 8–13

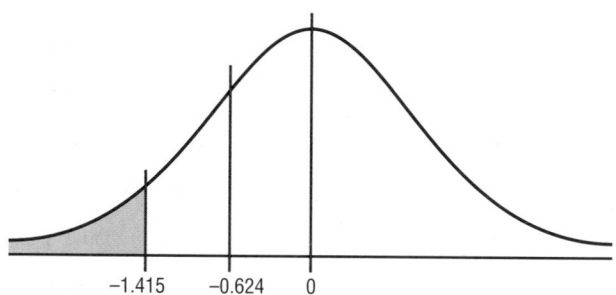

Step 5 There is not enough evidence to support the educator's claim that the average salary of substitute teachers in Allegheny County is less than $60 per day.

The P-values for the t test can be found by using Table F; however, specific P-values for t tests cannot be obtained from the table since only selected values of α (for example, 0.01, 0.05) are given. To find specific P-values for t tests, you would need a table similar to Table E for each degree of freedom. Since this is not practical, only *intervals* can be found for P-values. Examples 8–14 to 8–16 show how to use Table F to determine intervals for P-values for the t test.

Example 8–14

Find the P-value when the t test value is 2.056, the sample size is 11, and the test is right-tailed.

Solution

To get the P-value, look across the row with 10 degrees of freedom (d.f. = $n - 1$) in Table F and find the two values that 2.056 falls between. They are 1.812 and 2.228. Since this is a right-tailed test, look up to the row labeled One tail, α and find the two α values corresponding to 1.812 and 2.228. They are 0.05 and 0.025, respectively. See Figure 8–24.

Figure 8–24

Finding the *P*-Value for
Example 8–14

Confidence intervals	50%	80%	90%	95%	98%	99%
One tail, α	0.25	0.10	0.05	0.025	0.01	0.005
d.f. Two tails, α	0.50	0.20	0.10	0.05	0.02	0.01
1	1.000	3.078	6.314	12.706	31.821	63.657
2	0.816	1.886	2.920	4.303	6.965	9.925
3	0.765	1.638	2.353	3.182	4.541	5.841
4	0.741	1.533	2.132	2.776	3.747	4.604
5	0.727	1.476	2.015	2.571	3.365	4.032
6	0.718	1.440	1.943	2.447	3.143	3.707
7	0.711	1.415	1.895	2.365	2.998	3.499
8	0.706	1.397	1.860	2.306	2.896	3.355
9	0.703	1.383	1.833	2.262	2.821	3.250
10	0.700	1.372	1.812	2.228	2.764	3.169
11	0.697	1.363	1.796	2.201	2.718	3.106
12	0.695	1.356	1.782	2.179	2.681	3.055
13	0.694	1.350	1.771	2.160	2.650	3.012
14	0.692	1.345	1.761	2.145	2.624	2.977
15	0.691	1.341	1.753	2.131	2.602	2.947
(z) ∞	0.674	1.282	1.645	1.960	2.326	2.576

*2.056 falls between 1.812 and 2.228.

Hence, the *P*-value would be contained in the interval $0.025 < P\text{-value} < 0.05$. This means that the *P*-value is between 0.025 and 0.05. If α were 0.05, you would reject the null hypothesis since the *P*-value is less than 0.05. But if α were 0.01, you would not reject the null hypothesis since the *P*-value is greater than 0.01. (Actually, it is greater than 0.025.)

Example 8–15

Find the *P*-value when the *t* test value is 2.983, the sample size is 6, and the test is two-tailed.

Solution

To get the *P*-value, look across the row with d.f. = 5 and find the two values that 2.983 falls between. They are 2.571 and 3.365. Then look up to the row labeled Two tails, α to find the corresponding α values.

In this case, they are 0.05 and 0.02. Hence the *P*-value is contained in the interval $0.02 < P\text{-value} < 0.05$. This means that the *P*-value is between 0.02 and 0.05. In this case, if $\alpha = 0.05$, the null hypothesis can be rejected since *P*-value < 0.05; but if $\alpha = 0.01$, the null hypothesis cannot be rejected since *P*-value > 0.01 (actually *P*-value > 0.02).

Note: **Since many of you will be using calculators or computer programs that give the specific *P*-value for the *t* test and other tests presented later in this textbook, these specific values, in addition to the intervals, will be given for the answers to the examples and exercises.**

The *P*-value obtained from a calculator for Example 8–14 is 0.033. The *P*-value obtained from a calculator for Example 8–15 is 0.031.

To test hypotheses using the *P*-value method, follow the same steps as explained in Section 8–2. These steps are repeated here.

Step 1   State the hypotheses and identify the claim.

Step 2   Compute the test value.

Step 3   Find the *P*-value.

Step 4   Make the decision.

Step 5   Summarize the results.

This method is shown in Example 8–16.

| **Example 8–16** | **Jogger's Oxygen Uptake** |

A physician claims that joggers' maximal volume oxygen uptake is greater than the average of all adults. A sample of 15 joggers has a mean of 40.6 milliliters per kilogram (ml/kg) and a standard deviation of 6 ml/kg. If the average of all adults is 36.7 ml/kg, is there enough evidence to support the physician's claim at $\alpha = 0.05$?

Solution

Step 1   State the hypotheses and identify the claim.

$$H_0: \mu = 36.7 \quad \text{and} \quad H_1: \mu > 36.7 \text{ (claim)}$$

Step 2   Compute the test value. The test value is

$$t = \frac{\bar{X} - \mu}{s/\sqrt{n}} = \frac{40.6 - 36.7}{6/\sqrt{15}} = 2.517$$

Step 3   Find the *P*-value. Looking across the row with d.f. = 14 in Table F, you see that 2.517 falls between 2.145 and 2.624, corresponding to $\alpha = 0.025$ and $\alpha = 0.01$ since this is a right-tailed test. Hence, *P*-value > 0.01 and *P*-value < 0.025, or 0.01 < *P*-value < 0.025. That is, the *P*-value is somewhere between 0.01 and 0.025. (The *P*-value obtained from a calculator is 0.012.)

Step 4   Reject the null hypothesis since *P*-value < 0.05 (that is, *P*-value < α).

Step 5   There is enough evidence to support the claim that the joggers' maximal volume oxygen uptake is greater than 36.7 ml/kg.

Interesting Fact

The area of Alaska contains $\frac{1}{6}$ of the total area of the United States.

Students sometimes have difficulty deciding whether to use the *z* test or *t* test. The rules are the same as those pertaining to confidence intervals.

1. If σ is known, use the *z* test. The variable must be normally distributed if $n < 30$.

2. If σ is unknown but $n \geq 30$, use the *t* test.

3. If σ is unknown and $n < 30$, use the *t* test. (The population must be approximately normally distributed.)

These rules are summarized in Figure 8–25.

Speaking of
Statistics

Can Sunshine Relieve Pain?

A study conducted at the University of Pittsburgh showed that hospital patients in rooms with lots of sunlight required less pain medication the day after surgery and during their total stay in the hospital than patients who were in darker rooms.

Patients in the sunny rooms averaged 3.2 milligrams of pain reliever per hour for their total stay as opposed to 4.1 milligrams per hour for those in darker rooms. This study compared two groups of patients. Although no statistical tests were mentioned in the article, what statistical test do you think the researchers used to compare the groups?

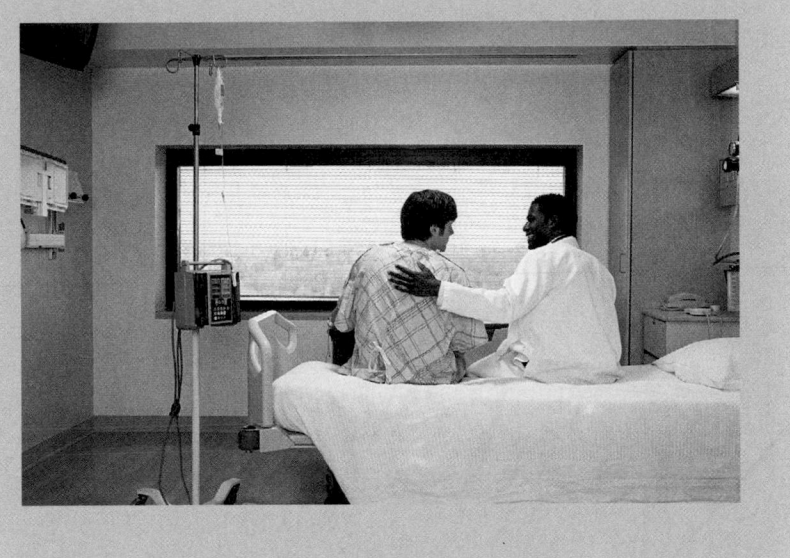

Figure 8–25

Using the z or t Test

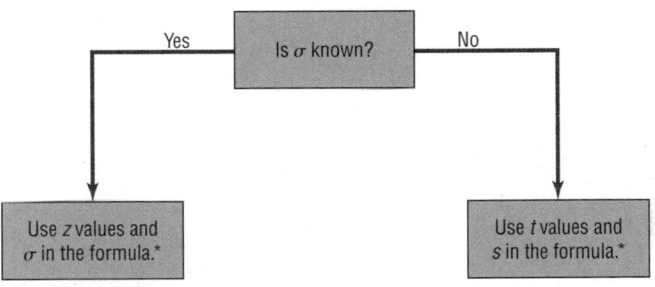

*If *n* < 30, the variable must be normally distributed.

Applying the Concepts **8–3**

How Much Nicotine Is in Those Cigarettes?

A tobacco company claims that its best-selling cigarettes contain at most 40 mg of nicotine. This claim is tested at the 1% significance level by using the results of 15 randomly selected cigarettes. The mean is 42.6 mg and the standard deviation is 3.7 mg. Evidence suggests that nicotine is normally distributed. Information from a computer output of the hypothesis test is listed.

Sample mean = 42.6   *P*-value = 0.008
Sample standard deviation = 3.7   Significance level = 0.01
Sample size = 15   Test statistic *t* = 2.72155
Degrees of freedom = 14   Critical value *t* = 2.62610

1. What are the degrees of freedom?
2. Is this a *z* or *t* test?
3. Is this a comparison of one or two samples?
4. Is this a right-tailed, left-tailed, or two-tailed test?

5. From observing the *P*-value, what would you conclude?

6. By comparing the test statistic to the critical value, what would you conclude?

7. Is there a conflict in this output? Explain.

8. What has been proved in this study?

See page 469 for the answers.

Exercises 8–3

1. In what ways is the *t* distribution similar to the standard normal distribution? In what ways is the *t* distribution different from the standard normal distribution?

2. What are the degrees of freedom for the *t* test?

3. Find the critical value (or values) for the *t* test for each.

   a. $n = 10$, $\alpha = 0.05$, right-tailed $+1.833$
   b. $n = 18$, $\alpha = 0.10$, two-tailed ± 1.740
   c. $n = 6$, $\alpha = 0.01$, left-tailed -3.365
   d. $n = 9$, $\alpha = 0.025$, right-tailed $+2.306$
   e. $n = 15$, $\alpha = 0.05$, two-tailed ± 2.145
   f. $n = 23$, $\alpha = 0.005$, left-tailed -2.819
   g. $n = 28$, $\alpha = 0.01$, two-tailed ± 2.771
   h. $n = 17$, $\alpha = 0.02$, two-tailed ± 2.583

4. (ans) Using Table F, find the *P*-value interval for each test value.

   a. $t = 2.321$, $n = 15$, right-tailed
   b. $t = 1.945$, $n = 28$, two-tailed
   c. $t = -1.267$, $n = 8$, left-tailed
   d. $t = 1.562$, $n = 17$, two-tailed
   e. $t = 3.025$, $n = 24$, right-tailed
   f. $t = -1.145$, $n = 5$, left-tailed
   g. $t = 2.179$, $n = 13$, two-tailed
   h. $t = 0.665$, $n = 10$, right-tailed

For Exercises 5 through 18, perform each of the following steps.

   a.   State the hypotheses and identify the claim.
   b.   Find the critical value(s).
   c.   Find the test value.
   d.   Make the decision.
   e.   Summarize the results.

Use the traditional method of hypothesis testing unless otherwise specified.

Assume that the population is approximately normally distributed.

5. **Veterinary Expenses of Cat Owners** According to the American Pet Products Manufacturers Association, cat owners spend an average of $179 annually in routine veterinary visits. A random sample of local cat owners revealed that 10 randomly selected owners spent an average of $205 with $s = \$26$. Is there a significant statistical difference at $\alpha = 0.01$?

Source: www.hsus.org/pets

6. **Park Acreage** A state executive claims that the average number of acres in western Pennsylvania state parks is less than 2000 acres. A random sample of five parks is selected, and the number of acres is shown. At $\alpha = 0.01$, is there enough evidence to support the claim?

959	1187	493	6249	541

Source: *Pittsburgh Tribune-Review.*

7. **Cell Phone Call Lengths** The average local cell phone call length was reported to be 2.27 minutes. A random sample of 20 phone calls showed an average of 2.98 minutes in length with a standard deviation of 0.98 minute. At $\alpha = 0.05$ can it be concluded that the average differs from the population average?

Source: *World Almanac.*

8. **Commute Time to Work** A survey of 15 large U.S. cities finds that the average commute time one way is 25.4 minutes. A chamber of commerce executive feels that the commute in his city is less and wants to publicize this. He randomly selects 25 commuters and finds the average is 22.1 minutes with a standard deviation of 5.3 minutes. At $\alpha = 0.10$, is he correct?

Source: *New York Times Almanac.*

9. **Heights of Tall Buildings** A researcher estimates that the average height of the buildings of 30 or more stories in a large city is at least 700 feet. A random sample of 10 buildings is selected, and the heights in feet are shown. At $\alpha = 0.025$, is there enough evidence to reject the claim?

485	511	841	725	615
520	535	635	616	582

Source: *Pittsburgh Tribune-Review.*

10. **Exercise and Reading Time Spent by Men** Men spend an average of 29 minutes per day on weekends and holidays exercising and playing sports. They spend an average of 23 minutes per day reading. A random sample of 25 men resulted in a mean of 35 minutes exercising with a standard deviation of 6.9 minutes and

an average of 20.5 minutes reading with $s = 7.2$ minutes. At $\alpha = 0.05$ for both, is there sufficient evidence that these two results differ from the national means?

Source: *Time* magazine.

11. Television Viewing by Teens Teens are reported to watch the fewest total hours of television per week of all the demographic groups. The average television viewing for teens on Sunday from 1:00 to 7:00 P.M. is 1 hour 13 minutes. A random sample of local teens disclosed the following times for Sunday afternoon television viewing. At $\alpha = 0.01$ can it be concluded that the average is greater than the national viewing time? (*Note:* Change all times to minutes.)

2:30	2:00	1:30	3:20
1:00	2:15	1:50	2:10
1:30	2:30		

Source: *World Almanac.*

12. Internet Visits A U.S. Web Usage Snapshot indicated a monthly average of 36 Internet visits per user from home. A random sample of 24 Internet users yielded a sample mean of 42.1 visits with a standard deviation of 5.3. At the 0.01 level of significance can it be concluded that this differs from the national average?

Source: *New York Times Almanac.*

13. Cost of Making a Movie During a recent year the average cost of making a movie was $54.8 million. This year, a random sample of 15 recent action movies had an average production cost of $62.3 million with a variance of $90.25 million. At the 0.05 level of significance, can it be concluded that it costs more than average to produce an action movie?

Source: *New York Times Almanac.*

14. Chocolate Chip Cookie Calories The average 1-ounce chocolate chip cookie contains 110 calories. A random sample of 15 different brands of 1-ounce chocolate chip cookies resulted in the following calorie amounts. At the $\alpha = 0.01$ level, is there sufficient evidence that the average calorie content is greater than 110 calories?

100	125	150	160	185	125	155	145	160
100	150	140	135	120	110			

Source: *The Doctor's Pocket Calorie, Fat, and Carbohydrate Counter.*

15. Cell Phone Bills The average monthly cell phone bill was reported to be $50.07 by the U.S. Wireless Industry. Random sampling of a large cell phone company found the following monthly cell phone charges:

55.83	49.88	62.98	70.42
60.47	52.45	49.20	50.02
58.60	51.29		

At the 0.05 level of significance can it be concluded that the average phone bill has increased?

Source: *World Almanac.*

16. Water Consumption The *Old Farmer's Almanac* stated that the average consumption of water per person per day was 123 gallons. To test the hypothesis that this figure may no longer be true, a researcher randomly selected 16 people and found that they used on average 119 gallons per day and $s = 5.3$. At $\alpha = 0.05$, is there enough evidence to say that the *Old Farmer's Almanac* figure might no longer be correct? Use the *P*-value method.

17. Doctor Visits A report by the Gallup Poll stated that on average a woman visits her physician 5.8 times a year. A researcher randomly selects 20 women and obtained these data.

3	2	1	3	7	2	9	4	6	6
8	0	5	6	4	2	1	3	4	1

At $\alpha = 0.05$ can it be concluded that the average is still 5.8 visits per year? Use the *P*-value method.

18. Number of Jobs The U.S. Bureau of Labor and Statistics reported that a person between the ages of 18 and 34 has had an average of 9.2 jobs. To see if this average is correct, a researcher selected a sample of 8 workers between the ages of 18 and 34 and asked how many different places they had worked. The results were as follows:

8	12	15	6	1	9	13	2

At $\alpha = 0.05$ can it be concluded that the mean is 9.2? Use the *P*-value method. Give one reason why the respondents might not have given the exact number of jobs that they have worked.

19. Teaching Assistants' Stipends A random sample of stipends of teaching assistants in economics is listed. Is there sufficient evidence at the $\alpha = 0.05$ level to conclude that the average stipend differs from $15,000? The stipends listed (in dollars) are for the academic year.

14,000	18,000	12,000	14,356	13,185
13,419	14,000	11,981	17,604	12,283
16,338	15,000			

Source: *Chronicle of Higher Education.*

20. Average Family Size The average family size was reported as 3.18. A random sample of families in a particular school district resulted in the following family sizes:

5	4	5	4	4	3	6	4	3	3	5
6	3	3	2	7	4	5	2	2	2	3
5	2									

At $\alpha = 0.05$, does the average family size differ from the national average?

Source: *New York Times Almanac.*

Technology *Step by Step*

MINITAB
Step by Step

Hypothesis Test for the Mean and the *t* Distribution

This relates to Example 8–13. Test the claim that the average salary for substitute teachers is less than $60 per day.

1. Enter the data into C1 of a MINITAB worksheet. Do not use the dollar sign. Name the column **Salary.**

2. Select **Stat>Basic Statistics>1-Sample t.**

3. Choose C1 Salary as the variable.

4. Click inside the text box for Test mean, and enter the hypothesized value of **60.**

5. Click [Options].

6. The Alternative should be less than.

7. Click [OK] twice.

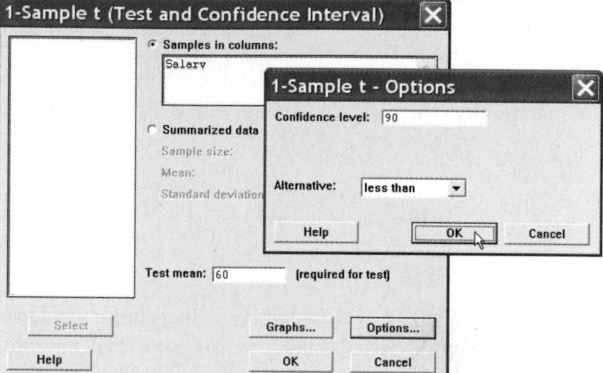

In the session window, the *P*-value for the test is 0.276.

One-Sample T: Salary

```
Test of mu = 60 vs < 60
```

Variable	N	Mean	StDev	SE Mean	90% Upper Bound	T	P
Salary	8	58.8750	5.0832	1.7972	61.4179	-0.63	0.276

We cannot reject H_0. There is not enough evidence in the sample to conclude the mean salary is less than $60.

TI-83 Plus or TI-84 Plus
Step by Step

Hypothesis Test for the Mean and the *t* Distribution (Data)

1. Enter the data values into L_1.
2. Press **STAT** and move the cursor to TESTS.
3. Press **2** for T-Test.
4. Move the cursor to Data and press **ENTER.**
5. Type in the appropriate values.
6. Move the cursor to the appropriate alternative hypothesis and press **ENTER.**
7. Move the cursor to Calculate and press **ENTER.**

Hypothesis Test for the Mean and the *t* Distribution (Statistics)

1. Press **STAT** and move the cursor to TESTS.
2. Press **2** for T-Test.
3. Move the cursor to Stats and press **ENTER.**
4. Type in the appropriate values.
5. Move the cursor to the appropriate alternative hypothesis and press **ENTER.**
6. Move the cursor to Calculate and press **ENTER.**

Excel
Step by Step

Hypothesis Test for the Mean: *t* Test

Excel does not have a procedure to conduct a hypothesis test for the mean. However, you may conduct the test of the mean using the MegaStat Add-in available on your CD. If you have not installed this add-in, do so, following the instructions from the Chapter 1 Excel Step by Step.

Example XL8–2

This example relates to Example 8–13 from the text. At the 10% significance level, test the claim that $\mu < 60$. The MegaStat *t* test uses the *P*-value method. Therefore, it is not necessary to enter a significance level.

1. Enter the data into column A of a new worksheet.
2. From the toolbar, select Add-Ins, **MegaStat>Hypothesis Tests>Mean vs. Hypothesized Value.** *Note:* You may need to open MegaStat from the MegaStat.xls file on your computer's hard drive.
3. Select data input and type **A1:A8** as the Input Range.
4. Type **60** for the Hypothesized mean and select the "less than" Alternative.
5. Select *t* test and click [OK].

The result of the procedure is shown next.

Hypothesis Test: Mean vs. Hypothesized Value

```
60.000  Hypothesized value
58.875  Mean data
 5.083  Standard deviation
 1.797  Standard error
     8  n
     7  d.f.

-0.63  t
0.2756  P-value (one-tailed, lower)
```

8–4

Objective 7

Test proportions, using the *z* test.

z Test for a Proportion

Many hypothesis-testing situations involve proportions. Recall from Chapter 7 that a *proportion* is the same as a percentage of the population.

These data were obtained from *The Book of Odds* by Michael D. Shook and Robert L. Shook (New York: Penguin Putnam, Inc.):

- 59% of consumers purchase gifts for their fathers.
- 85% of people over 21 said they have entered a sweepstakes.
- 51% of Americans buy generic products.
- 35% of Americans go out for dinner once a week.

A hypothesis test involving a population proportion can be considered as a binomial experiment when there are only two outcomes and the probability of a success does not change from trial to trial. Recall from Section 5–3 that the mean is $\mu = np$ and the standard deviation is $\sigma = \sqrt{npq}$ for the binomial distribution.

Since a normal distribution can be used to approximate the binomial distribution when $np \geq 5$ and $nq \geq 5$, the standard normal distribution can be used to test hypotheses for proportions.

Formula for the *z* Test for Proportions

$$z = \frac{\hat{p} - p}{\sqrt{pq/n}}$$

where

$$\hat{p} = \frac{X}{n} \quad \text{(sample proportion)}$$

p = population proportion

n = sample size

The formula is derived from the normal approximation to the binomial and follows the general formula

$$\text{Test value} = \frac{(\text{observed value}) - (\text{expected value})}{\text{standard error}}$$

We obtain $\hat{p}$ from the sample (i.e., observed value), p is the expected value (i.e., hypothesized population proportion), and $\sqrt{pq/n}$ is the standard error.

The formula $z = \dfrac{\hat{p} - p}{\sqrt{pq/n}}$ can be derived from the formula $z = \dfrac{X - \mu}{\sigma}$ by substituting $\mu = np$ and $\sigma = \sqrt{npq}$ and then dividing both numerator and denominator by n. Some algebra is used. See Exercise 23 in this section.

Assumptions for Testing a Proportion

1. The sample is a random sample.
2. The conditions for a binomial experiment are satisfied. (See Chapter 5.)
3. $np \geq 5$ and $nq \geq 5$.

The steps for hypothesis testing are the same as those shown in Section 8–3. Table E is used to find critical values and P-values.

Examples 8–17 to 8–19 show the traditional method of hypothesis testing. Example 8–20 shows the P-value method.

Sometimes it is necessary to find $\hat{p}$, as shown in Examples 8–17, 8–19, and 8–20, and sometimes $\hat{p}$ is given in the exercise. See Example 8–18.

Example 8–17

People Who Are Trying to Avoid Trans Fats

A dietitian claims that 60% of people are trying to avoid trans fats in their diets. She randomly selected 200 people and found that 128 people stated that they were trying to avoid trans fats in their diets. At $\alpha = 0.05$, is there enough evidence to reject the dietitian's claim?

Source: Based on a survey by the Gallup Poll.

Solution

Step 1 State the hypothesis and identify the claim.

$$H_0: p = 0.60 \text{ (claim)} \qquad \text{and} \qquad H_1: p \neq 0.60$$

Step 2 Find the critical values. Since $\alpha = 0.05$ and the test value is two-tailed, the critical values are ± 1.96.

Step 3 Compute the test value. First, it is necessary to find $\hat{p}$.

$$\hat{p} = \frac{X}{n} = \frac{128}{200} = 0.64 \qquad p = 0.60 \qquad q = 1 - 0.60 = 0.40$$

Substitute in the formula.

$$Z = \frac{\hat{p} - p}{\sqrt{pq/n}} = \frac{0.64 - 0.60}{\sqrt{(0.60)(0.40)/200}} = 1.15$$

Step 4 Make the decision. Do not reject the null hypothesis since the test value falls outside the critical region, as shown in Figure 8–26.

Figure 8–26

Critical and Test Values
for Example 8–17

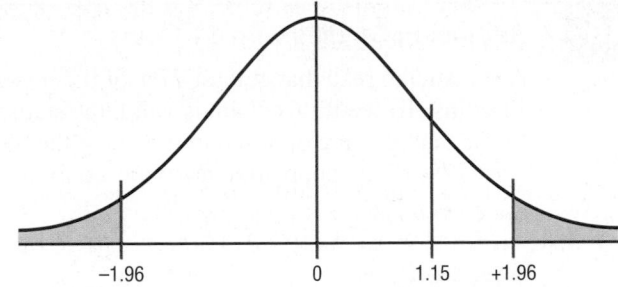

Step 5   Summarize the results. There is not enough evidence to reject the claim that 60% of people are trying to avoid trans fats in their diets.

Example 8–18   **Family and Medical Leave Act**

The Family and Medical Leave Act provides job protection and unpaid time off from work for a serious illness or birth of a child. In 2000, 60% of the respondents of a survey stated that it was very easy to get time off for these circumstances. A researcher wishes to see if the percentage who said that it was very easy to get time off has changed. A sample of 100 people who used the leave said that 53% found it easy to use the leave. At $\alpha = 0.01$, has the percentage changed?

Source: Department of Labor.

Solution

Step 1   State the hypotheses and identify the claim.

$H_0: p = 0.60$ and $H_1: p \neq 0.60$ (claim)

Step 2   Find the critical value(s). Since $\alpha = 0.01$ and this test is two-tailed, the critical values are ± 2.58.

Step 3   Compute the test value. It is not necessary to find $\hat{p}$ since it is given in the exercise; $\hat{p} = 53\%$. Substitute in the formula and evaluate.

$p = 0.6$ and $q = 1 - p = 1 - 0.6 = 0.4$

$$z = \frac{\hat{p} - p}{\sqrt{pq/n}} = \frac{0.53 - 0.60}{\sqrt{(0.6)(0.4)/100}} = -1.43$$

Step 4   Make the decision. Do not reject the null hypothesis, since the test value falls in the noncritical region, as shown in Figure 8–27.

Figure 8–27

Critical and Test Values
for Example 8–18

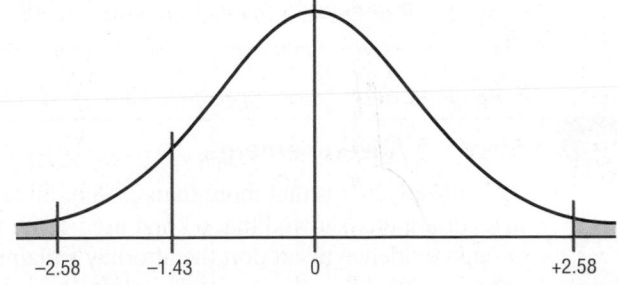

Step 5   Summarize the results. There is not enough evidence to support the claim that the percentage of those using the medical leave said that it was easy to get has changed.

Example 8–19   Replacing $1 Bills with $1 Coins

A statistician read that at least 77% of the population oppose replacing $1 bills with $1 coins. To see if this claim is valid, the statistician selected a sample of 80 people and found that 55 were opposed to replacing the $1 bills. At $\alpha = 0.01$, test the claim that at least 77% of the population are opposed to the change.

Source: *USA TODAY*.

Solution

Step 1   State the hypotheses and identify the claim.

$$H_0\colon p = 0.77 \text{ (claim)} \qquad \text{and} \qquad H_1\colon p < 0.77$$

Step 2   Find the critical value(s). Since $\alpha = 0.01$ and the test is left-tailed, the critical value is -2.33.

Step 3   Compute the test value.

$$\hat{p} = \frac{X}{n} = \frac{55}{80} = 0.6875$$

$$p = 0.77 \qquad \text{and} \qquad q = 1 - 0.77 = 0.23$$

$$z = \frac{\hat{p} - p}{\sqrt{pq/n}} = \frac{0.6875 - 0.77}{\sqrt{(0.77)(0.23)/80}} = -1.75$$

Step 4   Do not reject the null hypothesis, since the test value does not fall in the critical region, as shown in Figure 8–28.

Figure 8–28

Critical and Test Values for Example 8–19

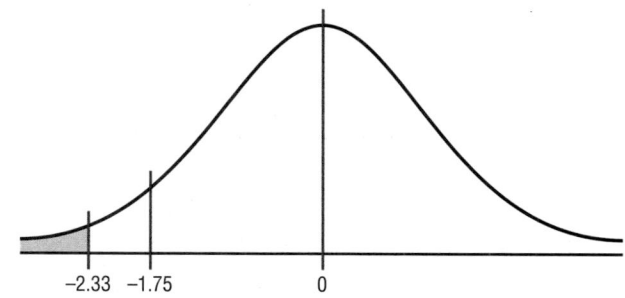

−2.33 −1.75   0

Step 5   There is not enough evidence to reject the claim that at least 77% of the population oppose replacing $1 bills with $1 coins.

Example 8–20   Attorney Advertisements

An attorney claims that more than 25% of all lawyers advertise. A sample of 200 lawyers in a certain city showed that 63 had used some form of advertising. At $\alpha = 0.05$, is there enough evidence to support the attorney's claim? Use the *P*-value method.

Solution

Step 1   State the hypotheses and identify the claim.

$$H_0\colon p = 0.25 \qquad \text{and} \qquad H_1\colon p > 0.25 \text{ (claim)}$$

Step 2 Compute the test value.

$$\hat{p} = \frac{X}{n} = \frac{63}{200} = 0.315$$

$$p = 0.25 \qquad \text{and} \qquad q = 1 - 0.25 = 0.75$$

$$z = \frac{\hat{p} - p}{\sqrt{pq/n}} = \frac{0.315 - 0.25}{\sqrt{(0.25)(0.75)/200}} = 2.12$$

Step 3 Find the *P*-value. The area under the curve for $z = 2.12$ is 0.9830. Subtracting the area from 1.0000, you get $1.0000 - 0.9830 = 0.0170$. The *P*-value is 0.0170.

Step 4 Reject the null hypothesis, since $0.0170 < 0.05$ (that is, *P*-value < 0.05). See Figure 8–29.

Figure 8–29

P-Value and α Value for Example 8–20

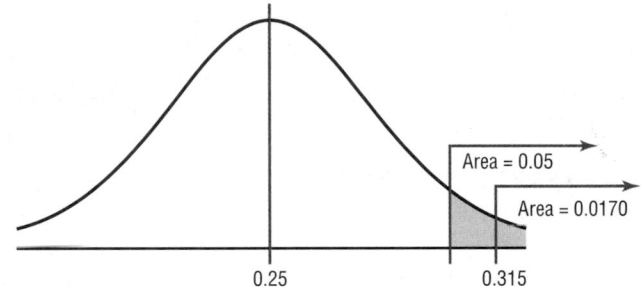

Step 5 There is enough evidence to support the attorney's claim that more than 25% of the lawyers use some form of advertising.

Applying the Concepts **8–4**

Quitting Smoking

Assume you are part of a research team that compares products designed to help people quit smoking. Condor Consumer Products Company would like more specific details about the study to be made available to the scientific community. Review the following and then answer the questions about how you would have conducted the study.

New StopSmoke

No method has been proved more effective. StopSmoke provides significant advantages over all other methods. StopSmoke is simpler to use, and it requires no weaning. StopSmoke is also significantly less expensive than the leading brands. StopSmoke's superiority has been proved in two independent studies.

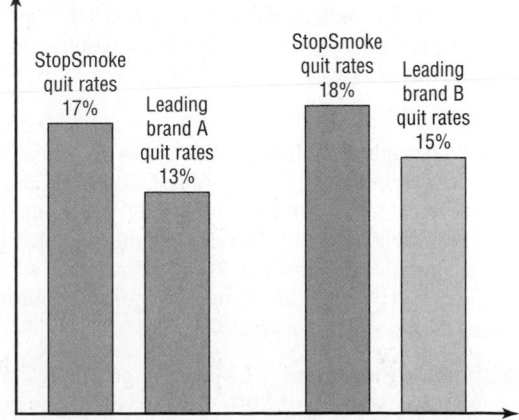

1. What were the statistical hypotheses?
2. What were the null hypotheses?
3. What were the alternative hypotheses?
4. Were any statistical tests run?
5. Were one- or two-tailed tests run?
6. What were the levels of significance?

7. If a type I error was committed, explain what it would have been.

8. If a type II error was committed, explain what it would have been.

9. What did the studies prove?

10. Two statements are made about significance. One states that StopSmoke provides significant advantages, and the other states that StopSmoke is significantly less expensive than other leading brands. Are they referring to statistical significance? What other type of significance is there?

See page 469 for the answers.

See page 469 for the answers.

Exercises 8–4

1. Give three examples of proportions. Answers will vary.

2. Why is a proportion considered a binomial variable?

3. When you are testing hypotheses by using proportions, what are the necessary requirements? $np \geq 5$ and $nq \geq 5$

4. What are the mean and the standard deviation of a proportion? $\mu = np; \sigma = \sqrt{pq/n}$

For Exercises 5 through 15, perform each of the following steps.

a. State the hypotheses and identify the claim.

b. Find the critical value(s).

c. Compute the test value.

d. Make the decision.

e. Summarize the results.

Use the traditional method of hypothesis testing unless otherwise specified.

5. **Home Ownership** A recent survey found that 68.6% of the population own their homes. In a random sample of 150 heads of households, 92 responded that they owned their homes. At the $\alpha = 0.01$ level of significance, does that suggest a difference from the national proportion?
Source: *World Almanac.*

6. **Stocks and Mutual Fund Ownership** It has been found that 50.3% of U.S. households own stocks and mutual funds. A random sample of 300 heads of households indicated that 171 owned some type of stock. At what level of significance would you conclude that this was a significant difference?
Source: www.census.gov

7. **Overweight Children** Health issues due to being overweight affect all age groups. Of children and adolescents 6–11 years of age, 18.8% are found to be overweight. A school district randomly sampled 130 in this age group and found that 20 were considered overweight. At $\alpha = 0.05$ is this less than the national proportion?
Source: *New York Times Almanac.*

8. **Female Physicians** The percentage of physicians who are women is 27.9%. In a survey of physicians employed by a large university health system, 45 of 120 randomly selected physicians were women. Is there sufficient evidence at the 0.05 level of significance to conclude that the proportion of women physicians at the university health system exceeds 27.9%?
Source: *New York Times Almanac.*

9. **Traveling Overseas** Of U.S. residents traveling overseas, 47% were women and 53% were men. A random sample of 500 travelers on a large airline revealed that of those 500, 263 were women. Does this differ from the national percentage at the 0.05 level of significance?
Source: *World Almanac.*

10. **Undergraduate Enrollment** It has been found that 85.6% of all enrolled college and university students in the United States are undergraduates. A random sample of 500 enrolled college students in a particular state revealed that 420 of them were undergraduates. Is there sufficient evidence to conclude that the proportion differs from the national percentage? Use $\alpha = 0.05$.
Source: *Time Almanac.*

11. **Moviegoers** The largest group of moviegoers by age is the 40- to 59-year-old age group. This group constitutes 32% of the movie-going population. A theater complex randomly surveyed the customers over a three-week period and found that out of 423 surveyed, 170 were 40 to 59 years of age. At the 0.01 level of significance does this differ from the stated proportion?
Source: *MPAA Study.*

12. **Exercise to Reduce Stress** A survey by *Men's Health* magazine stated that 14% of men said they used exercise to reduce stress. Use $\alpha = 0.10$. A random sample of 100 men was selected, and 10 said that they used exercise to relieve stress. Use the P-value method to test the claim. Could the results be generalized to all adult Americans?

13. **After-School Snacks** In the *Journal of the American Dietetic Association,* it was reported that 54% of kids said that they had a snack after school. A random sample of 60 kids was selected, and 36 said that they had a snack after school. Use $\alpha = 0.01$ and the P-value method to test the claim. On the basis of the results, should parents be concerned about their children eating a healthy snack?

14. Natural Gas Heat The Energy Information Administration reported that 51.7% of homes in the United States were heated by natural gas. A random sample of 200 homes found that 115 were heated by natural gas. Does the evidence support the claim, or has the percentage changed? Use $\alpha = 0.05$ and the P-value method. What could be different if the sample were taken in a different geographic area?

15. Youth Smoking Researchers suspect that 18% of all high school students smoke at least one pack of cigarettes a day. At Wilson High School, with an enrollment of 300 students, a study found that 50 students smoked at least one pack of cigarettes a day. At $\alpha = 0.05$, test the claim that 18% of all high school students smoke at least one pack of cigarettes a day. Use the P-value method.

16. Television Set Ownership According to Nielsen Media Research, of all the U.S. households that owned at least one television set, 83% had two or more sets. A local cable company canvassing the town to promote a new cable service found that of the 300 households visited, 240 had two or more television sets. At $\alpha = 0.05$ is there sufficient evidence to conclude that the proportion is less than the one in the report?

Source: *World Almanac.*

17. Borrowing Library Books For Americans using library services, the American Library Association (ALA)

claims that 67% borrow books. A library director feels that this is not true so he randomly selects 100 borrowers and finds that 82 borrowed books. Can he show that the ALA claim is incorrect? Use $\alpha = 0.05$.

Source: American Library Association; *USA TODAY.*

18. Doctoral Students' Salaries Nationally, at least 60% of Ph.D. students have paid assistantships. A college dean feels that this is not true in his state, so he randomly selects 50 Ph.D. students and finds that 26 have assistantships. At $\alpha = 0.05$, is the dean correct?

Source: U.S. Department of Education, *Chronicle of Higher Education.*

19. Football Injuries A report by the NCAA states that 57.6% of football injuries occur during practices. A head trainer claims that this is too high for his conference, so he randomly selects 36 injuries and finds that 17 occurred during practices. Is his claim correct, at $\alpha = 0.05$?

Source: *NCAA Sports Medicine Handbook.*

20. Foreign Languages Spoken in Homes Approximately 19.4% of the U.S. population 5 years old and older speaks a language other than English at home. In a large metropolitan area it was found that out of 400 randomly selected residents over 5 years of age, 94 spoke a language other than English at home. Is there sufficient evidence to conclude that the proportion is higher than the national proportion? You choose the level of significance.

Source: www.census.gov

Extending the Concepts

When np or nq is not 5 or more, the binomial table (Table B in Appendix C) must be used to find critical values in hypothesis tests involving proportions.

21. Coin Tossing A coin is tossed 9 times and 3 heads appear. Can you conclude that the coin is not balanced? Use $\alpha = 0.10$. [*Hint:* Use the binomial table and find $2P(X \le 3)$ with $p = 0.5$ and $n = 9$.] No

22. First-Class Airline Passengers In the past, 20% of all airline passengers flew first class. In a sample of 15 passengers, 5 flew first class. At $\alpha = 0.10$, can you conclude that the proportions have changed?

23. Show that $z = \dfrac{\hat{p} - p}{\sqrt{pq/n}}$ can be derived from $z = \dfrac{X - \mu}{\sigma}$ by substituting $\mu = np$ and $\sigma = \sqrt{npq}$ and dividing both numerator and denominator by n.

Technology *Step by Step*

MINITAB
Step by Step

Hypothesis Test for One Proportion and the z Distribution

MINITAB will calculate the test statistic and P-value for a test of a proportion, given the statistics from a sample or given the raw data. For example, test the claim that 40% of all telephone customers have call-waiting service, when $n = 100$ and $\hat{p} = 37\%$. Use $\alpha = 0.01$.

1. Select **Stat>Basic Statistics>1 Proportion.**
2. Click on the button for Summarized data. There are no data to enter in the worksheet.
3. Click in the box for Number of trials and enter **100.**
4. In the Number of events box enter **37.**
5. Click on [Options].
6. Type the complement of α, **99** for the confidence level.

7. Very important! Check the box for Use test and interval based on normal distribution.

8. Click [OK] twice.

The results for the confidence interval will be displayed in the session window. Since the *P*-value of 0.540 is greater than $\alpha = 0.01$, the null hypothesis cannot be rejected.

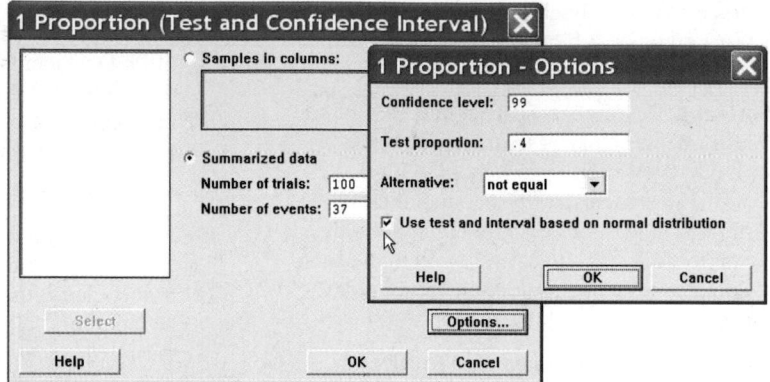

Test and CI for One Proportion

```
Test of p = 0.4 vs p not = 0.4
Sample   X    N   Sample p          99%  CI          Z-Value  P-Value
1        37  100  0.370000  (0.245638,  0.494362)     -0.61    0.540
```

There is not enough evidence to conclude that the proportion is different from 40%.

TI-83 Plus or TI-84 Plus
Step by Step

Hypothesis Test for the Proportion

1. Press **STAT** and move the cursor to TESTS.

2. Press **5** for 1-PropZTest.

3. Type in the appropriate values.

4. Move the cursor to the appropriate alternative hypothesis and press **ENTER**.

5. Move the cursor to Calculate and press **ENTER**.

Example TI8–3

This pertains to the previous example. Test the claim that $p = 40\%$, given $n = 100$ and $\hat{p} = 0.37$.

 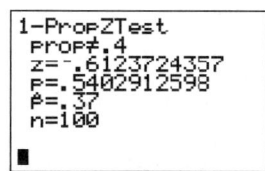

The test statistic is $z = -0.6123724357$, and the *P*-value is 0.5402912598.

Excel
Step by Step

Hypothesis Test for the Proportion: *z* Test

Excel does not have a procedure to conduct a hypothesis test for the population proportion. However, you may conduct the test of the proportion, using the MegaStat Add-in available on your CD. If you have not installed this add-in, do so, following the instructions from the Chapter 1 Excel Step by Step.

Example XL8–4

This example relates to the previous example. At the 1% significance level, test the claim that $p = 0.40$. The MegaStat test of the population proportion uses the *P*-value method. Therefore, it is not necessary to enter a significance level.

1. From the toolbar, select Add-Ins, **MegaStat>Hypothesis Tests>Proportion vs. Hypothesized Value.** *Note:* You may need to open MegaStat from the MegaStat.xls file on your computer's hard drive.

2. Type **0.37** for the Observed proportion, *p*.

3. Type **0.40** for the Hypothesized proportion, *p*.

4. Type **100** for the sample size, *n*.

5. Select the "not equal" Alternative.

6. Click [OK].

The result of the procedure is shown next.

Hypothesis Test for Proportion vs. Hypothesized Value

Observed	Hypothesized	
0.37	0.4	p (as decimal)
37/100	40/100	p (as fraction)
37.	40.	X
100	100	n
	0.049	standard error
	−0.61	z
	0.5403	p-value (two-tailed)

8–5 χ^2 **Test for a Variance or Standard Deviation**

Objective 8

Test variances or standard deviations, using the chi-square test.

In Chapter 7, the chi-square distribution was used to construct a confidence interval for a single variance or standard deviation. This distribution is also used to test a claim about a single variance or standard deviation.

To find the area under the chi-square distribution, use Table G in Appendix C. There are three cases to consider:

1. Finding the chi-square critical value for a specific α when the hypothesis test is right-tailed.

2. Finding the chi-square critical value for a specific α when the hypothesis test is left-tailed.

3. Finding the chi-square critical values for a specific α when the hypothesis test is two-tailed.

Example 8–21

Find the critical chi-square value for 15 degrees of freedom when $\alpha = 0.05$ and the test is right-tailed.

Solution

The distribution is shown in Figure 8–30.

Figure 8–30

Chi-Square Distribution for Example 8–21

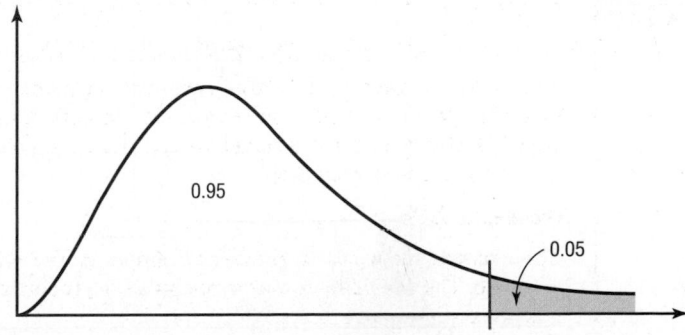

Find the α value at the top of Table G, and find the corresponding degrees of freedom in the left column. The critical value is located where the two columns meet—in this case, 24.996. See Figure 8–31.

Figure 8–31

Locating the Critical Value in Table G for Example 8–21

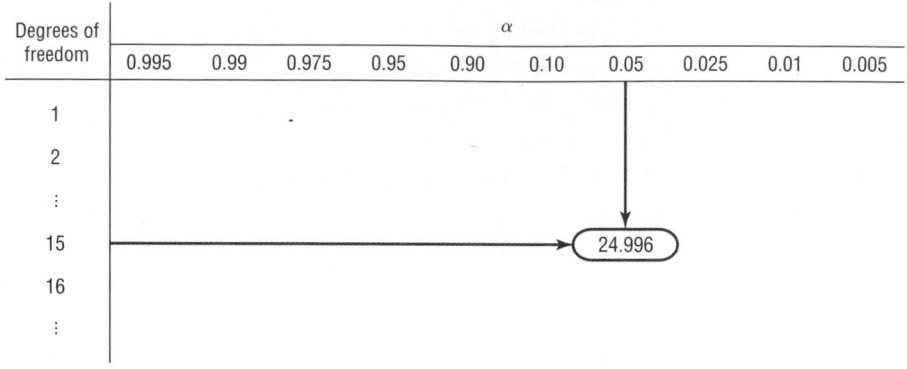

Example 8–22

Find the critical chi-square value for 10 degrees of freedom when $\alpha = 0.05$ and the test is left-tailed.

Solution

This distribution is shown in Figure 8–32.

Figure 8–32

Chi-Square Distribution for Example 8–22

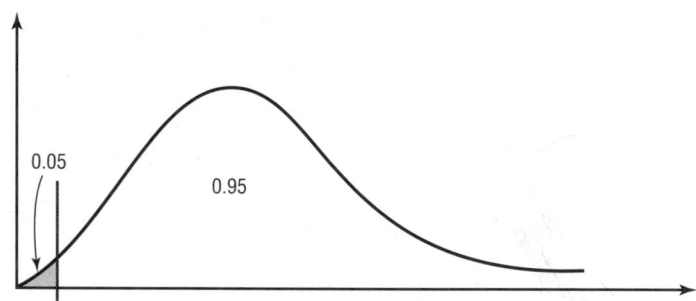

When the test is left-tailed, the α value must be subtracted from 1, that is, $1 - 0.05 = 0.95$. The left side of the table is used, because the chi-square table gives the area to the right of the critical value, and the chi-square statistic cannot be negative. The table is set up so that it gives the values for the area to the right of the critical value. In this case, 95% of the area will be to the right of the value.

For 0.95 and 10 degrees of freedom, the critical value is 3.940. See Figure 8–33.

Figure 8–33

Locating the Critical Value in Table G for Example 8–22

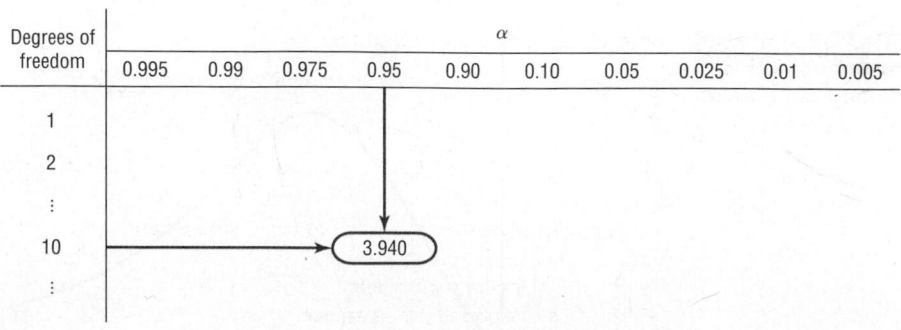

Example 8–23

Find the critical chi-square values for 22 degrees of freedom when $\alpha = 0.05$ and a two-tailed test is conducted.

Solution

When a two-tailed test is conducted, the area must be split, as shown in Figure 8–34. Note that the area to the right of the larger value is 0.025 (0.05/2 or $\alpha/2$), and the area to the right of the smaller value is 0.975 ($1.00 - 0.05/2$ or $1 - \alpha/2$).

Figure 8–34

Chi-Square Distribution for Example 8–23

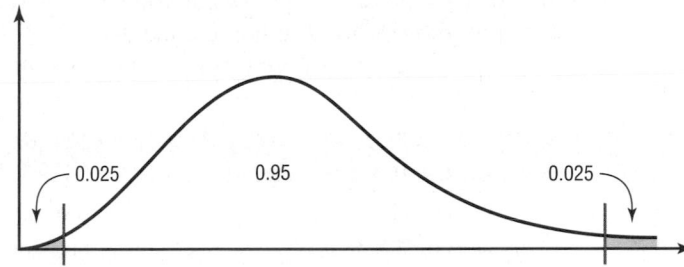

Remember that chi-square values cannot be negative. Hence, you must use α values in the table of 0.025 and 0.975. With 22 degrees of freedom, the critical values are 36.781 and 10.982, respectively.

After the degrees of freedom reach 30, Table G gives values only for multiples of 10 (40, 50, 60, etc.). When the exact degrees of freedom sought are not specified in the table, the closest smaller value should be used. For example, if the given degrees of freedom are 36, use the table value for 30 degrees of freedom. This guideline keeps the type I error equal to or below the α value.

When you are testing a claim about a single variance using the **chi-square test,** there are three possible test situations: right-tailed test, left-tailed test, and two-tailed test.

If a researcher believes the variance of a population to be greater than some specific value, say, 225, then the researcher states the hypotheses as

$$H_0: \sigma^2 = 225 \qquad \text{and} \qquad H_1: \sigma^2 > 225$$

and conducts a right-tailed test.

If the researcher believes the variance of a population to be less than 225, then the researcher states the hypotheses as

$$H_0: \sigma^2 = 225 \qquad \text{and} \qquad H_1: \sigma^2 < 225$$

and conducts a left-tailed test.

Finally, if a researcher does not wish to specify a direction, she or he states the hypotheses as

$$H_0: \sigma^2 = 225 \qquad \text{and} \qquad H_1: \sigma^2 \neq 225$$

and conducts a two-tailed test.

Formula for the Chi-Square Test for a Single Variance
$$\chi^2 = \frac{(n-1)s^2}{\sigma^2}$$ with degrees of freedom equal to $n-1$ and where n = sample size s^2 = sample variance σ^2 = population variance

You might ask, Why is it important to test variances? There are several reasons. First, in any situation where consistency is required, such as in manufacturing, you would like to have the smallest variation possible in the products. For example, when bolts are manufactured, the variation in diameters due to the process must be kept to a minimum, or the nuts will not fit them properly. In education, consistency is required on a test. That is, if the same students take the same test several times, they should get approximately the same grades, and the variance of each of the student's grades should be small. On the other hand, if the test is to be used to judge learning, the overall standard deviation of all the grades should be large so that you can differentiate those who have learned the subject from those who have not learned it.

Three assumptions are made for the chi-square test, as outlined here.

Unusual Stat

About 20% of cats owned in the United States are overweight.

Assumptions for the Chi-Square Test for a Single Variance

1. The sample must be randomly selected from the population.
2. The population must be normally distributed for the variable under study.
3. The observations must be independent of one another.

The traditional method for hypothesis testing follows the same five steps listed earlier. They are repeated here.

Step 1 State the hypotheses and identify the claim.

Step 2 Find the critical value(s).

Step 3 Compute the test value.

Step 4 Make the decision.

Step 5 Summarize the results.

Examples 8–24 through 8–26 illustrate the traditional hypothesis-testing procedure for variances.

Example 8–24

Variation of Test Scores

An instructor wishes to see whether the variation in scores of the 23 students in her class is less than the variance of the population. The variance of the class is 198. Is there enough evidence to support the claim that the variation of the students is less than the population variance ($\sigma^2 = 225$) at $\alpha = 0.05$? Assume that the scores are normally distributed.

Solution

Step 1 State the hypotheses and identify the claim.

$$H_0: \sigma^2 = 225 \qquad \text{and} \qquad H_1: \sigma^2 < 225 \text{ (claim)}$$

Step 2 Find the critical value. Since this test is left-tailed and $\alpha = 0.05$, use the value $1 - 0.05 = 0.95$. The degrees of freedom are $n - 1 = 23 - 1 = 22$. Hence, the critical value is 12.338. Note that the critical region is on the left, as shown in Figure 8–35.

Figure 8–35

Critical Value for
Example 8–24

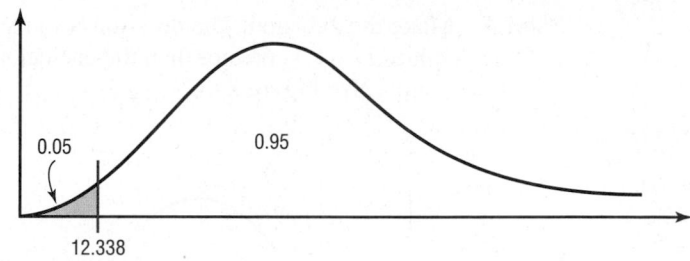

Step 3 Compute the test value.

$$\chi^2 = \frac{(n-1)s^2}{\sigma^2} = \frac{(23-1)(198)}{225} = 19.36$$

Step 4 Make the decision. Since the test value 19.36 falls in the noncritical region, as shown in Figure 8–36, the decision is to not reject the null hypothesis.

Figure 8–36

Critical and Test Values
for Example 8–24

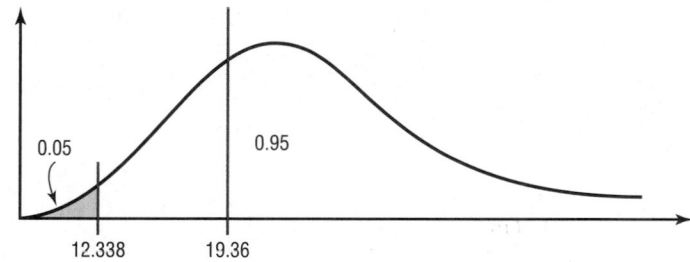

Step 5 Summarize the results. There is not enough evidence to support the claim that the variation in test scores of the instructor's students is less than the variation in scores of the population.

Example 8–25

Outpatient Surgery

 A hospital administrator believes that the standard deviation of the number of people using outpatient surgery per day is greater than 8. A random sample of 15 days is selected. The data are shown. At $\alpha = 0.10$, is there enough evidence to support the administrator's claim? Assume the variable is normally distributed.

25	30	5	15	18
42	16	9	10	12
12	38	8	14	27

Solution

Step 1 State the hypotheses and identify the claim.

$H_0: \sigma = 8$ and $H_1: \sigma > 8$ (claim)

Since the standard deviation is given, it should be squared to get the variance.

Step 2 Find the critical value. Since this test is right-tailed with d.f. of $15 - 1 = 14$ and $\alpha = 0.10$, the critical value is 21.064.

Step 3 Compute the test value. Since raw data are given, the standard deviation of the sample must be found by using the formula in Chapter 3 or your calculator. It is $s = 11.2$.

$$\chi^2 = \frac{(n-1)s^2}{\sigma^2} = \frac{(15-1)(11.2)^2}{64} = 27.44$$

Step 4 Make the decision. The decision is to reject the null hypothesis since the test value, 27.44, is greater than the critical value, 21.064, and falls in the critical region. See Figure 8–37.

Figure 8–37

Critical and Test Value for Example 8–25

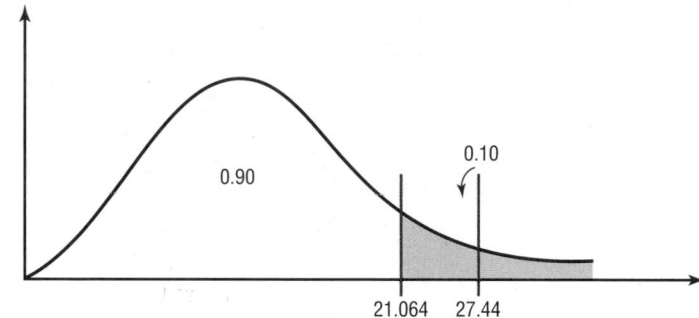

Step 5 Summarize the results. There is enough evidence to support the claim that the standard deviation is greater than 8.

Example 8–26

Nicotine Content of Cigarettes

A cigarette manufacturer wishes to test the claim that the variance of the nicotine content of its cigarettes is 0.644. Nicotine content is measured in milligrams, and assume that it is normally distributed. A sample of 20 cigarettes has a standard deviation of 1.00 milligram. At $\alpha = 0.05$, is there enough evidence to reject the manufacturer's claim?

Solution

Step 1 State the hypotheses and identify the claim.

$$H_0: \sigma^2 = 0.644 \text{ (claim)} \qquad \text{and} \qquad H_1: \sigma^2 \neq 0.644$$

Step 2 Find the critical values. Since this test is a two-tailed test at $\alpha = 0.05$, the critical values for 0.025 and 0.975 must be found. The degrees of freedom are 19; hence, the critical values are 32.852 and 8.907, respectively. The critical or rejection regions are shown in Figure 8–38.

Figure 8–38

Critical Values for Example 8–26

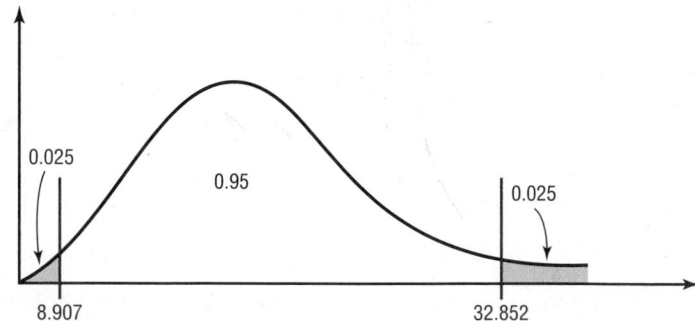

Step 3 Compute the test value.

$$\chi^2 = \frac{(n-1)s^2}{\sigma^2} = \frac{(20-1)(1.0)^2}{0.644} = 29.5$$

Since the standard deviation s is given in the problem, it must be squared for the formula.

Step 4 Make the decision. Do not reject the null hypothesis, since the test value falls between the critical values ($8.907 < 29.5 < 32.852$) and in the noncritical region, as shown in Figure 8–39.

Figure 8–39

Critical and Test Values for Example 8–26

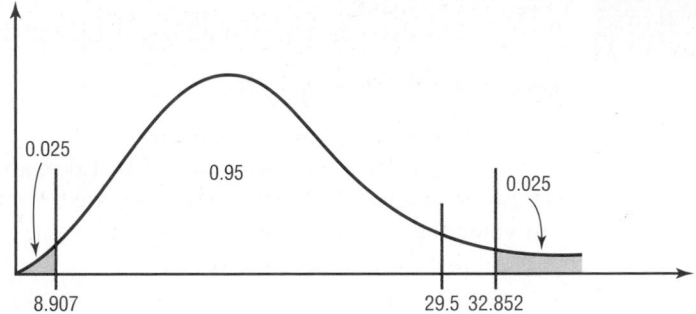

Step 5 Summarize the results. There is not enough evidence to reject the manufacturer's claim that the variance of the nicotine content of the cigarettes is equal to 0.644.

Approximate P-values for the chi-square test can be found by using Table G in Appendix C. The procedure is somewhat more complicated than the previous procedures for finding P-values for the z and t tests since the chi-square distribution is not exactly symmetric and χ^2 values cannot be negative. As we did for the t test, we will determine an *interval* for the P-value based on the table. Examples 8–27 through 8–29 show the procedure.

Example 8–27

Find the P-value when $\chi^2 = 19.274$, $n = 8$, and the test is right-tailed.

Solution

To get the P-value, look across the row with d.f. = 7 in Table G and find the two values that 19.274 falls between. They are 18.475 and 20.278. Look up to the top row and find the α values corresponding to 18.475 and 20.278. They are 0.01 and 0.005, respectively. See Figure 8–40. Hence the P-value is contained in the interval $0.005 < P\text{-value} < 0.01$. (The P-value obtained from a calculator is 0.007.)

Figure 8–40

P-Value Interval for Example 8–27

Degrees of freedom	α									
	0.995	0.99	0.975	0.95	0.90	0.10	0.05	0.025	0.01	0.005
1	—	—	0.001	0.004	0.016	2.706	3.841	5.024	6.635	7.879
2	0.010	0.020	0.051	0.103	0.211	4.605	5.991	7.378	9.210	10.597
3	0.072	0.115	0.216	0.352	0.584	6.251	7.815	9.348	11.345	12.838
4	0.207	0.297	0.484	0.711	1.064	7.779	9.488	11.143	13.277	14.860
5	0.412	0.554	0.831	1.145	1.610	9.236	11.071	12.833	15.086	16.750
6	0.676	0.872	1.237	1.635	2.204	10.645	12.592	14.449	16.812	18.548
7	0.989	1.239	1.690	2.167	2.833	12.017	11.067	16.013	18.475	20.278
8	1.344	1.646	2.180	2.733	3.490	13.362	15.507	17.535	20.090	21.955
9	1.735	2.088	2.700	3.325	4.168	14.684	16.919	19.023	21.666	23.589
10	2.156	2.558	3.247	3.940	4.865	15.987	18.307	20.483	23.209	25.188
⋮	⋮	⋮	⋮	⋮	⋮	⋮	⋮	⋮	⋮	⋮
100	67.328	70.065	74.222	77.929	82.358	118.498	124.342	129.561	135.807	140.169

*19.274 falls between 18.475 and 20.278

| Example 8–28 | Find the P-value when $\chi^2 = 3.823$, $n = 13$, and the test is left-tailed. |

Solution

To get the P-value, look across the row with d.f. = 12 and find the two values that 3.823 falls between. They are 3.571 and 4.404. Look up to the top row and find the values corresponding to 3.571 and 4.404. They are 0.99 and 0.975, respectively. When the χ^2 test value falls on the left side, each of the values must be subtracted from 1 to get the interval that P-value falls between.

$$1 - 0.99 = 0.01 \quad \text{and} \quad 1 - 0.975 = 0.025$$

Hence the P-value falls in the interval

$$0.01 < P\text{-value} < 0.025$$

(The P-value obtained from a calculator is 0.014.)

When the χ^2 test is two-tailed, both interval values must be doubled. If a two-tailed test were being used in Example 8–28, then the interval would be $2(0.01) < P$-value $< 2(0.025)$, or $0.02 < P$-value < 0.05.

The P-value method for hypothesis testing for a variance or standard deviation follows the same steps shown in the preceding sections.

Step 1 State the hypotheses and identify the claim.

Step 2 Compute the test value.

Step 3 Find the P-value.

Step 4 Make the decision.

Step 5 Summarize the results.

Example 8–29 shows the P-value method for variances or standard deviations.

| Example 8–29 | Car Inspection Times |

A researcher knows from past studies that the standard deviation of the time it takes to inspect a car is 16.8 minutes. A sample of 24 cars is selected and inspected. The standard deviation is 12.5 minutes. At $\alpha = 0.05$, can it be concluded that the standard deviation has changed? Use the P-value method.

Solution

Step 1 State the hypotheses and identify the claim.

$$H_0: \sigma = 16.8 \quad \text{and} \quad H_1: \sigma \neq 16.8 \text{ (claim)}$$

Step 2 Compute the test value.

$$\chi^2 = \frac{(n-1)s^2}{\sigma^2} = \frac{(24-1)(12.5)^2}{(16.8)^2} = 12.733$$

Step 3 Find the P-value. Using Table G with d.f. = 23, the value 12.733 falls between 11.689 and 13.091, corresponding to 0.975 and 0.95, respectively. Since these values are found on the left side of the distribution, each value must be subtracted from 1. Hence $1 - 0.975 = 0.025$ and $1 - 0.95 = 0.05$. Since this is a two-tailed test, the area must be doubled to obtain the P-value interval. Hence $0.05 < P$-value < 0.10, or somewhere between 0.05 and 0.10. (The P-value obtained from a calculator is 0.085.)

Step 4   Make the decision. Since $\alpha = 0.05$ and the *P*-value is between 0.05 and 0.10, the decision is to not reject the null hypothesis since *P*-value $> \alpha$.

Step 5   Summarize the results. There is not enough evidence to support the claim that the standard deviation has changed.

Applying the Concepts 8–5

Testing Gas Mileage Claims

Assume that you are working for the Consumer Protection Agency and have recently been getting complaints about the highway gas mileage of the new Dodge Caravans. Chrysler Corporation agrees to allow you to randomly select 40 of its new Dodge Caravans to test the highway mileage. Chrysler claims that the Caravans get 28 mpg on the highway. Your results show a mean of 26.7 and a standard deviation of 4.2. You support Chrysler's claim.

1. Show why you support Chrysler's claim by listing the *P*-value from your output. After more complaints, you decide to test the variability of the miles per gallon on the highway. From further questioning of Chrysler's quality control engineers, you find they are claiming a standard deviation of 2.1.
2. Test the claim about the standard deviation.
3. Write a short summary of your results and any necessary action that Chrysler must take to remedy customer complaints.
4. State your position about the necessity to perform tests of variability along with tests of the means.

See pages 469 and 470 for the answers.

---

1. Using Table G, find the critical value(s) for each, show the critical and noncritical regions, and state the appropriate null and alternative hypotheses. Use $\sigma^2 = 225$.

   a. $\alpha = 0.05, n = 18$, right-tailed
   b. $\alpha = 0.10, n = 23$, left-tailed
   c. $\alpha = 0.05, n = 15$, two-tailed
   d. $\alpha = 0.10, n = 8$, two-tailed
   e. $\alpha = 0.01, n = 17$, right-tailed
   f. $\alpha = 0.025, n = 20$, left-tailed
   g. $\alpha = 0.01, n = 13$, two-tailed
   h. $\alpha = 0.025, n = 29$, left-tailed

2. **(ans)** Using Table G, find the *P*-value interval for each χ^2 test value.

   a. $\chi^2 = 29.321, n = 16$, right-tailed
   b. $\chi^2 = 10.215, n = 25$, left-tailed
   c. $\chi^2 = 24.672, n = 11$, two-tailed
   d. $\chi^2 = 23.722, n = 9$, right-tailed
   e. $\chi^2 = 13.974, n = 28$, two-tailed
   f. $\chi^2 = 10.571, n = 19$, left-tailed
   g. $\chi^2 = 12.144, n = 6$, two-tailed
   h. $\chi^2 = 8.201, n = 23$, two-tailed

For Exercises 3 through 9, assume that the variables are normally or approximately normally distributed. Use the traditional method of hypothesis testing unless otherwise specified.

 3. **Calories in Pancake Syrup** A nutritionist claims that the standard deviation of the number of calories in 1 tablespoon of the major brands of pancake syrup is 60. A sample of major brands of syrup is selected, and the number of calories is shown. At $\alpha = 0.10$, can the claim be rejected?

53	210	100	200	100	220
210	100	240	200	100	210
100	210	100	210	100	60

Source: Based on information from *The Complete Book of Food Counts* by Corrine T. Netzer, Dell Publishers, New York.

4. **High Temperatures in January** Daily weather observations for southwestern Pennsylvania for the first three weeks of January show daily high temperatures as follows: 55, 44, 51, 59, 62, 60, 46, 51, 37, 30, 46, 51, 53, 57, 57, 39, 28, 37, 35, and 28 degrees Fahrenheit. The normal standard deviation in high temperatures for this time period is usually no more than 8 degrees. A meteorologist believes that with the unusual trend in

temperatures the standard deviation is greater. At $\alpha = 0.05$, can we conclude that the standard deviation is greater than 8 degrees?

Source: www.wunderground.com

5. Stolen Aircraft Test the claim that the standard deviation of the number of aircraft stolen each year in the United States is less than 15 if a sample of 12 years had a standard deviation of 13.6. Use $\alpha = 0.05$.

Source: Aviation Crime Prevention Institute.

6. Carbohydrates in Fast Foods The number of carbohydrates found in a random sample of fast-food entrees is listed below. Is there sufficient evidence to conclude that the variance differs from 100? Use the 0.05 level of significance.

53	46	39	39	30
47	38	73	43	41

Source: Fast Food Explorer (www.fatcalories.com).

7. Transferring Phone Calls The manager of a large company claims that the standard deviation of the time (in minutes) that it takes a telephone call to be transferred to the correct office in her company is 1.2 minutes or less. A sample of 15 calls is selected, and the calls are timed. The standard deviation of the sample is 1.8 minutes. At $\alpha = 0.01$, test the claim that the standard deviation is less than or equal to 1.2 minutes. Use the P-value method.

8. Soda Bottle Content A machine fills 12-ounce bottles with soda. For the machine to function properly, the standard deviation of the sample must be less than or equal to 0.03 ounce. A sample of 8 bottles is selected, and the number of ounces of soda in each bottle is given. At $\alpha = 0.05$, can we reject the claim that the machine is functioning properly? Use the P-value method.

12.03	12.10	12.02	11.98
12.00	12.05	11.97	11.99

9. High-Potassium Foods Potassium is important to good health in keeping fluids and minerals balanced and blood pressure low. High-potassium foods are those that contain more than 200 mg per serving. The amounts of potassium for a random sample are shown. At $\alpha = 0.10$ is the standard deviation of the potassium content greater than 100?

781	467	508	530
707	535	498	400

Source: www.drugs.com

10. Exam Grades A statistics professor is used to having a variance in his class grades of no more than 100. He feels that his current group of students

is different, and so he examines a random sample of midterm grades (listed below.) At $\alpha = 0.05$, can it be concluded that the variance in grades exceeds 100?

92.3	89.4	76.9	65.2	49.1
96.7	69.5	72.8	67.5	52.8
88.5	79.2	72.9	68.7	75.8

11. Tornado Deaths A researcher claims that the standard deviation of the number of deaths annually from tornadoes in the United States is less than 35. If a sample of 11 randomly selected years had a standard deviation of 32, is the claim believable? Use $\alpha = 0.05$.

Source: National Oceanic and Atmospheric Administration.

12. Interstate Speeds It has been reported that the standard deviation of the speeds of drivers on Interstate 75 near Findlay, Ohio, is 8 miles per hour for all vehicles. A driver feels from experience that this is very low. A survey is conducted, and for 50 drivers the standard deviation is 10.5 miles per hour. At $\alpha = 0.05$, is the driver correct?

13. College Room and Board Costs Room and board fees for a random sample of independent religious colleges are listed below.

7460	7959	7650	8120	7220
8768	7650	8400	7860	6782
8754	7443	9500	9100	

Estimate the standard deviation in costs based on $s \approx R/4$. Is there sufficient evidence to conclude that the sample standard deviation differs from this estimated amount? Use $\alpha = 0.05$.

Source: World Almanac.

14. Heights of Volcanoes A sample of heights (in feet) of active volcanoes in North America, outside of Alaska, is listed below. Is there sufficient evidence that the standard deviation in heights of volcanoes outside Alaska is less than the standard deviation in heights of Alaskan volcanoes, which is 2385.9 feet? Use $\alpha = 0.05$.

10,777	8159	11,240	10,456
14,163	8363		

Source: Time Almanac.

15. Manufactured Machine Parts A manufacturing process produces machine parts with measurements the standard deviation of which must be no more than 0.52 mm. A random sample of 20 parts in a given lot revealed a standard deviation in measurement of 0.568 mm. Is there sufficient evidence at $\alpha = 0.05$ to conclude that the standard deviation of the parts is outside the required guidelines?

Technology *Step by Step*

MINITAB
Step by Step

Hypothesis Test for Variance

For Example 8–25, test the administrator's claim that the standard deviation is greater than 8. There is no menu item to calculate the test statistic and *P*-value directly.

Calculate the Standard Deviation and Sample Size

1. Enter the data into a column of MINITAB. Name the column **OutPatients.**
2. The standard deviation and sample size will be calculated and stored.
   a) Select **Calc>Column Statistics.**
   b) Check the button for Standard deviation. You can only do one of these statistics at a time.
   c) Use OutPatients for the Input variable.
   d) Store the result in s, then click [OK].
3. Select **Edit>Edit Last Dialog Box,** then do three things:
   a) Change the Statistic option from Standard Deviation to N nonmissing.
   b) Type **n** in the text box for Store the result.
   c) Click [OK].

Calculate the Chi-Square Test Statistic

4. Select **Calc>Calculator.**

   a) In the text box for Store result in variable: type in **K3.** The chi-square value will be stored in a constant so it can be used later.

   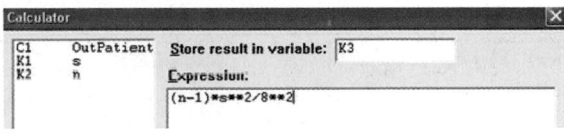

   b) In the expression, type in the formula as shown. The double asterisk is the symbol used for a power.
   c) Click [OK]. The chi-square value of 27.44 will be stored in K3.

Calculate the *P*-Value

   d) Select **Calc>Probability Distributions>Chi-Square.**
   e) Click the button for Cumulative probability.
   f) Type in **14** for Degrees of freedom.
   g) Click in the text box for Input constant and type **K3.**
   h) Type in **K4** for Optional storage.
   i) Click [OK]. Now K4 contains the area to the left of the chi-square test statistic.

   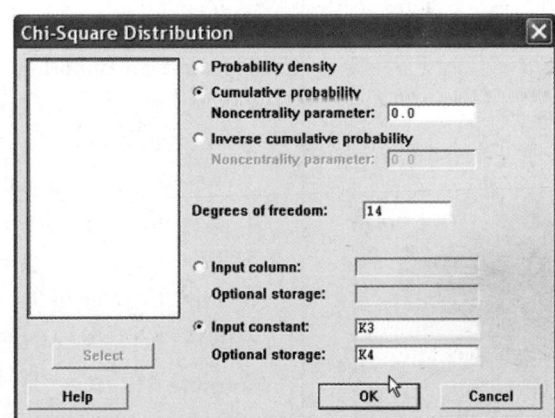

   Subtract the cumulative area from 1 to find the area on the right side of the chi-square test statistic. This is the *P*-value for a right-tailed test.

   j) Select **Calc>Calculator.**
   k) In the text box for Store result in variable, type in P-Value.
   l) The expression 1 − K4 calculates the complement of the cumulative area.
   m) Click [OK].

   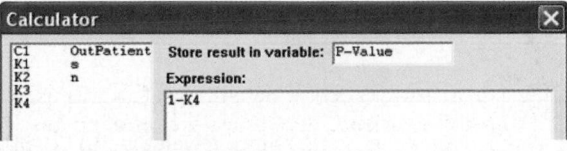

The result will be shown in the first row of C2, 0.0168057. Since the P-value is less than α, reject the null hypothesis. The standard deviation in the sample is 11.2, the point estimate for the true standard deviation σ.

TI-83 Plus or TI-84 Plus
Step by Step

The TI-83 Plus and TI-84 Plus do not have a built-in hypothesis test for the variance or standard deviation. However, the downloadable program named SDHYP is available on your CD and Online Learning Center. Follow the instructions with your CD for downloading the program.

Performing a Hypothesis Test for the Variance and Standard Deviation (Data)

1. Enter the values into L_1.
2. Press **PRGM,** move the cursor to the program named SDHYP, and press **ENTER** twice.
3. Press **1** for Data.
4. Type L_1 for the list and press **ENTER.**
5. Type the number corresponding to the type of alternative hypothesis.
6. Type the value of the hypothesized variance and press **ENTER.**
7. Press **ENTER** to clear the screen.

Example TI8–4

This pertains to Example 8–25 in the text. Test the claim that $\sigma > 8$ for these data.

25 30 5 15 18 42 16 9 10 12 12 38 8 14 27

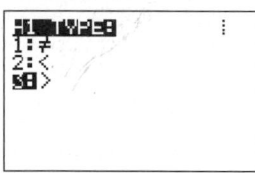 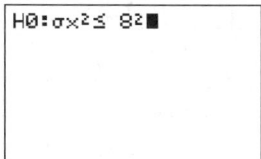 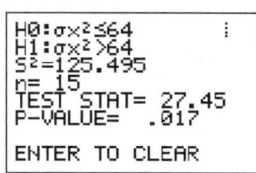

Since P-value $= 0.017 < 0.1$, we reject H_0 and conclude H_1. Therefore, there is enough evidence to support the claim that the standard deviation of the number of people using outpatient surgery is greater than 8.

Performing a Hypothesis Test for the Variance and Standard Deviation (Statistics)

1. Press **PRGM,** move the cursor to the program named SDHYP, and press **ENTER** twice.
2. Press **2** for Stats.
3. Type the sample standard deviation and press **ENTER.**
4. Type the sample size and press **ENTER.**
5. Type the number corresponding to the type of alternative hypothesis.
6. Type the value of the hypothesized variance and press **ENTER.**
7. Press **ENTER** to clear the screen.

Example TI8–5

This pertains to Example 8–26 in the text. Test the claim that $\sigma^2 = 0.644$, given $n = 20$ and $s = 1$.

 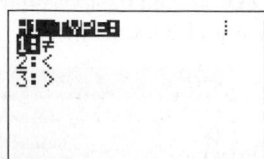 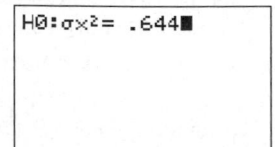 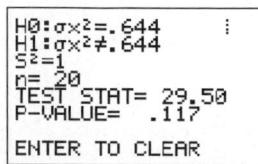

Since P-value $= 0.117 > 0.05$, we do not reject H_0 and do not conclude H_1. Therefore, there is not enough evidence to reject the manufacturer's claim that the variance of the nicotine content of the cigarettes is equal to 0.644.

Excel
Step by Step

Hypothesis Test for the Variance: Chi-Square Test

Excel does not have a procedure to conduct a hypothesis test for the variance. However, you may conduct the test of the variance using the MegaStat Add-in available on your CD. If you have not installed this add-in, do so, following the instructions from the Chapter 1 Excel Step by Step.

Example XL8–4

This example relates to Example 8–26 from the text. At the 5% significance level, test the claim that $\sigma^2 = 0.644$. The MegaStat chi-square test of the population variance uses the P-value method. Therefore, it is not necessary to enter a significance level.

1. Type a label for the variable: Nicotine in cell A1.
2. Type the observed variance: **1** in cell A2.
3. Type the sample size: **20** in cell A3.
4. From the toolbar, select Add-Ins, **MegaStat>Hypothesis Tests>Chi-Square Variance Test.** *Note:* You may need to open MegaStat from the MegaStat.xls file on your computer's hard drive.
5. Select summary input.
6. Type **A1:A3** for the Input Range.
7. Type **0.644** for the Hypothesized variance and select the "not equal" Alternative.
8. Click [OK].

The result of the procedure is shown next.

Chi-Square Variance Test

```
 0.64  Hypothesized variance
 1.00  Observed variance of nicotine
   20  n
   19  d.f.
29.50  Chi-square

0.1169  P-value (two-tailed)
```

8–6   Additional Topics Regarding Hypothesis Testing

In hypothesis testing, there are several other concepts that might be of interest to students in elementary statistics. These topics include the relationship between hypothesis testing and confidence intervals, and some additional information about the type II error.

Objective 9
Test hypotheses, using confidence intervals.

Confidence Intervals and Hypothesis Testing

There is a relationship between confidence intervals and hypothesis testing. When the null hypothesis is rejected in a hypothesis-testing situation, the confidence interval for the mean using the same level of significance *will not* contain the hypothesized mean. Likewise, when the null hypothesis is not rejected, the confidence interval computed using the same level of significance *will* contain the hypothesized mean. Examples 8–30 and 8–31 show this concept for two-tailed tests.

Example 8–30   Sugar Production

Sugar is packed in 5-pound bags. An inspector suspects the bags may not contain 5 pounds. A sample of 50 bags produces a mean of 4.6 pounds and a standard deviation of 0.7 pound. Is there enough evidence to conclude that the bags do not contain 5 pounds as stated at $\alpha = 0.05$? Also, find the 95% confidence interval of the true mean.

Solution

Now H_0: $\mu = 5$ and H_1: $\mu \neq 5$ (claim). The critical values are $+2.010$ and -2.010. The test value is

$$t = \frac{\bar{X} - \mu}{s/\sqrt{n}} = \frac{4.6 - 5.0}{0.7/\sqrt{50}} = \frac{-0.4}{0.099} = -4.04$$

Since $-4.04 < -2.010$, the null hypothesis is rejected. There is enough evidence to support the claim that the bags do not weigh 5 pounds.

The 95% confidence for the mean is given by

$$\bar{X} - t_{\alpha/2}\frac{s}{\sqrt{n}} < \mu < \bar{X} + t_{\alpha/2}\frac{s}{\sqrt{n}}$$

$$4.6 - (2.010)\left(\frac{0.7}{\sqrt{50}}\right) < \mu < 4.6 + (2.010)\left(\frac{0.7}{\sqrt{50}}\right)$$

$$4.4 < \mu < 4.8$$

Notice that the 95% confidence interval of μ does *not* contain the hypothesized value $\mu = 5$. Hence, there is agreement between the hypothesis test and the confidence interval.

Example 8–31

Hog Weights

A researcher claims that adult hogs fed a special diet will have an average weight of 200 pounds. A sample of 10 hogs has an average weight of 198.2 pounds and a standard deviation of 3.3 pounds. At $\alpha = 0.05$, can the claim be rejected? Also, find the 95% confidence interval of the true mean.

Solution

Now H_0: $\mu = 200$ pounds (claim) and H_1: $\mu \neq 200$ pounds. The t test must be used since σ is unknown. It is assumed that hog weights are normally distributed. The critical values at $\alpha = 0.05$ with 9 degrees of freedom are $+2.262$ and -2.262. The test value is

$$t = \frac{\bar{X} - \mu}{s/\sqrt{n}} = \frac{198.2 - 200}{3.3/\sqrt{10}} = \frac{-1.8}{1.0436} = -1.72$$

Thus, the null hypothesis is not rejected. There is not enough evidence to reject the claim that the weight of the adult hogs is 200 pounds.

The 95% confidence interval of the mean is

$$\bar{X} - t_{\alpha/2}\frac{s}{\sqrt{n}} < \mu < \bar{X} + t_{\alpha/2}\frac{s}{\sqrt{n}}$$

$$198.2 - (2.262)\left(\frac{3.3}{\sqrt{10}}\right) < \mu < 198.2 + (2.262)\left(\frac{3.3}{\sqrt{10}}\right)$$

$$198.2 - 2.361 < \mu < 198.2 + 2.361$$

$$195.8 < \mu < 200.6$$

The 95% confidence interval does contain the hypothesized mean $\mu = 200$. Again there is agreement between the hypothesis test and the confidence interval.

In summary, then, when the null hypothesis is rejected at a significance level of α, the confidence interval computed at the $1 - \alpha$ level will not contain the value of the mean that is stated in the null hypothesis. On the other hand, when the null hypothesis is not rejected, the confidence interval computed at the same significance level will contain the value of the mean stated in the null hypothesis. These results are true for other hypothesis-testing situations and are not limited to means tests.

The relationship between confidence intervals and hypothesis testing presented here is valid for two-tailed tests. The relationship between one-tailed hypothesis tests and one-sided or one-tailed confidence intervals is also valid; however, this technique is beyond the scope of this textbook.

Type II Error and the Power of a Test

Recall that in hypothesis testing, there are two possibilities: Either the null hypothesis H_0 is true, or it is false. Furthermore, on the basis of the statistical test, the null hypothesis is either rejected or not rejected. These results give rise to four possibilities, as shown in Figure 8–41. This figure is similar to Figure 8–2.

Objective 10

Explain the relationship between type I and type II errors and the power of a test.

As stated previously, there are two types of errors: type I and type II. A type I error can occur only when the null hypothesis is rejected. By choosing a level of significance, say, of 0.05 or 0.01, the researcher can determine the probability of committing a type I error. For example, suppose that the null hypothesis was H_0: $\mu = 50$, and it was rejected. At the 0.05 level (one tail), the researcher has only a 5% chance of being wrong, i.e., of rejecting a true null hypothesis.

On the other hand, if the null hypothesis is not rejected, then either it is true or a type II error has been committed. A type II error occurs when the null hypothesis is indeed false, but is not rejected. The probability of committing a type II error is denoted as β.

The value of β is not easy to compute. It depends on several things, including the value of α, the size of the sample, the population standard deviation, and the actual difference between the hypothesized value of the parameter being tested and the true parameter. The researcher has control over two of these factors, namely, the selection of α and the size of the sample. The standard deviation of the population is sometimes known or can be estimated. The major problem, then, lies in knowing the actual difference between the hypothesized parameter and the true parameter. If this difference were known, then the value of the parameter would be known; and if the parameter were known, then there would be no need to do any hypothesis testing. Hence, the value of β cannot be computed. But this does not mean that it should be ignored. What the researcher usually does is to try to minimize the size of β or to maximize the size of $1 - \beta$, which is called the **power of a test.**

The power of a statistical test measures the sensitivity of the test to detect a real difference in parameters if one actually exists. The power of a test is a probability and, like all probabilities, can have values ranging from 0 to 1. The higher the power, the more sensitive the test is to detecting a real difference between parameters if there is a difference.

Figure 8–41

Possibilities in Hypothesis Testing

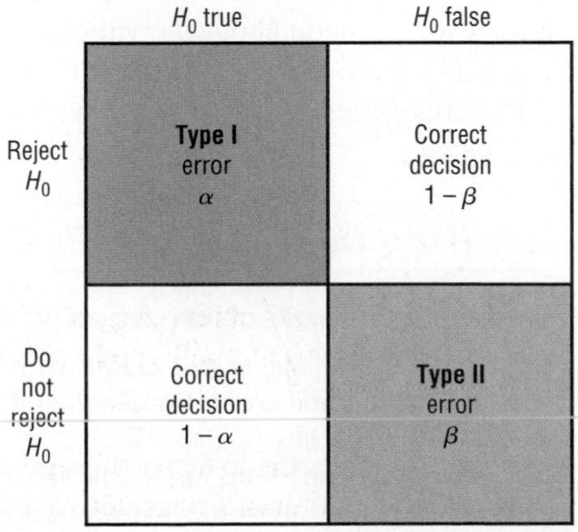

In other words, the closer the power of a test is to 1, the better the test is for rejecting the null hypothesis if the null hypothesis is, in fact, false.

The power of a test is equal to $1 - \beta$, that is, 1 minus the probability of committing a type II error. The power of the test is shown in the upper right-hand block of Figure 8–41. If somehow it were known that $\beta = 0.04$, then the power of a test would be $1 - 0.04 = 0.96$, or 96%. In this case, the probability of rejecting the null hypothesis when it is false is 96%.

As stated previously, the power of a test depends on the probability of committing a type II error, and since β is not easily computed, the power of a test cannot be easily computed. (See the Critical Thinking Challenges on page 468.)

However, there are some guidelines that can be used when you are conducting a statistical study concerning the power of a test. When you are conducting a statistical study, use the test that has the highest power for the data. There are times when the researcher has a choice of two or more statistical tests to test the hypotheses. The tests with the highest power should be used. It is important, however, to remember that statistical tests have assumptions that need to be considered.

If these assumptions cannot be met, then another test with lower power should be used. The power of a test can be increased by increasing the value of α. For example, instead of using $\alpha = 0.01$, use $\alpha = 0.05$. Recall that as α increases, β decreases. So if β is decreased, then $1 - \beta$ will increase, thus increasing the power of the test.

Another way to increase the power of a test is to select a larger sample size. A larger sample size would make the standard error of the mean smaller and consequently reduce β. (The derivation is omitted.)

These two methods should not be used at the whim of the researcher. Before α can be increased, the researcher must consider the consequences of committing a type I error. If these consequences are more serious than the consequences of committing a type II error, then α should not be increased.

Likewise, there are consequences to increasing the sample size. These consequences might include an increase in the amount of money required to do the study and an increase in the time needed to tabulate the data. When these consequences result, increasing the sample size may not be practical.

There are several other methods a researcher can use to increase the power of a statistical test, but these methods are beyond the scope of this book.

One final comment is necessary. When the researcher fails to reject the null hypothesis, this does not mean that there is not enough evidence to support alternative hypotheses. It may be that the null hypothesis is false, but the statistical test has too low a power to detect the real difference; hence, one can conclude only that in this study, there is not enough evidence to reject the null hypothesis.

The relationship among α, β, and the power of a test can be analyzed in greater detail than the explanation given here. However, it is hoped that this explanation will show you that there is no magic formula or statistical test that can guarantee foolproof results when a decision is made about the validity of H_0. Whether the decision is to reject H_0 or not to reject H_0, there is in either case a chance of being wrong. The goal, then, is to try to keep the probabilities of type I and type II errors as small as possible.

Applying the Concepts **8–6**

Consumer Protection Agency Complaints

Hypothesis testing and testing claims with confidence intervals are two different approaches that lead to the same conclusion. In the following activities, you will compare and contrast those two approaches.

Assume you are working for the Consumer Protection Agency and have recently been getting complaints about the highway gas mileage of the new Dodge Caravans. Chrysler

Corporation agrees to allow you to randomly select 40 of its new Dodge Caravans to test the highway mileage. Chrysler claims that the vans get 28 mpg on the highway. Your results show a mean of 26.7 and a standard deviation of 4.2. You are not certain if you should create a confidence interval or run a hypothesis test. You decide to do both at the same time.

1. Draw a normal curve, labeling the critical values, critical regions, test statistic, and population mean. List the significance level and the null and alternative hypotheses.

2. Draw a confidence interval directly below the normal distribution, labeling the sample mean, error, and boundary values.

3. Explain which parts from each approach are the same and which parts are different.

4. Draw a picture of a normal curve and confidence interval where the sample and hypothesized means are equal.

5. Draw a picture of a normal curve and confidence interval where the lower boundary of the confidence interval is equal to the hypothesized mean.

6. Draw a picture of a normal curve and confidence interval where the sample mean falls in the left critical region of the normal curve.

See page 470 for the answers.

Exercises 8–6

1. **Weekly Earnings for Leisure and Hospitality Workers** The average weekly earnings in the leisure and hospitality industry group for a recent year was $273. A random sample of 40 workers showed weekly average earnings of $285 with the population standard deviation equal to 58. At the 0.05 level of significance can it be concluded that the mean differs from $273? Find a 95% confidence interval for the weekly earnings and show that it supports the results of the hypothesis test.

Source: *New York Times Almanac.*

2. **One-Way Airfares** The average one-way airfare from Pittsburgh to Washington, D.C., is $236. A random sample of 20 one-way fares during a particular month had a mean of $210 with a standard deviation of $43. At $\alpha = 0.02$, is there sufficient evidence to conclude a difference from the stated mean? Use the sample statistics to construct a 98% confidence interval for the true mean one-way airfare from Pittsburgh to Washington, D.C., and compare your interval to the results of the test. Do they support or contradict one another?

Source: www.fedstats.gov

3. **IRS Audits** The IRS examined approximately 1% of individual tax returns for a specific year, and the average recommended additional tax per return was $19,150. Based on a random sample of 50 returns, the mean additional tax was $17,020. If the population standard deviation is $4080, is there sufficient evidence to conclude that the mean differs from $19,150 at

$\alpha = 0.05$? Does a 95% confidence interval support this result?

Source: *New York Times Almanac.*

4. **Canoe Trip Times** The average time it takes a person in a one-person canoe to complete a certain river course is 47 minutes. Because of rapid currents in the spring, a group of 10 people traverse the course in an average of 42 minutes. The standard deviation, known from previous trips, is 7 minutes. Test the claim that this group's time was different because of the strong currents. Use $\alpha = 0.10$. Find the 90% confidence interval of the true mean. Does the confidence interval interpretation agree with the results of the hypothesis test? Explain. Assume that the variable is normally distributed.

5. **Working at Home** Workers with a formal arrangement with their employer to be paid for time worked at home worked an average of 19 hours per week. A random sample of 15 mortgage brokers indicated that they worked a mean of 21.3 hours per week with a standard deviation of 6.5 hours. At $\alpha = 0.05$, is there sufficient evidence to conclude a difference? Construct a 95% confidence interval for the true mean number of paid working hours at home. Compare the results of your confidence interval to the conclusion of your hypothesis test and discuss the implications.

Source: www.bls.gov

6. **Newspaper Reading Times** A survey taken several years ago found that the average time a person spent reading the local daily newspaper was 10.8 minutes.

The standard deviation of the population was 3 minutes. To see whether the average time had changed since the newspaper's format was revised, the newspaper editor surveyed 36 individuals. The average time that the 36 people spent reading the paper was 12.2 minutes. At $\alpha = 0.02$, is there a change in the average time an individual spends reading the newspaper? Find the 98% confidence interval of the mean. Do the results agree? Explain.

7. What is meant by the power of a test?

8. How is the power of a test related to the type II error?

9. How can the power of a test be increased?

Summary

This chapter introduces the basic concepts of hypothesis testing. A statistical hypothesis is a conjecture about a population. There are two types of statistical hypotheses: the null and the alternative hypotheses. The null hypothesis states that there is no difference, and the alternative hypothesis specifies a difference. To test the null hypothesis, researchers use a statistical test. Many test values are computed by using

$$\text{Test value} = \frac{(\text{observed value}) - (\text{expected value})}{\text{standard error}}$$

- Researchers compute a test value from the sample data to decide whether the null hypothesis should be rejected. Statistical tests can be one-tailed or two-tailed, depending on the hypotheses.

   The null hypothesis is rejected when the difference between the population parameter and the sample statistic is said to be significant. The difference is significant when the test value falls in the critical region of the distribution. The critical region is determined by α, the level of significance of the test. The level is the probability of committing a type I error. This error occurs when the null hypothesis is rejected when it is true. Three generally agreed upon significance levels are 0.10, 0.05, and 0.01. A second kind of error, the type II error, can occur when the null hypothesis is not rejected when it is false. (8–1)

- There are two common methods used to test hypotheses; they are the traditional method and the *P*-value method. (8–2)

- All hypothesis-testing situations using the traditional method should include the following steps:

   1. State the null and alternative hypotheses and identify the claim.
   2. State an alpha level and find the critical value(s).
   3. Compute the test value.
   4. Make the decision to reject or not reject the null hypothesis.
   5. Summarize the results.

- All hypothesis-testing situations using the *P*-value method should include the following steps:

   1. State the hypotheses and identify the claim.
   2. Compute the test value.
   3. Find the *P*-value.
   4. Make the decision.
   5. Summarize the results.

- The *z* test is used to test a mean when the population standard deviation is known. When the sample size is less than 30, the population values need to be normally distributed. (8–2)

- When the population standard deviation is not known, researchers use a *t* test to test a claim about a mean. If the sample size is less than 30, the population values need to be normally or approximately normally distributed. (8–3)
- The *z* test can be used to test a claim about a population when $np \geq 5$ and $nq \geq 5$. (8–4)
- A single variance can be tested by using the chi-square test. (8–5)
- There is a relationship between confidence intervals and hypothesis testing. When the null hypothesis is rejected, the confidence interval for the mean using the same level of significance will not contain the hypothesized mean. When the null hypothesis is not rejected, the confidence interval, using the same level of significance, will contain the hypothesized mean. (8–6)
- The power of a statistical test measures the sensitivity of the test to detect a real difference in parameters if one actually exists. $1 - \beta$ is called the power of a test. (8–6)

Important Terms

α (alpha) 406

alternative
hypothesis 401

β (beta) 406

chi-square test 447

critical or rejection
region 406

critical value 406

hypothesis testing 400

left-tailed test 406

level of significance 406

noncritical or nonrejection
region 406

null hypothesis 401

one-tailed test 406

power of a test 459

P-value 418

research hypothesis 402

right-tailed test 406

statistical hypothesis 401

statistical test 404

test value 404

t test 427

two-tailed test 408

type I error 405

type II error 405

z test 413

Important Formulas

Formula for the *z* test for means:

$$z = \frac{\bar{X} - \mu}{\sigma/\sqrt{n}} \quad \text{if } n < 30, \text{ variable must be normally distributed}$$

Formula for the *t* test for means:

$$t = \frac{\bar{X} - \mu}{s/\sqrt{n}} \quad \text{if } n < 30, \text{ variable must be normally distributed}$$

Formula for the *z* test for proportions:

$$z = \frac{\hat{p} - p}{\sqrt{pq/n}}$$

Formula for the chi-square test for variance or standard deviation:

$$\chi^2 = \frac{(n - 1)s^2}{\sigma^2}$$

Review Exercises

For Exercises 1 through 19, perform each of the following steps.

a. State the hypotheses and identify the claim.
b. Find the critical value(s).
c. Compute the test value.
d. Make the decision.
e. Summarize the results.

Use the traditional method of hypothesis testing unless otherwise specified.

1. **High Temperatures in the United States** A meteorologist claims that the average of the highest temperatures in the United States is 98°. A random sample of 50 cities is selected, and the highest temperatures are

recorded. The data are shown. At $\alpha = 0.05$, can the claim be rejected? Assume $\sigma = 7.71$. (8–2)

97	94	96	105	99
96	80	95	101	97
101	87	88	97	94
98	95	88	94	94
99	99	98	96	96
97	98	99	92	97
99	108	97	98	114
91	96	102	99	102
100	93	88	102	99
98	80	95	101	61

Source: *The World Almanac & Book of Facts.*

2. Travel Times to Work Based on information from the U.S. Census Bureau, the mean travel time to work in minutes for all workers 16 years old and older was 25.3 minutes. A large company with offices in several states randomly sampled 100 of its workers to ascertain their commuting times. The sample mean was 23.9 minutes, and the population standard deviation is 6.39 minutes. At the 0.01 level of significance can it be concluded that the mean commuting time is less for this particular company? (8–2)

Source: factfinder.census.gov

3. Debt of College Graduates A random sample of the average debt (in dollars) at graduation from 30 of the top 100 public colleges and universities is listed below. Is there sufficient evidence at $\alpha = 0.01$ to conclude that the population mean debt at graduation is less than $18,000? Assume $\sigma = 2605$. (8–2)

16,012	15,784	16,597	18,105	12,665	14,734
17,225	16,953	15,309	15,297	14,437	14,835
13,607	13,374	19,410	18,385	22,312	16,656
20,142	17,821	12,701	22,400	15,730	17,673
18,978	13,661	12,580	14,392	16,000	15,176

Source: www.Kiplinger.com

4. Time Until Indigestion Relief An advertisement claims that Fasto Stomach Calm will provide relief from indigestion in less than 10 minutes. For a test of the claim, 35 individuals were given the product; the average time until relief was 9.25 minutes. From past studies, the standard deviation of the population is known to be 2 minutes. Can you conclude that the claim is justified? Find the *P*-value and let $\alpha = 0.05$. (8–2)

5. Monthly Home Rent The average monthly rent for a one-bedroom home in San Francisco is $1229. A random sample of 15 one-bedroom homes about 15 miles outside of San Francisco had a mean rent of $1350. The population standard deviation is $250. At $\alpha = 0.05$, can we conclude that the monthly rent outside San Francisco differs from that in the city? (8–2)

Source: *New York Times Almanac.*

6. Salaries for Actuaries Nationwide, the average salary of actuaries who achieve the rank of Fellow is $150,000.

An insurance executive wants to see how this compares with Fellows within his company. He checks the salaries of eight Fellows and finds the average salary to be $155,500 with a standard deviation of $15,000. Can he conclude that Fellows in his company make more than the national average, using $\alpha = 0.05$? (8–3)

Source: BeAnActuary.org

7. Weights of Men's Soccer Shoes Is lighter better? A random sample of men's soccer shoes from an international catalog had the following weights (in ounces).

10.8	9.8	8.8	9.6	9.9
10	8.4	9.6	10	9.4
9.8	9.4	9.8		

At $\alpha = 0.05$ can it be concluded that the average weight is less than 10 ounces? (8–3)

8. Whooping Crane Eggs Once down to about 15, the world's only wild flock of whooping cranes now numbers a record 237 birds in its Texas Coastal Bend wintering ground (www.SunHerald.com). The average whooping crane egg weighs 208 grams. A new batch of eggs was recently weighed, and their weights are listed below. At $\alpha = 0.01$, is there sufficient evidence to conclude that the weight is greater than 208 grams? (8–3)

210	208.5	211.6	212	210.3
210.2	209	206.4	209.7	

Source: http://www.pwrc.usgs.gov/cranes.htm

9. Union Membership Nationwide 13.7% of employed wage and salary workers are union members (down from 20.1% in 1983). A random sample of 300 local wage and salary workers showed that 50 belonged to a union. At $\alpha = 0.05$, is there sufficient evidence to conclude that the proportion of union membership differs from 13.7%? (8–4)

Source: *Time Almanac.*

10. Federal Prison Populations Nationally 60.2% of federal prisoners are serving time for drug offenses. A warden feels that in his prison the percentage is even higher. He surveys 400 inmates' records and finds that 260 of the inmates are drug offenders. At $\alpha = 0.05$, is he correct? (8–4)

Source: *New York Times Almanac.*

11. Free School Lunches It has been reported that 59.3% of U.S. school lunches served are free or at a reduced price. A random sample of 300 children in a large metropolitan area indicated that 156 of them received lunch free or at a reduced price. At the 0.01 level of significance, is there sufficient evidence to conclude that the proportion is less than 59.3%? (8–4)

Source: www.fns.usda.gov

12. MP3 Ownership An MP3 manufacturer claims that 65% of teenagers 13 to 16 years old have their own MP3 player. A researcher wishes to test the claim and selects a random sample of 80 teenagers. She finds that 57 have their own MP3 players. At $\alpha = 0.05$, should the claim be rejected? Use the *P*-value method. (8–4)

13. **Alcohol and Tobacco Use by High School Students** The use of both alcohol and tobacco by high school seniors has declined in the last 30 years. Alcohol use is down from 68.2 to 43.1%, and the use of cigarettes by high school seniors has decreased from 36.7 to 20.4%. A random sample of 300 high school seniors from a large region indicated that 18% had used cigarettes during the 30 days prior to the survey. At the 0.05 level of significance does this differ from the national proportion? (8–4)

Source: *New York Times Almanac.*

14. **Times of Videos** A film editor feels that the standard deviation for the number of minutes in a video is 3.4 minutes. A sample of 24 videos has a standard deviation of 4.2 minutes. At $\alpha = 0.05$, is the sample standard deviation different from what the editor hypothesized? (8–5)

15. **Fuel Consumption** The standard deviation of the fuel consumption of a certain automobile is hypothesized to be greater than or equal to 4.3 miles per gallon. A sample of 20 automobiles produced a standard deviation of 2.6 miles per gallon. Is the standard deviation really less than previously thought? Use $\alpha = 0.05$ and the *P*-value method. (8–5)

 16. **Movie Admission Prices** The average movie admission price for a recent year was $7.18. The population variance was 3.81. A random sample of 15 theater admission prices had a mean of $8.02 with a

standard deviation of 2.08. At $\alpha = 0.05$ is there sufficient evidence to conclude a difference from the population variance? (8–5)

Source: *New York Times Almanac.*

 17. **Games Played by NBA Scoring Leaders** A random sample of the number of games played by individual NBA scoring leaders is found below. Is there sufficient evidence to conclude that the variance in games played differs from 40? Use $\alpha = 0.05$. (8–5)

72	79	80	74	82
79	82	78	60	75

Source: *Time Almanac.*

18. **Tire Inflation** To see whether people are keeping their car tires inflated to the correct level of 35 pounds per square inch (psi), a tire company manager selects a sample of 36 tires and checks the pressure. The mean of the sample is 33.5 psi, and the population standard deviation is 3 psi. Are the tires properly inflated? Use $\alpha = 0.10$. Find the 90% confidence interval of the mean. Do the results agree? Explain. (8–6)

19. **Plant Leaf Lengths** A biologist knows that the average length of a leaf of a certain full-grown plant is 4 inches. The standard deviation of the population is 0.6 inch. A sample of 20 leaves of that type of plant given a new type of plant food had an average length of 4.2 inches. Is there reason to believe that the new food is responsible for a change in the growth of the leaves? Use $\alpha = 0.01$. Find the 99% confidence interval of the mean. Do the results concur? Explain. Assume that the variable is approximately normally distributed. (8–6)

Statistics Today

How Much Better Is Better?—Revisited

Now that you have learned the techniques of hypothesis testing presented in this chapter, you realize that the difference between the sample mean and the population mean must be *significant* before you can conclude that the students really scored above average. The superintendent should follow the steps in the hypothesis-testing procedure and be able to reject the null hypothesis before announcing that his students scored higher than average.

Data Analysis

The Data Bank is found in Appendix D, or on the World Wide Web by following links from
www.mhhe.com/math/stats/bluman/

1. From the Data Bank, select a random sample of at least 30 individuals, and test one or more of the following hypotheses by using the *z* test. Use $\alpha = 0.05$.

   a. For serum cholesterol, H_0: $\mu = 220$ milligram percent (mg%).

   b. For systolic pressure, H_0: $\mu = 120$ millimeters of mercury (mm Hg).

   c. For IQ, H_0: $\mu = 100$.

   d. For sodium level, H_0: $\mu = 140$ milliequivalents per liter (mEq/l).

2. Select a random sample of 15 individuals and test one or more of the hypotheses in Exercise 1 by using the *t* test. Use $\alpha = 0.05$.

3. Select a random sample of at least 30 individuals, and using the z test for proportions, test one or more of the following hypotheses. Use $\alpha = 0.05$.

   a. For educational level, H_0: $p = 0.50$ for level 2.

   b. For smoking status, H_0: $p = 0.20$ for level 1.

   c. For exercise level, H_0: $p = 0.10$ for level 1.

   d. For gender, H_0: $p = 0.50$ for males.

4. Select a sample of 20 individuals and test the hypothesis H_0: $\sigma^2 = 225$ for IQ level. Use $\alpha = 0.05$.

5. Using the data from Data Set XIII, select a sample of 10 hospitals and test H_0: $\mu = 250$ and H_1: $\mu < 250$ for the number of beds. Use $\alpha = 0.05$.

6. Using the data obtained in Exercise 5, test the hypothesis H_0: $\sigma \geq 150$. Use $\alpha = 0.05$.

Chapter Quiz

Determine whether each statement is true or false. If the statement is false, explain why.

1. No error is committed when the null hypothesis is rejected when it is false. True

2. When you are conducting the t test, the population must be approximately normally distributed. True

3. The test value separates the critical region from the noncritical region. False

4. The values of a chi-square test cannot be negative. True

5. The chi-square test for variances is always one-tailed. False

Select the best answer.

6. When the value of α is increased, the probability of committing a type I error is

   a. Decreased

   b. Increased

   c. The same

   d. None of the above

7. If you wish to test the claim that the mean of the population is 100, the appropriate null hypothesis is

   a. $\bar{X} = 100$

   b. $\mu \geq 100$

   c. $\mu \leq 100$

   d. $\mu = 100$

8. The degrees of freedom for the chi-square test for variances or standard deviations are

   a. 1

   b. n

   c. $n - 1$

   d. None of the above

9. For the t test, one uses _____ instead of σ.

   a. n

   b. s

   c. χ^2

   d. t

Complete the following statements with the best answer.

10. Rejecting the null hypothesis when it is true is called a(n) _____ error. Type I

11. The probability of a type II error is referred to as _____. β

12. A conjecture about a population parameter is called a(n) _____. Statistical hypothesis

13. To test the claim that the mean is greater than 87, you would use a(n) _____-tailed test. Right

14. The degrees of freedom for the t test are ___$n - 1$___.

For the following exercises where applicable:

   a. State the hypotheses and identify the claim.

   b. Find the critical value(s).

   c. Compute the test value.

   d. Make the decision.

   e. Summarize the results.

Use the traditional method of hypothesis testing unless otherwise specified.

15. **Ages of Professional Women** A sociologist wishes to see if it is true that for a certain group of professional women, the average age at which they have their first child is 28.6 years. A random sample of 36 women is selected, and their ages at the birth of their first child are recorded. At $\alpha = 0.05$, does the evidence refute the sociologist's assertion? Assume $\sigma = 4.18$.

32	28	26	33	35	34
29	24	22	25	26	28
28	34	33	32	30	29
30	27	33	34	28	25
24	33	25	37	35	33
34	36	38	27	29	26

16. **Home Closing Costs** A real estate agent believes that the average closing cost of purchasing a new home is $6500 over the purchase price. She selects 40 new home sales at random and finds that the average closing costs are $6600. The standard deviation of the population is $120. Test her belief at $\alpha = 0.05$.

17. Chewing Gum Use A recent study stated that if a person chewed gum, the average number of sticks of gum he or she chewed daily was 8. To test the claim, a researcher selected a random sample of 36 gum chewers and found the mean number of sticks of gum chewed per day was 9. The standard deviation of the population is 1. At $\alpha = 0.05$, is the number of sticks of gum a person chews per day actually greater than 8?

18. Hotel Rooms A travel agent claims that the average of the number of rooms in hotels in a large city is 500. At $\alpha = 0.01$ is the claim realistic? The data for a sample of six hotels are shown.

713   300   292   311   598   401   618

Give a reason why the claim might be deceptive.

19. Heights of Models In a New York modeling agency, a researcher wishes to see if the average height of female models is really less than 67 inches, as the chief claims. A sample of 20 models has an average height of 65.8 inches. The standard deviation of the sample is 1.7 inches. At $\alpha = 0.05$, is the average height of the models really less than 67 inches? Use the P-value method.

20. Experience of Taxi Drivers A taxi company claims that its drivers have an average of at least 12.4 years' experience. In a study of 15 taxi drivers, the average experience was 11.2 years. The standard deviation was 2. At $\alpha = 0.10$, is the number of years' experience of the taxi drivers really less than the taxi company claimed?

21. Ages of Robbery Victims A recent study in a small city stated that the average age of robbery victims was 63.5 years. A sample of 20 recent victims had a mean of 63.7 years and a standard deviation of 1.9 years. At $\alpha = 0.05$, is the average age higher than originally believed? Use the P-value method.

22. First-Time Marriages A magazine article stated that the average age of women who are getting married for the first time is 26 years. A researcher decided to test this hypothesis at $\alpha = 0.02$. She selected a sample of 25 women who were recently married for the first time and found the average was 25.1 years. The standard deviation was 3 years. Should the null hypothesis be rejected on the basis of the sample?

23. Survey on Vitamin Usage A survey in *Men's Health* magazine reported that 39% of cardiologists said that they took vitamin E supplements. To see if this is still true, a researcher randomly selected 100 cardiologists and found that 36 said that they took vitamin E supplements. At $\alpha = 0.05$ test the claim that 39% of the cardiologists took vitamin E supplements. A recent study said that taking too much vitamin E might be

harmful. How might this study make the results of the previous study invalid?

24. Breakfast Survey A dietitian read in a survey that at least 55% of adults do not eat breakfast at least 3 days a week. To verify this, she selected a random sample of 80 adults and asked them how many days a week they skipped breakfast. A total of 50% responded that they skipped breakfast at least 3 days a week. At $\alpha = 0.10$, test the claim.

25. Caffeinated Beverage Survey A Harris Poll found that 35% of people said that they drink a caffeinated beverage to combat midday drowsiness. A recent survey found that 19 out of 48 people stated that they drank a caffeinated beverage to combat midday drowsiness. At $\alpha = 0.02$ is the claim of the percentage found in the Harris Poll believable?

26. Radio Ownership A magazine claims that 75% of all teenage boys have their own radios. A researcher wished to test the claim and selected a random sample of 60 teenage boys. She found that 54 had their own radios. At $\alpha = 0.01$, should the claim be rejected?

27. Find the P-value for the z test in Exercise 15. P-value = 0.0324

28. Find the P-value for the z test in Exercise 16. P-value < 0.0001

29. Pages in Romance Novels A copyeditor thinks the standard deviation for the number of pages in a romance novel is greater than 6. A sample of 25 novels has a standard deviation of 9 pages. At $\alpha = 0.05$, is it higher, as the editor hypothesized?

30. Seed Germination Times It has been hypothesized that the standard deviation of the germination time of radish seeds is 8 days. The standard deviation of a sample of 60 radish plants' germination times was 6 days. At $\alpha = 0.01$, test the claim.

31. Pollution By-products The standard deviation of the pollution by-products released in the burning of 1 gallon of gas is 2.3 ounces. A sample of 20 automobiles tested produced a standard deviation of 1.9 ounces. Is the standard deviation really less than previously thought? Use $\alpha = 0.05$.

32. Strength of Wrapping Cord A manufacturer claims that the standard deviation of the strength of wrapping cord is 9 pounds. A sample of 10 wrapping cords produced a standard deviation of 11 pounds. At $\alpha = 0.05$, test the claim. Use the P-value method.

33. Find the 90% confidence interval of the mean in Exercise 15. Is μ contained in the interval? $28.9 < \mu < 31.2$; no

34. Find the 95% confidence interval for the mean in Exercise 16. Is μ contained in the interval? $\$6562.81 < \mu < \$6637.19$; no

Critical Thinking Challenges

The power of a test $(1 - \beta)$ can be calculated when a specific value of the mean is hypothesized in the alternative hypothesis; for example, let H_0: $\mu = 50$ and let H_1: $\mu = 52$. To find the power of a test, it is necessary to find the value of β. This can be done by the following steps:

Step 1   For a specific value of α find the corresponding value of $\overline{X}$, using $z = \dfrac{\overline{X} - \mu}{\sigma/\sqrt{n}}$, where μ is the hypothesized value given in H_0. Use a right-tailed test.

Step 2   Using the value of $\overline{X}$ found in step 1 and the value of μ in the alternative hypothesis,

find the area corresponding to z in the formula $z = \dfrac{\overline{X} - \mu}{\sigma/\sqrt{n}}$.

Step 3   Subtract this area from 0.5000. This is the value of β.

Step 4   Subtract the value of β from 1. This will give you the power of a test. See Figure 8–42.

1. Find the power of a test, using the hypotheses given previously and $\alpha = 0.05$, $\sigma = 3$, and $n = 30$.

2. Select several other values for μ in H_1 and compute the power of the test. Generalize the results.

Figure 8–42

Relationship Among α, β, and the Power of a Test

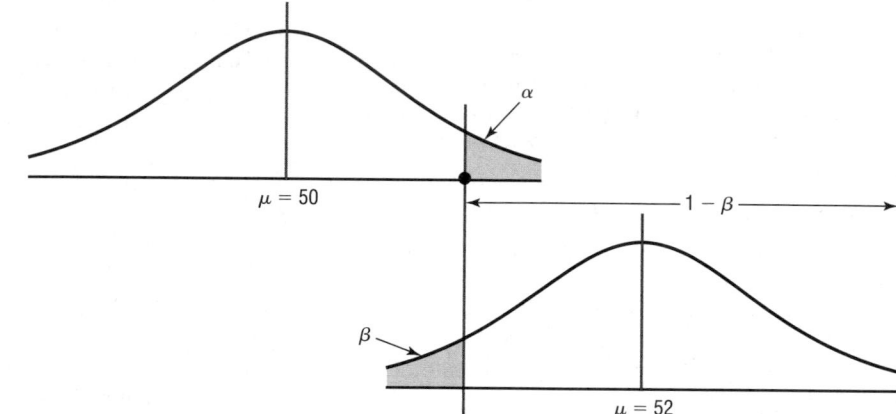

Data Projects

Use a significance level of 0.05 for all tests below.

1. **Business and Finance** Use the Dow Jones Industrial stocks in data project 1 of Chapter 7 as your data set. Find the gain or loss for each stock over the last quarter. Test the claim that the mean is that the stocks broke even (no gain or loss indicates a mean of 0).

2. **Sports and Leisure** Use the most recent NFL season for your data. For each team, find the quarterback rating for the number one quarterback. Test the claim that the mean quarterback rating for a number one quarterback is more than 80.

3. **Technology** Use your last month's itemized cell phone bill for your data. Determine the percentage of your text messages that were outgoing. Test the claim that a majority of your text messages were outgoing. Determine the mean, median, and standard deviation for the length of a call. Test the claim that the mean length

of a call is longer than the value for you found for the median length.

4. **Health and Wellness** Use the data collected in data project 4 of Chapter 7 for this exercise. Test the claim that the mean body temperature is less than 98.6 degrees Fahrenheit.

5. **Politics and Economics** Use the most recent results of the Presidential primary elections for both parties. Determine what percentage of voters in your state voted for the eventual Democratic nominee for President and what percentage voted for the eventual Republican nominee. Test the claim that a majority of your state favored the candidate who won the nomination for each party.

6. **Your Class** Use the data collected in data project 6 of Chapter 7 for this exercise. Test the claim that the mean BMI for a student is more than 25.

Answers to Applying the Concepts

Section 8–1   Eggs and Your Health

1. The study was prompted by claims that linked foods high in cholesterol to high blood serum cholesterol.

2. The population under study is people in general.

3. A sample of 500 subjects was collected.

4. The hypothesis was that eating eggs did not increase blood serum cholesterol.

5. Blood serum cholesterol levels were collected.

6. Most likely but we are not told which test.

7. The conclusion was that eating a moderate amount of eggs will not significantly increase blood serum cholesterol level.

Section 8–2   Car Thefts

1. The hypotheses are H_0: $\mu = 44$ and H_1: $\mu \neq 44$.

2. This sample can be considered large for our purposes.

3. The variable needs to be normally distributed.

4. We will use a z distribution.

5. Since we are interested in whether the car theft rate has changed, we use a two-tailed test.

6. Answers may vary. At the $\alpha = 0.05$ significance level, the critical values are $z = \pm 1.96$.

7. The sample mean is $\overline{X} = 55.97$, and the population standard deviation is 30.30. Our test statistic is
$z = \frac{55.97 - 44}{30.30/\sqrt{36}} = 2.37$.

8. Since $2.37 > 1.96$, we reject the null hypothesis.

9. There is enough evidence to conclude that the car theft rate has changed.

10. Answers will vary. Based on our sample data, it appears that the car theft rate has changed from 44 vehicles per 10,000 people. In fact, the data indicate that the car theft rate has increased.

11. Based on our sample, we would expect 55.97 car thefts per 10,000 people, so we would expect $(55.97)(5) = 279.85$, or about 280, car thefts in the city.

Section 8–3   How Much Nicotine Is in Those Cigarettes?

1. We have $15 - 1 = 14$ degrees of freedom.

2. This is a t test.

3. We are only testing one sample.

4. This is a right-tailed test, since the hypotheses of the tobacco company are H_0: $\mu = 40$ and H_1: $\mu > 40$.

5. The P-value is 0.008, which is less than the significance level of 0.01. We reject the tobacco company's claim.

6. Since the test statistic (2.72) is greater than the critical value (2.62), we reject the tobacco company's claim.

7. There is no conflict in this output, since the results based on the P-value and on the critical value agree.

8. Answers will vary. It appears that the company's claim is false and that there is more than 40 mg of nicotine in its cigarettes.

Section 8–4   Quitting Smoking

1. The statistical hypotheses were that StopSmoke helps more people quit smoking than the other leading brands.

2. The null hypotheses were that StopSmoke has the same effectiveness as or is not as effective as the other leading brands.

3. The alternative hypotheses were that StopSmoke helps more people quit smoking than the other leading brands. (The alternative hypotheses are the statistical hypotheses.)

4. No statistical tests were run that we know of.

5. Had tests been run, they would have been one-tailed tests.

6. Some possible significance levels are 0.01, 0.05, and 0.10.

7. A type I error would be to conclude that StopSmoke is better when it really is not.

8. A type II error would be to conclude that StopSmoke is not better when it really is.

9. These studies proved nothing. Had statistical tests been used, we could have tested the effectiveness of StopSmoke.

10. Answers will vary. One possible answer is that more than likely the statements are talking about practical significance and not statistical significance, since we have no indication that any statistical tests were conducted.

Section 8–5   Testing Gas Mileage Claims

1. The hypotheses are H_0: $\mu = 28$ and H_1: $\mu < 28$. The value of our test statistic is $t = -1.96$, and the associated P-value is 0.0287. We would reject Chrysler's claim that the Dodge Caravans are getting 28 mpg.

2. The hypotheses are H_0: $\sigma = 2.1$ and H_1: $\sigma > 2.1$. The value of our test statistic is $\chi^2 = \frac{(n-1)s^2}{\sigma^2} = \frac{(39)4.2^2}{2.1^2} = 156$, and the associated P-value is approximately zero. We would reject Chrysler's claim that the standard deviation is 2.1 mpg.

3. Answers will vary. It is recommended that Chrysler lower its claim about the highway miles per gallon of the Dodge Caravans. Chrysler should also try to reduce variability in miles per gallon and provide confidence intervals for the highway miles per gallon.

4. Answers will vary. There are cases when a mean may be fine, but if there is a lot of variability about the mean, there will be complaints (due to the lack of consistency).

Section 8–6   Consumer Protection Agency Complaints

1. Answers will vary.
2. Answers will vary.
3. Answers will vary.
4. Answers will vary.
5. Answers will vary.
6. Answers will vary.

9

Testing the Difference Between Two Means, Two Proportions, and Two Variances

Objectives

After completing this chapter, you should be able to

1 Test the difference between sample means, using the z test.

2 Test the difference between two means for independent samples, using the t test.

3 Test the difference between two means for  dependent samples.

4 Test the difference between two proportions.

5 Test the difference between two variances or standard deviations.

Outline

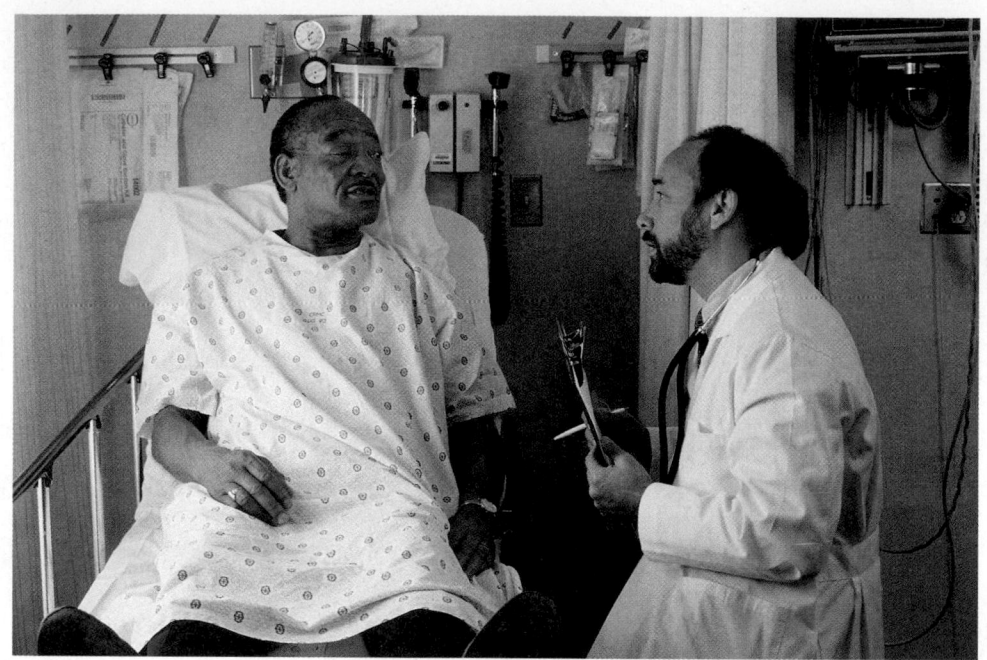

**Statistics
Today**

To Vaccinate or Not to Vaccinate? Small or Large?

Influenza is a serious disease among the elderly, especially those living in nursing homes. Those residents are more susceptible to influenza than elderly persons living in the community because the former are usually older and more debilitated, and they live in a closed environment where they are exposed more so than community residents to the virus if it is introduced into the home. Three researchers decided to investigate the use of vaccine and its value in determining outbreaks of influenza in small nursing homes.

These researchers surveyed 83 licensed homes in seven counties in Michigan. Part of the study consisted of comparing the number of people being vaccinated in small nursing homes (100 or fewer beds) with the number in larger nursing homes (more than 100 beds). Unlike the statistical methods presented in Chapter 8, these researchers used the techniques explained in this chapter to compare two sample proportions to see if there was a significant difference in the vaccination rates of patients in small nursing homes compared to those in large nursing homes. See Statistics Today—Revisited at the end of the chapter.

Source: Nancy Arden, Arnold S. Monto, and Suzanne E. Ohmit, "Vaccine Use and the Risk of Outbreaks in a Sample of Nursing Homes During an Influenza Epidemic," *American Journal of Public Health* 85, no. 3, pp. 399–401. Copyright by the American Public Health Association.

Introduction

The basic concepts of hypothesis testing were explained in Chapter 8. With the z, t, and χ^2 tests, a sample mean, variance, or proportion can be compared to a specific population mean, variance, or proportion to determine whether the null hypothesis should be rejected.

There are, however, many instances when researchers wish to compare two sample means, using experimental and control groups. For example, the average lifetimes of two different brands of bus tires might be compared to see whether there is any difference in tread wear. Two different brands of fertilizer might be tested to see whether one is better than the other for growing plants. Or two brands of cough syrup might be tested to see whether one brand is more effective than the other.

In the comparison of two means, the same basic steps for hypothesis testing shown in Chapter 8 are used, and the *z* and *t* tests are also used. When comparing two means by using the *t* test, the researcher must decide if the two samples are *independent* or *dependent*. The concepts of independent and dependent samples will be explained in Sections 9–2 and 9–3.

The *z* test can be used to compare two proportions, as shown in Section 9–4. Finally, two variances can be compared by using an *F* test as shown in Section 9–5.

<table><tr><td>**9–1**</td></tr></table>

Testing the Difference Between Two Means: Using the *z* Test

Objective 1

Test the difference between sample means, using the *z* test.

Suppose a researcher wishes to determine whether there is a difference in the average age of nursing students who enroll in a nursing program at a community college and those who enroll in a nursing program at a university. In this case, the researcher is not interested in the average age of all beginning nursing students; instead, he is interested in *comparing* the means of the two groups. His research question is, Does the mean age of nursing students who enroll at a community college differ from the mean age of nursing students who enroll at a university? Here, the hypotheses are

$$H_0: \mu_1 = \mu_2$$
$$H_1: \mu_1 \neq \mu_2$$

where

μ_1 = mean age of all beginning nursing students at the community college
μ_2 = mean age of all beginning nursing students at the university

Another way of stating the hypotheses for this situation is

$$H_0: \mu_1 - \mu_2 = 0$$
$$H_1: \mu_1 - \mu_2 \neq 0$$

If there is no difference in population means, subtracting them will give a difference of zero. If they are different, subtracting will give a number other than zero. Both methods of stating hypotheses are correct; however, the first method will be used in this book.

Assumptions for the *z* Test to Determine the Difference Between Two Means

1. Both samples are random samples.
2. The samples must be independent of each other. That is, there can be no relationship between the subjects in each sample.
3. The standard deviations of both populations must be known, and if the sample sizes are less than 30, the populations must be normally or approximately normally distributed.

The theory behind testing the difference between two means is based on selecting pairs of samples and comparing the means of the pairs. The population means need not be known.

All possible pairs of samples are taken from populations. The means for each pair of samples are computed and then subtracted, and the differences are plotted. If both populations have the same mean, then most of the differences will be zero or close to zero.

Figure 9–1

**Differences of Means
of Pairs of Samples**

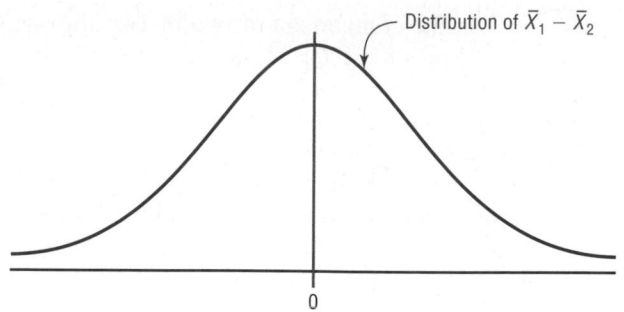

Occasionally, there will be a few large differences due to chance alone, some positive and others negative. If the differences are plotted, the curve will be shaped like a normal distribution and have a mean of zero, as shown in Figure 9–1.

The variance of the difference $\bar{X}_1 - \bar{X}_2$ is equal to the sum of the individual variances of $\bar{X}_1$ and $\bar{X}_2$. That is,

$$\sigma^2_{\bar{X}_1 - \bar{X}_2} = \sigma^2_{\bar{X}_1} + \sigma^2_{\bar{X}_2}$$

where $\sigma^2_{\bar{X}_1} = \dfrac{\sigma^2_1}{n_1}$ and $\sigma^2_{\bar{X}_2} = \dfrac{\sigma^2_2}{n_2}$

So the standard deviation of $\bar{X}_1 - \bar{X}_2$ is

$$\sqrt{\frac{\sigma^2_1}{n_1} + \frac{\sigma^2_2}{n_2}}$$

Formula for the z Test for Comparing Two Means from Independent Populations

$$z = \frac{(\bar{X}_1 - \bar{X}_2) - (\mu_1 - \mu_2)}{\sqrt{\dfrac{\sigma^2_1}{n_1} + \dfrac{\sigma^2_2}{n_2}}}$$

This formula is based on the general format of

$$\text{Test value} = \frac{(\text{observed value}) - (\text{expected value})}{\text{standard error}}$$

where $\bar{X}_1 - \bar{X}_2$ is the observed difference, and the expected difference $\mu_1 - \mu_2$ is zero when the null hypothesis is $\mu_1 = \mu_2$, since that is equivalent to $\mu_1 - \mu_2 = 0$. Finally, the standard error of the difference is

$$\sqrt{\frac{\sigma^2_1}{n_1} + \frac{\sigma^2_2}{n_2}}$$

In the comparison of two sample means, the difference may be due to chance, in which case the null hypothesis will not be rejected and the researcher can assume that the means of the populations are basically the same. The difference in this case is not significant. See Figure 9–2(a). On the other hand, if the difference is significant, the null hypothesis is rejected and the researcher can conclude that the population means are different. See Figure 9–2(b).

Figure 9–2

Hypothesis-Testing Situations in the Comparison of Means

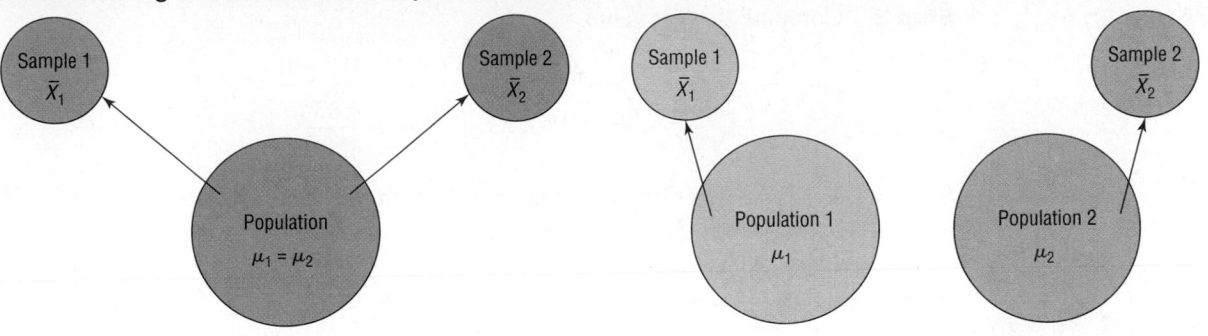

(a) Difference is not significant

Do not reject H_0: $\mu_1 = \mu_2$ since $\bar{X}_1 - \bar{X}_2$ is not significant.

(b) Difference is significant

Reject H_0: $\mu_1 = \mu_2$ since $\bar{X}_1 - \bar{X}_2$ is significant.

These tests can also be one-tailed, using the following hypotheses:

Right-tailed			**Left-tailed**		
H_0: $\mu_1 = \mu_2$	or	H_0: $\mu_1 - \mu_2 = 0$	H_0: $\mu_1 = \mu_2$	or	H_0: $\mu_1 - \mu_2 = 0$
H_1: $\mu_1 > \mu_2$		H_1: $\mu_1 - \mu_2 > 0$	H_1: $\mu_1 < \mu_2$		H_1: $\mu_1 - \mu_2 < 0$

The same critical values used in Section 8–2 are used here. They can be obtained from Table E in Appendix C.

If σ_1^2 and σ_2^2 are not known, the researcher can use the variances from each sample s_1^2 and s_2^2, but a t test must be used. This will be explained in Section 9–2.

The basic format for hypothesis testing using the traditional method is reviewed here.

Step 1 State the hypotheses and identify the claim.

Step 2 Find the critical value(s).

Step 3 Compute the test value.

Step 4 Make the decision.

Step 5 Summarize the results.

Example 9–1

Hotel Room Cost

A survey found that the average hotel room rate in New Orleans is \$88.42 and the average room rate in Phoenix is \$80.61. Assume that the data were obtained from two samples of 50 hotels each and that the standard deviations of the populations are \$5.62 and \$4.83, respectively. At $\alpha = 0.05$, can it be concluded that there is a significant difference in the rates?

Source: *USA TODAY.*

Solution

Step 1 State the hypotheses and identify the claim.

$$H_0: \mu_1 = \mu_2 \quad \text{and} \quad H_1: \mu_1 \neq \mu_2 \text{ (claim)}$$

Step 2 Find the critical values. Since $\alpha = 0.05$, the critical values are $+1.96$ and -1.96.

Step 3 Compute the test value.

$$z = \frac{(\bar{X}_1 - \bar{X}_2) - (\mu_1 - \mu_2)}{\sqrt{\dfrac{\sigma_1^2}{n_1} + \dfrac{\sigma_2^2}{n_2}}} = \frac{(88.42 - 80.61) - 0}{\sqrt{\dfrac{5.62^2}{50} + \dfrac{4.83^2}{50}}} = 7.45$$

Step 4 Make the decision. Reject the null hypothesis at $\alpha = 0.05$, since $7.45 > 1.96$. See Figure 9–3.

Figure 9–3

Critical and Test Values
for Example 9–1

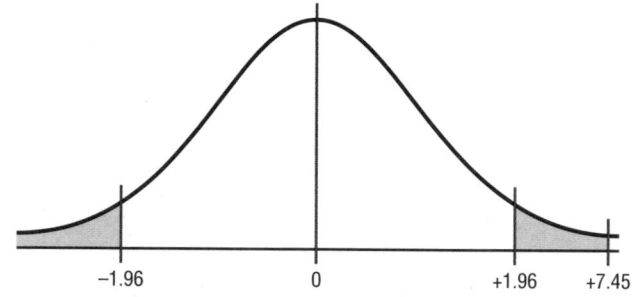

$-1.96 \qquad 0 \qquad +1.96 \quad +7.45$

Step 5 Summarize the results. There is enough evidence to support the claim that the means are not equal. Hence, there is a significant difference in the rates.

The *P*-values for this test can be determined by using the same procedure shown in Section 8–2. For example, if the test value for a two-tailed test is 1.40, then the *P*-value obtained from Table E is 0.1616. This value is obtained by looking up the area for $z = 1.40$, which is 0.9192. Then 0.9192 is subtracted from 1.0000 to get 0.0808. Finally, this value is doubled to get 0.1616 since the test is two-tailed. If $\alpha = 0.05$, the decision would be to not reject the null hypothesis, since *P*-value $> \alpha$.

The *P*-value method for hypothesis testing for this chapter also follows the same format as stated in Chapter 8. The steps are reviewed here.

Step 1 State the hypotheses and identify the claim.

Step 2 Compute the test value.

Step 3 Find the *P*-value.

Step 4 Make the decision.

Step 5 Summarize the results.

Example 9–2 illustrates these steps.

Example 9–2

College Sports Offerings

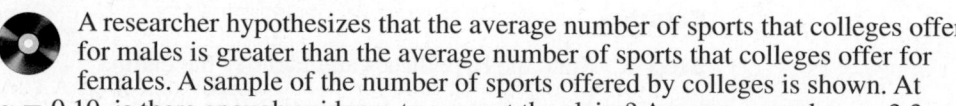

A researcher hypothesizes that the average number of sports that colleges offer for males is greater than the average number of sports that colleges offer for females. A sample of the number of sports offered by colleges is shown. At $\alpha = 0.10$, is there enough evidence to support the claim? Assume σ_1 and $\sigma_2 = 3.3$.

	Males					**Females**			
6	11	11	8	15	6	8	11	13	8
6	14	8	12	18	7	5	13	14	6
6	9	5	6	9	6	5	5	7	6
6	9	18	7	6	10	7	6	5	5
15	6	11	5	5	16	10	7	8	5
9	9	5	5	8	7	5	5	6	5
8	9	6	11	6	9	18	13	7	10
9	5	11	5	8	7	8	5	7	6
7	7	5	10	7	11	4	6	8	7
10	7	10	8	11	14	12	5	8	5

Source: *USA TODAY.*

Solution

Step 1   State the hypotheses and identify the claim.

$$H_0: \mu_1 = \mu_2 \qquad \text{and} \qquad H_1: \mu_1 > \mu_2 \text{ (claim)}$$

Step 2   Compute the test value. Using a calculator or the formula in Chapter 3, find the mean for each data set.

For the males $\overline{X}_1 = 8.6$ and $\sigma_1 = 3.3$

For the females $\overline{X}_2 = 7.9$ and $\sigma_2 = 3.3$

Substitute in the formula.

$$z = \frac{(\overline{X}_1 - \overline{X}_2) - (\mu_1 - \mu_2)}{\sqrt{\dfrac{\sigma_1^2}{n_1} + \dfrac{\sigma_2^2}{n_2}}} = \frac{(8.6 - 7.9) - 0}{\sqrt{\dfrac{3.3^2}{50} + \dfrac{3.3^2}{50}}} = 1.06*$$

Step 3   Find the *P*-value. For $z = 1.06$, the area is 0.8554, and $1.0000 - 0.8554 = 0.1446$, or a *P*-value of 0.1446.

Step 4   Make the decision. Since the *P*-value is larger than α (that is, $0.1446 > 0.10$), the decision is to not reject the null hypothesis. See Figure 9–4.

Step 5   Summarize the results. There is not enough evidence to support the claim that colleges offer more sports for males than they do for females.

Figure 9–4

P-Value and α Value for Example 9–2

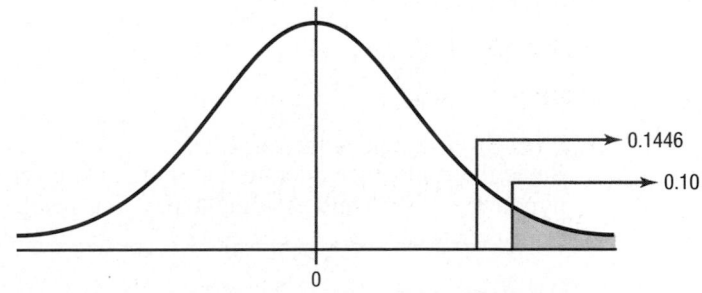

Note: Calculator results may differ due to rounding.

Sometimes, the researcher is interested in testing a specific difference in means other than zero. For example, he or she might hypothesize that the nursing students at a community college are, on average, 3.2 years older than those at a university. In this case, the hypotheses are

$$H_0: \mu_1 - \mu_2 = 3.2 \quad \text{and} \quad H_1: \mu_1 - \mu_2 > 3.2$$

The formula for the z test is still

$$z = \frac{(\bar{X}_1 - \bar{X}_2) - (\mu_1 - \mu_2)}{\sqrt{\dfrac{\sigma_1^2}{n_1} + \dfrac{\sigma_2^2}{n_2}}}$$

where $\mu_1 - \mu_2$ is the hypothesized difference or expected value. In this case, $\mu_1 - \mu_2 = 3.2$.

Confidence intervals for the difference between two means can also be found. When you are hypothesizing a difference of zero, if the confidence interval contains zero, the null hypothesis is not rejected. If the confidence interval does not contain zero, the null hypothesis is rejected.

Confidence intervals for the difference between two means can be found by using this formula:

Formula for the z Confidence Interval for Difference Between Two Means

$$(\bar{X}_1 - \bar{X}_2) - z_{\alpha/2}\sqrt{\frac{\sigma_1^2}{n_1} + \frac{\sigma_2^2}{n_2}} < \mu_1 - \mu_2 < (\bar{X}_1 - \bar{X}_2) + z_{\alpha/2}\sqrt{\frac{\sigma_1^2}{n_1} + \frac{\sigma_2^2}{n_2}}$$

Example 9–3

Find the 95% confidence interval for the difference between the means for the data in Example 9–1.

Solution

Substitute in the formula, using $z_{\alpha/2} = 1.96$.

$$(\bar{X}_1 - \bar{X}_2) - z_{\alpha/2}\sqrt{\frac{\sigma_1^2}{n_1} + \frac{\sigma_2^2}{n_2}} < \mu_1 - \mu_2$$

$$< (\bar{X}_1 - \bar{X}_2) + z_{\alpha/2}\sqrt{\frac{\sigma_1^2}{n_1} + \frac{\sigma_2^2}{n_2}}$$

$$(88.42 - 80.61) - 1.96\sqrt{\frac{5.62^2}{50} + \frac{4.83^2}{50}} < \mu_1 - \mu_2$$

$$< (88.42 - 80.61) + 1.96\sqrt{\frac{5.62^2}{50} + \frac{4.83^2}{50}}$$

$$7.81 - 2.05 < \mu_1 - \mu_2 < 7.81 + 2.05$$

$$5.76 < \mu_1 - \mu_2 < 9.86$$

Since the confidence interval does not contain zero, the decision is to reject the null hypothesis, which agrees with the previous result.

Applying the Concepts 9–1

Home Runs

For a sports radio talk show, you are asked to research the question whether more home runs are hit by players in the National League or by players in the American League. You decide to use the home run leaders from each league for a 40-year period as your data. The numbers are shown.

National League

47	49	73	50	65	70	49	47	40	43
46	35	38	40	47	39	49	37	37	36
40	37	31	48	48	45	52	38	38	36
44	40	48	45	45	36	39	44	52	47

American League

47	57	52	47	48	56	56	52	50	40
46	43	44	51	36	42	49	49	40	43
39	39	22	41	45	46	39	32	36	32
32	32	37	33	44	49	44	44	49	32

Using the data given, answer the following questions.

1. Define a population.
2. What kind of sample was used?
3. Do you feel that it is representative?
4. What are your hypotheses?
5. What significance level will you use?
6. What statistical test will you use?
7. What are the test results? (Assume $\sigma_1 = 8.8$ and $\sigma_2 = 7.8$.)
8. What is your decision?
9. What can you conclude?
10. Do you feel that using the data given really answers the original question asked?
11. What other data might be used to answer the question?

See pages 530 and 531 for the answers.

Exercises 9–1

1. Explain the difference between testing a single mean and testing the difference between two means.

2. When a researcher selects all possible pairs of samples from a population in order to find the difference between the means of each pair, what will be the shape of the distribution of the differences when the original distributions are normally distributed? What will be the mean of the distribution? What will be the standard deviation of the distribution?

3. What two assumptions must be met when you are using the z test to test differences between two means? Can the sample standard deviations s_1 and s_2 be used in place of the population standard deviations σ_1 and σ_2?

4. Show two different ways to state that the means of two populations are equal. $H_0: \mu_1 = \mu_2$ or $H_0: \mu_1 - \mu_2 = 0$

For Exercises 5 through 17, perform each of the following steps.

 a. State the hypotheses and identify the claim.
 b. Find the critical value(s).
 c. Compute the test value.
 d. Make the decision.
 e. Summarize the results.

Use the traditional method of hypothesis testing unless otherwise specified.

 5. **Lengths of Major U.S. Rivers** A researcher wishes to see if the average length of the major rivers in the United States is the same as the average length of the major rivers in Europe. The data (in miles) of a sample of rivers are shown. At $\alpha = 0.01$, is there enough evidence to reject the claim? Assume $\sigma_1 = 450$ and $\sigma_2 = 474$.

United States			Europe		
729	560	434	481	724	820
329	332	360	532	357	505
450	2315	865	1776	1122	496
330	410	1036	1224	634	230
329	800	447	1420	326	626
600	1310	652	877	580	210
1243	605	360	447	567	252
525	926	722	824	932	600
850	310	430	634	1124	1575
532	375	1979	565	405	2290
710	545	259	675	454	
300	470	425			

Source: *The World Almanac and Book of Facts.*

6. **Teachers' Salaries** California and New York lead the list of average teachers' salaries. The California yearly average is $64,421 while teachers in New York make an average annual salary of $62,332. Random samples of 45 teachers from each state yielded the following.

	California	New York
Sample mean	64,510	62,900
Population standard deviation	8,200	7,800

At $\alpha = 0.10$ is there a difference in means of the salaries?

Source: *World Almanac.*

7. **Commuting Times** The Bureau of the Census reports that the average commuting time for citizens of both Baltimore, Maryland, and Miami, Florida, is approximately 29 minutes. To see if their commuting times appear to be any different in the winter, random samples of 40 drivers were surveyed in each city and the average commuting time for the month of January was calculated for both cities. The results are provided below. At the 0.05 level of significance, can it be concluded that the commuting times are different in the winter?

	Miami	Baltimore
Sample size	40	40
Sample mean	28.5 min	35.2 min
Population standard deviation	7.2 min	9.1 min

Source: www.census.gov

8. **Heights of 9-Year-Olds** At age 9 the average weight (21.3 kg) and the average height (124.5 cm) for both boys and girls are exactly the same. A random sample of 9-year-olds yielded these results. Estimate the mean difference in height between boys and girls with 95% confidence. Does your interval support the given claim?

	Boys	Girls
Sample size	60	50
Mean height, cm	123.5	126.2
Population variance	98	120

Source: www.healthepic.com

9. **Length of Hospital Stays** The average length of "short hospital stays" for men is slightly longer than that for women, 5.2 days versus 4.5 days. A random sample of recent hospital stays for both men and women revealed the following. At $\alpha = 0.01$, is there sufficient evidence to conclude that the average hospital stay for men is longer than the average hospital stay for women?

	Men	Women
Sample size	32	30
Sample mean	5.5 days	4.2 days
Population standard deviation	1.2 days	1.5 days

Source: www.cdc.gov/nchs

10. **Home Prices** A real estate agent compares the selling prices of homes in two municipalities in southwestern Pennsylvania to see if there is a difference. The results of the study are shown. Is there enough evidence to reject the claim that the average cost of a home in both locations is the same? Use $\alpha = 0.01$.

Scott	Ligonier
$\bar{X}_1 = \$93,430^*$	$\bar{X}_2 = \$98,043^*$
$\sigma_1 = \$5602$	$\sigma_2 = \$4731$
$n_1 = 35$	$n_2 = 40$

*Based on information from RealSTATs.

11. **Women Science Majors** In a study of women science majors, the following data were obtained on two groups, those who left their profession within a few months after graduation (leavers) and those who remained in their profession after they graduated (stayers). Test the claim that those who stayed had a higher science grade point average than those who left. Use $\alpha = 0.05$.

Leavers	Stayers
$\bar{X}_1 = 3.16$	$\bar{X}_2 = 3.28$
$\sigma_1 = 0.52$	$\sigma_2 = 0.46$
$n_1 = 103$	$n_2 = 225$

Source: Paula Rayman and Belle Brett, "Women Science Majors: What Makes a Difference in Persistence after Graduation?" *The Journal of Higher Education.*

12. **ACT Scores** A survey of 1000 students nationwide showed a mean ACT score of 21.4. A survey of 500 Ohio scores showed a mean of 20.8. If the population standard deviation in each case is 3, can we conclude that Ohio is below the national average? Use $\alpha = 0.05$.

Source: Report of WFIN radio.

13. **Per Capita Income** The average per capita income for Wisconsin is reported to be $37,314, and for South Dakota it is $37,375—almost the same thing. A random sample of 50 workers from each state indicated the following sample statistics.

	Wisconsin	South Dakota
Size	50	50
Mean	$40,275	$38,750
Population standard deviation	$10,500	$12,500

At $\alpha = 0.05$ can we conclude a difference in means of the personal incomes?

Source: *New York Times Almanac.*

14. **Monthly Social Security Benefits** The average monthly Social Security benefit in 2004 for retired workers was $954.90 and for disabled workers was $894.10. Researchers used data from the Social Security records to test the claim that the difference in monthly benefits between the two groups was greater than $30.

Based on the following information, can the researchers' claim be supported at the 0.05 level of significance?

	Retired	Disabled
Sample size	60	60
Mean benefit	$960.50	$902.89
Population standard deviation	$98	$101

Source: *New York Times Almanac.*

15. **Self-Esteem Scores** In the study cited in Exercise 11, the researchers collected the data shown here on a self-esteem questionnaire. At $\alpha = 0.05$, can it be concluded that there is a difference in the self-esteem scores of the two groups? Use the *P*-value method.

Leavers	Stayers
$\bar{X}_1 = 3.05$	$\bar{X}_2 = 2.96$
$\sigma_1 = 0.75$	$\sigma_2 = 0.75$
$n_1 = 103$	$n_2 = 225$

Source: Paula Rayman and Belle Brett, "Women Science Majors: What Makes a Difference in Persistence after Graduation?" *The Journal of Higher Education.*

 16. **Ages of College Students** The dean of students wants to see whether there is a significant difference in ages of resident students and commuting students. She selects a sample of 50 students from each group. The ages are shown here. At $\alpha = 0.05$, decide if there is enough evidence to reject the claim of no difference in the ages of the two groups. Use the *P*-value method. Assume $\sigma_1 = 3.68$ and $\sigma_2 = 4.7$.

Resident students

22	25	27	23	26	28	26	24
25	20	26	24	27	26	18	19
18	30	26	18	18	19	32	23
19	19	18	29	19	22	18	22
26	19	19	21	23	18	20	18
22	21	19	21	21	22	18	20
19	23						

Commuter students

18	20	19	18	22	25	24	35
23	18	23	22	28	25	20	24
26	30	22	22	22	21	18	20
19	26	35	19	19	18	19	32
29	23	21	19	36	27	27	20
20	21	18	19	23	20	19	19
20	25						

17. **Problem-Solving Ability** Two groups of students are given a problem-solving test, and the results are compared. Find the 90% confidence interval of the true difference in means.

Mathematics majors	Computer science majors
$\bar{X}_1 = 83.6$	$\bar{X}_2 = 79.2$
$\sigma_1 = 4.3$	$\sigma_2 = 3.8$
$n_1 = 36$	$n_2 = 36$
	$2.8 < \mu_1 - \mu_2 < 6.0$

18. **Credit Card Debt** The average credit card debt for a recent year was $9205. Five years earlier the average credit card debt was $6618. Assume sample sizes of 35 were used and the population standard deviations of both samples were $1928. Is there enough evidence to believe that the average credit card debt has increased? Use $\alpha = 0.05$. Give a possible reason as to why or why not the debt was increased.

Source: CardWeb.com

19. **Literacy Scores** Adults aged 16 or older were assessed in three types of literacy in 2003: prose, document, and quantitative. The scores in document literacy were the same for 19- to 24-year-olds and for 40- to 49-year-olds. A random sample of scores from a later year showed the following statistics.

Age group	Mean score	Population standard deviation	Sample size
19–24	280	56.2	40
40–49	315	52.1	35

Construct a 95% confidence interval for the true difference in mean scores for these two groups. What does your interval say about the claim that there is no difference in mean scores?

Source: www.nces.ed.gov

20. **Battery Voltage** Two brands of batteries are tested, and their voltage is compared. The data follow. Find the 95% confidence interval of the true difference in the means. Assume that both variables are normally distributed.

$0.3 < \mu_1 - \mu_2 < 0.5$

Brand X	Brand Y
$\bar{X}_1 = 9.2$ volts	$\bar{X}_2 = 8.8$ volts
$\sigma_1 = 0.3$ volt	$\sigma_2 = 0.1$ volt
$n_1 = 27$	$n_2 = 30$

Extending the Concepts

21. **Exam Scores at Private and Public Schools** A researcher claims that students in a private school have exam scores that are at most 8 points higher than those of students in public schools. Random samples of 60 students from each type of school are selected and given an exam. The results are shown. At $\alpha = 0.05$, test the claim.

Private school	Public school
$\bar{X}_1 = 110$	$\bar{X}_2 = 104$
$\sigma_1 = 15$	$\sigma_2 = 15$
$n_1 = 60$	$n_2 = 60$

22. Sale Prices for Houses The average sales price of new one-family houses in the Midwest is $250,000 and in the South is $253,400. A random sample of 40 houses in each region was examined with the following results. At the 0.05 level of significance can it be concluded that the difference in mean sales price for the two regions is greater than $3400?

	South	Midwest
Sample size	40	40
Sample mean	261,500	248,200
Population standard deviation	10,500	12,000

Source: *New York Times Almanac.*

23. Average Earnings for College Graduates The average earnings of year-round full-time workers with bachelor's degrees or more is $88,641 for men and $58,000 for women—a difference of slightly over $30,000 a year. One hundred of each were sampled, resulting in a sample mean of $90,200 for men, and the population standard deviation is $15,000, and a mean of $57,800 for women, and the population standard deviation is $12,800. At the 0.01 level of significance can it be concluded that the difference in means is not $30,000?

Source: *New York Times Almanac.*

Technology *Step by Step*

TI-83 Plus or TI-84 Plus
Step by Step

Hypothesis Test for the Difference Between Two Means and *z* Distribution (Data)

1. Enter the data values into L_1 and L_2.
2. Press **STAT** and move the cursor to TESTS.
3. Press **3** for 2-SampZTest.
4. Move the cursor to Data and press **ENTER**.
5. Type in the appropriate values.
6. Move the cursor to the appropriate alternative hypothesis and press **ENTER**.
7. Move the cursor to Calculate and press **ENTER**.

Hypothesis Test for the Difference Between Two Means and *z* Distribution (Statistics)

1. Press **STAT** and move the cursor to TESTS.
2. Press **3** for 2-SampZTest.
3. Move the cursor to Stats and press **ENTER**.
4. Type in the appropriate values.
5. Move the cursor to the appropriate alternative hypothesis and press **ENTER**.
6. Move the cursor to Calculate and press **ENTER**.

Confidence Interval for the Difference Between Two Means and *z* Distribution (Data)

1. Enter the data values into L_1 and L_2.
2. Press **STAT** and move the cursor to TESTS.
3. Press **9** for 2-SampZInt.
4. Move the cursor to Data and press **ENTER**.
5. Type in the appropriate values.
6. Move the cursor to Calculate and press **ENTER**.

Confidence Interval for the Difference Between Two Means and *z* Distribution (Statistics)

1. Press **STAT** and move the cursor to TESTS.
2. Press **9** for 2-SampZInt.
3. Move the cursor to Stats and press **ENTER**.
4. Type in the appropriate values.
5. Move the cursor to Calculate and press **ENTER**.

Excel
Step by Step

z Test for the Difference Between Two Means

Excel has a two-sample z test included in the Data Analysis Add-in. To perform a z test for the difference between the means of two populations, given two independent samples, do this:

1. Enter the first sample data set into column A.
2. Enter the second sample data set into column B.
3. If the population variances are not known but $n \geq 30$ for both samples, use the formulas =VAR(A1:An) and =VAR(B1:Bn), where An and Bn are the last cells with data in each column, to find the variances of the sample data sets.
4. Select the **Data** tab from the toolbar. Then select **Data Analysis**.
5. In the Analysis Tools box, select z test: Two sample for Means.
6. Type the ranges for the data in columns A and B and type a value (usually 0) for the **Hypothesized Mean Difference.**
7. If the population variances are known, type them for Variable 1 and Variable 2. Otherwise, use the sample variances obtained in step 3.
8. Specify the confidence level **Alpha.**
9. Specify a location for the output, and click [**OK**].

Example XL9–1

Test the claim that the two population means are equal, using the sample data provided here, at $\alpha = 0.05$. Assume the population variances are $\sigma_A^2 = 10.067$ and $\sigma_B^2 = 7.067$.

Set A	10	2	15	18	13	15	16	14	18	12	15	15	14	18	16
Set B	5	8	10	9	9	11	12	16	8	8	9	10	11	7	6

The two-sample z test dialog box is shown (before the variances are entered); the results appear in the table that Excel generates. Note that the P-value and critical z value are provided for both the one-tailed test and the two-tailed test. The P-values here are expressed in scientific notation: $7.09045E{-}06 = 7.09045 \times 10^{-6} = 0.00000709045$. Because this value is less than 0.05, we reject the null hypothesis and conclude that the population means are not equal.

Two-Sample z Test
Dialog Box

z-Test Two Sample for Means	? X
Input	OK
Variable 1 Range:	$A$1:$A$15		Cancel
Variable 2 Range:	$B$1:$B$15		Help
Hypothesized Mean Difference:	0	
Variable 1 Variance (known):		
Variable 2 Variance (known):		
☐ Labels	
Alpha: 0.05	
Output options	
⦿ Output Range:	e3	
○ New Worksheet Ply:		
○ New Workbook	

z-Test: Two Sample for Means

	Variable 1	Variable 2
Mean	14.06666667	9.266666667
Known Variance	10.067	7.067
Observations	15	15
Hypothesized Mean Difference	0	
z	4.491149228	
P(Z<=z) one-tail	3.54522E-06	
z Critical one-tail	1.644853	
P(Z<=z) two-tail	7.09045E-06	
z Critical two-tail	1.959961082	

9–2 Testing the Difference Between Two Means of Independent Samples: Using the *t* Test

Objective 2

Test the difference between two means for independent samples, using the *t* test.

In Section 9–1, the *z* test was used to test the difference between two means when the population standard deviations were known and the variables were normally or approximately normally distributed, or when both sample sizes were greater than or equal to 30. In many situations, however, these conditions cannot be met—that is, the population standard deviations are not known. In these cases, a *t* test is used to test the difference between means when the two samples are independent and when the samples are taken from two normally or approximately normally distributed populations. Samples are **independent samples** when they are not related. Also it will be assumed that the variances are not equal.

Formula for the *t* Test—For Testing the Difference Between Two Means—Independent Samples

Variances are assumed to be unequal

$$t = \frac{(\bar{X}_1 - \bar{X}_2) - (\mu_1 - \mu_2)}{\sqrt{\dfrac{s_1^2}{n_1} + \dfrac{s_2^2}{n_2}}}$$

where the degrees of freedom are equal to the smaller of $n_1 - 1$ or $n_2 - 1$.

The formula

$$t = \frac{(\bar{X}_1 - \bar{X}_2) - (\mu_1 - \mu_2)}{\sqrt{\dfrac{s_1^2}{n_1} + \dfrac{s_2^2}{n_2}}}$$

follows the format of

$$\text{Test value} = \frac{(\text{observed value}) - (\text{expected value})}{\text{standard error}}$$

where $\bar{X}_1 - \bar{X}_2$ is the observed difference between sample means and where the expected value $\mu_1 - \mu_2$ is equal to zero when no difference between population means is hypothesized. The denominator $\sqrt{s_1^2/n_1 + s_2^2/n_2}$ is the standard error of the difference between two means. Since mathematical derivation of the standard error is somewhat complicated, it will be omitted here.

> **Assumptions for the *t* Test for Two Independent Means When σ_1 and σ_2 Are Unknown**
>
> 1. The samples are random samples.
> 2. The sample data are independent of one another.
> 3. When the sample sizes are less than 30, the populations must be normally or approximately normally distributed.

Example 9–4   **Farm Sizes**

The average size of a farm in Indiana County, Pennsylvania, is 191 acres. The average size of a farm in Greene County, Pennsylvania, is 199 acres. Assume the data were obtained from two samples with standard deviations of 38 and 12 acres, respectively, and sample sizes of 8 and 10, respectively. Can it be concluded at $\alpha = 0.05$ that the average size of the farms in the two counties is different? Assume the populations are normally distributed.

Source: *Pittsburgh Tribune-Review.*

Solution

Step 1   State the hypotheses and identify the claim for the means.

$$H_0: \mu_1 = \mu_2 \qquad \text{and} \qquad H_1: \mu_1 \neq \mu_2 \text{ (claim)}$$

Step 2   Find the critical values. Since the test is two-tailed, since $\alpha = 0.05$, and since the variances are unequal, the degrees of freedom are the smaller of $n_1 - 1$ or $n_2 - 1$. In this case, the degrees of freedom are $8 - 1 = 7$. Hence, from Table F, the critical values are $+2.365$ and -2.365.

Step 3   Compute the test value. Since the variances are unequal, use the first formula.

$$t = \frac{(\bar{X}_1 - \bar{X}_2) - (\mu_1 - \mu_2)}{\sqrt{\dfrac{s_1^2}{n_1} + \dfrac{s_2^2}{n_2}}} = \frac{(191 - 199) - 0}{\sqrt{\dfrac{38^2}{8} + \dfrac{12^2}{10}}} = -0.57$$

Step 4   Make the decision. Do not reject the null hypothesis, since $-0.57 > -2.365$. See Figure 9–5.

Figure 9–5

Critical and Test Values for Example 9–4

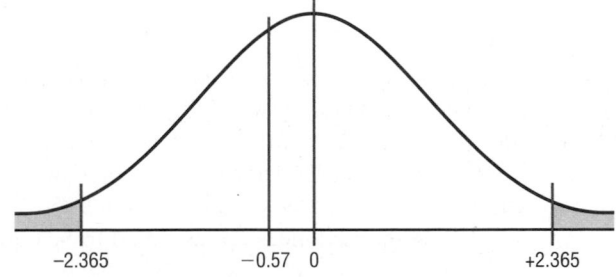

-2.365 $-0.57 \; 0$ $+2.365$

Step 5   Summarize the results. There is not enough evidence to support the claim that the average size of the farms is different.

When raw data are given in the exercises, use your calculator or the formulas in Chapter 3 to find the means and variances for the data sets. Then follow the procedures shown in this section to test the hypotheses.

Confidence intervals can also be found for the difference between two means with this formula:

Confidence Intervals for the Difference of Two Means: Independent Samples

Variances assumed to be unequal:

$$(\bar{X}_1 - \bar{X}_2) - t_{\alpha/2}\sqrt{\frac{s_1^2}{n_1} + \frac{s_2^2}{n_2}} < \mu_1 - \mu_2 < (\bar{X}_1 - \bar{X}_2) + t_{\alpha/2}\sqrt{\frac{s_1^2}{n_1} + \frac{s_2^2}{n_2}}$$

d.f. = smaller value of $n_1 - 1$ or $n_2 - 1$

Example 9–5

Find the 95% confidence interval for the data in Example 9–4.

Solution

Substitute in the formula.

$$(\bar{X}_1 - \bar{X}_2) - t_{\alpha/2}\sqrt{\frac{s_1^2}{n_1} + \frac{s_2^2}{n_2}} < \mu_1 - \mu_2$$

$$< (\bar{X}_1 - \bar{X}_2) + t_{\alpha/2}\sqrt{\frac{s_1^2}{n_1} + \frac{s_2^2}{n_2}}$$

$$(191 - 199) - 2.365\sqrt{\frac{38^2}{8} + \frac{12^2}{10}} < \mu_1 - \mu_2$$

$$< (191 - 199) + 2.365\sqrt{\frac{38^2}{8} + \frac{12^2}{10}}$$

$$-41.02 < \mu_1 - \mu_2 < 25.02$$

Since 0 is contained in the interval, the decision is to not reject the null hypothesis $H_0\colon \mu_1 = \mu_2$.

In many statistical software packages, a different method is used to compute the degrees of freedom for this t test. They are determined by the formula

$$\text{d.f.} = \frac{(s_1^2/n_1 + s_2^2/n_2)^2}{(s_1^2/n_1)^2/(n_1 - 1) + (s_2^2/n_2)^2/(n_2 - 1)}$$

This formula will not be used in this textbook.

There are actually two different options for the use of t tests. *One option is used when the variances of the populations are not equal, and the other option is used when the variances are equal.* To determine whether two sample variances are equal, the researcher can use an F test, as shown in Section 9–5.

When the variances are assumed to be equal, this formula is used and

$$t = \frac{(\bar{X}_1 - \bar{X}_2) - (\mu_1 - \mu_2)}{\sqrt{\dfrac{(n_1 - 1)s_1^2 + (n_2 - 1)s_2^2}{n_1 + n_2 - 2}}\sqrt{\dfrac{1}{n_1} + \dfrac{1}{n_2}}}$$

follows the format of

$$\text{Test value} = \frac{(\text{observed value}) - (\text{expected value})}{\text{standard error}}$$

For the numerator, the terms are the same as in the previously given formula. However, a note of explanation is needed for the denominator of the second test statistic. Since both populations are assumed to have the same variance, the standard error is computed with what is called a pooled estimate of the variance. A **pooled estimate of the variance** is a weighted average of the variance using the two sample variances and the *degrees of freedom* of each variance as the weights. Again, since the algebraic derivation of the standard error is somewhat complicated, it is omitted.

Note, however, that not all statisticians are in agreement about using the *F* test before using the *t* test. Some believe that conducting the *F* and *t* tests at the same level of significance will change the overall level of significance of the *t* test. Their reasons are beyond the scope of this textbook. Because of this, we will assume that $\sigma_1 \neq \sigma_2$ in this textbook.

Applying the Concepts **9–2**

Too Long on the Telephone

A company collects data on the lengths of telephone calls made by employees in two different divisions. The mean and standard deviation for the sales division are 10.26 and 8.56, respectively. The mean and standard deviation for the shipping and receiving division are 6.93 and 4.93, respectively. A hypothesis test was run, and the computer output follows.

Degrees of freedom = 56
Confidence interval limits = −0.18979, 6.84979
Test statistic $t = 1.89566$
Critical value $t = -2.0037, 2.0037$
P-value = 0.06317
Significance level − 0.05

1. Are the samples independent or dependent?
2. Which number from the output is compared to the significance level to check if the null hypothesis should be rejected?
3. Which number from the output gives the probability of a type I error that is calculated from the sample data?
4. Was a right-, left-, or two-tailed test done? Why?
5. What are your conclusions?
6. What would your conclusions be if the level of significance were initially set at 0.10?

See page 531 for the answers.

Exercises 9–2

For these exercises, perform each of these steps. Assume that all variables are normally or approximately normally distributed.

   a. State the hypotheses and identify the claim.
   b. Find the critical value(s).
   c. Compute the test value.
   d. Make the decision.
   e. Summarize the results.

Use the traditional method of hypothesis testing unless otherwise specified.
   For these exercises assume the variances are unequal.

1. **Bestseller Books** The mean for the number of weeks 15 *New York Times* hard-cover fiction books spent on the

bestseller list is 22 weeks. The standard deviation is 6.17 weeks. The mean for the number of weeks 15 *New York Times* hard-cover nonfiction books spent on the list is 28 weeks. The standard deviation is 13.2 weeks. At $\alpha = 0.10$, can we conclude that there is a difference in the mean times for the number of weeks the books were on the bestseller lists?

2. **Tax-Exempt Properties** A tax collector wishes to see if the mean values of the tax-exempt properties are different for two cities. The values of the tax-exempt properties for the two samples are shown. The data are given in millions of dollars. A $\alpha = 0.05$, is there enough evidence to support the tax collector's claim that the means are different?

City A				City B			
113	22	14	8	82	11	5	15
25	23	23	30	295	50	12	9
44	11	19	7	12	68	81	2
31	19	5	2	20	16	4	5

3. **Noise Levels in Hospitals** The mean noise level of 20 areas designated as "casualty doors" was 63.1 dBA, and the standard deviation is 4.1 dBA. The mean noise level for 24 areas designated as operating theaters was 56.3 dBA, and the standard deviation was 7.5 dBA. At $\alpha = 0.05$, can it be concluded that there is a difference in the means?

4. **Ages of Gamblers** The mean age of a sample of 25 people who were playing the slot machines is 48.7 years, and the standard deviation is 6.8 years. The mean age of a sample of 35 people who were playing roulette is 55.3 with a standard deviation of 3.2 years. Can it be concluded at $\alpha = 0.05$ that the mean age of those playing the slot machines is less than those playing roulette?

5. **Carbohydrates in Candies** The number of grams of carbohydrates contained in 1-ounce servings of randomly selected chocolate and nonchocolate candy is listed here. Is there sufficient evidence to conclude that the difference in the means is significant? Use $\alpha = 0.10$.

Chocolate:	29	25	17	36	41	25	32	29
	38	34	24	27	29			
Nonchocolate:	41	41	37	29	30	38	39	10
	29	55	29					

Source: *The Doctor's Pocket Calorie, Fat, and Carbohydrate Counter.*

6. **Teacher Salaries** A researcher claims that the mean of the salaries of elementary school teachers is greater than the mean of the salaries of secondary school teachers in a large school district. The mean of the salaries of a sample of 26 elementary school teachers is $48,256, and the sample standard deviation is $3,912.40. The mean of the salaries of a sample of 24 secondary school teachers is $45,633. The standard deviation is $5,533. At $\alpha = 0.05$, can it be concluded that the mean of the salaries of the elementary school teachers is greater than the mean of the salaries of the secondary school teachers? Use the P-value method.

7. **Weights of Running Shoes** The weights in ounces of a sample of running shoes for men and women are shown. Test the claim that the means are different. Use the P-value method with $\alpha = 0.05$.

Men		Women		
10.4	12.6	10.6	10.2	8.8
11.1	14.7	9.6	9.5	9.5
10.8	12.9	10.1	11.2	9.3
11.7	13.3	9.4	10.3	9.5
12.8	14.5	9.8	10.3	11.0

8. **Weights of Vacuum Cleaners** Upright vacuum cleaners have either a hard body type or a soft body type. Shown are the weights in pounds of a sample of each type. At $\alpha = 0.05$, can it be concluded that the means of the weights are different?

Hard body types				Soft body types			
21	17	17	20	24	13	11	13
16	17	15	20	12	15		
23	16	17	17				
13	15	16	18				
18							

9. Find the 95% confidence interval for the difference of the means in Exercise 3 of this section.
$$3.066 < \mu_1 - \mu_2 < 10.534$$

10. Find the 95% confidence interval for the difference of the means in Exercise 8 of this section.
$$-2.481 < \mu_1 - \mu_2 < 7.971$$

11. **Hours Spent Watching Television** According to Nielsen Media Research, children (ages 2–11) spend an average of 21 hours 30 minutes watching television per week while teens (ages 12–17) spend an average of 20 hours 40 minutes. Based on the sample statistics obtained below, is there sufficient evidence to conclude a difference in average television watching times between the two groups? Use $\alpha = 0.01$.

	Children	Teens
Sample mean	22.45	18.50
Sample variance	16.4	18.2
Sample size	15	15

Source: *Time Almanac.*

12. **NFL Salaries** An agent claims that there is no difference between the pay of safeties and linebackers in the NFL. A survey of 15 safeties found an average salary of $501,580, and a survey of 15 linebackers found an average salary of $513,360. If the standard deviation in the first sample is $20,000 and the standard deviation in the second sample is $18,000, is the agent correct? Use $\alpha = 0.05$.

Source: *NFL Players Assn./USA TODAY.*

13. **Cyber School Enrollment** The data show the number of students attending cyber charter schools in Allegheny County and the number of students attending cyber schools in counties surrounding Allegheny County. At $\alpha = 0.01$ is there enough evidence to support the claim that the average number of students in school districts in Allegheny County who attend cyber schools is greater than those who attend cyber schools in school districts outside Allegheny County? Give a factor that should be considered in interpreting this answer.

Allegheny County						Outside Allegheny County					
25	75	38	41	27	32	57	25	38	14	10	29

Source: *Pittsburgh Tribune-Review.*

14. Ages of Homes Whiting, Indiana, leads the "Top 100 Cities with the Oldest Houses" list with the average age of houses being 66.4 years. Farther down the list resides Franklin, Pennsylvania, with an average house age of 59.4 years. Researchers selected a random sample of 20 houses in each city and obtained the following statistics. At $\alpha = 0.05$, can it be concluded that the houses in Whiting are older? Use the *P*-value method.

	Whiting	Franklin
Mean age	62.1 years	55.6 years
Standard deviation	5.4 years	3.9 years

Source: www.city-data.com

15. Hospital Stays for Maternity Patients Health Care Knowledge Systems reported that an insured woman spends on average 2.3 days in the hospital for a routine childbirth, while an uninsured woman spends on average 1.9 days. Assume two samples of 16 women each were used in both samples. The standard deviation of the first sample is equal to 0.6 day, and the standard deviation of the second sample is 0.3 day. At $\alpha = 0.01$, test the claim that the means are equal. Find the 99% confidence interval for the differences of the means. Use the *P*-value method.

Source: Michael D. Shook and Robert L. Shook, *The Book of Odds.*

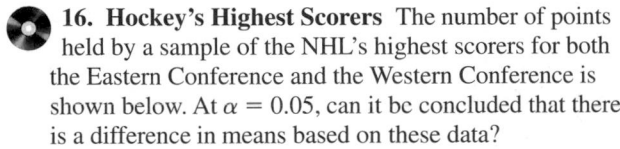 **16. Hockey's Highest Scorers** The number of points held by a sample of the NHL's highest scorers for both the Eastern Conference and the Western Conference is shown below. At $\alpha = 0.05$, can it be concluded that there is a difference in means based on these data?

Eastern Conference				Western Conference			
83	60	75	58	77	59	72	58
78	59	70	58	37	57	66	55
62	61	59		61			

Source: www.foxsports.com

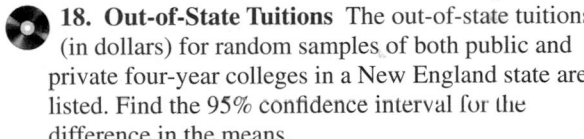 **17. Medical School Enrollments** A random sample of enrollments from medical schools that specialize in research and from those that are noted for primary care is listed. Find the 90% confidence interval for the difference in the means. $9.87 < \mu_1 - \mu_2 < 219.6$

Research				Primary care			
474	577	605	663	783	605	427	728
783	467	670	414	546	474	371	107
813	443	565	696	442	587	293	277
692	694	277	419	662	555	527	320
884							

Source: *U.S. News & World Report Best Graduate Schools.*

18. Out-of-State Tuitions The out-of-state tuitions (in dollars) for random samples of both public and private four-year colleges in a New England state are listed. Find the 95% confidence interval for the difference in the means.

Private		Public	
13,600	13,495	7,050	9,000
16,590	17,300	6,450	9,758
23,400	12,500	7,050	7,871
		16,100	

Source: *New York Times Almanac.* $\$1789.70 < \mu_1 - \mu_2 < \$12,425.41$

Technology *Step by Step*

MINITAB
Step by Step

Test the Difference Between Two Means: Independent Samples*

MINITAB will calculate the test statistic and *P*-value for differences between the means for two populations when the population standard deviations are unknown.

For Example 9–2, is the average number of sports for men higher than the average number for women?

1. Enter the data for Example 9–2 into C1 and C2. Name the columns **MaleS** and **FemaleS.**

2. Select **Stat>Basic Statistics>2-Sample t.**

3. Click the button for Samples in different columns.

There is one sample in each column.

4. Click in the box for First:. Double-click C1 MaleS in the list.

**MINITAB does not calculate a *z* test statistic. This statistic can be used instead.*

5. Click in the box for Second:, then double-click C2 FemaleS in the list. Do not check the box for Assume equal variances. MINITAB will use the large sample formula. The completed dialog box is shown.

6. Click [Options].

   a) Type in **90** for the Confidence level and **0** for the Test mean.

   b) Select greater than for the Alternative. This option affects the *P*-value. It must be correct.

7. Click [OK] twice. Since the *P*-value is greater than the significance level, $0.172 > 0.1$, do not reject the null hypothesis.

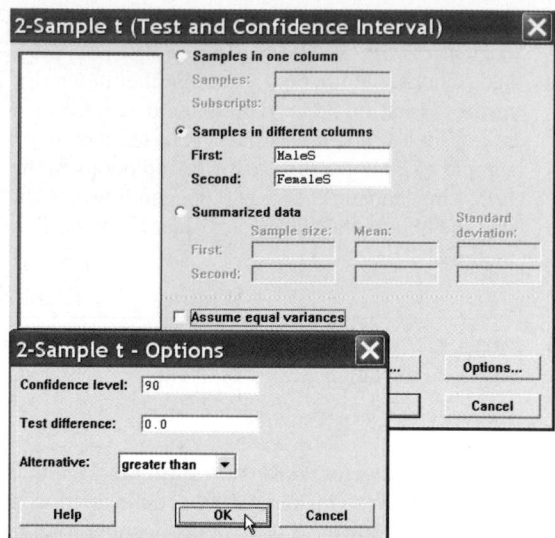

Two-Sample t-Test and CI: MaleS, FemaleS

```
Two-sample  t  for  MaleS  vs FemaleS
            N    Mean   StDev  SE Mean
MaleS      50    8.56   3.26    0.46
FemaleS    50    7.94   3.27    0.46
Difference = mu (MaleS) - mu (FemaleS)
Estimate for difference: 0.620000
90% lower bound for difference: -0.221962
t-Test of difference = 0 (vs >): t-Value = 0.95 P-Value = 0.172 DF = 97
```

TI-83 Plus or TI-84 Plus
Step by Step

Hypothesis Test for the Difference Between Two Means and *t* Distribution (Statistics)

1. Press **STAT** and move the cursor to TESTS.
2. Press **4** for 2-SampTTest.
3. Move the cursor to Stats and press **ENTER**.
4. Type in the appropriate values.
5. Move the cursor to the appropriate alternative hypothesis and press **ENTER**.
6. On the line for Pooled, move the cursor to No (standard deviations are assumed not equal) and press **ENTER**.
7. Move the cursor to Calculate and press **ENTER**.

Confidence Interval for the Difference Between Two Means and *t* Distribution (Data)

1. Enter the data values into L_1 and L_2.
2. Press **STAT** and move the cursor to TESTS.
3. Press **0** for 2-SampTInt.
4. Move the cursor to Data and press **ENTER**.
5. Type in the appropriate values.
6. On the line for Pooled, move the cursor to No (standard deviations are assumed not equal) and press **ENTER**.
7. Move the cursor to Calculate and press **ENTER**.

Confidence Interval for the Difference Between Two Means and *t* Distribution (Statistics)

1. Press **STAT** and move the cursor to TESTS.
2. Press **0** for 2-SampTInt.

3. Move the cursor to Stats and press **ENTER.**
4. Type in the appropriate values.
5. On the line for Pooled, move the cursor to No (standard deviations are assumed not equal) and press **ENTER.**
6. Move the cursor to Calculate and press **ENTER.**

Excel
Step by Step

Testing the Difference Between Two Means:
Independent Samples

Excel has a two-sample *t* test included in the Data Analysis Add-in. The following example shows how to perform a *t* test for the difference between two means.

Example XL9–2

Test the claim that there is no difference between population means based on these sample data. Assume the population variances are not equal. Use $\alpha = 0.05$.

Set A	32	38	37	36	36	34	39	36	37	42
Set B	30	36	35	36	31	34	37	33	32	

1. Enter the 10-number data set A into column A.
2. Enter the 9-number data set B into column B.
3. Select the Data tab from the toolbar. Then select Data Analysis.
4. In the Data Analysis box, under Analysis Tools select *t*-test: Two-Sample Assuming Unequal Variances, and click [OK].
5. In Input, type in the Variable 1 Range: **A1:A10** and the Variable 2 Range: **B1:B9.**
6. Type **0** for the Hypothesized Mean Difference.
7. Type **0.05** for Alpha.
8. In Output options, type D9 for the Output Range, then click [OK].

Two-Sample *t* Test in Excel

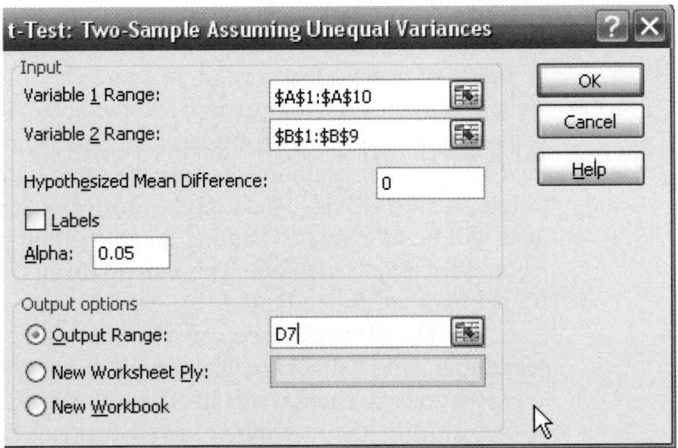

t-Test: Two-Sample Assuming Unequal Variances		
	Variable 1	*Variable 2*
Mean	36.7	33.77777778
Variance	7.344444444	5.944444444
Observations	10	9
Hypothesized Mean Difference	0	
df	17	
t Stat	2.474205364	
P(T<=t) one-tail	0.012095	
t Critical one-tail	1.739606716	
P(T<=t) two-tail	0.024189999	
t Critical two-tail	2.109815559	

Note: You may need to increase the column width to see all the results. To do this:

1. Highlight the columns D, E, and F.
2. Select **Format>AutoFit** Column Width.

The output reports both one- and two-tailed *P*-values.

Testing the Difference Between Two Means: Dependent Samples

Objective 3

Test the difference between two means for dependent samples.

In Section 9–2, the *t* test was used to compare two sample means when the samples were independent. In this section, a different version of the *t* test is explained. This version is used when the samples are dependent. Samples are considered to be **dependent samples** when the subjects are paired or matched in some way.

For example, suppose a medical researcher wants to see whether a drug will affect the reaction time of its users. To test this hypothesis, the researcher must pretest the subjects in the sample first. That is, they are given a test to ascertain their normal reaction times. Then after taking the drug, the subjects are tested again, using a posttest. Finally, the means of the two tests are compared to see whether there is a difference. Since the same subjects are used in both cases, the samples are *related;* subjects scoring high on the pretest will generally score high on the posttest, even after consuming the drug. Likewise, those scoring lower on the pretest will tend to score lower on the posttest. To take this effect into account, the researcher employs a *t* test, using the differences between the pretest values and the posttest values. Thus only the gain or loss in values is compared.

Here are some other examples of dependent samples. A researcher may want to design an SAT preparation course to help students raise their test scores the second time they take the SAT. Hence, the differences between the two exams are compared. A medical specialist may want to see whether a new counseling program will help subjects lose weight. Therefore, the preweights of the subjects will be compared with the postweights.

Besides samples in which the same subjects are used in a pre-post situation, there are other cases where the samples are considered dependent. For example, students might be matched or paired according to some variable that is pertinent to the study; then one student is assigned to one group, and the other student is assigned to a second group. For instance, in a study involving learning, students can be selected and paired according to their IQs. That is, two students with the same IQ will be paired. Then one will be assigned to one sample group (which might receive instruction by computers), and the other student will be assigned to another sample group (which might receive instruction by the lecture discussion method). These assignments will be done randomly. Since a student's IQ is important to learning, it is a variable that should be controlled. By matching subjects on IQ, the researcher can eliminate the variable's influence, for the most part. Matching, then, helps to reduce type II error by eliminating extraneous variables.

Two notes of caution should be mentioned. First, when subjects are matched according to one variable, the matching process does not eliminate the influence of other variables. Matching students according to IQ does not account for their mathematical ability or their familiarity with computers. Since not all variables influencing a study can be controlled, it is up to the researcher to determine which variables should be used in matching. Second, when the same subjects are used for a pre-post study, sometimes the knowledge that they are participating in a study can influence the results. For example, if people are placed in a special program, they may be more highly motivated to succeed simply because they have been selected to participate; the program itself may have little effect on their success.

When the samples are dependent, a special *t* test for dependent means is used. This test employs the difference in values of the matched pairs. The hypotheses are as follows:

Two-tailed	Left-tailed	Right-tailed
$H_0: \mu_D = 0$	$H_0: \mu_D = 0$	$H_0: \mu_D = 0$
$H_1: \mu_D \neq 0$	$H_1: \mu_D < 0$	$H_1: \mu_D > 0$

where μ_D is the symbol for the expected mean of the difference of the matched pairs. The general procedure for finding the test value involves several steps.

First, find the differences of the values of the pairs of data.

$$D = X_1 - X_2$$

Second, find the mean $\overline{D}$ of the differences, using the formula

$$\overline{D} = \frac{\Sigma D}{n}$$

where n is the number of data pairs. Third, find the standard deviation s_D of the differences, using the formula

$$s_D = \sqrt{\frac{n\Sigma D^2 - (\Sigma D)^2}{n(n-1)}}$$

Fourth, find the estimated standard error $s_{\overline{D}}$ of the differences, which is

$$s_{\overline{D}} = \frac{s_D}{\sqrt{n}}$$

Finally, find the test value, using the formula

$$t = \frac{\overline{D} - \mu_D}{s_D/\sqrt{n}} \qquad \text{with d.f.} = n - 1$$

The formula in the final step follows the basic format of

$$\text{Test value} = \frac{(\text{observed value}) - (\text{expected value})}{\text{standard error}}$$

where the observed value is the mean of the differences. The expected value μ_D is zero if the hypothesis is $\mu_D = 0$. The standard error of the difference is the standard deviation of the difference, divided by the square root of the sample size. Both populations must be normally or approximately normally distributed. Example 9–6 illustrates the hypothesis-testing procedure in detail.

Assumptions for the *t* Test for Two Means When the Samples Are Dependent

1. The sample or samples are random.
2. The sample data are dependent.
3. When the sample size or sample sizes are less than 30, the population or populations must be normally or approximately normally distributed.

Example 9–6

Bank Deposits

A sample of nine local banks shows their deposits (in billions of dollars) 3 years ago and their deposits (in billions of dollars) today. At $\alpha = 0.05$, can it be concluded that the average in deposits for the banks is greater today than it was 3 years ago? Use $\alpha = 0.05$.

Source: SNL Financial.

Bank	1	2	3	4	5	6	7	8	9
3 years ago	11.42	8.41	3.98	7.37	2.28	1.10	1.00	0.9	1.35
Today	16.69	9.44	6.53	5.58	2.92	1.88	1.78	1.5	1.22

Solution

Step 1   State the hypothesis and identify the claim. Since we are interested to see if there has been an increase in deposits, the deposits 3 years ago must be less than the deposits today; hence, the differences must be significantly less 3 years ago than they are today. Hence the mean of the differences must be less than zero.

$$H_0: \mu_D = 0 \quad \text{and} \quad H_1: \mu_D < 0 \text{ (claim)}$$

Step 2   Find the critical value. The degrees of freedom are $n - 1$, or $9 - 1 = 8$. The critical value for a left-tailed test with $\alpha = 0.05$ is -1.860.

Step 3   Compute the test value.

a. Make a table.

3 years ago (X_1)	Now (X_2)	A $D = X_1 - X_2$	B $D^2 = (X_1 - X_2)^2$
11.42	16.69		
8.41	9.44		
3.98	6.53		
7.37	5.58		
2.28	2.92		
1.10	1.88		
1.00	1.78		
0.90	1.50		
1.35	1.22		

b. Find the differences and place the results in column A.

$$
\begin{aligned}
11.42 - 16.69 &= -5.27 \\
8.41 - 9.44 &= -1.03 \\
3.98 - 6.53 &= -2.55 \\
7.37 - 5.58 &= +1.79 \\
2.28 - 2.92 &= -0.64 \\
1.10 - 1.88 &= -0.78 \\
1.00 - 1.78 &= -0.78 \\
0.9 - 1.50 &= -0.60 \\
1.35 - 1.22 &= +0.13 \\
\hline
\Sigma D &= \quad 9.73
\end{aligned}
$$

c. Find the means of the differences.

$$\overline{D} = \frac{\Sigma D}{n} = \frac{-9.73}{9} = -1.081$$

d. Square the differences and place the results in Column B.

$$
\begin{aligned}
(-5.27)^2 &= 27.7729 \\
(-1.03)^2 &= 1.0609 \\
(-2.55)^2 &= 6.5025 \\
(+1.79)^2 &= 3.2041 \\
(-0.64)^2 &= 0.4096 \\
(-0.78)^2 &= 0.6084 \\
(-0.78)^2 &= 0.6084 \\
(-0.60)^2 &= 0.3600 \\
(+0.13)^2 &= 0.1690 \\
\hline
\Sigma D^2 &= 40.5437
\end{aligned}
$$

The completed table is shown next.

3 years ago (X_1)	Now (X_2)	A $D = X_1 - X_2$	B $D^2 = (X_1 - X_2)^2$
11.42	16.69	−5.27	27.7299
8.41	9.44	−1.03	1.0609
3.98	6.53	−2.55	6.5025
7.37	5.58	+1.79	3.2041
2.28	2.92	−0.64	0.4096
1.10	1.88	−0.78	0.6084
1.00	1.78	−0.78	0.6084
0.90	1.58	−0.60	0.3600
1.35	1.22	+0.13	0.1690
		$\Sigma D = 9.73$	$\Sigma D^2 = 40.5437$

e. Find the standard deviation of the differences.

$$s_D = \sqrt{\frac{n\Sigma D^2 - (\Sigma D)^2}{n(n-1)}}$$

$$= \sqrt{\frac{9(40.5437) - (9.73)^2}{9(9-1)}}$$

$$= \sqrt{\frac{270.2204}{72}}$$

$$= 1.937$$

f. Find the test value.

$$t = \frac{\overline{D} - \mu_D}{s_D/\sqrt{n}} = \frac{-1.081 - 0}{1.937/\sqrt{9}} = -1.67$$

Step 4   Make the decision. Do not reject the null hypothesis since the test value, −1.67, is greater than the critical value, −1.860. See Figure 9–6.

Figure 9–6

Critical and Test Values for Example 9–6

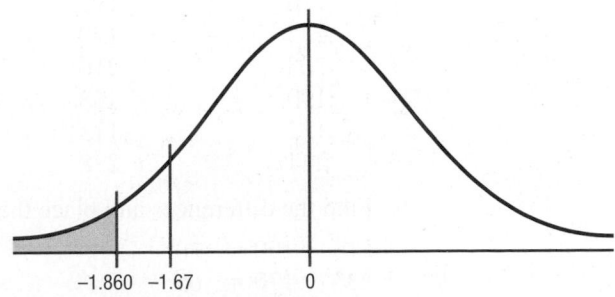

−1.860 −1.67   0

Step 5   Summarize the results. There is not enough evidence to show that the deposits have increased over the last 3 years.

The formulas for this t test are summarized next.

Formulas for the t Test for Dependent Samples

$$t = \frac{\overline{D} - \mu_D}{s_D / \sqrt{n}}$$

with d.f. $= n - 1$ and where

$$\overline{D} = \frac{\Sigma D}{n} \qquad \text{and} \qquad s_D = \sqrt{\frac{n \Sigma D^2 - (\Sigma D)^2}{n(n - 1)}}$$

Example 9–7

Cholesterol Levels

A dietitian wishes to see if a person's cholesterol level will change if the diet is supplemented by a certain mineral. Six subjects were pretested, and then they took the mineral supplement for a 6-week period. The results are shown in the table. (Cholesterol level is measured in milligrams per deciliter.) Can it be concluded that the cholesterol level has been changed at $\alpha = 0.10$? Assume the variable is approximately normally distributed.

Subject	1	2	3	4	5	6
Before (X_1)	210	235	208	190	172	244
After (X_2)	190	170	210	188	173	228

Solution

Step 1 State the hypotheses and identify the claim. If the diet is effective, the before cholesterol levels should be different from the after levels.

H_0: $\mu_D = 0$ and H_1: $\mu_D \neq 0$ (claim)

Step 2 Find the critical value. The degrees of freedom are 5. At $\alpha = 0.10$, the critical values are ± 2.015.

Step 3 Compute the test value.

  a. Make a table.

Before (X_1)	After (X_2)	A $D = X_1 - X_2$	B $D^2 = (X_1 - X_2)^2$
210	190		
235	170		
208	210		
190	188		
172	173		
244	228		

  b. Find the differences and place the results in column A.

$$210 - 190 = 20$$
$$235 - 170 = 65$$
$$208 - 210 = -2$$
$$190 - 188 = 2$$
$$172 - 173 = -1$$
$$244 - 228 = \underline{16}$$
$$\Sigma D = 100$$

    c. Find the mean of the differences.

$$\overline{D} = \frac{\Sigma D}{n} = \frac{100}{6} = 16.7$$

    d. Square the differences and place the results in column B.

$$(20)^2 = 400$$
$$(65)^2 = 4225$$
$$(-2)^2 = 4$$
$$(2)^2 = 4$$
$$(-1)^2 = 1$$
$$(16)^2 = \underline{256}$$
$$\Sigma D^2 = 4890$$

Then complete the table as shown.

Before (X_1)	After (X_2)	A $D = X_1 - X_2$	B $D^2 = (X_1 - X_2)^2$
210	190	20	400
235	170	65	4225
208	210	−2	4
190	188	2	4
172	173	−1	1
244	228	16	256
		$\Sigma D = 100$	$\Sigma D^2 = 4890$

    e. Find the standard deviation of the differences.

$$s_D = \sqrt{\frac{n\Sigma D^2 - (\Sigma D)^2}{n(n-1)}}$$
$$= \sqrt{\frac{6 \cdot 4890 - 100^2}{6(6-1)}}$$
$$= \sqrt{\frac{29{,}340 - 10{,}000}{30}}$$
$$= 25.4$$

    f. Find the test value.

$$t = \frac{\overline{D} - \mu_D}{s_D/\sqrt{n}} = \frac{16.7 - 0}{25.4/\sqrt{6}} = 1.610$$

Step 4   Make the decision. The decision is to not reject the null hypothesis, since the test value 1.610 is in the noncritical region, as shown in Figure 9–7.

Figure 9–7

Critical and Test Values
for Example 9–7

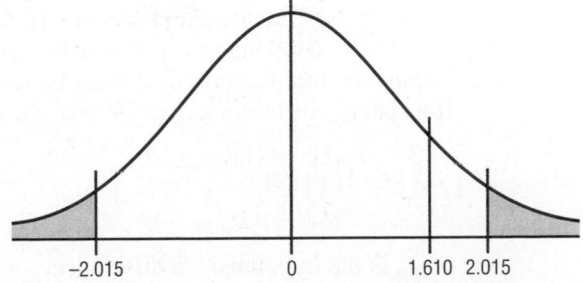

-2.015   0   1.610 2.015

Step 5   Summarize the results. There is not enough evidence to support the claim that the mineral changes a person's cholesterol level.

The steps for this *t* test are summarized in the Procedure Table.

Procedure Table

Testing the Difference Between Means for Dependent Samples

Step 1   State the hypotheses and identify the claim.

Step 2   Find the critical value(s).

Step 3   Compute the test value.

a. Make a table, as shown.

X_1	X_2	A $D = X_1 - X_2$	B $D^2 = (X_1 - X_2)^2$
$\vdots$	$\vdots$		
		$\Sigma D =$ _____	$\Sigma D^2 =$ _____

b. Find the differences and place the results in column A.

$$D = X_1 - X_2$$

c. Find the mean of the differences.

$$\overline{D} = \frac{\Sigma D}{n}$$

d. Square the differences and place the results in column B. Complete the table.

$$D^2 = (X_1 - X_2)^2$$

e. Find the standard deviation of the differences.

$$s_D = \sqrt{\frac{n\Sigma D^2 - (\Sigma D)^2}{n(n-1)}}$$

f. Find the test value.

$$t = \frac{\overline{D} - \mu_D}{s_D/\sqrt{n}} \quad \text{with d.f.} = n - 1$$

Step 4   Make the decision.

Step 5   Summarize the results.

Unusual Stat

About 4% of Americans spend at least one night in jail each year.

The *P*-values for the *t* test are found in Table F. For a two-tailed test with d.f. = 5 and *t* = 1.610, the *P*-value is found between 1.476 and 2.015; hence, 0.10 < *P*-value < 0.20. Thus, the null hypothesis cannot be rejected at $\alpha = 0.10$.

If a specific difference is hypothesized, this formula should be used

$$t = \frac{\overline{D} - \mu_D}{s_D/\sqrt{n}}$$

where μ_D is the hypothesized difference.

For example, if a dietitian claims that people on a specific diet will lose an average of 3 pounds in a week, the hypotheses are

$$H_0: \mu_D = 3 \qquad \text{and} \qquad H_1: \mu_D \neq 3$$

The value 3 will be substituted in the test statistic formula for μ_D.

Confidence intervals can be found for the mean differences with this formula.

Confidence Interval for the Mean Difference

$$\overline{D} - t_{\alpha/2}\frac{s_D}{\sqrt{n}} < \mu_D < \overline{D} + t_{\alpha/2}\frac{s_D}{\sqrt{n}}$$

d.f. $= n - 1$

Example 9–8

Find the 90% confidence interval for the data in Example 9–7.

Solution

Substitute in the formula.

$$\overline{D} - t_{\alpha/2}\frac{s_D}{\sqrt{n}} < \mu_D < \overline{D} + t_{\alpha/2}\frac{s_D}{\sqrt{n}}$$

$$16.7 - 2.015 \cdot \frac{25.4}{\sqrt{6}} < \mu_D < 16.7 + 2.015 \cdot \frac{25.4}{\sqrt{6}}$$

$$16.7 - 20.89 < \mu_D < 16.7 + 20.89$$

$$-4.19 < \mu_D < 37.59$$

Since 0 is contained in the interval, the decision is to not reject the null hypothesis $H_0: \mu_D = 0$.

Speaking of Statistics

Can Video Games Save Lives?

Can playing video games help doctors perform surgery? The answer is yes. A study showed that surgeons who played video games for at least 3 hours each week made about 37% fewer mistakes and finished operations 27% faster than those who did not play video games.

The type of surgery that they performed is called *laparoscopic* surgery, where the surgeon inserts a tiny video camera into the body and uses a joystick to maneuver the surgical instruments while watching the results on a television monitor. This study compares two groups and uses proportions. What statistical test do you think was used to compare the percentages? (See Section 9–4.)

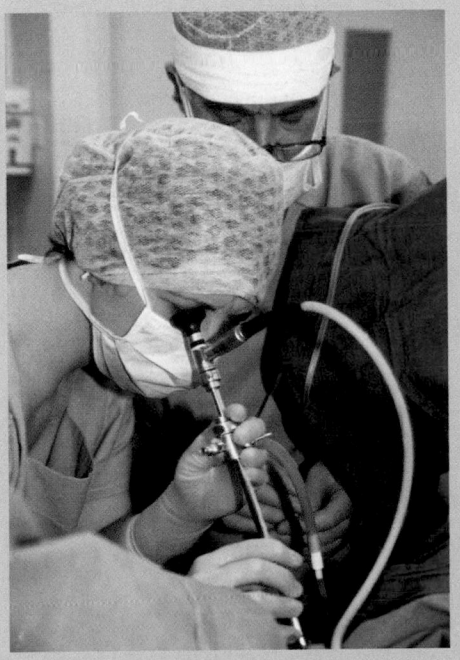

Applying the Concepts **9–3**

Air Quality

As a researcher for the EPA, you have been asked to determine if the air quality in the United States has changed over the past 2 years. You select a random sample of 10 metropolitan areas and find the number of days each year that the areas failed to meet acceptable air quality standards. The data are shown.

Year 1	18	125	9	22	138	29	1	19	17	31
Year 2	24	152	13	21	152	23	6	31	34	20

Source: *The World Almanac and Book of Facts.*

Based on the data, answer the following questions.

1. What is the purpose of the study?
2. Are the samples independent or dependent?
3. What hypotheses would you use?
4. What is (are) the critical value(s) that you would use?
5. What statistical test would you use?
6. How many degrees of freedom are there?
7. What is your conclusion?
8. Could an independent means test have been used?
9. Do you think this was a good way to answer the original question?

See page 531 for the answers.

Exercises 9–3

1. Classify each as independent or dependent samples.
   a. Heights of identical twins   Dependent
   b. Test scores of the same students in English and psychology   Dependent
   c. The effectiveness of two different brands of aspirin   Independent
   d. Effects of a drug on reaction time, measured by a before-and-after test   Dependent
   e. The effectiveness of two different diets on two different groups of individuals   Independent

For Exercises 2 through 10, perform each of these steps. Assume that all variables are normally or approximately normally distributed.
   a. State the hypotheses and identify the claim.
   b. Find the critical value(s).
   c. Compute the test value.
   d. Make the decision.
   e. Summarize the results.

Use the traditional method of hypothesis testing unless otherwise specified.

2. **Retention Test Scores** A sample of non-English majors at a selected college was used in a study to see if the student retained more from reading a 19th-century novel or by watching it in DVD form. Each student was assigned one novel to read and a different one to watch, and then they were given a 20-point written quiz on each novel. The test results are shown below. At $\alpha = 0.05$, can it be concluded that the book scores are higher than the DVD scores?

Book	90	80	90	75	80	90	84
DVD	85	72	80	80	70	75	80

3. **Improving Study Habits** As an aid for improving students' study habits, nine students were randomly selected to attend a seminar on the importance of education in life. The table shows the number of hours each student studied per week before and after the

seminar. At $\alpha = 0.10$, did attending the seminar increase the number of hours the students studied per week?

Before	9	12	6	15	3	18	10	13	7
After	9	17	9	20	2	21	15	22	6

4. Obstacle Course Times An obstacle course was set up on a campus, and 10 volunteers were given a chance to complete it while they were being timed. They then sampled a new energy drink and were given the opportunity to run the course again. The "before" and "after" times in seconds are shown below. Is there sufficient evidence at $\alpha = 0.05$ to conclude that the students did better the second time? Discuss possible reasons for your results.

Student	1	2	3	4	5	6	7	8
Before	67	72	80	70	78	82	69	75
After	68	70	76	65	75	78	65	68

5. Sleep Report Students in a statistics class were asked to report the number of hours they slept on weeknights and on weekends. At $\alpha = 0.05$, is there sufficient evidence that there is a difference in the mean number of hours slept?

Student	1	2	3	4	5	6	7	8
Hours, Sun.–Thurs.	8	5.5	7.5	8	7	6	6	8
Hours, Fri.–Sat.	4	7	10.5	12	11	9	6	9

6. PGA Golf Scores At a recent PGA tournament (the Honda Classic at Palm Beach Gardens, Florida) the following scores were posted for eight randomly selected golfers for two consecutive days. At $\alpha = 0.05$, is there evidence of a difference in mean scores for the two days?

Golfer	1	2	3	4	5	6	7	8
Thursday	67	65	68	68	68	70	69	70
Friday	68	70	69	71	72	69	70	70

Source: *Washington Observer-Reporter.*

7. Reducing Errors in Grammar A composition teacher wishes to see whether a new grammar program

will reduce the number of grammatical errors her students make when writing a two-page essay. The data are shown here. At $\alpha = 0.025$, can it be concluded that the number of errors has been reduced?

Student	1	2	3	4	5	6
Errors before	12	9	0	5	4	3
Errors after	9	6	1	3	2	3

8. Overweight Dogs A veterinary nutritionist developed a diet for overweight dogs. The total volume of food consumed remains the same, but one-half of the dog food is replaced with a low-calorie "filler" such as canned green beans. Six overweight dogs were randomly selected from her practice and were put on this program. Their initial weights were recorded, and then they were weighed again after 4 weeks. At the 0.05 level of significance can it be concluded that the dogs lost weight?

Before	42	53	48	65	40	52
After	39	45	40	58	42	47

9. Pulse Rates of Identical Twins A researcher wanted to compare the pulse rates of identical twins to see whether there was any difference. Eight sets of twins were selected. The rates are given in the table as number of beats per minute. At $\alpha = 0.01$, is there a significant difference in the average pulse rates of twins? Find the 99% confidence interval for the difference of the two. Use the *P*-value method.

Twin A	87	92	78	83	88	90	84	93
Twin B	83	95	79	83	86	93	80	86

10. A random sample of six music students played a short song, and the number of mistakes each student made was recorded. After they practiced the song 5 times, the number of mistakes each student made was recorded. The data are shown. At $\alpha = 0.05$, can it be concluded that there was a decrease in the mean number of mistakes?

Student	A	B	C	D	E	F
Before	10	6	8	8	13	8
After	4	2	2	7	8	9

Extending the Concepts

11. Instead of finding the mean of the differences between X_1 and X_2 by subtracting $X_1 - X_2$, you can find it by finding the means of X_1 and X_2 and then subtracting the means. Show that these two procedures will yield the same results.

Technology *Step by Step*

MINITAB
Step by Step

Test the Difference Between Two Means: Dependent Samples

A physical education director claims by taking a special vitamin, a weight lifter can increase his strength. Eight athletes are selected and given a test of strength, using the standard bench press. After 2 weeks of regular training, supplemented with the vitamin, they are tested again. Test the effectiveness of the vitamin regimen at $\alpha = 0.05$. Each value in these data represents the maximum number of pounds the athlete can bench-press. Assume that the variable is approximately normally distributed.

Athlete	1	2	3	4	5	6	7	8
Before (X_1)	210	230	182	205	262	253	219	216
After (X_2)	219	236	179	204	270	250	222	216

1. Enter the data into C1 and C2. Name the columns **Before** and **After.**

2. Select **Stat>Basic Statistics>Paired t.**

3. Double-click C1 Before for First sample.

4. Double-click C2 After for Second sample. The second sample will be subtracted from the first. The differences are not stored or displayed.

5. Click [Options].

6. Change the Alternative to less than.

7. Click [OK] twice.

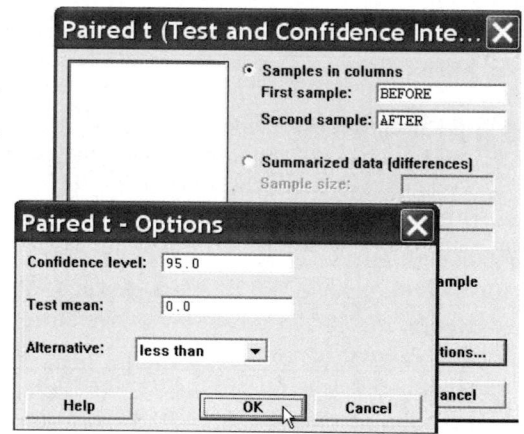

Paired t-Test and CI: BEFORE, AFTER

```
Paired t for BEFORE - AFTER
              N     Mean      StDev    SE Mean
BEFORE        8   222.125    25.920     9.164
AFTER         8   224.500    27.908     9.867
Difference    8   -2.37500   4.83846    1.71065

95% upper bound for mean difference: 0.86597
t-Test of mean difference = 0 (vs < 0) : t-Value = -1.39  P-Value = 0.104.
```

Since the *P*-value is 0.104, do not reject the null hypothesis. The sample difference of -2.38 in the strength measurement is not statistically significant.

TI-83 Plus or TI-84 Plus
Step by Step

Hypothesis Test for the Difference Between Two Means: Dependent Samples

1. Enter the data values into L_1 and L_2.

2. Move the cursor to the top of the L_3 column so that L_3 is highlighted.

3. Type $L_1 - L_2$, then press **ENTER.**

4. Press **STAT** and move the cursor to TESTS.

5. Press 2 for TTest.

6. Move the cursor to Data and press **ENTER.**

7. Type in the appropriate values, using **0** for μ_0 and $\mathbf{L_3}$ for the list.

8. Move the cursor to the appropriate alternative hypothesis and press **ENTER.**

9. Move the cursor to Calculate and press **ENTER.**

Confidence Interval for the Difference Between Two Means: Dependent Samples

1. Enter the data values into $\mathsf{L_1}$ and $\mathsf{L_2}$.

2. Move the cursor to the top of the $\mathsf{L_3}$ column so that $\mathsf{L_3}$ is highlighted.

3. Type $\mathbf{L_1 - L_2}$, then press **ENTER.**

4. Press **STAT** and move the cursor to TESTS.

5. Press **8** for TInterval.

6. Move the cursor to Stats and press **ENTER.**

7. Type in the appropriate values, using $\mathbf{L_3}$ for the list.

8. Move the cursor to Calculate and press **ENTER.**

Excel
Step by Step

Testing the Difference Between Two Means: Dependent Samples
Example XL9–3

Test the claim that there is no difference between population means based on these sample paired data. Use $\alpha = 0.05$.

Set A	33	35	28	29	32	34	30	34
Set B	27	29	36	34	30	29	28	24

1. Enter the 8-number data set A into column A.

2. Enter the 8-number data set B into column B.

3. Select the Data tab from the toolbar. Then select Data Analysis.

4. In the Data Analysis box, under Analysis Tools select *t*-test: Paired Two Sample for Means, and click [OK].

5. In Input, type in the Variable 1 Range: **A1:A8** and the Variable 2 Range: **B1:B8.**

6. Type **0** for the Hypothesized Mean Difference.

7. Type **0.05** for Alpha.

8. In Output options, type **D5** for the Output Range, then click [OK].

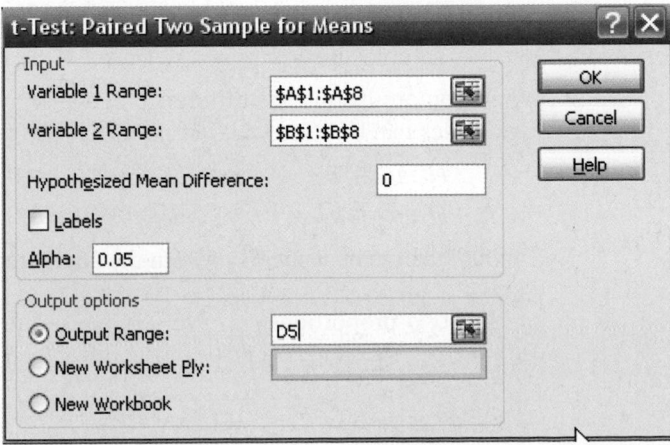

t-Test: Paired Two Sample for Means		
	Variable 1	Variable 2
Mean	31.875	29.625
Variance	6.696428571	14.55357143
Observations	8	8
Pearson Correlation	-0.757913399	
Hypothesized Mean Difference	0	
df	7	
t Stat	1.057517468	
P(T<=t) one-tail	0.1626994	
t Critical one-tail	1.894578604	
P(T<=t) two-tail	0.3253988	
t Critical two-tail	2.364624251	

Note: You may need to increase the column width to see all the results. To do this:

1. Highlight the columns D, E, and F.

2. Select **Format>AutoFit** Column Width.

The output shows a *P*-value of 0.3253988 for the two-tailed case. This value is greater than the alpha level of 0.05, so we fail to reject the null hypothesis.

9–4 Testing the Difference Between Proportions

Objective 4

Test the difference between two proportions.

The *z* test with some modifications can be used to test the equality of two proportions. For example, a researcher might ask, Is the proportion of men who exercise regularly less than the proportion of women who exercise regularly? Is there a difference in the percentage of students who own a personal computer and the percentage of nonstudents who own one? Is there a difference in the proportion of college graduates who pay cash for purchases and the proportion of non-college graduates who pay cash?

Recall from Chapter 7 that the symbol $\hat{p}$ ("p hat") is the sample proportion used to estimate the population proportion, denoted by *p*. For example, if in a sample of 30 college students, 9 are on probation, then the sample proportion is $\hat{p} = \frac{9}{30}$, or 0.3. The population proportion *p* is the number of all students who are on probation, divided by the number of students who attend the college. The formula for $\hat{p}$ is

$$\hat{p} = \frac{X}{n}$$

where

X = number of units that possess the characteristic of interest
n = sample size

When you are testing the difference between two population proportions p_1 and p_2, the hypotheses can be stated thus, if no difference between the proportions is hypothesized.

$$H_0: p_1 = p_2 \qquad H_0: p_1 - p_2 = 0$$
$$\text{or}$$
$$H_1: p_1 \neq p_2 \qquad H_1: p_1 - p_2 \neq 0$$

Similar statements using $<$ or $>$ in the alternate hypothesis can be formed for one-tailed tests.

For two proportions, $\hat{p}_1 = X_1/n_1$ is used to estimate p_1 and $\hat{p}_2 = X_2/n_2$ is used to estimate p_2. The standard error of the difference is

$$\sigma_{\hat{p}_1 - \hat{p}_2} = \sqrt{\sigma_{\hat{p}_1}^2 + \sigma_{\hat{p}_2}^2} = \sqrt{\frac{p_1 q_1}{n_1} + \frac{p_2 q_2}{n_2}}$$

where $\sigma_{p_1}^2$ and $\sigma_{p_2}^2$ are the variances of the proportions, $q_1 = 1 - p_1$, $q_2 = 1 - p_2$, and n_1 and n_2 are the respective sample sizes.

Since p_1 and p_2 are unknown, a weighted estimate of p can be computed by using the formula

$$\bar{p} = \frac{n_1 \hat{p}_1 + n_2 \hat{p}_2}{n_1 + n_2}$$

and $\bar{q} = 1 - \bar{p}$. This weighted estimate is based on the hypothesis that $p_1 = p_2$. Hence, $\bar{p}$ is a better estimate than either $\hat{p}_1$ or $\hat{p}_2$, since it is a combined average using both $\hat{p}_1$ and $\hat{p}_2$.

Since $\hat{p}_1 = X_1/n_1$ and $\hat{p}_2 = X_2/n_2$, $\bar{p}$ can be simplified to

$$\bar{p} = \frac{X_1 + X_2}{n_1 + n_2}$$

Finally, the standard error of the difference in terms of the weighted estimate is

$$\sigma_{\hat{p}_1 - \hat{p}_2} = \sqrt{\bar{p}\,\bar{q}\left(\frac{1}{n_1} + \frac{1}{n_2}\right)}$$

The formula for the test value is shown next.

Formula for the z Test for Comparing Two Proportions

$$z = \frac{(\hat{p}_1 - \hat{p}_2) - (p_1 - p_2)}{\sqrt{\bar{p}\,\bar{q}\left(\frac{1}{n_1} + \frac{1}{n_2}\right)}}$$

where

$$\bar{p} = \frac{X_1 + X_2}{n_1 + n_2} \qquad \hat{p}_1 = \frac{X_1}{n_1}$$

$$\bar{q} = 1 - \bar{p} \qquad \hat{p}_2 = \frac{X_2}{n_2}$$

This formula follows the format

$$\text{Test value} = \frac{(\text{observed value}) - (\text{expected value})}{\text{standard error}}$$

Assumptions for the z Test for Two Proportions

1. The samples must be random samples.
2. The sample data are independent of one another.
3. For both samples $np \geq 5$ and $nq \geq 5$.

Example 9-9

Vaccination Rates in Nursing Homes

In the nursing home study mentioned in the chapter-opening Statistics Today, the researchers found that 12 out of 34 small nursing homes had a resident vaccination rate of less than 80%, while 17 out of 24 large nursing homes had a vaccination rate of less than 80%. At $\alpha = 0.05$, test the claim that there is no difference in the proportions of the small and large nursing homes with a resident vaccination rate of less than 80%.

Source: Nancy Arden, Arnold S. Monto, and Suzanne E. Ohmit, "Vaccine Use and the Risk of Outbreaks in a Sample of Nursing Homes During an Influenza Epidemic," *American Journal of Public Health.*

Solution

Let $\hat{p}_1$ be the proportion of the small nursing homes with a vaccination rate of less than 80% and $\hat{p}_2$ be the proportion of the large nursing homes with a vaccination rate of less than 80%. Then

$$\hat{p}_1 = \frac{X_1}{n_1} = \frac{12}{34} = 0.35 \qquad \text{and} \qquad \hat{p}_2 = \frac{X_2}{n_2} = \frac{17}{24} = 0.71$$

$$\bar{p} = \frac{X_1 + X_2}{n_1 + n_2} = \frac{12 + 17}{34 + 24} = \frac{29}{58} = 0.5$$

$$\bar{q} = 1 - \bar{p} = 1 - 0.5 = 0.5$$

Now, follow the steps in hypothesis testing.

Step 1 State the hypotheses and identify the claim.

$$H_0\colon p_1 = p_2 \text{ (claim)} \qquad \text{and} \qquad H_1\colon p_1 \neq p_2$$

Step 2 Find the critical values. Since $\alpha = 0.05$, the critical values are $+1.96$ and -1.96.

Step 3 Compute the test value.

$$z = \frac{(\hat{p}_1 - \hat{p}_2) - (p_1 - p_2)}{\sqrt{\bar{p}\,\bar{q}\left(\dfrac{1}{n_1} + \dfrac{1}{n_2}\right)}}$$

$$= \frac{(0.35 - 0.71) - 0}{\sqrt{(0.5)(0.5)\left(\dfrac{1}{34} + \dfrac{1}{24}\right)}} = \frac{-0.36}{0.1333} = -2.7$$

Step 4 Make the decision. Reject the null hypothesis, since $-2.7 < -1.96$. See Figure 9–8.

Figure 9–8

Critical and Test Values for Example 9–9

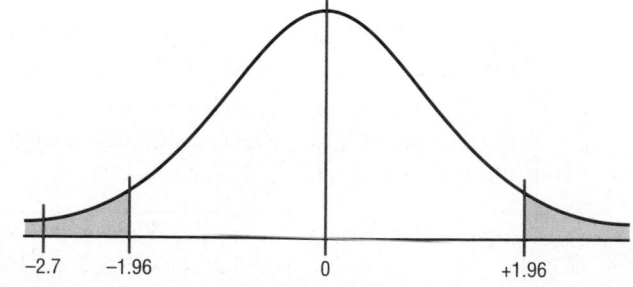

Step 5 Summarize the results. There is enough evidence to reject the claim that there is no difference in the proportions of small and large nursing homes with a resident vaccination rate of less than 80%.

Example 9-10

Texting While Driving

A survey of 1000 drivers this year showed that 29% of the people send text messages while driving. Last year a survey of 1000 drivers showed that 17% of those send text messages while driving. At $\alpha = 0.01$, can it be concluded that there has been an increase in the number of drivers who text while driving?

Source: FindLaw.com

Solution

You are given the percentages $\hat{p}_1 = 17\%$ or 0.17 and $\hat{p}_2 = 29\%$ or 0.29. To compute $\bar{p}$, you must find X_1 and X_2.

$$X_1 = \hat{p}_1 n_1 = 0.29\,(1000) = 290$$

$$X_2 = \hat{p}_2 n_2 = 0.17\,(1000) = 170$$

$$\bar{p} = \frac{X_1 + X_2}{n_1 + n_2} = \frac{290 + 170}{1000 + 1000} = \frac{460}{2000} = 0.23$$

$$\bar{q} = 1 - \bar{p} = 1 - 0.23 = 0.77$$

Step 1 State the hypotheses and identify the claim.

$$H_0: p_1 = p_2 \quad \text{and} \quad H_1: p_1 > p_2 \text{ (claim)}$$

Step 2 Find the critical value. Since $\alpha = 0.01$, the critical value is $z = 2.33$.

Step 3 Compute the test value.

$$z = \frac{(\hat{p}_1 - \hat{p}_2) - (p_1 - p_2)}{\sqrt{\bar{p}\,\bar{q}\left(\dfrac{1}{n_1} + \dfrac{1}{n_2}\right)}}$$

$$= \frac{(0.29 - 0.17) - 0}{\sqrt{(0.23)(0.77)\left(\dfrac{1}{1000} + \dfrac{1}{1000}\right)}} = 6.38$$

Step 4 Make the decision. Reject the null hypothesis since $6.38 > 2.33$.

Figure 9-9

Critical and Test Values for Example 9-10

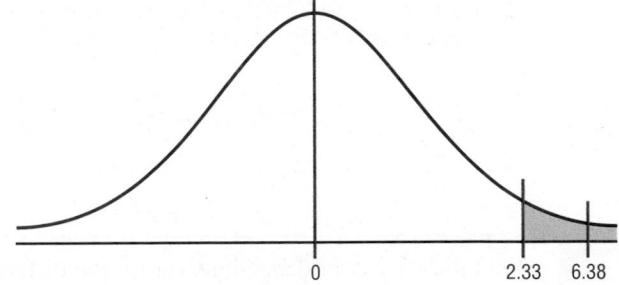

0   2.33   6.38

Step 5 Summarize the results. There is enough evidence to say that the proportion of drivers who send text messages is larger today than it was last year.

The *P*-value for the difference of proportions can be found from Table E, as shown in Section 9-1. For Example 9-10, 6.38 is beyond 3.49; hence, the null hypothesis can be rejected since the *P*-value is less than 0.001.

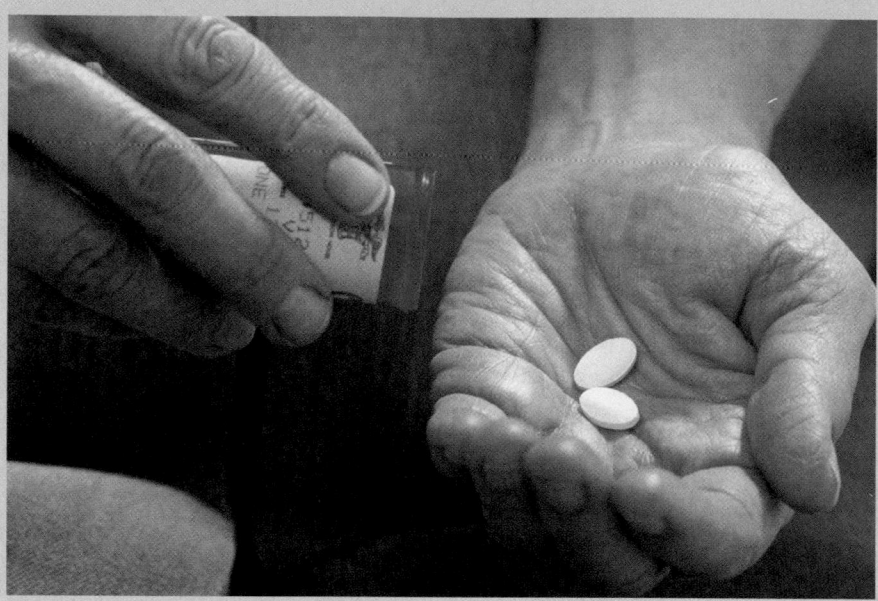

Speaking of
Statistics

Is More Expensive Better?

An article in the *Journal of the American Medical Association* explained a study done on placebo pain pills. Researchers randomly assigned 82 healthy people to two groups. The individuals in the first group were given sugar pills, but they were told that the pills were a new, fast-acting opioid pain reliever similar to codeine and that they were listed at $2.50 each. The individuals in the other group received the same sugar pills but were told that the pills had been marked down to 10¢ each.

Each group received electrical shocks before and after taking the pills. They were then asked if the pills reduced the pain. Eighty-five percent of the group who were told that the pain pills cost $2.50 said that they were effective, while 61% of the group who received the supposedly discounted pills said that they were effective.

State possible null and alternative hypotheses for this study. What statistical test could be used in this study? What might be the conclusion of the study?

The formula for the confidence interval for the difference between two proportions is shown next.

Confidence Interval for the Difference Between Two Proportions

$$(\hat{p}_1 - \hat{p}_2) - z_{\alpha/2}\sqrt{\frac{\hat{p}_1\hat{q}_1}{n_1} + \frac{\hat{p}_2\hat{q}_2}{n_2}} < p_1 - p_2 < (\hat{p}_1 - \hat{p}_2) + z_{\alpha/2}\sqrt{\frac{\hat{p}_1\hat{q}_1}{n_1} + \frac{\hat{p}_2\hat{q}_2}{n_2}}$$

Example 9–11

Find the 95% confidence interval for the difference of proportions for the data in Example 9–9.

Solution

$$\hat{p}_1 = \frac{12}{34} = 0.35 \qquad \hat{q}_1 = 0.65$$

$$\hat{p}_2 = \frac{17}{24} = 0.71 \qquad \hat{q}_2 = 0.29$$

Substitute in the formula.

$$(\hat{p}_1 - \hat{p}_2) - z_{\alpha/2}\sqrt{\frac{\hat{p}_1\hat{q}_1}{n_1} + \frac{\hat{p}_2\hat{q}_2}{n_2}} < p_1 - p_2$$

$$< (\hat{p}_1 - \hat{p}_2) + z_{\alpha/2}\sqrt{\frac{\hat{p}_1\hat{q}_1}{n_1} + \frac{\hat{p}_2\hat{q}_2}{n_2}}$$

$$(0.35 - 0.71) - 1.96\sqrt{\frac{(0.35)(0.65)}{34} + \frac{(0.71)(0.29)}{24}}$$

$$< p_1 - p_2 < (0.35 - 0.71) + 1.96\sqrt{\frac{(0.35)(0.65)}{34} + \frac{(0.71)(0.29)}{24}}$$

$$-0.36 - 0.242 < p_1 - p_2 < -0.36 + 0.242$$

$$-0.602 < p_1 - p_2 < -0.118$$

Since 0 is not contained in the interval, the decision is to reject the null hypothesis H_0: $p_1 = p_2$.

Applying the Concepts 9–4

Smoking and Education

You are researching the hypothesis that there is no difference in the percent of public school students who smoke and the percent of private school students who smoke. You find these results from a recent survey.

School	Percent who smoke
Public	32.3
Private	14.5

Based on these figures, answer the following questions.

1. What hypotheses would you use if you wanted to compare percentages of the public school students who smoke with the private school students who smoke?

2. What critical value(s) would you use?

3. What statistical test would you use to compare the two percentages?

4. What information would you need to complete the statistical test?

5. Suppose you found that 1000 individuals in each group were surveyed. Could you perform the statistical test?

6. If so, complete the test and summarize the results.

See page 531 for the answers.

Exercises 9–4

1a. Find the proportions $\hat{p}$ and $\hat{q}$ for each.

 a. $n = 48, X = 34 \quad \hat{p} = \frac{34}{48}, \hat{q} = \frac{14}{48}$
 b. $n = 75, X = 28 \quad \hat{p} = \frac{28}{75}, \hat{q} = \frac{47}{75}$
 c. $n = 100, X = 50 \quad \hat{p} = \frac{50}{100}, \hat{q} = \frac{50}{100}$
 d. $n = 24, X = 6 \quad \hat{p} = \frac{6}{24}, \hat{q} = \frac{18}{24}$
 e. $n = 144, X = 12 \quad \hat{p} = \frac{12}{144}, \hat{q} = \frac{132}{144}$

1b. Find each X, given $\hat{p}$.

 a. $\hat{p} = 0.16, n = 100 \quad 16$
 b. $\hat{p} = 0.08, n = 50 \quad 4$
 c. $\hat{p} = 6\%, n = 800 \quad 48$
 d. $\hat{p} = 52\%, n = 200 \quad 104$
 e. $\hat{p} = 20\%, n = 150 \quad 30$

2. Find $\bar{p}$ and $\bar{q}$ for each.

 a. $X_1 = 60, n_1 = 100, X_2 = 40, n_2 = 100 \quad \bar{p} = 0.5; \bar{q} = 0.5$
 b. $X_1 = 22, n_1 = 50, X_2 = 18, n_2 = 30 \quad \bar{p} = 0.5; \bar{q} = 0.5$
 c. $X_1 = 18, n_1 = 60, X_2 = 20, n_2 = 80 \quad \bar{p} = 0.27; \bar{q} = 0.73$
 d. $X_1 = 5, n_1 = 32, X_2 = 12, n_2 = 48 \quad \bar{p} = 0.2125; \bar{q} = 0.7875$
 e. $X_1 = 12, n_1 = 75, X_2 = 15, n_2 = 50 \quad \bar{p} = 0.216; \bar{q} = 0.784$

For Exercises 3 through 14, perform these steps.

 a. State the hypotheses and identify the claim.
 b. Find the critical value(s).
 c. Compute the test value.
 d. Make the decision.
 e. Summarize the results.

Use the traditional method of hypothesis testing unless otherwise specified.

3. Married People In a specific year 53.7% of men in the United States were married and 50.3% of women were married. Random samples of 300 men and 300 women found that 178 men and 139 women were married (not necessarily to each other.) At the 0.05 level of significance can it be concluded that the proportion of men who were married is greater than the proportion of women who were married?

Source: *New York Times Almanac.*

4. Undergraduate Financial Aid A study is conducted to determine if the percent of women who receive financial aid in undergraduate school is different from the percent of men who receive financial aid in undergraduate school. A random sample of undergraduates revealed these results. At $\alpha = 0.01$, is there significant evidence to reject the null hypothesis?

	Women	Men
Sample size	250	300
Number receiving aid	200	180

Source: U.S. Department of Education, National Center for Education Statistics.

5. High School Graduation Rates The overall U.S. public high school graduation rate is 73.4%. For Pennsylvania it is 83.5% and for Idaho 80.5%—a difference of 3%. Random samples of 1200 students from each state indicated that 980 graduated in Pennsylvania and 940 graduated in Idaho. At the 0.05 level of significance can it be concluded that there is a difference in the proportions of graduating students?

Source: *World Almanac.*

6. Animal Bites of Postal Workers In Cleveland, a sample of 73 mail carriers showed that 10 had been bitten by an animal during one week. In Philadelphia, in a sample of 80 mail carriers, 16 had received animal bites. Is there a significant difference in the proportions? Use $\alpha = 0.05$. Find the 95% confidence interval for the difference of the two proportions.

7. Lecture versus Computer-Assisted Instruction A survey found that 83% of the men questioned preferred computer-assisted instruction to lecture and 75% of the women preferred computer-assisted instruction to lecture. There were 100 individuals in each sample. At $\alpha = 0.05$, test the claim that there is no difference in the proportion of men and the proportion of women who favor computer-assisted instruction over lecture. Find the 95% confidence interval for the difference of the two proportions.

8. Leisure Time In a sample of 50 men, 44 said that they had less leisure time today than they had 10 years ago. In a sample of 50 women, 48 women said that they had less leisure time than they had 10 years ago. At $\alpha = 0.10$ is there a difference in the proportions? Find the 90% confidence interval for the difference of the two proportions. Does the confidence interval contain 0? Give a reason why this information would be of interest to a researcher.

Source: Based on statistics from Market Directory.

9. Desire to Be Rich In a sample of 80 Americans, 44 wished that they were rich. In a sample of 90 Europeans, 41 wished that they were rich. At $\alpha = 0.01$, is there a difference in the proportions? Find the 99% confidence interval for the difference of the two proportions.

10. Seat Belt Use In a sample of 200 men, 130 said they used seat belts. In a sample of 300 women, 63 said they used seat belts. Test the claim that men are more safety-conscious than women, at $\alpha = 0.01$. Use the P-value method.

11. Dog Ownership A survey found that in a sample of 75 families, 26 owned dogs. A survey done 15 years ago found that in a sample of 60 families, 26 owned dogs. At $\alpha = 0.05$ has the proportion of dog owners changed over the 15-year period? Find the 95% confidence interval of the true difference in the proportions. Does the confidence interval contain 0? Why would this fact be important to a researcher?

Source: Based on statistics from the American Veterinary Medical Association.

12. **Bullying** Bullying is a problem at any age but especially for students aged 12 to 18. A study showed that 7.2% of all students in this age bracket reported being bullied at school during the past six months with 6th grade having the highest incidence at 13.9% and 12th grade the lowest at 2.2%. To see if there is a difference between public and private schools, 200 students were randomly selected from each. At the 0.05 level of significance, can a difference be concluded?

	Private	Public
Sample size	200	200
No. bullied	13	16

Source: www.nces.ed.gov

13. **Survey on Inevitability of War** A sample of 200 teenagers shows that 50 believe that war is inevitable, and a sample of 300 people over age 60 shows that 93 believe war is inevitable. Is the proportion of teenagers who believe war is inevitable different from the proportion of people over age 60 who do? Use $\alpha = 0.01$. Find the 99% confidence interval for the difference of the two proportions.

14. **Hypertension** It has been found that 26% of men 20 years and older suffer from hypertension (high blood pressure) and 31.5% of women are hypertensive. A random sample of 150 of each gender was selected from recent hospital records, and the following results were obtained. Can you conclude that a higher percentage of women have high blood pressure? Use $\alpha = 0.05$.

Men   43 patients had high blood pressure
Women   52 patients had high blood pressure

Source: www.nchs.gov

15. **Partisan Support of Salary Increase Bill** Find the 99% confidence interval for the difference in the population proportions for the data of a study in which 80% of the 150 Republicans surveyed favored the bill for a salary increase and 60% of the

200 Democrats surveyed favored the bill for a salary increase. $0.077 < p_1 - p_2 < 0.323$

16. **Airlines On-Time Arrivals** The percentages of on-time arrivals for major U.S. airlines range from 68.6 to 91.1. Two regional airlines were surveyed with the following results. At $\alpha = 0.01$ is there a difference in proportions?

	Airline A	Airline B
No. of flights	300	250
No. of on-time flights	213	185

Source: *New York Times Almanac.*

17. **Senior Workers** It seems that people are choosing or finding it necessary to work later in life. Random samples of 200 men and 200 women age 65 or older were selected, and 80 men and 59 women were found to be working. At $\alpha = 0.01$, can it be concluded that the proportions are different?

Source: Based on www.census.gov

18. **Smoking Survey** National statistics show that 23% of men smoke and 18.5% of women do. A random sample of 180 men indicated that 50 were smokers, and of 150 women surveyed, 39 indicated that they smoked. Construct a 98% confidence interval for the true difference in proportions of male and female smokers. Comment on your interval—does it support the claim that there is a difference? $-0.0961 < p_1 - p_2 < 0.1319$

Source: www.nchs.gov

19. **College Education** The percentages of adults 25 years of age and older who have completed 4 or more years of college are 23.6% for females and 27.8% for males. A random sample of women and men who were 25 years old or older was surveyed with these results. Estimate the true difference in proportions with 95% confidence, and compare your interval with the *Almanac* statistics.

	Women	Men
Sample size	350	400
No. who completed 4 or more years	100	115

Source: *New York Times Almanac.*

Extending the Concepts

20. If there is a significant difference between p_1 and p_2 and between p_2 and p_3, can you conclude that there is a significant difference between p_1 and p_3?
No, p_1 could equal p_3.

Technology *Step by Step*

MINITAB
Step by Step

Test the Difference Between Two Proportions

For Example 9–9, test for a difference in the resident vaccination rates between small and large nursing homes.

1. This test does not require data. It doesn't matter what is in the worksheet.
2. Select **Stat>Basic Statistics>2 Proportions.**
3. Click the button for Summarized data.
4. Press **TAB** to move cursor to the first sample box for Trials.
   a) Enter **34, TAB,** then enter **12.**
   b) Press **TAB** or click in the second sample text box for Trials.
   c) Enter **24, TAB,** then enter **17.**
5. Click on [Options]. Check the box for Use pooled estimate of p for test. The Confidence level should be 95%, and the Test difference should be 0.
6. Click [OK] twice. The results are shown in the session window.

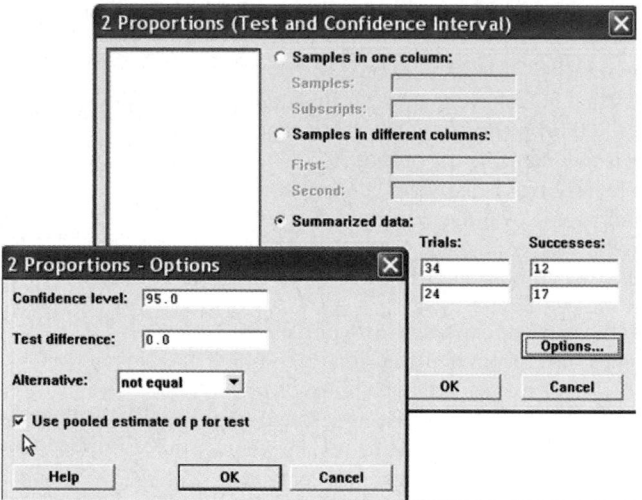

Test and CI for Two Proportions
```
Sample   X   N   Sample p
1       12  34   0.352941
2       17  24   0.708333

Difference = p (1) - p (2)
Estimate for difference:  -0.355392
95% CI for difference:  (-0.598025, -0.112759)
Test for difference = 0 (vs not = 0):  Z = -2.67 P-Value = 0.008
```

The *P*-value of the test is 0.008. Reject the null hypothesis. The difference is statistically significant. Of all small nursing homes 35%, compared to 71% of all large nursing homes, have an immunization rate of 80%. We can't tell why, only that there is a difference.

TI-83 Plus or TI-84 Plus
Step by Step

Hypothesis Test for the Difference Between Two Proportions

1. Press **STAT** and move the cursor to TESTS.
2. Press **6** for 2-PropZTEST.
3. Type in the appropriate values.
4. Move the cursor to the appropriate alternative hypothesis and press **ENTER.**
5. Move the cursor to Calculate and press **ENTER.**

Confidence Interval for the Difference Between Two Proportions

1. Press **STAT** and move the cursor to TESTS.
2. Press **B (ALPHA APPS)** for 2-PropZInt.
3. Type in the appropriate values.
4. Move the cursor to Calculate and press **ENTER.**

Excel
Step by Step

Testing the Difference Between Two Proportions

Excel does not have a procedure to test the difference between two population proportions. However, you may conduct this test using the MegaStat Add-in available on your CD. If you have not installed this add-in, do so, following the instructions from the Chapter 1 Excel Step by Step.

We will use the summary information from Example 9–9.

1. From the toolbar, select Add-Ins, **MegaStat>Hypothesis Tests>Compare Two Independent Proportions.** *Note:* You may need to open MegaStat from the MegaStat.xls file on your computer's hard drive.
2. Under Group 1, type **12** for *p* and **34** for *n*. Under Group 2, type **17** for *p* and **24** for *n*. MegaStat automatically changes *p* to *X* unless a decimal value less than 1 is typed in for these.
3. Type **0** for the Hypothesized difference and select the "not equal" Alternative, and click [OK].

Hypothesis Test for Two Independent Proportions

p_1	p_2		p_c	
0.3529	0.7083		0.5	p (as decimal)
12/34	17/24		29/58	p (as fraction)
12.	17.		29.	X
34	24		58	n

−0.3554	Difference
0.	Hypothesized difference
0.1333	Standard error
−2.67	z
0.0077	P-value (two-tailed)

9–5
Testing the Difference Between Two Variances

Objective 5
Test the difference between two variances or standard deviations.

In addition to comparing two means, statisticians are interested in comparing two variances or standard deviations. For example, is the variation in the temperatures for a certain month for two cities different?

In another situation, a researcher may be interested in comparing the variance of the cholesterol of men with the variance of the cholesterol of women. For the comparison of two variances or standard deviations, an *F* **test** is used. The *F* test should not be confused with the chi-square test, which compares a single sample variance to a specific population variance, as shown in Chapter 8.

If two independent samples are selected from two normally distributed populations in which the variances are equal ($\sigma_1^2 = \sigma_2^2$) and if the variances s_1^2 and s_2^2 are compared as $\frac{s_1^2}{s_2^2}$, the sampling distribution of the variances is called the *F* **distribution.**

Characteristics of the *F* Distribution

1. The values of *F* cannot be negative, because variances are always positive or zero.
2. The distribution is positively skewed.
3. The mean value of *F* is approximately equal to 1.
4. The *F* distribution is a family of curves based on the degrees of freedom of the variance of the numerator and the degrees of freedom of the variance of the denominator.

Figure 9–10 shows the shapes of several curves for the F distribution.

Figure 9–10

The *F* Family of Curves

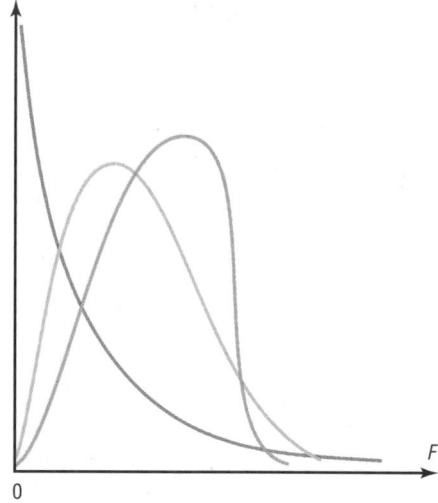

Formula for the *F* Test

$$F = \frac{s_1^2}{s_2^2}$$

where the larger of the two variances is placed in the numerator regardless of the subscripts. (See note on page 519.)

The F test has two terms for the degrees of freedom: that of the numerator, $n_1 - 1$, and that of the denominator, $n_2 - 1$, where n_1 is the sample size from which the larger variance was obtained.

When you are finding the F test value, *the larger of the variances is placed in the numerator of the F formula;* this is not necessarily the variance of the larger of the two sample sizes.

Table H in Appendix C gives the F critical values for $\alpha = 0.005, 0.01, 0.025, 0.05$, and 0.10 (each α value involves a separate table in Table H). These are one-tailed values; if a two-tailed test is being conducted, then the $\alpha/2$ value must be used. For example, if a two-tailed test with $\alpha = 0.05$ is being conducted, then the $0.05/2 = 0.025$ table of Table H should be used.

Example 9–12

Find the critical value for a right-tailed F test when $\alpha = 0.05$, the degrees of freedom for the numerator (abbreviated d.f.N.) are 15, and the degrees of freedom for the denominator (d.f.D.) are 21.

Solution

Since this test is right-tailed with $\alpha = 0.05$, use the 0.05 table. The d.f.N. is listed across the top, and the d.f.D. is listed in the left column. The critical value is found where the row and column intersect in the table. In this case, it is 2.18. See Figure 9–11.

Figure 9–11

Finding the Critical Value in Table H for Example 9–12

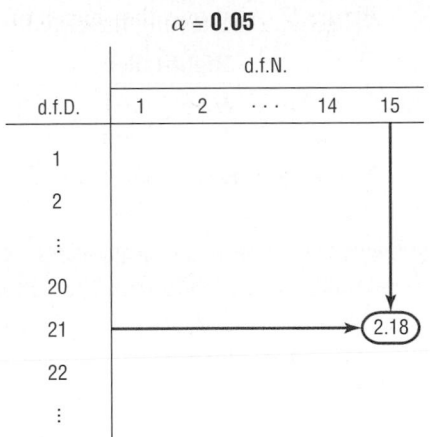

As noted previously, when the F test is used, the larger variance is always placed in the numerator of the formula. When you are conducting a two-tailed test, α is split; and even though there are two values, only the right tail is used. The reason is that the F test value is always greater than or equal to 1.

Example 9–13

Find the critical value for a two-tailed F test with $\alpha = 0.05$ when the sample size from which the variance for the numerator was obtained was 21 and the sample size from which the variance for the denominator was obtained was 12.

Solution

Since this is a two-tailed test with $\alpha = 0.05$, the $0.05/2 = 0.025$ table must be used. Here, d.f.N. $= 21 - 1 = 20$, and d.f.D. $= 12 - 1 = 11$; hence, the critical value is 3.23. See Figure 9–12.

Figure 9–12

Finding the Critical Value in Table H for Example 9–13

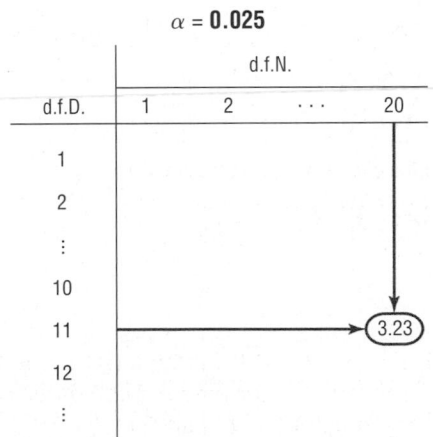

When the degree of freedom values cannot be found in the table, the closest value on the smaller side should be used. For example, if d.f.N. $= 14$, this value is between the given table values of 12 and 15; therefore, 12 should be used, to be on the safe side.

When you are testing the equality of two variances, these hypotheses are used:

Right-tailed	**Left-tailed**	**Two-tailed**
$H_0: \sigma_1^2 = \sigma_2^2$	$H_0: \sigma_1^2 = \sigma_2^2$	$H_0: \sigma_1^2 = \sigma_2^2$
$H_1: \sigma_1^2 > \sigma_2^2$	$H_1: \sigma_1^2 < \sigma_2^2$	$H_1: \sigma_1^2 \neq \sigma_2^2$

There are four key points to keep in mind when you are using the F test.

Notes for the Use of the F Test

1. The larger variance should always be placed in the numerator of the formula regardless of the subscripts. (See note on page 519.)

$$F = \frac{s_1^2}{s_2^2}$$

2. For a two-tailed test, the α value must be divided by 2 and the critical value placed on the right side of the F curve.

3. If the standard deviations instead of the variances are given in the problem, they must be squared for the formula for the F test.

4. When the degrees of freedom cannot be found in Table H, the closest value on the smaller side should be used.

Unusual Stat

Of all U.S. births, 2% are twins.

Assumptions for Testing the Difference Between Two Variances

1. The samples must be random samples.
2. The populations from which the samples were obtained must be normally distributed. (*Note:* The test should not be used when the distributions depart from normality.)
3. The samples must be independent of one another.

Remember also that in tests of hypotheses using the traditional method, these five steps should be taken:

Step 1   State the hypotheses and identify the claim.

Step 2   Find the critical value.

Step 3   Compute the test value.

Step 4   Make the decision.

Step 5   Summarize the results.

Example 9–14   **Heart Rates of Smokers**

A medical researcher wishes to see whether the variance of the heart rates (in beats per minute) of smokers is different from the variance of heart rates of people who do not smoke. Two samples are selected, and the data are as shown. Using $\alpha = 0.05$, is there enough evidence to support the claim?

Smokers	**Nonsmokers**
$n_1 = 26$	$n_2 = 18$
$s_1^2 = 36$	$s_2^2 = 10$

Solution

Step 1   State the hypotheses and identify the claim.

$$H_0: \sigma_1^2 = \sigma_2^2 \qquad \text{and} \qquad H_1: \sigma_1^2 \neq \sigma_2^2 \text{ (claim)}$$

Step 2   Find the critical value. Use the 0.025 table in Table H since $\alpha = 0.05$ and this is a two-tailed test. Here, d.f.N. $= 26 - 1 = 25$, and d.f.D. $= 18 - 1 = 17$. The critical value is 2.56 (d.f.N. $= 24$ was used). See Figure 9–13.

Figure 9–13

Critical Value for Example 9–14

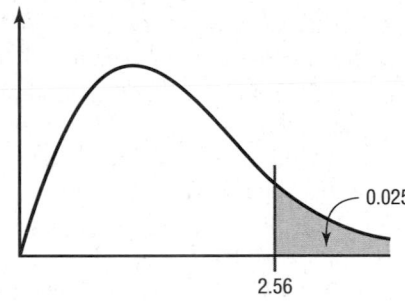

0.025

2.56

Step 3   Compute the test value.

$$F = \frac{s_1^2}{s_2^2} = \frac{36}{10} = 3.6$$

Step 4   Make the decision. Reject the null hypothesis, since $3.6 > 2.56$.

Step 5   Summarize the results. There is enough evidence to support the claim that the variance of the heart rates of smokers and nonsmokers is different.

Example 9–15

Waiting Time to See a Doctor

The standard deviation of the average waiting time to see a doctor for non-life-threatening problems in the emergency room at an urban hospital is 32 minutes. At a second hospital, the standard deviation is 28 minutes. If a sample of 16 patients was used in the first case and 18 in the second case, is there enough evidence to conclude at the 0.01 significance level that the standard deviation of the waiting times in the first hospital is greater than the standard deviation of the waiting times in the second hospital?

Solution

Step 1   State the hypotheses and identify the claim.

$$H_0: \sigma_1^2 = \sigma_2^2 \qquad \text{and} \qquad H_1: \sigma_1^2 > \sigma_2^2 \text{ (claim)}$$

Step 2   Find the critical value. Here, d.f.N. $= 16 - 1 = 15$, and d.f.D. $= 18 - 1 = 17$. From the 0.01 table, the critical value is 3.31.

Step 3   Compute the test value.

$$F = \frac{s_1^2}{s_2^2} = \frac{32^2}{28^2} = 1.31$$

Step 4   Do not reject the null hypothesis since $1.31 < 3.31$.

Step 5   Summarize the results. There is not enough evidence to support the claim that the standard deviation of the waiting times of the first hospital is greater than the standard deviation of the waiting times of the second hospital.

Finding P-values for the F test statistic is somewhat more complicated since it requires looking through all the F tables (Table H in Appendix C) using the specific d.f.N. and d.f.D. values. For example, suppose that a certain test has $F = 3.58$, d.f.N. $= 5$, and d.f.D. $= 10$. To find the P-value interval for $F = 3.58$, you must first find the corresponding F values for d.f.N. $= 5$ and d.f.D. $= 10$ for α equal to 0.005, 0.01, 0.025, 0.05, and 0.10 in Table H. Then make a table as shown.

α	0.10	0.05	0.025	0.01	0.005
F	2.52	3.33	4.24	5.64	6.87

Now locate the two F values that the test value 3.58 falls between. In this case, 3.58 falls between 3.33 and 4.24, corresponding to 0.05 and 0.025. Hence, the P-value for a right-tailed test for $F = 3.58$ falls between 0.025 and 0.05 (that is, $0.025 < P\text{-value} < 0.05$). For a right-tailed test, then, you would reject the null hypothesis at $\alpha = 0.05$ but not at $\alpha = 0.01$. The P-value obtained from a calculator is 0.0408. Remember that for a two-tailed test the values found in Table H for α must be doubled. In this case, $0.05 < P\text{-value} < 0.10$ for $F = 3.58$.

Once you understand the concept, you can dispense with making a table as shown and find the P-value directly from Table H.

Example 9–16   **Airport Passengers**

The CEO of an airport hypothesizes that the variance in the number of passengers for American airports is greater than the variance in the number of passengers for foreign airports. At $\alpha = 0.10$, is there enough evidence to support the hypothesis? The data in millions of passengers per year are shown for selected airports. Use the P-value method. Assume the variable is normally distributed.

American airports		Foreign airports	
36.8	73.5	60.7	51.2
72.4	61.2	42.7	38.6
60.5	40.1		

Source: Airports Council International.

Solution

Step 1   State the hypotheses and identify the claim.

$$H_0: \sigma_1^2 = \sigma_2^2 \quad \text{and} \quad H_1: \sigma_1^2 > \sigma_2^2 \text{ (claim)}$$

Step 2   Compute the test value. Using the formula in Chapter 3 or a calculator, find the variance for each group.

$$s_1^2 = 246.38 \quad \text{and} \quad s_2^2 = 95.87$$

Substitute in the formula and solve.

$$F = \frac{s_1^2}{s_2^2} = \frac{246.38}{95.87} = 2.57$$

Step 3   Find the P-value in Table H, using d.f.N. $= 5$ and d.f.D. $= 3$.

α	0.10	0.05	0.025	0.01	0.005
F	5.31	9.01	14.88	28.24	45.39

Since 2.57 is less than 5.31, the P-value is greater than 0.10. (The P-value obtained from a calculator is 0.234.)

Step 4   Make the decision. The decision is to not reject the null hypothesis since P-value > 0.10.

Step 5   Summarize the results. There is not enough evidence to support the claim that the variance in the number of passengers for American airports is greater than the variance in the number of passengers for foreign airports.

If the exact degrees of freedom are not specified in Table H, the closest smaller value should be used. For example, if $\alpha = 0.05$ (right-tailed test), d.f.N. = 18, and d.f.D. = 20, use the column d.f.N. = 15 and the row d.f.D. = 20 to get $F = 2.20$.

Note: It is not absolutely necessary to place the larger variance in the numerator when you are performing the F test. Critical values for left-tailed hypotheses tests can be found by interchanging the degrees of freedom and taking the reciprocal of the value found in Table H.

Also, you should use caution when performing the F test since the data can run contrary to the hypotheses on rare occasions. For example, if the hypotheses are $H_0: \sigma_1^2 \leq \sigma_2^2$ (written $H_0: \sigma_1^2 = \sigma_2^2$) and $H_1: \sigma_1^2 > \sigma_2^2$, but if $s_1^2 < s_2^2$, then the F test should not be performed and you would not reject the null hypothesis.

Applying the Concepts **9–5**

Variability and Automatic Transmissions

Assume the following data values are from the June 1996 issue of *Automotive Magazine*. An article compared various parameters of U.S.- and Japanese-made sports cars. This report centers on the price of an optional automatic transmission. Which country has the greater variability in the price of automatic transmissions? Input the data and answer the following questions.

Japanese cars		**U.S. cars**	
Nissan 300ZX	$1940	Dodge Stealth	$2363
Mazda RX7	1810	Saturn	1230
Mazda MX6	1871	Mercury Cougar	1332
Nissan NX	1822	Ford Probe	932
Mazda Miata	1920	Eagle Talon	1790
Honda Prelude	1730	Chevy Lumina	1833

1. What is the null hypothesis?
2. What test statistic is used to test for any significant differences in the variances?
3. Is there a significant difference in the variability in the prices between the Japanese cars and the U.S. cars?
4. What effect does a small sample size have on the standard deviations?
5. What degrees of freedom are used for the statistical test?
6. Could two sets of data have significantly different variances without having significantly different means?

See page 531 for the answers.

Exercises 9–5

1. When one is computing the F test value, what condition is placed on the variance that is in the numerator?
The variance in the numerator should be the larger of the two variances.

2. Why is the critical region always on the right side in the use of the F test? The larger variance is placed in the numerator of the formula; hence, $F \geq 1$.

3. What are the two different degrees of freedom associated with the F distribution?

4. What are the characteristics of the F distribution?

5. Using Table H, find the critical value for each.

   a. Sample 1: $s_1^2 = 128$, $n_1 = 23$
      Sample 2: $s_2^2 = 162$, $n_2 = 16$
      Two-tailed, $\alpha = 0.01$

   b. Sample 1: $s_1^2 = 37$, $n_1 = 14$
      Sample 2: $s_2^2 = 89$, $n_2 = 25$
      Right-tailed, $\alpha = 0.01$

   c. Sample 1: $s_1^2 = 232$, $n_1 = 30$
      Sample 2: $s_2^2 = 387$, $n_2 = 46$
      Two-tailed, $\alpha = 0.05$

   d. Sample 1: $s_1^2 = 164$, $n_1 = 21$
      Sample 2: $s_2^2 = 53$, $n_2 = 17$
      Two-tailed, $\alpha = 0.10$

   e. Sample 1: $s_1^2 = 92.8$, $n_1 = 11$
      Sample 2: $s_2^2 = 43.6$, $n_2 = 11$
      Right-tailed, $\alpha = 0.05$

6. **(ans)** Using Table H, find the P-value interval for each F test value.

   a. $F = 2.97$, d.f.N. = 9, d.f.D. = 14, right-tailed
   b. $F = 3.32$, d.f.N. = 6, d.f.D. = 12, two-tailed
   c. $F = 2.28$, d.f.N. = 12, d.f.D. = 20, right-tailed
   d. $F = 3.51$, d.f.N. = 12, d.f.D. = 21, right-tailed
   e. $F = 4.07$, d.f.N. = 6, d.f.D. = 10, two-tailed
   f. $F = 1.65$, d.f.N. = 19, d.f.D. = 28, right-tailed
   g. $F = 1.77$, d.f.N. = 28, d.f.D. = 28, right-tailed
   h. $F = 7.29$, d.f.N. = 5, d.f.D. = 8, two-tailed

For Exercises 7 through 20, perform the following steps. Assume that all variables are normally distributed.

   a. State the hypotheses and identify the claim.
   b. Find the critical value.
   c. Compute the test value.
   d. Make the decision.
   e. Summarize the results.

Use the traditional method of hypothesis testing unless otherwise specified.

7. **Ages of Hospital Patients** The average age of hospital inpatients has gradually increased to 52.5 years. Studies of two major health care systems found the following information. At the 0.05 level of significance is there sufficient evidence to conclude a difference between the two variances?

	System 1	System 2
Sample size	60	60
Sample mean	49.8	50.2
Sample standard deviation	5.4	7.6

Source: *New York Times Almanac.*

8. **Museum Attendance** A metropolitan children's museum open year-round wants to see if the variance in daily attendance differs between the summer and winter months. Random samples of 30 days each were selected and showed that in the winter months, the sample mean daily attendance was 300 with a standard deviation of 52, and the sample mean daily attendance for the summer months was 280 with a standard deviation of 65. At $\alpha = 0.05$ can we conclude a difference in variances?

9. **Wolf Pack Pups** Does the variance in average number of pups per pack differ between Montana and Idaho wolf packs? Random samples of packs were selected for each area, and the numbers of pups per pack were recorded. At the 0.05 level of significance, can a difference in variances be concluded?

Montana wolf packs	4	3	5	6	1	2	8	2
	3	1	7	6	5			
Idaho wolf packs	2	4	5	4	2	4	6	3
	1	4	2	1				

Source: www.fws.gov

10. **Noise Levels in Hospitals** In a hospital study, it was found that the standard deviation of the sound levels from 20 areas designated as "casualty doors" was 4.1 dBA and the standard deviation of 24 areas designated as operating theaters was 7.5 dBA. At $\alpha = 0.05$, can you substantiate the claim that there is a difference in the standard deviations?

Source: M. Bayo, A. Garcia, and A. Garcia, "Noise Levels in an Urban Hospital and Workers' Subjective Responses," *Archives of Environmental Health.*

11. **Calories in Ice Cream** The numbers of calories contained in $\frac{1}{2}$-cup servings of randomly selected flavors of ice cream from two national brands are listed here. At the 0.05 level of significance, is there sufficient evidence to conclude that the variance in the number of calories differs between the two brands?

Brand A		Brand B	
330	300	280	310
310	350	300	370
270	380	250	300
310	300	290	310

Source: *The Doctor's Pocket Calorie, Fat and Carbohydrate Counter.*

12. **Winter Temperatures** A random sample of daily high temperatures in January and February is listed below. At $\alpha = 0.05$ can it be concluded that there is a difference in variances in high temperature between the two months?

Jan.	31	31	38	24	24	42	22	43	35	42
Feb.	31	29	24	30	28	24	27	34	27	

13. **Population and Area** Cities were randomly selected from the list of the 50 largest cities in the United States (based on population). The areas of each

in square miles are indicated below. Is there sufficient evidence to conclude that the variance in area is greater for eastern cities than for western cities at $\alpha = 0.05$? At $\alpha = 0.01$?

Eastern		Western	
Atlanta, GA	132	Albuquerque, NM	181
Columbus, OH	210	Denver, CO	155
Louisville, KY	385	Fresno, CA	104
New York, NY	303	Las Vegas, NV	113
Philadelphia, PA	135	Portland, OR	134
Washington, DC	61	Seattle, WA	84
Charlotte, NC	242		

Source: *New York Times Almanac.*

14. Carbohydrates in Candy The number of grams of carbohydrates contained in 1-ounce servings of randomly selected chocolate and nonchocolate candy is listed here. Is there sufficient evidence to conclude that there is a difference between the variation in carbohydrate content for chocolate and nonchocolate candy? Use $\alpha = 0.10$.

Chocolate	29	25	17	36	41	25	32	29
	38	34	24	27	29			
Nonchocolate	41	41	37	29	30	38	39	10
	29	55	29					

Source: *The Doctor's Pocket Calorie, Fat and Carbohydrate Counter.*

15. Tuition Costs for Medical School The yearly tuition costs in dollars for random samples of medical schools that specialize in research and in primary care are listed. At $\alpha = 0.05$, can it be concluded that a difference between the variances of the two groups exists?

Research			Primary care		
30,897	34,280	31,943	26,068	21,044	30,897
34,294	31,275	29,590	34,208	20,877	29,691
20,618	20,500	29,310	33,783	33,065	35,000
21,274			27,297		

Source: *U.S. News & World Report Best Graduate Schools.*

16. County Size in Indiana and Iowa A researcher wishes to see if the variance of the areas in square miles for counties in Indiana is less than the variance of the areas for counties in Iowa. A random sample of counties is selected, and the data are shown. At $\alpha = 0.01$, can it be concluded that the variance of the areas for counties in Indiana is less than the variance of the areas for counties in Iowa?

Indiana				Iowa			
406	393	396	485	640	580	431	416
431	430	369	408	443	569	779	381
305	215	489	293	717	568	714	731
373	148	306	509	571	577	503	501
560	384	320	407	568	434	615	402

Source: *The World Almanac and Book of Facts.*

17. Heights of Tall Buildings Test the claim that the variance of heights of tall buildings in Denver is equal to the variance in heights of tall buildings in Detroit at $\alpha = 0.10$. The data are given in feet.

Denver			Detroit		
714	698	544	620	472	430
504	438	408	562	448	420
404			534	436	

Source: *The World Almanac and Book of Facts.*

18. Elementary School Teachers' Salaries A researcher claims that the variation in the salaries of elementary school teachers is greater than the variation in the salaries of secondary school teachers. A sample of the salaries of 30 elementary school teachers has a variance of $8324, and a sample of the salaries of 30 secondary school teachers has a variance of $2862. At $\alpha = 0.05$, can the researcher conclude that the variation in the elementary school teachers' salaries is greater than the variation in the secondary school teachers' salaries? Use the *P*-value method.

19. Weights of Running Shoes The weights in ounces of a sample of running shoes for men and women are shown. Calculate the variances for each sample, and test the claim that the variances are equal at $\alpha = 0.05$. Use the *P*-value method.

Men			Women		
11.9	10.4	12.6	10.6	10.2	8.8
12.3	11.1	14.7	9.6	9.5	9.5
9.2	10.8	12.9	10.1	11.2	9.3
11.2	11.7	13.3	9.4	10.3	9.5
13.8	12.8	14.5	9.8	10.3	11.0

20. Daily Stock Prices Two portfolios were randomly assembled from the New York Stock Exchange, and the daily stock prices are shown below. At the 0.05 level of significance, can it be concluded that a difference in variance in price exists between the two portfolios?

Portfolio A	36.44	44.21	12.21	59.60	55.44	39.42	51.29	48.68	41.59	19.49
Portfolio B	32.69	47.25	49.35	36.17	63.04	17.74	4.23	34.98	37.02	31.48

Source: *Washington Observer-Reporter.*

Technology *Step by Step*

MINITAB
Step by Step

Test for the Difference Between Two Variances

For Example 9–16, test the hypothesis that the variance in the number of passengers for American and foreign airports is different. Use the *P*-value approach.

American airports	Foreign airports
36.8	60.7
72.4	42.7
60.5	51.2
73.5	38.6
61.2	
40.1	

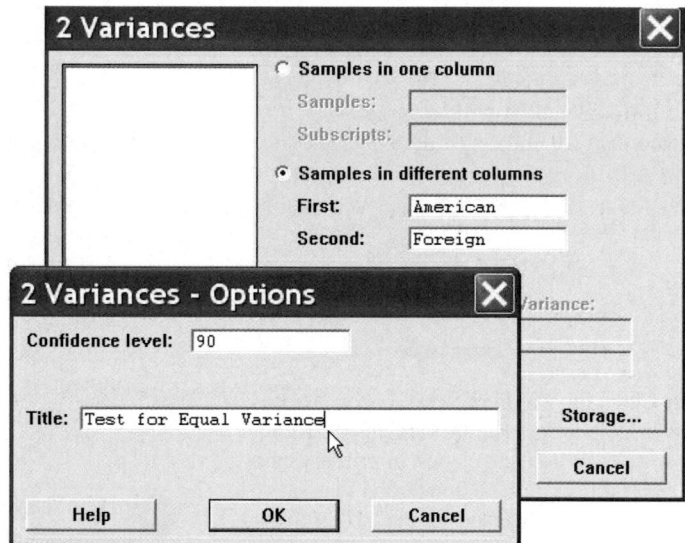

1. Enter the data into two columns of MINITAB.
2. Name the columns American and Foreign.
   a) Select **Stat>Basic Statistics>2-Variances.**
   b) Click the button for Samples in different columns.
   c) Click in the text box for First, then double-click C1 American.
   d) Double-click C2 Foreign, then click on [Options]. The dialog box is shown. Change the confidence level to **90** and type an appropriate title. In this dialog, we cannot specify a left- or right-tailed test.
3. Click [OK] twice. A graph window will open that includes a small window that says $F = 2.57$ and the *P*-value is 0.437. Divide this two-tailed *P*-value by 2 for a one-tailed test.

There is not enough evidence in the sample to conclude there is greater variance in the number of passengers in American airports compared to foreign airports.

TI-83 Plus or TI-84 Plus
Step by Step

Hypothesis Test for the Difference Between Two Variances (Data)

1. Enter the data values into L_1 and L_2.
2. Press **STAT** and move the cursor to TESTS.

3. Press **D (ALPHA X^{-1})** for 2-SampFTest.   (The TI-84 uses E)

4. Move the cursor to Data and press **ENTER**.

5. Type in the appropriate values.

6. Move the cursor to the appropriate alternative hypothesis and press **ENTER**.

7. Move the cursor to Calculate and press **ENTER**.

Hypothesis Test for the Difference Between Two Variances (Statistics)

1. Press **STAT** and move the cursor to TESTS.

2. Press **D (ALPHA X^{-1})** for 2-SampFTest.   (The TI-84 uses E)

3. Move the cursor to Stats and press **ENTER**.

4. Type in the appropriate values.

5. Move the cursor to the appropriate alternative hypothesis and press **ENTER**.

6. Move the cursor to Calculate and press **ENTER**.

Excel
Step by Step

F Test for the Difference Between Two Variances

Excel has a two-sample *F* test included in the Data Analysis Add-in. To perform an *F* test for the difference between the variances of two populations, given two independent samples, do this:

1. Enter the first sample data set into column A.

2. Enter the second sample data set into column B.

3. Select the Data tab from the toolbar. Then select Data Analysis.

4. In the Analysis Tools box, select F-test: Two-sample for Variances.

5. Type the ranges for the data in columns A and B.

6. Specify the confidence level Alpha.

7. Specify a location for the output, and click [OK].

Example XL9–4

 At $\alpha = 0.05$, test the hypothesis that the two population variances are equal, using the sample data provided here.

Set A	63	73	80	60	86	83	70	72	82
Set B	86	93	64	82	81	75	88	63	63

The results appear in the table that Excel generates, shown here. For this example, the output shows that the null hypothesis cannot be rejected at an α level of 0.05.

F-Test Two-Sample for Variances		
	Variable 1	*Variable 2*
Mean	74.33333333	77.22222222
Variance	82.75	132.9444444
Observations	9	9
df	8	8
F	0.622440451	
P(F<=f) one-tail	0.258814151	
F Critical one-tail	0.290858004	

Summary

Many times researchers are interested in comparing two parameters such as two means, two proportions, or two variances. These measures are obtained from two samples, then compared using a z test, t test, or an F test.

- If two sample means are compared, when the samples are independent and the population standard deviations are known, a z test is used. If the sample sizes are less than 30, the populations should be normally distributed. (9–1)
- If two means are compared when the samples are independent and the sample standard deviations are used, then a t test is used. Both variances are assumed to be unequal. (9–2)
- When the two samples are dependent or related, such as using the same subjects and comparing the means of before and after tests, then the t test for dependent samples is used. (9–3)
- Two proportions can be compared by using the z test for proportions. In this case, each of n_1p_1, n_1q_1, n_2p_2, and n_2q_2 must all be 5 or more. (9–4)
- Two variances can be compared by using an F test. The critical values for the F test are obtained from the F distribution. (9–5)
- Confidence intervals for differences between two parameters can also be found.

Important Terms

dependent samples 492

F distribution 513

F test 513

independent samples 484

pooled estimate of the variance 487

Important Formulas

Formula for the z test for comparing two means from independent populations; σ_1 and σ_2 are known:

$$z = \frac{(\bar{X}_1 - \bar{X}_2) - (\mu_1 - \mu_2)}{\sqrt{\frac{\sigma_1^2}{n_1} + \frac{\sigma_2^2}{n_2}}}$$

Formula for the confidence interval for difference of two means when σ_1 and σ_2 are known:

$$(\bar{X}_1 - \bar{X}_2) - z_{\alpha/2}\sqrt{\frac{\sigma_1^2}{n_1} + \frac{\sigma_2^2}{n_2}} < \mu_1 - \mu_2$$
$$< (\bar{X}_1 - \bar{X}_2) + z_{\alpha/2}\sqrt{\frac{\sigma_1^2}{n_1} + \frac{\sigma_2^2}{n_2}}$$

Formula for the t test for comparing two means (independent samples, variances not equal), σ_1 and σ_2 unknown:

$$t = \frac{(\bar{X}_1 - \bar{X}_2) - (\mu_1 - \mu_2)}{\sqrt{\frac{s_1^2}{n_1} + \frac{s_2^2}{n_2}}}$$

and d.f. = the smaller of $n_1 - 1$ or $n_2 - 1$.

Formula for the confidence interval for the difference of two means (independent samples, variances unequal), σ_1 and σ_2 unknown:

$$(\bar{X}_1 - \bar{X}_2) - t_{\alpha/2}\sqrt{\frac{s_1^2}{n_1} + \frac{s_2^2}{n_2}} < \mu_1 - \mu_2$$
$$< (\bar{X}_1 - \bar{X}_2) + t_{\alpha/2}\sqrt{\frac{s_1^2}{n_1} + \frac{s_2^2}{n_2}}$$

and d.f. = smaller of $n_1 - 1$ and $n_2 - 2$.

Formula for the t test for comparing two means from dependent samples:

$$t = \frac{\bar{D} - \mu_D}{s_D/\sqrt{n}}$$

where $\bar{D}$ is the mean of the differences

$$\bar{D} = \frac{\Sigma D}{n}$$

and s_D is the standard deviation of the differences

$$s_D = \sqrt{\frac{n\Sigma D^2 - (\Sigma D)^2}{n(n-1)}}$$

Formula for confidence interval for the mean of the difference for dependent samples:

$$\bar{D} - t_{\alpha/2}\frac{s_D}{\sqrt{n}} < \mu_D < \bar{D} + t_{\alpha/2}\frac{s_D}{\sqrt{n}}$$

and d.f. = $n - 1$.

Formula for the z test for comparing two proportions:

$$z = \frac{(\hat{p}_1 - \hat{p}_2) - (p_1 - p_2)}{\sqrt{\bar{p}\,\bar{q}\left(\dfrac{1}{n_1} + \dfrac{1}{n_2}\right)}}$$

where

$$\bar{p} = \frac{X_1 + X_2}{n_1 + n_2} \qquad \hat{p}_1 = \frac{X_1}{n_1}$$

$$\bar{q} = 1 - \bar{p} \qquad \hat{p}_2 = \frac{X_2}{n_2}$$

Formula for confidence interval for the difference of two proportions:

$$(\hat{p}_1 - \hat{p}_2) - z_{\alpha/2}\sqrt{\frac{\hat{p}_1\hat{q}_1}{n_1} + \frac{\hat{p}_2\hat{q}_2}{n_2}} < p_1 - p_2$$

$$< (\hat{p}_1 - \hat{p}_2) + z_{\alpha/2}\sqrt{\frac{\hat{p}_1\hat{q}_1}{n_1} + \frac{\hat{p}_2\hat{q}_2}{n_2}}$$

Formula for the F test for comparing two variances:

$$F = \frac{s_1^2}{s_2^2} \qquad \begin{aligned} \text{d.f.N.} &= n_1 - 1 \\ \text{d.f.D.} &= n_2 - 1 \end{aligned}$$

The larger variance is placed in the numerator.

Review Exercises

For each exercise, perform these steps. Assume that all variables are normally or approximately normally distributed.

   a. State the hypotheses and identify the claim.
   b. Find the critical value(s).
   c. Compute the test value.
   d. Make the decision.
   e. Summarize the results.

Use the traditional method of hypothesis testing unless otherwise specified.

1. Driving for Pleasure Two groups of drivers are surveyed to see how many miles per week they drive for pleasure trips. The data are shown. At $\alpha = 0.01$, can it be concluded that single drivers do more driving for pleasure trips on average than married drivers? Assume $\sigma_1 = 16.7$ and $\sigma_2 = 16.1$. (9–1)

Single drivers					Married drivers				
106	110	115	121	132	97	104	138	102	115
119	97	118	122	135	133	120	119	136	96
110	117	116	138	142	139	108	117	145	114
115	114	103	98	99	140	136	113	113	150
108	117	152	147	117	101	114	116	113	135
154	86	115	116	104	115	109	147	106	88
107	133	138	142	140	113	119	99	108	105

2. Average Earnings of College Graduates The average yearly earnings of male college graduates (with at least a bachelor's degree) are $58,500 for men aged 25 to 34. The average yearly earnings of female college graduates with the same qualifications are $49,339. Based on the results below, can it be concluded that there is a difference in mean earnings between male and female college graduates? Use the 0.01 level of significance. (9–1)

	Male	Female
Sample mean	$59,235	$52,487
Population standard deviation	8,945	10,125
Sample size	40	35

Source: *New York Times Almanac.*

3. Communication Times According to the Bureau of Labor Statistics' American Time Use Survey (ATUS), married persons spend an average of 8 minutes per day on phone calls, mail, and e-mail, while single persons spend an average of 14 minutes per day on these same tasks. Based on the following information, is there sufficient evidence to conclude that single persons spend, on average, a greater time each day communicating? Use the 0.05 level of significance. (9–2)

	Single	Married
Sample size	26	20
Sample mean	16.7 minutes	12.5 minutes
Sample variance	8.41	10.24

Source: *Time* magazine.

4. Average Temperatures The average temperatures for a 25-day period for Birmingham, Alabama, and Chicago, Illinois, are shown. Based on the samples, at $\alpha = 0.10$, can it be concluded that it is warmer in Birmingham? (9–2)

Birmingham					Chicago				
78	82	68	67	68	70	74	73	60	77
75	73	75	64	68	71	72	71	74	76
62	73	77	78	79	71	80	65	70	83
74	72	73	78	68	67	76	75	62	65
73	79	82	71	66	66	65	77	66	64

5. Teachers' Salaries A sample of 15 teachers from Rhode Island has an average salary of $35,270, with a standard deviation of $3256. A sample of 30 teachers from New York has an average salary of $29,512, with a standard deviation of $1432. Is there a significant difference in teachers' salaries between the two states? Use $\alpha = 0.02$. Find the 98% confidence interval for the difference of the two means. (9–2)

6. Soft Drinks in School The data show the amounts (in thousands of dollars) of the contracts for soft drinks in local school districts. At $\alpha = 0.10$ can it be concluded that there is a difference in the averages? Use the *P*-value method. Give a reason why the result would be of concern to a cafeteria manager. (9–2)

Pepsi						Coca-Cola		
46	120	80	500	100	59	420	285	57

Source: Local school districts.

7. High and Low Temperatures March is a month of variable weather in the Northeast. The chart below records the actual high and low temperatures for a selection of days in March from the weather report for Pittsburgh, Pennsylvania. At the 0.01 level of significance, is there sufficient evidence to conclude that there is more than a 10° difference between average highs and lows? (9–3)

Maximum	44	46	46	36	34	36	57	62	73	53
Minimum	27	34	24	19	19	26	33	57	46	26

Source: www.wunderground.com

8. Automobile Part Production In an effort to increase production of an automobile part, the factory manager decides to play music in the manufacturing area. Eight workers are selected, and the number of items each produced for a specific day is recorded. After one week of music, the same workers are monitored again. The data are given in the table. At $\alpha = 0.05$, can the manager conclude that the music has increased production? (9–3)

Worker	1	2	3	4	5	6	7	8
Before	6	8	10	9	5	12	9	7
After	10	12	9	12	8	13	8	10

9. Lay Teachers in Religious Schools A study found a slightly lower percentage of lay teachers in religious secondary schools than in elementary schools. A random sample of 200 elementary school and 200 secondary school teachers from religious schools in a large diocese found the following. At the 0.05 level of significance is there sufficient evidence to conclude a difference in proportions? (9–4)

	Elementary	Secondary
Sample size	200	200
Lay teachers	49	62

Source: *New York Times Almanac.*

10. Adopted Pets According to the 2005–2006 National Pet Owners Survey, only 16% of pet dogs were adopted from an animal shelter and 15% of pet cats were adopted. To test this difference in proportions of adopted pets, a survey was taken in a local region. Is there sufficient evidence to conclude that there is a difference in proportions? Use $\alpha = 0.05$. (9–4)

	Dogs	Cats
Number	180	200
Adopted	36	30

Source: www.hsus.org

11. Noise Levels in Hospitals In the hospital study cited previously, the standard deviation of the noise levels of the 11 intensive care units was 4.1 dBA, and the standard deviation of the noise levels of 24 nonmedical care areas, such as kitchens and machine rooms, was 13.2 dBA. At $\alpha = 0.10$, is there a significant difference between the standard deviations of these two areas? (9–5)

Source: M. Bayo, A. Garcia, and A. Garcia, "Noise Levels in an Urban Hospital and Workers' Subjective Responses," *Archives of Environmental Health.*

12. Heights of World Famous Cathedrals The heights (in feet) for a random sample of world famous cathedrals are listed below. In addition, the heights for a sample of the tallest buildings in the world are listed. Is there sufficient evidence at $\alpha = 0.05$ to conclude that there is a difference in the variances in height between the two groups? (9–5)

Cathedrals	72	114	157	56	83	108	90	151	
Tallest buildings	452	442	415	391	355	344	310	302	209

Source: www.infoplease.com

13. Paint Prices Two large home improvement stores advertise that they sell their paint at the same average price per gallon. A random sample of 25 cans from store Y had a standard deviation of $5.21, and store Z had a standard deviation of $4.08 based on a sample of 20 cans. At $\alpha = 0.05$ can we conclude that the variances are different? How much less would store Z's standard deviation have to be in order to conclude a difference? (9–5)

Statistics Today

To Vaccinate or Not to Vaccinate? Small or Large?—Revisited

Using a *z* test to compare two proportions, the researchers found that the proportion of residents in smaller nursing homes who were vaccinated (80.8%) was statistically greater than that of residents in large nursing homes who were vaccinated (68.7%). Using statistical methods presented in later chapters, they also found that the larger size of the nursing home and the lower frequency of vaccination were significant predictions of influenza outbreaks in nursing homes.

Data Analysis

The Data Bank is found in Appendix D, or on the World Wide Web by following links from
www.mhhe.com/math/stat/bluman/

1. From the Data Bank, select a variable and compare the mean of the variable for a random sample of at least 30 men with the mean of the variable for the random sample of at least 30 women. Use a z test.

2. Repeat the experiment in Exercise 1, using a different variable and two samples of size 15. Compare the means by using a t test.

3. Compare the proportion of men who are smokers with the proportion of women who are smokers. Use the data in the Data Bank. Choose random samples of size 30 or more. Use the z test for proportions.

4. Select two samples of 20 values from the data in Data Set IV in Appendix D. Test the hypothesis that the mean heights of the buildings are equal.

5. Using the same data obtained in Exercise 4, test the hypothesis that the variances are equal.

Chapter Quiz

Determine whether each statement is true or false. If the statement is false, explain why.

1. When you are testing the difference between two means, it is not important to distinguish whether the samples are independent of each other. False

2. If the same diet is given to two groups of randomly selected individuals, the samples are considered to be dependent. False

3. When computing the F test value, you always place the larger variance in the numerator of the fraction. True

4. Tests for variances are always two-tailed. False

Select the best answer.

5. To test the equality of two variances, you would use a(n) _____ test.
   a. z
   b. t
   c. Chi-square
   d. F

6. To test the equality of two proportions, you would use a(n) _____ test.
   a. z
   b. t
   c. Chi-square
   d. F

7. The mean value of the F is approximately equal to
   a. 0
   b. 0.5
   c. 1
   d. It cannot be determined.

8. What test can be used to test the difference between two sample means when the population variances are known?
   a. z
   b. t
   c. Chi-square
   d. F

Complete these statements with the best answer.

9. If you hypothesize that there is no difference between means, this is represented as H_0: _____. $\mu_1 = \mu_2$

10. When you are testing the difference between two means, the _____ test is used when the population variances are not known. t

11. When the t test is used for testing the equality of two means, the populations must be _____. Normal

12. The values of F cannot be _____. Negative

13. The formula for the F test for variances is _____. $\dfrac{s_1^2}{s_2^2}$

For each of these problems, perform the following steps.

 a. State the hypotheses and identify the claim.
 b. Find the critical value(s).
 c. Compute the test value.
 d. Make the decision.
 e. Summarize the results.

Use the traditional method of hypothesis testing unless otherwise specified.

14. **Cholesterol Levels** A researcher wishes to see if there is a difference in the cholesterol levels of two groups of men. A random sample of 30 men between the ages of 25 and 40 is selected and tested. The average level is 223. A second sample of 25 men between the ages of 41 and 56 is selected and tested. The average of this group is 229. The population standard deviation for both groups is 6. At $\alpha = 0.01$, is there a difference in the cholesterol levels between the two groups? Find the 99% confidence interval for the difference of the two means.

15. **Apartment Rental Fees** The data shown are the rental fees (in dollars) for two random samples of apartments in a large city. At $\alpha = 0.10$, can it be concluded that the average rental fee for apartments in the east is greater than the average rental fee in the west? Assume $\sigma_1 = 119$ and $\sigma_2 = 103$.

East					West				
495	390	540	445	420	525	400	310	375	750
410	550	499	500	550	390	795	554	450	370
389	350	450	530	350	385	395	425	500	550
375	690	325	350	799	380	400	450	365	425
475	295	350	485	625	375	360	425	400	475
275	450	440	425	675	400	475	430	410	450
625	390	485	550	650	425	450	620	500	400
685	385	450	550	425	295	350	300	360	400

Source: *Pittsburgh Post-Gazette.*

16. Prices of Low-Calorie Foods The average price of a sample of 12 bottles of diet salad dressing taken from different stores is $1.43. The standard deviation is $0.09. The average price of a sample of 16 low-calorie frozen desserts is $1.03. The standard deviation is $0.10. At $\alpha = 0.01$, is there a significant difference in price? Find the 99% confidence interval of the difference in the means.

17. Jet Ski Accidents The data shown represent the number of accidents people had when using jet skis and other types of wet bikes. At $\alpha = 0.05$, can it be concluded that the average number of accidents per year has increased from one period to the next?

1987–1991			1992–1996		
376	650	844	1650	2236	3002
1162	1513		4028	4010	

Source: *USA TODAY.*

18. Salaries of Chemists A sample of 12 chemists from Washington state shows an average salary of $39,420 with a standard deviation of $1659, while a sample of 26 chemists from New Mexico has an average salary of $30,215 with a standard deviation of $4116. Is there a significant difference between the two states in chemists' salaries at $\alpha = 0.02$? Find the 98% confidence interval of the difference in the means.

19. Family Incomes The average income of 15 families who reside in a large metropolitan East Coast city is $62,456. The standard deviation is $9652. The average income of 11 families who reside in a rural area of the Midwest is $60,213, with a standard deviation of $2009. At $\alpha = 0.05$, can it be concluded that the families who live in the cities have a higher income than those who live in the rural areas? Use the *P*-value method.

20. Mathematical Skills In an effort to improve the mathematical skills of 10 students, a teacher provides a weekly 1-hour tutoring session for the students. A pretest is given before the sessions, and a posttest is given after. The results are shown here. At $\alpha = 0.01$, can it be concluded that the sessions help to improve the students' mathematical skills?

Student	1	2	3	4	5	6	7	8	9	10
Pretest	82	76	91	62	81	67	71	69	80	85
Posttest	88	80	98	80	80	73	74	78	85	93

21. Egg Production To increase egg production, a farmer decided to increase the amount of time the lights in his hen house were on. Ten hens were selected, and the number of eggs each produced was recorded. After one week of lengthened light time, the same hens were monitored again. The data are given here. At $\alpha = 0.05$, can it be concluded that the increased light time increased egg production?

Hen	1	2	3	4	5	6	7	8	9	10
Before	4	3	8	7	6	4	9	7	6	5
After	6	5	9	7	4	5	10	6	9	6

22. Factory Worker Literacy Rates In a sample of 80 workers from a factory in city A, it was found that 5% were unable to read, while in a sample of 50 workers in city B, 8% were unable to read. Can it be concluded that there is a difference in the proportions of nonreaders in the two cities? Use $\alpha = 0.10$. Find the 90% confidence interval for the difference of the two proportions.

23. Male Head of Household A recent survey of 200 households showed that 8 had a single male as the head of household. Forty years ago, a survey of 200 households showed that 6 had a single male as the head of household. At $\alpha = 0.05$, can it be concluded that the proportion has changed? Find the 95% confidence interval of the difference of the two proportions. Does the confidence interval contain 0? Why is this important to know?

Source: Based on data from the U.S. Census Bureau.

24. Money Spent on Road Repair A politician wishes to compare the variances of the amount of money spent for road repair in two different counties. The data are given here. At $\alpha = 0.05$, is there a significant difference in the variances of the amounts spent in the two counties? Use the *P*-value method.

County A	County B
$s_1 = \$11,596$	$s_2 = \$14,837$
$n_1 = 15$	$n_2 = 18$

25. Heights of Basketball Players A researcher wants to compare the variances of the heights (in inches) of four-year college basketball players with those of players in junior colleges. A sample of 30 players from each type of school is selected, and the variances of the heights for each type are 2.43 and 3.15, respectively. At $\alpha = 0.10$, is there a significant difference between the variances of the heights in the two types of schools?

Critical Thinking Challenges

1. The study cited in the article entitled "Only the Timid Die Young" stated that "Timid rats were 60% more likely to die at any given time than were their outgoing brothers." Based on the results, answer the following questions.

a. Why were rats used in the study?

b. What are the variables in the study?

c. Why were infants included in the article?

d. What is wrong with extrapolating the results to humans?

e. Suggest some ways humans might be used in a study of this type.

ONLY THE TIMID DIE YOUNG

DO OVERACTIVE STRESS HORMONES DAMAGE HEALTH?

ABOUT 15 OUT OF 100 CHILDREN ARE BORN SHY, BUT ONLY THREE WILL BE SHY AS ADULTS.

FEARFUL TYPES MAY MEET THEIR maker sooner, at least among rats. Researchers have for the first time connected a personality trait—fear of novelty—to an early death.

Sonia Cavigelli and Martha McClintock, psychologists at the University of Chicago, presented unfamiliar bowls, tunnels and bricks to a group of young male rats. Those hesitant to explore the mystery objects were classified as "neophobic."

The researchers found that the neophobic rats produced high levels of stress hormones, called glucocorticoids—typically involved in the fight-or-flight stress response—when faced with strange situations. Those rats continued to have high levels of the hormones at random times throughout their lives, indicating that timidity is a fixed and stable trait. The team then set out to examine the cumulative effects of this personality trait on the rats' health.

Timid rats were 60 percent more likely to die at any given time than were their outgoing brothers. The causes of death were similar for both groups. "One hypothesis as to why the neophobic rats died earlier is that the stress hormones negatively affected their immune system," Cavigelli says. Neophobes died, on average, three months before their rat brothers, a significant gap, considering that most rats lived only two years.

Shyness—the human equivalent of neophobia—can be detected in infants as young as 14 months. Shy people also produce more stress hormones than "average," or thrill-seeking humans. But introverts don't necessarily stay shy for life, as rats apparently do. Jerome Kagan, a professor of psychology at Harvard University, has found that while 15 out of every 100 children will be born with a shy temperament, only three will appear shy as adults. None, however, will be extroverts.

Extrapolating from the doomed fate of neophobic rats to their human counterparts is difficult. "But it means that something as simple as a personality trait could have physiological consequences," Cavigelli says.

—*Carlin Flora*

Reprinted with permission from *Psychology Today Magazine* (Copyright © 2004, Sussex Publishers, LLC).

2. Based on the study presented in the article entitled "Sleeping Brain, Not at Rest," answer these questions.

a. What were the variables used in the study?

b. How were they measured?

c. Suggest a statistical test that might have been used to arrive at the conclusion.

d. Based on the results, what would you suggest for students preparing for an exam?

SLEEPING BRAIN, NOT AT REST

Regions of the brain that have spent the day learning sleep more heavily at night.

In a study published in the journal *Nature*, Giulio Tononi, a psychiatrist at the University of Wisconsin–Madison, had subjects perform a simple point-and-click task with a computer adjusted so that its cursor didn't track in the right direction. Afterward, the subjects' brain waves were recorded while they slept, then examined for "slow wave" activity, a kind of deep sleep.

Compared with people who'd completed the same task with normal cursors, Tononi's subjects showed elevated slow wave activity in brain areas associated with spatial orientation, indicating that their brains were adjusting to the day's learning by making cellular-level changes. In the morning, Tononi's subjects performed their tasks better than they had before going to sleep.

—*Richard A. Love*

Reprinted with permission from *Psychology Today Magazine* (Copyright © 2004, Sussex Publishers, LLC).

Data Projects

Use a significance level of 0.05 for all tests below.

1. **Business and Finance** Use the data collected in data project 1 of Chapter 2 to complete this problem. Test the claim that the mean earnings per share for Dow Jones stocks are greater than for NASDAQ stocks.

2. **Sports and Leisure** Use the data collected in data project 2 of Chapter 7 regarding home runs for this problem. Test the claim that the mean number of home runs hit by the American League sluggers is the same as the mean for the National League.

3. **Technology** Use the cell phone data collected for data project 2 in Chapter 8 to complete this problem. Test the claim that the mean length for outgoing calls is the same as that for incoming calls. Test the claim that the standard deviation for outgoing calls is more than that for incoming calls.

4. **Health and Wellness** Use the data regarding BMI that were collected in data project 6 of Chapter 7 to complete this problem. Test the claim that the mean BMI for males is the same as that for females. Test the claim that the standard deviation for males is the same as that for females.

5. **Politics and Economics** Use data from the last Presidential election to categorize the 50 states as "red" or "blue" based on who was supported for President in that state, the Democratic or Republican candidate. Use the data collected in data project 5 of Chapter 2 regarding income. Test the claim that the mean incomes for red states and blue states are equal.

6. **Your Class** Use the data collected in data project 6 of Chapter 2 regarding heart rates. Test the claim that the heart rates after exercise are more variable than the heart rates before exercise.

Answers to Applying the Concepts

Section 9–1   Home Runs

1. The population is all home runs hit by major league baseball players.

2. A cluster sample was used.

3. Answers will vary. While this sample is not representative of all major league baseball players per se, it does allow us to compare the leaders in each league.

4. $H_0: \mu_1 = \mu_2$ and $H_1: \mu_1 \neq \mu_2$

5. Answers will vary. Possible answers include the 0.05 and 0.01 significance levels.

6. We will use the z test for the difference in means.

7. Our test statistic is $z = \dfrac{44.75 - 42.88}{\sqrt{\dfrac{8.8^2}{40} + \dfrac{7.8^2}{40}}} = 1.01$, and our P-value is 0.3124.

8. We fail to reject the null hypothesis.

9. There is not enough evidence to conclude that there is a difference in the number of home runs hit by National League versus American League baseball players.

10. Answers will vary. One possible answer is that since we do not have a random sample of data from each league, we cannot answer the original question asked.

11. Answers will vary. One possible answer is that we could get a random sample of data from each league from a recent season.

Section 9–2   Too Long on the Telephone

1. These samples are independent.

2. We compare the P-value of 0.06317 to the significance level to check if the null hypothesis should be rejected.

3. The P-value of 0.06317 also gives the probability of a type I error.

4. Since two critical values are shown, we know that a two-tailed test was done.

5. Since the P-value of 0.06317 is greater than the significance value of 0.05, we fail to reject the null hypothesis and find that we do not have enough evidence to conclude that there is a difference in the lengths of telephone calls made by employees in the two divisions of the company.

6. If the significance level had been 0.10, we would have rejected the null hypothesis, since the P-value would have been less than the significance level.

Section 9–3   Air Quality

1. The purpose of the study is to determine if the air quality in the United States has changed over the past 2 years.

2. These are dependent samples, since we have two readings from each of 10 metropolitan areas.

3. The hypotheses we will test are $H_0: \mu_D = 0$ and $H_1: \mu_D \neq 0$.

4. We will use the 0.05 significance level and critical values of $t = \pm 2.262$.

5. We will use the t test for dependent samples.

6. There are $10 - 1 = 9$ degrees of freedom.

7. Our test statistic is $t = \dfrac{-6.7 - 0}{11.27/\sqrt{10}} = -1.879$. We fail to reject the null hypothesis and find that there is not enough evidence to conclude that the air quality in the United States has changed over the past 2 years.

8. No, we could not use an independent means test since we have two readings from each metropolitan area.

9. Answers will vary. One possible answer is that there are other measures of air quality that we could have examined to answer the question.

Section 9–4   Smoking and Education

1. Our hypotheses are $H_0: p_1 = p_2$ and $H_1: p_1 \neq p_2$.

2. At the 0.05 significance level, our critical values are $z = \pm 1.96$.

3. We will use the z test for the difference between proportions.

4. To complete the statistical test, we would need the sample sizes.

5. Knowing the sample sizes were 1000, we can now complete the test.

6. Our test statistic is $z = \dfrac{0.323 - 0.145}{\sqrt{(0.234)(0.766)\left(\dfrac{1}{1000} + \dfrac{1}{1000}\right)}} =$
9.40, and our P-value is very close to zero. We reject the null hypothesis and find that there is enough evidence to conclude that there is a difference in the proportions of high school graduates and college graduates who smoke.

Section 9–5   Variability and Automatic Transmissions

1. The null hypothesis is that the variances are the same: $H_0: \sigma_1^2 = \sigma_2^2$ ($H_1: \sigma_1^2 \neq \sigma_2^2$).

2. We will use an F test.

3. The value of the test statistic is $F = \dfrac{s_1^2}{s_2^2} = \dfrac{514.8^2}{77.7^2} = 43.92$, and the P-value is 0.0008. There is a significant difference in the variability of the prices between the two countries.

4. Small sample sizes are highly impacted by outliers.

5. The degrees of freedom for the numerator and denominator are both 5.

6. Yes, two sets of data can center on the same mean but have very different standard deviations.

Hypothesis-Testing Summary 1

1. Comparison of a sample mean with a specific population mean.

   Example: $H_0: \mu = 100$

   a. Use the z test when σ is known:

   $$z = \frac{\bar{X} - \mu}{\sigma/\sqrt{n}}$$

   b. Use the t test when σ is unknown:

   $$t = \frac{\bar{X} - \mu}{s/\sqrt{n}} \qquad \text{with d.f.} = n - 1$$

2. Comparison of a sample variance or standard deviation with a specific population variance or standard deviation.

   Example: $H_0: \sigma^2 = 225$

   Use the chi-square test:

   $$\chi^2 = \frac{(n-1)s^2}{\sigma^2} \qquad \text{with d.f.} = n - 1$$

3. Comparison of two sample means.

   Example: $H_0: \mu_1 = \mu_2$

   a. Use the z test when the population variances are known:

   $$z = \frac{(\bar{X}_1 - \bar{X}_2) - (\mu_1 - \mu_2)}{\sqrt{\dfrac{\sigma_1^2}{n_1} + \dfrac{\sigma_2^2}{n_2}}}$$

   b. Use the t test for independent samples when the population variances are unknown and assume the sample variances are unequal:

   $$t = \frac{(\bar{X}_1 - \bar{X}_2) - (\mu_1 - \mu_2)}{\sqrt{\dfrac{s_1^2}{n_1} + \dfrac{s_2^2}{n_2}}}$$

   with d.f. = the smaller of $n_1 - 1$ or $n_2 - 1$.

   Formula for the t test for comparing two means (independent samples, variances equal):

   $$t = \frac{(\bar{X}_1 - \bar{X}_2) - (\mu_1 - \mu_2)}{\sqrt{\dfrac{(n_1-1)s_1^2 + (n_2-1)s_2^2}{n_1 + n_2 - 2}}\sqrt{\dfrac{1}{n_1} + \dfrac{1}{n_2}}}$$

   with d.f. = $n_1 + n_2 - 2$.

   c. Use the t test for means for dependent samples:

   Example: $H_0: \mu_D = 0$

   $$t = \frac{\bar{D} - \mu_D}{s_D/\sqrt{n}} \qquad \text{with d.f.} = n - 1$$

   where n = number of pairs.

4. Comparison of a sample proportion with a specific population proportion.

   Example: $H_0: p = 0.32$

   Use the z test:

   $$z = \frac{X - \mu}{\sigma} \qquad \text{or} \qquad z = \frac{\hat{p} - p}{\sqrt{pq/n}}$$

5. Comparison of two sample proportions.

   Example: $H_0: p_1 = p_2$

   Use the z test:

   $$z = \frac{(\hat{p}_1 - \hat{p}_2) - (p_1 - p_2)}{\sqrt{\bar{p}\,\bar{q}\left(\dfrac{1}{n_1} + \dfrac{1}{n_2}\right)}}$$

   where

   $$\bar{p} = \frac{X_1 + X_2}{n_1 + n_2} \qquad \hat{p}_1 = \frac{X_1}{n_1}$$

   $$\bar{q} = 1 - \bar{p} \qquad \hat{p}_2 = \frac{X_2}{n_2}$$

6. Comparison of two sample variances or standard deviations.

   Example: $H_0: \sigma_1^2 = \sigma_2^2$

   Use the F test:

   $$F = \frac{s_1^2}{s_2^2}$$

   where

   s_1^2 = larger variance   d.f.N. = $n_1 - 1$
   s_2^2 = smaller variance   d.f.D. = $n_2 - 1$

CHAPTER 10

Correlation and Regression

Objectives

After completing this chapter, you should be able to

1 Draw a scatter plot for a set of ordered pairs.

2 Compute the correlation coefficient.

3 Test the hypothesis H_0: $\rho = 0$.

4 Compute the equation of the regression line.

5 Compute the coefficient of determination.

6 Compute the standard error of the estimate.

7 Find a prediction interval.

8 Be familiar with the concept of multiple regression.

Outline

**Statistics
Today**

Do Dust Storms Affect Respiratory Health?

Southeast Washington state has a long history of seasonal dust storms. Several researchers decided to see what effect, if any, these storms had on the respiratory health of the people living in the area. They undertook (among other things) to see if there was a relationship between the amount of dust and sand particles in the air when the storms occur and the number of hospital emergency room visits for respiratory disorders at three community hospitals in southeast Washington. Using methods of correlation and regression, which are explained in this chapter, they were able to determine the effect of these dust storms on local residents. See Statistics Today—Revisited at the end of the chapter.

Source: B. Hefflin, B. Jalaludin, N. Cobb, C. Johnson, L. Jecha, and R. Etzel, "Surveillance for Dust Storms and Respiratory Diseases in Washington State," *Archives of Environmental Health* 49, no. 3 (May–June), pp. 170–74. Reprinted with permission of the Helen Dwight Reid Education Foundation. Published by Heldref Publications, 1319 18th St. N.W., Washington, D.C. 20036-1802.

Introduction

In Chapters 7 and 8, two areas of inferential statistics—confidence intervals and hypothesis testing—were explained. Another area of inferential statistics involves determining whether a relationship exists between two or more numerical or quantitative variables. For example, a businessperson may want to know whether the volume of sales for a given month is related to the amount of advertising the firm does that month. Educators are interested in determining whether the number of hours a student studies is related to the student's score on a particular exam. Medical researchers are interested in questions such as, Is caffeine related to heart damage? or Is there a relationship between a person's age and his or her blood pressure? A zoologist may want to know whether the birth weight of a certain animal is related to its life span. These are only a few of the many questions that can be answered by using the techniques of correlation and regression analysis. **Correlation** is a statistical method used to determine whether a linear relationship between variables exists. **Regression** is a statistical method used to describe the nature of the relationship between variables, that is, positive or negative, linear or nonlinear.

The purpose of this chapter is to answer these questions statistically:

1. Are two or more variables linearly related?
2. If so, what is the strength of the relationship?

3. What type of relationship exists?

4. What kind of predictions can be made from the relationship?

To answer the first two questions, statisticians use a numerical measure to determine whether two or more variables are linearly related and to determine the strength of the relationship between or among the variables. This measure is called a *correlation coefficient.* For example, there are many variables that contribute to heart disease, among them lack of exercise, smoking, heredity, age, stress, and diet. Of these variables, some are more important than others; therefore, a physician who wants to help a patient must know which factors are most important.

To answer the third question, you must ascertain what type of relationship exists. There are two types of relationships: *simple* and *multiple.* In a **simple relationship,** there are two variables—an **independent variable,** also called an explanatory variable or a predictor variable, and a **dependent variable,** also called a response variable. A simple relationship analysis is called *simple regression,* and there is one independent variable that is used to predict the dependent variable. For example, a manager may wish to see whether the number of years the salespeople have been working for the company has anything to do with the amount of sales they make. This type of study involves a simple relationship, since there are only two variables—years of experience and amount of sales.

In a **multiple relationship,** called *multiple regression,* two or more independent variables are used to predict one dependent variable. For example, an educator may wish to investigate the relationship between a student's success in college and factors such as the number of hours devoted to studying, the student's GPA, and the student's high school background. This type of study involves several variables.

Simple relationships can also be positive or negative. A **positive relationship** exists when both variables increase or decrease at the same time. For instance, a person's height and weight are related; and the relationship is positive, since the taller a person is, generally, the more the person weighs. In a **negative relationship,** as one variable increases, the other variable decreases, and vice versa. For example, if you measure the strength of people over 60 years of age, you will find that as age increases, strength generally decreases. The word *generally* is used here because there are exceptions.

Finally, the fourth question asks what type of predictions can be made. Predictions are made in all areas and daily. Examples include weather forecasting, stock market analyses, sales predictions, crop predictions, gasoline price predictions, and sports predictions. Some predictions are more accurate than others, due to the strength of the relationship. That is, the stronger the relationship is between variables, the more accurate the prediction is.

10–1   Scatter Plots and Correlation

Objective 1

Draw a scatter plot for a set of ordered pairs.

In simple correlation and regression studies, the researcher collects data on two numerical or quantitative variables to see whether a relationship exists between the variables. For example, if a researcher wishes to see whether there is a relationship between number of hours of study and test scores on an exam, she must select a random sample of students, determine the hours each studied, and obtain their grades on the exam. A table can be made for the data, as shown here.

Student	Hours of study x	Grade y (%)
A	6	82
B	2	63
C	1	57
D	5	88
E	2	68
F	3	75

As stated previously, the two variables for this study are called the independent variable and the dependent variable. The independent variable is the variable in regression that can be controlled or manipulated. In this case, the number of hours of study is the independent variable and is designated as the x variable. The dependent variable is the variable in regression that cannot be controlled or manipulated. The grade the student received on the exam is the dependent variable, designated as the y variable. The reason for this distinction between the variables is that you assume that the grade the student earns *depends* on the number of hours the student studied. Also, you assume that, to some extent, the student can regulate or *control* the number of hours he or she studies for the exam.

The determination of the x and y variables is not always clear-cut and is sometimes an arbitrary decision. For example, if a researcher studies the effects of age on a person's blood pressure, the researcher can generally assume that age affects blood pressure. Hence, the variable *age* can be called the *independent variable,* and the variable *blood pressure* can be called the *dependent variable.* On the other hand, if a researcher is studying the attitudes of husbands on a certain issue and the attitudes of their wives on the same issue, it is difficult to say which variable is the independent variable and which is the dependent variable. In this study, the researcher can arbitrarily designate the variables as independent and dependent.

The independent and dependent variables can be plotted on a graph called a *scatter plot.* The independent variable x is plotted on the horizontal axis, and the dependent variable y is plotted on the vertical axis.

A **scatter plot** is a graph of the ordered pairs (x, y) of numbers consisting of the independent variable x and the dependent variable y.

The scatter plot is a visual way to describe the nature of the relationship between the independent and dependent variables. The scales of the variables can be different, and the coordinates of the axes are determined by the smallest and largest data values of the variables.

The procedure for drawing a scatter plot is shown in Examples 10–1 through 10–3.

Example 10–1   Car Rental Companies

 Construct a scatter plot for the data shown for car rental companies in the United States for a recent year.

Company	Cars (in ten thousands)	Revenue (in billions)
A	63.0	$7.0
B	29.0	3.9
C	20.8	2.1
D	19.1	2.8
E	13.4	1.4
F	8.5	1.5

Source: *Auto Rental News.*

Solution

Step 1   Draw and label the x and y axes.

Step 2   Plot each point on the graph, as shown in Figure 10–1.

Figure 10–1

Scatter Plot for
Example 10–1

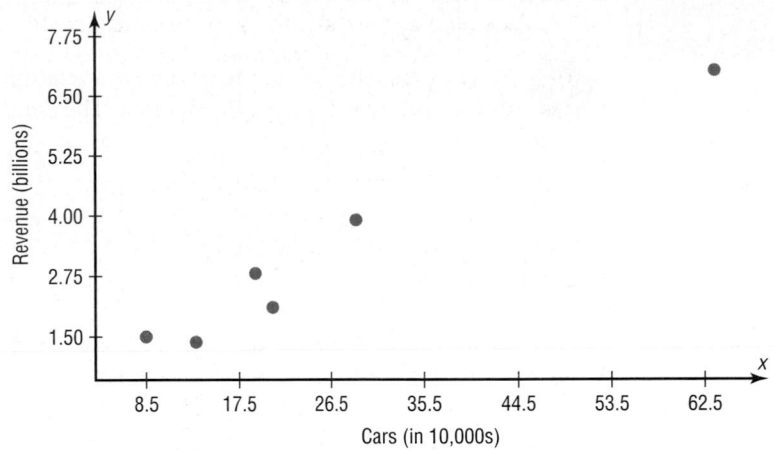

Example 10–2

Absences and Final Grades

 Construct a scatter plot for the data obtained in a study on the number of absences and the final grades of seven randomly selected students from a statistics class. The data are shown here.

Student	Number of absences x	Final grade y (%)
A	6	82
B	2	86
C	15	43
D	9	74
E	12	58
F	5	90
G	8	78

Solution

Step 1 Draw and label the x and y axes.

Step 2 Plot each point on the graph, as shown in Figure 10–2.

Figure 10–2

Scatter Plot for
Example 10–2

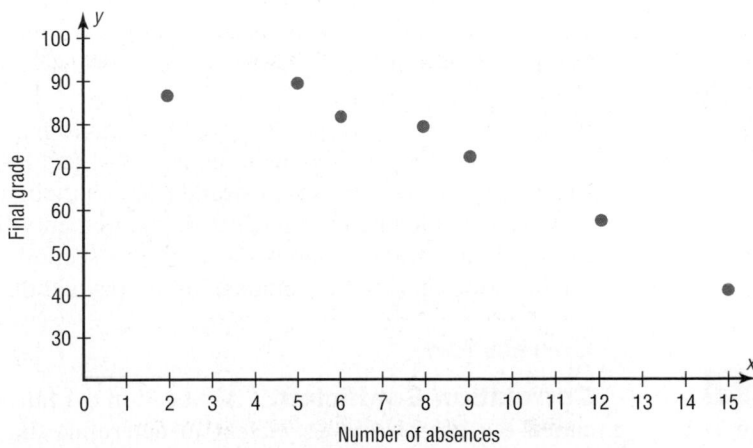

Example 10–3

Age and Wealth

A researcher wishes to see if there is a relationship between the ages and net worth of the wealthiest people in America. The data for a specific year are shown.

Person	Age x	Net wealth y ($ billions)
A	73	16
B	65	26
C	53	50
D	54	21.5
E	79	40
F	69	16
G	61	19.6
H	65	19

Source: *Forbes magazine.*

Solution

Step 1 Draw and label the *x* and *y* axes.

Step 2 Plot each point on the graph, as shown in Figure 10–3.

Figure 10–3

Scatter Plot for Example 10–3

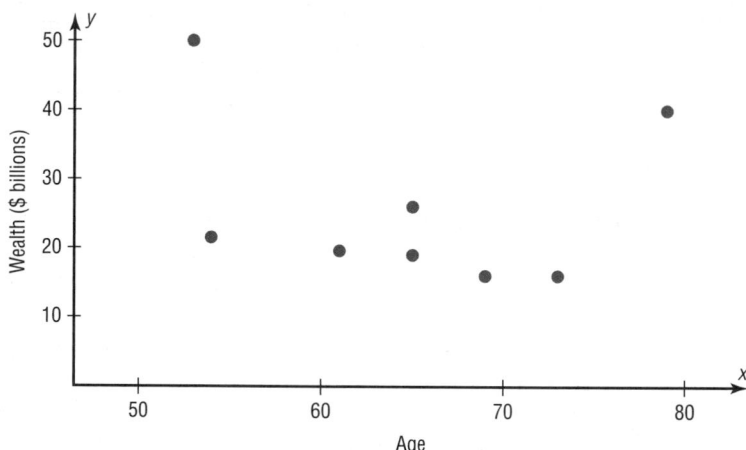

After the plot is drawn, it should be analyzed to determine which type of relationship, if any, exists. For example, the plot shown in Figure 10–1 suggests a positive relationship, since as the number of cars rented increases, revenue tends to increase also. The plot of the data shown in Figure 10–2 suggests a negative relationship, since as the number of absences increases, the final grade decreases. Finally, the plot of the data shown in Figure 10–3 shows no specific type of relationship, since no pattern is discernible.

Note that the data shown in Figures 10–1 and 10–2 also suggest a linear relationship, since the points seem to fit a straight line, although not perfectly. Sometimes a scatter plot, such as the one in Figure 10–4, shows a curvilinear relationship between the data. In this situation, the methods shown in this section and in Section 10–2 cannot be used. Methods for curvilinear relationships are beyond the scope of this book.

Correlation

Objective 2

Compute the correlation coefficient.

Correlation Coefficient As stated in the Introduction, statisticians use a measure called the *correlation coefficient* to determine the strength of the linear relationship between two variables. There are several types of correlation coefficients. The one

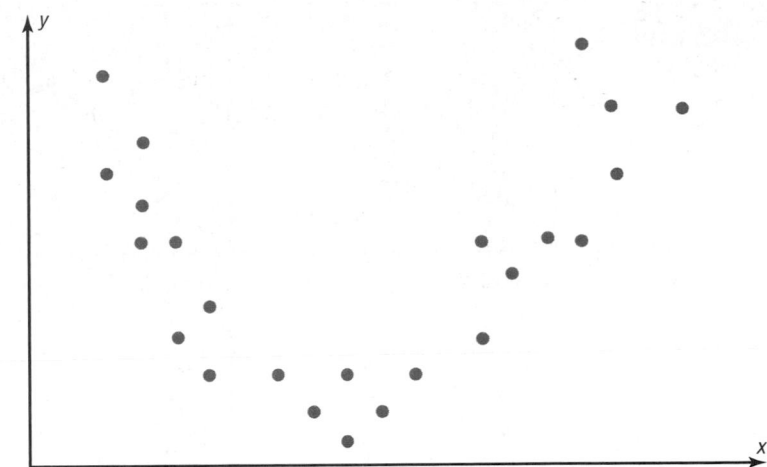

Figure 10–4

Scatter Plot
Suggesting a
Curvilinear
Relationship

explained in this section is called the **Pearson product moment correlation coefficient (PPMC),** named after statistician Karl Pearson, who pioneered the research in this area.

> The **correlation coefficient** computed from the sample data measures the strength and direction of a linear relationship between two quantitative variables. The symbol for the sample correlation coefficient is r. The symbol for the population correlation coefficient is ρ (Greek letter rho).

The *range of the correlation coefficient* is from -1 to $+1$. If there is a *strong positive linear relationship* between the variables, the value of r will be close to $+1$. If there is a *strong negative linear relationship* between the variables, the value of r will be close to -1. When there is no linear relationship between the variables or only a weak relationship, the value of r will be close to 0. See Figure 10–5.

The graphs in Figure 10–6 show the relationship between the correlation coefficients and their corresponding scatter plots. Notice that as the value of the correlation coefficient increases from 0 to $+1$ (parts a, b, and c), data values become closer to an increasingly strong relationship. As the value of the correlation coefficient decreases from 0 to -1 (parts d, e, and f), the data values also become closer to a straight line. Again this suggests a stronger relationship.

There are several ways to compute the value of the correlation coefficient. One method is to use the formula shown here.

Formula for the Correlation Coefficient r

$$r = \frac{n(\Sigma xy) - (\Sigma x)(\Sigma y)}{\sqrt{[n(\Sigma x^2) - (\Sigma x)^2][n(\Sigma y^2) - (\Sigma y)^2]}}$$

where n is the number of data pairs.

Figure 10–5

Range of Values for the
Correlation Coefficient

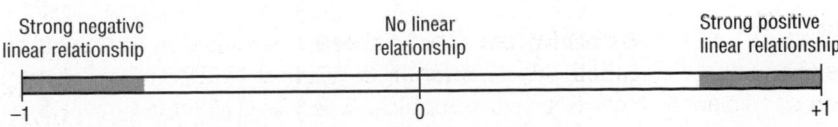

Strong negative linear relationship   No linear relationship   Strong positive linear relationship

-1 0 $+1$

Figure 10–6

Relationship Between
the Correlation
Coefficient and the
Scatter Plot

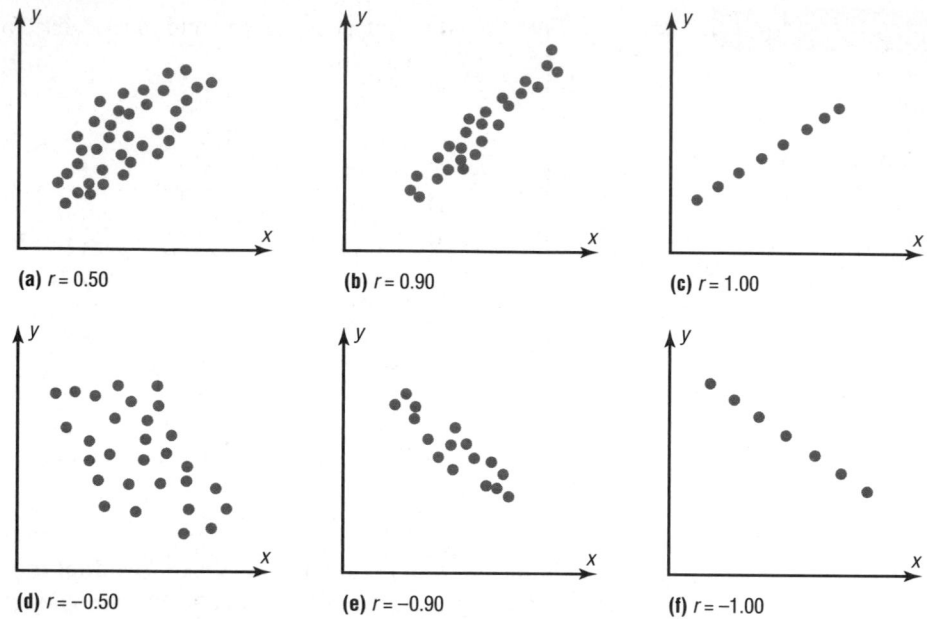

(a) $r = 0.50$ (b) $r = 0.90$ (c) $r = 1.00$

(d) $r = -0.50$ (e) $r = -0.90$ (f) $r = -1.00$

Assumptions for the Correlation Coefficient

1. The sample is a random sample.
2. The data pairs fall approximately on a straight line and are measured at the interval or ratio level.
3. The variables have a joint normal distribution. (This means that given any specific value of x, the y values are normally distributed; and given any specific value of y, the x values are normally distributed.)

Rounding Rule for the Correlation Coefficient   Round the value of r to three decimal places.

The formula looks somewhat complicated, but using a table to compute the values, as shown in Example 10–4, makes it somewhat easier to determine the value of r.

There are no units associated with r, and the value of r will remain unchanged if the x and y values are switched.

Example 10–4

Car Rental Companies

Compute the correlation coefficient for the data in Example 10–1.

Solution

Step 1   Make a table as shown here.

Company	Cars x (in ten thousands)	Revenue y (in billions)	xy	x^2	y^2
A	63.0	7.0			
B	29.0	3.9			
C	20.8	2.1			
D	19.1	2.8			
E	13.4	1.4			
F	8.5	1.5			

Step 2 Find the values of xy, x^2, and y^2 and place these values in the corresponding columns of the table.

The completed table is shown.

Company	Cars x (in 10,000s)	Revenue y (in billions)	xy	x^2	y^2
A	63.0	7.0	441.00	3969.00	49.00
B	29.0	3.9	113.10	841.00	15.21
C	20.8	2.1	43.68	432.64	4.41
D	19.1	2.8	53.48	364.81	7.84
E	13.4	1.4	18.76	179.56	1.96
F	8.5	1.5	12.75	72.25	2.25
	$\Sigma x = 153.8$	$\Sigma y = 18.7$	$\Sigma xy = 682.77$	$\Sigma x^2 = 5859.26$	$\Sigma y^2 = 80.67$

Step 3 Substitute in the formula and solve for r.

$$r = \frac{n(\Sigma xy) - (\Sigma x)(\Sigma y)}{\sqrt{[n(\Sigma x^2) - (\Sigma x)^2][n(\Sigma y^2) - (\Sigma y)^2]}}$$

$$= \frac{(6)(682.77) - (153.8)(18.7)}{\sqrt{[(6)(5859.26) - (153.8)^2][(6)(80.67) - (18.7)^2]}} = 0.982$$

The correlation coefficient suggests a strong relationship between the number of cars a rental agency has and its annual revenue.

Example 10–5

Absences and Final Grades

Compute the value of the correlation coefficient for the data obtained in the study of the number of absences and the final grade of the seven students in the statistics class given in Example 10–2.

Solution

Step 1 Make a table.

Step 2 Find the values of xy, x^2, and y^2; place these values in the corresponding columns of the table.

Student	Number of absences x	Final grade y (%)	xy	x^2	y^2
A	6	82	492	36	6,724
B	2	86	172	4	7,396
C	15	43	645	225	1,849
D	9	74	666	81	5,476
E	12	58	696	144	3,364
F	5	90	450	25	8,100
G	8	78	624	64	6,084
	$\Sigma x = 57$	$\Sigma y = 511$	$\Sigma xy = 3745$	$\Sigma x^2 = 579$	$\Sigma y^2 = 38,993$

Step 3 Substitute in the formula and solve for r.

$$r = \frac{n(\Sigma xy) - (\Sigma x)(\Sigma y)}{\sqrt{[n(\Sigma x^2) - (\Sigma x)^2][n(\Sigma y^2) - (\Sigma y)^2]}}$$

$$= \frac{(7)(3745) - (57)(511)}{\sqrt{[(7)(579) - (57)^2][(7)(38,993) - (511)^2]}} = -0.944$$

The value of r suggests a strong negative relationship between a student's final grade and the number of absences a student has. That is, the more absences a student has, the lower is his or her grade.

Example 10–6 | **Age and Wealth**

Compute the value of the correlation coefficient for the data given in Example 10–3 for the age and wealth of the richest persons in the United States.

Solution

Step 1 Make a table.

Step 2 Find the values of xy, x^2, and y^2, and place these values in the corresponding columns of the table.

Person	Age x	Net wealth y	xy	x^2	y^2
A	73	16	1,168	5,329	256
B	65	26	1,690	4,225	676
C	53	50	2,650	2,809	2,500
D	54	21.5	1,161	2,916	462.25
E	79	40	3,160	6,241	1,600
F	69	16	1,104	4,761	256
G	61	19.6	1,195.6	3,721	384.16
H	65	19	1,235	4,225	361

$\Sigma x = 519$ $\Sigma y = 208.1$ $\Sigma xy = 13,363.6$ $\Sigma x^2 = 34,227$ $\Sigma y^2 = 6,495.41$

Step 3 Substitute in the formula and solve for r.

$$r = \frac{n(\Sigma xy) - (\Sigma x)(\Sigma y)}{\sqrt{[n(\Sigma x^2) - (\Sigma x)^2][n(\Sigma y^2) - (\Sigma y)^2]}}$$

$$= \frac{8(13,363.6) - (519)(208.1)}{\sqrt{[8(34,227) - (519)^2][8(6495.41) - (208.1)^2]}}$$

$$= \frac{-1095.1}{\sqrt{(4455)(8657.67)}}$$

$$= \frac{-1095.1}{6210.469}$$

$$= -0.176$$

The value of r indicates a very weak negative relationship between the variables.

In Example 10–4, the value of r was high (close to 1.00); in Example 10–6, the value of r was much lower (close to 0). This question then arises, When is the value of r due to chance, and when does it suggest a significant linear relationship between the variables? This question will be answered next.

Objective 3

Test the hypothesis H_0: $\rho = 0$.

The Significance of the Correlation Coefficient As stated before, the range of the correlation coefficient is between -1 and $+1$. When the value of r is near $+1$ or -1, there is a strong linear relationship. When the value of r is near 0, the linear relationship is weak or nonexistent. Since the value of r is computed from data obtained from samples, there are two possibilities when r is not equal to zero: either the value of r is high enough to conclude that there is a significant linear relationship between the variables, or the value of r is due to chance.

To make this decision, you use a hypothesis-testing procedure. The traditional method is similar to the one used in previous chapters.

Step 1   State the hypotheses.

Step 2   Find the critical values.

Step 3   Compute the test value.

Step 4   Make the decision.

Step 5   Summarize the results.

The population correlation coefficient is computed from taking all possible (x, y) pairs; it is designated by the Greek letter ρ (rho). The sample correlation coefficient can then be used as an estimator of ρ if the following assumptions are valid.

1. The variables x and y are *linearly* related.

2. The variables are *random* variables.

3. The two variables have a *bivariate normal distribution*.

A biviarate normal distribution means that for the pairs of (x, y) data values, the corresponding y values have a bell-shaped distribution for any given x value, and the x values for any given y value have a bell-shaped distribution.

Formally defined, the **population correlation coefficient** ρ is the correlation computed by using all possible pairs of data values (x, y) taken from a population.

In hypothesis testing, one of these is true:

$H_0\colon \rho = 0$ This null hypothesis means that there is no correlation between the x and y variables in the population.

$H_1\colon \rho \neq 0$ This alternative hypothesis means that there is a significant correlation between the variables in the population.

When the null hypothesis is rejected at a specific level, it means that there is a significant difference between the value of r and 0. When the null hypothesis is not rejected, it means that the value of r is not significantly different from 0 (zero) and is probably due to chance.

Several methods can be used to test the significance of the correlation coefficient. Three methods will be shown in this section. The first uses the t test.

Formula for the t Test for the Correlation Coefficient

$$t = r\sqrt{\frac{n-2}{1-r^2}}$$

with degrees of freedom equal to $n - 2$.

Although hypothesis tests can be one-tailed, most hypotheses involving the correlation coefficient are two-tailed. Recall that ρ represents the population correlation coefficient. Also, if there is no linear relationship, the value of the correlation coefficient will be 0. Hence, the hypotheses will be

$$H_0\colon \rho = 0 \qquad \text{and} \qquad H_1\colon \rho \neq 0$$

You do not have to identify the claim here, since the question will always be whether there is a significant linear relationship between the variables.

The two-tailed critical values are used. These values are found in Table F in Appendix C. Also, when you are testing the significance of a correlation coefficient, both variables x and y must come from normally distributed populations.

Example 10–7

Test the significance of the correlation coefficient found in Example 10–4. Use $\alpha = 0.05$ and $r = 0.982$.

Solution

Step 1 State the hypotheses.

$$H_0: \rho = 0 \qquad \text{and} \qquad H_1: \rho \neq 0$$

Step 2 Find the critical values. Since $\alpha = 0.05$ and there are $6 - 2 = 4$ degrees of freedom, the critical values obtained from Table F are ± 2.776, as shown in Figure 10–7.

Figure 10–7

Critical Values for Example 10–7

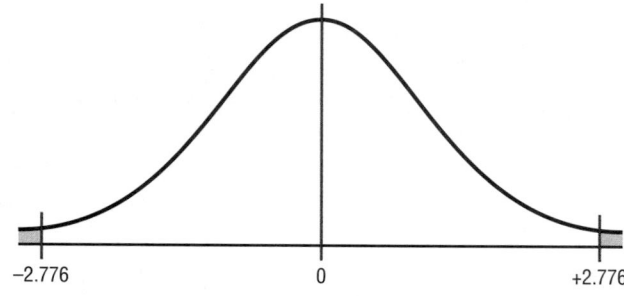

Step 3 Compute the test value.

$$t = r\sqrt{\frac{n - 2}{1 - r^2}} = 0.982\sqrt{\frac{6 - 2}{1 - (0.982)^2}} = 10.4$$

Step 4 Make the decision. Reject the null hypothesis, since the test value falls in the critical region, as shown in Figure 10–8.

Figure 10–8

Test Value for Example 10–7

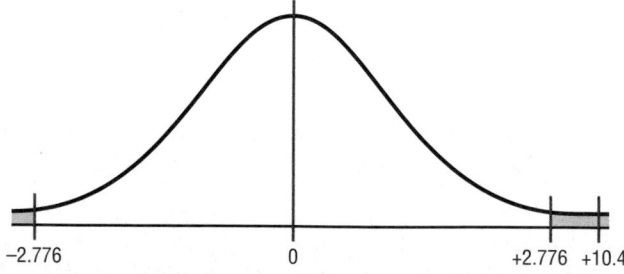

Step 5 Summarize the results. There is a significant relationship between the number of cars a rental agency owns and its annual income.

The second method that can be used to test the significance of r is the P-value method. The method is the same as that shown in Chapters 8 and 9. It uses the following steps.

Step 1 State the hypotheses.

Step 2 Find the test value. (In this case, use the t test.)

Step 3   Find the *P*-value. (In this case, use Table F.)

Step 4   Make the decision.

Step 5   Summarize the results.

Consider an example where $t = 4.059$ and d.f. $= 4$. Using Table F with d.f. $= 4$ and the row Two tails, the value 4.059 falls between 3.747 and 4.604; hence, $0.01 <$ *P*-value < 0.02. (The *P*-value obtained from a calculator is 0.015.) That is, the *P*-value falls between 0.01 and 0.02. The decision, then, is to reject the null hypothesis since *P*-value < 0.05.

The third method of testing the significance of *r* is to use Table I in Appendix C. This table shows the values of the correlation coefficient that are significant for a specific α level and a specific number of degrees of freedom. For example, for 7 degrees of freedom and $\alpha = 0.05$, the table gives a critical value of 0.666. Any value of *r* greater than $+0.666$ or less than -0.666 will be significant, and the null hypothesis will be rejected. See Figure 10–9. When Table I is used, you need not compute the *t* test value. Table I is for two-tailed tests only.

Figure 10–9

Finding the Critical Value from Table I

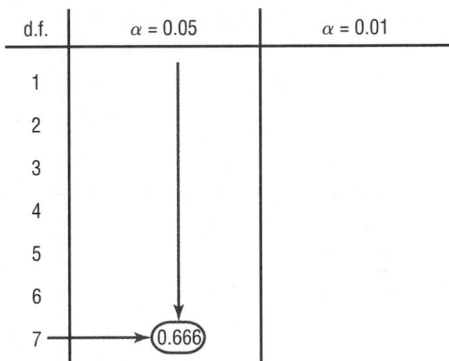

Example 10–8

Using Table I, test the significance at $\alpha = 0.01$ of the correlation coefficient $r = -0.176$, obtained in Example 10–6.

Solution

$$H_0: \rho = 0 \qquad \text{and} \qquad H_1: \rho \neq 0$$

Since the sample size is 8, there are $n - 2$, or $8 - 2 = 6$, degrees of freedom. When $\alpha = 0.01$ and d.f. $= 6$, the value obtained from Table I is 0.834. For a significant relationship, a value of *r* greater than $+0.834$ or less than -0.834 is needed. Since the value of $r = -0.176$ is greater than -0.834, the null hypothesis is not rejected. Hence there is not enough evidence to say that there is a significant linear relationship between age and wealth. See Figure 10–10.

Figure 10–10

Rejection and Nonrejection Regions for Example 10–8

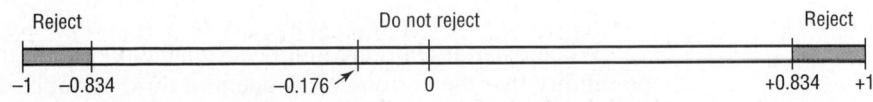

Correlation and Causation   Researchers must understand the nature of the linear relationship between the independent variable x and the dependent variable y. When a hypothesis test indicates that a significant linear relationship exists between the variables, researchers must consider the possibilities outlined next.

Possible Relationships Between Variables

When the null hypothesis has been rejected for a specific α value, any of the following five possibilities can exist.

1. *There is a direct cause-and-effect relationship between the variables.* That is, x causes y. For example, water causes plants to grow, poison causes death, and heat causes ice to melt.

2. *There is a reverse cause-and-effect relationship between the variables.* That is, y causes x. For example, suppose a researcher believes excessive coffee consumption causes nervousness, but the researcher fails to consider that the reverse situation may occur. That is, it may be that an extremely nervous person craves coffee to calm his or her nerves.

3. *The relationship between the variables may be caused by a third variable.* For example, if a statistician correlated the number of deaths due to drowning and the number of cans of soft drink consumed daily during the summer, he or she would probably find a significant relationship. However, the soft drink is not necessarily responsible for the deaths, since both variables may be related to heat and humidity.

4. *There may be a complexity of interrelationships among many variables.* For example, a researcher may find a significant relationship between students' high school grades and college grades. But there probably are many other variables involved, such as IQ, hours of study, influence of parents, motivation, age, and instructors.

5. *The relationship may be coincidental.* For example, a researcher may be able to find a significant relationship between the increase in the number of people who are exercising and the increase in the number of people who are committing crimes. But common sense dictates that any relationship between these two values must be due to coincidence.

When two variables are highly correlated, item 3 in the box states that there exists a possibility that the correlation is due to a third variable. If this is the case and the third variable is unknown to the researcher or not accounted for in the study, it is called a

lurking variable. An attempt should be made by the researcher to identify such variables and to use methods to control their influence.

It is important to restate the fact that even if the correlation between two variables is high, it does not necessarily mean causation. There are other possibilities, such as lurking variables or just a coincidental relationship. See the Speaking of Statistics article on page 548.

Also, you should be cautious when the data for one or both of the variables involve averages rather than individual data. It is not wrong to use averages, but the results cannot be generalized to individuals since averaging tends to smooth out the variability among individual data values. The result could be a higher correlation than actually exists.

Thus, when the null hypothesis is rejected, the researcher must consider all possibilities and select the appropriate one as determined by the study. Remember, correlation does not necessarily imply causation.

Applying the Concepts **10–1**

Stopping Distances

In a study on speed control, it was found that the main reasons for regulations were to make traffic flow more efficient and to minimize the risk of danger. An area that was focused on in the study was the distance required to completely stop a vehicle at various speeds. Use the following table to answer the questions.

MPH	Braking distance (feet)
20	20
30	45
40	81
50	133
60	205
80	411

Assume MPH is going to be used to predict stopping distance.

1. Which of the two variables is the independent variable?

2. Which is the dependent variable?

3. What type of variable is the independent variable?

4. What type of variable is the dependent variable?

5. Construct a scatter plot for the data.

6. Is there a linear relationship between the two variables?

7. Redraw the scatter plot, and change the distances between the independent-variable numbers. Does the relationship look different?

8. Is the relationship positive or negative?

9. Can braking distance be accurately predicted from MPH?

10. List some other variables that affect braking distance.

11. Compute the value of r.

12. Is r significant at $\alpha = 0.05$?

See page 589 for the answers.

Statistics

Coffee Not Disease Culprit, Study Says

In correlation and regression studies, it is difficult to control all variables. This study shows some of the consequences when researchers overlook certain aspects in studies. Suggest ways that the extraneous variables might be controlled in future studies.

NEW YORK (AP)—Two new studies suggest that coffee drinking, even up to $5\frac{1}{2}$ cups per day, does not increase the risk of heart disease, and other studies that claim to have found increased risks might have missed the true culprits, a researcher says.

"It might not be the coffee cup in one hand, it might be the cigarette or coffee roll in the other," said Dr. Peter W. F. Wilson, the author of one of the new studies.

He noted in a telephone interview Thursday that many coffee drinkers, particularly heavy coffee drinkers, are smokers. And one of the new studies found that coffee drinkers had excess fat in their diets.

The findings of the new studies conflict sharply with a study reported in November 1985 by Johns Hopkins University scientists in Baltimore.

The Hopkins scientists found that coffee drinkers who consumed five or more cups of coffee per day had three times the heart-disease risk of non-coffee drinkers.

The reason for the discrepancy appears to be that many of the coffee drinkers in the Hopkins study also smoked—and it was the smoking that increased their heart-disease risk, said Wilson.

Wilson, director of laboratories for the Framingham Heart Study in Framingham, Mass., said Thursday at a conference sponsored by the American Heart Association in Charleston, S.C., that he had examined the coffee intake of 3,937 participants in the Framingham study during 1956–66 and an additional 2,277 during the years 1972–1982.

In contrast to the subjects in the Hopkins study, most of these coffee drinkers consumed two or three cups per day, Wilson said. Only 10 percent drank six or more cups per day.

He then looked at blood cholesterol levels and heart and blood vessel disease in the two groups. "We ran these analyses for coronary heart disease, heart attack, sudden death and stroke and in absolutely every analysis, we found no link with coffee," Wilson said.

He found that coffee consumption was linked to a significant decrease in total blood cholesterol in men, and to a moderate increase in total cholesterol in women.

Source: Reprinted with permission of the Associated Press.

Exercises 10–1

1. What is meant by the statement that two variables are related? Two variables are related when a discernible pattern exists between them.

2. How is a linear relationship between two variables measured in statistics? Explain.

3. What is the symbol for the sample correlation coefficient? The population correlation coefficient? r, ρ (rho)

4. What is the range of values for the correlation coefficient? The range of r is from -1 to $+1$.

5. What is meant when the relationship between the two variables is called positive? Negative?

6. Give examples of two variables that are positively correlated and two that are negatively correlated. Answers will vary.

7. Give an example of a correlation study, and identify the independent and dependent variables. Answers will vary.

8. What is the diagram of the independent and dependent variables called? Why is drawing this diagram important? The diagram is called a scatter plot. It shows the nature of the relationship.

9. What is the name of the correlation coefficient used in this section? Pearson product moment correlation coefficient

10. What statistical test is used to test the significance of the correlation coefficient? t test

11. When two variables are correlated, can the researcher be sure that one variable causes the other? Why or why not? There are many other possibilities, such as chance or relationship to a third variable.

For Exercises 12 through 27, perform the following steps.

  a. Draw the scatter plot for the variables.

  b. Compute the value of the correlation coefficient.

c. State the hypotheses.

d. Test the significance of the correlation coefficient at $\alpha = 0.05$, using Table I.

e. Give a brief explanation of the type of relationship.

12. Gas Tax and Fuel Use The data below indicate the state gas tax in cents per gallon and the fuel use per registered vehicle (in gallons). Is there a significant relationship between these two variables?

Tax	21.5	23	18	24.5	26.4	19
Usage	1062	631	920	686	736	684

(The information in this exercise will be used for Exercise 12 in Section 10–2.)

Source: *World Almanac.*

13. Commercial Movie Releases The yearly data have been published showing the number of releases for each of the commercial movie studios and the gross receipts for those studios thus far. Based on these data, can it be concluded that there is a relationship between the number of releases and the gross receipts?

No. of releases *x*	361	270	306	22	35	10	8	12	21
Gross receipts *y* (million $)	3844	1962	1371	1064	334	241	188	154	125

(The information in this exercise will be used for Exercises 13 and 36 in Section 10–2 and Exercises 15 and 19 in Section 10–3.)

Source: www.showbizdata.com

14. Forest Fires and Acres Burned An environmentalist wants to determine the relationships between the numbers (in thousands) of forest fires over the year and the number (in hundred thousands) of acres burned. The data for 8 recent years are shown. Describe the relationship.

Number of fires *x*	72	69	58	47	84	62	57	45
Number of acres burned *y*	62	42	19	26	51	15	30	15

Source: National Interagency Fire Center.

(The information in this exercise will be used for Exercise 14 in Section 10–2 and Exercises 16 and 20 in Section 10–3.)

15. Alumni Contributions The director of an alumni association for a small college wants to determine whether there is any type of relationship between the amount of an alumnus's contribution (in dollars) and the years the alumnus has been out of school. The data follow. (The information is used for Exercises 15, 36, and 37 in Section 10–2 and Exercises 17 and 21 in Section 10–3.)

Years *x*	1	5	3	10	7	6
Contribution *y*	500	100	300	50	75	80

16. State Debt and Per Capita Tax An economics student wishes to see if there is a relationship between the amount of state debt per capita and the amount of tax per capita at the state level. Based on the following data, can she or he conclude that per capita state debt and per capita state taxes are related? Both amounts are in dollars and represent five randomly selected states. (The information in this exercise will be used for Exercises 16 and 37 in Section 10–2 and Exercises 18 and 22 in Section 10–3.)

Per capita debt *x*	1924	907	1445	1608	661
Per capita tax *y*	1685	1838	1734	1842	1317

Source: *World Almanac.*

17. School Districts and Secondary Schools A random sample of states yielded the following numbers of local school districts and the corresponding numbers of secondary schools. Is there a significant relationship between the data?

School districts	53	19	24	17	95	68
Secondary schools	50	27	187	84	143	216

Source: *World Almanac.*

(The information in this exercise will be used for Exercise 17 of Section 10–2.)

18. Triples and Home Runs The data below show the number of three-base hits (triples) and the number of home runs hit during the season by a random sample of MLB teams. Is there a significant relationship between the data?

Triples	25	23	51	19	20	43
Home runs	212	199	144	160	149	122

Source: *New York Times Almanac.*

(The information in this exercise will be used for Exercises 18 and 38 in Section 10–2.)

19. Egg Production Recent agricultural data showed the number of eggs produced and the price received per dozen for a given year. Based on the following data for a random selection of states, can it be concluded that a relationship exists between the number of eggs produced and the price per dozen? (The information in this exercise will be used for Exercise 19 in Section 10–2.)

No. of eggs (millions) *x*	957	1332	1163	1865	119	273
Price per dozen (dollars) *y*	0.770	0.697	0.617	0.652	1.080	1.420

Source: *World Almanac.*

20. Emergency Calls and Temperature An emergency service wishes to see whether a relationship exists between the outside temperature and the number of emergency calls it receives for a 7-hour period. The data are shown. (The information in this exercise will be used for Exercises 20 and 38 in Section 10–2.)

Temperature x	68	74	82	88	93	99	101
No. of calls y	7	4	8	10	11	9	13

21. Faculty and Students The number of faculty and the number of students are shown for a random selection of small colleges. Is there a significant relationship between the two variables? Switch x and y and repeat the process. Which do you think is really the independent variable?

Faculty	99	110	113	116	138	174	220
Students	1353	1290	1091	1213	1384	1283	2075

Source: *World Almanac.*

(The information in this exercise will be used for Exercises 21 and 36 in Section 10–2.)

22. Precipitation and Snow/Sleet For a random selection of U.S. cities, the following data show the number of days for which the precipitation is greater than or equal to 0.01 inch and the number of days for which there is at least 1 inch of snow and/or sleet. Is there a significant linear relationship between the variables?

Precipitation ≥ 0.01 inch	61	111	140	116	88	136
Snow/sleet ≥ 1 in	2	15	21	8	11	13

Source: *World Almanac.*

(The information in this exercise will be used for Exercise 22 in Section 10–2.)

23. Average Temperature and Precipitation The average normal daily temperature (in degrees Fahrenheit) and the corresponding average monthly precipitation (in inches) for the month of June are shown here for seven randomly selected cities in the United States. Determine if there is a relationship between the two variables. (The information in this exercise will be used for Exercise 23 in Section 10–2.)

Avg. daily temp. x	86	81	83	89	80	74	64
Avg. mo. precip. y	3.4	1.8	3.5	3.6	3.7	1.5	0.2

Source: *New York Times Almanac.*

24. NHL Assists and Total Points A random sample of scoring leaders from the NHL showed the following numbers of assists and total points. Based on these data, can it be concluded that there is a significant relationship between the two?

Assists	26	29	32	34	36	37	40
Total points	48	68	66	69	76	67	84

Source: *Associated Press.*

(The information in this exercise will be used for Exercise 24 in Section 10–2.)

25. Fat Grams and Secondary Schools The numbers of fat calories and grams of saturated fat in a number of fast-food nonbreakfast entrees are shown below. Is there sufficient evidence to conclude a significant relationship between the two variables?

Fat calories	190	220	270	360	460	540
Sat. fat (g)	9	8	13	17	23	27

Source: www.fatcalories.com

(The information in this exercise will be used in Exercise 25 in Section 10–2.)

26. Tall Buildings An architect wants to determine the relationship between the heights (in feet) of a building and the number of stories in the building. The data for a sample of 10 buildings in Pittsburgh are shown. Explain the relationship.

Stories x	64	54	40	31	45	38	42	41	37	40
Height y	841	725	635	616	615	582	535	520	511	485

Source: *World Almanac Book of Facts.*

(The information in this exercise will be used for Exercise 26 of Section 10–2.)

27. Hospital Beds A hospital administrator wants to see if there is a relationship between the number of licensed beds and the number of staffed beds in local hospitals. The data for a specific day are shown. Describe the relationship.

Licensed beds x	144	32	175	185	208	100	169
Staffed beds y	112	32	162	141	103	80	118

Source: *Pittsburgh Tribune-Review.*

(The information in this exercise will be used for Exercise 28 of this section and Exercise 27 in Section 10–2.)

Extending the Concepts

28. One of the formulas for computing r is

$$r = \frac{\Sigma(x - \bar{x})(y - \bar{y})}{(n - 1)(s_x)(s_y)}$$

Using the data in Exercise 27, compute r with this formula. Compare the results.

 29. Compute r for the data set shown. Explain the reason for this value of r. Now, interchange the values of x and y and compute r again. Compare this value with the previous one. Explain the results of the comparison.

x	1	2	3	4	5
y	3	5	7	9	11

 30. Compute r for the following data and test the hypothesis $H_0\colon \rho = 0$. Draw the scatter plot; then explain the results.

x	−3	−2	−1	0	1	2	3
y	9	4	1	0	1	4	9

10–2   Regression

Objective 4

Compute the equation of the regression line.

In studying relationships between two variables, collect the data and then construct a scatter plot. The purpose of the scatter plot, as indicated previously, is to determine the nature of the relationship. The possibilities include a positive linear relationship, a negative linear relationship, a curvilinear relationship, or no discernible relationship. After the scatter plot is drawn, the next steps are to compute the value of the correlation coefficient and to test the significance of the relationship. If the value of the correlation coefficient is significant, the next step is to determine the equation of the **regression line,** which is the data's line of best fit. (*Note:* Determining the regression line when r is not significant and then making predictions using the regression line are meaningless.) The purpose of the regression line is to enable the researcher to see the trend and make predictions on the basis of the data.

Line of Best Fit

Figure 10–11 shows a scatter plot for the data of two variables. It shows that several lines can be drawn on the graph near the points. Given a scatter plot, you must be able to draw the *line of best fit. Best fit* means that the sum of the squares of the vertical distances from

Figure 10–11

Scatter Plot with Three Lines Fit to the Data

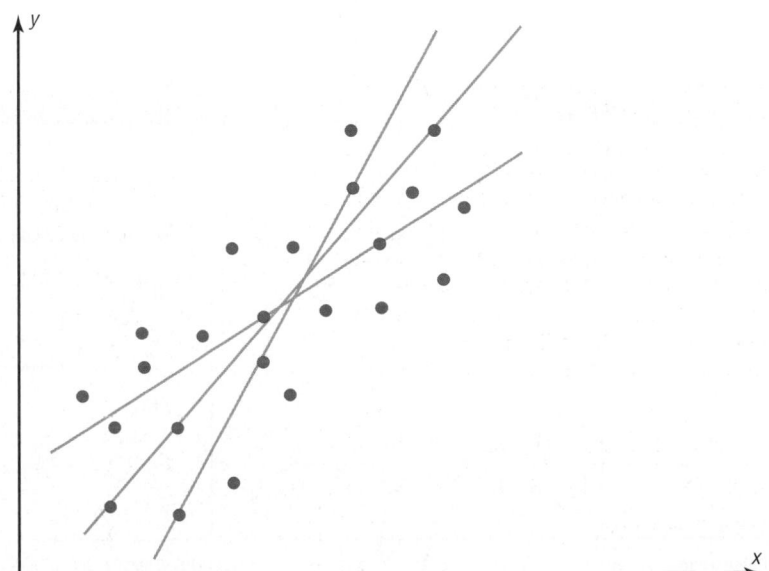

Figure 10–12

Line of Best Fit for a
Set of Data Points

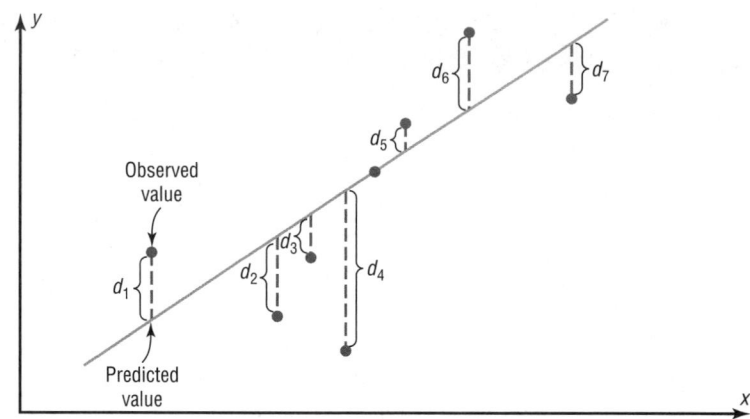

Historical Notes

Francis Galton drew
the line of best fit
visually. An assistant
of Karl Pearson's
named G. Yule
devised the
mathematical
solution using the
least-squares
method, employing
a mathematical
technique developed
by Adrien-Marie
Legendre about
100 years earlier.

each point to the line is at a minimum. The reason you need a line of best fit is that the values of y will be predicted from the values of x; hence, the closer the points are to the line, the better the fit and the prediction will be. See Figure 10–12. When r is positive, the line slopes upward and to the right. When r is negative, the line slopes downward from left to right.

Determination of the Regression Line Equation

In algebra, the equation of a line is usually given as $y = mx + b$, where m is the slope of the line and b is the y intercept. (Students who need an algebraic review of the properties of a line should refer to Appendix A, Section A–3, before studying this section.) In statistics, the equation of the regression line is written as $y' = a + bx$, where a is the y' intercept and b is the slope of the line. See Figure 10–13.

There are several methods for finding the equation of the regression line. Two formulas are given here. *These formulas use the same values that are used in computing the value of the correlation coefficient.* The mathematical development of these formulas is beyond the scope of this book.

Figure 10–13

A Line as Represented in Algebra and in Statistics

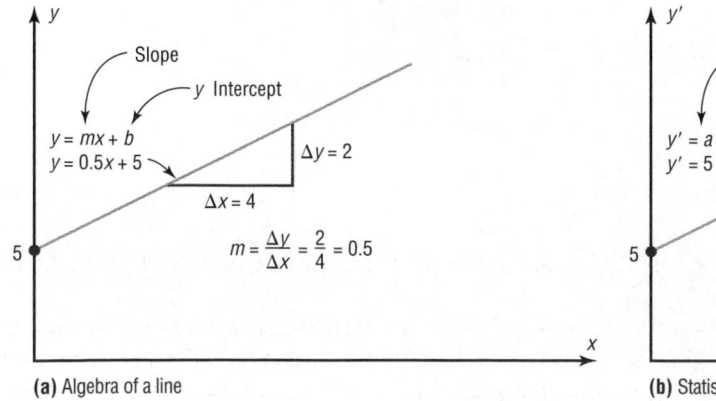

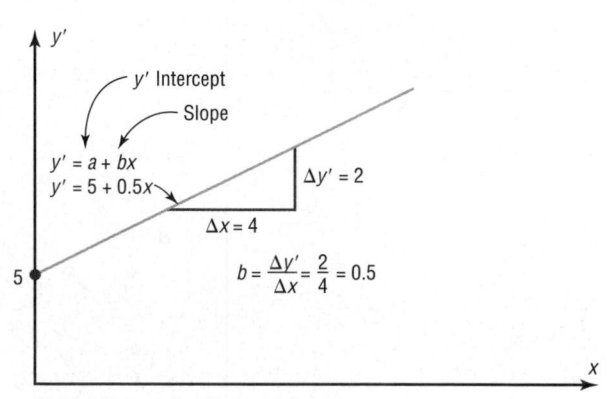

(a) Algebra of a line

(b) Statistical notation for a regression line

Formulas for the Regression Line $y' = a + bx$

$$a = \frac{(\Sigma y)(\Sigma x^2) - (\Sigma x)(\Sigma xy)}{n(\Sigma x^2) - (\Sigma x)^2}$$

$$b = \frac{n(\Sigma xy) - (\Sigma x)(\Sigma y)}{n(\Sigma x^2) - (\Sigma x)^2}$$

where a is the y' intercept and b is the slope of the line.

Rounding Rule for the Intercept and Slope   Round the values of a and b to three decimal places.

| Example 10–9 | Car Rental Companies |

Find the equation of the regression line for the data in Example 10–4, and graph the line on the scatter plot of the data.

Solution

The values needed for the equation are $n = 6$, $\Sigma x = 153.8$, $\Sigma y = 18.7$, $\Sigma xy = 682.77$, and $\Sigma x^2 = 5859.26$. Substituting in the formulas, you get

$$a = \frac{(\Sigma y)(\Sigma x^2) - (\Sigma x)(\Sigma xy)}{n(\Sigma x^2) - (\Sigma x)^2} = \frac{(18.7)(5859.26) - (153.8)(682.77)}{(6)(5859.26) - (153.8)^2} = 0.396$$

$$b = \frac{n(\Sigma xy) - (\Sigma x)(\Sigma y)}{n(\Sigma x^2) - (\Sigma x)^2} = \frac{6(682.77) - (153.8)(18.7)}{(6)(5859.26) - (153.8)^2} = 0.106$$

Hence, the equation of the regression line $y' = a + bx$ is

$$y' = 0.396 + 0.106x$$

To graph the line, select any two points for x and find the corresponding values for y. Use any x values between 10 and 60. For example, let $x = 15$. Substitute in the equation and find the corresponding y' value.

$$y' = 0.396$$
$$= 0.396 + 0.106(15)$$
$$= 1.986$$

Let $x = 40$; then

$$y' = 0.396 + 0.106x$$
$$= 0.396 + 0.106(40)$$
$$= 4.636$$

Then plot the two points (15, 1.986) and (40, 4.636) and draw a line connecting the two points. See Figure 10–14.

Note: When you draw the regression line, it is sometimes necessary to *truncate* the graph (see Chapter 2). This is done when the distance between the origin and the first labeled coordinate on the x axis is not the same as the distance between the rest of the

Figure 10–14

Regression Line for Example 10–9

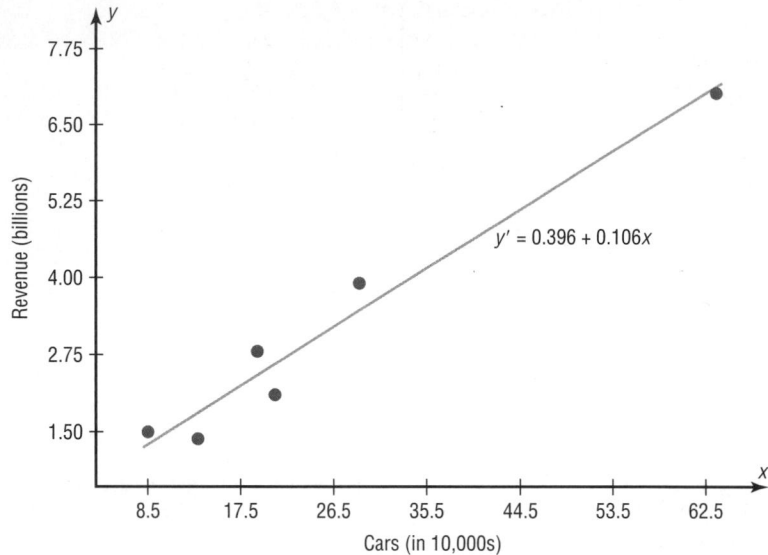

labeled x coordinates or the distance between the origin and the first labeled y' coordinate is not the same as the distance between the other labeled y' coordinates. When the x axis or the y axis has been truncated, do not use the y' intercept value to graph the line. When you graph the regression line, always select x values between the smallest x data value and the largest x data value.

Example 10–10

Absences and Final Grades

Find the equation of the regression line for the data in Example 10–5, and graph the line on the scatter plot.

Solution

The values needed for the equation are $n = 7$, $\Sigma x = 57$, $\Sigma y = 511$, $\Sigma xy = 3745$, and $\Sigma x^2 = 579$. Substituting in the formulas, you get

$$a = \frac{(\Sigma y)(\Sigma x^2) - (\Sigma x)(\Sigma xy)}{n(\Sigma x^2) - (\Sigma x)^2} = \frac{(511)(579) - (57)(3745)}{(7)(579) - (57)^2} = 102.493$$

$$b = \frac{n(\Sigma xy) - (\Sigma x)(\Sigma y)}{n(\Sigma x^2) - (\Sigma x)^2} = \frac{(7)(3745) - (57)(511)}{(7)(579) - (57)^2} = -3.622$$

Hence, the equation of the regression line $y' = a + bx$ is

$$y' = 102.493 - 3.622x$$

The graph of the line is shown in Figure 10–15.

Historical Note

In 1795, Adrien-Marie Legendre (1752–1833) measured the meridian arc on the earth's surface from Barcelona, Spain, to Dunkirk, England. This measure was used as the basis for the measure of the meter. Legendre developed the least-squares method around the year 1805.

The sign of the correlation coefficient and the sign of the slope of the regression line will always be the same. That is, if r is positive, then b will be positive; if r is negative, then b will be negative. The reason is that the numerators of the formulas are the same and determine the signs of r and b, and the denominators are always positive. The regression line will always pass through the point whose x coordinate is the mean of the x values and whose y coordinate is the mean of the y values, that is, $(\bar{x}, \bar{y})$.

Figure 10-15

Regression Line for
Example 10-10

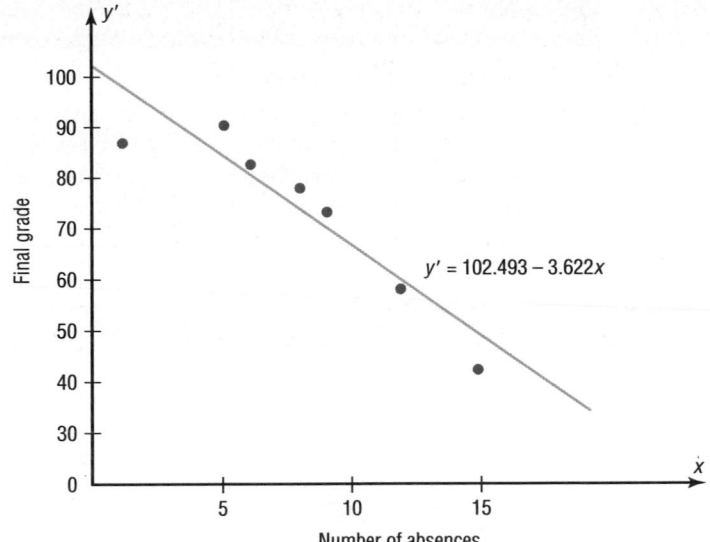

$$y' = 102.493 - 3.622x$$

Final grade (y-axis)

Number of absences (x-axis)

The regression line can be used to make predictions for the dependent variable. The method for making predictions is shown in Example 10-11.

Example 10-11   Car Rental Companies

Use the equation of the regression line to predict the income of a car rental agency that has 200,000 automobiles.

Solution

Since the x values are in 10,000s, divide 200,000 by 10,000 to get 20, and then substitute 20 for x in the equation.

$$y' = 0.396 + 0.106x$$
$$= 0.396 + 0.106(20)$$
$$= 2.516$$

Hence, when a rental agency has 200,000 automobiles, its revenue will be approximately $2.516 billion.

The value obtained in Example 10-11 is a point prediction, and with point predictions, no degree of accuracy or confidence can be determined. More information on prediction is given in Section 10-3.

The magnitude of the change in one variable when the other variable changes exactly 1 unit is called a **marginal change.** The value of slope b of the regression line equation represents the marginal change. For example, in Example 10-9 the slope of the regression line is 0.106, which means for each increase of 10,000 cars, the value of y changes 0.106 unit ($106 million) on average.

When r is not significantly different from 0, the best predictor of y is the mean of the data values of y. For valid predictions, the value of the correlation coefficient must be significant. Also, two other assumptions must be met.

Assumptions for Valid Predictions in Regression

1. The sample is a random sample.
2. For any specific value of the independent variable x, the value of the dependent variable y must be normally distributed about the regression line. See Figure 10–16(a).
3. The standard deviation of each of the dependent variables must be the same for each value of the independent variable. See Figure 10–16(b).

Figure 10–16

Assumptions for Predictions

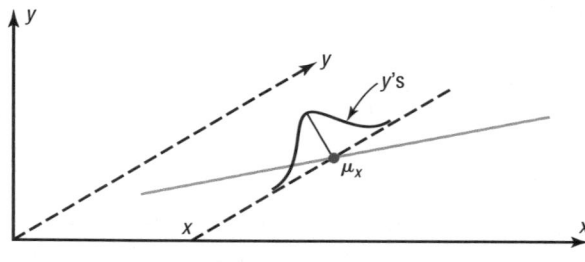

(a) Dependent variable y normally distributed

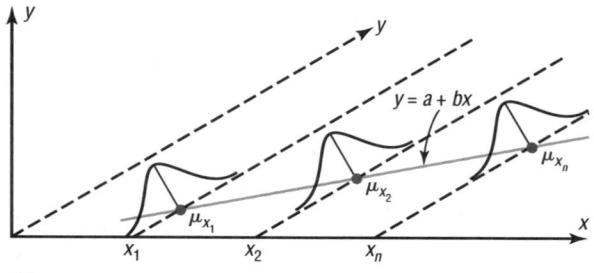

(b) $\sigma_1 = \sigma_2 = \cdots = \sigma_n$

Extrapolation, or making predictions beyond the bounds of the data, must be interpreted cautiously. For example, in 1979, some experts predicted that the United States would run out of oil by the year 2003. This prediction was based on the current consumption and on known oil reserves at that time. However, since then, the automobile industry has produced many new fuel-efficient vehicles. Also, there are many as yet undiscovered oil fields. Finally, science may someday discover a way to run a car on something as unlikely but as common as peanut oil. In addition, the price of a gallon of gasoline was predicted to reach $10 a few years later. Fortunately this has not come to pass. *Remember that when predictions are made, they are based on present conditions or on the premise that present trends will continue.* This assumption may or may not prove true in the future.

The steps for finding the value of the correlation coefficient and the regression line equation are summarized in this Procedure Table:

Interesting Fact

It is estimated that wearing a motorcycle helmet reduces the risk of a fatal accident by 30%.

Procedure Table

Finding the Correlation Coefficient and the Regression Line Equation

Step 1 Make a table, as shown in step 2.

Step 2 Find the values of xy, x^2, and y^2. Place them in the appropriate columns and sum each column.

x	y	xy	x^2	y^2
.	.	.	.	.
.	.	.	.	.
.	.	.	.	.
$\Sigma x =$	$\Sigma y =$	$\Sigma xy =$	$\Sigma x^2 =$	$\Sigma y^2 =$

Procedure Table (*Continued*)

Step 3   Substitute in the formula to find the value of r.

$$r = \frac{n(\Sigma xy) - (\Sigma x)(\Sigma y)}{\sqrt{[n(\Sigma x^2) - (\Sigma x)^2][n(\Sigma y^2) - (\Sigma y)^2]}}$$

Step 4   When r is significant, substitute in the formulas to find the values of a and b for the regression line equation $y' = a + bx$.

$$a = \frac{(\Sigma y)(\Sigma x^2) - (\Sigma x)(\Sigma xy)}{n(\Sigma x^2) - (\Sigma x)^2} \qquad b = \frac{n(\Sigma xy) - (\Sigma x)(\Sigma y)}{n(\Sigma x^2) - (\Sigma x)^2}$$

A scatter plot should be checked for outliers. An outlier is a point that seems out of place when compared with the other points (see Chapter 3). Some of these points can affect the equation of the regression line. When this happens, the points are called **influential points** or **influential observations.**

When a point on the scatter plot appears to be an outlier, it should be checked to see if it is an influential point. An influential point tends to "pull" the regression line toward the point itself. To check for an influential point, the regression line should be graphed with the point included in the data set. Then a second regression line should be graphed that excludes the point from the data set. If the position of the second line is changed considerably, the point is said to be an influential point. Points that are outliers in the x direction tend to be influential points.

Researchers should use their judgment as to whether to include influential observations in the final analysis of the data. If the researcher feels that the observation is not necessary, then it should be excluded so that it does not influence the results of the study. However, if the researcher feels that it is necessary, then he or she may want to obtain additional data values whose x values are near the x value of the influential point and then include them in the study.

Explain that to me. "

© Dave Carpenter. King Features Syndicate.

Applying the Concepts 10–2

Stopping Distances Revisited

In a study on speed and braking distance, researchers looked for a method to estimate how fast a person was traveling before an accident by measuring the length of the skid marks. An area that was focused on in the study was the distance required to completely stop a vehicle at various speeds. Use the following table to answer the questions.

MPH	Braking distance (feet)
20	20
30	45
40	81
50	133
60	205
80	411

Assume MPH is going to be used to predict stopping distance.

1. Find the linear regression equation.
2. What does the slope tell you about MPH and the braking distance? How about the y intercept?
3. Find the braking distance when MPH = 45.
4. Find the braking distance when MPH = 100.
5. Comment on predicting beyond the given data values.

See page 590 for the answers.

Exercises 10–2

1. What two things should be done before one performs a regression analysis?

2. What are the assumptions for regression analysis?

3. What is the general form for the regression line used in statistics? $y' = a + bx$

4. What is the symbol for the slope? For the y intercept? b, a

5. What is meant by the *line of best fit?*

6. When all the points fall on the regression line, what is the value of the correlation coefficient? r would equal +1 or −1.

7. What is the relationship between the sign of the correlation coefficient and the sign of the slope of the regression line? When r is positive, b will be positive. When r is negative, b will be negative.

8. As the value of the correlation coefficient increases from 0 to 1, or decreases from 0 to −1, how do the points of the scatter plot fit the regression line? They would be clustered closer to the line.

9. How is the value of the correlation coefficient related to the accuracy of the predicted value for a specific value of x? The closer r is to +1 or −1, the more accurate the predicted value will be.

10. If the value of r is not significant, what can be said about the regression line?

11. When the value of r is not significant, what value should be used to predict y? When r is not significant, the mean of the y values should be used to predict y.

For Exercises 12 through 27, use the same data as for the corresponding exercises in Section 10–1. For each exercise, find the equation of the regression line and find the y' value for the specified x value. Remember that no regression should be done when r is not significant.

12. **Gas Tax and Fuel Use** The gas tax and fuel use are shown.

Tax	21.5	23	18	24.5	26.4	19
Usage	1062	631	920	686	736	684

Find y' when $x = \$0.25$. Not significant so no regression should be done.

13. **Commercial Movie Releases** New movie releases per studio and gross receipts are as follows:

No. of releases	361	270	306	22	35	10	8	12	21
Gross receipts (million $)	3844	1962	1371	1064	334	241	188	154	125

Find y' when $x = 200$ new releases. $y' = 181.661 + 7.319x$; $y' = 1645.5$ (million $)

14. Forest Fires and Acres Burned Number of fires and number of acres burned are as follows:

Fires x	72	69	58	47	84	62	57	45
Acres y	62	41	19	26	51	15	30	15

Find y' when $x = 60$. $y' = -31.46 + 1.036x$; 30.7

15. Years and contribution data are as follows:

Years x	1	5	3	10	7	6
Contribution y, \$	500	100	300	50	75	80

Find y' when $x = 4$ years. $y' = 453.176 - 50.439x$; 251.42

16. State Debt and Per Capita Taxes Data for per capita state debt and per capita state tax are as follows:

Per capita debt	1924	907	1445	1608	661
Per capita tax	1685	1838	1734	1842	1317

Find y' when $x = \$1500$ in per capita debt.
Not significant so no regression should be done.

17. School Districts and Secondary Schools The number of school districts and the number of secondary schools in the district are shown.

School districts	53	19	24	17	95	68
Secondary schools	50	27	187	84	143	216

Find y' when $x = 70$. Since r is not significant, no regression should be done.

18. Triples and Home Runs The number of triples and the number of home runs obtained by a selected sample of MLB players are shown.

Triples	25	23	51	19	20	43
Home runs	212	199	144	160	149	122

Find y' when $x = 33$. Since r is not significant, no regression should be done.

19. Egg Production Number of eggs and price per dozen are shown.

No. of eggs (million)	957	1332	1163	1865	119	273
Price per dozen (\$)	0.770	0.697	0.617	0.652	1.080	1.420

Find y' when $x = 1600$ million eggs.
$y' = 1.252 - 0.000398x$; $y' = 0.615$ per dozen

20. Emergency Calls and Temperature Temperature in degrees Fahrenheit and number of emergency calls are shown.

Temperature x	68	74	82	88	93	99	101
No. of calls y	7	4	8	10	11	9	13

Find y' when $x = 80°F$. $y' = -7.544 + 0.190x$; 7.656, or 8 calls

21. Faculty and Students The number of faculty and the number of students in a random selection of small colleges are shown.

Faculty	99	110	113	116	138	174	220
Students	1353	1290	1091	1213	1384	1283	2075

Now find the equation of the regression line when x and y are interchanged. $y' = -14.974 + 0.111x$

22. Precipitation and Snowfall/Sleet The number of days of precipitation and snowfall/sleet are shown.

Precipitation	61	111	140	116	88	136
Snow/sleet	2	15	21	8	11	13

Find y' when $x = 100$ days.
$y' = -7.327 + 0.175x$; 10.173 in

23. Average Temperature and Precipitation Temperatures (in degrees Fahrenheit) and precipitation (in inches) are as follows:

Avg. daily temp. x	86	81	83	89	80	74	64
Avg. mo. precip. y	3.4	1.8	3.5	3.6	3.7	1.5	0.2

Find y' when $x = 70°F$. $y' = -8.994 + 0.1448x$; 1.1

24. NHL Assists and Total Points The number of assists and the total number of points for a sample of NHL scoring leaders are shown.

Assists	26	29	32	34	36	37	40
Total points	48	68	66	69	76	67	84

Find y' when $x = 30$ assists. $y' = 2.693 + 1.962x$; 62

25. Fat Calories and Fat Grams The number of fat calories and the number of saturated fat grams for a random selection of breakfast entrees are shown.

Fat calories	190	220	270	360	460	540
Sat. fat (g)	9	8	13	17	23	27

Find y' when $x = 400$ fat calories. $y' = -2.417 + 0.055x$; 19.6 grams

26. Tall Buildings Stories and heights of buildings data follow:

Stories x	64	54	40	31	45	38	42	41	37	40
Heights y	841	725	635	616	615	582	535	520	511	485

Find y' when $x = 44$. $y' = 206.399 + 9.262x$; 613.9

27. Hospital Beds Licensed beds and staffed beds data follow:

Licensed beds x	144	32	175	185	208	100	169
Staffed beds y	112	32	162	141	103	80	118

Find y' when $x = 44$. $y' = 22.659 + 0.582x$; 48.267

For Exercises 28 through 33, do a complete regression analysis by performing these steps.

 a. Draw a scatter plot.
 b. Compute the correlation coefficient.
 c. State the hypotheses.
 d. Test the hypotheses at $\alpha = 0.05$. Use Table I.
 e. Determine the regression line equation.
 f. Plot the regression line on the scatter plot.
 g. Summarize the results.

28. Fireworks and Injuries These data were obtained for the years 1993 through 1998 and indicate the number of fireworks (in millions) used and the related injuries. Predict the number of injuries if 100 million fireworks are used during a given year.

Fireworks in use x	67.6	87.1	117	115	118	113
Related injuries y	12,100	12,600	12,500	10,900	7800	7000

Source: National Council of Fireworks Safety, American Pyrotechnic Assoc.

29. **Farm Acreage** Is there a relationship between the number of farms in a state and the acreage per farm? A random selection of states across the country, both eastern and western, produced the following results. Can a relationship between these two variables be concluded?

No. of farms (thousands) x	77	52	20.8	49	28	58.2
Acreage per farm y	347	173	173	218	246	132

Source: *World Almanac.*

30. **SAT Scores** Educational researchers desired to find out if a relationship exists between the average SAT verbal score and the average SAT mathematical score. Several states were randomly selected, and their SAT average scores are recorded below. Is there sufficient evidence to conclude a relationship between the two scores?

Verbal x	526	504	594	585	503	589
Math y	530	522	606	588	517	589

Source: *World Almanac.*

31. **Coal Production** These data were obtained from a sample of counties in southwestern Pennsylvania and indicate the number (in thousands) of tons of bituminous coal produced in each county and the number of employees working in coal production in each county. Predict the amount of coal produced for a county that has 500 employees.

No. of employees x	110	731	1031	20	118	1162	103	752
Tons y	227	5410	5328	147	729	8095	635	6157

32. **Television Viewers** A television executive selects 10 television shows and compares the average number of viewers the show had last year with the average number of viewers this year. The data (in millions) are shown. Describe the relationship.

Viewers last year x	26.6	17.85	20.3	16.8	20.8
Viewers this year y	28.9	19.2	26.4	13.7	20.2
Viewers last year x	16.7	19.1	18.9	16.0	15.8
Viewers this year y	18.8	25.0	21.0	16.8	15.3

Source: Nielsen Media Research.

33. **Absences and Final Grades** An educator wants to see how the number of absences for a student in her class affects the student's final grade. The data obtained from a sample are shown.

No. of absences x	10	12	2	0	8	5
Final grade y	70	65	96	94	75	82

For Exercises 34 and 35, do a complete regression analysis and test the significance of r at $\alpha = 0.05$, using the P-value method.

34. **Father's and Son's Weights** A physician wishes to know whether there is a relationship between a father's weight (in pounds) and his newborn son's weight (in pounds). The data are given here.

Father's weight x	176	160	187	210	196	142	205	215
Son's weight y	6.6	8.2	9.2	7.1	8.8	9.3	7.4	8.6

35. **Age and Net Worth** Is a person's age related to his or her net worth? A sample of 10 billionaires is selected, and the person's age and net worth are compared. The data are given here.

Age x	56	39	42	60	84	37	68	66	73	55
Net worth (billion $) y	18	14	12	14	11	10	10	7	7	5

Source: The Associated Press.

Extending the Concepts

36. For Exercises 13, 15, and 21 in Section 10–1, find the mean of the x and y variables. Then substitute the mean of the x variable into the corresponding regression line equations found in Exercises 13, 15, and 21 in this section and find y'. Compare the value of y' with $\bar{y}$ for each exercise. Generalize the results.

37. The y intercept value a can also be found by using the equation

$$a = \bar{y} - b\bar{x}$$

Verify this result by using the data in Exercises 15 and 16 of Sections 10–1 and 10–2. 453.173; regression should not be done.

38. The value of the correlation coefficient can also be found by using the formula

$$r = \frac{bs_x}{s_y}$$

where s_x is the standard deviation of the x values and s_y is the standard deviation of the y values. Verify this result for Exercises 18 and 20 of Section 10–1. $r = -0.543; 0.812$

Technology *Step by Step*

MINITAB
Step by Step

Create a Scatter Plot

1. These instructions use the following data:

x	6	2	15	9	12	5	8
y	82	86	43	74	58	90	78

Enter the data into three columns. The subject column is optional (see step 6b).

2. Name the columns **C1 Subject, C2 Age,** and **C3 Pressure.**

3. Select **Graph>Scatterplot,** then select Simple and click [OK].

4. Double-click on C3 Pressure for the [Y] variable and C2 Age for the predictor [X] variable.

5. Click [Data View]. The Data Display should be Symbols. If not, click the option box to select it. Click [OK].

6. Click [Labels].

   a) Type **Pressure vs. Age** in the text box for Titles/Footnotes, then type **Your Name** in the box for Subtitle 1.

   b) *Optional:* Click the tab for Data Labels, then click the option to Use labels from column.

   c) Select C1 Subject.

7. Click [OK] twice.

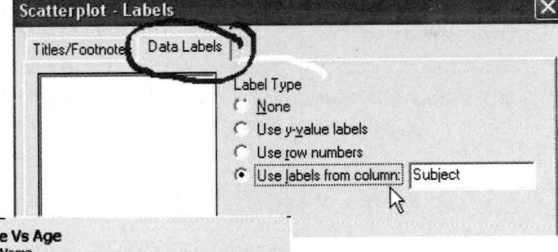

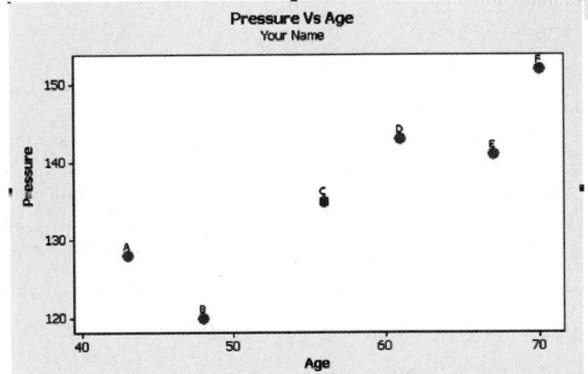

Calculate the Correlation Coefficient

8. Select **Stat>Basic Statistics>Correlation.**

9. Double-click C3 Pressure, then double-click C2 Age. The box for Display p-values should be checked.

10. Click [OK]. The correlation coefficient will be displayed in the session window, $r = +0.897$ with a *P*-value of 0.015.

Determine the Equation of the Least-Squares Regression Line

11. Select **Stat>Regression>Regression.**

12. Double-click Pressure in the variable list to select it for the Response variable Y.

13. Double-click C2 Age in the variable list to select it for the Predictors variable X.

14. Click on [Storage], then check the boxes for Residuals and Fits.

15. Click [OK] twice.

The session window will contain the regression analysis as shown.

In the worksheet two new columns will be added with the fitted values and residuals. Summary: The scatter plot and correlation coefficient confirm a strong positive linear correlation between pressure and age. The null hypothesis would be rejected at a significance level of 0.015. The equation of the regression equation is pressure = 81.0 + 0.964 (age).

↓	C1-T	C2	C3	C4	C5
	Subject	Age	Pressure	RESI1	FITS1
1	A	43	128	5.48353	122.516
2	B	48	120	-7.33838	127.338
3	C	56	135	-0.05343	135.053
4	D	61	143	3.12467	139.875
5	E	67	141	-4.66162	145.662
6	F	70	152	3.44524	148.555

Regression Analysis: Pressure versus Age

```
The regression equation is
Pressure = 81.0 + 0.964 Age

Predictor         Coef       SE Coef           T          P
Constant         81.05         13.88        5.84      0.004
Age             0.9644        0.2381        4.05      0.015
S = 5.641     R-Sq = 80.4%      R-Sq (adj) = 75.5%

Analysis of Variance
Source            DF            SS           MS          F          P
Regression         1        522.21       522.21      16.41      0.015
Residual Error     4        127.29        31.82
Total              5        649.50
```

TI-83 Plus or TI-84 Plus
Step by Step

Correlation and Regression

To graph a scatter plot:

1. Enter the x values in L_1 and the y values in L_2.
2. Make sure the Window values are appropriate. Select an Xmin slightly less than the smallest x data value and an Xmax slightly larger than the largest x data value. Do the same for Ymin and Ymax. Also, you may need to change the Xscl and Yscl values, depending on the data.
3. Press **2nd [STAT PLOT] 1** for Plot 1. The other y functions should be turned off.
4. Move the cursor to On and press **ENTER** on the Plot 1 menu.
5. Move the cursor to the graphic that looks like a scatter plot next to Type (first graph), and press **ENTER.** Make sure the X list is L_1, and the Y list is L_2.
6. Press **GRAPH.**

Example TI10–1

Draw a scatter plot for the following data.

x	43	48	56	61	67	70
y	128	120	135	143	141	152

The input and output screens are shown.

Input	Input	Output
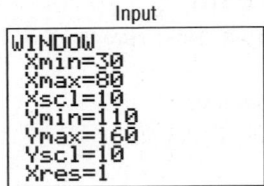	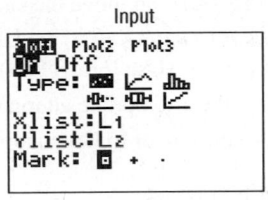	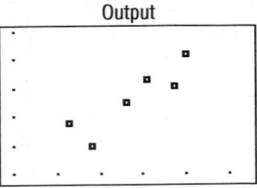

To find the equation of the regression line:

1. Press **STAT** and move the cursor to Calc.

2. Press **8** for LinReg(a+bx) then **ENTER**. The values for *a* and *b* will be displayed.

In order to have the calculator compute and display the correlation coefficient and coefficient of determination as well as the equation of the line, you must set the diagnostics display mode to on. Follow these steps:

1. Press **2nd [CATALOG]**.

2. Use the arrow keys to scroll down to DiagnosticOn.

3. Press **ENTER** to copy the command to the home screen.

4. Press **ENTER** to execute the command.

You will have to do this only once. Diagnostic display mode will remain on until you perform a similar set of steps to turn it off.

Example TI10–2

Find the equation of the regression line for the data in Example TI10–1. The input and output screens are shown.

Input	Output
	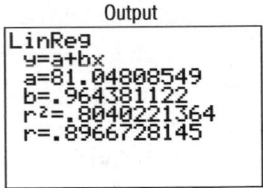

The equation of the regression line is $y' = 81.04808549 + 0.964381122x$.

To plot the regression line on the scatter plot:

1. Press **Y=** and **CLEAR** to clear any previous equations.

2. Press **VARS** and then **5** for Statistics.

3. Move the cursor to EQ and press **1** for RegEQ. The line will be in the Y= screen.

4. Press **GRAPH**.

Example TI10–3

Draw the regression line found in Example TI10–2 on the scatter plot.

The output screens are shown.

Output	Output
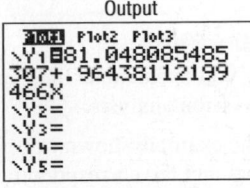	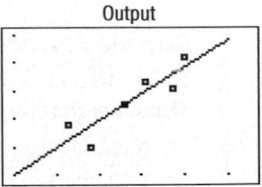

To test the significance of b and ρ:

1. Press **STAT** and move the cursor to TESTS.
2. Press **E (ALPHA SIN)** for LinRegTTest. Make sure the Xlist is L_1, the Ylist is L_2, and the Freq is 1.   (Use F for TI-84)
3. Select the appropriate alternative hypothesis.
4. Move the cursor to Calculate and press **ENTER.**

Example TI10–4

Test the hypothesis H_0: $\rho = 0$ for the data in Example TI 10–1. Use $\alpha = 0.05$.

Input	Output	Output
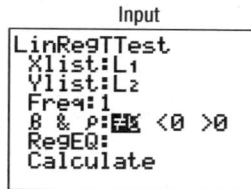	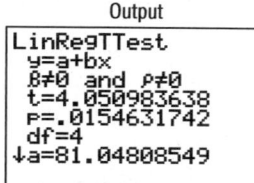	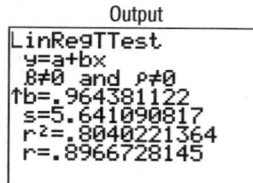

In this case, the t test value is 4.050983638. The P-value is 0.0154631742, which is significant. The decision is to reject the null hypothesis at $\alpha = 0.05$, since $0.0154631742 < 0.05$; $r = 0.8966728145$, $r^2 = 0.8040221364$.

There are two other ways to store the equation for the regression line in Y_1 for graphing.

1. Type Y_1 after the LinReg(a+bx) command.
2. Type Y_1 in the RegEQ: spot in the LinRegTTest.

To get Y_1 do this:

Press **VARS** for variables, move cursor to Y-VARS, press **1** for Function, press **1** for Y_1.

Excel
Step by Step

Scatter Plots

Creating a scatter plot is straightforward when you use the Chart Wizard.

1. You must have at least two columns of data to use the Scatter Plot option.
2. Highlight the data to be plotted. Select the Insert tab from the toolbar. Then select the Scatter chart and the first type (Scatter with only markers).
3. By left-clicking anywhere on the chart, you automatically bring up the Chart Tools group on the toolbar. The Chart Tools menu includes three additional tabs for editing your chart: Design, Layout, and Format.
4. You can add titles to your chart and to the axes by selecting the Layout tab, then selecting the appropriate option from the Labels group.

Correlation Coefficient

The CORREL function in Excel returns the correlation coefficient without regression analysis.

1. Enter the data in columns A and B.
2. Select a blank cell, and then select the Formulas tab from the toolbar.
3. Select Insert Function fx icon from the toolbar.
4. Select the Statistical function category and select the CORREL function.
5. Enter the data range **A1:AN,** where **N** is the number of sample data pairs for the first variable in Array1. Enter the data range **B1:BN** for the second variable in Array2, and then click [OK].

Correlation and Regression

This procedure will allow you to calculate the Pearson product moment correlation coefficient without performing a regression analysis.

1. Enter the data from the example shown in a new worksheet. Enter the six values for the x numbers in column A and the corresponding y numbers in column B.

Example

x	43	48	56	61	67	70
y	128	120	135	143	141	152

2. Select Data from the toolbar. Then select Data Analysis. Under Analysis Tools, select Correlation.

3. In the Correlation dialog box, type **A1:B6** for the Input Range and check the Grouped By: Columns option.

4. Under Output options, select Output Range, and type **D2**. Then click [OK].

This procedure will allow you to conduct a regression analysis and compute the correlation coefficient. Use the data from Example 10–2.

1. Select the Data tab on the toolbar, then **Data Analysis>Regression.**

2. In the Regression dialog box, type **B1:B6** in the Input Y Range and type **A1:A6** in the Input X Range.

3. Under Output options, select Output Range, and type **D6.** Then click [OK].

Note: To see all of the decimal places for the statistics in the Summary Output, expand the width of columns D to L.

1. Highlight columns D through L.

2. Select the Home tab, and then select Format Autofit Column Width.

SUMMARY OUTPUT

Regression Statistics

Multiple R	0.896672815
R Square	0.804022136
Adjusted R Square	0.75502767
Standard Error	5.641090817
Observations	6

ANOVA

	df	SS	MS	F	Significance F
Regression	1	522.2123776	522.2123776	16.41046844	0.015463174
Residual	4	127.2876224	31.82190561		
Total	5	649.5			

	Coefficients	Standard Error	t Stat	P-value	Lower 95%	Upper 95%	Lower 95.0%	Upper 95.0%
Intercept	81.04808549	13.88088081	5.838828717	0.004289034	42.50858191	119.5875891	42.50858191	119.5875891
X Variable 1	0.964381122	0.238060977	4.050983638	0.015463174	0.303417888	1.625344356	0.303417888	1.625344356

10–3 Coefficient of Determination and Standard Error of the Estimate

The previous sections stated that if the correlation coefficient is significant, the equation of the regression line can be determined. Also, for various values of the independent variable x, the corresponding values of the dependent variable y can be predicted. Several other measures are associated with the correlation and regression techniques. They include the coefficient of determination, the standard error of the estimate, and the prediction interval. But before these concepts can be explained, the different types of variation associated with the regression model must be defined.

Types of Variation for the Regression Model

Consider the following hypothetical regression model.

x	1	2	3	4	5
y	10	8	12	16	20

The equation of the regression line is $y' = 4.8 + 2.8x$, and $r = 0.919$. The sample y values are 10, 8, 12, 16, and 20. The predicted values, designated by y', for each x can be found by substituting each x value into the regression equation and finding y'. For example, when $x = 1$,

$$y' = 4.8 + 2.8x = 4.8 + (2.8)(1) = 7.6$$

Now, for each x, there is an observed y value and a predicted y' value; for example, when $x = 1$, $y = 10$, and $y' = 7.6$. Recall that the closer the observed values are to the predicted values, the better the fit is and the closer r is to $+1$ or -1.

The *total variation* $\Sigma(y - \bar{y})^2$ is the sum of the squares of the vertical distances each point is from the mean. The total variation can be divided into two parts: that which is attributed to the relationship of x and y and that which is due to chance. The variation obtained from the relationship (i.e., from the predicted y' values) is $\Sigma(y' - \bar{y})^2$ and is called the *explained variation*. Most of the variations can be explained by the relationship. The closer the value r is to $+1$ or -1, the better the points fit the line and the closer $\Sigma(y' - \bar{y})^2$ is to $\Sigma(y - \bar{y})^2$. In fact, if all points fall on the regression line, $\Sigma(y' - \bar{y})^2$ will equal $\Sigma(y - \bar{y})^2$, since y' is equal to y in each case.

On the other hand, the variation due to chance, found by $\Sigma(y - y')^2$, is called the *unexplained variation*. This variation cannot be attributed to the relationship. When the unexplained variation is small, the value of r is close to $+1$ or -1. If all points fall on the regression line, the unexplained variation $\Sigma(y - y')^2$ will be 0. Hence, the *total variation* is equal to the sum of the explained variation and the unexplained variation. That is,

$$\Sigma(y - \bar{y})^2 = \Sigma(y' - \bar{y})^2 + \Sigma(y - y')^2$$

These values are shown in Figure 10–17. For a single point, the differences are called *deviations*. For the hypothetical regression model given earlier, for $x = 1$ and $y = 10$, you get $y' = 7.6$ and $\bar{y} = 13.2$.

The procedure for finding the three types of variation is illustrated next.

Step 1   Find the predicted y' values.

For $x = 1$ $y' = 4.8 + 2.8x = 4.8 + (2.8)(1) = 7.6$

For $x = 2$ $y' = 4.8 + (2.8)(2) = 10.4$

For $x = 3$ $y' = 4.8 + (2.8)(3) = 13.2$

For $x = 4$ $y' = 4.8 + (2.8)(4) = 16.0$

For $x = 5$ $y' = 4.8 + (2.8)(5) = 18.8$

Figure 10–17

Deviations for the Regression Equation

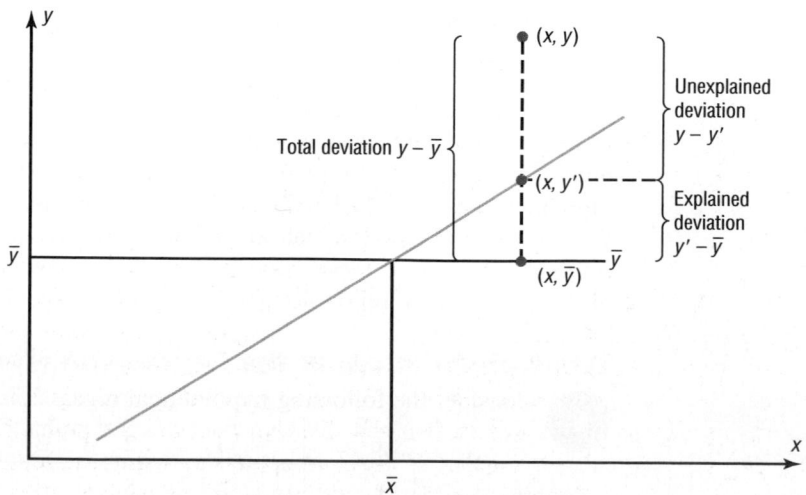

Hence, the values for this example are as follows:

x	y	y'
1	10	7.6
2	8	10.4
3	12	13.2
4	16	16.0
5	20	18.8

Step 2   Find the mean of the y values.

$$\bar{y} = \frac{10 + 8 + 12 + 16 + 20}{5} = 13.2$$

Step 3   Find the total variation $\Sigma(y - \bar{y})^2$.

$$(10 - 13.2)^2 = 10.24$$
$$(8 - 13.2)^2 = 27.04$$
$$(12 - 13.2)^2 = 1.44$$
$$(16 - 13.2)^2 = 7.84$$
$$(20 - 13.2)^2 = \underline{46.24}$$
$$\Sigma(y - \bar{y})^2 = 92.8$$

Step 4   Find the explained variation $\Sigma(y' - \bar{y})^2$.

$$(7.6 - 13.2)^2 = 31.36$$
$$(10.4 - 13.2)^2 = 7.84$$
$$(13.2 - 13.2)^2 = 0.00$$
$$(16 - 13.2)^2 = 7.84$$
$$(18.8 - 13.2)^2 = \underline{31.36}$$
$$\Sigma(y' - \bar{y})^2 = 78.4$$

Step 5   Find the unexplained variation $\Sigma(y - y')^2$.

$$(10 - 7.6)^2 = 5.76$$
$$(8 - 10.4)^2 = 5.76$$
$$(12 - 13.2)^2 = 1.44$$
$$(16 - 16)^2 = 0.00$$
$$(20 - 18.8)^2 = \underline{1.44}$$
$$\Sigma(y - y')^2 = 14.4$$

Notice that

Total variation = explained variation + unexplained variation
$$92.8 = 78.4 + 14.4$$

Note: The values $(y - y')$ are called *residuals*. A **residual** is the difference between the actual value of y and the predicted value y' for a given x value. The mean of the residuals is always zero. As stated previously, the regression line determined by the formulas in Section 10–2 is the line that best fits the points of the scatter plot. The sum of the squares of the residuals computed by using the regression line is the smallest possible value. For this reason, a regression line is also called a **least-squares line.**

Residual Plots

As previously stated, the values $y - y'$ are called residuals (sometimes called the *prediction errors*). These values can be plotted with the x values, and the plot, called a **residual plot,** can be used to determine how well the regression line can be used to make predictions.

The residuals for the previous example are calculated as shown.

x	y	y'	$y - y'$ = **residual**
1	10	7.6	$10 - 7.6 = 2.4$
2	8	10.4	$8 - 10.4 = -2.4$
3	12	13.2	$12 - 13.2 = -1.2$
4	16	16	$16 - 16 = 0$
5	20	18.8	$20 - 18.8 = 1.2$

The x values are plotted using the horizontal axis, and the residuals are plotted using the vertical axis. Since the mean of the residuals is always zero, a horizontal line with a y coordinate of zero is placed on the y axis as shown in Figure 10–18.

Plot the x and residual values as shown in Figure 10–18.

x	1	2	3	4	5
$y - y'$	2.4	-2.4	-1.2	0	1.2

Figure 10–18

Residual Plot

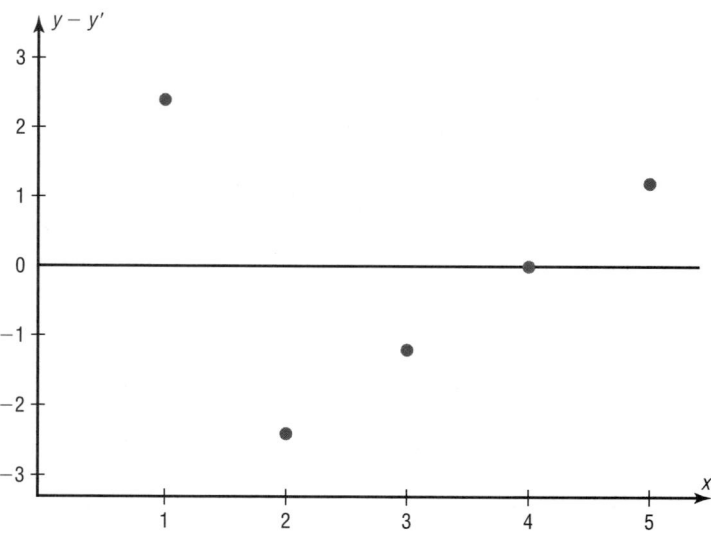

To interpret a residual plot, you need to determine if the residuals form a pattern. Figure 10–19 shows four examples of residual plots. If the residual values are more or less evenly distributed about the line, as shown in Figure 10–19(a), then the relationship between x and y is linear and the regression line can be used to make predictions. This means that the standard deviations of each of the dependent variables must be the same for each value of the independent variable. This is called the *homoscedasticity assumption.* See assumption 3 on page 556.

Figure 10–19(b) shows that the variance of the residuals increases as the values of x increase. This means that the regression line is not suitable for predictions.

Figure 10–19(c) shows a curvilinear relationship between the x values and the residual values; hence, the regression line is not suitable for making predictions.

Figure 10–19(d) shows that as the x values increase, the residuals increase and become more dispersed. This means that the regression line is not suitable for making predictions.

The residual plot in Figure 10–18 shows that the regression line $y' = 4.8 + 2.8x$ is somewhat questionable for making predictions due to a small sample size.

Figure 10–19

Examples of
Residual Plots

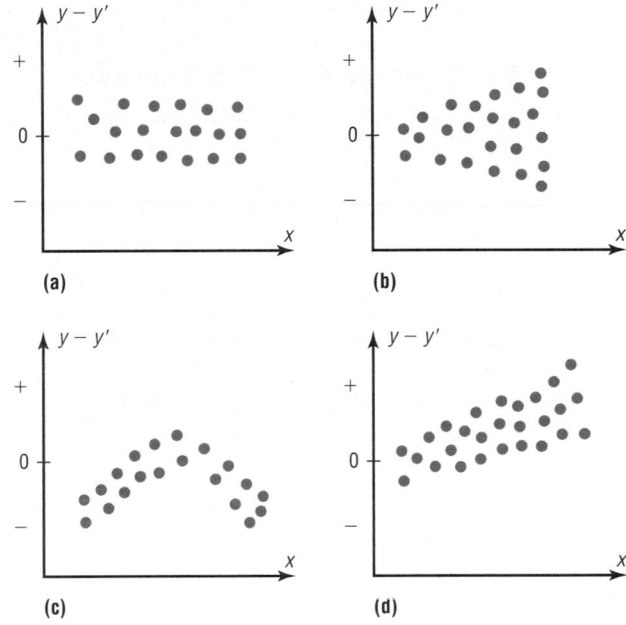

(a)   (b)

(c)   (d)

Objective 5

Compute the coefficient of determination.

Coefficient of Determination

The *coefficient of determination* is the ratio of the explained variation to the total variation and is denoted by r^2. That is,

$$r^2 = \frac{\text{explained variation}}{\text{total variation}}$$

For the example, $r^2 = 78.4/92.8 = 0.845$. The term r^2 is usually expressed as a percentage. So in this case, 84.5% of the total variation is explained by the regression line using the independent variable.

Another way to arrive at the value for r^2 is to square the correlation coefficient. In this case, $r = 0.919$ and $r^2 = 0.845$, which is the same value found by using the variation ratio.

The **coefficient of determination** is a measure of the variation of the dependent variable that is explained by the regression line and the independent variable. The symbol for the coefficient of determination is r^2.

Of course, it is usually easier to find the coefficient of determination by squaring r and converting it to a percentage. Therefore, if $r = 0.90$, then $r^2 = 0.81$, which is equivalent to 81%. This result means that 81% of the variation in the dependent variable is accounted for by the variations in the independent variable. The rest of the variation, 0.19, or 19%, is unexplained. This value is called the *coefficient of nondetermination* and is found by subtracting the coefficient of determination from 1. As the value of r approaches 0, r^2 decreases more rapidly. For example, if $r = 0.6$, then $r^2 = 0.36$, which means that only 36% of the variation in the dependent variable can be attributed to the variation in the independent variable.

Coefficient of Nondetermination
$1.00 - r^2$

Standard Error of the Estimate

When a y' value is predicted for a specific x value, the prediction is a point prediction. However, a prediction interval about the y' value can be constructed, just as a confidence interval was constructed for an estimate of the population mean. The prediction interval uses a statistic called the *standard error of the estimate.*

The **standard error of the estimate,** denoted by s_{est}, is the standard deviation of the observed y values about the predicted y' values. The formula for the standard error of the estimate is

$$s_{est} = \sqrt{\frac{\Sigma(y - y')^2}{n - 2}}$$

The standard error of the estimate is similar to the standard deviation, but the mean is not used. As can be seen from the formula, the standard error of the estimate is the square root of the unexplained variation—that is, the variation due to the difference of the observed values and the expected values—divided by $n - 2$. So the closer the observed values are to the predicted values, the smaller the standard error of the estimate will be.

Example 10–12 shows how to compute the standard error of the estimate.

Example 10–12   Copy Machine Maintenance Costs

A researcher collects the following data and determines that there is a significant relationship between the age of a copy machine and its monthly maintenance cost. The regression equation is $y' = 55.57 + 8.13x$. Find the standard error of the estimate.

Machine	Age x (years)	Monthly cost y
A	1	$ 62
B	2	78
C	3	70
D	4	90
E	4	93
F	6	103

Solution

Step 1 Make a table, as shown.

x	y	y'	$y - y'$	$(y - y')^2$
1	62			
2	78			
3	70			
4	90			
4	93			
6	103			

Step 2   Using the regression line equation $y' = 55.57 + 8.13x$, compute the predicted values y' for each x and place the results in the column labeled y'.

$$x = 1 \qquad y' = 55.57 + (8.13)(1) = 63.70$$
$$x = 2 \qquad y' = 55.57 + (8.13)(2) = 71.83$$
$$x = 3 \qquad y' = 55.57 + (8.13)(3) = 79.96$$
$$x = 4 \qquad y' = 55.57 + (8.13)(4) = 88.09$$
$$x = 6 \qquad y' = 55.57 + (8.13)(6) = 104.35$$

Step 3   For each y, subtract y' and place the answer in the column labeled $y - y'$.

$$62 - 63.70 = -1.70 \qquad 90 - 88.09 = 1.91$$
$$78 - 71.83 = 6.17 \qquad 93 - 88.09 = 4.91$$
$$70 - 79.96 = -9.96 \qquad 103 - 104.35 = -1.35$$

Step 4   Square the numbers found in step 3 and place the squares in the column labeled $(y - y')^2$.

Step 5   Find the sum of the numbers in the last column. The completed table is shown.

x	y	y'	$y - y'$	$(y - y')^2$
1	62	63.70	−1.70	2.89
2	78	71.83	6.17	38.0689
3	70	79.96	−9.96	99.2016
4	90	88.09	1.91	3.6481
4	93	88.09	4.91	24.1081
6	103	104.35	−1.35	1.8225
				169.7392

Step 6   Substitute in the formula and find s_{est}.

$$s_{est} = \sqrt{\frac{\Sigma(y - y')^2}{n - 2}} = \sqrt{\frac{169.7392}{6 - 2}} = 6.51$$

In this case, the standard deviation of observed values about the predicted values is 6.51.

The standard error of the estimate can also be found by using the formula

$$s_{est} = \sqrt{\frac{\Sigma y^2 - a\,\Sigma y - b\,\Sigma xy}{n - 2}}$$

Example 10–13

Find the standard error of the estimate for the data for Example 10–12 by using the preceding formula. The equation of the regression line is $y' = 55.57 + 8.13x$.

Solution

Step 1   Make a table.

Step 2   Find the product of x and y values, and place the results in the third column.

Step 3   Square the y values, and place the results in the fourth column.

Step 4   Find the sums of the second, third, and fourth columns. The completed table is shown here.

x	y	xy	y^2
1	62	62	3,844
2	78	156	6,084
3	70	210	4,900
4	90	360	8,100
4	93	372	8,649
6	103	618	10,609
	$\Sigma y = 496$	$\Sigma xy = 1778$	$\Sigma y^2 = 42,186$

Step 5   From the regression equation $y' = 55.57 + 8.13x$, $a = 55.57$, and $b = 8.13$.

Step 6   Substitute in the formula and solve for s_{est}.

$$s_{est} = \sqrt{\frac{\Sigma y^2 - a\,\Sigma y - b\,\Sigma xy}{n - 2}}$$

$$= \sqrt{\frac{42,186 - (55.57)(496) - (8.13)(1778)}{6 - 2}} = 6.48$$

This value is close to the value found in Example 10–12. The difference is due to rounding.

Objective **7**

Find a prediction interval.

Prediction Interval

The standard error of the estimate can be used for constructing a **prediction interval** (similar to a confidence interval) about a y' value.

     When a specific value x is substituted into the regression equation, the y' that you get is a point estimate for y. For example, if the regression line equation for the age of a machine and the monthly maintenance cost is $y' = 55.57 + 8.13x$ (Example 10–12), then the predicted maintenance cost for a 3-year-old machine would be $y' = 55.57 + 8.13(3)$, or $79.96. Since this is a point estimate, you have no idea how accurate it is. But you can construct a prediction interval about the estimate. By selecting an α value, you can achieve a $(1 - \alpha) \cdot 100\%$ confidence that the interval contains the actual mean of the y values that correspond to the given value of x.

     The reason is that there are possible sources of prediction errors in finding the regression line equation. One source occurs when finding the standard error of the estimate s_{est}. Two others are errors made in estimating the slope and the y' intercept, since the equation of the regression line will change somewhat if different random samples are used when calculating the equation.

Formula for the Prediction Interval about a Value y'

$$y' - t_{\alpha/2}s_{est}\sqrt{1 + \frac{1}{n} + \frac{n(x - \overline{X})^2}{n\,\Sigma x^2 - (\Sigma x)^2}} < y < y' + t_{\alpha/2}s_{est}\sqrt{1 + \frac{1}{n} + \frac{n(x - \overline{X})^2}{n\,\Sigma x^2 - (\Sigma x)^2}}$$

with d.f. $= n - 2$.

| Example 10–14 | For the data in Example 10–12, find the 95% prediction interval for the monthly maintenance cost of a machine that is 3 years old. |

Solution

Step 1 Find Σx, Σx^2, and $\overline{X}$.

$$\Sigma x = 20 \qquad \Sigma x^2 = 82 \qquad \overline{X} = \frac{20}{6} = 3.3$$

Step 2 Find y' for $x = 3$.

$$y' = 55.57 + 8.13x$$
$$= 55.57 + 8.13(3) = 79.96$$

Step 3 Find s_{est}.

$$s_{est} = 6.48$$

as shown in Example 10–13.

Step 4 Substitute in the formula and solve: $t_{\alpha/2} = 2.776$, d.f. $= 6 - 2 = 4$ for 95%.

$$y' - t_{\alpha/2}s_{est}\sqrt{1 + \frac{1}{n} + \frac{n(x - \overline{X})^2}{n\,\Sigma x^2 - (\Sigma x)^2}} < y < y'$$

$$+ t_{\alpha/2}s_{est}\sqrt{1 + \frac{1}{n} + \frac{n(x - \overline{X})^2}{n\,\Sigma x^2 \quad (\Sigma x)^2}}$$

$$79.96 - (2.776)(6.48)\sqrt{1 + \frac{1}{6} + \frac{6(3 - 3.3)^2}{6(82) - (20)^2}} < y < 79.96$$

$$+ (2.776)(6.48)\sqrt{1 + \frac{1}{6} + \frac{6(3 - 3.3)^2}{6(82) - (20)^2}}$$

$$79.96 - (2.776)(6.48)(1.08) < y < 79.96 + (2.776)(6.48)(1.08)$$
$$79.96 - 19.43 < y < 79.96 + 19.43$$
$$60.53 < y < 99.39$$

Hence, you can be 95% confident that the interval $60.53 < y < 99.39$ contains the actual value of y.

Applying the Concepts 10–3

Interpreting Simple Linear Regression

Answer the questions about the following computer-generated information.

Linear correlation coefficient $r = 0.794556$
Coefficient of determination $= 0.631319$
Standard error of estimate $= 12.9668$
Explained variation $= 5182.41$
Unexplained variation $= 3026.49$
Total variation $= 8208.90$
Equation of regression line $y' = 0.725983X + 16.5523$
Level of significance $= 0.1$
Test statistic $= 0.794556$
Critical value $= 0.378419$

1. Are both variables moving in the same direction?

2. Which number measures the distances from the prediction line to the actual values?

3. Which number is the slope of the regression line?

4. Which number is the *y* intercept of the regression line?

5. Which number can be found in a table?

6. Which number is the allowable risk of making a type I error?

7. Which number measures the variation explained by the regression?

8. Which number measures the scatter of points about the regression line?

9. What is the null hypothesis?

10. Which number is compared to the critical value to see if the null hypothesis should be rejected?

11. Should the null hypothesis be rejected?

See page 590 for the answers.

Exercises 10–3

1. What is meant by the *explained variation?* How is it computed? Explained variation is the variation due to the relationship. It is computed by $\Sigma(y' - \overline{y})^2$.

2. What is meant by the *unexplained variation?* How is it computed? Unexplained variation is the variation due to chance. It is computed by $\Sigma(y - y')^2$.

3. What is meant by the *total variation?* How is it computed?

4. Define the coefficient of determination.

5. How is the coefficient of determination found?

6. Define the coefficient of nondetermination. It is the percent of the variation in *y* that is not due to the variation in *x*.

7. How is the coefficient of nondetermination found? The coefficient of nondetermination is found by subtracting r^2 from 1.

For Exercises 8 through 13, find the coefficients of determination and nondetermination and explain the meaning of each.

8. $r = 0.80$ $R^2 = 0.64$; 64% of the variation of *y* is due to the variation of *x*; 36% is due to chance.

9. $r = 0.75$ $R^2 = 0.5625$; 56.25% of the variation of *y* is due to the variation of *x*; 43.75% is due to chance.

10. $r = 0.35$ $R^2 = 0.1225$; 12.25% of the variation of *y* is due to the variation of *x*; 87.75% is due to chance.

11. $r = 0.42$ $R^2 = 0.1764$; 17.64% of the variation of *y* is due to the variation of *x*; 82.36% is due to chance.

12. $r = 0.18$ $R^2 = 0.0324$; 3.24% of the variation of *y* is due to the variation of *x*; 96.76% is due to chance.

13. $r = 0.91$ $R^2 = 0.8281$; 82.81% of the variation of *y* is due to the variation of *x*; 17.19% is due to chance.

14. Define the standard error of the estimate for regression. When can the standard error of the estimate be used to construct a prediction interval about a value y'?

15. Compute the standard error of the estimate for Exercise 13 in Section 10–1. The regression line equation was found in Exercise 13 in Section 10–2. 629.4862

16. Compute the standard error of the estimate for Exercise 14 in Section 10–1. The regression line equation was found in Exercise 14 in Section 10–2. 12.03* (TI value 12.06)

17. Compute the standard error of the estimate for Exercise 15 in Section 10–1. The regression line equation was found in Exercise 15 in Section 10–2. 94.22*

18. Compute the standard error of the estimate for Exercise 16 in Section 10–1. The regression line equation was found in Exercise 16 in Section 10–2. The standard error should not be calculated.

19. For the data in Exercises 13 in Sections 10–1 and 10–2 and 15 in Section 10–3, find the 90% prediction interval when *x* = 200 new releases. 365.88 < *y'* < 2925.04*

20. For the data in Exercises 14 in Sections 10–1 and 10–2 and 16 in Section 10–3, find the 95% prediction interval when *x* = 60. The prediction interval should not be calculated.

21. For the data in Exercises 15 in Sections 10–1 and 10–2 and 17 in Section 10–3, find the 90% prediction interval when *x* = 4 years. $30.46 < *y* < $472.38*

22. For the data in Exercises 16 in Sections 10–1 and 10–2 and 18 in Section 10–3, find the 98% prediction interval when *x* = 47 years. The prediction interval should not be calculated.

*Answers may vary due to rounding.

<table>
<tr><td>

10–4

Objective 8

Be familiar with the concept of multiple regression.

</td><td>

Multiple Regression (Optional)

The previous sections explained the concepts of simple linear regression and correlation. In simple linear regression, the regression equation contains one independent variable x and one dependent variable y' and is written as

$$y' = a + bx$$

where a is the y' intercept and b is the slope of the regression line.

In **multiple regression,** there are several independent variables and one dependent variable, and the equation is

$$y' = a + b_1 x_1 + b_2 x_2 + \cdots + b_k x_k$$

where $x_1, x_2, \ldots, x_k$ are the independent variables.

For example, suppose a nursing instructor wishes to see whether there is a relationship between a student's grade point average, age, and score on the state board nursing examination. The two independent variables are GPA (denoted by x_1) and age (denoted by x_2). The instructor will collect the data for all three variables for a sample of nursing students. Rather than conduct two separate simple regression studies, one using the GPA and state board scores and another using ages and state board scores, the instructor can conduct one study using multiple regression analysis with two independent variables—GPA and ages—and one dependent variable—state board scores.

A multiple regression correlation R can also be computed to determine if a significant relationship exists between the independent variables and the dependent variable. Multiple regression analysis is used when a statistician thinks there are several independent variables contributing to the variation of the dependent variable. This analysis then can be used to increase the accuracy of predictions for the dependent variable over one independent variable alone.

Two other examples for multiple regression analysis are when a store manager wants to see whether the amount spent on advertising and the amount of floor space used for a display affect the amount of sales of a product, and when a sociologist wants to see whether the amount of time children spend watching television and playing video games is related to their weight. Multiple regression analysis can also be conducted by using

</td></tr>
</table>

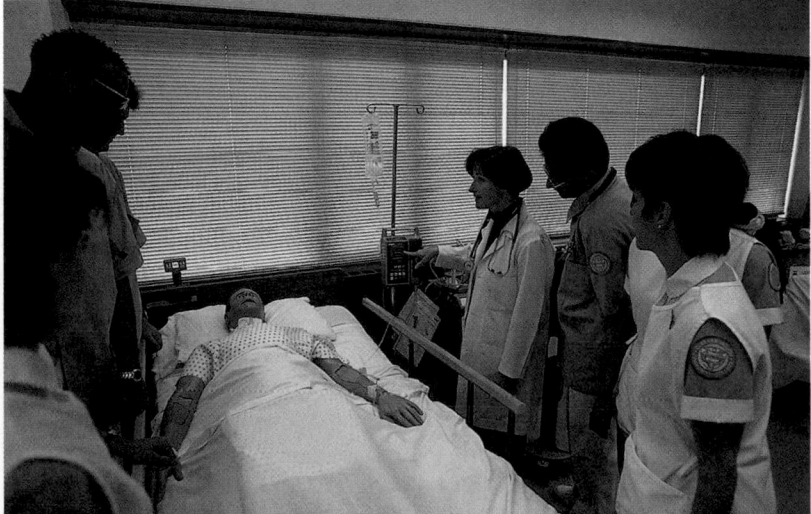

<table>
<tr><td>

Unusual Stats

The most popular single-digit number played by people who purchase lottery tickets is 7.

</td></tr>
</table>

In this study, researchers found a correlation between the cleanliness of the homes children are raised in and the years of schooling completed and earning potential for those children. What interfering variables were controlled? How might these have been controlled? Summarize the conclusions of the study.

SUCCESS

HOME SMART HOME

KIDS WHO GROW UP IN A CLEAN HOUSE FARE BETTER AS ADULTS

Good-bye, GPA. So long, SATs. New research suggests that we may be able to predict children's future success from the level of cleanliness in their homes.

A University of Michigan study presented at the annual meeting of the American Economic Association uncovered a surprising correlation: children raised in clean homes were later found to have completed more school and to have higher earning potential than those raised in dirty homes. The clean homes may indicate a family that values organization and similarly helpful skills at school and work, researchers say.

Cleanliness ratings for about 5,000 households were assessed between 1968 and 1972, and respondents were interviewed 25 years later to determine educational achievement and professional earnings of the young adults who had grown up there, controlling for variables such as race, socio-economic status and level of parental education. The data showed that those raised in homes rated "clean" to "very clean" had completed an average of 1.6 more years of school than those raised in "not very clean" or "dirty" homes. Plus, the first group's annual wages averaged about $3,100 more than the second's.

But don't buy stock in Mr. Clean and Pine Sol just yet. "We're not advocating that everyone go out and clean their homes right this minute," explains Rachel Dunifon, a University of Michigan doctoral candidate and a researcher on the study. Rather, the main implication of the study, Dunifon says, is that there is significant evidence that non-cognitive factors, such as organization and efficiency, play a role in determining academic and financial success.

— Jackie Fisherman

more than two independent variables, denoted by $x_1, x_2, x_3, \ldots, x_m$. Since these computations are quite complicated and for the most part would be done on a computer, this chapter will show the computations for two independent variables only.

For example, the nursing instructor wishes to see whether a student's grade point average and age are related to the student's score on the state board nursing examination. She selects five students and obtains the following data.

Student	GPA x_1	Age x_2	State board score y
A	3.2	22	550
B	2.7	27	570
C	2.5	24	525
D	3.4	28	670
E	2.2	23	490

The multiple regression equation obtained from the data is

$$y' = -44.81 + 87.64x_1 + 14.533x_2$$

If a student has a GPA of 3.0 and is 25 years old, her predicted state board score can be computed by substituting these values in the equation for x_1 and x_2, respectively, as shown.

$$y' = -44.81 + 87.64(3.0) + 14.533(25)$$
$$= 581.44 \text{ or } 581$$

Hence, if a student has a GPA of 3.0 and is 25 years old, the student's predicted state board score is 581.

The Multiple Regression Equation

A multiple regression equation with two independent variables (x_1 and x_2) and one dependent variable has the form

$$y' = a + b_1x_1 + b_2x_2$$

A multiple regression equation with three independent variables (x_1, x_2, and x_3) and one dependent variable has the form

$$y' = a + b_1x_1 + b_2x_2 + b_3x_3$$

General Form of the Multiple Regression Equation

The general form of the multiple regression equation with k independent variables is

$$y' = a + b_1x_1 + b_2x_2 + \cdots + b_kx_k$$

The x's are the independent variables. The value for a is more or less an intercept, although a multiple regression equation with two independent variables constitutes a plane rather than a line. The b's are called *partial regression coefficients*. Each b represents the amount of change in y' for one unit of change in the corresponding x value when the other x values are held constant. In the example just shown, the regression equation was $y' = -44.81 + 87.64x_1 + 14.533x_2$. In this case, for each unit of change in the student's GPA, there is a change of 87.64 units in the state board score with the student's age x_2 being held constant. And for each unit of change in x_2 (the student's age), there is a change of 14.533 units in the state board score with the GPA held constant.

Assumptions for Multiple Regression

The assumptions for multiple regression are similar to those for simple regression.

1. For any specific value of the independent variable, the values of the y variable are normally distributed. (This is called the *normality* assumption.)
2. The variances (or standard deviations) for the y variables are the same for each value of the independent variable. (This is called the *equal-variance* assumption.)
3. There is a linear relationship between the dependent variable and the independent variables. (This is called the *linearity* assumption.)
4. The independent variables are not correlated. (This is called the *nonmulticollinearity* assumption.)
5. The values for the y variables are independent. (This is called the *independence* assumption.)

In multiple regression, as in simple regression, the strength of the relationship between the independent variables and the dependent variable is measured by a correlation coefficient. This **multiple correlation coefficient** is symbolized by R. The value of R can range from 0 to $+1$; R can never be negative. The closer to $+1$, the stronger the relationship; the closer to 0, the weaker the relationship. The value of R takes into account all the independent variables and can be computed by using the values of the individual correlation coefficients. The formula for the multiple correlation coefficient when there are two independent variables is shown next.

Formula for the Multiple Correlation Coefficient

The formula for R is

$$R = \sqrt{\frac{r_{yx_1}^2 + r_{yx_2}^2 - 2r_{yx_1} \cdot r_{yx_2} \cdot r_{x_1x_2}}{1 - r_{x_1x_2}^2}}$$

where r_{yx_1} is the value of the correlation coefficient for variables y and x_1; r_{yx_2} is the value of the correlation coefficient for variables y and x_2; and $r_{x_1x_2}$ is the value of the correlation coefficient for variables x_1 and x_2.

In this case, R is 0.989, as shown in Example 10–15. The multiple correlation coefficient is always higher than the individual correlation coefficients. For this specific example, the multiple correlation coefficient is higher than the two individual correlation coefficients computed by using grade point average and state board scores ($r_{yx_1} = 0.845$) or age and state board scores ($r_{yx_2} = 0.791$). *Note:* $r_{x_1x_2} = 0.371$.

Example 10–15

State Board Scores

For the data regarding state board scores, find the value of R.

Solution

The values of the correlation coefficients are

$$r_{yx_1} = 0.845$$
$$r_{yx_2} = 0.791$$
$$r_{x_1x_2} = 0.371$$

Substituting in the formula, you get

$$R = \sqrt{\frac{r_{yx_1}^2 + r_{yx_2}^2 - 2r_{yx_1} \cdot r_{yx_2} \cdot r_{x_1x_2}}{1 - r_{x_1x_2}^2}}$$

$$= \sqrt{\frac{(0.845)^2 + (0.791)^2 - 2(0.845)(0.791)(0.371)}{1 - 0.371^2}}$$

$$= \sqrt{\frac{0.8437569}{0.862359}} = \sqrt{0.9784288} = 0.989$$

Hence, the correlation between a student's grade point average and age with the student's score on the nursing state board examination is 0.989. In this case, there is a strong relationship among the variables; the value of R is close to 1.00.

As with simple regression, R^2 is the *coefficient of multiple determination,* and it is the amount of variation explained by the regression model. The expression $1 - R^2$ represents the amount of unexplained variation, called the *error* or *residual variation.* Since $R = 0.989$, $R^2 = 0.978$ and $1 - R^2 = 1 - 0.978 = 0.022$.

Testing the Significance of R

An F test is used to test the significance of R. The hypotheses are

$$H_0: \rho = 0 \quad \text{and} \quad H_1: \rho \neq 0$$

where ρ represents the population correlation coefficient for multiple correlation.

F Test for Significance of R

The formula for the F test is

$$F = \frac{R^2/k}{(1 - R^2)/(n - k - 1)}$$

where n is the number of data groups $(x_1, x_2, \ldots, y)$ and k is the number of independent variables.

The degrees of freedom are d.f.N. $= n - k$ and d.f.D. $= n - k - 1$.

Example 10–16   State Board Scores

Test the significance of the R obtained in Example 10–15 at $\alpha = 0.05$.

Solution

$$F = \frac{R^2/k}{(1 - R^2)/(n - k - 1)}$$

$$= \frac{0.978/2}{(1 - 0.978)/(5 - 2 - 1)} = \frac{0.489}{0.011} = 44.45$$

The critical value obtained from Table H with $\alpha = 0.05$, d.f.N. $= 3$, and d.f.D. $= 5 - 2 - 1 = 2$ is 19.16. Hence, the decision is to reject the null hypothesis and conclude that there is a significant relationship among the student's GPA, age, and score on the nursing state board examination.

Adjusted R^2

Since the value of R^2 is dependent on n (the number of data pairs) and k (the number of variables), statisticians also calculate what is called an **adjusted R^2,** denoted by R^2_{adj}. This is based on the number of degrees of freedom.

Formula for the Adjusted R^2

The formula for the adjusted R^2 is

$$R^2_{\text{adj}} = 1 - \frac{(1 - R^2)(n - 1)}{n - k - 1}$$

The adjusted R^2 is smaller than R^2 and takes into account the fact that when n and k are approximately equal, the value of R may be artificially high, due to sampling error rather than a true relationship among the variables. This occurs because the chance variations of all the variables are used in conjunction with one another to derive the regression equation. Even if the individual correlation coefficients for each independent variable and the dependent variable were all zero, the multiple correlation coefficient due to sampling error could be higher than zero.

Hence, both R^2 and R^2_{adj} are usually reported in a multiple regression analysis.

Example 10–17

State Board Scores

Calculate the adjusted R^2 for the data in Example 10–16. The value for R is 0.989.

Solution

$$R^2_{adj} = 1 - \frac{(1 - R^2)(n - 1)}{n - k - 1}$$
$$= 1 - \frac{(1 - 0.989^2)(5 - 1)}{5 - 2 - 1}$$
$$= 1 - 0.043758$$
$$= 0.956$$

In this case, when the number of data pairs and the number of independent variables are accounted for, the adjusted multiple coefficient of determination is 0.956.

Applying the Concepts 10–4

More Math Means More Money

In a study to determine a person's yearly income 10 years after high school, it was found that the two biggest predictors are number of math courses taken and number of hours worked per week during a person's senior year of high school. The multiple regression equation generated from a sample of 20 individuals is

$$y' = 6000 + 4540x_1 + 1290x_2$$

Let x_1 represent the number of mathematics courses taken and x_2 represent hours worked. The correlation between income and mathematics courses is 0.63. The correlation between income and hours worked is 0.84, and the correlation between mathematics courses and hours worked is 0.31. Use this information to answer the following questions.

1. What is the dependent variable?
2. What are the independent variables?
3. What are the multiple regression assumptions?
4. Explain what 4540 and 1290 in the equation tell us.
5. What is the predicted income if a person took 8 math classes and worked 20 hours per week during her or his senior year in high school?
6. What does a multiple correlation coefficient of 0.77 mean?
7. Compute R^2.
8. Compute the adjusted R^2.
9. Would the equation be considered a good predictor of income?
10. What are your conclusions about the relationship among courses taken, hours worked, and yearly income?

See page 590 for the answers.

Exercises 10–4

1. Explain the similarities and differences between simple linear regression and multiple regression.

2. What is the general form of the multiple regression equation? What does a represent? What do the b's represent? $y' = a + b_1x_1 + b_2x_2 + \cdots + b_kx_k$; a is the slope and the b's are the partial regression coefficients.

3. Why would a researcher prefer to conduct a multiple regression study rather than separate regression studies using one independent variable and the dependent variable? The relationship would include all variables in one equation.

4. What are the assumptions for multiple regression? Normality, equal variance, linearity, nonmulticollinearity, and independence

5. How do the values of the individual correlation coefficients compare to the value of the multiple correlation coefficient? They will all be smaller.

6. Age, GPA, and Income A researcher has determined that a significant relationship exists among an employee's age x_1, grade point average x_2, and income y. The multiple regression equation is $y' = -34,127 + 132x_1 + 20,805x_2$. Predict the income of a person who is 32 years old and has a GPA of 3.4. $40,834

7. Assembly Line Work A manufacturer found that a significant relationship exists among the number of hours an assembly line employee works per shift x_1, the total number of items produced x_2, and the number of defective items produced y. The multiple regression equation is $y' = 9.6 + 2.2x_1 - 1.08x_2$. Predict the number of defective items produced by an employee who has worked 9 hours and produced 24 items. 3.48 or 3

8. Special Occasion Cakes A pastry chef who specializes in special occasion cakes uses the following equation to help calculate the price of a cake: $y = -26.279 + 14.855x_1 + 3.1035x_2 + 0.73079x_3$, where x_1 is the number of layers desired, x_2 the number of servings

needed, and x_3 the amount of filling mix used. Calculate the price of a three-layer cake to serve 48 people using 40 ounces of filling. $196.49

9. Aspects of Students' Academic Behavior A college statistics professor is interested in the relationship among various aspects of students' academic behavior and their final grade in the class. She found a significant relationship between the number of hours spent studying statistics per week, the number of classes attended per semester, the number of assignments turned in during the semester, and the student's final grade. This relationship is described by the multiple regression equation $y' = -14.9 + 0.93359x_1 + 0.99847x_2 + 5.3844x_3$. Predict the final grade for a student who studies statistics 8 hours per week (x_1), attends 34 classes (x_2), and turns in 11 assignments (x_3). 85.75 (grade) or 86

10. Age, Cholesterol, and Sodium A medical researcher found a significant relationship among a person's age x_1, cholesterol level x_2, sodium level of the blood x_3, and systolic blood pressure y. The regression equation is $y' = 97.7 + 0.691x_1 + 219x_2 - 299x_3$. Predict the systolic blood pressure of a person who is 35 years old and has a cholesterol level of 194 milligrams per deciliter (mg/dl) and a sodium blood level of 142 milliequivalents per liter (mEq/l). 149.885 or 150

11. Explain the meaning of the multiple correlation coefficient R.

12. What is the range of values R can assume? 0 to 1

13. Define R^2 and R^2_{adj}. R^2 is the coefficient of multiple determination. R^2_{adj} is adjusted for sample size and number of predictors.

14. What are the hypotheses used to test the significance of R? $H_0: \rho = 0$ and $H_1: \rho \neq 0$

15. What test is used to test the significance of R? F test

16. What is the meaning of the adjusted R^2? Why is it computed?

MINITAB
Step by Step

Multiple Regression

In Example 10–15, is there a correlation between a student's score and her or his age and grade point average?

1. Enter the data for the example into three columns of MINITAB. Name the columns **GPA**, **AGE**, and **SCORE**.

2. Click **Stat>Regression> Regression.**

3. Double-click on C3 SCORE, the response variable.

4. Double-click C1 GPA, then C2 AGE.

5. Click on [Storage].

   a) Check the box for Residuals.

   b) Check the box for Fits.

6. Click [OK] twice.

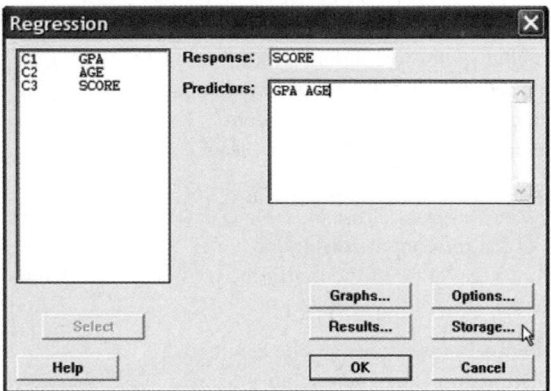

Regression Analysis: SCORE versus GPA, AGE

```
The regression equation is
SCORE = -44.8 + 87.6 GPA + 14.5 Age

Predictor     Coef   SE Coef      T       P
Constant    -44.81     69.25  -0.65   0.584
GPA          87.64     15.24   5.75   0.029
AGE         14.533     2.914   4.99   0.038
S = 14.0091   R-Sq = 97.9%   R-Sq(adj) = 95.7%

Analysis of Variance
Source          DF       SS       MS       F       P
Regression       2  18027.5   9013.7   45.93   0.021
Residual Error   2    392.5    196.3
Total            4  18420.0
```

The test statistic and P-value are 45.93 and 0.021, respectively. Since the P-value is less than α, reject the null hypothesis. There is enough evidence in the sample to conclude the scores are related to age and grade point average.

TI-83 Plus or TI-84 Plus
Step by Step

The TI-83 Plus and the TI-84 Plus do not have a built-in function for multiple regression. However, the downloadable program named MULREG is available on your CD and Online Learning Center. Follow the instructions with your CD for downloading the program.

Finding a Multiple Regression Equation

1. Enter the sets of data values into L_1, L_2, L_3, etc. Make note of which lists contain the independent variables and which list contains the dependent variable as well as how many data values are in each list.

2. Press **PRGM,** move the cursor to the program named MULREG, and press **ENTER** twice.

3. Type the number of independent variables and press **ENTER.**

4. Type the number of cases for each variable and press **ENTER.**

5. Type the name of the list that contains the data values for the first independent variable and press **ENTER.** Repeat this for all independent variables and the dependent variable.

6. The program will show the regression coefficients.

7. Press **ENTER** to see the values of R^2 and adjusted R^2.

8. Press **ENTER** to see the values of the F test statistics and the P-value.

Find the multiple regression equation for these data used in this section:

Student	GPA x_1	Age x_2	State board score y
A	3.2	22	550
B	2.7	27	570
C	2.5	24	525
D	3.4	28	670
E	2.2	23	490

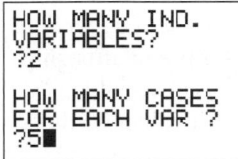 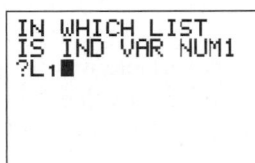 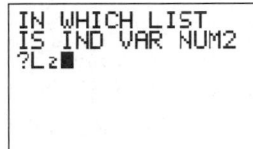 

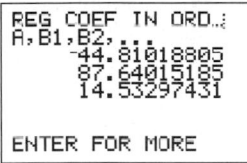 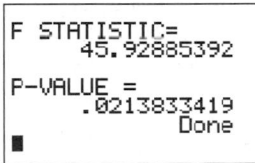

Excel
Step by Step

Multiple Regression

These instructions use data from the nursing examination example discussed at the beginning of Section 10–4.

1. Enter the data from the example into three separate columns of a new worksheet—GPAs in cells A1:A5, ages in cells B1:B5, and scores in cells C1:C5.

2. Select the Data tab on the toolbar, then **Data Analysis>Regression.**

3. In the Regression dialog box, type C1:C5 for the Input Y Range and type A1:B5 for the Input X range.

4. Type D2 for the Output Range and click [OK].

Regression Statistics								
Multiple R	0.989288203							
R Square	0.978691148							
Adjusted R Square	0.957382297							
Standard Error	14.00908721							
Observations	5							
ANOVA								
	df	*SS*	*MS*	*F*	*Significance F*			
Regression	2	18027.49095	9013.745475	45.92885435	0.021308852			
Residual	2	392.5090492	196.2545246					
Total	4	18420						
	Coefficients	*Standard Error*	*t Stat*	*P-value*	*Lower 95%*	*Upper 95%*	*Lower 95.0%*	*Upper 95.0%*
Intercept	-44.81018805	69.24686663	-0.647107808	0.583915745	-342.7554078	253.1350317	-342.7554078	253.1350317
X Variable 1	87.64015185	15.23718666	5.751727913	0.028922601	22.07982906	153.2004746	22.07982906	153.2004746
X Variable 2	14.53297431	2.913737536	4.987743106	0.037924877	1.996173546	27.06977507	1.996173546	27.06977507

The session window shows the correlation coefficient for each pair of variables. The multiple correlation coefficient is significant at 0.021. Ninety-six percent of the variation from the mean is explained by the regression equation. The regression equation is SCORE = −44.8 + 87.6*GPA + 1.45*AGE.

Summary

- Many relationships among variables exist in the real world. One way to determine whether a linear relationship exists is to use the statistical techniques known as correlation and regression. The strength and direction of a linear relationship are measured by the value of the correlation coefficient. It can assume values between and including +1 and −1. The closer the value of the correlation coefficient is to +1 or −1, the stronger the linear relationship is between the variables. A value of +1 or −1 indicates a perfect linear relationship. A positive relationship between two variables means that for small values of the independent variable, the values of the dependent variable will be small, and that for large values of the independent variable, the values of the dependent variable will be large. A negative relationship between two variables means that for small values of the independent variable, the values of the dependent variable will be large, and that for large values of the independent variable, the values of the dependent variable will be small. (10–1)

- Remember that a significant relationship between two variables does not necessarily mean that one variable is a direct cause of the other variable. In some cases this is true, but other possibilities that should be considered include a complex relationship involving other (perhaps unknown) variables, a third variable interacting with both variables, or a relationship due solely to chance. (10–1)

- Relationships can be linear or curvilinear. To determine the shape, you draw a scatter plot of the variables. If the relationship is linear, the data can be approximated by a straight line, called the *regression line,* or the *line of best fit.* The closer the value of r is to +1 or −1, the more closely the points will fit the line. (10–2)

- A residual plot can be used to determine if the regression line equation can be used for predictions. (10–3)

- The coefficient of determination is a better indicator of the strength of a linear relationship than the correlation coefficient. It is better because it identifies the percentage of variation of the dependent variable that is directly attributable to the variation of the independent variable. The coefficient of determination is obtained by squaring the correlation coefficient and converting the result to a percentage. (10–3)

- Another statistic used in correlation and regression is the standard error of the estimate, which is an estimate of the standard deviation of the y values about the predicted y' values. The standard error of the estimate can be used to construct a prediction interval about a specific value point estimate y' of the mean of the y values for a given value of x. (10–3)

- In addition, relationships can be multiple. That is, there can be two or more independent variables and one dependent variable. A coefficient of correlation and a regression equation can be found for multiple relationships, just as they can be found for simple relationships. (10–4)

"At this point in my report, I'll ask all of you to follow me to the conference room directly below us!"

Source: Cartoon by Bradford Veley, Marquette, Michigan. Reprinted with permission.

Important Terms

adjusted R^2 579

coefficient of determination 569

correlation 534

correlation coefficient 539

dependent variable 535

extrapolation 556

independent variable 535

influential point or observation 557

least-squares line 567

lurking variable 547

marginal change 555

multiple correlation coefficient 578

multiple regression 575

multiple relationship 535

negative relationship 535

Pearson product moment correlation coefficient 539

population correlation coefficient 543

positive relationship 535

prediction interval 572

regression 534

regression line 551

residual 567

residual plot 568

scatter plot 536

simple relationship 535

standard error of the estimate 570

Important Formulas

Formula for the correlation coefficient:

$$r = \frac{n(\Sigma xy) - (\Sigma x)(\Sigma y)}{\sqrt{[n(\Sigma x^2) - (\Sigma x)^2][n(\Sigma y^2) - (\Sigma y)^2]}}$$

Formula for the t test for the correlation coefficient:

$$t = r\sqrt{\frac{n-2}{1-r^2}} \qquad \text{d.f.} = n - 2$$

The regression line equation:

$$y' = a + bx$$

where

$$a = \frac{(\Sigma y)(\Sigma x^2) - (\Sigma x)(\Sigma xy)}{n(\Sigma x^2) - (\Sigma x)^2}$$

$$b = \frac{n(\Sigma xy) - (\Sigma x)(\Sigma y)}{n(\Sigma x^2) - (\Sigma x)^2}$$

Formula for the standard error of the estimate:

$$s_{est} = \sqrt{\frac{\Sigma(y - y')^2}{n-2}}$$

or

$$s_{est} = \sqrt{\frac{\Sigma y^2 - a\,\Sigma y - b\,\Sigma xy}{n-2}}$$

Formula for the prediction interval for a value y':

$$y' - t_{\alpha/2}s_{est}\sqrt{1 + \frac{1}{n} + \frac{n(x - \bar{X})^2}{n\,\Sigma x^2 - (\Sigma x)^2}} < y$$

$$< y' + t_{\alpha/2}s_{est}\sqrt{1 + \frac{1}{n} + \frac{n(x - \bar{X})^2}{n\,\Sigma x^2 - (\Sigma x)^2}}$$

$$\text{d.f.} = n - 2$$

Formula for the multiple correlation coefficient:

$$R = \sqrt{\frac{r_{yx_1}^2 + r_{yx_2}^2 - 2r_{yx_1} \cdot r_{yx_2} \cdot r_{x_1x_2}}{1 - r_{x_1x_2}^2}}$$

Formula for the F test for the multiple correlation coefficient:

$$F = \frac{R^2/k}{(1 - R^2)/(n - k - 1)}$$

with d.f.N $= n - k$ and d.f.D $= n - k - 1$.

Formula for the adjusted R^2:

$$R_{adj}^2 = 1 - \frac{(1 - R^2)(n - 1)}{n - k - 1}$$

Review Exercises

For Exercises 1 through 7, do a complete regression analysis by performing the following steps.

a. Draw the scatter plot.

b. Compute the value of the correlation coefficient.

c. Test the significance of the correlation coefficient at $\alpha = 0.01$, using Table I.

d. Determine the regression line equation.

e. Plot the regression line on the scatter plot.

f. Predict y' for a specific value of x.

1. Passengers and Airline Fares The U.S. Department of Transportation Office of Aviation Analysis provides the weekly average number of passengers per flight and the average one-way fare in dollars for common commercial routes. Randomly selected flights are listed below with the reported data. Is there evidence of a relationship between these two variables? (10–1)(10–2)

Flight	Avg. no. of passengers x	Avg. one-way fare y
Pittsburgh–Washington, DC	310	$236
Chicago–Pittsburgh	1388	105
Cincinnati–New York City	750	339
Denver–Phoenix	3019	96
Denver–Los Angeles	2151	176
Houston–Philadelphia	1104	180

Source: www.fedstats.gov

2. Elementary and Secondary Schools School district information was examined for a random selection of states. The data below show the number of elementary schools and the number of secondary schools for each particular state. Is there a significant relationship between the variables? Predict the number of secondary schools when the number of elementary schools is 300. (10–1)(10–2)

Elementary	201	766	148	218	519	396	274
Secondary	50	280	27	41	108	82	63

Source: *World Almanac.*

3. Touchdowns and QB Ratings Listed below are the number of touchdown passes thrown in the season and the quarterback rating for a random sample of NFL quarterbacks. Is there a significant linear relationship between the variables? (10–1)(10–2)

TDs	34	21	15	22	34	26	23
QB rating	106	89	82	81	96	91	86

Source: *New York Times Almanac.*

4. Driver's Age and Accidents A study is conducted to determine the relationship between a driver's age and the number of accidents he or she has over a 1-year period. The data are shown here. (This information will be used for Exercise 8.) If there is a significant relationship, predict the number of accidents of a driver who is 28. (10–1)(10–2)

Driver's age x	16	24	18	17	23	27	32
No. of accidents y	3	2	5	2	0	1	1

5. Typing Speed and Word Processing A researcher desires to know whether the typing speed of a secretary (in words per minute) is related to the time (in hours) that it takes the secretary to learn to use a new word processing program. The data are shown.

Speed x	48	74	52	79	83	56	85	63	88	74	90	92
Time y	7	4	8	3.5	2	6	2.3	5	2.1	4.5	1.9	1.5

If there is a significant relationship, predict the time it will take the average secretary who has a typing speed of 72 words per minute to learn the word processing program. (This information will be used for Exercises 9 and 11.) (10–1)(10–2)

6. Protein and Diastolic Blood Pressure A study was conducted with vegetarians to see whether the number of grams of protein each ate per day was related to diastolic blood pressure. The data are given here. (This information will be used for Exercises 10 and 12.) If there is a significant relationship, predict the diastolic pressure of a vegetarian who consumes 8 grams of protein per day. (10–1)(10–2)

Grams x	4	6.5	5	5.5	8	10	9	8.2	10.5
Pressure y	73	79	83	82	84	92	88	86	95

7. Medical Specialties and Gender Although more and more women are becoming physicians each year, it is well known that men outnumber women in many specialties. Randomly selected specialties are listed below with the numbers of male and female physicians in each. Can it be concluded that there is a significant relationship between the two variables? Predict the number of male specialists when there are 2000 female specialists. (10–1)(10–2)

Specialty	Female x	Male y
Dermatology	3,482	6,506
Emergency medicine	5,098	20,429
Neurology	2,895	10,088
Pediatric cardiology	459	1,241
Radiology	1,218	7,574
Forensic pathology	181	399
Radiation oncology	968	3,215

Source: *World Almanac.*

8. For Exercise 4, find the standard error of the estimate. (10–3) 1.417* For calculation purposes only. No regression should be done.

9. For Exercise 5, find the standard error of the estimate. (10–3) 0.468* (TI value 0.513)

10. For Exercise 6, find the standard error of the estimate. (10–3) 2.89 (TI value 2.845)

11. For Exercise 5, find the 90% prediction interval for time when the speed is 72 words per minute. (10–3) $3.34 < y < 5.10$*

12. For Exercise 6, find the 95% prediction interval for pressure when the number of grams is 8. (10–3) $79 < y < 93$

13. (Opt.) A study found a significant relationship among a person's years of experience on a particular job x_1, the number of workdays missed per month x_2, and the person's age y. The regression equation is $y' = 12.8 + 2.09x_1 + 0.423x_2$. Predict a person's age if he or she has been employed for 4 years and has missed 2 workdays a month. (10–4) 22.01*

14. (Opt.) Find R when $r_{yx_1} = 0.681$ and $r_{yx_2} = 0.872$ and $r_{x_1x_2} = 0.746$. (10–4) $R = 0.873$

15. (Opt.) Find R^2_{adj} when $R = 0.873$, $n = 10$, and $k = 3$. (10–4) $R^2_{adj} = 0.643^*$

*Answers may vary due to rounding.

Statistics Today

Do Dust Storms Affect Respiratory Health?–Revisited

The researchers correlated the dust pollutant levels in the atmosphere and the number of daily emergency room visits for several respiratory disorders, such as bronchitis, sinusitis, asthma, and pneumonia. Using the Pearson correlation coefficient, they found overall a significant but low correlation, $r = 0.13$, for bronchitis visits only. However, they found a much higher correlation value for sinusitis, P-value $= 0.08$, when pollutant levels exceeded maximums set by the Environmental Protection Agency (EPA). In addition, they found statistically significant correlation coefficients $r = 0.94$ for sinusitis visits and $r = 0.74$ for upper-respiratory-tract infection visits 2 days after the dust pollutants exceeded the maximum levels set by the EPA.

Data Analysis

The Data Bank is found in Appendix D, or on the World Wide Web by following links from

www.mhhe.com/math/stat/bluman/

1. From the Data Bank, choose two variables that might be related: for example, IQ and educational level; age and cholesterol level; exercise and weight; or weight and systolic pressure. Do a complete correlation and regression analysis by performing the following steps. Select a random sample of at least 10 subjects.

   a. Draw a scatter plot.

   b. Compute the correlation coefficient.

   c. Test the hypothesis H_0: $\rho = 0$.

   d. Find the regression line equation.

   e. Summarize the results.

2. Repeat Exercise 1, using samples of values of 10 or more obtained from Data Set V in Appendix D. Let $x =$ the number of suspensions and $y =$ the enrollment size.

3. Repeat Exercise 1, using samples of 10 or more values obtained from Data Set XIII. Let $x =$ the number of beds and $y =$ the number of personnel employed.

Chapter Quiz

Determine whether each statement is true or false. If the statement is false, explain why.

1. A negative relationship between two variables means that for the most part, as the x variable increases, the y variable increases. False

2. A correlation coefficient of -1 implies a perfect linear relationship between the variables. True

3. Even if the correlation coefficient is high (near $+1$) or low (near -1), it may not be significant. True

4. When the correlation coefficient is significant, you can assume x causes y. False

5. It is not possible to have a significant correlation by chance alone. False

6. In multiple regression, there are several dependent variables and one independent variable. False

Select the best answer.

7. The strength of the linear relationship between two quantitative variables is determined by the value of

   (a.) r c. x
   b. a d. s_{est}

8. To test the significance of r, a(n) _____ test is used.

   (a.) t c. χ^2
   b. F d. None of the above

9. The test of significance for r has _____ degrees of freedom.

   a. 1   c. $n - 1$
   b. n (d.) $n - 2$

10. The equation of the regression line used in statistics is

   a. $x = a + by$ (c.) $y' = a + bx$
   b. $y = bx + a$ d. $x = ay + b$

11. The coefficient of determination is

   a. r c. a
   (b.) r^2 d. b

Complete the following statements with the best answer.

12. A statistical graph of two quantitative variables is called a(n) _____. Scatter plot

13. The x variable is called the _____ variable. Independent

14. The range of r is from _____ to _____. $-1, +1$

15. The sign of r and _____ will always be the same. b (slope)

16. The regression line is called the _____ . Line of best fit

17. If all the points fall on a straight line, the value of r will be _____ or _____ . $+1, -1$

For Exercises 18 through 21, do a complete regression analysis.

    a. Draw the scatter plot.

    b. Compute the value of the correlation coefficient.

    c. Test the significance of the correlation coefficient at $\alpha = 0.05$.

    d. Determine the regression line equation.

    e. Plot the regression line on the scatter plot.

    f. Predict y' for a specific value of x.

18. Prescription Drug Prices A medical researcher wants to determine the relationship between the price per dose of prescription drugs in the United States and the price of the same dose in Australia. The data are shown. Describe the relationship.

U.S. price x	3.31	3.16	2.27	3.13	2.54	1.98	2.22
Australian price y	1.29	1.75	0.82	0.83	1.32	0.84	0.82

19. Age and Driving Accidents A study is conducted to determine the relationship between a driver's age and the number of accidents he or she has over a 1-year period. The data are shown here. If there is a significant relationship, predict the number of accidents of a driver who is 64.

Driver's age x	63	65	60	62	66	67	59
No. of accidents y	2	3	1	0	3	1	4

20. Age and Cavities A researcher desires to know if the age of a child is related to the number of cavities he or she has. The data are shown here. If there is a

significant relationship, predict the number of cavities for a child of 11.

Age of child x	6	8	9	10	12	14
No. of cavities y	2	1	3	4	6	5

21. Fat and Cholesterol A study is conducted with a group of dieters to see if the number of grams of fat each consumes per day is related to cholesterol level. The data are shown here. If there is a significant relationship, predict the cholesterol level of a dieter who consumes 8.5 grams of fat per day.

Fat grams x	6.8	5.5	8.2	10	8.6	9.1	8.6	10.4
Cholesterol level y	183	201	193	283	222	250	190	218

22. For Exercise 20, find the standard error of the estimate. 1.129*

23. For Exercise 21, find the standard error of the estimate. 29.5* For calculation purposes only. No regression should be done.

24. For Exercise 20, find the 90% prediction interval of the number of cavities for a 7-year-old. $0 < y < 5$*

25. For Exercise 21, find the 95% prediction interval of the cholesterol level of a person who consumes 10 grams of fat. 217.5 (average of y' values is used since there is no significant relationship)

26. (Opt.) A study was conducted, and a significant relationship was found among the number of hours a teenager watches television per day x_1, the number of hours the teenager talks on the telephone per day x_2, and the teenager's weight y. The regression equation is $y' = 98.7 + 3.82x_1 + 6.51x_2$. Predict a teenager's weight if she averages 3 hours of TV and 1.5 hours on the phone per day. 119.9*

27. (Opt.) Find R when $r_{yx_1} = 0.561$ and $r_{yx_2} = 0.714$ and $r_{x_1x_2} = 0.625$. $R = 0.729$*

28. (Opt.) Find R^2_{adj} when $R = 0.774$, $n = 8$, and $k = 2$. $R^2_{adj} = 0.439$*

These answers may vary due to the method of calculation or rounding.

Critical Thinking Challenges

Product Sales When the points in a scatter plot show a curvilinear trend rather than a linear trend, statisticians have methods of fitting curves rather than straight lines to the data, thus obtaining a better fit and a better prediction model. One type of curve that can be used is the logarithmic regression curve. The data shown are the number of items of a new product sold over a period of 15 months at a certain store. Notice that sales rise during the beginning months and then level off later on.

Month x	1	3	6	8	10	12	15
No. of items sold y	10	12	15	19	20	21	21

1. Draw the scatter plot for the data.

2. Find the equation of the regression line.

3. Describe how the line fits the data.

4. Using the log key on your calculator, transform the x values into log x values.

5. Using the log x values instead of the x values, find the equation of a and b for the regression line.

6. Next, plot the curve $y = a + b \log x$ on the graph.

7. Compare the line $y = a + bx$ with the curve $y = a + b \log x$ and decide which one fits the data better.

8. Compute r, using the x and y values; then compute r, using the log x and y values. Which is higher?

9. In your opinion, which (the line or the logarithmic curve) would be a better predictor for the data? Why?

Data Projects

Use a significance level of 0.05 for all tests below.

1. **Business and Finance** Use the stocks in data project 1 of Chapter 2 identified as the Dow Jones Industrials as the sample. For each, note the current price and the amount of the last year's dividends. Are the two variables linearly related? How much variability in amount of dividend is explainable by the price?

2. **Sports and Leisure** For each team in major league baseball note the number of wins the team had last year and the number of home runs by its best home run hitter. Is the number of wins linearly related to the number of home runs hit? How much variability in total wins is explained by home runs hit? Write a regression equation to determine how many wins you would expect a team to have, knowing their top home run output.

3. **Technology** Use the data collected in data project 3 of Chapter 2 for this problem. For the data set note the length of the song and the year it was released. Is there a linear relationship between the length of a song and the year it was released? Is the sign on the correlation coefficient positive or negative? What does

the sign on the coefficient indicate about the relationship?

4. **Health and Wellness** Use a fast-food restaurant to compile your data. For each menu item note its fat grams and its total calories. Is there a linear relationship between the two variables? How much variance in total calories is explained by fat grams? Write a regression equation to determine how many total calories you would expect in an item, knowing its fat grams.

5. **Politics and Economics** For each state find its average SAT Math score, SAT English score, and average household income. Which has the strongest linear relationship, SAT Math and SAT English, SAT Math and income, or SAT English and income?

6. **Your Class** Use the data collected in data project 6 of Chapter 2 regarding heart rates. Is there a linear relationship between the heart rates before and after exercise? How much of the variability in heart rate after exercise is explainable by heart rate before exercise? Write a regression equation to determine what heart rate after exercise you would expect for a person, given the person's heart rate before exercise.

Answers to Applying the Concepts

Section 10–1   Stopping Distances

1. The independent variable is miles per hour (mph).

2. The dependent variable is braking distance (feet).

3. Miles per hour is a continuous quantitative variable.

4. Braking distance is a continuous quantitative variable.

5. A scatter plot of the data is shown.

Scatter plot of braking distance vs. mph

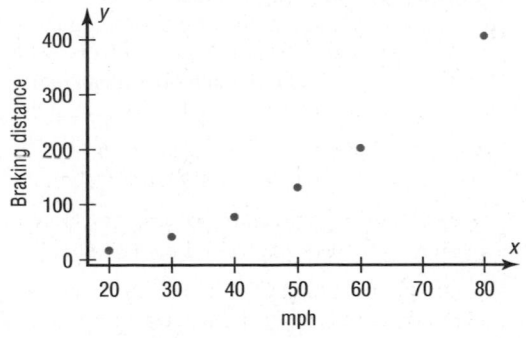

6. There might be a linear relationship between the two variables, but there is a bit of a curve in the data.

7. Changing the distances between the mph increments will change the appearance of the relationship.

8. There is a positive relationship between the two variables—higher speeds are associated with longer braking distances.

9. The strong relationship between the two variables suggests that braking distance can be accurately predicted from mph. We might still have some concern about the curve in the data.

10. Answers will vary. Some other variables that might affect braking distance include road conditions, driver response time, and condition of the brakes.

11. The correlation coefficient is $r = 0.966$.

12. The value for $r = 0.966$ is significant at $\alpha = 0.05$. This confirms the strong positive relationship between the variables.

Section 10–2   Stopping Distances Revisited

1. The linear regression equation is
$y' = -151.90 + 6.4514x$

2. The slope says that for each additional mile per hour a car is traveling, we expect the stopping distance to increase by 6.45 feet, on average. The y intercept is the braking distance we would expect for a car traveling 0 mph—this is meaningless in this context, but is an important part of the model.

3. $y' = -151.90 + 6.4514(45) = 138.4$ The braking distance for a car traveling 45 mph is approximately 138 feet.

4. $y' = -151.90 + 6.4514(100) = 493.2$ The braking distance for a car traveling 100 mph is approximately 493 feet.

5. It is not appropriate to make predictions of braking distance for speeds outside of the given data values (for example, the 100 mph above) because we know nothing about the relationship between the two variables outside of the range of the data.

Section 10–3   Interpreting Simple Linear Regression

1. Both variables are moving in the same direction. In others words, the two variables are positively associated. This is so because the correlation coefficient is positive.

2. The unexplained variation of 3026.49 measures the distances from the prediction line to the actual values.

3. The slope of the regression line is 0.725983.

4. The y intercept is 16.5523.

5. The critical value of 0.378419 can be found in a table.

6. The allowable risk of making a type I error is 0.10, the level of significance.

7. The variation explained by the regression is 0.631319, or about 63.1%.

8. The average scatter of points about the regression line is 12.9668, the standard error of the estimate.

9. The null hypothesis is that there is no correlation, $H_0: \rho = 0$.

10. We compare the test statistic of 0.794556 to the critical value to see if the null hypothesis should be rejected.

11. Since $0.794556 > 0.378419$, we reject the null hypothesis and find that there is enough evidence to conclude that the correlation is not equal to zero.

Section 10–4   More Math Means More Money

1. The dependent variable is yearly income 10 years after high school.

2. The independent variables are number of math courses taken and number of hours worked per week during the senior year of high school.

3. Multiple regression assumes that the independent variables are not highly correlated.

4. We expect a person's yearly income 10 years after high school to be $4540 more, on average, for each additional math course taken, all other variables held constant. We expect a person's yearly income 10 years after high school to be $1290 more, on average, for each additional hour worked per week during the senior year of high school, all other variables held constant.

5. $y' = 6000 + 4540(8) + 1290(20) = 68,120$. The predicted yearly income 10 years after high school is $68,120.

6. The multiple correlation coefficient of 0.77 means that there is a fairly strong positive relationship between the independent variables (number of math courses and hours worked during senior year of high school) and the dependent variable (yearly income 10 years after high school).

7. $R^2 = (0.77)^2 = 0.5929$

8. $R^2_{adj} = 1 - \dfrac{(1 - R^2)(n - 1)}{n - k - 1}$

   $= 1 - \dfrac{(1 - 0.5929)(20 - 1)}{20 - 2 - 1}$

   $= 1 - \dfrac{(0.4071)(19)}{17} = 0.5450$

9. The equation appears to be a fairly good predictor of income, since 54.5% of the variation in yearly income 10 years after high school is explained by the regression model.

10. Answers will vary. One possible answer is that yearly income 10 years after high school increases with more math classes and more hours of work during the senior year of high school. The number of math classes has a higher coefficient, so more math does mean more money!

11

Other Chi-Square Tests

Objectives

After completing this chapter, you should be able to

1 Test a distribution for goodness of fit, using chi-square.

2 Test two variables for independence, using chi-square.

3 Test proportions for homogeneity, using chi-square.

Statistics and Heredity

An Austrian monk, Gregor Mendel (1822–1884), studied genetics, and his principles are the foundation for modern genetics. Mendel used his spare time to grow a variety of peas at the monastery. One of his many experiments involved crossbreeding peas that had smooth yellow seeds with peas that had wrinkled green seeds. He noticed that the results occurred with regularity. That is, some of the offspring had smooth yellow seeds, some had smooth green seeds, some had wrinkled yellow seeds, and some had wrinkled green seeds. Furthermore, after several experiments, the percentages of each type seemed to remain approximately the same. Mendel formulated his theory based on the assumption of dominant and recessive traits and tried to predict the results. He then crossbred his peas and examined 556 seeds over the next generation.

Finally, he compared the actual results with the theoretical results to see if his theory was correct. To do this, he used a "simple" chi-square test, which is explained in this chapter. See Statistics Today—Revisited at the end of this chapter.

Source: J. Hodges, Jr., D. Krech, and R. Crutchfield, *Stat Lab, An Empirical Introduction to Statistics* (New York: McGraw-Hill), pp. 228–229. Used with permission.

Introduction

The chi-square distribution was used in Chapters 7 and 8 to find a confidence interval for a variance or standard deviation and to test a hypothesis about a single variance or standard deviation.

It can also be used for tests concerning *frequency distributions,* such as "If a sample of buyers is given a choice of automobile colors, will each color be selected with the same frequency?" The chi-square distribution can be used to test the *independence* of two variables, for example, "Are senators' opinions on gun control independent of party affiliations?" That is, do the Republicans feel one way and the Democrats feel differently, or do they have the same opinion?

Finally, the chi-square distribution can be used to test the *homogeneity of proportions*. For example, is the proportion of high school seniors who attend college immediately after graduating the same for the northern, southern, eastern, and western parts of the United States?

This chapter explains the chi-square distribution and its applications. In addition to the applications mentioned here, chi-square has many other uses in statistics.

11–1   Test for Goodness of Fit

Objective 1

Test a distribution for goodness of fit, using chi-square.

In addition to being used to test a single variance, the chi-square statistic can be used to see whether a frequency distribution fits a specific pattern. For example, to meet customer demands, a manufacturer of running shoes may wish to see whether buyers show a preference for a specific style. A traffic engineer may wish to see whether accidents occur more often on some days than on others, so that she can increase police patrols accordingly. An emergency service may want to see whether it receives more calls at certain times of the day than at others, so that it can provide adequate staffing.

When you are testing to see whether a frequency distribution fits a specific pattern, you can use the chi-square **goodness-of-fit test.** For example, suppose as a market analyst you wished to see whether consumers have any preference among five flavors of a new fruit soda. A sample of 100 people provided these data:

Karl Pearson (1857–1936) first used the chi-square distribution as a goodness-of-fit test for data. He developed many types of descriptive graphs and gave them unusual names such as stigmograms, topograms, stereograms, and radiograms.

Historical Note

Cherry	Strawberry	Orange	Lime	Grape
32	28	16	14	10

If there were no preference, you would expect each flavor to be selected with equal frequency. In this case, the equal frequency is $100/5 = 20$. That is, *approximately* 20 people would select each flavor.

Since the frequencies for each flavor were obtained from a sample, these actual frequencies are called the **observed frequencies.** The frequencies obtained by calculation (as if there were no preference) are called the **expected frequencies.** A completed table for the test is shown.

Frequency	Cherry	Strawberry	Orange	Lime	Grape
Observed	32	28	16	14	10
Expected	20	20	20	20	20

The observed frequencies will almost always differ from the expected frequencies due to sampling error; that is, the values differ from sample to sample. But the question is: Are these differences significant (a preference exists), or are they due to chance? The chi-square goodness-of-fit test will enable the researcher to determine the answer.

Before computing the test value, you must state the hypotheses. The null hypothesis should be a statement indicating that there is no difference or no change. For this example, the hypotheses are as follows:

H_0: Consumers show no preference for flavors of the fruit soda.

H_1: Consumers show a preference.

In the goodness-of-fit test, the degrees of freedom are equal to the number of categories minus 1. For this example, there are five categories (cherry, strawberry, orange, lime, and grape); hence, the degrees of freedom are $5 - 1 = 4$. This is so because the number of subjects in each of the first four categories is free to vary. But in order for the sum to be 100—the total number of subjects—the number of subjects in the last category is fixed.

Formula for the Chi-Square Goodness-of-Fit Test

$$\chi^2 = \sum \frac{(O - E)^2}{E}$$

with degrees of freedom equal to the number of categories minus 1, and where
O = observed frequency
E = expected frequency

Two assumptions are needed for the goodness-of-fit test. These assumptions are given next.

Assumptions for the Chi-Square Goodness-of-Fit Test

1. The data are obtained from a random sample.
2. The expected frequency for each category must be 5 or more.

This test is a right-tailed test, since when the $O - E$ values are squared, the answer will be positive or zero. This formula is explained in Example 11–1.

Example 11–1

Fruit Soda Flavor Preference

Is there enough evidence to reject the claim that there is no preference in the selection of fruit soda flavors, using the data shown previously? Let $\alpha = 0.05$.

Solution

Step 1 State the hypotheses and identify the claim.

H_0: Consumers show no preference for flavors (claim).

H_1: Consumers show a preference.

Step 2 Find the critical value. The degrees of freedom are $5 - 1 = 4$, and $\alpha = 0.05$. Hence, the critical value from Table G in Appendix C is 9.488.

Step 3 Compute the test value by subtracting the expected value from the corresponding observed value, squaring the result and dividing by the expected value, and finding the sum. The expected value for each category is 20, as shown previously.

$$\chi^2 = \sum \frac{(O - E)^2}{E}$$

$$= \frac{(32 - 20)^2}{20} + \frac{(28 - 20)^2}{20} + \frac{(16 - 20)^2}{20} + \frac{(14 - 20)^2}{20} + \frac{(10 - 20)^2}{20}$$

$$= 18.0$$

Step 4 Make the decision. The decision is to reject the null hypothesis, since $18.0 > 9.488$, as shown in Figure 11–1.

Figure 11–1

Critical and Test Values
for Example 11–1

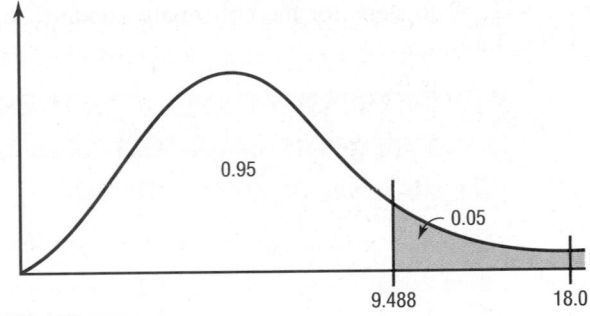

Step 5   Summarize the results. There is enough evidence to reject the claim that
consumers show no preference for the flavors.

To get some idea of why this test is called the goodness-of-fit test, examine graphs
of the observed values and expected values. See Figure 11–2. From the graphs, you can
see whether the observed values and expected values are close together or far apart.

Figure 11–2

Graphs of the
Observed and
Expected Values for
Soda Flavors

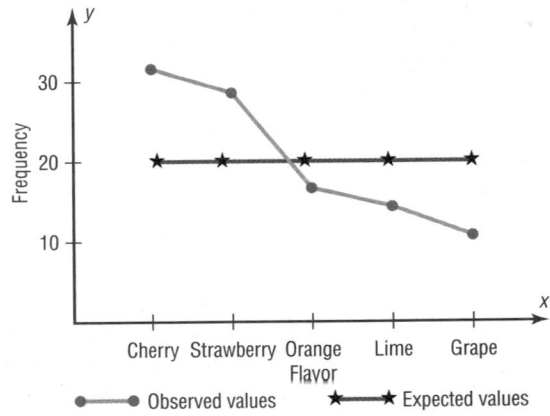

When the observed values and expected values are close together, the chi-square test
value will be small. Then the decision will be to not reject the null hypothesis—hence,
there is "a good fit." See Figure 11–3(a). When the observed values and the expected values are far apart, the chi-square test value will be large. Then the null hypothesis will be
rejected—hence, there is "not a good fit." See Figure 11–3(b).

Figure 11–3

Results of the
Goodness-of-Fit Test

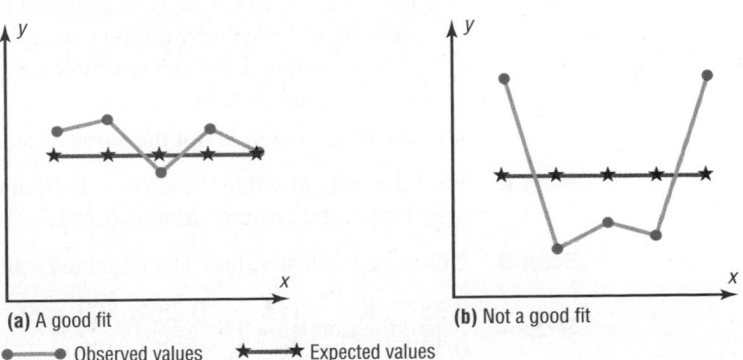

The steps for the chi-square goodness-of-fit test are summarized in this Procedure Table.

Procedure Table

The Chi-Square Goodness-of-Fit Test

Step 1 State the hypotheses and identify the claim.

Step 2 Find the critical value. The test is always right-tailed.

Step 3 Compute the test value.

Find the sum of the $\dfrac{(O - E)^2}{E}$ values.

Step 4 Make the decision.

Step 5 Summarize the results.

When there is perfect agreement between the observed and the expected values, $\chi^2 = 0$. Also, χ^2 can never be negative. Finally, the test is right-tailed because "H_0: Good fit" and "H_1: Not a good fit" mean that χ^2 will be small in the first case and large in the second case.

Example 11–2

Retired Senior Executives Return to Work

The Russel Reynold Association surveyed retired senior executives who had returned to work. They found that after returning to work, 38% were employed by another organization, 32% were self-employed, 23% were either freelancing or consulting, and 7% had formed their own companies. To see if these percentages are consistent with those of Allegheny County residents, a local researcher surveyed 300 retired executives who had returned to work and found that 122 were working for another company, 85 were self-employed, 76 were either freelancing or consulting, and 17 had formed their own companies. At $\alpha = 0.10$, test the claim that the percentages are the same for those people in Allegheny County.

Source: Michael L. Shook and Robert D. Shook, *The Book of Odds*.

Solution

Step 1 State the hypotheses and identify the claim.

H_0: The retired executives who returned to work are distributed as follows: 38% are employed by another organization, 32% are self-employed, 23% are either freelancing or consulting, and 7% have formed their own companies (claim).

H_1: The distribution is not the same as stated in the null hypothesis.

Step 2 Find the critical value. Since $\alpha = 0.10$ and the degrees of freedom are $4 - 1 = 3$, the critical value is 6.251.

Step 3 Compute the test value. The expected values are computed as follows:

$0.38 \times 300 = 114$ $0.23 \times 300 = 69$

$0.32 \times 300 = 96$ $0.07 \times 300 = 21$

$$\chi^2 = \sum \frac{(O - E)^2}{E}$$

$$= \frac{(122 - 114)^2}{114} + \frac{(85 - 96)^2}{96} + \frac{(76 - 69)^2}{69} + \frac{(17 - 21)^2}{21}$$

$$= 3.2939$$

Step 4 Make the decision. Since $3.2939 < 6.251$, the decision is not to reject the null hypothesis. See Figure 11–4.

Figure 11–4

Critical and Test Values for Example 11–2

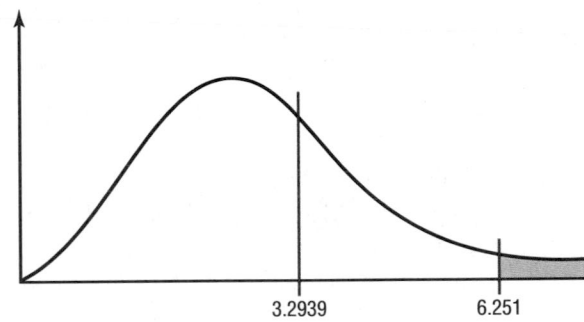

3.2939   6.251

Step 5 Summarize the results. There is not enough evidence to reject the claim. It can be concluded that the percentages are not significantly different from those given in the null hypothesis.

Example 11–3

Firearm Deaths

A researcher read that firearm-related deaths for people aged 1 to 18 were distributed as follows: 74% were accidental, 16% were homicides, and 10% were suicides. In her district, there were 68 accidental deaths, 27 homicides, and 5 suicides during the past year. At $\alpha = 0.10$, test the claim that the percentages are equal.

Source: Centers for Disease Control and Prevention.

Solution

Step 1 State the hypotheses and identify the claim:

H_0: The deaths due to firearms for people aged 1 through 18 are distributed as follows: 74% accidental, 16% homicides, and 10% suicides (claim).

H_1: The distribution is not the same as stated in the null hypothesis.

Step 2 Find the critical value. Since $\alpha = 0.10$ and the degrees of freedom are $3 - 1 = 2$, the critical value is 4.605.

Step 3 Compute the test value. The expected values are as follows:

$0.74 \times 100 = 74$
$0.16 \times 100 = 16$
$0.10 \times 100 = 10$

$$\chi^2 = \sum \frac{(O - E)^2}{E}$$

$$= \frac{(68 - 74)^2}{74} + \frac{(27 - 16)^2}{16} + \frac{(5 - 10)^2}{10}$$

$$= 10.549$$

Step 4 Reject the null hypothesis, since 10.549 > 4.605, as shown in Figure 11–5.

Figure 11–5

Critical and Test Values
for Example 11–3

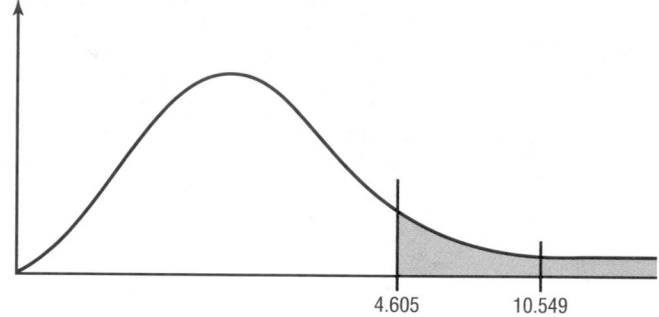

Step 5 Summarize the results. There is enough evidence to reject the claim that the distribution is 74% accidental, 16% homicides, and 10% suicides.

The *P*-value method of hypothesis testing can also be used for the chi-square tests explained in this chapter. The *P*-values for chi-square are found in Table G in Appendix C. The method used to find the *P*-value for a chi-square test value is the same as the method shown in Section 8–5. The *P*-value for $\chi^2 = 3.2939$ with d.f. $= 3$ (for the data in Example 11–2) is greater than 0.10 since 6.251 is the value in Table G for $\alpha = 0.10$. (The *P*-value obtained from a calculator is 0.348.) Hence *P*-value > 0.10. The decision is to not reject the null hypothesis, which is consistent with the decision made in Example 11–2 using the traditional method of hypothesis testing.

For use of the chi-square goodness-of-fit test, statisticians have determined that the expected frequencies should be at least 5, as stated in the assumptions. The reasoning is as follows: The chi-square distribution is continuous, whereas the goodness-of-fit test is discrete. However, the continuous distribution is a good approximation and can be used when the expected value for each class is at least 5. If an expected frequency of a class is less than 5, then that class can be combined with another class so that the expected frequency is 5 or more.

Test of Normality (Optional)

The chi-square goodness-of-fit test can be used to test a variable to see if it is normally distributed. The null hypotheses are

H_0: The variable is normally distributed.

H_1: The variable is not normally distributed.

The procedure is somewhat complicated. It involves finding the expected frequencies for each class of a frequency distribution by using the standard normal distribution. Then the actual frequencies (i.e., observed frequencies) are compared to the expected frequencies, using the chi-square goodness-of-fit test. If the observed frequencies are close in value to the expected frequencies, the chi-square test value will be small, and the null hypothesis cannot be rejected. In this case, it can be concluded that the variable is approximately normally distributed.

On the other hand, if there is a large difference between the observed frequencies and the expected frequencies, the chi-square test value will be larger, and the null hypothesis can be rejected. In this case, it can be concluded that the variable is not normally distributed. Example 11–4 illustrates the procedure for the chi-square test of normality. To find the areas in the examples, you might want to review Section 6–2.

Example 11–4 shows how to do the calculations.

| Example 11–4 | **Test of Normality** |

Use chi-square to determine if the variable shown in the frequency distribution is normally distributed. Use $\alpha = 0.05$.

Boundaries	Frequency
89.5–104.5	24
104.5–119.5	62
119.5–134.5	72
134.5–149.5	26
149.5–164.5	12
164.5–179.5	4
	200

Solution

H_0: The variable is normally distributed.

H_1: The variable is not normally distributed.

First find the mean and standard deviation of the variable. Then find the area under the standard normal distribution, using z values and Table E for each class. Find the expected frequencies for each class by multiplying the area by 200. Finally, find the chi-square test value by using the formula $\chi^2 = \sum \dfrac{(O - E)^2}{E}$.

Boundaries	f	X_m	$f \cdot X_m$	$f \cdot X_m^2$
89.5–104.5	24	97	2,328	225,816
104.5–119.5	62	112	6,944	777,728
119.5–134.5	72	127	9,144	1,161,288
134.5–149.5	26	142	3,692	524,264
149.5–164.5	12	157	1,884	295,788
164.5–179.5	4	172	688	118,336
	200		24,680	3,103,220

$$\bar{X} = \frac{24,680}{200} = 123.4$$

$$s = \sqrt{\frac{200(3,103,220) - 24,680^2}{200(199)}} = \sqrt{290} = 17.03$$

The area to the left of $x = 104.5$ is found as

$$z = \frac{104.5 - 123.4}{17.03} = -1.11$$

The area for $z < -1.11$ is 0.1335.

    The area between 104.5 and 119.5 is found as

$$z = \frac{119.5 - 123.4}{17.03} = -0.23$$

The area for $-1.11 < z < -0.23$ is $0.4090 - 0.1335 = 0.2755$.

    The area between 119.5 and 134.5 is found as

$$z = \frac{134.5 - 123.4}{17.03} = 0.65$$

The area for $-0.23 < z < 0.65$ is $0.7422 - 0.4090 = 0.3332$.

    The area between 134.5 and 149.5 is found as

$$z = \frac{149.5 - 123.4}{17.03} = 1.53$$

The area for $0.65 < z < 1.53$ is $0.9370 - 0.7422 = 0.1948$.

The area between 149.5 and 164.5 is found as

$$z = \frac{164.5 - 123.4}{17.03} = 2.41$$

The area for $1.53 < z < 2.41$ is $0.9920 - 0.9370 = 0.0550$.
The area to the right of $x = 164.5$ is found as

$$z = \frac{164.5 - 123.4}{17.03} = 2.41$$

The area is $1.0000 - 0.9920 = 0.0080$.
The expected frequencies are found by

$$0.1335 \cdot 200 = 26.7$$
$$0.2755 \cdot 200 = 55.1$$
$$0.3332 \cdot 200 = 66.64$$
$$0.1948 \cdot 200 = 38.96$$
$$0.0550 \cdot 200 = 11.0$$
$$0.0080 \cdot 200 = 1.6$$

Note: Since the expected frequency for the last category is less than 5, it can be combined with the previous category.
The χ^2 is found by

O	24	62	72	26	16
E	26.7	55.1	66.64	38.96	12.6

$$\chi^2 = \frac{(24 - 26.7)^2}{26.7} + \frac{(62 - 55.1)^2}{55.1} + \frac{(72 - 66.64)^2}{66.64} + \frac{(26 - 38.96)^2}{38.96}$$
$$+ \frac{(16 - 12.6)^2}{12.6}$$
$$= 6.797$$

The critical value in this test has the degrees of freedom equal to the number of categories -3 since one degree of freedom is lost for each parameter that is estimated. In this case, the mean and standard deviation have been estimated so two additional degrees of freedom are needed.

The C.V. with d.f. $= 2$ and $\alpha = 0.05$ is 5.991, so the null hypothesis is rejected. Hence, the distribution can be considered not normally distributed.

Note: At $\alpha = 0.01$, the C.V. $= 9.210$ and the null hypotheses would not be rejected. So it is important to decide which critical value should be used.

Applying the Concepts **11–1**

Never the Same Amounts

M&M/Mars, the makers of Skittles candies, states that the flavor blend is 20% for each flavor. Skittles is a combination of lemon, lime, orange, strawberry, and grape flavored candies. The following data list the results of four randomly selected bags of Skittles and their flavor blends. Use the data to answer the questions.

Bag	Green	Orange	Red	Purple	Yellow
1	7	20	10	7	14
2	20	5	5	13	17
3	4	16	13	21	4
4	12	9	16	3	17
Total	43	50	44	44	52

(column group header: Flavor)

1. Are the variables quantitative or qualitative?

2. What type of test can be used to compare the observed values to the expected values?

3. Perform a chi-square test on the total values.

4. What hypotheses did you use?

5. What were the degrees of freedom for the test? What is the critical value?

6. What is your conclusion?

See page 627 for the answers.

Exercises 11–1

1. How does the goodness-of-fit test differ from the chi-square variance test?

2. How are the degrees of freedom computed for the goodness-of-fit test? The degrees of freedom are the number of categories minus 1.

3. How are the expected values computed for the goodness-of-fit test?

4. When the expected frequencies are less than 5 for a specific class, what should be done so that you can use the goodness-of-fit test? The categories should be combined with other categories.

For Exercises 5 through 19, perform these steps.

a. State the hypotheses and identify the claim.

b. Find the critical value.

c. Compute the test value.

d. Make the decision.

e. Summarize the results.

Use the traditional method of hypothesis testing unless otherwise specified.

5. **Home-Schooled Student Activities** Students who are home-schooled often attend their local schools to participate in various types of activities such as sports or musical ensembles. According to the government, 82% of home-schoolers receive their education entirely at home, while 12% attend school up to 9 hours per week and 6% spend from 9 to 25 hours per week at school. A survey of 85 students who are home-schooled revealed the following information about where they receive their education.

Entirely at home	Up to 9 hours	9 to 25 hours
50	25	10

At $\alpha = 0.05$, is there sufficient evidence to conclude that the proportions differ from those stated by the government?
Source: www.nces.ed.gov

6. **Combatting Midday Drowsiness** A researcher wishes to see if the five ways (drinking decaffeinated beverages, taking a nap, going for a walk, eating a sugary snack, other) people use to combat midday drowsiness are equally distributed among office workers. A sample of 60 office workers is selected, and the following data are obtained. At $\alpha = 0.10$ can it be concluded that there is no preference? Why would the results be of interest to an employer?

Method	Beverage	Nap	Walk	Snack	Other
Number	21	16	10	8	5

Source: Based on information from Harris Interactive.

7. **Music Sales** In a recent year, 77.8% of recorded music sales were full-length CDs, 12.8% digital downloads, 3.8% singles, and the rest a mixture of other formats. A survey of college students and their recent music purchases indicated that out of 200 purchases, 48 were downloads, 135 were full-length CDs, 10 were singles, and the rest fit into the "other" category. At the 0.05 level of significance is there sufficient evidence to conclude that at least one of the proportions differs from its original value?
Source: New York Times Almanac.

8. **On-Time Performance by Airlines** According to the Bureau of Transportation statistics, on-time performance by the airlines is described as follows:

Action	% of Time
On time	70.8
National Aviation System delay	8.2
Aircraft arriving late	9.0
Other (because of weather and other conditions)	12.0

Records of 200 flights for a major airline company showed that 125 planes were on time; 40 were delayed because of weather, 10 because of a National Aviation System delay, and the rest because of arriving late. At $\alpha = 0.05$, do these results differ from the government's statistics?
Source: www.transtats.bts.gov

9. **Genetically Modified Food** An ABC News poll asked adults whether they felt genetically modified food was safe to eat. Thirty-five percent felt it was safe, 52% felt it was not safe, and 13% had no opinion.

A random sample of 120 adults was asked the same question at a local county fair. Forty people felt that genetically modified food was safe, 60 felt that it was not safe, and 20 had no opinion. At the 0.01 level of significance, is there sufficient evidence to conclude that the proportions differ from those reported in the survey?

Source: ABCNews.com Poll, www.pollingreport.com

10. Truck Colors In a recent year, the most popular colors for light trucks were white, 30%; black, 17%; red, 14%; silver, 12%; gray, 11%; blue, 8%; and other, 8%. A survey of light truck owners in a particular area revealed the following. At $\alpha = 0.05$ do the proportions differ from those stated?

White	Black	Red	Silver	Gray	Blue	Other
45	32	30	30	22	15	6

Source: World Almanac.

11. Assessment of Mathematics Students As part of the Mathematics Assessment, eighth-graders were asked about the frequency with which they used calculators while taking tests or quizzes. The results for national public schools were as follows: never, 28%; sometimes, 51%, and always, 21%. A random sample of 140 eighth-grade students in a large urban school district indicated that 30 said never, 78 said sometimes, and 32 said always. At $\alpha = 0.05$ do these proportions differ from the national report?

Source: Nationsreportcard.gov

12. Ages of Head Start Program Students The Head Start Program provides a wide range of services to low-income children up to the age of 5 and their families. Its goals are to provide services to improve social and learning skills and to improve health and nutrition status so that the participants can begin school on an equal footing with their more advantaged peers. The distribution of ages for participating children is as follows: 4% five-year-olds, 52% four-year-olds, 34% three-year-olds, and 10% under 3 years. When the program was assessed in a particular region, it was found that of the 200 participants, 20 were 5 years old, 120 were 4 years old, 40 were 3 years old, and 20 were under 3 years. Is there sufficient evidence at $\alpha = 0.05$ that the proportions differ from the program's? Use the P-value method.

Source: New York Times Almanac/www.fedstats.dhhs.gov

13. Payment Preference A USA TODAY Snapshot states that 53% of adult shoppers prefer to pay cash for purchases, 30% use checks, 16% use credit cards, and 1% have no preference. The owner of a large store randomly selected 800 shoppers and asked their payment preferences. The results were that 400 paid cash,

210 paid by check, 170 paid with a credit card, and 20 had no preference. At $\alpha = 0.01$, test the claim that the owner's customers have the same preferences as those surveyed.

Source: USA TODAY.

14. College Degree Recipients A survey of 800 recent degree recipients found that 155 received associate degrees; 450, bachelor degrees; 20, first professional degrees; 160, master degrees; and 15, doctorates. Is there sufficient evidence to conclude that at least one of the proportions differs from a report which stated that 23.3% were associate degrees; 51.1%, bachelor degrees; 3%, first professional degrees; 20.6%, master degrees; and 2%, doctorates? Use $\alpha = 0.05$.

Source: New York Times Almanac.

15. Internet Users A survey was targeted at determining if educational attainment affected Internet use. Randomly selected shoppers at a busy mall were asked if they used the Internet and their highest level of education attained. The results are listed below. Is there sufficient evidence at the 0.05 level of significance that the proportion of Internet users differs for any of the groups?

Graduated college +	Attended college	Did not attend
44	41	40

Source: www.infoplease.com

16. Education Level and Health Insurance A researcher wishes to see if the number of adults who do not have health insurance is equally distributed among three categories (less than 12 years of education, 12 years of education, more than 12 years of education). A sample of 60 adults who do not have health insurance is selected, and the results are shown. At $\alpha = 0.05$ can it be concluded that the frequencies are not equal? Use the P-value method. If the null hypothesis is rejected, give a possible reason for this.

Category	Less than 12 years	12 years	More than 12 years
Frequency	29	20	11

Source: U.S. Census Bureau.

17. Paying for Prescriptions A medical researcher wishes to determine if the way people pay for their medical prescriptions is distributed as follows: 60% personal funds, 25% insurance, 15% Medicare. A sample of 50 people found that 32 paid with their own money, 10 paid using insurance, and 8 paid using Medicare. At $\alpha = 0.05$ is the assumption correct? Use the P-value method. What would be an implication of the results?

Source: U.S. Health Care Financing.

Extending the Concepts

18. Tossing Coins Three coins are tossed 72 times, and the number of heads is shown. At $\alpha = 0.05$, test the null hypothesis that the coins are balanced and randomly tossed. (*Hint:* Use the binomial distribution.)

No. of heads	0	1	2	3
Frequency	3	10	17	42

19. State Lottery Numbers Select a three-digit state lottery number over a period of 50 days. Count the number of times each digit, 0 through 9, occurs. Test the claim, at $\alpha = 0.05$, that the digits occur at random.
Answers will vary.

Step by Step

MINITAB
Step by Step

Chi-Square Test for Goodness of Fit

For Example 11–1, is there a preference for flavor of soda? There is no menu command to do this directly. Use the calculator.

1. Enter the observed counts into C1 and the expected counts into C2. Name the columns O and E.

2. Select **Calc>Calculator.**

   a) Type **K1** in the Store result in variable.

   b) In the Expression box type the formula **SUM((O-E)**2/E).**

   c) Click [OK]. The chi-square test statistic will be displayed in the constant K1.

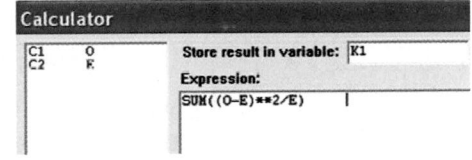

   d) Click the Project Manager icon, then navigate to the Worksheet 1>Constants. K1 is unnamed and equal to 18.

Calculate the *P*-Value

3. Select **Calc>Probability Distributions.**

4. Click on Chi-square.

   a) Click the button for Cumulative probability.

   b) In the box for Degrees of freedom type **4.**

   c) Click the button for Input constant, then click in the text box and select K1 in the variable list.

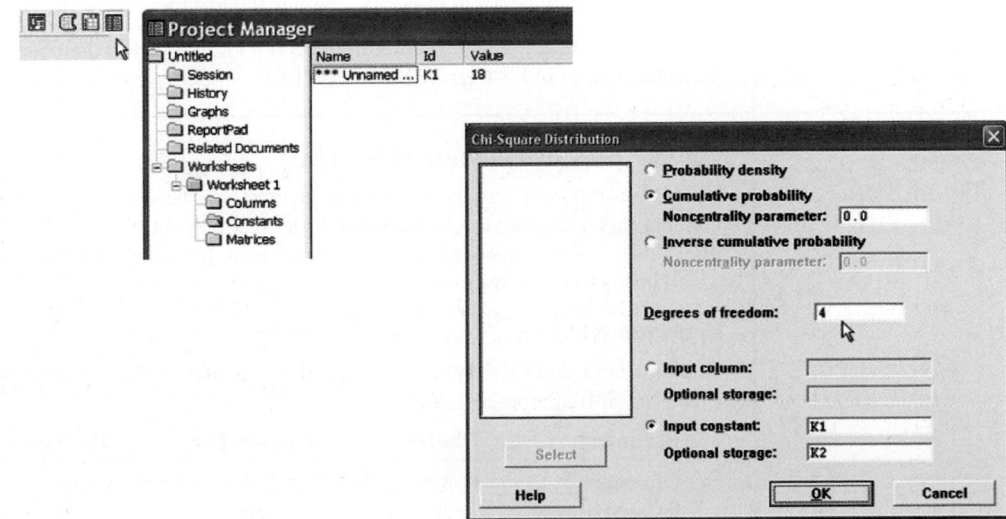

d) In the text box, Optional storage type **K2.** This is the area to the left of the test statistic. To calculate the *P*-value, we need the complement—the area to the right.

e) Select **Calc>Calculator,** then type **K3** for the storage variable and **1 − K2** for the expression.

f) Click [OK]. In the Project Manager you will see K3 = 0.00121341. This is the *P*-value for the test.

Reject the null hypothesis. There is enough evidence in the sample to conclude there is not 20% of each flavor.

TI-83 Plus or TI-84 Plus
Step by Step

Goodness-of-Fit Test
Example TI11–1

This pertains to Example 11–1 from the text. At the 5% significance level, test the claim that there is no preference in the selection of fruit soda flavors for the data.

Frequency	Cherry	Strawberry	Orange	Lime	Grape
Observed	32	28	16	14	10
Expected	20	20	20	20	20

To calculate the test statistic:

1. Enter the observed frequencies in L_1 and the expected frequencies in L_2.
2. Press **2nd [QUIT]** to return to the home screen.
3. Press **2nd [LIST]**, move the cursor to MATH, and press **5** for sum(.
4. Type $(L_1 − L_2)^2/L_2)$, then press **ENTER.**

To calculate the *P*-value:
Press **2nd [DISTR]** then press **7** to get χ^2cdf(.   (Use 8 on the TI-84.)
For this *P*-value, the χ^2cdf(command has form χ^2cdf(test statistic, ∞, degrees of freedom).
Use E99 for ∞. Type **2nd [EE]** to get the small E.
For this example use χ^2cdf(18, E99,4):

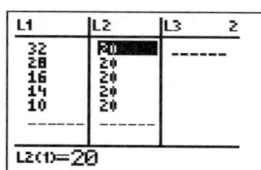 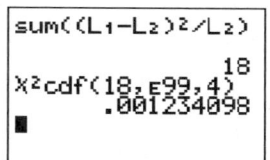

Since *P*-value = 0.001234098 < 0.05 = significance level, reject H_0 and conclude H_1. Therefore, there is enough evidence to reject the claim that consumers show no preference for soda flavors.

Note: On the newer TI-84 calculator, there is a GOF test. Put the observed values in L1 and the expected values in L2 then TESTS-ALPHA D, enter df, and then calculate.

Excel
Step by Step

Chi-Square Goodness-of-Fit Test

Excel does not have a procedure to conduct the goodness-of-fit test. However, you may conduct this test using the MegaStat Add-in available on your CD. If you have not installed this add-in, do so, following the instructions from the Chapter 1 Excel Step by Step.
    This example pertains to Example 11–1 from the text.

Example XL11–1

Test the claim that there is no preference for soda flavor. Use a significance level of $\alpha = 0.05$. The table of frequencies is shown below.

Frequency	Cherry	Strawberry	Orange	Lime	Grape
Observed	32	28	16	14	10
Expected	20	20	20	20	20

1. Enter the observed frequencies in row 1 (cells: A1 to E1) of a new worksheet.
2. Enter the expected frequencies in row 2 (cells: A2 to E2).
3. From the toolbar, select Add-Ins, **MegaStat>Chi-Square/Crosstab>Goodness of Fit Test.** *Note:* You may need to open MegaStat from the MegaStat.xls file on your computer's hard drive.
4. In the dialog box, type **A1:E1** for the Observed values and **A2:E2** for the Expected values. Then click [OK].

Goodness-of-Fit Test

Observed	Expected	$O - E$	$(O - E)^2/E$	% of chisq
32	20.000	12.000	7.200	40.00
28	20.000	8.000	3.200	17.78
16	20.000	−4.000	0.800	4.44
14	20.000	−6.000	1.800	10.00
10	20.000	−10.000	5.000	27.78
100	100.000	0.000	18.000	100.00

18.00	chi-square
4	df
0.0012	*P*-value

Since the *P*-value is less than the significance level, the null hypothesis is rejected and thus the claim of no preference is supported.

Chi-Square Test for Normality

Example XL11–2

This example refers to Example 11–4. At the 5% significance level, determine if the variable is normally distributed.

Start with the table of observed and expected values:

Observed	24	62	72	26	16
Expected	26.7	55.1	66.64	38.96	12.6

1. Enter the Observed values in row 1 of a new worksheet.
2. Enter the Expected values in row 2.

Note: You may include labels for the observed and expected values in cells A1 and A2, respectively.

3. Select the Formulas tab, then Insert Function.
4. In the Insert Function dialog box, select the Statistical category and the CHITEST function.
5. Type **B1:F1** for the Actual Range and **B2:F2** for the Expected Range, then click [OK].

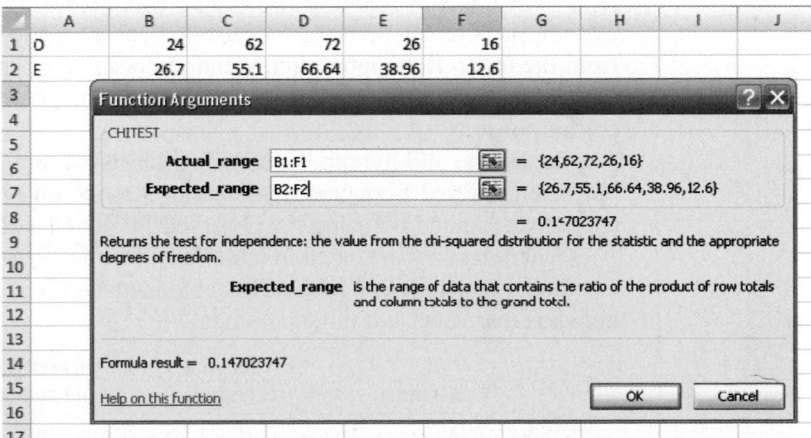

The *P*-value of 0.1470 is greater than the significance level of 0.05. So we do not reject the null hypothesis. Thus, the distribution of the variable is approximately normal.

Tests Using Contingency Tables

When data can be tabulated in table form in terms of frequencies, several types of hypotheses can be tested by using the chi-square test.

Two such tests are the independence of variables test and the homogeneity of proportions test. The test of independence of variables is used to determine whether two variables are independent of or related to each other when a single sample is selected. The test of homogeneity of proportions is used to determine whether the proportions for a variable are equal when several samples are selected from different populations. Both tests use the chi-square distribution and a contingency table, and the test value is found in the same way. The independence test will be explained first.

Objective **2**

Test two variables for independence, using chi-square.

Test for Independence

The chi-square **independence test** can be used to test the independence of two variables. For example, suppose a new postoperative procedure is administered to a number of patients in a large hospital. The researcher can ask the question, Do the doctors feel differently about this procedure from the nurses, or do they feel basically the same way? Note that the question is not whether they prefer the procedure but whether there is a difference of opinion between the two groups.

To answer this question, a researcher selects a sample of nurses and doctors and tabulates the data in table form, as shown.

Group	Prefer new procedure	Prefer old procedure	No preference
Nurses	100	80	20
Doctors	50	120	30

As the survey indicates, 100 nurses prefer the new procedure, 80 prefer the old procedure, and 20 have no preference; 50 doctors prefer the new procedure, 120 like the old procedure, and 30 have no preference. Since the main question is whether there is a difference in opinion, the null hypothesis is stated as follows:

H_0: The opinion about the procedure is *independent* of the profession.

The alternative hypothesis is stated as follows:

H_1: The opinion about the procedure is *dependent* on the profession.

If the null hypothesis is not rejected, the test means that both professions feel basically the same way about the procedure and the differences are due to chance. If the null hypothesis is rejected, the test means that one group feels differently about the procedure from the other. Remember that rejection does *not* mean that one group favors the procedure and the other does not. Perhaps both groups favor it or both dislike it, but in different proportions.

To test the null hypothesis by using the chi-square independence test, you must compute the expected frequencies, assuming that the null hypothesis is true. These frequencies are computed by using the observed frequencies given in the table.

When data are arranged in table form for the chi-square independence test, the table is called a **contingency table.** The table is made up of R rows and C columns. The table here has two rows and three columns.

Group	Prefer new procedure	Prefer old procedure	No preference
Nurses	100	80	20
Doctors	50	120	30

Note that row and column headings do not count in determining the number of rows and columns.

A contingency table is designated as an $R \times C$ (rows by columns) table. In this case, $R = 2$ and $C = 3$; hence, this table is a 2×3 contingency table. Each block in the table is called a *cell* and is designated by its row and column position. For example, the cell with a frequency of 80 is designated as $C_{1,2}$, or row 1, column 2. The cells are shown below.

	Column 1	Column 2	Column 3
Row 1	$C_{1,1}$	$C_{1,2}$	$C_{1,3}$
Row 2	$C_{2,1}$	$C_{2,2}$	$C_{2,3}$

The degrees of freedom for any contingency table are (rows $-$ 1) times (columns $-$ 1); that is, d.f. $= (R-1)(C-1)$. In this case, $(2-1)(3-1) = (1)(2) = 2$. The reason for this formula for d.f. is that all the expected values except one are free to vary in each row and in each column.

Using the previous table, you can compute the expected frequencies for each block (or cell), as shown next.

1. Find the sum of each row and each column, and find the grand total, as shown.

Group	Prefer new procedure	Prefer old procedure	No preference	Total
Nurses	100	80	20	Row 1 sum ⤳ 200
Doctors	+50	+120	+30	Row 2 sum ⤳ 200
Total	150	200	50	400
	Column 1 sum	Column 2 sum	Column 3 sum	Grand total

2. For each cell, multiply the corresponding row sum by the column sum and divide by the grand total, to get the expected value:

$$\text{Expected value} = \frac{\text{row sum} \times \text{column sum}}{\text{grand total}}$$

For example, for $C_{1,2}$, the expected value, denoted by $E_{1,2}$, is (refer to the previous tables)

$$E_{1,2} = \frac{(200)(200)}{400} = 100$$

For each cell, the expected values are computed as follows:

$$E_{1,1} = \frac{(200)(150)}{400} = 75 \qquad E_{1,2} = \frac{(200)(200)}{400} = 100 \qquad E_{1,3} = \frac{(200)(50)}{400} = 25$$

$$E_{2,1} = \frac{(200)(150)}{400} = 75 \qquad E_{2,2} = \frac{(200)(200)}{400} = 100 \qquad E_{2,3} = \frac{(200)(50)}{400} = 25$$

The expected values can now be placed in the corresponding cells along with the observed values, as shown.

Group	Prefer new procedure	Prefer old procedure	No preference	Total
Nurses	100 (75)	80 (100)	20 (25)	200
Doctors	50 (75)	120 (100)	30 (25)	200
Total	150	200	50	400

The rationale for the computation of the expected frequencies for a contingency table uses proportions. For $C_{1,1}$ a total of 150 out of 400 people prefer the new procedure. And since there are 200 nurses, you would expect, if the null hypothesis were true, $(150/400)(200)$, or 75, of the nurses to be in favor of the new procedure.

The formula for the test value for the independence test is the same as the one used for the goodness-of-fit test. It is

$$\chi^2 = \sum \frac{(O - E)^2}{E}$$

For the previous example, compute the $(O - E)^2/E$ values for each cell, and then find the sum.

$$\chi^2 = \sum \frac{(O - E)^2}{E}$$
$$= \frac{(100 - 75)^2}{75} + \frac{(80 - 100)^2}{100} + \frac{(20 - 25)^2}{25} + \frac{(50 - 75)^2}{75}$$
$$+ \frac{(120 - 100)^2}{100} + \frac{(30 - 25)^2}{25}$$
$$= 26.67$$

The final steps are to make the decision and summarize the results. This test is always a right-tailed test, and the degrees of freedom are $(R - 1)(C - 1) = (2 - 1)(3 - 1) = 2$. If $\alpha = 0.05$, the critical value from Table G is 5.991. Hence, the decision is to reject the null hypothesis since $26.67 > 5.991$. See Figure 11–6.

Figure 11–6

Critical and Test Values for the Postoperative Procedures Example

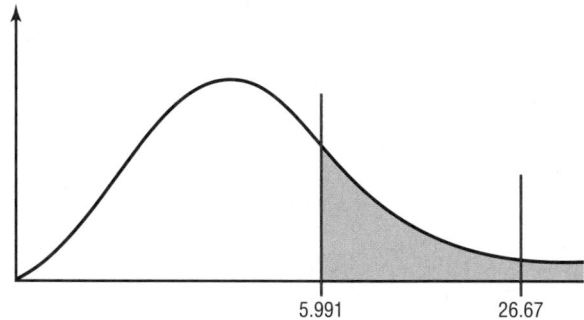

5.991   26.67

The conclusion is that there is enough evidence to support the claim that opinion is related to (dependent on) profession—that is, that the doctors and nurses differ in their opinions about the procedure.

Examples 11–5 and 11–6 illustrate the procedure for the chi-square test of independence.

Example 11–5

Hospitals and Infections

A researcher wishes to see if there is a relationship between the hospital and the number of patient infections. A sample of 3 hospitals was selected, and the number of infections for a specific year has been reported. The data are shown next.

Hospital	Surgical site infections	Pneumonia infections	Bloodstream infections	Total
A	41	27	51	119
B	36	3	40	79
C	169	106	109	384
Total	246	136	200	582

Source: Pennsylvania Health Care Cost Containment Council.

At $\alpha = 0.05$ can it be concluded that the number of infections is related to the hospital where they occurred?

Solution

Step 1 State the hypothesis and identify the claim.

   H_0: The number of infections is independent of the hospital.

   H_1: The number of infections is dependent on the hospital (claim).

Step 2 Find the critical value. The critical value at $\alpha = 0.05$ with $(3 - 1)(3 - 1) = (2)(2) = 4$ degrees of freedom is 9.488.

Step 3 Compute the test value. First find the expected values.

$$E_{1,1} = \frac{(119)(246)}{582} = 50.30 \qquad E_{1,2} = \frac{(119)(136)}{582} = 27.81 \qquad E_{1,3} = \frac{(119)(200)}{582} = 40.89$$

$$E_{2,1} = \frac{(79)(246)}{582} = 33.39 \qquad E_{2,2} = \frac{(79)(136)}{582} = 18.46 \qquad E_{2,3} = \frac{(79)(200)}{582} = 27.15$$

$$E_{3,1} = \frac{(384)(246)}{582} = 162.31 \qquad E_{3,2} = \frac{(384)(136)}{582} = 89.73 \qquad E_{3,3} = \frac{(384)(200)}{582} = 131.96$$

The completed table is shown.

Hospital	Surgical site infections	Pneumonia infections	Bloodstream infections	Total
A	41 (50.30)	27 (27.81)	51 (40.89)	119
B	36 (33.39)	3 (18.46)	40 (27.15)	79
C	169 (162.31)	106 (89.73)	109 (131.96)	384
Total	246	136	200	582

Then substitute in the formula and evaluate.

$$\chi^2 = \sum \frac{(O - E)^2}{E}$$

$$= \frac{(41 - 50.30)^2}{50.30} + \frac{(27 - 27.81)^2}{27.81} + \frac{(51 - 40.89)^2}{40.89}$$

$$+ \frac{(36 - 33.39)^2}{33.39} + \frac{(3 - 18.46)^2}{18.46} + \frac{(40 - 27.15)^2}{27.15}$$

$$+ \frac{(169 - 162.31)^2}{162.31} + \frac{(106 - 89.73)^2}{89.73} + \frac{(109 - 131.96)^2}{131.96}$$

$$= 1.719 + 0.024 + 2.500 + 0.204 + 12.948 + 6.082$$

$$+ 0.276 + 2.950 + 3.995$$

$$= 30.698$$

Step 4 Make the decision. The decision is to reject the null hypothesis since $30.518 > 9.488$. See Figure 11–7.

Figure 11–7

Critical and Test Values for Example 11–5

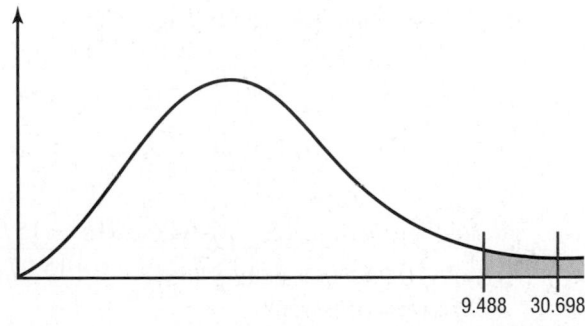

9.488 30.698

Step 5 Summarize the results. There is enough evidence to support the claim that the number of infections is related to the hospital where they occurred.

Example 11–6 **Alcohol and Gender**

A researcher wishes to determine whether there is a relationship between the gender of an individual and the amount of alcohol consumed. A sample of 68 people is selected, and the following data are obtained.

| | Alcohol consumption | | | |
Gender	Low	Moderate	High	Total
Male	10	9	8	27
Female	13	16	12	41
Total	23	25	20	68

At $\alpha = 0.10$, can the researcher conclude that alcohol consumption is related to gender?

Solution

Step 1 State the hypotheses and identify the claim.

H_0: The amount of alcohol that a person consumes is independent of the individual's gender.

H_1: The amount of alcohol that a person consumes is dependent on the individual's gender (claim).

Step 2 Find the critical value. The critical value is 4.605, since the degrees of freedom are $(2 - 1)(3 - 1) = 2$.

Step 3 Compute the test value. First, compute the expected values.

$$E_{1,1} = \frac{(27)(23)}{68} = 9.13 \qquad E_{1,2} = \frac{(27)(25)}{68} = 9.93 \qquad E_{1,3} = \frac{(27)(20)}{68} = 7.94$$

$$E_{2,1} = \frac{(41)(23)}{68} = 13.87 \qquad E_{2,2} = \frac{(41)(25)}{68} = 15.07 \qquad E_{2,3} = \frac{(41)(20)}{68} = 12.06$$

The completed table is shown.

| | Alcohol consumption | | | |
Gender	Low	Moderate	High	Total
Male	10 (9.13)	9 (9.93)	8 (7.94)	27
Female	13 (13.87)	16 (15.07)	12 (12.06)	41
Total	23	25	20	68

Then the test value is

$$\chi^2 = \sum \frac{(O - E)^2}{E}$$

$$= \frac{(10 - 9.13)^2}{9.13} + \frac{(9 - 9.93)^2}{9.93} + \frac{(8 - 7.94)^2}{7.94}$$

$$+ \frac{(13 - 13.87)^2}{13.87} + \frac{(16 - 15.07)^2}{15.07} + \frac{(12 - 12.06)^2}{12.06}$$

$$= 0.283$$

Step 4 Make the decision. The decision is to not reject the null hypothesis, since $0.283 < 4.605$. See Figure 11–8.

Figure 11–8

Critical and Test Values for Example 11–6

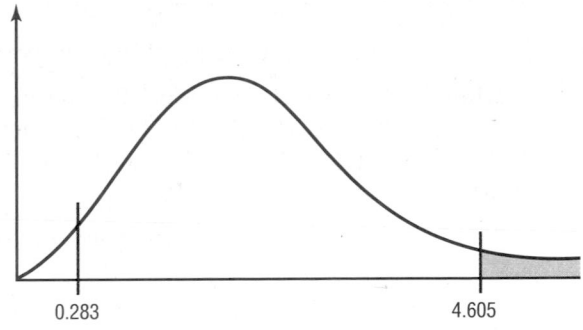

Step 5 Summarize the results. There is not enough evidence to support the claim that the amount of alcohol a person consumes is dependent on the individual's gender.

Test for Homogeneity of Proportions

Objective 3

Test proportions for homogeneity, using chi-square.

The second chi-square test that uses a contingency table is called the **homogeneity of proportions test.** In this situation, samples are selected from several different populations, and the researcher is interested in determining whether the proportions of elements that have a common characteristic are the same for each population. The sample sizes are specified in advance, making either the row totals or column totals in the contingency table known before the samples are selected. For example, a researcher may select a sample of 50 freshmen, 50 sophomores, 50 juniors, and 50 seniors and then find the proportion of students who are smokers in each level. The researcher will then compare the proportions for each group to see if they are equal. The hypotheses in this case would be

H_0: $p_1 = p_2 = p_3 = p_4$

H_1: At least one proportion is different from the others.

If the researcher does not reject the null hypothesis, it can be assumed that the proportions are equal and the differences in them are due to chance. Hence, the proportion of students who smoke is the same for grade levels freshmen through senior. When the null hypothesis is rejected, it can be assumed that the proportions are not all equal. The computational procedure is the same as that for the test of independence shown in Example 11–7.

Interesting Facts

Water is the most critical nutrient in your body. It is needed for just about everything that happens. Water is lost fast: 2 cups daily is lost just exhaling, 10 cups through normal waste and body cooling, and 1 to 2 quarts per hour running, biking, or working out.

Example 11–7

Money and Happiness

A psychologist selected 100 people from each of four income groups and asked them if they were "very happy." The percent for each group who responded yes and the number from the survey are shown in the table. At $\alpha = 0.05$ test the claim that there is no difference in the proportions.

Household income	Less than $30,000 (24%)	$30,000–$74,999 (33%)	$75,000–$99,999 (38%)	$100,000 or more (49%)	Total
Yes	24	33	38	49	144
No	76	67	62	51	256
	100	100	100	100	400

Source: Based on information from Princeton Survey Research Associates International.

Solution

Step 1 State the hypotheses and identify the claim.

H_0: $p_1 = p_2 = p_3 = p_4$ (claim)

H_1: At least one proportion differs from the others.

Step 2 Find the critical value. The formula for the degrees of freedom is the same as before: $(R - 1)(C - 1) = (2 - 1)(4 - 1) = 1(3) = 3$. The critical value is 7.815.

Step 3 Compute the test value. Since we want to test the claim that the proportions are equal, we use the expected value as $\frac{1}{4} \cdot 400 = 100$ and the formula

$$\chi^2 = \sum \frac{(O - E)^2}{E}.$$

$$E_{1,1} = \frac{(144)(100)}{400} = 36 \qquad E_{1,2} = \frac{(144)(100)}{400} = 36 \qquad E_{1,3} = \frac{(144)(100)}{400} = 36 \qquad E_{1,4} = \frac{(144)(100)}{400} = 36$$

$$E_{2,1} = \frac{(256)(100)}{400} = 64 \qquad E_{2,2} = \frac{(256)(100)}{400} = 64 \qquad E_{2,3} = \frac{(256)(100)}{400} = 64 \qquad E_{2,4} = \frac{(256)(100)}{400} = 64$$

The completed table is shown.

Household income	Less than $30,000 (24%)	$30,000–$74,999 (33%)	$75,000–$99,999 (38%)	$100,000 or more (49%)	Total
Yes	24 (36)	33 (36)	38 (36)	49 (36)	144
No	76 (64)	67 (64)	62 (64)	51 (64)	256
	100	100	100	100	400

$$\chi^2 = \sum \frac{(O - E)^2}{E}$$

$$= \frac{(24 - 36)^2}{36} + \frac{(33 - 36)^2}{36} + \frac{(38 - 36)^2}{36} + \frac{(49 - 36)^2}{36}$$

$$+ \frac{(76 - 64)^2}{64} + \frac{(67 - 64)^2}{64} + \frac{(62 - 62)^2}{64} + \frac{(51 - 64)^2}{64}$$

$$= 4 + 0.25 + 0.1111 + 4.6944 + 2.25 + 0.1406 + 0.0625 + 2.6406$$

$$= 14.149$$

Step 4 Make the decision. Reject the null hypothesis since $14.149 > 7.815$. See Figure 11–9.

Figure 11–9

Critical and Test Values for Example 11–7

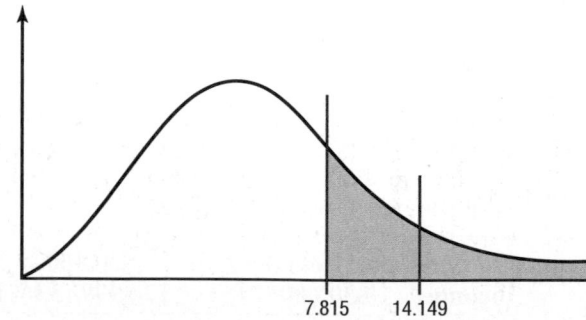

7.815   14.149

Step 5 Summarize the results. There is enough evidence to reject the claim that there is no difference in the proportions. Hence the incomes seem to make a difference in the proportions.

When the degrees of freedom for a contingency table are equal to 1—that is, the table is a 2×2 table—some statisticians suggest using the *Yates correction for continuity*. The formula for the test is then

$$\chi^2 = \sum \frac{(|O - E| - 0.5)^2}{E}$$

Since the chi-square test is already conservative, most statisticians agree that the Yates correction is not necessary. (See Exercise 33 in Exercises 11–2.)

The steps for the chi-square independence and homogeneity tests are summarized in this Procedure Table.

Procedure Table

The Chi-Square Independence and Homogeneity Tests

Step 1   State the hypotheses and identify the claim.

Step 2   Find the critical value in the right tail. Use Table G.

Step 3   Compute the test value. To compute the test value, first find the expected values. For each cell of the contingency table, use the formula

$$E = \frac{(\text{row sum})(\text{column sum})}{\text{grand total}}$$

to get the expected value. To find the test value, use the formula

$$\chi^2 = \sum \frac{(O - E)^2}{E}$$

Step 4   Make the decision.

Step 5   Summarize the results.

The assumptions for the two chi-square tests are given next.

Assumptions for the Chi-Square Independence and Homogeneity Tests

1. The data are obtained from a random sample.
2. The expected value in each cell must be 5 or more.

If the expected values are not 5 or more, combine categories.

Applying the Concepts **11–2**

Satellite Dishes in Restricted Areas

The Senate is expected to vote on a bill to allow the installation of satellite dishes of any size in deed-restricted areas. The House had passed a similar bill. An opinion poll was taken to see if how a person felt about satellite dish restrictions was related to his or her age. A chi-square test was run, creating the following computer-generated information.

Degrees of freedom d.f. = 6
Test statistic $\chi^2 = 61.25$
Critical value C.V. = 12.6
P-value = 0.00
Significance level = 0.05

	18–29	30–49	50–64	65 and up
For	96 (79.5)	96 (79.5)	90 (79.5)	36 (79.5)
Against	201 (204.75)	189 (204.75)	195 (204.75)	234 (204.75)
Don't know	3 (15.75)	15 (15.75)	15 (15.75)	30 (15.75)

1. Which number from the output is compared to the significance level to check if the null hypothesis should be rejected?

2. Which number from the output gives the probability of a type I error that is calculated from your sample data?

3. Was a right-, left-, or two-tailed test run? Why?

4. Can you tell how many rows and columns there were by looking at the degrees of freedom?

5. Does increasing the sample size change the degrees of freedom?

6. What are your conclusions? Look at the observed and expected frequencies in the table to draw some of your own specific conclusions about response and age.

7. What would your conclusions be if the level of significance were initially set at 0.10?

8. Does chi-square tell you which cell's observed and expected frequencies are significantly different?

See page 627 for the answers.

Exercises 11–2

1. How is the chi-square independence test similar to the goodness-of-fit test? How is it different?

2. How are the degrees of freedom computed for the independence test? d.f. = (rows − 1)(columns − 1)

3. Generally, how would the null and alternative hypotheses be stated for the chi-square independence test?

4. What is the name of the table used in the independence test? Contingency table

5. How are the expected values computed for each cell in the table? The expected values are computed as (row total × column total) ÷ grand total.

6. Explain how the chi-square independence test differs from the chi-square homogeneity of proportions test.

7. How are the null and alternative hypotheses stated for the test of homogeneity of proportions?

For Exercises 8 through 31, perform the following steps.

  a. State the hypotheses and identify the claim.
  b. Find the critical value.
  c. Compute the test value.
  d. Make the decision.
  e. Summarize the results.

Use the traditional method of hypothesis testing unless otherwise specified.

8. **Ethnicity and Movie Admissions** Are movie admissions related to ethnicity? A 2007 study indicated the following numbers of admissions (in thousands) for two different years. At the 0.05 level of significance can it be concluded that movie attendance by year was dependent upon ethnicity?

	Caucasian	Hispanic	African-American	Other
2006	936	240	195	101
2007	909	297	150	115

Source: *MPAA Study 2007.*

9. **Endangered or Threatened Species** Can you conclude a relationship between the class of vertebrate and whether it is endangered or threatened? Use the 0.05 level of significance. Is there a different result for the 0.01 level of significance?

	Mammal	Bird	Reptile	Amphibian	Fish
Endangered	68	76	14	13	76
Threatened	13	15	23	10	61

Source: www.infoplease.com

10. **Women in the Military** This table lists the numbers of officers and enlisted personnel for women in the military. At $\alpha = 0.05$, is there sufficient evidence to conclude that a relationship exists between rank and branch of the Armed Forces?

	Officers	Enlisted
Army	10,791	62,491
Navy	7,816	42,750
Marine Corps	932	9,525
Air Force	11,819	54,344

Source: *New York Times Almanac.*

11. Composition of State Legislatures Is the composition of state legislatures in the House of Representatives related to the specific state? Use $\alpha = 0.05$.

	Democrats	Republicans
Pennsylvania	100	103
Ohio	39	59
West Virginia	75	25
Maryland	106	35

Source: *New York Times Almanac.*

12. Population and Age Is the size of the population by age related to the state that it's in? Use $\alpha = 0.05$. (Population values are in thousands.)

	Under 5	5–17	18–24	25–44	45–64	65+
Pennsylvania	721	2140	1025	3515	2702	1899
Ohio	740	2104	1065	3359	2487	1501

Source: *New York Times Almanac.*

13. Medal Counts for the Olympics The 2010 Winter Olympics final medal counts for the top four nations are shown below. At the 0.10 level of significance can it be concluded that the type of medal won was dependent upon the competing country?

	Gold	Silver	Bronze
United States	9	15	13
Germany	10	13	7
Canada	14	7	5
Norway	9	8	6

14. Congressional Representatives Four states were randomly selected, and their members in the U.S. House of Representatives (111th Congress) are noted below. At $\alpha = 0.10$ can it be concluded that there is a dependent relationship between the state and the political party affiliation of their representatives?

	California	Florida	Illinois	Texas
Democrat	33	10	12	12
Republican	19	15	7	20

Source: *New York Times Almanac.*

15. Student Majors at Colleges The table below shows the number of students (in thousands) participating in various programs at both two-year and four-year institutions. At $\alpha = 0.05$, can it be concluded that there is a relationship between program of study and type of institution?

	Two-year	Four-year
Agriculture and related sciences	36	52
Criminal justice	210	231
Foreign languages and literature	28	59
Mathematics and statistics	28	63

Source: *Time Almanac.*

16. Organ Transplantation Listed below is information regarding organ transplantation for three different years. Based on these data, is there sufficient evidence at $\alpha = 0.01$ to conclude that a relationship exists between year and type of transplant?

Year	Heart	Kidney/Pancreas	Lung
2003	2056	870	1085
2004	2016	880	1173
2005	2127	903	1408

Source: www.infoplease.com

17. Weekend Furniture Sales A large furniture retailer with stores in three cities had the following results from a special weekend sale. At $\alpha = 0.05$ is there sufficient evidence that the type of furniture sold was dependent upon the store?

	Recliner	Sofa	Loveseat
Store 22A	15	12	18
Store 22B	20	10	12
Store 22C	10	10	10

18. Record CDs Sold Are the sales of CDs (in thousands) by genre related to the year in which the sales occurred? Use the 0.05 level of significance.

Year	Classical	Jazz	Soundtracks
2005	15,875	17,139	22,849
2004	18,686	18,794	27,367

Source: *Time Almanac.*

19. Choice of Exercise Equipment Is the choice of exercise equipment dependent upon gender? Recent records from a large gym indicated the following equipment usage. At the 0.05 level of significance, is there a relationship?

	Treadmill	Elliptical	Bike
Male	120	60	75
Female	100	95	82

20. Effectiveness of New Drug To test the effectiveness of a new drug, a researcher gives one group of individuals the new drug and another group a placebo. The results of the study are shown here. At $\alpha = 0.10$, can the researcher conclude that the drug is effective? Use the P-value method.

Medication	Effective	Not effective
Drug	32	9
Placebo	12	18

21. Recreational Reading and Gender A book publisher wishes to determine whether there is a difference in the type of book selected by males and females for recreational reading. A random sample provides the data given here. At $\alpha = 0.05$, test the claim that the type of book selected is independent of the gender of the individual. Use the *P*-value method.

Type of book

Gender	Mystery	Romance	Self-help
Male	243	201	191
Female	135	149	202

22. Foreign Language Speaking Dorms A local college recently made the news by offering foreign language–speaking dorm rooms to its students. When questioned at another school, 50 students from each class responded as shown. At $\alpha = 0.05$, is there sufficient evidence to conclude that the proportions of students favoring foreign language–speaking dorms are not the same for each class?

	Freshmen	Sophomores	Juniors	Seniors
Yes (favor)	10	15	20	22
No	40	35	30	28

23. Youth Physical Fitness According to a recent survey, 64% of Americans between the ages of 6 and 17 cannot pass a basic fitness test. A physical education instructor wishes to determine if the percentages of such students in different schools in his school district are the same. He administers a basic fitness test to 120 students in each of four schools. The results are shown here. At $\alpha = 0.05$, test the claim that the proportions who pass the test are equal.

	Southside	West End	East Hills	Jefferson
Passed	49	38	46	34
Failed	71	82	74	86
Total	120	120	120	120

Source: *The Harper's Index Book.*

24. Participation in Market Research Survey An advertising firm has decided to ask 92 customers at each of three local shopping malls if they are willing to take part in a market research survey. According to previous studies, 38% of Americans refuse to take part in such surveys. The results are shown here. At $\alpha = 0.01$, test the claim that the proportions of those who are willing to participate are equal.

	Mall A	Mall B	Mall C
Will participate	52	45	36
Will not participate	40	47	56
Total	92	92	92

Source: *The Harper's Index Book.*

25. Workforce Distribution A researcher wishes to see if the proportions of workers for each type of job have changed during the last 10 years. A sample of

100 workers is selected, and the results are shown. At $\alpha = 0.05$, test the claim that the proportions have not changed. Can the results be generalized to the population of the United States?

	Services	Manu-facturing	Government	Other
10 years ago	33	13	11	3
Now	18	12	8	2
Total	51	25	19	5

Source: Pennsylvania Department of Labor and Industry.

26. Mothers Working Outside the Home According to a recent survey, 59% of Americans aged 8 to 17 would prefer that their mother work outside the home, regardless of what she does now. A school district psychologist decided to select three samples of 60 students each in elementary, middle, and high school to see how the students in her district felt about the issue. At $\alpha = 0.10$, test the claim that the proportions of the students who prefer that their mother have a job are equal.

	Elementary	Middle	High
Prefers mother work	29	38	51
Prefers mother not work	31	22	9
Total	60	60	60

Source: Daniel Weiss, *100% American.*

27. Volunteer Practices of Students The Bureau of Labor Statistics reported information on volunteers by selected characteristics. They found that 24.4% of the population aged 16 to 24 volunteers a median number of 36 hours per year. A survey of 75 students in each age group revealed the following data on volunteer practices. At $\alpha = 0.05$, can it be concluded that the proportions of volunteers are the same for each group?

	Age				
	18	19	20	21	22
Yes (volunteer)	19	18	23	31	13
No	56	57	52	44	62

Source: *Time Almanac.*

28. Fathers in the Delivery Room On average, 79% of American fathers are in the delivery room when their children are born. A physician's assistant surveyed 300 first-time fathers to determine if they had been in the delivery room when their children were born. The results are shown here. At $\alpha = 0.05$, is there enough evidence to reject the claim that the proportions of those who were in the delivery room at the time of birth are the same?

	Hos-pital A	Hos-pital B	Hos-pital C	Hos-pital D
Present	66	60	57	56
Not present	9	15	18	19
Total	75	75	75	75

Source: Daniel Weiss, *100% American.*

29. Injuries on Monkey Bars A children's playground equipment manufacturer read in a survey that 55% of all U.S. playground injuries occur on the monkey bars. The manufacturer wishes to investigate playground injuries in four different parts of the country to determine if the proportions of accidents on the monkey bars are equal. The results are shown here. At $\alpha = 0.05$, test the claim that the proportions are equal. Use the *P*-value method.

Accidents	North	South	East	West
On monkey bars	15	18	13	16
Not on monkey bars	15	12	17	14
Total	30	30	30	30

Source: Michael D. Shook and Robert L. Shook, *The Book of Odds.*

30. Thanksgiving Travel According to the American Automobile Association, 31 million Americans travel over the Thanksgiving holiday. To determine whether to stay open or not, a national restaurant chain surveyed 125 customers at each of four locations to see if they would be traveling over the holiday. The results are shown here. At $\alpha = 0.10$, test the claim that the proportions of Americans who will travel over the Thanksgiving holiday are equal. Use the *P*-value method.

	Loca-tion A	Loca-tion B	Loca-tion C	Loca-tion D
Will travel	37	52	46	49
Will not travel	88	73	79	76
Total	125	125	125	125

Source: Michael D. Shook and Robert L. Shook, *The Book of Odds.*

31. Grocery Lists The vice president of a large supermarket chain wished to determine if her customers made a list before going grocery shopping. She surveyed 288 customers in three stores. The results are shown here. At $\alpha = 0.10$, test the claim that the proportions of the customers in the three stores who made a list before going shopping are equal.

	Store A	Store B	Store C
Made list	77	74	68
No list	19	22	28
Total	96	96	96

Source: Daniel Weiss, *100% American.*

Extending the Concepts

32. For a 2×2 table, a, b, c, and d are the observed values for each cell, as shown.

a	b
c	d

The chi-square test value can be computed as

$$\chi^2 = \frac{n(ad - bc)^2}{(a + b)(a + c)(c + d)(b + d)}$$

where $n = a + b + c + d$. Compute the χ^2 test value by using the above formula and the formula $\Sigma(O - E)^2/E$, and compare the results for the following table. Both answers are the same. $\chi^2 = 1.70$

12	15
9	23

33. For the contingency table shown in Exercise 32, compute the chi-square test value by using the Yates correction (page 613) for continuity. $\chi^2 = 1.075$

34. When the chi-square test value is significant and there is a relationship between the variables, the strength of this relationship can be measured by using the *contingency coefficient*. The formula for the contingency coefficient is

$$C = \sqrt{\frac{\chi^2}{\chi^2 + n}}$$

where χ^2 is the test value and n is the sum of frequencies of the cells. The contingency coefficient will always be less than 1. Compute the contingency coefficient for Exercises 8 and 20. 0.1277; 0.361

Speaking of Statistics

Does Color Affect Your Appetite?

It has been suggested that color is related to appetite in humans. For example, if the walls in a restaurant are painted certain colors, it is thought that the customer will eat more food. A study was done at the University of Illinois and the University of Pennsylvania. When people were given six varieties of jellybeans mixed in a bowl or separated by color, they ate about twice as many from the bowl with the mixed jellybeans as from the bowls that were separated by color.

It is thought that when the jellybeans were mixed, people felt that it offered a greater variety of choices, and the variety of choices increased their appetites.

In this case one variable—color—is categorical, and the other variable—amount of jellybeans eaten—is numerical. Could a chi-square goodness-of-fit test be used here? If so, suggest how it could be set up.

Technology *Step by Step*

MINITAB
Step by Step

Tests Using Contingency Tables

Examples

A sociologist wishes to see whether the number of years of college a person has completed is related to her or his place of residence. A sample of 88 people is taken and classified as shown.

Location	No college	Four-year degree	Advanced degree
Urban	15	12	8
Suburban	8	15	9
Rural	6	8	7
Total	29	35	24

At $\alpha = 0.05$, can the sociologist conclude that a person's location is dependent on the number of years of college?

Calculate the Chi-Square Test Statistic and *P*-Value

1. Enter the observed frequencies for the example shown above into three columns of MINITAB. Name the columns but not the rows. Exclude totals. The complete worksheet is shown.

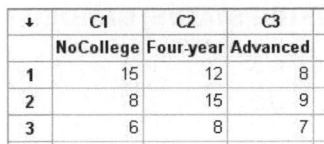

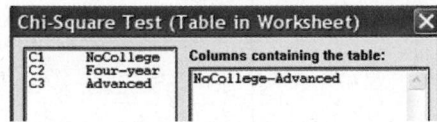

2. Select **Stat>Tables>Chi-Square Test.**

3. Drag the mouse over the three columns in the list.

4. Click [Select]. The three columns will be placed in the Columns box as a sequence, NoCollege through Advanced.

5. Click [OK].

The chi-square test statistic 3.006 has a *P*-value of 0.557. Do not reject the null hypothesis.

There is no relationship between level of education and place of residence.

Chi-Square Test: NoCollege, Four-year, Advanced
```
Expected counts are printed below observed counts
Chi-Square contributions are printed below expected counts

        NoCollege  Four-year   Advanced   Total
    1          15         12          8      35
            11.53      13.92       9.55
            1.041      0.265      0.250

    2           8         15          9      32
            10.55      12.73       8.73
            0.614      0.406      0.009

    3           6          8          7      21
             6.92       8.35       5.73
             0.122      0.015      0.283

Total          29         35         24      88
Chi-Sq = 3.006, DF = 4, P-Value = 0.557
```

Construct a Contingency Table and Calculate the Chi-Square Test Statistic
In Chapter 4 we learned how to construct a contingency table by using gender and smoking status in the Data Bank file described in Appendix D. Are smoking status and gender related? Who is more likely to smoke, men or women?

1. Use **File>Open Worksheet** to open the Data Bank file. Remember do *not* click the file icon.

2. Select **Stat>Tables>Cross Tabulation and Chi-Square.**

3. Double-click Smoking Status for rows and Gender for columns.

4. The Display option for Counts should be checked.

5. Click [Chi-Square].

   a) Check Chi-Square analysis.

   b) Check Expected cell counts.

6. Click [OK] twice.

In the session window the contingency table and the chi-square analysis will be displayed.

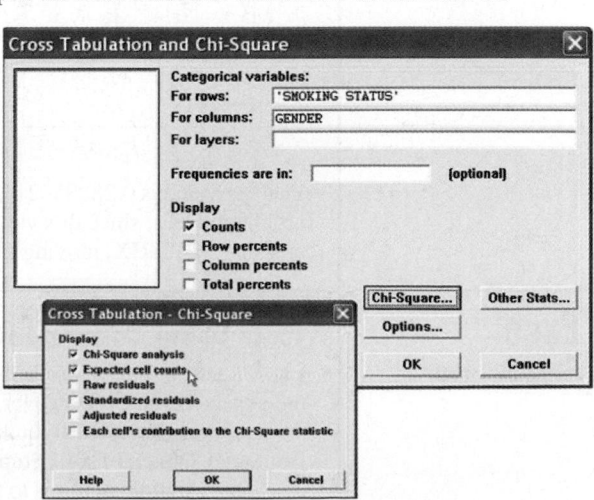

Tabulated statistics: SMOKING STATUS, GENDER

```
Rows: SMOKING STATUS   Columns: Gender

                F       M       All

0              25      22       47
            23.50   23.50    47.00

1              18      19       37
            18.50   18.50    37.00

2               7       9       16
             8.00    8.00    16.00

All            50      50      100
            50.00   50.00   100.00
Cell Contents:          Count
                        Expected count


Pearson Chi-Square = 0.469, DF = 2, P-Value = 0.791
```

There is not enough evidence to conclude that smoking is related to gender.

TI-83 Plus or TI-84 Plus
Step by Step

Chi-Square Test for Independence

1. Press **2nd** [X^{-1}] for **MATRIX** and move the cursor to Edit, then press **ENTER.**
2. Enter the number of rows and columns. Then press **ENTER.**
3. Enter the values in the matrix as they appear in the contingency table.
4. Press **STAT** and move the cursor to TESTS. Press **C (ALPHA PRGM)** for χ^2-Test. Make sure the observed matrix is [A] and the expected matrix is [B].
5. Move the cursor to Calculate and press **ENTER.**

Example TI11–2

Using the data shown from Example 11–6, test the claim of independence at $\alpha = 0.10$.

$$
\begin{array}{ccc}
10 & 9 & 8 \\
13 & 16 & 12
\end{array}
$$

Input
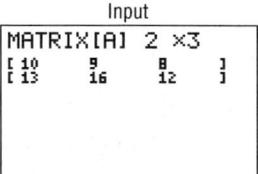

Input
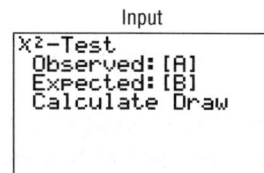

Output
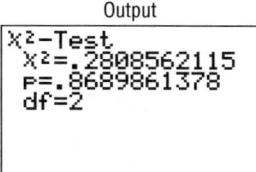

The test value is 0.2808562115. The P-value is 0.8689861378. The decision is to not reject the null hypothesis, since this value is greater than 0.10. You can find the expected values by pressing **MATRIX,** moving the cursor to [B], and pressing **ENTER** twice.

Excel
Step by Step

Tests Using Contingency Tables

Excel does not have a procedure to conduct tests using contingency tables without including the expected values. However, you may conduct such tests using the MegaStat Add-in available on your CD. If you have not installed this add-in, do so, following the instructions from the Chapter 1 Excel Step by Step.

This example pertains to the example shown in the previous MINITAB section.

Example XL11–3

Using a significance level $\alpha = 0.05$, determine whether the number of years of college a person has completed is related to residence.

1. Enter the location variable labels in column A, beginning at cell A2.

2. Enter the categories for the number of years of college in cells B1, C1, and D1, respectively.

3. Enter the observed values in the appropriate block (cell).

4. From the toolbar, select Add-Ins, **MegaStat>Chi-Square/Crosstab>Contingency Table.** *Note:* You may need to open MegaStat from the MegaStat.xls file on your computer's hard drive.

5. In the dialog box, type **A1:D4** for the Input range.

6. Check chi-square from the Output Options.

7. Click [OK].

Chi-Square Contingency Table Test for Independence

	None	4-year	Advanced	Total
Urban	15	12	8	35
Suburban	8	15	9	32
Rural	6	8	7	21
Total	29	35	24	88

```
3.01 chi-square
   4 df
.5569 P-value
```

The results of the test indicate that at the 5% level of significance, there is not enough evidence to conclude that a person's location is dependent on number of years of college.

Summary

- Three uses of the chi-square distribution were explained in this chapter. It can be used as a goodness-of-fit test to determine whether the frequencies of a distribution are the same as the hypothesized frequencies. For example, is the number of defective parts produced by a factory the same each day? This test is always a right-tailed test. (11–1)

- The test of independence is used to determine whether two variables are related or are independent. This test uses a contingency table and is always a right-tailed test. An example of its use is a test to determine whether the attitudes of urban residents about the recycling of trash differ from the attitudes of rural residents. (11–2)

- Finally, the homogeneity of proportions test is used to determine if several proportions are all equal when samples are selected from different populations. (11–2)
  The chi-square distribution is also used for other types of statistical hypothesis tests, such as the Kruskal-Wallis test, which is explained in Chapter 13.

Important Terms

contingency table 606

expected frequency 593

goodness-of-fit test 593

homogeneity of proportions test 611

independence test 606

observed frequency 593

Important Formulas

Formula for the chi-square test for goodness of fit:

$$\chi^2 = \sum \frac{(O - E)^2}{E}$$

with degrees of freedom equal to the number of categories minus 1 and where

> O = observed frequency
>
> E = expected frequency

Formula for the chi-square independence and homogeneity of proportions tests:

$$\chi^2 = \sum \frac{(O - E)^2}{E}$$

with degrees of freedom equal to (rows $-$ 1) times (columns $-$ 1). Formula for the expected value for each cell:

$$E = \frac{\textbf{(row sum)(column sum)}}{\textbf{grand total}}$$

Review Exercises

For Exercises 1 through 10, follow these steps.

> *a.* State the hypotheses and identify the claim.
> *b.* Find the critical value(s).
> *c.* Compute the test value.
> *d.* Make the decision.
> *e.* Summarize the results.

Use the traditional method of hypothesis testing unless otherwise specified.

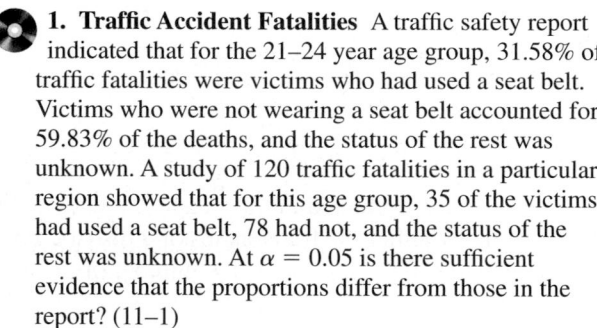

1. Traffic Accident Fatalities A traffic safety report indicated that for the 21–24 year age group, 31.58% of traffic fatalities were victims who had used a seat belt. Victims who were not wearing a seat belt accounted for 59.83% of the deaths, and the status of the rest was unknown. A study of 120 traffic fatalities in a particular region showed that for this age group, 35 of the victims had used a seat belt, 78 had not, and the status of the rest was unknown. At $\alpha = 0.05$ is there sufficient evidence that the proportions differ from those in the report? (11–1)

Source: *New York Times Almanac.*

2. Displaced Workers The reasons that workers in the 25–54 year old category were displaced are listed below.

Plant closed/moved	44.8%
Insufficient work	25.2%
Position eliminated	30%

A random sample of 180 displaced workers (in this age category) found that 40 lost their jobs due to their position being eliminated, 53 due to insufficient work, and the rest due to the company being closed or moving. At the 0.01 level of significance are these proportions different from those from the U.S. Department of Labor? (11–1)

Source: *BLS-World Almanac.*

3. Tire Labeling The federal government has proposed labeling tires by fuel efficiency to save fuel and cut emissions. A survey was taken to see who would use these labels. At $\alpha = 0.10$, is the gender of the individual related to whether or not a person would use these labels? The data from a sample are shown here. (11–1)

Gender	Yes	No	Undecided
Men	114	30	6
Women	136	16	8

Source: *USA TODAY.*

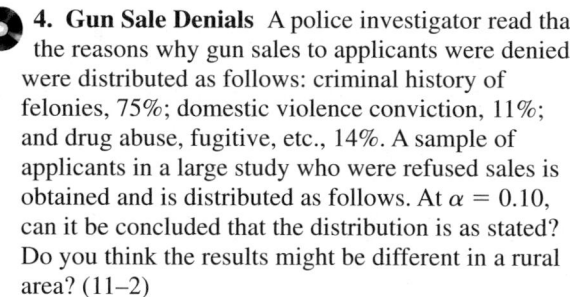

4. Gun Sale Denials A police investigator read that the reasons why gun sales to applicants were denied were distributed as follows: criminal history of felonies, 75%; domestic violence conviction, 11%; and drug abuse, fugitive, etc., 14%. A sample of applicants in a large study who were refused sales is obtained and is distributed as follows. At $\alpha = 0.10$, can it be concluded that the distribution is as stated? Do you think the results might be different in a rural area? (11–2)

Reason	Criminal history	Domestic violence	Drug abuse etc.
Number	120	42	38

Source: Based on FBI statistics.

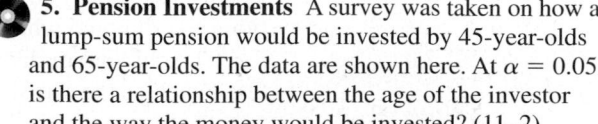

5. Pension Investments A survey was taken on how a lump-sum pension would be invested by 45-year-olds and 65-year-olds. The data are shown here. At $\alpha = 0.05$, is there a relationship between the age of the investor and the way the money would be invested? (11–2)

	Large company stock funds	Small company stock funds	Inter-national stock funds	CDs or money market funds	Bonds
Age 45	20	10	10	15	45
Age 65	42	24	24	6	24

Source: *USA TODAY.*

Statistics Today

Statistics and Heredity—Revisited

Using probability, Mendel predicted the following:

	Smooth		Wrinkled	
	Yellow	Green	Yellow	Green
Expected	0.5625	0.1875	0.1875	0.0625

The observed results were these:

	Smooth		Wrinkled	
	Yellow	Green	Yellow	Green
Observed	0.5666	0.1942	0.1816	0.0556

Using chi-square tests on the data, Mendel found that his predictions were accurate in most cases (i.e., a good fit), thus supporting his theory. He reported many highly successful experiments. Mendel's genetic theory is simple but useful in predicting the results of hybridization.

A Fly in the Ointment

Although Mendel's theory is basically correct, an English statistician named R. A. Fisher examined Mendel's data some 50 years later. He found that the observed (actual) results agreed too closely with the expected (theoretical) results and concluded that the data had in some way been falsified. The results were too good to be true. Several explanations have been proposed, ranging from deliberate misinterpretation to an assistant's error, but no one can be sure how this happened.

---

6. Tornadoes According to records from the Storm Prediction Center, the following numbers of tornadoes occurred in the first quarter of each of the years 2003–2006. Is there sufficient evidence to conclude that a relationship exists between the month and year in which the tornadoes occurred? Use $\alpha = 0.05$. (11–2)

	2006	2005	2004	2003
January	48	33	3	0
February	12	10	9	18
March	113	62	50	43

Source: National Weather Service Storm Prediction Center.

7. Employment of High School Females A guidance counselor wishes to determine if the proportions of female high school students in his school district who have jobs are equal to the national average of 36%. He surveys 80 female students, ages 16 through 18, to determine if they work. The results are shown. At $\alpha = 0.01$, test the claim that the proportions of female students who work are equal. Use the P-value method. (11–2)

	16-year-olds	17-year-olds	18-year-olds
Work	45	31	38
Don't work	35	49	42
Total	80	80	80

Source: Michael D. Shook and Robert L. Shook, *The Book of Odds*.

8. Risk of Injury The risk of injury is higher for males compared to females (57% versus 43%). A hospital emergency room supervisor wishes to determine if the proportions of injuries to males in his hospital are the same for each of four months. He surveys 100 injuries treated in his ER for each month. The results are shown here. At $\alpha = 0.05$, can he reject the claim that the proportions of injuries for males are equal for each of the four months? (11–2)

	May	June	July	August
Male	51	47	58	63
Female	49	53	42	37
Total	100	100	100	100

Source: Michael D. Shook and Robert L. Shook, *The Book of Odds*.

9. Health Insurance Coverage Based on the following data showing the numbers of people (in thousands) with and without health insurance, can it be concluded at the 0.01 level of significance that the proportion with or without health insurance is related to the state chosen? (11–2)

	With	Without
Arkansas	552	123
Montana	793	146
North Dakota	553	61
Wyoming	447	70

Source: *New York Times Almanac.*

10. Cardiovascular Procedures Is the frequency of cardiovascular procedure related to gender? The following data were obtained for selected procedures for a recent year. At $\alpha = 0.10$ is there sufficient evidence to conclude a dependent relationship between gender and procedure? (11–2)

	Coronary artery stent	Coronary artery bypass	Pacemaker
Men	425	320	198
Women	227	123	219

Source: *New York Times Almanac.*

Data Analysis

The Data Bank is located in Appendix D, or on the World Wide Web by following links from www.mhhe.com/math/stat/bluman

1. Select a sample of 40 individuals from the Data Bank. Use the chi-square goodness-of-fit test to see if the marital status of individuals is equally distributed.

2. Use the chi-square test of independence to test the hypothesis that smoking is independent of gender. Use a sample of at least 75 people.

3. Using the data from Data Set X in Appendix D, classify the data as 1–3, 4–6, 7–9, etc. Use the chi-square goodness-of-fit test to see if the number of times each ball is drawn is equally distributed.

Chapter Quiz

Determine whether each statement is true or false. If the statement is false, explain why.

1. The chi-square test of independence is always two-tailed.
   False
2. The test values for the chi-square goodness-of-fit test and the independence test are computed by using the same formula. True

3. When the null hypothesis is rejected in the goodness-of-fit test, it means there is close agreement between the observed and expected frequencies. False

Select the best answer.

4. The values of the chi-square variable cannot be

    a. Positive   c. Negative
    b. 0   d. None of the above

5. The null hypothesis for the chi-square test of independence is that the variables are

    a. Dependent   c. Related
    b. Independent   d. Always 0

6. The degrees of freedom for the goodness-of-fit test are

    a. 0   c. Sample size − 1
    b. 1   d. Number of categories − 1

Complete the following statements with the best answer.

7. The degrees of freedom for a 4×3 contingency table are _____. 6

8. An important assumption for the chi-square test is that the observations must be _____. Independent

9. The chi-square goodness-of-fit test is always _____-tailed. Right

10. In the chi-square independence test, the expected frequency for each class must always be _____. At least 5

For Exercises 11 through 19, follow these steps.

    a. State the hypotheses and identify the claim.
    b. Find the critical value.
    c. Compute the test value.
    d. Make the decision.
    e. Summarize the results.

Use the traditional method of hypothesis testing unless otherwise specified.

11. **Job Loss Reasons** A survey of why people lost their jobs produced the following results. At $\alpha = 0.05$, test the claim that the number of responses is equally distributed. Do you think the results might be different if the study were done 10 years ago?

Reason	Company closing	Position abolished	Insufficient work
Number	26	18	28

Source: Based on information from U.S. Department of Labor.

12. **Consumption of Takeout Foods** A food service manager read that the place where people consumed takeout food is distributed as follows: home, 53%; car, 19%; work, 14%; other, 14%. A survey of 300 individuals showed the following results. At $\alpha = 0.01$, can it be concluded that the distribution is as stated? Where would a fast-food restaurant want to target its advertisements?

Place	Home	Car	Work	Other
Number	142	57	51	50

Source: Beef Industry Council.

13. **Television Viewing** A survey found that 62% of the respondents stated that they never watched the home shopping channels on cable television, 23% stated that they watched the channels rarely, 11% stated that they

watched them occasionally, and 4% stated that they watched them frequently. A group of 200 college students was surveyed, and 105 stated that they never watched the home shopping channels, 72 stated that they watched them rarely, 13 stated that they watched them occasionally, and 10 stated that they watched them frequently. At $\alpha = 0.05$, can it be concluded that the college students differ in their preference for the home shopping channels?

Source: Based on information obtained from *USA TODAY* Snapshots.

14. **Ways to Get to Work** The 2000 Census indicated the following percentages for means of commuting to work for workers over 15 years of age.

Alone	75.7
Carpooling	12.2
Public	4.7
Walked	2.9
Other	1.2
Worked at home	3.3

A random sample of workers found that 320 drove alone, 100 carpooled, 30 used public transportation, 20 walked, 10 used other forms of transportation, and 20 worked at home. Is there sufficient evidence to conclude that the proportions of workers using each type of transportation differ from those in the Census report? Use $\alpha = 0.05$.

Source: Census Bureau, *Washington Observer-Reporter.*

15. **Favorite Ice Cream Flavor** A survey of women and men asked what their favorite ice cream flavor was. The results are shown. At $\alpha = 0.05$, can it be concluded that the favorite flavor is independent of gender?

	Flavor			
	Vanilla	**Chocolate**	**Strawberry**	**Other**
Women	62	36	10	2
Men	49	37	5	9

 16. **Types of Pizzas Purchased** A pizza shop owner wishes to determine if the type of pizza a person selects is related to the age of the individual. The data obtained from a sample are shown. At $\alpha = 0.10$, is the age of the purchaser related to the type of pizza ordered? Use the *P*-value method.

	Type of pizza			
Age	**Plain**	**Pepperoni**	**Mushroom**	**Double cheese**
10–19	12	21	39	71
20–29	18	76	52	87
30–39	24	50	40	47
40–49	52	30	12	28

17. **Pennant Colors Purchased** A survey at a ballpark shows the following selection of pennants sold to fans. The data are presented here. At $\alpha = 0.10$, is the color of the pennant purchased independent of the gender of the individual?

	Blue	**Yellow**	**Red**
Men	519	659	876
Women	487	702	787

18. **Tax Credit Refunds** In a survey of children ages 8 through 11, these data were obtained as to what their parents should do with the money from a $400 tax credit.

	Keep it for themselves	**Give it to their children**	**Don't know**
Girls	162	132	6
Boys	147	147	6

At $\alpha = 0.10$, is there a relationship between the feelings of the children and the gender of the children?

Source: Based on information from *USA TODAY* Snapshot.

19. **Employment Satisfaction** A survey of 60 men and 60 women asked if they would be happy spending the rest of their careers with their present employers. The results are shown. At $\alpha = 0.10$, can it be concluded that the proportions are equal? If they are not equal, give a possible reason for the difference.

	Yes	**No**	**Undecided**
Men	40	15	5
Women	36	9	15

Source: Based on information from a Maritz Poll.

Critical Thinking Challenges

1. **Random Digits** Use your calculator or the MINITAB random number generator to generate 100 two-digit random numbers. Make a grouped frequency distribution, using the chi-square goodness-of-fit test to see if the distribution is random. To do this, use an expected frequency of 10 for each class. Can it be concluded that the distribution is random? Explain.

2. **Lottery Numbers** Simulate the state lottery by using your calculator or MINITAB to generate 100 three-digit random numbers. Group these numbers 100–199, 200–299, etc. Use the chi-square goodness-of-fit test to see if the numbers are random. The expected frequency for each class should be 10. Explain why.

3. Purchase a bag of M&M's candy and count the number of pieces of each color. Using the information as your sample, state a hypothesis for the distribution of colors, and compare your hypothesis to H_0: The distribution of colors of M&M's candy is 13% brown, 13% red, 14% yellow, 16% green, 20% orange, and 24% blue.

 **Data Projects**

Use a significance level of 0.05 for all tests below.

1. **Business and Finance** Many of the companies that produce multicolored candy will include on their website information about the production percentages for the various colors. Select a favorite multicolored candy. Find out what percentage of each color is produced. Open up a bag of the candy, noting how many of each color are in the bag (be careful to count them before you eat them). Is the bag distributed as expected based on the production percentages? If no production percentages can be found, test to see if the colors are uniformly distributed.

2. **Sports and Leisure** Use a local (or favorite) basketball, football, baseball, and hockey team as the data set. For the most recently completed season, note the teams' home record for wins and losses. Test to see whether home field advantage is independent of sport.

3. **Technology** Use the data collected in data project 3 of Chapter 2 regarding song genres. Do the data indicate that songs are uniformly distributed among the genres?

4. **Health and Wellness** Research the percentages of each blood type that the Red Cross states are in the population. Now use your class as a sample. For each student note the blood type. Is the distribution of blood types in your class as expected based on the Red Cross percentages?

5. **Politics and Economics** Research the distribution (by percent) of registered Republicans, Democrats, and Independents in your state. Use your class as a sample. For each student, note the party affiliation. Is the distribution as expected based on the percentages for your state? What might be problematic about using your class as a sample for this exercise?

6. **Your Class** Conduct a classroom poll to determine which of the following sports each student likes best: baseball, football, basketball, hockey, or NASCAR. Also, note the gender of the individual. Is preference for sport independent of gender?

Answers to Applying the Concepts

Section 11–1   Never the Same Amounts

1. The variables are qualitative and we have the counts for each category.

2. We can use a chi-square goodness-of-fit test.

3. There are a total of 233 candies, so we would expect 46.6 of each color. Our test statistic is $\chi^2 = 1.442$.

4. H_0: The colors are equally distributed.
   H_1: The colors are not equally distributed.

5. There are $5 - 1 = 4$ degrees of freedom for the test. The critical value depends on the choice of significance level. At the 0.05 significance level, the critical value is 9.488.

6. Since $1.442 < 9.488$, we fail to reject the null hypothesis. There is not enough evidence to conclude that the colors are not equally distributed.

Section 11–2   Satellite Dishes in Restricted Areas

1. We compare the P-value to the significance level of 0.05 to check if the null hypothesis should be rejected.

2. The P-value gives the probability of a type I error.

3. This is a right-tailed test, since chi-square tests of independence are always right-tailed.

4. You cannot tell how many rows and columns there were just by looking at the degrees of freedom.

5. Increasing the sample size does not increase the degrees of freedom, since the degrees of freedom are based on the number of rows and columns.

6. We will reject the null hypothesis. There are a number of cells where the observed and expected frequencies are quite different.

7. If the significance level were initially set at 0.10, we would still reject the null hypothesis.

8. No, the chi-square value does not tell us which cells have observed and expected frequencies that are very different.

12

Analysis of Variance

Objectives

After completing this chapter, you should be able to

1 Use the one-way ANOVA technique to determine if there is a significant difference among three or more means.

2 Determine which means differ, using the Scheffé or Tukey test if the null hypothesis is rejected in the ANOVA.

3 Use the two-way ANOVA technique to determine if there is a significant difference in the main effects or interaction.

Outline

Source: C. Luus, G. Wells, and J. Turtle, "Child Eyewitnesses: Seeing Is Believing," *Journal of Applied Psychology* 80, no. 2, pp. 317–26.

Statistics Today

Is Seeing Really Believing?

Many adults look on the eyewitness testimony of children with skepticism. They believe that young witnesses' testimony is less accurate than the testimony of adults in court cases. Several statistical studies have been done on this subject.

In a preliminary study, three researchers selected fourteen 8-year-olds, fourteen 12-year-olds, and fourteen adults. The researchers showed each group the same video of a crime being committed. The next day, each witness responded to direct and cross-examination questioning. Then the researchers, using statistical methods explained in this chapter, were able to determine if there were differences in the accuracy of the testimony of the three groups on direct examination and on cross-examination. The statistical methods used here differ from the ones explained in Chapter 9 because there are three groups rather than two. See Statistics Today—Revisited at the end of this chapter.

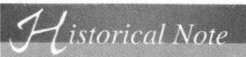

Historical Note

The methods of analysis of variance were developed by R. A. Fisher in the early 1920s.

Introduction

The F test, used to compare two variances as shown in Chapter 9, can also be used to compare three or more means. This technique is called *analysis of variance,* or *ANOVA.* It is used to test claims involving three or more means. (*Note:* The F test can also be used to test the equality of two means. But since it is equivalent to the t test in this case, the t test is usually used instead of the F test when there are only two means.) For example, suppose a researcher wishes to see whether the means of the time it takes three groups of students to solve a computer problem using Fortran, Basic, and Pascal are different. The researcher will use the ANOVA technique for this test. The z and t tests should not be used when three or more means are compared, for reasons given later in this chapter.

For three groups, the F test can only show whether a difference exists among the three means. It cannot reveal where the difference lies—that is, between $\overline{X}_1$ and $\overline{X}_2$, or $\overline{X}_1$ and $\overline{X}_3$, or $\overline{X}_2$ and $\overline{X}_3$. If the F test indicates that there is a difference among the means, other statistical tests are used to find where the difference exists. The most commonly used tests are the Scheffé test and the Tukey test, which are also explained in this chapter.

The analysis of variance that is used to compare three or more means is called a *one-way analysis of variance* since it contains only one variable. In the previous example, the variable is the type of computer language used. The analysis of variance can be extended to studies involving two variables, such as type of computer language used and mathematical background of the students. These studies involve a *two-way analysis of variance.* Section 12–3 explains the two-way analysis of variance.

12–1

Objective 1

Use the one-way ANOVA technique to determine if there is a significant difference among three or more means.

One-Way Analysis of Variance

When an F test is used to test a hypothesis concerning the means of three or more populations, the technique is called **analysis of variance** (commonly abbreviated as **ANOVA**). At first glance, you might think that to compare the means of three or more samples, you can use the t test, comparing two means at a time. But there are several reasons why the t test should not be done.

First, when you are comparing two means at a time, the rest of the means under study are ignored. With the F test, all the means are compared simultaneously. Second, when you are comparing two means at a time and making all pairwise comparisons, the probability of rejecting the null hypothesis when it is true is increased, since the more t tests that are conducted, the greater is the likelihood of getting significant differences by chance alone. Third, the more means there are to compare, the more t tests are needed. For example, for the comparison of 3 means two at a time, 3 t tests are required. For the comparison of 5 means two at a time, 10 tests are required. And for the comparison of 10 means two at a time, 45 tests are required.

Assumptions for the F Test for Comparing Three or More Means

1. The populations from which the samples were obtained must be normally or approximately normally distributed.
2. The samples must be independent of one another.
3. The variances of the populations must be equal.

Even though you are comparing three or more means in this use of the F test, *variances* are used in the test instead of means.

With the F test, two different estimates of the population variance are made. The first estimate is called the **between-group variance,** and it involves finding the variance of the means. The second estimate, the **within-group variance,** is made by computing the variance using all the data and is not affected by differences in the means. If there is no difference in the means, the between-group variance estimate will be approximately equal to the within-group variance estimate, and the F test value will be approximately equal to 1. The null hypothesis will not be rejected. However, when the means differ significantly, the between-group variance will be much larger than the within-group variance; the F test value will be significantly greater than 1; and the null hypothesis will be rejected. Since variances are compared, this procedure is called *analysis of variance* (ANOVA).

For a test of the difference among three or more means, the following hypotheses should be used:

H_0: $\mu_1 = \mu_2 = \cdots = \mu_k$
H_1: At least one mean is different from the others.

As stated previously, a significant test value means that there is a high probability that this difference in means is not due to chance, but it does not indicate where the difference lies.

The degrees of freedom for this F test are d.f.N. $= k - 1$, where k is the number of groups, and d.f.D. $= N - k$, where N is the sum of the sample sizes of the groups $N = n_1 + n_2 + \cdots + n_k$. The sample sizes need not be equal. The F test to compare means is always right-tailed.

Examples 12–1 and 12–2 illustrate the computational procedure for the ANOVA technique for comparing three or more means, and the steps are summarized in the Procedure Table shown after the examples.

Example 12–1	**Lowering Blood Pressure**

A researcher wishes to try three different techniques to lower the blood pressure of individuals diagnosed with high blood pressure. The subjects are randomly assigned to three groups; the first group takes medication, the second group exercises, and the third group follows a special diet. After four weeks, the reduction in each person's blood pressure is recorded. At $\alpha = 0.05$, test the claim that there is no difference among the means. The data are shown.

Medication	**Exercise**	**Diet**
10	6	5
12	8	9
9	3	12
15	0	8
13	2	4
$\overline{X}_1 = 11.8$	$\overline{X}_2 = 3.8$	$\overline{X}_3 = 7.6$
$s_1^2 = 5.7$	$s_2^2 = 10.2$	$s_3^2 = 10.3$

Solution

Step 1   State the hypotheses and identify the claim.

H_0: $\mu_1 = \mu_2 = \mu_3$ (claim)

H_1: At least one mean is different from the others.

Step 2   Find the critical value. Since $k = 3$ and $N = 15$,

d.f.N. $= k - 1 = 3 - 1 = 2$

d.f.D. $= N - k = 15 - 3 = 12$

The critical value is 3.89, obtained from Table H in Appendix C with $\alpha = 0.05$.

Step 3   Compute the test value, using the procedure outlined here.

     a. Find the mean and variance of each sample (these values are shown below the data).

     b. Find the grand mean. The *grand mean,* denoted by $\overline{X}_{GM}$, is the mean of all values in the samples.

$$\overline{X}_{GM} = \frac{\Sigma X}{N} = \frac{10 + 12 + 9 + \cdots + 4}{15} = \frac{116}{15} = 7.73$$

When samples are equal in size, find $\overline{X}_{GM}$ by summing the $\overline{X}$'s and dividing by k, where $k =$ the number of groups.

c. Find the between-group variance, denoted by s_B^2.

$$s_B^2 = \frac{\sum n_i(\overline{X}_i - \overline{X}_{GM})^2}{k - 1}$$

$$= \frac{5(11.8 - 7.73)^2 + 5(3.8 - 7.73)^2 + 5(7.6 - 7.73)^2}{3 - 1}$$

$$= \frac{160.13}{2} = 80.07$$

Note: This formula finds the variance among the means by using the sample sizes as weights and considers the differences in the means.

d. Find the within-group variance, denoted by s_W^2.

$$s_W^2 = \frac{\sum(n_i - 1)s_i^2}{\sum(n_i - 1)}$$

$$= \frac{(5 - 1)(5.7) + (5 - 1)(10.2) + (5 - 1)(10.3)}{(5 - 1) + (5 - 1) + (5 - 1)}$$

$$= \frac{104.80}{12} = 8.73$$

Note: This formula finds an overall variance by calculating a weighted average of the individual variances. It does not involve using differences of the means.

e. Find the F test value.

$$F = \frac{s_B^2}{s_W^2} = \frac{80.07}{8.73} = 9.17$$

Step 4   Make the decision. The decision is to reject the null hypothesis, since $9.17 > 3.89$.

Step 5   Summarize the results. There is enough evidence to reject the claim and conclude that at least one mean is different from the others.

The numerator of the fraction obtained in step 3, part c, of the computational procedure is called the **sum of squares between groups,** denoted by SS_B. The numerator of the fraction obtained in step 3, part d, of the computational procedure is called the **sum of squares within groups,** denoted by SS_W. This statistic is also called the *sum of squares for the error.* SS_B is divided by d.f.N. to obtain the between-group variance. SS_W is divided by $N - k$ to obtain the within-group or error variance. These two variances are sometimes called **mean squares,** denoted by MS_B and MS_W. These terms are used to summarize the analysis of variance and are placed in a summary table, as shown in Table 12–1.

Table 12–1	**Analysis of Variance Summary Table**			
Source	Sum of squares	d.f.	Mean square	F
Between	SS_B	$k - 1$	MS_B	
Within (error)	SS_W	$N - k$	MS_W	
Total				

In the table,

$$SS_B = \text{sum of squares between groups}$$

$$SS_W = \text{sum of squares within groups}$$

$$k = \text{number of groups}$$

$$N = n_1 + n_2 + \cdots + n_k = \text{sum of sample sizes for groups}$$

$$MS_B = \frac{SS_B}{k-1}$$

$$MS_W = \frac{SS_W}{N-k}$$

$$F = \frac{MS_B}{MS_W}$$

The totals are obtained by adding the corresponding columns. For Example 12–1, the **ANOVA summary table** is shown in Table 12–2.

Table 12–2	Analysis of Variance Summary Table for Example 12–1			
Source	Sum of squares	d.f.	Mean square	F
Between	160.13	2	80.07	9.17
Within (error)	104.80	12	8.73	
Total	264.93	14		

Most computer programs will print out an ANOVA summary table.

Example 12–2

Employees at Toll Road Interchanges

 A state employee wishes to see if there is a significant difference in the number of employees at the interchanges of three state toll roads. The data are shown. At $\alpha = 0.05$, can it be concluded that there is a significant difference in the average number of employees at each interchange?

Pennsylvania Turnpike	Greensburg Bypass/ Mon-Fayette Expressway	Beaver Valley Expressway
7	10	1
14	1	12
32	1	1
19	0	9
10	11	1
11	1	11
$\bar{X}_1 = 15.5$	$\bar{X}_2 = 4.0$	$\bar{X}_3 = 5.8$
$s_1^2 = 81.9$	$s_2^2 = 25.6$	$s_3^2 = 29.0$

Source: Pennsylvania Turnpike Commission.

Solution

Step 1 State the hypotheses and identify the claim.

$$H_0: \mu_1 = \mu_2 = \mu_3$$

H_1: At least one mean is different from the others (claim).

Step 2 Find the critical value. Since $k = 3$, $N = 18$, and $\alpha = 0.05$,
d.f.N. $= k - 1 = 3 - 1 = 2$
d.f.D. $= N - k = 18 - 3 = 15$
The critical value is 3.68.

Step 3 Compute the test value.

a. Find the mean and variance of each sample (these values are shown below the data columns in the example).

b. Find the grand mean.

$$\bar{X}_{GM} = \frac{\Sigma X}{N} = \frac{7 + 14 + 32 + \cdots + 11}{18} = \frac{152}{18} = 8.4$$

c. Find the between-group variance.

$$s_B^2 = \frac{\Sigma n_i(\bar{X}_i - \bar{X}_{GM})^2}{k - 1}$$
$$= \frac{6(15.5 - 8.4)^2 + 6(4 - 8.4)^2 + 6(5.8 - 8.4)^2}{3 - 1}$$
$$= \frac{459.18}{2} = 229.59$$

d. Find the within-group variance.

$$s_W^2 = \frac{\Sigma(n_i - 1)s_i^2}{\Sigma(n_i - 1)}$$
$$= \frac{(6 - 1)(81.9) + (6 - 1)(25.6) + (6 - 1)(29.0)}{(6 - 1) + (6 - 1) + (6 - 1)}$$
$$= \frac{682.5}{15} = 45.5$$

e. Find the F test value.

$$F = \frac{s_B^2}{s_W^2} = \frac{229.59}{45.5} = 5.05$$

Step 4 Make the decision. Since $5.05 > 3.68$, the decision is to reject the null hypothesis.

Step 5 Summarize the results. There is enough evidence to support the claim that there is a difference among the means. The ANOVA summary table for this example is shown in Table 12–3.

Table 12–3 Analysis of Variance Summary Table for Example 12-2

Source	Sum of squares	d.f.	Mean square	F
Between	459.18	2	229.59	5.05
Within	682.5	15	45.5	
Total	1141.68	17		

The steps for computing the F test value for the ANOVA are summarized in this Procedure Table.

Procedure Table

Finding the F Test Value for the Analysis of Variance

Step 1 Find the mean and variance of each sample.

$$(\overline{X}_1, s_1^2), (\overline{X}_2, s_2^2), \ldots, (\overline{X}_k, s_k^2)$$

Step 2 Find the grand mean.

$$\overline{X}_{GM} = \frac{\Sigma X}{N}$$

Step 3 Find the between-group variance.

$$s_B^2 = \frac{\Sigma n_i(\overline{X}_i - \overline{X}_{GM})^2}{k - 1}$$

Step 4 Find the within-group variance.

$$s_W^2 = \frac{\Sigma(n_i - 1)s_i^2}{\Sigma(n_i - 1)}$$

Step 5 Find the F test value.

$$F = \frac{s_B^2}{s_W^2}$$

The degrees of freedom are

d.f.N. $= k - 1$

where k is the number of groups, and

d.f.D. $= N - k$

where N is the sum of the sample sizes of the groups

$$N = n_1 + n_2 + \cdots + n_k$$

The P-values for ANOVA are found by using the procedure shown in Section 9–2. For Example 12–2, find the two α values in the tables for the F distribution (Table H), using d.f.N. $= 2$ and d.f.D. $= 15$, where $F = 5.05$ falls between. In this case, 5.05 falls between 4.77 and 6.36, corresponding, respectively, to $\alpha = 0.025$ and $\alpha = 0.01$; hence, $0.01 < P\text{-value} < 0.025$. Since the P-value is between 0.01 and 0.025 and since $P\text{-value} < 0.05$ (the originally chosen value for α), the decision is to reject the null hypothesis. (The P-value obtained from a calculator is 0.021.)

When the null hypothesis is rejected in ANOVA, it only means that at least one mean is different from the others. To locate the difference or differences among the means, it is necessary to use other tests such as the Tukey or the Scheffé test.

Applying the Concepts **12–1**

Colors That Make You Smarter

The following set of data values was obtained from a study of people's perceptions on whether the color of a person's clothing is related to how intelligent the person looks. The subjects rated the person's intelligence on a scale of 1 to 10. Group 1 subjects were randomly shown people with clothing in shades of blue and gray. Group 2 subjects were randomly shown people with clothing in shades of brown and yellow. Group 3 subjects were randomly shown people with clothing in shades of pink and orange. The results follow.

Group 1	Group 2	Group 3
8	7	4
7	8	9
7	7	6
7	7	7
8	5	9
8	8	8
6	5	5
8	8	8
8	7	7
7	6	5
7	6	4
8	6	5
8	6	4

1. Use ANOVA to test for any significant differences between the means.
2. What is the purpose of this study?
3. Explain why separate t tests are not accepted in this situation.

See page 668 for the answers.

Exercises 12–1

1. What test is used to compare three or more means?

2. State three reasons why multiple t tests cannot be used to compare three or more means.

3. What are the assumptions for ANOVA?

4. Define between-group variance and within-group variance.

5. What is the F test formula for comparing three or more means? $F = \dfrac{s_B^2}{s_W^2}$

6. State the hypotheses used in the ANOVA test.

7. When there is no significant difference among three or more means, the value of F will be close to what number?
One

For Exercises 8 through 19, assume that all variables are normally distributed, that the samples are independent, and that the population variances are equal. Also, for each exercise, perform the following steps.

   a. State the hypotheses and identify the claim.

   b. Find the critical value.

   c. Compute the test value.

   d. Make the decision.

   e. Summarize the results, and explain where the differences in the means are.

Use the traditional method of hypothesis testing unless otherwise specified.

8. Sodium Contents of Foods The amount of sodium (in milligrams) in one serving for a random sample of three different kinds of foods is listed here. At the 0.05 level of significance, is there sufficient evidence to conclude that a difference in mean sodium amounts exists among condiments, cereals, and desserts?

Condiments	Cereals	Desserts
270	260	100
130	220	180
230	290	250
180	290	250
80	200	300
70	320	360
200	140	300
		160

Source: *The Doctor's Pocket Calorie, Fat, and Carbohydrate Counter.*

9. Hybrid Vehicles A study was done before the recent surge in gasoline prices to compare the cost to drive 25 miles for different types of hybrid vehicles. The cost of a gallon of gas at the time of the study was approximately $2.50. Based on the information given below for different models of hybrid cars, trucks, and SUVs, is there sufficient evidence to conclude a difference in the mean cost to drive 25 miles? Use $\alpha = 0.05$. (The information in this exercise will be used in Exercise 3 in Section 12–2.)

Hybrid cars	Hybrid SUVs	Hybrid trucks
2.10	2.10	3.62
2.70	2.42	3.43
1.67	2.25	
1.67	2.10	
1.30	2.25	

Source: www.fueleconomy.com

10. Healthy Eating Americans appear to be eating healthier. Between 1970 and 2007 the per capita consumption of broccoli increased 1000% from 0.5 to 5.5 pounds. A nutritionist followed a group of people randomly assigned to one of three groups and noted their monthly broccoli intake (in pounds). At $\alpha = 0.05$ is there a difference in means?

Group A	Group B	Group C
2.0	2.0	3.7
1.5	1.5	2.5
0.75	4.0	4.0
1.0	3.0	5.1
1.3	2.5	3.8
3.0	2.0	2.9

Source: World Almanac.

11. Lengths of Suspension Bridges The lengths (in feet) of a random sample of suspension bridges

in the United States, Europe, and Asia are shown. At $\alpha = 0.05$, is there sufficient evidence to conclude that there is a difference in mean lengths?

United States	Europe	Asia
4260	5238	6529
3500	4626	4543
2300	4347	3668
2000	3300	3379
1850		2874

Source: New York Times Almanac.

12. Weight Gain of Athletes A researcher wishes to see whether there is any difference in the weight gains of athletes following one of three special diets. Athletes are randomly assigned to three groups and placed on the diet for 6 weeks. The weight gains (in pounds) are shown here. At $\alpha = 0.05$, can the researcher conclude that there is a difference in the diets?

Diet A	Diet B	Diet C
3	10	8
6	12	3
7	11	2
4	14	5
	8	
	6	

A computer printout for this problem is shown. Use the P-value method and the information in this printout to test the claim. (The information in this exercise will be used in Exercise 4 of Section 12–2.)

Computer Printout for Exercise 12

```
ANALYSIS OF VARIANCE SOURCE TABLE
Source        df    Sum of Squares   Mean Square      F      P-value

Bet Groups     2        101.095        50.548       7.740    0.00797
W/I Groups    11         71.833         6.530

Total         13        172.929

DESCRIPTIVE STATISTICS
Condit         N           Means        St Dev

diet A         4          5.000         1.826
diet B         6         10.167         2.858
diet C         4          4.500         2.646
```

13. Expenditures per Pupil The per-pupil costs (in thousands of dollars) for cyber charter school tuition for school districts in three areas of southwestern Pennsylvania are shown. At $\alpha = 0.05$, is there a difference in the means? If so, give a possible reason for the difference. (The information in this exercise will be used in Exercise 5 of Section 12–2.)

Area I	Area II	Area III
6.2	7.5	5.8
9.3	8.2	6.4
6.8	8.5	5.6
6.1	8.2	7.1
6.7	7.0	3.0
6.9	9.3	3.5

Source: Tribune-Review.

14. Cell Phone Bills The average local cell phone monthly bill is $50.07. A random sample of monthly bills from three different providers is listed below. At $\alpha = 0.05$ is there a difference in mean bill amounts among providers?

Provider X	Provider Y	Provider Z
48.20	105.02	59.27
60.59	85.73	65.25
72.50	61.95	70.27
55.62	75.69	42.19
89.47	82.11	52.34

Source: *World Almanac.*

15. Number of Farms The numbers (in thousands) of farms per state found in three sections of the country are listed next. Test the claim at $\alpha = 0.05$ that the mean number of farms is the same across these three geographic divisions.

Eastern third	Middle third	Western third
48	95	29
57	52	40
24	64	40
10	64	68
38		

Source: *New York Times Almanac.*

16. Annual Child Care Costs Annual child care costs for infants are considerably higher than for older children. At $\alpha = 0.05$, can you conclude a difference in mean infant day care costs for different regions of the United States? (Annual costs per infant are given in dollars.) (The information in this exercise will be used in Exercise 6 of Section 12–2.)

New England	Midwest	Southwest
10,390	9,449	7,644
7,592	6,985	9,691
8,755	6,677	5,996
9,464	5,400	5,386
7,328	8,372	

Source: www.naccrra.org (National Association of Child Care Resources and Referral Agencies: "Breaking the Piggy Bank").

17. Microwave Oven Prices A research organization tested microwave ovens. At $\alpha = 0.10$, is there a significant difference in the average prices of the three types of oven?

Watts		
1000	900	800
270	240	180
245	135	155
190	160	200
215	230	120
250	250	140
230	200	180
	200	140
	210	130

A computer printout for this exercise is shown. Use the P-value method and the information in this printout to test the claim. (The information in this exercise will be used in Exercise 7 of Section 12–2.)

Computer Printout for Exercise 17

```
ANALYSIS OF VARIANCE SOURCE TABLE
Source         df      Sum of Squares    Mean Square        F      P-value

Bet Groups      2      21729.735         10864.867      10.118    0.00102
W/I Groups     19      20402.083          1073.794

Total          21      42131.818

DESCRIPTIVE STATISTICS
Condit          N                 Means         St Dev

  1000          6               233.333         28.23
   900          8               203.125         39.36
   800          8               155.625         28.21
```

18. Calories in Fast-Food Sandwiches Three popular fast-food restaurant franchises specializing in burgers were surveyed to find out the number of calories in their frequently ordered sandwiches. At the 0.05 level of significance can it be concluded that a difference in mean number of calories per burger exists? The information in this exercise will be used for Exercise 8 in Section 12–2.

FF#1	FF#2	FF#3
970	1010	740
880	970	540
840	920	510
710	850	510
	820	

Source: www.fatcalories.com

19. Basketball Scores for College Teams Below are randomly selected scores for winning college basketball teams in each of three regions for a particular weekend. At the 0.10 level of significance is there sufficient evidence that there is a difference in mean scores by region?

East	Midwest	South
68	78	62
75	79	74
90	65	71
85	67	70
84	60	72
67	79	72
85	57	64
75	74	75

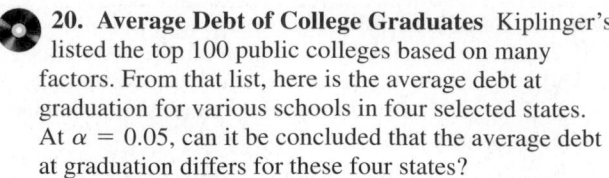

20. Average Debt of College Graduates Kiplinger's listed the top 100 public colleges based on many factors. From that list, here is the average debt at graduation for various schools in four selected states. At $\alpha = 0.05$, can it be concluded that the average debt at graduation differs for these four states?

New York	Virginia	California	Pennsylvania
14,734	14,524	13,171	18,105
16,000	15,176	14,431	17,051
14,347	12,665	14,689	16,103
14,392	12,591	13,788	22,400
12,500	18,385	15,297	17,976

Source: www.Kiplinger.com

Technology *Step by Step*

MINITAB
Step by Step

One-Way Analysis of Variance (ANOVA)

Which treatment is most effective in lowering cholesterol—medication, diet, or exercise?

1. Enter the data for Example 12–1 into columns of MINITAB.
2. Name the columns **Medication, Exercise,** and **Diet.**
3. Select **Stat>ANOVA>One-Way (Unstacked).**
4. Drag the mouse over the three columns in the list box and then click [Select].
5. Click [OK]. In the session window the ANOVA table will be displayed, showing the test statistic $F = 9.17$ whose P-value is 0.004.

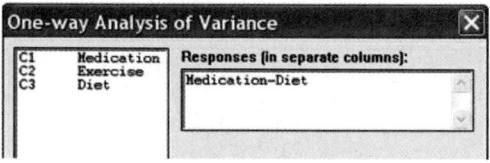

One-Way ANOVA: Medication, Exercise, Diet

```
Source   DF      SS      MS      F       P
Factor    2   160.13   80.07   9.17   0.004
Error    12   104.80    8.73
Total    14   264.93
```

```
                                  Individual 95% CIs For Mean
                                  Based on Pooled StDev
Level        N    Mean   StDev   -------+---------+---------+---------+--
Medication   5  11.800   2.387                        (--------*-------)
Exercise     5   3.800   3.194   (-------*-------)
Diet         5   7.600   3.209         (-------*-------)
                                  -------+---------+---------+---------+--
                                     3.5       7.0      10.5      14.0
Pooled StDev = 2.955
```

Reject the null hypothesis. There is enough evidence to conclude that there is a difference between the treatments. Section 12–2 will explain.

TI-83 Plus or TI-84 Plus
Step by Step

One-Way Analysis of Variance (ANOVA)

1. Enter the data into L_1, L_2, L_3, etc.
2. Press **STAT** and move the cursor to TESTS.
3. Press **F (ALPHA COS)** for ANOVA(.   (Use H for the TI-84.)
4. Type each list followed by a comma. End with) and press **ENTER.**

Example TI12–1

Test the claim H_0: $\mu_1 = \mu_2 = \mu_3$ at $\alpha = 0.05$ for these data from Example 12–1.

Medication	Exercise	Diet
10	6	5
12	8	9
9	3	12
15	0	8
13	2	4

Input

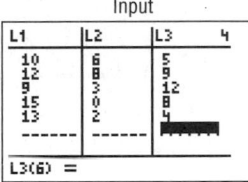

Input

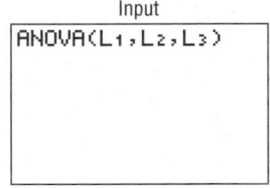

Output

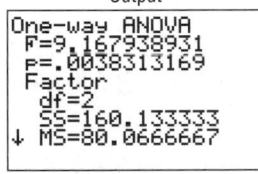

Output

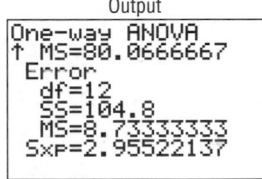

The F test value is 9.167938931. The P-value is 0.0038313169, which is significant at $\alpha = 0.05$. The factor variable has

    d.f. = 2
    SS = 160.133333
    MS = 80.0666667

The error has

    d.f. = 12
    SS = 104.8
    MS = 8.73333333

Excel
Step by Step

One-Way Analysis of Variance (ANOVA)

Example XL12–1

1. Enter the data below in columns A, B, and C.

9	8	12
6	7	15
15	12	18
4	3	9
3	5	10

2. From the toolbar, select Data, then Data Analysis.
3. Select Anova: Single Factor under Analysis tools, then [OK].

4. In the Anova: Single Factor dialog box, type **A1:C5** for the Input Range.

5. Check Grouped By: Columns.

6. Type **0.05** for the Alpha level.

7. Under Output options, check Output Range and type **E2**.

8. Click [OK].

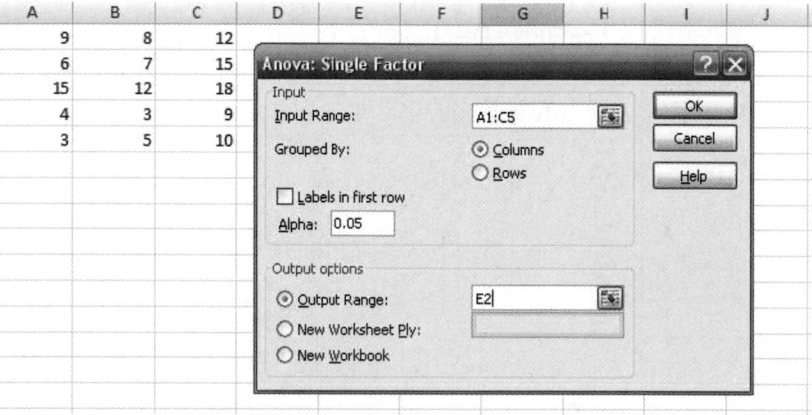

The results of the ANOVA are shown below.

Anova: Single Factor

SUMMARY

Groups	Count	Sum	Average	Variance
Column 1	5	37	7.4	23.3
Column 2	5	35	7	11.5
Column 3	5	64	12.8	13.7

ANOVA

Source of Variation	SS	df	MS	F	P-value	F crit
Between Groups	104.9333333	2	52.46666667	3.245360825	0.074707518	3.885293835
Within Groups	194	12	16.16666667			
Total	298.9333333	14				

12–2 The Scheffé Test and the Tukey Test

When the null hypothesis is rejected using the F test, the researcher may want to know where the difference among the means is. Several procedures have been developed to determine where the significant differences in the means lie after the ANOVA procedure has been performed. Among the most commonly used tests are the *Scheffé test* and the *Tukey test*.

Objective 2

Determine which means differ, using the Scheffé or Tukey test if the null hypothesis is rejected in the ANOVA.

Scheffé Test

To conduct the **Scheffé test,** you must compare the means two at a time, using all possible combinations of means. For example, if there are three means, the following comparisons must be done:

$$\overline{X}_1 \text{ versus } \overline{X}_2 \qquad \overline{X}_1 \text{ versus } \overline{X}_3 \qquad \overline{X}_2 \text{ versus } \overline{X}_3$$

Formula for the Scheffé Test

$\mathcal{U}$*nusual Stat*

According to the *British Medical Journal*, the body's circadian rhythms produce drowsiness during the midafternoon, matched only by the 2:00 A.M. to 7:00 A.M. period for sleep-related traffic accidents.

$$F_S = \frac{(\bar{X}_i - \bar{X}_j)^2}{s_W^2[(1/n_i) + (1/n_j)]}$$

where $\bar{X}_i$ and $\bar{X}_j$ are the means of the samples being compared, n_i and n_j are the respective sample sizes, and s_W^2 is the within-group variance.

To find the critical value F' for the Scheffé test, multiply the critical value for the F test by $k - 1$:

$$F' = (k - 1)(\text{C.V.})$$

There is a significant difference between the two means being compared when F_S is greater than F'. Example 12–3 illustrates the use of the Scheffé test.

Example 12–3

Using the Scheffé test, test each pair of means in Example 12–1 to see whether a specific difference exists, at $\alpha = 0.05$.

Solution

a. For $\bar{X}_1$ versus $\bar{X}_2$,

$$F_S = \frac{(\bar{X}_1 - \bar{X}_2)^2}{s_W^2[(1/n_1) + (1/n_2)]} = \frac{(11.8 - 3.8)^2}{8.73[(1/5) + (1/5)]} = 18.33$$

b. For $\bar{X}_2$ versus $\bar{X}_3$,

$$F_S = \frac{(\bar{X}_2 - \bar{X}_3)^2}{s_W^2[(1/n_2) + (1/n_3)]} = \frac{(3.8 - 7.6)^2}{8.73[(1/5) + (1/5)]} = 4.14$$

c. For $\bar{X}_1$ versus $\bar{X}_3$,

$$F_S = \frac{(\bar{X}_1 - \bar{X}_3)^2}{s_W^2[(1/n_1) + (1/n_3)]} = \frac{(11.8 - 7.6)^2}{8.73[(1/5) + (1/5)]} = 5.05$$

The critical value for the analysis of variance for Example 12–1 was 3.89, found by using Table H with $\alpha = 0.05$, d.f.N. $= k - 1 = 2$, and d.f.D. $= N - k = 12$. In this case, it is multiplied by $k - 1$ as shown.

The critical value for F' at $\alpha = 0.05$, with d.f.N. $= 2$ and d.f.D. $= 12$, is

$$F' = (k - 1)(\text{C.V.}) = (3 - 1)(3.89) = 7.78$$

Since only the F test value for part *a* ($\bar{X}_1$ versus $\bar{X}_2$) is greater than the critical value, 7.78, the only significant difference is between $\bar{X}_1$ and $\bar{X}_2$, that is, between medication and exercise.

On occasion, when the F test value is greater than the critical value, the Scheffé test may not show any significant differences in the pairs of means. This result occurs because the difference may actually lie in the average of two or more means when compared with the other mean. The Scheffé test can be used to make these types of comparisons, but the technique is beyond the scope of this book.

This study involved three groups. The results showed that patients in all three groups felt better after 2 years. State possible null and alternative hypotheses for this study. Was the null hypothesis rejected? Explain how the statistics could have been used to arrive at the conclusion.

HEALTH

TRICKING KNEE PAIN

You sign up for a clinical trial of arthroscopic surgery used to relieve knee pain caused by arthritis. Youre sedated and wake up with tiny incisions. Soon your bum knee feels better. Two years later you find out you had "placebo" surgery. In a study at the Houston VA Medical Center, researchers divided 180 patients into three groups: two groups had damaged cartilage removed, while the third got simulated surgery. Yet an equal number of patients in all groups felt better after two years. Some 650,000 people have the surgery annually, but they're wasting their money, says Dr. Nelda P. Wray, who led the study. And the patients who got fake surgery? "They aren't angry at us," she says. "They still report feeling better."

— STEPHEN P. WILLIAMS

Tukey Test

The **Tukey test** can also be used after the analysis of variance has been completed to make pairwise comparisons between means when the groups have the same sample size. The symbol for the test value in the Tukey test is q.

Formula for the Tukey Test

$$q = \frac{\bar{X}_i - \bar{X}_j}{\sqrt{s_W^2/n}}$$

where $\bar{X}_i$ and $\bar{X}_j$ are the means of the samples being compared, n is the size of the samples, and s_W^2 is the within-group variance.

When the absolute value of q is greater than the critical value for the Tukey test, there is a significant difference between the two means being compared. The procedures for finding q and the critical value from Table N in Appendix C for the Tukey test are shown in Example 12–4.

Example 12–4

Using the Tukey test, test each pair of means in Example 12–1 to see whether a specific difference exists, at $\alpha = 0.05$.

Solution

a. For $\bar{X}_1$ versus $\bar{X}_2$,

$$q = \frac{\bar{X}_1 - \bar{X}_2}{\sqrt{s_W^2/n}} = \frac{11.8 - 3.8}{\sqrt{8.73/5}} = \frac{8}{1.32} = 6.06$$

b. For $\bar{X}_1$ versus $\bar{X}_3$,

$$q = \frac{\bar{X}_1 - \bar{X}_3}{\sqrt{s_W^2/n}} = \frac{11.8 - 7.6}{\sqrt{8.73/5}} = \frac{4.2}{1.32} = 3.18$$

c. For $\bar{X}_2$ versus $\bar{X}_3$,

$$q = \frac{\bar{X}_2 - \bar{X}_3}{\sqrt{s_W^2/n}} = \frac{3.8 - 7.6}{\sqrt{8.73/5}} = \frac{-3.8}{1.32} = -2.88$$

To find the critical value for the Tukey test, use Table N in Appendix C. The number of means k is found in the row at the top, and the degrees of freedom for s_W^2 are found in the left column (denoted by v). Since $k = 3$, d.f. $= 12$, and $\alpha = 0.05$, the critical value is 3.77. See Figure 12–1. Hence, the only q value that is greater in absolute value than the critical value is the one for the difference between $\bar{X}_1$ and $\bar{X}_2$. The conclusion, then, is that there is a significant difference in means for medication and exercise. These results agree with the Scheffé analysis.

Figure 12–1

Finding the Critical Value in Table N for the Tukey Test (Example 12–4)

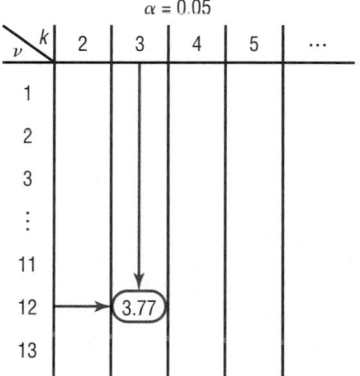

You might wonder why there are two different tests that can be used after the ANOVA. Actually, there are several other tests that can be used in addition to the Scheffé and Tukey tests. It is up to the researcher to select the most appropriate test. The Scheffé test is the most general, and it can be used when the samples are of different sizes. Furthermore, the Scheffé test can be used to make comparisons such as the average of $\bar{X}_1$ and $\bar{X}_2$ compared with $\bar{X}_3$. However, the Tukey test is more powerful than the Scheffé test for making pairwise comparisons for the means. A rule of thumb for pairwise comparisons is to use the Tukey test when the samples are equal in size and the Scheffé test when the samples differ in size. This rule will be followed in this textbook.

Applying the Concepts 12–2

Colors That Make You Smarter

The following set of data values was obtained from a study of people's perceptions on whether the color of a person's clothing is related to how intelligent the person looks. The subjects rated the person's intelligence on a scale of 1 to 10. Group 1 subjects were randomly shown people with clothing in shades of blue and gray. Group 2 subjects were randomly shown people with clothing in shades of brown and yellow. Group 3 subjects were randomly shown people with clothing in shades of pink and orange. The results follow.

Group 1	Group 2	Group 3
8	7	4
7	8	9
7	7	6
7	7	7
8	5	9
8	8	8
6	5	5
8	8	8
8	7	7
7	6	5
7	6	4
8	6	5
8	6	4

1. Use the Tukey test to test all possible pairwise comparisons.
2. Are there any contradictions in the results?
3. Explain why separate *t* tests are not accepted in this situation.
4. When would Tukey's test be preferred over the Scheffé method? Explain.

See page 668 for the answers.

Exercises 12–2

1. What two tests can be used to compare two means when the null hypothesis is rejected using the one-way ANOVA *F* test? The Scheffé and Tukey tests are used.

2. Explain the difference between the two tests used to compare two means when the null hypothesis is rejected using the one-way ANOVA *F* test.

For Exercises 3 through 9, the null hypothesis was rejected. Use the Scheffé test when sample sizes are unequal or the Tukey test when sample sizes are equal, to test the differences between the pairs of means. Assume all variables are normally distributed, samples are independent, and the population variances are equal.

3. Exercise 9 in Section 12–1.

4. Exercise 12 in Section 12–1.

5. Exercise 13 in Section 12–1.

6. Exercise 16 in Section 12–1. No further testing should be done.

7. Exercise 17 in Section 12–1.

8. Exercise 18 in Section 12–1.

9. Exercise 20 in Section 12–1.

For Exercises 10 through 13, do a complete one-way ANOVA. If the null hypothesis is rejected, use either the Scheffé or Tukey test to see if there is a significant difference in the pairs of means. Assume all assumptions are met.

 10. Weights of Digital Cameras The data consist of the weights in ounces of three different types of digital camera. Use $\alpha = 0.05$ to see if the means are equal.

2–3 Megapixels	4–5 Megapixels	6–8 Megapixels
6	14	19
8	11	27
7	15	21
11	24	23
4	17	24
8	10	33

 11. Fiber Content of Foods The number of grams of fiber per serving for a random sample of three different kinds of foods is listed. Is there sufficient evidence at the 0.05 level of significance to conclude that there is a difference in mean fiber content among breakfast cereals, fruits, and vegetables?

Breakfast cereals	Fruits	Vegetables
3	5.5	10
4	2	1.5
6	4.4	3.5
4	1.6	2.7
10	3.8	2.5
5	4.5	6.5
6	2.8	4
8		3
5		

Source: *The Doctor's Pocket Calorie, Fat, and Carbohydrate Counter.*

12. **Per-Pupil Expenditures** The expenditures (in dollars) per pupil for states in three sections of the country are listed. Using $\alpha = 0.05$, can you conclude that there is a difference in means?

Eastern third	Middle third	Western third
4946	6149	5282
5953	7451	8605
6202	6000	6528
7243	6479	6911
6113		

Source: *New York Times Almanac.*

 13. **Weekly Unemployment Benefits** The average weekly unemployment benefit for the entire United States is $297. Three states are randomly selected, and a sample of weekly unemployment benefits is recorded for each. At $\alpha = 0.05$ is there sufficient evidence to conclude a difference in means? If so, perform the appropriate test to find out where the difference exists.

Florida	Pennsylvania	Maine
200	300	250
187	350	195
192	295	275
235	362	260
260	280	220
175	340	290

Source: *World Almanac.*

12–3 Two-Way Analysis of Variance

Objective 3

Use the two-way ANOVA technique to determine if there is a significant difference in the main effects or interaction.

The analysis of variance technique shown previously is called a **one-way ANOVA** since there is only *one independent variable.* The **two-way ANOVA** is an extension of the one-way analysis of variance; it involves *two independent variables.* The independent variables are also called **factors.**

The two-way analysis of variance is quite complicated, and many aspects of the subject should be considered when you are using a research design involving a two-way ANOVA. For the purposes of this textbook, only a brief introduction to the subject will be given.

In doing a study that involves a two-way analysis of variance, the researcher is able to test the effects of two independent variables or factors on one *dependent variable.* In addition, the interaction effect of the two variables can be tested.

For example, suppose a researcher wishes to test the effects of two different types of plant food and two different types of soil on the growth of certain plants. The two independent variables are the type of plant food and the type of soil, while the dependent variable is the plant growth. Other factors, such as water, temperature, and sunlight, are held constant.

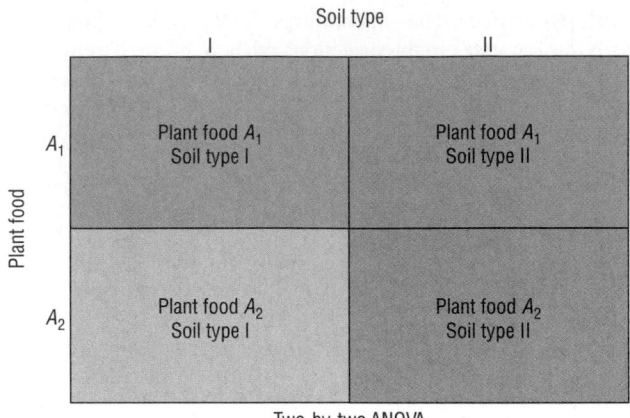

Figure 12–2

Treatment Groups for the Plant Food–Soil Type Experiment

To conduct this experiment, the researcher sets up four groups of plants. See Figure 12–2. Assume that the plant food type is designated by the letters A_1 and A_2 and the soil type by the Roman numerals I and II. The groups for such a two-way ANOVA are sometimes called **treatment groups.** The four groups are

Group 1   Plant food A_1, soil type I
Group 2   Plant food A_1, soil type II
Group 3   Plant food A_2, soil type I
Group 4   Plant food A_2, soil type II

The plants are assigned to the groups at random. This design is called a 2×2 (read "two-by-two") design, since each variable consists of two **levels,** that is, two different treatments.

The two-way ANOVA enables the researcher to test the effects of the plant food and the soil type in a single experiment rather than in separate experiments involving the plant food alone and the soil type alone. Furthermore, the researcher can test an additional hypothesis about the effect of the *interaction* of the two variables—plant food and soil type—on plant growth. For example, is there a difference between the growth of plants using plant food A_1 and soil type II and the growth of plants using plant food A_2 and soil type I? When a difference of this type occurs, the experiment is said to have a significant **interaction effect.** That is, the types of plant food affect the plant growth differently in different soil types. When the interaction effect is statistically significant the researcher should not consider the effects of the individual factors without considering the interaction effect.

There are many different kinds of two-way ANOVA designs, depending on the number of levels of each variable. Figure 12–3 shows a few of these designs. As stated previously, the plant food–soil type experiment uses a 2×2 ANOVA.

The design in Figure 12–3(a) is called a 3×2 design, since the factor in the rows has three levels and the factor in the columns has two levels. Figure 12–3(b) is a 3×3 design, since each factor has three levels. Figure 12–3(c) is a 4×3 design.

The two-way ANOVA design has several null hypotheses. There is one for each independent variable and one for the interaction. In the plant food–soil type problem, the hypotheses are as follows:

1. H_0: There is no interaction effect between type of plant food used and type of soil used on plant growth.

    H_1: There is an interaction effect between food type and soil type on plant growth.

2. H_0: There is no difference in means of heights of plants grown using different foods.

    H_1: There is a difference in means of heights of plants grown using different foods.

Figure 12–3

Some Types of
Two-Way ANOVA
Designs

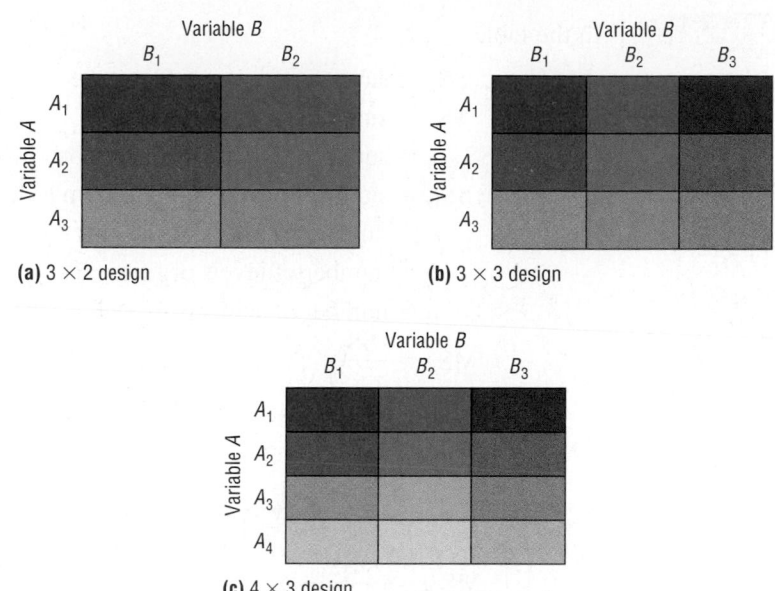

(a) 3 × 2 design

(b) 3 × 3 design

(c) 4 × 3 design

3. H_0: There is no difference in means of heights of plants grown in different soil types.

   H_1: There is a difference in means of heights of plants grown in different soil types.

The first set of hypotheses concerns the interaction effect; the second and third sets test the effects of the independent variables, which are sometimes called the **main effects.**

As with the one-way ANOVA, a between-group variance estimate is calculated, and a within-group variance estimate is calculated. An F test is then performed for each of the independent variables and the interaction. The results of the two-way ANOVA are summarized in a two-way table, as shown in Table 12–4 for the plant experiment.

Table 12–4   ANOVA Summary Table for Plant Food and Soil Type

Source	Sum of squares	d.f.	Mean square	F
Plant food				
Soil type				
Interaction				
Within (error)				
Total				

In general, the two-way **ANOVA summary table** is set up as shown in Table 12–5.

Table 12–5   ANOVA Summary Table

Source	Sum of squares	d.f.	Mean square	F
A	SS_A	$a - 1$	MS_A	F_A
B	SS_B	$b - 1$	MS_B	F_B
$A \times B$	$SS_{A \times B}$	$(a - 1)(b - 1)$	$MS_{A \times B}$	$F_{A \times B}$
Within (error)	SS_W	$ab(n - 1)$	MS_W	
Total				

In the table,

$$SS_A = \text{sum of squares for factor } A$$
$$SS_B = \text{sum of squares for factor } B$$
$$SS_{A \times B} = \text{sum of squares for interaction}$$
$$SS_W = \text{sum of squares for error term (within-group)}$$
$$a = \text{number of levels of factor } A$$
$$b = \text{number of levels of factor } B$$
$$n = \text{number of subjects in each group}$$

$$MS_A = \frac{SS_A}{a - 1}$$

$$MS_B = \frac{SS_B}{b - 1}$$

$$MS_{A \times B} = \frac{SS_{A \times B}}{(a - 1)(b - 1)}$$

$$MS_W = \frac{SS_W}{ab(n - 1)}$$

$$F_A = \frac{MS_A}{MS_W} \qquad \text{with d.f.N.} = a - 1, \text{d.f.D.} = ab(n - 1)$$

$$F_B = \frac{MS_B}{MS_W} \qquad \text{with d.f.N.} = b - 1, \text{d.f.D.} = ab(n - 1)$$

$$F_{A \times B} = \frac{MS_{A \times B}}{MS_W} \qquad \text{with d.f.N.} = (a - 1)(b - 1), \text{d.f.D.} = ab(n - 1)$$

The assumptions for the two-way analysis of variance are basically the same as those for the one-way ANOVA, except for sample size.

Assumptions for the Two-Way ANOVA

1. The populations from which the samples were obtained must be normally or approximately normally distributed.
2. The samples must be independent.
3. The variances of the populations from which the samples were selected must be equal.
4. The groups must be equal in sample size.

The computational procedure for the two-way ANOVA is quite lengthy. For this reason, it will be omitted in Example 12–5, and only the two-way ANOVA summary table will be shown. The table used in Example 12–5 is similar to the one generated by most computer programs. You should be able to interpret the table and summarize the results.

Example 12–5

Gasoline Consumption

A researcher wishes to see whether the type of gasoline used and the type of automobile driven have any effect on gasoline consumption. Two types of gasoline, regular and high-octane, will be used, and two types of automobiles, two-wheel- and four-wheel-drive, will be used in each group. There will be two automobiles in each group, for a total of eight automobiles used. Using a two-way analysis of variance, the researcher will perform the following steps.

Step 1 State the hypotheses.

Step 2 Find the critical value for each F test, using $\alpha = 0.05$.

Step 3 Complete the summary table to get the test value.

Step 4 Make the decision.

Step 5 Summarize the results.

 The data (in miles per gallon) are shown here, and the summary table is given in Table 12–6.

	Type of automobile	
Gas	**Two-wheel-drive**	**Four-wheel-drive**
Regular	26.7 25.2	28.6 29.3
High-octane	32.3 32.8	26.1 24.2

Table 12–6   ANOVA Summary Table for Example 12–5

Source	SS	d.f.	MS	F
Gasoline A	3.920			
Automobile B	9.680			
Interaction (A × B)	54.080			
Within (error)	3.300			
Total	70.980			

Solution

Step 1 State the hypotheses. The hypotheses for the interaction are these:

H_0: There is no interaction effect between type of gasoline used and type of automobile a person drives on gasoline consumption.

H_1: There is an interaction effect between type of gasoline used and type of automobile a person drives on gasoline consumption.

The hypotheses for the gasoline types are

H_0: There is no difference between the means of gasoline consumption for two types of gasoline.

H_1: There is a difference between the means of gasoline consumption for two types of gasoline.

The hypotheses for the types of automobile driven are

H_0: There is no difference between the means of gasoline consumption for two-wheel-drive and four-wheel-drive automobiles.

H_1: There is a difference between the means of gasoline consumption for two-wheel-drive and four-wheel-drive automobiles.

Step 2 Find the critical values for each F test. In this case, each independent variable, or factor, has two levels. Hence, a 2×2 ANOVA table is used. Factor A is designated as the gasoline type. It has two levels, regular and high-octane; therefore, $a = 2$. Factor B is designated as the automobile type. It also has two levels; therefore, $b = 2$. The degrees of freedom for each factor are as follows:

$$\text{Factor } A: \quad \text{d.f.N.} = a - 1 = 2 - 1 = 1$$

$$\text{Factor } B: \quad \text{d.f.N.} = b - 1 = 2 - 1 = 1$$

$$\text{Interaction } (A \times B): \quad \text{d.f.N.} = (a - 1)(b - 1)$$

$$= (2 - 1)(2 - 1) = 1 \cdot 1 = 1$$

$$\text{Within (error)}: \quad \text{d.f.D.} = ab(n - 1)$$

$$= 2 \cdot 2(2 - 1) = 4$$

where n is the number of data values in each group. In this case, $n = 2$.

The critical value for the F_A test is found by using $\alpha = 0.05$, d.f.N. = 1, and d.f.D. = 4. In this case, $F_A = 7.71$. The critical value for the F_B test is found by using $\alpha = 0.05$, d.f.N. = 1, and d.f.D. = 4; also F_B is 7.71. Finally, the critical value for the $F_{A \times B}$ test is found by using d.f.N. = 1 and d.f.D. = 4; it is also 7.71.

Note: If there are different levels of the factors, the critical values will not all be the same. For example, if factor A has three levels and factor b has four levels, and if there are two subjects in each group, then the degrees of freedom are as follows:

$$\text{d.f.N.} = a - 1 = 3 - 1 = 2 \qquad \text{factor } A$$

$$\text{d.f.N.} = b - 1 = 4 - 1 = 3 \qquad \text{factor } B$$

$$\text{d.f.N.} = (a - 1)(b - 1) = (3 - 1)(4 - 1)$$

$$= 2 \cdot 3 = 6 \qquad \text{factor } A \times B$$

$$\text{d.f.N.} = ab(n - 1) = 3 \cdot 4(2 - 1) = 12 \qquad \text{within (error) factor}$$

Step 3 Complete the ANOVA summary table to get the test values. The mean squares are computed first.

$$\text{MS}_A = \frac{\text{SS}_A}{a - 1} = \frac{3.920}{2 - 1} = 3.920$$

$$\text{MS}_B = \frac{\text{SS}_B}{b - 1} = \frac{9.680}{2 - 1} = 9.680$$

$$\text{MS}_{A \times B} = \frac{\text{SS}_{A \times B}}{(a - 1)(b - 1)} = \frac{54.080}{(2 - 1)(2 - 1)} = 54.080$$

$$\text{MS}_W = \frac{\text{SS}_W}{ab(n - 1)} = \frac{3.300}{4} = 0.825$$

The F values are computed next.

$$F_A = \frac{\text{MS}_A}{\text{MS}_W} = \frac{3.920}{0.825} = 4.752 \qquad \text{d.f.N.} = a - 1 = 1 \qquad \text{d.f.D.} = ab(n - 1) = 4$$

$$F_B = \frac{\text{MS}_B}{\text{MS}_W} = \frac{9.680}{0.825} = 11.733 \qquad \text{d.f.N.} = b - 1 = 1 \qquad \text{d.f.D.} = ab(n - 1) = 4$$

$$F_{A \times B} = \frac{\text{MS}_{A \times B}}{\text{MS}_W} = \frac{54.080}{0.825} = 65.552 \qquad \text{d.f.N.} = (a - 1)(b - 1) = 1 \qquad \text{d.f.D.} = ab(n - 1) = 4$$

The completed ANOVA table is shown in Table 12–7.

Table 12–7	Completed ANOVA Summary Table for Example 12–5			
Source	**SS**	**d.f.**	**MS**	**F**
Gasoline *A*	3.920	1	3.920	4.752
Automobile *B*	9.680	1	9.680	11.733
Interaction ($A \times B$)	54.080	1	54.080	65.552
Within (error)	3.300	4	0.825	
Total	70.980	7		

Step 4   Make the decision. Since $F_B = 11.733$ and $F_{A \times B} = 65.552$ are greater than the critical value 7.71, the null hypotheses concerning the type of automobile driven and the interaction effect should be rejected. Since the interaction effect is statistically significant no decision should be made about the automobile type without further investigation.

Step 5   Summarize the results. Since the null hypothesis for the interaction effect was rejected, it can be concluded that the combination of type of gasoline and type of automobile does affect gasoline consumption.

Interesting Fact

Some birds can fly as high as 5 miles.

In the preceding analysis, the effect of the type of gasoline used and the effect of the type of automobile driven are called the *main effects*. If there is no significant interaction effect, the main effects can be interpreted independently. However, if there is a significant interaction effect, the main effects must be interpreted cautiously.

To interpret the results of a two-way analysis of variance, researchers suggest drawing a graph, plotting the means of each group, analyzing the graph, and interpreting the results. In Example 12–5, find the means for each group or cell by adding the data values in each cell and dividing by *n*. The means for each cell are shown in the chart here.

Type of automobile

Gas	Two-wheel-drive	Four-wheel-drive
Regular	$\bar{X} = \dfrac{26.7 + 25.2}{2} = 25.95$	$\bar{X} = \dfrac{28.6 + 29.3}{2} = 28.95$
High-octane	$\bar{X} = \dfrac{32.3 + 32.8}{2} = 32.55$	$\bar{X} = \dfrac{26.1 + 24.2}{2} = 25.15$

The graph of the means for each of the variables is shown in Figure 12–4. In this graph, the lines cross each other. When such an intersection occurs and the interaction is significant, the interaction is said to be a **disordinal interaction.** When there is a disordinal interaction, you should not interpret the main effects without considering the interaction effect.

The other type of interaction that can occur is an *ordinal interaction*. Figure 12–5 shows a graph of means in which an ordinal interaction occurs between two variables. The lines do not cross each other, nor are they parallel. If the *F* test value for the interaction is significant and the lines do not cross each other, then the interaction is said to be an **ordinal interaction** and the main effects can be interpreted independently of each other.

Finally, when there is no significant interaction effect, the lines in the graph will be parallel or approximately parallel. When this situation occurs, the main effects can be interpreted independently of each other because there is no significant interaction.

Figure 12–4

Graph of the Means
of the Variables in
Example 12–5

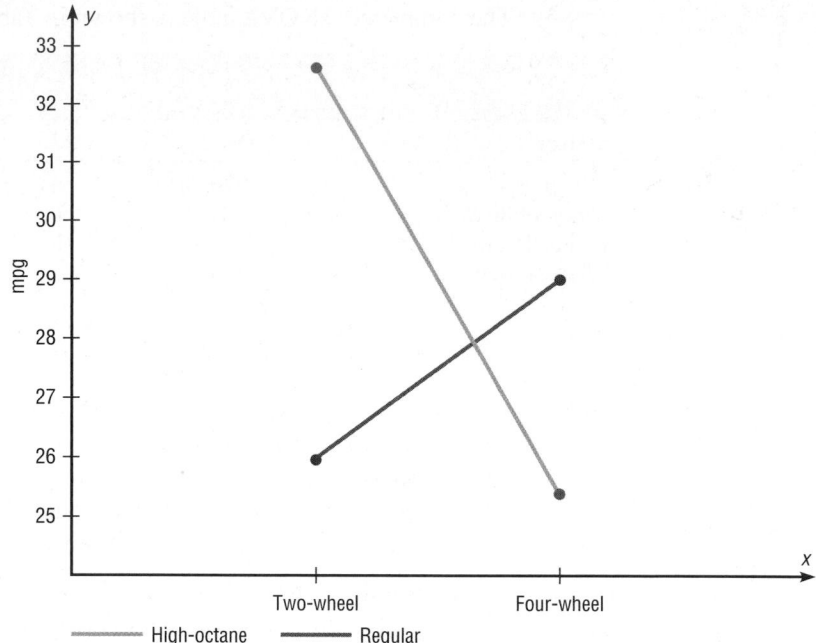

Figure 12–5

Graph of Two Variables
Indicating an Ordinal
Interaction

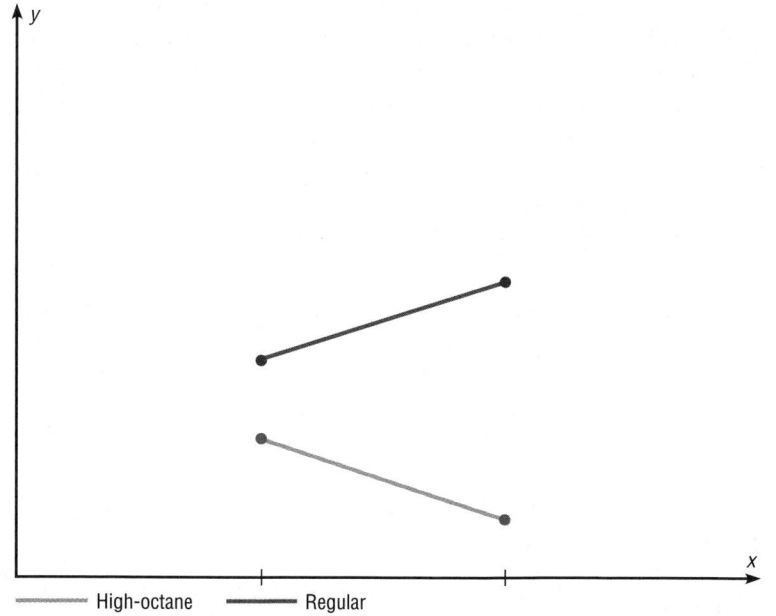

Figure 12–6 shows the graph of two variables when the interaction effect is not significant; the lines are parallel.

Example 12–5 was an example of a 2×2 two-way analysis of variance, since each independent variable had two levels. For other types of variance problems, such as a 3×2 or a 4×3 ANOVA, interpretation of the results can be quite complicated. Procedures using tests such as the Tukey and Scheffé tests for analyzing the cell means exist and are similar to the tests shown for the one-way ANOVA, but they are beyond the scope of this textbook. Many other designs for analysis of variance are available to researchers, such as three-factor designs and repeated-measure designs; they are also beyond the scope of this book.

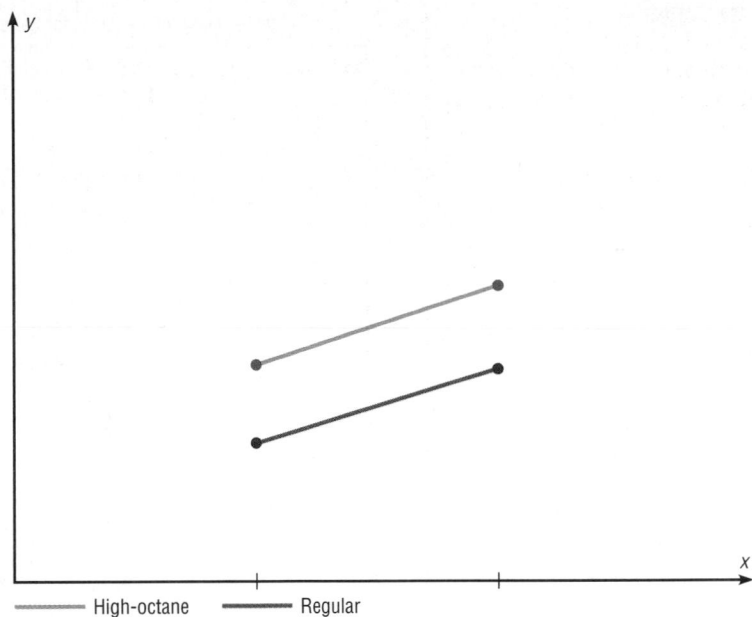

Figure 12–6

Graph of Two Variables Indicating No Interaction

———— High-octane   ———— Regular

In summary, the two-way ANOVA is an extension of the one-way ANOVA. The former can be used to test the effects of two independent variables and a possible interaction effect on a dependent variable.

Applying the Concepts **12–3**

Automobile Sales Techniques

The following outputs are from the result of an analysis of how car sales are affected by the experience of the salesperson and the type of sales technique used. Experience was broken up into four levels, and two different sales techniques were used. Analyze the results and draw conclusions about level of experience with respect to the two different sales techniques and how they affect car sales.

Two-Way Analysis of Variance

```
Analysis of Variance for Sales
Source         DF        SS        MS
Experience      3     3414.0    1138.0
Presentation    1        6.0       6.0
Interaction     3      414.0     138.0
Error          16      838.0      52.4
Total          23     4672.0

                            Individual 95% CI
Experience    Mean    -----+---------+---------+---------+------
1             62.0    (-----*-----)
2             63.0     (-----*-----)
3             78.0                     (-----*-----)
4             91.0                                  (-----*-----)
                      -----+---------+---------+---------+------
                        60.0      70.0      80.0      90.0

                            Individual 95% CI
Presentation  Mean    ------+---------+---------+---------+-----
1             74.0       (-----------------*-----------------)
2             73.0    (-----------------*-----------------)
                      ------+---------+---------+---------+-----
```

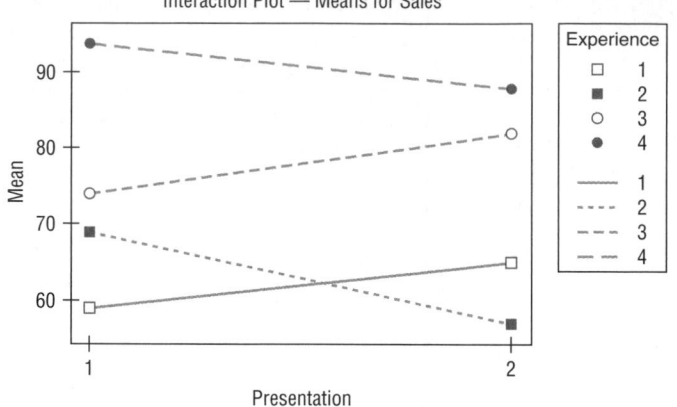

See page 668 for the answers.

Exercises 12–3

1. How does the two-way ANOVA differ from the one-way ANOVA?

2. Explain what is meant by *main effects* and *interaction effect*.

3. How are the values for the mean squares computed?

4. How are the *F* test values computed?

5. In a two-way ANOVA, variable *A* has three levels and variable *B* has two levels. There are five data values in each cell. Find each degrees-of-freedom value.

   a. d.f.N. for factor *A* For factor *A*, $\text{d.f.}_A = 2$

   b. d.f.N. for factor *B* For factor *B*, $\text{d.f.}_B = 1$

   c. d.f.N. for factor *A* × *B* $\text{d.f.}_{A \times B} = 2$

   d. d.f.D. for the within (error) factor $\text{d.f.}_{\text{within}} = 24$

6. In a two-way ANOVA, variable *A* has six levels and variable *B* has five levels. There are seven data values in each cell. Find each degrees-of-freedom value.

   a. d.f.N. for factor *A* 5

   b. d.f.N. for factor *B* 4

   c. d.f.N. for factor *A* × *B* 20

   d. d.f.D. for the within (error) factor 180

7. What are the two types of interactions that can occur in the two-way ANOVA? The two types of interactions that can occur are ordinal and disordinal.

8. When can the main effects for the two-way ANOVA be interpreted independently?

9. Describe what the graph of the variables would look like for each situation in a two-way ANOVA experiment.

   a. No interaction effect occurs.

   b. An ordinal interaction effect occurs.

   c. A disordinal interaction effect occurs.

For Exercises 10 through 15, perform these steps. Assume that all variables are normally or approximately normally distributed, that the samples are independent, and that the population variances are equal.

   a. State the hypotheses.

   b. Find the critical value for each *F* test.

   c. Complete the summary table and find the test value.

   d. Make the decision.

   e. Summarize the results. (Draw a graph of the cell means if necessary.)

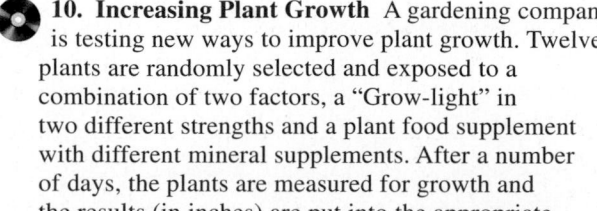 **10. Increasing Plant Growth** A gardening company is testing new ways to improve plant growth. Twelve plants are randomly selected and exposed to a combination of two factors, a "Grow-light" in two different strengths and a plant food supplement with different mineral supplements. After a number of days, the plants are measured for growth and the results (in inches) are put into the appropriate boxes.

	Grow-light 1	Grow-light 2
Plant food A	9.2, 9.4, 8.9	8.5, 9.2, 8.9
Plant food B	7.1, 7.2, 8.5	5.5, 5.8, 7.6

Can an interaction between the two factors be concluded? Is there a difference in mean growth with respect to light? With respect to plant food? Use $\alpha = 0.05$.

11. Environmentally Friendly Air Freshener As a new type of environmentally friendly, natural air freshener is being developed, it is tested to see whether the effects of temperature and humidity affect the length of time that the scent is effective. The numbers of days that the air freshener had a significant level of scent are

listed below for two temperature and humidity levels. Can an interaction between the two factors be concluded? Is there a difference in mean length of effectiveness with respect to humidity? With respect to temperature? Use $\alpha = 0.05$.

	Temperature 1	Temperature 2
Humidity 1	35, 25, 26	35, 31, 37
Humidity 2	28, 22, 21	23, 19, 18

12. Home-Building Times A contractor wishes to see whether there is a difference in the time (in days) it takes two subcontractors to build three different types of homes. At $\alpha = 0.05$, analyze the data shown here, using a two-way ANOVA. See below for raw data.

Data for Exercise 12

	Home type		
Subcontractor	**I**	**II**	**III**
A	25, 28, 26, 30, 31	30, 32, 35, 29, 31	43, 40, 42, 49, 48
B	15, 18, 22, 21, 17	21, 27, 18, 15, 19	23, 25, 24, 17, 13

ANOVA Summary Table for Exercise 12

Source	SS	d.f.	MS	F
Subcontractor	1672.553			
Home type	444.867			
Interaction	313.267			
Within	328.800			
Total	2759.487			

13. Durability of Paint A pigment laboratory is testing both dry additives and solution-based additives to see their effect on the durability rating (a number from 1 to 10) of a finished paint product. The paint to be tested is divided into four equal quantities, and a different combination of the two additives is added to one-fourth of each quantity. After a prescribed number of hours, the durability rating is obtained for each of the 16 samples, and the results are recorded below in the appropriate space.

	Dry additive 1	Dry additive 2
Solution additive A	9, 8, 5, 6	4, 5, 8, 9
Solution additive B	7, 7, 6, 8	10, 8, 6, 7

Can an interaction be concluded between the dry and solution additives? Is there a difference in mean durability rating with respect to dry additive used? With respect to solution additive? Use $\alpha = 0.05$.

14. Types of Outdoor Paint Two types of outdoor paint, enamel and latex, were tested to see how long (in months) each lasted before it began to crack, flake, and peel. They were tested in four geographic locations in the United States to study the effects of climate on the paint. At $\alpha = 0.01$, analyze the data shown, using a two-way ANOVA shown below. Each group contained five test panels. See below for raw data.

Data for Exercise 14

	Geographic location			
Type of paint	**North**	**East**	**South**	**West**
Enamel	60, 53, 58, 62, 57	54, 63, 62, 71, 76	80, 82, 62, 88, 71	62, 76, 55, 48, 61
Latex	36, 41, 54, 65, 53	62, 61, 77, 53, 64	68, 72, 71, 82, 86	63, 65, 72, 71, 63

ANOVA Summary Table for Exercise 14

Source	SS	d.f.	MS	F
Paint type	12.1			
Location	2501.0			
Interaction	268.1			
Within	2326.8			
Total	5108.0			

15. Age and Sales A company sells three items: swimming pools, spas, and saunas. The owner decides to see whether the age of the sales representative and the type of item affect monthly sales. At $\alpha = 0.05$, analyze the data shown, using a two-way ANOVA. Sales are given in hundreds of dollars for a randomly selected month, and five salespeople were selected for each group.

ANOVA Summary Table for Exercise 15

Source	SS	d.f.	MS	F
Age	168.033			
Product	1,762.067			
Interaction	7,955.267			
Within	2,574.000			
Total	12,459.367			

Data for Exercise 15

Age of salesperson	Product		
	Pool	Spa	Sauna
Over 30	56, 23, 52, 28, 35	43, 25, 16, 27, 32	47, 43, 52, 61, 74
30 or under	16, 14, 18, 27, 31	58, 62, 68, 72, 83	15, 14, 22, 16, 27

Technology *Step by Step*

MINITAB
Step by Step

Two-Way Analysis of Variance

For Example 12–5, how do gasoline type and vehicle type affect gasoline mileage?

1. Enter the data into three columns of a worksheet. The data for this analysis have to be "stacked" as shown.

   a) All the gas mileage data are entered in a single column named MPG.

   b) The second column contains codes identifying the gasoline type, a 1 for regular or a 2 for high-octane.

   c) The third column will contain codes identifying the type of automobile, 1 for two-wheel-drive or 2 for four-wheel-drive.

2. Select **Stat>ANOVA>Two-Way.**

   a) Double-click MPG in the list box.

   b) Double-click GasCode as Row factor.

   c) Double-click TypeCode as Column factor.

   d) Check the boxes for Display means, then click [OK].

The session window will contain the results.

↓	C1	C2	C3
	MPG	**GasCode**	**TypeCode**
1	26.7	1	1
2	25.2	1	1
3	32.3	2	1
4	32.8	2	1
5	28.6	1	2
6	29.3	1	2
7	26.1	2	2
8	24.2	2	2

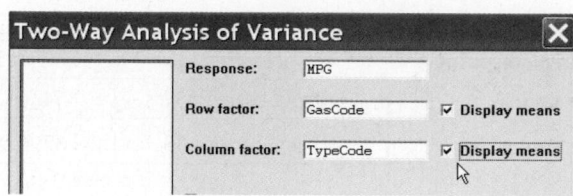

Two-Way ANOVA: MPG versus GasCode, TypeCode

```
Source         DF    SS       MS        F       P
GasCode        1    3.92     3.920    4.75    0.095
TypeCode       1    9.68     9.680   11.73    0.027
Interaction    1   54.08    54.080   65.55    0.001
Error          4    3.30     0.825
Total          7   70.98
```

```
                      Individual 95% CIs For Mean Based on
                      Pooled StDev
GasCode   Mean    --------+--------+--------+--------+-
1         27.45   (-----------*-----------)
2         28.85               (------------*------------)
                  --------+--------+--------+--------+-
                     27.0     28.0     29.0     30.0
                      Individual 95% CIs For Mean Based on
                      Pooled StDev
TypeCode  Mean    -----+---------+---------+---------+----
1         29.25                     (---------*---------)
2         27.05   (---------*-----------)
                  -----+---------+---------+---------+----
                     26.4      27.6      28.8      30.0
```

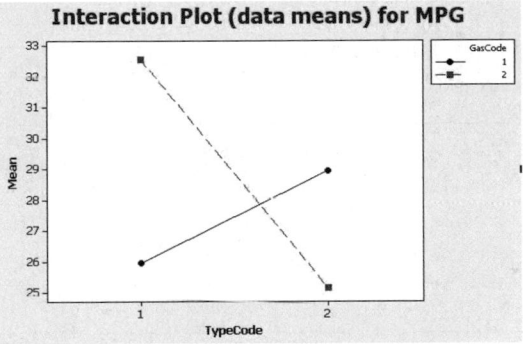

Interaction Plot (data means) for MPG

Plot Interactions

3. Select **Stat>ANOVA>Interactions Plot.**

   a) Double-click MPG for the response variable and GasCodes and TypeCodes for the factors.

   b) Click [OK].

Intersecting lines indicate a significant interaction of the two independent variables.

TI-83 Plus or TI-84 Plus
Step by Step

The TI-83 Plus and TI-84 Plus do not have a built-in function for two-way analysis of variance. However, the downloadable program named TWOWAY is available on your CD and Online Learning Center. Follow the instructions with your CD for downloading the program.

Performing a Two-Way Analysis of Variance

1. Enter the data values of the dependent variable into L_1 and the coded values for the levels of the factors into L_2 and L_3.

2. Press **PRGM,** move the cursor to the program named TWOWAY, and press **ENTER** twice.

3. Type L_1 for the list that contains the dependent variable and press **ENTER.**

4. Type L_2 for the list that contains the coded values for the first factor and press **ENTER.**

5. Type L_3 for the list that contains the coded values for the second factor and press **ENTER.**

6. The program will show the statistics for the first factor.

7. Press **ENTER** to see the statistics for the second factor.

8. Press **ENTER** to see the statistics for the interaction.

9. Press **ENTER** to see the statistics for the error.

10. Press **ENTER** to clear the screen.

Example TI12–2

Perform a two-way analysis of variance for the gasoline data (Example 12–5 in the text). The gas mileages are the data values for the dependent variable. Factor A is the type of gasoline (1 for regular, 2 for high-octane). Factor B is the type of automobile (1 for two-wheel-drive, 2 for four-wheel-drive).

Gas mileages (L_1)	Type of gasoline (L_2)	Type of automobile (L_3)
26.7	1	1
25.2	1	1
32.3	2	1
32.8	2	1
28.6	1	2
29.3	1	2
26.1	2	2
24.2	2	2

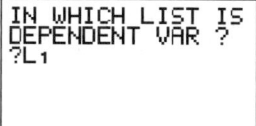 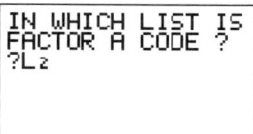 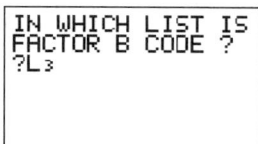

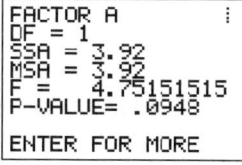 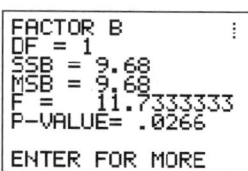 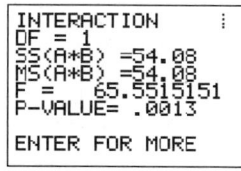 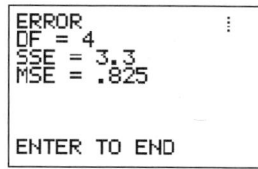

Excel
Step by Step

Two-Way Analysis of Variance (ANOVA)

This example pertains to Example 12–5 from the text.

Example XL12–2

A researcher wishes to see if type of gasoline used and type of automobile driven have any effect on gasoline consumption. Use $\alpha = 0.05$.

1. Enter the data exactly as shown in the figure below in an Excel worksheet.

	A	B	C
1		2-wheel Drive	4-wheel drive
2	Regular	26.7	28.6
3		25.2	29.3
4	Hi-octane	32.3	26.1
5		32.8	24.2

2. From the toolbar, select Data, then Data Analysis.
3. Select Anova: Two-Factor With Replication under Analysis tools, then [OK].
4. In the Anova: Single Factor dialog box, type **A1:C5** for the Input Range.
5. Type **2** for the Rows per sample.
6. Type **0.05** for the Alpha level.
7. Under Output options, check Output Range and type **E2**.
8. Click [OK].

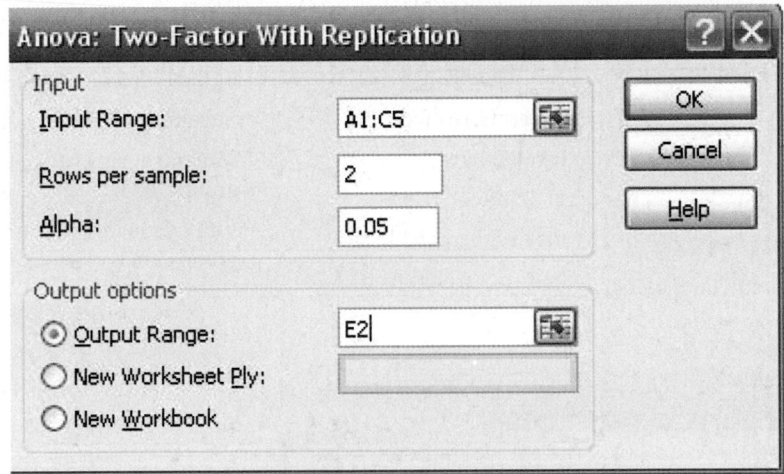

The two-way ANOVA table is shown below.

ANOVA						
Source of Variation	*SS*	*df*	*MS*	*F*	*P-value*	*F crit*
Sample	3.92	1	3.92	4.751515152	0.094766001	7.708647421
Columns	9.68	1	9.68	11.73333333	0.026647909	7.708647421
Interaction	54.08	1	54.08	65.55151515	0.00126491	7.708647421
Within	3.3	4	0.825			
Total	70.98	7				

Summary

- The F test, as shown in Chapter 9, can be used to compare two sample variances to determine whether they are equal. It can also be used to compare three or more means. When three or more means are compared, the technique is called analysis of variance (ANOVA). The ANOVA technique uses two estimates of the population variance. The between-group variance is the variance of the sample means; the within-group variance is the overall variance of all the values. When there is no significant difference among the means, the two estimates will be approximately equal and the F test value will be close to 1. If there is a significant difference among the means, the between-group variance estimate will be larger than the within-group variance estimate and a significant test value will result. (12–1)

- If there is a significant difference among means, the researcher may wish to see where this difference lies. Several statistical tests can be used to compare the sample means after the ANOVA technique has been done. The most common are the Scheffé test and the Tukey test. When the sample sizes are the same, the Tukey test can be used. The Scheffé test is more general and can be used when the sample sizes are equal or not equal. (12–2)

- When there is one independent variable, the analysis of variance is called a one-way ANOVA. When there are two independent variables, the analysis of variance is called a two-way ANOVA. The two-way ANOVA enables the researcher to test the effects of two independent variables and a possible interaction effect on one dependent variable. If an interaction effect is found to be statistically significant, the researcher must investigate further to find out if the main effects can be examined. (12–3)

Important Terms

analysis of variance (ANOVA) 631

ANOVA summary table 634

between-group variance 631

disordinal interaction 653

factors 647

interaction effect 648

level 648

main effect 649

mean square 633

one-way ANOVA 647

ordinal interaction 653

Scheffé test 642

sum of squares between groups 633

sum of squares within groups 633

treatment groups 648

Tukey test 644

two-way ANOVA 647

within-group variance 631

Important Formulas

Formulas for the ANOVA test:

$$\overline{X}_{GM} = \frac{\Sigma X}{N}$$

$$F = \frac{s_B^2}{s_W^2}$$

where

$$s_B^2 = \frac{\Sigma n_i(\overline{X}_i - \overline{X}_{GM})^2}{k - 1} \qquad s_W^2 = \frac{\Sigma(n_i - 1)s_i^2}{\Sigma(n_i - 1)}$$

d.f.N. $= k - 1$ $\qquad N = n_1 + n_2 + \cdots + n_k$

d.f.D. $= N - k$ $\qquad k =$ **number of groups**

Formulas for the Scheffé test:

$$F_s = \frac{(\overline{X}_i - \overline{X}_j)^2}{s_W^2[(1/n_i) + (1/n_j)]} \qquad \text{and} \qquad F' = (k - 1)(\text{C.V.})$$

Formula for the Tukey test:

$$q = \frac{\overline{X}_i - \overline{X}_j}{\sqrt{s_W^2/n}}$$

d.f.N. $= k$ $\qquad$ and $\qquad$ **d.f.D.** $=$ **degrees of freedom for s_W^2**

Formulas for the two-way ANOVA:

$$\text{MS}_A = \frac{\text{SS}_A}{a - 1} \qquad\qquad F_A = \frac{\text{MS}_A}{\text{MS}_W} \qquad \begin{aligned}&\textbf{d.f.N.} = a - 1\\ &\textbf{d.f.D.} = ab(n - 1)\end{aligned}$$

$$\text{MS}_B = \frac{\text{SS}_B}{b - 1} \qquad\qquad F_B = \frac{\text{MS}_B}{\text{MS}_W} \qquad \begin{aligned}&\textbf{d.f.N.} = b - 1\\ &\textbf{d.f.D.} = ab(n - 1)\end{aligned}$$

$$\text{MS}_{A \times B} = \frac{\text{SS}_{A \times B}}{(a - 1)(b - 1)} \qquad F_{A \times B} = \frac{\text{MS}_{A \times B}}{\text{MS}_W} \qquad \begin{aligned}&\textbf{d.f.N.} = (a - 1)(b - 1)\\ &\textbf{d.f.D.} = ab(n - 1)\end{aligned}$$

$$\text{MS}_W = \frac{\text{SS}_W}{ab(n - 1)}$$

Review Exercises

If the null hypothesis is rejected in Exercises 1 through 7, use the Scheffé test when the sample sizes are unequal to test the differences between the means, and use the Tukey test when the sample sizes are equal. For these exercises, perform these steps.

a. State the hypotheses and identify the claim.

b. Find the critical value(s).

c. Compute the test value.

d. Make the decision.

e. Summarize the results, and explain where the differences in means are.

Use the traditional method of hypothesis testing unless otherwise specified.

 1. Lengths of Various Types of Bridges The data represent the lengths in feet of three types of bridges in the United States. At $\alpha = 0.01$, test the claim that there is no significant difference in the means of the lengths of the types of bridges. (12–1)(12–2)

Simple truss	Segmented concrete	Continuous plate
745	820	630
716	750	573
700	790	525
650	674	510
647	660	480
625	640	460
608	636	451
598	620	450
550	520	450
545	450	425
534	392	420
528	370	360

Source: *World Almanac and Book of Facts.*

2. Number of State Parks The numbers of state parks found in selected states in three different regions of the country are listed below. At $\alpha = 0.05$ can it be concluded that the average number of state parks differs by region? (12–1)(12–2)

South	West	New England
51	28	94
64	44	72
35	24	14
24	31	52
47	40	

Source: *Time Almanac.*

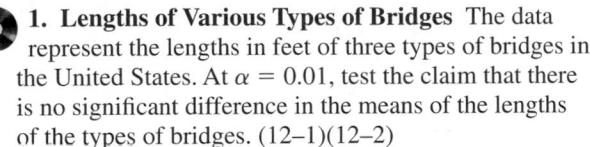 **3. Carbohydrates in Cereals** The number of carbohydrates per serving in randomly selected cereals from three manufacturers is shown. At the 0.05

level of significance, is there sufficient evidence to conclude a difference in the average number of carbohydrates? (12–1)(12–2)

Manufacturer 1	Manufacturer 2	Manufacturer 3
25	23	24
26	44	39
24	24	28
26	24	25
26	36	23
41	27	32
26	25	
43		

Source: *The Doctor's Pocket Calorie, Fat, and Carbohydrate Counter.*

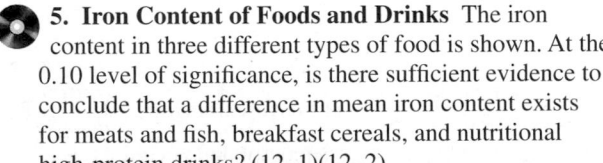 **4. Grams of Fat per Serving of Pizza** The number of grams of fat per serving for three different kinds of pizza from several manufacturers is listed below. At the 0.01 level of significance, is there sufficient evidence that a difference exists in mean fat content? (12–1)(12–2)

Cheese	Pepperoni	Supreme/Deluxe
18	20	16
11	17	27
19	15	17
20	18	17
16	23	12
21	23	27
16	21	20

Source: *The Doctor's Pocket Calorie, Fat, and Carbohydrate Counter.*

5. Iron Content of Foods and Drinks The iron content in three different types of food is shown. At the 0.10 level of significance, is there sufficient evidence to conclude that a difference in mean iron content exists for meats and fish, breakfast cereals, and nutritional high-protein drinks? (12–1)(12–2)

Meats and fish	Breakfast cereals	Nutritional drinks
3.4	8	3.6
2.5	2	3.6
5.5	1.5	4.5
5.3	3.8	5.5
2.5	3.8	2.7
1.3	6.8	3.6
2.7	1.5	6.3
	4.5	

Source: *The Doctor's Pocket Calorie, Fat, and Carbohydrate Counter.*

6. Temperatures in January The average January high temperatures (in degrees Fahrenheit) for selected tourist

Is Seeing Really Believing?—Revisited

To see if there were differences in the testimonies of the witnesses in the three age groups, the witnesses responded to 17 questions, 10 on direct examination and 7 on cross-examination. These were then scored for accuracy. An analysis of variance test with age as the independent variable was used to compare the total number of questions answered correctly by the groups. The results showed no significant differences among the age groups for the direct examination questions. However, there was a significant difference among the groups on the cross-examination questions. Further analysis showed the 8-year-olds were significantly less accurate under cross-examination compared to the other two groups. The 12-year-old and adult eyewitnesses did not differ in the accuracy of their cross-examination responses.

cities on different continents are listed below. Is there sufficient evidence to conclude a difference in mean temperatures for the three areas? Use the 0.05 level of significance. (12–1)(12–2)

Europe	Central and South America	Asia
41	87	89
38	75	35
36	66	83
56	84	67
50	75	48

Source: *Time Almanac.*

7. **School Incidents Involving Police Calls** A researcher wishes to see if there is a difference in the average number of times local police were called in school incidents. Samples of school districts were selected, and the numbers of incidents for a specific year were reported. At $\alpha = 0.05$, is there a difference in the means? If so, suggest a reason for the difference. (12–1)(12–2)

County A	County B	County C	County D
13	16	15	11
11	33	12	31
2	12	19	3
2	2		
2			

Source: U.S. Department of Education.

8. **Review Preparation for Statistics** A statistics instructor wanted to see if student participation in review preparation methods led to higher examination scores. Five students were randomly selected and placed in each test group for a three-week unit on statistical inference. Everyone took the same examination at the end of the unit, and the resulting scores are shown below. Is there sufficient evidence at $\alpha = 0.05$ to conclude an interaction between the two factors? Is there sufficient evidence to conclude a difference in mean scores based on formula delivery system? Is there sufficient evidence to conclude a difference in mean scores based on the review organization technique? (12–3)

	Formulas provided	Student-made formula cards
Student-led review	89, 76, 80, 90, 75	94, 86, 80, 79, 82
Instructor-led review	75, 80, 68, 65, 79	88, 78, 85, 65, 72

9. **Effects of Different Types of Diets** A medical researcher wishes to test the effects of two different diets and two different exercise programs on the glucose level in a person's blood. The glucose level is measured in milligrams per deciliter (mg/dl). Three subjects are randomly assigned to each group. Analyze the data shown here, using a two-way ANOVA with $\alpha = 0.05$. (12–3)

Exercise program	Diet A	B
I	62, 64, 66	58, 62, 53
II	65, 68, 72	83, 85, 91

ANOVA Summary Table for Exercise 9

Source	SS	d.f.	MS	F
Exercise	816.750			
Diet	102.083			
Interaction	444.083			
Within	108.000			
Total	1470.916			

Data Analysis

The Data Bank is found in Appendix D, or on the World Wide Web by following links from
www.mhhe.com/math/stat/bluman

1. From the Data Bank, select a random sample of subjects, and test the hypothesis that the mean cholesterol levels of the nonsmokers, less-than-one-pack-a-day smokers, and one-pack-plus smokers are equal. Use an ANOVA test. If the null hypothesis is rejected, conduct the Scheffé test to find where the difference is. Summarize the results.

2. Repeat Exercise 2 for the mean IQs of the various educational levels of the subjects.

3. Using the Data Bank, randomly select 12 subjects and randomly assign them to one of the four groups in the following classifications.

	Smoker	Nonsmoker
Male		
Female		

Use one of these variables—weight, cholesterol, or systolic pressure—as the dependent variable, and perform a two-way ANOVA on the data. Use a computer program to generate the ANOVA table.

Chapter Quiz

Determine whether each statement is true or false. If the statement is false, explain why.

1. In analysis of variance, the null hypothesis should be rejected only when there is a significant difference among all pairs of means. False

2. The F test does not use the concept of degrees of freedom. False

3. When the F test value is close to 1, the null hypothesis should be rejected. False

4. The Tukey test is generally more powerful than the Scheffé test for pairwise comparisons. True

Select the best answer.

5. Analysis of variance uses the _____ test.

   a. z c. χ^2
   b. t (d.) F

6. The null hypothesis in ANOVA is that all the means are _____.

   (a.) Equal   c. Variable
   b. Unequal   d. None of the above

7. When you conduct an F test, _____ estimates of the population variance are compared.

   (a.) Two   c. Any number of
   b. Three   d. No

8. If the null hypothesis is rejected in ANOVA, you can use the _____ test to see where the difference in the means is found.

   a. z or t (c.) Scheffé or Tukey
   b. F or χ^2 d. Any of the above

Complete the following statements with the best answer.

9. When three or more means are compared, you use the _____ technique. ANOVA

10. If the null hypothesis is rejected in ANOVA, the _____ test should be used when sample sizes are equal. Tukey

11. In a two-way ANOVA, you can test _____ main hypotheses and one interactive hypothesis. Two

For Exercises 12 through 16 use the traditional method of hypothesis testing unless otherwise specified.

 12. **Voters in Presidential Elections** In a recent Presidential election, a sample of the percentage of voters who voted is shown. At $\alpha = 0.05$, is there a difference in the mean percentage of voters who voted?

Northeast	Southeast	Northwest	Southwest
65.3	54.8	60.5	42.3
59.9	61.8	61.0	61.2
66.9	49.6	74.0	54.7
64.2	58.6	61.4	56.7

Source: Committee for the Study of the American Electorate.

13. **Ages of Late-Night TV Talk Show Viewers** A media researcher wanted to see if there was a difference in the ages of viewers of three late-night television talk shows. Three samples of viewers were selected, and the ages of the viewers are shown. At $\alpha = 0.01$, is there a difference in the means of the ages of the viewers? Why is the average age of a viewer important to a television show writer?

David Letterman	Jay Leno	Conan O'Brien
53	48	40
46	51	36
48	57	35
42	46	42
35	38	39

Source: Based on information from Nielsen Media Research.

14. Prices of Body Soap A consumer group desired to compare the mean price for 12-ounce bottles of liquid body soap from two nationwide brands and one store brand. Four different bottles of each were selected at a large discount drug store, and the prices are noted below. At the 0.05 level of significance is there sufficient evidence to conclude a difference in mean prices? If so, perform the appropriate test to find out where.

Brand X	Brand Y	Store brand
5.99	8.99	4.99
6.99	7.99	3.99
8.59	6.29	5.29
6.49	7.29	4.49

15. Air Pollution A lot of different factors contribute to air pollution. One particular factor, particulate matter, was measured for prominent cities of three continents. Particulate matter includes smoke, soot, dust, and liquid droplets from combustion such that the particle is less than 10 microns in diameter and thus capable of reaching deep into the respiratory system. The measurements are listed below. At the 0.05 level of significance is there sufficient evidence to conclude a difference in means? If so, perform the appropriate test to find out where.

Asia	Europe	Africa
79	34	33
104	35	16
40	30	43
73	43	

Source: *World Almanac.*

16. Alumni Gift Solicitation Several students volunteered for an alumni phone-a-thon to solicit

alumni gifts. The number of calls made by randomly selected students from each class is listed. At $\alpha = 0.05$, is there sufficient evidence to conclude a difference in means?

Freshmen	Sophomores	Juniors	Seniors
25	17	20	20
29	25	24	25
32	20	25	26
15	26	30	32
18	30	15	19
26	28	18	20
35			

17. Diets and Exercise Programs A researcher conducted a study of two different diets and two different exercise programs. Three randomly selected subjects were assigned to each group for one month. The values indicate the amount of weight each lost.

Exercise program \ Diet	A	B
I	5, 6, 4	8, 10, 15
II	3, 4, 8	12, 16, 11

Answer the following questions for the information in the printout shown below.

a. What procedure is being used? Two-way ANOVA
b. What are the names of the two variables? Diet and exercise program
c. How many levels does each variable contain? 2
d. What are the hypotheses for the study?
e. What are the *F* values for the hypotheses? State which are significant, using the *P*-values.
f. Based on the answers to part *e*, which hypotheses can be rejected? Reject the null hypothesis for the diets.

Computer Printout for Problem 17

```
Datafile: NONAME.SST    Procedure: Two-way ANOVA

TABLE OF MEANS:
                          DIET
                   A .....      B .....       Row Mean
EX PROG I .....    5.000        11.000        8.000
       II .....    5.000        13.000        9.000
       Col Mean    5.000        12.000
       Tot Mean    8.500

  SOURCE TABLE:
       Source      df    Sums of Squares   Mean Square    F Ratio    p-value
         DIET      1          147.000        147.000       21.000     0.00180
      EX PROG      1            3.000          3.000        0.429     0.53106
  DIET X EX P      1            3.000          3.000        0.429     0.53106
       Within      8           56.000          7.000
        Total     11          209.000
```

Critical Thinking Challenges

Adult Children of Alcoholics

Shown here are the abstract and two tables from a research study entitled "Adult Children of Alcoholics: Are They at Greater Risk for Negative Health Behaviors?" by Arlene E. Hall. Based on the abstract and the tables, answer these questions.

1. What was the purpose of the study?

2. How many groups were used in the study?

3. By what means were the data collected?

4. What was the sample size?

5. What type of sampling method was used?

6. How might the population be defined?

7. What may have been the hypothesis for the ANOVA part of the study?

8. Why was the one-way ANOVA procedure used, as opposed to another test, such as the t test?

9. What part of the ANOVA table did the conclusion "ACOAs had significantly lower wellness scores (WS) than non-ACOAs" come from?

10. What level of significance was used?

11. In the following excerpts from the article, the researcher states that

> . . . using the Tukey-HSD procedure revealed a significant difference between ACOAs and non-ACOAs, $p = 0.05$, but no significant difference was found between ACOAs and Unsures or between non-ACOAs and Unsures.

Using Tables 12–8 and 12–9 and the means, explain why the Tukey test would have enabled the researcher to draw this conclusion.

Abstract The purpose of the study was to examine and compare the health behaviors of adult children of alcoholics (ACOAs) and their non-ACOA peers within a university population. Subjects were 980 undergraduate students from a major university in the East. Three groups (ACOA, non-ACOA, and Unsure) were identified from subjects' responses to three direct questions regarding parental drinking behaviors. A questionnaire was used to collect data for the study. Included were questions related to demographics, parental drinking behaviors, and the College Wellness Check (WS), a health risk appraisal designed especially for college students (Dewey & Cabral, 1986). Analysis of variance procedures revealed that ACOAs had significantly lower wellness scores (WS) than non-ACOAs. Chi-square analyses of the individual variables revealed that ACOAs and non-ACOAs were significantly different on 15 of the 50 variables of the WS. A discriminant analysis procedure revealed the similarities between Unsure subjects and ACOA subjects. The results provide valuable information regarding ACOAs in a nonclinical setting and contribute to our understanding of the influences related to their health risk behaviors.

Table 12–8	Means and Standard Deviations for the Wellness Scores (WS) Group by ($N = 945$)		
Group	**N**	**$\overline{X}$**	**S.D.**
ACOAs	143	69.0	13.6
Non-ACOAs	746	73.2	14.5
Unsure	56	70.1	14.0
Total	945	212.3	42.1

Table 12–9	ANOVA of Group Means for the Wellness Scores (WS)			
Source	**d.f.**	**SS**	**MS**	**F**
Between groups	2	2,403.5	1,201.7	5.9*
Within groups	942	193,237.4	205.1	
Total	944	195,640.8		

*$p < 0.01$

Source: Arlene E. Hall, "Adult Children of Alcoholics: Are They at Greater Risk for Negative Health Behaviors?" *Journal of Health Education* 12, no. 4, pp. 232–238.

Data Projects

Use a significance level of 0.05 for all tests below.

1. **Business and Finance** Select 10 stocks at random from the Dow Jones Industrials, the NASDAQ, and the S&P 500. For each, note the gain or loss in the last quarter. Use analysis of variance to test the claim that stocks from all three groups have had equal performance.

2. **Sports and Leisure** Use total earnings data for movies that were released in the previous year. Sort them by rating (G, PG, PG13, and R). Is the mean revenue for movies the same regardless of rating?

3. **Technology** Use the data collected in data project 3 of Chapter 2 regarding song lengths. Consider only three genres. For example, use rock, alternative, and hip hop/rap. Conduct an analysis of variance to determine if the mean song lengths for the genres are the same.

4. **Health and Wellness** Select 10 cereals from each of the following categories: cereal targeted at children, cereal targeted at dieters, and cereal that fits neither of the previous categories. For each cereal note its calories per cup (this may require some computation since serving sizes vary for cereals). Use analysis of variance to test the claim that the calorie content of these different types of cereals is the same.

5. **Politics and Economics** Conduct an anonymous survey to obtain your data. Ask the participants to identify which of the following categories describes them best: registered Republican, Democrat, Independent, or not registered to vote. Also ask them to give their age. Use an analysis of variance to determine whether there is a difference in mean age between the different political designations.

6. **Your Class** Split the class into four groups, those whose favorite type of music is rock, whose favorite is country, whose favorite is rap or hip hop, and whose favorite is another type of music. Make a list of the ages of students for each of the four groups. Use analysis of variance to test the claim that the means for all four groups are equal.

Answers to Applying the Concepts

Section 12–1   Colors That Make You Smarter

1. The ANOVA produces a test statistic of $F = 3.06$, with a P-value of 0.059. We would fail to reject the null hypothesis and find that there is not enough evidence to conclude that the color of a person's clothing is related to people's perceptions of how intelligent the person looks.

2. Answers will vary. One possible answer is that the purpose of the study was to determine if the color of a person's clothing is related to people's perceptions of how intelligent the person looks.

3. We would have to perform three separate t tests, which would inflate the error rate.

Section 12–2   Colors That Make You Smarter

1. Tukey's pairwise comparisons show no significant difference in the three pairwise comparisons of the means.

2. This agrees with the nonsignificant results of the general ANOVA test conducted in Applying the Concepts 12–1.

3. The t tests should not be used since they would inflate the error rate.

4. We prefer the Tukey test over the Scheffé test when the samples are all the same size.

Section 12–3   Automobile Sales Techniques

There is no significant difference between levels 1 and 2 of experience. Level 3 and level 4 salespersons did significantly better than those at levels 1 and 2, with level 4 showing the best results, on average. If type of presentation is taken into consideration, the interaction plot shows a significant difference. The best combination seems to be level 4 experience with presentation style 1.

Hypothesis-Testing Summary 2*

7. Test of the significance of the correlation coefficient.

Example: $H_0: \rho = 0$

Use a t test:

$$t = r\sqrt{\frac{n-2}{1-r^2}} \quad \text{with d.f.} = n - 2$$

8. Formula for the F test for the multiple correlation coefficient.

Example: $H_0: \rho = 0$

$$F = \frac{R^2/k}{(1-R^2)/(n-k-1)}$$

$$\text{d.f.N.} = n - k \qquad \text{d.f.D.} = n - k - 1$$

9. Comparison of a sample distribution with a specific population.

Example: H_0: There is no difference between the two distributions.

Use the chi-square goodness-of-fit test:

$$\chi^2 = \sum \frac{(O-E)^2}{E}$$

d.f. = no. of categories − 1

10. Comparison of the independence of two variables.

Example: H_0: Variable A is independent of variable B.

Use the chi-square independence test:

$$\chi^2 = \sum \frac{(O-E)^2}{E}$$

$$\text{d.f.} = (R-1)(C-1)$$

11. Test for homogeneity of proportions.

Example: $H_0: p_1 = p_2 = p_3$

Use the chi-square test:

$$\chi^2 = \sum \frac{(O-E)^2}{E}$$

$$\text{d.f.} = (R-1)(C-1)$$

12. Comparison of three or more sample means.

Example: $H_0: \mu_1 = \mu_2 = \mu_3$

Use the analysis of variance test:

$$F = \frac{s_B^2}{s_W^2}$$

where

$$s_B^2 = \frac{\sum n_i(\bar{X}_i - \bar{X}_{GM})^2}{k-1}$$

$$s_W^2 = \frac{\sum(n_i - 1)s_i^2}{\sum(n_i - 1)}$$

d.f.N. = $k - 1$ $N = n_1 + n_2 + \cdots + n_k$

d.f.D. = $N - k$ k = number of groups

13. Test when the F value for the ANOVA is significant. Use the Scheffé test to find what pairs of means are significantly different.

$$F_s = \frac{(\bar{X}_i - \bar{X}_j)^2}{s_W^2[(1/n_i) + (1/n_j)]}$$

$$F' = (k-1)(\text{C.V.})$$

Use the Tukey test to find which pairs of means are significantly different.

$$q = \frac{\bar{X}_i - X_j}{\sqrt{s_W^2/n}} \qquad \begin{array}{l} \text{d.f.N.} = k \\ \text{d.f.D.} = \text{degrees of freedom for } s_W^2 \end{array}$$

14. Test for the two-way ANOVA.

Example:

H_0: There is no significant difference for the main effects.

H_1: There is no significant difference for the interaction effect.

$$MS_A = \frac{SS_A}{a-1}$$

$$MS_B = \frac{SS_B}{b-1}$$

$$MS_{A\times B} = \frac{SS_{A\times B}}{(a-1)(b-1)}$$

$$MS_W = \frac{SS_W}{ab(n-1)}$$

$$F_A = \frac{MS_A}{MS_W} \qquad \begin{array}{l} \text{d.f.N.} = a - 1 \\ \text{d.f.D.} = ab(n-1) \end{array}$$

$$F_B = \frac{MS_B}{MS_W} \qquad \begin{array}{l} \text{d.f.N.} = (b-1) \\ \text{d.f.D.} = ab(n-1) \end{array}$$

$$F_{A\times B} = \frac{MS_{A\times B}}{MS_W} \qquad \begin{array}{l} \text{d.f.N.} = (a-1)(b-1) \\ \text{d.f.D.} = ab(n-1) \end{array}$$

*This summary is a continuation of Hypothesis-Testing Summary 1, at the end of Chapter 9.

CHAPTER 13

Nonparametric Statistics

Objectives

After completing this chapter, you should be able to

1. State the advantages and disadvantages of nonparametric methods.

2. Test hypotheses, using the sign test.

3. Test hypotheses, using the Wilcoxon rank sum test.

4. Test hypotheses, using the signed-rank test.

5. Test hypotheses, using the Kruskal-Wallis test.

6. Compute the Spearman rank correlation coefficient.

7. Test hypotheses, using the runs test.

Outline

Statistics Today

Too Much or Too Little?

Suppose a manufacturer of ketchup wishes to check the bottling machines to see if they are functioning properly. That is, are they dispensing the right amount of ketchup per bottle? A 40-ounce bottle is currently used. Because of the natural variation in the manufacturing process, the amount of ketchup in a bottle will not always be exactly 40 ounces. Some bottles will contain less than 40 ounces, and others will contain more than 40 ounces. To see if the variation is due to chance or to a malfunction in the manufacturing process, a runs test can be used. The runs test is a nonparametric statistical technique. See Statistics Today—Revisited at the end of this chapter. This chapter explains such techniques, which can be used to help the manufacturer determine the answer to the question.

Introduction

Statistical tests, such as the z, t, and F tests, are called parametric tests. **Parametric tests** are statistical tests for population parameters such as means, variances, and proportions that involve assumptions about the populations from which the samples were selected. One assumption is that these populations are normally distributed. But what if the population in a particular hypothesis-testing situation is *not* normally distributed? Statisticians have developed a branch of statistics known as **nonparametric statistics** or **distribution-free statistics** to use when the population from which the samples are selected is not normally distributed. Nonparametric statistics can also be used to test hypotheses that do not involve specific population parameters, such as μ, σ, or p.

For example, a sportswriter may wish to know whether there is a relationship between the rankings of two judges on the diving abilities of 10 Olympic swimmers. In another situation, a sociologist may wish to determine whether men and women enroll at random for a specific drug rehabilitation program. The statistical tests used in these situations are nonparametric or distribution-free tests. The term *nonparametric* is used for both situations.

The nonparametric tests explained in this chapter are the sign test, the Wilcoxon rank sum test, the Wilcoxon signed-rank test, the Kruskal-Wallis test, and the runs test.

In addition, the Spearman rank correlation coefficient, a statistic for determining the relationship between ranks, is explained.

13–1

Advantages and Disadvantages of Nonparametric Methods

As stated previously, nonparametric tests and statistics can be used in place of their parametric counterparts (z, t, and F) when the assumption of normality cannot be met. However, you should not assume that these statistics are a better alternative than the parametric statistics. There are both advantages and disadvantages in the use of nonparametric methods.

Advantages

Objective 1

State the advantages and disadvantages of nonparametric methods.

There are five advantages that nonparametric methods have over parametric methods:

1. They can be used to test population parameters when the variable is not normally distributed.
2. They can be used when the data are nominal or ordinal.
3. They can be used to test hypotheses that do not involve population parameters.
4. In some cases, the computations are easier than those for the parametric counterparts.
5. They are easy to understand.

Disadvantages

There are three disadvantages of nonparametric methods:

1. They are *less sensitive* than their parametric counterparts when the assumptions of the parametric methods are met. Therefore, larger differences are needed before the null hypothesis can be rejected.
2. They tend to use *less information* than the parametric tests. For example, the sign test requires the researcher to determine only whether the data values are above or below the median, not how much above or below the median each value is.
3. They are *less efficient* than their parametric counterparts when the assumptions of the parametric methods are met. That is, larger sample sizes are needed to overcome the loss of information. For example, the nonparametric sign test is about 60% as efficient as its parametric counterpart, the z test. Thus, a sample size of 100 is needed for use of the sign test, compared with a sample size of 60 for use of the z test to obtain the same results.

Since there are both advantages and disadvantages to the nonparametric methods, the researcher should use caution in selecting these methods. If the parametric assumptions can be met, the parametric methods are preferred. However, when parametric assumptions cannot be met, the nonparametric methods are a valuable tool for analyzing the data.

The basic assumption for nonparametric statistics is that the sample or samples are randomly obtained. When two or more samples are used, they must be independent of each other unless otherwise stated.

Ranking

Many nonparametric tests involve the **ranking** of data, that is, the positioning of a data value in a data array according to some rating scale. Ranking is an ordinal variable. For example, suppose a judge decides to rate five speakers on an ascending scale of 1 to 10, with 1 being the best and 10 being the worst, for categories such as voice, gestures, logical presentation, and platform personality. The ratings are shown in the chart.

Speaker	A	B	C	D	E
Rating	8	6	10	3	1

Interesting Fact

Older men have the biggest ears. James Heathcote, M.D., says, "On average, our ears seem to grow 0.22 millimeter a year. This is roughly a centimeter during the course of 50 years."

The rankings are shown next.

Speaker	E	D	B	A	C
Rating	1	3	6	8	10
Ranking	1	2	3	4	5

Since speaker E received the lowest score, 1 point, he or she is ranked first. Speaker D received the next-lower score, 3 points; he or she is ranked second; and so on.

What happens if two or more speakers receive the same number of points? Suppose the judge awards points as follows:

Speaker	A	B	C	D	E
Rating	8	6	10	6	3

The speakers are then ranked as follows:

Speaker	E	D	B	A	C
Rating	3	6	6	8	10
Ranking	1	Tie for 2nd and 3rd		4	5

When there is a tie for two or more places, the average of the ranks must be used. In this case, each would be ranked as

$$\frac{2 + 3}{2} = \frac{5}{2} = 2.5$$

Hence, the rankings are as follows:

Speaker	E	D	B	A	C
Rating	3	6	6	8	10
Ranking	1	2.5	2.5	4	5

Many times, the data are already ranked, so no additional computations must be done. For example, if the judge does not have to award points but can simply select the speakers who are best, second-best, third-best, and so on, then these ranks can be used directly.

P-values can also be found for nonparametric statistical tests, and the *P*-value method can be used to test hypotheses that use nonparametric tests. For this chapter, the *P*-value method will be limited to some of the nonparametric tests that use the standard normal distribution or the chi-square distribution.

Applying the Concepts **13–1**

Ranking Data

The following table lists the percentages of patients who experienced side effects from a drug used to lower a person's cholesterol level.

Side effect	Percent
Chest pain	4.0
Rash	4.0
Nausea	7.0
Heartburn	5.4
Fatigue	3.8
Headache	7.3
Dizziness	10.0
Chills	7.0
Cough	2.6

Rank each value in the table.

See page 717 for the answer.

1. What is meant by *nonparametric statistics?*

2. When should nonparametric statistics be used?

3. List the advantages and disadvantages of nonparametric statistics.

For Exercises 4 through 10, rank each set of data.

4. 3, 8, 6, 1, 4, 10, 7

5. 22, 66, 32, 43, 65, 43, 71, 34

6. 83, 460, 582, 177, 241

7. 19.4, 21.8, 3.2, 23.1, 5.9, 10.3, 11.1

8. 10.9, 20.2, 43.9, 9.5, 17.6, 5.6, 32.6, 0.85, 17.6

9. 28, 50, 52, 11, 71, 36, 47, 88, 41, 50, 71, 50

10. 90.6, 47.0, 82.2, 9.27, 327.0, 52.9, 18.0, 145.0, 34.5, 9.54

13–2

The Sign Test

Single-Sample Sign Test

Objective 2

Test hypotheses, using the sign test.

The simplest nonparametric test, the **sign test** for single samples, is used to test the value of a median for a specific sample. When using the sign test, the researcher hypothesizes the specific value for the median of a population; then he or she selects a sample of data and compares each value with the conjectured median. If the data value is above the conjectured median, it is assigned a plus sign. If it is below the conjectured median, it is assigned a minus sign. And if it is exactly the same as the conjectured median, it is assigned a 0. Then the numbers of plus and minus signs are compared. If the null hypothesis is true, the number of plus signs should be approximately equal to the number of minus signs. If the null hypothesis is not true, there will be a disproportionate number of plus or minus signs.

Test Value for the Sign Test

The test value is the smaller number of plus or minus signs.

For example, if there are 8 positive signs and 3 negative signs, the test value is 3. When the sample size is 25 or less, Table J in Appendix C is used to determine the critical value. For a specific α, if the test value is less than or equal to the critical value obtained from the table, the null hypothesis should be rejected. The values in Table J are obtained from the binomial distribution. The derivation is omitted here.

Example 13–1

Snow Cone Sales

A convenience store owner hypothesizes that the median number of snow cones she sells per day is 40. A random sample of 20 days yields the following data for the number of snow cones sold each day.

18	43	40	16	22
30	29	32	37	36
39	34	39	45	28
36	40	34	39	52

At $\alpha = 0.05$, test the owner's hypothesis.

Solution

Step 1   State the hypotheses and identify the claim.

H_0: median $= 40$ (claim)   and H_1: median $\neq 40$

Step 2   Find the critical value. Compare each value of the data with the median. If the value is greater than the median, replace the value with a plus sign. If it is less than the median, replace it with a minus sign. And if it is equal to the median, replace it with a 0. The completed table follows.

$-$	$+$	0	$-$	$-$
$-$	$-$	$-$	$-$	$-$
$-$	$-$	$-$	$+$	$-$
$-$	0	$-$	$-$	$+$

Refer to Table J in Appendix C, using $n = 18$ (the total number of plus and minus signs; omit the zeros) and $\alpha = 0.05$ for a two-tailed test; the critical value is 4. See Figure 13–1.

Figure 13–1

Finding the Critical Value in Table J for Example 13–1

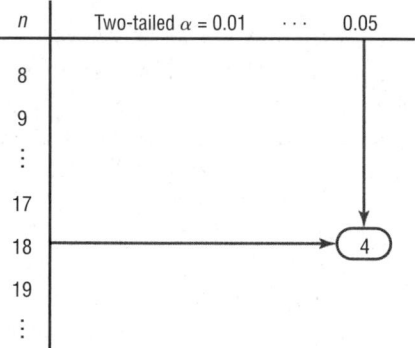

Step 3   Compute the test value. Count the number of plus and minus signs obtained in step 2, and use the smaller value as the test value. Since there are 3 plus signs and 15 minus signs, 3 is the test value.

Step 4   Make the decision. Compare the test value 3 with the critical value 4. If the test value is less than or equal to the critical value, the null hypothesis is rejected. In this case, the null hypothesis is rejected since $3 < 4$.

Step 5   Summarize the results. There is enough evidence to reject the claim that the median number of snow cones sold per day is 40.

When the sample size is 26 or more, the normal approximation can be used to find the test value. The formula is given. The critical value is found in Table E in Appendix C.

Formula for the z Test Value in the Sign Test When $n \geq 26$

$$z = \frac{(X + 0.5) - (n/2)}{\sqrt{n}/2}$$

where

$X =$ smaller number of $+$ or $-$ signs
$n =$ sample size

Example 13–2	Age of Foreign-Born Residents

Based on information from the U.S. Census Bureau, the median age of foreign-born U.S. residents is 36.4 years. A researcher selects a sample of 50 foreign-born U.S. residents in his area and finds that 21 are older than 36.4 years. At $\alpha = 0.05$, test the claim that the median age of the residents is at least 36.4 years.

Solution

Step 1 State the hypotheses and identify the claim.

$$H_0: \text{MD} = 36.4 \text{ (claim)} \qquad \text{and} \qquad H_1: \text{MD} < 36.4$$

Step 2 Find the critical value. Since $\alpha = 0.05$ and $n = 50$, and since this is a left-tailed test, the critical value is -1.65, obtained from Table E.

Step 3 Compute the test value.

$$z = \frac{(X + 0.5) - (n/2)}{\sqrt{n}/2} = \frac{(21 + 0.5) - (50/2)}{\sqrt{50}/2} = \frac{-3.5}{3.5355} = -0.99$$

Step 4 Make the decision. Since the test value of -0.99 is greater than -1.65, the decision is to not reject the null hypothesis.

Step 5 Summarize the results. There is not enough evidence to reject the claim that the median age of the residents is at least 36.4.

In Example 13–2, the sample size was 50, and 21 residents are older than 36.4. So $50 - 21$, or 29, residents are not older than 36.4. The value of X corresponds to the smaller of the two numbers 21 and 29. In this case, $X = 21$ is used in the formula; since 21 is the smaller of the two numbers, the value of X is 21.

Suppose a researcher hypothesized that the median age of houses in a certain municipality was 40 years. In a random sample of 100 houses, 68 were older than 40 years. Then the value used for X in the formula would be $100 - 68$, or 32, since it is the smaller of the two numbers 68 and 32. When 40 is subtracted from the age of a house older than 40 years, the answer is positive. When 40 is subtracted from the age of a house that is less than 40 years old, the result is negative. There would be 68 positive signs and 32 negative signs (assuming that no house was exactly 40 years old). Hence, 32 would be used for X, since it is the smaller of the two values.

Paired-Sample Sign Test

The sign test can also be used to test sample means in a comparison of two dependent samples, such as a before-and-after test. Recall that when dependent samples are taken from normally distributed populations, the t test is used (Section 9–4). When the condition of normality cannot be met, the nonparametric sign test can be used, as shown in Example 13–3.

Example 13–3	Ear Infections in Swimmers

A medical researcher believed the number of ear infections in swimmers can be reduced if the swimmers use earplugs. A sample of 10 people was selected, and the number of infections for a four-month period was recorded. During the first two months, the swimmers did not use the earplugs; during the second two months, they did. At the beginning of the second two-month period, each swimmer was examined to make sure that no infections were present. The data are shown here. At $\alpha = 0.05$, can the researcher conclude that using earplugs reduced the number of ear infections?

Number of ear infections

Swimmer	Before, X_B	After, X_A
A	3	2
B	0	1
C	5	4
D	4	0
E	2	1
F	4	3
G	3	1
H	5	3
I	2	2
J	1	3

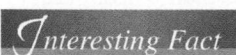

Room temperature is generally considered 72° since at this temperature a clothed person's body heat is allowed to escape at a rate that is most comfortable to him or her.

Solution

Step 1 State the hypotheses and identify the claim.

H_0: The number of ear infections will not be reduced.

H_1: The number of ear infections will be reduced (claim).

Step 2 Find the critical value. Subtract the after values X_A from the before values X_B and indicate the difference by a positive or negative sign or 0, according to the value, as shown in the table.

Swimmer	Before, X_B	After, X_A	Sign of difference
A	3	2	+
B	0	1	−
C	5	4	+
D	4	0	+
E	2	1	+
F	4	3	+
G	3	1	+
H	5	3	+
I	2	2	0
J	1	3	−

From Table J, with $n = 9$ (the total number of positive and negative signs; the 0 is not counted) and $\alpha = 0.05$ (one-tailed), at most 1 negative sign is needed to reject the null hypothesis because 1 is the smallest entry in the $\alpha = 0.05$ column of Table J.

Step 3 Compute the test value. Count the number of positive and negative signs found in step 2, and use the smaller value as the test value. There are 2 negative signs, so the test value is 2.

Step 4 Make the decision. There are 2 negative signs. The decision is to not reject the null hypothesis. The reason is that with $n = 9$, C.V. = 1 and $1 < 2$.

Step 5 Summarize the results. There is not enough evidence to support the claim that the use of earplugs reduced the number of ear infections.

When conducting a one-tailed sign test, the researcher must scrutinize the data to determine whether they support the null hypothesis. If the data support the null hypothesis, there is no need to conduct the test. In Example 13–3, the null hypothesis states that the number of ear infections will not be reduced. The data would support the null hypothesis if there were more negative signs than positive signs. The reason is that the before values X_B in most cases would be smaller than the after values X_A, and the $X_B - X_A$ values would be negative more often than positive. This would indicate that there is not enough evidence to reject the null hypothesis. The researcher would stop here, since there is no need to continue the procedure.

On the other hand, if the number of ear infections were reduced, the X_B values, for the most part, would be larger than the X_A values, and the $X_B - X_A$ values would most often be positive, as in Example 13–3. Hence, the researcher would continue the procedure. A word of caution is in order, and a little reasoning is required.

When the sample size is 26 or more, the normal approximation can be used in the same manner as in Example 13–2. The steps for conducting the sign test for single or paired samples are given in the Procedure Table.

Procedure Table

Sign Test for Single and Paired Samples

Step 1 State the hypotheses and identify the claim.

Step 2 Find the critical value(s). For the single-sample test, compare each value with the conjectured median. If the value is larger than the conjectured median, replace it with a positive sign. If it is smaller than the conjectured median, replace it with a negative sign.

For the paired-sample sign test, subtract the after values from the before values, and indicate the difference with a positive or negative sign or 0, according to the value. Use Table J and n = total number of positive and negative signs.

Check the data to see whether they support the null hypothesis. If they do, do not reject the null hypothesis. If not, continue with step 3.

Step 3 Compute the test value. Count the numbers of positive and negative signs found in step 2, and use the smaller value as the test value.

Step 4 Make the decision. Compare the test value with the critical value in Table J. If the test value is less than or equal to the critical value, reject the null hypothesis.

Step 5 Summarize the results.

Note: If the sample size n is 26 or more, use Table E and the following formula for the test value:

$$z = \frac{(X + 0.5) - (n/2)}{\sqrt{n}/2}$$

where

X = smaller number of + or − signs

n = sample size

Applying the Concepts **13–2**

Clean Air

An environmentalist suggests that the median of the number of days per month that a large city failed to meet the EPA acceptable standards for clean air is 11 days per month. A random sample of 20 months shows the number of days per month that the air quality was below the EPA's standards.

15	14	1	9	0	3	3	1	10	8
6	16	21	22	3	19	16	5	23	13

1. What is the claim?

2. What test would you use to test the claim? Why?

3. What would the hypotheses be?

4. Select a value for α and find the corresponding critical value.

5. What is the test value?

6. What is your decision?

7. Summarize the results.

8. Could a parametric test be used?

See page 717 for the answers.

Exercises 13–2

1. Why is the sign test the simplest nonparametric test to use? The sign test uses only positive or negative signs.

2. What population parameter can be tested with the sign test? The median

3. In the sign test, what is used as the test value when $n < 26$? The smaller number of positive or negative signs

4. When $n \geq 26$, what is used in place of Table J for the sign test? The normal approximation

For Exercises 5 through 20, perform these steps.

  a. State the hypotheses and identify the claim.
  b. Find the critical value(s).
  c. Compute the test value.
  d. Make the decision.
  e. Summarize the results.

Use the traditional method of hypothesis testing unless otherwise specified.

5. **Ages When Married** The median age at first marriage for men in the United States in 2008 was 27.6 years. Alumni officers at a large university contacted recent newlyweds to see if their median age was different. Their ages (in years) at marriage are listed below. At $\alpha = 0.05$ can it be concluded that the median age for these alumni is different?

31.8	39.9	34.1	22.9
29.2	33.9	34.0	36.9
33.8	36.2	26.1	35.1
23.1	25.2	32.6	26.3

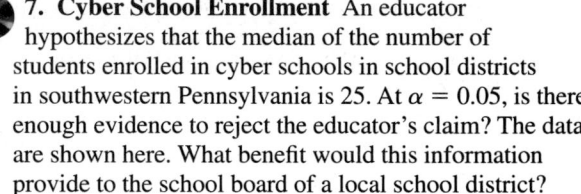

6. **Game Attendance** An athletic director suggests the median number for the paid attendance at 20 local football games is 3000. The data for a sample are shown. At $\alpha = 0.05$, is there enough evidence to reject the claim? If you were printing the programs for the games, would you use this figure as a guide?

6210	3150	2700	3012	4875
3540	6127	2581	2642	2573
2792	2800	2500	3700	6030
5437	2758	3490	2851	2720

Source: *Pittsburgh Post Gazette.*

7. **Cyber School Enrollment** An educator hypothesizes that the median of the number of students enrolled in cyber schools in school districts in southwestern Pennsylvania is 25. At $\alpha = 0.05$, is there enough evidence to reject the educator's claim? The data are shown here. What benefit would this information provide to the school board of a local school district?

12	41	26	14	4
38	27	27	9	11
17	11	66	5	14
8	35	16	25	17

Source: *Pittsburgh Tribune-Review.*

8. **Weekly Earnings of Women** According to the Women's Bureau of the U.S. Department of Labor, the occupation with the highest median weekly earnings among women is pharmacist with median weekly earnings of $1603. Based on the weekly earnings listed below from a sample of female pharmacists, can it be concluded that the median is less than $1603? Use $\alpha = 0.05$.

1550	1355	1777
1430	1570	1701
2465	1655	1484
1429	1829	1812
1217	1501	1449

9. **Natural Gas Costs** For a specific year, the median price of natural gas was $10.86 per 1000 cubic feet. A researcher wishes to see if there is enough evidence to reject the claim. Out of 42 households, 18 paid less than $10.86 per 1000 cubic feet for natural gas. Test the claim at $\alpha = 0.05$. How could a prospective home buyer use this information?

Source: Based on information from the Energy Information Administration.

10. Family Income The median U.S. family income is believed to be $63,211. In a survey of families in a particular neighborhood, it was found that out of 40 families surveyed, 10 had incomes below $63,211. At the 0.05 level of significance is there sufficient evidence to conclude that the median income is not $63,211?

11. Number of Faculty for Proprietary Schools An educational researcher believes that the median number of faculty for proprietary (for-profit) colleges and universities is 150. The data provided list the number of faculty at a selected number of proprietary colleges and universities. At the 0.05 level of significance, is there sufficient evidence to reject his claim?

372	111	165	95	191	83	136	149	37	119
142	136	137	171	122	133	133	342	126	64
61	100	225	127	92	140	140	75	108	96
138	318	179	243	109					

Source: *World Almanac.*

12. Television Viewers A researcher read that the median age for viewers of the Carson Daly show is 39. To test the claim, 75 viewers were surveyed, and 27 were under the age of 39. At $\alpha = 0.02$ test the claim. Give one reason why an advertiser might like to know the results of this study.

Source: Nielsen Media Research.

13. Students' Opinions on Lengthening the School Year One hundred students are asked if they favor increasing the school year by 20 days. The responses are 62 no, 36 yes, and 2 undecided. At $\alpha = 0.10$, test the hypothesis that 50% of the students are against extending the school year. Use the *P*-value method.

14. Deaths due to Severe Weather A meteorologist suggests that the median number of deaths per year from tornadoes in the United States is 60. The number of deaths for a sample of 11 years is shown. At $\alpha = 0.05$ is there enough evidence to reject the claim? If you took proper safety precautions during a tornado, would you feel relatively safe?

| 53 | 39 | 39 | 67 | 69 | 40 |
| 25 | 33 | 30 | 130 | 94 | |

Source: NOAA.

15. Diet Medication and Weight A study was conducted to see whether a certain diet medication had an effect on the weights (in pounds) of eight women. Their weights were taken before and six weeks after daily administration of the medication. The data are shown here. At $\alpha = 0.05$, can you conclude that the medication had an effect (increase or decrease) on the weights of the women?

Subject	A	B	C	D	E	F	G	H
Weight before	187	163	201	158	139	143	198	154
Weight after	178	162	188	156	133	150	175	150

16. Exam Scores A statistics professor wants to investigate the relationship between a student's midterm examination score and the score on the final. Eight students were selected, and their scores on the two examinations are noted below. At the 0.10 level of significance, is there sufficient evidence to conclude that there is a difference in scores?

Student	1	2	3	4	5	6	7	8
Midterm	75	92	68	85	65	80	75	80
Final	82	90	79	95	70	83	72	79

17. Increasing Supervisory Skills A large corporation sent several of its prospective supervisors to a two-day seminar in identifying and increasing supervisory skills. Participants were given a pretest at the start of the seminar and a posttest at the conclusion. Their scores are listed below. At $\alpha = 0.05$ can it be concluded that the training program was effective?

Employee	1	2	3	4	5	6	7	8
Pretest	70	65	73	72	80	77	69	68
Posttest	68	72	75	70	83	82	72	75

18. Effects of a Pill on Appetite A researcher wishes to test the effects of a pill on a person's appetite. Twelve subjects are allowed to eat a meal of their choice, and their caloric intake is measured. The next day, the same subjects take the pill and eat a meal of their choice. The caloric intake of the second meal is measured. The data are shown here. At $\alpha = 0.02$, can the researcher conclude that the pill had an effect on a person's appetite?

Subject	1	2	3	4	5	6	7
Meal 1	856	732	900	1321	843	642	738
Meal 2	843	721	872	1341	805	531	740

Subject	8	9	10	11	12
Meal 1	1005	888	756	911	998
Meal 2	900	805	695	878	914

19. Television Viewers A researcher wishes to determine if the number of viewers for 10 returning television shows has not changed since last year. The data are given in millions of viewers. At $\alpha = 0.01$, test the claim that the number of viewers has not changed. Depending on your answer, would a television executive plan to air these programs for another year?

Show	1	2	3	4	5	6
Last year	28.9	26.4	20.8	25.0	21.0	19.2
This year	26.6	20.5	20.2	19.1	18.9	17.8

Show	7	8	9	10
Last year	13.7	18.8	16.8	15.3
This year	16.8	16.7	16.0	15.8

Source: Based on information from Nielsen Media Research.

20. Routine Maintenance and Defective Parts A manufacturer believes that if routine maintenance (cleaning and oiling of machines) is increased to once a day rather than once a week, the number of defective parts produced by the machines will decrease. Nine machines are selected, and the number of defective parts produced over a 24-hour operating period is counted. Maintenance is then increased to once a day for a week, and the number of defective parts each machine produces is again counted over a 24-hour operating period. The data are shown here. At $\alpha = 0.01$, can the manufacturer conclude that increased maintenance reduces the number of defective parts manufactured by the machines?

Machine	1	2	3	4	5	6	7	8	9
Before	6	18	5	4	16	13	20	9	3
After	5	16	7	4	18	12	14	7	1

Extending the Concepts

Confidence Interval for the Median

The confidence interval for the median of a set of values less than or equal to 25 in number can be found by ordering the data from smallest to largest, finding the median, and using Table J. For example, to find the 95% confidence interval of the true median for 17, 19, 3, 8, 10, 15, 1, 23, 2, 12, order the data:

1, 2, 3, 8, 10, 12, 15, 17, 19, 23

From Table J, select $n = 10$ and $\alpha = 0.05$, and find the critical value. Use the two-tailed row. In this case, the critical value is 1. Add 1 to this value to get 2. In the ordered list, count from the left two numbers and from the right two numbers, and use these numbers to get the confidence interval, as shown:

1, 2, 3, 8, 10, 12, 15, 17, 19, 23

$2 \leq MD \leq 19$

Always add 1 to the number obtained from the table before counting. For example, if the critical value is 3, then count 4 values from the left and right.

For Exercises 21 through 25, find the confidence interval of the median, indicated in parentheses, for each set of data.

21. 3, 12, 15, 18, 16, 15, 22, 30, 25, 4, 6, 9 (95%)
    $6 \leq \text{median} \leq 22$
22. 101, 115, 143, 106, 100, 142, 157, 163, 155, 141, 145, 153, 152, 147, 143, 115, 164, 160, 147, 150 (90%)
    MD = 146; $141 \leq MD \leq 153$
23. 8.2, 7.1, 6.3, 5.2, 4.8, 9.3, 7.2, 9.3, 4.5, 9.6, 7.8, 5.6, 4.7, 4.2, 9.5, 5.1 (98%) $4.7 \leq \text{median} \leq 9.3$

24. 1, 8, 2, 6, 10, 15, 24, 33, 56, 41, 58, 54, 5, 3, 42, 31, 15, 65, 21 (99%) MD = 21; $5 \leq MD \leq 54$

25. 12, 15, 18, 14, 17, 19, 25, 32, 16, 47, 14, 23, 27, 42, 33, 35, 39, 41, 21, 19 (95%) $17 \leq \text{median} \leq 33$

Technology *Step by Step*

MINITAB
Step by Step

The Sign Test

1. Type the data for Example 13–1 into a column of MINITAB. Name the column SnowCones.

2. Select **Stat>Nonparametrics> 1-Sample Sign Test.**

3. Double-click SnowCones in the list box.

4. Click on Test median, then enter the hypothesized value of **40.**

5. Click [OK]. In the session window the *P*-value is 0.0075.

The Paired-Sample Sign Test

1. Enter the data for Example 13–3 into a worksheet; only the Before and After columns are necessary. Calculate a column with the differences to begin the process.

2. Select **Calc>Calculator.**

3. Type **D** in the box for Store result in variable.

4. Move to the Expression box, then click on Before, the subtraction sign, and After. The completed entry is shown.

5. Click [OK].

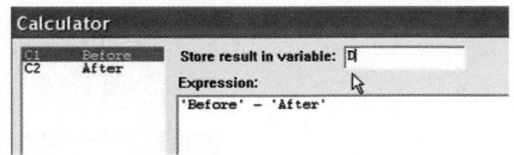

MINITAB will calculate the differences and store them in the first available column with the name "D." Use the instructions for the Sign Test on the differences D with a hypothesized value of zero.

Sign Test for Median: D

Sign test of median = 0.00000 versus not = 0.00000

	N	Below	Equal	Above	P	Median
D	10	2	1	7	0.1797	1.000

The P-value is 0.1797. Do not reject the null hypothesis.

Excel
Step by Step

The Sign Test

Excel does not have a procedure to conduct the sign test. However, you may conduct this test by using the MegaStat Add-in available on your CD. If you have not installed this add-in, do so, following the instructions from the Chapter 1 Excel Step by Step.

1. Enter the data from Example 13–1 into column A of a new worksheet.
2. From the toolbar, select Add-Ins, **MegaStat>Nonparametric Tests>Sign Test.** *Note:* You may need to open MegaStat from the MegaStat.xls file on your computer's hard drive.
3. Type **A1:A20** for the Input range.
4. Type **40** for the Hypothesized value, and select the "not equal" Alternative.
5. Click [OK].

The P-value is 0.0075. Reject the null hypothesis.

13–3

Objective 3

Test hypotheses, using the Wilcoxon rank sum test.

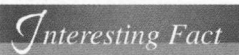

One in four married women now earns more than her husband.

The Wilcoxon Rank Sum Test

The sign test does not consider the magnitude of the data. For example, whether a value is 1 point or 100 points below the median, it will receive a negative sign. And when you compare values in the pretest/posttest situation, the magnitude of the differences is not considered. The Wilcoxon tests consider differences in magnitudes by using ranks.

The two tests considered in this section and in Section 13–4 are the **Wilcoxon rank sum test,** which is used for independent samples, and the **Wilcoxon signed-rank test,** which is used for dependent samples. Both tests are used to compare distributions. The parametric equivalents are the z and t tests for independent samples (Sections 9–1 and 9–3) and the t test for dependent samples (Section 9–4). For the parametric tests, as stated previously, the samples must be selected from approximately normally distributed populations, but the only assumption for the Wilcoxon signed-rank tests is that the population of differences has a symmetric distribution.

In the Wilcoxon tests, the values of the data for both samples are combined and then ranked. If the null hypothesis is true—meaning that there is no difference in the population distributions—then the values in each sample should be ranked approximately the same. Therefore, when the ranks are summed for each sample, the sums should be approximately equal, and the null hypothesis will not be rejected. If there is a large difference in the sums of the ranks, then the distributions are not identical, and the null hypothesis will be rejected.

The first test to be considered is the Wilcoxon rank sum test for independent samples. For this test, both sample sizes must be greater than or equal to 10. The formulas needed for the test are given next.

Formula for the Wilcoxon Rank Sum Test When Samples Are Independent

$$z = \frac{R - \mu_R}{\sigma_R}$$

where

$$\mu_R = \frac{n_1(n_1 + n_2 + 1)}{2}$$

$$\sigma_R = \sqrt{\frac{n_1 n_2 (n_1 + n_2 + 1)}{12}}$$

R = sum of ranks for smaller sample size (n_1)

n_1 = smaller of sample sizes

n_2 = larger of sample sizes

$n_1 \geq 10$ and $n_2 \geq 10$

Note that if both samples are the same size, either size can be used as n_1.

Example 13–4 illustrates the Wilcoxon rank sum test for independent samples.

Example 13–4

Times to Complete an Obstacle Course

Two independent samples of army and marine recruits are selected, and the time in minutes it takes each recruit to complete an obstacle course is recorded, as shown in the table. At $\alpha = 0.05$, is there a difference in the times it takes the recruits to complete the course?

Army	15	18	16	17	13	22	24	17	19	21	26	28	Mean = 19.67
Marines	14	9	16	19	10	12	11	8	15	18	25		Mean = 14.27

Solution

Step 1   State the hypotheses and identify the claim.

H_0: There is no difference in the times it takes the recruits to complete the obstacle course.

H_1: There is a difference in the times it takes the recruits to complete the obstacle course (claim).

Step 2   Find the critical value. Since $\alpha = 0.05$ and this test is a two-tailed test, use the z values of $+1.96$ and -1.96 from Table E.

Step 3   Compute the test value.

a. Combine the data from the two samples, arrange the combined data in order, and rank each value. Be sure to indicate the group.

Time	8	9	10	11	12	13	14	15	15	16	16	17
Group	M	M	M	M	M	A	M	A	M	A	M	A
Rank	1	2	3	4	5	6	7	8.5	8.5	10.5	10.5	12.5

Time	17	18	18	19	19	21	22	24	25	26	28
Group	A	M	A	A	M	A	A	A	M	A	A
Rank	12.5	14.5	14.5	16.5	16.5	18	19	20	21	22	23

b. Sum the ranks of the group with the smaller sample size. (*Note:* If both groups have the same sample size, either one can be used.) In this case, the sample size for the marines is smaller.

$$R = 1 + 2 + 3 + 4 + 5 + 7 + 8.5 + 10.5 + 14.5 + 16.5 + 21$$
$$= 93$$

c. Substitute in the formulas to find the test value.

$$\mu_R = \frac{n_1(n_1 + n_2 + 1)}{2} = \frac{(11)(11 + 12 + 1)}{2} = 132$$

$$\sigma_R = \sqrt{\frac{n_1 n_2(n_1 + n_2 + 1)}{12}} = \sqrt{\frac{(11)(12)(11 + 12 + 1)}{12}}$$
$$= \sqrt{264} = 16.2$$

$$z = \frac{R - \mu_R}{\sigma_R} = \frac{93 - 132}{16.2} = -2.41$$

Step 4   Make the decision. The decision is to reject the null hypothesis, since $-2.41 < -1.96$.

Step 5   Summarize the results. There is enough evidence to support the claim that there is a difference in the times it takes the recruits to complete the course.

The steps for the Wilcoxon rank sum test are given in the Procedure Table.

Procedure Table

Wilcoxon Rank Sum Test

Step 1   State the hypotheses and identify the claim.

Step 2   Find the critical value(s). Use Table E.

Step 3   Compute the test value.

a. Combine the data from the two samples, arrange the combined data in order, and rank each value.

b. Sum the ranks of the group with the smaller sample size. (*Note:* If both groups have the same sample size, either one can be used.)

c. Use these formulas to find the test value.

$$\mu_R = \frac{n_1(n_1 + n_2 + 1)}{2}$$

$$\sigma_R = \sqrt{\frac{n_1 n_2(n_1 + n_2 + 1)}{12}}$$

$$z = \frac{R - \mu_R}{\sigma_R}$$

where R is the sum of the ranks of the data in the smaller sample and n_1 and n_2 are each greater than or equal to 10.

Step 4   Make the decision.

Step 5   Summarize the results.

Applying the Concepts **13–3**

School Lunch

A nutritionist decided to see if there was a difference in the number of calories served for lunch in elementary and secondary schools. She selected a random sample of eight elementary schools and another random sample of eight secondary schools in Pennsylvania. The data are shown.

Elementary	Secondary
648	694
589	730
625	750
595	810
789	860
727	702
702	657
564	761

1. Are the samples independent or dependent?
2. What are the hypotheses?
3. What nonparametric test would you use to test the claim?
4. What critical value would you use?
5. What is the test value?
6. What is your decision?
7. What is the corresponding parametric test?
8. What assumption would you need to meet to use the parametric test?
9. If this assumption were not met, would the parametric test yield the same results?

See page 717 for the answers.

Exercises 13–3

1. What are the minimum sample sizes for the Wilcoxon rank sum test? n_1 and n_2 are each greater than or equal to 10.

2. What are the parametric equivalent tests for the Wilcoxon rank sum tests? The t test for independent samples

3. What distribution is used for the Wilcoxon rank sum test? The standard normal distribution

For Exercises 4 through 11, use the Wilcoxon rank sum test. Assume that the samples are independent. Also perform each of these steps.

  a. State the hypotheses and identify the claim.
  b. Find the critical value(s).
  c. Compute the test value.
  d. Make the decision.
  e. Summarize the results.

Use the traditional method of hypothesis testing unless otherwise specified.

 4. Lengths of Prison Sentences A random sample of men and women in prison was asked to give the length

of sentence each received for a certain type of crime. At $\alpha = 0.05$, test the claim that there is no difference in the sentence received by each gender. The data (in months) are shown here.

Males	8	12	6	14	22	27	32	24	26
Females	7	5	2	3	21	26	30	9	4

Males	19	15	13		
Females	17	23	12	11	16

5. **Technology Proficiency Test** The following are scores from a technology proficiency test required of all new incoming students at a particular college. Use the Wilcoxon rank sum test to see if there is a difference in scores between freshmen and transfer students at the 0.05 level of significance.

Freshmen	40	32	40	32	47	39	38	39	29	35	30
Transfers	38	43	35	45	37	36	36	33	46	44	41

6. Lifetimes of Handheld Video Games To test the claim that there is no difference in the lifetimes of two brands of handheld video games, a researcher selects a sample of 11 video games of each brand. The lifetimes (in months) of each brand are shown here. At $\alpha = 0.01$, can the researcher conclude that there is a difference in the distributions of lifetimes for the two brands?

Brand A	42	34	39	42	22	47	51	34	41	39	28
Brand B	29	39	38	43	45	49	53	38	44	43	32

 7. Stopping Distances of Automobiles A researcher wishes to see if the stopping distance for midsize automobiles is different from the stopping distance for compact automobiles at a speed of 70 miles per hour. The data are shown. At $\alpha = 0.10$, test the claim that the stopping distances are the same. If one of your safety concerns is stopping distance, would it make a difference which type of automobile you purchase?

Automobile	1	2	3	4	5	6	7	8	9	10
Midsize	188	190	195	192	186	194	188	187	214	203
Compact	200	211	206	297	198	204	218	212	196	193

Source: Based on information from the National Highway Traffic Safety Administration.

 8. Winning Baseball Games For the years 1970–1993 the National League (NL) and the American League (AL) (major league baseball) were each divided into two divisions: East and West. Below is a sample of the number of games won by each league's Eastern Division. At $\alpha = 0.05$, is there sufficient evidence to conclude a difference in the number of wins?

NL	89	96	88	101	90	91	92	96	108	100	95
AL	108	86	91	97	100	102	95	104	95	89	88	101

Source: World Almanac.

9. Hunting Accidents A game commissioner wishes to see if the number of hunting accidents in counties in western Pennsylvania is different from the number of hunting accidents in counties in eastern Pennsylvania. A sample of counties from the two regions is selected, and the numbers of hunting accidents are shown. At $\alpha = 0.05$, is there a difference in the number of accidents in the two areas? If so, give a possible reason for the difference.

Western Pa.	10	21	11	11	9	17	13	8	15	17
Eastern Pa.	14	3	7	13	11	2	8	5	5	6

Source: Pennsylvania Game Commission.

10. Medical School Enrollments Samples of enrollments from medical schools that specialize in research and in primary care are listed below. At $\alpha = 0.05$, can it be concluded that there is a difference?

Research	474 577 605 663 813 443 565 696 692 217
Primary care	783 546 442 662 605 474 587 555 427 320 293

Source: U.S. News & World Report Best Graduate Schools.

11. Speed of Pain Relievers Volunteers were randomly assigned to one of two groups to test the speed with which a pain reliever brought relief. One group took the standard dose of extra-strength acetaminophen (group A) while the other group (group N) took a newly approved pain-relieving drug. The number of minutes until symptoms abated is listed for each member of each group. At $\alpha = 0.05$ can it be concluded that there is a difference in time until pain is relieved?

Group A	15	20	12	20	17	14	15	17	18	11
Group N	7	14	13	11	10	16	12	9	10	9

Technology *Step by Step*

MINITAB
Step by Step

Wilcoxon Rank Sum Test (Mann-Whitney)

1. Enter the data for Example 13–4 into two columns of a worksheet.

2. Name the columns **Army** and **Marines**.

3. Select **Stat>Nonparametric>Mann-Whitney.**

4. Double-click Army for the First Sample.

5. Double-click Marines for the Second Sample.

6. Click [OK].

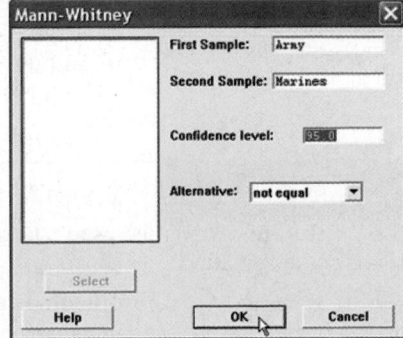

Mann-Whitney Test and CI: Army, Marines

```
            N  Median
Army       12  18.500
Marines    11  14.000
```

```
Point estimate for ETA1-ETA2 is 6.000
95.5 Percent CI for ETA1-ETA2 is (1.003, 9.998)
W = 183.0
Test of ETA1 = ETA2 vs ETA1 not = ETA2 is significant at 0.0178
The test is significant at 0.0177 (adjusted for ties)
```

The P-value for the test is 0.0177. Reject the null hypothesis. There is a significant difference in the times it takes the recruits to complete the course.

Excel
Step by Step

The Wilcoxon Mann-Whitney Test

Excel does not have a procedure to conduct the Mann-Whitney rank sum test. However, you may conduct this test by using the MegaStat Add-in available on your CD. If you have not installed this add-in, do so, following the instructions from the Chapter 1 Excel Step by Step.

1. Enter the data from Example 13–4 into columns A and B of a new worksheet.

2. From the toolbar, select Add-Ins, **MegaStat>Nonparametric Tests>Wilcoxon-Mann/ Whitney Test.** *Note:* You may need to open MegaStat from the MegaStat.xls file on your computer's hard drive.

3. Type **A1:A12** in the box for Group 1.

4. Type **B1:B11** in the box for Group 2.

5. Check the option labeled Correct for ties, and select the "not equal" Alternative.

6. Click [OK].

Wilcoxon Mann-Whitney Test

n	Sum of ranks	
12	183	Group 1
11	93	Group 2
23	276	Total

```
          144.00 Expected value
           16.23 Standard deviation
            2.37 z, corrected for ties
          0.0177 P-value (two-tailed)
```

The P-value is 0.0177. Reject the null hypothesis.

13–4
The Wilcoxon Signed-Rank Test

Objective 4

Test hypotheses, using the signed-rank test.

When the samples are dependent, as they would be in a before-and-after test using the same subjects, the Wilcoxon signed-rank test can be used in place of the t test for dependent samples. Again, this test does not require the condition of normality. Table K is used to find the critical values.

The procedure for this test is shown in Example 13–5.

Example 13–5

Shoplifting Incidents

In a large department store, the owner wishes to see whether the number of shoplifting incidents per day will change if the number of uniformed security officers is doubled. A sample of 7 days before security is increased and 7 days after the increase shows the number of shoplifting incidents.

	Number of shoplifting incidents	
Day	**Before**	**After**
Monday	7	5
Tuesday	2	3
Wednesday	3	4
Thursday	6	3
Friday	5	1
Saturday	8	6
Sunday	12	4

Is there enough evidence to support the claim, at $\alpha = 0.05$, that there is a difference in the number of shoplifting incidents before and after the increase in security?

Solution

Step 1　State the hypotheses and identify the claim.

H_0: There is no difference in the number of shoplifting incidents before and after the increase in security.

H_1: There is a difference in the number of shoplifting incidents before and after the increase in security (claim).

Step 2　Find the critical value from Table K. Since $n = 7$ and $\alpha = 0.05$ for this two-tailed test, the critical value is 2. See Figure 13–2.

Figure 13–2

Finding the Critical Value in Table K for Example 13–5

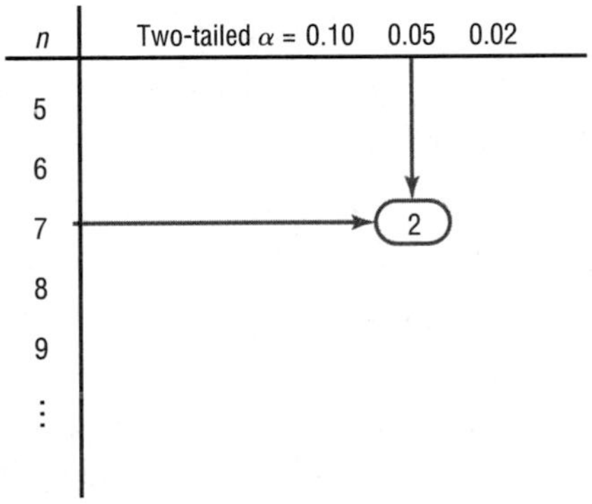

Step 3　Find the test value.

a. Make a table as shown here.

Day	Before, X_B	After, X_A	Difference $D = X_B - X_A$	Absolute value $|D|$	Rank	Signed rank
Mon.	7	5				
Tues.	2	3				
Wed.	3	4				
Thurs.	6	3				
Fri.	5	1				
Sat.	8	6				
Sun.	12	4				

b. Find the differences (before minus after), and place the values in the Difference column.

$$7 - 5 = 2 \qquad 6 - 3 = 3 \qquad 8 - 6 = 2$$
$$2 - 3 = -1 \qquad 5 - 1 = 4 \qquad 12 - 4 = 8$$
$$3 - 4 = -1$$

c. Find the absolute value of each difference, and place the results in the Absolute value column. (*Note:* The absolute value of any number except 0 is the positive value of the number. Any differences of 0 should be ignored.)

$$|2| = 2 \qquad |3| = 3 \qquad |2| = 2$$
$$|-1| = 1 \qquad |4| = 4 \qquad |8| = 8$$
$$|-1| = 1$$

d. Rank each absolute value from lowest to highest, and place the rankings in the Rank column. In the case of a tie, assign the values that rank plus 0.5.

Value	2	1	1	3	4	2	8
Rank	3.5	1.5	1.5	5	6	3.5	7

e. Give each rank a plus or minus sign, according to the sign in the Difference column. The completed table is shown here.

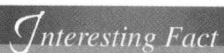

Day	Before, X_B	After, X_A	Difference $D = X_B - X_A$	Absolute value $|D|$	Rank	Signed rank
Mon.	7	5	2	2	3.5	+3.5
Tues.	2	3	−1	1	1.5	−1.5
Wed.	3	4	−1	1	1.5	−1.5
Thurs.	6	3	3	3	5	+5
Fri.	5	1	4	4	6	+6
Sat.	8	6	2	2	3.5	+3.5
Sun.	12	4	8	8	7	+7

f. Find the sum of the positive ranks and the sum of the negative ranks separately.

Positive rank sum $\qquad (+3.5) + (+5) + (+6) + (+3.5) + (+7) = +25$
Negative rank sum $\qquad (-1.5) + (-1.5) \qquad\qquad\qquad\qquad = -3$

g. Select the smaller of the absolute values of the sums ($|-3|$), and use this absolute value as the test value w_s. In this case, $w_s = |-3| = 3$.

Step 4 Make the decision. Reject the null hypothesis if the test value is less than or equal to the critical value. In this case, $3 > 2$; hence, the decision is not to reject the null hypothesis.

Step 5 Summarize the results. There is not enough evidence to support the claim that there is a difference in the number of shoplifting incidents. Hence, the security increase probably made no difference in the number of shoplifting incidents.

The rationale behind the signed-rank test can be explained by a diet example. If the diet is working, then the majority of the postweights will be smaller than the preweights. When the postweights are subtracted from the preweights, the majority of the signs will be positive, and the absolute value of the sum of the negative ranks will be small. This sum will probably be smaller than the critical value obtained from Table K, and the null hypothesis will be rejected. On the other hand, if the diet does not work, some people will gain weight, other people will lose weight, and still other people will remain about the same weight. In this case, the sum of the positive ranks and the absolute value of the sum of the negative ranks will be approximately equal and will be about one-half of the sum of the absolute value of all the ranks. In this case, the smaller of the absolute values of the two sums will still be larger than the critical value obtained from Table K, and the null hypothesis will not be rejected.

When $n \geq 30$, the normal distribution can be used to approximate the Wilcoxon distribution. The same critical values from Table E used for the z test for specific α values are used. The formula is

$$z = \frac{w_s - \dfrac{n(n+1)}{4}}{\sqrt{\dfrac{n(n+1)(2n+1)}{24}}}$$

where

$\quad n =$ number of pairs where difference is not 0

$\quad w_s =$ smaller sum in absolute value of signed ranks

The steps for the Wilcoxon signed-rank test are given in the Procedure Table.

Procedure Table

Wilcoxon Signed-Rank Test

Step 1   State the hypotheses and identify the claim.

Step 2   Find the critical value from Table K.

Step 3   Compute the test value.

    a. Make a table, as shown.

Before, X_B	After, X_A	Difference $D = X_B - X_A$	Absolute value $|D|$	Rank	Signed rank

    b. Find the differences (before $-$ after), and place the values in the Difference column.

    c. Find the absolute value of each difference, and place the results in the Absolute value column.

    d. Rank each absolute value from lowest to highest, and place the rankings in the Rank column.

    e. Give each rank a positive or negative sign, according to the sign in the Difference column.

    f. Find the sum of the positive ranks and the sum of the negative ranks separately.

    g. Select the smaller of the absolute values of the sums, and use this absolute value as the test value w_s.

Step 4   Make the decision. Reject the null hypothesis if the test value is less than or equal to the critical value.

Step 5   Summarize the results.

    Note: When $n \geq 30$, use Table E and the test value

$$z = \frac{w_s - \dfrac{n(n+1)}{4}}{\sqrt{\dfrac{n(n+1)(2n+1)}{24}}}$$

where

$\quad n =$ number of pairs where difference is not 0

$\quad w_s =$ smaller sum in absolute value of signed ranks

Applying the Concepts **13–4**

Pain Medication

A researcher decides to see how effective a pain medication is. Eight subjects were asked to determine the severity of their pain by using a scale of 1 to 10, with 1 being very minor and 10 being very severe. Then each was given the medication, and after 1 hour, they were asked to rate the severity of their pain, using the same scale.

Subject	1	2	3	4	5	6	7	8
Before	8	6	2	3	4	6	2	7
After	6	5	3	1	2	6	1	6

1. What is the purpose of the study?
2. Are the samples independent or dependent?
3. What are the hypotheses?
4. What nonparametric test could be used to test the claim?
5. What significance level would you use?
6. What is your decision?
7. What parametric test could you use?
8. Would the results be the same?

See page 717 for the answers.

Exercises 13–4

1. What is the parametric equivalent test for the Wilcoxon signed-rank test? The *t* test for dependent samples

For Exercises 2 and 3, find the sum of the signed ranks. Assume that the samples are dependent. State which sum is used as the test value.

2.

Pretest	65	103	79	92	72	91	76	95
Posttest	72	105	64	95	78	92	76	93

3.

Pretest	108	97	115	162	156	105	153
Posttest	110	97	103	168	143	112	141

For Exercises 4 through 8, use Table K to determine whether the null hypothesis should be rejected.

4. $w_s = 62$, $n = 21$, $\alpha = 0.05$, two-tailed test
C.V. = 59; do not reject

5. $w_s = 18$, $n = 15$, $\alpha = 0.02$, two-tailed test
C.V. = 20; reject

6. $w_s = 53$, $n = 20$, $\alpha = 0.05$, two-tailed test
C.V. = 52; do not reject

7. $w_s = 102$, $n = 28$, $\alpha = 0.01$, one-tailed test
C.V. = 102; reject

8. $w_s = 33$, $n = 18$, $\alpha = 0.01$, two-tailed test
C.V. = 28; do not reject

9. Drug Prices Eight drugs were selected, and the prices for the human doses and the animal doses for the same amounts were compared. At $\alpha = 0.05$, can it be concluded that the prices for the animal doses are significantly less than the prices for the human doses? If the null hypothesis is rejected, give one reason why animal doses might cost less than human doses.

Human dose	0.67	0.64	1.20	0.51	0.87	0.74	0.50	1.22
Animal dose	0.13	0.18	0.42	0.25	0.57	0.57	0.49	1.28

Source: House Committee on Government Reform.

10. Property Assessments Use the sign test to test the hypothesis that the assessed value has changed between 2006 and 2010. Use $\alpha = 0.05$. Do you think land values in a large city would be normally distributed?

Ward	A	B	C	D	E	F	G	H	I	J	K
2006	184	414	22	99	116	49	24	50	282	25	141
2010	161	382	22	190	120	52	28	50	297	40	148

11. Weight Loss Through Diet Eight subjects were weighed before and after a new three-week "healthy" diet. At the 0.05 level of significance, can it be concluded that a difference in weight resulted? (Weights are in pounds.)

Subject	A	B	C	D	E	F	G	H
Before	150	195	188	197	204	175	160	180
After	152	190	185	191	200	170	162	179

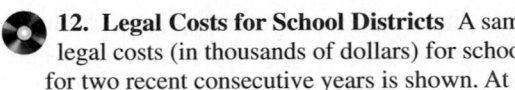

12. Legal Costs for School Districts A sample of legal costs (in thousands of dollars) for school districts for two recent consecutive years is shown. At $\alpha = 0.05$, is there a difference in the costs?

Year 1	108	36	65	108	87	94	10	40
Year 2	138	28	67	181	97	126	18	67

Source: *Pittsburgh Tribune-Review.*

13. Drug Prices A researcher wishes to compare the prices for prescription drugs in the United States with those in Canada. The same drugs and dosages were compared in each country. At $\alpha = 0.05$, can it be concluded that the drugs in Canada are cheaper?

Drug	1	2	3	4	5	6
United States	3.31	2.27	2.54	3.13	23.40	3.16
Canada	1.47	1.07	1.34	1.34	21.44	1.47

Drug	7	8	9	10
United States	1.98	5.27	1.96	1.11
Canada	1.07	3.39	2.22	1.13

Source: IMS Health and other sources.

Technology *Step by Step*

MINITAB
Step by Step

Wilcoxon Signed-Rank Test

Test the median value for the differences of two dependent samples. Use Example 13–5.

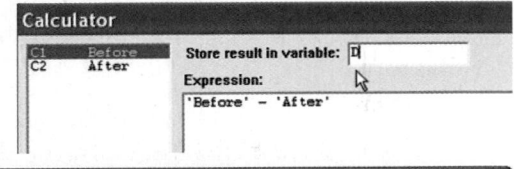

1. Enter the data into two columns of a worksheet. Name the columns **Before** and **After.**

2. Calculate the differences, using **Calc>Calculator.**

3. Type **D** in the box for Store result in variable.

4. In the expression box, type **Before − After.**

5. Click [OK].

6. Select **Stat>Nonparametric> 1-Sample Wilcoxon**

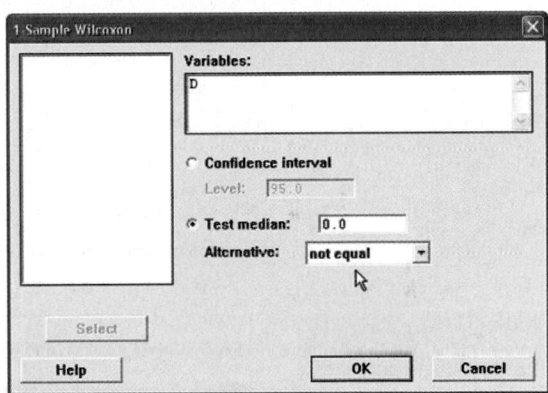

7. Select C3 for the Variable.

8. Click on Test median. The value should be 0.

9. Click [OK].

Wilcoxon Signed-Rank Test: D

```
Test of median = 0.000000 versus median not = 0.000000
              N
         for   Wilcoxon              Estimated
     N  Test   Statistic      P       Median
D    7    7         25.0   0.076       2.250
```

The *P*-value of the test is 0.076. Do not reject the null hypothesis.

13–5

Objective 5

Test hypotheses, using the Kruskal-Wallis test.

The Kruskal-Wallis Test

The analysis of variance uses the F test to compare the means of three or more populations. The assumptions for the ANOVA test are that the populations are normally distributed and that the population variances are equal. When these assumptions cannot be met, the nonparametric **Kruskal-Wallis test,** sometimes called the *H test,* can be used to compare three or more means.

In this test, each sample size must be 5 or more. In these situations, the distribution can be approximated by the chi-square distribution with $k - 1$ degrees of freedom, where k = number of groups. This test also uses ranks. The formula for the test is given next.

In the Kruskal-Wallis test, you consider all the data values as a group and then rank them. Next, the ranks are separated and the H formula is computed. This formula approximates the variance of the ranks. If the samples are from different populations, the sums of the ranks will be different and the H value will be large; hence, the null hypothesis will be rejected if the H value is large enough. If the samples are from the same population, the sums of the ranks will be approximately the same and the H value will be small; therefore, the null hypothesis will not be rejected. This test is always a right-tailed test. The chi-square table, Table G, with d.f. $= k - 1$, should be used for critical values.

Formula for the Kruskal-Wallis Test

$$H = \frac{12}{N(N + 1)}\left(\frac{R_1^2}{n_1} + \frac{R_2^2}{n_2} + \cdots + \frac{R_k^2}{n_k}\right) - 3(N + 1)$$

where

R_1 = sum of ranks of sample 1
n_1 = size of sample 1
R_2 = sum of ranks of sample 2
n_2 = size of sample 2
.
.
.
R_k = sum of ranks of sample k
n_k = size of sample k
$N = n_1 + n_2 + \cdots + n_k$
k = number of samples

Example 13–6 illustrates the procedure for conducting the Kruskal-Wallis test.

Example 13–6

Hospital Infections

A researcher wishes to see if the total number of infections that occurred in three groups of hospitals is the same. The data are shown in the table. At $\alpha = 0.05$ is there enough evidence to reject the claim that the number of infections in the three groups of hospitals is the same?

Group A	Group B	Group C
557	476	105
315	232	110
920	80	167
178	116	155

Source: Pennsylvania Health Care Cost Containment Council.

Solution

Step 1 State the hypotheses and identify the claim.

H_0: There is no difference in the number of infections in the three groups of hospitals (claim).

H_1: There is a difference in the number of infections in the three groups of hospitals.

Step 2   Find the critical value. Use the chi-square table (Table G) with d.f. $= k - 1$, where $k =$ the number of groups. With $\alpha = 0.05$ and d.f. $= 3 - 1 = 2$, the critical value is 5.991.

Step 3   Compute the test value.

   a. Arrange all the data from the lowest value to the highest value and rank each value.

Amount	Group	Rank
80	B	1
105	C	2
110	C	3
116	B	4
155	C	5
167	C	6
178	A	7
232	B	8
315	A	9
476	B	10
557	A	11
920	A	12

   b. Find the sum of the ranks for each group.

Group A	$7 + 9 + 11 + 12 = 39$
Group B	$1 + 4 + 8 + 10 = 23$
Group C	$2 + 3 + 5 + 6 = 16$

   c. Substitute in the formula.

$$H = \frac{12}{N(N + 1)}\left(\frac{R_1^2}{n_1} + \frac{R_2^2}{n_2} + \frac{R_3^2}{n_3}\right) - 3(N + 1)$$

where

$$N = 12 \qquad R_1 = 39 \qquad R_2 = 23 \qquad R_3 = 16$$
$$n_1 = n_2 = n_3 = 4$$

Therefore,

$$H = \frac{12}{12(12 + 1)}\left(\frac{39^2}{4} + \frac{23^2}{4} + \frac{16^2}{4}\right) - 3(12 + 1)$$
$$= 5.346$$

Step 4   Make the decision. Since 5.346 is less than the critical value of 5.991, the decision is to not reject the null hypothesis.

Step 5   Summarize the results. There is not enough evidence to reject the claim that there is no difference in the number of infections in the groups of hospitals. Hence the differences are not significant at $\alpha = 0.05$.

The steps for the Kruskal-Wallis test are given in the Procedure Table.

Procedure Table

Kruskal-Wallis Test

Step 1   State the hypotheses and identify the claim.

Step 2   Find the critical value. Use the chi-square table, Table G, with d.f. $= k - 1$ (k = number of groups).

Step 3   Compute the test value.
  - *a.* Arrange the data from lowest to highest and rank each value.
  - *b.* Find the sum of the ranks of each group.
  - *c.* Substitute in the formula

$$H = \frac{12}{N(N+1)}\left(\frac{R_1^2}{n_1} + \frac{R_2^2}{n_2} + \cdots + \frac{R_k^2}{n_k}\right) - 3(N+1)$$

where
$$N = n_1 + n_2 + \cdots + n_k$$
$$R_k = \text{sum of ranks for } k\text{th group}$$
$$k = \text{number of groups}$$

Step 4   Make the decision.

Step 5   Summarize the results.

Applying the Concepts 13–5

Heights of Waterfalls

You are doing research for an article on the waterfalls on our planet. You want to make a statement about the heights of waterfalls on three continents. Three samples of waterfall heights (in feet) are shown.

North America	Africa	Asia
600	406	330
1200	508	830
182	630	614
620	726	1100
1170	480	885
442	2014	330

1. What questions are you trying to answer?

2. What nonparametric test would you use to find the answer?

3. What are the hypotheses?

4. Select a significance level and run the test. What is the H value?

5. What is your conclusion?

6. What is the corresponding parametric test?

7. What assumptions would you need to make to conduct this test?

See page 718 for the answers.

Exercises 13–5

For Exercises 1 through 11, perform these steps.

a. State the hypotheses and identify the claim.
b. Find the critical value.
c. Compute the test value.
d. Make the decision.
e. Summarize the results.

Use the traditional method of hypothesis testing unless otherwise specified.

1. Calories in Cereals Samples of four different cereals show the following numbers of calories for the suggested servings of each brand. At $\alpha = 0.05$, is there a difference in the number of calories for the different brands?

Brand A	Brand B	Brand C	Brand D
112	110	109	106
120	118	116	122
135	123	125	130
125	128	130	117
108	102	128	116
121	101	132	114

2. Mathematics Literacy Scores Through the Organization for Economic Cooperation and Development (OECD), 15-year-olds are tested in member countries in mathematics, reading, and science literacy. Below are listed total mathematics literacy scores (i.e., both genders) for selected countries in different parts of the world. Test, using the Kruskal-Wallis test, to see if there is a difference in means at $\alpha = 0.05$.

Western Hemisphere	Europe	Eastern Asia
527	520	523
406	510	547
474	513	547
381	548	391
411	496	549

Source: www.nces.ed.gov

3. Lawnmower Costs A researcher wishes to compare the prices of three types of lawnmowers. At $\alpha = 0.10$, can it be concluded that there is a difference in the prices? Based on your answer, do you feel that the cost should be a factor in determining which type of lawnmower a person would purchase?

Gas-powered self-propelled	Gas-powered push	Electric
290	320	188
325	360	245
210	200	470
300	229	395
330	160	

4. Sodium Content of Microwave Dinners Three brands of microwave dinners were advertised as low in sodium. Samples of the three different brands show the following milligrams of sodium. At $\alpha = 0.05$, is there a difference in the amount of sodium among the brands?

Brand A	Brand B	Brand C
810	917	893
702	912	790
853	952	603
703	958	744
892	893	623
732		743
713		609
613		

5. Unemployment Benefits In Chapter 12 we did this exercise assuming that the populations were normally distributed and that the population variances were equal. Assume that this is not the case. Using the Kruskal-Wallis test, is the outcome affected? Do you think unemployment benefits are normally distributed? Test for a difference in means at $\alpha = 0.05$.

Florida	Pennsylvania	Maine
200	300	250
187	350	195
192	295	275
235	362	260
260	280	220
175	340	290

6. Job Offers for Chemical Engineers A recent study recorded the number of job offers received by newly graduated chemical engineers at three colleges. The data are shown here. At $\alpha = 0.05$, is there a difference in the average number of job offers received by the graduates at the three colleges?

College A	College B	College C
6	2	10
8	1	12
7	0	9
5	3	13
6	6	4

7. Expenditures for Pupils The expenditures in dollars per pupil for states in three sections of the country are listed below. At $\alpha = 0.05$, can it be concluded that there is a difference in spending between regions?

Eastern third	Middle third	Western third
6701	9854	7584
6708	8414	5474
9186	7279	6622
6786	7311	9673
9261	6947	7353

Source: *New York Times Almanac.*

8. Printer Costs An electronics store manager wishes to compare the costs (in dollars) of three types of computer printers. The data are shown. At $\alpha = 0.05$, can it be concluded that there is a difference in the prices? Based on your answer, do you think that a certain type of printer generally costs more than the other types?

Inkjet printers	Multifunction printers	Laser printers
149	98	192
199	119	159
249	149	198
239	249	198
99	99	229
79	199	

9. Number of Crimes per Week In a large city, the number of crimes per week in five precincts is recorded for five weeks. The data are shown here. At $\alpha = 0.01$, is there a difference in the number of crimes?

Precinct 1	Precinct 2	Precinct 3	Precinct 4	Precinct 5
105	87	74	56	103
108	86	83	43	98
99	91	78	52	94
97	93	74	58	89
92	82	60	62	88

10. Amounts of Caffeine in Beverages The amounts of caffeine in a regular (small) serving of assorted beverages are listed below. If someone wants to limit caffeine intake, does it really matter which beverage she or he chooses? Is there a difference in caffeine content at $\alpha = 0.05$?

Teas	Coffees	Colas
70	120	35
40	80	48
30	160	55
25	90	43
40	140	42

Source: *Doctor's Pocket Calorie, Fat & Carbohydrate Counter.*

11. Maximum Speeds of Animals A human is said to be able to reach a maximum speed of 27.89 miles per hour. The maximum speeds of various types of other animals are listed below. Based on these particular groupings is there evidence of a difference in speeds? Use the 0.05 level of significance.

Predatory mammals	Deerlike animals	Domestic animals
70	50	47.5
50	35	39.35
43	32	35
42	30	30
40	61	11

MINITAB
Step by Step

Kruskal-Wallis Test

Example: Milliequivalents of Potassium in Breakfast Drinks

A researcher tests three different brands of breakfast drinks to see how many milliequivalents of potassium per quart each contains. These data are obtained.

Brand A	Brand B	Brand C
4.7	5.3	6.3
3.2	6.4	8.2
5.1	7.3	6.2
5.2	6.8	7.1
5.0	7.2	6.6

At $\alpha = 0.05$, is there enough evidence to reject the hypothesis that all brands contain the same amount of potassium?

The data for this test must be "stacked." All the numeric data must be in one column, and the second column identifies the brand.

1. Stack the data for the example into two columns of a worksheet.

   a) First, enter all the potassium amounts into one column.

   b) Name this column **Potassium.**

   c) Enter code **A, B,** or **C** for the brand into the next column.

   d) Name this column **Brand.**

↓	C1 Potassium	C2-T Brand
1	4.7	A
2	3.2	A
3	5.1	A
4	5.2	A
5	5.0	A
6	5.3	B
7	6.4	B
8	7.3	B
9	6.8	B
10	7.2	B
11	6.3	C
12	8.2	C
13	6.2	C
14	7.1	C
15	6.6	C

The worksheet is shown.

2. Select **Stat>Nonparametric>Kruskal-Wallis.**
3. Double-click C1 Potassium to select it for Response.

This variable must be quantitative so the column for Brand will not be available in the list until the cursor is in the Factor text box.

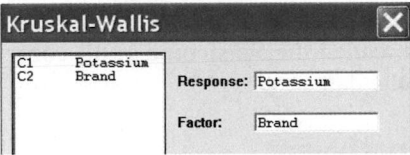

4. Select C2 Brand for Factor.
5. Click [OK].

Kruskal-Wallis Test: Potassium versus Brand
```
Kruskal-Wallis Test on Potassium
Brand    N   Median   Ave Rank       Z
A        5    5.000        3.0   -3.06
B        5    6.800       10.6    1.59
C        5    6.600       10.4    1.47
Overall 15                8.0
H = 9.38   DF = 2   P = 0.009
```

The value $H = 9.38$ has a P-value of 0.009. Reject the null hypothesis.

Excel
Step by Step

The Kruskal-Wallis Test

Excel does not have a procedure to conduct the Kruskal-Wallis test. However, you may conduct this test by using the MegaStat Add-in available on your CD. If you have not installed this add-in, do so, following the instructions from the Chapter 1 Excel Step by Step.

1. Enter the data from previous example into columns A, B, and C of a new worksheet.
2. From the toolbar, select Add-Ins, **MegaStat>Nonparametric Tests>Kruskal-Wallis Test.** *Note:* You may need to open MegaStat from the MegaStat.xls file on your computer's hard drive.
3. Type **A1:C5** in the box for Input range.
4. Check the option labeled Correct for ties, and select the "not equal" Alternative.
5. Click [OK].

Kruskal-Wallis Test

Median	n	Avg. rank	
5.00	5	3.00	Group 1
6.80	5	10.60	Group 2
6.60	5	10.40	Group 3
6.30	15		Total

```
                 9.380   H
                     2   d.f.
                0.0092   P-value
      Multiple comparison values for avg. ranks
         6.77(0.05)            8.30(0.01)
```

The P-value is 0.0092. Reject the null hypothesis.

13–6

The Spearman Rank Correlation Coefficient and the Runs Test

The techniques of regression and correlation were explained in Chapter 10. To determine whether two variables are linearly related, you use the Pearson product moment correlation coefficient. Its values range from $+1$ to -1. One assumption for testing the hypothesis that $\rho = 0$ for the Pearson coefficient is that the populations from which the samples are obtained are normally distributed. If this requirement cannot be met, the nonparametric equivalent, called the **Spearman rank correlation coefficient** (denoted by r_s), can be used when the data are ranked.

Rank Correlation Coefficient

The computations for the rank correlation coefficient are simpler than those for the Pearson coefficient and involve ranking each set of data. The difference in ranks is found, and r_s is computed by using these differences. If both sets of data have the same ranks, r_s will be $+1$. If the sets of data are ranked in exactly the opposite way, r_s will be -1. If there is no relationship between the rankings, r_s will be near 0.

Objective 6

Compute the Spearman rank correlation coefficient.

Formula for Computing the Spearman Rank Correlation Coefficient

$$r_s = 1 - \frac{6 \sum d^2}{n(n^2 - 1)}$$

where
  d = difference in ranks
  n = number of data pairs

This formula is algebraically equivalent to the formula for r given in Chapter 10, except that ranks are used instead of raw data.

The computational procedure is shown in Example 13–7. For a test of the significance of r_s, Table L is used for values of n up to 30. For larger values, the normal distribution can be used. (See Exercises 24 through 28 in the exercise section.)

Example 13–7

Bank Branches and Deposits

A researcher wishes to see if there is a relationship between the number of branches a bank has and the total number of deposits (in billions of dollars) the bank receives. A sample of eight regional banks is selected, and the number of branches and the amount of deposits are shown in the table. At $\alpha = 0.05$ is there a significant linear correlation between the number of branches and the amount of the deposits?

Bank	Number of branches	Deposits (in billions)
A	209	$23
B	353	31
C	19	7
D	201	12
E	344	26
F	132	5
G	401	24
H	126	5

Source: SNL Financial.

Solution

Step 1 State the hypotheses.

$H_0: \rho = 0$ and $H_1: \rho \neq 0$

Step 2 Find the critical value. Use Table L to find the value for $n = 8$ and $\alpha = 0.05$. It is 0.738. See Figure 13–3.

Figure 13–3

Finding the Critical Value in Table L for Example 13–7

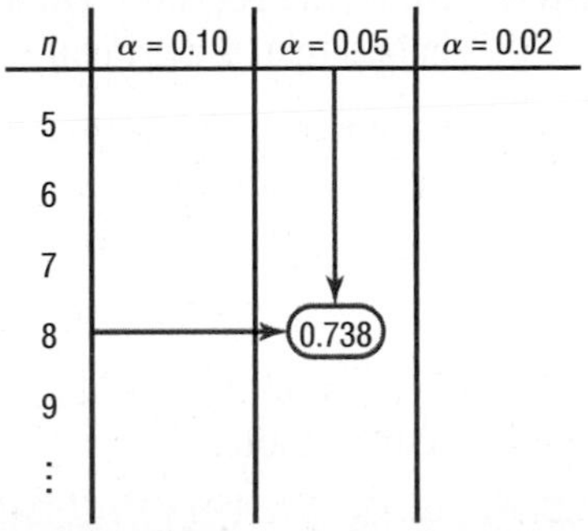

n	$\alpha = 0.10$	$\alpha = 0.05$	$\alpha = 0.02$
5			
6			
7			
8		0.738	
9			
⋮			

Step 3 Find the test value.

a. Rank each data set as shown in the table.

Bank	Branches	Rank	Deposits	Rank
A	209	4	23	4
B	353	2	31	1
C	19	8	7	6
D	201	5	12	5
E	344	3	26	2
F	132	6	5	7
G	401	1	24	3
H	126	7	4	8

b. Let X_1 be the rank of the branches and X_2 be the rank of the deposits.

c. Subtract the ranking $(X_1 - X_2)$.

$4 - 4 = 0$ $2 - 1 = 1$ $8 - 6 = 2$ etc.

d. Square the differences.

$0^2 = 0$ $1^2 = 1$ $2^2 = 4$ etc.

e. Find the sum of the squares

$0 + 1 + 4 + 0 + 1 + 1 + 4 + 1 = 12$

The results can be summarized in a table as shown.

X_1	X_2	$d = X_1 - X_2$	d^2
4	4	0	0
2	1	1	1
8	6	2	4
5	5	0	0
3	2	1	1
6	7	-1	1
1	3	-2	4
7	8	-1	1
			$\Sigma d^2 = 12$

f. Substitute in the formula for r_s.

$$r_s = 1 - \frac{6 \, \Sigma d^2}{n(n^2 - 1)} \quad \text{where } n = \text{number of pairs}$$

$$r_s = 1 - \frac{6 \cdot 12}{6(6^2 - 1)} = 1 - \frac{72}{210} = 0.657$$

Step 4 Make the decision. Do not reject the null hypothesis since $r_s = 0.657$, which is less than the critical value of 0.738.

Step 5 Summarize the results. There is not enough evidence to say that there is a linear relationship between the number of branches a bank has and the deposits of the bank.

The steps for finding and testing the Spearman rank correlation coefficient are given in the Procedure Table.

Procedure Table

Finding and Testing the Spearman Rank Correlation Coefficient

Step 1 State the hypotheses.

Step 2 Rank each data set.

Step 3 Subtract the rankings $(X_1 - X_2)$.

Step 4 Square the differences.

Step 5 Find the sum of the squares.

Step 6 Substitute in the formula.

$$r_s = 1 - \frac{6 \, \Sigma d^2}{n(n^2 - 1)}$$

where
d = difference in ranks
n = number of pairs of data

Step 7 Find the critical value.

Step 8 Make the decision.

Step 9 Summarize the results.

Objective **7**

Test hypotheses, using the runs test.

The Runs Test

When samples are selected, you assume that they are selected at random. How do you know if the data obtained from a sample are truly random? Before the answer to this question is given, consider the following situations for a researcher interviewing 20 people for a survey. Let their gender be denoted by M for male and F for female. Suppose the participants were chosen as follows:

      Situation 1   M M M M M M M M M M F F F F F F F F F F

It does not look as if the people in this sample were selected at random, since 10 males were selected first, followed by 10 females.

    Consider a different selection:

      Situation 2   F M F M F M F M F M F M F M F M F M F M

In this case, it seems as if the researcher selected a female, then a male, etc. This selection is probably not random either.

Finally, consider the following selection:

Situation 3   F F F M M F M F M M F F M M F F M M M F

This selection of data looks as if it may be random, since there is a mix of males and females and no apparent pattern to their selection.

Rather than try to guess whether the data of a sample have been selected at random, statisticians have devised a nonparametric test to determine randomness. This test is called the **runs test.**

A **run** is a succession of identical letters preceded or followed by a different letter or no letter at all, such as the beginning or end of the succession.

For example, the first situation presented has two runs:

Run 1:   M M M M M M M M M M

Run 2:   F F F F F F F F F F

The second situation has 20 runs. (Each letter constitutes one run.) The third situation has 11 runs.

Run 1:	F F F	Run 5:	F	Run 9:	F F
Run 2:	M M	Run 6:	M M	Run 10:	M M M
Run 3:	F	Run 7:	F F	Run 11:	F
Run 4:	M	Run 8:	M M		

Example 13–8

Determine the number of runs in each sequence.

a. M M F F F M F F

b. H T H H H

c. A B A A A B B A B B B

Solution

a. There are four runs, as shown.

M M	F F F	M	F F
1	2	3	4

b. There are three runs, as shown.

H	T	H H H
1	2	3

c. There are six runs, as shown.

A	B	A A A	B B	A	B B B
1	2	3	4	5	6

The test for randomness considers the number of runs rather than the frequency of the letters. For example, for data to be selected at random, there should not be too few or too many runs, as in situations 1 and 2. The runs test does not consider the questions of how many males or females were selected or how many of each are in a specific run.

To determine whether the number of runs is within the random range, use Table M in Appendix C. The values are for a two-tailed test with $\alpha = 0.05$. For a sample of 12 males and 8 females, the table values shown in Figure 13–4 mean that any number of runs from 7 to 15 would be considered random. If the number of runs is 6 or less or 16 or more, the sample is probably not random, and the null hypothesis should be rejected.

Example 13–9 shows the procedure for conducting the runs test by using letters as data. Example 13–10 shows how the runs test can be used for numerical data.

Figure 13–4

Finding the Critical Value in Table M

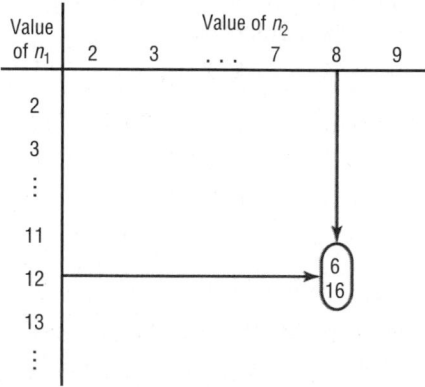

Example 13–9

Gender of Train Passengers

On a commuter train, the conductor wishes to see whether the passengers enter the train at random. He observes the first 25 people, with the following sequence of males (M) and females (F).

$$F\,F\,F\,M\,M\,F\,F\,F\,F\,M\,F\,M\,M\,M\,F\,F\,F\,F\,M\,M\,F\,F\,F\,M\,M$$

Test for randomness at $\alpha = 0.05$.

Solution

Step 1 State the hypotheses and identify the claim.

H_0: The passengers board the train at random, according to gender (claim).

H_1: The null hypothesis is not true.

Step 2 Find the number of runs. Arrange the letters according to runs of males and females, as shown.

Run	Gender
1	F F F
2	M M
3	F F F F
4	M
5	F
6	M M M
7	F F F F
8	M M
9	F F F
10	M M

There are 15 females (n_1) and 10 males (n_2).

Step 3 Find the critical value. Find the number of runs in Table M for $n_1 = 15$, $n_2 = 10$, and $\alpha = 0.05$. The values are 7 and 18. *Note:* In this situation the critical value is found after the number of runs is determined.

Step 4 Make the decision. Compare these critical values with the number of runs. Since the number of runs is 10 and 10 is between 7 and 18, do not reject the null hypothesis.

Step 5 Summarize the results. There is not enough evidence to reject the hypothesis that the passengers board the train at random according to gender.

Example 13–10

Ages of Drug Program Participants

Twenty people enrolled in a drug abuse program. Test the claim that the ages of the people, according to the order in which they enroll, occur at random, at $\alpha = 0.05$. The data are 18, 36, 19, 22, 25, 44, 23, 27, 27, 35, 19, 43, 37, 32, 28, 43, 46, 19, 20, 22.

Solution

Step 1 State the hypotheses and identify the claim.

H_0: The ages of the people, according to the order in which they enroll in a drug program, occur at random (claim).

H_1: The null hypothesis is not true.

Step 2 Find the number of runs.

a. Find the median of the data. Arrange the data in ascending order.

18 19 19 19 20 22 22 23 25 27 27

28 32 35 36 37 43 43 44 46

The median is 27.

b. Replace each number in the original sequence with an A if it is above the median and with a B if it is below the median. Eliminate any numbers that are equal to the median.

B A B B B A B A B A A A A A A B B B

c. Arrange the letters according to runs.

Run	Letters
1	B
2	A
3	B B B
4	A
5	B
6	A
7	B
8	A A A A A A
9	B B B

Step 3 Find the critical value. Table M shows that with $n_1 = 9$, $n_2 = 9$, and $\alpha = 0.05$, the number of runs should be between 5 and 15.

Step 4 Make the decision. Since there are 9 runs and 9 falls between 5 and 15, the null hypothesis is not rejected.

Step 5 Summarize the results. There is not enough evidence to reject the hypothesis that the ages of the people who enroll occur at random.

The steps for the runs test are given in the Procedure Table.

Procedure Table

The Runs Test

Step 1 State the hypotheses and identify the claim.

Step 2 Find the number of runs.
  Note: When the data are numerical, find the median. Then compare each data value with the median and classify it as above or below the median. Other methods such as odd-even can also be used. (Discard any value that is equal to the median.)

Step 3 Find the critical value. Use Table M.

Step 4 Make the decision. Compare the actual number of runs with the critical value.

Step 5 Summarize the results.

Applying the Concepts **13–6**

Tall Trees

As a biologist, you wish to see if there is a relationship between the heights of tall trees and their diameters. You find the following data for the diameter (in inches) of the tree at 4.5 feet from the ground and the corresponding heights (in feet).

Diameter (in.)	Height (ft)
1024	261
950	321
451	219
505	281
761	159
644	83
707	191
586	141
442	232
546	108

Source: *The World Almanac and Book of Facts.*

1. What question are you trying to answer?
2. What type of nonparametric analysis could be used to answer the question?
3. What would be the corresponding parametric test that could be used?
4. Which test do you think would be better?
5. Perform both tests and write a short statement comparing the results.

See page 718 for the answer.

Exercises 13–6

For Exercises 1 through 4, find the critical value from Table L for the rank correlation coefficient, given sample size *n* and α. Assume that the test is two-tailed.

1. $n = 14$, $\alpha = 0.01$ 0.716
2. $n = 28$, $\alpha = 0.02$ 0.488
3. $n = 10$, $\alpha = 0.05$ 0.648
4. $n = 9$, $\alpha = 0.01$ 0.833

For Exercises 5 through 14, perform these steps.

  a. Find the Spearman rank correlation coefficient.
  b. State the hypotheses.
  c. Find the critical value. Use $\alpha = 0.05$.
  d. Make the decision.
  e. Summarize the results.

Use the traditional method of hypothesis testing unless otherwise specified.

5. Mathematics Achievement Test Scores The National Assessment of Educational Progress (U.S. Department of Education) tests mathematics, reading, and science achievement in grades 4 and 8. A random sample of states is selected, and their mathematics achievement scores are noted for fourth- and eighth-graders. At $\alpha = 0.05$ can a linear relationship be concluded between the data?

Grade 4	89	84	80	89	88	77	80
Grade 8	81	75	66	76	80	59	74

Source: *World Almanac.*

6. Subway and Commuter Rail Passengers Six cities are selected, and the number of daily passenger trips (in thousands) for subways and commuter rail service is obtained. At $\alpha = 0.05$, is there a relationship between the variables? Suggest one reason why the transportation authority might use the results of this study.

City	1	2	3	4	5	6
Subway	845	494	425	313	108	41
Rail	39	291	142	103	33	39

Source: American Public Transportation Association.

7. Motion Picture Releases and Gross Revenue In Chapter 10 it was demonstrated that there was a significant linear relationship between the numbers of releases that a motion picture studio put out and its gross receipts for the year. Is there a relationship between the two at the 0.05 level of significance?

No. of releases	361	270	306	22	35	10	8	12	21
Receipts	2844	1967	1371	1064	667	241	188	154	125

Source: www.showbizdata.com

8. Hospitals and Nursing Homes Find the Spearman rank correlation coefficient for the following data, which represent the number of hospitals and nursing homes in each of seven randomly selected states. At the 0.05 level of significance, is there enough evidence to conclude that there is a correlation between the two?

Hospitals	107	61	202	133	145	117	108
Nursing homes	230	134	704	376	431	538	373

Source: *World Almanac.*

9. Calories and Cholesterol in Fast-Food Sandwiches Use the Spearman rank correlation coefficient to see if there is a linear relationship between these two sets of data, representing the number of calories and the amount of cholesterol in fast-food sandwiches.

Calories	580	580	270	470	420	415	330	430
Cholesterol (mg)	205	225	285	270	185	215	185	220

Source: www.fatcalories.com

10. Book Publishing The data below show the number of books published in six different subject areas for the years 1980 and 2004. Use $\alpha = 0.05$ to see if there is a relationship between the two data sets. Do you think the same relationship will hold true 20 years from now? (In case you're curious, the subjects represented are agriculture, home economics, literature, music, science, and sports and recreation.)

1980	461	879	1686	357	3109	971
2004	1065	3639	4671	2764	8509	4806

Source: *New York Times Almanac.*

11. Gasoline Costs Shown is a comparison between the average gasoline prices charged by a gasoline station and a car rental company for 10 cities in the United States before the recent surge in gasoline prices. At $\alpha = 0.05$, is there a relationship between the prices? How might a person who travels a lot and rents an automobile use the information obtained from this study?

Car rental agency price	5.12	5.27	5.29	5.18	5.59
Gas station price	2.09	1.96	2.29	1.94	2.20
Car rental agency price	5.30	5.83	5.46	5.12	5.15
Gas station price	2.20	2.40	2.12	2.15	2.11

Source: AAA Oil Price Information Service and car rental agencies.

12. Motor Vehicle Thefts and Burglaries Is there a relationship between the number of motor vehicle (MV) thefts and the number of burglaries (per 100,000 population) for different metropolitan areas? Use $\alpha = 0.05$.

MV theft	220.5	499.4	285.6	159.2	104.3	444
Burglary	913.6	909.2	803.6	520.9	477.8	993.7

Source: *New York Times Almanac.*

13. Cyber School Enrollments Shown are the number of students enrolled in cyber school for five randomly selected school districts and the per-pupil costs for the cyber school education. At $\alpha = 0.10$, is there a relationship between the two variables? How might this information be useful to school administrators?

Number of students	10	6	17	8	11
Per-pupil cost	7200	9393	7385	4500	8203

Source: *Pittsburgh Tribune-Review.*

14. Drug Prices Shown are the price for a human dose of several prescription drugs and the price for an equivalent dose for animals. At $\alpha = 0.10$, is there a relationship between the variables?

Humans	0.67	0.64	1.20	0.51	0.87	0.74	0.50	1.22
Animals	0.13	0.18	0.42	0.25	0.57	0.57	0.49	1.28

Source: House Committee on Government Reform.

15. A school dentist wanted to test the claim, at $\alpha = 0.05$, that the number of cavities in fourth-grade students is random. Forty students were checked, and the number of cavities each had is shown here. Test for randomness of the values above or below the median.

0	4	6	0	6	2	5	3	1	5	1
2	2	1	3	7	3	6	0	2	6	0
2	3	1	5	2	1	3	0	2	3	7
3	1	5	1	1	2	2				

16. Daily Lottery Numbers Listed below are the daily numbers (daytime drawing) for the Pennsylvania State Lottery for February 2007. Using O for odd and E for even, test for randomness at $\alpha = 0.05$.

270	054	373	204	908	121	121
804	116	467	357	926	626	247
783	554	406	272	508	764	890
441	964	606	568	039	370	583

Source: www.palottery.com

17. Cola Orders Many eating facilities serve one brand of soft drinks only, but the College Corner Café serves two different brands. On a Friday night here are the orders for cola. Test for randomness at the 0.05 level of significance.

P	P	C	C	C	P	C	P	P	C	P
P	P	C	P	C	P	C	C	C	C	P
C	C	P	P	P	P	C				

18. Random Numbers Random? A calculator generated these integers randomly. Apply the runs test to see if

you can reject the hypothesis that the numbers are truly random. Use $\alpha = 0.05$.

1	1	1	1	1	1	2	1	1	1	1
2	2	1	2	1	2	2	1	2	1	1
2	1	1								

19. Concert Seating As students, faculty, friends, and family arrived for the Spring Wind Ensemble Concert at Shafer Auditorium, they were asked whether they were going to sit in the balcony (B) or on the ground floor (G). Use the responses listed below and test for randomness at $\alpha = 0.05$.

B B G G B B G B B B B B B G B B
G G B B B B G G G G B G B B B G G

20. Twenty shoppers are in a checkout line at a grocery store. At $\alpha = 0.05$, test for randomness of their gender: male (M) or female (F). The data are shown here.

F M M F F M F M M F

F M M M F F F F F M

21. Employee Absences A supervisor records the number of employees absent over a 30-day period. Test for randomness, at $\alpha = 0.05$.

27	6	19	24	18	12	15	17	18	20
0	9	4	12	3	2	7	7	0	5
32	16	38	31	27	15	5	9	4	10

22. Skiing Conditions A ski lodge manager observes the weather for the month of February. If his customers are able to ski, he records S; if weather conditions do not permit skiing, he records N. Test for randomness, at $\alpha = 0.05$.

S S S S S N N N N N N N

N S S S N N S S S S S S S

23. Tossing a Coin Toss a coin 30 times and record the outcomes (H or T). Test the results for randomness at $\alpha = 0.05$. Repeat the experiment a few times and compare your results. Answers will vary.

Extending the Concepts

When $n \geq 30$, the formula $r = \dfrac{\pm z}{\sqrt{n-1}}$ can be used to find the critical values for the rank correlation coefficient. For example, if $n = 40$ and $\alpha = 0.05$ for a two-tailed test,

$$r = \frac{\pm 1.96}{\sqrt{40 - 1}} = \pm 0.314$$

Hence, any r_s greater than or equal to $+0.314$ or less than or equal to -0.314 is significant.

For Exercises 24 through 28, find the critical r value for each (assume that the test is two-tailed).

24. $n = 50$, $\alpha = 0.05$ ±0.28

25. $n = 30$, $\alpha = 0.01$ ±0.479

26. $n = 35$, $\alpha = 0.02$ ±0.400

27. $n = 60$, $\alpha = 0.10$ ±0.215

28. $n = 40$, $\alpha = 0.01$ ±0.413

Technology *Step by Step*

MINITAB
Step by Step

Runs Test for Randomness

1. Sequence is important! Enter the data down C1 in the same order they were collected. Do not sort them! Use the data from Example 13–10.

2. Calculate the median and store it as a constant.

   a) Select **Calc>Column Statistics.**

   b) Check the option for Median.

   c) Use C1 Age for the Input Variable.

   d) Type the name of the constant MedianAge in the Store result in text box.

   e) Click [OK].

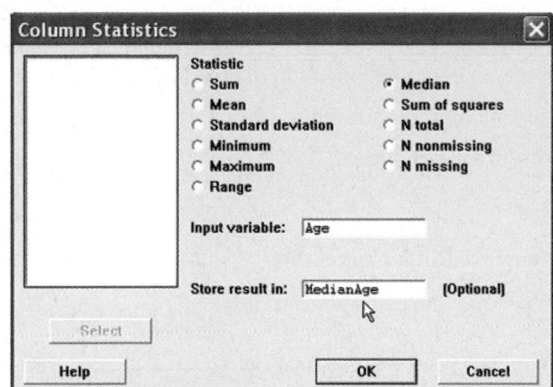 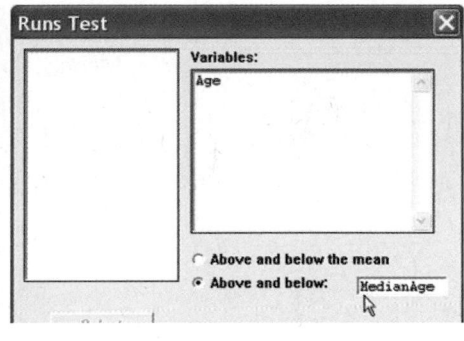

3. Select **Stat>Nonparametric>Runs Test.**

4. Select C1 Age as the variable.

5. Click the button for Above and below, then select MedianAge in the text box.

6. Click [OK]. The results will be displayed in the session window.

Runs Test: Age
```
Runs test for Age
Runs above and below K = 27

The observed number of runs = 9
The expected number of runs = 10.9
9 observations above K, 11 below
* N is small, so the following approximation may be invalid.
P-value = 0.378
```

The *P*-value is 0.378. Do not reject the null hypothesis.

Excel
Step by Step

Spearman Rank Correlation Coefficient

Example: Textbook Ratings

Two students were asked to rate eight different textbooks for a specific course on an ascending scale from 0 to 20 points. Points were assigned for each of several categories, such as reading level, use of illustrations, and use of color. At $\alpha = 0.05$, test the hypothesis that there is a significant linear correlation between the two students' ratings. The data are shown in the following table.

Textbook	Student 1's rating	Student 2's rating
A	4	4
B	10	6
C	18	20
D	20	14
E	12	16
F	2	8
G	5	11
H	9	7

Excel does not have a procedure to compute the Spearman rank correlation coefficient. However, you may compute this statistic by using the MegaStat Add-in available on your CD. If you have not installed this add-in, do so, following the instructions from the Chapter 1 Excel Step by Step.

1. Enter the rating scores from the example into columns A and B of a new worksheet.

2. From the toolbar, select Add-Ins, **MegaStat>Nonparametric Tests>Spearman Coefficient of Rank Correlation.** *Note:* You may need to open MegaStat from the MegaStat.xls file on your computer's hard drive.

3. Type **A1:B8** in the box for Input range.

4. Check the Correct for ties option.

5. Click [OK].

Spearman Coefficient of Rank Correlation

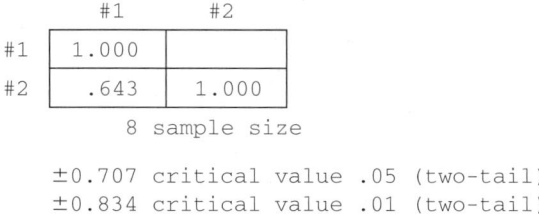

	#1	#2
#1	1.000	
#2	.643	1.000

8 sample size

±0.707 critical value .05 (two-tail)
±0.834 critical value .01 (two-tail)

Since the correlation coefficient 0.643 is less than the critical value, there is not enough evidence to reject the null hypothesis of a nonzero correlation between the variables.

Summary

- In many research situations, the assumptions (particularly that of normality) for the use of parametric statistics cannot be met. Also, some statistical studies do not involve parameters such as means, variances, and proportions. For both situations, statisticians have developed nonparametric statistical methods, also called *distribution-free methods.* (13–1)

- There are several advantages to the use of nonparametric methods. The most important one is that no knowledge of the population distributions is required. Other advantages include ease of computation and understanding. The major disadvantage is that they are less efficient than their parametric counterparts when the assumptions for the parametric methods are met. In other words, larger sample sizes are needed to get results as accurate as those given by their parametric counterparts. (13–1)

- This list gives the nonparametric statistical tests presented in this chapter, along with their parametric counterparts.

Nonparametric test	Parametric test	Condition
Single-sample sign test (13–2)	z or t test	One sample
Paired-sample sign test (13–2)	z or t test	Two dependent samples
Wilcoxon rank sum test (13–3)	z or t test	Two independent samples
Wilcoxon signed-rank test (13–4)	t test	Two dependent samples
Kruskal-Wallis test (13–5)	ANOVA	Three or more independent samples
Spearman rank correlation coefficient (13–6)	Pearson's correlation coefficient	Relationships between variables
Runs test (13–6)	None	Randomness

- When the assumptions of the parametric tests can be met, the parametric tests should be used instead of their nonparametric counterparts.

Important Terms

distribution-free statistics 672

Kruskal-Wallis test 693

nonparametric statistics 672

parametric tests 672

ranking 673

run 703

runs test 703

sign test 675

Spearman rank correlation coefficient 700

Wilcoxon rank sum test 683

Wilcoxon signed-rank test 683

Important Formulas

Formula for the z test value in the sign test:

$$z = \frac{(X + 0.5) - (n/2)}{\sqrt{n}/2}$$

where

n = sample size (greater than or equal to 26)

X = smaller number of positive or negative signs

Formula for the Wilcoxon rank sum test:

$$z = \frac{R - \mu_R}{\sigma_R}$$

where

$$\mu_R = \frac{n_1(n_1 + n_2 + 1)}{2}$$

$$\sigma_R = \sqrt{\frac{n_1 n_2(n_1 + n_2 + 1)}{12}}$$

R = sum of ranks for smaller sample size (n_1)

n_1 = smaller of sample sizes

n_2 = larger of sample sizes

$n_1 \geq 10$ and $n_2 \geq 10$

Formula for the Wilcoxon signed-rank test:

$$z = \frac{w_s - \frac{n(n + 1)}{4}}{\sqrt{\frac{n(n + 1)(2n + 1)}{24}}}$$

where

n = number of pairs where difference is not 0 and $n \geq 30$

w_s = smaller sum in absolute value of signed ranks

Formula for the Kruskal-Wallis test:

$$H = \frac{12}{N(N + 1)}\left(\frac{R_1^2}{n_1} + \frac{R_2^2}{n_2} + \cdots + \frac{R_k^2}{n_k}\right) - 3(N + 1)$$

where

R_1 = sum of ranks of sample 1

n_1 = size of sample 1

R_2 = sum of ranks of sample 2

n_2 = size of sample 2

$\vdots$

R_k = sum of ranks of sample k

n_k = size of sample k

$N = n_1 + n_2 + \cdots + n_k$

k = number of samples

Formula for the Spearman rank correlation coefficient:

$$r_s = 1 - \frac{6 \Sigma d^2}{n(n^2 - 1)}$$

where

d = difference in ranks

n = number of data pairs

Review Exercises

For Exercises 1 through 13, follow this procedure:

  a. State the hypotheses and identify the claim.

  b. Find the critical value(s).

  c. Compute the test value.

  d. Make the decision.

  e. Summarize the results.

Use the traditional method of hypothesis testing unless otherwise specified.

  1. **Ages of City Residents** The median age for the total population of the state of Maine is 41.2, the highest in the nation. The mayor of a particular city believes that his population is considerably "younger" and that the median age there is 36 years. At $\alpha = 0.05$, is there sufficient evidence to reject his claim? The data here represent a random selection of persons from the household population of the city.

40	56	42	72	12	22
25	43	39	48	50	37
18	35	15	30	52	45
10	24	25	39	29	19
30	60	38	42	41	61

Source: www.factfinder.census.gov

  2. **Lifetime of Truck Tires** A tire manufacturer claims that the median lifetime of a certain brand of truck tires is 40,000 miles. A sample of 30 tires shows that 12 lasted longer than 40,000 miles. Is there enough evidence to reject the claim at $\alpha = 0.05$? Use the sign test.

  3. **Grocery Store Repricing** A grocery store chain has decided to help customers save money by instituting "temporary repricing" to help cut costs. Nine products from the sale flyer are featured below with their regular price and their "temporary" new price. Using the paired-sample sign test and $\alpha = 0.05$, is there evidence of a difference in price? Comment on your results.

Old	2.59	0.69	1.29	3.10	1.89	2.05	1.58	2.75	1.99
New	2.09	0.70	1.18	2.95	1.59	1.75	1.32	2.19	1.99

   4. **Record High Temperatures** Shown here are the record high temperatures for Dawson Creek in British Columbia, Canada, and for Whitehorse in Yukon, Canada, for 12 months. Using the Wilcoxon rank sum test at $\alpha = 0.05$, do you find a difference in the record high temperatures? Use the *P*-value method.

Dawson Creek	52	60	57	71	86	89	94	93	88	80	66	52
White-horse	47	50	51	69	86	89	91	86	80	66	51	47

Source: Jack Williams, *The USA TODAY Weather Almanac.*

  5. **Hours Worked by Student Employees** Student employees are a major part of most college campus employment venues. Two major departments that participate in student hiring are listed below with the number of hours worked by students for a month. At the 0.10 level of significance, is there sufficient evidence to conclude a difference? Is the conclusion the same for the 0.05 level of significance?

Athletics	20	24	17	12	18	22	25	30	15	19	
Library	35	28	24	20	25	18	22	26	31	21	19

   6. **Fuel Efficiency of Automobiles** Twelve automobiles were tested to see how many miles per gallon each one obtained. Under similar driving conditions, they were tested again, using a special additive. The data are shown here. At $\alpha = 0.05$, did the additive improve gas mileage? Use the Wilcoxon signed-rank test.

Before		After	
13.6	18.3	22.6	23.7
18.2	19.5	21.9	20.8
16.1	18.2	25.3	25.3
15.3	16.7	28.6	27.2
19.2	21.3	15.2	17.2
18.8	17.2	16.3	18.5

  7. **Lunch Costs** Full-time employees in a large city were asked how much they spent on a typical weekday lunch and how much they spent on the weekend. The amounts are listed below. At $\alpha = 0.05$, is there sufficient evidence to conclude a difference in the amounts spent?

Weekday	7.00	5.50	4.50	10.00	6.75	5.00	6.00
Weekend	6.00	10.00	7.00	12.00	8.50	7.00	8.00

  8. **Breaking Strengths of Ropes** Samples of three types of ropes are tested for breaking strength. The data (in pounds) are shown here. At $\alpha = 0.05$, is there a difference in the breaking strength of the ropes? Use the Kruskal-Wallis test.

Cotton	Nylon	Hemp
230	356	506
432	303	527
505	361	581
487	405	497
451	432	459
380	378	507
462	361	562
531	399	571
366	372	499
372	363	475
453	306	505
488	304	561
462	318	532
467	322	501

Statistics Today

Too Much or Too Little?—Revisited

In this case, the manufacturer would select a sequence of bottles and see how many bottles contained more than 40 ounces, denoted by plus, and how many bottles contained less than 40 ounces, denoted by minus. The sequence could then be analyzed according to the number of runs, as explained in Section 13–6. If the sequence were not random, then the machine would need to be checked to see if it was malfunctioning. Another method that can be used to see if machines are functioning properly is *statistical quality control*. This method is beyond the scope of this book.

9. **Beach Temperatures for July** The National Oceanographic Data Center provides useful data for vacation planning. Below are listed beach temperatures in the month of July for various U.S. coastal areas. Using the 0.05 level of significance, can it be concluded that there is a difference in temperatures? Omit the Southern Pacific temperatures and repeat the procedure. Is the conclusion the same?

Southern Pacific	Western Gulf	Eastern Gulf	Southern Atlantic
67	86	87	76
68	86	87	81
66	84	86	82
69	85	86	84
63	79	85	80
62	85	84	86
		85	87

Source: www.nodc.noaa.gov

 10. **Homework Exercises and Exam Scores** A statistics instructor wishes to see whether there is a relationship between the number of homework exercises a student completes and her or his exam score. The data are shown here. Using the Spearman rank correlation coefficient, test the hypothesis that there is no relationship at $\alpha = 0.05$.

Homework problems	63	55	58	87	89	52	46	75	105
Exam score	85	71	75	98	93	63	72	89	100

11. Shown below is the average number of viewers for 10 television shows for two consecutive years. At $\alpha = 0.05$, is there a relationship between the number of viewers?

Last year	28.9	26.4	20.8	25.0	21.0	19.2
This year	26.6	20.5	20.2	19.1	18.9	17.8

Last year	13.7	18.8	16.8	15.3
This year	16.8	16.7	16.0	15.8

12. **Book Arrangements** A bookstore has a display of sale books arranged on shelves in the store window. A combination of hardbacks (H) and paperbacks (P) is arranged as follows. Test for randomness at $\alpha = 0.05$.

H H H P P P P P H P H P H P H H H H H P P P P P
H H P P P H P P P P P P

13. **Exam Scores** An instructor wishes to see whether grades of students who finish an exam occur at random. Shown here are the grades of 30 students in the order that they finished an exam. (Read from left to right across each row, and then proceed to the next row.) Test for randomness, at $\alpha = 0.05$.

87	93	82	77	64	98
100	93	88	65	72	73
56	63	85	92	95	91
88	63	72	79	55	53
65	68	54	71	73	72

Data Analysis

The Data Bank is found in Appendix D, or on the World Wide Web by following links from www.mhhe.com/math/stat/bluman

1. From the Data Bank, choose a sample and use the sign test to test one of the following hypotheses.

   a. For serum cholesterol, test H_0: median = 220 milligram percent (mg%).

   b. For systolic pressure, test H_0: median = 120 millimeters of mercury (mm Hg).

   c. For IQ, test H_0: median = 100.

   d. For sodium level, test H_0: median = 140 mEq/l.

2. From the Data Bank, select a sample of subjects. Use the Kruskal-Wallis test to see if the sodium levels of smokers and nonsmokers are equal.

3. From the Data Bank select a sample of 50 subjects. Use the Wilcoxon rank sum test to see if the means of the sodium levels of the males differ from those of the females.

Chapter Quiz

Determine whether each statement is true or false. If the statement is false, explain why.

1. Nonparametric statistics cannot be used to test the difference between two means. False

2. Nonparametric statistics are more sensitive than their parametric counterparts. False

3. Nonparametric statistics can be used to test hypotheses about parameters other than means, proportions, and standard deviations. True

4. Parametric tests are preferred over their nonparametric counterparts, if the assumptions can be met. True

Select the best answer.

5. The _____ test is used to test means when samples are dependent and the normality assumption cannot be met.

   ⓐ. Wilcoxon signed-rank   c. Sign
   b. Wilcoxon rank sum   d. Kruskal-Wallis

6. The Kruskal-Wallis test uses the _____ distribution.

   a. z   ⓒ Chi-square
   b. t   d. F

7. The nonparametric counterpart of ANOVA is the
   _____.

   a. Wilcoxon signed-rank test
   b. Sign test
   c. Runs test
   ⓓ. None of the above

8. To see if two rankings are related, you can use the
   _____.

   a. Runs test
   ⓑ. Spearman correlation coefficient
   c. Sign test
   d. Kruskal-Wallis test

Complete the following statements with the best answer.

9. When the assumption of normality cannot be met, you can use _____ tests. Nonparametric

10. When data are _____ or _____ in nature, nonparametric methods are used. Nominal, ordinal

11. To test to see whether a median was equal to a specific value, you would use the _____ test. Sign

12. Nonparametric tests are less _____ than their parametric counterparts. Sensitive

For the following exercises, use the traditional method of hypothesis testing unless otherwise specified.

 13. Home Prices The median price for an existing home in 2009 was $177,500. A random sample of homes for sale listed by a local realtor indicated homes available for the following prices. Test the claim that the median is not $177,500. Use $\alpha = 0.05$.

184,500	174,900	155,000	210,000	235,500	399,900
355,900	182,500	229,900	199,900	169,900	219,900

Source: *World Almanac.*

14. Lifetimes of Batteries A battery manufacturer claims that the median lifetime of a certain brand of heavy-duty battery is 1200 hours. A sample of 25 batteries shows that 15 lasted longer than 1200 hours. Test the claim at $\alpha = 0.05$. Use the sign test.

 15. Weights of Turkeys A special diet is fed to adult turkeys to see if they will gain weight. The before and after weights (in pounds) are given here. Use the paired-sample sign test at $\alpha = 0.05$ to see if there is weight gain.

Before	28	24	29	30	32	33	25	26	28
After	30	29	31	32	32	35	29	25	31

 16. Charity Donations Two teams of 10 members each solicited donations for their participation in a charity walk for blood cancer research. The teams received the following amounts. At $\alpha = 0.05$ can it be concluded that there is a difference in amounts?

Team A	100	50	65	50	60	75	100	150	108	120
Team B	135	90	80	140	155	60	200	58	70	72

17. Textbook Costs Samples of students majoring in law and nursing are selected, and the amount each spent on textbooks for the spring semester is recorded here, in dollars. Using the Wilcoxon rank sum test at $\alpha = 0.10$, is there a difference in the amount spent by each group?

Law	167	158	162	106	98	206	112	121
Nursing	98	198	209	168	157	126	104	122

Law	133	145	151	199
Nursing	111	138	116	201

 18. Student Grade Point Averages The grade point average of a group of students was recorded for one month. During the next nine-week grading period, the students attended a workshop on study skills. Their GPAs were recorded at the end of the grading period, and the data appear here. Using the Wilcoxon signed-rank test at $\alpha = 0.05$, can it be concluded that the GPA increased?

Before	3.0	2.9	2.7	2.5	2.1	2.6	1.9	2.0
After	3.2	3.4	2.9	2.5	3.0	3.1	2.4	2.8

19. Sodium Content of Fast-Food Sandwiches
Sometimes calories and cholesterol are not the only considerations in healthy eating. Below are listed the sodium contents (in mg) for sandwiches from three popular fast-food restaurants. Use $\alpha = 0.05$.

No. 1	No. 2	No. 3
2940	2010	1130
3720	1850	1190
3180	1980	1220
2260	1640	1640
2780	1440	1240

Source: www.fatcalories.com

20. Medication and Reaction Times Three different groups of monkeys were fed three different medications for one month to see if the medication has any effect on reaction time. Each monkey was then taught to repeat a series of steps to receive a reward. The number of trials it took each to receive the reward is shown here. At $\alpha = 0.05$, does the medication have an effect on reaction time? Use the Kruskal-Wallis test. Use the P-value method.

Med. 1	8	7	11	14	8	6	5
Med. 2	3	4	6	7	9	3	4
Med. 3	8	14	13	7	5	9	12

21. Drug Prices Is there a relationship between the prescription drug prices in Canada and Great Britain? Use $\alpha = 0.10$.

Canada	1.47	1.07	1.34	1.34	1.47	1.07	3.39	1.11	1.13
Great Britain	1.67	1.08	1.67	0.82	1.73	0.95	2.86	0.41	1.70

Source: USA TODAY.

22. Funding and Enrollment for Head Start Students Is there a relationship between the amount of money (in millions of dollars) spent on the Head Start Program by the states and the number of students enrolled (in thousands)? Use $\alpha = 0.10$.

Funding	100	50	22	88	49	219
Enrollment	16	7	3	14	8	31

Source: Gannet News Service.

23. Birth Registry At the state registry of vital statistics, the birth certificates issued for females (F) and males (M) were tallied. At $\alpha = 0.05$, test for randomness. The data are shown here.

M M F F F F F F F F M M M M F F
M F M F M M M F F F

24. Output of Motors The output in revolutions per minute (rpm) of 10 motors was obtained. The motors were tested again under similar conditions after they had been reconditioned. The data are shown here. At $\alpha = 0.05$, did the reconditioning improve the motors' performance? Use the Wilcoxon signed-rank test.

Before	413	701	397	602	405	512	450	487	388	351
After	433	712	406	650	450	550	450	500	402	415

25. State Lottery Numbers A statistician wishes to determine if a state's lottery numbers are selected at random. The winning numbers selected for the month of February are shown here. Test for randomness at $\alpha = 0.05$.

321	909	715	700	487	808	509	606	943	761
200	123	367	012	444	576	409	128	567	908
103	407	890	193	672	867	003	578		

Critical Thinking Challenges

1. Tolls for Bridge Two commuters ride to work together in one car. To decide who pays the toll for a bridge on the way to work, they flip a coin and the loser pays. Explain why over a period of one year, one person might have to pay the toll 5 days in a row. There is no toll on the return trip. (*Hint:* You may want to use random numbers.)

2. Olympic Medals Shown in the next column are the type and number of medals each country won in the 2000 Summer Olympic Games. You are to rank the countries from highest to lowest. Gold medals are highest, followed by silver, followed by bronze. There are many different ways to rank objects and events. Here are several suggestions.

a. Rank the countries according to the total medals won.
b. List some advantages and disadvantages of this method.
c. Rank each country separately for the number of gold medals won, then for the number of silver medals won, and then for the number of bronze medals won. Then rank the countries according to the sum of the *ranks* for the categories.
d. Are the rankings of the countries the same as those in step *a*? Explain any differences.
e. List some advantages and disadvantages of this method of ranking.
f. A third way to rank the countries is to assign a weight to each medal. In this case, assign 3 points

for each gold medal, 2 points for each silver medal, and 1 point for each bronze medal the country won. Multiply the number of medals by the weights for each medal and find the sum. For example, since Austria won 2 gold medals, 1 silver medal, and 0 bronze medals, its rank sum is $(2 \times 3) + (1 \times 2) + (0 \times 1) = 8$. Rank the countries according to this method.

g. Compare the ranks using this method with those using the other two methods. Are the rankings the same or different? Explain.

h. List some advantages and disadvantages of this method.

i. Select two of the rankings, and run the Spearman rank correlation test to see if they differ significantly.

Summer Olympic Games 2000 Final Medal Standings

Country	Gold	Silver	Bronze
Austria	2	1	0
Canada	3	3	8
Germany	14	17	26
Italy	13	8	13
Norway	4	3	3
Russia	32	28	28
Switzerland	1	6	2
United States	40	24	33

Source: Reprinted with permission from the *World Almanac and Book of Facts*. World Almanac Education Group Inc.

 Data Projects

Use a significance level of 0.05 for all tests below.

1. **Business and Finance** Monitor the price of a stock over a five-week period. Note the amount of gain or loss per day. Test the claim that the median is 0. Perform a runs test to see if the distribution of gains and losses is random.

2. **Sports and Leisure** Watch a basketball game, baseball game, or football game. For baseball, monitor an inning's pitches for balls and strikes (all fouls and balls in play also count as strikes). For football monitor a series of plays for runs versus passing plays. For basketball monitor one team's shots for misses versus made shots. For the collected data, conduct a runs test to see if the distribution is random.

3. **Technology** Use the data collected in data project 3 of Chapter 2 regarding song lengths. Consider only three genres. For example, use rock, alternative, and hip

hop/rap. Conduct a Kruskal-Wallis test to determine if the mean song lengths for the genres are the same.

4. **Health and Wellness** Have everyone in class take her or his pulse during the first minute of class. Have everyone take his or her pulse again 30 minutes into class. Conduct a paired-sample sign test to determine if there is a difference in pulse rates.

5. **Politics and Economics** Find the ranking for each state for its mean SAT Mathematics scores, its mean SAT English score, and its mean for income. Conduct a rank correlation analysis using Math and English, Math and income, and English and income. Which pair has the strongest relationship?

6. **Your Class** Have everyone in class take his or her temperature on a healthy day. Test the claim that the median body temperature is 98.6°F.

Hypothesis-Testing Summary 3*

15. Test to see whether the median of a sample is a specific value when $n \geq 26$.

Example: H_0: median $= 100$

Use the sign test:

$$z = \frac{(X + 0.5) - (n/2)}{\sqrt{n}/2}$$

16. Test to see whether two independent samples are obtained from populations that have identical distributions.

Example: H_0: There is no difference in the ages of the subjects.

Use the Wilcoxon rank sum test:

$$z = \frac{R - \mu_R}{\sigma_R}$$

where

$$\mu_R = \frac{n_1(n_1 + n_2 + 1)}{2}$$

$$\sigma_R = \sqrt{\frac{n_1 n_2 (n_1 + n_2 + 1)}{12}}$$

*This summary is a continuation of Hypothesis-Testing Summary 2 at the end of Chapter 12.

17. Test to see whether two dependent samples have identical distributions.

    Example: H_0: There is no difference in the effects of a tranquilizer on the number of hours a person sleeps at night.

    Use the Wilcoxon signed-rank test:

    $$z = \frac{w_s - \dfrac{n(n+1)}{4}}{\sqrt{\dfrac{n(n+1)(2n+1)}{24}}}$$

    when $n \geq 30$.

18. Test to see whether three or more samples come from identical populations.

    Example: H_0: There is no difference in the weights of the three groups.

    Use the Kruskal-Wallis test:

    $$H = \frac{12}{N(N+1)}\left(\frac{R_1^2}{n_1} + \frac{R_2^2}{n_2} + \cdots + \frac{R_k^2}{n_k}\right) - 3(N+1)$$

19. Rank correlation coefficient.

    $$r_s = 1 - \frac{6\,\Sigma d^2}{n(n^2 - 1)}$$

20. Test for randomness: Use the runs test.

Answers to Applying the Concepts

Section 13–1   Ranking Data

Percent	2.6	3.8	4.0	4.0	5.4	7.0	7.0	7.3	10.0
Rank	1	2	3.5	3.5	5	6.5	6.5	8	9

Section 13–2   Clean Air

1. The claim is that the median number of days that a large city failed to meet EPA standards is 11 days per month.

2. We will use the sign test, since we do not know anything about the distribution of the variable and we are testing the median.

3. H_0: median = 11 and H_1: median > 11.

4. If $\alpha = 0.05$, then the critical value is 5.

5. The test value is 9.

6. Since $9 > 5$, do not reject the null hypothesis.

7. There is not enough evidence to conclude that the median is not 11 days per month.

8. We cannot use a parametric test in this situation.

Section 13–3   School Lunch

1. The samples are independent since two different random samples were selected.

2. H_0: There is no difference in the number of calories served for lunch in elementary and secondary schools.

   H_1: There is a difference in the number of calories served for lunch in elementary and secondary schools.

3. We will use the Wilcoxon rank sum test.

4. The critical value is ± 1.96 if we use $\alpha = 0.05$.

5. The test statistic is $z = -2.15$.

6. Since $-2.15 < -1.96$, we reject the null hypothesis and conclude that there is a difference in the number of calories served for lunch in elementary and secondary schools.

7. The corresponding parametric test is the two-sample t test.

8. We would need to know that the samples were normally distributed to use the parametric test.

9. Since t tests are robust against variations from normality, the parametric test would yield the same results.

Section 13–4   Pain Medication

1. The purpose of the study is to see how effective a pain medication is.

2. These are dependent samples, since we have before and after readings on the same subjects.

3. H_0: The severity of pain after is the same as the severity of pain before the medication was administered.

   H_1: The severity of pain after is less than the severity of pain before the medication was administered.

4. We will use the Wilcoxon signed-rank test.

5. We will choose to use a significance level of 0.05.

6. The test statistic is $w_s = 2.5$. The critical value is 4. Since $2.5 < 4$, we reject the null hypothesis. There is enough evidence to conclude that the severity of pain after is less than the severity of pain before the medication was administered.

7. The parametric test that could be used is the t test for small dependent samples.

8. The results for the parametric test would be the same.

Section 13–5 Heights of Waterfalls

1. We are investigating the heights of waterfalls on three continents.

2. We will use the Kruskal-Wallis test.

3. H_0: There is no difference in the heights of waterfalls on the three continents.

   H_1: There is a difference in the heights of waterfalls on the three continents.

4. We will use the 0.05 significance level. The critical value is 5.991. Our test statistic is $H = 0.01$.

5. Since $0.01 < 5.991$, we fail to reject the null hypothesis. There is not enough evidence to conclude that there is a difference in the heights of waterfalls on the three continents.

6. The corresponding parametric test is analysis of variance (ANOVA).

7. To perform an ANOVA, the population must be normally distributed, the samples must be independent of each other, and the variances of the samples must be equal.

Section 13–6 Tall Trees

1. The biologist is trying to see if there is a relationship between the heights and diameters of tall trees.

2. We will use a Spearman rank correlation analysis.

3. The corresponding parametric test is the Pearson product moment correlation analysis.

4. Answers will vary.

5. The Pearson correlation coefficient is $r = 0.329$. The associated P-value is 0.353. We would fail to reject the null hypothesis that the correlation is zero. The Spearman's rank correlation coefficient is $r_s = 0.115$. We would reject the null hypothesis, at the 0.05 significance level, if $r_s > 0.648$. Since $0.115 < 0.648$, we fail to reject the null hypothesis that the correlation is zero. Both the parametric and nonparametric tests find that the correlation is not statistically significantly different from zero—it appears that no linear relationship exists between the heights and diameters of tall trees.

Sampling and Simulation

Objectives

After completing this chapter, you should be able to

1 Demonstrate a knowledge of the four basic sampling methods.

2 Recognize faulty questions on a survey and other factors that can bias responses.

3 Solve problems, using simulation techniques.

Outline

The Monty Hall Problem

On the game show *Let's Make A Deal,* host Monty Hall gave a contestant a choice of three doors. A valuable prize was behind one door, and nothing was behind the other two doors. When the contestant selected one door, host Monty Hall opened one of the other doors that the contestant didn't select and that had no prize behind it. (Monty Hall knew in advance which door had the prize.) Then he asked the contestant if he or she wanted to change doors or keep the one that the contestant originally selected. Now the question is, Should the contestant switch doors, or does it really matter? This chapter will show you how you can solve this problem by simulation. For the answer, see Statistics Today—Revisited at the end of the chapter.

Introduction

Most people have heard of Gallup and Nielsen. These and other pollsters gather information about the habits and opinions of the U.S. people. Such survey firms, and the U.S. Census Bureau, gather information by selecting samples from well-defined populations. Recall from Chapter 1 that the subjects in the sample should be a subgroup of the subjects in the population. Sampling methods often use what are called *random numbers* to select samples.

Since many statistical studies use surveys and questionnaires, some information about these is presented in Section 14–2.

Random numbers are also used in *simulation techniques.* Instead of studying a real-life situation, which may be costly or dangerous, researchers create a similar situation in a laboratory or with a computer. Then, by studying the simulated situation, researchers can gain the necessary information about the real-life situation in a less expensive or safer manner. This chapter will explain some common methods used to obtain samples as well as the techniques used in simulations.

14–1

Common Sampling Techniques

Objective 1

Demonstrate a knowledge of the four basic sampling methods.

In Chapter 1, a *population* was defined as all subjects (human or otherwise) under study. Since some populations can be very large, researchers cannot use every single subject, so a sample must be selected. A *sample* is a subgroup of the population. Any subgroup of the population, technically speaking, can be called a sample. However, for researchers to make valid inferences about population characteristics, the sample must be random.

> For a sample to be a **random sample,** every member of the population must have an equal chance of being selected.

When a sample is chosen at random from a population, it is said to be an **unbiased sample.** That is, the sample, for the most part, is representative of the population. Conversely, if a sample is selected incorrectly, it may be a biased sample. Samples are said to be **biased samples** when some type of systematic error has been made in the selection of the subjects.

A sample is used to get information about a population for several reasons:

1. *It saves the researcher time and money.*

2. *It enables the researcher to get information that he or she might not be able to obtain otherwise.* For example, if a person's blood is to be analyzed for cholesterol, a researcher cannot analyze every single drop of blood without killing the person. Or if the breaking strength of cables is to be determined, a researcher cannot test to destruction every cable manufactured, since the company would not have any cables left to sell.

3. *It enables the researcher to get more detailed information about a particular subject.* If only a few people are surveyed, the researcher can conduct in-depth interviews by spending more time with each person, thus getting more information about the subject. This is not to say that the smaller the sample, the better; in fact, the opposite is true. In general, larger samples—if correct sampling techniques are used—give more reliable information about the population.

It would be ideal if the sample were a perfect miniature of the population in all characteristics. This ideal, however, is impossible to achieve, because there are so many human traits (height, weight, IQ, etc.). The best that can be done is to select a sample that will be representative with respect to *some* characteristics, preferably those pertaining to the study. For example, if one-half of the population subjects are female, then approximately one-half of the sample subjects should be female. Likewise, other characteristics, such as age, socioeconomic status, and IQ, should be represented proportionately. To obtain unbiased samples, statisticians have developed several basic sampling methods. The most common methods are *random, systematic, stratified,* and *cluster sampling.* Each method will be explained in detail in this section.

In addition to the basic methods, there are other methods used to obtain samples. Some of these methods are also explained in this section.

Random Sampling

A random sample is obtained by using methods such as random numbers, which can be generated from calculators, computers, or tables. In *random sampling,* the basic requirement is that, for a sample of size *n,* all possible samples of this size have an equal chance of being selected from the population. But before the correct method of obtaining a random sample is explained, several incorrect methods commonly used by various researchers and agencies to gain information are discussed.

One incorrect method commonly used is to ask "the person on the street." News reporters use this technique quite often. Selecting people haphazardly on the street does not meet the requirement for simple random sampling, since not all possible samples of a specific size have an equal chance of being selected. Many people will be at home or at work when the interview is being conducted and therefore do not have a chance of being selected.

Another incorrect technique is to ask a question by either radio or television and have the listeners or viewers call the station to give their responses or opinions. Again, this sample is not random, since only those who feel strongly for or against the issue may respond and people may not have heard or seen the program. A third erroneous method is to ask people to respond by mail. Again, only those who are concerned and who have the time are likely to respond.

These methods do not meet the requirement of random sampling, since not all possible samples of a specific size have an equal chance of being selected. To meet this requirement, researchers can use one of two methods. The first method is to number each element of the population and then place the numbers on cards. Place the cards in a hat or fishbowl, mix them, and then select the sample by drawing the cards. When using this procedure, researchers must ensure that the numbers are well mixed. On occasion, when this procedure is used, the numbers are not mixed well, and the numbers chosen for the sample are those that were placed in the bowl last.

The second and preferred way of selecting a random sample is to use random numbers. Figure 14–1 shows a table of two-digit random numbers generated by a computer. A more detailed table of random numbers is found in Table D of Appendix C.

The theory behind random numbers is that each digit, 0 through 9, has an equal probability of occurring. That is, in every sequence of 10 digits, each digit has a probability of $\frac{1}{10}$ of occurring. This does not mean that in every sequence of 10 digits, you will find each digit. Rather, it means that on the average, each digit will occur once. For example, the digit 2 may occur 3 times in a sequence of 10 digits, but in later sequences, it may not occur at all, thus averaging to a probability of $\frac{1}{10}$.

To obtain a sample by using random numbers, number the elements of the population sequentially and then select each person by using random numbers. This process is shown in Example 14–1.

Random samples can be selected with or without replacement. If the same member of the population cannot be used more than once in the study, then the sample is selected without replacement. That is, once a random number is selected, it cannot be used later.

Figure 14–1												
Table of Random Numbers	79	41	71	93	60	35	04	67	96	04	79	10	86
	26	52	53	13	43	50	92	09	87	21	83	75	17
	18	13	41	30	56	20	37	74	49	56	45	46	83
	19	82	02	69	34	27	77	34	24	93	16	77	00
	14	57	44	30	93	76	32	13	55	29	49	30	77
	29	12	18	50	06	33	15	79	50	28	50	45	45
	01	27	92	67	93	31	97	55	29	21	64	27	29
	55	75	65	68	65	73	07	95	66	43	43	92	16
	84	95	95	96	62	30	91	64	74	83	47	89	71
	62	62	21	37	82	62	19	44	08	64	34	50	11
	66	57	28	69	13	99	74	31	58	19	47	66	89
	48	13	69	97	29	01	75	58	05	40	40	18	29
	94	31	73	19	75	76	33	18	05	53	04	51	41
	00	06	53	98	01	55	08	38	49	42	10	44	38
	46	16	44	27	80	15	28	01	64	27	89	03	27
	77	49	85	95	62	93	25	39	63	74	54	82	85
	81	96	43	27	39	53	85	61	12	90	67	96	02
	40	46	15	73	23	75	96	68	13	99	49	64	11

Note: In the explanations and examples of the sampling procedures, a small population will be used, and small samples will be selected from this population. Small populations are used for illustrative purposes only, because the entire population could be included with little difficulty. In real life, however, researchers must usually sample from very large populations, using the procedures shown in this chapter.

| Example 14–1 | Television Show Interviews |

Suppose a researcher wants to produce a television show featuring in-depth interviews with state governors on the subject of capital punishment. Because of time constraints, the 60-minute program will have room for only 10 governors. The researcher wishes to select the governors at random. Select a random sample of 10 states from 50.

Note: This answer is not unique.

Solution

Step 1 Number each state from 1 to 50, as shown. In this case, they are numbered alphabetically.

01. Alabama	14. Indiana	27. Nebraska	40. South Carolina
02. Alaska	15. Iowa	28. Nevada	41. South Dakota
03. Arizona	16. Kansas	29. New Hampshire	42. Tennessee
04. Arkansas	17. Kentucky	30. New Jersey	43. Texas
05. California	18. Louisiana	31. New Mexico	44. Utah
06. Colorado	19. Maine	32. New York	45. Vermont
07. Connecticut	20. Maryland	33. North Carolina	46. Virginia
08. Delaware	21. Massachusetts	34. North Dakota	47. Washington
09. Florida	22. Michigan	35. Ohio	48. West Virginia
10. Georgia	23. Minnesota	36. Oklahoma	49. Wisconsin
11. Hawaii	24. Mississippi	37. Oregon	50. Wyoming
12. Idaho	25. Missouri	38. Pennsylvania	
13. Illinois	26. Montana	39. Rhode Island	

Step 2 Using the random numbers shown in Figure 14–1, find a starting point. To find a starting point, you generally close your eyes and place your finger anywhere on the table. In this case, the first number selected was 27 in the fourth column. Going down the column and continuing on to the next column, select the first 10 numbers. They are 27, 95, 27, 73, 60, 43, 56, 34, 93, and 06. See Figure 14–2. (Note that 06 represents 6.)

Figure 14–2

Selecting a Starting Point and 10 Numbers from the Random Number Table

79	41	71	93	60 ✔	35	04	67	96	04	79	10	86
26	52	53	13	43 ✔	50	92	09	87	21	83	75	17
18	13	41	30	56 ✔	20	37	74	49	56	45	46	83
19	82	02	69	34 ✔	27	77	34	24	93	16	77	00
14	57	44	30	93 ✔	76	32	13	55	29	49	30	77
29	12	18	50	06 ✔	33	15	79	50	28	50	45	45
01	27	92	67	93	31	97	55	29	21	64	27	29
55	75	65	68	65	73	07	95	66	43	43	92	16
84	95	95	96	62	30	91	64	74	83	47	89	71
62	62	21	37	82	62	19	44	08	64	34	50	11
66	57	28	69	13	99	74	31	58	19	47	66	89
48	13	69	97	29	01	75	58	05	40	40	18	29
94	31	73	19	75	76	33	18	05	53	04	51	41
00	06	53	*Start here	01	55	08	38	49	42	10	44	38
46	16	44	㉗ ✔	80	15	28	01	64	27	89	03	27
77	49	85	95 ✔	62	93	25	39	63	74	54	82	85
81	96	43	27 ✔	39	53	85	61	12	90	67	96	02
40	46	15	73 ✔	23	75	96	68	13	99	49	64	11

Now, refer to the list of states and identify the state corresponding to each number. The sample consists of the following states:

27	Nebraska	43	Texas
95		56	
27	Nebraska	34	North Dakota
73		93	
60		06	Colorado

Step 3 Since the numbers 95, 73, 60, 56, and 93 are too large, they are disregarded. And since 27 appears twice, it is also disregarded the second time. Now, you must select six more random numbers between 1 and 50 and omit duplicates, since this sample will be selected without replacement. Make this selection by continuing down the column and moving over to the next column until a total of 10 numbers is selected. The final 10 numbers are 27, 43, 34, 06, 13, 29, 01, 39, 23, and 35. See Figure 14–3.

Figure 14–3

The Final 10 Numbers Selected

79	41	71	93	60	(35)	04	67	96	04	79	10	86
26	52	53	13	(43)	50	92	09	87	21	83	75	17
18	13	41	30	56	20	37	74	49	56	45	46	83
19	82	02	69	(34)	27	77	34	24	93	16	77	00
14	57	44	30	93	76	32	13	55	29	49	30	77
29	12	18	50	(06)	33	15	79	50	28	50	45	45
01	27	92	67	93	31	97	55	29	21	64	27	29
55	75	65	68	65	73	07	95	66	43	43	92	16
84	95	95	96	62	30	91	64	74	83	47	89	71
62	62	21	37	82	62	19	44	08	64	34	50	11
66	57	28	69	(13)	99	74	31	58	19	47	66	89
48	13	69	97	(29)	01	75	58	05	40	40	18	29
94	31	73	19	75	76	33	18	05	53	04	51	41
00	06	53	98	(01)	55	08	38	49	42	10	44	38
46	16	44	(27)	80	15	28	01	64	27	89	03	27
77	49	85	95	62	93	25	39	63	74	54	82	85
81	96	43	27	(39)	53	85	61	12	90	67	96	02
40	46	15	73	(23)	75	96	68	13	99	49	64	11

These numbers correspond to the following states:

27	Nebraska	29	New Hampshire
43	Texas	01	Alabama
34	North Dakota	39	Rhode Island
06	Colorado	23	Minnesota
13	Illinois	35	Ohio

Thus, the governors of these 10 states will constitute the sample.

Random sampling has one limitation. If the population is extremely large, it is time-consuming to number and select the sample elements. Also, notice that the random numbers in the table are two-digit numbers. If three digits are needed, then the first digit from the next column can be used, as shown in Figure 14–4. Table D in Appendix C gives five-digit random numbers.

Should We Be Afraid of Lightning?

The National Weather Service collects various types of data about the weather. For example, each year in the United States about 400 million lightning strikes occur. On average, 400 people are struck by lightning, and 85% of those struck are men. About 100 of these people die. The cause of most of these deaths is not burns, even though temperatures as high as 54,000°F are reached, but heart attacks. The lightning strike short-circuits the body's autonomic nervous system, causing the heart to stop beating. In some instances, the heart will restart on its own. In other cases, the heart victim will need emergency resuscitation.

The most dangerous places to be during a thunderstorm are open fields, golf courses, under trees, and near water, such as a lake or swimming pool. It's best to be inside a building during a thunderstorm although there's no guarantee that the building won't be struck by lightning. Are these statistics descriptive or inferential? Why do you think more men are struck by lightning than women? Should you be afraid of lightning?

Figure 14–4

Method for Selecting Three-Digit Numbers

79	41	71	93	60	35	04	67	96	04	79	10	86
26	52	53	13	43	50	92	09	87	21	83	75	17
18	13	41	30	56	20	37	74	49	56	45	46	83
19	82	02	69	34	27	77	34	24	93	16	77	00
14	57	44	30	93	76	32	13	55	29	49	30	77
29	12	18	50	06	33	15	79	50	28	50	45	45
01	27	92	67	93	31	97	55	29	21	64	27	29
55	75	65	68	65	73	07	95	66	43	43	92	16
84	95	95	96	62	30	91	64	74	83	47	89	71
62	62	21	37	82	62	19	44	08	64	34	50	11
66	57	28	69	13	99	74	31	58	19	47	66	89
48	13	69	97	29	01	75	58	05	40	40	18	29
94	31	73	19	75	76	33	18	05	53	04	51	41
00	06	53	98	01	55	08	38	49	42	10	44	38
46	16	44	27	80	15	28	01	64	27	89	03	27
77	49	85	95	62	93	25	39	63	74	54	82	85
81	96	43	27	39	53	85	61	12	90	67	96	02
40	46	15	73	23	75	96	68	13	99	49	64	11

Use one column and part of the next column for three digits, that is, 404.

Systematic Sampling

A **systematic sample** is a sample obtained by numbering each element in the population and then selecting every third or fifth or tenth, etc., number from the population to be included in the sample. This is done after the first number is selected at random.

The procedure of systematic sampling is illustrated in Example 14–2.

| Example 14–2 | Television Show Interviews |

Using the population of 50 states in Example 14–1, select a systematic sample of 10 states.

Solution

Step 1 Number the population units as shown in Example 14–1.

Step 2 Since there are 50 states and 10 are to be selected, the rule is to select every fifth state. This rule was determined by dividing 50 by 10, which yields 5.

Step 3 Using the table of random numbers, select the first digit (from 1 to 5) at random. In this case, 4 was selected.

Step 4 Select every fifth number on the list, starting with 4. The numbers include the following:

1 2 3 ④ 5 6 7 8 ⑨ 10 11 12 13 ⑭ · · ·

The selected states are as follows:

4	Arkansas	29	New Hampshire
9	Florida	34	North Dakota
14	Indiana	39	Rhode Island
19	Maine	44	Utah
24	Mississippi	49	Wisconsin

The advantage of systematic sampling is the ease of selecting the sample elements. Also, in many cases, a numbered list of the population units may already exist. For example, the manager of a factory may have a list of employees who work for the company, or there may be an in-house telephone directory.

When doing systematic sampling, you must be careful how the items are arranged on the list. For example, if each unit were arranged, say, as

1. Husband
2. Wife
3. Husband
4. Wife

then the selection of the starting number could produce a sample of all males or all females, depending on whether the starting number is even or odd and whether the number to be added is even or odd. As another example, if the list were arranged in order of heights of individuals, you would get a different average from two samples if the first were selected by using a small starting number and the second by using a large starting number.

Stratified Sampling

A **stratified sample** is a sample obtained by dividing the population into subgroups, called *strata*, according to various homogeneous characteristics and then selecting members from each stratum for the sample.

For example, a population may consist of males and females who are smokers or nonsmokers. The researcher will want to include in the sample people from each group—that is, males who smoke, males who do not smoke, females who smoke, and females who do not smoke. To accomplish this selection, the researcher divides the population into four subgroups and then selects a random sample from each subgroup. This method ensures that the sample is representative on the basis of the characteristics of gender and smoking. Of course, it may not be representative on the basis of other characteristics.

Example 14–3

Using the population of 20 students shown in Figure 14–5, select a sample of eight students on the basis of gender (male/female) and grade level (freshman/sophomore) by stratification.

Figure 14–5

Population of Students for Example 14–3

1. Ald, Peter	M	Fr	11. Martin, Janice	F	Fr
2. Brown, Danny	M	So	12. Meloski, Gary	M	Fr
3. Bear, Theresa	F	Fr	13. Oeler, George	M	So
4. Carson, Susan	F	Fr	14. Peters, Michele	F	So
5. Collins, Carolyn	F	Fr	15. Peterson, John	M	Fr
6. Davis, William	M	Fr	16. Smith, Nancy	F	Fr
7. Hogan, Michael	M	Fr	17. Thomas, Jeff	M	So
8. Jones, Lois	F	So	18. Toms, Debbie	F	So
9. Lutz, Harry	M	So	19. Unger, Roberta	F	So
10. Lyons, Larry	M	So	20. Zibert, Mary	F	So

Solution

Step 1 Divide the population into two subgroups, consisting of males and females, as shown in Figure 14–6.

Figure 14–6

Population Divided into Subgroups by Gender

Males			**Females**		
1. Ald, Peter	M	Fr	1. Bear, Theresa	F	Fr
2. Brown, Danny	M	So	2. Carson, Susan	F	Fr
3. Davis, William	M	Fr	3. Collins, Carolyn	F	Fr
4. Hogan, Michael	M	Fr	4. Jones, Lois	F	So
5. Lutz, Harry	M	So	5. Martin, Janice	F	Fr
6. Lyons, Larry	M	So	6. Peters, Michele	F	So
7. Meloski, Gary	M	Fr	7. Smith, Nancy	F	Fr
8. Oeler, George	M	So	8. Toms, Debbie	F	So
9. Peterson, John	M	Fr	9. Unger, Roberta	F	So
10. Thomas, Jeff	M	So	10. Zibert, Mary	F	So

Step 2 Divide each subgroup further into two groups of freshmen and sophomores, as shown in Figure 14–7.

Figure 14–7

Each Subgroup Divided into Subgroups by Grade Level

Group 1			**Group 2**		
1. Ald, Peter	M	Fr	1. Bear, Theresa	F	Fr
2. Davis, William	M	Fr	2. Carson, Susan	F	Fr
3. Hogan, Michael	M	Fr	3. Collins, Carolyn	F	Fr
4. Meloski, Gary	M	Fr	4. Martin, Janice	F	Fr
5. Peterson, John	M	Fr	5. Smith, Nancy	F	Fr

Group 3			**Group 4**		
1. Brown, Danny	M	So	1. Jones, Lois	F	So
2. Lutz, Harry	M	So	2. Peters, Michele	F	So
3. Lyons, Larry	M	So	3. Toms, Debbie	F	So
4. Oeler, George	M	So	4. Unger, Roberta	F	So
5. Thomas, Jeff	M	So	5. Zibert, Mary	F	So

Step 3 Determine how many students need to be selected from each subgroup to have a proportional representation of each subgroup in the sample. There are four groups, and since a total of eight students is needed for the sample, two students must be selected from each subgroup.

Step 4 Select two students from each group by using random numbers. In this case, the random numbers are as follows:

Group 1	Students 5 and 4	Group 2	Students 5 and 2
Group 3	Students 1 and 3	Group 4	Students 3 and 4

The stratified sample then consists of the following people:

Peterson, John	M	Fr	Smith, Nancy	F	Fr
Meloski, Gary	M	Fr	Carson, Susan	F	Fr
Brown, Danny	M	So	Toms, Debbie	F	So
Lyons, Larry	M	So	Unger, Roberta	F	So

The major advantage of stratification is that it ensures representation of all population subgroups that are important to the study. There are two major drawbacks to stratification, however. First, if there are many variables of interest, dividing a large population into representative subgroups requires a great deal of effort. Second, if the variables are somewhat complex or ambiguous (such as beliefs, attitudes, or prejudices), it is difficult to separate individuals into the subgroups according to these variables.

Cluster Sampling

A **cluster sample** is a sample obtained by selecting a preexisting or natural group, called a *cluster,* and using the members in the cluster for the sample.

For example, many studies in education use already existing classes, such as the seventh grade in Wilson Junior High School. The voters of a certain electoral district might be surveyed to determine their preferences for a mayoral candidate in the upcoming election. Or the residents of an entire city block might be polled to ascertain the percentage of households that have two or more incomes. In cluster sampling, researchers may use all units of a cluster if that is feasible, or they may select only part of a cluster to use as a sample. This selection is done by random methods.

There are three advantages to using a cluster sample instead of other types of samples: (1) A cluster sample can reduce costs, (2) it can simplify fieldwork, and (3) it is convenient. For example, in a dental study involving X-raying fourth-grade students' teeth to see how many cavities each child had, it would be a simple matter to select a single classroom and bring the X-ray equipment to the school to conduct the study. If other sampling methods were used, researchers might have to transport the machine to several different schools or transport the pupils to the dental office.

The major disadvantage of cluster sampling is that the elements in a cluster may not have the same variations in characteristics as elements selected individually from a population. The reason is that groups of people may be more homogeneous (alike) in specific clusters such as neighborhoods or clubs. For example, the people who live in a certain neighborhood tend to have similar incomes, drive similar cars, live in similar houses, and, for the most part, have similar habits.

Speaking of
Statistics

In this study, the researchers found that subjects did better on fill-in-the-blank questions than on multiple-choice questions. Do you agree with the professor's statement, "Trusting your first impulse is your best strategy?" Explain your answer.

TESTS

Is That Your Final Answer?

Beating game shows takes more than smarts: Contestants must also overcome self-doubt and peer pressure. Two new studies suggest today's hottest game shows are particularly challenging because the very mechanisms employed to help contestants actually lead them astray.

Multiple-choice questions are one such offender, as alternative answers seem to make test-takers ignore gut instincts. To learn why, researchers at Southern Methodist University (SMU) gave two identical tests: one using multiple-choice questions and the other fill-in-the-blank. The results, recently published in the *Journal of Educational Psychology*, show that test-takers were incorrect more often when given false alternatives, and that the longer they considered those alternatives, the more credible the answers looked.

"If you sit and stew, you forget that you know the right answer," says Alan Brown, Ph.D., a psychology professor at SMU. "Trusting your first impulse is your best strategy."

Audiences can also be trouble, says Jennifer Butler, Ph.D., a Wittenberg University psychology professor. Her recent study in the *Journal of Personality and Social Psychology* found that contestants who see audience participation as peer pressure slow down to avoid making embarrassing mistakes. But this strategy backfires, as more contemplation produces more wrong answers. Worse, Butler says, if perceived peer pressure grows unbearable, contestants may opt out of answering at all, "thinking that it's better to stop than to have your once supportive audience come to believe you're an idiot."

— *Sarah Smith*

Source: Reprinted with permission from *Psychology Today Magazine*, (Copyright © 2000 Sussex Publishers, LLC.).

Interesting Fact

Folks in extra-large aerobics classes— those with 70 to 90 participants—show up more often and are more fond of their classmates than exercisers in sessions of 18 to 26 people, report researchers at the University of Arizona.

Other Types of Sampling Techniques

In addition to the four basic sampling methods, other methods are sometimes used. In **sequence sampling,** which is used in quality control, successive units taken from production lines are sampled to ensure that the products meet certain standards set by the manufacturing company.

In **double sampling,** a very large population is given a questionnaire to determine those who meet the qualifications for a study. After the questionnaires are reviewed, a second, smaller population is defined. Then a sample is selected from this group.

In **multistage sampling,** the researcher uses a combination of sampling methods. For example, suppose a research organization wants to conduct a nationwide survey for a new product being manufactured. A sample can be obtained by using the following combination of methods. First the researchers divide the 50 states into four or five regions (or clusters). Then several states from each region are selected at random. Next the states are divided into various areas by using large cities and small towns. Samples of these areas are then selected. Next, each city and each town are divided into districts or wards. Finally, streets in these wards are selected at random, and the families living on these streets are given samples of the product to test and are asked to report the results. This hypothetical example illustrates a typical multistage sampling method.

The steps for conducting a sample survey are given in the Procedure Table.

Procedure Table

Conducting a Sample Survey

Step 1 Decide what information is needed.

Step 2 Determine how the data will be collected (phone interview, mail survey, etc.).

Step 3 Select the information-gathering instrument or design the questionnaire if one is not available.

Step 4 Set up a sampling list, if possible.

Step 5 Select the best method for obtaining the sample (random, systematic, stratified, cluster, or other).

Step 6 Conduct the survey and collect the data.

Step 7 Tabulate the data.

Step 8 Conduct the statistical analysis.

Step 9 Report the results.

Applying the Concepts 14–1

The White or Wheat Bread Debate

Read the following study and answer the questions.

A baking company selected 36 women weighing different amounts and randomly assigned them to four different groups. The four groups were white bread only, brown bread only, low-fat white bread only, and low-fat brown bread only. Each group could eat only the type of bread assigned to the group. The study lasted for eight weeks. No other changes in any of the women's diets were allowed. A trained evaluator was used to check for any differences in the women's diets. The results showed that there were no differences in weight gain between the groups over the eight-week period.

1. Did the researchers use a population or a sample for their study?
2. Based on who conducted this study, would you consider the study to be biased?
3. Which sampling method do you think was used to obtain the original 36 women for the study (random, systematic, stratified, or clustered)?
4. Which sampling method would you use? Why?
5. How would you collect a random sample for this study?
6. Does random assignment help representativeness the same as random selection does? Explain.

See page 750 for the answers.

Exercises 14–1

1. Name the four basic sampling techniques. Random, systematic, stratified, cluster
2. Why are samples used in statistics?
3. What is the basic requirement for a sample? A sample must be randomly selected.
4. Why should random numbers be used when you are selecting a random sample?
5. List three incorrect methods that are often used to obtain a sample.

6. What is the principle behind random numbers? Over the long run each digit, 0 through 9, will occur with the same probability.
7. List the advantages and disadvantages of random sampling.
8. List the advantages and disadvantages of systematic sampling.
9. List the advantages and disadvantages of stratified sampling.

10. List the advantages and disadvantages of cluster sampling.

Use Figure 14–8 to answer Exercises 11 through 14.

11. Population and Area of U.S. Cities Using the table of random numbers, select 10 cities and find the sample mean (average) of the population, the area in square miles,

Figure 14–8			
The 50 Largest Cities in the United States (Based on the 2000 Census)			
City	Population	Area (sq. mi.)	Avg. annual rainfall (in.)
1. Albuquerque, NM	448,607	127.2	8.12
2. Atlanta, GA	416,474	131.2	48.61
3. Austin, TX	656,562	232	31.50
4. Baltimore, MD	651,154	80.3	43.39
5. Boston, MA	589,141	47.2	43.81
6. Charlotte, NC	540,828	152.14	43.16
7. Chicago, IL	2,896,016	228.1	33.34
8. Cleveland, OH	478,403	79	35.40
9. Colorado Springs, CO	360,890	183.2	16.24
10. Columbus, OH	711,470	186.8	36.97
11. Dallas, TX	1,188,580	331.4	34.16
12. Denver, CO	554,636	106.8	15.31
13. Detroit, MI	951,270	135.6	30.97
14. El Paso, TX	563,662	239.7	7.82
15. Fort Worth, TX	534,694	258.5	29.45
16. Fresno, CA	427,652	99.4	10
17. Honolulu, HI	371,657	25.3	23.47
18. Houston, TX	1,953,631	572.7	44.77
19. Indianapolis, IN	791,926	352	39.12
20. Jacksonville, FL	735,617	840	52.77
21. Kansas City, MO	441,545	316.4	29.27
22. Las Vegas, NV	478,434	83.3	4
23. Long Beach, CA	461,522	49.8	12
24. Los Angeles, CA	3,694,820	465.9	14.85
25. Memphis, TN	650,100	264.1	51.57
26. Mesa, AZ	396,375	124.62	7.52
27. Miami, FL	362,470	34.3	57.55
28. Milwaukee, WI	596,974	95.8	30.94
29. Minneapolis, MN	382,618	55.1	26.36
30. Nashville, TN	569,891	479.5	48.49
31. New Orleans, LA	484,674	199.4	59.74
32. New York City, NY	8,008,278	301.5	44.12
33. Oakland, CA	399,484	53.9	18.03
34. Oklahoma City, OK	506,132	604	30.89
35. Omaha, NE	390,007	99.3	30.34
36. Philadelphia, PA	1,517,550	136	41.42
37. Phoenix, AZ	1,321,045	375	7.11
38. Portland, OR	529,121	113.9	37.39
39. Sacramento, CA	407,018	97.3	17.87
40. San Antonio, TX	1,144,646	304.5	29.13
41. San Diego, CA	1,223,400	329	9.32
42. San Francisco, CA	776,733	46.4	19.71
43. San Jose, CA	894,943	169.2	13.86
44. Seattle, WA	563,374	83.6	38.85
45. St. Louis, MO	348,189	61.4	33.91
46. Tucson, AZ	486,699	125	11.14
47. Tulsa, OK	393,049	186.1	38.77
48. Virginia Beach, VA	425,257	225.9	45.22
49. Washington, DC	572,059	62.7	39
50. Wichita, KS	344,284	140.2	29

and the average annual rainfall. Compare these sample means with the population means. Answers will vary.

12. **Rainfall in U.S. Cities** Select a sample of 10 cities by the systematic method. Compute the sample means of the population, area, and average annual rainfall. Compare to the population means. Answers will vary.

13. **Wind Speeds** Select a cluster sample of 10 cities and calculate the average rainfall. Compare with the population mean. Answers will vary.

14. Are there any characteristics of these data that might create problems in sampling? Answers will vary.

Record Highest Temperatures by State (°F)

Alabama 112	Alaska 100	Arizona 128	Arkansas 120
California 134	Colorado 118	Connecticut 106	Delaware 110
Florida 109	Georgia 112	Hawaii 100	Idaho 118
Illinois 117	Indiana 116	Iowa 118	Kansas 121
Kentucky 114	Louisiana 114	Maine 105	Maryland 109
Massachusetts 107	Michigan 112	Minnesota 114	Mississippi 115
Missouri 118	Montana 117	Nebraska 118	Nevada 125
New Hampshire 106	New Jersey 110	New Mexico 122	New York 108
North Carolina 110	North Dakota 121	Ohio 113	Oklahoma 120
Oregon 119	Pennsylvania 111	Rhode Island 104	South Carolina 111
South Dakota 120	Tennessee 113	Texas 120	Utah 117
Vermont 105	Virginia 110	Washington 118	West Virginia 112
Wisconsin 114	Wyoming 115		

Use the above data for Exercises 15 and 16.

15. Which method of sampling might be good for this set of data? Choose one to select 10 states and calculate the sample mean. Compare with the population mean. Answers will vary.

16. **Record High Temperatures** Choose a different method to select 10 states and compute the sample mean high temperature. Compare with your answer in Exercise 15 and with the population mean. Do you see

any features of this data set that might affect the results of obtaining a sample mean? Answers will vary.

17. **Electoral Votes** Select a systematic sample of 10 states and compute the mean number of electoral votes for the sample. Compare this mean with the population mean. Answers will vary.

18. **Electoral Votes** Divide the 50 states into five subgroups by geographic location, using a map of the

---

Figure 14–9

States and Number of Electoral Votes for Each (for Exercises 17 through 19)

1. Alabama	9	14. Indiana	12	27. Nebraska	5	40. South Carolina	8
2. Alaska	3	15. Iowa	8	28. Nevada	4	41. South Dakota	3
3. Arizona	7	16. Kansas	7	29. New Hampshire	4	42. Tennessee	11
4. Arkansas	6	17. Kentucky	9	30. New Jersey	16	43. Texas	29
5. California	47	18. Louisiana	10	31. New Mexico	5	44. Utah	5
6. Colorado	8	19. Maine	4	32. New York	36	45. Vermont	3
7. Connecticut	8	20. Maryland	10	33. North Carolina	13	46. Virginia	12
8. Delaware	3	21. Massachusetts	13	34. North Dakota	3	47. Washington	10
9. Florida	21	22. Michigan	20	35. Ohio	23	48. West Virginia	6
10. Georgia	12	23. Minnesota	10	36. Oklahoma	8	49. Wisconsin	11
11. Hawaii	4	24. Mississippi	7	37. Oregon	7	50. Wyoming	3
12. Idaho	4	25. Missouri	11	38. Pennsylvania	25		
13. Illinois	24	26. Montana	4	39. Rhode Island	4		

United States. Each subgroup should include 10 states. The subgroups should be northeast, southeast, central, northwest, and southwest. Select two states from each subgroup, and find the mean number of electoral votes for the sample. Compare these means with the population mean. Answers will vary.

19. **Electoral Votes** Select a cluster of 10 states and compute the mean number of electoral votes for the sample. Compare this mean with the population mean. Answers will vary.

20. Many research studies described in newspapers and magazines do not report the sample size or the sampling method used. Try to find a research article that gives this information; state the sampling method that was used and the sample size. Answers will vary.

"Now think carefully. The answer you give will represent the opinion of millions of Americans."

Source: *The Saturday Evening Post*, BFL&MS, Inc.

Technology *Step by Step*

MINITAB
Step by Step

Select a Random Sample with Replacement

A simple random sample selected with replacement allows some values to be used more than once, duplicates. In the first example, a random sample of integers will be selected with replacement.

1. Select **Calc>Random Data>Integer.**

2. Type **10** for rows of data.

3. Type the name of a column, Random1, in the box for Store in column(s).

4. Type **1** for Minimum and **50** for Maximum, then click [OK].

A sample of 10 integers between 1 and 50 will be displayed in the first column of the worksheet. Every list will be different.

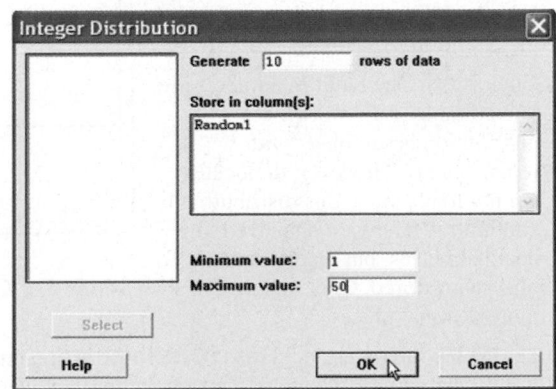

Select a Random Sample Without Replacement

To sample without replacement, make a list of integers and sample from the columns.

1. Select **Calc>Make Patterned Data>Simple Set of Numbers.**

2. Type **Integers** in the text box for Store patterned data in.

3. Type **1** for Minimum and **50** for Maximum. Leave 1 for steps and click [OK]. A list of the integers from 1 to 50 will be created in the worksheet.

4. Select **Calc>Random Data>Sample from columns.**

5. Sample **10** for the number of rows and Integers for the name of the column.

6. Type Random2 as the name of the new column. Be sure to leave the option for Sample with replacement unchecked.

7. Click [OK]. The new sample will be in the worksheet. There will be no duplicates.

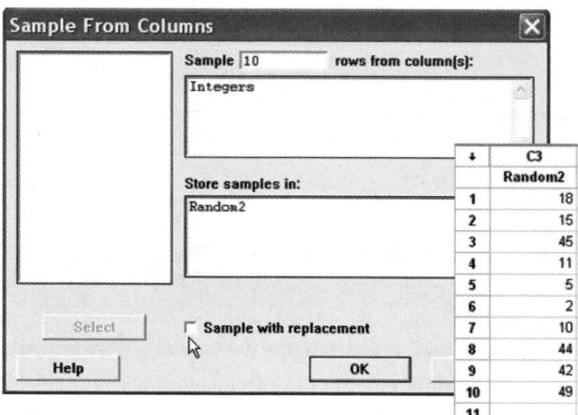

Select a Random Sample from a Normal Distribution

No data are required in the worksheet.

1. Select **Calc>Random Data>Normal . . .**

2. Type **50** for the number of rows.

3. Press TAB or click in the box for Store in columns. Type in RandomNormal.

4. Type in **500** for the Mean and **75** for the Standard deviation.

5. Click [OK]. The random numbers are in a column of the worksheet. The distribution is sampled "with replacement." However, duplicates are not likely since this distribution is continuous. They are displayed to 3 decimal places, but many more places are stored. Click in any cell such as row 5 of C4 RandomNormal, and you will see more decimal places.

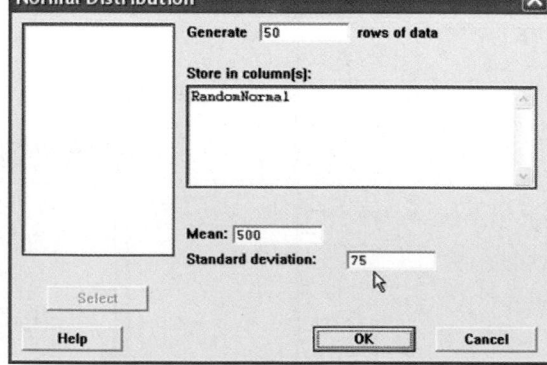

6. To display the list, select **Data>Display data,** then select C1 RandomNormal and click [OK]. They are displayed in the same order they were selected, but going across not down.

TI-83 Plus or TI-84 Plus
Step by Step

Generate Random Numbers

To generate random numbers from 0 to 1 by using the TI-83 Plus or TI-84 Plus:

1. Press **MATH** and move the cursor to PRB and press **1** for rand, then press **ENTER**. The calculator will generate a random decimal from 0 to 1.
2. To generate additional random numbers press **ENTER**.

To generate a list of random integers between two specific values:

1. Press **MATH** and move the cursor to PRB.
2. Press **5** for randInt(.
3. Enter the lowest value followed by a comma, then the largest value followed by a comma, then the number of random numbers desired followed by). Press **ENTER**.

Example: Generate five three-digit random numbers.

Enter **0, 999, 5**) at the randInt(as shown.

The calculator will generate five three-digit random numbers. Use the arrow keys to view the entire list.

```
rand
            .9435974025
randInt(0,999,5)
  {908 146 514 40…
```

Excel
Step by Step

Generate Random Numbers

The Data Analysis Add-In in Excel has a feature to generate random numbers from a specified probability distribution. For this example, a list of 50 random real numbers will be generated from a uniform distribution. The real numbers will then be rounded to integers between 1 and 50.

1. Open a new worksheet and select the Data tab, then **Data Analysis>Random Number Generation** from Analysis Tools. Click [OK].
2. In the dialog box, type **1** for the Number of Variables. Leave the Number of Random Numbers box empty.
3. For Distribution, select Uniform.
4. In the Parameters box, type **1** for the lower bound and **51** for the upper bound.
5. You may type in an integer value between **1** and **51** for the Random Seed. For this example, type **3** for the Random Seed.
6. Select Output Range and type in **A1:A50.**
7. Click [OK].

To convert the random numbers to a list of integers:

8. Select cell B1 and select the Formulas tab, and then the Insert Function icon.

9. Select the Math & Trig Function category and scroll to the Function name INT to convert the data in column A to integer values.

Note: The INT function rounds the argument (input) down to the nearest integer.

10. Type **A1** for the Number in the INT dialog box. Click [OK].

11. While cell B1 is selected in the worksheet, move the pointer to the lower right-hand corner of the cell until a thick plus sign appears. Right-click on the mouse and drag the plus down to cell **B50;** then release the mouse key.

12. The numbers from column A should have been rounded to integers in column B.

Here is a sample of the data produced from the preceding procedure.

1.073244	1
11.98056	11
15.18195	15
14.87219	14
11.72878	11
36.97674	36
28.01193	28
36.86383	36
42.53111	42
19.56746	19

14–2   Surveys and Questionnaire Design

Objective 2

Recognize faulty questions on a survey and other factors that can bias responses.

Many statistical studies obtain information from *surveys*. A survey is conducted when a sample of individuals is asked to respond to questions about a particular subject. There are two types of surveys: interviewer-administered and self-administered. Interviewer-administered surveys require a person to ask the questions. The interview can be conducted face to face in an office, on a street, or in the mall, or via telephone.

Self-administered surveys can be done by mail or in a group setting such as a classroom.

When analyzing the results of surveys, you should be very careful about the interpretations. The way a question is phrased can influence the way people respond. For example, when a group of people were asked if they favored a waiting period and background check before guns could be sold, 91% of the respondents were in favor of it and 7% were against it. However, when asked if there should be a national gun registration program costing about 20% of all dollars spent on crime control, only 33% of the respondents were in favor of it and 61% were against it.

As you can see, by phrasing questions in different ways, different responses can be obtained, since the purpose of a national gun registry would include a waiting period and a background check.

When you are writing questions for a questionnaire, it is important to avoid these common mistakes.

1. *Asking biased questions.* By asking questions in a certain way, the researcher can lead the respondents to answer in the way he or she wants them to. For example, asking a question such as "Are you going to vote for candidate Jones even though the latest survey indicates that he will lose the election?" instead of "Are you going to vote for candidate Jones?" may dissuade some people from answering in the affirmative.

2. *Using confusing words.* In this case, the participant misinterprets the meaning of the words and answers the questions in a biased way. For example, the question "Do you think people would live longer if they were on a diet?" could be misinterpreted since there are many different types of diets—weight loss diets, low-salt diets, medically prescribed diets, etc.

3. *Asking double-barreled questions.* Sometimes questions contain compound sentences that require the participant to respond to two questions at the same time. For example, the question "Are you in favor of a special tax to provide national health care for the citizens of the United States?" asks two questions: "Are you in favor of a national health care program?" and "Do you favor a tax to support it?"

4. *Using double negatives in questions.* Questions with double negatives can be confusing to the respondents. For example, the question "Do you feel that it is not appropriate to have areas where people cannot smoke?" is very confusing since *not* is used twice in the sentence.

5. *Ordering questions improperly.* By arranging the questions in a certain order, the researcher can lead the participant to respond in a way that he or she may otherwise not have done. For example, a question might ask the respondent, "At what age should an elderly person not be permitted to drive?" A later question might ask the respondent to list some problems of elderly people. The respondent may indicate that transportation is a problem based on reading the previous question.

Other factors can also bias a survey. For example, the participant may not know anything about the subject of the question but will answer the question anyway to avoid being considered uninformed. For example, many people might respond yes or no to the following question: "Would you be in favor of giving pensions to the widows of unknown soldiers?" In this case, the question makes no sense since if the soldiers were unknown, their widows would also be unknown.

Many people will make responses on the basis of what they think the person asking the questions wants to hear. For example, if a question states, "How often do you lie?" people may *understate* the incidences of their lying.

Participants will, in some cases, respond differently to questions depending on whether their identity is known. This is especially true if the questions concern sensitive issues such as income, sexuality, and abortion. Researchers try to ensure confidentiality (i.e., keeping the respondent's identity secret) rather than anonymity (soliciting unsigned responses); however, many people will be suspicious in either case.

Still other factors that could bias a survey include the time and place of the survey and whether the questions are open-ended or closed ended. The time and place where a survey is conducted can influence the results. For example, if a survey on airline safety is conducted immediately after a major airline crash, the results may differ from those obtained in a year in which no major airline disasters occurred.

Finally, the type of questions asked influences the responses. In this case, the concern is whether the question is open-ended or closed-ended.

An *open-ended question* would be one such as "List three activities that you plan to spend more time on when you retire." A *closed-ended question* would be one such as "Select three activities that you plan to spend more time on after you retire: traveling; eating out; fishing and hunting; exercising; visiting relatives."

One problem with a closed-ended question is that the respondent is forced to choose the answers that the researcher gives and cannot supply his or her own. But there is also a problem with open-ended questions in that the results may be so varied that attempting to summarize them might be difficult, if not impossible. Hence, you should be aware of what types of questions are being asked before you draw any conclusions from the survey.

There are several other things to consider when you are conducting a study that uses questionnaires. For example, a pilot study should be done to test the design and usage of the questionnaire (i.e., the *validity* of the questionnaire). The pilot study helps the researcher to pretest the questionnaire to determine if it meets the objectives of the study. It also helps the researcher to rewrite any questions that may be misleading, ambiguous, etc.

Unusual Stat

Of people who are struck by lightning, 85% are men.

If the questions are being asked by an interviewer, some training should be given to that person. If the survey is being done by mail, a cover letter and clear directions should accompany the questionnaire.

Questionnaires help researchers to gather needed statistical information for their studies; however, much care must be given to proper questionnaire design and usage; otherwise, the results will be unreliable.

Applying the Concepts **14–2**

Smoking Bans and Profits

Assume you are a restaurant owner and are concerned about the recent bans on smoking in public places. Will your business lose money if you do not allow smoking in your restaurant? You decide to research this question and find two related articles in regional newspapers. The first article states that randomly selected restaurants in Derry, Pennsylvania, that have completely banned smoking have lost 25% of their business. In that study, a survey was used and the owners were asked how much business they thought they lost. The survey was conducted by an anonymous group. It was reported in the second article that there had been a modest increase in business among restaurants that banned smoking in that same area. Sales receipts were collected and analyzed against last year's profits. The second survey was conducted by the Restaurants Business Association.

1. How has the public smoking ban affected restaurant business in Derry, Pennsylvania?
2. Why do you think the surveys reported conflicting results?
3. Should surveys based on anecdotal responses be allowed to be published?
4. Can the results of a sample be representative of a population and still offer misleading information?
5. How critical is measurement error in survey sampling?

See pages 750 and 751 for the answers.

Exercises 14–2

Exercises 1 through 8 include questions that contain a flaw. Identify the flaw and rewrite the question, following the guidelines presented in this section.

1. Which type of artificial sweetener do you think is the least unhealthy? Flaw—biased; it's confusing.

2. Do you like the mayor? Flaw—the purpose of the question is unclear. You could like him personally but not politically.

3. Do you approve of the mayor's political agenda? Flaw—the question is too broad.

4. Do you approve of the mayor's position on the new soft drink tax? Flaw—none. The question is good if the respondent knows the mayor's position; otherwise his position needs to be stated.

5. How long have you studied for this examination? Flaw—confusing words. How many hours did you study for this exam?

6. Which artificial sweetener do you prefer? Possible order problem—ask first, "Do you use artificial sweetener regularly?"

7. If a plane were to crash on the border of New York and New Jersey, where should the survivors be buried?

8. Are you in favor of imposing a tax on tobacco to pay for health care related to diseases caused by smoking? Flaw—none.

9. Find a study that uses a questionnaire. Select any questions that you feel are improperly written. Answers will vary.

10. Many television and radio stations have a phone vote poll. If there is one in your area, select a specific day and write a brief paragraph stating the question of the day and state if it could be misleading in any way. Answers will vary.

14–3   Simulation Techniques and the Monte Carlo Method

Many real-life problems can be solved by employing simulation techniques.

A **simulation technique** uses a probability experiment to mimic a real-life situation.

Instead of studying the actual situation, which might be too costly, too dangerous, or too time-consuming, scientists and researchers create a similar situation but one that is less expensive, less dangerous, or less time-consuming. For example, NASA uses space shuttle flight simulators so that its astronauts can practice flying the shuttle. Most video games use the computer to simulate real-life sports such as boxing, wrestling, baseball, and hockey.

Simulation techniques go back to ancient times when the game of chess was invented to simulate warfare. Modern techniques date to the mid-1940s when two physicists, John Von Neumann and Stanislaw Ulam, developed simulation techniques to study the behavior of neutrons in the design of atomic reactors.

Mathematical simulation techniques use probability and random numbers to create conditions similar to those of real-life problems. Computers have played an important role in simulation techniques, since they can generate random numbers, perform experiments, tally the outcomes, and compute the probabilities much faster than human beings. The basic simulation technique is called the *Monte Carlo method*. This topic is discussed next.

The Monte Carlo Method

Objective 3

Solve problems, using simulation techniques.

The **Monte Carlo method** is a simulation technique using random numbers. Monte Carlo simulation techniques are used in business and industry to solve problems that are extremely difficult or involve a large number of variables. The steps for simulating real-life experiments in the Monte Carlo method are as follows:

1. List all possible outcomes of the experiment.
2. Determine the probability of each outcome.
3. Set up a correspondence between the outcomes of the experiment and the random numbers.
4. Select random numbers from a table and conduct the experiment.
5. Repeat the experiment and tally the outcomes.
6. Compute any statistics and state the conclusions.

Before examples of the complete simulation technique are given, an illustration is needed for step 3 (set up a correspondence between the outcomes of the experiment and the random numbers). Tossing a coin, for instance, can be simulated by using random numbers as follows: Since there are only two outcomes, heads and tails, and since each outcome has a probability of $\frac{1}{2}$, the odd digits (1, 3, 5, 7, and 9) can be used to represent a head, and the even digits (0, 2, 4, 6, and 8) can represent a tail.

Suppose a random number 8631 is selected. This number represents four tosses of a single coin and the results T, T, H, H. Or this number could represent one toss of four coins with the same results.

An experiment of rolling a single die can also be simulated by using random numbers. In this case, the digits 1, 2, 3, 4, 5, and 6 can represent the number of spots that appear on the face of the die. The digits 7, 8, 9, and 0 are ignored, since they cannot be rolled.

Figure 14–10

Spinner with Four Numbers

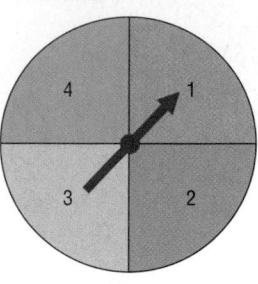

When two dice are rolled, two random digits are needed. For example, the number 26 represents a 2 on the first die and a 6 on the second die. The random number 37 represents a 3 on the first die, but the 7 cannot be used, so another digit must be selected. As another example, a three-digit daily lotto number can be simulated by using three-digit random numbers. Finally, a spinner with four numbers, as shown in Figure 14–10, can be simulated by letting the random numbers 1 and 2 represent 1 on the spinner, 3 and 4 represent 2 on the spinner, 5 and 6 represent 3 on the spinner, and 7 and 8 represent 4 on the spinner, since each number has a probability of $\frac{1}{4}$ of being selected. The random numbers 9 and 0 are ignored in this situation.

Many real-life games, such as bowling and baseball, can be simulated by using random numbers, as shown in Figure 14–11.

Figure 14–11

Example of Simulation of a Game

Source: Albert Shuylte, "Simulated Bowling Game," Student Math Notes, March 1986. Published by the National Council of Teachers of Mathematics. Reprinted with permission.

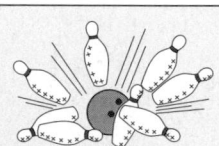

Simulated Bowling Game

Let's use the random digit table to simulate a bowling game. Our game is much simpler than commercial simulation games.

First Ball		Second Ball			
		2-Pin Split		No split	
Digit	Results	Digit	Results	Digit	Results
1–3	Strike	1	Spare	1–3	Spare
4–5	2-pin split	2–8	Leave one pin	4–6	Leave 1 pin
6–7	9 pins down	9–0	Miss both pins	7–8	*Leave 2 pins
8	8 pins down			9	+Leave 3 pins
9	7 pins down			0	Leave all pins
0	6 pins down				

*If there are fewer than 2 pins, result is a spare.
+If there are fewer than 3 pins, those pins are left.

Here's how to score bowling:
1. There are 10 frames to a **game** or **line**.
2. You roll two balls for each frame, unless you knock all the pins down with the first ball (a **strike**).
3. Your score for a frame is the sum of the pins knocked down by the two balls, if you don't knock down all 10.
4. If you knock all 10 pins down with two balls (a **spare**, shown as ⟋), your score is 10 pins plus the number knocked down with the next ball.
5. If you knock all 10 pins down with the first ball (a **strike**, shown as ⊠), your score is 10 pins plus the number knocked down by the next **two** balls.
6. A **split** (shown as ◯) is when there is a big space between the remaining pins. Place in the circle the number of pins remaining after the second ball.
7. A **miss** is shown as —.

Here is how one person simulated a bowling game using the random digits 7 2 7 4 8 2 2 3 6 1 6 0 4 6 1 5 5, chosen in that order from the table.

Frame

	1	2	3	4	5	6	7	8	9	10
Digit(s)	7/2	7/4	8/2	2	3	6/1	6/0	4/6	1	5/5
Bowling result	9⟋ 19	9— 28	8⟋ 48	⊠ 77	⊠ 97	9⟋ 116	9— 125	8 ① 134	⊠ 153	8 ① 162	162

Now you try several.

Frame

	1	2	3	4	5	6	7	8	9	10
Digit(s)										
Bowling result										

	1	2	3	4	5	6	7	8	9	10
Digit(s)										
Bowling result										

If you wish to, you can change the probabilities in the simulation to better reflect *your* actual bowling ability.

Example 14–4	Snoring

According to the CDC, the chance that a person snores while sleeping is 20%. Use random numbers to simulate a sample of 20 people and identify those who snore.

Solution

Now 20% is $\frac{20}{100} = \frac{1}{5}$, so one out of every five people snores while sleeping. Using random digits, select 20 single numbers and assign 1 and 2 as people who snore and 3 through 9 and 0 as people who do not snore. (*Note:* You can use any two digits for those who snore.) Then the 1s and 2s represent people who snore; 0 and 3 through 9 represent those who do not snore.

Example 14–5	Outcomes of a Tennis Game

Using random numbers, simulate the outcomes of a tennis game between Bill and Mike, with the additional condition that Bill is twice as good as Mike.

Solution

Since Bill is twice as good as Mike, he will win approximately two games for every one Mike wins; hence, the probability that Bill wins will be $\frac{2}{3}$, and the probability that Mike wins will be $\frac{1}{3}$. The random digits 1 through 6 can be used to represent a game Bill wins; the random digits 7, 8, and 9 can be used to represent Mike's wins. The digit 0 is disregarded. Suppose they play five games, and the random number 86314 is selected. This number means that Bill won games 2, 3, 4, and 5 and Mike won the first game. The sequence is

8	6	3	1	4
M	B	B	B	B

More complex problems can be solved by using random numbers, as shown in Examples 14–6 to 14–8.

Example 14–6	Rolling a Die

A die is rolled until a 6 appears. Using simulation, find the average number of rolls needed. Try the experiment 20 times.

Solution

Step 1 List all possible outcomes. They are 1, 2, 3, 4, 5, 6.

Step 2 Assign the probabilities. Each outcome has a probability of $\frac{1}{6}$.

Step 3 Set up a correspondence between the random numbers and the outcome. Use random numbers 1 through 6. Omit the numbers 7, 8, 9, and 0.

Step 4 Select a block of random numbers, and count each digit 1 through 6 until the first 6 is obtained. For example, the block 857236 means that it takes 4 rolls to get a 6.

8	5	7	2	3	6
	↑		↑	↑	↑
	5		2	3	6

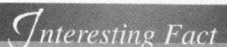

Step 5   Repeat the experiment 19 more times and tally the data as shown.

Trial	Random number	Number of rolls
1	8 5 7 2 3 6	4
2	2 1 0 4 8 0 1 5 1 1 0 1 5 3 6	11
3	2 3 3 6	4
4	2 4 1 3 0 4 8 3 6	7
5	4 2 1 6	4
6	3 7 5 2 0 3 9 8 7 5 8 1 8 3 7 1 6	9
7	7 7 9 2 1 0 6	3
8	9 9 5 6	2
9	9 6	1
10	8 9 5 7 9 1 4 3 4 2 6	7
11	8 5 4 7 5 3 6	5
12	2 8 9 1 8 6	3
13	6	1
14	0 9 4 2 9 9 3 9 6	4
15	1 0 3 6	3
16	0 7 1 1 9 9 7 3 3 6	5
17	5 1 0 8 5 1 2 7 6	6
18	0 2 3 6	3
19	0 1 0 1 1 5 4 0 9 2 3 3 3 6	10
20	5 2 1 6	4
		Total 96

Step 6   Compute the results and draw a conclusion. In this case, you must find the average.

$$\overline{X} = \frac{\Sigma X}{n} = \frac{96}{20} = 4.8$$

Hence, the average is about 5 rolls.

Note: The theoretical average obtained from the expected value formula is 6. If this experiment is done many times, say 1000 times, the results should be closer to the theoretical results.

Example 14–7

Selecting a Key

A person selects a key at random from four keys to open a lock. Only one key fits. If the first key does not fit, she tries other keys until one fits. Find the average of the number of keys a person will have to try to open the lock. Try the experiment 25 times.

Solution

Assume that each key is numbered from 1 through 4 and that key 2 fits the lock. Naturally, the person doesn't know this, so she selects the keys at random. For the simulation, select a sequence of random digits, using only 1 through 4, until the digit 2 is reached. The trials are shown here.

Trial	Random digit (key)	Number	Trial	Random digit (key)	Number
1	2	1	14	2	1
2	2	1	15	4 2	2
3	1 2	2	16	1 3 2	3
4	1 4 3 2	4	17	1 2	2
5	3 2	2	18	2	1
6	3 1 4 2	4	19	3 4 2	3
7	4 2	2	20	2	1
8	4 3 2	3	21	2	1
9	4 2	2	22	2	1
10	2	1	23	4 2	2
11	4 2	2	24	4 3 1 2	4
12	3 1 2	3	25	3 1 2	3
13	3 1 2	3			Total 54

Next, find the average:

$$\overline{X} = \frac{\Sigma X}{n} = \frac{1 + 1 + \cdots + 3}{25} = \frac{54}{25} = 2.16$$

The theoretical average is 2.5. Again, only 25 repetitions were used; more repetitions should give a result closer to the theoretical average.

Example 14–8

Selecting a Monetary Bill

A box contains five \$1 bills, three \$5 bills, and two \$10 bills. A person selects a bill at random. What is the expected value of the bill? Perform the experiment 25 times.

Solution

Step 1 List all possible outcomes. They are \$1, \$5, and \$10.

Step 2 Assign the probabilities to each outcome:

$$P(\$1) = \tfrac{5}{10} \qquad P(\$5) = \tfrac{3}{10} \qquad P(\$10) = \tfrac{2}{10}$$

Step 3 Set up a correspondence between the random numbers and the outcomes. Use random numbers 1 through 5 to represent a \$1 bill being selected, 6 through 8 to represent a \$5 bill being selected, and 9 and 0 to represent a \$10 bill being selected.

Steps 4 and 5 Select 25 random numbers and tally the results.

Number	Results (\$)
4 5 8 2 9	1, 1, 5, 1, 10
2 5 6 4 6	1, 1, 5, 1, 5
9 1 8 0 3	10, 1, 5, 10, 1
8 4 0 6 0	5, 1, 10, 5, 10
9 6 9 4 3	10, 5, 10, 1, 1

Step 6 Compute the average:

$$\overline{X} = \frac{\Sigma X}{n} = \frac{\$1 + \$1 + \$5 + \cdots + \$1}{25} = \frac{\$116}{25} = \$4.64$$

Hence, the average (expected value) is \$4.64.

Recall that using the expected value formula $E(X) = \Sigma[X \cdot P(X)]$ gives a theoretical average of

$$E(X) = \Sigma[X \cdot P(X)] = (0.5)(\$1) + (0.3)(\$5) + (0.2)(\$10) = \$4.00$$

Remember that simulation techniques do not give exact results. The more times the experiment is performed, though, the closer the actual results should be to the theoretical results. (Recall the law of large numbers.)

The steps for solving problems using the Monte Carlo method are summarized in the Procedure Table.

Procedure Table

Simulating Experiments Using the Monte Carlo Method

Step 1 List all possible outcomes of the experiment.

Step 2 Determine the probability of each outcome.

Step 3 Set up a correspondence between the outcomes of the experiment and the random numbers.

Step 4 Select random numbers from a table and conduct the experiment.

Step 5 Repeat the experiment and tally the outcomes.

Step 6 Compute any statistics and state the conclusions.

Applying the Concepts **14–3**

Simulations

Answer the following questions:

1. Define simulation technique.
2. Have simulation techniques been used for very many years?
3. Is it cost-effective to do simulation testing on some things such as airplanes or automobiles?
4. Why might simulation testing be better than real-life testing? Give examples.
5. When did physicists develop computer simulation techniques to study neutrons?
6. When could simulations be misleading or harmful? Give examples.
7. Could simulations have prevented previous disasters such as the Hindenburg or the 1986 Space Shuttle disaster?
8. What discipline is simulation theory based on?

See page 751 for the answers.

Exercises 14–3

1. Define simulation techniques.

2. Give three examples of simulation techniques.
   Answers will vary.

3. Who is responsible for the development of modern simulation techniques? John Von Neumann and Stanislaw Ulam

4. What role does the computer play in simulation?

5. What are the steps in the simulation of an experiment?

6. What purpose do random numbers play in simulation? Random numbers can be used to ensure the outcomes occur with appropriate probability.

7. What happens when the number of repetitions is increased? When the repetitions increase, there is a higher probability that the simulation will yield more precise answers.

For Exercises 8 through 13, explain how each experiment can be simulated by using random numbers.

8. **Foreign-Born Residents** Almost 16% of Texas residents are foreign-born. Explain how to select a sample of 40 based on this scenario.

   Source: factfinder.census.gov

9. **Stay-at-Home Parents** Fewer than one-half of all mothers are stay-at-home parents. Recent statistics indicate that 68.1% of all mothers with children under age 18 are in the labor force. Explain how to create a simulation to represent this situation.

   Source: *New York Times Almanac.*

10. **Playing Basketball** Two basketball players have a free-throw contest—one is a 70% shooter and the other is a 75% shooter. They each shoot 20 shots in groups of 5 shots each. Use a calculator to simulate the contest and find out who wins. (Repeat a number of times and compare your answers.)

11. **Television Set Ownership** Thirty-five percent of U.S. households with at least one television set have premium cable service. Explain how to simulate this with random numbers. Use your method to select a random sample of 100 households and test the hypothesis that *p* does not equal 35%.

12. **Matching Pennies** Two players match pennies. Use the odd digits to represent a match and the even digits to represent a nonmatch.

13. **Odd Man Out** Three players play odd man out. (Three coins are tossed; if all three match, the game is repeated and no one wins. If two players match, the third person wins all three coins.) Let an odd number represent heads and an even number represent tails. Then each person selects a digit at random.

For Exercises 14 through 21, use random numbers to simulate the experiments. The number in parentheses is the number of times the experiment should be repeated.

14. **Tossing a Coin** A coin is tossed until four heads are obtained. Find the average number of tosses necessary. (50) Answers will vary.

15. **Rolling a Die** A die is rolled until all faces appear at least once. Find the average number of tosses. (30) Answers will vary.

16. **Prizes in Caramel Corn Boxes** A caramel corn company gives four different prizes, one in each box. They are placed in the boxes at random. Find the average number of boxes a person needs to buy to get all four prizes. (40) Answers will vary.

17. **Keys to a Door** The probability that a door is locked is 0.6, and there are five keys, one of which will unlock the door. The experiment consists of choosing one key at random and seeing if you can open the door. Repeat the experiment 50 times and calculate the empirical probability of opening the door. Compare your result to the theoretical probability for this experiment. Answers will vary.

18. **Lottery Winner** To win a certain lotto, a person must spell the word *big*. Sixty percent of the tickets contain the letter b, 30% contain the letter i, and 10% contain the letter g. Find the average number of tickets a person must buy to win the prize. (30) Answers will vary.

19. **Clay Pigeon Shooting** Two shooters shoot clay pigeons. Gail has an 80% accuracy rate and Paul has a 60% accuracy rate. Paul shoots first. The first person who hits the target wins. Find the probability that each wins. (30). Answers will vary.

20. In Exercise 19, find the average number of shots fired. (30) Answers will vary.

21. **Basketball Foul Shots** A basketball player has a 60% success rate for shooting foul shots. If she gets two shots, find the probability that she will make one or both shots. (50). Answers will vary.

22. Which would be easier to simulate with random numbers, baseball or soccer? Explain. Answers will vary.

23. Explain how cards can be used to generate random numbers. Answers will vary.

24. Explain how a pair of dice can be used to generate random numbers. Answers will vary.

Summary

- To obtain information and make inferences about a large population, researchers select a sample. A sample is a subgroup of the population. Using a sample rather than a population, researchers can save time and money, get more detailed information, and get information that otherwise would be impossible to obtain. (14–1)

- The four most common methods researchers use to obtain samples are random, systematic, stratified, and cluster sampling methods. In random sampling, some type of random method (usually random numbers) is used to obtain the sample. In systematic sampling, the researcher selects every *k*th person or item after selecting the first one at random. In stratified sampling, the population is divided into subgroups according to various characteristics, and elements are then selected at random from the subgroups. In cluster sampling, the researcher selects an intact group to use as a sample. When the population is large, multistage sampling (a combination of methods) is used to obtain a subgroup of the population. (14–1)

• Researchers must use caution when conducting surveys and designing questionnaires; otherwise, conclusions obtained from these will be inaccurate. Guidelines were presented in Section 14–2. (14–2)

• Most sampling methods use random numbers, which can also be used to simulate many real-life problems or situations. The basic method of simulation is known as the Monte Carlo method. The purpose of simulation is to duplicate situations that are too dangerous, too costly, or too time-consuming to study in real life. Most simulation techniques can be done on the computer or calculator, since they can rapidly generate random numbers, count the outcomes, and perform the necessary computations. (14–3)

Sampling and simulation are two techniques that enable researchers to gain information that might otherwise be unobtainable.

Important Terms

biased sample 721	Monte Carlo method 739	sequence sampling 729	systematic sample 725
cluster sample 728	multistage sampling 729	simulation technique 739	unbiased sample 721
double sampling 729	random sample 721	stratified sample 726	

Review Exercises

Wind Speed of Hurricanes

The 2005 Atlantic hurricane season was notable for many reasons, among them the most named storms and the most hurricanes. Use Figure 14–12 to answer questions 1 through 4.

Figure 14–12

2005 Hurricane Season

Name	Max. Wind	Classification	Name	Max. Wind	Classification
Arlene	70	Storm	Ophelia	85	Hurricane
Bret	40	S	Philippe	80	H
Cindy	75	H	Rita	175	H
Dennis	150	H	Stan	80	H
Emily	160	H	Unnamed	50	S
Franklin	70	S	Tammy	50	S
Gert	45	S	Vince	75	H
Harvey	65	S	Wilma	175	H
Irene	105	H	Alpha	50	S
Jose	50	S	Beta	115	H
Katrina	175	H	Gamma	55	S
Lee	40	S	Delta	70	S
Maria	115	H	Epsilon	85	H
Nate	90	H	Zeta	65	S

1. **Hurricanes** Select a random sample of eight storms by using random numbers, and find the average maximum wind speed. Compare with the population mean. Answers will vary.

2. **Hurricanes** Select a systematic sample of eight storms and calculate the average maximum wind speed. Compare with the population mean. Answers will vary.

3. Hurricanes Select a cluster of 10 storms. Compute the sample means wind speeds. Compare these sample means with the population means. Answers will vary.

4. Hurricanes Divide the 28 storms into 4 subgroups. Then select a sample of three storms from each group. Compute the means for wind speeds. Compare these means to the population mean. Answers will vary.

Composition of State Legislatures

State	Senate	House	State	Senate	House
Alabama	35	105	Montana	50	100
Alaska	20	40	Nebraska	Unicameral—49	
Arizona	30	60	Nevada	21	42
Arkansas	35	100	New Hampshire	24	400
California	40	80	New Jersey	40	80
Colorado	35	65	New Mexico	42	70
Connecticut	36	151	New York	62	150
Delaware	21	41	North Carolina	50	120
Florida	40	120	North Dakota	47	94
Georgia	56	180	Ohio	33	99
Hawaii	25	51	Oklahoma	48	101
Idaho	35	70	Oregon	30	60
Illinois	59	118	Pennsylvania	50	203
Indiana	50	100	Rhode Island	38	75
Iowa	50	100	South Carolina	46	124
Kansas	40	125	South Dakota	35	70
Kentucky	38	100	Tennessee	33	99
Louisiana	39	105	Texas	31	150
Maine	35	151	Utah	29	75
Maryland	47	141	Vermont	30	150
Massachusetts	40	160	Virginia	40	100
Michigan	38	110	Washington	49	98
Minnesota	67	134	West Virginia	34	100
Mississippi	52	122	Wisconsin	33	99
Missouri	34	163	Wyoming	30	60

Use the above data to answer the following questions.

5. Senators and Representatives Select random samples of 10 states and find the mean number of state senators for this sample. Compare this mean with the population mean. Repeat for state representatives. Answers will vary.

6. Senators and Representatives Select a systematic sample of 10 states and compute the mean number of state senators. Compare with the population mean. Repeat for state representatives. Answers will vary.

7. Senators and Representatives Divide the 50 states into five subgroups by geographic location, using a map of the United States. Each subgroup (northeast, southeast, central, northwest, and southwest) should include 10 states. Select two from each subgroup and find the mean number of state senators (representatives) for this sample. Compare with the population means. Answers will vary.

8. Senators and Representatives Select a cluster of 10 states and compute the mean number of state senators

(representatives) for the sample. Compare with the population means. Answers will vary.

For Exercises 9 through 13, explain how to simulate each experiment by using random numbers.

9. A baseball player strikes out 40% of the time.

10. An airline overbooks 15% of the time.

11. Two players roll a die. The higher number wins.

12. Player 1 rolls two dice. Player 2 rolls one die. If the number on the single die matches one number of the player who rolled the two dice, player 2 wins. Otherwise, player 1 wins.

13. Rock, Paper, Scissors Two players play rock, paper, scissors. The rules are as follows: Since paper covers rock, paper wins. Since rock breaks scissors, rock wins. Since scissors cut paper, scissors win. Each person selects rock, paper, or scissors by random numbers and then compares results.

For Exercises 14 through 18, use random numbers to simulate the experiments. The number in parentheses is the number of times the experiment should be repeated.

14. **Football** A football is placed on the 10-yard line, and a team has four downs to score a touchdown. The team can move the ball only 0 to 5 yards per play. Find the average number of times the team will score a touchdown. (30) Answers will vary.

15. In Exercise 14, find the average number of plays it will take to score a touchdown. Ignore the four-downs rule and keep playing until a touchdown is scored. (30) Answers will vary.

16. **Rolling a Die** Four dice are rolled 50 times. Find the average of the sum of the number of spots that will appear. (50) Answers will vary.

17. **Field Goals** A field goal kicker is successful in 60% of his kicks inside the 35-yard line. Find the probability of kicking three field goals in a row. (50) Answers will vary.

18. **Making a Sale** A sales representative finds that there is a 30% probability of making a sale by visiting the potential customer personally. For every 20 calls, find the probability of making three sales in a row. (50) Answers will vary.

For Exercises 19 through 22, explain what is wrong with each question. Rewrite each one following the guidelines in this chapter.

19. How often do you run red lights? Flaw—asking a biased question. Have you ever driven through a red light?

20. Do you think students who are not failing should not be tutored? Flaw—using a double negative. Do you think students who are not failing should be given tutoring if they request it?

21. Do you think all automobiles should have heavy-duty bumpers, even though it will raise the price of the cars by $500? Flaw—asking a double-barreled question. Do you think all automobiles should have heavy-duty bumpers?

22. Explain the difference between an open-ended question and a closed-ended question. Answers will vary.

Data Analysis

The Data Bank is found in Appendix D.

1. From the Data Bank, choose a variable. Select a random sample of 20 individuals, and find the mean of the data.

2. Select a systematic sample of 20 individuals, and using the same variable as in Exercise 1, find the mean.

3. Select a cluster sample of 20 individuals, and using the same variable as in Exercise 1, find the mean.

4. Stratify the data according to marital status and gender, and sample 20 individuals. Compute the mean of the sample variable selected in Exercise 1 (use four groups of five individuals).

5. Compare all four means and decide which one is most appropriate. (*Hint:* Find the population mean.)

Chapter Quiz

Determine whether each statement is true or false. If the statement is false, explain why.

1. When researchers are sampling from large populations, such as adult citizens living in the United States, they may use a combination of sampling techniques to ensure representativeness. True

2. Simulation techniques using random numbers are a substitute for performing the actual statistical experiment. True

3. When researchers perform simulation experiments, they do not need to use random numbers since they can make up random numbers. False

4. Random samples are said to be unbiased. True

Select the best answer.

5. When all subjects under study are used, the group is called a _____.
   - *a.* Population
   - *b.* Large group
   - *c.* Sample
   - *d.* Study group

6. When a population is divided into subgroups with similar characteristics and then a sample is obtained, this method is called _____ sampling.
   - *a.* Random
   - *b.* Systematic
   - *c.* Stratified
   - *d.* Cluster

7. Interviewing selected people at a local supermarket can be considered an example of _____ sampling.
   - *a.* Random
   - *b.* Systematic
   - *c.* Convenience
   - *d.* Stratified

Complete the following statements with the best answer.

8. In general, when you conduct sampling, the _____ the sample, the more representative it will be. Larger

9. When samples are not representative, they are said to be _____. Biased

10. When all residents of a street are interviewed for a survey, the sampling method used is _____. Cluster

The Monty Hall Problem—Revisited

It appears that it does not matter whether the contestant switches doors because he is given a choice of two doors, and the chance of winning the prize is 1 out of 2, or $\frac{1}{2}$. This reasoning, however, is incorrect. Consider the three possibilities for the prize. It could be behind door A, B, or C. Also consider the fact that the contestant has selected door A. Now the three situations look like this:

	Door		
Case	**A**	**B**	**C**
1	Prize	Empty	Empty
2	Empty	Prize	Empty
3	Empty	Empty	Prize

In case 1, the contestant selected door A, and if the contestant switched after being shown that there was no prize behind either door B or door C, he'd lose. In case 2, the contestant selected door A, and Monty will open door C, so if the contestant switched, he would win the prize. In case 3, the contestant selected door A, and Monty will open door B, so if the contestant switched, he would win the prize. Hence, by switching, the probability of winning is $\frac{2}{3}$ and the probability of losing is $\frac{1}{3}$. The same reasoning can be used no matter which door you select.

You can simulate this problem by using three cards, say, an ace (prize) and two other cards. Have a person arrange the cards in a row and let you select a card. After the person turns over one of the cards (a nonace), then switch. Keep track of the number of times you win. You can also play this game on the Internet by going to the website http://www.stat.sc.edu/~west/javahtml/LetsMakeaDeal.html.

Use Figure 14–12 in the Review Exercises (page 746) for Exercises 11 through 14.

11. Select a random sample of 12 people, and find the mean of the blood pressures of the individuals. Compare this with the population mean. Answers will vary.

12. Select a systematic sample of 12 people, and compute the mean of their blood pressures. Compare this with the population mean. Answers will vary.

13. Divide the individuals into subgroups of six males and six females. Find the means of their blood pressures. Compare these means with the population mean. Answers will vary.

14. Select a cluster of 12 people, and find the mean of their blood pressures. Compare this with the population mean. Answers will vary.

For Exercises 15 through 19, explain how each could be simulated by using random numbers.

15. A chess player wins 45% of his games.

16. A travel agency has a 5% cancellation rate.

17. Two players select a card from a deck with no face cards. The player who gets the higher card wins.

18. One player rolls two dice. The other player selects a card from a deck. Face cards count as 11 for a jack, 12 for a queen, and 13 for a king. The player with the higher total points wins.

19. Two players toss two coins. If they match, player 1 wins; otherwise, player 2 wins.

For Exercises 20 through 24, use random numbers to simulate the experiments. The number in parentheses is the number of times the experiment should be done.

20. **Phone Sales** A telephone solicitor finds that there is a 15% probability of selling her product over the phone. For every 20 calls, find the probability of making two sales in a row. (100) Answers will vary.

21. **Field Goals** A field goal kicker is successful in 65% of his kicks inside the 40-yard line. Find the probability of his kicking four field goals in a row. (40) Answers will vary.

22. **Tossing Coins** Two coins are tossed. Find the average number of times two tails will appear. (40) Answers will vary.

23. **Selecting Cards** A single card is drawn from a deck. Find the average number of times it takes to draw an ace. (30) Answers will vary.

24. **Bowling** A bowler finds that there is a 30% probability that he will make a strike. For every 15 frames he bowls, find the probability of making two strikes. (30) Answers will vary.

Critical Thinking Challenges

1. Explain why two different opinion polls might yield different results on a survey. Also, give an example of an opinion poll and explain how the data may have been collected.

2. Use a computer to generate random numbers to simulate the following real-life problem.

In a certain geographic region, 40% of the people have type O blood. On a certain day, the blood center needs 4 pints of type O blood. On average, how many donors are needed to obtain 4 pints of type O blood?

 ## Data Projects

1. **Business and Finance** A car salesperson has six automobiles on the car lot. Roll a die, using the numbers 1 through 6 to represent each car. If only one car can be sold on each day, how long will it take him to sell all the automobiles? In other words, see how many tosses of the die it will take to get the numbers 1 through 6.

2. **Sports and Leisure** Using the rules given in Figure 14–4 on page 725, play the simulated bowling game. Each game consists of 10 frames.

3. **Technology** In a carton of 12 iPods, three are defective. If four are sold on Saturday, find the probability that at least one will be defective. Use random numbers to simulate this exercise 50 times.

4. **Health and Wellness** Of people who go on a special diet, 25% will lose at least 10 pounds in 10 weeks. A drug manufacturer says that if people take its special herbal pill, that will increase the number of people who lose at least 10 pounds in 10 weeks. The company conducts an experiment, giving its pills to 20 people. Seven people lost at least 10 pounds in 10 weeks. The

drug manufacturer claims that the study "proves" the success of the herbal pills. Using random numbers, simulate the experiment 30 times, assuming the pills are ineffective. What can you conclude about the result that 7 out of 20 people lost at least 10 pounds?

5. **Politics and Economics** In Exercise Section 2–3, problem 2 shows the numbers of signers of the Declaration of Independence from each state. A student decides to write a paper on two of the signers, who are selected at random. What is the probability that both signers will be from the same state? Use random numbers to simulate the experiment, and perform the experiment 50 times.

6. **Your Class** Simulate the classical birthday problem given in the Critical Thinking Challenge 3 in Chapter 4. Select a sample size of 25 and generate random numbers between 1 and 365. Are there any two random numbers that are the same? Select a sample of 50. Are there any two random numbers that are the same? Repeat the experiments 10 times and explain your answers.

Answers to Applying the Concepts

Section 14–1   The White or Wheat Bread Debate

1. The researchers used a sample for their study.

2. Answers will vary. One possible answer is that we might have doubts about the validity of the study, since the baking company that conducted the experiment has an interest in the outcome of the experiment.

3. The sample was probably a convenience sample.

4. Answers will vary. One possible answer would be to use a simple random sample.

5. Answers will vary. One possible answer is that a list of women's names could be obtained from the city in which the women live. Then a simple random sample could be selected from this list.

6. The random assignment helps to spread variation among the groups. The random selection helps to generalize from the sample back to the population. These are two different issues.

Section 14–2   Smoking Bans and Profits

1. It is uncertain how public smoking bans affected restaurant business in Derry, Pennsylvania, since the survey results were conflicting.

2. Since the data were collected in different ways, the survey results were bound to have different answers. Perceptions of the owners will definitely be different from an analysis of actual sales receipts, particularly if the owners assumed that the public smoking bans would hurt business.

3. Answers will vary. One possible answer is that it would be difficult to not allow surveys based on anecdotal responses to be published. At the same time, it would be good for those publishing such survey results to comment on the limitations of these surveys.

4. We can get results from a representative sample that offer misleading information about the population.

5. Answers will vary. One possible answer is that measurement error is important in survey sampling in order to give ranges for the population parameters that are being investigated.

Section 14–3   Simulations

1. A simulation uses a probability experiment to mimic a real-life situation.

2. Simulation techniques date back to ancient times.

3. It is definitely cost-effective to run simulations for expensive items such as airplanes and automobiles.

4. Simulation testing is safer, faster, and less expensive than many real-life testing situations.

5. Computer simulation techniques were developed in the mid-1940s.

6. Answers will vary. One possible answer is that some simulations are far less harmful than conducting an actual study on the real-life situation of interest.

7. Answers will vary. Simulations could have possibly prevented disasters such as the Hindenburg or the 1986 Space Shuttle disaster. For example, data analysis after the Space Shuttle disaster showed that there was a decent chance that something would go wrong on that flight. See http://history.nasa.gov/sts511.html

8. Simulation theory is based in probability theory.

Appendix A

Algebra Review

A–1 Factorials

Definition and Properties of Factorials

The notation called factorial notation is used in probability. *Factorial notation* uses the exclamation point and involves multiplication. For example,

$$5! = 5 \cdot 4 \cdot 3 \cdot 2 \cdot 1 = 120$$
$$4! = 4 \cdot 3 \cdot 2 \cdot 1 = 24$$
$$3! = 3 \cdot 2 \cdot 1 = 6$$
$$2! = 2 \cdot 1 - 2$$
$$1! = 1$$

In general, a factorial is evaluated as follows:

$$n! = n(n - 1)(n - 2) \cdots 3 \cdot 2 \cdot 1$$

Note that the factorial is the product of n factors, with the number decreased by 1 for each factor.

One property of factorial notation is that it can be stopped at any point by using the exclamation point. For example,

$$5! = 5 \cdot 4! \qquad \text{since} \quad 4! = 4 \cdot 3 \cdot 2 \cdot 1$$
$$= 5 \cdot 4 \cdot 3! \qquad \text{since} \quad 3! = 3 \cdot 2 \cdot 1$$
$$= 5 \cdot 4 \cdot 3 \cdot 2! \qquad \text{since} \quad 2! = 2 \cdot 1$$
$$= 5 \cdot 4 \cdot 3 \cdot 2 \cdot 1$$

Thus,
$$n! = n(n - 1)!$$
$$= n(n - 1)(n - 2)!$$
$$= n(n - 1)(n - 2)(n - 3)! \qquad \text{etc.}$$

Another property of factorials is

$$0! = 1$$

This fact is needed for formulas.

Operations with Factorials

Factorials cannot be added or subtracted directly. They must be multiplied out. Then the products can be added or subtracted.

Example A–1

Evaluate $3! + 4!$.

Solution

$$3! + 4! = (3 \cdot 2 \cdot 1) + (4 \cdot 3 \cdot 2 \cdot 1)$$
$$= 6 + 24 = 30$$

Note: $3! + 4! \neq 7!$, since $7! = 5040$.

Example A–2

Evaluate $5! - 3!$.

Solution

$$5! - 3! = (5 \cdot 4 \cdot 3 \cdot 2 \cdot 1) - (3 \cdot 2 \cdot 1)$$
$$= 120 - 6 = 114$$

Note: $5! - 3! \neq 2!$, since $2! = 2$.

Factorials cannot be multiplied directly. Again, you must multiply them out and then multiply the products.

Example A–3

Evaluate $3! \cdot 2!$.

Solution

$$3! \cdot 2! = (3 \cdot 2 \cdot 1) \cdot (2 \cdot 1) = 6 \cdot 2 = 12$$

Note: $3! \cdot 2! \neq 6!$, since $6! = 720$.

Finally, factorials cannot be divided directly unless they are equal.

Example A–4

Evaluate $6! \div 3!$.

Solution

$$\frac{6!}{3!} = \frac{6 \cdot 5 \cdot 4 \cdot 3 \cdot 2 \cdot 1}{3 \cdot 2 \cdot 1} = \frac{720}{6} = 120$$

Note: $\dfrac{6!}{3!} \neq 2!$ since $2! = 2$

But $\dfrac{3!}{3!} = \dfrac{3 \cdot 2 \cdot 1}{3 \cdot 2 \cdot 1} = \dfrac{6}{6} = 1$

In division, you can take some shortcuts, as shown:

$$\frac{6!}{3!} = \frac{6 \cdot 5 \cdot 4 \cdot 3!}{3!} \qquad \text{and} \qquad \frac{3!}{3!} = 1$$
$$= 6 \cdot 5 \cdot 4 = 120$$

$$\frac{8!}{6!} = \frac{8 \cdot 7 \cdot 6!}{6!} \qquad \text{and} \qquad \frac{6!}{6!} = 1$$
$$= 8 \cdot 7 = 56$$

Another shortcut that can be used with factorials is cancellation, after factors have been expanded. For example,

$$\frac{7!}{(4!)(3!)} = \frac{7 \cdot 6 \cdot 5 \cdot 4!}{3 \cdot 2 \cdot 1 \cdot 4!}$$

Now cancel both instances of 4!. Then cancel the $3 \cdot 2$ in the denominator with the 6 in the numerator.

$$\frac{7 \cdot \overset{1}{\cancel{6}} \cdot 5 \cdot \cancel{4!}}{\underset{1}{\cancel{3}} \cdot \underset{1}{\cancel{2}} \cdot 1 \cdot \underset{1}{\cancel{4!}}} = 7 \cdot 5 = 35$$

Example A–5

Evaluate $10! \div (6!)(4!)$.

Solution

$$\frac{10!}{(6!)(4!)} = \frac{10 \cdot \overset{3}{\cancel{9}} \cdot \overset{1}{\cancel{8}} \cdot 7 \cdot \cancel{6!}}{\underset{1}{\cancel{4}} \cdot \underset{1}{\cancel{3}} \cdot \underset{1}{\cancel{2}} \cdot 1 \cdot \underset{1}{\cancel{6!}}} = 10 \cdot 3 \cdot 7 = 210$$

Exercises

Evaluate each expression.

A–1. 9!   362,880

A–2. 7!   5040

A–3. 5!   120

A–4. 0!   1

A–5. 1!   1

A–6. 3!   6

A–7. $\frac{12!}{9!}$ 1320

A–8. $\frac{10!}{2!}$ 1,814,400

A–9. $\frac{5!}{3!}$ 20

A–10. $\frac{11!}{7!}$ 7920

A–11. $\frac{9!}{(4!)(5!)}$ 126

A–12. $\frac{10!}{(7!)(3!)}$ 120

A–13. $\frac{8!}{(4!)(4!)}$ 70

A–14. $\frac{15!}{(12!)(3!)}$ 455

A–15. $\frac{10!}{(10!)(0!)}$ 1

A–16. $\frac{5!}{(3!)(2!)(1!)}$ 10

A–17. $\frac{8!}{(3!)(3!)(2!)}$ 560

A–18. $\frac{11!}{(7!)(2!)(2!)}$ 1980

A–19. $\frac{10!}{(3!)(2!)(5!)}$ 2520

A–20. $\frac{6!}{(2!)(2!)(2!)}$ 90

A–2   Summation Notation

In mathematics, the symbol Σ (Greek capital letter sigma) means to add or find the sum. For example, ΣX means to add the numbers represented by the variable X. Thus, when X represents 5, 8, 2, 4, and 6, then ΣX means $5 + 8 + 2 + 4 + 6 = 25$.

Sometimes, a subscript notation is used, such as

$$\sum_{i=1}^{5} X_i$$

This notation means to find the sum of five numbers represented by X, as shown:

$$\sum_{i=1}^{5} X_i = X_1 + X_2 + X_3 + X_4 + X_5$$

When the number of values is not known, the unknown number can be represented by n, such as

$$\sum_{i=1}^{n} X_i = X_1 + X_2 + X_3 + \cdots + X_n$$

There are several important types of summation used in statistics. The notation ΣX^2 means to square each value before summing. For example, if the values of the X's are 2, 8, 6, 1, and 4, then

$$\Sigma X^2 = 2^2 + 8^2 + 6^2 + 1^2 + 4^2$$
$$= 4 + 64 + 36 + 1 + 16 = 121$$

The notation $(\Sigma X)^2$ means to find the sum of X's and then square the answer. For instance, if the values for X are 2, 8, 6, 1, and 4, then

$$(\Sigma X)^2 = (2 + 8 + 6 + 1 + 4)^2$$
$$= (21)^2 = 441$$

Another important use of summation notation is in finding the mean (shown in Section 3–1). The mean $\bar{X}$ is defined as

$$\bar{X} = \frac{\Sigma X}{n}$$

For example, to find the mean of 12, 8, 7, 3, and 10, use the formula and substitute the values, as shown:

$$\bar{X} = \frac{\Sigma X}{n} = \frac{12 + 8 + 7 + 3 + 10}{5} = \frac{40}{5} = 8$$

The notation $\Sigma(X - \bar{X})^2$ means to perform the following steps.

STEP 1 Find the mean.

STEP 2 Subtract the mean from each value.

STEP 3 Square the answers.

STEP 4 Find the sum.

Example A–6

Find the value of $\Sigma(X - \bar{X})^2$ for the values 12, 8, 7, 3, and 10 of X.

Solution

STEP 1 Find the mean.

$$\bar{X} = \frac{12 + 8 + 7 + 3 + 10}{5} = \frac{40}{5} = 8$$

STEP 2 Subtract the mean from each value.

$$12 - 8 = 4 \qquad 7 - 8 = -1 \qquad 10 - 8 = 2$$
$$8 - 8 = 0 \qquad 3 - 8 = -5$$

STEP 3 Square the answers.

$$4^2 = 16 \qquad (-1)^2 = 1 \qquad 2^2 = 4$$
$$0^2 = 0 \qquad (-5)^2 = 25$$

STEP 4 Find the sum.

$$16 + 0 + 1 + 25 + 4 = 46$$

Example A–7

Find $\Sigma(X - \bar{X})^2$ for the following values of X: 5, 7, 2, 1, 3, 6.

Solution

Find the mean.

$$\bar{X} = \frac{5 + 7 + 2 + 1 + 3 + 6}{6} = \frac{24}{6} = 4$$

Then the steps in Example A–6 can be shortened as follows:

$$\begin{aligned}
\Sigma(X - \bar{X})^2 &= (5 - 4)^2 + (7 - 4)^2 + (2 - 4)^2 \\
&\quad + (1 - 4)^2 + (3 - 4)^2 + (6 - 4)^2 \\
&= 1^2 + 3^2 + (-2)^2 + (-3)^2 \\
&\quad + (-1)^2 + 2^2 \\
&= 1 + 9 + 4 + 9 + 1 + 4 = 28
\end{aligned}$$

Exercises

For each set of values, find ΣX, ΣX^2, $(\Sigma X)^2$, and $\Sigma(X - \bar{X})^2$.

A–21. 9, 17, 32, 16, 8, 2, 9, 7, 3, 18 121; 2181; 14,641; 716.9

A–22. 4, 12, 9, 13, 0, 6, 2, 10 56; 550; 3136; 158

A–23. 5, 12, 8, 3, 4 32; 258; 1024; 53.2

A–24. 6, 2, 18, 30, 31, 42, 16, 5 150; 4270; 22,500; 1457.5

A–25. 80, 76, 42, 53, 77 328; 22,678; 107,584; 1161.2

A–26. 123, 132, 216, 98, 146, 114
829; 123,125; 687,241; 8584.8333

A–27. 53, 72, 81, 42, 63, 71, 73, 85, 98, 55
693; 50,511; 480,249; 2486.1

A–28. 43, 32, 116, 98, 120 409; 40,333; 167,281; 6876.80

A–29. 12, 52, 36, 81, 63, 74 318; 20,150; 101,124; 3296

A–30. −9, −12, 18, 0, −2, −15 −20; 778; 400; 711.3334

A–3 The Line

The following figure shows the *rectangular coordinate system,* or *Cartesian plane.* This figure consists of two axes: the horizontal axis, called the x axis, and the vertical axis, called the y axis. Each axis has numerical scales. The point of intersection of the axes is called the *origin.*

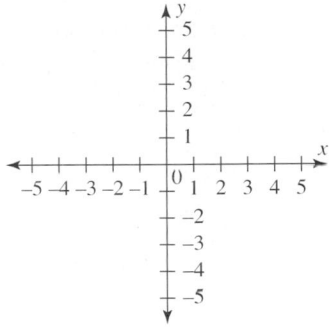

Points can be graphed by using coordinates. For example, the notation for point $P(3, 2)$ means that the x coordinate is 3 and the y coordinate is 2. Hence, P is located at the intersection of $x = 3$ and $y = 2$, as shown.

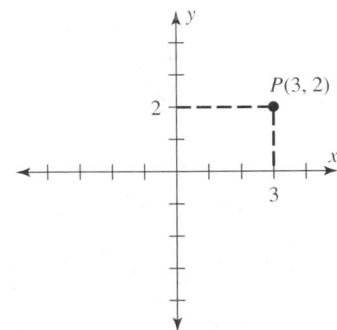

Other points, such as $Q(-5, 2)$, $R(4, 1)$, and $S(-3, -4)$, can be plotted, as shown in the next figure.

When a point lies on the y axis, the x coordinate is 0, as in $(0, 6)(0, -3)$, etc. When a point lies on the x axis, the y coordinate is 0, as in $(6, 0)(-8, 0)$, etc., as shown at the top of the next page.

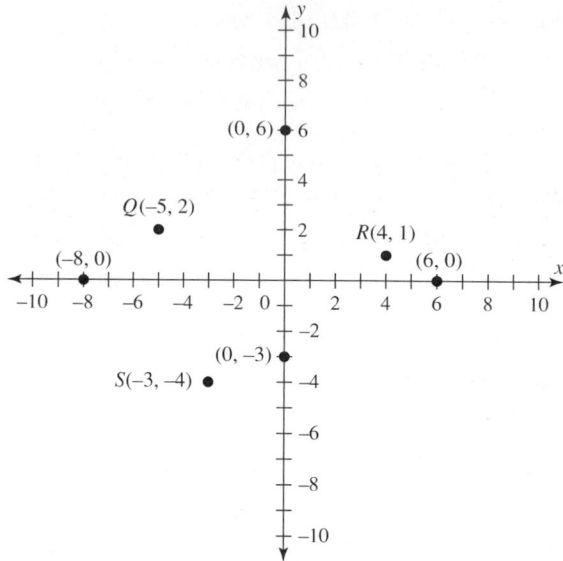

Two points determine a line. There are two properties of a line: its slope and its equation. The *slope m* of a line is determined by the ratio of the rise (called Δy) to the run (Δx).

$$m = \frac{\text{rise}}{\text{run}} = \frac{\Delta y}{\Delta x}$$

For example, the slope of the line shown below is $\frac{3}{2}$, or 1.5, since the height Δy is 3 units and the run Δx is 2 units.

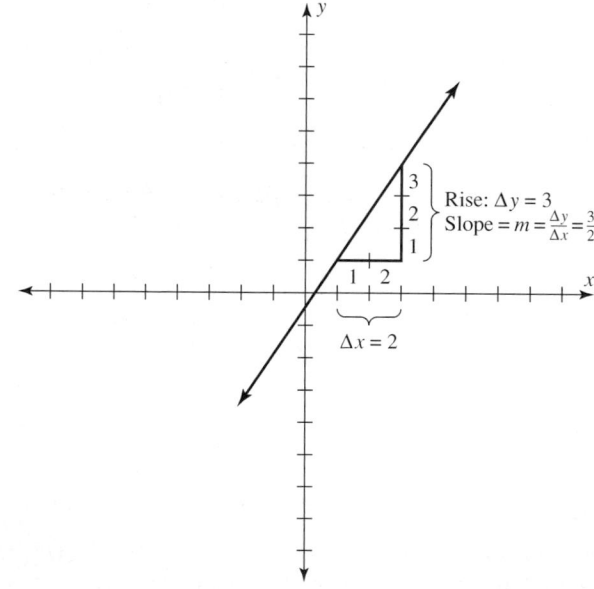

The slopes of lines can be positive, negative, or zero. A line going uphill from left to right has a positive slope. A line going downhill from left to right has a negative slope. And a line that is horizontal has a slope of zero.

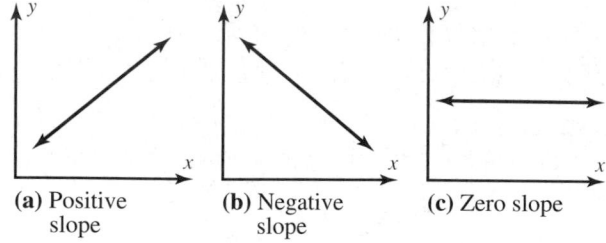

(a) Positive slope **(b)** Negative slope **(c)** Zero slope

A point *b* where the line crosses the *x* axis is called the *x intercept* and has the coordinates (*b*, 0). A point *a* where the line crosses the *y* axis is called the *y intercept* and has the coordinates (0, *a*).

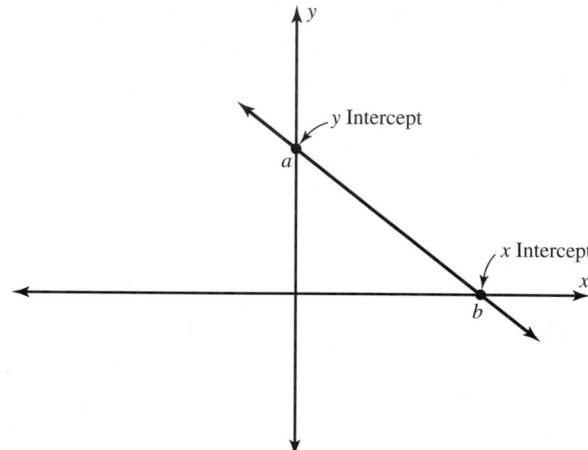

Every line has a unique equation of the form $y = a + bx$. For example, the equations

$$y = 5 + 3x$$
$$y = 8.6 + 3.2x$$
$$y = 5.2 - 6.1x$$

all represent different, unique lines. The number represented by *a* is the *y* intercept point; the number represented by *b* is the slope. The line whose equation is $y = 3 + 2x$ has a *y* intercept at 3 and a slope of 2, or $\frac{2}{1}$. This line can be shown as in the following graph.

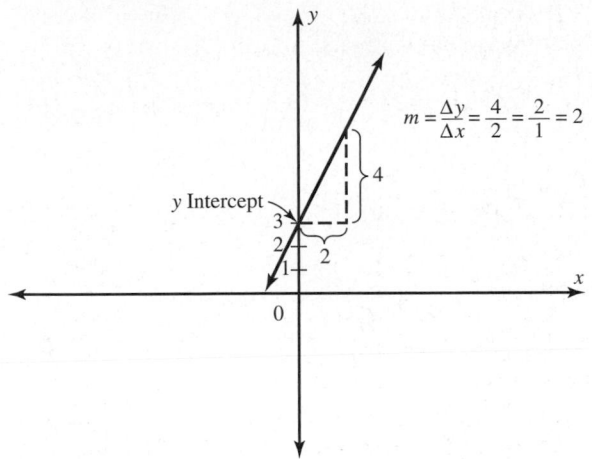

If two points are known, then the graph of the line can be plotted. For example, to find the graph of a line passing through the points $P(2, 1)$ and $Q(3, 5)$, plot the points and connect them as shown below.

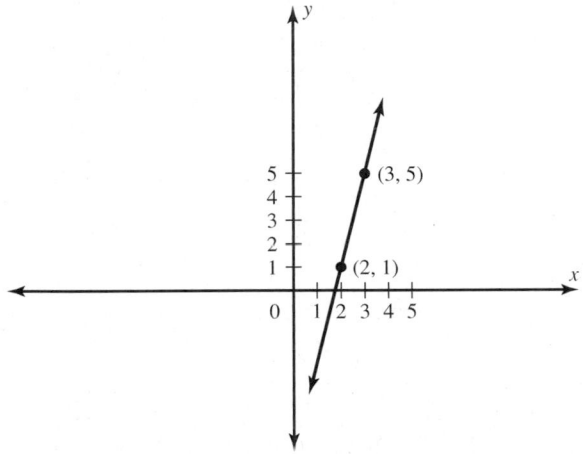

Given the equation of a line, you can graph the line by finding two points and then plotting them.

Example A–8

Plot the graph of the line whose equation is $y = 3 + 2x$.

Solution

Select any number as an x value, and substitute it in the equation to get the corresponding y value. Let $x = 0$.

Then

$$y = 3 + 2x = 3 + 2(0) = 3$$

Hence, when $x = 0$, then $y = 3$, and the line passes through the point $(0, 3)$.

Now select any other value of x, say, $x = 2$.

$$y = 3 + 2x = 3 + 2(2) = 7$$

Hence, a second point is $(2, 7)$. Then plot the points and graph the line.

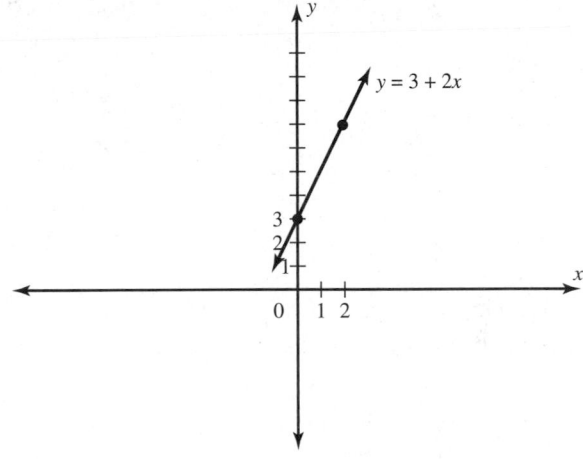

Exercises

Plot the line passing through each set of points.

A–31. $P(3, 2)$, $Q(1, 6)$ **A–34.** $P(-1, -2)$, $Q(-7, 8)$

A–32. $P(0, 5)$, $Q(8, 0)$ **A–35.** $P(6, 3)$, $Q(10, 3)$

A–33. $P(-2, 4)$, $Q(3, 6)$

Find at least two points on each line, and then graph the line containing these points.

A–36. $y = 5 + 2x$ **A–39.** $y = -2 - 2x$

A–37. $y = -1 + x$ **A–40.** $y = 4 - 3x$

A–38. $y = 3 + 4x$

Writing the Research Report

After conducting a statistical study, a researcher must write a final report explaining how the study was conducted and giving the results. The formats of research reports, theses, and dissertations vary from school to school; however, they tend to follow the general format explained here.

Front Materials

The front materials typically include the following items:

Title page
Copyright page
Acknowledgments
Table of contents
Table of appendixes
List of tables
List of figures

Chapter 1: Nature and Background of the Study

This chapter should introduce the reader to the nature of the study and present some discussion on the background. It should contain the following information:

Introduction
Statement of the problem
Background of the problem
Rationale for the study
Research questions and/or hypotheses
Assumptions, limitations, and delimitations
Definitions of terms

Chapter 2: Review of Literature

This chapter should explain what has been done in previous research related to the study. It should contain the following information:

Prior research
Related literature

Chapter 3: Methodology

This chapter should explain how the study was conducted. It should contain the following information:

Development of questionnaires, tests, survey instruments, etc.
Definition of the population
Sampling methods used
How the data were collected
Research design used
Statistical tests that will be used to analyze the data

Chapter 4: Analysis of Data

This chapter should explain the results of the statistical analysis of the data. It should state whether the null hypothesis should be rejected. Any statistical tables used to analyze the data should be included here.

Chapter 5: Summary, Conclusions, and Recommendations

This chapter summarizes the results of the study and explains any conclusions that have resulted from the statistical analysis of the data. The researchers should cite and explain any shortcomings of the study. Recommendations obtained from the study should be included here, and further studies should be suggested.

Bayes' Theorem

Objective B-1

Find the probability of an event, using Bayes' theorem.

Given two dependent events A and B, the previous formulas for conditional probability allow you to find $P(A \text{ and } B)$, or $P(B|A)$. Related to these formulas is a rule developed by the English Presbyterian minister Thomas Bayes (1702–1761). The rule is known as **Bayes' theorem.**

It is possible, given the outcome of the second event in a sequence of two events, to determine the probability of various possibilities for the first event. In Example 4–31, there were two boxes, each containing red balls and blue balls. A box was selected and a ball was drawn. The example asked for the probability that the ball selected was red. Now a different question can be asked: If the ball is red, what is the probability it came from box 1? In this case, the outcome is known, a red ball was selected, and you are asked to find the probability that it is a result of a previous event, that it came from box 1. Bayes' theorem can enable

you to compute this probability and can be explained by using tree diagrams.

The tree diagram for the solution of Example 4–31 is shown in Figure B–1, along with the appropriate notation and the corresponding probabilities. In this case, A_1 is the event of selecting box 1, A_2 is the event of selecting box 2, R is the event of selecting a red ball, and B is the event of selecting a blue ball.

To answer the question "If the ball selected is red, what is the probability that it came from box 1?" two formulas

$$P(B|A) = \frac{P(A \text{ and } B)}{P(A)} \tag{1}$$

$$P(A \text{ and } B) = P(A) \cdot P(B|A) \tag{2}$$

can be used. The notation that will be used is that of Example 4–31, shown in Figure B–1. Finding the probability that box 1 was selected given that the ball selected was red can be written symbolically as $P(A_1|R)$. By formula 1,

$$P(A_1|R) = \frac{P(R \text{ and } A_1)}{P(R_1)}$$

Note: $P(R \text{ and } A_1) = P(A_1 \text{ and } R)$.

Figure B–1

Tree Diagram for
Example 4–31

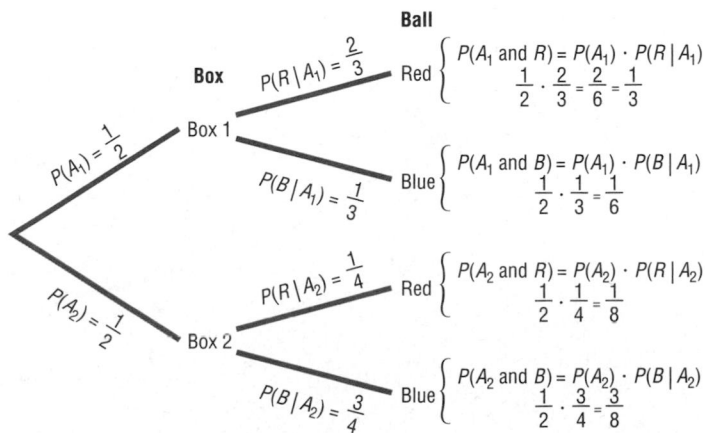

$$A\text{-}9$$

By formula 2,

$$P(A_1 \text{ and } R) = P(A_1) \cdot P(R|A_1)$$

and

$$P(R) = P(A_1 \text{ and } R) + P(A_2 \text{ and } R)$$

as shown in Figure B–1; $P(R)$ was found by adding the products of the probabilities of the branches in which a red ball was selected. Now,

$$P(A_1 \text{ and } R) = P(A_1) \cdot P(R|A_1)$$
$$P(A_2 \text{ and } R) = P(A_2) \cdot P(R|A_2)$$

Substituting these values in the original formula for $P(A_1|R)$, you get

$$P(A_1|R) = \frac{P(A_1) \cdot P(R|A_1)}{P(A_1) \cdot P(R|A_1) + P(A_2) \cdot P(R|A_2)}$$

Refer to Figure B–1. The numerator of the fraction is the product of the top branch of the tree diagram, which consists of selecting a red ball and selecting box 1. And the denominator is the sum of the products of the two branches of the tree where the red ball was selected.

Using this formula and the probability values shown in Figure B–1, you can find the probability that box 1 was selected given that the ball was red, as shown.

$$P(A_1|R) = \frac{P(A_1) \cdot P(R|A_1)}{P(A_1) \cdot P(R|A_1) + P(A_2) \cdot P(R|A_2)}$$

$$= \frac{\frac{1}{2} \cdot \frac{2}{3}}{\frac{1}{2} \cdot \frac{2}{3} + \frac{1}{2} \cdot \frac{1}{4}} = \frac{\frac{1}{3}}{\frac{1}{3} + \frac{1}{8}} = \frac{\frac{1}{3}}{\frac{8}{24} + \frac{3}{24}} = \frac{\frac{1}{3}}{\frac{11}{24}}$$

$$= \frac{1}{3} \div \frac{11}{24} = \frac{1}{\cancel{3}} \cdot \frac{\overset{8}{\cancel{24}}}{11} = \frac{8}{11}$$

This formula is a simplified version of Bayes' theorem.

Before Bayes' theorem is stated, another example is shown.

Example B–1

A shipment of two boxes, each containing six telephones, is received by a store. Box 1 contains one defective phone, and box 2 contains two defective phones. After the boxes are unpacked, a phone is selected and found to be defective. Find the probability that it came from box 2.

Solution

STEP 1   Select the proper notation. Let A_1 represent box 1 and A_2 represent box 2. Let D represent a defective phone and ND represent a phone that is not defective.

STEP 2   Draw a tree diagram and find the corresponding probabilities for each branch. The probability of selecting box 1 is $\frac{1}{2}$, and the probability of selecting box 2 is $\frac{1}{2}$. Since there is one defective phone in box 1, the probability of selecting it is $\frac{1}{6}$. The probability of selecting a nondefective phone from box 1 is $\frac{5}{6}$.

     Since there are two defective phones in box 2, the probability of selecting a defective phone from box 2 is $\frac{2}{6}$, or $\frac{1}{3}$; and the probability of selecting a nondefective phone is $\frac{4}{6}$, or $\frac{2}{3}$. The tree diagram is shown in Figure B–2.

STEP 3   Write the corresponding formula. Since the example is asking for the probability that, given a defective phone, it came from box 2, the corresponding formula is as shown.

$$P(A_2|D) = \frac{P(A_2) \cdot P(D|A_2)}{P(A_1) \cdot P(D|A_1) + P(A_2) \cdot P(D|A_2)}$$

$$= \frac{\frac{1}{2} \cdot \frac{2}{6}}{\frac{1}{2} \cdot \frac{1}{6} + \frac{1}{2} \cdot \frac{2}{6}} = \frac{\frac{1}{6}}{\frac{1}{12} + \frac{2}{12}} = \frac{\frac{1}{6}}{\frac{3}{12}}$$

$$= \frac{1}{6} \div \frac{3}{12} = \frac{1}{\cancel{6}} \cdot \frac{\overset{2}{\cancel{12}}}{3} = \frac{2}{3}$$

Figure B–2

Tree Diagram for Example B–1

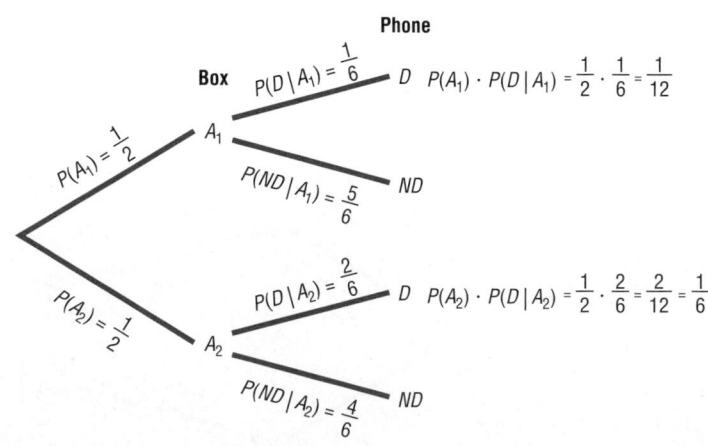

Bayes' theorem can be generalized to events with three or more outcomes and formally stated as in the next box.

Bayes' theorem For two events A and B, where event B follows event A, event A can occur in $A_1, A_2, \ldots, A_n$ mutually exclusive ways, and event B can occur in $B_1, B_2, \ldots, B_m$ mutually exclusive ways,

$$P(A_1|B_1) = \frac{P(A_1) \cdot P(B_1|A_1)}{[P(A_1) \cdot P(B_1|A_1) + P(A_2) \cdot P(B_1|A_2)} $$
$$+ \cdots + P(A_n) \cdot P(B_1|A_n)]$$

for any specific events A_1 and B_1.

The numerator is the product of the probabilities on the branch of the tree that consists of outcomes A_1 and B_1. The denominator is the sum of the products of the probabilities of the branches containing B_1 and A_1, B_1 and A_2, ... , B_1 and A_n.

Example B–2

On a game show, a contestant can select one of four boxes. Box 1 contains one $100 bill and nine $1 bills. Box 2 contains two $100 bills and eight $1 bills. Box 3 contains three $100 bills and seven $1 bills. Box 4 contains five $100 bills and five $1 bills. The contestant selects a box at random and selects a bill from the box at random. If a $100 bill is selected, find the probability that it came from box 4.

Solution

STEP 1 Select the proper notation. Let B_1, B_2, B_3, and B_4 represent the boxes and 100 and 1 represent the values of the bills in the boxes.

STEP 2 Draw a tree diagram and find the corresponding probabilities. The probability of selecting each box is $\frac{1}{4}$, or 0.25. The probabilities of selecting the $100 bill from each box, respectively, are $\frac{1}{10} = 0.1$, $\frac{2}{10} = 0.2$, $\frac{3}{10} = 0.3$, and $\frac{5}{10} = 0.5$. The tree diagram is shown in Figure B–3.

STEP 3 Using Bayes' theorem, write the corresponding formula. Since the example asks for the probability that box 4 was selected, given that $100 was obtained, the corresponding formula is as follows:

$$P(B_4|100) = \frac{P(B_4) \cdot P(100|B_4)}{[P(B_1) \cdot P(100|B_1) + P(B_2) \cdot P(100|B_2)}$$
$$+ P(B_3) \cdot P(100|B_3) + P(B_4) \cdot P(100|B_4)]$$

$$= \frac{0.125}{0.025 + 0.05 + 0.075 + 0.125}$$

$$= \frac{0.125}{0.275} = 0.455$$

Figure B–3

Tree Diagram for Example B–2

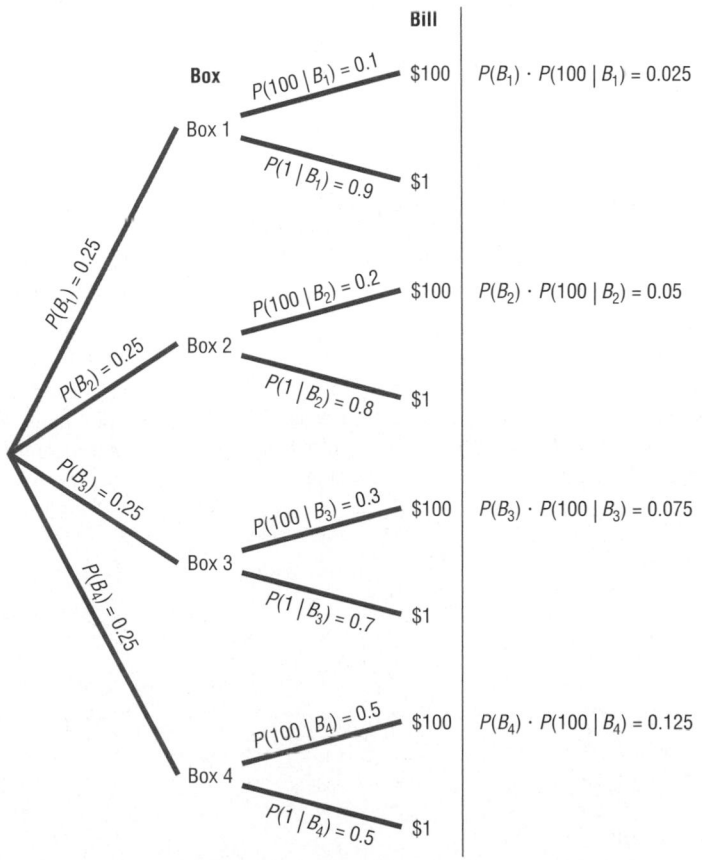

In Example B–2, the original probability of selecting box 4 was 0.25. However, once additional information was obtained—and the condition was considered that a $100 bill was selected—the revised probability of selecting box 4 became 0.455.

Bayes' theorem can be used to revise probabilities of events once additional information becomes known. Bayes' theorem is used as the basis for a branch of statistics called *Bayesian decision making,* which includes the use of subjective probabilities in making statistical inferences.

Exercises

B–1. An appliance store purchases electric ranges from two companies. From company A, 500 ranges are purchased and 2% are defective. From company B, 850 ranges are purchased and 2% are defective. Given that a range is defective, find the probability that it came from company B. 0.65

B–2. Two manufacturers supply blankets to emergency relief organizations. Manufacturer A supplies 3000 blankets, and 4% are irregular in workmanship. Manufacturer B supplies 2400 blankets, and 7% are found to be irregular. Given that a blanket is irregular, find the probability that it came from manufacturer B. 0.579

B–3. A test for a certain disease is found to be 95% accurate, meaning that it will correctly diagnose the disease in 95 out of 100 people who have the ailment. For a certain segment of the population, the incidence of the disease is 9%. If a person tests positive, find the probability that the person actually has the disease. The test is also 95% accurate for a negative result. 0.653

B–4. Using the test in Exercise B–3, if a person tests negative for the disease, find the probability that the person actually has the disease. Remember, 9% of the population has the disease. 0.005

B–5. A corporation has three methods of training employees. Because of time, space, and location, it sends 20% of its employees to location A, 35% to location B, and 45% to location C. Location A has an 80% success rate. That is, 80% of the employees who complete the course will pass the licensing

exam. Location B has a 75% success rate, and location C has a 60% success rate. If a person has passed the exam, find the probability that the person went to location B. 0.379

B–6. In Exercise B–5, if a person failed the exam, find the probability that the person went to location C. 0.585

B–7. A store purchases baseball hats from three different manufacturers. In manufacturer A's box, there are 12 blue hats, 6 red hats, and 6 green hats. In manufacturer B's box, there are 10 blue hats, 10 red hats, and 4 green hats. In manufacturer C's box, there are 8 blue hats, 8 red hats, and 8 green hats. A box is selected at random, and a hat is selected at random from that box. If the hat is red, find the probability that it came from manufacturer A's box. $\frac{1}{4}$

B–8. In Exercise B–7, if the hat selected is green, find the probability that it came from manufacturer B's box. $\frac{2}{9}$

B–9. A driver has three ways to get from one city to another. There is an 80% probability of encountering a traffic jam on route 1, a 60% probability on route 2, and a 30% probability on route 3. Because of other factors, such as distance and speed limits, the driver uses route 1 fifty percent of the time and routes 2 and 3 each 25% of the time. If the driver calls the dispatcher to inform him that she is in a traffic jam, find the probability that she has selected route 1. 0.64

B–10. In Exercise B–9, if the driver did not encounter a traffic jam, find the probability that she selected route 3. 0.467

B–11. A store owner purchases telephones from two companies. From company A, 350 telephones are purchased and 2% are defective. From company B, 1050 telephones are purchased and 4% are defective. Given that a phone is defective, find the probability that it came from company B. 0.857

B–12. Two manufacturers supply food to a large cafeteria. Manufacturer A supplies 2400 cans of soup, and 3% are found to be dented. Manufacturer B supplies 3600 cans, and 1% are found to be dented. Given that a can of soup is dented, find the probability that it came from manufacturer B. 0.33

Alternate Approach to the Standard Normal Distribution

The following procedure may be used to replace the cumulative area to the left procedure shown in Section 6–1. This method determines areas from the mean where $z = 0$.

Finding Areas Under the Standard Normal Distribution

For the solution of problems using the standard normal distribution, a four-step procedure may be used with the use of the Procedure Table shown.

STEP 1 Sketch the normal curve and label.

STEP 2 Shade the area desired.

STEP 3 Find the figure that matches the shaded area from the following procedure table.

STEP 4 Follow the directions given in the appropriate block of the procedure table to get the desired area.

Note: Table B–1 gives the area between 0 and any z score to the right of 0, and all areas are positive.

There are seven basic types of problems and all seven are summarized in the Procedure Table, with appropriate examples.

Procedure Table

Examples

1. Between 0 and any z score:
Look up the z score in the table to get the area.

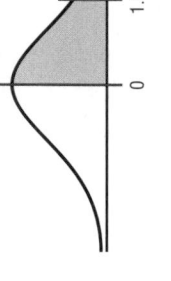

B–3–1: Find the area between $z = 0$ and $z = 1.23$.

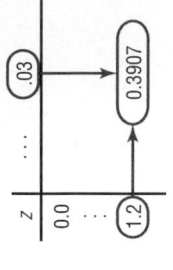

Look up area from $z = 0$ to $z = 1.23$ on Table B–1, as shown below.

The area between $z = 0$ and $z = 1.23$ is 0.3907.

2. In any tail:
  a. Look up the z score to get the area.
  b. Subtract the area from 0.5000.

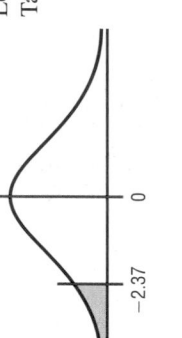

B–3–2: Find the area to the left of $z = -2.37$.

Look up area from $z = 0$ to $z = 2.37$ on Table B–1, as shown below.

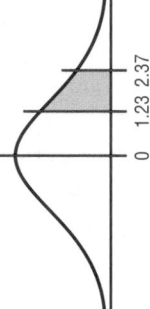

The area from $z = 0$ to $z = 2.37$ is the same as the area from $z = 0$ to $z = -2.37$.
Therefore, the area to the left of $z = -2.37 = 0.5000 - 0.4911 = 0.0089$.

3. Between two z scores on the same side of the mean:
  a. Look up both z scores to get the areas.
  b. Subtract the smaller area from the larger area.

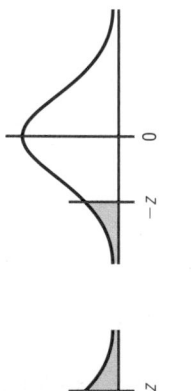

B–3–3: Find the area between $z = 1.23$ and $z = 2.37$.

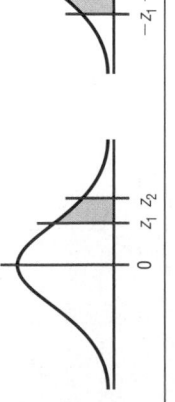

Look up areas for $z = 0$ to $z = 1.23$ and $z = 0$ to $z = 2.37$, as shown in B–3–1 and B–3–2, respectively.

The area between $z = 1.23$ and $z = 2.37 = 0.4911 - 0.3907 = 0.1004$.

4. Between two z scores on opposite sides of the mean:
  a. Look up both z scores to get the areas.
  b. Add the areas.

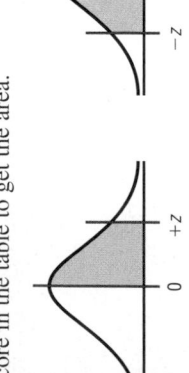

B–3–4: Find the area between $z = -1.23$ and $z = 2.37$.

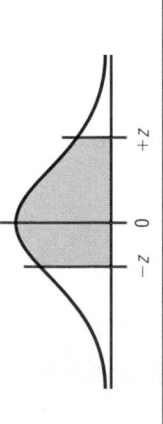

Look up areas for $z = 0$ to $z = 1.23$ and $z = 0$ to $z = 2.37$, as shown in B–3–1 and B–3–2, respectively.

The area between $z = -1.23$ and $z = 2.37 = 0.3907 + 0.4911 = 0.8818$.

5. To the left of any z score, where z is greater than the mean:

a. Look up the z score to get the area.

b. Add 0.5000 to the area.

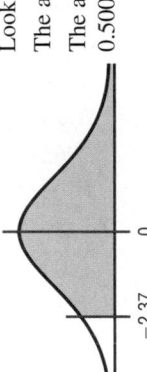

B–3–5: Find the area to the left of $z = 2.37$.

Look up area for $z = 2.37$, as shown in B–3–2.

The area to the left of $z = 0$ is 0.5000.

The area to the left of $z = 2.37 = 0.4911 + 0.5000 = 0.9911$.

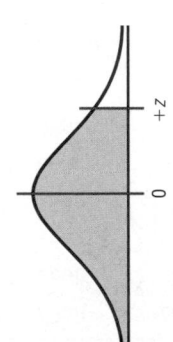

6. To the right of any z score, where z is less than the mean:

a. Look up the area in the table to get the area.

b. Add 0.5000 to the area.

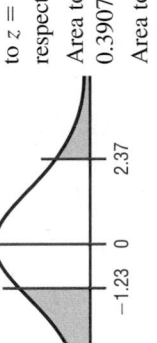

B–3–6: Find the area to the right of $z = -2.37$.

Look up area for $z = 2.37$, as shown in B–3–2.

The area to the right of $z = 0$ is 0.5000.

The area to the right of $z = -2.37 = 0.4911 + 0.5000 = 0.9911$.

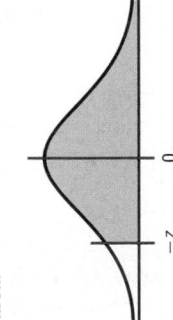

7. In any two tails:

a. Look up the z scores in the table to get the areas.

b. Subtract both areas from 0.5000.

c. Add the answers.

B–3–7: Find the area to the left of $z = -1.23$ and to the right of $z = 2.37$.

Look up areas for $z = 0$ to $z = 1.23$ and $z = 0$ to $z = 2.37$, as shown in B–3–1 and B–3–2, respectively.

Area to the left of $z = -1.23 = 0.5000 - 0.3907 = 0.1093$.

Area to the right of $z = 2.37 = 0.5000 - 0.4911 = 0.0089$.

The area to the left of $z = -1.23$ and to the right of $z = 2.37 = 0.1093 + 0.0089 = 0.1182$.

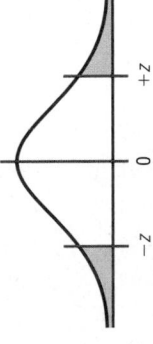

Table B-1	The Standard Normal Distribution								
z	.00	.01	.02	.03	.04	.05	.06	.07	.08	.09
0.0	.0000	.0040	.0080	.0120	.0160	.0199	.0239	.0279	.0319	.0359
0.1	.0398	.0438	.0478	.0517	.0557	.0596	.0636	.0675	.0714	.0753
0.2	.0793	.0832	.0871	.0910	.0948	.0987	.1026	.1064	.1103	.1141
0.3	.1179	.1217	.1255	.1293	.1331	.1368	.1406	.1443	.1480	.1517
0.4	.1554	.1591	.1628	.1664	.1700	.1736	.1772	.1808	.1844	.1879
0.5	.1915	.1950	.1985	.2019	.2054	.2088	.2123	.2157	.2190	.2224
0.6	.2257	.2291	.2324	.2357	.2389	.2422	.2454	.2486	.2517	.2549
0.7	.2580	.2611	.2642	.2673	.2704	.2734	.2764	.2794	.2823	.2852
0.8	.2881	.2910	.2939	.2967	.2995	.3023	.3051	.3078	.3106	.3133
0.9	.3159	.3186	.3212	.3238	.3264	.3289	.3315	.3340	.3365	.3389
1.0	.3413	.3438	.3461	.3485	.3508	.3531	.3554	.3577	.3599	.3621
1.1	.3643	.3665	.3686	.3708	.3729	.3749	.3770	.3790	.3810	.3830
1.2	.3849	.3869	.3888	.3907	.3925	.3944	.3962	.3980	.3997	.4015
1.3	.4032	.4049	.4066	.4082	.4099	.4115	.4131	.4147	.4162	.4177
1.4	.4192	.4207	.4222	.4236	.4251	.4265	.4279	.4292	.4306	.4319
1.5	.4332	.4345	.4357	.4370	.4382	.4394	.4406	.4418	.4429	.4441
1.6	.4452	.4463	.4474	.4484	.4495	.4505	.4515	.4525	.4535	.4545
1.7	.4554	.4564	.4573	.4582	.4591	.4599	.4608	.4616	.4625	.4633
1.8	.4641	.4649	.4656	.4664	.4671	.4678	.4686	.4693	.4699	.4706
1.9	.4713	.4719	.4726	.4732	.4738	.4744	.4750	.4756	.4761	.4767
2.0	.4772	.4778	.4783	.4788	.4793	.4798	.4803	.4808	.4812	.4817
2.1	.4821	.4826	.4830	.4834	.4838	.4842	.4846	.4850	.4854	.4857
2.2	.4861	.4864	.4868	.4871	.4875	.4878	.4881	.4884	.4887	.4890
2.3	.4893	.4896	.4898	.4901	.4904	.4906	.4909	.4911	.4913	.4916
2.4	.4918	.4920	.4922	.4925	.4927	.4929	.4931	.4932	.4934	.4936
2.5	.4938	.4940	.4941	.4943	.4945	.4946	.4948	.4949	.4951	.4952
2.6	.4953	.4955	.4956	.4957	.4959	.4960	.4961	.4962	.4963	.4964
2.7	.4965	.4966	.4967	.4968	.4969	.4970	.4971	.4972	.4973	.4974
2.8	.4974	.4975	.4976	.4977	.4977	.4978	.4979	.4979	.4980	.4981
2.9	.4981	.4982	.4982	.4983	.4984	.4984	.4985	.4985	.4986	.4986
3.0	.4987	.4987	.4987	.4988	.4988	.4989	.4989	.4989	.4990	.4990
3.1	.4990	.4991	.4991	.4991	.4992	.4992	.4992	.4992	.4993	.4993
3.2	.4993	.4993	.4994	.4994	.4994	.4994	.4994	.4995	.4995	.4995
3.3	.4995	.4995	.4995	.4996	.4996	.4996	.4996	.4996	.4996	.4997
3.4	.4997	.4997	.4997	.4997	.4997	.4997	.4997	.4997	.4997	.4998

For z values greater than 3.49, use 0.4999.

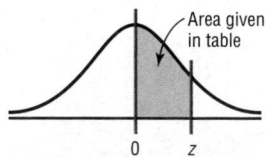

Appendix C

Tables

Table A	Factorials
n	$n!$
0	1
1	1
2	2
3	6
4	24
5	120
6	720
7	5,040
8	40,320
9	362,880
10	3,628,800
11	39,916,800
12	479,001,600
13	6,227,020,800
14	87,178,291,200
15	1,307,674,368,000
16	20,922,789,888,000
17	355,687,428,096,000
18	6,402,373,705,728,000
19	121,645,100,408,832,000
20	2,432,902,008,176,640,000

Table B		The Binomial Distribution										
							p					
n	*x*	0.05	0.1	0.2	0.3	0.4	0.5	0.6	0.7	0.8	0.9	0.95
2	0	0.902	0.810	0.640	0.490	0.360	0.250	0.160	0.090	0.040	0.010	0.002
	1	0.095	0.180	0.320	0.420	0.480	0.500	0.480	0.420	0.320	0.180	0.095
	2	0.002	0.010	0.040	0.090	0.160	0.250	0.360	0.490	0.640	0.810	0.902
3	0	0.857	0.729	0.512	0.343	0.216	0.125	0.064	0.027	0.008	0.001	
	1	0.135	0.243	0.384	0.441	0.432	0.375	0.288	0.189	0.096	0.027	0.007
	2	0.007	0.027	0.096	0.189	0.288	0.375	0.432	0.441	0.384	0.243	0.135
	3		0.001	0.008	0.027	0.064	0.125	0.216	0.343	0.512	0.729	0.857
4	0	0.815	0.656	0.410	0.240	0.130	0.062	0.026	0.008	0.002		
	1	0.171	0.292	0.410	0.412	0.346	0.250	0.154	0.076	0.026	0.004	
	2	0.014	0.049	0.154	0.265	0.346	0.375	0.346	0.265	0.154	0.049	0.014
	3		0.004	0.026	0.076	0.154	0.250	0.346	0.412	0.410	0.292	0.171
	4			0.002	0.008	0.026	0.062	0.130	0.240	0.410	0.656	0.815
5	0	0.774	0.590	0.328	0.168	0.078	0.031	0.010	0.002			
	1	0.204	0.328	0.410	0.360	0.259	0.156	0.077	0.028	0.006		
	2	0.021	0.073	0.205	0.309	0.346	0.312	0.230	0.132	0.051	0.008	0.001
	3	0.001	0.008	0.051	0.132	0.230	0.312	0.346	0.309	0.205	0.073	0.021
	4			0.006	0.028	0.077	0.156	0.259	0.360	0.410	0.328	0.204
	5				0.002	0.010	0.031	0.078	0.168	0.328	0.590	0.774
6	0	0.735	0.531	0.262	0.118	0.047	0.016	0.004	0.001			
	1	0.232	0.354	0.393	0.303	0.187	0.094	0.037	0.010	0.002		
	2	0.031	0.098	0.246	0.324	0.311	0.234	0.138	0.060	0.015	0.001	
	3	0.002	0.015	0.082	0.185	0.276	0.312	0.276	0.185	0.082	0.015	0.002
	4		0.001	0.015	0.060	0.138	0.234	0.311	0.324	0.246	0.098	0.031
	5			0.002	0.010	0.037	0.094	0.187	0.303	0.393	0.354	0.232
	6				0.001	0.004	0.016	0.047	0.118	0.262	0.531	0.735
7	0	0.698	0.478	0.210	0.082	0.028	0.008	0.002				
	1	0.257	0.372	0.367	0.247	0.131	0.055	0.017	0.004			
	2	0.041	0.124	0.275	0.318	0.261	0.164	0.077	0.025	0.004		
	3	0.004	0.023	0.115	0.227	0.290	0.273	0.194	0.097	0.029	0.003	
	4		0.003	0.029	0.097	0.194	0.273	0.290	0.227	0.115	0.023	0.004
	5			0.004	0.025	0.077	0.164	0.261	0.318	0.275	0.124	0.041
	6				0.004	0.017	0.055	0.131	0.247	0.367	0.372	0.257
	7					0.002	0.008	0.028	0.082	0.210	0.478	0.698
8	0	0.663	0.430	0.168	0.058	0.017	0.004	0.001				
	1	0.279	0.383	0.336	0.198	0.090	0.031	0.008	0.001			
	2	0.051	0.149	0.294	0.296	0.209	0.109	0.041	0.010	0.001		
	3	0.005	0.033	0.147	0.254	0.279	0.219	0.124	0.047	0.009		
	4		0.005	0.046	0.136	0.232	0.273	0.232	0.136	0.046	0.005	
	5			0.009	0.047	0.124	0.219	0.279	0.254	0.147	0.033	0.005
	6			0.001	0.010	0.041	0.109	0.209	0.296	0.294	0.149	0.051
	7				0.001	0.008	0.031	0.090	0.198	0.336	0.383	0.279
	8					0.001	0.004	0.017	0.058	0.168	0.430	0.663

Table B	(continued)										
						p					
n	x	0.05	0.1	0.2	0.3	0.4	0.5	0.6	0.7	0.8	0.9	0.95
9	0	0.630	0.387	0.134	0.040	0.010	0.002					
	1	0.299	0.387	0.302	0.156	0.060	0.018	0.004				
	2	0.063	0.172	0.302	0.267	0.161	0.070	0.021	0.004			
	3	0.008	0.045	0.176	0.267	0.251	0.164	0.074	0.021	0.003		
	4	0.001	0.007	0.066	0.172	0.251	0.246	0.167	0.074	0.017	0.001	
	5		0.001	0.017	0.074	0.167	0.246	0.251	0.172	0.066	0.007	0.001
	6			0.003	0.021	0.074	0.164	0.251	0.267	0.176	0.045	0.008
	7				0.004	0.021	0.070	0.161	0.267	0.302	0.172	0.063
	8					0.004	0.018	0.060	0.156	0.302	0.387	0.299
	9						0.002	0.010	0.040	0.134	0.387	0.630
10	0	0.599	0.349	0.107	0.028	0.006	0.001					
	1	0.315	0.387	0.268	0.121	0.040	0.010	0.002				
	2	0.075	0.194	0.302	0.233	0.121	0.044	0.011	0.001			
	3	0.010	0.057	0.201	0.267	0.215	0.117	0.042	0.009	0.001		
	4	0.001	0.011	0.088	0.200	0.251	0.205	0.111	0.037	0.006		
	5		0.001	0.026	0.103	0.201	0.246	0.201	0.103	0.026	0.001	
	6			0.006	0.037	0.111	0.205	0.251	0.200	0.088	0.011	0.001
	7			0.001	0.009	0.042	0.117	0.215	0.267	0.201	0.057	0.010
	8				0.001	0.011	0.044	0.121	0.233	0.302	0.194	0.075
	9					0.002	0.010	0.040	0.121	0.268	0.387	0.315
	10						0.001	0.006	0.028	0.107	0.349	0.599
11	0	0.569	0.314	0.086	0.020	0.004						
	1	0.329	0.384	0.236	0.093	0.027	0.005	0.001				
	2	0.087	0.213	0.295	0.200	0.089	0.027	0.005	0.001			
	3	0.014	0.071	0.221	0.257	0.177	0.081	0.023	0.004			
	4	0.001	0.016	0.111	0.220	0.236	0.161	0.070	0.017	0.002		
	5		0.002	0.039	0.132	0.221	0.226	0.147	0.057	0.010		
	6			0.010	0.057	0.147	0.226	0.221	0.132	0.039	0.002	
	7			0.002	0.017	0.070	0.161	0.236	0.220	0.111	0.016	0.001
	8				0.004	0.023	0.081	0.177	0.257	0.221	0.071	0.014
	9				0.001	0.005	0.027	0.089	0.200	0.295	0.213	0.087
	10					0.001	0.005	0.027	0.093	0.236	0.384	0.329
	11							0.004	0.020	0.086	0.314	0.569
12	0	0.540	0.282	0.069	0.014	0.002						
	1	0.341	0.377	0.206	0.071	0.017	0.003					
	2	0.099	0.230	0.283	0.168	0.064	0.016	0.002				
	3	0.017	0.085	0.236	0.240	0.142	0.054	0.012	0.001			
	4	0.002	0.021	0.133	0.231	0.213	0.121	0.042	0.008	0.001		
	5		0.004	0.053	0.158	0.227	0.193	0.101	0.029	0.003		
	6			0.016	0.079	0.177	0.226	0.177	0.079	0.016		
	7			0.003	0.029	0.101	0.193	0.227	0.158	0.053	0.004	
	8			0.001	0.008	0.042	0.121	0.213	0.231	0.133	0.021	0.002
	9				0.001	0.012	0.054	0.142	0.240	0.236	0.085	0.017
	10					0.002	0.016	0.064	0.168	0.283	0.230	0.099
	11						0.003	0.017	0.071	0.206	0.377	0.341
	12							0.002	0.014	0.069	0.282	0.540

Table B	(continued)										
						p						
n	*x*	**0.05**	**0.1**	**0.2**	**0.3**	**0.4**	**0.5**	**0.6**	**0.7**	**0.8**	**0.9**	**0.95**
13	0	0.513	0.254	0.055	0.010	0.001						
	1	0.351	0.367	0.179	0.054	0.011	0.002					
	2	0.111	0.245	0.268	0.139	0.045	0.010	0.001				
	3	0.021	0.100	0.246	0.218	0.111	0.035	0.006	0.001			
	4	0.003	0.028	0.154	0.234	0.184	0.087	0.024	0.003			
	5		0.006	0.069	0.180	0.221	0.157	0.066	0.014	0.001		
	6		0.001	0.023	0.103	0.197	0.209	0.131	0.044	0.006		
	7			0.006	0.044	0.131	0.209	0.197	0.103	0.023	0.001	
	8			0.001	0.014	0.066	0.157	0.221	0.180	0.069	0.006	
	9				0.003	0.024	0.087	0.184	0.234	0.154	0.028	0.003
	10				0.001	0.006	0.035	0.111	0.218	0.246	0.100	0.021
	11					0.001	0.010	0.045	0.139	0.268	0.245	0.111
	12						0.002	0.011	0.054	0.179	0.367	0.351
	13							0.001	0.010	0.055	0.254	0.513
14	0	0.488	0.229	0.044	0.007	0.001						
	1	0.359	0.356	0.154	0.041	0.007	0.001					
	2	0.123	0.257	0.250	0.113	0.032	0.006	0.001				
	3	0.026	0.114	0.250	0.194	0.085	0.022	0.003				
	4	0.004	0.035	0.172	0.229	0.155	0.061	0.014	0.001			
	5		0.008	0.086	0.196	0.207	0.122	0.041	0.007			
	6		0.001	0.032	0.126	0.207	0.183	0.092	0.023	0.002		
	7			0.009	0.062	0.157	0.209	0.157	0.062	0.009		
	8			0.002	0.023	0.092	0.183	0.207	0.126	0.032	0.001	
	9				0.007	0.041	0.122	0.207	0.196	0.086	0.008	
	10				0.001	0.014	0.061	0.155	0.229	0.172	0.035	0.004
	11					0.003	0.022	0.085	0.194	0.250	0.114	0.026
	12					0.001	0.006	0.032	0.113	0.250	0.257	0.123
	13						0.001	0.007	0.041	0.154	0.356	0.359
	14							0.001	0.007	0.044	0.229	0.488
15	0	0.463	0.206	0.035	0.005							
	1	0.366	0.343	0.132	0.031	0.005						
	2	0.135	0.267	0.231	0.092	0.022	0.003					
	3	0.031	0.129	0.250	0.170	0.063	0.014	0.002				
	4	0.005	0.043	0.188	0.219	0.127	0.042	0.007	0.001			
	5	0.001	0.010	0.103	0.206	0.186	0.092	0.024	0.003			
	6		0.002	0.043	0.147	0.207	0.153	0.061	0.012	0.001		
	7			0.014	0.081	0.177	0.196	0.118	0.035	0.003		
	8			0.003	0.035	0.118	0.196	0.177	0.081	0.014		
	9			0.001	0.012	0.061	0.153	0.207	0.147	0.043	0.002	
	10				0.003	0.024	0.092	0.186	0.206	0.103	0.010	0.001
	11				0.001	0.007	0.042	0.127	0.219	0.188	0.043	0.005
	12					0.002	0.014	0.063	0.170	0.250	0.129	0.031
	13						0.003	0.022	0.092	0.231	0.267	0.135
	14							0.005	0.031	0.132	0.343	0.366
	15								0.005	0.035	0.206	0.463

Table B	(continued)										
							p					
n	*x*	0.05	0.1	0.2	0.3	0.4	0.5	0.6	0.7	0.8	0.9	0.95
16	0	0.440	0.185	0.028	0.003							
	1	0.371	0.329	0.113	0.023	0.003						
	2	0.146	0.275	0.211	0.073	0.015	0.002					
	3	0.036	0.142	0.246	0.146	0.047	0.009	0.001				
	4	0.006	0.051	0.200	0.204	0.101	0.028	0.004				
	5	0.001	0.014	0.120	0.210	0.162	0.067	0.014	0.001			
	6		0.003	0.055	0.165	0.198	0.122	0.039	0.006			
	7			0.020	0.101	0.189	0.175	0.084	0.019	0.001		
	8			0.006	0.049	0.142	0.196	0.142	0.049	0.006		
	9			0.001	0.019	0.084	0.175	0.189	0.101	0.020		
	10				0.006	0.039	0.122	0.198	0.165	0.055	0.003	
	11				0.001	0.014	0.067	0.162	0.210	0.120	0.014	0.001
	12					0.004	0.028	0.101	0.204	0.200	0.051	0.006
	13					0.001	0.009	0.047	0.146	0.246	0.142	0.036
	14						0.002	0.015	0.073	0.211	0.275	0.146
	15							0.003	0.023	0.113	0.329	0.371
	16								0.003	0.028	0.185	0.440
17	0	0.418	0.167	0.023	0.002							
	1	0.374	0.315	0.096	0.017	0.002						
	2	0.158	0.280	0.191	0.058	0.010	0.001					
	3	0.041	0.156	0.239	0.125	0.034	0.005					
	4	0.008	0.060	0.209	0.187	0.080	0.018	0.002				
	5	0.001	0.017	0.136	0.208	0.138	0.047	0.008	0.001			
	6		0.004	0.068	0.178	0.184	0.094	0.024	0.003			
	7		0.001	0.027	0.120	0.193	0.148	0.057	0.009			
	8			0.008	0.064	0.161	0.185	0.107	0.028	0.002		
	9			0.002	0.028	0.107	0.185	0.161	0.064	0.008		
	10				0.009	0.057	0.148	0.193	0.120	0.027	0.001	
	11				0.003	0.024	0.094	0.184	0.178	0.068	0.004	
	12				0.001	0.008	0.047	0.138	0.208	0.136	0.017	0.001
	13					0.002	0.018	0.080	0.187	0.209	0.060	0.008
	14						0.005	0.034	0.125	0.239	0.156	0.041
	15						0.001	0.010	0.058	0.191	0.280	0.158
	16							0.002	0.017	0.096	0.315	0.374
	17								0.002	0.023	0.167	0.418

Table B	(continued)										
						p					
n	x	0.05	0.1	0.2	0.3	0.4	0.5	0.6	0.7	0.8	0.9	0.95
18	0	0.397	0.150	0.018	0.002							
	1	0.376	0.300	0.081	0.013	0.001						
	2	0.168	0.284	0.172	0.046	0.007	0.001					
	3	0.047	0.168	0.230	0.105	0.025	0.003					
	4	0.009	0.070	0.215	0.168	0.061	0.012	0.001				
	5	0.001	0.022	0.151	0.202	0.115	0.033	0.004				
	6		0.005	0.082	0.187	0.166	0.071	0.015	0.001			
	7		0.001	0.035	0.138	0.189	0.121	0.037	0.005			
	8			0.012	0.081	0.173	0.167	0.077	0.015	0.001		
	9			0.003	0.039	0.128	0.185	0.128	0.039	0.003		
	10			0.001	0.015	0.077	0.167	0.173	0.081	0.012		
	11				0.005	0.037	0.121	0.189	0.138	0.035	0.001	
	12				0.001	0.015	0.071	0.166	0.187	0.082	0.005	
	13					0.004	0.033	0.115	0.202	0.151	0.022	0.001
	14					0.001	0.012	0.061	0.168	0.215	0.070	0.009
	15						0.003	0.025	0.105	0.230	0.168	0.047
	16						0.001	0.007	0.046	0.172	0.284	0.168
	17							0.001	0.013	0.081	0.300	0.376
	18								0.002	0.018	0.150	0.397
19	0	0.377	0.135	0.014	0.001							
	1	0.377	0.285	0.068	0.009	0.001						
	2	0.179	0.285	0.154	0.036	0.005						
	3	0.053	0.180	0.218	0.087	0.017	0.002					
	4	0.011	0.080	0.218	0.149	0.047	0.007	0.001				
	5	0.002	0.027	0.164	0.192	0.093	0.022	0.002				
	6		0.007	0.095	0.192	0.145	0.052	0.008	0.001			
	7		0.001	0.044	0.153	0.180	0.096	0.024	0.002			
	8			0.017	0.098	0.180	0.144	0.053	0.008			
	9			0.005	0.051	0.146	0.176	0.098	0.022	0.001		
	10			0.001	0.022	0.098	0.176	0.146	0.051	0.005		
	11				0.008	0.053	0.144	0.180	0.098	0.071		
	12				0.002	0.024	0.096	0.180	0.153	0.044	0.001	
	13				0.001	0.008	0.052	0.145	0.192	0.095	0.007	
	14					0.002	0.022	0.093	0.192	0.164	0.027	0.002
	15					0.001	0.007	0.047	0.149	0.218	0.080	0.011
	16						0.002	0.017	0.087	0.218	0.180	0.053
	17							0.005	0.036	0.154	0.285	0.179
	18							0.001	0.009	0.068	0.285	0.377
	19								0.001	0.014	0.135	0.377

Table B		(concluded)										
							p					
n	x	0.05	0.1	0.2	0.3	0.4	0.5	0.6	0.7	0.8	0.9	0.95
20	0	0.358	0.122	0.012	0.001							
	1	0.377	0.270	0.058	0.007							
	2	0.189	0.285	0.137	0.028	0.003						
	3	0.060	0.190	0.205	0.072	0.012	0.001					
	4	0.013	0.090	0.218	0.130	0.035	0.005					
	5	0.002	0.032	0.175	0.179	0.075	0.015	0.001				
	6		0.009	0.109	0.192	0.124	0.037	0.005				
	7		0.002	0.055	0.164	0.166	0.074	0.015	0.001			
	8			0.022	0.114	0.180	0.120	0.035	0.004			
	9			0.007	0.065	0.160	0.160	0.071	0.012			
	10			0.002	0.031	0.117	0.176	0.117	0.031	0.002		
	11				0.012	0.071	0.160	0.160	0.065	0.007		
	12				0.004	0.035	0.120	0.180	0.114	0.022		
	13				0.001	0.015	0.074	0.166	0.164	0.055	0.002	
	14					0.005	0.037	0.124	0.192	0.109	0.009	
	15					0.001	0.015	0.075	0.179	0.175	0.032	0.002
	16						0.005	0.035	0.130	0.218	0.090	0.013
	17						0.001	0.012	0.072	0.205	0.190	0.060
	18							0.003	0.028	0.137	0.285	0.189
	19								0.007	0.058	0.270	0.377
	20								0.001	0.012	0.122	0.358

Note: All values of 0.0005 or less are omitted.

Source: J. Freund and G. Simon, *Modern Elementary Statistics,* Table "The Binomial Distribution," © 1992 Prentice-Hall, Inc. Reproduced by permission of Pearson Education, Inc.

Table C	The Poisson Distribution

	λ									
x	**0.1**	**0.2**	**0.3**	**0.4**	**0.5**	**0.6**	**0.7**	**0.8**	**0.9**	**1.0**
0	.9048	.8187	.7408	.6703	.6065	.5488	.4966	.4493	.4066	.3679
1	.0905	.1637	.2222	.2681	.3033	.3293	.3476	.3595	.3659	.3679
2	.0045	.0164	.0333	.0536	.0758	.0988	.1217	.1438	.1647	.1839
3	.0002	.0011	.0033	.0072	.0126	.0198	.0284	.0383	.0494	.0613
4	.0000	.0001	.0003	.0007	.0016	.0030	.0050	.0077	.0111	.0153
5	.0000	.0000	.0000	.0001	.0002	.0004	.0007	.0012	.0020	.0031
6	.0000	.0000	.0000	.0000	.0000	.0000	.0001	.0002	.0003	.0005
7	.0000	.0000	.0000	.0000	.0000	.0000	.0000	.0000	.0000	.0001

	λ									
x	**1.1**	**1.2**	**1.3**	**1.4**	**1.5**	**1.6**	**1.7**	**1.8**	**1.9**	**2.0**
0	.3329	.3012	.2725	.2466	.2231	.2019	.1827	.1653	.1496	.1353
1	.3662	.3614	.3543	.3452	.3347	.3230	.3106	.2975	.2842	.2707
2	.2014	.2169	.2303	.2417	.2510	.2584	.2640	.2678	.2700	.2707
3	.0738	.0867	.0998	.1128	.1255	.1378	.1496	.1607	.1710	.1804
4	.0203	.0260	.0324	.0395	.0471	.0551	.0636	.0723	.0812	.0902
5	.0045	.0062	.0084	.0111	.0141	.0176	.0216	.0260	.0309	.0361
6	.0008	.0012	.0018	.0026	.0035	.0047	.0061	.0078	.0098	.0120
7	.0001	.0002	.0003	.0005	.0008	.0011	.0015	.0020	.0027	.0034
8	.0000	.0000	.0001	.0001	.0001	.0002	.0003	.0005	.0006	.0009
9	.0000	.0000	.0000	.0000	.0000	.0000	.0001	.0001	.0001	.0002

	λ									
x	**2.1**	**2.2**	**2.3**	**2.4**	**2.5**	**2.6**	**2.7**	**2.8**	**2.9**	**3.0**
0	.1225	.1108	.1003	.0907	.0821	.0743	.0672	.0608	.0550	.0498
1	.2572	.2438	.2306	.2177	.2052	.1931	.1815	.1703	.1596	.1494
2	.2700	.2681	.2652	.2613	.2565	.2510	.2450	.2384	.2314	.2240
3	.1890	.1966	.2033	.2090	.2138	.2176	.2205	.2225	.2237	.2240
4	.0992	.1082	.1169	.1254	.1336	.1414	.1488	.1557	.1622	.1680
5	.0417	.0476	.0538	.0602	.0668	.0735	.0804	.0872	.0940	.1008
6	.0146	.0174	.0206	.0241	.0278	.0319	.0362	.0407	.0455	.0504
7	.0044	.0055	.0068	.0083	.0099	.0118	.0139	.0163	.0188	.0216
8	.0011	.0015	.0019	.0025	.0031	.0038	.0047	.0057	.0068	.0081
9	.0003	.0004	.0005	.0007	.0009	.0011	.0014	.0018	.0022	.0027
10	.0001	.0001	.0001	.0002	.0002	.0003	.0004	.0005	.0006	.0008
11	.0000	.0000	.0000	.0000	.0000	.0001	.0001	.0001	.0002	.0002
12	.0000	.0000	.0000	.0000	.0000	.0000	.0000	.0000	.0000	.0001

	λ									
x	**3.1**	**3.2**	**3.3**	**3.4**	**3.5**	**3.6**	**3.7**	**3.8**	**3.9**	**4.0**
0	.0450	.0408	.0369	.0334	.0302	.0273	.0247	.0224	.0202	.0183
1	.1397	.1304	.1217	.1135	.1057	.0984	.0915	.0850	.0789	.0733
2	.2165	.2087	.2008	.1929	.1850	.1771	.1692	.1615	.1539	.1465
3	.2237	.2226	.2209	.2186	.2158	.2125	.2087	.2046	.2001	.1954
4	.1734	.1781	.1823	.1858	.1888	.1912	.1931	.1944	.1951	.1954

Table C	(continued)								
					λ				
x	3.1	3.2	3.3	3.4	3.5	3.6	3.7	3.8	3.9	4.0
5	.1075	.1140	.1203	.1264	.1322	.1377	.1429	.1477	.1522	.1563
6	.0555	.0608	.0662	.0716	.0771	.0826	.0881	.0936	.0989	.1042
7	.0246	.0278	.0312	.0348	.0385	.0425	.0466	.0508	.0551	.0595
8	.0095	.0111	.0129	.0148	.0169	.0191	.0215	.0241	.0269	.0298
9	.0033	.0040	.0047	.0056	.0066	.0076	.0089	.0102	.0116	.0132
10	.0010	.0013	.0016	.0019	.0023	.0028	.0033	.0039	.0045	.0053
11	.0003	.0004	.0005	.0006	.0007	.0009	.0011	.0013	.0016	.0019
12	.0001	.0001	.0001	.0002	.0002	.0003	.0003	.0004	.0005	.0006
13	.0000	.0000	.0000	.0000	.0001	.0001	.0001	.0001	.0002	.0002
14	.0000	.0000	.0000	.0000	.0000	.0000	.0000	.0000	.0000	.0001

					λ				
x	4.1	4.2	4.3	4.4	4.5	4.6	4.7	4.8	4.9	5.0
0	.0166	.0150	.0136	.0123	.0111	.0101	.0091	.0082	.0074	.0067
1	.0679	.0630	.0583	.0540	.0500	.0462	.0427	.0395	.0365	.0337
2	.1393	.1323	.1254	.1188	.1125	.1063	.1005	.0948	.0894	.0842
3	.1904	.1852	.1798	.1743	.1687	.1631	.1574	.1517	.1460	.1404
4	.1951	.1944	.1933	.1917	.1898	.1875	.1849	.1820	.1789	.1755
5	.1600	.1633	.1662	.1687	.1708	.1725	.1738	.1747	.1753	.1755
6	.1093	.1143	.1191	.1237	.1281	.1323	.1362	.1398	.1432	.1462
7	.0640	.0686	.0732	.0778	.0824	.0869	.0914	.0959	.1002	.1044
8	.0328	.0360	.0393	.0428	.0463	.0500	.0537	.0575	.0614	.0653
9	.0150	.0168	.0188	.0209	.0232	.0255	.0280	.0307	.0334	.0363
10	.0061	.0071	.0081	.0092	.0104	.0118	.0132	.0147	.0164	.0181
11	.0023	.0027	.0032	.0037	.0043	.0049	.0056	.0064	.0073	.0082
12	.0008	.0009	.0011	.0014	.0016	.0019	.0022	.0026	.0030	.0034
13	.0002	.0003	.0004	.0005	.0006	.0007	.0008	.0009	.0011	.0013
14	.0001	.0001	.0001	.0001	.0002	.0002	.0003	.0003	.0004	.0005
15	.0000	.0000	.0000	.0000	.0001	.0001	.0001	.0001	.0001	.0002

					λ				
x	5.1	5.2	5.3	5.4	5.5	5.6	5.7	5.8	5.9	6.0
0	.0061	.0055	.0050	.0045	.0041	.0037	.0033	.0030	.0027	.0025
1	.0311	.0287	.0265	.0244	.0225	.0207	.0191	.0176	.0162	.0149
2	.0793	.0746	.0701	.0659	.0618	.0580	.0544	.0509	.0477	.0446
3	.1348	.1293	.1239	.1185	.1133	.1082	.1033	.0985	.0938	.0892
4	.1719	.1681	.1641	.1600	.1558	.1515	.1472	.1428	.1383	.1339

Table C	(continued)

					λ					
x	5.1	5.2	5.3	5.4	5.5	5.6	5.7	5.8	5.9	6.0
5	.1753	.1748	.1740	.1728	.1714	.1697	.1678	.1656	.1632	.1606
6	.1490	.1515	.1537	.1555	.1571	.1584	.1594	.1601	.1605	.1606
7	.1086	.1125	.1163	.1200	.1234	.1267	.1298	.1326	.1353	.1377
8	.0692	.0731	.0771	.0810	.0849	.0887	.0925	.0962	.0998	.1033
9	.0392	.0423	.0454	.0486	.0519	.0552	.0586	.0620	.0654	.0688
10	.0200	.0220	.0241	.0262	.0285	.0309	.0334	.0359	.0386	.0413
11	.0093	.0104	.0116	.0129	.0143	.0157	.0173	.0190	.0207	.0225
12	.0039	.0045	.0051	.0058	.0065	.0073	.0082	.0092	.0102	.0113
13	.0015	.0018	.0021	.0024	.0028	.0032	.0036	.0041	.0046	.0052
14	.0006	.0007	.0008	.0009	.0011	.0013	.0015	.0017	.0019	.0022
15	.0002	.0002	.0003	.0003	.0004	.0005	.0006	.0007	.0008	.0009
16	.0001	.0001	.0001	.0001	.0001	.0002	.0002	.0002	.0003	.0003
17	.0000	.0000	.0000	.0000	.0000	.0000	.0001	.0001	.0001	.0001

					λ					
x	6.1	6.2	6.3	6.4	6.5	6.6	6.7	6.8	6.9	7.0
0	.0022	.0020	.0018	.0017	.0015	.0014	.0012	.0011	.0010	.0009
1	.0137	.0126	.0116	.0106	.0098	.0090	.0082	.0076	.0070	.0064
2	.0417	.0390	.0364	.0340	.0318	.0296	.0276	.0258	.0240	.0223
3	.0848	.0806	.0765	.0726	.0688	.0652	.0617	.0584	.0552	.0521
4	.1294	.1249	.1205	.1162	.1118	.1076	.1034	.0992	.0952	.0912
5	.1579	.1549	.1519	.1487	.1454	.1420	.1385	.1349	.1314	.1277
6	.1605	.1601	.1595	.1586	.1575	.1562	.1546	.1529	.1511	.1490
7	.1399	.1418	.1435	.1450	.1462	.1472	.1480	.1486	.1489	.1490
8	.1066	.1099	.1130	.1160	.1188	.1215	.1240	.1263	.1284	.1304
9	.0723	.0757	.0791	.0825	.0858	.0891	.0923	.0954	.0985	.1014
10	.0441	.0469	.0498	.0528	.0558	.0588	.0618	.0649	.0679	.0710
11	.0245	.0265	.0285	.0307	.0330	.0353	.0377	.0401	.0426	.0452
12	.0124	.0137	.0150	.0164	.0179	.0194	.0210	.0227	.0245	.0264
13	.0058	.0065	.0073	.0081	.0089	.0098	.0108	.0119	.0130	.0142
14	.0025	.0029	.0033	.0037	.0041	.0046	.0052	.0058	.0064	.0071
15	.0010	.0012	.0014	.0016	.0018	.0020	.0023	.0026	.0029	.0033
16	.0004	.0005	.0005	.0006	.0007	.0008	.0010	.0011	.0013	.0014
17	.0001	.0002	.0002	.0002	.0003	.0003	.0004	.0004	.0005	.0006
18	.0000	.0001	.0001	.0001	.0001	.0001	.0001	.0002	.0002	.0002
19	.0000	.0000	.0000	.0000	.0000	.0000	.0000	.0001	.0001	.0001

Table C	(continued)									
					λ					
x	7.1	7.2	7.3	7.4	7.5	7.6	7.7	7.8	7.9	8.0
0	.0008	.0007	.0007	.0006	.0006	.0005	.0005	.0004	.0004	.0003
1	.0059	.0054	.0049	.0045	.0041	.0038	.0035	.0032	.0029	.0027
2	.0208	.0194	.0180	.0167	.0156	.0145	.0134	.0125	.0116	.0107
3	.0492	.0464	.0438	.0413	.0389	.0366	.0345	.0324	.0305	.0286
4	.0874	.0836	.0799	.0764	.0729	.0696	.0663	.0632	.0602	.0573
5	.1241	.1204	.1167	.1130	.1094	.1057	.1021	.0986	.0951	.0916
6	.1468	.1445	.1420	.1394	.1367	.1339	.1311	.1282	.1252	.1221
7	.1489	.1486	.1481	.1474	.1465	.1454	.1442	.1428	.1413	.1396
8	.1321	.1337	.1351	.1363	.1373	.1382	.1388	.1392	.1395	.1396
9	.1042	.1070	.1096	.1121	.1144	.1167	.1187	.1207	.1224	.1241
10	.0740	.0770	.0800	.0829	.0858	.0887	.0914	.0941	.0967	.0993
11	.0478	.0504	.0531	.0558	.0585	.0613	.0640	.0667	.0695	.0722
12	.0283	.0303	.0323	.0344	.0366	.0388	.0411	.0434	.0457	.0481
13	.0154	.0168	.0181	.0196	.0211	.0227	.0243	.0260	.0278	.0296
14	.0078	.0086	.0095	.0104	.0113	.0123	.0134	.0145	.0157	.0169
15	.0037	.0041	.0046	.0051	.0057	.0062	.0069	.0075	.0083	.0090
16	.0016	.0019	.0021	.0024	.0026	.0030	.0033	.0037	.0041	.0045
17	.0007	.0008	.0009	.0010	.0012	.0013	.0015	.0017	.0019	.0021
18	.0003	.0003	.0004	.0004	.0005	.0006	.0006	.0007	.0008	.0009
19	.0001	.0001	.0001	.0002	.0002	.0002	.0003	.0003	.0003	.0004
20	.0000	.0000	.0001	.0001	.0001	.0001	.0001	.0001	.0001	.0002
21	.0000	.0000	.0000	.0000	.0000	.0000	.0000	.0000	.0001	.0001
					λ					
x	8.1	8.2	8.3	8.4	8.5	8.6	8.7	8.8	8.9	9.0
0	.0003	.0003	.0002	.0002	.0002	.0002	.0002	.0002	.0001	.0001
1	.0025	.0023	.0021	.0019	.0017	.0016	.0014	.0013	.0012	.0011
2	.0100	.0092	.0086	.0079	.0074	.0068	.0063	.0058	.0054	.0050
3	.0269	.0252	.0237	.0222	.0208	.0195	.0183	.0171	.0160	.0150
4	.0544	.0517	.0491	.0466	.0443	.0420	.0398	.0377	.0357	.0337
5	.0882	.0849	.0816	.0784	.0752	.0722	.0692	.0663	.0635	.0607
6	.1191	.1160	.1128	.1097	.1066	.1034	.1003	.0972	.0941	.0911
7	.1378	.1358	.1338	.1317	.1294	.1271	.1247	.1222	.1197	.1171
8	.1395	.1392	.1388	.1382	.1375	.1366	.1356	.1344	.1332	.1318
9	.1256	.1269	.1280	.1290	.1299	.1306	.1311	.1315	.1317	.1318

Table C	(continued)								
					λ				
x	8.1	8.2	8.3	8.4	8.5	8.6	8.7	8.8	8.9	9.0
10	.1017	.1040	.1063	.1084	.1104	.1123	.1140	.1157	.1172	.1186
11	.0749	.0776	.0802	.0828	.0853	.0878	.0902	.0925	.0948	.0970
12	.0505	.0530	.0555	.0579	.0604	.0629	.0654	.0679	.0703	.0728
13	.0315	.0334	.0354	.0374	.0395	.0416	.0438	.0459	.0481	.0504
14	.0182	.0196	.0210	.0225	.0240	.0256	.0272	.0289	.0306	.0324
15	.0098	.0107	.0116	.0126	.0136	.0147	.0158	.0169	.0182	.0194
16	.0050	.0055	.0060	.0066	.0072	.0079	.0086	.0093	.0101	.0109
17	.0024	.0026	.0029	.0033	.0036	.0040	.0044	.0048	.0053	.0058
18	.0011	.0012	.0014	.0015	.0017	.0019	.0021	.0024	.0026	.0029
19	.0005	.0005	.0006	.0007	.0008	.0009	.0010	.0011	.0012	.0014
20	.0002	.0002	.0002	.0003	.0003	.0004	.0004	.0005	.0005	.0006
21	.0001	.0001	.0001	.0001	.0001	.0002	.0002	.0002	.0002	.0003
22	.0000	.0000	.0000	.0000	.0001	.0001	.0001	.0001	.0001	.0001

					λ				
x	9.1	9.2	9.3	9.4	9.5	9.6	9.7	9.8	9.9	10.0
0	.0001	.0001	.0001	.0001	.0001	.0001	.0001	.0001	.0001	.0000
1	.0010	.0009	.0009	.0008	.0007	.0007	.0006	.0005	.0005	.0005
2	.0046	.0043	.0040	.0037	.0034	.0031	.0029	.0027	.0025	.0023
3	.0140	.0131	.0123	.0115	.0107	.0100	.0093	.0087	.0081	.0076
4	.0319	.0302	.0285	.0269	.0254	.0240	.0226	.0213	.0201	.0189
5	.0581	.0555	.0530	.0506	.0483	.0460	.0439	.0418	.0398	.0378
6	.0881	.0851	.0822	.0793	.0764	.0736	.0709	.0682	.0656	.0631
7	.1145	.1118	.1091	.1064	.1037	.1010	.0982	.0955	.0928	.0901
8	.1302	.1286	.1269	.1251	.1232	.1212	.1191	.1170	.1148	.1126
9	.1317	.1315	.1311	.1306	.1300	.1293	.1284	.1274	.1263	.1251
10	.1198	.1210	.1219	.1228	.1235	.1241	.1245	.1249	.1250	.1251
11	.0991	.1012	.1031	.1049	.1067	.1083	.1098	.1112	.1125	.1137
12	.0752	.0776	.0799	.0822	.0844	.0866	.0888	.0908	.0928	.0948
13	.0526	.0549	.0572	.0594	.0617	.0640	.0662	.0685	.0707	.0729
14	.0342	.0361	.0380	.0399	.0419	.0439	.0459	.0479	.0500	.0521
15	.0208	.0221	.0235	.0250	.0265	.0281	.0297	.0313	.0330	.0347
16	.0118	.0127	.0137	.0147	.0157	.0168	.0180	.0192	.0204	.0217
17	.0063	.0069	.0075	.0081	.0088	.0095	.0103	.0111	.0119	.0128
18	.0032	.0035	.0039	.0042	.0046	.0051	.0055	.0060	.0065	.0071
19	.0015	.0017	.0019	.0021	.0023	.0026	.0028	.0031	.0034	.0037

Table C	(continued)

					λ					
x	**9.1**	**9.2**	**9.3**	**9.4**	**9.5**	**9.6**	**9.7**	**9.8**	**9.9**	**10.0**
20	.0007	.0008	.0009	.0010	.0011	.0012	.0014	.0015	.0017	.0019
21	.0003	.0003	.0004	.0004	.0005	.0006	.0006	.0007	.0008	.0009
22	.0001	.0001	.0002	.0002	.0002	.0002	.0003	.0003	.0004	.0004
23	.0000	.0001	.0001	.0001	.0001	.0001	.0001	.0001	.0002	.0002
24	.0000	.0000	.0000	.0000	.0000	.0000	.0000	.0001	.0001	.0001

					λ					
x	**11**	**12**	**13**	**14**	**15**	**16**	**17**	**18**	**19**	**20**
0	.0000	.0000	.0000	.0000	.0000	.0000	.0000	.0000	.0000	.0000
1	.0002	.0001	.0000	.0000	.0000	.0000	.0000	.0000	.0000	.0000
2	.0010	.0004	.0002	.0001	.0000	.0000	.0000	.0000	.0000	.0000
3	.0037	.0018	.0008	.0004	.0002	.0001	.0000	.0000	.0000	.0000
4	.0102	.0053	.0027	.0013	.0006	.0003	.0001	.0001	.0000	.0000
5	.0224	.0127	.0070	.0037	.0019	.0010	.0005	.0002	.0001	.0001
6	.0411	.0255	.0152	.0087	.0048	.0026	.0014	.0007	.0004	.0002
7	.0646	.0437	.0281	.0174	.0104	.0060	.0034	.0018	.0010	.0005
8	.0888	.0655	.0457	.0304	.0194	.0120	.0072	.0042	.0024	.0013
9	.1085	.0874	.0661	.0473	.0324	.0213	.0135	.0083	.0050	.0029
10	.1194	.1048	.0859	.0663	.0486	.0341	.0230	.0150	.0095	.0058
11	.1194	.1144	.1015	.0844	.0663	.0496	.0355	.0245	.0164	.0106
12	.1094	.1144	.1099	.0984	.0829	.0661	.0504	.0368	.0259	.0176
13	.0926	.1056	.1099	.1060	.0956	.0814	.0658	.0509	.0378	.0271
14	.0728	.0905	.1021	.1060	.1024	.0930	.0800	.0655	.0514	.0387
15	.0534	.0724	.0885	.0989	.1024	.0992	.0906	.0786	.0650	.0516
16	.0367	.0543	.0719	.0866	.0960	.0992	.0963	.0884	.0772	.0646
17	.0237	.0383	.0550	.0713	.0847	.0934	.0963	.0936	.0863	.0760
18	.0145	.0256	.0397	.0554	.0706	.0830	.0909	.0936	.0911	.0844
19	.0084	.0161	.0272	.0409	.0557	.0699	.0814	.0887	.0911	.0888
20	.0046	.0097	.0177	.0286	.0418	.0559	.0692	.0798	.0866	.0888
21	.0024	.0055	.0109	.0191	.0299	.0426	.0560	.0684	.0783	.0846
22	.0012	.0030	.0065	.0121	.0204	.0310	.0433	.0560	.0676	.0769
23	.0006	.0016	.0037	.0074	.0133	.0216	.0320	.0438	.0559	.0669
24	.0003	.0008	.0020	.0043	.0083	.0144	.0226	.0328	.0442	.0557
25	.0001	.0004	.0010	.0024	.0050	.0092	.0154	.0237	.0336	.0446
26	.0000	.0002	.0005	.0013	.0029	.0057	.0101	.0164	.0246	.0343
27	.0000	.0001	.0002	.0007	.0016	.0034	.0063	.0109	.0173	.0254
28	.0000	.0000	.0001	.0003	.0009	.0019	.0038	.0070	.0117	.0181
29	.0000	.0000	.0001	.0002	.0004	.0011	.0023	.0044	.0077	.0125

Table C	(concluded)									
	λ									
x	11	12	13	14	15	16	17	18	19	20
30	.0000	.0000	.0000	.0001	.0002	.0006	.0013	.0026	.0049	.0083
31	.0000	.0000	.0000	.0000	.0001	.0003	.0007	.0015	.0030	.0054
32	.0000	.0000	.0000	.0000	.0001	.0001	.0004	.0009	.0018	.0034
33	.0000	.0000	.0000	.0000	.0000	.0001	.0002	.0005	.0010	.0020
34	.0000	.0000	.0000	.0000	.0000	.0000	.0001	.0002	.0006	.0012
35	.0000	.0000	.0000	.0000	.0000	.0000	.0000	.0001	.0003	.0007
36	.0000	.0000	.0000	.0000	.0000	.0000	.0000	.0001	.0002	.0004
37	.0000	.0000	.0000	.0000	.0000	.0000	.0000	.0000	.0001	.0002
38	.0000	.0000	.0000	.0000	.0000	.0000	.0000	.0000	.0000	.0001
39	.0000	.0000	.0000	.0000	.0000	.0000	.0000	.0000	.0000	.0001

Reprinted with permission from W. H. Beyer, *Handbook of Tables for Probability and Statistics,* 2nd ed. Copyright CRC Press, Boca Raton, Fla., 1986.

Table D		Random Numbers											
10480	15011	01536	02011	81647	91646	67179	14194	62590	36207	20969	99570	91291	90700
22368	46573	25595	85393	30995	89198	27982	53402	93965	34095	52666	19174	39615	99505
24130	48360	22527	97265	76393	64809	15179	24830	49340	32081	30680	19655	63348	58629
42167	93093	06243	61680	07856	16376	39440	53537	71341	57004	00849	74917	97758	16379
37570	39975	81837	16656	06121	91782	60468	81305	49684	60672	14110	06927	01263	54613
77921	06907	11008	42751	27756	53498	18602	70659	90655	15053	21916	81825	44394	42880
99562	72905	56420	69994	98872	31016	71194	18738	44013	48840	63213	21069	10634	12952
96301	91977	05463	07972	18876	20922	94595	56869	69014	60045	18425	84903	42508	32307
89579	14342	63661	10281	17453	18103	57740	84378	25331	12566	58678	44947	05584	56941
85475	36857	43342	53988	53060	59533	38867	62300	08158	17983	16439	11458	18593	64952
28918	69578	88231	33276	70997	79936	56865	05859	90106	31595	01547	85590	91610	78188
63553	40961	48235	03427	49626	69445	18663	72695	52180	20847	12234	90511	33703	90322
09429	93969	52636	92737	88974	33488	36320	17617	30015	08272	84115	27156	30613	74952
10365	61129	87529	85689	48237	52267	67689	93394	01511	26358	85104	20285	29975	89868
07119	97336	71048	08178	77233	13916	47564	81056	97735	85977	29372	74461	28551	90707
51085	12765	51821	51259	77452	16308	60756	92144	49442	53900	70960	63990	75601	40719
02368	21382	52404	60268	89368	19885	55322	44819	01188	65255	64835	44919	05944	55157
01011	54092	33362	94904	31273	04146	18594	29852	71585	85030	51132	01915	92747	64951
52162	53916	46369	58586	23216	14513	83149	98736	23495	64350	94738	17752	35156	35749
07056	97628	33787	09998	42698	06691	76988	13602	51851	46104	88916	19509	25625	58104
48663	91245	85828	14346	09172	30168	90229	04734	59193	22178	30421	61666	99904	32812
54164	58492	22421	74103	47070	25306	76468	26384	58151	06646	21524	15227	96909	44592
32639	32363	05597	24200	13363	38005	94342	28728	35806	06912	17012	64161	18296	22851
29334	27001	87637	87308	58731	00256	45834	15398	46557	41135	10367	07684	36188	18510
02488	33062	28834	07351	19731	92420	60952	61280	50001	67658	32586	86679	50720	94953
81525	72295	04839	96423	24878	82651	66566	14778	76797	14780	13300	87074	79666	95725
29676	20591	68086	26432	46901	20849	89768	81536	86645	12659	92259	57102	80428	25280
00742	57392	39064	66432	84673	40027	32832	61362	98947	96067	64760	64584	96096	98253
05366	04213	25669	26422	44407	44048	37937	63904	45766	66134	75470	66520	34693	90449
91921	26418	64117	94305	26766	25940	39972	22209	71500	64568	91402	42416	07844	69618
00582	04711	87917	77341	42206	35126	74087	99547	81817	42607	43808	76655	62028	76630
00725	69884	62797	56170	86324	88072	76222	36086	84637	93161	76038	65855	77919	88006
69011	65797	95876	55293	18988	27354	26575	08625	40801	59920	29841	80150	12777	48501
25976	57948	29888	88604	67917	48708	18912	82271	65424	69774	33611	54262	85963	03547
09763	83473	73577	12908	30883	18317	28290	35797	05998	41688	34952	37888	38917	88050
91567	42595	27958	30134	04024	86385	29880	99730	55536	84855	29080	09250	79656	73211
17955	56349	90999	49127	20044	59931	06115	20542	18059	02008	73708	83517	36103	42791
46503	18584	18845	49618	02304	51038	20655	58727	28168	15475	56942	53389	20562	87338
92157	89634	94824	78171	84610	82834	09922	25417	44137	48413	25555	21246	35509	20468
14577	62765	35605	81263	39667	47358	56873	56307	61607	49518	89656	20103	77490	18062
98427	07523	33362	64270	01638	92477	66969	98420	04880	45585	46565	04102	46880	45709
34914	63976	88720	82765	34476	17032	87589	40836	32427	70002	70663	88863	77775	69348
70060	28277	39475	46473	23219	53416	94970	25832	69975	94884	19661	72828	00102	66794
53976	54914	06990	67245	68350	82948	11398	42878	80287	88267	47363	46634	06541	97809
76072	29515	40980	07391	58745	25774	22987	80059	39911	96189	41151	14222	60697	59583
90725	52210	83974	29992	65831	38857	50490	83765	55657	14361	31720	57375	56228	41546
64364	67412	33339	31926	14883	24413	59744	92351	97473	89286	35931	04110	23726	51900
08962	00358	31662	25388	61642	34072	81249	35648	56891	69352	48373	45578	78547	81788
95012	68379	93526	70765	10593	04542	76463	54328	02349	17247	28865	14777	62730	92277
15664	10493	20492	38391	91132	21999	59516	81652	27195	48223	46751	22923	32261	85653

Reprinted with permission from W. H. Beyer, *Handbook of Tables for Probability and Statistics,* 2nd ed. Copyright CRC Press, Boca Raton, Fla., 1986.

Table E	The Standard Normal Distribution

Cumulative Standard Normal Distribution

z	.00	.01	.02	.03	.04	.05	.06	.07	.08	.09
−3.4	.0003	.0003	.0003	.0003	.0003	.0003	.0003	.0003	.0003	.0002
−3.3	.0005	.0005	.0005	.0004	.0004	.0004	.0004	.0004	.0004	.0003
−3.2	.0007	.0007	.0006	.0006	.0006	.0006	.0006	.0005	.0005	.0005
−3.1	.0010	.0009	.0009	.0009	.0008	.0008	.0008	.0008	.0007	.0007
−3.0	.0013	.0013	.0013	.0012	.0012	.0011	.0011	.0011	.0010	.0010
−2.9	.0019	.0018	.0018	.0017	.0016	.0016	.0015	.0015	.0014	.0014
−2.8	.0026	.0025	.0024	.0023	.0023	.0022	.0021	.0021	.0020	.0019
−2.7	.0035	.0034	.0033	.0032	.0031	.0030	.0029	.0028	.0027	.0026
−2.6	.0047	.0045	.0044	.0043	.0041	.0040	.0039	.0038	.0037	.0036
−2.5	.0062	.0060	.0059	.0057	.0055	.0054	.0052	.0051	.0049	.0048
−2.4	.0082	.0080	.0078	.0075	.0073	.0071	.0069	.0068	.0066	.0064
−2.3	.0107	.0104	.0102	.0099	.0096	.0094	.0091	.0089	.0087	.0084
−2.2	.0139	.0136	.0132	.0129	.0125	.0122	.0119	.0116	.0113	.0110
−2.1	.0179	.0174	.0170	.0166	.0162	.0158	.0154	.0150	.0146	.0143
−2.0	.0228	.0222	.0217	.0212	.0207	.0202	.0197	.0192	.0188	.0183
−1.9	.0287	.0281	.0274	.0268	.0262	.0256	.0250	.0244	.0239	.0233
−1.8	.0359	.0351	.0344	.0336	.0329	.0322	.0314	.0307	.0301	.0294
−1.7	.0446	.0436	.0427	.0418	.0409	.0401	.0392	.0384	.0375	.0367
−1.6	.0548	.0537	.0526	.0516	.0505	.0495	.0485	.0475	.0465	.0455
−1.5	.0668	.0655	.0643	.0630	.0618	.0606	.0594	.0582	.0571	.0559
−1.4	.0808	.0793	.0778	.0764	.0749	.0735	.0721	.0708	.0694	.0681
−1.3	.0968	.0951	.0934	.0918	.0901	.0885	.0869	.0853	.0838	.0823
−1.2	.1151	.1131	.1112	.1093	.1075	.1056	.1038	.1020	.1003	.0985
−1.1	.1357	.1335	.1314	.1292	.1271	.1251	.1230	.1210	.1190	.1170
−1.0	.1587	.1562	.1539	.1515	.1492	.1469	.1446	.1423	.1401	.1379
−0.9	.1841	.1814	.1788	.1762	.1736	.1711	.1685	.1660	.1635	.1611
−0.8	.2119	.2090	.2061	.2033	.2005	.1977	.1949	.1922	.1894	.1867
−0.7	.2420	.2389	.2358	.2327	.2296	.2266	.2236	.2206	.2177	.2148
−0.6	.2743	.2709	.2676	.2643	.2611	.2578	.2546	.2514	.2483	.2451
−0.5	.3085	.3050	.3015	.2981	.2946	.2912	.2877	.2843	.2810	.2776
−0.4	.3446	.3409	.3372	.3336	.3300	.3264	.3228	.3192	.3156	.3121
−0.3	.3821	.3783	.3745	.3707	.3669	.3632	.3594	.3557	.3520	.3483
−0.2	.4207	.4168	.4129	.4090	.4052	.4013	.3974	.3936	.3897	.3859
−0.1	.4602	.4562	.4522	.4483	.4443	.4404	.4364	.4325	.4286	.4247
−0.0	.5000	.4960	.4920	.4880	.4840	.4801	.4761	.4721	.4681	.4641

For z values less than −3.49, use 0.0001.

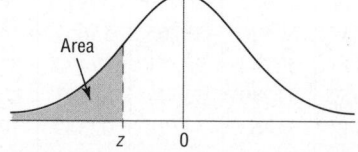

Area

z   0

Table E	(*continued*)

Cumulative Standard Normal Distribution

z	.00	.01	.02	.03	.04	.05	.06	.07	.08	.09
0.0	.5000	.5040	.5080	.5120	.5160	.5199	.5239	.5279	.5319	.5359
0.1	.5398	.5438	.5478	.5517	.5557	.5596	.5636	.5675	.5714	.5753
0.2	.5793	.5832	.5871	.5910	.5948	.5987	.6026	.6064	.6103	.6141
0.3	.6179	.6217	.6255	.6293	.6331	.6368	.6406	.6443	.6480	.6517
0.4	.6554	.6591	.6628	.6664	.6700	.6736	.6772	.6808	.6844	.6879
0.5	.6915	.6950	.6985	.7019	.7054	.7088	.7123	.7157	.7190	.7224
0.6	.7257	.7291	.7324	.7357	.7389	.7422	.7454	.7486	.7517	.7549
0.7	.7580	.7611	.7642	.7673	.7704	.7734	.7764	.7794	.7823	.7852
0.8	.7881	.7910	.7939	.7967	.7995	.8023	.8051	.8078	.8106	.8133
0.9	.8159	.8186	.8212	.8238	.8264	.8289	.8315	.8340	.8365	.8389
1.0	.8413	.8438	.8461	.8485	.8508	.8531	.8554	.8577	.8599	.8621
1.1	.8643	.8665	.8686	.8708	.8729	.8749	.8770	.8790	.8810	.8830
1.2	.8849	.8869	.8888	.8907	.8925	.8944	.8962	.8980	.8997	.9015
1.3	.9032	.9049	.9066	.9082	.9099	.9115	.9131	.9147	.9162	.9177
1.4	.9192	.9207	.9222	.9236	.9251	.9265	.9279	.9292	.9306	.9319
1.5	.9332	.9345	.9357	.9370	.9382	.9394	.9406	.9418	.9429	.9441
1.6	.9452	.9463	.9474	.9484	.9495	.9505	.9515	.9525	.9535	.9545
1.7	.9554	.9564	.9573	.9582	.9591	.9599	.9608	.9616	.9625	.9633
1.8	.9641	.9649	.9656	.9664	.9671	.9678	.9686	.9693	.9699	.9706
1.9	.9713	.9719	.9726	.9732	.9738	.9744	.9750	.9756	.9761	.9767
2.0	.9772	.9778	.9783	.9788	.9793	.9798	.9803	.9808	.9812	.9817
2.1	.9821	.9826	.9830	.9834	.9838	.9842	.9846	.9850	.9854	.9857
2.2	.9861	.9864	.9868	.9871	.9875	.9878	.9881	.9884	.9887	.9890
2.3	.9893	.9896	.9898	.9901	.9904	.9906	.9909	.9911	.9913	.9916
2.4	.9918	.9920	.9922	.9925	.9927	.9929	.9931	.9932	.9934	.9936
2.5	.9938	.9940	.9941	.9943	.9945	.9946	.9948	.9949	.9951	.9952
2.6	.9953	.9955	.9956	.9957	.9959	.9960	.9961	.9962	.9963	.9964
2.7	.9965	.9966	.9967	.9968	.9969	.9970	.9971	.9972	.9973	.9974
2.8	.9974	.9975	.9976	.9977	.9977	.9978	.9979	.9979	.9980	.9981
2.9	.9981	.9982	.9982	.9983	.9984	.9984	.9985	.9985	.9986	.9986
3.0	.9987	.9987	.9987	.9988	.9988	.9989	.9989	.9989	.9990	.9990
3.1	.9990	.9991	.9991	.9991	.9992	.9992	.9992	.9992	.9993	.9993
3.2	.9993	.9993	.9994	.9994	.9994	.9994	.9994	.9995	.9995	.9995
3.3	.9995	.9995	.9995	.9996	.9996	.9996	.9996	.9996	.9996	.9997
3.4	.9997	.9997	.9997	.9997	.9997	.9997	.9997	.9997	.9997	.9998

For z values greater than 3.49, use 0.9999.

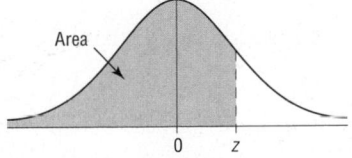

Table F	The *t* Distribution					
	Confidence intervals	**80%**	**90%**	**95%**	**98%**	**99%**
	One tail, α	**0.10**	**0.05**	**0.025**	**0.01**	**0.005**
d.f.	**Two tails, α**	**0.20**	**0.10**	**0.05**	**0.02**	**0.01**
1		3.078	6.314	12.706	31.821	63.657
2		1.886	2.920	4.303	6.965	9.925
3		1.638	2.353	3.182	4.541	5.841
4		1.533	2.132	2.776	3.747	4.604
5		1.476	2.015	2.571	3.365	4.032
6		1.440	1.943	2.447	3.143	3.707
7		1.415	1.895	2.365	2.998	3.499
8		1.397	1.860	2.306	2.896	3.355
9		1.383	1.833	2.262	2.821	3.250
10		1.372	1.812	2.228	2.764	3.169
11		1.363	1.796	2.201	2.718	3.106
12		1.356	1.782	2.179	2.681	3.055
13		1.350	1.771	2.160	2.650	3.012
14		1.345	1.761	2.145	2.624	2.977
15		1.341	1.753	2.131	2.602	2.947
16		1.337	1.746	2.120	2.583	2.921
17		1.333	1.740	2.110	2.567	2.898
18		1.330	1.734	2.101	2.552	2.878
19		1.328	1.729	2.093	2.539	2.861
20		1.325	1.725	2.086	2.528	2.845
21		1.323	1.721	2.080	2.518	2.831
22		1.321	1.717	2.074	2.508	2.819
23		1.319	1.714	2.069	2.500	2.807
24		1.318	1.711	2.064	2.492	2.797
25		1.316	1.708	2.060	2.485	2.787
26		1.315	1.706	2.056	2.479	2.779
27		1.314	1.703	2.052	2.473	2.771
28		1.313	1.701	2.048	2.467	2.763
29		1.311	1.699	2.045	2.462	2.756
30		1.310	1.697	2.042	2.457	2.750
32		1.309	1.694	2.037	2.449	2.738
34		1.307	1.691	2.032	2.441	2.728
36		1.306	1.688	2.028	2.434	2.719
38		1.304	1.686	2.024	2.429	2.712
40		1.303	1.684	2.021	2.423	2.704
45		1.301	1.679	2.014	2.412	2.690
50		1.299	1.676	2.009	2.403	2.678
55		1.297	1.673	2.004	2.396	2.668
60		1.296	1.671	2.000	2.390	2.660
65		1.295	1.669	1.997	2.385	2.654
70		1.294	1.667	1.994	2.381	2.648
75		1.293	1.665	1.992	2.377	2.643
80		1.292	1.664	1.990	2.374	2.639
90		1.291	1.662	1.987	2.368	2.632
100		1.290	1.660	1.984	2.364	2.626
500		1.283	1.648	1.965	2.334	2.586
1000		1.282	1.646	1.962	2.330	2.581
(z) ∞		1.282[a]	1.645[b]	1.960	2.326[c]	2.576[d]

[a]This value has been rounded to 1.28 in the textbook.
[b]This value has been rounded to 1.65 in the textbook.
[c]This value has been rounded to 2.33 in the textbook.
[d]This value has been rounded to 2.58 in the textbook.

Source: Adapted from W. H. Beyer, *Handbook of Tables for Probability and Statistics,* 2nd ed., CRC Press, Boca Raton, Fla., 1986. Reprinted with permission.

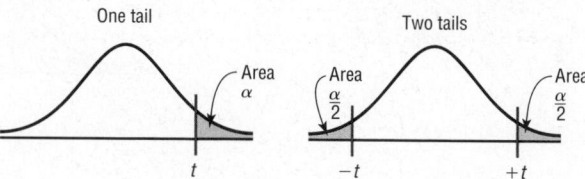

| Table G | The Chi-Square Distribution | | | | | | | | |

Degrees of freedom	α									
	0.995	0.99	0.975	0.95	0.90	0.10	0.05	0.025	0.01	0.005
1	—	—	0.001	0.004	0.016	2.706	3.841	5.024	6.635	7.879
2	0.010	0.020	0.051	0.103	0.211	4.605	5.991	7.378	9.210	10.597
3	0.072	0.115	0.216	0.352	0.584	6.251	7.815	9.348	11.345	12.838
4	0.207	0.297	0.484	0.711	1.064	7.779	9.488	11.143	13.277	14.860
5	0.412	0.554	0.831	1.145	1.610	9.236	11.071	12.833	15.086	16.750
6	0.676	0.872	1.237	1.635	2.204	10.645	12.592	14.449	16.812	18.548
7	0.989	1.239	1.690	2.167	2.833	12.017	14.067	16.013	18.475	20.278
8	1.344	1.646	2.180	2.733	3.490	13.362	15.507	17.535	20.090	21.955
9	1.735	2.088	2.700	3.325	4.168	14.684	16.919	19.023	21.666	23.589
10	2.156	2.558	3.247	3.940	4.865	15.987	18.307	20.483	23.209	25.188
11	2.603	3.053	3.816	4.575	5.578	17.275	19.675	21.920	24.725	26.757
12	3.074	3.571	4.404	5.226	6.304	18.549	21.026	23.337	26.217	28.299
13	3.565	4.107	5.009	5.892	7.042	19.812	22.362	24.736	27.688	29.819
14	4.075	4.660	5.629	6.571	7.790	21.064	23.685	26.119	29.141	31.319
15	4.601	5.229	6.262	7.261	8.547	22.307	24.996	27.488	30.578	32.801
16	5.142	5.812	6.908	7.962	9.312	23.542	26.296	28.845	32.000	34.267
17	5.697	6.408	7.564	8.672	10.085	24.769	27.587	30.191	33.409	35.718
18	6.265	7.015	8.231	9.390	10.865	25.989	28.869	31.526	34.805	37.156
19	6.844	7.633	8.907	10.117	11.651	27.204	30.144	32.852	36.191	38.582
20	7.434	8.260	9.591	10.851	12.443	28.412	31.410	34.170	37.566	39.997
21	8.034	8.897	10.283	11.591	13.240	29.615	32.671	35.479	38.932	41.401
22	8.643	9.542	10.982	12.338	14.042	30.813	33.924	36.781	40.289	42.796
23	9.262	10.196	11.689	13.091	14.848	32.007	35.172	38.076	41.638	44.181
24	9.886	10.856	12.401	13.848	15.659	33.196	36.415	39.364	42.980	45.559
25	10.520	11.524	13.120	14.611	16.473	34.382	37.652	40.646	44.314	46.928
26	11.160	12.198	13.844	15.379	17.292	35.563	38.885	41.923	45.642	48.290
27	11.808	12.879	14.573	16.151	18.114	36.741	40.113	43.194	46.963	49.645
28	12.461	13.565	15.308	16.928	18.939	37.916	41.337	44.461	48.278	50.993
29	13.121	14.257	16.047	17.708	19.768	39.087	42.557	45.722	49.588	52.336
30	13.787	14.954	16.791	18.493	20.599	40.256	43.773	46.979	50.892	53.672
40	20.707	22.164	24.433	26.509	29.051	51.805	55.758	59.342	63.691	66.766
50	27.991	29.707	32.357	34.764	37.689	63.167	67.505	71.420	76.154	79.490
60	35.534	37.485	40.482	43.188	46.459	74.397	79.082	83.298	88.379	91.952
70	43.275	45.442	48.758	51.739	55.329	85.527	90.531	95.023	100.425	104.215
80	51.172	53.540	57.153	60.391	64.278	96.578	101.879	106.629	112.329	116.321
90	59.196	61.754	65.647	69.126	73.291	107.565	113.145	118.136	124.116	128.299
100	67.328	70.065	74.222	77.929	82.358	118.498	124.342	129.561	135.807	140.169

Source: Owen, *Handbook of Statistical Tables,* Table A–4 "Chi-Square Distribution Table," © 1962 by Addison-Wesley Publishing Company, Inc. Copyright renewal © 1990. Reproduced by permission of Pearson Education, Inc.

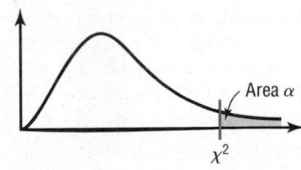

Table H The F Distribution

$\alpha = 0.005$

d.f.D.: degrees of freedom, denominator	1	2	3	4	5	6	7	8	9	10	12	15	20	24	30	40	60	120	∞
							d.f.N.: degrees of freedom, numerator												
1	16,211	20,000	21,615	22,500	23,056	23,437	23,715	23,925	24,091	24,224	24,426	24,630	24,836	24,940	25,044	25,148	25,253	25,359	25,465
2	198.5	199.0	199.2	199.2	199.3	199.3	199.4	199.4	199.4	199.4	199.4	199.4	199.4	199.5	199.5	199.5	199.5	199.5	199.5
3	55.55	49.80	47.47	46.19	45.39	44.84	44.43	44.13	43.88	43.69	43.39	43.08	42.78	42.62	42.47	42.31	42.15	41.99	41.83
4	31.33	26.28	24.26	23.15	22.46	21.97	21.62	21.35	21.14	20.97	20.70	20.44	20.17	20.03	19.89	19.75	19.61	19.47	19.32
5	22.78	18.31	16.53	15.56	14.94	14.51	14.20	13.96	13.77	13.62	13.38	13.15	12.90	12.78	12.66	12.53	12.40	12.27	12.14
6	18.63	14.54	12.92	12.03	11.46	11.07	10.79	10.57	10.39	10.25	10.03	9.81	9.59	9.47	9.36	9.24	9.12	9.00	8.88
7	16.24	12.40	10.88	10.05	9.52	9.16	8.89	8.68	8.51	8.38	8.18	7.97	7.75	7.65	7.53	7.42	7.31	7.19	7.08
8	14.69	11.04	9.60	8.81	8.30	7.95	7.69	7.50	7.34	7.21	7.01	6.81	6.61	6.50	6.40	6.29	6.18	6.06	5.95
9	13.61	10.11	8.72	7.96	7.47	7.13	6.88	6.69	6.54	6.42	6.23	6.03	5.83	5.73	5.62	5.52	5.41	5.30	5.19
10	12.83	9.43	8.08	7.34	6.87	6.54	6.30	6.12	5.97	5.85	5.66	5.47	5.27	5.17	5.07	4.97	4.86	4.75	4.64
11	12.23	8.91	7.60	6.88	6.42	6.10	5.86	5.68	5.54	5.42	5.24	5.05	4.86	4.76	4.65	4.55	4.44	4.34	4.23
12	11.75	8.51	7.23	6.52	6.07	5.76	5.52	5.35	5.20	5.09	4.91	4.72	4.53	4.43	4.33	4.23	4.12	4.01	3.90
13	11.37	8.19	6.93	6.23	5.79	5.48	5.25	5.08	4.94	4.82	4.64	4.46	4.27	4.17	4.07	3.97	3.87	3.76	3.65
14	11.06	7.92	6.68	6.00	5.56	5.26	5.03	4.86	4.72	4.60	4.43	4.25	4.06	3.96	3.86	3.76	3.66	3.55	3.44
15	10.80	7.70	6.48	5.80	5.37	5.07	4.85	4.67	4.54	4.42	4.25	4.07	3.88	3.79	3.69	3.58	3.48	3.37	3.26
16	10.58	7.51	6.30	5.64	5.21	4.91	4.69	4.52	4.38	4.27	4.10	3.92	3.73	3.64	3.54	3.44	3.33	3.22	3.11
17	10.38	7.35	6.16	5.50	5.07	4.78	4.56	4.39	4.25	4.14	3.97	3.79	3.61	3.51	3.41	3.31	3.21	3.10	2.98
18	10.22	7.21	6.03	5.37	4.96	4.66	4.44	4.28	4.14	4.03	3.86	3.68	3.50	3.40	3.30	3.20	3.10	2.99	2.87
19	10.07	7.09	5.92	5.27	4.85	4.56	4.34	4.18	4.04	3.93	3.76	3.59	3.40	3.31	3.21	3.11	3.00	2.89	2.78
20	9.94	6.99	5.82	5.17	4.76	4.47	4.26	4.09	3.96	3.85	3.68	3.50	3.32	3.22	3.12	3.02	2.92	2.81	2.69
21	9.83	6.89	5.73	5.09	4.68	4.39	4.18	4.01	3.88	3.77	3.60	3.43	3.24	3.15	3.05	2.95	2.84	2.73	2.61
22	9.73	6.81	5.65	5.02	4.61	4.32	4.11	3.94	3.81	3.70	3.54	3.36	3.18	3.08	2.98	2.88	2.77	2.66	2.55
23	9.63	6.73	5.58	4.95	4.54	4.26	4.05	3.88	3.75	3.64	3.47	3.30	3.12	3.02	2.92	2.82	2.71	2.60	2.48
24	9.55	6.66	5.52	4.89	4.49	4.20	3.99	3.83	3.69	3.59	3.42	3.25	3.06	2.97	2.87	2.77	2.66	2.55	2.43
25	9.48	6.60	5.46	4.84	4.43	4.15	3.94	3.78	3.64	3.54	3.37	3.20	3.01	2.92	2.82	2.72	2.61	2.50	2.38
26	9.41	6.54	5.41	4.79	4.38	4.10	3.89	3.73	3.60	3.49	3.33	3.15	2.97	2.87	2.77	2.67	2.56	2.45	2.33
27	9.34	6.49	5.36	4.74	4.34	4.06	3.85	3.69	3.56	3.45	3.28	3.11	2.93	2.83	2.73	2.63	2.52	2.41	2.29
28	9.28	6.44	5.32	4.70	4.30	4.02	3.81	3.65	3.52	3.41	3.25	3.07	2.89	2.79	2.69	2.59	2.48	2.37	2.25
29	9.23	6.40	5.28	4.66	4.26	3.98	3.77	3.61	3.48	3.38	3.21	3.04	2.86	2.76	2.66	2.56	2.45	2.33	2.24
30	9.18	6.35	5.24	4.62	4.23	3.95	3.74	3.58	3.45	3.34	3.18	3.01	2.82	2.73	2.63	2.52	2.42	2.30	2.18
40	8.83	6.07	4.98	4.37	3.99	3.71	3.51	3.35	3.22	3.12	2.95	2.78	2.60	2.50	2.40	2.30	2.18	2.06	1.93
60	8.49	5.79	4.73	4.14	3.76	3.49	3.29	3.13	3.01	2.90	2.74	2.57	2.39	2.29	2.19	2.08	1.96	1.83	1.69
120	8.18	5.54	4.50	3.92	3.55	3.28	3.09	2.93	2.81	2.71	2.54	2.37	2.19	2.09	1.98	1.87	1.75	1.61	1.43
∞	7.88	5.30	4.28	3.72	3.35	3.09	2.90	2.74	2.62	2.52	2.36	2.19	2.00	1.90	1.79	1.67	1.53	1.36	1.00

Table H *(continued)*

$\alpha = 0.01$

d.f.D.: degrees of freedom, denominator	d.f.N.: degrees of freedom, numerator																		
	1	2	3	4	5	6	7	8	9	10	12	15	20	24	30	40	60	120	∞
1	4052	4999.5	5403	5625	5764	5859	5928	5982	6022	6056	6106	6157	6209	6235	6261	6287	6313	6339	6366
2	98.50	99.00	99.17	99.25	99.30	99.33	99.36	99.37	99.39	99.40	99.42	99.43	99.45	99.46	99.47	99.47	99.48	99.49	99.50
3	34.12	30.82	29.46	28.71	28.24	27.91	27.67	27.49	27.35	27.23	27.05	26.87	26.69	26.60	26.50	26.41	26.32	26.22	26.13
4	21.20	18.00	16.69	15.98	15.52	15.21	14.98	14.80	14.66	14.55	14.37	14.20	14.02	13.93	13.84	13.75	13.65	13.56	13.46
5	16.26	13.27	12.06	11.39	10.97	10.67	10.46	10.29	10.16	10.05	9.89	9.72	9.55	9.47	9.38	9.29	9.20	9.11	9.02
6	13.75	10.92	9.78	9.15	8.75	8.47	8.26	8.10	7.98	7.87	7.72	7.56	7.40	7.31	7.23	7.14	7.06	6.97	6.88
7	12.25	9.55	8.45	7.85	7.46	7.19	6.99	6.84	6.72	6.62	6.47	6.31	6.16	6.07	5.99	5.91	5.82	5.74	5.65
8	11.26	8.65	7.59	7.01	6.63	6.37	6.18	6.03	5.91	5.81	5.67	5.52	5.36	5.28	5.20	5.12	5.03	4.95	4.86
9	10.56	8.02	6.99	6.42	6.06	5.80	5.61	5.47	5.35	5.26	5.11	4.96	4.81	4.73	4.65	4.57	4.48	4.40	4.31
10	10.04	7.56	6.55	5.99	5.64	5.39	5.20	5.06	4.94	4.85	4.71	4.56	4.41	4.33	4.25	4.17	4.08	4.00	3.91
11	9.65	7.21	6.22	5.67	5.32	5.07	4.89	4.74	4.63	4.54	4.40	4.25	4.10	4.02	3.94	3.86	3.78	3.69	3.60
12	9.33	6.93	5.95	5.41	5.06	4.82	4.64	4.50	4.39	4.30	4.16	4.01	3.86	3.78	3.70	3.62	3.54	3.45	3.36
13	9.07	6.70	5.74	5.21	4.86	4.62	4.44	4.30	4.19	4.10	3.96	3.82	3.66	3.59	3.51	3.43	3.34	3.25	3.17
14	8.86	6.51	5.56	5.04	4.69	4.46	4.28	4.14	4.03	3.94	3.80	3.66	3.51	3.43	3.35	3.27	3.18	3.09	3.00
15	8.68	6.36	5.42	4.89	4.56	4.32	4.14	4.00	3.89	3.80	3.67	3.52	3.37	3.29	3.21	3.13	3.05	2.96	2.87
16	8.53	6.23	5.29	4.77	4.44	4.20	4.03	3.89	3.78	3.69	3.55	3.41	3.26	3.18	3.10	3.02	2.93	2.84	2.75
17	8.40	6.11	5.18	4.67	4.34	4.10	3.93	3.79	3.68	3.59	3.46	3.31	3.16	3.08	3.00	2.92	2.83	2.75	2.65
18	8.29	6.01	5.09	4.58	4.25	4.01	3.84	3.71	3.60	3.51	3.37	3.23	3.08	3.00	2.92	2.84	2.75	2.66	2.57
19	8.18	5.93	5.01	4.50	4.17	3.94	3.77	3.63	3.52	3.43	3.30	3.15	3.00	2.92	2.84	2.76	2.67	2.58	2.49
20	8.10	5.85	4.94	4.43	4.10	3.87	3.70	3.56	3.46	3.37	3.23	3.09	2.94	2.86	2.78	2.69	2.61	2.52	2.42
21	8.02	5.78	4.87	4.37	4.04	3.81	3.64	3.51	3.40	3.31	3.17	3.03	2.88	2.80	2.72	2.64	2.55	2.46	2.36
22	7.95	5.72	4.82	4.31	3.99	3.76	3.59	3.45	3.35	3.26	3.12	2.98	2.83	2.75	2.67	2.58	2.50	2.40	2.31
23	7.88	5.66	4.76	4.26	3.94	3.71	3.54	3.41	3.30	3.21	3.07	2.93	2.78	2.70	2.62	2.54	2.45	2.35	2.26
24	7.82	5.61	4.72	4.22	3.90	3.67	3.50	3.36	3.26	3.17	3.03	2.89	2.74	2.66	2.58	2.49	2.40	2.31	2.21
25	7.77	5.57	4.68	4.18	3.85	3.63	3.46	3.32	3.22	3.13	2.99	2.85	2.70	2.62	2.54	2.45	2.36	2.27	2.17
26	7.72	5.53	4.64	4.14	3.82	3.59	3.42	3.29	3.18	3.09	2.96	2.81	2.66	2.58	2.50	2.42	2.33	2.23	2.13
27	7.68	5.49	4.60	4.11	3.78	3.56	3.39	3.26	3.15	3.06	2.93	2.78	2.63	2.55	2.47	2.38	2.29	2.20	2.10
28	7.64	5.45	4.57	4.07	3.75	3.53	3.36	3.23	3.12	3.03	2.90	2.75	2.60	2.52	2.44	2.35	2.26	2.17	2.06
29	7.60	5.42	4.54	4.04	3.73	3.50	3.33	3.20	3.09	3.00	2.87	2.73	2.57	2.49	2.41	2.33	2.23	2.14	2.03
30	7.56	5.39	4.51	4.02	3.70	3.47	3.30	3.17	3.07	2.98	2.84	2.70	2.55	2.47	2.39	2.30	2.21	2.11	2.01
40	7.31	5.18	4.31	3.83	3.51	3.29	3.12	2.99	2.89	2.80	2.66	2.52	2.37	2.29	2.20	2.11	2.02	1.92	1.80
60	7.08	4.98	4.13	3.65	3.34	3.12	2.95	2.82	2.72	2.63	2.50	2.35	2.20	2.12	2.03	1.94	1.84	1.73	1.60
120	6.85	4.79	3.95	3.48	3.17	2.96	2.79	2.66	2.56	2.47	2.34	2.19	2.03	1.95	1.86	1.76	1.66	1.53	1.38
∞	6.63	4.61	3.78	3.32	3.02	2.80	2.64	2.51	2.41	2.32	2.18	2.04	1.88	1.79	1.70	1.59	1.47	1.32	1.00

Table H (continued)

α = 0.025

d.f.D.: degrees of freedom, denominator	d.f.N.: degrees of freedom, numerator																		
	1	2	3	4	5	6	7	8	9	10	12	15	20	24	30	40	60	120	∞
1	647.8	799.5	864.2	899.6	921.8	937.1	948.2	956.7	963.3	968.6	976.7	984.9	993.1	997.2	1001	1006	1010	1014	1018
2	38.51	39.00	39.17	39.25	39.30	39.33	39.36	39.37	39.39	39.40	39.41	39.43	39.45	39.46	39.46	39.47	39.48	39.49	39.50
3	17.44	16.04	15.44	15.10	14.88	14.73	14.62	14.54	14.47	14.42	14.34	14.25	14.17	14.12	14.08	14.04	13.99	13.95	13.90
4	12.22	10.65	9.98	9.60	9.36	9.20	9.07	8.98	8.90	8.84	8.75	8.66	8.56	8.51	8.46	8.41	8.36	8.31	8.26
5	10.01	8.43	7.76	7.39	7.15	6.98	6.85	6.76	6.68	6.62	6.52	6.43	6.33	6.28	6.23	6.18	6.12	6.07	6.02
6	8.81	7.26	6.60	6.23	5.99	5.82	5.70	5.60	5.52	5.46	5.37	5.27	5.17	5.12	5.07	5.01	4.96	4.90	4.85
7	8.07	6.54	5.89	5.52	5.29	5.12	4.99	4.90	4.82	4.76	4.67	4.57	4.47	4.42	4.36	4.31	4.25	4.20	4.14
8	7.57	6.06	5.42	5.05	4.82	4.65	4.53	4.43	4.36	4.30	4.20	4.10	4.00	3.95	3.89	3.84	3.78	3.73	3.67
9	7.21	5.71	5.08	4.72	4.48	4.32	4.20	4.10	4.03	3.96	3.87	3.77	3.67	3.61	3.56	3.51	3.45	3.39	3.33
10	6.94	5.46	4.83	4.47	4.24	4.07	3.95	3.85	3.78	3.72	3.62	3.52	3.42	3.37	3.31	3.26	3.20	3.14	3.08
11	6.72	5.26	4.63	4.28	4.04	3.88	3.76	3.66	3.59	3.53	3.43	3.33	3.23	3.17	3.12	3.06	3.00	2.94	2.88
12	6.55	5.10	4.47	4.12	3.89	3.73	3.61	3.51	3.44	3.37	3.28	3.18	3.07	3.02	2.96	2.91	2.85	2.79	2.72
13	6.41	4.97	4.35	4.00	3.77	3.60	3.48	3.39	3.31	3.25	3.15	3.05	2.95	2.89	2.84	2.78	2.72	2.66	2.60
14	6.30	4.86	4.24	3.89	3.66	3.50	3.38	3.29	3.21	3.15	3.05	2.95	2.84	2.79	2.73	2.67	2.61	2.55	2.49
15	6.20	4.77	4.15	3.80	3.58	3.41	3.29	3.20	3.12	3.06	2.96	2.86	2.76	2.70	2.64	2.59	2.52	2.46	2.40
16	6.12	4.69	4.08	3.73	3.50	3.34	3.22	3.12	3.05	2.99	2.89	2.79	2.68	2.63	2.57	2.51	2.45	2.38	2.32
17	6.04	4.62	4.01	3.66	3.44	3.28	3.16	3.06	2.98	2.92	2.82	2.72	2.62	2.56	2.50	2.44	2.38	2.32	2.25
18	5.98	4.56	3.95	3.61	3.38	3.22	3.10	3.01	2.93	2.87	2.77	2.67	2.56	2.50	2.44	2.38	2.32	2.26	2.19
19	5.92	4.51	3.90	3.56	3.33	3.17	3.05	2.96	2.88	2.82	2.72	2.62	2.51	2.45	2.39	2.33	2.27	2.20	2.13
20	5.87	4.46	3.86	3.51	3.29	3.13	3.01	2.91	2.84	2.77	2.68	2.57	2.46	2.41	2.35	2.29	2.22	2.16	2.09
21	5.83	4.42	3.82	3.48	3.25	3.09	2.97	2.87	2.80	2.73	2.64	2.53	2.42	2.37	2.31	2.25	2.18	2.11	2.04
22	5.79	4.38	3.78	3.44	3.22	3.05	2.93	2.84	2.76	2.70	2.60	2.50	2.39	2.33	2.27	2.21	2.14	2.08	2.00
23	5.75	4.35	3.75	3.41	3.18	3.02	2.90	2.81	2.73	2.67	2.57	2.47	2.36	2.30	2.24	2.18	2.11	2.04	1.97
24	5.72	4.32	3.72	3.38	3.15	2.99	2.87	2.78	2.70	2.64	2.54	2.44	2.33	2.27	2.21	2.15	2.08	2.01	1.94
25	5.69	4.29	3.69	3.35	3.13	2.97	2.85	2.75	2.68	2.61	2.51	2.41	2.30	2.24	2.18	2.12	2.05	1.98	1.91
26	5.66	4.27	3.67	3.33	3.10	2.94	2.82	2.73	2.65	2.59	2.49	2.39	2.28	2.22	2.16	2.09	2.03	1.95	1.88
27	5.63	4.24	3.65	3.31	3.08	2.92	2.80	2.71	2.63	2.57	2.47	2.36	2.25	2.19	2.13	2.07	2.00	1.93	1.85
28	5.61	4.22	3.63	3.29	3.06	2.90	2.78	2.69	2.61	2.55	2.45	2.34	2.23	2.17	2.11	2.05	1.98	1.91	1.83
29	5.59	4.20	3.61	3.27	3.04	2.88	2.76	2.67	2.59	2.53	2.43	2.32	2.21	2.15	2.09	2.03	1.96	1.89	1.81
30	5.57	4.18	3.59	3.25	3.03	2.87	2.75	2.65	2.57	2.51	2.41	2.31	2.20	2.14	2.07	2.01	1.94	1.87	1.79
40	5.42	4.05	3.46	3.13	2.90	2.74	2.62	2.53	2.45	2.39	2.29	2.18	2.07	2.01	1.94	1.88	1.80	1.74	1.64
60	5.29	3.93	3.34	3.01	2.79	2.63	2.51	2.41	2.33	2.27	2.17	2.06	1.94	1.88	1.82	1.74	1.67	1.58	1.48
120	5.15	3.80	3.23	2.89	2.67	2.52	2.39	2.30	2.22	2.16	2.05	1.94	1.82	1.76	1.69	1.61	1.53	1.43	1.31
∞	5.02	3.69	3.12	2.79	2.57	2.41	2.29	2.19	2.11	2.05	1.94	1.83	1.71	1.64	1.57	1.48	1.39	1.27	1.00

Table H (continued)

$\alpha = 0.05$

d.f.N.: degrees of freedom, numerator

d.f.D.: degrees of freedom, denominator	1	2	3	4	5	6	7	8	9	10	12	15	20	24	30	40	60	120	∞
1	161.4	199.5	215.7	224.6	230.2	234.0	236.8	238.9	240.5	241.9	243.9	245.9	248.0	249.1	250.1	251.1	252.2	253.3	254.3
2	18.51	19.00	19.16	19.25	19.30	19.33	19.35	19.37	19.38	19.40	19.41	19.43	19.45	19.45	19.46	19.47	19.48	19.49	19.50
3	10.13	9.55	9.28	9.12	9.01	8.94	8.89	8.85	8.81	8.79	8.74	8.70	8.66	8.64	8.62	8.59	8.57	8.55	8.53
4	7.71	6.94	6.59	6.39	6.26	6.16	6.09	6.04	6.00	5.96	5.91	5.86	5.80	5.77	5.75	5.72	5.69	5.66	5.63
5	6.61	5.79	5.41	5.19	5.05	4.95	4.88	4.82	4.77	4.74	4.68	4.62	4.56	4.53	4.50	4.46	4.43	4.40	4.36
6	5.99	5.14	4.76	4.53	4.39	4.28	4.21	4.15	4.10	4.06	4.00	3.94	3.87	3.84	3.81	3.77	3.74	3.70	3.67
7	5.59	4.74	4.35	4.12	3.97	3.87	3.79	3.73	3.68	3.64	3.57	3.51	3.44	3.41	3.38	3.34	3.30	3.27	3.23
8	5.32	4.46	4.07	3.84	3.69	3.58	3.50	3.44	3.39	3.35	3.28	3.22	3.15	3.12	3.08	3.04	3.01	2.97	2.93
9	5.12	4.26	3.86	3.63	3.48	3.37	3.29	3.23	3.18	3.14	3.07	3.01	2.94	2.90	2.86	2.83	2.79	2.75	2.71
10	4.96	4.10	3.71	3.48	3.33	3.22	3.14	3.07	3.02	2.98	2.91	2.85	2.77	2.74	2.70	2.66	2.62	2.58	2.54
11	4.84	3.98	3.59	3.36	3.20	3.09	3.01	2.95	2.90	2.85	2.79	2.72	2.65	2.61	2.57	2.53	2.49	2.45	2.40
12	4.75	3.89	3.49	3.26	3.11	3.00	2.91	2.85	2.80	2.75	2.69	2.62	2.54	2.51	2.47	2.43	2.38	2.34	2.30
13	4.67	3.81	3.41	3.18	3.03	2.92	2.83	2.77	2.71	2.67	2.60	2.53	2.46	2.42	2.38	2.34	2.30	2.25	2.21
14	4.60	3.74	3.34	3.11	2.96	2.85	2.76	2.70	2.65	2.60	2.53	2.46	2.39	2.35	2.31	2.27	2.22	2.18	2.13
15	4.54	3.68	3.29	3.06	2.90	2.79	2.71	2.64	2.59	2.54	2.48	2.40	2.33	2.29	2.25	2.20	2.16	2.11	2.07
16	4.49	3.63	3.24	3.01	2.85	2.74	2.66	2.59	2.54	2.49	2.42	2.35	2.28	2.24	2.19	2.15	2.11	2.06	2.01
17	4.45	3.59	3.20	2.96	2.81	2.70	2.61	2.55	2.49	2.45	2.38	2.31	2.23	2.19	2.15	2.10	2.06	2.01	1.96
18	4.41	3.55	3.16	2.93	2.77	2.66	2.58	2.51	2.46	2.41	2.34	2.27	2.19	2.15	2.11	2.06	2.02	1.97	1.92
19	4.38	3.52	3.13	2.90	2.74	2.63	2.54	2.48	2.42	2.38	2.31	2.23	2.16	2.11	2.07	2.03	1.98	1.93	1.88
20	4.35	3.49	3.10	2.87	2.71	2.60	2.51	2.45	2.39	2.35	2.28	2.20	2.12	2.08	2.04	1.99	1.95	1.90	1.84
21	4.32	3.47	3.07	2.84	2.68	2.57	2.49	2.42	2.37	2.32	2.25	2.18	2.10	2.05	2.01	1.96	1.92	1.87	1.81
22	4.30	3.44	3.05	2.82	2.66	2.55	2.46	2.40	2.34	2.30	2.23	2.15	2.07	2.03	1.98	1.94	1.89	1.84	1.78
23	4.28	3.42	3.03	2.80	2.64	2.53	2.44	2.37	2.32	2.27	2.20	2.13	2.05	2.01	1.96	1.91	1.86	1.81	1.76
24	4.26	3.40	3.01	2.78	2.62	2.51	2.42	2.36	2.30	2.25	2.18	2.11	2.03	1.98	1.94	1.89	1.84	1.79	1.73
25	4.24	3.39	2.99	2.76	2.60	2.49	2.40	2.34	2.28	2.24	2.16	2.09	2.01	1.96	1.92	1.87	1.82	1.77	1.71
26	4.23	3.37	2.98	2.74	2.59	2.47	2.39	2.32	2.27	2.22	2.15	2.07	1.99	1.95	1.90	1.85	1.80	1.75	1.69
27	4.21	3.35	2.96	2.73	2.57	2.46	2.37	2.31	2.25	2.20	2.13	2.06	1.97	1.93	1.88	1.84	1.79	1.73	1.67
28	4.20	3.34	2.95	2.71	2.56	2.45	2.36	2.29	2.24	2.19	2.12	2.04	1.96	1.91	1.87	1.82	1.77	1.71	1.65
29	4.18	3.33	2.93	2.70	2.55	2.43	2.35	2.28	2.22	2.18	2.10	2.03	1.94	1.90	1.85	1.81	1.75	1.70	1.64
30	4.17	3.32	2.92	2.69	2.53	2.42	2.33	2.27	2.21	2.16	2.09	2.01	1.93	1.89	1.84	1.79	1.74	1.68	1.62
40	4.08	3.23	2.84	2.61	2.45	2.34	2.25	2.18	2.12	2.08	2.00	1.92	1.84	1.79	1.74	1.69	1.64	1.58	1.51
60	4.00	3.15	2.76	2.53	2.37	2.25	2.17	2.10	2.04	1.99	1.92	1.84	1.75	1.70	1.65	1.59	1.53	1.47	1.39
120	3.92	3.07	2.68	2.45	2.29	2.17	2.09	2.02	1.96	1.91	1.83	1.75	1.66	1.61	1.55	1.50	1.43	1.35	1.25
∞	3.84	3.00	2.60	2.37	2.21	2.10	2.01	1.94	1.88	1.83	1.75	1.67	1.57	1.52	1.46	1.39	1.32	1.22	1.00

Table H (concluded)

$\alpha = 0.10$

d.f.N.: degrees of freedom, numerator

d.f.D.: degrees of freedom, denominator	1	2	3	4	5	6	7	8	9	10	12	15	20	24	30	40	60	120	∞
1	39.86	49.50	53.59	55.83	57.24	58.20	58.91	59.44	59.86	60.19	60.71	61.22	61.74	62.00	62.26	62.53	62.79	63.06	63.33
2	8.53	9.00	9.16	9.24	9.29	9.33	9.35	9.37	9.38	9.39	9.41	9.42	9.44	9.45	9.46	9.47	9.47	9.48	9.49
3	5.54	5.46	5.39	5.34	5.31	5.28	5.27	5.25	5.24	5.23	5.22	5.20	5.18	5.18	5.17	5.16	5.15	5.14	5.13
4	4.54	4.32	4.19	4.11	4.05	4.01	3.98	3.95	3.94	3.92	3.90	3.87	3.84	3.83	3.82	3.80	3.79	3.78	3.76
5	4.06	3.78	3.62	3.52	3.45	3.40	3.37	3.34	3.32	3.30	3.27	3.24	3.21	3.19	3.17	3.16	3.14	3.12	3.10
6	3.78	3.46	3.29	3.18	3.11	3.05	3.01	2.98	2.96	2.94	2.90	2.87	2.84	2.82	2.80	2.78	2.76	2.74	2.72
7	3.59	3.26	3.07	2.96	2.88	2.83	2.78	2.75	2.72	2.70	2.67	2.63	2.59	2.58	2.56	2.54	2.51	2.49	2.47
8	3.46	3.11	2.92	2.81	2.73	2.67	2.62	2.59	2.56	2.54	2.50	2.46	2.42	2.40	2.38	2.36	2.34	2.32	2.29
9	3.36	3.01	2.81	2.69	2.61	2.55	2.51	2.47	2.44	2.42	2.38	2.34	2.30	2.28	2.25	2.23	2.21	2.18	2.16
10	3.29	2.92	2.73	2.61	2.52	2.46	2.41	2.38	2.35	2.32	2.28	2.24	2.20	2.18	2.16	2.13	2.11	2.08	2.06
11	3.23	2.86	2.66	2.54	2.45	2.39	2.34	2.30	2.27	2.25	2.21	2.17	2.12	2.10	2.08	2.05	2.03	2.00	1.97
12	3.18	2.81	2.61	2.48	2.39	2.33	2.28	2.24	2.21	2.19	2.15	2.10	2.06	2.04	2.01	1.99	1.96	1.93	1.90
13	3.14	2.76	2.56	2.43	2.35	2.28	2.23	2.20	2.16	2.14	2.10	2.05	2.01	1.98	1.96	1.93	1.90	1.88	1.85
14	3.10	2.73	2.52	2.39	2.31	2.24	2.19	2.15	2.12	2.10	2.05	2.01	1.96	1.94	1.91	1.89	1.86	1.83	1.80
15	3.07	2.70	2.49	2.36	2.27	2.21	2.16	2.12	2.09	2.06	2.02	1.97	1.92	1.90	1.87	1.85	1.82	1.79	1.76
16	3.05	2.67	2.46	2.33	2.24	2.18	2.13	2.09	2.06	2.03	1.99	1.94	1.89	1.87	1.84	1.81	1.78	1.75	1.72
17	3.03	2.64	2.44	2.31	2.22	2.15	2.10	2.06	2.03	2.00	1.96	1.91	1.86	1.84	1.81	1.78	1.75	1.72	1.69
18	3.01	2.62	2.42	2.29	2.20	2.13	2.08	2.04	2.00	1.98	1.93	1.89	1.84	1.81	1.78	1.75	1.72	1.69	1.66
19	2.99	2.61	2.40	2.27	2.18	2.11	2.06	2.02	1.98	1.96	1.91	1.86	1.81	1.79	1.76	1.73	1.70	1.67	1.63
20	2.97	2.59	2.38	2.25	2.16	2.09	2.04	2.00	1.96	1.94	1.89	1.84	1.79	1.77	1.74	1.71	1.68	1.64	1.61
21	2.96	2.57	2.36	2.23	2.14	2.08	2.02	1.98	1.95	1.92	1.87	1.83	1.78	1.75	1.72	1.69	1.66	1.62	1.59
22	2.95	2.56	2.35	2.22	2.13	2.06	2.01	1.97	1.93	1.90	1.86	1.81	1.76	1.73	1.70	1.67	1.64	1.60	1.57
23	2.94	2.55	2.34	2.21	2.11	2.05	1.99	1.95	1.92	1.89	1.84	1.80	1.74	1.72	1.69	1.66	1.62	1.59	1.55
24	2.93	2.54	2.33	2.19	2.10	2.04	1.98	1.94	1.91	1.88	1.83	1.78	1.73	1.70	1.67	1.64	1.61	1.57	1.53
25	2.92	2.53	2.32	2.18	2.09	2.02	1.97	1.93	1.89	1.87	1.82	1.77	1.72	1.69	1.66	1.63	1.59	1.56	1.52
26	2.91	2.52	2.31	2.17	2.08	2.01	1.96	1.92	1.88	1.86	1.81	1.76	1.71	1.68	1.65	1.61	1.58	1.54	1.50
27	2.90	2.51	2.30	2.17	2.07	2.00	1.95	1.91	1.87	1.85	1.80	1.75	1.70	1.67	1.64	1.60	1.57	1.53	1.49
28	2.89	2.50	2.29	2.16	2.06	2.00	1.94	1.90	1.87	1.84	1.79	1.74	1.69	1.66	1.63	1.59	1.56	1.52	1.48
29	2.89	2.50	2.28	2.15	2.06	1.99	1.93	1.89	1.86	1.83	1.78	1.73	1.68	1.65	1.62	1.58	1.55	1.51	1.47
30	2.88	2.49	2.28	2.14	2.05	1.98	1.93	1.88	1.85	1.82	1.77	1.72	1.67	1.64	1.61	1.57	1.54	1.50	1.46
40	2.84	2.44	2.23	2.09	2.00	1.93	1.87	1.83	1.79	1.76	1.71	1.66	1.61	1.57	1.54	1.51	1.47	1.42	1.38
60	2.79	2.39	2.18	2.04	1.95	1.87	1.82	1.77	1.74	1.71	1.66	1.60	1.54	1.51	1.48	1.44	1.40	1.35	1.29
120	2.75	2.35	2.13	1.99	1.90	1.82	1.77	1.72	1.68	1.65	1.60	1.55	1.48	1.45	1.41	1.37	1.32	1.26	1.19
∞	2.71	2.30	2.08	1.94	1.85	1.77	1.72	1.67	1.63	1.60	1.55	1.49	1.42	1.38	1.34	1.30	1.24	1.17	1.00

From M. Merrington and C. M. Thompson (1943). Table of Percentage Points of the Inverted Beta (*F*) Distribution. *Biometrika 33*, pp. 74–87. Reprinted with permission from Biometrika.

Table I	Critical Values for the PPMC

Reject H_0: $\rho = 0$ if the absolute value of r is greater than the value given in the table. The values are for a two-tailed test; d.f. $= n - 2$.

d.f.	$\alpha = 0.05$	$\alpha = 0.01$
1	0.999	0.999
2	0.950	0.999
3	0.878	0.959
4	0.811	0.917
5	0.754	0.875
6	0.707	0.834
7	0.666	0.798
8	0.632	0.765
9	0.602	0.735
10	0.576	0.708
11	0.553	0.684
12	0.532	0.661
13	0.514	0.641
14	0.497	0.623
15	0.482	0.606
16	0.468	0.590
17	0.456	0.575
18	0.444	0.561
19	0.433	0.549
20	0.423	0.537
25	0.381	0.487
30	0.349	0.449
35	0.325	0.418
40	0.304	0.393
45	0.288	0.372
50	0.273	0.354
60	0.250	0.325
70	0.232	0.302
80	0.217	0.283
90	0.205	0.267
100	0.195	0.254

Source: From *Biometrika Tables for Statisticians,* vol. 1 (1962), p. 138. Reprinted with permission.

Table J	Critical Values for the Sign Test

Reject the null hypothesis if the smaller number of positive or negative signs is less than or equal to the value in the table.

n	One-tailed, $\alpha = 0.005$ / Two-tailed, $\alpha = 0.01$	$\alpha = 0.01$ / $\alpha = 0.02$	$\alpha = 0.025$ / $\alpha = 0.05$	$\alpha = 0.05$ / $\alpha = 0.10$
8	0	0	0	1
9	0	0	1	1
10	0	0	1	1
11	0	1	1	2
12	1	1	2	2
13	1	1	2	3
14	1	2	3	3
15	2	2	3	3
16	2	2	3	4
17	2	3	4	4
18	3	3	4	5
19	3	4	4	5
20	3	4	5	5
21	4	4	5	6
22	4	5	5	6
23	4	5	6	7
24	5	5	6	7
25	5	6	6	7

Note: Table J is for one-tailed or two-tailed tests. The term n represents the total number of positive and negative signs. The test value is the number of less frequent signs.

Source: Table 1, p. 560, from "The Statistical Sign Test" by W. J. Dixon and A. M. Mood, vol. 41. no. 236 (Dec. 1946), pp. 557–566.

Table K — Critical Values for the Wilcoxon Signed-Rank Test

Reject the null hypothesis if the test value is less than or equal to the value given in the table.

	One-tailed, $\alpha = 0.05$	$\alpha = 0.025$	$\alpha = 0.01$	$\alpha = 0.005$
n	Two-tailed, $\alpha = 0.10$	$\alpha = 0.05$	$\alpha = 0.02$	$\alpha = 0.01$
5	1			
6	2	1		
7	4	2	0	
8	6	4	2	0
9	8	6	3	2
10	11	8	5	3
11	14	11	7	5
12	17	14	10	7
13	21	17	13	10
14	26	21	16	13
15	30	25	20	16
16	36	30	24	19
17	41	35	28	23
18	47	40	33	28
19	54	46	38	32
20	60	52	43	37
21	68	59	49	43
22	75	66	56	49
23	83	73	62	55
24	92	81	69	61
25	101	90	77	68
26	110	98	85	76
27	120	107	93	84
28	130	117	102	92
29	141	127	111	100
30	152	137	120	109

Source: From *Some Rapid Approximate Statistical Procedures,* Copyright 1949, 1964 Lerderle Laboratories, American Cyanamid Co., Wayne, N.J. Reprinted with permission.

Table L — Critical Values for the Rank Correlation Coefficient

Reject H_0: $\rho = 0$ if the absolute value of r_s is greater than the value given in the table.

n	$\alpha = 0.10$	$\alpha = 0.05$	$\alpha = 0.02$	$\alpha = 0.01$
5	0.900	—	—	—
6	0.829	0.886	0.943	—
7	0.714	0.786	0.893	0.929
8	0.643	0.738	0.833	0.881
9	0.600	0.700	0.783	0.833
10	0.564	0.648	0.745	0.794
11	0.536	0.618	0.709	0.818
12	0.497	0.591	0.703	0.780
13	0.475	0.566	0.673	0.745
14	0.457	0.545	0.646	0.716
15	0.441	0.525	0.623	0.689
16	0.425	0.507	0.601	0.666
17	0.412	0.490	0.582	0.645
18	0.399	0.476	0.564	0.625
19	0.388	0.462	0.549	0.608
20	0.377	0.450	0.534	0.591
21	0.368	0.438	0.521	0.576
22	0.359	0.428	0.508	0.562
23	0.351	0.418	0.496	0.549
24	0.343	0.409	0.485	0.537
25	0.336	0.400	0.475	0.526
26	0.329	0.392	0.465	0.515
27	0.323	0.385	0.456	0.505
28	0.317	0.377	0.488	0.496
29	0.311	0.370	0.440	0.487
30	0.305	0.364	0.432	0.478

Source: From N. L. Johnson and F. C. Leone, *Statistical and Experimental Design,* vol. I (1964), p. 412. Reprinted with permission from the Institute of Mathematical Statistics.

Table M Critical Values for the Number of Runs

This table gives the critical values at $\alpha = 0.05$ for a two-tailed test. Reject the null hypothesis if the number of runs is less than or equal to the smaller value or greater than or equal to the larger value.

Value of n_1	Value of n_2: 2	3	4	5	6	7	8	9	10	11	12	13	14	15	16	17	18	19	20
2	1	1	1	1	1	1	1	1	1	1	2	2	2	2	2	2	2	2	2
	6	6	6	6	6	6	6	6	6	6	6	6	6	6	6	6	6	6	6
3	1	1	1	1	2	2	2	2	2	2	2	2	2	3	3	3	3	3	3
	6	8	8	8	8	8	8	8	8	8	8	8	8	8	8	8	8	8	8
4	1	1	1	2	2	2	3	3	3	3	3	3	3	3	4	4	4	4	4
	6	8	9	9	9	10	10	10	10	10	10	10	10	10	10	10	10	10	10
5	1	1	2	2	3	3	3	3	3	4	4	4	4	4	4	4	5	5	5
	6	8	9	10	10	11	11	12	12	12	12	12	12	12	12	12	12	12	12
6	1	2	2	3	3	3	3	4	4	4	4	5	5	5	5	5	5	6	6
	6	8	9	10	11	12	12	13	13	13	13	14	14	14	14	14	14	14	14
7	1	2	2	3	3	3	4	4	5	5	5	5	5	6	6	6	6	6	6
	6	8	10	11	12	13	13	14	14	14	14	15	15	15	16	16	16	16	16
8	1	2	3	3	3	4	4	5	5	5	6	6	6	6	6	7	7	7	7
	6	8	10	11	12	13	14	14	15	15	16	16	16	16	17	17	17	17	17
9	1	2	3	3	4	4	5	5	5	6	6	6	7	7	7	7	8	8	8
	6	8	10	12	13	14	14	15	16	16	16	17	17	18	18	18	18	18	18
10	1	2	3	3	4	5	5	5	6	6	7	7	7	7	8	8	8	8	9
	6	8	10	12	13	14	15	16	16	17	17	18	18	18	19	19	19	20	20
11	1	2	3	4	4	5	5	6	6	7	7	7	8	8	8	9	9	9	9
	6	8	10	12	13	14	15	16	17	17	18	19	19	19	20	20	20	21	21
12	2	2	3	4	4	5	6	6	7	7	7	8	8	8	9	9	9	10	10
	6	8	10	12	13	14	16	16	17	18	19	19	20	20	21	21	21	22	22
13	2	2	3	4	5	5	6	6	7	7	8	8	9	9	9	10	10	10	10
	6	8	10	12	14	15	16	17	18	19	19	20	20	21	21	22	22	23	23
14	2	2	3	4	5	5	6	7	7	8	8	9	9	9	10	10	10	11	11
	6	8	10	12	14	15	16	17	18	19	20	20	21	22	22	23	23	23	24
15	2	3	3	4	5	6	6	7	7	8	8	9	9	10	10	11	11	11	12
	6	8	10	12	14	15	16	18	18	19	20	21	22	22	23	23	24	24	25
16	2	3	4	4	5	6	6	7	8	8	9	9	10	10	11	11	11	12	12
	6	8	10	12	14	16	17	18	19	20	21	21	22	23	23	24	25	25	25
17	2	3	4	4	5	6	7	7	8	9	9	10	10	11	11	11	12	12	13
	6	8	10	12	14	16	17	18	19	20	21	22	23	23	24	25	25	26	26
18	2	3	4	5	5	6	7	8	8	9	9	10	10	11	11	12	12	13	13
	6	8	10	12	14	16	17	18	19	20	21	22	23	24	25	25	26	26	27
19	2	3	4	5	6	6	7	8	8	9	10	10	11	11	12	12	13	13	13
	6	8	10	12	14	16	17	18	20	21	22	23	23	24	25	26	26	27	27
20	2	3	4	5	6	6	7	8	9	9	10	10	11	12	12	13	13	13	14
	6	8	10	12	14	16	17	18	20	21	22	23	24	25	25	26	27	27	28

Source: Adapted from C. Eisenhardt and F. Swed, "Tables for Testing Randomness of Grouping in a Sequence of Alternatives," *The Annals of Statistics,* vol. 14 (1943), pp. 83–86. Reprinted with permission of the Institute of Mathematical Statistics and of the Benjamin/Cummings Publishing Company, in whose publication, *Elementary Statistics,* 3rd ed. (1989), by Mario F. Triola, this table appears.

| Table N | Critical Values for the Tukey Test |

α = 0.01

k / v	2	3	4	5	6	7	8	9	10	11	12	13	14	15	16	17	18	19	20
1	90.03	135.0	164.3	185.6	202.2	215.8	227.2	237.0	245.6	253.2	260.0	266.2	271.8	277.0	281.8	286.3	290.4	294.3	298.0
2	14.04	19.02	22.29	24.72	26.63	28.20	29.53	30.68	31.69	32.59	33.40	34.13	34.81	35.43	36.00	36.53	37.03	37.50	37.95
3	8.26	10.62	12.17	13.33	14.24	15.00	15.64	16.20	16.69	17.13	17.53	17.89	18.22	18.52	18.81	19.07	19.32	19.55	19.77
4	6.51	8.12	9.17	9.96	10.58	11.10	11.55	11.93	12.27	12.57	12.84	13.09	13.32	13.53	13.73	13.91	14.08	14.24	14.40
5	5.70	6.98	7.80	8.42	8.91	9.32	9.67	9.97	10.24	10.48	10.70	10.89	11.08	11.24	11.40	11.55	11.68	11.81	11.93
6	5.24	6.33	7.03	7.56	7.97	8.32	8.61	8.87	9.10	9.30	9.48	9.65	9.81	9.95	10.08	10.21	10.32	10.43	10.54
7	4.95	5.92	6.54	7.01	7.37	7.68	7.94	8.17	8.37	8.55	8.71	8.86	9.00	9.12	9.24	9.35	9.46	9.55	9.65
8	4.75	5.64	6.20	6.62	6.96	7.24	7.47	7.68	7.86	8.03	8.18	8.31	8.44	8.55	8.66	8.76	8.85	8.94	9.03
9	4.60	5.43	5.96	6.35	6.66	6.91	7.13	7.33	7.49	7.65	7.78	7.91	8.03	8.13	8.23	8.33	8.41	8.49	8.57
10	4.48	5.27	5.77	6.14	6.43	6.67	6.87	7.05	7.21	7.36	7.49	7.60	7.71	7.81	7.91	7.99	8.08	8.15	8.23
11	4.39	5.15	5.62	5.97	6.25	6.48	6.67	6.84	6.99	7.13	7.25	7.36	7.46	7.56	7.65	7.73	7.81	7.88	7.95
12	4.32	5.05	5.50	5.84	6.10	6.32	6.51	6.67	6.81	6.94	7.06	7.17	7.26	7.36	7.44	7.52	7.59	7.66	7.73
13	4.26	4.96	5.40	5.73	5.98	6.19	6.37	6.53	6.67	6.79	6.90	7.01	7.10	7.19	7.27	7.35	7.42	7.48	7.55
14	4.21	4.89	5.32	5.63	5.88	6.08	6.26	6.41	6.54	6.66	6.77	6.87	6.96	7.05	7.13	7.20	7.27	7.33	7.39
15	4.17	4.84	5.25	5.56	5.80	5.99	6.16	6.31	6.44	6.55	6.66	6.76	6.84	6.93	7.00	7.07	7.14	7.20	7.26
16	4.13	4.79	5.19	5.49	5.72	5.92	6.08	6.22	6.35	6.46	6.56	6.66	6.74	6.82	6.90	6.97	7.03	7.09	7.15
17	4.10	4.74	5.14	5.43	5.66	5.85	6.01	6.15	6.27	6.38	6.48	6.57	6.66	6.73	6.81	6.87	6.94	7.00	7.05
18	4.07	4.70	5.09	5.38	5.60	5.79	5.94	6.08	6.20	6.31	6.41	6.50	6.58	6.65	6.73	6.79	6.85	6.91	6.97
19	4.05	4.67	5.05	5.33	5.55	5.73	5.89	6.02	6.14	6.25	6.34	6.43	6.51	6.58	6.65	6.72	6.78	6.84	6.89
20	4.02	4.64	5.02	5.29	5.51	5.69	5.84	5.97	6.09	6.19	6.28	6.37	6.45	6.52	6.59	6.65	6.71	6.77	6.82
24	3.96	4.55	4.91	5.17	5.37	5.54	5.69	5.81	5.92	6.02	6.11	6.19	6.26	6.33	6.39	6.45	6.51	6.56	6.61
30	3.89	4.45	4.80	5.05	5.24	5.40	5.54	5.65	5.76	5.85	5.93	6.01	6.08	6.14	6.20	6.26	6.31	6.36	6.41
40	3.82	4.37	4.70	4.93	5.11	5.26	5.39	5.50	5.60	5.69	5.76	5.83	5.90	5.96	6.02	6.07	6.12	6.16	6.21
60	3.76	4.28	4.59	4.82	4.99	5.13	5.25	5.36	5.45	5.53	5.60	5.67	5.73	5.78	5.84	5.89	5.93	5.97	6.01
120	3.70	4.20	4.50	4.71	4.87	5.01	5.12	5.21	5.30	5.37	5.44	5.50	5.56	5.61	5.66	5.71	5.75	5.79	5.83
∞	3.64	4.12	4.40	4.60	4.76	4.88	4.99	5.08	5.16	5.23	5.29	5.35	5.40	5.45	5.49	5.54	5.57	5.61	5.65

Table N *(continued)*

$\alpha = 0.05$

k \ v	2	3	4	5	6	7	8	9	10	11	12	13	14	15	16	17	18	19	20
1	17.97	26.98	32.82	37.08	40.41	43.12	45.40	47.36	49.07	50.59	51.96	53.20	54.33	55.36	56.32	57.22	58.04	58.83	59.56
2	6.08	8.33	9.80	10.88	11.74	12.44	13.03	13.54	13.99	14.39	14.75	15.08	15.38	15.65	15.91	16.14	16.37	16.57	16.77
3	4.50	5.91	6.82	7.50	8.04	8.48	8.85	9.18	9.46	9.72	9.95	10.15	10.35	10.53	10.69	10.84	10.98	11.11	11.24
4	3.93	5.04	5.76	6.29	6.71	7.05	7.35	7.60	7.83	8.03	8.21	8.37	8.52	8.66	8.79	8.91	9.03	9.13	9.23
5	3.64	4.60	5.22	5.67	6.03	6.33	6.58	6.80	6.99	7.17	7.32	7.47	7.60	7.72	7.83	7.93	8.03	8.12	8.21
6	3.46	4.34	4.90	5.30	5.63	5.90	6.12	6.32	6.49	6.65	6.79	6.92	7.03	7.14	7.24	7.34	7.43	7.51	7.59
7	3.34	4.16	4.68	5.06	5.36	5.61	5.82	6.00	6.16	6.30	6.43	6.55	6.66	6.76	6.85	6.94	7.02	7.10	7.17
8	3.26	4.04	4.53	4.89	5.17	5.40	5.60	5.77	5.92	6.05	6.18	6.29	6.39	6.48	6.57	6.65	6.73	6.80	6.87
9	3.20	3.95	4.41	4.76	5.02	5.24	5.43	5.59	5.74	5.87	5.98	6.09	6.19	6.28	6.36	6.44	6.51	6.58	6.64
10	3.15	3.88	4.33	4.65	4.91	5.12	5.30	5.46	5.60	5.72	5.83	5.93	6.03	6.11	6.19	6.27	6.34	6.40	6.47
11	3.11	3.82	4.26	4.57	4.82	5.03	5.20	5.35	5.49	5.61	5.71	5.81	5.90	5.98	6.06	6.13	6.20	6.27	6.33
12	3.08	3.77	4.20	4.51	4.75	4.95	5.12	5.27	5.39	5.51	5.61	5.71	5.80	5.88	5.95	6.02	6.09	6.15	6.21
13	3.06	3.73	4.15	4.45	4.69	4.88	5.05	5.19	5.32	5.43	5.53	5.63	5.71	5.79	5.86	5.93	5.99	6.05	6.11
14	3.03	3.70	4.11	4.41	4.64	4.83	4.99	5.13	5.25	5.36	5.46	5.55	5.64	5.71	5.79	5.85	5.91	5.97	6.03
15	3.01	3.67	4.08	4.37	4.59	4.78	4.94	5.08	5.20	5.31	5.40	5.49	5.57	5.65	5.72	5.78	5.85	5.90	5.96
16	3.00	3.65	4.05	4.33	4.56	4.74	4.90	5.03	5.15	5.26	5.35	5.44	5.52	5.59	5.66	5.73	5.79	5.84	5.90
17	2.98	3.63	4.02	4.30	4.52	4.70	4.86	4.99	5.11	5.21	5.31	5.39	5.47	5.54	5.61	5.67	5.73	5.79	5.84
18	2.97	3.61	4.00	4.28	4.49	4.67	4.82	4.96	5.07	5.17	5.27	5.35	5.43	5.50	5.57	5.63	5.69	5.74	5.79
19	2.96	3.59	3.98	4.25	4.47	4.65	4.79	4.92	5.04	5.14	5.23	5.31	5.39	5.46	5.53	5.59	5.65	5.70	5.75
20	2.95	3.58	3.96	4.23	4.45	4.62	4.77	4.90	5.01	5.11	5.20	5.28	5.36	5.43	5.49	5.55	5.61	5.66	5.71
24	2.92	3.53	3.90	4.17	4.37	4.54	4.68	4.81	4.92	5.01	5.10	5.18	5.25	5.32	5.38	5.44	5.49	5.55	5.59
30	2.89	3.49	3.85	4.10	4.30	4.46	4.60	4.72	4.82	4.92	5.00	5.08	5.15	5.21	5.27	5.33	5.38	5.43	5.47
40	2.86	3.44	3.79	4.04	4.23	4.39	4.52	4.63	4.73	4.82	4.90	4.98	5.04	5.11	5.16	5.22	5.27	5.31	5.36
60	2.83	3.40	3.74	3.98	4.16	4.31	4.44	4.55	4.65	4.73	4.81	4.88	4.94	5.00	5.06	5.11	5.15	5.20	5.24
120	2.80	3.36	3.68	3.92	4.10	4.24	4.36	4.47	4.56	4.64	4.71	4.78	4.84	4.90	4.95	5.00	5.04	5.09	5.13
∞	2.77	3.31	3.63	3.86	4.03	4.17	4.29	4.39	4.47	4.55	4.62	4.68	4.74	4.80	4.85	4.89	4.93	4.97	5.01

Table N *(concluded)*

$\alpha = 0.10$

k / v	2	3	4	5	6	7	8	9	10	11	12	13	14	15	16	17	18	19	20
1	8.93	13.44	16.36	18.49	20.15	21.51	22.64	23.62	24.48	25.24	25.92	26.54	27.10	27.62	28.10	28.54	28.96	29.35	29.71
2	4.13	5.73	6.77	7.54	8.14	8.63	9.05	9.41	9.72	10.01	10.26	10.49	10.70	10.89	11.07	11.24	11.39	11.54	11.68
3	3.33	4.47	5.20	5.74	6.16	6.51	6.81	7.06	7.29	7.49	7.67	7.83	7.98	8.12	8.25	8.37	8.48	8.58	8.68
4	3.01	3.98	4.59	5.03	5.39	5.68	5.93	6.14	6.33	6.49	6.65	6.78	6.91	7.02	7.13	7.23	7.33	7.41	7.50
5	2.85	3.72	4.26	4.66	4.98	5.24	5.46	5.65	5.82	5.97	6.10	6.22	6.34	6.44	6.54	6.63	6.71	6.79	6.86
6	2.75	3.56	4.07	4.44	4.73	4.97	5.17	5.34	5.50	5.64	5.76	5.87	5.98	6.07	6.16	6.25	6.32	6.40	6.47
7	2.68	3.45	3.93	4.28	4.55	4.78	4.97	5.14	5.28	5.41	5.53	5.64	5.74	5.83	5.91	5.99	6.06	6.13	6.19
8	2.63	3.37	3.83	4.17	4.43	4.65	4.83	4.99	5.13	5.25	5.36	5.46	5.56	5.64	5.72	5.80	5.87	5.93	6.00
9	2.59	3.32	3.76	4.08	4.34	4.54	4.72	4.87	4.99	5.13	5.23	5.33	5.42	5.51	5.58	5.66	5.72	5.79	5.85
10	2.56	3.27	3.70	4.02	4.26	4.47	4.64	4.78	4.91	5.03	5.13	5.23	5.32	5.40	5.47	5.54	5.61	5.67	5.73
11	2.54	3.23	3.66	3.96	4.20	4.40	4.57	4.71	4.84	4.95	5.05	5.15	5.23	5.31	5.38	5.45	5.51	5.57	5.63
12	2.52	3.20	3.62	3.92	4.16	4.35	4.51	4.65	4.78	4.89	4.99	5.08	5.16	5.24	5.31	5.37	5.44	5.49	5.55
13	2.50	3.18	3.59	3.88	4.12	4.30	4.46	4.60	4.72	4.83	4.93	5.02	5.10	5.18	5.25	5.31	5.37	5.43	5.48
14	2.49	3.16	3.56	3.85	4.08	4.27	4.42	4.56	4.68	4.79	4.88	4.97	5.05	5.12	5.19	5.26	5.32	5.37	5.43
15	2.48	3.14	3.54	3.83	4.05	4.23	4.39	4.52	4.64	4.75	4.84	4.93	5.01	5.08	5.15	5.21	5.27	5.32	5.38
16	2.47	3.12	3.52	3.80	4.03	4.21	4.36	4.49	4.61	4.71	4.81	4.89	4.97	5.04	5.11	5.17	5.23	5.28	5.33
17	2.46	3.11	3.50	3.78	4.00	4.18	4.33	4.46	4.58	4.68	4.77	4.86	4.93	5.01	5.07	5.13	5.19	5.24	5.30
18	2.45	3.10	3.49	3.77	3.98	4.16	4.31	4.44	4.55	4.65	4.75	4.83	4.90	4.98	5.04	5.10	5.16	5.21	5.26
19	2.45	3.09	3.47	3.75	3.97	4.14	4.29	4.42	4.53	4.63	4.72	4.80	4.88	4.95	5.01	5.07	5.13	5.18	5.23
20	2.44	3.08	3.46	3.74	3.95	4.12	4.27	4.40	4.51	4.61	4.70	4.78	4.85	4.92	4.99	5.05	5.10	5.16	5.20
24	2.42	3.05	3.42	3.69	3.90	4.07	4.21	4.34	4.44	4.54	4.63	4.71	4.78	4.85	4.91	4.97	5.02	5.07	5.12
30	2.40	3.02	3.39	3.65	3.85	4.02	4.16	4.28	4.38	4.47	4.56	4.64	4.71	4.77	4.83	4.89	4.94	4.99	5.03
40	2.38	2.99	3.35	3.60	3.80	3.96	4.10	4.21	4.32	4.41	4.49	4.56	4.63	4.69	4.75	4.81	4.86	4.90	4.95
60	2.36	2.96	3.31	3.56	3.75	3.91	4.04	4.16	4.25	4.34	4.42	4.49	4.56	4.62	4.67	4.73	4.78	4.82	4.86
120	2.34	2.93	3.28	3.52	3.71	3.86	3.99	4.10	4.19	4.28	4.35	4.42	4.48	4.54	4.60	4.65	4.69	4.74	4.78
∞	2.33	2.90	3.24	3.48	3.66	3.81	3.93	4.04	4.13	4.21	4.28	4.35	4.41	4.47	4.52	4.57	4.61	4.65	4.69

Source: "Tables of Range and Studentized Range," *Annals of Mathematical Statistics*, vol. 31, no. 4. Reprinted with permission of the Institute of Mathematical Sciences.

Appendix D

Data Bank

Data Bank Values

This list explains the values given for the categories in the Data Bank.

1. "Age" is given in years.

2. "Educational level" values are defined as follows:

   0 = no high school degree   2 = college graduate
   1 = high school graduate   3 = graduate degree

3. "Smoking status" values are defined as follows:

   0 = does not smoke
   1 = smokes less than one pack per day
   2 = smokes one or more than one pack per day

4. "Exercise" values are defined as follows:

   0 = none   2 = moderate
   1 = light   3 = heavy

5. "Weight" is given in pounds.

6. "Serum cholesterol" is given in milligram percent (mg%).

7. "Systolic pressure" is given in millimeters of mercury (mm Hg).

8. "IQ" is given in standard IQ test score values.

9. "Sodium" is given in milliequivalents per liter (mEq/1).

10. "Gender" is listed as male (M) or female (F).

11. "Marital status" values are defined as follows:

    M = married   S = single
    W = widowed   D = divorced

Data Bank

ID number	Age	Educational level	Smoking status	Exercise	Weight	Serum cholesterol	Systolic pressure	IQ	Sodium	Gender	Marital status
01	27	2	1	1	120	193	126	118	136	F	M
02	18	1	0	1	145	210	120	105	137	M	S
03	32	2	0	0	118	196	128	115	135	F	M
04	24	2	0	1	162	208	129	108	142	M	M
05	19	1	2	0	106	188	119	106	133	F	S
06	56	1	0	0	143	206	136	111	138	F	W
07	65	1	2	0	160	240	131	99	140	M	W
08	36	2	1	0	215	215	163	106	151	M	D
09	43	1	0	1	127	201	132	111	134	F	M
10	47	1	1	1	132	215	138	109	135	F	D

Data Bank *(continued)*

ID number	Age	Educational level	Smoking status	Exercise	Weight	Serum cholesterol	Systolic pressure	IQ	Sodium	Gender	Marital status
11	48	3	1	2	196	199	148	115	146	M	D
12	25	2	2	3	109	210	115	114	141	F	S
13	63	0	1	0	170	242	149	101	152	F	D
14	37	2	0	3	187	193	142	109	144	M	M
15	40	0	1	1	234	208	156	98	147	M	M
16	25	1	2	1	199	253	135	103	148	M	S
17	72	0	0	0	143	288	156	103	145	F	M
18	56	1	1	0	156	164	153	99	144	F	D
19	37	2	0	2	142	214	122	110	135	M	M
20	41	1	1	1	123	220	142	108	134	F	M
21	33	2	1	1	165	194	122	112	137	M	S
22	52	1	0	1	157	205	119	106	134	M	D
23	44	2	0	1	121	223	135	116	133	F	M
24	53	1	0	0	131	199	133	121	136	F	M
25	19	1	0	3	128	206	118	122	132	M	S
26	25	1	0	0	143	200	118	103	135	M	M
27	31	2	1	1	152	204	120	119	136	M	M
28	28	2	0	0	119	203	118	116	138	F	M
29	23	1	0	0	111	240	120	105	135	F	S
30	47	2	1	0	149	199	132	123	136	F	M
31	47	2	1	0	179	235	131	113	139	M	M
32	59	1	2	0	206	260	151	99	143	M	W
33	36	2	1	0	191	201	148	118	145	M	D
34	59	0	1	1	156	235	142	100	132	F	W
35	35	1	0	0	122	232	131	106	135	F	M
36	29	2	0	2	175	195	129	121	148	M	M
37	43	3	0	3	194	211	138	129	146	M	M
38	44	1	2	0	132	240	130	109	132	F	S
39	63	2	2	1	188	255	156	121	145	M	M
40	36	2	1	1	125	220	126	117	140	F	S
41	21	1	0	1	109	206	114	102	136	F	M
42	31	2	0	2	112	201	116	123	133	F	M
43	57	1	1	1	167	213	141	103	143	M	W
44	20	1	2	3	101	194	110	111	125	F	S
45	24	2	1	3	106	188	113	114	127	F	D
46	42	1	0	1	148	206	136	107	140	M	S
47	55	1	0	0	170	257	152	106	130	F	M
48	23	0	0	1	152	204	116	95	142	M	M
49	32	2	0	0	191	210	132	115	147	M	M
50	28	1	0	1	148	222	135	100	135	M	M
51	67	0	0	0	160	250	141	116	146	F	W
52	22	1	1	1	109	220	121	103	144	F	M
53	19	1	1	1	131	231	117	112	133	M	S
54	25	2	0	2	153	212	121	119	149	M	D
55	41	3	2	2	165	236	130	131	152	M	M

Data Bank (concluded)

ID number	Age	Educational level	Smoking status	Exercise	Weight	Serum cholesterol	Systolic pressure	IQ	Sodium	Gender	Marital status
56	24	2	0	3	112	205	118	100	132	F	S
57	32	2	0	1	115	187	115	109	136	F	S
58	50	3	0	1	173	203	136	126	146	M	M
59	32	2	1	0	186	248	119	122	149	M	M
60	26	2	0	1	181	207	123	121	142	M	S
61	36	1	1	0	112	188	117	98	135	F	D
62	40	1	1	0	130	201	121	105	136	F	D
63	19	1	1	1	132	237	115	111	137	M	S
64	37	2	0	2	179	228	141	127	141	F	M
65	65	3	2	1	212	220	158	129	148	M	M
66	21	1	2	2	99	191	117	103	131	F	S
67	25	2	2	1	128	195	120	121	131	F	S
68	68	0	0	0	167	210	142	98	140	M	W
69	18	1	1	2	121	198	123	113	136	F	S
70	26	0	1	1	163	235	128	99	140	M	M
71	45	1	1	1	185	229	125	101	143	M	M
72	44	3	0	0	130	215	128	128	137	F	M
73	50	1	0	0	142	232	135	104	138	F	M
74	63	0	0	0	166	271	143	103	147	F	W
75	48	1	0	3	163	203	131	103	144	M	M
76	27	2	0	3	147	186	118	114	134	M	M
77	31	3	1	1	152	228	116	126	138	M	D
78	28	2	0	2	112	197	120	123	133	F	M
79	36	2	1	2	190	226	123	121	147	M	M
80	43	3	2	0	179	252	127	131	145	M	D
81	21	1	0	1	117	185	116	105	137	F	S
82	32	2	1	0	125	193	123	119	135	F	M
83	29	2	1	0	123	192	131	116	131	F	D
84	49	2	2	1	185	190	129	127	144	M	M
85	24	1	1	1	133	237	121	114	129	M	M
86	36	2	0	2	163	195	115	119	139	M	M
87	34	1	2	0	135	199	133	117	135	F	M
88	36	0	0	1	142	216	138	88	137	F	M
89	29	1	1	1	155	214	120	98	135	M	S
90	42	0	0	2	169	201	123	96	137	M	D
91	41	1	1	1	136	214	133	102	141	F	D
92	29	1	1	0	112	205	120	102	130	F	M
93	43	1	1	0	185	208	127	100	143	M	M
94	61	1	2	0	173	248	142	101	141	M	M
95	21	1	1	3	106	210	111	105	131	F	S
96	56	0	0	0	149	232	142	103	141	F	M
97	63	0	1	0	192	193	163	95	147	M	M
98	74	1	0	0	162	247	151	99	151	F	W
99	35	2	0	1	151	251	147	113	145	F	M
100	28	2	0	3	161	199	129	116	138	M	M

Data Set I Record Temperatures

Record high temperatures by state in degrees Fahrenheit

112	100	128	120	134
118	106	110	115	109
112	100	118	117	116
118	121	114	114	105
109	107	112	114	115
118	117	118	125	106
110	122	108	110	121
113	120	119	111	104
111	120	113	120	117
105	110	118	112	114

Record low temperatures by state in degrees Fahrenheit

-27	-80	-40	-29	-45
-61	-32	-17	-66	-2
-17	12	-60	-36	-36
-47	-40	-37	-16	-48
-40	-35	-51	-60	-19
-40	-70	-47	-50	-47
-39	-50	-52	-34	-60
-19	-27	-54	-42	-25
-50	-58	-32	-23	-69
	-30	-48	-37	-55

Data Set II Identity Theft Complaints

The data values show the number of complaints of identity theft for 50 selected cities in the year 2002.

2609	1202	2730	483	655
626	393	1268	279	663
817	1165	551	2654	592
128	189	424	585	78
1836	154	248	239	5888
574	75	226	28	205
176	372	84	229	15
148	117	22	211	31
77	41	200	35	30
88	20	84	465	136

Source: Federal Trade Commission.

Data Set III Length of Major North American Rivers

729	610	325	392	524
1459	450	465	605	330
950	906	329	290	1000
600	1450	862	532	890
407	525	720	1243	850
649	730	352	390	420
710	340	693	306	250
470	724	332	259	2340
560	1060	774	332	3710

Data Set III Length of Major North American Rivers (continued)

2315	2540	618	1171	460
431	800	605	410	1310
500	790	531	981	460
926	375	1290	1210	1310
383	380	300	310	411
1900	434	420	545	569
425	800	865	380	445
538	1038	424	350	377
540	659	652	314	360
301	512	500	313	610
360	430	682	886	447
338	485	625	722	525
800	309	435		

Data Set IV Heights (in Feet) of 80 Tallest Buildings in New York City

1250	861	1046	952	552
915	778	856	850	927
729	745	757	752	814
750	697	743	739	750
700	670	716	707	730
682	648	687	687	705
650	634	664	674	685
640	628	630	653	673
625	620	628	645	650
615	592	620	630	630
595	580	614	618	629
587	575	590	609	615
575	572	580	588	603
574	563	575	577	587
565	555	562	570	576
557	570	555	561	574

Heights (in Feet) of 25 Tallest Buildings in Calgary, Alberta

689	530	460	410
645	525	449	410
645	507	441	408
626	500	435	407
608	469	435	
580	468	432	
530	463	420	

Data Set V   School Suspensions

The data values show the number of suspensions and the number of students enrolled in 40 local school districts in southwestern Pennsylvania.

Suspensions	Enrollment	Suspensions	Enrollment
37	1316	63	1588
29	1337	500	6046
106	4904	5	3610
47	5301	117	4329
51	1380	13	1908
46	1670	8	1341
65	3446	71	5582
223	1010	57	1869
10	795	16	1697
60	2094	60	2269
15	926	51	2307
198	1950	48	1564
56	3005	20	4147
72	4575	80	3182
110	4329	43	2982
6	3238	15	3313
37	3064	187	6090
26	2638	182	4874
140	4949	76	8286
39	3354	37	539
42	3547		

Source: U.S. Department of Education, *Pittsburgh Tribune-Review.*

Data Set VI   Acreage of U.S. National Parks, in Thousands of Acres

41	66	233	775	169
36	338	223	46	64
183	4724	61	1449	7075
1013	3225	1181	308	77
520	77	27	217	5
539	3575	650	462	1670
2574	106	52	52	236
505	913	94	75	265
402	196	70	13	132
28	7656	2220	760	143

Source: The Universal Almanac.

Data Set VII   Acreage Owned by 35 Municipalities in Southwestern Pennsylvania

384	44	62	218	250
198	60	306	105	600
10	38	87	227	340
48	70	58	223	3700
22	78	165	150	160
130	120	100	234	1200
4200	402	180	200	200

Source: Pittsburgh Tribune-Review.

Data Set VIII   Oceans of the World

Ocean	Area (thousands of square miles)	Maximum depth (feet)
Arctic	5,400	17,881
Caribbean Sea	1,063	25,197
Mediterranean Sea	967	16,470
Norwegian Sea	597	13,189
Gulf of Mexico	596	14,370
Hudson Bay	475	850
Greenland Sea	465	15,899
North Sea	222	2,170
Black Sea	178	7,360
Baltic Sea	163	1,440
Atlantic Ocean	31,830	30,246
South China Sea	1,331	18,241
Sea of Okhotsk	610	11,063
Bering Sea	876	13,750
Sea of Japan	389	12,280
East China Sea	290	9,126
Yellow Sea	161	300
Pacific Ocean	63,800	36,200
Arabian Sea	1,492	19,029
Bay of Bengal	839	17,251
Red Sea	169	7,370
Indian Ocean	28,360	24,442

Source: The Universal Almanac.

Data Set IX   Commuter and Rapid Rail Systems in the United States

System	Stations	Miles	Vehicles operated
Long Island RR	134	638.2	947
N.Y. Metro North	108	535.9	702
New Jersey Transit	158	926.0	582
Chicago RTA	117	417.0	358
Chicago & NW Transit	62	309.4	277
Boston Amtrak/MBTA	101	529.8	291
Chicago, Burlington, Northern	27	75.0	139
NW Indiana CTD	18	134.8	39
New York City TA	469	492.9	4923
Washington Metro Area TA	70	162.1	534
Metro Boston TA	53	76.7	368
Chicago TA	137	191.0	924
Philadelphia SEPTA	76	75.8	300
San Francisco BART	34	142.0	415
Metro Atlantic RTA	29	67.0	136
New York PATH	13	28.6	282
Miami/Dade Co TA	21	42.2	82
Baltimore MTA	12	26.6	48
Philadelphia PATCO	13	31.5	102
Cleveland RTA	18	38.2	30
New York, Staten Island RT	22	28.6	36

Source: The Universal Almanac.

Data Set X Keystone Jackpot Analysis*

Ball	Times drawn	Ball	Times drawn	Ball	Times drawn
1	11	12	10	23	7
2	5	13	11	24	8
3	10	14	5	25	13
4	11	15	8	26	11
5	7	16	14	27	7
6	13	17	8	28	10
7	8	18	11	29	11
8	10	19	10	30	5
9	16	20	7	31	7
10	12	21	11	32	8
11	10	22	6	33	11

*Times each number has been selected in the regular drawings of the Pennsylvania Lottery.

Source: Copyright *Pittsburgh Post-Gazette,* all rights reserved. Reprinted with permission.

Data Set XI Pages in Statistics Books

The data values represent the number of pages found in statistics textbooks.

616	578	569	511	468
493	564	801	483	847
525	881	757	272	703
741	556	500	668	967
608	465	739	669	651
495	613	774	274	542
739	488	601	727	556
589	724	731	662	680
589	435	742	567	574
733	576	526	443	478
586	282			

Source: Allan G. Bluman.

Data Set XII Fifty Top Grossing Movies–2000

The data values represent the gross income in millions of dollars for the 50 top movies for the year 2000.

253.4	123.3	90.2	61.3	57.3
215.4	122.8	90.0	61.3	57.2
186.7	117.6	89.1	60.9	56.9
182.6	115.8	77.1	60.8	56.0
161.3	113.7	73.2	60.6	53.3
157.3	113.3	71.2	60.1	53.3
157.0	109.7	70.3	60.0	51.9
155.4	106.8	69.7	59.1	50.9
137.7	101.6	68.5	58.3	50.8
126.6	90.6	68.4	58.1	50.2

Source: Reprinted with permission from the *World Almanac and Book of Facts.* Copyright © K-III Reference Corporation. All rights reserved.

Data Set XIII Hospital Data*

Number	Number of beds	Admissions	Payroll ($000)	Personnel
1	235	6,559	18,190	722
2	205	6,237	17,603	692
3	371	8,915	27,278	1,187
4	342	8,659	26,722	1,156
5	61	1,779	5,187	237
6	55	2,261	7,519	247
7	109	2,102	5,817	245
8	74	2,065	5,418	223
9	74	3,204	7,614	326
10	137	2,638	7,862	362
11	428	18,168	70,518	2,461
12	260	12,821	40,780	1,422
13	159	4,176	11,376	465
14	142	3,952	11,057	450
15	45	1,179	3,370	145
16	42	1,402	4,119	211
17	92	1,539	3,520	158
18	28	503	1,172	72
19	56	1,780	4,892	195
20	68	2,072	6,161	243
21	206	9,868	30,995	1,142
22	93	3,642	7,912	305
23	68	1,558	3,929	180
24	330	7,611	33,377	1,116
25	127	4,716	13,966	498
26	87	2,432	6,322	240
27	577	19,973	60,934	1,822
28	310	11,055	31,362	981
29	49	1,775	3,987	180
30	449	17,929	53,240	1,899
31	530	15,423	50,127	1,669
32	498	15,176	49,375	1,549
33	60	565	5,527	251
34	350	11,793	34,133	1,207
35	381	13,133	49,641	1,731
36	585	22,762	71,232	2,608
37	286	8,749	28,645	1,194
38	151	2,607	12,737	377
39	98	2,518	10,731	352
40	53	1,848	4,791	185
41	142	3,658	11,051	421
42	73	3,393	9,712	385
43	624	20,410	72,630	2,326
44	78	1,107	4,946	139
45	85	2,114	4,522	221

(continued)

Data Set XIII Hospital Data* (continued)

Number	Number of beds	Admissions	Payroll ($000)	Personnel
46	120	3,435	11,479	417
47	84	1,768	4,360	184
48	667	22,375	74,810	2,461
49	36	1,008	2,311	131
50	598	21,259	113,972	4,010
51	1,021	40,879	165,917	6,264
52	233	4,467	22,572	558
53	205	4,162	21,766	527
54	80	469	8,254	280
55	350	7,676	58,341	1,525
56	290	7,499	57,298	1,502
57	890	31,812	134,752	3,933
58	880	31,703	133,836	3,914
59	67	2,020	8,533	280
60	317	14,595	68,264	2,772
61	123	4,225	12,161	504
62	285	7,562	25,930	952
63	51	1,932	6,412	472
64	34	1,591	4,393	205
65	194	5,111	19,367	753
66	191	6,729	21,889	946
67	227	5,862	18,285	731
68	172	5,509	17,222	680
69	285	9,855	27,848	1,180
70	230	7,619	29,147	1,216
71	206	7,368	28,592	1,185
72	102	3,255	9,214	359
73	76	1,409	3,302	198
74	540	396	22,327	788
75	110	3,170	9,756	409
76	142	4,984	13,550	552
77	380	335	11,675	543
78	256	8,749	23,132	907
79	235	8,676	22,849	883
80	580	1,967	33,004	1,059
81	86	2,477	7,507	309
82	102	2,200	6,894	225
83	190	6,375	17,283	618
84	85	3,506	8,854	380
85	42	1,516	3,525	166
86	60	1,573	15,608	236
87	485	16,676	51,348	1,559
88	455	16,285	50,786	1,537
89	266	9,134	26,145	939
90	107	3,497	10,255	431
91	122	5,013	17,092	589

Data Set XIII Hospital Data* (continued)

Number	Number of beds	Admissions	Payroll ($000)	Personnel
92	36	519	1,526	80
93	34	615	1,342	74
94	37	1,123	2,712	123
95	100	2,478	6,448	265
96	65	2,252	5,955	237
97	58	1,649	4,144	203
98	55	2,049	3,515	152
99	109	1,816	4,163	194
100	64	1,719	3,696	167
101	73	1,682	5,581	240
102	52	1,644	5,291	222
103	326	10,207	29,031	1,074
104	268	10,182	28,108	1,030
105	49	1,365	4,461	215
106	52	763	2,615	125
107	106	4,629	10,549	456
108	73	2,579	6,533	240
109	163	201	5,015	260
110	32	34	2,880	124
111	385	14,553	52,572	1,724
112	95	3,267	9,928	366
113	339	12,021	54,163	1,607
114	50	1,548	3,278	156
115	55	1,274	2,822	162
116	278	6,323	15,697	722
117	298	11,736	40,610	1,606
118	136	2,099	7,136	255
119	97	1,831	6,448	222
120	369	12,378	35,879	1,312
121	288	10,807	29,972	1,263
122	262	10,394	29,408	1,237
123	94	2,143	7,593	323
124	98	3,465	9,376	371
125	136	2,768	7,412	390
126	70	824	4,741	208
127	35	883	2,505	142
128	52	1,279	3,212	158

*This information was obtained from a sample of hospitals in a selected state. The hospitals are identified by number instead of name.

Glossary

adjusted R^2 used in multiple regression when n and k are approximately equal, to provide a more realistic value of R^2

alpha the probability of a type I error, represented by the Greek letter α

alternative hypothesis a statistical hypothesis that states a difference between a parameter and a specific value or states that there is a difference between two parameters

analysis of variance (ANOVA) a statistical technique used to test a hypothesis concerning the means of three or more populations

ANOVA summary table the table used to summarize the results of an ANOVA test

Bayes' theorem a theorem that allows you to compute the revised probability of an event that occurred before another event when the events are dependent

beta the probability of a type II error, represented by the Greek letter β

between-group variance a variance estimate using the means of the groups or between the groups in an F test

biased sample a sample for which some type of systematic error has been made in the selection of subjects for the sample

bimodal a data set with two modes

binomial distribution the outcomes of a binomial experiment and the corresponding probabilities of these outcomes

binomial experiment a probability experiment in which each trial has only two outcomes, there are a fixed number of trials, the outcomes of the trials are independent, and the probability of success remains the same for each trial

boxplot a graph used to represent a data set when the data set contains a small number of values

categorical frequency distribution a frequency distribution used when the data are categorical (nominal)

central limit theorem a theorem that states that as the sample size increases, the shape of the distribution of the sample means taken from the population with mean μ and standard deviation σ will approach a normal distribution; the distribution will have a mean μ and a standard deviation $\sigma/\sqrt{n}$

Chebyshev's theorem a theorem that states that the proportion of values from a data set that fall within k standard deviations of the mean will be at least $1 - 1/k^2$, where k is a number greater than 1

chi-square distribution a probability distribution obtained from the values of $(n - 1)s^2/\sigma^2$ when random samples are selected from a normally distributed population whose variance is σ^2

class boundaries the upper and lower values of a class for a grouped frequency distribution whose values have one additional decimal place more than the data and end in the digit 5

class midpoint a value for a class in a frequency distribution obtained by adding the lower and upper class boundaries (or the lower and upper class limits) and dividing by 2

class width the difference between the upper class boundary and the lower class boundary for a class in a frequency distribution

classical probability the type of probability that uses sample spaces to determine the numerical probability that an event will happen

cluster sample a sample obtained by selecting a preexisting or natural group, called a cluster, and using the members in the cluster for the sample

coefficient of determination a measure of the variation of the dependent variable that is explained by the regression line and the independent variable; the ratio of the explained variation to the total variation

coefficient of variation the standard deviation divided by the mean with the result expressed as a percentage

combination a selection of objects without regard to order

complement of an event the set of outcomes in the sample space that are not among the outcomes of the event itself

compound event an event that consists of two or more outcomes or simple events

conditional probability the probability that an event B occurs after an event A has already occurred

confidence interval a specific interval estimate of a parameter determined by using data obtained from a sample and the specific confidence level of the estimate

confidence level the probability that a parameter lies within the specified interval estimate of the parameter

confounding variable a variable that influences the outcome variable but cannot be separated from the other variables that influence the outcome variable

consistent estimator an estimator whose value approaches the value of the parameter estimated as the sample size increases

contingency table data arranged in table form for the chi-square independence test, with R rows and C columns

continuous variable a variable that can assume all values between any two specific values; a variable obtained by measuring

control group a group in an experimental study that is not given any special treatment

convenience sample sample of subjects used because they are convenient and available

correction for continuity a correction employed when a continuous distribution is used to approximate a discrete distribution

correlation a statistical method used to determine whether a linear relationship exists between variables

correlation coefficient a statistic or parameter that measures the strength and direction of a linear relationship between two variables

critical or **rejection region** the range of values of the test value that indicates that there is a significant difference and the null hypothesis should be rejected in a hypothesis test

critical value (C.V.) a value that separates the critical region from the noncritical region in a hypothesis test

cumulative frequency the sum of the frequencies accumulated up to the upper boundary of a class in a frequency distribution

data measurements or observations for a variable

data array a data set that has been ordered

data set a collection of data values

data value or **datum** a value in a data set

decile a location measure of a data value; it divides the distribution into 10 groups

degrees of freedom the number of values that are free to vary after a sample statistic has been computed; used when a distribution (such as the t distribution) consists of a family of curves

dependent events events for which the outcome or occurrence of the first event affects the outcome or occurrence of the second event in such a way that the probability is changed

dependent samples samples in which the subjects are paired or matched in some way; i.e., the samples are related

dependent variable a variable in correlation and regression analysis that cannot be controlled or manipulated

descriptive statistics a branch of statistics that consists of the collection, organization, summarization, and presentation of data

discrete variable a variable that assumes values that can be counted

disordinal interaction an interaction between variables in ANOVA, indicated when the graphs of the lines connecting the mean intersect

distribution-free statistics *see* nonparametric statistics

double sampling a sampling method in which a very large population is given a questionnaire to determine those who meet the qualifications for a study; the questionnaire is reviewed, a second smaller population is defined, and a sample is selected from this group

empirical probability the type of probability that uses frequency distributions based on observations to determine numerical probabilities of events

empirical rule a rule that states that when a distribution is bell-shaped (normal), approximately 68% of the data values will fall within 1 standard deviation of the mean; approximately 95% of the data values will fall within 2 standard deviations of the mean; and approximately 99.7% of the data values will fall within 3 standard deviations of the mean

equally likely events the events in the sample space that have the same probability of occurring

estimation the process of estimating the value of a parameter from information obtained from a sample

estimator a statistic used to estimate a parameter

event outcome of a probability experiment

expected frequency the frequency obtained by calculation (as if there were no preference) and used in the chi-square test

expected value the theoretical average of a variable that has a probability distribution

experimental study a study in which the researcher manipulates one of the variables and tries to determine how the manipulation influences other variables

explanatory variable a variable that is being manipulated by the researcher to see if it affects the outcome variable

exploratory data analysis the act of analyzing data to determine what information can be obtained by using stem and leaf plots, medians, interquartile ranges, and boxplots

extrapolation use of the equation for the regression line to predict y' for a value of x that is beyond the range of the data values of x

F distribution the sampling distribution of the variances when two independent samples are selected from two normally distributed populations in which the variances are equal and the variances s_1^2 and s_2^2 are compared as $s_1^2 \div s_2^2$

F test a statistical test used to compare two variances or three or more means

factors the independent variables in ANOVA tests

finite population correction factor a correction factor used to correct the standard error of the mean when the sample size is greater than 5% of the population size

five-number summary five specific values for a data set that consist of the lowest and highest values, Q_1 and Q_3, and the median

frequency the number of values in a specific class of a frequency distribution

frequency distribution an organization of raw data in table form, using classes and frequencies

frequency polygon a graph that displays the data by using lines that connect points plotted for the frequencies at the midpoints of the classes

goodness-of-fit test a chi-square test used to see whether a frequency distribution fits a specific pattern

grouped frequency distribution a distribution used when the range is large and classes of several units in width are needed

Hawthorne effect an effect on an outcome variable caused by the fact that subjects of the study know that they are participating in the study

histogram a graph that displays the data by using vertical bars of various heights to represent the frequencies of a distribution

homogeneity of proportions test a test used to determine the equality of three or more proportions

hypergeometric distribution the distribution of a variable that has two outcomes when sampling is done without replacement

hypothesis testing a decision-making process for evaluating claims about a population

independence test a chi-square test used to test the independence of two variables when data are tabulated in table form in terms of frequencies

independent events events for which the probability of the first occurring does not affect the probability of the second occurring

independent samples samples that are not related

independent variable a variable in correlation and regression analysis that can be controlled or manipulated

inferential statistics a branch of statistics that consists of generalizing from samples to populations, performing hypothesis testing, determining relationships among variables, and making predictions

influential observation an observation that when removed from the data values would markedly change the position of the regression line

interaction effect the effect of two or more variables on each other in a two-way ANOVA study

interquartile range $Q_3 - Q_1$

interval estimate a range of values used to estimate a parameter

interval level of measurement a measurement level that ranks data and in which precise differences between units of measure exist. *See also* nominal, ordinal, and ratio levels of measurement

Kruskal-Wallis test a nonparametric test used to compare three or more means

law of large numbers when a probability experiment is repeated a large number of times, the relative frequency probability of an outcome will approach its theoretical probability

least-squares line another name for the regression line

left-tailed test a test used on a hypothesis when the critical region is on the left side of the distribution

level a treatment in ANOVA for a variable

level of significance the maximum probability of committing a type I error in hypothesis testing

lower class limit the lower value of a class in a frequency distribution that has the same decimal place value as the data

lurking variable a variable that influences the relationship between x and y, but was not considered in the study

main effect the effect of the factors or independent variables when there is a nonsignificant interaction effect in a two-way ANOVA study

marginal change the magnitude of the change in the dependent variable when the independent variable changes 1 unit

maximum error of estimate the maximum likely difference between the point estimate of a parameter and the actual value of the parameter

mean the sum of the values, divided by the total number of values

mean square the variance found by dividing the sum of the squares of a variable by the corresponding degrees of freedom; used in ANOVA

measurement scales a type of classification that tells how variables are categorized, counted, or measured; the four types of scales are nominal, ordinal, interval, and ratio

median the midpoint of a data array

midrange the sum of the lowest and highest data values, divided by 2

modal class the class with the largest frequency

mode the value that occurs most often in a data set

Monte Carlo method a simulation technique using random numbers

multimodal a data set with three or more modes

multinomial distribution a probability distribution for an experiment in which each trial has more than two outcomes

multiple correlation coefficient a measure of the strength of the relationship between the independent variables and the dependent variable in a multiple regression study

multiple regression a study that seeks to determine if several independent variables are related to a dependent variable

multiple relationship a relationship in which many variables are under study

multistage sampling a sampling technique that uses a combination of sampling methods

mutually exclusive events probability events that cannot occur at the same time

negative relationship a relationship between variables such that as one variable increases, the other variable decreases, and vice versa

negatively skewed or **left-skewed distribution** a distribution in which the majority of the data values fall to the right of the mean

nominal level of measurement a measurement level that classifies data into mutually exclusive (nonoverlapping) exhaustive categories in which no order or ranking can be imposed on them. *See also* interval, ordinal, and ratio levels of measurement

noncritical or **nonrejection region** the range of values of the test value that indicates that the difference was probably due to chance and the null hypothesis should not be rejected

nonparametric statistics a branch of statistics for use when the population from which the samples are selected is not normally distributed and for use in testing hypotheses that do not involve specific population parameters

nonrejection region *see* noncritical region

normal distribution a continuous, symmetric, bell-shaped distribution of a variable

normal quantile plot graphical plot used to determine whether a variable is approximately normally distributed

null hypothesis a statistical hypothesis that states that there is no difference between a parameter and a specific value or that there is no difference between two parameters

observational study a study in which the researcher merely observes what is happening or what has happened in the past and draws conclusions based on these observations

observed frequency the actual frequency value obtained from a sample and used in the chi-square test

ogive a graph that represents the cumulative frequencies for the classes in a frequency distribution

one-tailed test a test that indicates that the null hypothesis should be rejected when the test statistic value is in the critical region on one side of the mean

one-way ANOVA a study used to test for differences among means for a single independent variable when there are three or more groups

open-ended distribution a frequency distribution that has no specific beginning value or no specific ending value

ordinal interaction an interaction between variables in ANOVA, indicated when the graphs of the lines connecting the means do not intersect

ordinal level of measurement a measurement level that classifies data into categories that can be ranked; however, precise differences between the ranks do not exist. *See also* interval, nominal, and ratio levels of measurement

outcome the result of a single trial of a probability experiment

outcome variable a variable that is studied to see if it has changed significantly due to the manipulation of the explanatory variable

outlier an extreme value in a data set; it is omitted from a boxplot

parameter a characteristic or measure obtained by using all the data values for a specific population

parametric tests statistical tests for population parameters such as means, variances, and proportions that involve assumptions about the populations from which the samples were selected

Pareto chart chart that uses vertical bars to represent frequencies for a categorical variable

Pearson product moment correlation coefficient (PPMCC) a statistic used to determine the strength of a relationship when the variables are normally distributed

Pearson's index of skewness value used to determine the degree of skewness of a variable

percentile a location measure of a data value; it divides the distribution into 100 groups

permutation an arrangement of *n* objects in a specific order

pie graph a circle that is divided into sections or wedges according to the percentage of frequencies in each category of the distribution

point estimate a specific numerical value estimate of a parameter

Poisson distribution a probability distribution used when *n* is large and *p* is small and when the independent variables occur over a period of time

pooled estimate of the variance a weighted average of the variance using the two sample variances and their respective degrees of freedom as the weights

population the totality of all subjects possessing certain common characteristics that are being studied

population correlation coefficient the value of the correlation coefficient computed by using all possible pairs of data values (*x, y*) taken from a population

positive relationship a relationship between two variables such that as one variable increases, the other variable increases or as one variable decreases, the other decreases

positively skewed or **right-skewed distribution** a distribution in which the majority of the data values fall to the left of the mean

power of a test the probability of rejecting the null hypothesis when it is false

prediction interval a confidence interval for a predicted value *y*

probability the chance of an event occurring

probability distribution the values a random variable can assume and the corresponding probabilities of the values

probability experiment a chance process that leads to well-defined results called outcomes

proportion a part of a whole, represented by a fraction, a decimal, or a percentage

***P*-value** the actual probability of getting the sample mean value if the null hypothesis is true

qualitative variable a variable that can be placed into distinct categories, according to some characteristic or attribute

quantiles values that separate the data set into approximately equal groups

quantitative variable a variable that is numerical in nature and that can be ordered or ranked

quartile a location measure of a data value; it divides the distribution into four groups

quasi-experimental study a study that uses intact groups rather than random assignment of subjects to groups

random sample a sample obtained by using random or chance methods; a sample for which every member of the population has an equal chance of being selected

random variable a variable whose values are determined by chance

range the highest data value minus the lowest data value

range rule of thumb dividing the range by 4, given an approximation of the standard deviation

ranking the positioning of a data value in a data array according to some rating scale

ratio level of measurement a measurement level that possesses all the characteristics of interval measurement and a true zero; it also has true ratios between different units of measure. *See also* interval, nominal, and ordinal levels of measurement

raw data data collected in original form

regression a statistical method used to describe the nature of the relationship between variables, that is, a positive or negative, linear or nonlinear relationship

regression line the line of best fit of the data

rejection region *see* critical region

relative frequency graph a graph using proportions instead of raw data as frequencies

relatively efficient estimator an estimator that has the smallest variance from among all the statistics that can be used to estimate a parameter

residual the difference between the actual value of *y* and the predicted value *y′* for a specific value of *x*

residual plot plot of the *x* values and the residuals to determine how well the regression line can be used to make predictions

resistant statistic a statistic that is not affected by an extremely skewed distribution

right-tailed test a test used on a hypothesis when the critical region is on the right side of the distribution

run a succession of identical letters preceded by or followed by a different letter or no letter at all, such as the beginning or end of the succession

runs test a nonparametric test used to determine whether data are random

sample a group of subjects selected from the population

sample space the set of all possible outcomes of a probability experiment

sampling distribution of sample means a distribution obtained by using the means computed from random samples taken from a population

sampling error the difference between the sample measure and the corresponding population measure due to the fact that the sample is not a perfect representation of the population

scatter plot a graph of the independent and dependent variables in regression and correlation analysis

Scheffé test a test used after ANOVA, if the null hypothesis is rejected, to locate significant differences in the means

sequence sampling a sampling technique used in quality control in which successive units are taken from production lines and tested to see whether they meet the standards set by the manufacturing company

sign test a nonparametric test used to test the value of the median for a specific sample or to test sample means in a comparison of two dependent samples

simple event an outcome that results from a single trial of a probability experiment

simple relationship a relationship in which only two variables are under study

simulation techniques techniques that use probability experiments to mimic real-life situations

Spearman rank correlation coefficient the nonparametric equivalent to the correlation coefficient, used when the data are ranked

standard deviation the square root of the variance

standard error of the estimate the standard deviation of the observed y values about the predicted y' values in regression and correlation analysis

standard error of the mean the standard deviation of the sample means for samples taken from the same population

standard normal distribution a normal distribution for which the mean is equal to 0 and the standard deviation is equal to 1

standard score the difference between a data value and the mean, divided by the standard deviation

statistic a characteristic or measure obtained by using the data values from a sample

statistical hypothesis a conjecture about a population parameter, which may or may not be true

statistical test a test that uses data obtained from a sample to make a decision about whether the null hypothesis should be rejected

statistics the science of conducting studies to collect, organize, summarize, analyze, and draw conclusions from data

stem and leaf plot a data plot that uses part of a data value as the stem and part of the data value as the leaf to form groups or classes

stratified sample a sample obtained by dividing the population into subgroups, called strata, according to various homogeneous characteristics and then selecting members from each stratum

subjective probability the type of probability that uses a probability value based on an educated guess or estimate, employing opinions and inexact information

sum of squares between groups a statistic computed in the numerator of the fraction used to find the between-group variance in ANOVA

sum of squares within groups a statistic computed in the numerator of the fraction used to find the within-group variance in ANOVA

symmetric distribution a distribution in which the data values are uniformly distributed about the mean

systematic sample a sample obtained by numbering each element in the population and then selecting every kth number from the population to be included in the sample

t distribution a family of bell-shaped curves based on degrees of freedom, similar to the standard normal distribution with the exception that the variance is greater than 1; used when you are testing small samples and when the population standard deviation is unknown

t test a statistical test for the mean of a population, used when the population is normally distributed and the population standard deviation is unknown

test value the numerical value obtained from a statistical test, computed from (observed value − expected value) ÷ standard error

time series graph a graph that represents data that occur over a specific time

treatment group a group in an experimental study that has received some type of treatment

treatment groups the groups used in an ANOVA study

tree diagram a device used to list all possibilities of a sequence of events in a systematic way

Tukey test a test used to make pairwise comparisons of means in an ANOVA study when samples are the same size

two-tailed test a test that indicates that the null hypothesis should be rejected when the test value is in either of the two critical regions

two-way ANOVA a study used to test the effects of two or more independent variables and the possible interaction between them

type I error the error that occurs if you reject the null hypothesis when it is true

type II error the error that occurs if you do not reject the null hypothesis when it is false

unbiased estimator an estimator whose value approximates the expected value of a population parameter, used for the variance or standard deviation when the sample size is less than 30; an estimator whose expected value or mean must be equal to the mean of the parameter being estimated

unbiased sample a sample chosen at random from the population that is, for the most part, representative of the population

ungrouped frequency distribution a distribution that uses individual data and has a small range of data

uniform distribution a distribution whose values are evenly distributed over its range

upper class limit the upper value of a class in a frequency distribution that has the same decimal place value as the data

variable a characteristic or attribute that can assume different values

variance the average of the squares of the distance that each value is from the mean

Venn diagram a diagram used as a pictorial representative for a probability concept or rule

weighted mean the mean found by multiplying each value by its corresponding weight and dividing by the sum of the weights

Wilcoxon rank sum test a nonparametric test used to test independent samples and compare distributions

Wilcoxon signed-rank test a nonparametric test used to test dependent samples and compare distributions

within-group variance a variance estimate using all the sample data for an F test; it is not affected by differences in the means

z **distribution** *see* standard normal distribution

z **score** *see* standard score

z **test** a statistical test for means and proportions of a population, used when the population is normally distributed and the population standard deviation is known

z **value** same as z score

Glossary of Symbols

a	y intercept of a line	F_S	Scheffé test value
α	Probability of a type I error	GM	Geometric mean
b	Slope of a line	H	Kruskal-Wallis test value
β	Probability of a type II error	H_0	Null hypothesis
C	Column frequency	H_1	Alternative hypothesis
cf	Cumulative frequency	HM	Harmonic mean
$_nC_r$	Number of combinations of n objects taking r objects at a time	k	Number of samples
		λ	Number of occurrences for the Poisson distribution
C.V.	Critical value	s_D	Standard deviation of the differences
CVar	Coefficient of variation	s_{est}	Standard error of estimate
D	Difference; decile	SS_B	Sum of squares between groups
$\bar{D}$	Mean of the differences	SS_W	Sum of squares within groups
d.f.	Degrees of freedom	s_B^2	Between-group variance
d.f.N.	Degrees of freedom, numerator	s_W^2	Within-group variance
d.f.D.	Degrees of freedom, denominator	t	t test value
E	Event; expected frequency; maximum error of estimate	$t_{\alpha/2}$	Two-tailed t critical value
		μ	Population mean
$\bar{E}$	Complement of an event	μ_D	Mean of the population differences
e	Euler's constant ≈ 2.7183	$\mu_{\bar{X}}$	Mean of the sample means
$E(X)$	Expected value	w	Class width; weight
f	Frequency	r	Sample correlation coefficient
F	F test value; failure	R	Multiple correlation coefficient
F'	Critical value for the Scheffé test	r^2	Coefficient of determination
MD	Median	ρ	Population correlation coefficient
MR	Midrange	r_S	Spearman rank correlation coefficient
MS_B	Mean square between groups	S	Sample space; success
MS_W	Mean square within groups (error)	s	Sample standard deviation
n	Sample size	s^2	Sample variance
N	Population size	σ	Population standard deviation
$n(E)$	Number of ways E can occur	σ^2	Population variance
$n(S)$	Number of outcomes in the sample space	$\sigma_{\bar{X}}$	Standard error of the mean
O	Observed frequency	Σ	Summation notation
P	Percentile; probability	w_s	Smaller sum of signed ranks, Wilcoxon signed-rank test
p	Probability; population proportion		
$\hat{p}$	Sample proportion	X	Data value; number of successes for a binomial distribution
$\bar{p}$	Weighted estimate of p		
$P(B|A)$	Conditional probability	$\bar{X}$	Sample mean
$P(E)$	Probability of an event E	x	Independent variable in regression
$P(\bar{E})$	Probability of the complement of E	$\bar{X}_{GM}$	Grand mean
$_nP_r$	Number of permutations of n objects taking r objects at a time	X_m	Midpoint of a class
		χ^2	Chi-square
π	Pi ≈ 3.14	y	Dependent variable in regression
Q	Quartile	y'	Predicted y value
q	$1-p$; test value for Tukey test	z	z test value or z score
$\hat{q}$	$1-\hat{p}$	$z_{\alpha/2}$	Two-tailed z critical value
$\bar{q}$	$1-\bar{p}$	!	Factorial
R	Range; rank sum		

Appendix F

Bibliography

Aczel, Amir D. *Complete Business Statistics,* 3rd ed. Chicago: Irwin, 1996.

Beyer, William H. *CRC Handbook of Tables for Probability and Statistics,* 2nd ed. Boca Raton, Fla.: CRC Press, 1986.

Brase, Charles, and Corrinne P. Brase. *Understanding Statistics,* 5th ed. Lexington, Mass.: D.C. Heath, 1995.

Chao, Lincoln L. *Introduction to Statistics.* Monterey, Calif.: Brooks/Cole, 1980.

Daniel, Wayne W., and James C. Terrell. *Business Statistics,* 4th ed. Boston: Houghton Mifflin, 1986.

Edwards, Allan L. *An Introduction to Linear Regression and Correlation,* 2nd ed. New York: Freeman, 1984.

Eves, Howard. *An Introduction to the History of Mathematics,* 3rd ed. New York: Holt, Rinehart and Winston, 1969.

Famighetti, Robert, ed. *The World Almanac and Book of Facts 1996.* New York: Pharos Books, 1995.

Freund, John E., and Gary Simon. *Statistics—A First Course,* 6th ed. Englewood Cliffs, N.J.: Prentice-Hall, 1995.

Gibson, Henry R. *Elementary Statistics.* Dubuque, Iowa: Wm. C. Brown Publishers, 1994.

Glass, Gene V., and Kenneth D. Hopkins. *Statistical Methods in Education and Psychology,* 2nd ed. Englewood Cliffs, N.J.: Prentice-Hall, 1984.

Guilford, J. P. *Fundamental Statistics in Psychology and Education,* 4th ed. New York: McGraw-Hill, 1965.

Haack, Dennis G. *Statistical Literacy: A Guide to Interpretation.* Boston: Duxbury Press, 1979.

Hartwig, Frederick, with Brian Dearing. *Exploratory Data Analysis.* Newbury Park, Calif.: Sage Publications, 1979.

Henry, Gary T. *Graphing Data: Techniques for Display and Analysis.* Thousand Oaks, Calif.: Sage Publications, 1995.

Isaac, Stephen, and William B. Michael. *Handbook in Research and Evaluation,* 2nd ed. San Diego: EdITS, 1990.

Johnson, Robert. *Elementary Statistics,* 6th ed. Boston: PWS–Kent, 1992.

Kachigan, Sam Kash. *Statistical Analysis.* New York: Radius Press, 1986.

Khazanie, Ramakant. *Elementary Statistics in a World of Applications,* 3rd ed. Glenview, Ill.: Scott, Foresman, 1990.

Kuzma, Jan W. *Basic Statistics for the Health Sciences.* Mountain View, Calif.: Mayfield, 1984.

Lapham, Lewis H., Michael Pollan, and Eric Ethridge. *The Harper's Index Book.* New York: Henry Holt, 1987.

Lipschultz, Seymour. *Schaum's Outline of Theory and Problems of Probability.* New York: McGraw-Hill, 1968.

Marascuilo, Leonard A., and Maryellen McSweeney. *Nonparametric and Distribution-Free Methods for the Social Sciences.* Monterey, Calif.: Brooks/Cole, 1977.

Marzillier, Leon F. *Elementary Statistics.* Dubuque, Iowa: Wm. C. Brown Publishers, 1990.

Mason, Robert D., Douglas A. Lind, and William G. Marchal. *Statistics: An Introduction.* New York: Harcourt Brace Jovanovich, 1988.

MINITAB. *MINITAB Reference Manual.* State College, Pa.: MINITAB, Inc., 1994.

Minium, Edward W. *Statistical Reasoning in Psychology and Education.* New York: Wiley, 1970.

Moore, David S. *The Basic Practice of Statistics.* New York: W. H. Freeman and Co., 1995.

Moore, David S., and George P. McCabe. *Introduction to the Practice of Statistics,* 3rd ed. New York: W. H. Freeman, 1999.

Newmark, Joseph. *Statistics and Probability in Modern Life.* New York: Saunders, 1988.

Pagano, Robert R. *Understanding Statistics,* 3rd ed. New York: West, 1990.

Phillips, John L., Jr. *How to Think about Statistics.* New York: Freeman, 1988.

Reinhardt, Howard E., and Don O. Loftsgaarden. *Elementary Probability and Statistical Reasoning.* Lexington, Mass.: Heath, 1977.

Roscoe, John T. *Fundamental Research Statistics for the Behavioral Sciences,* 2nd ed. New York: Holt, Rinehart and Winston, 1975.

Rossman, Allan J. *Workshop Statistics, Discovery with Data.* New York: Springer, 1996.

Runyon, Richard P., and Audrey Haber. *Fundamentals of Behavioral Statistics,* 6th ed. New York: Random House, 1988.

Shulte, Albert P., 1981 yearbook editor, and James R. Smart, general yearbook editor. *Teaching Statistics and Probability, 1981 Yearbook.* Reston, Va.: National Council of Teachers of Mathematics, 1981.

Smith, Gary. *Statistical Reasoning.* Boston: Allyn and Bacon, 1985.

Spiegel, Murray R. *Schaum's Outline of Theory and Problems of Statistics.* New York: McGraw-Hill, 1961.

Texas Instruments. *TI-83 Graphing Calculator Guidebook.* Temple, Tex.: Texas Instruments, 1996.

Triola, Mario G. *Elementary Statistics,* 7th ed. Reading, Mass.: Addison-Wesley, 1998.

Wardrop, Robert L. *Statistics: Learning in the Presence of Variation.* Dubuque, Iowa: Wm. C. Brown Publishers, 1995.

Warwick, Donald P., and Charles A. Lininger. *The Sample Survey: Theory and Practice.* New York: McGraw-Hill, 1975.

Weiss, Daniel Evan. *100% American.* New York: Poseidon Press, 1988.

Williams, Jack. *The USA Today Weather Almanac 1995.* New York: Vintage Books, 1994.

Wright, John W., ed. *The Universal Almanac 1995.* Kansas City, Mo.: Andrews & McMeel, 1994.

Appendix G

Photo Credits

Chapter 1
Opener(both): © Getty RF; p.2: © Getty RF; p. 10: © Banana Stock Ltd RF; p. 11: © Ingram Publishing/Super Stock RF.

Chapter 2
Opener: © Corbis RF; p. 36: © Corbis RF; p. 81: © Getty RF.

Chapter 3
Opener: © Comstock/Jupiter Images RF; p. 104: © Getty RF; p. 105: © Image 100 RF; p. 109: © Getty RF.

Chapter 4
Opener: © Corbis RF; p. 182: © Getty RF; p. 230: © The McGraw-Hill Companies, Inc./Evelyn Jo Hebert, photographer; p. 237: © Corbis RF; p. 240: © Corbis RF.

Chapter 5
Opener: © Alamy RF; p. 252: © Fotosearch RF; p. 256: © Brand X/Punchstock RF; p. 270: © Getty RF.

Chapter 6
Opener: Library of Congress; p. 300, 318: © Corbis RF.

Chapter 7
Opener: USDA; p. 356: © Corbis RF; p. 381:© Brand X Pictures/Getty RF; p. 386: © Corbis RF.

Chapter 8
Opener: © Dr. Parvinder Sethi; p. 400: © PhotoDisc/Punchstock RF; p. 433: © Jupiterimages/Imagesource RF; p. 458: © Getty RF.

Chapter 9
Opener(both): © SuperStock RF; p. 472: © Corbis RF; p. 499: © Antonio Reeve/Photo Researchers; p. 508: © Comstcok/PictureQuest RF.

Chapter 10
Opener: © Getty RF; p. 534, 546: © Getty RF; p. 573: © Michael Kagan.

Chapter 11
Opener: © Comstock RF; p. 592, 618: © Getty RF; p. 626: © The McGraw-Hill Companies, Inc./Jill Braaten, photographer.

Chapter 12
Opener: © Getty RF; p. 630: © Brand X RF; p. 647: Photo by Jeff Vanuga, USDA Natural Resources Conservation Service.

Chapter 13
Opener: © Getty RF; p. 672: © The McGraw-Hill Companies, Inc./Andrew Resek, photographer.

Chapter 14
Opener: © SuperStock RF; p. 720: Courtesy of Hastos-Hall Productions; p. 725: © Getty RF.

Instructor's Section

Outline

Teaching Tips

Chapter 1

It is important to emphasize that statistical studies use *random variables* and the values of the variables are called *data*. Since data can be used in different ways, statistics can be divided into two main branches, *descriptive statistics* and *inferential statistics*. The two branches can be illustrated to the class with examples selected from newspapers and magazines.

A very important requirement of a statistical study is to define the population and select a random *sample*. These concepts should be introduced, and then, if desired, students can be referred to Chapter 14, which contains more information on the selection process. It should be pointed out that inferential statistics is based on *probability* theory.

Since statisticians use data, the various types of data and the measurement levels should be explained. Again, real-life examples can be selected from newspapers and magazine articles.

Two types of statistical studies are explained here. Students should be aware of the advantages and disadvantages of observational and experimental studies. Newspaper and magazine articles can be used to illustrate each type of study. A brief explanation of the uses and misuses of statistics is presented in Chapter 1.

Chapter 2

It is important to emphasize that the reason data are organized into a frequency distribution is to enable the researcher to make sense out of seemingly random occurrences. Once data are organized, they can be studied for various patterns and information. Also, a frequency distribution is used to draw various statistical graphs and is used in computing descriptive statistical measures, such as means and standard deviations.

When you are teaching the histogram, frequency polygon, and ogive, explain to students that the histogram and ogive use the class boundaries on the x axis and the frequency polygon uses class midpoints. Stress that the points for the ogive are plotted at the *upper class* boundaries, except for the first point whose frequency is zero. Instructors wishing to teach scatter plots along with the other graphs in this chapter can teach Section 10–1 here.

Stem and leaf plots have been moved from Chapter 3 to Chapter 2.

Chapter 3

The purpose of Chapter 3 is to explain the basic descriptive measures that are used in statistics. They can be divided into three groups:

1. Measures of average (mean, median, and mode).
2. Measures of variation (range, variance, and standard deviation).
3. Measures of position (percentiles, deciles, and quartiles).

Before teaching this chapter, the instructor may wish to teach the summation notation section found in Appendix A: Algebra Review. In addition, students should be made aware that the measures of central tendency will usually be different for the same data, and that each measure has a specific purpose.

Students should know that there are differences between the population variance and the unbiased estimate of the variance. For those students who have statistical calculators, the two different keys should be explained.

Students sometimes have difficulty understanding what is meant by the standard deviation. It helps to explain that for many data sets, most of the values fall within 2 standard deviations on either side of the mean. A better approximation is given by Chebyshev's theorem.

Section 3–3 explains position measures. Note that there are several different ways to compute the percentile ranks for individual data. The method presented is consistent with the computation of the median. Procedures for finding percentiles for grouped data have been omitted in this textbook; instead, graphic methods are used. The relationship between the cumulative frequency graph and the percentile graph should be pointed out to students.

Finally, a graphic technique called the boxplot can be used to describe a data set that is too small to be represented by a histogram. Two boxplots drawn on the same axes can also be used to compare two data sets.

Chapter 4

Students should be made aware that probability is used as a basis for inferential statistics. It is important to emphasize the concept of a sample space and the basic probability rules and to distinguish among three types of probabilities. The addition, multiplication, and complementary rules build on the basic probability rules.

This chapter includes the counting rules. These rules are helpful in determining the number of outcomes in a sample space in order to compute probabilities. Section 4–5 shows how to use the counting rules along with the probability rules.

In teaching this chapter, most professors will want to review the factorial notation given in Appendix A unless students have covered it in a previous course. Also emphasize the difference between doing something when repetitions are allowed and when they are not allowed, and the difference between a permutation and a combination. Some students will have difficulty making this distinction at the end of the chapter. Explain that when order is important, such as in license plates, ID numbers, or street addresses, you should use the multiplication rule or the permutation rules.

Chapter 5

It is important to teach the binomial distribution (Sections 5–1 through 5–3). These are necessary prerequisites for later sections on hypothesis testing. Section 5–4, on the multinomial, Poisson, and hypergeometric distributions, can be omitted at the instructor's discretion.

Chapter 6

In Chapter 6, it is important to emphasize the characteristics of the theoretical normal distribution and to show that it can be used as a model for real-life variables that are approximately normally distributed. Using these properties and the Procedure Table, students can solve all problems involving finding areas under the normal distribution. Applications of the normal distribution to real-life variables involve transforming these variables to z values and using the Procedure Table.

Section 6–4 explains the central limit theorem, which will lead to hypothesis testing in Chapter 8.

Section 6–5 explains how the normal distribution can be used to approximate a variable that has the binomial distribution.

Since many applications of statistics require a distribution to be normal, several methods can be used to determine whether a distribution is normal. It should be emphasized that no real-world distribution is perfectly normal.

Chapter 7

Many statistical procedures involve making estimates. To reinforce this, the instructor can have students bring in examples of estimates from newspaper and magazine articles. In most cases, the estimate given will look like a point estimate; however, most editors omit the confidence interval. For example, if an estimate is 7 ± 2 years, it will be reported as 7 years.

Be sure to point out that the more confident you wish to be, the larger the interval should be. There are ways to reduce the size of the interval without changing the level of confidence, for example, by increasing sample size.

Chapter 8

The hypothesis-testing procedure is difficult for many students, since it involves many different concepts. It is important to explain that you can never be 100% sure of the correctness of the results when samples are used.

In teaching this chapter, it is helpful to explain generally what is happening, then follow this explanation by using specific examples. Be sure to emphasize that the claim could be either the null or alternative hypothesis and relate the level of significance to the central limit theorem.

The two methods of hypothesis testing are explained in this chapter. Both methods use five basic steps. With the advent of computers and calculators that can compute the P-value quickly, this method is being used more frequently than in the past. Explain that when one is finding P-values from Tables F, G, and H, only approximate interval values can be used. In the examples and exercises, the specific P-value obtained from a calculator has been given with the interval.

Since P-value intervals for the t and chi-square tests are somewhat difficult to find from the table, you may want to have students use technology for these exercises or use the traditional method for hypotheses testing.

It is important to show students how to use the chi-square table and to emphasize that the variance test can be one- or two-tailed. The one-tailed test can be either right or left. When testing a right-tailed hypothesis, students should use the right side of the table and the specific value, for example, 0.05. When testing a left-tailed hypothesis, the students should use the left side of the table and find the $1 - \alpha$ value. For example, if $\alpha = 0.05$, the $1 - 0.05$, or 0.95, column should be used.

For a two-tailed test, the area must be split. For example, if $\alpha = 0.05$, you should use the 0.025 and 0.975 columns.

Chapter 9

Once students have mastered Chapter 8, this chapter becomes a continuation of the concepts of hypothesis testing. One difficulty students encounter is that there are five different formulas for testing differences. It is important to emphasize the different situations and which test is appropriate in each case.

Section 9–2 explains the F test for comparing two variances. This section is somewhat difficult for students. This section can be taught with the ANOVA material in Chapter 12; however, the students would need to be told whether the variances are equal for the exercises in Section 9–3.

Chapter 10

It is important to stress that many real-life variables are related in some way. Using correlation and regression techniques, statisticians can determine the nature of the relationship. Examples of relationships found in newspapers and magazines can be used in a discussion of the topics involved, such as the strength and type of the relationship. Students should understand that if the graph is cut off, the y intercept value may not be appropriate when the regression line is plotted.

A brief explanation of extrapolation, lurking variables, marginal change, least-squares line, residuals, and influential observations has been added to this chapter.

Chapter 11

Emphasize that the variance test shown in Chapter 8 can be one- or two-tailed, but the goodness-of-fit test and the independence test are always right-tailed.

Chapter 12

It is important to make students aware that they should use the analysis of variance when testing the equality of three or more means. Also emphasize that the ANOVA will not specify where the difference lies. After a significant F, they can perform the Scheffé or Tukey test to determine where the difference lies. There are many other tests that can be conducted after the F test, but they are beyond the scope of this book.

Chapter 13

The important point to emphasize in this chapter is that the parametric statistics require the assumption of normality of the data. When this assumption cannot be met, the corresponding counterpart, nonparametric statistics, can be used to test similar hypotheses. In most cases, however, larger sample sizes are needed to obtain the same results as with the parametric counterparts. Students should know that there is not complete agreement among statisticians as to the use of nonparametric statistics.

Chapter 14

There are two main topics in Chapter 14: sampling and simulation. Both topics use random numbers. For sampling, the instructor can explain the basic methods and have students actually select samples from a small population, then compute the mean of the variable being sampled. The instructor can have each student read his or her mean aloud to see how close it is to the actual population mean. This is an excellent introduction to the central limit theorem presented in Chapter 6. The mean of the sample means can be computed, and if the class is large, the students will see how close this mean is to the population mean.

The purpose of the second half of this chapter is to enable students to understand the nature of statistical simulation. Again using random numbers and sampling, the students will see how close the means found by simulation techniques are to the theoretical means computed mathematically.

In addition to the two main topics, a third topic on surveys and questionnaires has been added. This topic will help increase the statistical literacy of the students.

Video Resource Guide

The following guide is for two statistics video programs: *Against All Odds: Inside Statistics* (AAO), an Annenberg/CPB Project; and *Decision through Data* (DTD), COMAP, Inc. These video series are available to qualified adopters. Please contact your local sales representative for more information about this program.

Chapter 1
AAO: Programs 1, 14
DTD: Units 1, 17, 18

Chapter 2
AAO: Program 2
DTD: Unit 3

Chapter 3
AAO: Program 3
DTD: Units 2, 4, 5, 6

Chapter 4
AAO: Program 15

Chapter 5
AAO: Programs 16, 17

Chapter 6
AAO: Programs 4, 5, 18
DTD: Units 7, 8, 19

Chapter 7
AAO: Program 19
DTD: Unit 20

Chapter 8
AAO: Programs 20, 21, 23
DTD: Unit 21

Chapter 9
AAO: Program 22

Chapter 10
AAO: Programs 7, 8, 9, 11, 25
DTD: Units 9, 11, 12, 13, 14, 16

Chapter 11
AAO: Program 24

Chapter 14
AAO: Program 14
DTD: Unit 17

Selected Answers*

Chapter 1

Review Exercises

1. Descriptive statistics describe the data set. Inferential statistics use the data to draw conclusions about the population.

2. Probability deals with events that occur by chance. It is used in gambling and insurance.

3. Answers will vary.

4. A population is the totality of all subjects possessing certain common characteristics that are being studied.

5. A sample is a subset or portion of the population that we actually do study to find out information about the population. Samples are used to save time and money when the population is large and when the units must be destroyed to gain information.

6. *a.* Inferential *e.* Inferential
   b. Descriptive *f.* Inferential
   c. Descriptive *g.* Descriptive
   d. Descriptive *h.* Inferential

7. *a.* Ratio *f.* Ratio
   b. Ordinal *g.* Ordinal
   c. Interval *h.* Ratio
   d. Ratio *i.* Ratio
   e. Ratio *j.* Nominal

8. *a.* Qualitative *e.* Quantitative
   b. Quantitative *f.* Quantitative
   c. Quantitative *g.* Quantitative
   d. Qualitative

9. *a.* Discrete *e.* Continuous
   b. Continuous *f.* Discrete
   c. Discrete *g.* Continuous
   d. Continuous

10. *a.* 35.5–36.5
    b. 105.35–105.45
    c. 72.55–72.65
    d. 5.265–5.275
    e. 4.5–5.5

11. Random, systematic, stratified, cluster

12. *a.* Cluster *c.* Random *e.* Stratified
    b. Systematic *d.* Systematic

13. Answers will vary. 14. Answers will vary.

15. Answers will vary. 16. Answers will vary.

17. *a.* Experimental *c.* Observational
    b. Observational *d.* Experimental

18. *a.* Independent variable: type of pill received; dependent variable: number of respiratory infections.
    b. Independent variable: color of automobile; dependent variable: running red lights.
    c. Independent variable: level of hostility; dependent variable: cholesterol level.
    d. Independent variable: type of diet; dependent variable: blood pressure.

19. Possible answers:
    a. Workplace of subjects, smoking habits, etc.
    b. Gender, age, etc.
    c. Diet, type of job, etc.
    d. Exercise, heredity, age, etc.

20. Only 20 people were used in the study.

21. The only time claims can be proved is when the entire population is used.

22. It is meaningless since there is no definition of "the road less traveled." Also, there is no way to know that for *every* 100 women, 91 would say that they have taken "the road less traveled."

23. Since the results are not typical, the advertisers selected only a few people for whom the weight loss product worked extremely well.

24. There is no mention of how this conclusion was obtained.

25. "74% more calories" than what? No comparison group is stated.

26. Since the word *may* is used, there is no guarantee that the product will help fight cancer.

27. What is meant by "24 hours of acid control"?

28. No. There are many other factors that contribute to criminal behavior.

29. Possible answer: It could be the amount of caffeine in the coffee or tea. It could have been the brewing method.

30. Answers will vary.

31. Answers will vary.

32. Answers will vary.

Chapter Quiz

1. True 2. True
3. True 4. False
5. True 6. True
7. False 8. *c*
9. *b* 10. *d*
11. *a* 12. *c*
13. *a* 14. Descriptive, inferential

*Answers may vary due to rounding or use of technology.

15. Gambling, insurance   16. Population

17. Sample

18. *a.* Saves time   *c.* Use when population is infinite
    b. Saves money

19. *a.* Random   *c.* Cluster
    b. Systematic   *d.* Stratified

20. Quasi-experimental   21. Random

22. *a.* Descriptive   *d.* Inferential
    b. Inferential   *e.* Inferential
    c. Descriptive

23. *a.* Nominal   *d.* Interval
    b. Ratio   *e.* Ratio
    c. Ordinal

24. *a.* Continuous   *d.* Continuous
    b. Discrete   *e.* Discrete
    c. Continuous

25. *a.* 31.5–32.5 minutes
    b. 0.475–0.485 millimeter
    c. 6.15–6.25 inches
    d. 18.5–19.5 pounds
    e. 12.05–12.15 quarts

Chapter 2

Exercises 2–1

1. To organize data in a meaningful way, to determine the shape of the distribution, to facilitate computational procedures for statistics, to make it easier to draw charts and graphs, to make comparisons among different sets of data

2. Categorical, ungrouped, grouped

3. *a.* 31.5–38.5, 35, 7
   b. 85.5–104.5, 95, 19
   c. 894.5–905.5, 900, 11
   d. 12.25–13.55, 12.9, 1.3
   e. 3.175–4.965, 4.07, 1.79

4. 5–20; class width should be an odd number so that the midpoints of the classes are in the same place value as the data.

5. *a.* Class width is not uniform.
   b. Class limits overlap, and class width is not uniform.
   c. A class has been omitted.
   d. Class width is not uniform.

6. An open-ended frequency distribution has either a first class with no lower limit or a last class with no upper limit. They are necessary to accommodate all the data.

7.

Class	Tally	Frequency	Percent
A	////	4	10
M	𝚼𝚼𝚼 𝚼𝚼𝚼 𝚼𝚼𝚼 𝚼𝚼𝚼 𝚼𝚼𝚼 ///	28	70
H	𝚼𝚼𝚼 /	6	15
S	//	2	5
		40	100

8.

Limits	Boundaries	*f*
21–27	20.5–27.5	6
28–34	27.5–34.5	9
35–41	34.5–41.5	5
42–48	41.5–48.5	7
49–55	48.5–55.5	3
		30

	cf
Less than 20.5	0
Less than 27.5	6
Less than 34.5	15
Less than 41.5	20
Less than 48.5	27
Less than 55.5	30

9.

Limits	Boundaries	*f*
165–185	164.5–185.5	4
186–206	185.5–206.5	6
207–227	206.5–227.5	15
228–248	227.5–248.5	13
249–269	248.5–269.5	9
270–290	269.5–290.5	1
291–311	290.5–311.5	1
312–332	311.5–332.5	1
		50

A peak occurs in class 207–227 (206.5–227.5). There are no gaps in the distribution, and there is one value in each of the three highest classes.

	cf
Less than 164.5	0
Less than 185.5	4
Less than 206.5	10
Less than 227.5	25
Less than 248.5	38
Less than 269.5	47
Less than 290.5	48
Less than 311.5	49
Less than 332.5	50

10.

Limits	Boundaries	f
54–62	53.5–62.5	7
63–71	62.5–71.5	6
72–80	71.5–80.5	8
81–89	80.5–89.5	4
90–98	89.5–98.5	1
99–107	98.5–107.5	3
108–116	107.5–116.5	1
		30

	cf
Less than 53.5	0
Less than 62.5	7
Less than 71.5	13
Less than 80.5	21
Less than 89.5	25
Less than 98.5	26
Less than 107.5	29
Less than 116.5	30

11.

Limits	Boundaries	f
746–752	745.5–752.5	4
753–759	752.5–759.5	6
760–766	759.5–766.5	8
767–773	766.5–773.5	9
774–780	773.5–780.5	3
		30

	cf
Less than 745.5	0
Less than 752.5	4
Less than 759.5	10
Less than 766.5	18
Less than 773.5	27
Less than 780.5	30

12.

Limits	Boundaries	f
5,427–17,733	5,426.5–17,733.5	17
17,734–30,040	17,733.5–30,040.5	1
30,041–42,347	30,040.5–42,347.5	1
42,348–54,654	42,347.5–54,654.5	1
54,655–66,961	54,654.5–66,961.5	1
66,962–79,268	66,961.5–79,268.5	1
79,269–91,575	79,268.5–91,575.5	3
		25

The majority of the data values fall in the lowest class. There are no gaps in the distribution.

	cf
Less than 5,426.5	0
Less than 17,733.5	17
Less than 30,040.5	18
Less than 42,347.5	19
Less than 54,654.5	20
Less than 66,961.5	21
Less than 79,268.5	22
Less than 91,575.5	25

13.

Limits	Boundaries	f
27–33	26.5–33.5	7
34–40	33.5–40.5	14
41–47	40.5–47.5	15
48–54	47.5–54.5	11
55–61	54.5–61.5	3
62–68	61.5–68.5	3
69–75	68.5–75.5	2
		55

	cf
Less than 26.5	0
Less than 33.5	7
Less than 40.5	21
Less than 47.5	36
Less than 54.5	47
Less than 61.5	50
Less than 68.5	53
Less than 75.5	55

14.

Limits	Boundaries	f
0–10	−0.5–10.5	7
11–21	10.5–21.5	6
22–32	21.5–32.5	2
33–43	32.5–43.5	0
44–54	43.5–54.5	1
		16

	cf
Less than −0.5	0
Less than 10.5	7
Less than 21.5	13
Less than 32.5	15
Less than 43.5	15
Less than 54.5	16

15.

Limits	Boundaries	f
6–132	5.5–132.5	16
133–259	132.5–259.5	3
260–386	259.5–386.5	0
387–513	386.5–513.5	0
514–640	513.5–640.5	1
		20

The lowest class has the most data values, 16, and the next class has 3 values. There is one extremely large data value, 635, and it is in the last class, 514–640 (513.5–640.5).

	cf
Less than 5.5	0
Less than 132.5	16
Less than 259.5	19
Less than 386.5	19
Less than 513.5	19
Less than 640.5	20

16.

Limits	Boundaries	f
140–230	139.5–230.5	11
231–321	230.5–321.5	5
322–412	321.5–412.5	4
413–503	412.5–503.5	4
504–594	503.5–594.5	4
595–685	594.5–685.5	1
686–776	685.5–776.5	0
777–867	776.5–867.5	1
		30

	cf
Less than 139.5	0
Less than 230.5	11
Less than 321.5	16
Less than 412.5	20
Less than 503.5	24
Less than 594.5	28
Less than 685.5	29
Less than 776.5	29
Less than 867.5	30

17. $H = 123$ $L = 77$
Range $= 123 - 77 = 46$
Width $= 46 \div 7 = 6.6$ or 7

Limits	Boundaries	f
77–83	76.5–83.5	1
84–90	83.5–90.5	1
91–97	90.5–97.5	6
98–104	97.5–104.5	14
105–111	104.5–111.5	8
112–118	111.5–118.5	1
119–125	118.5–125.5	1
		32

	cf
Less than 76.5	0
Less than 83.5	1
Less than 90.5	2
Less than 97.5	8
Less than 104.5	22
Less than 111.5	30
Less than 118.5	31
Less than 125.5	32

18. $H = 31.5$ $L = 7.5$
Range $= 31.5 - 7.5 = 24$
Width $= 24 \div 5 = 4.8$ or 5

Limits	Boundaries	f
7.5–12.4	7.45–12.45	1
12.5–17.4	12.45–17.45	4
17.5–22.4	17.45–22.45	10
22.5–27.4	22.45–27.45	6
27.5–32.4	27.45–32.45	4
		25

	cf
Less than 7.45	0
Less than 12.45	1
Less than 17.45	5
Less than 22.45	15
Less than 27.45	21
Less than 32.45	25

19. The percents sum to 101. They should sum to 100% unless rounding was used.

Exercises 2–2

1. Eighty applicants do not need to enroll in the developmental programs.

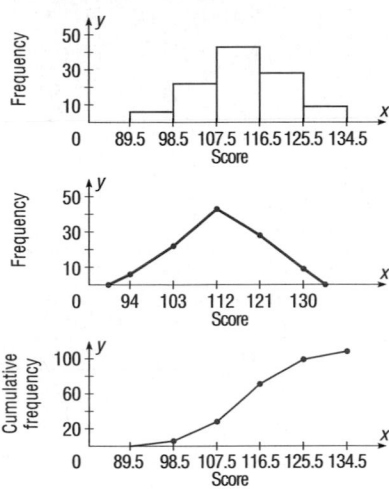

2.

Limits	Boundaries	f
70–116	69.5–116.5	5
117–163	116.5–163.5	9
164–210	163.5–210.5	6
211–257	210.5–257.5	6
258–304	257.5–304.5	0
305–351	304.5–351.5	1
352–398	351.5–398.5	1
		28

	cf
Less than 69.5	0
Less than 116.5	5
Less than 163.5	14
Less than 210.5	20
Less than 257.5	26
Less than 304.5	26
Less than 351.5	27
Less than 398.5	28

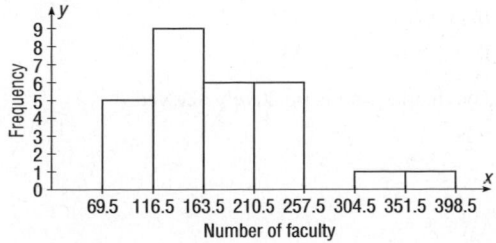

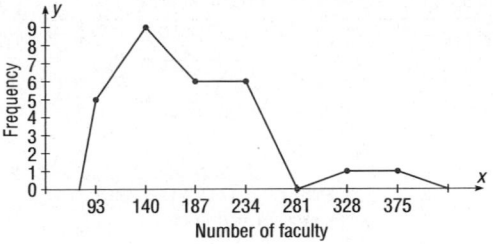

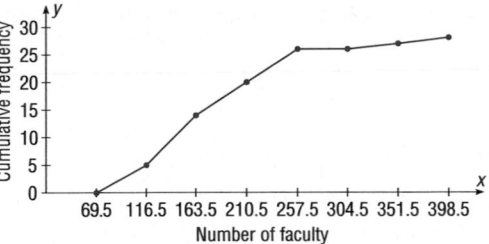

$\frac{12}{28} = 0.429$ or 42.9% have 180 or more. The histogram and frequency polygon are positively skewed.

3.

Limits	Boundaries	f
3–45	2.5–45.5	19
46–88	45.5–88.5	19
89–131	88.5–131.5	10
132–174	131.5–174.5	1
175–217	174.5–217.5	0
218–260	217.5–260.5	1
		50

	cf
Less than 2.5	0
Less than 45.5	19
Less than 88.5	38
Less than 131.5	48
Less than 174.5	49
Less than 217.5	49
Less than 260.5	50

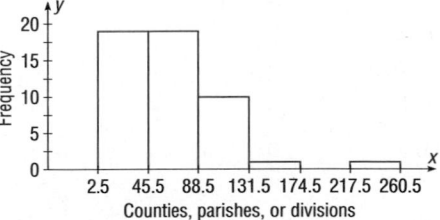

The distribution is positively skewed.

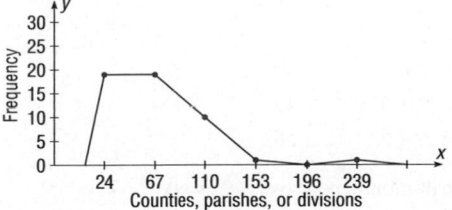

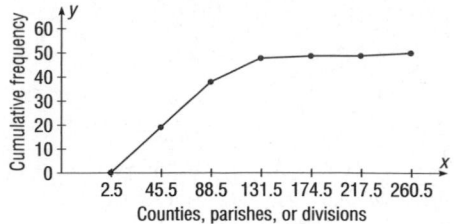

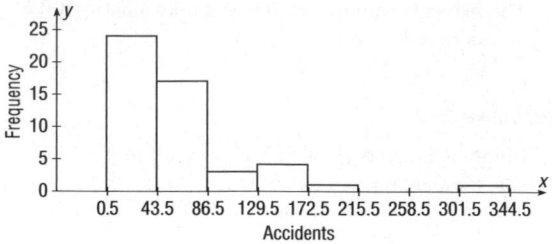

4.

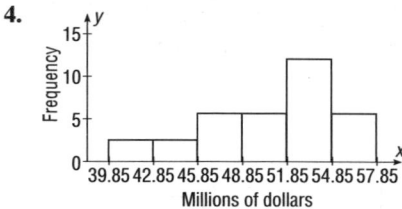

The distribution is negatively or left-skewed.

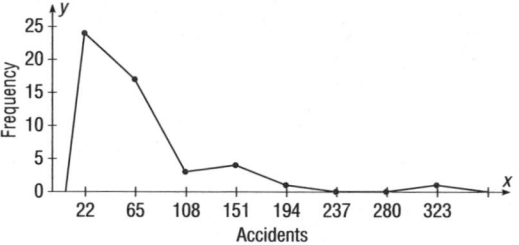

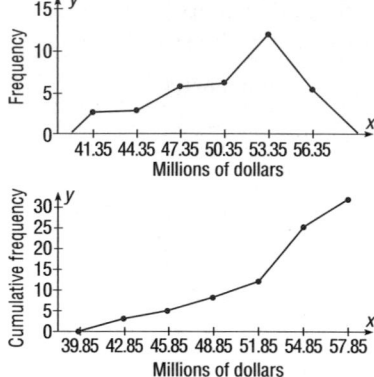

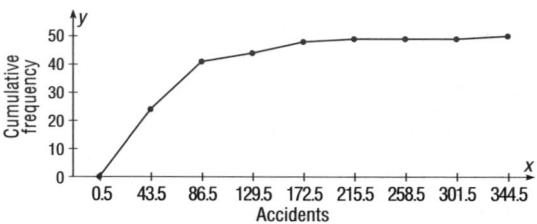

6.

Limits	Boundaries	f
6–8	5.5–8.5	12
9–11	8.5–11.5	16
12–14	11.5–14.5	3
15–17	14.5–17.5	1
18–20	17.5–20.5	0
21–23	20.5–23.5	0
24–26	23.5–26.5	1
		33

5.

Limits	Boundaries	f
1–43	0.5–43.5	24
44–86	43.5–86.5	17
87–129	86.5–129.5	3
130–172	129.5–172.5	4
173–215	172.5–215.5	1
216–258	215.5–258.5	0
259–301	258.5–301.5	0
302–344	301.5–344.5	1

	cf
Less than 0.5	0
Less than 43.5	24
Less than 86.5	41
Less than 129.5	44
Less than 172.5	48
Less than 215.5	49
Less than 258.5	49
Less than 301.5	49
Less than 344.5	50

	cf
Less than 5.5	0
Less than 8.5	12
Less than 11.5	28
Less than 14.5	31
Less than 17.5	32
Less than 20.5	32
Less than 23.5	32
Less than 26.5	33

Yes. The distribution is positively skewed.

The distribution is positively skewed.

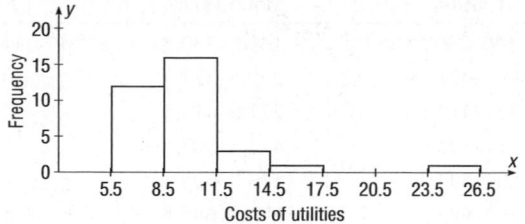

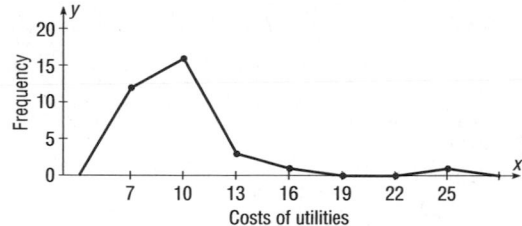

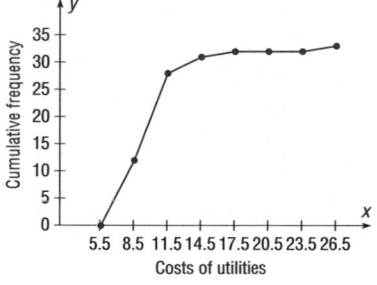

7.

Limits	Boundaries	f(1998)	f(2003)
0–22	−0.5–22.5	18	26
23–45	22.5–45.5	7	1
46–68	45.5–68.5	3	0
69–91	68.5–91.5	1	1
92–114	91.5–114.5	1	0
115–137	114.5–137.5	0	1
138–160	137.5–160.5	0	1
		30	30

Both distributions are positively skewed, but the data are somewhat more spread out in the first three classes in 1998 than in 2003, and there are two large data values in the 2003 data.

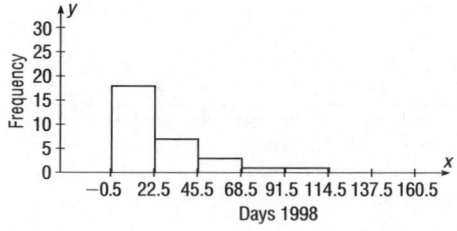

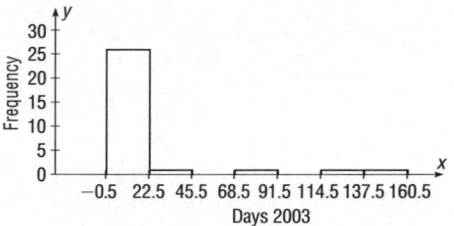

8. The data values fall somewhat on the left side of the distribution. The histogram is right-skewed. There are no gaps in the histogram.

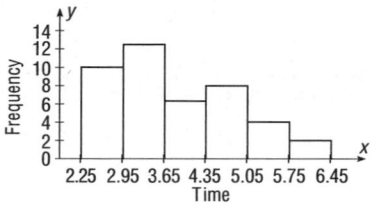

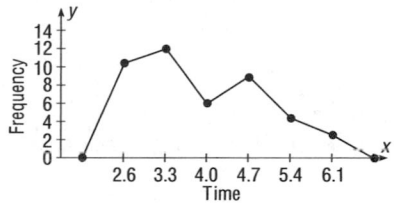

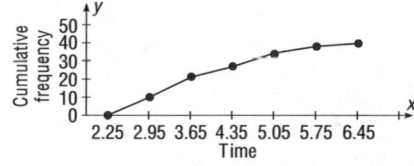

9.

Limits	Boundaries	f
83.1–90.0	83.05–90.05	3
90.1–97.0	90.05–97.05	5
97.1–104.0	97.05–104.05	6
104.1–111.0	104.05–111.05	7
111.1–118.0	111.05–118.05	3
118.1–125.0	118.05–125.05	1
		25

	cf
Less than 83.05	0
Less than 90.05	3
Less than 97.05	8
Less than 104.05	14
Less than 111.05	21
Less than 118.05	24
Less than 125.05	25

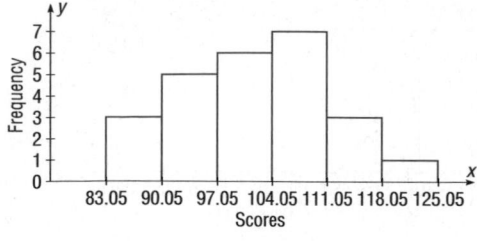

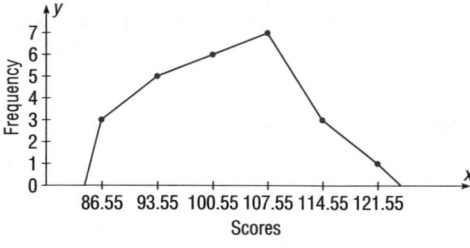

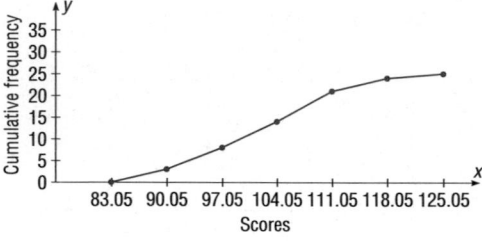

10. The distribution of math percentages is more bell-shaped than the distribution of reading percentages, and its peak in the class of 32.5–37.5 is not as high as the peak of the reading percentages.

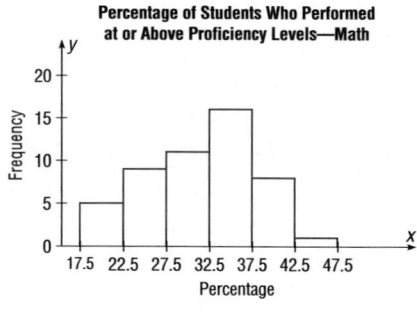

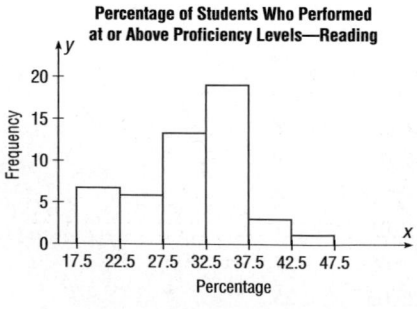

11.

Limits	Boundaries	f
140–230	139.5–230.5	11
231–321	230.5–321.5	5
322–412	321.5–412.5	4
413–503	412.5–503.5	4
504–594	503.5–594.5	4
595–685	594.5–685.5	1
686–776	685.5–776.5	0
777–867	776.5–867.5	1
		30

	cf
Less than 139.5	0
Less than 230.5	11
Less than 321.5	16
Less than 412.5	20
Less than 503.5	24
Less than 594.5	28
Less than 685.5	29
Less than 776.5	29
Less than 867.5	30

The distribution is positively skewed.

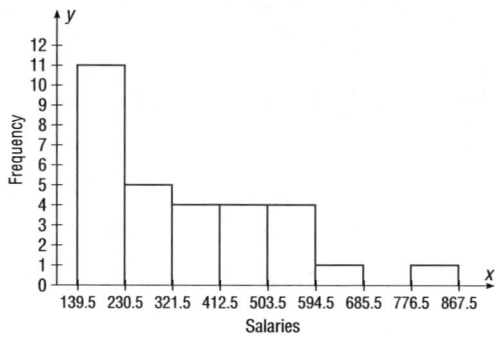

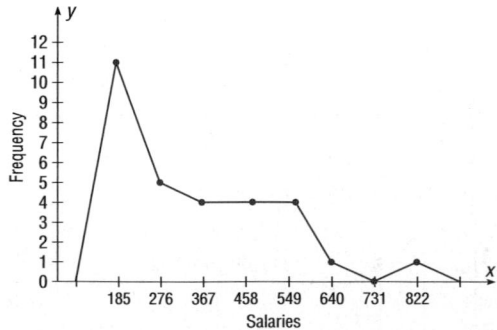

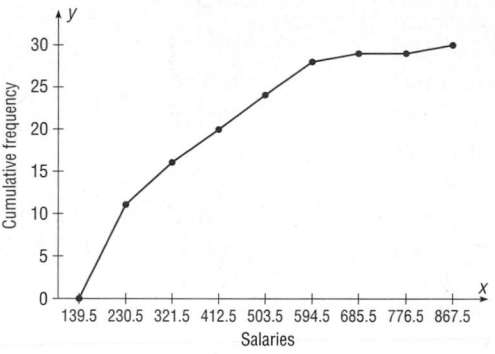

	crf
Less than 0.5	0.00
Less than 43.5	0.48
Less than 86.5	0.82
Less than 129.5	0.88
Less than 172.5	0.96
Less than 215.5	0.98
Less than 258.5	0.98
Less than 301.5	0.98
Less than 344.5	1.00

Of the states 82% have fewer than 87 accidents per year.

12. The histograms show that the distances of the home runs McGwire hit are more variable (spread out) than those hit by Sosa.

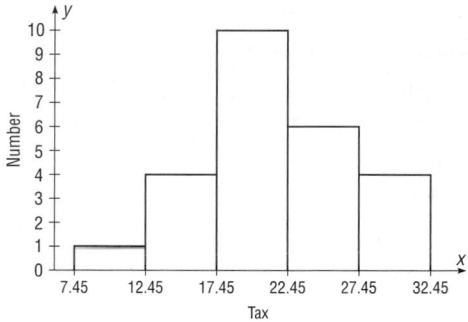

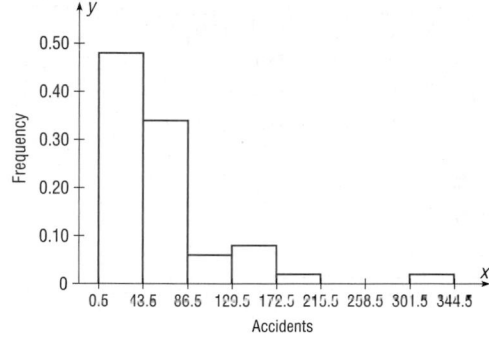

13. The proportion of applicants who need to enroll in the developmental program is about 0.26.

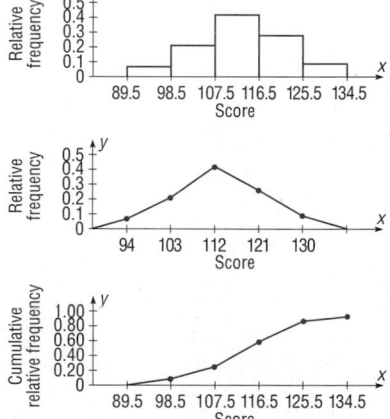

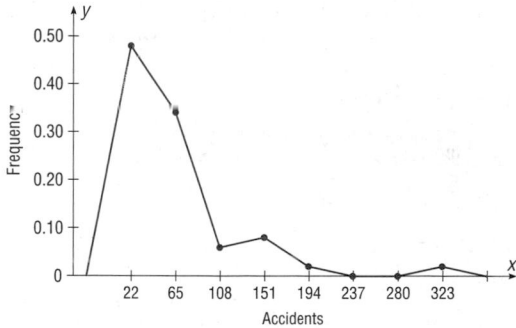

14.

Limits	Boundaries	rf
1–43	0.5–43.5	0.48
44–86	43.5–86.5	0.34
87–129	86.5–129.5	0.06
130–172	129.5–172.5	0.08
173–215	172.5–215.5	0.02
216–258	215.5–258.5	0.00
259–301	258.5–301.5	0.00
302–344	301.5–344.5	0.02

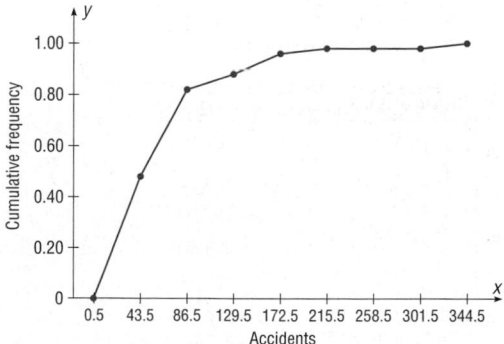

15.

Class boundaries	rf
79.5–108.5	0.17
108.5–137.5	0.28
137.5–166.5	0.04
166.5–195.5	0.20
195.5–224.5	0.22
224.5–253.5	0.04
253.5–282.5	0.04
	0.99

	crf
Less than 79.5	0.00
Less than 108.5	0.17
Less than 137.5	0.45
Less than 166.5	0.49
Less than 195.5	0.69
Less than 224.5	0.91
Less than 253.5	0.95
Less than 282.5	0.99*

*Due to rounding.

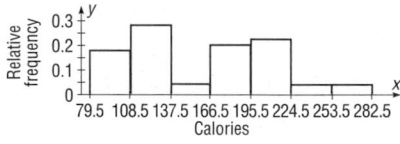

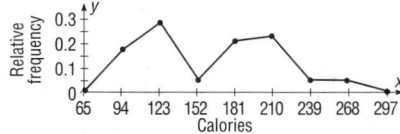

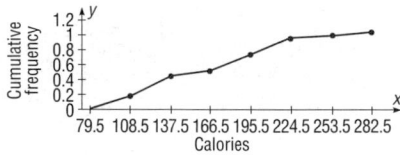

The histogram has two peaks.

16.

Class boundaries	rf
11.5–19.5	0.175
19.5–27.5	0.425
27.5–35.5	0.250
35.5–43.5	0.100
43.5–51.5	0.025
51.5–59.5	0.025
	1.000

	cf
Less than 11.5	0
Less than 19.5	0.175
Less than 27.5	0.600
Less than 35.5	0.850
Less than 43.5	0.950
Less than 51.5	0.975
Less than 59.5	1.000

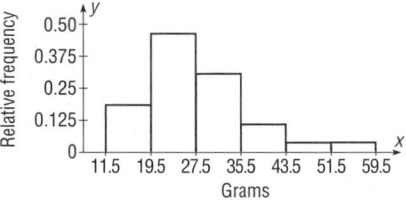

The histogram is positively skewed.

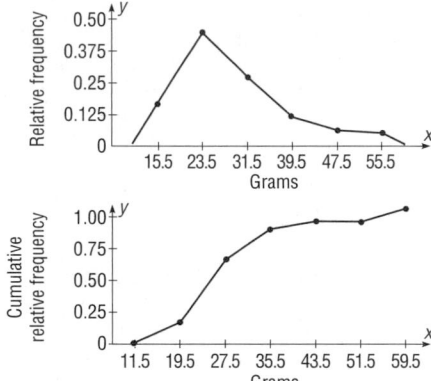

17.

Class boundaries	rf
−0.5–27.5	0.87
27.5–55.5	0.03
55.5–83.5	0.00
83.5–111.5	0.03
111.5–139.5	0.00
139.5–167.5	0.03
167.5–195.5	0.03
	0.99

	crf
Less than −0.5	0.00
Less than 27.5	0.87
Less than 55.5	0.90
Less than 83.5	0.90
Less than 111.5	0.93
Less than 139.5	0.93
Less than 167.5	0.96
Less than 195.5	0.99

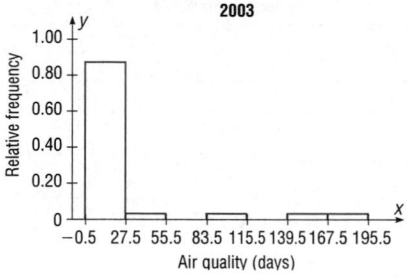

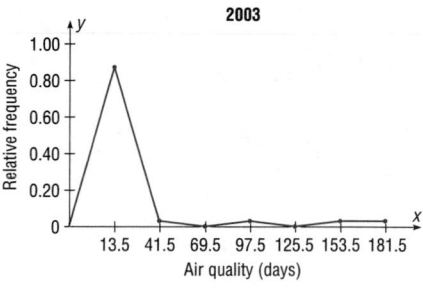

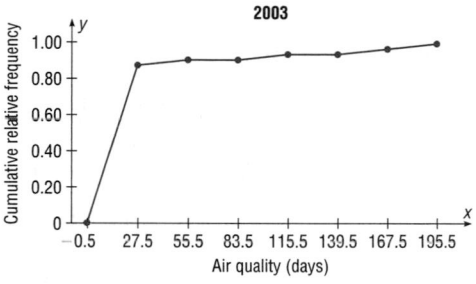

18. Based on the histograms, the older dogs have longer reaction times to the stimulus. Also, the spread (variability) is somewhat smaller for the older dogs.

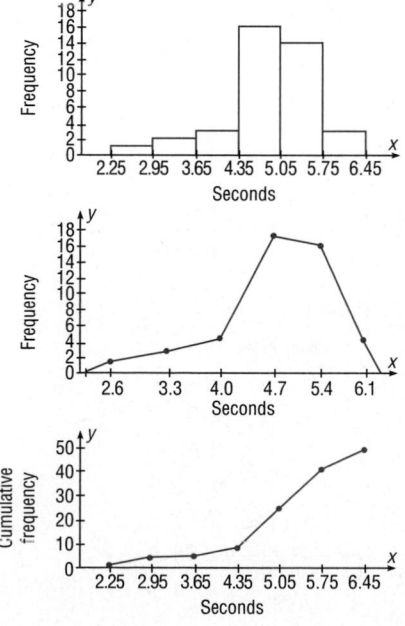

19. *a.*

Limits	Boundaries	Midpoints	*f*
22–24	21.5–24.5	23	1
25–27	24.5–27.5	26	3
28–30	27.5–30.5	29	0
31–33	30.5–33.5	32	6
34–36	33.5–36.5	35	5
37–39	36.5–39.5	38	3
40–42	39.5–42.5	41	2

	cf
Less than 21.5	0
Less than 24.5	1
Less than 27.5	4
Less than 30.5	4
Less than 33.5	10
Less than 36.5	15
Less than 39.5	18
Less than 42.5	20

b.

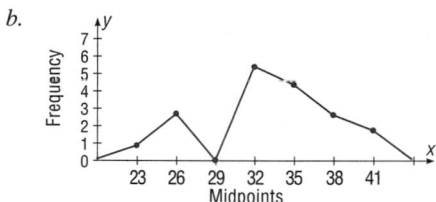

c.

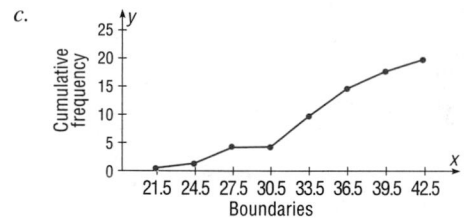

20. *a.* 0   *b.* 14   *c.* 10   *d.* 16

Exercises 2–3

1.

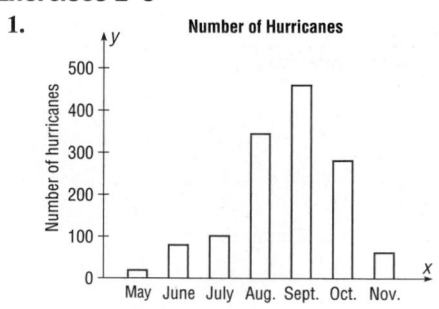

2.

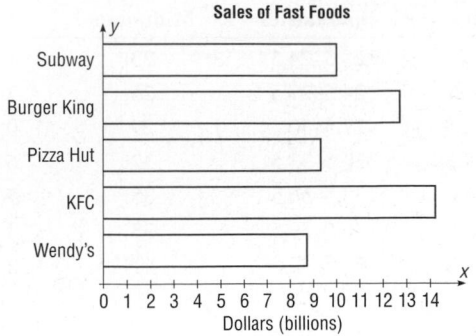

Sales of Fast Foods

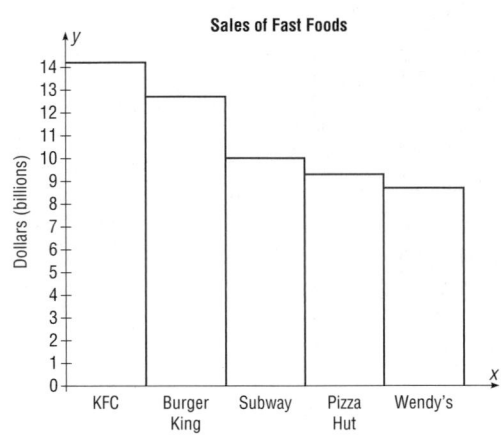

Sales of Fast Foods

3.

Calories Burned While Exercising

4.

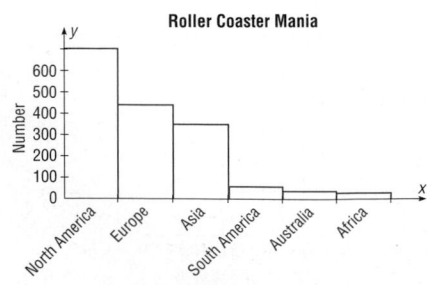

Roller Coaster Mania

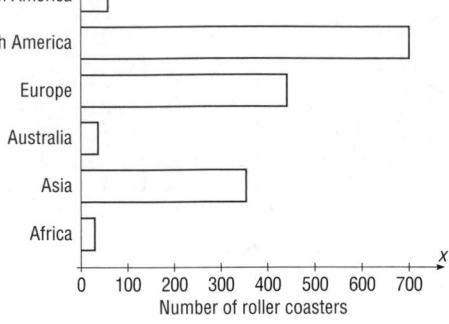

Roller Coaster Mania

5.

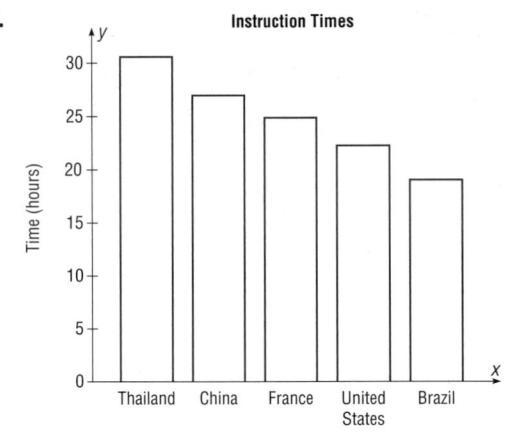

Instruction Times

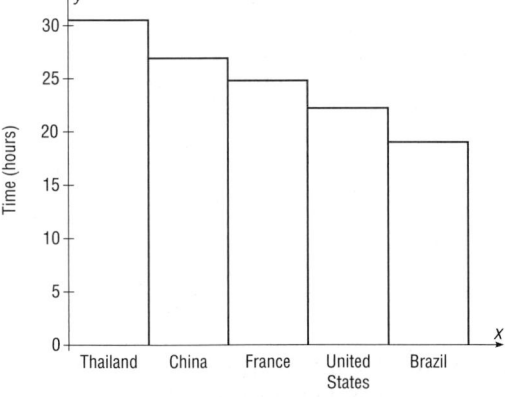

Instruction Times

6. The sales are increasing.

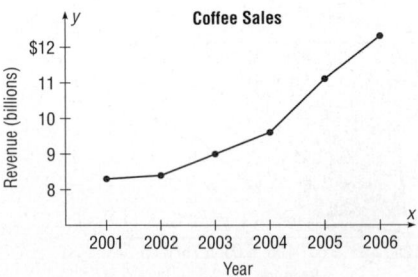

Coffee Sales

7.

Safety Record of U.S. Airlines

8.

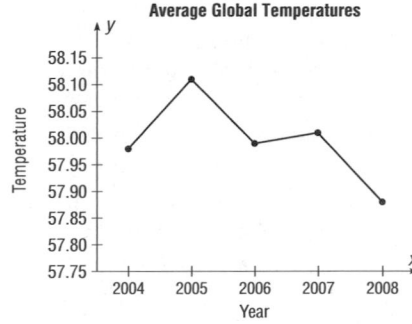

Average Global Temperatures

After a slight increase in 2005, the average temperature has declined somewhat in the following years.

9. The atmospheric concentration of carbon dioxide has been steadily increasing over the years.

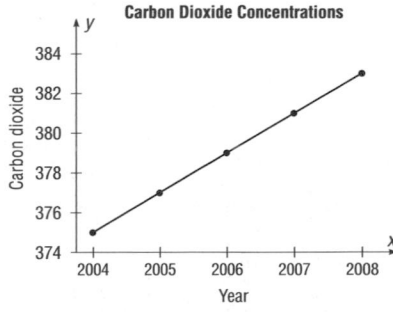

Carbon Dioxide Concentrations

10. About one-third of the travelers visit friends or relatives, and the fewest travel for personal business.

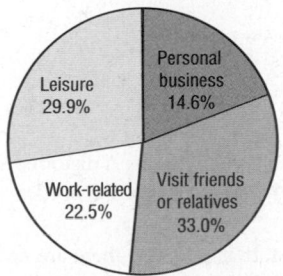

Reasons for Travel

11.

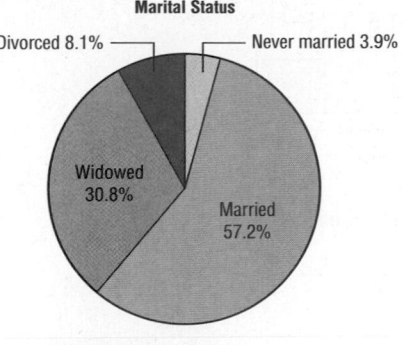

Marital Status

Educational Attainment

H. S. graduate
Some college
Bachelor's/advanced degree
Less than 9th grade
Grades 9–12 but no diploma

12.

White	19%	68.4°
Silver	18%	64.8°
Black	16%	57.6°
Red	13%	46.8°
Gray	12%	43.2°
Blue	12%	43.2°
Other	10%	36.0°

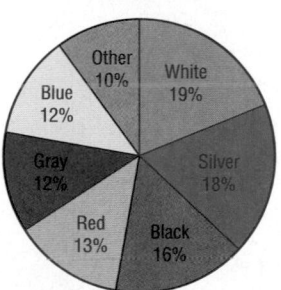

Popular Vehicle Colors

13. The pie graph better represents the data since we are looking at parts of a whole.

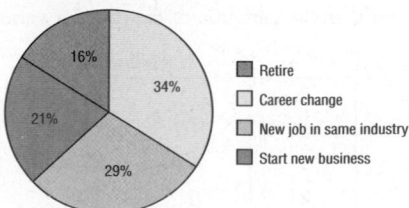

Workers Who Switch Jobs

Retire
Career change
New job in same industry
Start new business

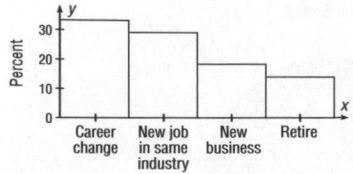

14. *a.* Time series graph

    b. Pie graph

    c. Pareto chart

    d. Pie graph

    e. Time series graph

    f. Pareto chart

15. The distribution is somewhat symmetric and unimodal and has a peak in the 50s.

```
4 | 2 3
4 | 6 6 7 7 8 9 9
5 | 0 1 1 1 1 1 2 2 4 4 4 4 4
5 | 5 5 5 5 6 6 6 7 7 7 7 8
6 | 0 1 1 1 2 4 4
6 | 5 8 9
```

16.
```
10 | 0 0 0 0 0
11 | 0 0 5
12 | 0 0 0 0 0 0
13 | 0 0 0 0 0
14 | 0 0 0 0 5 5 5
15 | 0 0 0
16 | 0 0 0 0 0
17 | 0
18 | 0 0
19 | 0
```

17.

Variety 1		Variety 2
2	1	3 8
3 0	2	5
9 8 8 5 2	3	6 8
3 3 1	4	1 2 5 5
9 9 8 5 3 3 2 1 0	5	0 3 5 5 6 7 9
	6	2 2

The distributions are somewhat similar in their shapes; however, the variation of the data for variety 2 is slightly larger than the variation of the data for variety 1.

18.

Math		Reading
9 9 9 7 5 5 2	5	
9 8 6 3 2 1	6	1 1 5 6 6 7 9
6 4 3 3 2	7	0 0 1 6 6 6 7 7 7 8
	8	0

19. Answers will vary.

20. The United States has many more launches than Japan. The number of space launches by Japan is relatively stable for the period. The number of launches for the United States dropped in 1995 and then increased after that.

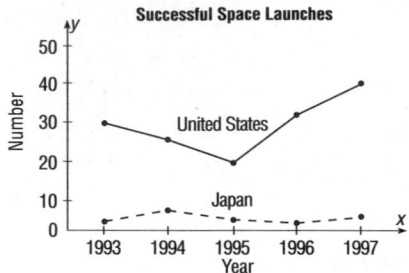

21. Production of both veal and lamb is decreasing with the exception of 1990, where both show an increase.

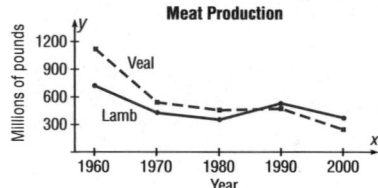

22. A Pareto chart is most appropriate.

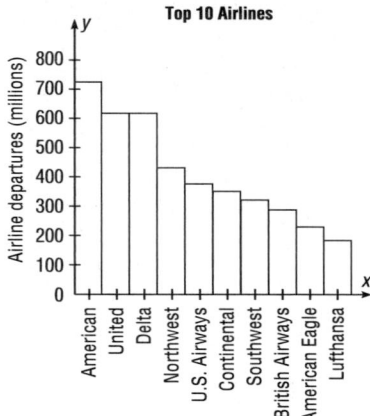

23.

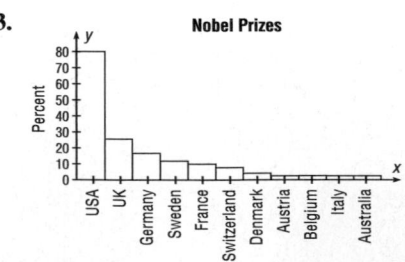

24. The bottle for 2004 is much wider, giving a distorted view of the difference since only the heights of the bottles should be compared.

25. The values on the *y* axis start at 3.5. Also there are no data values shown for the years 2004 through 2011.

Review Exercises

1.

Class	f
Newspaper	10
Television	16
Radio	12
Internet	12
	50

2. The graph shows that the percentage of the people who receive their news by television is larger than the percentage who receive their news by other means.

How People Receive News

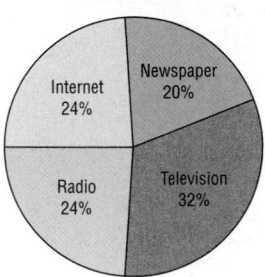

Internet 24%
Newspaper 20%
Radio 24%
Television 32%

3.

Class	f
Baseballs	4
Golf balls	5
Tennis balls	6
Soccer balls	5
Footballs	5
	25

4. The percentage of tennis balls sold was the largest of any group.

Ball Sales

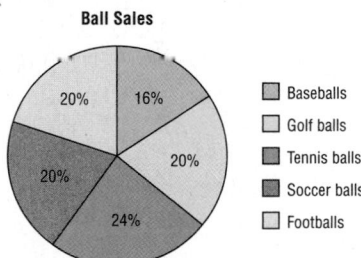

16%
20%
20%
24%
20%

- Baseballs
- Golf balls
- Tennis balls
- Soccer balls
- Footballs

5.

Class	f
11	1
12	2
13	2
14	2
15	1
16	2
17	4
18	2
19	2
20	1
21	0
22	1
	20

	cf
Less than 10.5	0
Less than 11.5	1
Less than 12.5	3
Less than 13.5	5
Less than 14.5	7
Less than 15.5	8
Less than 16.5	10
Less than 17.5	14
Less than 18.5	16
Less than 19.5	18
Less than 20.5	19
Less than 21.5	19
Less than 22.5	20

6. The distribution is somewhat uniform with the exception of the class 16.5–17.5 where it is peaked. There is a gap for the class 20.5–21.5.

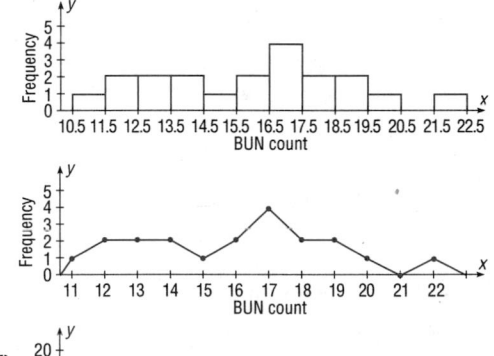

7.

Class limits	Class boundaries	f
15–19	14.5–19.5	3
20–24	19.5–24.5	18
25–29	24.5–29.5	18
30–34	29.5–34.5	8
35–39	34.5–39.5	3
		50

	cf
Less than 14.5	0
Less than 19.5	3
Less than 24.5	21
Less than 29.5	39
Less than 34.5	47
Less than 39.5	50

8.

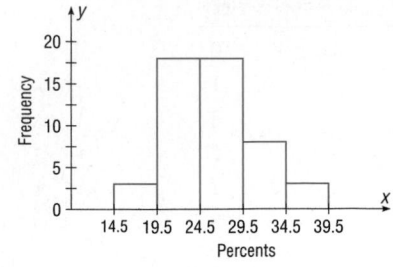

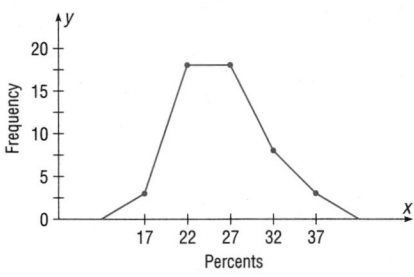

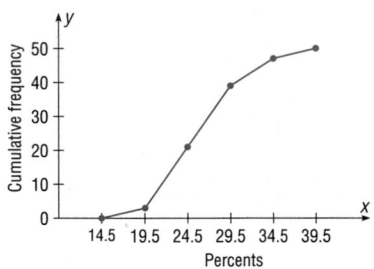

9.

Class limits	Class boundaries	f
170–188	169.5–188.5	11
189–207	188.5–207.5	9
208–226	207.5–226.5	4
227–245	226.5–245.5	5
246–264	245.5–264.5	0
265–283	264.5–283.5	0
284–302	283.5–302.5	0
303–321	302.5–321.5	1
		30

	cf
Less than 169.5	0
Less than 188.5	11
Less than 207.5	20
Less than 226.5	24
Less than 245.5	29
Less than 264.5	29
Less than 283.5	29
Less than 302.5	29
Less than 321.5	30

10. The distribution is peaked in the class of $169.50–$188.50. Most of the data cluster on the left side of the distribution. There is a large gap, and the value 320 may be an outlier.

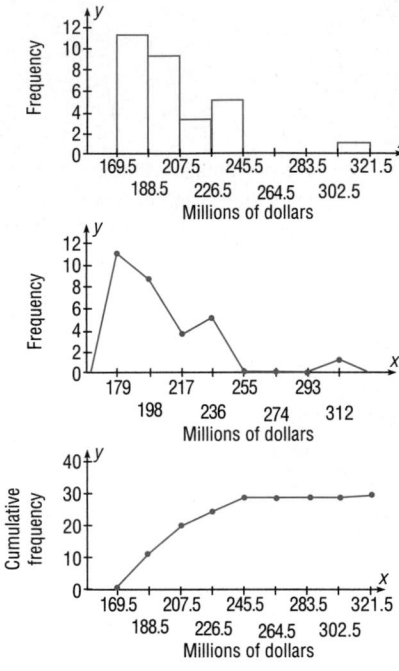

11.

Limits	Boundaries	rf
51–59	50.5–59.5	0.125
60–68	59.5–68.5	0.300
69–77	68.5–77.5	0.275
78–86	77.5–86.5	0.200
87–95	86.5–95.5	0.050
96–104	95.5–104.5	0.050
		1.000

	crf
Less than 50.5	0.000
Less than 59.5	0.125
Less than 68.5	0.425
Less than 77.5	0.700
Less than 86.5	0.900
Less than 95.5	0.950
Less than 104.5	1.000

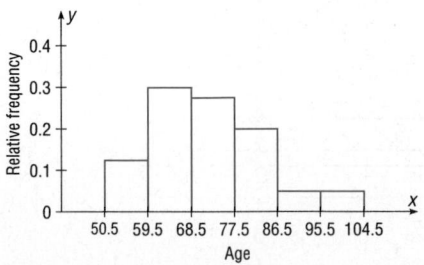

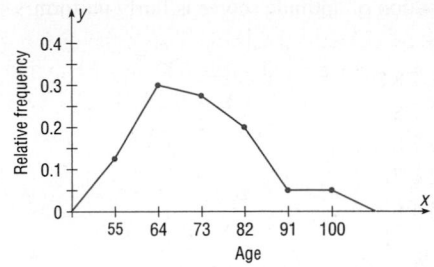

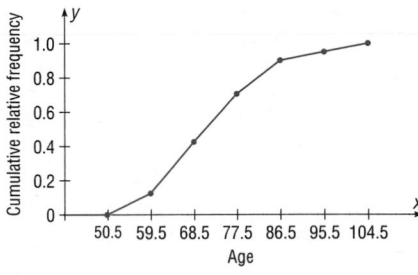

12.

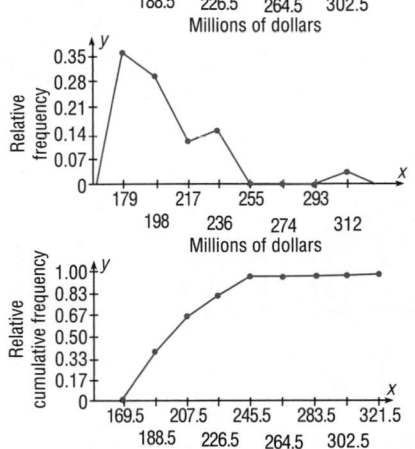

13.

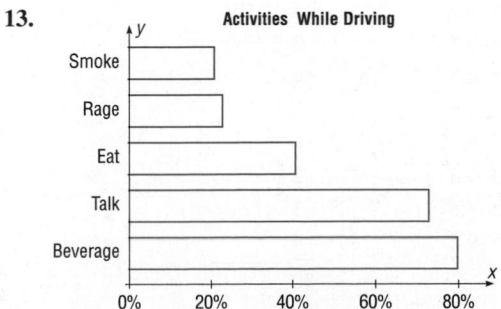

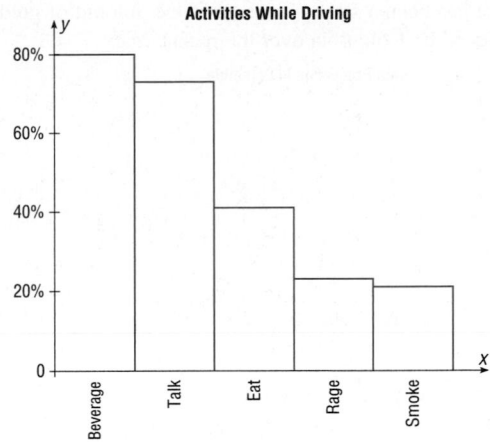

Activities While Driving

14.

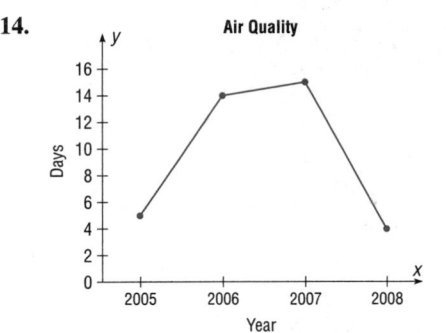

15. The bank failures increased in 2002 from 4 to 11, then dropped until 2008, when they increased to 28. The year 2009 brought an increase to 98.

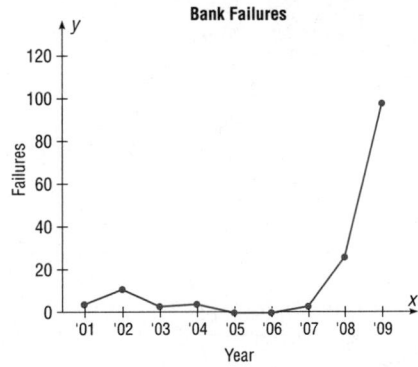

16.

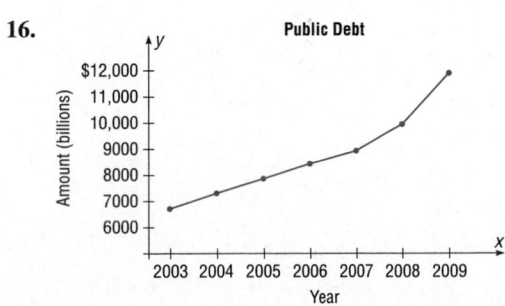

17. There has been a steady increase in the amount of gold produced by Colombia over the recent years.

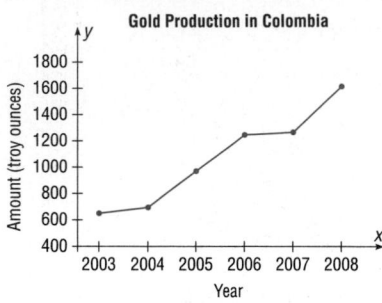

18.

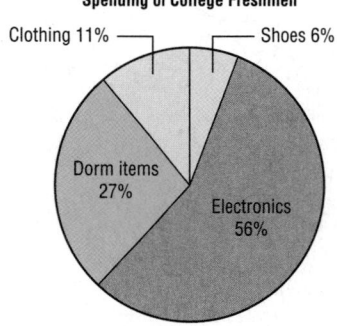

19.

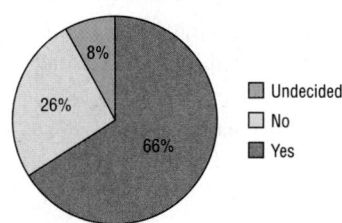

20.
```
2 | 9 9
3 | 2 4 5 6 8 8
4 | 1 2 3 7 7
5 | 1 3 5 8
6 | 2 2 2 3 7
7 | 2 3
```

21.
```
10 | 2 8 8
11 | 3
12 |
13 |
14 | 2 4
15 |
16 |
17 | 6 6 6
18 | 4 9
19 | 2
20 | 5 9
21 | 0
```

22. The distribution of aptitude scores is fairly uniform.
```
20 | 0 4 9
21 | 0 1 2 7 8 8
22 | 2 7 7 7 8
23 | 0 1 3 7 8
24 | 1 2 2 3 7
25 | 1 1 3 4 6
26 | 0
```

Chapter Quiz

1. False
2. False
3. False
4. True
5. True
6. False
7. False
8. *c*
9. *c*
10. *b*
11. *b*
12. Categorical, ungrouped, grouped
13. 5, 20
14. Categorical
15. Time series
16. Stem and leaf plot
17. Vertical or *y*

18.

Class	*f*
H	6
A	5
M	6
C	8
	25

19.

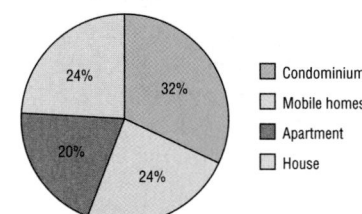

20.

Class boundaries	*f*
0.5–1.5	1
1.5–2.5	5
2.5–3.5	3
3.5–4.5	4
4.5–5.5	2
5.5–6.5	6
6.5–7.5	2
7.5–8.5	3
8.5–9.5	4
	30

	cf
Less than 0.5	0
Less than 1.5	1
Less than 2.5	6
Less than 3.5	9
Less than 4.5	13
Less than 5.5	15
Less than 6.5	21
Less than 7.5	23
Less than 8.5	26
Less than 9.5	30

21.

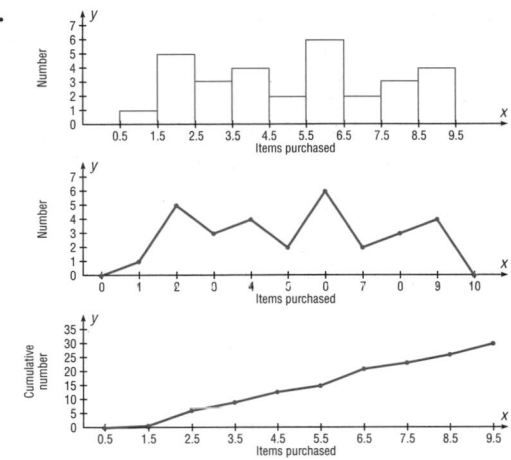

22.

Class limits	f	Class boundaries
27–90	13	26.5–90.5
91–154	2	90.5–154.5
155–218	0	154.5–218.5
219–282	5	218.5–282.5
283–346	0	282.5–346.5
347–410	2	346.5–410.5
411–474	0	410.5–474.5
475–538	1	474.5–538.5
539–602	2	538.5–602.5
	25	

23. The distribution is positively skewed with one more than one-half of the data values in the lowest class.

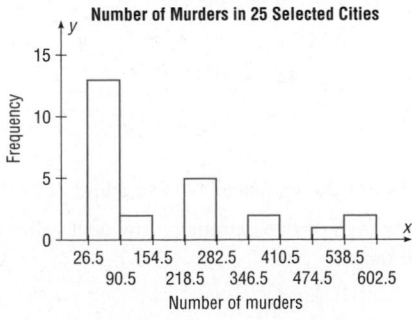

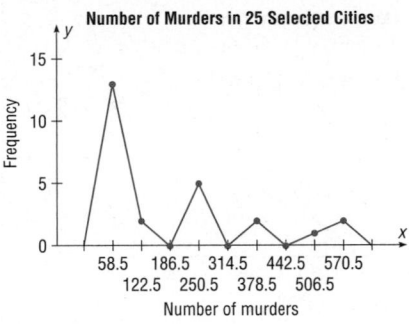

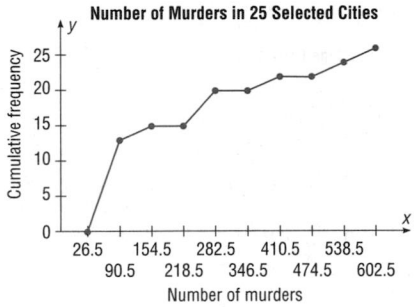

24.

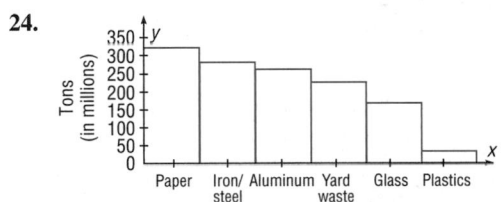

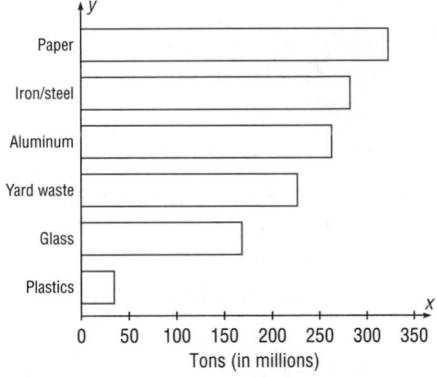

25.

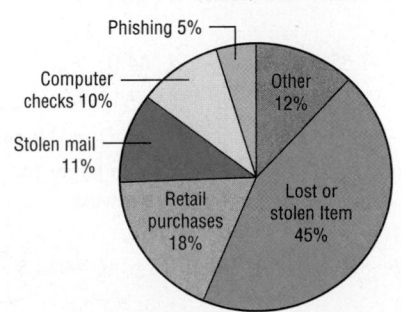

26.

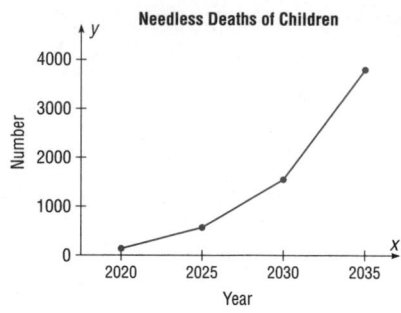

Needless Deaths of Children

Based on these data, it appears that the population is declining.

27.

1	5 9
2	6 8
3	1 5 8 8 9
4	1 7 8
5	3 3 4
6	2 3 7 8
7	6 9
8	6 8 9
9	8

Chapter 3

Exercises 3–1

1. *a.* 3.724 *b.* 3.73 *c.* 3.74 and 3.70 *d.* 3.715

2. *a.* 3174.6 *b.* 1479 *c.* No mode *d.* 5012.5

3. *a.* 68.1 *b.* 68 *c.* 42, 62, 64, 66, 72, 74
d. 64.5

4. Observers:
a. 380.4 *b.* 365 *c.* No mode
d. 393. These values are higher.

Visits:

a. 276.9 *b.* 206.5 *c.* No mode *d.* 374

5. *a.* 9422.2 *b.* 8988 *c.* 7552, 12,568, 8632
d. 9434. Claim seems a little high.

6. *a.* 19 *b.* 10 *c.* 7
d. 28.5 (Isn't it cool that Albert Einstein is on this list?)

7. *a.* 6.63 *b.* 6.45 *c.* 5.4, 6.2, 6.4, 7.2
d. 6.7; answers will vary

8. 24.42; 23.45; 16.9, 17.2, 18, 19.1, 24, 25.2, 31.7; 32.1.
It appears that the mean and median are good measures
of the average.

9. *a.* 46.78 *b.* 47.65 *c.* None *d.* 44.05

10. New England: *a.* 2451.5 *b.* 1453.5 *c.* No mode
Northwest: *a.* 569.8 *b.* 396 *c.* No mode
The measures of central tendency are much larger for
New England compared to those for Northwest.

11. 2004:
a. 8421.2 *b.* 8197 *c.* No mode *d.* 9984.5
1990:
a. 9810 *b.* 9214.5 *c.* No mode *d.* 13345.5

12. *a.* 5 *b.* 3.5–6.5

13. *a.* 17.68 *b.* 2.48–7.48 and 17.51–22.51.
Group mean is less.

14. *a.* 19.7 *b.* 17.5–22.5

15. *a.* 6.5 *b.* 0.8–4.4. Probably not—data are
"top heavy."

16. Younger dogs $\overline{X} = 3.83$; modal class 2.95–3.65
Older dogs $\overline{X} = 4.85$; modal class 4.35–5.05
The means are different.

17. *a.* 26.7 *b.* 24.2–28.6

18. *a.* 42.9 *b.* 32–42

19. *a.* 34.1 *b.* 0.5–19.5

20. *a.* 180.3 *b.* 177–185

21. *a.* 23.7 *b.* 21.5–24.5

22. *a.* 14.6 *b.* 0–10

23. 44.8; 40.5–47.5

24. *a.* 64.4 *b.* 3–45 and 46–88

25. *a.* 1804.6 *b.* 1013–1345

26. $9866.67

27. 2.896 **28.** 35.4%

29. $545,666.67 **30.** 83.2

31. 82.7

32. *a.* Mode *c.* Median *e.* Mean
b. Median *d.* Mode *f.* Median

33. *a.* Median *c.* Mode *e.* Mode
b. Mean *d.* Mode *f.* Mean

34. Roman letters, $\overline{X}$; Greek letters, μ

35. Both could be true since one may be using the mean for
the average salary and the other may be using the mode
for the average.

36. 320 **37.** 6

38. *a.* 40 *b.* 20 *c.* 300 *d.* 3
e. The results will be the same as if you add, subtract,
multiply, and divide the mean by 10.

39. *a.* 36 mph *b.* 30.77 mph *c.* $16.67

40. *a.* 25.5% *b.* 5.7% *c.* 8.4% *d.* 3.2%

41. 5.48 **42.** 4.31

Exercises 3–2

1. The square root of the variance is the standard deviation.

2. One extremely high or one extremely low data value will
influence the range.

3. σ^2; σ **4.** s^2; s

5. When the sample size is less than 30, the formula for the variance of the sample will underestimate the population variance.

6. No, *a* has the smallest variation; *c* has the biggest variation.

7. 48; 254.7; 15.9 (rounded to 16)   The data vary widely.

8. 62; 332.4; 18.2; using the range rule of thumb, $s \approx 15.5$. This is close to the actual standard deviation of 18.2.

9.

	Temp. (°F)	Precip. (inches)
Range	32	4
Variance	147.7	1.89
Standard deviation	12.15	1.373

The temperatures are more variable.

10. The surface area for the Western states is more variable since the standard deviation is 16,178.4 as compared to 6440.2 for the Eastern states.

11. Houston: $\overline{X} = 55.8$, $s = 8.88$, CVar $= 15.91\%$. Pittsburgh: $\overline{X} = 41.5$, $s = 9.42$, CVar $= 22.7\%$. Pittsburgh is more variable.

12. Europe: $\overline{X} = 34{,}637$, $s = 7609.8$, CVar $= 21.97\%$. Asia: $\overline{X} = 16{,}326.3$, $s = 8054.5$, CVar $= 49.33\%$. Asia is more variable.

13. $s \approx R/4$ so $s \approx 5$ years.

14. *a.* 22   *b.* 35.5   *c.* 5.96

15. *a.* 160   *b.* 1984.5   *c.* 44.5

16. *a.* 2721   *b.* 355,427.6   *c.* 596.2

17. *a.* 46   *b.* 77.48   *c.* 8.8

18. NL: $s^2 = 0.00004$, $s = 0.0066$
AL: $s^2 = 0.0000476$, $s = 0.0069$

19. 133.6; 11.6   20. 25.7; 5.1

21. 27,941.46; 167.2   22. 0.847; 0.920

23. 167.2; 12.93   24. 134.3; 11.6

25. 211.2; 14.5; no, the variability of the lifetimes of the batteries is quite large.

26. Younger dogs: 1.1; 1.0; older dogs: 0.6; 0.8; the variability of the reaction times of the younger dogs is greater than the variability of that of the older dogs.

27. 11.7; 3.4

28. 20.9%; 22.5%. The factory workers' data are more variable.

29. United States: $\overline{X} = 3386.6$, $s = 693.9$, CVar $= 20.49\%$. World: $\overline{X} = 4997.8$, $s = 803.2$, CVar $= 16.07\%$. The United States is more variable.

30. 13.1%; 15.2%. The waiting time for people who are discharged is more variable.

31. 23.1%; 12.9%; age is more variable.

32. *a.* 75%   *b.* 56%

33. *a.* 96%   *b.* 93.75%

34. At least 93.75%

35. Between 164 and 316 calories

36. Between 84 and 276 minutes

37. Between 385 and 895 pounds

38. Between $149,300 and $343,300

39. 86%   40. At least 84%

41. 16%

42. *a.* No more than 12.5%   *b.* 2.5%

43. All the data values fall within 2 standard deviations of the mean.

44. 93.3%   All but two data values fall within 2 standard deviations of the mean.

45. 56%; 75%; 84%; 88.89%; 92%

46. *a.* 15.81   *c.* 15.81   *e.* 3.16
b. 15.81   *d.* 79.06
f. The standard deviation is unchanged when a specific number is added to or subtracted from each data value. If each data value is multiplied by a number, the standard deviation increases by the number times the original standard deviation. For division, the standard deviation is divided by the number.
g. When the same number is added to or subtracted from each data value, the mean will increase or decrease by that number, but the standard deviation will remain unchanged. When each data value is multiplied by the same number, the mean or standard deviation will be equal to that number times the original mean or standard deviation. When each data value is divided by the same number, the mean or standard deviation will be equal to the original mean or standard deviation divided by that number.

47. 4.36

48. *a.* 2, positively skewed
b. -2.25, negatively skewed
c. 0, symmetric
d. 0.3, positively skewed

49. It must be an incorrect data value, since it is beyond the range using the formula $s\sqrt{n-1}$.

Exercises 3–3

1. A z score tells how many standard deviations the data value is above or below the mean.

2. A percentile rank indicates the percentage of data values that fall below the specific rank.

3. A percentile is a relative measurement of position; a percentage is an absolute measure of the part to the total.

4. A quartile is a relative measure of position obtained by dividing the data set into quarters.

5. $Q_1 = P_{25}$; $Q_2 = P_{50}$; $Q_3 = P_{75}$

6. A decile is a relative measure of position obtained by dividing the data set into tenths.

7. $D_1 = P_{10}$; $D_2 = P_{20}$; $D_3 = P_{30}$; etc.

8. P_{50}; Q_2; D_5

9. Canada -0.40, Italy 1.47, United States -1.91

10. Byrd: $z = 2.30$ Sununu: $z = -1.70$

11. *a.* 0.6 *b.* -1.2 *c.* 2.4 *d.* -2.2 *e.* 0.2

12. *a.* $74,566   *b.* $43,966   *c.* $54,166   *d.* $79,666
    e. $37,846

13. Neither; $z = 1.5$ for each

14. 0.64; 0.95. The student from the university has a higher relative debt.

15. *a.* -0.93 *b.* -0.85 *c.* -1.4; score in part *b* is highest

16. *a.* $5806 *b.* $6563   *c.* $7566   *d.* $8563

17. *a.* 24th   *b.* 67th   *c.* 48th   *d.* 88th

18. *a.* 6   *b.* 24   *c.* 68   *d.* 76   *e.* 94

19. *a.* 234   *b.* 251   *c.* 263   *d.* 274   *e.* 284

20. *a.* 375   *b.* 389   *c.* 433   *d.* 477   *e.* 504

21. *a.* 13th   *b.* 40th   *c.* 54th   *d.* 76th   *e.* 92nd

22. 94th; 72nd; 61st; 17th; 83rd; 50th; 39th; 28th; 6th

23. 597

24. 7th; 21st; 36th; 50th; 64th; 79th; 93rd

25. 47

26. 5th; 15th; 25th; 35th; 45th; 55th; 65th; 75th; 85th; 95th

27. 2.1

28. 8th; 25th; 42nd; 58th; 75th; 92nd

29. 12

30. *a.* 3   *c.* None   *e.* 145
    b. 54   *d.* None   *f.* None

31. *a.* 12; 20.5; 32; 22; 20   *b.* 62; 94; 99; 80.5; 37

Exercises 3–4

1. 6, 8, 19, 32, 54; 24

2. 7, 11.5, 19, 35, 48; 23.5

3. 188, 192, 339, 437, 589; 245

4. 147, 156, 273, 543, 632; 387

5. 14.6, 15.05, 16.3, 19, 19.8; 3.95

6. 2.2, 3.7, 4.6, 9.4, 9.7; 5.7

7. 11, 3, 8, 5, 9, 4

8. 325, 200, 275, 225, 300, 75

9. 95, 55, 70, 65, 90, 25

10. 6000, 2000, 4000, 3000, 5000; 2000

11.
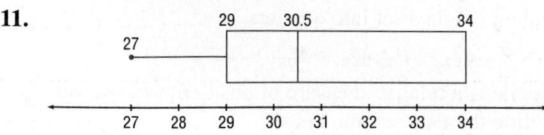

12. The distribution is slightly left-skewed.

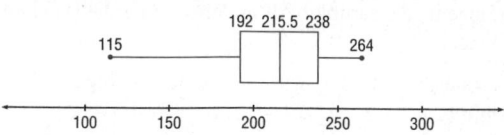

13. The graph of the data is somewhat positively skewed.

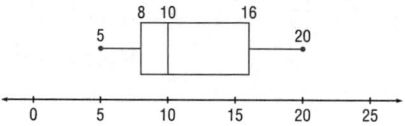

14. The graph of the data is somewhat positively skewed.

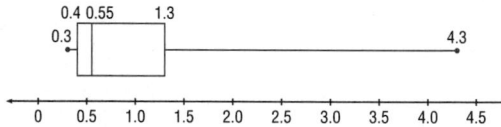

15. Based on the median, the data are left-skewed. Based on the lines, the data are right-skewed.

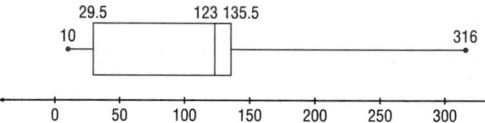

16. The range and variation of the capacity of the dams in South America are considerably larger than those of the United States.

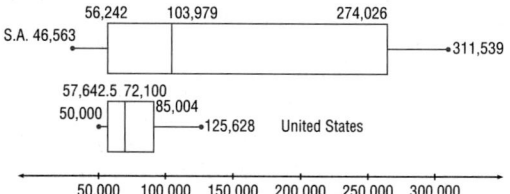

17. *a.* May: 391.7   *b.* 2003: 289.8

c.

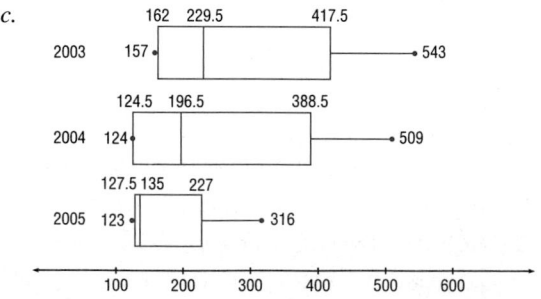

18.

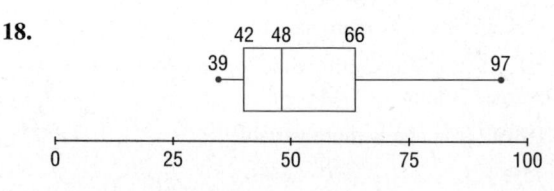

There are no outliers.

Review Exercises

1. $\bar{X} = 27.2$, MD $= 19$, mode $= 17$, MR $= 38$, $R = 42$, $S^2 = 239.96$, $S = 15.5$

2. Attacks:
   $\bar{X} = 63.6$; MD $= 64$; No mode; MR $= 64$; $R = 14$; $S^2 = 26.80$; $S = 5.2$, C.Var $= 8.18\%$

   Deaths:
   $\bar{X} = 4$, MD $= 4$, mode $= 4$, MR $= 4$, $R = 6$, $S^2 = 4.50$, $S = 2.1$, C.Var $= 52.5\%$ Deaths are more variable.

3. *a.* 7.3   *b.* 7–9   *c.* 10.0   *d.* 3.2

4. $\bar{X} = 531.5$, modal class 505–531, $s^2 = 1360.8$, $s = 36.9$, skewed right

5. *a.* 55.5   *b.* 57.5–72.5   *c.* 566.1   *d.* 23.8

6. *a.* 18.5   *b.* 19–21   *c.* 17.7   *d.* 4.2

7. 1.43 viewers

8. $4700.00

9. 6

10. 31.25%; 18.6%; the number of books is more variable

11. Magazine variance: 0.214; year variance: 0.417; years are more variable

12. *a.* 59th; 32nd; 41st; 77th; 50th; 14th; 5th; 86th; 68th; 23rd; 95th

    b. 40th percentile: 16

    The distribution is generally symmetric.

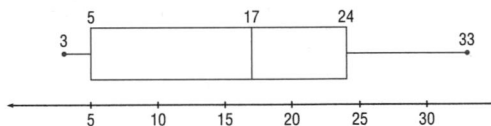

13. *a.*

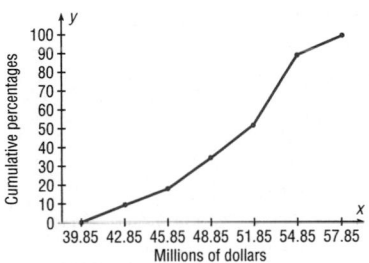

    b. 50, 53, 55

    c. 10th; 26th; 78th

14. *a.* 400   *c.* None

    b. None   *d.* None

15. $0.26–$0.38

16. *a.* Nothing because $k = 1$

    b. At most ¼ or 25%   *c.* At most 7.3%

17. 56%   18. 83%

19. 88.89%

20. *a.* -0.5 *b.* -0.8

    The test in part *a* is better.

21. The range is much larger.

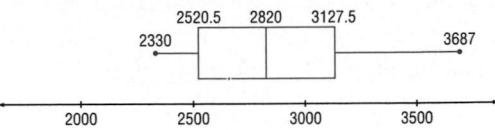

22. The employees worked more hours before Christmas than after Christmas. Also, the range and variability of the distribution of hours are greater before Christmas.

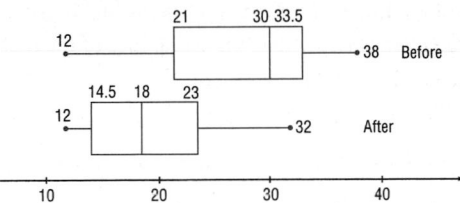

23. $23.7 - 35.7$

Chapter Quiz

1. True   2. True

3. False   4. False

5. False   6. False

7. False   8. False

9. False   10. *c*

11. *c*   12. *a* and *b*

13. *b*   14. *d*

15. *b*   16. Statistic

17. Parameters, statistics   18. Standard deviation

19. σ 20. Midrange

21. Positively   22. Outlier

23. *a.* 15.3   *c.* 15, 16, and 17   *e.* 6   *g.* 1.9

    b. 15.5   *d.* 15   *f.* 3.57

24. *a.* 6.4   *b.* 6–8   *c.* 11.6   *d.* 3.4

25. *a.* 51.4   *b.* 35.5–50.5   *c.* 451.5   *d.* 21.2

26. *a.* 8.2   *b.* 7–9   *c.* 21.6   *d.* 4.6

27. 1.6   28. 4.5

29. 0.33; 0.162; newspapers   30. 0.3125; 0.229; brands

31. -0.75; -1.67; science

32. *a.* 0.5   *b.* 1.6   *c.* 15, *c* is highest

33. *a.* 56.25; 43.75; 81.25; 31.25; 93.75; 18.75; 6.25; 68.75

    b. 0.9

    c.

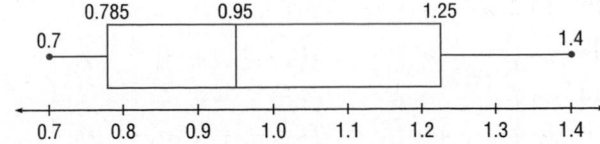

34. *a.*

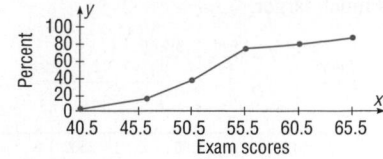

    b. 47; 55; 64

    c. 56th, 6th, 99th percentiles

35. The cost of prebuy gas is much less than that of the return without filling gas. The variability of the return without filling gas is larger than the variability of the prebuy gas.

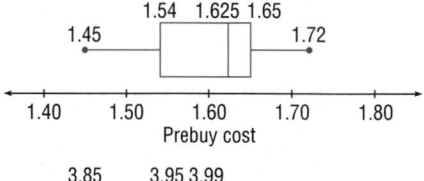

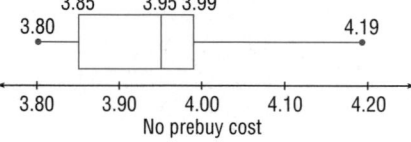

36. 16%, 97.5%

Chapter 4

Exercises 4–1

1. A probability experiment is a chance process that leads to well-defined outcomes.

2. The set of all possible outcomes of a probability experiment is called a sample space.

3. An outcome is the result of a single trial of a probability experiment, but an event can consist of more than one outcome.

4. Equally likely events have the same probability of occurring.

5. The range of values is 0 to 1 inclusive.

6. 1

7. 0

8. 1

9. 0.80   Since the probability that it won't rain is 80%, you could leave your umbrella at home and be fairly safe.

10. *c, d, e, h*

11. *a.* Empirical   *d.* Classical   *f.* Empirical
    b. Classical   *e.* Empirical   *g.* Subjective
    c. Empirical

12. *a.* $\frac{1}{6}$ *c.* $\frac{1}{2}$ *e.* 1   *g.* $\frac{1}{3}$
    b. 0   *d.* $\frac{2}{3}$ *f.* $\frac{2}{3}$

13. *a.* $\frac{1}{9}$ *b.* $\frac{2}{9}$ *c.* $\frac{1}{6}$ *d.* $\frac{13}{18}$ *e.* $\frac{1}{6}$

14. *a.* $\frac{1}{13}$ *c.* $\frac{1}{52}$ *e.* $\frac{4}{13}$ *g.* $\frac{1}{26}$ *i.* $\frac{1}{2}$
    b. $\frac{1}{4}$ *d.* $\frac{2}{13}$ *f.* $\frac{1}{52}$ *h.* $\frac{1}{26}$ *j.* $\frac{1}{2}$

15. *a.* 0.1   *b.* 0.2   *c.* 0.8

16. *a.* $\frac{4}{25}$ *b.* $\frac{19}{25}$

17. *a.* 0.43   *b.* 0.52   *c.* 0.17

18. 0.428

19. *a.* 0.04   *b.* 0.52   *c.* 0.4

20. *a.* 0.38   *b.* 0.62   *c.* 0.74

21. *a.* $\frac{1}{8}$ *b.* $\frac{1}{4}$ *c.* $\frac{3}{4}$ *d.* $\frac{3}{4}$

22. $\frac{2}{9}$ **23.** $\frac{1}{9}$

24. *a.* 0.295   *b.* 0.419   *c.* 0.093

25. *a.* 27%   *b.* 33%   *c.* 67%   *d.* 14%

26. *a.* 0.7   *b.* 1   *c.* 0

27. 0.662

28. *a.* 0.61   *b.* 0.63   *c.* 0.08

29. *a.* Sample space

	1	2	3	4	5	6
1	1	2	3	4	5	6
2	2	4	6	8	10	12
3	3	6	9	12	15	18
4	4	8	12	16	20	24
5	5	10	15	20	25	30
6	6	12	18	24	30	36

    b. $\frac{5}{12}$

    c. $\frac{17}{36}$

30. 0.6

31.

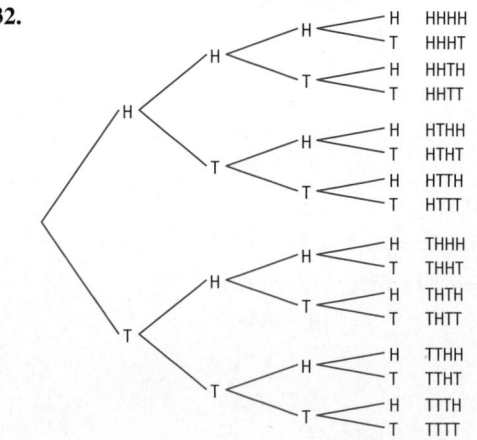

32.

33.

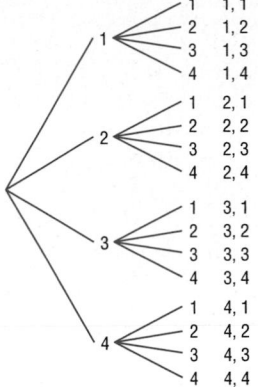

34.

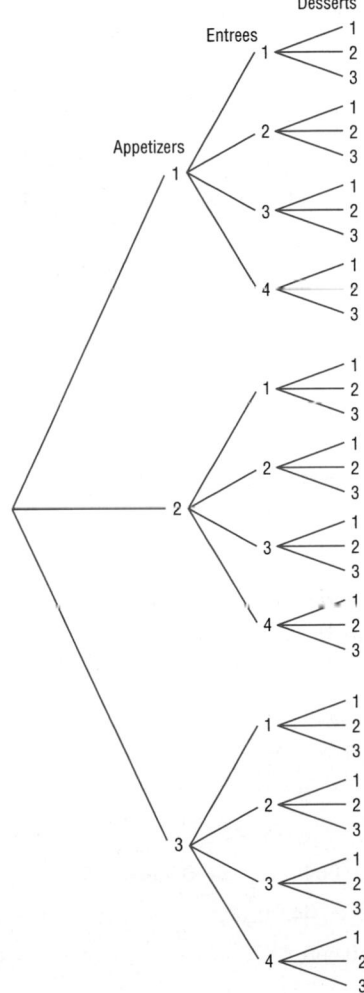

35.

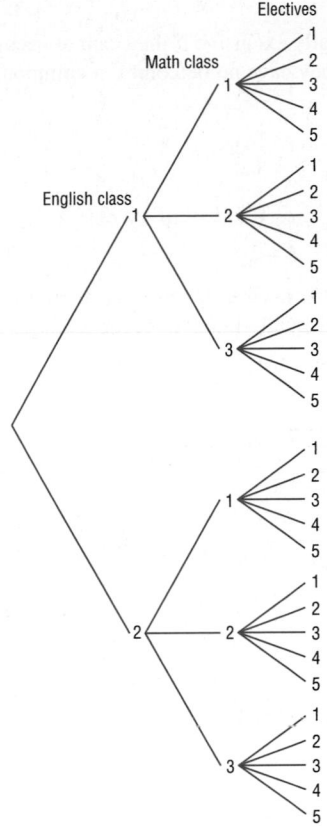

36.

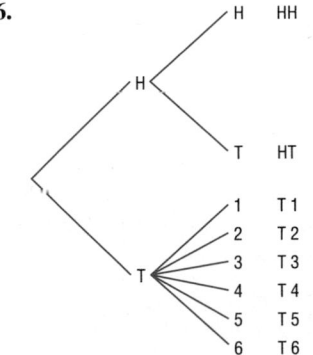

37. *a.* 0.08 *b.* 0.01 *c.* 0.35 *d.* 0.36

38. Probably

39. The statement is probably not based on empirical probability and is probably not true.

40. *a.* $\frac{6}{25}$ *b.* $\frac{2}{5}$ *c.* $\frac{9}{25}$ *d.* $\frac{12}{25}$ *e.* $\frac{1}{5}$

41. Answers will vary.

42. Approximately $\frac{1}{4}$, $\frac{1}{2}$, and $\frac{1}{4}$, respectively

43. *a.* 1:5, 5:1 *d.* 1:1, 1:1 *g.* 1:1, 1:1
  b. 1:1, 1:1 *e.* 1:12, 12:1
  c. 1:3, 3:1 *f.* 1:3, 3:1

Exercises 4–2

1. Two events are mutually exclusive if they cannot occur at the same time (i.e., they have no outcomes in common). Examples will vary.

2. *a.* No   *c.* Yes   *e.* No   *g.* Yes
   b. No   *d.* No   *f.* Yes

3. *a.* 0.707   *b.* 0.589   *c.* 0.011   *d.* 0.731

4. $\frac{26}{31}$ 5. $\frac{11}{19}$

6. $\frac{23}{49}$; the probability of the event is slightly less than 0.5, which makes it about equally likely to occur or not to occur.

7. *a.* $\frac{7}{11}$ *b.* $\frac{6}{11}$ *c.* $\frac{7}{11}$ *d.* $\frac{1}{2}$

8. 0.10 or 10%   9. 0.55

10. *a.* $\frac{4}{13}$ *b.* $\frac{1}{2}$ *c.* $\frac{7}{13}$

11. *a.* $\frac{6}{7}$ *b.* $\frac{4}{7}$ *c.* 1

12. *a.* 0.5   *b.* 0.7692   *c.* 0.6154

13. *a.* 0.058   *b.* 0.942   *c.* 0.335

14. *a.* 0.072   *b.* 0.229   *c.* 0.4856

15. *a.* 0.056   *b.* 0.004   *c.* 0.076

16. *a.* 0.301   *b.* 0.592   *c.* 0.412

17. *a.* $\frac{7}{58}$ *b.* $\frac{16}{29}$ *c.* $\frac{12}{29}$

18. *a.* $\frac{467}{1392}$ *b.* $\frac{47}{58}$ *c.* $\frac{833}{1392}$

19. *a.* $\frac{1}{15}$ *b.* $\frac{1}{3}$ *c.* $\frac{5}{6}$ *d.* $\frac{5}{6}$ *e.* $\frac{1}{3}$

20. *a.* 0.2813   *b.* 0.9375   *c.* 0.5625

21. *a.* $\frac{5}{12}$ *b.* $\frac{1}{8}$ *c.* $\frac{2}{3}$ *d.* $\frac{23}{24}$

22. *a.* $\frac{29}{100}$ *b.* $\frac{2}{3}$ *c.* $\frac{157}{300}$

23. *a.* $\frac{3}{13}$ *b.* $\frac{3}{4}$ *c.* $\frac{19}{52}$ *d.* $\frac{7}{13}$ *e.* $\frac{15}{26}$

24. *a.* $\frac{1}{3}$ *b.* $\frac{1}{3}$ *c.* $\frac{1}{4}$ *d.* Choice *c* is least likely to occur.

25. 0.318   26. *a.* $\frac{1}{36}$ *b.* $\frac{1}{36}$

27. 0.06   28. 0.10

29. 0.30   30. No. $P(A \cap B) \neq 0$

Exercises 4–3

1. *a.* Independent   *e.* Independent
   b. Dependent   *f.* Dependent
   c. Dependent   *g.* Dependent
   d. Dependent   *h.* Independent

2. 0.007; the event is very unlikely to occur since its probability is very small.

3. *a.* 0.009   *b.* 0.227

4. 7.3%

5. 0.00194 The event is highly unlikely since the probability is small.

6. 6.3%

7. 0.5139

8. *a.* 0.807   *b.* 0.194

9. 0.179

10. *a.* $\frac{1}{17}$ *b.* $\frac{4}{17}$ *c.* $\frac{1}{221}$

11. *a.* 0.003   *b.* 0.636   *c.* 0.997

12. $\frac{1}{15}$

13. *a.* $\frac{1}{270,725}$ *b.* $\frac{11}{4165}$ *c.* $\frac{46}{833}$

14. $\frac{5}{28}$ 15. $\frac{1}{56}$

16. $\frac{3}{29}$ unlikely

17. $\frac{38}{87}$ Number 20 is more likely to occur.

18. 0.116

19. *a.* 0.167   *b.* 0.406   *c.* 0.691

20. $\frac{49}{72}$

21. 0.03

22. 0.78   0.0192

23. 0.071   24. 0.1157

25. $\frac{1}{7}$ 26. 89%

27. 0.2   28. 0.9

29. 68.4%

30. *a.* 0.7143   *b.* 0.4348   *c.* 0.1558

31. *a.* 0.06   *b.* 0.4353   *c.* 0.35
    d. 0.1667

32. *a.* 0.498   *b.* 0.109
    c. No. $P(\text{path}|\text{female}) \neq P(\text{path})$

33. *a.* 0.327   *b.* 0.119   *c.* No. $P(G|\text{U.S.}) \neq P(G)$

34. *a.* 0.0954   *b.* 0.9046   *c.* 0.1601

35. *a.* 0.0197   *b.* 0.611

36. 0.231

37. *a.* 0.1717   *b.* 0.8283

38. *a.* 0.157   *b.* 0.097   *c.* 0.903

39. 0.574   40. 0.202

41. 0.9869

42. *a.* 0.216   *b.* 0.064   *c.* 0.936

43. $\frac{14,498}{20,825}$ 44. 0.131

45. *a.* 0.332   *b.* 0.668

46. 96.8%

47. $\frac{31}{32}$

48. 0.111; the event is very unlikely to occur since the probability is only about 11%.

49. 0.665 It will happen almost 67% of the time. It's somewhat likely.

50. *a.* 0.011   *b.* 0.022   *c.* 0.978

51. $\frac{7}{8}$

52. 0.678; yes the event is a little more likely to occur than not since the probability is about 68%.

53. No, since $P(A \cap B) = 0$ and does not equal $P(A) \cdot P(B)$.

54. No, since $P(C|D) \neq P(C)$.

55. Enrollment and meeting with DW and meeting with MH are dependent. Since meeting with MH has a low probability and meeting with LP has no effect, all students, if possible, should meet with DW.

56. *a.* The events are dependent, and the commercial hurts sales since the probability that a person buys the product is less than 0.35.

b. The events are independent; hence, the commercial has no effect.

c. The events are dependent, and the commercial helps since the probability that a person buys the product is higher than 0.35.

Exercises 4–4

1. 100,000; 30,240   **2.** 362,880

3. 720   **4.** 362,880

5. 5040 ways   **6.** 120; 12

7. 3,628,000   **8.** 27,600; 35,152

9. 1000; 72   **10.** 1296; 360

11. 600   **12.** 8

13. *a.* 40,320   *c.* 1   *e.* 2520   *g.* 60   *i.* 120
    b. 3,628,800   *d.* 1   *f.* 11,880   *h.* 1   *j.* 30

14. 40,320   **15.** 24

16. 210   **17.** 7315

18. 30,240   **19.** 840

20. 120   **21.** 151,200

22. 286; 378 (count 0)   **23.** 5,527,200

24. 330   **25.** 495; 11,880

26. 2520

27. *a.* 10   *c.* 35   *e.* 15   *g.* 1   *i.* 66
    b. 56   *d.* 15   *f.* 1   *h.* 36   *j.* 4

28. 22,100   **29.** 120

30. 41,580   **31.** 210

32. 120   **33.** 15,504

34. 462   **35.** 43,758; 12,870

36. 166,320   **37.** 495; 210; 420

38. 14,400   **39.** 475

40. 194,040   **41.** 2970

42. 53,130

43. $_7C_2$ is 21 combinations + 7 double tiles = 28

44. 191,100   **45.** 330

46. 1287   **47.** 194,040

48. 24,310

49. 125,970   **50.** 120

51. 1,860,480   **52.** 3080

53. 136   **54.** 45

55. 120   **56.** 300

57. 200   **58.** 126

59. 336   **60.** 15

61. 2; 6; $(n-1)!$

62. *a.* 48   *b.* 60   *c.* 72

63. *a.* 4   *b.* 36   *c.* 624   *d.* 3744

Exercises 4–5

1. $\frac{11}{221}$

2. *a.* $\frac{1}{2530}$ *b.* $\frac{38}{253}$ *c.* $\frac{969}{2530}$ *d.* $\frac{114}{253}$

3. *a.* $\frac{4}{35}$ *b.* $\frac{1}{35}$ *c.* $\frac{12}{35}$ *d.* $\frac{18}{35}$

4. 0.0659; 0.1810; 0.0289

5. *a.* 0.129   *b.* 0.107   *c.* 0.0908

6. *a.* $\frac{14}{55}$ *b.* $\frac{28}{55}$ *c.* $\frac{1}{55}$

7. $\frac{1}{1225}$ **8.** $\frac{18}{12,495} = \frac{6}{4165}$

9. *a.* 0.120   *b.* 0.296   *c.* 0.182

10. *a.* 0.002   *b.* 0.246   *c.* 0.751   *d.* 0.249

11. *a.* 0.3216   *b.* 0.1637   *c.* 0.5146
    d. It probably got lost in the wash!

12. $\frac{882}{2431}$ **13.** $\frac{5}{72}$

14. *a.* 0.0003   *b.* 0.089   *c.* 0.053   *d.* 0.496

15. $\frac{1}{60}$

16. *a.* $\frac{4}{2,598,960}$ *b.* $\frac{36}{2,598,960}$ *c.* $\frac{624}{2,598,960}$

17. 0.727

Review Exercises

1. *a.* 0.167   *b.* 0.667   *c.* 0.5

2. *a.* $\frac{1}{4}$ *b.* $\frac{11}{26}$ *c.* $\frac{1}{52}$ *d.* $\frac{1}{13}$ *e.* $\frac{1}{2}$

3. *a.* 0.7   *b.* 0.5

4. *a.* 0.333   *b.* 0.444

5. $\frac{13}{60}$

6. *a.* $\frac{9}{35}$ *b.* $\frac{23}{35}$ *c.* $\frac{19}{35}$ *d.* $\frac{19}{35}$

7. 0.19

8. *a.* $\frac{1}{4}$ *b.* $\frac{1}{6}$ *c.* $\frac{1}{4}$ *d.* $\frac{1}{4}$
    e. 0   *f.* 1

9. 0.98   **10.** 0.24

11. *a.* 0.0001   *b.* 0.402   *c.* 0.598

12. 0.1350; 0.0039

13. *a.* $\frac{2}{17}$ *b.* $\frac{11}{850}$ *c.* $\frac{1}{5525}$

14. *a.* $\frac{1}{26}$ *b.* $\frac{1}{4}$ *c.* $\frac{1}{8}$

15. *a.* 0.603   *b.* 0.340   *c.* 0.324   *d.* 0.379

16. *a.* $\frac{57}{104}$ *b.* $\frac{10}{13}$ *c.* $\frac{3}{4}$

17. 0.4   **18.** 5.8%

19. 0.51

20. 0.0417; impossible; 0.25; 0.625

21. 57.3%   **22.** 0.058

23. *a.* $\frac{19}{44}$ *b.* $\frac{1}{4}$

24. 0.558; 0.442

25. 0.718   **26.** 55.6%

27. 175,760,000; 78,624,000; 88,583,040

28. 220   **29.** 350

30. 40,320   **31.** 45

32. 800

33. 100! (Answers may vary regarding calculator.)

34. 30   **35.** 495

36. 286   **37.** 15,504

38. 60   **39.** 175,760,000; 0.0000114

40. $\frac{1}{8}$ **41.** 0.097

42. 0.000772; 0.0000006

43.

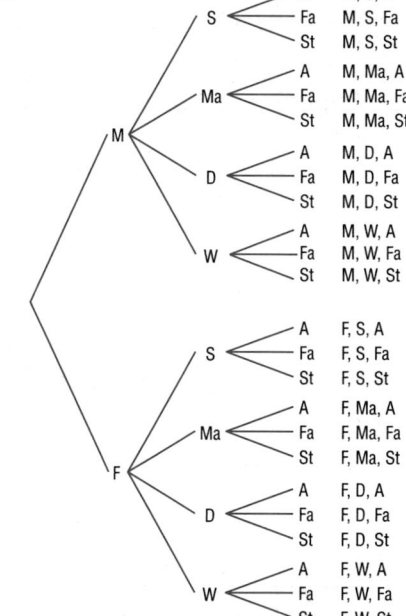

17. *b*   **18.** Sample space

19. 0, 1   **20.** 0

21. 1   **22.** Mutually exclusive

23. *a.* $\frac{1}{13}$ *b.* $\frac{1}{13}$ *c.* $\frac{4}{13}$

24. *a.* $\frac{1}{4}$ *b.* $\frac{4}{13}$ *c.* $\frac{1}{52}$ *d.* $\frac{1}{13}$ *e.* $\frac{1}{2}$

25. *a.* $\frac{12}{31}$ *b.* $\frac{12}{31}$ *c.* $\frac{27}{31}$ *d.* $\frac{24}{31}$

26. *a.* $\frac{11}{36}$ *b.* $\frac{5}{18}$ *c.* $\frac{11}{36}$ *d.* $\frac{1}{3}$ *e.* 0   *f.* $\frac{11}{12}$

27. 0.68   **28.** 0.002

29. *a.* $\frac{253}{9996}$ *b.* $\frac{33}{66,640}$ *c.* 0

30. 0.54   **31.** 0.53

32. 0.81   **33.** 0.056

34. *a.* $\frac{1}{2}$ *b.* $\frac{3}{7}$

35. 0.99   **36.** 0.518

37. 0.9999886   **38.** 2646

39. 40,320   **40.** 1365

41. 1,188,137,600; 710,424,000

42. 720   **43.** 33,554,432

44. 56   **45.** $\frac{1}{4}$

46. $\frac{3}{14}$ **47.** $\frac{12}{55}$

48.

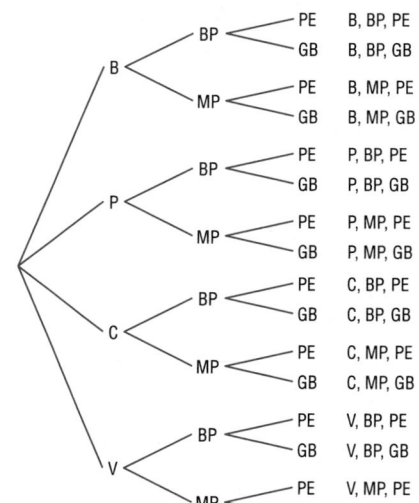

Chapter Quiz

1. False   2. False
3. True   4. False
5. False   6. False
7. True   8. False
9. *b*   10. *b* and *d*
11. *d*   12. *b*
13. *c*   14. *b*
15. *d*   16. *b*

Chapter 5

Exercises 5–1

1. A random variable is a variable whose values are determined by chance. Examples will vary.

2. If the values that a random variable can assume are countable, then the variable is called discrete; otherwise, it is called a continuous random variable.

3. The number of commercials a radio station plays during each hour. The number of times a student uses his or her calculator during a mathematics exam. The number of leaves on a specific type of tree. (Answers will vary.)

4. The weights of strawberries grown in a specific plot; the heights of all seniors at a specific college; the times it takes students to complete a mathematics exam. (Answers will vary.)

5. A probability distribution is a distribution that consists of the values a random variable can assume along with the corresponding probabilities of these values. (Examples will vary.)

6. No. The sum of the probabilities of the events does not equal 1.

7. No; probabilities cannot be negative, and the sum of the probabilities is not 1.

8. No. Probabilities cannot be negative.

9. Yes   10. Yes

11. No. A probability cannot be greater than 1.

12. Continuous   13. Discrete

14. Discrete   15. Continuous

16. Continuous   17. Discrete

18. Continuous

19.

X	0	1	2	3
$P(X)$	$\frac{6}{15}$	$\frac{5}{15}$	$\frac{3}{15}$	$\frac{1}{15}$

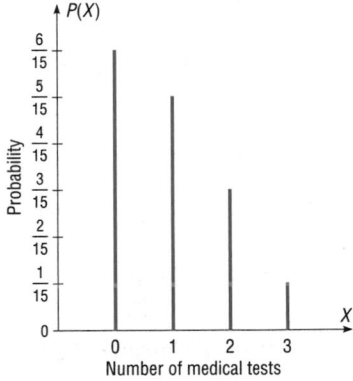

20.

X	\$5000	\$7000	\$9000
$P(X)$	$\frac{1}{2}$	$\frac{3}{8}$	$\frac{1}{8}$

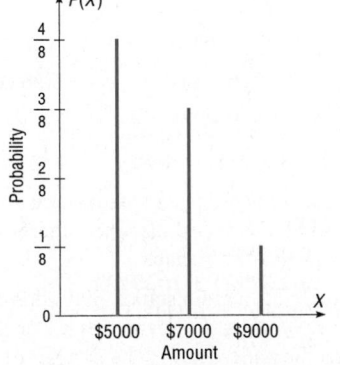

21.

X	2	3	5	7
$P(X)$	0.35	0.41	0.15	0.09

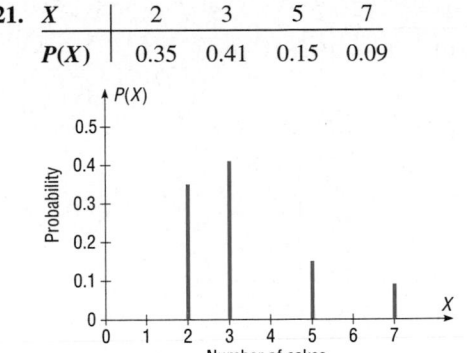

22.

X	0	1	2	3	4
$P(X)$	0.15	0.25	0.3	0.25	0.05

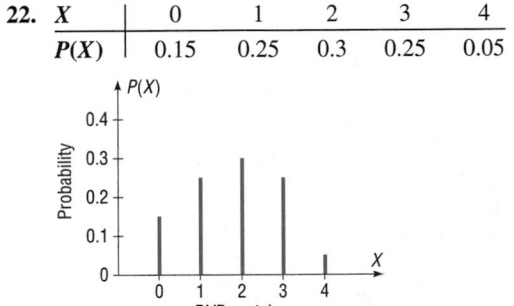

23.

X	1	2	3	4	5	6
$P(X)$	$\frac{1}{2}$	$\frac{1}{6}$	$\frac{1}{12}$	$\frac{1}{12}$	$\frac{1}{12}$	$\frac{1}{12}$

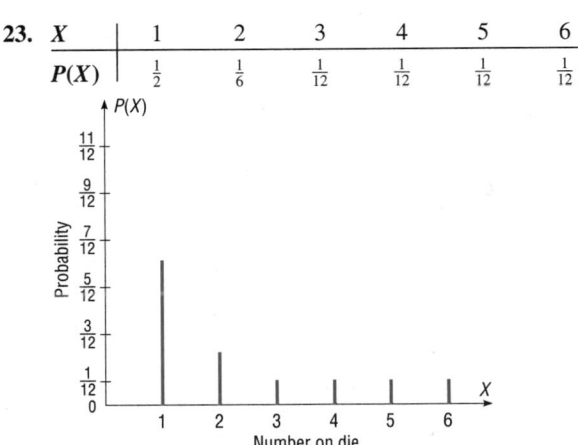

24.

X	1	2	3	4	5
$P(X)$	0.32	0.12	0.23	0.18	0.15

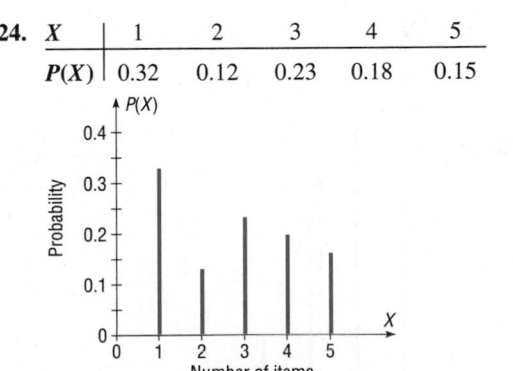

25.

X	2	3	4	5
P(X)	0.01	0.34	0.62	0.03

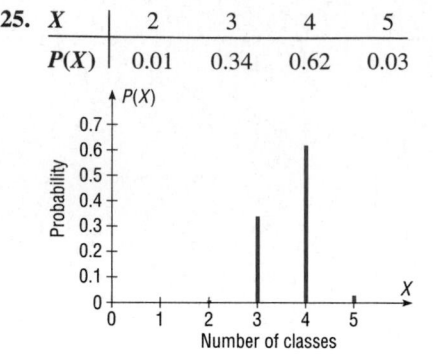

26.

X	0	1	2	3
P(X)	0.22	0.33	0.37	0.08

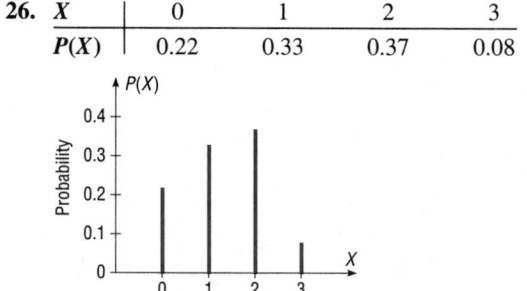

27.

X	1	5	10	20
P(X)	$\frac{3}{11}$	$\frac{2}{11}$	$\frac{5}{11}$	$\frac{1}{11}$

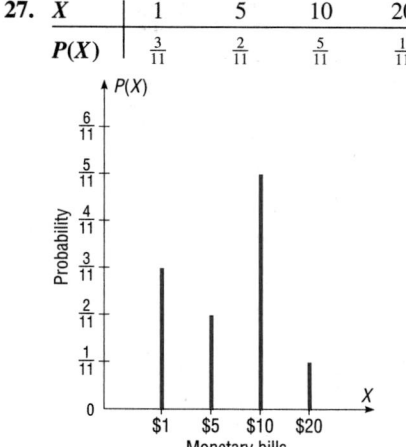

28.

X	0	1	2	3	4
P(X)	$\frac{1}{16}$	$\frac{1}{4}$	$\frac{3}{8}$	$\frac{1}{4}$	$\frac{1}{16}$

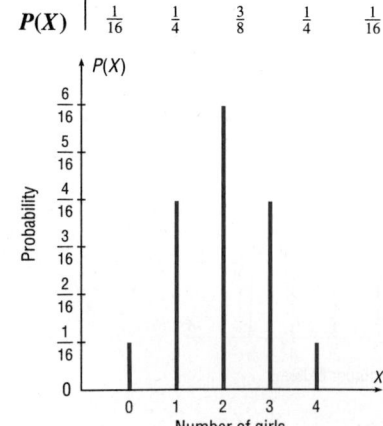

29.

X	1	2	3	4
P(X)	$\frac{1}{4}$	$\frac{1}{4}$	$\frac{3}{8}$	$\frac{1}{8}$

30.

X	2	3	4	5	6	7	8	9	10	11	12
P(X)	$\frac{1}{36}$	$\frac{1}{18}$	$\frac{1}{12}$	$\frac{1}{9}$	$\frac{5}{36}$	$\frac{1}{6}$	$\frac{5}{36}$	$\frac{1}{9}$	$\frac{1}{12}$	$\frac{1}{18}$	$\frac{1}{36}$

31.

X	1	2	3
P(X)	$\frac{1}{6}$	$\frac{1}{3}$	$\frac{1}{2}$

Yes

32.

X	0.2	0.3	0.5
P(X)	0.2	0.3	0.5

Yes

33.

X	3	4	7
P(X)	$\frac{3}{6}$	$\frac{4}{6}$	$\frac{7}{6}$

No, the sum of the probabilities is greater than 1.

34.

X	0.1	0.02	0.04
P(X)	0.2	0.12	0.14

No, the sum of the probabilities is less than 1.

35.

X	1	2	4
P(X)	$\frac{1}{7}$	$\frac{2}{7}$	$\frac{4}{7}$

Yes

36.

X	0	1	2
P(X)	0	$\frac{1}{3}$	$\frac{1}{2}$

No, the sum of the probabilities is less than 1.

Exercises 5–2

1. 0.17; 0.321; 0.567   **2.** 20.8; 1.6; 1.3; 104 suits

3. 1.3, 0.9, 1. No, on average, each person has about 1 credit card.

4. 7.4; 1.84; 1.356   **5.** 5.4; 2.94; 1.71; 0.027

6. 1.3; 1.81; 1.35   **7.** 6.6; 1.3; 1.1

8. 1.9; 0.6; 0.8; 2 diaries   **9.** 9.4; 5.24; 2.289; 0.25

10. 37.1; 1.3; 1.1; it could happen (perhaps on a Super Bowl Sunday), but it is highly unlikely.

11. \$260   **12.** \$7200

13. \$0.83   **14.** −33.3 cents; no

15. −\$1.00   **16.** −\$2.00

17. −\$0.50, −\$0.52   **18.** \$265.70

19. *a.* −5.26 cents   *c.* −5.26 cents   *e.* −5.26 cents
b. −5.26 cents   *d.* −5.26 cents

20. 7; 5.8; 2.4

21. 10.5

22. $\sigma^2 = \Sigma(X - \mu)^2 \times P(X)$
$\sigma^2 = \Sigma(X^2 - 2\mu X + \mu^2)P(X)$
$\sigma^2 = \Sigma X^2 \times P(X) - 2\mu\Sigma XP(X) + \mu^2\Sigma P(X)$
$\sigma^2 = \Sigma X^2 \times P(X) - 2\mu \times \mu + \mu^2(1)$
$\sigma^2 = \Sigma X^2 \times P(X) - 2\mu^2 + \mu^2$
$\sigma^2 = \Sigma X^2 \times P(X) - \mu^2$

23. Answers will vary.   **24.** Answers will vary.

25. Answers will vary.

26. $1.56 with the cost of a stamp = $0.44

Exercises 5–3

1. *a.* Yes   *c.* Yes   *e.* No   *g.* Yes   *i.* No
   b. Yes   *d.* No   *f.* Yes   *h.* Yes   *j.* Yes

2. *a.* 0.420   *c.* 0.590   *e.* 0.000   *g.* 0.418   *i.* 0.246
   b. 0.346   *d.* 0.251   *f.* 0.250   *h.* 0.176

3. *a.* 0.0005   *c.* 0.342   *e.* 0.173
   b. 0.131   *d.* 0.007

4. *a.* 0.021 (TI 0.0214)   *b.* 0.001 (TI 0.001158)
   c. 0 (TI 0.0000003)
   d. Since the probability of each event becomes less likely, the probabilities become smaller.

5. 0.021; no, it's only about a 2% chance.

6. 0.000; the probability is extremely small.

7. *a.* 0.124   *b.* 0.912   *c.* 0.017

8. *a.* 0.925   *b.* 0.998   *c.* 0.337

9. 0.071

10. *a.* 0.025   *b.* 0.215   *c.* 0.162

11. *a.* 0.346   *b.* 0.913   *c.* 0.663   *d.* 0.683

12. *a.* 0.047   *b.* 0.065   *c.* 0.821

13. *a.* 0.242   *b.* 0.547   *c.* 0.306

14. *a.* 75; 18.8; 4.3   *e.* 100; 90; 9.5
   b. 90; 63; 7.9   *f.* 125; 93.8; 9.7
   c. 10; 5; 2.2   *g.* 20; 12; 3.5
   d. 8; 1.6; 1.3   *h.* 6; 5; 2.2

15. 8; 7.9; 2.8   **16.** 5; 2.5; 1.58

17. 9; 8.73; 2.95   **18.** 166; 28.2; 5.3

19. 210; 165.9; 12.9   **20.** 102; 15.3; 3.912

21. 0.199

22. 0.217

23. 0.559

24. 64; 43.52; 6.597

25. 0.177

26. 0.018

27. 0.246

28. Yes. $P(3) = 0.216$. This implies that $p = 0.6$ and then $q = 0.4$. $P(0)$, $P(1)$, and $P(2)$ all check out.

29.

X	0	1	2	3
P(X)	0.125	0.375	0.375	0.125

30.

Exercises 5–4

1. *a.* 0.135   *c.* 0.0096   *e.* 0.0112
   b. 0.0324   *d.* 0.18

2. 0.0016   **3.** 0.0025

4. 0.1   **5.** $\frac{1}{108}$

6. 0.002

7. *a.* 0.1563   *c.* 0.0504   *e.* 0.1241
   b. 0.1465   *d.* 0.071

8. 0.1606

9. *a.* 0.0183   *b.* 0.0733   *c.* 0.1465   *d.* 0.7619

10. 0.3033   **11.** 0.3554

12. 0.2642   **13.** 0.0498

14. 0.2205   **15.** 0.1563

16. 0.0004   **17.** 0.117

18. 0.252   **19.** 0.321

20. 0.712   **21.** 0.597

Review Exercises

1. Yes

2. No. The sum of the probabilities does not equal 1.

3. No; the sum of the probabilities is greater than 1.

4.

5. *a.* 0.35   *b.* 1.55; 1.8075; 1.3444

6.

X	$0.01	$0.10	$0.25	$0.50
P(X)	$\frac{1}{2}$	$\frac{3}{10}$	$\frac{1}{10}$	$\frac{1}{10}$

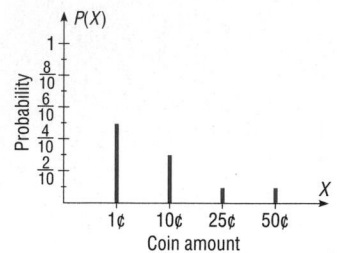

7.

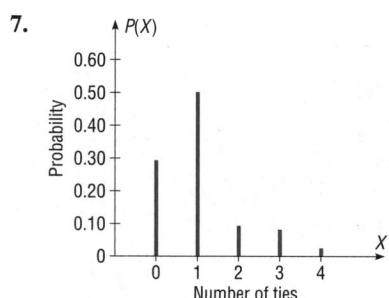

8. 2.1; 1.4; 1.2   **9.** 7.22; 2.1716; 1.47

10. 2.1; 1.5; 1.2   **11.** 24.2; 1.5; 1.2

12. $8100   **13.** $2.15

14. $4.92

15. *a.* 0.008   *b.* 0.724   *c.* 0.0002   *d.* 0.276

16. 0.2639; 0.155

17. 120; 24; 4.9   **18.** 189; 69.93; 8.3624

19. 0.886   **20.** 135; 98.6; 9.9

21. 0.190   **22.** 61.5; 46.371; 6.8096

23. 0.0193   **24.** 0.007

25. 0.050   **26.** 0.1203

27. *a.* 0.5543   *b.* 0.8488   *c.* 0.4457

28. 0.0504   **29.** 0.27

30. 0.21   **31.** 0.0862

Chapter Quiz

1. True   **2.** False

3. False   **4.** True

5. chance   **6.** $n \cdot p$

7. 1   **8.** *c*

9. *c*   **10.** *d*

11. No, since $\Sigma P(X) > 1$ **12.** Yes

13. Yes   **14.** Yes

15.

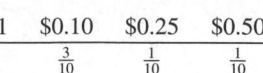

16.

X	0	1	2	3	4
P(X)	0.02	0.3	0.48	0.13	0.07

17. 2.0; 1.3; 1.1   **18.** 32.2; 1.1; 1.0

19. 5.2   **20.** $9.65

21. 0.124

22. *a.* 0.075   *b.* 0.872   *c.* 0.125

23. 240; 48; 6.9   **24.** 9; 7.9; 2.8

25. 0.008   **26.** 0.0003

27. 0.061   **28.** 0.122

29. *a.* 0.5470   *b.* 0.9863   *c.* 0.4529

30. 0.128

31. *a.* 0.160   *b.* 0.42   *c.* 0.07

Chapter 6

Exercises 6–1

1. The characteristics of the normal distribution are as follows:

   a. It is bell-shaped.

   b. It is symmetric about the mean.

   c. Its mean, median, and mode are equal.

   d. It is continuous.

   e. It never touches the *x* axis.

   f. The area under the curve is equal to 1.

   g. It is unimodal.

2. Many variables are normally distributed, and the distribution can be used to describe these variables.

3. 1 or 100%

4. 50% of the area lies below the mean, and 50% of the area lies above the mean.

5. 68%; 95%; 99.7%

6. 0.4616

7. 0.2734

8. 0.1255

9. 0.4808

10. 0.0222

11. 0.3859

12. 0.2266

13. 0.0823

14. 0.0806

15. 0.1094

16. 0.1909

17. 0.0258

18. 0.4634

19. 0.0482

20. 0.9049

21. 0.9826

22. 0.9726

23. 0.5675

24. 0.0684

25. 0.3574

26. 0.4750

27. 0.2486

28. 0.3907

29. 0.4236

30. 0.2061

31. 0.0023

32. 0.0384

33. 0.0934

34. 0.5199

35. 0.9522 (TI: 0.9521)

36. 0.0550

37. 0.0706 (TI: 0.0707)

38. 0.9236

39. 0.9222

40. 1.32

41. $z = -1.39$ (TI: -1.3885)

42. 1.98

43. $z = -2.08$ (TI: -2.0792)

44. 1.84

45. -1.26 (TI: -1.2602)

46. *a.* 0.12   *b.* 0.52   *c.* 1.18

47. *a.* -2.28 (TI: -2.2801)
    b. -0.92 (TI: -0.91995)
    c. -0.27 (TI: -0.26995)

48. $z = \pm0.64$

49. *a.* $z = +1.96$ and $z = -1.96$ (TI: ±1.95996)
    b. $z = +1.65$ and $z = -1.65$, approximately (TI: ±1.64485)
    c. $z = +2.58$ and $z = -2.58$, approximately (TI: ±2.57583)

50. 0.6745; 0.8416; 1.41

51. 0.6827; 0.9545; 0.9973; they are very close.

52. 1.16

53. 2.10

54. -0.75

55. -1.45 and 0.11

56. 1.175

57. $y = \dfrac{e^{-X^2/2}}{\sqrt{2\pi}}$

58.

X	-2	-1.5	-1	-0.5	0	0.5	1	1.5	2
y	0.05	0.13	0.24	0.35	0.4	0.35	0.24	0.13	0.05

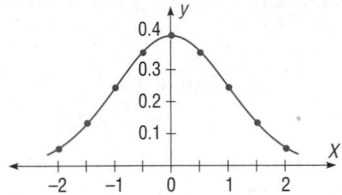

Exercises 6–2

1. 0.0022

2. *a.* 0.3031   *b.* 0.9131
   c. Not too happy—it's really at the bottom of the heap! (prob. = 0.0016)

3. *a.* 0.2005 (TI: 0.2007)   *b.* 0.4315 (TI: 0.4316)

4. 1146; 0.0307

5. *a.* 0.3023   *b.* 0.0062

6. *a.* 0.4602   *b.* 0.0031   *c.* 0.6676

7. *a.* 0.3557 (TI: 0.3547)
   b. 0.8389 (TI: 0.8391)

8. *a.* 0.0749   *b.* 0.2385

9. 0.0262; 0.0001; would want to know why it had only been driven less than 6000 miles (TI: 0.0260; 0.0002)

10. 0.2061; 0.1251

11. *a.* 0.9803 (TI: 0.9801)
    b. 0.2514 (TI: 0.2511)
    c. 0.3434 (TI: 0.3430)

12. 35.1 cents

13. *a.* 0.3057   *b.* 0.5688
    c. The person could assume it will be between the mean time plus or minus 2 standard deviations of the mean.

14. Less than 0.0001

15. *a.* 0.3281   *b.* 0.4002   *c.* Not usually

16. Men: $104,053   Women: $94,698

17. 0.0080 or 0.8%. A temperature of 63° is unlikely since the probability is about 0.8%.

18. $722.99 and $861.01

19. The maximum size is 1927.76 square feet; the minimum size is 1692.24 square feet. (TI: 1927.90 maximum, 1692.10 minimum)

20. $227,100 to $265,500

21. 0.006; $821

22. 92.99 or 93

23. The maximum price is $9222, and the minimum price is $7290. (TI: $7288.14 minimum, $9223.86 maximum)

24. 4.05

25. 6.7; 4.05 (TI: for 10%, 6.657; for 30%, 4.040)

26. *a.* 588 *b.* 183

27. $18,840.48 (TI: $18,869.48)

28. 0.0968; 0.6641

29. 18.6 months 30. 71.6 or 72

31. *a.* $\mu = 120, \sigma = 20$ *b.* $\mu = 15, \sigma = 2.5$
    c. $\mu = 30, \sigma = 5$

32. No. Any subgroup would not be a perfect representation of the seniors; therefore, the mean and standard deviation would be different.

33. There are several mathematics tests that can be used.

34. No. The shape of the distribution would be the same.

35. 3.125 36. 95.68

37. $\mu = 45, \sigma = 1.34$

38. 77 and up   A
    68–76   B
    52–67   C
    44–51   D
    0–43   F

39. Not normal

40. Not normal

41. Not normal

42. Not normal

Exercises 6–3

1. The distribution is called the sampling distribution of sample means.

2. The sample is not a perfect representation of the population. The difference is due to what is called sampling error.

3. The mean of the sample means is equal to the population mean.

4. The standard error of the mean: $\sigma_{\bar{x}} = \sigma / \sqrt{n}$.

5. The distribution will be approximately normal when the sample size is large.

6. $z = \dfrac{X - \mu}{\sigma}$ 7. $z = \dfrac{\bar{X} - \mu}{\sigma/\sqrt{n}}$

8. 0.7135

9. *a.* 0.0026 (TI: 0.0026)
   b. 0.8212 (TI: 0.8201)
   c. 0.1787 (TI: 0.1799)

10. *a.* 0.2389 *b.* 0.0375

11. 0.2673

12. *a.* 0.0778 *b.* 0.3446

13. 0.0427; 0.9572 (TI: 0.0423; 0.9577)

14. Yes—the probability of such is less than 0.0001.

15. *a.* 0.3859 (TI: 0.3875) *b.* 0.1841 (TI: 0.1831)
    c. Individual values are more variable than means.

16. 0.1357

17. 0.4176 (TI: 0.4199)

18. *a.* 0.0051 *b.* 0.3632

19. 0.1254 (TI: 0.12769)

20. 0.9850; Less than 0.0001

21. *a.* 0.4052 or 40.52% *b.* 0.0901 or 9.01%
    c. Yes, the probability is slightly more than 40%.
    d. It's possible since the probability is about 9%.

22. *a.* 0.1255 *b.* 0.4608
    c. Means are less variable than individual data.

23. *a.* 0.3707 (TI: 0.3694) *b.* 0.0475 (TI: 0.04779)

24. *a.* 0.1567 *b.* 0.4963

25. 0.0174 No—the central limit theorem applies.

26. 0.0025

27. 0.0143 28. 1963.10 pounds

29. $\sigma_{\bar{x}} = 1.5, n = 25$ 30. 400

Exercises 6–4

1. When p is approximately 0.5, as n increases, the shape of the binomial distribution becomes similar to that of the normal distribution. The conditions are that $n \cdot p$ and $n \cdot q$ are both ≥ 5. The correction is necessary because the normal distribution is continuous and the binomial distribution is discrete.

2. *a.* 0.0811 *c.* 0.1052 *e.* 0.2327
   b. 0.0516 *d.* 0.1711 *f.* 0.9988

3. *a.* Yes *c.* No *e.* Yes
   b. No *d.* Yes *f.* No

4. 0.9970 5. 0.8577

6. 0.1003 7. 0.9875

8. 0.0984 9. 0.3936

10. 0.0985 11. 0.0087

12. 0.2005

13. 0.9951; yes (TI: 0.9950)

14. 0.0559

15. *a.* $n \geq 50$ *c.* $n \geq 10$ *e.* $n \geq 50$
    b. $n \geq 17$ *d.* $n \geq 25$

Review Exercises

1. *a.* 0.4744 *e.* 0.2139 *h.* 0.9131
   b. 0.1443 *f.* 0.8284 *i.* 0.0183
   c. 0.0590 *g.* 0.0233 *j.* 0.9535
   d. 0.8329 (TI: 0.8330)

2. *a.* 0.4808   *e.* 0.6391   *i.* 0.9732
   b. 0.0336   *f.* 0.0485   *j.* 0.9616
   c. 0.9219   *g.* 0.0212
   d. 0.0617   *h.* 0.8830

3. 0.1131; $4872 and $5676
   (TI: $4869.31 minimum, $5678.69 maximum)

4. *a.* 0.1587   *b.* 0.0013

5. *a.* 0.3621 or 36.21%   *b.* 0.1190 or 11.9%
   c. 0.0606 or 6.06%

6. 0.0239; 0.1654

7. $130.92

8. Not normal

9. Not normal

10. *a.* 0.7054 (TI: 0.7057)   *b.* 0.8869 (TI: 0.8868)

11. *a.* 0.0143 (TI: 0.0142)   *b.* 0.9641

12. 0.0023; yes, since the probability is less than 1%.

13. 0.5234

14. 0.0496

15. 0.7123; 0.9999 (TI: 0.7139; 0.9999)

16. 0.0668

17. 0.0465

Chapter Quiz

1. False   2. True
3. True   4. True
5. False   6. False
7. *a*   8. *a*
9. *b*   10. *b*
11. *c*   12. 0.5
13. Sampling error
14. The population mean
15. Standard error of the mean
16. 5   17. 5%
18. *a.* 0.4332   *d.* 0.1029   *g.* 0.0401   *j.* 0.9131
    b. 0.3944   *e.* 0.2912   *h.* 0.8997
    c. 0.0344   *f.* 0.8284   *i.* 0.017
19. *a.* 0.4846   *d.* 0.0188   *g.* 0.0089   *j.* 0.8461
    b. 0.4693   *e.* 0.7461   *h.* 0.9582
    c. 0.9334   *f.* 0.0384   *i.* 0.9788
20. *a.* 0.7734   *b.* 0.0516   *c.* 0.3837
    d. Any rainfall above 65 inches could be considered an extremely wet year since this value is 2 standard deviations above the mean.
21. *a.* 0.0668   *b.* 0.0228   *c.* 0.4649   *d.* 0.0934
22. *a.* 0.4525   *b.* 0.3707   *c.* 0.3707   *d.* 0.019

23. *a.* 0.0013   *b.* 0.5   *c.* 0.0081   *d.* 0.5511
24. *a.* 0.0037   *b.* 0.0228   *c.* 0.5   *d.* 0.3232
25. 8.804 centimeters
26. 121.24 is the lowest acceptable score.
27. 0.015   28. 0.9738
29. 0.0495; no   30. 0.0455 or 4.55%
31. 0.8577   32. 0.0495
33. Not normal
34. Approximately normal

Chapter 7
Exercises 7–1

1. A point estimate of a parameter specifies a particular value, such as $\mu = 87$; an interval estimate specifies a range of values for the parameter, such as $84 < \mu < 90$. The advantage of an interval estimate is that a specific confidence level (say 95%) can be selected, and one can be 95% confident that the interval contains the parameter that is being estimated.

2. The standard deviation of the population must be known, or it must be estimated or specified in terms of E. Sample size must be specified, and the degree of confidence must be selected.

3. The margin of error is the likely range of values to the right or left of the statistic that may contain the parameter.

4. A 95% confidence interval means one can be 95% confident that the confidence interval will contain the parameter being estimated.

5. A good estimator should be unbiased, consistent, and relatively efficient.

6. $\overline{X}$

7. For one to be able to determine sample size, the margin of error and the degree of confidence must be specified and the population standard deviation must be known.

8. No, as long as it is much larger than the sample size needed.

9. *a.* 2.58   *c.* 1.96   *e.* 1.88
   b. 2.33   *d.* 1.65

10. $295.15 < \mu < 397.35$

11. *a.* 16.6 hours   *b.* $15.7 < \mu < 17.5$
    c. $15.4 < \mu < 17.8$
    d. The 99% confidence interval is larger since you want to be 99% confident that the mean is contained in the interval rather than 95% confident.

12. $2.55 < \mu < 3.09$

13. $1.72 < \mu < 1.88$; lower

14. *a.* 7.2 jobs   *b.* $6.6 < \mu < 7.8$ *c.* $6.4 < \mu < 8.0$

    d. The 95% confidence is smaller since there is less of a chance that the mean is contained in the interval as opposed to the 99% confidence interval.

15. $145{,}030 < \mu < 154{,}970$

16. $34.3 < \mu < 52.7$

17. $4913 < \mu < 5087$; 4000 hours does not seem reasonable since it is outside the interval.

18. $\$3840 < \mu < \$4134$; $3800

19. $59.5 < \mu < 62.9$ **20.** $172.74 < \mu < 208.66$

21. 123 subjects   **22.** $57.4 < \mu < 58.6$

23. 44 subjects   **24.** 12

25. 240 exams

26. $37.71 < \mu < 38.89$; the 90% interval

Exercises 7–2

1. The characteristics of the *t* distribution are as follows: It is bell-shaped, it is symmetric about the mean, and it never touches the *x* axis. The mean, median, and mode are equal to 0 and are located at the center of the distribution. The variance is greater than 1. The *t* distribution is a family of curves based on degrees of freedom. As a sample size increases, the *t* distribution approaches the standard normal distribution.

2. The degrees of freedom are the number of values free to vary after a sample statistic has been computed.

3. The *t* distribution should be used when σ is unknown.

4. *a.* 2.898   *c.* 2.624   *e.* 2.093

    b. 2.074   *d.* 1.833

5. $21.8 < \mu < 30.4$

6. $205.2 < \mu < 230.2$. Assume the variable is normally distributed.

7. $\overline{X} = 33.4$; $s = 28.7$; $21.2 < \mu < 45.6$; the point estimate is 33.4, and it is close to 32. Also, the interval does indeed contain $\mu = 32$. The data value 132 is unusually large (an outlier). The mean may not be the best estimate in this case.

8. $38.70 < \mu < 48.28$. Assume normal distribution; yes.

9. $496.8 < \mu < 650.8$. No, 625 homicides would not be considered high since it would be inside the 99% confidence interval.

10. $25.8 < \mu < 33.9$. Assume normal distribution.

11. $13.5 < \mu < 15.1$; about 30 minutes.

12. $13.6 < \mu < 16.4$; 16.4 miles per hour.

13. $17.87 < \mu < 20.53$. Assume normal distribution; it's higher.

14. $9.7 < \mu < 16.5$

15. $28.4 < \mu < 38.0$

16. $38.4 < \mu < 44.8$

17. $32.0 < \mu < 71$. Assume normal distribution.

18. $84.2 < \mu < 87.8$. He probably used a maximum pulse rate of 88 on average.

19. Answers will vary.

20. $8.8 < \mu < 58.2$

21. $\overline{X} = 2.175$; $s = 0.585$; $\mu > \$1.95$ means one can be 95% confident that the mean revenue is greater than $1.95; $\mu < \$2.40$ means one can be 95% confident that the mean revenue is less than $2.40.

Exercises 7–3

1. *a.* 0.5, 0.5   *c.* 0.46, 0.54   *e.* 0.45, 0.55

    b. 0.45, 0.55   *d.* 0.58, 0.42

2. *a.* $\hat{p} = 0.25$, $\hat{q} = 0.75$ *d.* $\hat{p} = 0.55$, $\hat{q} = 0.45$

    b. $\hat{p} = 0.42$, $\hat{q} = 0.58$ *e.* $\hat{p} = 0.12$, $\hat{q} = 0.88$

    c. $\hat{p} = 0.68$, $\hat{q} = 0.32$

3. $0.365 < p < 0.415$

4. $0.388 < p < 0.492$. It is probably higher because of increased awareness in a college town.

5. $0.092 < p < 0.153$; 11% is contained in the confidence interval.

6. $0.233 < p < 0.401$ **7.** $0.797 < p < 0.883$

8. $0.400 < p < 0.463$ **9.** $0.596 < p < 0.704$

10. $0.721 < p < 0.819$

11. $0.125 < p < 0.375$. No, since 0.28 is contained in the interval.

12. $0.188 < p < 0.288$; yes

13. $0.419 < p < 0.481$

14. $0.529 < p < 0.591$

15. 385; 601

16. *a.* 225   *b.* 273

17. 801 homes; 1068 homes

18. 318   **19.** 1089

20. 1893   **21.** 95%

22. 96%

Exercises 7–4

1. Chi-square

2. The variable must be normally distributed.

3. *a.* 3.816; 21.920   *d.* 0.412; 16.750

    b. 10.117; 30.144   *e.* 26.509; 55.758

    c. 13.844; 41.923

4. $15.1 < \sigma^2 < 40.5$
$3.9 < \sigma < 6.4$

5. $56.6 < \sigma^2 < 236.3$; $7.5 < \sigma < 15.4$

6. $5.0 < \sigma^2 < 204.0$
$2.2 < \sigma < 14.3$

7. Use $\sigma = r \div 4$
$1,593,756 < \sigma^2 < 16,537,507$; $1262.4 < \sigma < 4066.6$;
$8,469,845 < \sigma^2 < 87,886,811$; $2910.3 < \sigma < 9374.8$

8. $3.5 < \sigma^2 < 9.3$
$1.9 < \sigma < 3.0$

9. $604 < \sigma^2 < 5837$; $24.6 < \sigma < 76.4$

10. $259.343 < \sigma^2 < 772.724$
$16.104 < \sigma < 27.798$

11. $130,136 < \sigma^2 < 413,084$
$361 < \sigma < 643$

12. $6.8 < \sigma^2 < 140$
$2.6 < \sigma < 11.8$

13. $16.2 < \sigma < 19.8$

Review Exercises

1. $13.99 < \mu < 25.27$ (or $14 < \mu < 25$)
(TI: $14.005 < \mu < 25.255$)

2. 7.5; $7.46 < \mu < 7.54$

3. 28

4. \$23.45; $22.79 < \mu < \$24.11$

5. $76.9 < \mu < 88.3$. Assume normal distribution.

6. $25 < \mu < 31$ **7.** $0.409 < p < 0.471$

8. $0.395 < p < 0.445$ **9.** $0.343 < p < 0.457$

10. $0.414 < p < 0.531$ **11.** 460

12. 842 children; 1068 children

13. $0.218 < \sigma < 0.435$. Yes. It seems that there is a large standard deviation.

14. $1.5 < \sigma^2 < 5.3$ **15.** $5.1 < \sigma^2 < 18.3$

16. $28.6 < \sigma^2 < 334.2$; $5.3 < \sigma < 18.3$

Chapter Quiz

1. True   **2.** True

3. False   **4.** True

5. *b*   **6.** *a*

7. *b*

8. Unbiased, consistent, relatively efficient

9. Margin of error

10. Point   **11.** 90; 95; 99

12. \$121.60; $\$119.85 < \mu < \$123.35$

13. \$44.80; $\$43.15 < \mu < \$46.45$

14. 4150; $3954 < \mu < 4346$

15. $45.7 < \mu < 51.5$

16. $418 < \mu < 458$ **17.** $26 < \mu < 36$

18. 180   **19.** 25

20. $0.374 < p < 0.486$ **21.** $0.295 < p < 0.425$

22. $0.342 < p < 0.547$ **23.** 545

24. $7 < \sigma < 13$

25. $30.9 < \sigma^2 < 78.2$ **26.** $1.8 < \sigma < 3.2$
$5.6 < \sigma < 8.8$

Chapter 8

Note: For Chapters 8–13, specific *P*-values are given in parentheses after the *P*-value intervals. When the specific *P*-value is extremely small, it is not given.

Exercises 8–1

1. The null hypothesis states that there is no difference between a parameter and a specific value or that there is no difference between two parameters. The alternative hypothesis states that there is a specific difference between a parameter and a specific value or that there is a difference between two parameters. Examples will vary.

2. A type I error occurs when the null hypothesis is rejected when it is true. A type II error occurs when the null hypothesis is not rejected when it is false. They are related in that decreasing the probability of one type of error increases the probability of the other type of error.

3. A statistical test uses the data obtained from a sample to make a decision about whether the null hypothesis should be rejected.

4. A one-tailed test indicates the null hypothesis should be rejected when the test statistic value is in the critical region on one side of the mean. A two-tailed test indicates the null hypothesis should be rejected when the test statistic value is in either critical region on either side of the mean.

5. The critical region is the range of values of the test statistic that indicates that there is a significant difference and the null hypothesis should be rejected. The noncritical region is the range of values of the test statistic that indicates that the difference was probably due to chance and the null hypothesis should not be rejected.

6. H_0 represents the null hypothesis; H_1 represents the alternative hypothesis.

7. α, β

8. When the difference between the sample mean and the hypothesized population mean is large, then the difference is said to be significant and probably not due to chance.

9. A one-tailed test should be used when a specific direction, such as greater than or less than, is being hypothesized; when no direction is specified, a two-tailed test should be used.

10. The steps in hypothesis testing are as follows.

　　a. State the hypotheses and identify the claim.

　　b. Find the critical value(s).

　　c. Compute the test statistic value.

　　d. Make the decision.

　　e. Summarize the results.

11. Hypotheses can be proved true only when the entire population is used to compute the test statistic. In most cases, this is impossible.

12. *a.* ± 1.96　　　*d.* $+2.33$　　　*g.* $+1.65$　　　*i.* -1.75

　　b. -2.33　　　*e.* -1.65　　　*h.* ± 2.58　　　*j.* $+2.05$

　　c. $+2.58$　　　*f.* -2.05

13. *a.* H_0: $\mu = 24.6$ and H_1: $\mu \neq 24.6$

　　b. H_0: $\mu = \$51,497$ and H_1: $\mu \neq \$51,497$

　　c. H_0: $\mu = 25.4$ and H_1: $\mu > 25.4$

　　d. H_0: $\mu = 88$ and H_1: $\mu < 88$

　　e. H_0: $\mu = 70$ and H_1: $\mu < 70$

　　f. H_0: $\mu = \$79.95$ and H_1: $\mu \neq \$79.95$

　　g. H_0: $\mu = 8.2$ and H_1: $\mu \neq 8.2$

Exercises 8–2

1. H_0: $\mu = 305$; H_1: $\mu > 305$ (claim); C.V. $= 1.65$; $z = 4.71$; reject. There is enough evidence to support the claim that the mean depth is greater than 305 feet. It might be due to warmer temperatures or more rainfall.

2. H_0: $\mu = \$3262$ and H_1: $\mu < \$3262$ (claim); C.V. $= -1.65$; $z = -1.72$; reject. Yes. There is enough evidence to support the claim that the average credit card debt is less than $3262.

3. H_0: $\mu = \$24$ billion and H_1: $\mu > \$24$ billion (claim); C.V. $= 1.65$; $z = 1.85$; reject. There is enough evidence to support the claim that the average revenue is greater than $24 billion.

4. H_0: $\mu = 8.5$; H_1: $\mu \neq 8.5$ (claim); C.V. $= \pm 1.96$; $z = 2.17$; reject. There is enough evidence to support the claim that there is a difference.

5. H_0: $\mu = 30.9$; H_1: $\mu \neq 30.9$ (claim); C.V. $= \pm 2.58$; $z = 1.89$; do not reject. There is not enough evidence to support the claim that the mean has changed.

6. H_0: $\mu = 3000$ and H_1: $\mu > 3000$ (claim); C.V. $= 1.65$; $z = 1.61$; do not reject. No. There is not enough evidence to say that the average production has increased.

7. H_0: $\mu = 29$ and H_1: $\mu \neq 29$ (claim); C.V. $= \pm 1.96$; $z = 0.944$; do not reject. There is not enough evidence to say that the average height differs from 29 inches.

8. H_0: $\mu = 59,593$; H_1: $\mu < 59,593$ (claim); C.V. $= -2.33$; $z = -2.90$; reject H_0. There is sufficient evidence at $\alpha = 0.01$ to conclude that the state employees earn less than the federal employees.

9. H_0: $\mu = \$8121$; H_1: $\mu > \$8121$ (claim); C.V. $= 2.33$; $z = 1.93$; do not reject. There is not enough evidence to support the claim that the mean is greater than $8121.

10. H_0: $\mu = \$60,000$ (claim) and H_1: $\mu \neq \$60,000$; C.V. $= \pm 1.96$; $z = 1.78$; do not reject. There is not enough evidence to reject the claim that the average price of a home is $60,000.

11. H_0: $\mu = 500$; H_1: $\mu \neq 500$ (claim); C.V. $= \pm 2.58$; $z = -4.04$; reject H_0. There is sufficient evidence to conclude that the mean differs from 500.

12. H_0: $\mu = \$10,337$; H_1: $\mu \neq \$10,337$ (claim); C.V. $= \pm 1.65$ at 0.10, ± 1.96 at 0.05, and ± 2.58 at 0.01; $z = 3.62$; reject at 0.10, 0.05, and 0.01. There is enough evidence to support the claim that the mean expenditure has changed.

13. H_0: $\mu = 60.35$; H_1: $\mu < 60.35$ (claim); C.V. $= -1.65$; $z = -4.82$; reject H_0. There is sufficient evidence to conclude that the state senators are younger.

14. The *P*-value is the actual probability of getting the sample mean if the null hypothesis is true.

15. *a.* Do not reject.　　　　*d.* Reject.

　　b. Reject.　　　　　　　　*e.* Reject.

　　c. Do not reject.

16. H_0: $\mu = 52$ (claim) and H_1: $\mu \neq 52$; $z = 8.69$; *P*-value < 0.01; reject. There is enough evidence to reject the claim that the mean is 52. The researcher's claim is not valid.

17. H_0: $\mu = 264$ and H_1: $\mu < 264$ (claim); $z = -2.53$; *P*-value $= 0.0057$; reject. There is enough evidence to support the claim that the average stopping distance is less than 264 ft. (TI: *P*-value $= 0.0056$)

18. H_0: $\mu = 40$ and H_1: $\mu < 40$ (claim); $z = -2.45$; *P*-value $= 0.0069$ (TI: *P*-value $= 0.0070$); reject. There is enough evidence to support the claim that the average number of pages copied is less than 40.

19. H_0: $\mu = 546$ and H_1: $\mu < 546$ (claim); $z = -2.4$; *P*-value $= 0.0082$. Yes, it can be concluded that the number of calories burned is less than originally thought. (TI: *P*-value $= 0.0082$)

20. H_0: $\mu = 800$ (claim) and H_1: $\mu \neq 800$; $z = -2.61$; *P*-value $= 0.0090$; reject. There is enough evidence to reject the null hypothesis that the breaking strength is 800 pounds.

21. H_0: $\mu = 444$; H_1: $\mu \neq 444$; $z = -1.70$; *P*-value $= 0.0892$; do not reject H_0. There is insufficient evidence at $\alpha = 0.05$ to conclude that the average size differs from 444 acres. (TI: *P*-value $= 0.0886$)

22. H_0: $\mu = 65$ (claim) and H_1: $\mu \neq 65$; $z = -1.21$; *P*-value $= 0.2262$ (TI: *P*-value $= 0.2278$); do not reject. There is not enough evidence to reject the hypothesis that the average acreage is 65 acres.

23. H_0: $\mu = 30,000$ (claim) and H_1: $\mu \neq 30,000$; $z = 1.71$; *P*-value $= 0.0872$; reject. There is enough evidence to reject the claim that the customers are adhering to the recommendation. Yes, the 0.10 level is appropriate. (TI: *P*-value $= 0.0868$)

24. H_0: $\mu = 60$ (claim) and H_1: $\mu \neq 60$; $z = -0.03$; P-value $= 0.976$; since P-value > 0.05, do not reject. There is not enough evidence to reject the claim that the average number of tickets issued is 60.

25. H_0: $\mu = 10$ and H_1: $\mu < 10$ (claim); $z = -8.67$; P-value < 0.0001; since P-value < 0.05, reject. Yes, there is enough evidence to support the claim that the average number of days missed per year is less than 10. (TI: P-value $= 0$)

26. Reject the claim at $\alpha = 0.05$ but not at $\alpha = 0.01$. There is no contradiction, since the value of α should be chosen before the test is conducted.

27. H_0: $\mu = 8.65$ (claim) and H_1: $\mu \neq 8.65$; C.V. $= \pm 1.96$; $z = -1.35$; do not reject. Yes; there is not enough evidence to reject the claim that the average hourly wage of the employees is \$8.65.

Exercises 8–3

1. It is bell-shaped, it is symmetric about the mean, and it never touches the x axis. The mean, median, and mode are all equal to 0, and they are located at the center of the distribution. The t distribution differs from the standard normal distribution in that it is a family of curves and the variance is greater than 1; and as the degrees of freedom increase, the t distribution approaches the standard normal distribution.

2. The degrees of freedom are the number of values that are free to vary after a sample statistic has been computed. They tell the researcher which specific curve to use when a distribution consists of a family of curves.

3. *a.* $+1.833$ *c.* -3.365 *e.* ± 2.145 *g.* ± 2.771
    b. ± 1.740 *d.* $+2.306$ *f.* -2.819 *h.* ± 2.583

4. Specific P-values are in parentheses.
    a. $0.01 < P$-value < 0.025 (0.018)
    b. $0.05 < P$-value < 0.10 (0.062)
    c. $0.10 < P$-value < 0.25 (0.123)
    d. $0.10 < P$-value < 0.20 (0.138)
    e. P-value < 0.005 (0.003)
    f. $0.10 < P$-value < 0.25 (0.158)
    g. P-value $= 0.05$ (0.05)
    h. P-value > 0.25 (0.261)

5. H_0: $\mu = 179$; H_1: $\mu \neq 179$ (claim); C.V. $= \pm 3.250$; d.f. $= 9$; $t = 3.162$; do not reject H_0. There is insufficient evidence to conclude that the mean differs from \$179.

6. H_0: $\mu = 2000$ and H_1: $\mu < 2000$ (claim); C.V. $= -3.747$; d.f. $= 4$; $t = -0.104$; do not reject. There is not enough evidence to support the claim that the average number of acres is less than 2000.

7. H_0: $\mu = 2.27$; H_1: $\mu \neq 2.27$ (claim); C.V. $= \pm 2.093$; d.f. $= 19$; $t = 3.240$; reject. There is enough evidence to support the claim that the average time differs from 2.27.

8. H_0: $\mu = 25.4$ and H_1: $\mu < 25.4$ (claim); C.V. $= -1.318$; d.f. $= 24$; $t = -3.11$; reject. Yes. There is enough evidence to support the claim that the average commuting time is less than 25.4 minutes.

9. H_0: $\mu = 700$ (claim) and H_1: $\mu < 700$; C.V. $= -2.262$; d.f. $= 9$; $t = -2.71$; reject. There is enough evidence to reject the claim that the average height of the buildings is at least 700 feet.

10. Exercise: H_0: $\mu = 29$; H_1: $\mu \neq 29$ (claim); C.V. $= \pm 2.064$; d.f. $= 24$; $t = 4.348$; reject H_0. There is sufficient evidence to conclude that the mean exercise time differs from 29 minutes per day.
    Reading: H_0: $\mu = 23$; H_1: $\mu \neq 23$ (claim); C.V. $= \pm 2.064$; d.f. $= 24$; $t = -1.736$; do not reject H_0. There is insufficient evidence to conclude that the mean time spent reading differs from 23 minutes per day.

11. H_0: $\mu = 73$; H_1: $\mu > 73$ (claim); C.V. $= 2.821$; d.f. $= 9$; $t = 4.063$; reject. There is enough evidence to support the claim that the average is greater than the national average.

12. H_0: $\mu = 36$; H_1: $\mu \neq 36$ (claim); C.V. $= \pm 2.807$; d.f. $= 23$; $t = 5.638$; reject. There is enough evidence to support the claim that the mean is not 36 visits.

13. H_0: $\mu = \$54.8$ million and H_1: $\mu > \$54.8$ million (claim); C.V. $= 1.761$; d.f. $= 14$; $t = 3.058$; reject. Yes. There is enough evidence to support the claim that the average cost of an action movie is greater than \$54.8 million.

14. H_0: $\mu = 110$ and H_1: $\mu > 110$ (claim); C.V. $= 2.624$; d.f. $= 14$; $t = 4.389$; reject. Yes. There is enough evidence to support the claim that the average calorie content is greater than 110 calories.

15. H_0: $\mu = \$50.07$, H_1: $\mu > \$50.07$ (claim); C.V. $= 1.833$; d.f. $= 9$; $t = 2.741$; reject. There is enough evidence to support the claim that the average phone bill has increased.

16. H_0: $\mu = 123$ and H_1: $\mu \neq 123$ (claim); d.f. $= 15$; $t = -3.02$; P-value < 0.01 (0.0086); reject. There is enough evidence to support the hypothesis that the mean has changed. The *Old Farmer's Almanac* figure may have changed.

17. H_0: $\mu = 5.8$ and H_1: $\mu \neq 5.8$ (claim); d.f. $= 19$; $t = -3.462$; P-value < 0.01; reject. There is enough evidence to support the claim that the mean number of times has changed. (TI: P-value $= 0.0026$)

18. H_0: $\mu = 9.2$ (claim) and H_1: $\mu \neq 9.2$; d.f. $= 7$; $t = -0.531$; P-value > 0.50 (0.612); do not reject. There is not enough evidence to reject the claim that the mean is 9.2. One reason why a person may not give the exact number of past jobs is that he or she may have forgotten about a particular job.

19. H_0: $\mu = \$15,000$ and H_1: $\mu \neq \$15,000$; d.f. $= 11$; $t = -1.10$; C.V. $= \pm 2.201$; do not reject. There is not enough evidence to conclude that the average stipend differs from \$15,000.

20. H_0: $\mu = 3.18$ and H_1: $\mu \neq 3.18$ (claim); C.V. $= \pm 2.069$; d.f. $= 23$; $t = 2.231$; reject. Yes. There is enough evidence to support the claim that the average family size is different from 3.18.

Exercises 8–4

1. Answers will vary.

2. The proportion of A items can be considered a success, whereas the proportion of items that are not included in A can be considered a failure. Hence there are two outcomes.

3. $np \geq 5$ and $nq \geq 5$

4. $\mu = np$; $\sigma = \sqrt{pq/n}$

5. H_0: $p = 0.686$; H_1: $p \neq 0.686$ (claim); C.V. $= \pm 2.58$; $z = -1.93$; do not reject H_0. There is insufficient evidence to conclude that the proportion differs.

6. H_0: $p = 0.503$; H_1: $p \neq 0.503$ (claim); $z = 2.32$; therefore, reject H_0 at any $\alpha \leq 0.025$.

7. H_0: $p = 0.188$; H_1: $p < 0.188$ (claim); C.V. $= -1.65$; $z = -1.00$; do not reject. There is not enough evidence to support the claim that the proportion is less than the national proportion.

8. H_0: $p = 0.279$ and H_1: $p > 0.279$ (claim); C.V. $= 1.65$; $z = 2.35$; reject. Yes. There is enough evidence to conclude that the proportion of women physicians exceeds 27.9%.

9. H_0: $p = 0.47$; H_1: $p \neq 0.47$ (claim); C.V. $= \pm 1.96$; $z = 2.51$; reject. There is enough evidence to support the claim that the proportion is different from the national proportion.

10. H_0: $p = 0.856$; H_1: $p \neq 0.856$ (claim); C.V. $= \pm 1.96$; $z = -1.02$; do not reject H_0. There is insufficient evidence to conclude that the proportion differs from the national rate.

11. H_0: $p = 0.32$; H_1: $p \neq 0.32$ (claim); C.V. $= \pm 2.58$; $z = 3.61$; reject. There is enough evidence to support the claim that the proportion is different than 32%.

12. H_0: $p = 0.14$ (claim) and H_1: $p \neq 0.14$; $z = -1.15$; P-value $= 0.250$; do not reject. There is not enough evidence to reject the claim that 14% of men use exercise to relieve stress. No, the results cannot be generalized to all adult Americans since only men were surveyed.

13. H_0: $p = 0.54$ (claim) and H_1: $p \neq 0.54$; $z = 0.93$; P-value $= 0.3524$; do not reject. There is not enough evidence to reject the claim that the proportion is 0.54. Yes, a healthy snack should be made available for children to eat after school. (TI: P-value $= 0.3511$)

14. H_0: $p = 0.517$ (claim) and H_1: $p \neq 0.517$; $z = 1.64$; P-value $= 0.101$; do not reject. There is not enough evidence to reject the claim that the proportion is 0.517. The evidence supports the claim.

    The percentage of homes heated by natural gas might be different.

15. H_0: $p = 0.18$ (claim) and H_1: $p > 0.18$; $z = -0.60$; P-value $= 0.5486$; since P-value > 0.05, do not reject. There is not enough evidence to reject the claim that 18%

of all high school students smoke at least a pack of cigarettes a day. (TI: P-value $= 0.5478$)

16. H_0: $p = 0.83$; H_1: $p < 0.83$ (claim); C.V. $= -1.65$; $z = -1.38$; do not reject. There is not enough evidence to support the claim that the proportion is less than 83%.

17. H_0: $p = 0.67$ and H_1: $p \neq 0.67$ (claim); C.V. $= \pm 1.96$; $z = 3.19$; reject. Yes. There is enough evidence to support the claim that the percentage is not 67%.

18. H_0: $p = 0.60$ and H_1: $p < 0.60$ (claim); C.V. $= -1.65$; $z = -1.15$; do not reject. There is not enough evidence to support the claim that the proportion is less than 0.60.

19. H_0: $p = 0.576$ and H_1: $p < 0.576$ (claim); C.V. $= -1.65$; $z = -1.26$; do not reject. There is not enough evidence to support the claim that the proportion is less than 0.576.

20. H_0: $p = 0.194$; H_1: $p > 0.194$ (claim); C.V. $= 1.65$; $z = 2.07$; reject H_0. There is sufficient evidence at $\alpha = 0.05$ to conclude that the proportion is higher than the national proportion.

21. No

22. H_0: $p = 0.20$ and H_1: $p \neq 0.20$ (claim). We have a binomial with $p = 0.20$, $n = 15$. Our P-value is $2 \cdot P(X > 5) = 2(0.061) = 0.122$. Do not reject H_0. There is not enough evidence to conclude that the proportions have changed.

23. $z = \dfrac{X - \mu}{\sigma}$

    $z = \dfrac{X - np}{\sqrt{npq}}$ since $\mu = np$ and $\sigma = \sqrt{npq}$

    $z = \dfrac{X/n - np/n}{\sqrt{npq}/n}$

    $z = \dfrac{X/n - np/n}{\sqrt{npq/n^2}}$

    $z = \dfrac{\hat{p} - p}{\sqrt{pq/n}}$ since $\hat{p} = X/n$

Exercises 8–5

1. a. H_0: $\sigma^2 = 225$ and H_1: $\sigma^2 > 225$; C.V. $= 27.587$; d.f. $= 17$

   b. H_0: $\sigma^2 = 225$ and H_1: $\sigma^2 < 225$; C.V. $= 14.042$; d.f. $= 22$

   c. H_0: $\sigma^2 = 225$ and H_1: $\sigma^2 \neq 225$; C.V. $= 5.629$; 26.119; d.f. $= 14$

   d. H_0: $\sigma^2 = 225$ and H_1: $\sigma^2 \neq 225$; C.V. $= 2.167$; 14.067; d.f. $= 7$

   e. H_0: $\sigma^2 = 225$ and H_1: $\sigma^2 > 225$; C.V. $= 32.000$; d.f. $= 16$

   f. H_0: $\sigma^2 = 225$ and H_1: $\sigma^2 < 225$; C.V. $= 8.907$; d.f. $= 19$

   g. H_0: $\sigma^2 = 225$ and H_1: $\sigma^2 \neq 225$; C.V. $= 3.074$; 28.299; d.f. $= 12$

   h. H_0: $\sigma^2 = 225$ and H_1: $\sigma^2 < 225$; C.V. $= 15.308$; d.f. $= 28$

2. *a.* $0.01 < P$-value < 0.025 (0.015)

    b. $0.005 < P$-value < 0.01 (0.006)

    c. $0.01 < P$-value < 0.02 (0.012)

    d. P-value < 0.005 (0.003)

    e. $0.02 < P$-value < 0.05 (0.037)

    f. $0.05 < P$-value < 0.10 (0.088)

    g. $0.05 < P$-value < 0.10 (0.066)

    h. P-value < 0.01 (0.007)

3. H_0: $\sigma = 60$ (claim) and H_1: $\sigma \neq 60$; C.V. $= 8.672$; 27.587; d.f. $= 17$; $\chi^2 = 19.707$; do not reject. There is not enough evidence to reject the claim that the standard deviation is 60.

4. H_0: $\sigma = 8$; H_1: $\sigma > 8$ (claim); C.V. $= 30.144$; $\chi^2 = 36.033$; reject H_0. There is sufficient evidence to conclude that the standard deviation is greater than 8 degrees.

5. H_0: $\sigma = 15$ and H_1: $\sigma < 15$ (claim); C.V. $= 4.575$; d.f. $= 11$; $\chi^2 = 9.0425$; do not reject. There is not enough evidence to support the claim that the standard deviation is less than 15.

6. H_0: $\sigma^2 = 100$; H_1: $\sigma^2 \neq 100$ (claim); C.V. $= 2.700$, 19.023; d.f. $= 9$; $\chi^2 = 12.189$; do not reject. There is not enough evidence to support the claim that the variance differs from 100.

7. H_0: $\sigma = 1.2$ (claim) and H_1: $\sigma > 1.2$; $\alpha = 0.01$; d.f. $= 14$; $\chi^2 = 31.5$; P-value < 0.005 (0.0047); since P-value < 0.01, reject. There is enough evidence to reject the claim that the standard deviation is less than or equal to 1.2 minutes.

8. H_0: $\sigma = 0.03$ (claim) and H_1: $\sigma > 0.03$; $\alpha = 0.05$; d.f. $= 7$; $\chi^2 = 14.381$; $0.025 < P$-value < 0.05 (0.045); since P-value < 0.05, reject. Yes, there is enough evidence to reject the claim that the standard deviation is less than or equal to 0.03 ounce.

9. H_0: $\sigma = 100$; H_1: $\sigma > 100$ (claim); C.V. $= 12.017$; d.f. $= 7$; $\chi^2 = 11.241$; do not reject. There is not enough evidence to support the claim that the standard deviation is greater than 100 mg.

10. H_0: $\sigma^2 = 100$; H_1: $\sigma^2 > 100$ (claim); C.V. $= 23.685$; $\chi^2 = 25.729$; reject H_0. There is sufficient evidence to conclude that the variance in grades exceeds 100.

11. H_0: $\sigma = 35$ and H_1: $\sigma < 35$ (claim); C.V. $= 3.940$; d.f. $= 10$; $\chi^2 = 8.359$; do not reject. There is not enough evidence to support the claim that the standard deviation is less than 35.

12. H_0: $\sigma = 8$ and H_1: $\sigma > 8$ (claim); C.V. $= 55.758$; d.f. $= 49$; $\chi^2 = 84.4$; reject. Yes. There is enough evidence to support the claim that the standard deviation is greater than 8.

13. H_0: $\sigma = 679.5$; H_1: $\sigma \neq 679.5$ (claim); C.V. $= 5.009$, 24.736; d.f. $= 13$; $\chi^2 = 16.723$; do not reject. There is not

enough evidence to support the claim that the sample standard deviation differs from the estimated standard deviation.

14. H_0: $\sigma = 2385.9$; H_1: $\sigma < 2385.9$ (claim); C.V. $= 1.145$; $\chi^2 = 4.231$; do not reject H_0. There is insufficient evidence to conclude that the standard deviation is less.

15. H_0: $\sigma = 0.52$; H_1: $\sigma > 0.52$ (claim); C.V. $= 30.144$; $\chi^2 = 22.670$; do not reject H_0. There is insufficient evidence to conclude that the standard deviation is outside the guidelines.

Exercises 8–6

1. H_0: $\mu = \$273$; H_1: $\mu \neq \$273$ (claim); C.V. $= \pm 1.96$; $z = 1.31$; $267.03 < \mu < 302.97$; do not reject. There is not enough evidence to support the claim that the mean has changed. The interval supports the result.

2. H_0: $\mu = \$236$; H_1: $\mu \neq \$236$ (claim); C.V. $= \pm 2.539$; d.f. $= 19$; $t = -2.704$; reject H_0. There is sufficient evidence to conclude that the mean cost differs from $236.

    98% C.I.: $185.59 < \mu < 234.41$. They support one another because $236 is outside the interval, implying a difference.

3. H_0: $\mu = \$19{,}150$; H_1: $\mu \neq \$19{,}150$ (claim); C.V. $= \pm 1.96$; $z = -3.69$; $15{,}889 < \mu < 18.151$; reject. There is enough evidence to support the claim that the mean differs from $19,150. Yes, the interval supports the results.

4. H_0: $\mu = 47$ and H_1: $\mu \neq 47$ (claim); C.V. $= \pm 1.65$; $z = -2.26$; reject; $38.35 < \mu < 45.65$. There is enough evidence to support the claim that the mean time has changed. The confidence interval does not contain the hypothesized mean 47.

5. H_0: $\mu = 19$; H_1: $\mu \neq 19$ (claim); C.V. $= \pm 2.145$; d.f. $= 14$; $t = 1.37$; do not reject H_0. There is insufficient evidence to conclude that the mean number of hours differs from 19. 95% C.I.: $17.7 < \mu < 24.9$. Because the mean ($\mu = 19$) is in the interval, there is no evidence to support the idea that a difference exists.

6. H_0: $\mu = 10.8$ (claim) and H_1: $\mu \neq 10.8$; C.V. $= \pm 2.33$; $z = 2.80$; reject; $11.035 < \mu < 13.365$. There is enough evidence to reject the claim that the average time a person spends reading a newspaper is 10.8 minutes. The confidence interval does not contain the hypothesized mean 10.8.

7. The power of a statistical test is the probability of rejecting the null hypothesis when it is false.

8. The power of a test is equal to $1 - \beta$, where β is the probability of a type II error.

9. The power of a test can be increased by increasing α or selecting a larger sample size.

Review Exercises

1. H_0: $\mu = 98°$ (claim) and H_1: $\mu \neq 98°$; C.V. = ± 1.96; $z = -2.02$; reject. There is enough evidence to reject the claim that the average high temperature in the United States is 98°.

2. H_0: $\mu = 25.3$; H_1: $\mu < 25.3$ (claim); C.V. = -2.33; $z = -2.19$; do not reject. There is not enough evidence to support the claim that the average time is less than 25.3 minutes.

3. H_0: $\mu = 18,000$; H_1: $\mu < 18,000$ (claim); C.V. = -2.33; test statistic $z = -3.58$; reject H_0. There is sufficient evidence to conclude that the mean debt is less than $18,000.

4. H_0: $\mu = 10$ and H_1: $\mu < 10$ (claim); $z = -2.22$; P-value = 0.0132; reject. There is enough evidence to support the claim that the average time is less than 10 minutes.

5. H_0: $\mu = 1229$; H_1: $\mu \neq 1229$ (claim); C.V. = ± 1.96; $z = 1.875$; do not reject H_0. There is insufficient evidence to conclude that the rent differs.

6. H_0: $\mu = \$150,000$ and H_1: $\mu > \$150,000$ (claim); C.V. = 1.895; d.f. = 7; $t = 1.04$; do not reject. There is not enough evidence to support the claim that the average salary is greater than $150,000.

7. H_0: $\mu = 10$; H_1: $\mu < 10$ (claim); C.V. = -1.782; d.f. = 12; $t = -2.230$; reject. There is enough evidence to support the claim that the mean weight is less than 10 ounces.

8. H_0: $\mu = 208$; H_1: $\mu > 208$ (claim); C.V. = 2.896; d.f. = 9; $t = 3.13$; reject H_0. There is sufficient evidence that the mean weight is greater than 208 g.

9. H_0: $p = 0.137$; H_1: $p \neq 0.137$ (claim); C.V. = ± 1.96; $z = 1.51$; do not reject H_0. There is insufficient evidence to conclude that the proportion of union membership differs from 13.7%.

10. H_0: $p = 0.602$ and H_1: $p > 0.602$ (claim); C.V. = 1.65; $z = 1.96$; reject. Yes. There is enough evidence to support the claim that the proportion is greater than 0.602.

11. H_0: $p = 0.593$; H_1: $p < 0.593$ (claim); C.V. = -2.33; $z = -2.57$; reject H_0. There is sufficient evidence to conclude that the proportion of free and reduced lunches is less than 59.3%.

12. H_0: $p = 0.65$ (claim) and H_1: $p \neq 0.65$; $z = 1.17$; P-value = 0.242; since P-value > 0.05, do not reject. There is not enough evidence to reject the claim that 65% of teenagers own their own MP3 players. (TI: P-value = 0.2412)

13. H_0: $p = 0.204$; H_1: $p \neq 0.204$ (claim); C.V. = ± 1.96; $z = -1.03$; do not reject. There is not enough evidence to support the claim that the proportion is different from the national proportion.

14. H_0: $\sigma = 3.4$ (claim) and H_1: $\sigma \neq 3.4$; C.V. = 11.689 and 38.076; d.f. = 23; $\chi^2 = 35.1$; do not reject. No, there is not enough evidence to reject the claim that the standard deviation is 3.4 minutes.

15. H_0: $\sigma = 4.3$ (claim) and H_1: $\sigma < 4.3$; d.f. = 19; $\chi^2 = 6.95$; $0.005 < P$-value < 0.01 (0.006); since P-value < 0.05, reject. Yes, there is enough evidence to reject the claim that the standard deviation is greater than or equal to 4.3 miles per gallon.

16. H_0: $\sigma^2 = 3.81$; H_1: $\sigma^2 \neq 3.81$ (claim); C.V. = 5.629, 26.119; d.f. = 14; $\chi^2 = 15.898$; do not reject. There is not enough evidence to support the claim that the variance is different than 3.81.

17. H_0: $\sigma^2 = 40$; H_1: $\sigma^2 \neq 40$ (claim); C.V. = 2.700 and 19.023; test statistic $\chi^2 = 9.68$; do not reject H_0. There is insufficient evidence to conclude that the variance in the number of games played differs from 40.

18. H_0: $\mu = 35$ (claim) and H_1: $\mu \neq 35$; C.V. = ± 1.65; $z = -3.00$; reject; $32.675 < \mu < 34.325$. No. There is enough evidence to reject the claim that the mean is 35 pounds. Yes, the results agree. The mean is not contained in the interval.

19. H_0: $\mu = 4$ and H_1: $\mu \neq 4$ (claim); C.V. = ± 2.58; $z = 1.49$; $3.85 < \mu < 4.55$; do not reject. There is not enough evidence to support the claim that the growth has changed.

Chapter Quiz

1. True
2. True
3. False
4. True
5. False
6. b
7. d
8. c
9. b
10. Type I
11. β
12. Statistical hypothesis
13. Right
14. $n - 1$

15. H_0: $\mu = 28.6$ (claim) and H_1: $\mu \neq 28.6$; $z = 2.15$; C.V. = ± 1.96; reject. There is enough evidence to reject the claim that the average age of the mothers is 28.6 years.

16. H_0: $\mu = \$6500$ (claim) and H_1: $\mu \neq \$6500$; $z = 5.27$; C.V. = ± 1.96; reject. There is enough evidence to reject the agent's claim.

17. H_0: $\mu = 8$ and H_1: $\mu > 8$ (claim); $z = 6$; C.V. = 1.65; reject. There is enough evidence to support the claim that the average is greater than 8.

18. H_0: $\mu = 500$ (claim) and H_1: $\mu \neq 500$; d.f. = 6; $t = -0.571$; C.V. = ± 3.707; do not reject. There is not enough evidence to reject the claim that the mean is 500.

19. H_0: $\mu = 67$ and H_1: $\mu < 67$ (claim); $t = -3.1568$; P-value < 0.005 (0.003); since P-value < 0.05, reject. There is enough evidence to support the claim that the average height is less than 67 inches.

20. H_0: $\mu = 12.4$ and H_1: $\mu < 12.4$ (claim); $t = -2.324$; C.V. $= -1.345$; reject. There is enough evidence to support the claim that the average is less than the company claimed.

21. H_0: $\mu = 63.5$ and H_1: $\mu > 63.5$ (claim); $t = 0.47075$; P-value > 0.25 (0.322); since P-value > 0.05, do not reject. There is not enough evidence to support the claim that the average is greater than 63.5.

22. H_0: $\mu = 26$ (claim) and H_1: $\mu \neq 26$; $t = -1.5$; C.V. $= \pm 2.492$; do not reject. There is not enough evidence to reject the claim that the average is 26.

23. H_0: $p = 0.39$ (claim) and H_1: $p \neq 0.39$; C.V. $= \pm 1.96$; $z = -0.62$; do not reject. There is not enough evidence to reject the claim that 39% took supplements. The study supports the results of the previous study.

24. H_0: $p = 0.55$ (claim) and H_1: $p < 0.55$; $z = -0.8989$; C.V. $= -1.28$; do not reject. There is not enough evidence to reject the survey's claim.

25. H_0: $p = 0.35$ (claim) and H_1: $p \neq 0.35$; C.V. $= \pm 2.33$; $z = 0.666$; do not reject. There is not enough evidence to reject the claim that the proportion is 35%.

26. H_0: $p = 0.75$ (claim) and H_1: $p \neq 0.75$; $z = 2.6833$; C.V. $= \pm 2.58$; reject. There is enough evidence to reject the claim.

27. P-value $= 0.0324$

28. P-value < 0.0001

29. H_0: $\sigma = 6$ and H_1: $\sigma > 6$ (claim); $\chi^2 = 54$; C.V. $= 36.415$; reject. There is enough evidence to support the claim.

30. H_0: $\sigma = 8$ (claim) and H_1: $\sigma \neq 8$; $\chi^2 = 33.2$; C.V. $= 27.991, 79.490$; do not reject. There is not enough evidence to reject the claim that $\sigma = 8$.

31. H_0: $\sigma = 2.3$ and H_1: $\sigma < 2.3$ (claim); $\chi^2 = 13$; C.V. $= 10.117$; do not reject. There is not enough evidence to support the claim that the standard deviation is less than 2.3.

32. H_0: $\sigma = 9$ (claim) and H_1: $\sigma \neq 9$; $\chi^2 = 13.4$; P-value > 0.20 (0.291); since P-value > 0.05, do not reject. There is not enough evidence to reject the claim that $\sigma = 9$.

33. $28.9 < \mu < 31.2$; no

34. $\$6562.81 < \mu < \$6637.19$; no

Chapter 9

Exercises 9–1

1. Testing a single mean involves comparing a sample mean to a specific value such as $\mu = 100$; testing the difference between two means involves comparing the means of two samples, such as $\mu_1 = \mu_2$.

2. When both samples are larger than or equal to 30, the distribution will be approximately normal. The mean of

the differences will be equal to zero. The standard deviation of the differences will be

$$\sqrt{\frac{\sigma_1^2}{n_1} + \frac{\sigma_2^2}{n_2}}$$

3. The populations must be independent of each other, and they must be normally distributed; s_1 and s_2 can be used in place of σ_1 and σ_2 when σ_1 and σ_2 are unknown, but a t test must be used.

4. H_0: $\mu_1 = \mu_2$ or H_0: $\mu_1 - \mu_2 = 0$

5. H_0: $\mu_1 = \mu_2$ (claim) and H_1: $\mu_1 \neq \mu_2$; C.V. $= \pm 2.58$; $z = -0.88$; do not reject. There is not enough evidence to reject the claim that the average lengths of the major rivers are the same. (TI: $z = -0.856$)

6. H_0: $\mu_1 = \mu_2$; H_1: $\mu_1 \neq \mu_2$ (claim); C.V. $= \pm 1.65$; $z = 0.95$; do not reject. There is not enough evidence to reject the claim that the means are different.

7. H_0: $\mu_1 = \mu_2$; H_1: $\mu_1 \neq \mu_2$ (claim); C.V. $= \pm 1.96$; $z = -3.65$; reject. There is sufficient evidence at $\alpha = 0.05$ to conclude that the commuting times differ in the winter.

8. $-1.2363 < \mu_1 - \mu_2 < 6.6363$. Yes, since the interval contains 0.

9. H_0: $\mu_1 = \mu_2$; H_1: $\mu_1 > \mu_2$ (claim); C.V. $= 2.33$; $z = 3.75$; reject. There is sufficient evidence at $\alpha = 0.01$ to conclude that the average hospital stay for men is longer.

10. H_0: $\mu_1 = \mu_2$ (claim) and H_1: $\mu_1 \neq \mu_2$; C.V. $= \pm 2.58$; $z = -3.82$; reject. There is enough evidence to reject the claim that the average costs of the homes in both locations are the same.

11. H_0: $\mu_1 = \mu_2$ and H_1: $\mu_1 < \mu_2$ (claim); C.V. $= -1.65$; $z = -2.01$; reject. There is enough evidence to support the claim that the stayers had a higher grade point average.

12. H_0: $\mu_1 = \mu_2$ and H_1: $\mu_1 > \mu_2$ (claim); C.V. $= 1.65$; $z = 3.65$; reject. There is enough evidence to support the claim that Ohio students are below the national average.

13. H_0: $\mu_1 = \mu_2$; H_1: $\mu_1 \neq \mu_2$ (claim); C.V. $= \pm 1.96$; $z = 0.66$; do not reject. There is not enough evidence to support the claim that there is a difference in the means.

14. H_0: $\mu_1 - \mu_2 = 30$; H_1: $\mu_1 - \mu_2 > 30$ (claim); C.V. $= 1.645$; $z = 1.52$; do not reject. There is insufficient evidence to conclude that the difference in benefits is greater than $30.

15. H_0: $\mu_1 = \mu_2$ and H_1: $\mu_1 \neq \mu_2$ (claim); $z = 1.01$; P-value $= 0.3124$; do not reject. There is not enough evidence to support the claim that there is a difference in self-esteem scores. (TI: P-value $= 0.3131$)

16. H_0: $\mu_1 = \mu_2$ (claim) and H_1: $\mu_1 \neq \mu_2$; $z = -0.76$; P-value $= 0.4472$; do not reject the null hypothesis.

There is not enough evidence to reject the claim that there is no difference in the ages.

17. $2.8 < \mu_1 - \mu_2 < 6.0$

18. $H_0: \mu_1 = \mu_2$ and $H_1: \mu_1 > \mu_2$ (claim); C.V. = 1.65; $z = 5.61$; reject. There is enough evidence to support the claim that the average credit card debt has increased. One possible reason for the increase could be that the price of the merchandise purchased has increased.

19. $10.48 < \mu_1 - \mu_2 < 59.52$. The interval provides evidence to reject the claim that there is no difference in mean scores because the interval for the difference is entirely positive. That is, 0 is not in the interval.

20. $0.3 < \mu_1 - \mu_2 < 0.5$

21. $H_0: \mu_1 - \mu_2 = 8$ (claim) and $H_1: \mu_1 - \mu_2 > 8$; C.V. = $+1.65$; $z = -0.73$; do not reject. There is not enough evidence to reject the claim that private school students have exam scores that are at most 8 points higher than those of students in public schools.

22. $H_0: \mu_1 - \mu_2 = \$3400$; $H_1: \mu_1 - \mu_2 > \$3400$ (claim); C.V. = 1.65; $z = 3.93$; reject. There is enough evidence to support the claim that the difference in the means of the sale prices is greater than $3400.

23. $H_0: \mu_1 - \mu_2 = \$30{,}000$; $H_1: \mu_1 - \mu_2 \neq \$30{,}000$ (claim); C.V. = ± 2.58; $z = 1.22$; do not reject. There is not enough evidence to support the claim that the difference in income is not $30,000.

Exercises 9–2

1. $H_0: \mu_1 = \mu_2$; $H_1: \mu_1 \neq \mu_2$ (claim); C.V. = ± 1.761; d.f. = 14; $t = -1.595$; do not reject. There is not enough evidence to support the claim that the means are different.

2. $H_0: \mu_1 = \mu_2$; $H_1: \mu_1 \neq \mu_2$ (claim); C.V. = ± 2.131; d.f. = 15; $t = -0.942$; do not reject. There is not enough evidence to support the claim that the means are different. (*Note:* In each data set there is a suspected outlier that may make the results suspect.)

3. $H_0: \mu_1 = \mu_2$; $H_1: \mu_1 \neq \mu_2$ (claim); C.V. = ± 2.093; d.f. = 19; $t = 3.811$; reject. There is enough evidence to support the claim that the mean noise levels are different.

4. $H_0: \mu_1 = \mu_2$; $H_1: \mu_1 < \mu_2$ (claim); C.V. = -1.711; d.f. = 24; $t = -4.509$; reject. There is enough evidence to support the claim that the mean age of those playing the slot machines is less than that of those playing roulette.

5. $H_0: \mu_1 = \mu_2$; $H_1: \mu_1 \neq \mu_2$ (claim); C.V. = ± 1.812; d.f. = 10; $t = -1.220$; do not reject. There is not enough evidence to support the claim that the means are not equal.

6. $H_0: \mu_1 = \mu_2$; $H_1: \mu_1 > \mu_2$ (claim); d.f. = 23; $t = 1.921$; the P-value for the t test is $0.025 < P$-value 0.05 (0.031), so the decision is to reject H_0 at 0.05. There is enough evidence to support the claim that the mean salary for the elementary school teachers is greater than the mean salary of the secondary school teachers.

7. $H_0: \mu_1 = \mu_2$; $H_1: \mu_1 \neq \mu_2$ (claim); d.f. = 9; $t = 5.103$; the P-value for the t test is P-value < 0.0001; reject. There is enough evidence to support the claim that the means are different.

8. $H_0: \mu_1 = \mu_2$; $H_1: \mu_1 \neq \mu_2$ (claim); C.V. = ± 2.571; d.f. = 5; $t = 1.351$; do not reject. There is not enough evidence to support the claim that the means are not equal.

9. $3.066 < \mu_1 - \mu_2 < 10.534$
(TI: Interval $3.18 < \mu_1 - \mu_2 < 10.42$)

10. $-2.481 < \mu_1 - \mu_2 < 7.971$
(TI: Interval $-2.24 < \mu_1 - \mu_2 < 7.73$)

11. $H_0: \mu_1 = \mu_2$; $H_1: \mu_1 \neq \mu_2$ (claim); C.V. = ± 2.977; d.f. = 14; $t = 2.60$; do not reject. There is insufficient evidence to conclude a difference in viewing times.

12. $H_0: \mu_1 = \mu_2$ (claim) and $H_1: \mu_1 \neq \mu_2$; C.V. = ± 2.145; $t = -1.70$; do not reject. There is not enough evidence to reject the claim that the means are equal.

13. $H_0: \mu_1 = \mu_2$ and $H_1: \mu_1 > \mu_2$ (claim); C.V. = 3.365; d.f. = 5; $t = 1.057$; do not reject. There is not enough evidence to support the claim that the average number of students attending cyber charter schools in Allegheny County is greater that the average number of students attending cyber charter schools in surrounding counties. One reason why caution should be used is that cyber charter schools are a relatively new concept.

14. $H_0: \mu_1 = \mu_2$; $H_1: \mu_1 > \mu_2$ (claim); $t = 4.36$; P-value 0.00 (0.00005) $< \alpha$; reject. There is sufficient evidence to conclude that the houses in Whiting are older. (TI: P-value = 0.000055)

15. $H_0: \mu_1 = \mu_2$ (claim) and $H_1: \mu_1 \neq \mu_2$; d.f. = 15; $t = 2.385$. The P-value for the t test is $0.02 < P$-value < 0.05 (0.026). Do not reject since P-value > 0.01. There is not enough evidence to reject the claim that the means are equal. $-0.09 < \mu_1 - \mu_2 < 0.89$ (TI: Interval $-0.07 < \mu_1 - \mu_2 < 0.87$)

16. $H_0: \mu_1 = \mu_2$; $H_1: \mu_1 \neq \mu_2$ (claim); C.V. = ± 2.306; $t = 1.17$; do not reject. There is insufficient evidence to conclude a difference in means.

17. $9.87 < \mu_1 - \mu_2 < 219.6$
(TI: Interval $13.23 < \mu_1 - \mu_2 < 216.24$)

18. $\$1789.70 < \mu_1 - \mu_2 < \$12{,}425.41$
(TI: Interval $\$2484.60 < \mu_1 - \mu_2 < \$11{,}731$)

Exercises 9–3

1. a. Dependent   d. Dependent
b. Dependent   e. Independent
c. Independent

2. $H_0: \mu_D = 0$; $H_1: \mu_D > 0$ (claim); C.V. = 1.943; $t = 2.812$; reject. There is sufficient evidence to conclude that the book scores are higher than DVD scores.

3. $H_0: \mu_D = 0$ and $H_1: \mu_D < 0$ (claim); C.V. = -1.397; d.f. = 8; $t = -2.8$; reject. There is enough evidence to support the claim that the seminar increased the number of hours students studied.

4. $H_0: \mu_D = 0$; $H_1: \mu_D > 0$ (claim); C.V. $= 1.895$; $t = 4.249$; reject. There is sufficient evidence to conclude that students did better the second time. Possible reasons: familiar with course; warmed up, etc.

5. $H_0: \mu_D = 0$ and $H_1: \mu_D \neq 0$ (claim); C.V. $= \pm 2.365$; d.f. $= 7$; $t = 1.6583$; do not reject. There is not enough evidence to support the claim that the means are different.

6. $H_0: \mu_D = 0$; $H_1: \mu_D \neq 0$ (claim); C.V. $= \pm 2.365$; $t = -2.411$; reject. There is sufficient evidence to conclude a difference in mean scores.

7. $H_0: \mu_D = 0$ and $H_1: \mu_D > 0$ (claim); C.V. $= 2.571$; d.f. $= 5$; $t = 2.24$; do not reject. There is not enough evidence to support the claim that the errors have been reduced.

8. $H_0: \mu_D = 0$; $H_1: \mu_D > 0$ (claim); C.V. $= 2.015$; $t = 3.060$; reject. There is enough evidence to support the claim that the dogs lost weight.

9. $H_0: \mu_D = 0$ and $H_1: \mu_D \neq 0$ (claim); d.f. $= 7$; $t = 0.978$; $0.20 < P\text{-value} < 0.50$ (0.361). Do not reject since $P\text{-value} > 0.01$. There is not enough evidence to support the claim that there is a difference in the pulse rates. $-3.23 < \mu_D < 5.73$

10. $H_0: \mu_D = 0$; $H_1: \mu_D > 0$ (claim); C.V. $= 2.015$; $t = 2.976$; reject. There is enough evidence to support the claim that the mean number of mistakes has decreased.

11. $\overline{X_1 - X_2} = \Sigma \dfrac{X_1 - X_2}{n} = \Sigma\left(\dfrac{X_1}{n} - \dfrac{X_2}{n}\right)$

$\qquad\qquad = \Sigma\dfrac{X_1}{n} - \Sigma\dfrac{X_2}{n} = \overline{X}_1 - \overline{X}_2$

Exercises 9–4

1a. *a.* $\hat{p} = \frac{34}{48}, \hat{q} = \frac{14}{48}$ *d.* $\hat{p} = \frac{6}{24}, \hat{q} = \frac{18}{24}$

     b. $\hat{p} = \frac{28}{75}, \hat{q} = \frac{47}{75}$ *e.* $\hat{p} = \frac{12}{144}, \hat{q} = \frac{132}{144}$

     c. $\hat{p} = \frac{50}{100}, \hat{q} = \frac{50}{100}$

1b. *a.* 16   *b.* 4   *c.* 48

     d. 104   *e.* 30

2. *a.* $\bar{p} = 0.5$; $\bar{q} = 0.5$

     b. $\bar{p} = 0.5$; $\bar{q} = 0.5$

     c. $\bar{p} = 0.27$; $\bar{q} = 0.73$

     d. $\bar{p} = 0.2125$; $\bar{q} = 0.7875$

     e. $\bar{p} = 0.216$; $\bar{q} = 0.784$

3. $\hat{p}_1 = 0.593$; $\hat{p}_2 = 0.463$; $\bar{p} = 0.528$; $\bar{q} = 0.472$; $H_0: p_1 = p_2$; $H_1: p_1 > p_2$ (claim); C.V. $= 1.65$; $z = 3.19$; reject. There is enough evidence to support the claim that the proportion of married men is greater than the proportion of married women.

4. $\hat{p}_1 = 0.60$; $\hat{p}_2 = 0.80$; $\bar{p} = 0.69$; $\bar{q} = 0.31$; $H_0: p_1 = p_2$ and $H_1: p_1 \neq p_2$ (claim); C.V. $= \pm 2.58$; $z = -5.053$; reject. There is enough evidence to support the claim that the proportions are different.

5. $\hat{p}_1 = 0.817$; $\hat{p}_2 = 0.783$; $\bar{p} = 0.8$; $\bar{q} = 0.2$; $H_0: p_1 = p_2$; $H_1: p_1 \neq p_2$ (claim); C.V. $= \pm 1.96$; $z = 2.04$; reject. There is enough evidence to support the claim that the proportions are different.

6. $\hat{p}_1 = \frac{10}{73} = 0.14$; $\hat{p}_2 = \frac{16}{80} = 0.20$; $\bar{p} = 0.17$; $\bar{q} = 0.83$; $H_0: p_1 = p_2$ and $H_1: p_1 \neq p_2$ (claim); C.V. $= \pm 1.96$; $z = -0.99$ (TI: $z = -1.04$); do not reject. No, there is not enough evidence to support the claim that there is a difference in the proportions. $-0.181 < p_1 - p_2 < 0.055$

7. $\hat{p}_1 = 0.83$; $\hat{p}_2 = 0.75$; $\bar{p} = 0.79$; $\bar{q} = 0.21$; $H_0: p_1 = p_2$ (claim) and $H_1: p_1 \neq p_2$; C.V. $= \pm 1.96$; $z = 1.39$; do not reject. There is not enough evidence to reject the claim that the proportions are equal. $-0.032 < p_1 - p_2 < 0.192$

8. $\hat{p}_1 = 0.88$; $\hat{p}_2 = 0.96$; $\bar{p} = 0.92$; $\bar{q} = 0.08$; $H_0: p_1 = p_2$ and $H_1: p_1 \neq p_2$ (claim); C.V. $= \pm 1.65$; $z = -1.47$; do not reject. (TI: Interval $-0.168 < p_1 - p_2 < 0.008$) There is not enough evidence to support the claim that there is a difference in the proportions. A researcher might want to find out why people feel that they have less leisure time now as opposed to 10 years ago. Are they working more? Raising a family? etc.

9. $\hat{p}_1 = 0.55$; $\hat{p}_2 = 0.45$; $\bar{p} = 0.5$; $\bar{q} = 0.5$; $H_0: p_1 = p_2$ and $H_1: p_1 \neq p_2$ (claim); C.V. $= \pm 2.58$; $z = 1.23$; do not reject. There is not enough evidence to support the claim that the proportions are different. $(-0.103 < p_1 - p_2 < 0.291)$

10. $\hat{p}_1 = \frac{130}{200} = 0.65$; $\hat{p}_2 = \frac{63}{300} = 0.21$; $\bar{p} = 0.386$; $\bar{q} = 0.614$; $H_0: p_1 = p_2$ and $H_1: p_1 > p_2$ (claim); $z = 9.90$; $P\text{-value} < 0.001$; reject since $P\text{-value} < 0.01$. There is enough evidence to support the claim that men are more safety conscious than women.

11. $\hat{p}_1 = 0.347$; $\hat{p}_2 = 0.433$; $\bar{p} = 0.385$; $\bar{q} = 0.615$; $H_0: p_1 = p_2$ and $H_1: p_1 \neq p_2$ (claim); C.V. $= \pm 1.96$; $z = -1.03$; do not reject. There is not enough evidence to say that the proportion of dog owners has changed $(-0.252 < p_1 - p_2 < 0.079)$. Yes, the confidence interval contains 0. This is another way to conclude that there is no difference in the proportions.

12. $\hat{p}_1 = 0.065$; $\hat{p}_2 = 0.08$; $\bar{p} = 0.0725$; $\bar{q} = 0.9275$; $H_0: p_1 = p_2$; $H_1: p_1 \neq p_2$ (claim); C.V. $= \pm 1.96$; $z = -0.58$; do not reject. There is insufficient evidence to conclude a difference.

13. $\hat{p}_1 = 0.25$; $\hat{p}_2 = 0.31$; $\bar{p} = 0.286$; $\bar{q} = 0.714$; $H_0: p_1 = p_2$ and $H_1: p_1 \neq p_2$ (claim); C.V. $= \pm 2.58$; $z = -1.45$; do not reject. There is not enough evidence to support the claim that the proportions are different. $-0.165 < p_1 - p_2 < 0.045$

14. $\hat{p}_1 = 0.287$; $\hat{p}_2 = 0.347$; $\bar{p} = 0.317$; $\bar{q} = 0.683$; $H_0: p_1 = p_2$; $H_1: p_1 < p_2$ (claim); C.V. $= -1.645$; $z = -1.12$; do not reject. There is insufficient evidence to conclude that the proportion of women is higher.

15. $0.077 < p_1 - p_2 < 0.323$

16. $\hat{p}_1 = 0.71$; $\hat{p}_2 = 0.74$; $\bar{p} = 0.724$; $\bar{q} = 0.276$; $H_0: p_1 = p_2$; $H_1: p_1 \neq p_2$ (claim); C.V. $= \pm 2.58$; $z = -0.78$; do not reject. There is not enough evidence to support the claim that the proportions are different.

17. $\hat{p}_1 = 0.4$; $\hat{p}_2 = 0.295$; $\bar{p} = 0.3475$; $\bar{q} = 0.6525$; $H_0: p_1 = p_2$; $H_1: p_1 \neq p_2$ (claim); C.V. $= \pm 2.58$; $z = 2.21$;

do not reject. There is not enough evidence to support the claim that the proportions are different.

18. $\hat{p}_1 = 0.278$; $\hat{p}_2 = 0.26$; $-0.0961 < p_1 - p_2 < 0.1319$. The interval does not support the claim that there is a difference because 0 is contained in the interval and thus allows for the possibility that no difference exists.

19. $-0.0631 < p_1 - p_2 < 0.0667$. It does agree with the *Almanac* statistics stating a difference of -0.042 since -0.042 is contained in the interval.

20. No, p_1 could equal p_3.

Exercises 9–5

1. The variance in the numerator should be the larger of the two variances.

2. The larger variance is placed in the numerator of the formula; hence, $F \geq 1$.

3. One degree of freedom is used for the variance associated with the numerator, and one is used for the variance associated with the denominator.

4. The characteristics of the F distribution are as follows:
  a. The values of F cannot be negative.
  b. The distribution is positively skewed.
  c. The mean value of the F distribution is approximately equal to 1.
  d. The F distribution is a family of curves based on the degrees of freedom.

5. *a.* d.f.N. = 15, d.f.D. = 22; C.V. = 3.36
  b. d.f.N. = 24, d.f.D. = 13; C.V. = 3.59
  c. d.f.N. = 45, d.f.D. = 29; C.V. = 2.03
  d. d.f.N. = 20, d.f.D. = 16; C.V. = 2.28
  e. d.f.N. = 10, d.f.D. = 10; C.V. = 2.98

6. Specific P-values are in parentheses.
  a. $0.025 < P\text{-value} < 0.05$ (0.033)
  b. $0.05 < P\text{-value} < 0.10$ (0.072)
  c. $P\text{-value} = 0.05$
  d. $0.005 < P\text{-value} < 0.01$ (0.006)
  e. $P\text{-value} = 0.05$
  f. $P\text{-value} > 0.10$ (0.112)
  g. $0.05 < P\text{-value} < 0.10$ (0.068)
  h. $0.01 < P\text{-value} < 0.02$ (0.015)

7. H_0: $\sigma_1^2 = \sigma_2^2$; H_1: $\sigma_1^2 \neq \sigma_2^2$ (claim); C.V. = ± 1.88; d.f.N. = 59; d.f.D. = 59; $F = 1.981$; reject. There is enough evidence to support the claim that the variances are not equal.

8. H_0: $\sigma_1^2 = \sigma_2^2$; H_1: $\sigma_1^2 \neq \sigma_2^2$ (claim); C.V. = 2.15; d.f.N. = 29; d.f.D. = 29; $F = 1.563$; do not reject. There is not enough evidence to support the claim that the variances are different.

9. H_0: $\sigma_1^2 = \sigma_2^2$; H_1: $\sigma_1^2 \neq \sigma_2^2$ (claim); C.V. = 3.430; d.f.N. = 12; d.f.D. = 11; $F = 2.085$; do not reject. There

is not enough evidence to support the claim that the variances are different.

10. H_0: $\sigma_1 = \sigma_2$ and H_1: $\sigma_1 \neq \sigma_2$ (claim); C.V. = 2.51; d.f.N. = 23; d.f.D. = 19; $F = 3.346$; reject. There is enough evidence to support the claim that the standard deviations are different.

11. H_0: $\sigma_1^2 = \sigma_2^2$ and H_1: $\sigma_1^2 \neq \sigma_2^2$ (claim); C.V. = 4.99; d.f.N. = 7; d.f.D. = 7; $F = 1$; do not reject. There is not enough evidence to support the claim that there is a difference in the variances.

12. H_0: $\sigma_1^2 = \sigma_2^2$; H_1: $\sigma_1^2 \neq \sigma_2^2$ (claim); C.V. = 4.36; d.f.N. = 9; d.f.D. = 8; $F = 6.187$; reject. There is enough evidence to support the claim that the variances are not equal.

13. H_0: $\sigma_1^2 = \sigma_2^2$; H_1: $\sigma_1^2 > \sigma_2^2$ (claim); C.V. = 4.950; $F = 9.801$; reject. There is sufficient evidence at $\alpha = 0.05$ to conclude that the variance in area is greater for Eastern cities. C.V. = 10.67; do not reject. There is insufficient evidence to conclude the variance is greater at $\alpha = 0.01$.

14. H_0: $\sigma_1^2 = \sigma_2^2$ and H_1: $\sigma_1^2 \neq \sigma_2^2$ (claim); C.V. = 2.75; d.f.N. = 10; d.f.D. = 12; $F = 2.9707$; reject. There is enough evidence to support the claim that the variances are not equal.

15. H_0: $\sigma_1^2 = \sigma_2^2$ and H_1: $\sigma_1^2 \neq \sigma_2^2$ (claim); C.V. = 4.03; d.f.N. = 9; d.f.D. = 9; $F = 1.1026$; do not reject. There is not enough evidence to support the claim that the variances are not equal.

16. H_0: $\sigma_1^2 = \sigma_2^2$ and H_1: $\sigma_1^2 < \sigma_2^2$ (claim); C.V. = 3.15; d.f.N. = 19; d.f.D. = 19; $F = 1.45$; do not reject. There is not enough evidence to support the claim that the variance of the areas for the counties in Indiana is less than the variance of the areas for the counties in Iowa.

17. H_0: $\sigma_1^2 = \sigma_2^2$ (claim) and H_1: $\sigma_1^2 \neq \sigma_2^2$; C.V. = 3.87; d.f.N. = 6; d.f.D. = 7; $F = 3.18$; do not reject. There is not enough evidence to reject the claim that the variances of the heights are equal.

18. H_0: $\sigma_1^2 = \sigma_2^2$ and H_1: $\sigma_1^2 > \sigma_2^2$ (claim); $F = 2.91$; d.f.N. = 29; d.f.D. = 29; $P\text{-value} < 0.005$ (0.003); reject. There is enough evidence to support the claim that the variation in the salaries of the elementary school teachers is greater than the variation in the salaries of the secondary school teachers.

19. H_0: $\sigma_1^2 = \sigma_2^2$ (claim) and H_1: $\sigma_1^2 \neq \sigma_2^2$; $F = 5.32$; d.f.N. = 14; d.f.D. = 14; $P\text{-value} < 0.01$ (0.004); reject. There is enough evidence to reject the claim that the variances of the weights are equal. The variance for men is 2.363 and the variance for women is 0.444.

20. H_0: $\sigma_1^2 = \sigma_2^2$; H_1: $\sigma_1^2 \neq \sigma_2^2$ (claim); C.V. = 4.03; $F = 1.178$; do not reject. There is insufficient evidence to conclude a difference in variances.

Review Exercises

1. H_0: $\mu_1 = \mu_2$ and H_1: $\mu_1 > \mu_2$ (claim); C.V. = 2.33; $z = 0.59$; do not reject. There is not enough evidence to support the claim that single drivers do more pleasure driving than married drivers.

2. H_0: $\mu_1 = \mu_2$; H_1: $\mu_1 \neq \mu_2$ (claim); C.V. $= \pm 2.58$; $z = 3.04$; reject. There is sufficient evidence to conclude a difference in mean earnings.

3. H_0: $\mu_1 = \mu_2$; H_1: $\mu_1 > \mu_2$ (claim); C.V. $= 1.729$; $t = 4.595$; reject. There is sufficient evidence to conclude that single persons spend a greater time communicating.

4. H_0: $\mu_1 = \mu_2$ and H_1: $\mu_1 > \mu_2$ (claim); C.V. $= 1.318$; $t = 1.324$; do not reject. There is not enough evidence to support the claim that it is warmer in Birmingham.

5. H_0: $\mu_1 = \mu_2$ and H_1: $\mu_1 \neq \mu_2$ (claim); C.V. $= \pm 2.624$; d.f. $= 14$; $t = 6.54$; reject. Yes, there is enough evidence to support the claim that there is a difference in the teachers' salaries. $\$3494.80 < \mu_1 - \mu_2 < \$8021.20$

6. H_0: $\mu_1 = \mu_2$ and H_1: $\mu_1 \neq \mu_2$ (claim); d.f. $= 2$; $t = -0.81$; do not reject. Since $p > 0.10$, there is not enough evidence to support the claim that the means are different. A cafeteria manager would want to know the results to make a decision on which beverage to serve.

7. H_0: $\mu_D = 10$; H_1: $\mu_D > 10$ (claim); C.V. $= 2.821$; $t = 3.249$; reject. There is sufficient evidence to conclude that the difference in temperature is greater than 10 degrees.

8. H_0: $\mu_D = 0$ and H_1: $\mu_D < 0$ (claim); C.V. $= -1.895$; d.f. $= 7$; $t = -2.73$; reject. There is enough evidence to support the claim that the music has increased production; however other things (e.g. experience) could have changed as well.

9. H_0: $p_1 = p_2$; H_1: $p_1 \neq p_2$ (claim); C.V. $= \pm 1.96$; $z = -1.45$; do not reject. There is not enough evidence to support the claim that the proportions are different.

10. $\hat{p}_1 = 0.2$; $\hat{p}_2 = 0.15$; $\bar{p} = 0.17$; $\bar{q} = 0.83$; H_0: $p_1 = p_2$; H_1: $p_1 \neq p_2$ (claim); C.V. $= \pm 1.96$; $z = 1.28$; do not reject. There is insufficient evidence to conclude that there is a difference in proportions.

11. H_0: $\sigma_1 = \sigma_2$ and H_1: $\sigma_1 \neq \sigma_2$ (claim); C.V. $= 2.77$; $\alpha = 0.10$; d.f.N. $= 23$; d.f.D. $= 10$; $F = 10.365$; reject. There is enough evidence to support the claim that there is a difference in the standard deviations.

12. H_0: $\sigma_1^2 = \sigma_2^2$; H_1: $\sigma_1^2 \neq \sigma_2^2$ (claim); C.V. $= 4.90$; $F = 4.623$; do not reject. There is insufficient evidence to conclude a difference in variances.

13. H_0: $\sigma_1^2 = \sigma_2^2$; H_1: $\sigma_1^2 \neq \sigma_2^2$ (claim); C.V. $= 2.45$; d.f.N. $= 24$; d.f.D. $= 19$; $F = 1.631$; do not reject. There is not enough evidence to support the claim that the standard deviations are different. Store Z's paint would have to have a standard deviation of $3.33.

Chapter Quiz

1. False
2. False
3. True
4. False
5. *d*
6. *a*
7. *c*
8. *a*

9. $\mu_1 = \mu_2$

10. *t*

11. Normal

12. Negative

13. $\dfrac{s_1^2}{s_2^2}$

14. H_0: $\mu_1 = \mu_2$ and H_1: $\mu \neq \mu_2$ (claim); $z = -3.69$; C.V. $= \pm 2.58$; reject. There is enough evidence to support the claim that there is a difference in the cholesterol levels of the two groups. $-10.2 < \mu_1 - \mu_2 < -1.8$

15. H_0: $\mu_1 = \mu_2$ and H_1: $\mu > \mu_2$ (claim); C.V. $= 1.28$; $z = 1.60$; reject. There is enough evidence to support the claim that the average rental fees for the apartments in the East are greater than the average rental fees for the apartments in the West.

16. H_0: $\mu_1 = \mu_2$ and H_1: $\mu_1 \neq \mu_2$ (claim); $t = 11.094$; C.V. $= \pm 2.779$; reject. There is enough evidence to support the claim that the average prices are different. $0.298 < \mu_1 - \mu_2 < 0.502$ (TI: Interval $0.2995 < \mu_1 - \mu_2 < 0.5005$)

17. H_0: $\mu_1 = \mu_2$ and H_1: $\mu_1 < \mu_2$ (claim); C.V. $= -1.860$; d.f. $= 8$; $t = -4.05$; reject. There is enough evidence to support the claim that accidents have increased.

18. H_0: $\mu_1 = \mu_2$ and H_1: $\mu_1 \neq \mu_2$ (claim); $t = 9.807$; C.V. $= \pm 2.718$; reject. There is enough evidence to support the claim that the salaries are different. $\$6653 < \mu_1 - \mu_2 < \$11,757$ (TI: Interval $\$6619 < \mu_1 - \mu_2 < \$11,491$)

19. H_0: $\mu_1 = \mu_2$ and H_1: $\mu_1 > \mu_2$ (claim); d.f. $= 10$; $t = 0.874$; $0.10 < P\text{-value} < 0.25$ (0.198); do not reject since $P\text{-value} > 0.05$. There is not enough evidence to support the claim that the incomes of city residents are greater than the incomes of rural residents.

20. H_0: $\mu_D = 0$ and H_1: $\mu_D < 0$ (claim); $t = -4.17$; C.V. $= -2.821$; reject. There is enough evidence to support the claim that the sessions improved math skills.

21. H_0: $\mu_D = 0$ and H_1: $\mu_D < 0$ (claim); $t = -1.71$; C.V. $= -1.833$; do not reject. There is not enough evidence to support the claim that egg production was increased.

22. H_0: $p_1 = p_2$ and H_1: $p_1 \neq p_2$ (claim); $z = -0.69$; C.V. $= \pm 1.65$; do not reject. There is not enough evidence to support the claim that the proportions are different. $-0.105 < p_1 - p_2 < 0.045$

23. H_0: $p_1 = p_2$ and H_1: $p_1 \neq p_2$ (claim); C.V. $= \pm 1.96$; $z = 0.544$; do not reject. There is not enough evidence to support the claim that the proportions have changed. $-0.026 < p_1 - p_2 < 0.0460$. Yes, the confidence interval contains 0; hence, the null hypothesis is not rejected.

24. H_0: $\sigma_1^2 = \sigma_2^2$ and H_1: $\sigma_1^2 \neq \sigma_2^2$ (claim); $F = 1.637$; d.f.N. $= 17$; d.f.D. $= 14$; $P\text{-value} > 0.20$ (0.357). Do not reject since $P\text{-value} > 0.05$. There is not enough evidence to support the claim that the variances are different.

25. H_0: $\sigma_1^2 = \sigma_2^2$ and H_1: $\sigma_1^2 \neq \sigma_2^2$ (claim); $F = 1.296$; C.V. $= 1.90$; do not reject. There is not enough evidence to support the claim that the variances are different.

Chapter 10

Exercises 10–1

1. Two variables are related when a discernible pattern exists between them.

2. Relationships are measured by the correlation coefficient. When r is near $+1$, there is a strong positive linear relationship between the variables. When r is near -1, there is a strong negative linear relationship. When r is near 0, there is no linear relationship between the variables.

3. r, ρ (rho)

4. The range of r is from -1 to $+1$.

5. A positive relationship means that as x increases, y increases. A negative relationship means that as x increases, y decreases.

6. Answers will vary.

7. Answers will vary.

8. The diagram is called a scatter plot. It shows the nature of the relationship.

9. Pearson product moment correlation coefficient

10. t test

11. There are many other possibilities, such as chance or relationship to a third variable.

12. H_0: $\rho = 0$; H_1: $\rho \neq 0$; $r = -0.367$; C.V. $= \pm 0.811$; do not reject. There is not a significant linear relationship between the gasoline tax and the fuel use per registered vehicle.

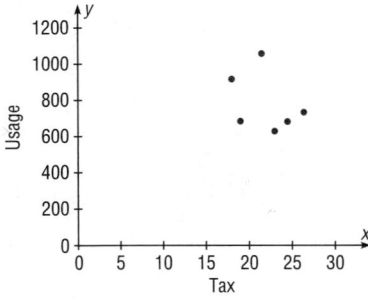

13. H_0: $\rho = 0$; H_1: $\rho \neq 0$; $r = 0.880$; C.V. $= \pm 0.666$; reject. There is sufficient evidence to conclude that a significant relationship exists between the number of releases and gross receipts.

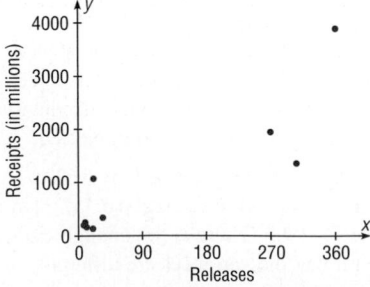

14. H_0: $\rho = 0$; H_1: $\rho \neq 0$; $r = 0.771$; C.V. $= \pm 0.707$; reject. There is a significant linear relationship between the number of forest fires and the number of acres burned.

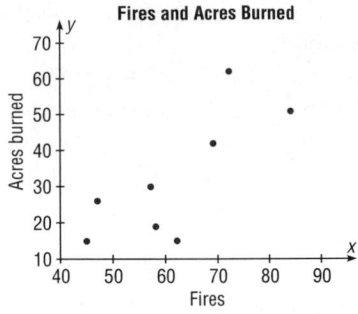

15. H_0: $\rho = 0$; H_1: $\rho \neq 0$; $r = -0.883$; C.V. $= \pm 0.811$; reject. There is a significant relationship between the number of years a person has been out of school and his or her contribution.

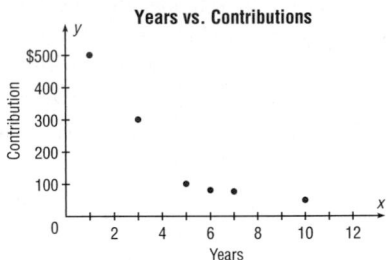

16. H_0: $\rho = 0$; H_1: $\rho \neq 0$; $r = 0.518$; C.V. $= \pm 0.878$; do not reject. There is insufficient evidence to conclude a relationship exists between per capita debt and tax.

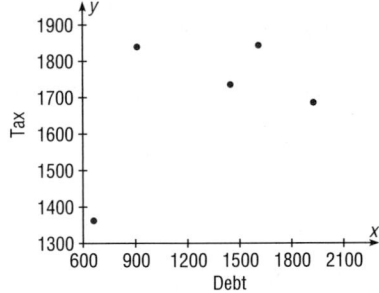

17. H_0: $\rho = 0$; H_1: $\rho \neq 0$; $r = 0.401$; C.V. $= \pm 0.811$; do not reject. There is not a significant linear relationship between the number of local school districts and the corresponding number of secondary schools.

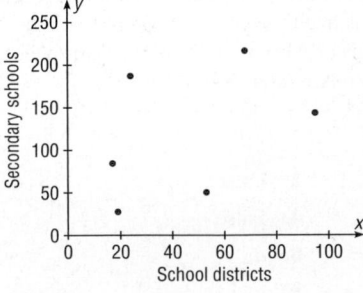

18. $H_0: \rho = 0$; $H_1: \rho \neq 0$; $r = -0.543$; C.V. $= \pm 0.811$; do not reject. There is not a significant linear relationship between the number of triples and the number of home runs.

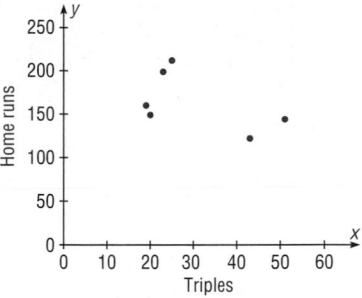

19. $H_0: \rho = 0$; $H_1: \rho \neq 0$; $r = -0.833$; C.V. $= \pm 0.811$; reject. There is sufficient evidence to conclude a relationship exists between the number of eggs produced and the price per dozen.

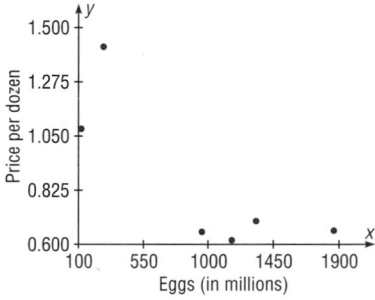

20. $H_0: \rho = 0$; $H_1: \rho \neq 0$; $r = 0.811$; C.V. $= \pm 0.754$; reject. There is a significant relationship between the temperature and the number of emergency calls received

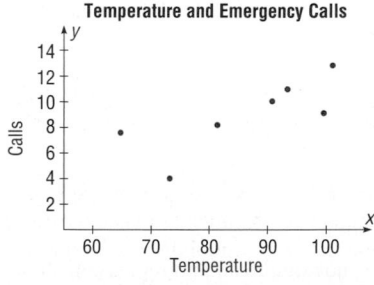

21. $H_0: \rho = 0$; $H_1: \rho \neq 0$; $r = 0.812$; C.V. $= \pm 0.754$; reject. There is a significant linear relationship between the number of faculty and the number of students at small colleges. When the values for x and y are switched, the

results are identical. The independent variable is most likely the number of students.

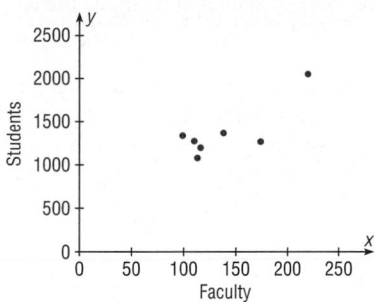

22. $H_0: \rho = 0$; $H_1: \rho \neq 0$; $r = 0.813$; C.V. $= \pm 0.811$; reject. There is a significant linear relationship between the precipitation and the amount of snow/sleet.

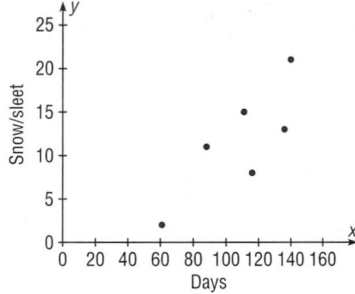

23. $H_0: \rho = 0$; $H_1: \rho \neq 0$; $r = 0.883$; C.V. $= \pm 0.754$; reject. There is a significant linear relationship between the average daily temperature and the average monthly precipitation.

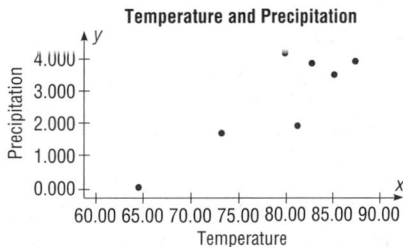

24. $H_0: \rho = 0$; $H_1: \rho \neq 0$; $r = 0.861$; C.V. $= \pm 0.754$; reject. There is a significant relationship between the number of assists and the total number of points.

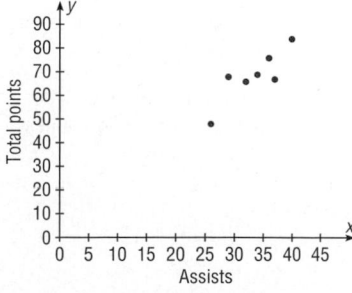

25. $H_0: \rho = 0$; $H_1: \rho \neq 0$; $r = 0.993$; C.V. $= \pm0.811$; reject. There is a significant linear relationship between the fat calories and the amount of saturated fat in the breakfast foods.

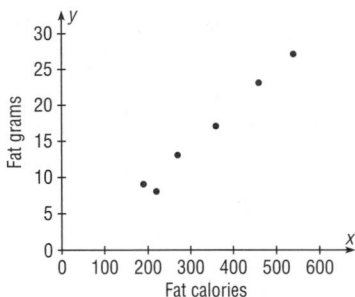

26. $H_0: \rho = 0$; $H_1: \rho \neq 0$; $r = 0.797$; C.V. $= \pm0.632$; reject. There is a significant linear relationship between the height of buildings and the number of stories these buildings contain.

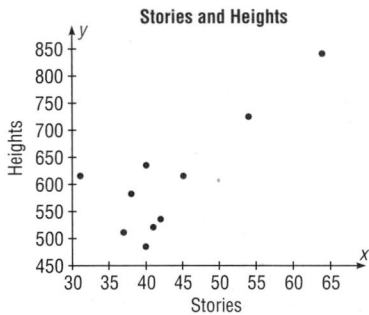

27. $H_0: \rho = 0$; $H_1: \rho \neq 0$; $r = 0.831$; C.V. $= \pm0.754$; reject. There is a significant linear relationship between the number of licensed beds in a hospital and the number of staffed beds.

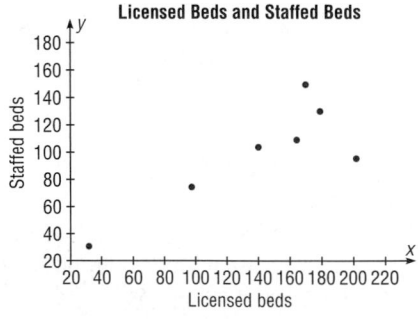

28. $r = 0.831$. The results are the same. (*Note:* There may be a slight difference due to rounding.)

29. $r = 1.00$: All values fall in a straight line. $r = 1.00$: The value of r between x and y is the same when x and y are interchanged.

30. $r = 0$. The relationship is nonlinear as shown.

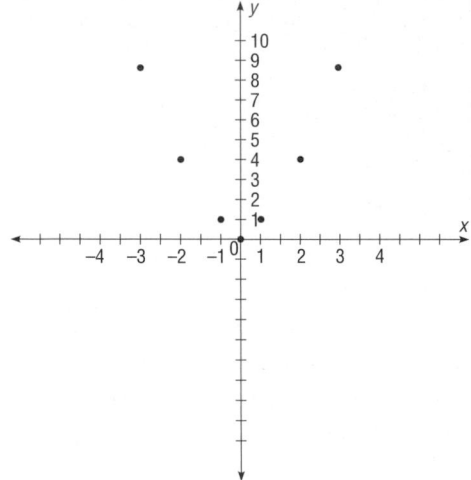

Exercises 10–2

1. A scatter plot should be drawn, and the value of the correlation coefficient should be tested to see whether it is significant.

2. 1. For any specific value of the independent variable x, the value of the dependent variable y must be normally distributed about the regression line.

   2. The standard deviation of each of the dependent variables must be the same for each value of the independent variable.

3. $y' = a + bx$

4. b, a

5. It is the line that is drawn through the points on the scatter plot such that the sum of the squares of the vertical distances from each point to the line is a minimum.

6. r would equal $+1$ or -1.

7. When r is positive, b will be positive. When r is negative, b will be negative.

8. They would be clustered closer to the line.

9. The closer r is to $+1$ or -1, the more accurate the predicted value will be.

10. If the value of r is not significant, no regression should be done. Any regression line is meaningless.

11. When r is not significant, the mean of the y values should be used to predict y.

12. Not significant so no regression should be done.

13. $y' = 181.661 + 7.319x$; $y' = 1645.5$ (million $)

14. $y' = -31.46 + 1.036x$; 30.7

15. $y' = 453.176 - 50.439x$; 251.42

16. Not significant so no regression should be done.

17. Since r is not significant, no regression should be done.

18. Since r is not significant, no regression should be done.

19. $y' = 1.252 - 0.000398x$; $y' = 0.615$ per dozen

20. $y' = -7.544 + 0.190x$; 7.656, or 8 calls

21. $y' = -14.974 + 0.111x$

22. $y' = -7.327 + 0.175x$; 10.173 in

23. $y' = -8.994 + 0.1448x$; 1.1

24. $y' = 2.693 + 1.962x$; 62

25. $y' = -2.417 + 0.055x$; 19.6 grams

26. $y' = 206.399 + 9.262x$; 613.9

27. $y' = 22.659 + 0.582x$; 48.267

28. $H_0: \rho = 0$; $H_1: \rho \neq 0$; $r = -0.514$; C.V. $= \pm 0.811$; do not reject. There is no significant relationship between the number of fireworks in use and the number of related injuries. No regression should be done.

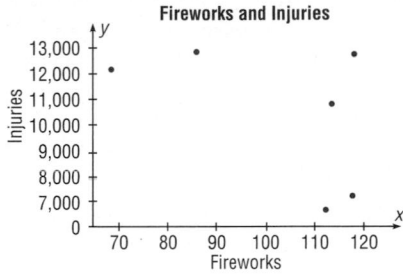

29. $H_0: \rho = 0$; $H_1: \rho \neq 0$; $r = 0.429$; C.V. $= \pm 0.811$; do not reject. There is insufficient evidence to conclude a relationship exists between number of farms and acreage.

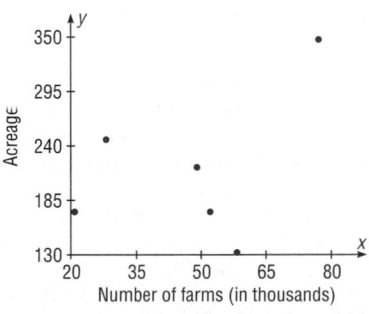

30. $H_0: \rho = 0$; $H_1: \rho \neq 0$; $r = 0.99$; C.V. $= \pm 0.811$; reject. There is sufficient evidence to conclude a relationship exists between verbal and mathematical scores. $y' = 63.472 + 0.900x$

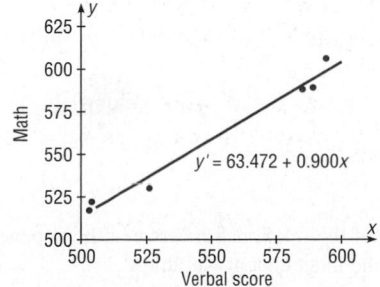

31. $H_0: \rho = 0$; $H_1: \rho \neq 0$; $r = 0.970$; C.V. $= \pm 0.707$; reject; $y' = -33.358 + 6.703x$; when $x = 500$, $y' = 3318.142$. There is a significant relationship between number of employees and tons of coal produced.

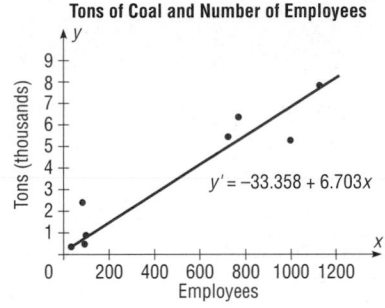

32. $H_0: \rho = 0$; $H_1: \rho \neq 0$; $r = 0.839$; C.V. $= \pm 0.632$; reject. There is a significant linear relationship between the number of viewers of last year's show and the number of viewers of the same shows this year. $y' = -3.668 + 1.281x$

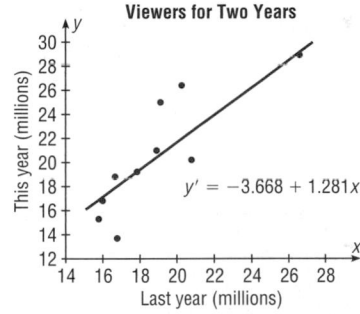

33. $H_0: \rho = 0$; $H_1: \rho \neq 0$; $r = -0.981$; C.V. $= \pm 0.811$; reject. There is a significant relationship between the number of absences and the final grade; $y' = 96.784 - 2.668x$.

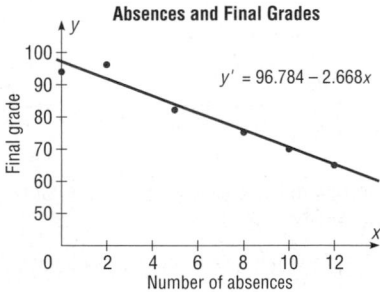

34. H_0: $\rho = 0$; H_1: $\rho \neq 0$; $r = -0.306$; d.f. = 6; $t = -0.787$; $0.20 < P$-value < 0.50 (0.462); do not reject since P-value > 0.05. There is no significant relationship between the weights of the fathers and sons. Since r is not significant, no regression analysis should be done.

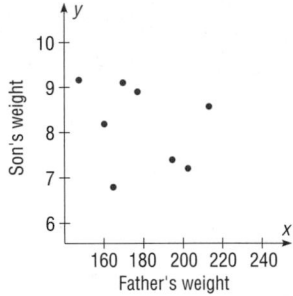

35. H_0: $\rho = 0$; H_1: $\rho \neq 0$; $r = -0.265$; P-value > 0.05 (0.459); do not reject. There is no significant linear relationship between the ages of billionaires and their net worth. No regression should be done.

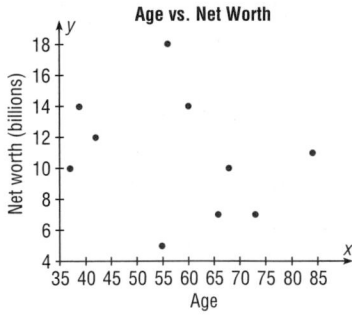

36. $\bar{y} = y' = 1031.44$; $\bar{y} = y' = 184$; $\bar{y} = y' = 136.6$; in all cases, $\bar{y} = y'$; hence, the regression line will always pass through the point $(\overline{X}, \overline{Y})$. Slight differences occur due to rounding.

37. 453.173; regression should not be done

38. $r = -0.543$; $r = 0.812$

Exercises 10–3

1. Explained variation is the variation due to the relationship. It is computed by $\Sigma(y' - \bar{y})^2$.

2. Unexplained variation is the variation due to chance. It is computed by $\Sigma(y - y')^2$.

3. Total variation is the sum of the squares of the vertical distances of the points from the mean. It is computed by $\Sigma(y - \bar{y})^2$.

4. The coefficient of determination is a measure of variation of the dependent variable that is explained by the regression line and the independent variable.

5. The coefficient of determination is found by squaring the value of the correlation coefficient.

6. It is the percent of the variation in y that is not due to the variation in x.

7. The coefficient of nondetermination is found by subtracting r^2 from 1.

8. $R^2 = 0.64$; 64% of the variation of y is due to the variation of x; 36% is due to chance.

9. $R^2 = 0.5625$; 56.25% of the variation of y is due to the variation of x; 43.75% is due to chance.

10. $R^2 = 0.1225$; 12.25% of the variation of y is due to the variation of x; 87.75% is due to chance.

11. $R^2 = 0.1764$; 17.64% of the variation of y is due to the variation of x; 82.36% is due to chance.

12. $R^2 = 0.0324$; 3.24% of the variation of y is due to the variation of x; 96.76% is due to chance.

13. $R^2 = 0.8281$; 82.81% of the variation of y is due to the variation of x; 17.19% is due to chance.

14. The standard error of the estimate is the standard deviation of the observed y values about the predicted y' values. It can be used when you are using the t distribution.

15. 629.4862

16. 12.03* (TI value 12.06)

17. 94.22*

18. The standard error should not be calculated.

19. $365.88 < y' < 2925.04$*

20. The prediction interval should not be calculated.

21. $\$30.46 < y < \$472.38$*

22. The prediction interval should not be calculated.

*Answers may vary due to rounding.

Exercises 10–4

1. Simple regression has one dependent variable and one independent variable. Multiple regression has one dependent variable and two or more independent variables.

2. $y' = a + b_1x_1 + b_2x_2 + \cdots + b_kx_k$; a is the slope and the b's are the partial regression coefficients.

3. The relationship would include all variables in one equation.

4. Normality, equal variance, linearity, nonmulticollinearity, and independence

5. They will all be smaller.   **6.** $40,834

7. 3.48 or 3

8. $196.49

9. 85.75 (grade) or 86

10. 149.885 or 150

11. R is the strength of the relationship between the dependent variable and all the independent variables.

12. 0 to 1

13. R^2 is the coefficient of multiple determination. R^2_{adj} is adjusted for sample size and number of predictors.

14. H_0: $\rho = 0$ and H_1: $\rho \neq 0$

15. F test

16. It is the adjusted coefficient of multiple determination. It is computed when sample size is small and is a better estimate since R^2 is larger when sample size is small. ($n \approx k$)

Review Exercises

1. H_0: $\rho = 0$; H_1: $\rho \neq 0$; $r = -0.686$; C.V. $= \pm 0.917$; do not reject. There is insufficient evidence to conclude that a relationship exists between number of passengers and one-way fare cost.

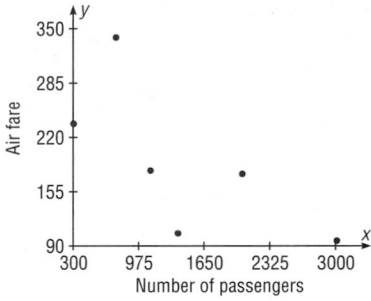

2. H_0: $\rho = 0$; H_1: $\rho \neq 0$; $r = 0.952$; C.V. $= +0.875$; reject. There is a significant linear relationship between the number of elementary schools and the number of secondary schools. $y' = -42.425 + 0.376x$; $y' = 70$ (rounded)

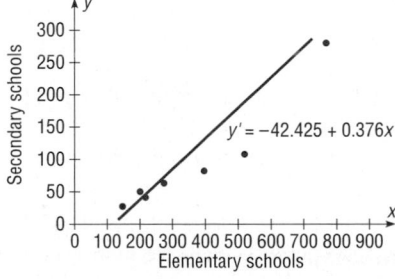

3. H_0: $\rho = 0$; H_1: $\rho \neq 0$; $r = 0.873$; C.V. $= \pm 0.875$; do not reject. There is not a significant linear relationship between the number of touchdowns and the quarterback's rating. No regression should be done.

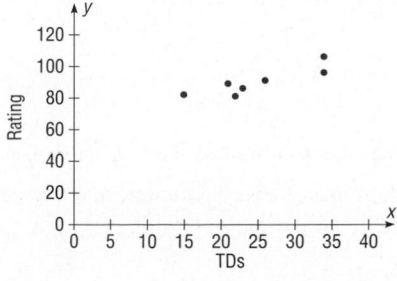

4. H_0: $\rho = 0$; H_1: $\rho \neq 0$; $r = -0.610$; C.V. $= \pm 0.875$; do not reject. There is not a significant relationship between age and the number of accidents a person has. No regression analysis should be done, since the null hypothesis has not been rejected.

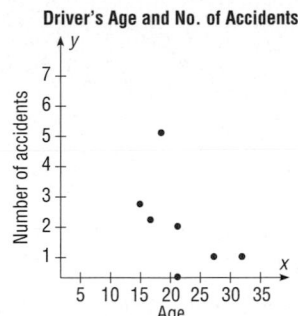

5. H_0: $\rho = 0$; H_1: $\rho \neq 0$; $r = -0.974$; C.V. $= \pm 0.708$; d.f. $= 10$; reject. There is a significant relationship between speed and time; $y' = 14.086 - 0.137x$; $y' = 4.222$.

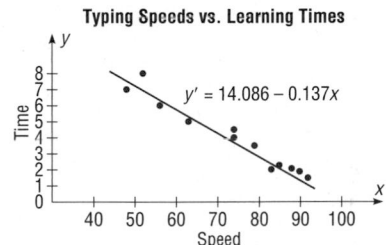

6. H_0: $\rho = 0$; H_1: $\rho \neq 0$; $r = 0.916$; C.V. $= \pm 0.798$; d.f. $= 7$; reject. There is a significant relationship between grams and pressure. $y' = 64.936 + 2.662x$; $y' = 86.232$

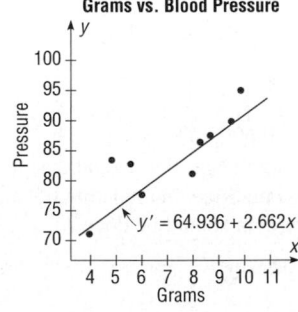

7. $H_0: \rho = 0$; $H_1: \rho \neq 0$; $r = 0.907$; C.V. $= \pm 0.875$; reject. There is sufficient evidence to conclude a relationship exists between the numbers of female physicians and male physicians in a given field. $y' = 102.846 + 3.408x$; $y' = 6919$

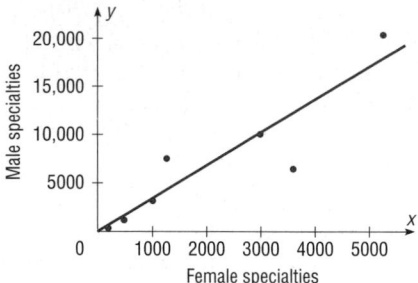

8. 1.417* For calculation purposes only. No regression should be done.

9. 0.468* (TI value 0.513)

10. 2.89 (TI value 2.845)

11. $3.34 < y < 5.10$*

12. $79 < y < 93$ **13.** 22.01*

14. $R = 0.873$ **15.** $R_{adj}^2 = 0.643$*

*Answers may vary due to rounding.

Chapter Quiz

1. False   **2.** True

3. True   **4.** False

5. False   **6.** False

7. a **8.** a

9. d **10.** c

11. b **12.** Scatter plot

13. Independent   **14.** $-1, +1$

15. b

16. Line of best fit

17. $+1, -1$

18. $H_0: \rho = 0$; $H_1: \rho \neq 0$; $r = 0.600$; C.V. $= \pm 0.754$; do not reject. There is no significant linear relationship between the price of the same drugs in the United States and in Australia. No regression should be done.

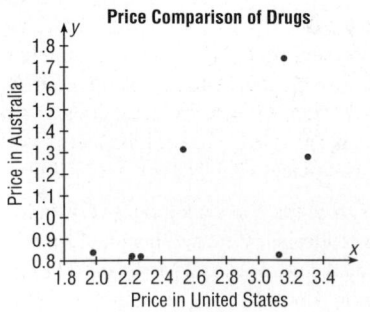

19. $H_0: \rho = 0$; $H_1: \rho \neq 0$; $r = -0.078$; C.V. $= \pm 0.754$; do not reject. No regression should be done.

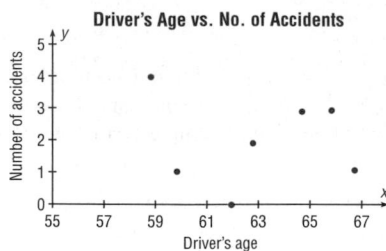

20. $H_0: \rho = 0$; $H_1: \rho \neq 0$; $r = 0.842$; C.V. $= \pm 0.811$; reject. $y' = -1.918 + 0.551x$; 4.14 or 4

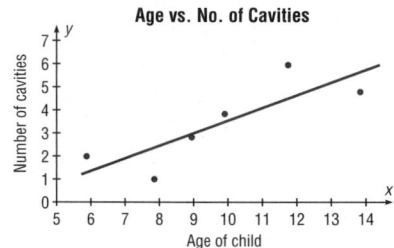

21. $H_0: \rho = 0$; $H_1: \rho \neq 0$; $r = 0.602$; C.V. $= \pm 0.707$; do not reject. No regression should be done.

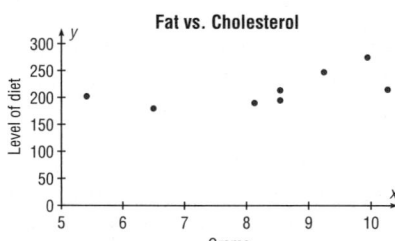

22. 1.129*

23. 29.5* For calculation purposes only. No regression should be done.

24. $0 < y < 5$*

25. 217.5 (average of y' values is used since there is no significant relationship)

26. 119.9*

27. $R = 0.729$*

28. $R_{adj}^2 = 0.439$*

*These answers may vary due to the method of calculation or rounding.

Chapter 11

Exercises 11–1

1. The variance test compares a sample variance with a hypothesized population variance; the goodness-of-fit test compares a distribution obtained from a sample with a hypothesized distribution.

2. The degrees of freedom are the number of categories minus 1.

3. The expected values are computed on the basis of what the null hypothesis states about the distribution.

4. The categories should be combined with other categories.

5. H_0: 82% of home-schooled students receive their education entirely at home, 12% attend school up to 9 hours per week, and 6% spend from 9 to 25 hours per week at school. H_1: The proportions differ from those stated in the null hypothesis (claim). C.V. = 5.991; $\chi^2 = 31.75$; reject. There is sufficient evidence to conclude that the proportions differ from those stated by the government.

6. H_0: The methods used by workers to combat midday drowsiness are equally distributed among the five categories (claim). H_1: The methods are not equally distributed among the five categories. C.V. = 7.779; d.f. = 4; $\chi^2 = 13.83$; reject. There is enough evidence to reject the claim that the methods used are equally distributed over the categories. An employer could plan ways to help workers. For example, the employer could install a beverage machine in the workplace.

7. H_0: The distribution of the recorded music sales were as follows: full-length CDs, 77.8%; digital downloads, 12.8%; singles, 3.8%; and other formats, 5.6%. H_1: The distribution is not the same as that stated in the null hypothesis (claim). C.V. = 7.815; $\chi^2 = 24.66$; reject. There is enough evidence to support the claim that the distribution is not the same as stated in the null hypothesis.

8. H_0: The performance of airlines is that 70.8% were on time, 7.8% were air carrier delayed, 8.2% were delayed by the National Aviation System, 9% were delayed by other aircraft arriving late, and 12% were delayed for other reasons. H_1: The proportions of delays are different from those stated in the null hypothesis (claim). C.V. = 7.815; $\chi^2 = 17.833$; reject. There is sufficient evidence to conclude that the proportions differ.

9. H_0: 35% feel that genetically modified food is safe to eat, 52% feel that genetically modified food is not safe to eat, and 13% have no opinion. H_1: The distribution is not the same as stated in the null hypothesis (claim). C.V. = 9.210; d.f. = 2; $\chi^2 = 1.4286$; do not reject. There is not enough evidence to support the claim that the proportions are different from those reported in the survey.

10. H_0: The distribution of truck colors is white, 30%; black, 17%; red, 14%; silver, 12%; gray, 11%; blue, 8%; other, 8%. H_1: The distribution differs from that stated in the

null hypothesis (claim). C.V. = 12.592; $\chi^2 = 10.914$; do not reject. There is not enough evidence to support the claim that the distribution differs from that stated in the null hypothesis.

11. H_0: The distribution of students who use calculators on tests is as follows: never, 28%; sometimes, 51%; and always, 21%. H_1: The distribution is not the same as stated in the null hypothesis (claim). C.V. = 5.991; $\chi^2 = 2.999$; do not reject. There is not enough evidence to support the claim that the distribution is different from the one stated in the null hypothesis.

12. H_0: The distribution for participating children is 4% five-year-olds, 52% four-year-olds, 34% three-year-olds, and 10% under 3 years of age. H_1: The distribution is not the same as stated in the null hypothesis (claim). $\chi^2 = 31.991$; P-value 0.00 < 0.05; reject. There is sufficient evidence to conclude that the proportions differ. (TI: P-value = 0.00000053)

13. H_0: The methods of payments of adult shoppers for purchases are distributed as follows: 53% pay cash, 30% use checks, 16% use credit cards, and 1% have no preference (claim). H_1: The distribution is not the same as stated in the null hypothesis. C.V. = 11.345; d.f. = 3; $\chi^2 = 36.8897$; reject. There is enough evidence to reject the claim that the distribution at the large store is the same as in the survey.

14. H_0: The distribution of degree recipients is as follows: associate degrees, 23.3%; bachelor degrees, 51.1%; professional degrees, 3%; master degrees, 20.6%; and doctoral degrees, 2%. H_1: The distribution of degree recipients is not the same as stated in the null hypothesis (claim). C.V. = 9.488; $\chi^2 = 10.311$; reject. There is enough evidence to support the claim that the distribution is different from what is stated in the null hypothesis.

15. H_0: The proportion of Internet users is the same for the groups. H_1: The proportion of Internet users is not the same for the groups (claim). C.V. = 5.991; $\chi^2 = 0.208$; do not reject. There is insufficient evidence to conclude that the proportions differ.

16. H_0: The number of people who do not have health insurance is equally distributed over the three educational categories. H_1: The number of people who do not have health insurance is not equally distributed over the three categories (claim). The d.f. = 2; $\alpha = 0.05$; $\chi^2 = 8.1$; reject at 0.05 since 0.01 < P-value < 0.025. There is enough evidence to support the claim that the number of people who don't have health insurance is not equally distributed over the three educational categories. Perhaps those with more education have better jobs that provide employee health insurance. (TI: P-value = 0.01742)

17. H_0: The distribution of the ways people pay for their prescriptions is as follows: 60% used personal funds, 25% used insurance, and 15% used Medicare (claim). H_1: The distribution is not the same as stated in the null

hypothesis. The d.f. $= 2$; $\alpha = 0.05$; $\chi^2 = 0.667$; do not reject since P-value > 0.05. There is not enough evidence to reject the claim that the distribution is the same as stated in the null hypothesis. An implication of the results is that the majority of people are using their own money to pay for medications. Maybe the medication should be less expensive to help out these people. (TI: P-value $= 0.7164$)

18. H_0: The coins are balanced and randomly tossed (claim). H_1: The coins are not balanced and are not randomly tossed. C.V. $= 7.815$; d.f. $= 3$; $\chi^2 = 139.4$; reject the null hypothesis. There is enough evidence to reject the claim that the coins are balanced and randomly tossed.

19. Answers will vary.

Exercises 11–2

1. The independence test and the goodness-of-fit test both use the same formula for computing the test value. However, the independence test uses a contingency table, whereas the goodness-of-fit test does not.

2. d.f. $=$ (rows $-$ 1)(columns $-$ 1)

3. H_0: The variables are independent (or not related). H_1: The variables are dependent (or related).

4. Contingency table

5. The expected values are computed as (row total $\times$ column total) $\div$ grand total.

6. The test of independence is used to determine whether two variables selected from a single sample are related. The test of homogeneity of proportions is used to determine whether proportions are equal.

7. H_0: $p_1 = p_2 = p_3 = p_4 = \cdots = p_n$. H_1: At least one proportion is different from the others.

8. H_0: The movie attendance by year is independent of the ethnicity of the movie goers. H_1: The movie attendance by year is dependent upon the ethnicity of the movie goers (claim). C.V. $= 7.815$; $\chi^2 = 13.222$; reject. There is sufficient evidence to support the claim that movie attendance by year is dependent upon the ethnicity of movie goers.

9. H_0: The number of endangered species is independent of the number of threatened species. H_1: The number of endangered species is dependent upon the number of threatened species (claim). C.V. $= 9.488$; $\chi^2 = 45.315$; reject. There is sufficient evidence to conclude a relationship. The result is not different at $\alpha = 0.01$.

10. H_0: The rank of women personnel is independent of the military branch of service. H_1: The rank of women personnel is dependent on the military branch of service (claim). C.V. $= 7.815$; d.f. $= 3$; $\chi^2 = 654.27$; reject. There is enough evidence to support the claim that the rank is dependent on the military branch of service.

11. H_0: The composition of the legislature (House of Representatives) is independent of the state. H_1: The composition of the legislature is dependent upon the state (claim). C.V. $= 7.815$; d.f. $= 3$; $\chi^2 = 48.7521$; reject. There is enough evidence to support the claim that the composition of the legislature is dependent upon the state.

12. H_0: The size of the population (by age) is independent of the state. H_1: The size of the population (by age) is dependent on the state (claim). C.V. $= 11.071$; d.f. $= 5$; $\chi^2 = 36.4656$; reject. There is enough evidence to support the claim that the size of the population (by age) is dependent on the state.

13. H_0: The type of Olympic medal won is independent of the country that won the medal. H_1: The type of medal won is dependent on the country that won the medal (claim). C.V. $= 9.236$; $\chi^2 = 6.651$; do not reject. There is not enough evidence to support the claim that the type of medal won is dependent on the country that won the medal.

14. H_0: The political party affiliation of congressional representatives is independent of the state of the representative. H_1: The political party affiliation of congressional representatives is dependent on the state of the representative (claim). C.V. $= 6.251$; $\chi^2 = 7.821$; reject. There is enough evidence to support the claim that the political party of the representative is dependent upon the state of the representative.

15. H_0: The program of study of a student is independent of the type of institution. H_1: The program of study of a student is dependent upon the type of institution (claim). C.V. $= 7.815$; $\chi^2 = 13.702$; reject. There is sufficient evidence to conclude that there is a relationship between program of study and type of institution.

16. H_0: The type of transplant is independent of the year in which the transplant was received. H_1: The type of transplant is dependent upon the year it was received (claim). C.V. $= 13.277$; $\chi^2 = 23.211$; reject. There is sufficient evidence to conclude that a relationship exists between year and type of transplant.

17. H_0: The type of furniture sold is independent of the store that sold the furniture. H_1: The type of furniture sold is dependent on the store that sold it (claim). C.V. $= 9.488$; $\chi^2 = 2.86$; do not reject. There is not enough evidence to support the claim that the type of furniture sold is dependent on the store that sold the furniture.

18. H_0: The genre of CDs sold is independent of the year in which the sale occurred. H_1: The genre of the CDs sold is dependent upon the year in which the sale occurred (claim). C.V. $= 5.991$; $\chi^2 = 42.939$; reject. There is sufficient evidence to conclude that the sales by genre are related to the year.

19. H_0: The choice of exercise equipment is independent of the gender of the individual using it. H_1: The choice of

exercise equipment is dependent upon the gender of the individual using it (claim). C.V. = 5.991; χ^2 = 9.139; reject. There is enough evidence to support the claim that the choice of exercise equipment is dependent upon the gender of the user.

20. H_0: The drug is not effective. H_1: The drug is effective (claim). α = 0.10; χ^2 = 10.643; d.f. = 1; P-value < 0.005 (0.001); reject since P-value < 0.10. There is enough evidence to support the claim that the drug is effective.

21. H_0: The type of book purchased by an individual is independent of the gender of the individual (claim). H_1: The type of book purchased by an individual is dependent on the gender of the individual. The d.f. = 2; α = 0.05; χ^2 = 19.43; P-value < 0.05; reject since P-value < 0.05. There is enough evidence to reject the claim that the type of book purchased by an individual is independent of the gender of the individual. (TI: P-value = 0.00006)

22. H_0: $p_1 = p_2 = p_3 = p_4$ (claim). H_1: At least one proportion is different. C.V. = 7.815; χ^2 = 7.788; do not reject. There is insufficient evidence to conclude that the proportions differ.

23. H_0: $p_1 = p_2 = p_3 = p_4$ (claim). H_1: At least one proportion is different. C.V. = 7.815; d.f. = 3; χ^2 = 5.317; do not reject. There is not enough evidence to reject the claim that the proportions are equal.

24. H_0: $p_1 = p_2 = p_3$ (claim). H_1: At least one proportion is different. C.V. = 9.210; d.f. = 2; χ^2 = 5.602; do not reject. There is not enough evidence to reject the claim that the proportions are equal.

25. H_0: $p_1 = p_2 = p_3 = p_4$ (claim). H_1: At least one of the proportions is different from the others. C.V. = 7.815; d.f. = 3; χ^2 = 1.172; do not reject. There is not enough evidence to reject the claim that the proportions are equal. Since the survey was done in Pennsylvania, it is doubtful that it can be generalized to the population of the United States.

26. H_0: $p_1 = p_2 = p_3$ (claim). H_1: At least one proportion is different. C.V. = 4.605; d.f. = 2; χ^2 = 18.06; reject. There is enough evidence to reject the claim that the proportions are equal.

27. H_0: $p_1 = p_2 = p_3 = p_4 = p_5$. H_1: At least one proportion is different. C.V. = 9.488; χ^2 = 12.028; reject. There is sufficient evidence to conclude that the proportions differ.

28. H_0: $p_1 = p_2 = p_3 = p_4$ (claim). H_1: At least one proportion is different. C.V. = 7.815; d.f. = 3; χ^2 = 5.0; do not reject. There is not enough evidence to reject the claim that the proportions are equal.

29. H_0: $p_1 = p_2 = p_3 = p_4$ (claim). H_1: At least one proportion is different. The d.f. = 3; χ^2 = 1.734; α = 0.05;

P-value > 0.10 (0.629); do not reject since P-value > 0.05. There is not enough evidence to reject the claim that the proportions are equal. (TI: P-value = 0.6291)

30. H_0: $p_1 = p_2 = p_3 = p_4$ (claim). H_1: At least one proportion is different. χ^2 = 4.334; α = 0.10; d.f. = 3; P-value > 0.10 (0.228); do not reject since P-value > 0.10. There is not enough evidence to reject the claim that the proportions are equal.

31. H_0: $p_1 = p_2 = p_3$ (claim). H_1: At least one proportion is different. C.V. = 4.605; d.f. = 2; χ^2 = 2.401; do not reject. There is not enough evidence to reject the claim that the proportions are equal.

32. Both answers are the same. χ^2 = 1.70

33. χ^2 = 1.075

34. 0.1277; 0.361

Review Exercises

1. H_0: The distribution of traffic fatalities were as follows: used seat belt, 31.58%; did not use seat belt, 59.83%; status unknown, 8.59%. H_1: The distribution is not as stated in the null hypothesis (claim). C.V. = 5.991; χ^2 = 1.819; do not reject. There is not enough evidence to support the claim that the distribution differs from the one stated in the null hypothesis.

2. H_0: The distribution of the reasons why workers were displaced is as follows: plant closed or moved, 44.8%; insufficient work, 25.2%; and position eliminated, 30%. H_1: The distribution of reasons why workers were displaced is not the same as stated in the null hypothesis (claim). C.V. = 9.210; χ^2 = 5.418; do not reject. There is not enough evidence to support the claim that the distribution is different from that stated in the null hypothesis.

3. H_0: Opinion is independent of gender. H_1: Opinion is dependent on gender (claim). C.V. = 4.605; d.f. = 2; χ^2 = 6.166; reject. There is enough evidence to support the claim that opinion is dependent on gender.

4. H_0: The distribution of denials for gun permits is as follows: 75% for criminal history, 11% for domestic violence, and 14% for other reasons. H_1: The distribution is not the same as stated in the null hypothesis. C.V. = 4.605; d.f. = 2; χ^2 = 27.75; reject. There is enough evidence to reject the claim that the distribution is as stated in the null hypothesis. Yes, the distribution may vary in different geographic locations.

5. H_0: The type of investment is independent of the age of the investor. H_1: The type of investment is dependent on the age of the investor (claim). C.V. = 9.488; d.f. = 4; χ^2 = 28.0; reject. There is enough evidence to support the claim that the type of investment is dependent on the age of the investor.

6. H_0: The month in which tornadoes occurred is independent of the year in which they occurred. H_1: The month in which tornadoes occurred is dependent upon the year in which they occurred (claim). C.V. = 12.592; $\chi^2 = 52.45$; reject. There is sufficient evidence to conclude that a relationship exists between the month and the year in which the tornadoes occurred.

7. H_0: $p_1 = p_2 = p_3$ (claim). H_1: At least one proportion is different. $\chi^2 = 4.912$; d.f. = 2; $\alpha = 0.01$; $0.05 < P$-value < 0.10 (0.086); do not reject since P-value > 0.01. There is not enough evidence to reject the claim that the proportions are equal.

8. H_0: $p_1 = p_2 = p_3 = p_4$ (claim). H_1: At least one proportion is different. C.V. = 7.815; d.f. = 3; $\chi^2 = 6.166$; do not reject. There is not enough evidence to reject the claim that the proportions are equal.

9. H_0: Health care coverage is independent of the state of residence of the individual. H_1: Health care coverage is related to the state of residence of the individual (claim). C.V. = 11.345; $\chi^2 = 18.993$; reject. There is sufficient evidence to say that health care coverage is related to the state of residence of the individual.

10. H_0: The incidence of the cardiovascular procedure is independent of the gender of the individual. H_1: The incidence of cardiovascular procedure is dependent on the gender of the individual (claim). C.V. = 4.605; $\chi^2 = 59.949$; reject. There is enough evidence to support the claim that the procedure is dependent on the gender of the individual.

Chapter Quiz

1. False
2. True
3. False
4. c
5. b
6. d
7. 6
8. Independent
9. Right
10. At least 5

11. H_0: The reasons why people lost their jobs are equally distributed (claim). H_1: The reasons why people lost their jobs are not equally distributed. C.V. = 5.991; d.f. = 2; $\chi^2 = 2.334$; do not reject. There is not enough evidence to reject the claim that the reasons why people lost their jobs are equally distributed. The results could have been different 10 years ago since different factors of the economy existed then.

12. H_0: Takeout food is consumed according to the following distribution: 53% at home, 19% in the car, 14% at work, and 14% at other places (claim). H_1: The distribution is different from that stated in the null hypothesis. C.V. = 11.345; d.f. = 3; $\chi^2 = 5.271$; do not reject. There is not enough evidence to reject the claim that the distribution is as stated. Fast-food restaurants may want to make their advertisements appeal to those who like to take their food home to eat.

13. H_0: College students show the same preference for shopping channels as those surveyed. H_1: College students show a different preference for shopping channels (claim). C.V. = 7.815; d.f. = 3; $\alpha = 0.05$; $\chi^2 = 21.789$; reject. There is enough evidence to support the claim that college students show a different preference for shopping channels.

14. H_0: The number of commuters is distributed as follows: 75.7%, alone; 12.2%, carpooling; 4.7%, public transportation; 2.9%, walking; 1.2%, other; and 3.3%, working at home. H_1: The proportion of workers using each type of transportation differs from the stated proportions. C.V. = 11.071; d.f. = 5; $\chi^2 = 41.269$; reject. There is enough evidence to support the claim that the distribution is different from the one stated in the null hypothesis.

15. H_0: Ice cream flavor is independent of the gender of the purchaser (claim). H_1: Ice cream flavor is dependent upon the gender of the purchaser. C.V. = 7.815; d.f. = 3; $\chi^2 = 7.198$; do not reject. There is not enough evidence to reject the claim that ice cream flavor is independent of the gender of the purchaser.

16. H_0: The type of pizza ordered is independent of the age of the individual who purchases it. H_1: The type of pizza ordered is dependent on the age of the individual who purchases it (claim). $\chi^2 = 107.3$; d.f. = 9; $\alpha = 0.10$; P-value < 0.005; reject since P-value < 0.10. There is enough evidence to support the claim that the pizza purchased is related to the age of the purchaser.

17. H_0: The color of the pennant purchased is independent of the gender of the purchaser (claim). H_1: The color of the pennant purchased is dependent on the gender of the purchaser. $\chi^2 = 5.632$; C.V. = 4.605; reject. There is enough evidence to reject the claim that the color of the pennant purchased is independent of the gender of the purchaser.

18. H_0: The opinion of the children on the use of the tax credit is independent of the gender of the children. H_1: The opinion of the children on the use of the tax credit is dependent upon the gender of the children (claim). C.V. = 4.605; d.f. = 2; $\chi^2 = 1.534$; do not reject. There is not enough evidence to support the claim that the opinion of the children on the use of the tax credit is dependent on their gender.

19. H_0: $p_1 = p_2 = p_3$ (claim). H_1: At least one proportion is different from the others. C.V. = 4.605; d.f. = 2; $\chi^2 = 6.711$; reject. There is enough evidence to reject the claim that the proportions are equal. It seems that more women are undecided about their jobs. Perhaps they want better income or greater chances of advancement.

Chapter 12

Exercises 12–1

1. The analysis of variance using the F test can be employed to compare three or more means.

2. *a.* Comparing two means at a time ignores all other means.
   b. The probability of a type I error is larger than α when multiple t tests are used.
   c. The more sample means, the more t tests are needed.

3. The populations from which the samples were obtained must be normally distributed. The samples must be independent of each other. The variances of the populations must be equal.

4. The between-group variance estimates the population variance using the means. The within-group variance estimates the population variance using all the data values.

5. $F = \dfrac{s_B^2}{s_W^2}$

6. $H_0: \mu_1 = \mu_2 = \cdots = \mu_k.$ H_1: At least one mean is different from the others.

7. One

8. $H_0: \mu_1 = \mu_2 = \mu_3.$ H_1: At least one mean is different from the others (claim). C.V. = 3.52; $\alpha = 0.05$; d.f.N. = 2; d.f.D. = 19; $F = 2.3985$; do not reject. There is not enough evidence to support the claim that at least one mean is different from the others.

9. $H_0: \mu_1 = \mu_2 = \mu_3.$ H_1: At least one of the means differs from the others. C.V. = 4.26; d.f.N. = 2; d.f.D. = 9; $F = 14.149$; reject. There is sufficient evidence to conclude at least one mean is different from the others.

10. $H_0: \mu_1 = \mu_2 = \mu_3.$ H_1: At least one mean differs from the others (claim). C.V. = 3.68; d.f.N. = 2; d.f.D. = 15; $F = 8.515$; reject. There is enough evidence to conclude that at least one mean differs from the others.

11. $H_0: \mu_1 = \mu_2 = \mu_3.$ H_1: At least one mean is different from the others (claim). C.V. = 3.98; $\alpha = 0.05$; d.f.N. = 2; d.f.D. = 11; $F = 2.7313$; do not reject. There is not enough evidence to support the claim that at least one mean is different from the others.

12. $H_0: \mu_1 = \mu_2 = \mu_3.$ H_1: At least one mean is different from the others (claim). $F = 7.740$; P-value = 0.00797; reject. There is enough evidence to conclude that at least one mean is different from the others.

13. $H_0: \mu_1 = \mu_2 = \mu_3.$ H_1: At least one mean is different from the others (claim). C.V. = 3.68; $\alpha = 0.05$; d.f.N. = 2; d.f.D. = 15; $F = 8.14$; reject. There is enough evidence to support the claim that at least one mean is different from the others.

14. $H_0: \mu_1 = \mu_2 = \mu_3.$ H_1: At least one mean differs from the others (claim). C.V. = 3.89; d.f.N. = 2; d.f.D. = 12; $F = 3.677$; do not reject. There is not enough evidence to support the claim that at least one mean differs from the others.

15. $H_0: \mu_1 = \mu_2 = \mu_3$ (claim). H_1: At least one mean is different from the others. C.V. = 4.10; $\alpha = 0.05$; d.f.N. = 2; d.f.D. = 10; $F = 3.9487$; do not reject. There is not enough evidence to reject the claim that the means are equal.

16. $H_0: \mu_1 = \mu_2 = \mu_3.$ H_1: At least one of the means differs from the others. C.V. = 3.98; d.f.N. = 2; d.f.D. = 11; $F = 1.3066$; do not reject. There is insufficient evidence to conclude that at least one mean is different from the others.

17. $H_0: \mu_1 = \mu_2 = \mu_3.$ H_1: At least one mean is different from the others (claim). $F = 10.118$; P-value = 0.00102; reject. There is enough evidence to conclude that at least one mean is different from the others.

18. $H_0: \mu_1 = \mu_2 = \mu_3.$ H_1: At least one mean differs from the others (claim). C.V. = 4.10; d.f.N. = 2; d.f.D. = 10; $F = 14.204$; reject. There is enough evidence to support the claim that at least one mean differs from the others.

19. $H_0: \mu_1 = \mu_2 = \mu_3.$ H_1: At least one mean differs from the others (claim). C.V. = 2.57; d.f.N. = 2; d.f.D. = 21; $F = 3.497$; reject. There is sufficient evidence to conclude at least one mean is different from the others.

20. $H_0: \mu_1 = \mu_2 = \mu_3 = \mu_4.$ H_1: At least one of the means differs from the others. C.V. = 3.24; d.f.N. = 3; d.f.D. = 16; $F = 5.543$; reject. There is sufficient evidence to conclude that at least one of the means differs from the others.

Exercises 12–2

1. The Scheffé and Tukey tests are used.

2. The Scheffé test is usually used when sample sizes are not the same. The Tukey test is usually used when the sample sizes are equal.

3. $F_{1\times2} = 2.059$; $F_{2\times3} = 17.640$; $F_{1\times3} = 27.929$. Scheffé test: C.V. = 8.52. There is sufficient evidence to conclude a difference in mean cost to drive 25 miles between hybrid cars and hybrid trucks and between hybrid SUVs and hybrid trucks.

4. Scheffé test: C.V. = 7.96; $\overline{X}_1$ versus $\overline{X}_2$: $F_S = 9.81$; $\overline{X}_1$ versus $\overline{X}_3$: $F_S = 0.077$; $\overline{X}_2$ versus $\overline{X}_3$: $F_S = 11.80$. There is a significant difference between $\overline{X}_1$ and $\overline{X}_2$, and $\overline{X}_2$ and $\overline{X}_3$.

5. Tukey test: C.V. = 3.29; $\overline{X}_1 = 7.0$; $\overline{X}_2 = 8.12$; $\overline{X}_3 = 5.23$; $\overline{X}_1$ versus $\overline{X}_2$, $q = -2.196$; $\overline{X}_1$ versus $\overline{X}_3$, $q = 3.47$; $\overline{X}_2$ versus $\overline{X}_3$, $q = -6.35$. There is a significant difference between $\overline{X}_1$ and $\overline{X}_3$, and $\overline{X}_2$ and $\overline{X}_3$. One reason for the difference might be that the students are enrolled in cyber schools with different fees.

6. No further testing should be done.

7. Scheffé test: C.V. = 5.22; $\overline{X}_1$ versus $\overline{X}_2$, $F = 2.91$; $\overline{X}_1$ versus $\overline{X}_3$, $F = 19.3$; $\overline{X}_2$ versus $\overline{X}_3$, $F = 8.40$. There is a significant difference between $\overline{X}_1$ and $\overline{X}_3$, and $\overline{X}_2$ and $\overline{X}_3$.

8. Scheffé test: C.V. = 8.20; $\bar{X}_1$ versus $\bar{X}_2$, $F_3 = 0.936$; $\bar{X}_1$ versus $\bar{X}_3$, $F = 15.557$; $\bar{X}_2$ versus $\bar{X}_3$, $F = 26.268$. There is a significant difference between $\bar{X}_1$ and $\bar{X}_3$ and $\bar{X}_2$ and $\bar{X}_3$.

9. Tukey test: C.V. = 3.08; $\bar{X}_1$ versus $\bar{X}_2$, $q = 3.262$; $\bar{X}_1$ versus $\bar{X}_3$, $q = 3.215$; $\bar{X}_2$ versus $\bar{X}_3$, $q = -0.047$. There is a significant difference between $\bar{X}_1$ and $\bar{X}_2$ and $\bar{X}_2$ and $\bar{X}_3$.

10. $H_0: \mu_1 = \mu_2 = \mu_3$ (claim). H_1: At least one mean is different from the others. C.V. = 3.68; $\alpha = 0.05$; d.f.N. = 2; d.f.D. = 15; $F = 23.94$; reject. There is enough evidence to reject the hypothesis that the means are equal. Tukey test: C.V. = 3.67; $\bar{X}_1 = 7.33$; $\bar{X}_2 = 15.17$; $\bar{X}_3 = 24.5$; $\bar{X}_1$ versus $\bar{X}_2$, $q = -4.45$; $\bar{X}_1$ versus $\bar{X}_3$, $q = -9.76$; $\bar{X}_2$ versus $\bar{X}_3$, $q = -5.30$. There is a significant difference between $\bar{X}_1$ and $\bar{X}_2$, and $\bar{X}_1$ and $\bar{X}_3$, and $\bar{X}_2$ and $\bar{X}_3$.

11. $H_0: \mu_1 = \mu_2 = \mu_3$. H_1: At least one mean is different from the others (claim). C.V. = 3.47; $\alpha = 0.05$; d.f.N. = 2; d.f.D. = 21; $F = 1.9912$; do not reject. There is not enough evidence to support the claim that at least one mean is different from the others.

12. $H_0: \mu_1 = \mu_2 = \mu_3$. H_1: At least one mean is different from the others (claim). C.V. = 4.10; $\alpha = 0.05$; d.f.N. = 2; d.f.D. = 10; $F = 0.6488$; do not reject. There is not enough evidence to support the claim that at least one mean is different from the others.

13. $H_0: \mu_1 = \mu_2 = \mu_3$. H_1: At least one mean differs from the others (claim). C.V. = 3.68; d.f.N. = 2; d.f.D. = 16; $F = 17.172$; reject. There is enough evidence to support the claim that at least one mean differs from the others. Tukey test: C.V. = 3.67; $\bar{X}_1$ versus $\bar{X}_2$, $q = -8.17$; $\bar{X}_1$ versus $\bar{X}_3$, $q = -2.91$; $\bar{X}_2$ versus $\bar{X}_3$, $q = 5.269$. There is a significant difference between $\bar{X}_1$ and $\bar{X}_2$ and between $\bar{X}_2$ and $\bar{X}_3$.

Exercises 12–3

1. The two-way ANOVA allows the researcher to test the effects of two independent variables and a possible interaction effect. The one-way ANOVA can test the effects of only one independent variable.

2. The main effects are the effects of the independent variables taken separately. The interaction effect occurs when one independent variable affects the dependent variable differently at different levels of the other independent variable.

3. The mean square values are computed by dividing the sum of squares by the corresponding degrees of freedom.

4. One computes the F test value by dividing the mean square for the variable by the mean square for the within (error) term.

5. a. For factor A, $\text{d.f.}_A = 2$ c. $\text{d.f.}_{A \times B} = 2$
   b. For factor B, $\text{d.f.}_B = 1$ d. $\text{d.f.}_{\text{within}} = 24$

6. a. 5   b. 4   c. 20   d. 180

7. The two types of interactions that can occur are ordinal and disordinal.

8. The main effects can be interpreted independently when the interaction effect is not significant or the interaction is ordinal.

9. a. The lines will be parallel or approximately parallel. They may also coincide.
   b. The lines will not intersect and they will not be parallel.
   c. The lines will intersect.

10. *Interaction:* H_0: There is no interaction effect between the strength of the Grow-light strength and the plant food supplement. H_1: There is an interaction effect between the Grow-light strength and the plant food supplement. *Plant food:* H_0: There is no difference in the mean growth with respect to the type of plant food supplement. H_1: There is a difference in the mean growth with respect to the type of plant food supplement. *Grow-light:* H_0: There is no difference in the mean growth with respect to the strength of the Grow-light. H_1: There is a difference in the mean growth with respect to the strength of the Grow-light. C.V. = 5.32; d.f.N. = 1; d.f.D. = 8; $F = 24.56$ for plant food. There is sufficient evidence to conclude there is a difference in the mean growth for the plant food. Plant light strength and the interaction have no effect.

ANOVA Summary

Source	SS	d.f.	MS	F	P-value
Plant food	12.8133	1	12.8133	24.56	0.001
Grow-light	1.9200	1	1.9200	3.68	0.091
Interaction	0.7500	1	0.7500	1.44	0.265
Within	4.1733	8	0.5217		
Total	19.6567	11			

11. *Interaction:* H_0: There is no interaction effect between the temperature and the level of humidity. H_1: There is an interactive effect between the temperature and the level of humidity. *Humidity:* H_0: There is no difference in mean length of effectiveness with respect to humidity. H_1: There is a difference in mean length of effectiveness with respect to humidity. *Temperature:* H_0: There is no difference in the mean length of effectiveness based on temperature. H_1: There is a difference in mean length of effectiveness based on temperature. C.V. = 5.318; d.f.N. = 1; d.f.D. = 8; $F = 18.383$ for humidity. There is sufficient evidence to conclude a difference in mean length of effectiveness based on the humidity level. The temperature and interaction effects are not significant.

ANOVA Summary Table for Exercise 11

Source of variation	SS	d.f.	MS	F	P-value
Humidity	280.3333	1	280.3333	18.383	0.003
Temperature	3	1	3	0.197	0.669
Interaction	65.33333	1	65.33333	4.284	0.0722
Within	122	8	15.25		
Total	470.6667	11			

12. H_0: There is no interaction effect between the subcontractors and the types of homes they build on the times it takes to build the homes. H_1: There is an interaction effect between the subcontractors and the types of homes they build on the times it takes to build the homes.

H_0: There is no difference in the means of the times it takes the subcontractors to build the homes. H_1: There is a difference in the means of the times it takes the subcontractors to build the homes.

H_0: There is no difference among the means of the times for the types of homes built. H_1: There is a difference among the means of the times for the types of homes built.

ANOVA Summary Table

Source	SS	d.f.	MS	F
Subcontractor	1672.553	1	1672.553	122.084
Home type	444.867	2	222.434	16.236
Interaction	313.267	2	156.634	11.433
Within	328.800	24	13.700	
Total	2759.487	29		

The critical values at $\alpha = 0.05$: for the subcontractor with d.f.N. = 1 and d.f.D. = 24, C.V. = 4.26; for the home type and interaction with d.f.N. = 2 and d.f.D. = 24, C.V. = 3.40. All F test values exceed the critical values, and all the null hypotheses are rejected. Since there is a significant interaction effect, the means of the cells must be computed and graphed to determine

the type of interaction. Cell means:

Subcontractor \ Home type	I	II	III
A	28	31.4	44.4
B	18.6	20.0	20.4

Since all three means for the home types for subcontractor A are greater than the three means for subcontractor B and the differences are not equal, there is an ordinal interaction. Hence, it can be concluded that there is a difference in means for the subcontractors and home types. In addition, there is a significant interaction between subcontractors and home types.

13. *Interaction:* H_0: There is no interaction effect on the durability rating between the dry additives and the solution-based additives. H_1: There is an interaction effect on the durability rating between the dry additives and the solution-based additives. *Solution-based additive:* H_0: There is no difference in the mean durability rating with respect to the solution-based additives. H_1: There is a difference in the mean durability rating with respect to the solution-based additives. *Dry additive:* H_0: There is no difference in the mean durability rating with respect to the dry additive. H_1: There is a difference in the mean durability rating with respect to the dry additive. C.V. = 4.75; d.f.N. = 1; d.f.D. = 12. There is not a significant interaction effect. Neither the solution additive nor the dry additive have a significant effect on mean durability.

ANOVA Summary Table for Exercise 13

Source	SS	d.f.	MS	F	P-value
Solution additive	1.563	1	1.563	0.497	0.494
Dry additive	0.063	1	0.063	0.020	0.898
Interaction	1.563	1	1.563	0.497	0.494
Within	37.750	12	3.146		
Total	40.939	15			

14. H_0: There is no interaction effect between the type of paint and the geographic location on the lifetimes of the paint. H_1: There is an interaction effect between the type of paint and the geographic location on the lifetimes of the paint.

H_0: There is no difference between the means of the lifetimes of the two types of paints. H_1: There is a difference between the means of the lifetimes of the two types of paints.

H_0: There is no difference among the means of the lifetimes of the paints used in different geographic locations.
H_1: There is a difference in the means of the lifetimes of the paints used in different geographic locations.

ANOVA Summary Table

Source	SS	d.f.	MS	F
Paint type	12.1	1	12.1	0.166
Location	2501.0	3	833.667	11.465
Interaction	268.1	3	89.367	1.229
Within	2326.8	32	72.713	
Total	5108.0	39		

The critical values for $\alpha = 0.01$: for the paint type with d.f.N. = 1 and d.f.D. = 32 (use 30), C.V. = 7.56; for the location and interaction with d.f.N. = 3 and d.f.D. = 32 (use 30), C.V. = 4.51.

There is not a significant interaction effect, so the main effects can be interpreted. There is a significant difference in the means for the geographic location, but not for the type of paint.

15. H_0: There is no interaction effect between the ages of the salespeople and the products they sell on the monthly sales. H_1: There is an interaction effect between the ages of the salespeople and the products they sell on the monthly sales.

H_0: There is no difference in the means of the monthly sales of the two age groups. H_1: There is a difference in the means of the monthly sales of the two age groups.

H_0: There is no difference among the means of the sales for the different products. H_1: There is a difference among the means of the sales for the different products.

ANOVA Summary Table

Source	SS	d.f.	MS	F
Age	168.033	1	168.033	1.567
Product	1,762.067	2	881.034	8.215
Interaction	7,955.267	2	3,977.634	37.087
Within	2,574.000	24	107.250	
Total	12,459.367	29		

At $\alpha = 0.05$, the critical values are: for age, d.f.N. = 1, d.f.D. = 24, C.V. = 4.26; for product and interaction, d.f.N. = 2 and d.f.D. = 24; C.V. = 3.40. There is a significant interaction between the age of the salesperson and the type of product sold, so no main effects should be interpreted without further study.

	Product		
Age	Pools	Spas	Saunas
Over 30	38.8	28.6	55.4
30 and under	21.2	68.6	18.8

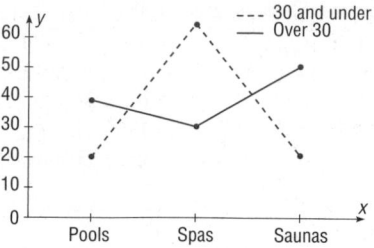

Since the lines cross, there is a disordinal interaction; hence, there is an interaction effect between the ages of salespeople and the type of products sold.

Review Exercises

1. H_0: $\mu_1 = \mu_2 = \mu_3$ (claim). H_1: At least one mean is different from the others. C.V. = 5.39; d.f.N. = 2; d.f.D. = 33; $\alpha = 0.01$; $F = 6.94$; reject. Tukey test: C.V. = 4.45; $\bar{X}_1$ versus $\bar{X}_2$: $q = 0.342$; $\bar{X}_1$ versus $\bar{X}_3$: $q = 4.72$; $\bar{X}_2$ versus $\bar{X}_3$: $q = 4.38$. There is a significant difference between $\bar{X}_1$ and $\bar{X}_3$.

2. H_0: $\mu_1 = \mu_2 = \mu_3$. H_1: At least one of the means differs from the others. C.V. = 3.982; d.f.N. = 2; d.f.D. = 11; $F = 1.580$; do not reject. There is insufficient evidence to conclude at least one mean differs from the others.

3. H_0: $\mu_1 = \mu_2 = \mu_3$. H_1: At least one mean is different from the others (claim). C.V. = 3.55; $\alpha = 0.05$; d.f.N. = 2; d.f.D. = 18; $F = 0.0408$; do not reject. There is not enough evidence to support the claim that at least one mean is different from the others.

4. H_0: $\mu_1 = \mu_2 = \mu_3$. H_1: At least one mean is different from the others (claim). C.V. = 6.01; $\alpha = 0.01$; d.f.N. = 2; d.f.D. = 18; $F = 0.6519$; do not reject. There is not enough evidence to support the claim that at least one mean is different from the others.

5. H_0: $\mu_1 = \mu_2 = \mu_3$. H_1: At least one mean is different from the others (claim). C.V. = 2.61; $\alpha = 0.10$; d.f.N. = 2; d.f.D. = 19; $F = 0.4876$; do not reject. There is not enough evidence to support the claim that at least one mean is different from the others.

6. H_0: $\mu_1 = \mu_2 = \mu_3$. H_1: At least one of the means differs from the others. C.V. = 3.89; d.f.N. = 2; d.f.D. = 12; $F = 6.320$; reject. There is sufficient evidence to conclude a difference in means. Tukey test: C.V. (from Table N) $F = 3.77$; $F_{1 \times 2} = -4.989$; $F_{2 \times 3} = 1.953$; $F_{1 \times 3} = -3.035$. There is sufficient evidence to conclude a difference in mean January high temperatures between Europe and Central and South America.

7. H_0: $\mu_1 = \mu_2 = \mu_3 = \mu_4$. H_1: At least one mean is different from the others (claim). C.V. = 3.59; $\alpha = 0.05$; d.f.N. = 3; d.f.D. = 11; $F = 0.182$; do not reject. There is not enough evidence to support the claim that at least one mean is different from the others.

8. *Interaction*: H_0: There is no interaction effect between type of formula delivery system and review organization.

H_1: There is an interaction effect between type of formula delivery system and review organization. *Review:* H_0: There is no difference in mean scores based on who leads the review. H_1: There is a difference in mean scores based on who leads the review. *Formulas:* H_0: There is no difference in mean scores based on who provides the

formulas. H_1: There is a difference in mean scores based on who provides the formulas.

C.V. = 4.49; d.f.N. = 1; d.f.D. = 16; $F = 5.244$ for review organization. There is sufficient evidence to conclude a difference in mean scores based on who leads the review. The formula and interaction effects are not significant.

ANOVA Summary Table for Exercise 8

Source of variation	SS	d.f.	MS	F	P-value
Sample	288.8	1	288.8	5.244	0.036
Columns	51.2	1	51.2	0.930	0.349
Interaction	5	1	5	0.091	0.767
Within	881.2	16	55.075		
Total	1226.2	19			

9. H_0: There is no interaction effect between the type of exercise program and the type of diet on a person's glucose level. H_1: There is an interaction effect between type of exercise program and the type of diet on a person's glucose level.

H_0: There is no difference in the means for the glucose levels of the people in the two exercise programs. H_1: There is a difference in the means for the glucose levels of the people in the two exercise programs.

H_0: There is no difference in the means for the glucose levels of the people in the two diet programs. H_1: There is a difference in the means for the glucose levels of the people in the two diet programs.

ANOVA Summary Table

Source	SS	d.f.	MS	F
Exercise	816.750	1	816.750	60.50
Diet	102.083	1	102.083	7.56
Interaction	444.083	1	444.083	32.90
Within	108.000	8	13.500	
Total	1470.916	11		

At $\alpha = 0.05$, d.f.N. = 1, d.f.D. = 8, and the critical value is 5.32 for each F_A, F_B, and $F_{A \times B}$. Hence, all three null hypotheses are rejected. The cell means should be calculated.

Exercise \ Diet	A	B
I	64.000	57.667
II	68.333	86.333

Since the means for exercise program I are both smaller than those for exercise program II and the vertical differences are not the same, the interaction is ordinal. Hence you can say that there is a difference for exercise and diet, and that an interaction effect is present.

Chapter Quiz

1. False
2. False
3. False
4. True
5. d
6. a
7. a
8. c
9. ANOVA
10. Tukey
11. Two

12. H_0: $\mu_1 = \mu_2 = \mu_3 = \mu_4$. H_1: At least one mean is different from the others (claim). C.V. = 3.49; $\alpha = 0.05$; d.f.N. = 3; d.f.D. = 12; $F = 3.23$; do not reject. There is not enough evidence to support the claim that there is a difference in the means.

13. H_0: $\mu_1 = \mu_2 = \mu_3$. H_1: At least one mean is different from the others (claim). C.V. = 6.93; $\alpha = 0.01$; d.f.N. = 2; d.f.D. = 12; $F = 3.49$. There is not enough evidence to support the claim that at least one mean is different from the others. Writers would want to target their material to the age group of the viewers.

14. H_0: $\mu_1 = \mu_2 = \mu_3$. H_1: At least one mean differs from the others (claim). C.V. = 4.26; d.f.N. = 2; d.f.D. = 9; $F = 10.025$; reject. There is enough evidence to conclude that at least one mean differs from the others. Tukey test: C.V. = 3.95; $\bar{X}_1$ versus $\bar{X}_2$, $q = -1.28$; $\bar{X}_1$ versus $\bar{X}_3$, $q = 4.74$; $\bar{X}_2$ versus $\bar{X}_3$, $q = 6.02$. There is a significant difference between $\bar{X}_1$ and $\bar{X}_3$ and between $\bar{X}_2$ and $\bar{X}_3$.

15. H_0: $\mu_1 = \mu_2 = \mu_3$. H_1: At least one mean differs from the others (claim). C.V. = 2.92; d.f.N. = 2; d.f.D. = 8; $F = 6.652$; reject. Scheffé test: C.V. = 8.918; $\bar{X}_1$ versus $\bar{X}_2$, $F_s = 9.32$; $\bar{X}_1$ versus $\bar{X}_3$, $F_s = 10.132$; $\bar{X}_2$ versus $\bar{X}_3$, $F_s = 0.1258$. There is a significant difference between $\bar{X}_1$ and $\bar{X}_2$ and between $\bar{X}_1$ and $\bar{X}_3$.

16. H_0: $\mu_1 = \mu_2 = \mu_3 = \mu_4$. H_1: At least one mean is different from the others (claim). C.V. = 3.07; $\alpha = 0.05$; d.f.N. = 3; d.f.D. = 21; $F = 0.4564$; do not reject. There is not enough evidence to support the claim that at least one mean is different from the others.

17. *a.* Two-way ANOVA

b. Diet and exercise program

c. 2

d. H_0: There is no interaction effect between the type of exercise program and the type of diet on a person's weight loss. H_1: There is an interaction effect between the type of exercise program and the type of diet on a person's weight loss.

H_0: There is no difference in the means of the weight losses of people in the exercise programs. H_1: There is a difference in the means of the weight losses of people in the exercise programs.

H_0: There is no difference in the means of the weight losses of people in the diet programs. H_1: There is a difference in the means of the weight losses of people in the diet programs.

e. Diet: $F = 21.0$, significant; exercise program: $F = 0.429$, not significant; interaction: $F = 0.429$, not significant

f. Reject the null hypothesis for the diets.

Chapter 13

Exercises 13–1

1. *Nonparametric* means hypotheses other than those using population parameters can be tested; *distribution-free* means no assumptions about the population distributions have to be satisfied.

2. When the assumptions for the parametric methods cannot be met, statisticians use nonparametric methods.

3. Nonparametric methods have the following advantages:

   a. They can be used to test population parameters when the variable is not normally distributed.

   b. They can be used when data are nominal or ordinal.

   c. They can be used to test hypotheses other than those involving population parameters.

   d. The computations are easier in some cases than the computations of the parametric counterparts.

   e. They are easier to understand.

   The disadvantages are as follows:

   a. They are less sensitive than their parametric counterparts.

   b. They tend to use less information than their parametric counterparts.

   c. They are less efficient than their parametric counterparts.

4.
Data	1	3	4	6	7	8	10
Rank	1	2	3	4	5	6	7

5.
Data	22	32	34	43	43	65	66	71
Rank	1	2	3	4.5	4.5	6	7	8

6.
Data	83	177	241	460	582
Rank	1	2	3	4	5

7.
Data	3.2	5.9	10.3	11.1	19.4	21.8	23.1
Rank	1	2	3	4	5	6	7

8.
Data	0.85	5.6	9.5	10.9	17.6	17.6	20.2	32.6	43.9
Rank	1	2	3	4	5.5	5.5	7	8	9

9.
Data	11	28	36	41	47	50	50	50	52	71	71	88
Rank	1	2	3	4	5	7	7	7	9	10.5	10.5	12

10.
Data	9.27	9.54	18.0	34.5	47.0	52.9	82.2	90.6	145.0	327.0
Rank	1	2	3	4	5	6	7	8	9	10

Exercises 13–2

1. The sign test uses only positive or negative signs.

2. The median

3. The smaller number of positive or negative signs

4. The normal approximation

5. H_0: median = 27.6 years and H_1: median $\neq$ 27.6 years (claim); test value = 5; C.V. = 3; do not reject. There is insufficient evidence to support the claim that the median is not 27.6 years.

6. H_0: median = 3000 (claim) and H_1: median $\neq$ 3000; test value = 10; C.V. = 5; do not reject. There is not enough evidence to reject the claim that the median is 3000. Yes, you could use 3000 as a guide.

7. H_0: median = 25 (claim) and H_1: median $\neq$ 25; test value = 7; C.V. = 4; do not reject. There is not enough evidence to reject the claim that the median is 25. School boards could use the median to plan for the costs of cyber school enrollments.

8. H_0: median = \$1603 and H_1: median $<$ \$1603 (claim); test value = 6; C.V. = 3; do not reject. There is not enough evidence to support the claim that the median is less than \$1603.

9. H_0: median = \$10.86 (claim) and H_1: median $\neq$ \$10.86; C.V. = ± 1.96; $z = -0.77$; do not reject. There is not enough evidence to reject the claim that the median is \$10.86. Home buyers could estimate the yearly cost of their gas bills.

10. H_0: median = \$63,211 and H_1: median $\neq$ \$63,211 (claim); $z = -3.00$; C.V. = ± 1.96; reject. There is sufficient evidence to support the claim that the median is not \$63,211.

11. H_0: the median number of faculty = 150 and H_1: the median $\neq$ 150; C.V. = ± 1.96; $z = -2.70$; reject. There is sufficient evidence at the 0.05 level of significance to reject the claim that the median number of faculty is 150.

12. H_0: median = 39 (claim) and H_1: median $\neq$ 39; C.V. = ± 2.33; $z = -2.31$; do not reject. There is not enough evidence to reject the claim that the median is 39. One reason that this information would be valuable is that the sponsors of the show would target viewers under 39.

13. H_0: median = 50 (claim) and H_1: median ≠ 50; $z = -2.3$; P-value = 0.0214; reject. There is enough evidence to reject the claim that 50% of the students are against extending the school year.

14. H_0: median = 60 (claim) and H_1: median ≠ 60; C.V. = 1; test value = 4; do not reject. There is not enough evidence to reject the claim that the median is 60. Yes, considering the number of tornadoes, the median of 60 is relatively small.

15. H_0: the medication has no effect on weight loss and H_1: the medication affects weight loss (claim); C.V. = 0; test value = 1; do not reject. There is not enough evidence to support the claim that the medication affects weight loss.

16. H_0: there is no difference between scores and H_1: there is a difference between scores; test value = 3; C.V. = 1; do not reject. There is insufficient evidence to conclude a difference in scores.

17. H_0: there is no difference in the test scores and H_1: there is an increase in the test scores (i.e., the program is effective) (claim); test value = 2; C.V. = 0; do not reject. There is insufficient evidence to support the claim that the program is effective.

18. H_0: the pill has no effect on the caloric intake of the person eating and H_1: the pill has an effect on the caloric intake of the person eating (claim); C.V. = 1; test value = 2; do not reject. There is not enough evidence to support the claim that the pill has an effect on caloric intake.

19. H_0: the number of viewers is the same as last year (claim) and H_1: the number of viewers is not the same as last year; C.V. = 0; test value = 2; do not reject. There is not enough evidence to reject the claim that the number of viewers is the same as last year.

20. H_0: increased maintenance does not reduce the number of defective parts a machine produces and H_1: increased maintenance reduces the number of defective parts a machine produces (claim); C.V. = 0; test value = 2; do not reject. There is not enough evidence to support the claim that increased maintenance reduces the number of defective parts manufactured by the machines.

21. 6 ≤ median ≤ 22

22. MD = 146; 141 ≤ MD ≤ 153

23. 4.7 ≤ median ≤ 9.3

24. MD = 21; 5 ≤ MD ≤ 54

25. 17 ≤ median ≤ 33

Exercises 13–3

1. n_1 and n_2 are each greater than or equal to 10.

2. The t test for independent samples

3. The standard normal distribution

4. H_0: there is no difference in the length of the sentences of the males and females (claim) and H_1: there is a difference in the length of the sentences of the males and females; C.V. = ±1.96; $z = 1.49$; do not reject. There is not enough evidence to reject the claim that there are no differences in the sentences received by the males and females.

5. H_0: there is no difference in the test scores and H_1: there is a difference in the test scores (claim); C.V. = ±1.96; $z = -1.215$; do not reject. There is not enough evidence to support the claim that there is a difference in the test scores.

6. H_0: there is no difference in the lifetimes of the two brands of video game (claim) and H_1: there is a difference in the lifetimes of the two brands of video game; C.V. = ±2.58; $z = -0.89$; do not reject. There is not enough evidence to reject the claim that there is no difference in the lifetimes of the two brands of video game.

7. H_0: there is no difference between the stopping distances of the two types of automobiles (claim) and H_1: there is a difference between the stopping distances of the two types of automobiles; C.V. = ±1.65; $z = -2.72$; reject. There is not enough evidence to reject the claim that there is no difference in the stopping distances of the automobiles. In this case, midsize cars have a smaller stopping distance.

8. H_0: there is no difference in the number of wins and H_1: there is a difference in the number of wins; $R = 125$; $\mu_R = 132$; $\sigma_R = 16.2481$; C.V. = ±1.96; $z = -0.431$; do not reject. There is insufficient evidence to conclude a difference in the number of wins.

9. H_0: there is no difference in the number of hunting accidents in the two geographic areas and H_1: there is a difference in the number of hunting accidents (claim); C.V. = ±1.96; $z = 2.57$; reject. There is enough evidence to support the claim that there is a difference in the number of accidents in the two areas. The number of accidents may be related to the number of hunters in the areas.

10. H_0: there is no difference in the size of enrollments and H_1: there is a difference in the size of enrollments; $R = 127$; $\mu_R = 110$; $\sigma_R = 14.2009$; C.V. = ±1.96; $z = 1.20$; do not reject. There is insufficient evidence to conclude a difference in enrollments.

11. H_0: there is no difference in the pain relief times of the drugs and H_1: there is a difference in the pain relief times of the drugs (claim); C.V. = ±1.96; $z = 2.91$; reject. There is enough evidence to support the claim that there is a difference in the pain relief times of the drugs.

Exercises 13–4

1. The t test for dependent samples

2. The sum of the positive ranks is 9.5. The sum of the negative ranks is 18.5. The test value is 9.5.

3. Sum of minus ranks is -6; sum of plus ranks is $+15$. The test value is 6.

4. C.V. $= 59$; do not reject

5. C.V. $= 20$; reject

6. C.V. $= 52$; do not reject

7. C.V. $= 102$; reject

8. C.V. $= 28$; do not reject

9. H_0: the human dose is equal to the animal dose and H_1: the human dose is more than the animal dose (claim); C.V. $= 6$; $w_s = 2$; reject. There is enough evidence to support the claim that the human dose costs more than the equivalent animal dose. One reason is that some people might not be inclined to pay a lot of money for their pets' medication.

10. H_0: there is no difference in the assessed values of the properties for the given two years and H_1: there is a difference in the assessed values of the properties for the given two years (claim); C.V. $= 11$; $w_s = 15$; do not reject. There is not enough evidence to support the claim that the assessed value of the properties has changed. The assessed property values would probably not be normally distributed.

11. H_0: there is no difference in the weights of the subjects and H_1: there is a difference in the weights of the subjects (claim); C.V. $= 4$; $w_s = 5$; do not reject. There is insufficient evidence to support the claim that the weights have changed.

12. H_0: there is no difference in legal costs and H_1: there is a difference in legal costs; $w_s = 2.5$; C.V. $= 4$; reject. There is sufficient evidence to conclude a difference in legal costs.

13. H_0: the prices of prescription drugs in the United States are equal to the prices in Canada and H_1: the drugs sold in Canada are cheaper; C.V. $= 11$; $w_s = 3$; reject. There is enough evidence to support the claim that the drugs are less expensive in Canada.

Exercises 13–5

1. H_0: there is no difference in the number of calories and H_1: there is a difference in the number of calories (claim); C.V. $= 7.815$; $H = 2.842$; do not reject. There is not enough evidence to support the claim that there is a difference in the number of calories.

2. H_0: there is no difference in the mathematical literacy scores of the individuals and H_1: there is a difference in the mathematical literacy of the individuals (claim); C.V. $= 5.991$; $H = 4.16$; do not reject. There is not enough evidence to support the claim that there is a difference in the mathematical literacy scores of the individuals.

3. H_0: there is no difference in the prices of the three types of lawnmowers and H_1: there is a difference in the prices of the three types of lawnmowers (claim); C.V. $= 4.605$; $H = 1.07$; do not reject. There is not enough evidence to support the claim that the prices are different. No, price

is not a factor. Results are suspect since one sample is less than 5.

4. H_0: there is no difference in the amounts of sodium in the different brands of microwave dinners and H_1: there is a difference in the amounts of sodium in the different brands of microwave dinners (claim); C.V. $= 5.991$; $H = 10.533$; reject. There is enough evidence to support the claim that there is a difference in the amounts of sodium in the different brands of microwave dinner.

5. H_0: there is no difference in the amounts of the benefits for the areas and H_1: there is a difference in the amount of the benefits for the areas (claim); C.V. $= 5.991$; $H = 12.43$; reject. There is significant evidence to support the claim that there is a difference in the amount of the benefits for the areas. The benefits are probably not normally distributed.

6. H_0: there is no difference in the number of job offers received by each group and H_1: there is a difference in the number of job offers received by each group (claim); C.V. $= 5.991$; $H = 8.54$; reject. There is enough evidence to support the claim that the number of job offers is different.

7. H_0: there is no difference in spending between regions and H_1: there is a difference in spending between regions; $H = 0.74$; C.V. $= 5.991$; do not reject. There is insufficient evidence to conclude a difference in spending.

8. H_0: there is no difference in the prices of the three types of printer and H_1: there is a difference in the prices of the three types of printer (claim); C.V. $= 5.991$; $H = 0.809$; do not reject. There is not enough evidence to support the claim that there is a difference in the prices of the printers. No, based on these samples, you cannot conclude that one type of printer generally costs more than another type.

9. H_0: there is no difference in the number of crimes in the five precincts and H_1: there is a difference in the number of crimes in the five precincts (claim); C.V. $= 13.277$; $H = 20.753$; reject. There is enough evidence to support the claim that there is a difference in the number of crimes in the five precincts.

10. H_0: there is no difference in caffeine content and H_1: there is a difference in caffeine content; $H = 9.98$; C.V. $= 5.991$; reject. There is sufficient evidence to conclude a difference in caffeine content.

11. H_0: there is no difference in speeds and H_1: there is a difference in speeds; $H = 3.815$; C.V. $= 5.991$; do not reject. There is insufficient evidence to conclude a difference in speeds.

Exercises 13–6

1. 0.716   2. 0.488

3. 0.648   4. 0.833

5. $r_s = 0.929$; H_0: $\rho = 0$ and H_1: $\rho \neq 0$; C.V. $= \pm 0.786$; reject. There is enough evidence to say that there is a relationship between the grade 4 achievement tests and the grade 8 achievement tests.

6. $r_s = 0.471$; H_0: $\rho = 0$ and H_1: $\rho \neq 0$; C.V. $= \pm 0.886$; do not reject. There is no significant linear relationship.

7. $r_s = 0.817$; H_0: $\rho = 0$ and H_1: $\rho \neq 0$; C.V. $= \pm 0.700$; reject. There is a significant relationship between the number of new releases and the gross receipts.

8. $r_s = 0.893$; H_0: $\rho = 0$ and H_1: $\rho \neq 0$; C.V. $= \pm 0.786$; reject. There is a significant relationship between the number of hospitals and the number of nursing homes in a state.

9. $r_s = 0.048$; H_0: $\rho = 0$ and H_1: $\rho \neq 0$; C.V. $= \pm 0.738$; do not reject. There is not enough evidence to say that a significant correlation exists between calories and the cholesterol amounts in fast-food sandwiches.

10. $r_s = 0.8857$; H_0: $\rho = 0$ and H_1: $\rho \neq 0$; C.V. $= \pm 0.886$. Very close! There is not a significant relationship between the number of books published in 1980 and in 2004 in the same subject area. Since r is not significant, no relationship can be predicted 20 years from now. Even if r is significant, you should not make a prediction for 20 years from now. That would be extrapolating.

11. $r_s = 0.624$; H_0: $\rho = 0$ and H_1: $\rho \neq 0$; C.V. $= \pm 0.700$; do not reject. There is no significant relationship between gasoline prices paid to the car rental agency and regular gasoline prices. One would wonder how the car rental agencies determine their prices.

12. $r_s = 0.714$; H_0: $\rho = 0$ and H_1: $\rho \neq 0$; C.V. $= \pm 0.886$; do not reject. There is not sufficient evidence to conclude a significant relationship between the number of motor vehicle thefts and burglaries.

13. $r_s = -0.10$; H_0: $\rho = 0$ and H_1: $\rho \neq 0$; C.V. $= \pm 0.900$; do not reject. There is no significant relationship between the number of cyber school students and the cost per pupil. In this case, the cost per pupil is different in each district.

14. $r_s = 0.542$; H_0: $\rho = 0$ and H_1: $\rho \neq 0$; C.V. $= \pm 0.643$; do not reject. There is no significant relationship between the costs of the drugs.

15. H_0: the number of cavities in a person occurs at random and H_1: the null hypothesis is not true. There are 21 runs; the expected number of runs is between 10 and 22. Therefore, do not reject the null hypothesis; the number of cavities in a person occurs at random.

16. H_0: the numbers occur at random and H_1: the null hypothesis is not true. There are 14 runs. Since the expected number of runs is between 8 and 20, do not reject. The numbers occur at random.

17. H_0: the purchases of soft drinks occur at random and H_1: the null hypothesis is not true. There are 16 runs, and the expected number of runs is between 9 and 22, so do not reject the null hypothesis. Hence the purchases of soft drinks occur at random.

18. H_0: the integers generated by a calculator occur at random and H_1: the null hypothesis is not true. There are 13 runs, and the expected number of runs is between 7 and 17, so the null hypothesis is not rejected. The integers occur at random.

19. H_0: the seating occurs at random and H_1: the null hypothesis is not true. There are 14 runs. Since the expected number of runs is between 10 and 23, do not reject. The seating occurs at random.

20. H_0: the gender of the shoppers in line at the grocery store is random (claim) and H_1: the null hypothesis is not true. There are 10 runs. Since the expected number of runs is between 6 and 16, the null hypothesis should not be rejected. There is not enough evidence to reject the hypothesis that the gender of the shoppers in line is random.

21. H_0: the number of absences of employees occurs at random over a 30-day period and H_1: the null hypothesis is not true. There are only 6 runs, and this value does not fall within the 9-to-21 range. Hence, the null hypothesis is rejected; the absences do not occur at random.

22. H_0: the days customers are able to ski occur at random (claim) and H_1: the null hypothesis is not true. There are 5 runs. Since this number is not between 9 and 20, the decision is to reject the null hypothesis. There is enough evidence to reject the claim that the days customers are able to ski occur at random.

23. Answers will vary.

24. ± 0.28

25. ± 0.479

26. ± 0.400

27. ± 0.215

28. ± 0.413

Review Exercises

1. H_0: median $= 36$ years and H_1: median $\neq 36$ years; $z = -0.548$; C.V. $= \pm 1.96$; do not reject. There is insufficient evidence to conclude that the median differs from 36.

2. H_0: median $= 40,000$ miles (claim) and H_1: median $\neq 40,000$ miles; $z = -0.913$; C.V. $= \pm 1.96$; do not reject. There is not enough evidence to reject the claim that the median is 40,000 miles.

3. H_0: there is no difference in prices and H_1: there is a difference in prices; test value $= 1$; C.V. $= 0$; do not reject. There is insufficient evidence to conclude a difference in prices. Comments: Examine what affects the result of this test.

4. H_0: there is no difference in the record high temperatures of the two cities and H_1: there is a difference in the record high temperatures of the two cities (claim); $z = -1.24$; P-value $= 0.2150$; do not reject. There is not enough evidence to support the claim that there is a difference in the record high temperatures of the two cities.

5. H_0: there is no difference in the hours worked and H_1: there is a difference in the hours worked; $R = 85$; $\mu_R = 110$; $\sigma_R = 14.2009$; $z = -1.76$; C.V. $= \pm 1.645$; reject. There is sufficient evidence to conclude a difference in the hours worked. C.V. $= \pm 1.96$; do not reject.

6. H_0: the additive did not improve the gas mileage and H_1: the additive did improve the gas mileage (claim); C.V. = 14; w_s = 14; reject. There is enough evidence to support the claim that the additive improved the gas mileage.

7. H_0: there is no difference in the amount spent and H_1: there is a difference in the amount spent; w_s = 1; C.V. = 2; reject. There is sufficient evidence of a difference in amount spent at the 0.05 level of significance.

8. H_0: there is no difference in the breaking strengths of the ropes and H_1: there is a difference in the breaking strengths of the ropes (claim); C.V. = 5.991; H = 28.02; reject. There is enough evidence to support the claim that there is a difference in the breaking strengths of the ropes.

9. H_0: there is no difference in beach temperatures and H_1: there is a difference in temperatures; H = 15.524; C.V. = 7.815; reject. There is sufficient evidence to conclude a difference in beach temperatures. (Without the Southern Pacific: H = 3.661; C.V. = 5.991; do not reject.)

10. r_s = 0.933; H_0: ρ = 0 and H_1: $\rho \neq 0$; C.V. = ± 0.700; reject. There is a significant relationship between the rankings.

11. r_s = 0.891; H_0: ρ = 0 and H_1: $\rho \neq 0$; C.V. = ± 0.648; reject. There is a significant relationship in the average number of people who are watching the television shows for both years.

12. H_0: the books are arranged at random and H_1: the null hypothesis is not true. There are 12 runs. Since the expected number of runs is between 10 and 22, do not reject. The books are arranged at random.

13. H_0: the grades of students who finish the exam occur at random and H_1: the null hypothesis is not true. Since there are 8 runs and this value does not fall in the 9-to-21 interval, the null hypothesis is rejected. The grades do not occur at random.

Chapter Quiz

1. False
2. False
3. True
4. True
5. *a*
6. *c*
7. *d*
8. *b*
9. Nonparametric
10. Nominal, ordinal
11. Sign
12. Sensitive

13. H_0: median = \$177,500; H_1: median $\neq$ \$177,500 (claim); C.V. = 2; test value = 3; do not reject. There is not enough evidence to say that the median is not \$177,500.

14. H_0: median = 1200 (claim) and H_1: median $\neq$ 1200. There are 10 minus signs. Do not reject since 10 is greater than the critical value 6. There is not enough evidence to reject the claim that the median is 1200.

15. H_0: there will be no change in the weight of the turkeys after the special diet and H_1: the turkeys will weigh more after the special diet (claim). There is 1 plus sign; hence, the null hypothesis is rejected. There is enough evidence to support the claim that the turkeys gained weight on the special diet.

16. H_0: there is no difference in the amounts of money received by the teams and H_1: there is a difference in the amounts of money each team received; C.V. = ± 1.96; z = -0.79; do not reject. There is not enough evidence to say that the amounts differ.

17. H_0: the distributions are the same and H_1: the distributions are different (claim); z = -0.14434; C.V. = ± 1.65; do not reject the null hypothesis. There is not enough evidence to support the claim that the distributions are different.

18. H_0: there is no difference in the GPA of the students before and after the workshop and H_1: there is a difference in the GPA of the students before and after the workshop (claim); test statistic = 0; C.V. = 2; reject the null hypothesis. There is enough evidence to support the claim that there is a difference in the GPAs of the students.

19. H_0: there is no difference in the amounts of sodium in the three sandwiches and H_1: there is a difference in the amounts of sodium in the sandwiches; C.V. = 5.991; H = 11.795; reject. There is enough evidence to conclude that there is a difference in the amounts of sodium in the sandwiches.

20. H_0: there is no difference in the reaction times of the monkeys and H_1: there is a difference in the reaction times of the monkeys (claim); H = 6.9; $0.025 < P\text{-value} < 0.05$ (0.032); reject the null hypothesis. There is enough evidence to support the claim that there is a difference in the reaction times of the monkeys.

21. r_s = 0.683; H_0: ρ = 0 and H_1: $\rho \neq 0$; C.V. = ± 0.600; reject. There is enough evidence to say that there is a significant relationship between the drug prices.

22. r_s = 0.943; H_0: ρ = 0 and H_1: $\rho \neq 0$; C.V. = ± 0.829; reject. There is a significant relationship between the amount of money spent on Head Start and the number of students enrolled in the program.

23. H_0: the births of babies occur at random according to gender and H_1: the null hypothesis is not true. There are 10 runs, and since this is between 8 and 19, the null hypothesis is not rejected. There is not enough evidence to reject the null hypothesis that the gender occurs at random.

24. H_0: there is no difference in the rpm of the motors before and after the reconditioning and H_1: there is a difference in the rpm of the motors before and after the reconditioning (claim); test statistic = 0; C.V. = 6; do not reject. There is not enough evidence to support the claim that there is a difference in the rpm of the motors before and after reconditioning.

25. H_0: the numbers occur at random and H_1: the null hypothesis is not true. There are 16 runs, and since this is between 9 and 21, the null hypothesis is not rejected. There is not enough evidence to reject the null hypothesis that the numbers occur at random.

Chapter 14

Exercises 14–1

1. Random, systematic, stratified, cluster

2. Samples can save the researcher time and money. They are used when the population is large or infinite. They are used when the original units are to be destroyed, such as in testing the breaking strength of ropes.

3. A sample must be randomly selected.

4. Random numbers are used to ensure every element of the population has the same chance of being selected.

5. Talking to people on the street, calling people on the phone, and asking your friends are three incorrect ways of obtaining a sample.

6. Over the long run each digit, 0 through 9, will occur with the same probability.

7. Random sampling has the advantage that each unit of the population has an equal chance of being selected. One disadvantage is that the units of the population must be numbered; if the population is large, this could be somewhat time-consuming.

8. Systematic sampling has an advantage in that once the first unit is selected, each succeeding unit selected has been determined. This saves time. A disadvantage would be if the list of units was arranged in some manner so that a bias would occur, such as selecting all men when the population consists of both men and women.

9. An advantage of stratified sampling is that it ensures representation for the groups used in stratification; however, it is virtually impossible to stratify the population so that all groups are represented.

10. Clusters are easy to use since they already exist, but it is difficult to justify that the clusters actually represent the population.

11–20. Answers will vary.

Exercises 14–2

1. Flaw—biased; it's confusing.

2. Flaw—the purpose of the question is unclear. You could like him personally but not politically.

3. Flaw—the question is too broad.

4. Flaw—none. The question is good if the respondent knows the mayor's position; otherwise his position needs to be stated.

5. Flaw—confusing words. How many hours did you study for this exam?

6. Possible order problem—ask first, "Do you use artificial sweetener regularly?"

7. Flaw—confusing words. If a plane were to crash on the border of New York and New Jersey, where should the victims be buried?

8. Flaw—none.

9. Answers will vary.

10. Answers will vary.

Exercises 14–3

1. Simulation involves setting up probability experiments that mimic the behavior of real-life events.

2. Answers will vary.

3. John Von Neumann and Stanislaw Ulam

4. Using the computer to simulate real-life situations can save time, since the computer can generate random numbers and keep track of the outcomes very quickly and easily.

5. The steps are as follows:
   a. List all possible outcomes.
   b. Determine the probability of each outcome.
   c. Set up a correspondence between the outcomes and the random numbers.
   d. Conduct the experiment by using random numbers.
   e. Repeat the experiment and tally the outcomes.
   f. Compute any statistics and state the conclusions.

6. Random numbers can be used to ensure the outcomes occur with appropriate probability.

7. When the repetitions increase, there is a higher probability that the simulation will yield more precise answers.

8. Use a table of random numbers. Select 40 random numbers. Numbers 01 through 16 mean the person is foreign-born.

9. Use three-digit random numbers; numbers 001 through 681 mean that the mother is in the labor force.

10. Select two-digit random numbers in groups of 5. For one person, 01 through 70 means a success. For the other person, 01 through 75 means a success.

11. Select 100 two-digit random numbers. Numbers 00 to 34 mean the household has at least one set with premium cable service. Numbers 35 to 99 mean the household does not have the service.

12. Use the odd digits to represent a match and the even digits to represent a nonmatch.

13. Let an odd number represent heads and an even number represent tails. Then each person selects a digit at random.

14–24. Answers will vary.

Review Exercises

1–8. Answers will vary.

9. Use one-digit random numbers 1 through 4 for a strikeout and 5 through 9 and 0 represent anything other than a strikeout.

10. Use two-digit random numbers: 01 through 15 represent an overbooked plane, and 16 through 99 and 00 represent a plane that is not overbooked.

11. In this case, a one-digit random number is selected. Numbers 1 through 6 represent the numbers on the face. Ignore 7, 8, 9, and 0 and select another number.

12. The first person selects a two-digit random number. Any two-digit random number that has a 7, 8, 9, or 0 is ignored, and another random number is selected. Player 1 selects a one-digit random number; any random number that is not 1 through 6 is ignored, and another one is selected.

13. Let the digits 1 through 3 represent rock, let 4 through 6 represent paper, let 7 through 9 represent scissors, and omit 0.

14–18. Answers will vary.

19. Flaw—asking a biased question. Have you ever driven through a red light?

20. Flaw—using a double negative. Do you think students who are not failing should be given tutoring if they request it?

21. Flaw—asking a double-barreled question. Do you think all automobiles should have heavy-duty bumpers?

22. Answers will vary.

Chapter Quiz

1. True   **2.** True

3. False   **4.** True

5. *a*   **6.** *c*

7. *c*   **8.** Larger

9. Biased   **10.** Cluster

11–14. Answers will vary.

15. Use two-digit random numbers: 01 through 45 means the player wins. Any other two-digit random number means the player loses.

16. Use two-digit random numbers: 01 through 05 means a cancellation. Any other two-digit random number means the person shows up.

17. The random numbers 01 through 10 represent the 10 cards in hearts. The random numbers 11 through 20 represent the 10 cards in diamonds. The random numbers 21 through 30 represent the 10 spades, and 31 through 40 represent the 10 clubs. Any number over 40 is ignored.

18. Use two-digit random numbers to represent the spots on the face of the dice. Ignore any two-digit random numbers with 7, 8, 9, or 0. For cards, use two-digit random numbers between 01 and 13.

19. Use two-digit random numbers. The first digit represents the first player, and the second digit represents the second player. If both numbers are odd or even, player 1 wins. If a digit is odd and the other digit is even, player 2 wins.

20–24. Answers will vary.

Appendix A

A–1. 362,880   **A–2.** 5040

A–3. 120   **A–4.** 1

A–5. 1   **A–6.** 6

A–7. 1320   **A–8.** 1,814,400

A–9. 20   **A–10.** 7920

A–11. 126   **A–12.** 120

A–13. 70   **A–14.** 455

A–15. 1   **A–16.** 10

A–17. 560   **A–18.** 1980

A–19. 2520   **A–20.** 90

A–21. 121; 2181; 14,641; 716.9

A–22. 56; 550; 3136; 158

A–23. 32; 258; 1024; 53.2

A–24. 150; 4270; 22,500; 1457.5

A–25. 328; 22,678; 107,584; 1161.2

A–26. 829; 123,125; 687,241; 8584.8333

A–27. 693; 50,511; 480,249; 2486.1

A–28. 409; 40,333; 167,281; 6876.80

A–29. 318; 20,150; 101,124; 3296

A–30. −20; 778; 400; 711.3334

A–31.

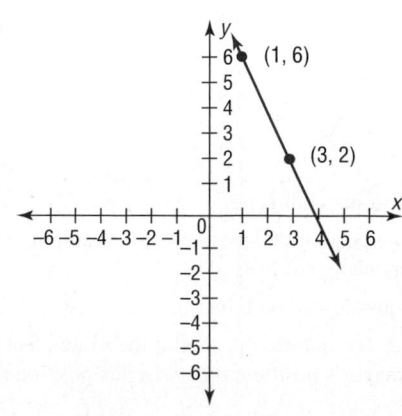

A–32.

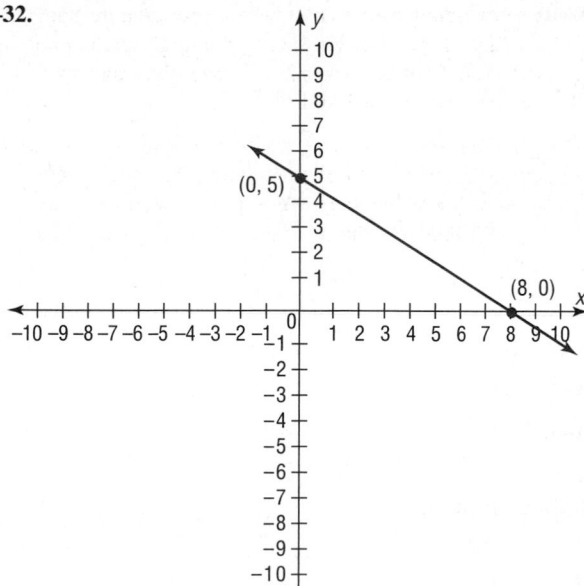

A–35.

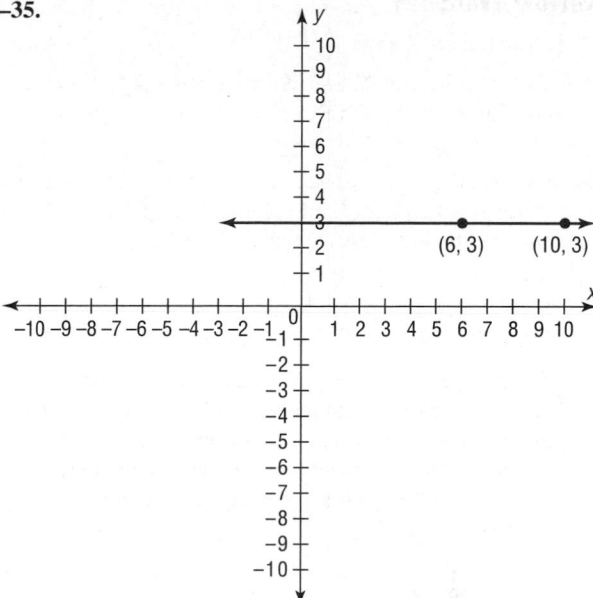

A–33.

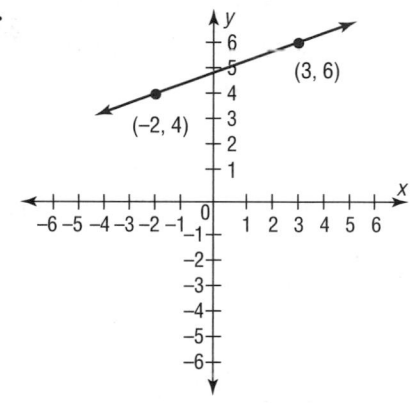

A–36.

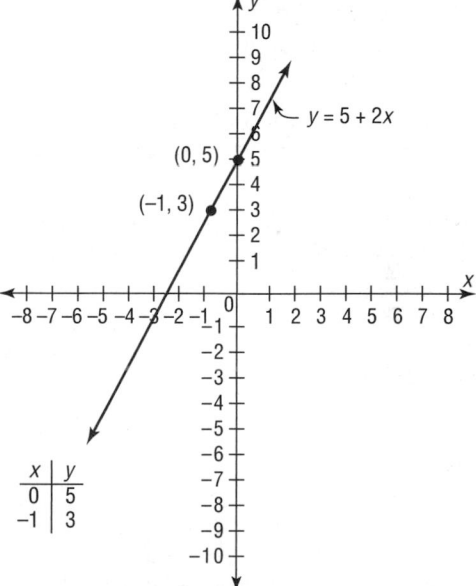

A–34.

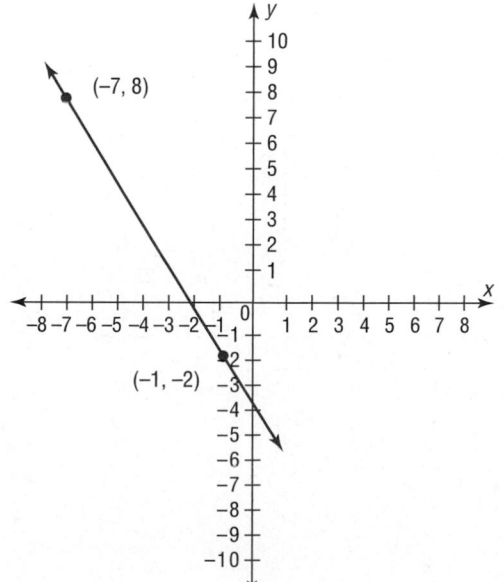

A–37.

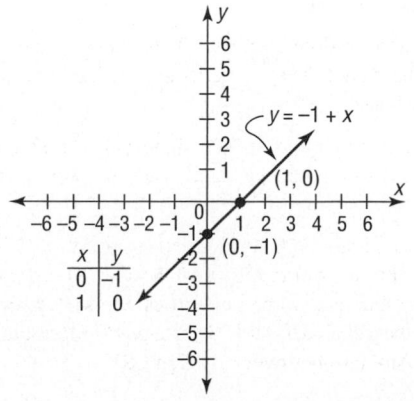

A–38.

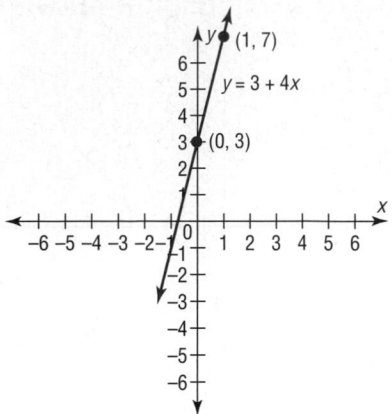

A–39.

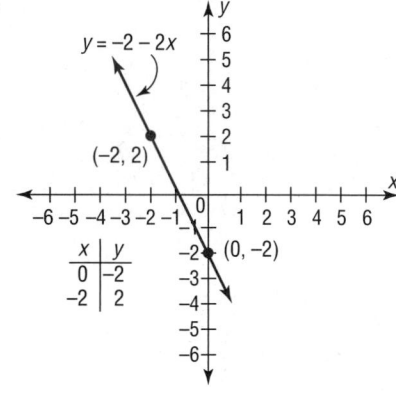

A–40.

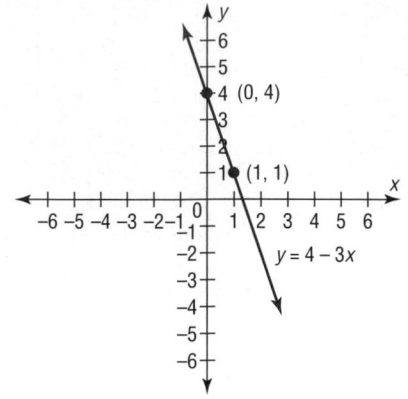

Appendix B–2

B–1.	0.65	**B–2.**	0.579
B–3.	0.653	**B–4.**	0.005
B–5.	0.379	**B–6.**	0.585
B–7.	$\frac{1}{4}$	**B–8.**	$\frac{2}{9}$
B–9.	0.64	**B–10.**	0.467
B–11.	0.857	**B–12.**	0.33

Index

Confidence interval, 358
  hypothesis testing, 457–459
  mean, 358–373
  means, difference of, 478, 486, 499
  median, 672
  proportion, 377–379
  proportions, differences, 508–509
  variances and standard deviations,
      385–390
Confidence level, 358
Confounding variable, 15
Consistent estimator, 357
Contingency coefficient, 617
Contingency table, 606–607
Continuous variable, 6–7, 253, 300
Control group, 14
Convenience sample, 12–13
Correction factor for
      continuity, 342
Correlation, 534, 538–547
Correlation coefficient, 539
  multiple, 578
  Pearson's product moment, 539
  population, 543
  Spearman's rank, 700–702
Critical region, 406
Critical value, 406
Cumulative frequency, 54
Cumulative frequency distribution, 42–43
Cumulative frequency graph, 54–56
Cumulative relative frequency, 57–58

D

Data, 3
Data array, 109
Data set, 3
Data value (datum), 3
Deciles, 151
Degrees of freedom, 370
Dependent events, 213
Dependent samples, 492
Dependent variable, 14, 535
Descriptive statistics, 4
Difference between two means, 473–479,
      484–487, 492–499
  assumptions for the test to
      determine, 473, 486, 493
  proportions, 504–509
Discrete probability distribution, 254
Discrete variable, 6, 253
Disordinal interaction, 653
Distribution-free statistics
      (nonparametric), 672

Distributions
  bell-shaped, 59, 301
  bimodal, 60, 111
  binomial, 270–276
  chi-square, 386–388
  F, 513
  frequency, 37
  hypergeometric, 286–289
  multinomial, 283–284
  negatively skewed, 60, 117, 301
  normal, 302–311
  Poisson, 284–286
  positively skewed, 60, 117, 301
  probability, 253–258
  sampling, 331–333
  standard normal, 304
  symmetrical, 59, 117, 301
Double sampling, 729

E

Empirical probability, 191–193
Empirical rule, 136
Equally likely events, 186
Estimation, 356
Estimator, properties of a good, 357
Event, simple, 185
Events
  complementary, 189–190
  compound, 189
  dependent, 213
  equally likely, 186
  independent, 211
  mutually exclusive, 199–200
Expectation, 264–266
Expected frequency, 593
Expected value, 264
Experimental study, 14
Explained variation, 566
Explanatory variable, 14
Exploratory data analysis (EDA),
      162–165
Extrapolation, 556

F

Factorial notation, 227
Factors, 647
F-distribution, characteristics of, 513
Finite population correction
      factor, 337
Five-number summary, 162
Frequency, 37

Table F	The *t* Distribution					
	Confidence intervals	**80%**	**90%**	**95%**	**98%**	**99%**
	One tail, α	**0.10**	**0.05**	**0.025**	**0.01**	**0.005**
d.f.	**Two tails, α**	**0.20**	**0.10**	**0.05**	**0.02**	**0.01**
1		3.078	6.314	12.706	31.821	63.657
2		1.886	2.920	4.303	6.965	9.925
3		1.638	2.353	3.182	4.541	5.841
4		1.533	2.132	2.776	3.747	4.604
5		1.476	2.015	2.571	3.365	4.032
6		1.440	1.943	2.447	3.143	3.707
7		1.415	1.895	2.365	2.998	3.499
8		1.397	1.860	2.306	2.896	3.355
9		1.383	1.833	2.262	2.821	3.250
10		1.372	1.812	2.228	2.764	3.169
11		1.363	1.796	2.201	2.718	3.106
12		1.356	1.782	2.179	2.681	3.055
13		1.350	1.771	2.160	2.650	3.012
14		1.345	1.761	2.145	2.624	2.977
15		1.341	1.753	2.131	2.602	2.947
16		1.337	1.746	2.120	2.583	2.921
17		1.333	1.740	2.110	2.567	2.898
18		1.330	1.734	2.101	2.552	2.878
19		1.328	1.729	2.093	2.539	2.861
20		1.325	1.725	2.086	2.528	2.845
21		1.323	1.721	2.080	2.518	2.831
22		1.321	1.717	2.074	2.508	2.819
23		1.319	1.714	2.069	2.500	2.807
24		1.318	1.711	2.064	2.492	2.797
25		1.316	1.708	2.060	2.485	2.787
26		1.315	1.706	2.056	2.479	2.779
27		1.314	1.703	2.052	2.473	2.771
28		1.313	1.701	2.048	2.467	2.763
29		1.311	1.699	2.045	2.462	2.756
30		1.310	1.697	2.042	2.457	2.750
32		1.309	1.694	2.037	2.449	2.738
34		1.307	1.691	2.032	2.441	2.728
36		1.306	1.688	2.028	2.434	2.719
38		1.304	1.686	2.024	2.429	2.712
40		1.303	1.684	2.021	2.423	2.704
45		1.301	1.679	2.014	2.412	2.690
50		1.299	1.676	2.009	2.403	2.678
55		1.297	1.673	2.004	2.396	2.668
60		1.296	1.671	2.000	2.390	2.660
65		1.295	1.669	1.997	2.385	2.654
70		1.294	1.667	1.994	2.381	2.648
75		1.293	1.665	1.992	2.377	2.643
80		1.292	1.664	1.990	2.374	2.639
90		1.291	1.662	1.987	2.368	2.632
100		1.290	1.660	1.984	2.364	2.626
500		1.283	1.648	1.965	2.334	2.586
1000		1.282	1.646	1.962	2.330	2.581
(*z*) ∞		1.282[a]	1.645[b]	1.960	2.326[c]	2.576[d]

[a]This value has been rounded to 1.28 in the textbook
[b]This value has been rounded to 1.65 in the textbook.
[c]This value has been rounded to 2.33 in the textbook.
[d]This value has been rounded to 2.58 in the textbook.

Source: Adapted from W. H. Beyer, *Handbook of Tables for Probability and Statistics,* 2nd ed., CRC Press, Boca Raton, Fla., 1986. Reprinted with permission.

One tail

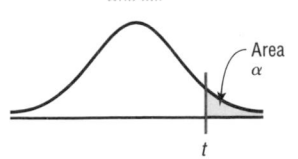

Two tails

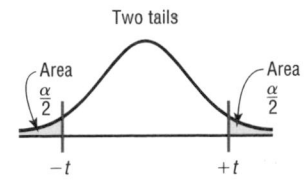

Glossary of Symbols

a	y intercept of a line	MR	Midrange
α	Probability of a type I error	MS_B	Mean square between groups
b	Slope of a line	MS_W	Mean square within groups (error)
β	Probability of a type II error	n	Sample size
C	Column frequency	N	Population size
cf	Cumulative frequency	$n(E)$	Number of ways E can occur
$_nC_r$	Number of combinations of n objects taking r objects at a time	$n(S)$	Number of outcomes in the sample space
C.V.	Critical value	O	Observed frequency
CVar	Coefficient of variation	P	Percentile; probability
D	Difference; decile	p	Probability; population proportion
$\overline{D}$	Mean of the differences	$\hat{p}$	Sample proportion
d.f.	Degrees of freedom	$\bar{p}$	Weighted estimate of p
d.f.N.	Degrees of freedom, numerator	$P(B|A)$	Conditional probability
d.f.D.	Degrees of freedom, denominator	$P(E)$	Probability of an event E
E	Event; expected frequency; maximum error of estimate	$P(\overline{E})$	Probability of the complement of E
$\overline{E}$	Complement of an event	$_nP_r$	Number of permutations of n objects taking r objects at a time
e	Euler's constant ≈ 2.7183	π	Pi ≈ 3.14
$E(X)$	Expected value	Q	Quartile
f	Frequency	q	$1 - p$; test value for Tukey test
F	F test value; failure	$\hat{q}$	$1 - \hat{p}$
F'	Critical value for the Scheffé test	$\bar{q}$	$1 - \bar{p}$
MD	Median	R	Range; rank sum
		F_S	Scheffé test value
		GM	Geometric mean